Mushrooms
Demystified

David Arora

Mushrooms Demystified

*A Comprehensive Guide
to the
Fleshy Fungi*

SECOND EDITION

Ten Speed Press
Berkeley

Copyright © 1979, 1986 by David Arora

All rights reserved. Published in the United States by Ten Speed Press, an imprint of the Crown Publishing Group, a division of Random House, Inc., New York.
www.crownpublishing.com
www.tenspeed.com

Ten Speed Press and the Ten Speed Press colophon are registered trademarks of Random House, Inc.

Grateful acknowledgement is made to Alan Snitow of the *San Francisco Bay Guardian* and to the Peoples Press, in whose pages some of the introductory material originally appeared.

Library of Congress Catalog Number 79-85123

The author welcomes information and feedback from readers.
You can write to him care of the publisher.

Library of Congress Cataloging-in-Publication Data

Arora, David
     Mushrooms demystified.

     Bibliography.
     Includes index.
     1.Mushrooms–California–Identification.
    2. Mushrooms–Pacific States–Identification.
    I. Title
    QK617 .A69 1986 589.2'223'09794 86-5917
    ISBN-13: 978-0-89815-169-5

Printed and bound in China

Cover design by Brenton Beck,
Fifth Street Design Associates.
Illustrations by Michael Cabaniss.
All photographs by David Arora
unless otherwise indicated.
Front cover photograph of *Agaricus augustus* by Bill Donaldson.

23 22 21

Second Edition

I dedicate this book
with love
to my mother and father,
whose admonitions to me as a teen-ager
to stay away from mushrooms
inspired me to get closer

PREFACE TO THE SECOND EDITION

THE first edition of *Mushrooms Demystified* was designed primarily for California. The second edition is useful throughout the United States and Canada. The distinction is not as great as it sounds because most of the mushrooms found in California are widely distributed; the text has undergone considerable change nevertheless. Prominent and distinctive species are included from all parts of the continent: New England, the South, Midwest, Southwest, Rocky Mountains, Pacific Northwest, and of course, California. In addition, the nomenclature has been updated, over 200 new color plates have been added, most of the original black-and-white photographs have been replaced, and certain groups such as the truffles are treated in greater detail. One thing that has not changed is the tone of the tome—I have once again made a special effort to keep the terminology simple and the language entertaining. The result is a bigger book, and I think, a better one.

Of course, in trying to become more things to more people, one runs the risk of becoming less to some. Despite a California slant (intended or not, every field guide has built-in geographical bias because every author has more field experience in some regions than others), the inclusion of species from across the country may disappoint Californians who found the limited scope of the first edition reassuring. The text and keys are longer because of the larger number of species covered, but they are not any more difficult to use. Simple geography, in fact, often facilitates recognition (e.g., *Suillus lakei* of western North America and *S. pictus*, its eastern counterpart).

I have, however, maintained a certain provincial flavor by providing detailed season and habitat information for the Central California Coast (hereafter referred to as "our area"). Southern Californians, northern Californians, eastern Californians, non-Californians, and worldly central Californians may find this disconcerting, but there are good reasons for doing so. The terrain, climate, and vegetation of North America are remarkably varied. Anyone who has used or written books of national scope knows that in striving to cover such a diverse area, one is often forced to present information in an extremely generalized, vague way. Just how useful is it to say of a particular mushroom, "found in woods across the country in spring, summer, fall, or winter," when in any one area it is likely to have a favorite host tree and limited fruiting period? Instead of providing information on habitat and season that is marginally useful to all readers, I have decided to supply information that is very useful to readers fortunate enough to live in or visit "our area," and marginally useful (i.e., as useful as that in any other national field guide) to the rest. Besides, any mushroom hunter worth her salt-and-butter quickly learns when and where mushrooms grow in her area.

I have tried to make *Mushrooms Demystified* as accessible as it is comprehensive. For the uninitiated, there are special chapters on mushroom terminology and classification, when and where to find mushrooms, how to collect and identify mushrooms, how to use the keys (which are the backbone of the book), mushroom cookery and mushroom toxins, and what the scientific (Latin) names of mushrooms mean. The book is comprehensive enough to interest intermediate and advanced mushroom hunters. However, beginners intimidated by its length can selectively ignore the more technical discussions and concentrate on learning the "Seventy Distinctive Mushrooms" listed on pp. 48-51.

In using this book it is important to realize that the identification of mushrooms differs in several key respects from the identification of, say, birds or wildflowers. For one thing, there is no book that "has them all." Over 2000 species are described, illustrated, or mentioned in this book—more than in any other North American field guide—but you will quickly discover that many of the mushrooms you find are not included in this book and many of the included mushrooms do not occur in your area (unless it is "our area"). This will be truer of some regions (e.g., the Southeast) than others, and there is no way it can be avoided—there are just too many mushrooms! As consolation I offer the words of Gary Lincoff (author of another national field guide): "It is better for a book to contain good descriptions of mushrooms that don't occur in your area than bad descriptions of ones that do!" As every author has a different set of experiences to share, I recommend that all mushroom hunters, even those in California, supplement this book with others, particularly regional guides (if they exist) because their limited scope enables them to present more specific information.

It will be many years before we have a complete inventory of North American mushrooms. More than anything this is a tribute to the elusive nature of mushrooms. They are difficult to study because they are ephemeral and unpredictable, and so much depends on being in the right place at the right time. It is also a comment on the fact that few people take mushrooms seriously. As a result, the documented distribution of the lesser-known mushrooms corresponds to the undocumented distribution of the better-known "mushroomologists." Furthermore, many mushroom species still await classification. While these factors make identification more difficult, they also add an element of suspense to the hunt. One is continually finding species that are new to science, or new to North America, or new to one's area, or at least new to oneself. Mycology (the study of fungi) is a field to which discerning amateurs can make significant contributions!

Another important point to realize is that many mushrooms cannot be positively identified unless you have access to a microscope and technical literature and know how to use them. Microscopic characteristics are *not* stressed in this book. This means that many of the species mentioned briefly are merely *suggestions* as to what an unidentified mushroom might be, and many of the species that are fully described are actually "complexes"—groups of closely related species whose exact identities are a matter for the specialist.

In other words, I've chosen to sacrifice a certain degree of exactitude by interpreting many species broadly. For instance, the description of *Agaricus silvicola* embraces a confusing group of white woodland mushrooms with a sweet odor and yellow-staining skin. Whether or not the "true" *A. silvicola* occurs in North America, I see no harm in applying the name to the "complex" *providing readers are made aware of the situation* (for instance, by appending the word "group" to the name) and any attendant risks. Purists may object to this approach, but there are many amateurs who would like to apply names to these "complexes" even if they are unequipped to make the subtler distinctions between species or varieties within a complex. Those who wish to verify identifications, look up species' authorities, or otherwise pursue the matter further can do so by making use of the literature listed in the bibliography. Students and professionals interested in the more unusual species I have collected can write to me care of the publisher—I will be happy to furnish the names of herbariums to which specimens have been sent.

I would also like to say a few words about the title of this book, since there have been varied reactions to it. One person has stated publicly that mushrooms cannot be demystified because there is nothing mystifying or mysterious about them; others have questioned the desirability of demystification and suggested that I call the second edition "Mushrooms *Re*mystified." I, for one, was first attracted to mushrooms because I found them mysterious. I *still* find them mysterious and I see no danger that this book or any other will deprive them of their mystery. Mushrooms, like other forms of life, are miracles—miracles which we can explain but not fully comprehend. Anyone who has studied anything knows that answers beget questions as surely as questions lead to answers, that the more one knows about a subject, the larger and more challenging it becomes. In addition to being an informative home and field companion, I hope *Mushrooms Demystified* will be an inspiration—to look more closely, ask more questions, seek more answers. Identification, after all, is not an end in itself, but a means toward acquiring a deeper knowledge and keener appreciation of our co-inhabitants on this planet.

The selection of color plates reflects my hope that this book will be more than a tool for identification. Many plates were chosen for their usefulness, but some were picked for their beauty, dramatic effect, or charm. (A book does little good if it does not invite one to take it off the shelf!) *Mushrooms Demystified* is also a vehicle for expressing my love for (and exasperation with) mushrooms and for people who love mushrooms, my respect for life and for people who respect life. Much of the pleasure in getting to know wild mushrooms is directly attributable to the "wild" companionship of fellow fungophiles, and I have paid tribute to these folks wherever possible.

Acknowledgments

Mushrooms Demystified is a compilation of information from many diverse sources. I have acted as assembler and interpreter, supplementing the information with my own knowledge, experience, keys, comments, and photographs, and binding it together in an accessible, useful, and cohesive (assuming the pages don't fall out) form. I get credit for writing the book; the many mycologists whose monographs and articles form its factual foundation do not. Yet compared to the vast fund of knowledge painstakingly accumulated by these mycologists, my own two cents' worth (15 years, actually) is insignificant. I have provided a bibliography of primary references, but in the tradition of other field guides, I have not made extensive use of footnotes or included species' authorities (the names of those who described them). I would like to express my appreciation, then, to all the authors listed in the bibliography plus any I have inadvertently omitted. I want them to know that they have my congratulations and respect, and I want my readers to know that this book would not have been possible without their dedicated efforts. If at times I seem irreverent in my discussions of mushrooms and mycologists, it is only because I am so by nature, and a sense of humor helps me to cope with the more exasperating aspects of mushroom scrutiny and book-writing.

I want to extend special thanks to those mycologists who directly contributed to this book by giving me their time and expertise, providing identifications of "problem" mushrooms, and offering suggestions on the manuscript. These mycologists are (in alphabetical order, with the areas in which they have helped): Prof. Joseph Ammirati of the University of Washington, *Cortinarius*; Chuck Barrows of Santa Fe, New Mexico, mushrooms of the Southwest; Prof. Howard Bigelow of the University of Massachusetts, *Clitocybe* and *Marasmius*; William Burk of the University of North Carolina, stinkhorns; Dennis Desjardin of San Francisco State University, *Marasmius* and *Collybia*; Prof. Robert Fogel of the University of Michigan, *Hymenogaster* and *Destuntzia;* Prof. Robert Gilbertson of the University of Arizona, resupinate (crust) fungi and mushrooms of southern Arizona; Bill Isaacs of Santa Fe, New Mexico, *Agaricus* and *Endoptychum*; Prof. David Jenkins of the University of Alabama-Birmingham, *Amanita*; Rick Kerrigan of the University of California-Santa Barbara, *Agaricus*; Prof. David Largent of Humboldt State University, *Leptonia, Nolanea, Entoloma,* and *Hygrophorus*; Herb Saylor of San Francisco State University, truffles, false truffles, and coral fungi; Prof. Robert Shaffer of the University of Michigan, *Russula*; Prof. Emeritus Alexander Smith of the University of Michigan, *Cortinarius* and many other mushrooms; Prof. James Trappe of the U.S.D.A. Forestry Sciences Laboratory, Corvallis, Oregon, truffles (especially *Tuber*) and false truffles; and Greg Wright of Claremont, California, southern Californian and southwestern mushrooms. Although these people have been of great assistance in the preparation of this book, any errors of fact are my responsibility, not theirs.

I am also very grateful to Prof. Isabelle Tavares of the University of California-Berkeley, for helping me to locate, examine, and photograph various Gasteromycetes in the herbarium there; Barbara Waaland and the biology department of the University of California-Santa Cruz, for providing laboratory facilities; John Anderson, Charles Prentiss, and John Lane of the Santa Cruz City Museum, for their longtime support of my studies; Michael Cabaniss of Santa Cruz, for his fine illustrations; and the many people who helped make this book as comprehensive as it is by graciously providing the photographs which I lacked: Chuck Barrows, Alan Bessette, Nancy Burnett, Bill Everson, Michael Fogden, Ray Gipson, Dan Harper, Richard Homola, Nancy Jarvis, Rick Kerrigan, Joel Leivick, Keith Muscutt, Herb Saylor, Phil Sharp, Bob Short, Bob Tally, Bob Winter, Greg Wright, and Joan Zeller.

Many people have played a vital role in the production of the book. I want to thank Phil Wood and George Young of Ten Speed Press for underwriting the entire project and having faith in it; Neil Cossman of ASAP Typography for three years of extraordinary patience and cooperation while the text was being tyepset and corrected, typeset and corercted, typeset and corrected; Hal Hershey of Hal Hershey Book Design & Production and Brenton Beck of Fifth Street Design, for their advice and expertise on the design and layout; and Ralph and Mildred Buchsbaum of Boxwood Press, whose initial interest in the first edition made the second edition possible.

Also deserving recognition are the many friends and strangers who have neipea me along in my research. Some have provided me with food, drink, housing, and companionship during my travels; others have supplied valuable information or opinions on mushrooms or clues to their whereabouts. Some have even risked their health or that of their loved ones in order to determine the edibility of untested mushrooms. Many have braved torrential downpours, suffered severe cases of poison oak, fought off swarms of mosquitoes and ticks, and trudged miles out of their way to bring me rare and interesting (as well as uncommonly delicious) mushrooms. To Ed Aguilar, Mine and Marc Doolittle, Joe and Nancy Haydock, Mark Hildebrand, Ginny and Rosalie Hunt, Luen Miller, Craig Mitchell, Ciro and Rose Marie Milazzo, Suzanne and Beanie Rainbow, Bob Sellers, Sue Willis, Bob Winter, Reggie, Willie, Duke, Max, Butch, Hank, and all the rest—thank you from the bottom of my basket!

I would also like to thank my parents, Harbans and Shirley Arora, for their encouragement in all phases of this lengthy project; my legion of ex-housemates for putting up with my habit of lapsing into Latin monologues at the dinner table and tolerating a chronic case of "There's too many fungi in the refrigerator!"; all those who contributed to the first edition (without which the second would not be possible); the many readers of the first edition whose letters have brought me such joy and satisfaction (readers of the second edition, take note!); and any and all contributors I have forgotten to mention (my propensity for forgetting the names of important people is equaled only by my propensity for remembering the names of obscure mushrooms).

Finally, I would like to express the deepest gratitude to my wife, Judith Scott Mattoon, who, I am pleased to say, did not type a single word of the manuscript nor provide countless hours of selfless assistance without which this book would not have been possible. This book, in fact, would have been possible without her, but our marriage would not. She has been a lovely and lively companion, steadfast friend, and ruthless editor on those occasions when my literary capacities deserted me. True, she'd rather forage for *Agaricus bisporus* in a grocery store than *Craterellus cornucopioides* in the forest, but I haven't given up on her yet!

David Arora
Santa Cruz
1985

CONTENTS

Fungophobia   1
What Is a Mushroom?   4
Mushrooms and the Environment   6
Names and Classification   8
Collecting Mushrooms   11
Identification and Terminology   14
How to Use the Keys   21
Questions About Mushrooms   23
    Which Mushrooms Are Good to Eat?   23
    What's the Difference Between a Mushroom and a Toadstool? 25
    When and Where Do Mushrooms Grow?   25
    Can People Harm Mushrooms By Picking Them?   26
    Can People Harm Themselves By Picking Mushrooms?   27
    Do Other Animals Eat Mushrooms?   28
    What Is the Nutritional Value of Mushrooms?   30
    What Is the Medicinal Value of Mushrooms?   30
    Can You Grow Wild Mushrooms?   30
    Hey, Maaaaaaaaan, Do Any Psilocybin Mushrooms
      Grow Around Here?   31
LBM's: Little Brown Mushrooms   32
Habitats   34
Seventy Distinctive Mushrooms (A quick-reference check list)   48
Key to the Major Groups of Fleshy Fungi   52

Basidiomycotina (Basidiomycetes)   57
  Hymenomycetes   57
    Agaricales (Agarics or Gilled Mushrooms)   58
      Russulaceae   63
      Hygrophoraceae   103
      Tricholomataceae   129
      Entolomataceae   238
      Pluteaceae   253
      Amanitaceae   262
      Lepiotaceae   293
      Agaricaceae   310
      Coprinaceae   341
      Strophariaceae   367
      Cortinariaceae   396
      Bolbitiaceae   466
      Paxillaceae   476
      Gomphidiaceae   481
      Boletaceae (Boletes)   488

Aphyllophorales 548
   Polyporaceae & Allies (Polypores and Bracket Fungi) 549
   Stereaceae & Allies (Crust and Parchment Fungi) 604
   Hydnaceae (Teeth Fungi) 611
   Clavariaceae (Coral and Club Fungi) 630
   Cantharellaceae (Chanterelles) 658
  Tremellales & Allies (Jelly Fungi) 669
Gasteromycetes 676
  Lycoperdales & Allies (Puffballs and Earthstars) 677
  Tulostomatales (Stalked Puffballs) 715
  Podaxales & Allies (Gastroid Agarics) 724
  Hymenogastrales & Allies (False Truffles) 739
  Phallales (Stinkhorns) 764
  Nidulariales (Bird's Nest Fungi) 778
Ascomycotina (Ascomycetes) 782
  Discomycetes 783
   Pezizales 783
    Morchellaceae (Morels and Allies) 784
    Helvellaceae (False Morels and Elfin Saddles) 796
    Pezizaceae & Allies (Cup Fungi) 817
   Tuberales (Truffles) 841
   Helotiales (Earth Tongues) 865
  Pyrenomycetes (Flask Fungi) 878

Mushroom Cookery 888
Mushroom Toxins 892
What It All Means (A Short Dictionary of Scientific Names) 899
Glossary 913
Bibliography: Suggested Readings and Primary References 919
General Index 926
Genus and Species Index 936

FUNGOPHOBIA

BRING home what looks like a wild onion for dinner, and no one gives it a second thought—despite the fact it might be a death camas you have, especially if you didn't bother to smell it. But bring home a wild mushroom for dinner, and watch the faces of your friends crawl with various combinations of fear, anxiety, loathing, and distrust! Appetites are suddenly and mysteriously misplaced, vague announcements are hurriedly mumbled as to dinner engagements elsewhere, until you're finally left alone to "enjoy" your meal in total silence.

For there are few things that strike as much fear in your average American as the mere mention of wild mushrooms or "toadstools." Like snakes, slugs, worms, and spiders, they're regarded as unearthly and unworthy, despicable and inexplicable—the vermin of the vegetable world. And yet, consider this: out of several thousand different kinds of wild mushrooms in North America, only five or six are deadly poisonous! And once you know what to look for, it's about as difficult to tell a deadly *Amanita* from a savory chanterelle as it is a lima bean from an artichoke.

This irrational fear of fungi is by no means a universal trait. The media and medical profession have done their part to perpetuate it, but they are certainly not responsible for its origin. To a large extent, we inherited our fungophobia from the British. William Delisle Hays, an astute Englishman writing in the 1800's, expressed it this way:

> (All mushrooms) . . . are lumped together in one sweeping condemnation. They are looked upon as vegetable vermin only made to be destroyed. No English eye can see their beauties, their office is unknown, their varieties not regarded. They are hardly allowed a place among nature's lawful children, but are considered something abnormal, worthless, and inexplicable. By precept and example children are taught from earliest infancy to despise, loathe, and avoid all kinds of "toadstools." The individual who desires to engage in the study of them must boldly face a good deal of scorn. He is laughed at for his strange taste among the better classes, and is actually regarded as a sort of idiot among the lower orders. No fad or hobby is esteemed so contemptible as that of "fungus-hunter," or "toadstool-eater."
>
> This popular sentiment, which we may coin the word "fungophobia" to express, is very curious. If it were human—that is, universal—one would be inclined to set it down as an instinct, and to revere it accordingly. But it is not human—it is merely British. It is so deep and intense a prejudice that it amounts to a national superstition . . .
>
> It is a striking instance of the confused popular notions of fungi in England that hardly any species have or ever had colloquial English names. They are all "toadstools," and therefore are thought unworthy of baptism. Can anything more fully demonstrate the existence of that deep-rooted prejudice called here "fungophobia"? . . .

A century later, in America, Alan Snitow echoes similar sentiments in the *San Francisco Bay Guardian*:

> I wander through Bay Area woods and fields looking for mushrooms. When I first started, I thought my new hobby offered final proof positive that I am —well, a weird person. I didn't know anyone else "into" mushrooms. My

friends' puzzled looks seemed to confirm my view that I must suffer from a form of repressed necrophilia, a perverse fascination with things rotted and decayed . . .

(That some mushrooms can be eaten helps) . . . to justify the quest, especially to friends who raise their eyebrows at the mention of fungi. If I am off looking for mushrooms, they think I have a rational purpose to my actions. Otherwise they would say: "He's off walking around in the rain again." Rather than pitying looks at the soggy human who returns, they see a slightly eccentric, but productive, worker, toiling to bring back something for the evening's table.

Mushrooms can be every bit as beautiful as birds, butterflies, shells, and flowers, yet we never think to describe them in such flattering terms. When novelists or poets want to conjure up an emotion of fear, loathing, total revulsion, and imminent decay, they inevitably drag in the mushrooms and toadstools— malignant instruments of death and disease that appear only in the dankest and most abominable of situations. Witness Shelley:

> And agarics and fungi, with mildew and mould
> Started like mist from the wet ground cold
> Pale, fleshy, as if the decaying dead
> With a spirit of growth had been animated . . .

And Sir Arthur Conan Doyle, bygone creator of Sherlock Holmes:

> . . . A sickly autumn shone upon the land. Wet and rotten leaves reeked and festered under the foul haze. The fields were spotted with monstrous fungi of a size and colour never matched before—scarlet and mauve and liver and black—it was as though the sick earth had burst into foul pustules. Mildew and lichen mottled the walls and with that filthy crop, death sprang also from the watersoaked earth.

And D.H. Lawrence, that chronic belittler of the British bourgeoisie:

> How beastly the bourgeois is
> especially the male of the species—

> Nicely groomed, like a mushroom
> standing there so sleek and erect and eyeable—
> and like a fungus, living on the remains of bygone life,
> sucking his life out of the dead leaves of greater life than his own.

> And even so, he's stale, he's been there too long,
> Touch him, and you'll find he's all gone inside
> just like an old mushroom, all wormy inside, and hollow
> under a smooth skin and an upright appearance.

> Full of seething, wormy, hollow feelings
> rather nasty—
> How beastly the bourgeois is!

> Standing in their thousands, these appearances, in damp England
> what a pity they can't all be kicked over
> like sickening toadstools, and left to melt back, swiftly
> into the soil of England.

And that prim American poet Emily Dickinson:

> Had nature any outcast face
> Could she a son condemn
> Had nature an Iscariot
> That mushroom—it is him.

Has any group of organisms been so unjustly maligned? Actually, Dickenson's limp effort should come as no surprise, since she was a virtual recluse. Prejudice is largely a measure of ignorance!

And yet, if you go to continental Europe, you'll find that fungophobia is the exception and not the rule. Most Europeans, especially those who live close to the woods, know which mushrooms to pick and how to cook them. They bestow upon each species an individual name and sell them in the markets. Many Americans, on the other hand, are completely oblivious to the fact that there is more than one fungus among us—those of recent European or Oriental ancestry being notable exceptions.

The farther east you go in Europe, the more passionate is the love for mushrooms. Which brings us to Russia. The Russians go absolutely bananas over fungus. Mushrooming is a commonplace tradition there, not the hallowed turf of the academic or connoisseur. Instead of talking about the weather, strangers often engage in polite conversation about how the mushroom season is progressing. And Russian children are raised on mushroom lore from earliest infancy. Many family names are derived from fungi: Bribov, Borovikov, Gruzdjov, Ryshikov, Opjonkin. Another one is Griboyedev, or "Mr. Mushroomeater." The poet Majokovsky was a mushroom addict. (Poetry, like mushroom hunting, is a great tradition there. A Russian poet draws 5,000 to a poetry reading —here you're lucky to draw 50.) Even Lenin is said to have been possessed by a *razh* or "mushroom passion."

In this country, it is only with the renewed interest in natural foods and the desire to return to the earth (and what's good for you) that mushrooms are being noticed again. Mycological societies are sprouting up in the major cities. And of course, business is capitalizing on the trend. Polka-dotted mushrooms have appeared in startling profusion on curtains and calendars, pottery and stationery, potholders and incense holders, bumper stickers and birthday cakes.

And yet, when it comes down to actually *eating* wild mushrooms, most Americans are still afraid. Instead they opt for something more familiar and not half as good, such as Grape Nuts or Malt Balls. Yet it stands to reason that if mushroom-eating were an inherently dangerous activity, it could not exist to the extent it does in Europe. And the mycological societies in America would be in dire need of new members, their ranks depleted annually by the insidious "Mushroom Menace." Like driving, swimming, walking, or breathing, mushroom-eating *is only made dangerous by those who approach it frivolously.*

If you treat mushrooms with discrimination and respect, you can learn to pick your own edible wild mushrooms without fear of confusing them with poisonous types—mushrooms which are nutritious, far more flavorful than the mass-produced cultivated variety, and best of all, free! It does, however, require time and effort—a willingness to plunge into the woods, to uncover their secrets, to learn their characteristics, to penetrate their haunts. That's what this book is about.

WHAT IS A MUSHROOM?

FUNGI are neither plants nor animals. They don't contain chlorophyll like green plants, and as a result cannot manufacture their own food. In this respect they resemble animals, because they feed themselves by digesting other organic matter. However, they lack the nervous system, specialized organs, and mobility characteristic of most animals. Furthermore, fungi reproduce by means of microscopic reproductive units called **spores.** These are far simpler in structure than seeds or eggs, and in fact, usually consist of only one cell.

The term **mushroom** is most often used to describe the reproductive structure **(fruiting body)** of a fungus. In this sense a mushroom, like a potato or persimmon, is *not* an organism, but a *part* of an organism. However, the term "mushroom" can also mean any fungus which produces a fleshy fruiting body (that is, one that has substance). By this definition, not all fungi qualify as mushrooms. Athlete's foot fungus, bread molds, water molds, yeasts, and mildews are examples of fungi which do *not* form fleshy fruiting bodies. The term "mushroom" can also be applied in a more restricted sense to those fleshy fungi like the cultivated mushroom whose fruiting bodies bear spores on radiating blades called **gills.**

Many fungi are exquisitely constructed, and their life cycles are among the most complex to be found. It is not the purpose of this book to explore their biology, but it *is* necessary to consider briefly how mushrooms grow and reproduce. All of the mushrooms in this book belong to two subdivisions of the true fungi. Most of them produce their spores on the *exterior* of microscopic club-shaped cells called **basidia** (singular: **basidium**), hence they are called **Basidiomycetes.** A smaller number produce their spores *inside* microscopic saclike mother cells called **asci** (singular: **ascus**), hence they are called **Ascomycetes.**

The fruiting bodies of the Basidiomycetes and Ascomycetes vary greatly in detail and design, but their function is always the same—they perpetuate their species by disseminating spores. A typical gilled mushroom (the most common type of fruiting body) is a straightforward structure consisting of a **cap, gills,** and (usually) a **stalk** (see diagram). A protective covering called a **veil** may also be present, and if so, will frequently form a **ring (annulus)** and/or a **volva** when it ruptures. The parts of a gilled mushroom are discussed in more detail on pages 14-18, and fruiting bodies of a radically different structure, such as puffballs, are illustrated and discussed in their respective chapters.

Spore formation. At left is a typical club-shaped basidium, with four small stalks (sterigmata) on which the spores form. At right is an ascus, inside of which spores (usually eight) form.

basidium   ascus

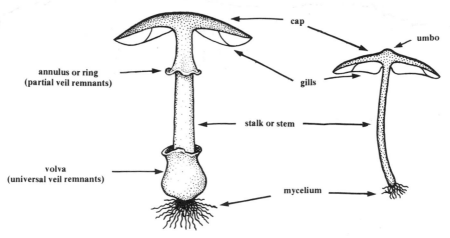

Parts of a gilled mushroom. Mature *Amanita* at left has cap, stalk, gills, annulus, and volva. The partial veil covers the gills when young and breaks to form a ring (annulus) on the stalk, while the universal veil at first envelops the entire fruiting body and breaks to form a volva (sack, collar, or series of concentric rings) at base of stalk. Development of fruiting body is shown on pp. 270-271. At right is a mature *Marasmius,* which has neither annulus nor volva, but often has an umbo (knob) on cap.

In a gilled mushroom, millions of spores are produced on basidia which line the gills. These spores are subsequently discharged and carried by air currents to new localities. Each is theoretically capable of germination, but only a small percentage land in a favorable environment. Spores germinate by sending out a **germ tube** which branches to form many threadlike cells called **hyphae.** When two spores of *different but compatible* strains (or "sexes") germinate in close proximity to each other, their hyphae merge to form hyphae with two nuclei (one from each parent). These hyphae grow rapidly, forming an intricate network of filaments called the **mycelium** or **spawn** (see photo on p. 43). The mycelium is the *vegetative portion* of the fungus. The tips of the mycelial hyphae liberate enzymes which digest food to support growth.

Once the mycelium has established itself and built up an adequate food reserve, it becomes capable of producing mushrooms. Under favorable conditions (for most species this means damp but not soggy, cool but not cold), some hyphae bundle together to form knots of tissue which gradually develop into fruiting bodies. When these fruiting bodies are differentiated but not fully developed (that is, after they have a cap and stalk but before the cap expands), they are called **buttons.** The stalk then elongates, pushing the cap above the surface of the ground (or other substrate). Finally, the cap opens and the veil (if present) breaks, exposing the gills on which spores form. The mature fruiting body is essentially a bundle of threadlike hyphae (each with two nuclei), but the filaments in the bundle terminate in either spore-producing cells (basidia), specialized sterile cells (**cystidia**), or unspecialized cells (**basidioles**, or in some cases, **paraphyses**). (See illustration on p. 6.)

Though they lack the sexual organs of plants and animals, mushrooms reproduce sexually, i.e., genes are recombined so that *offspring are not genetically identical to parents*. Gene exchange takes place in the basidium (or in Ascomycetes, a special cell called the **ascogonium**). The two parent nuclei fuse, doubling the chromosome number, then divide twice while replicating their chromosomes only once—thereby reducing the chromosome number to half

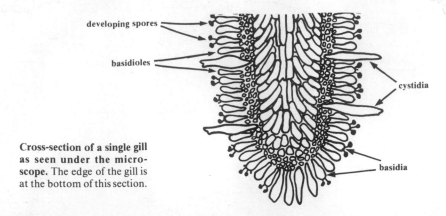

developing spores

basidioles

cystidia

Cross-section of a single gill as seen under the micro-scope. The edge of the gill is at the bottom of this section.

basidia

that after fusion. The four remaining nuclei migrate to the tip of the basidium, and walls form behind them to produce four spores—two of each strain ("sex"), each with one nucleus. With their subsequent discharge the life cycle is completed.

The above life cycle is typical of most Basidiomycetes, except that it is often complicated by the presence of more than two strains (just as our life cycle would be unimaginably complicated by the existence of more than two sexes!). Also, some mushrooms are capable of forming spores asexually.

MUSHROOMS AND THE ENVIRONMENT

IT IS the "role" of fungi to break things down, to give things back. One of the more obvious laws of nature is that existing life must die if new life is to flourish. Stale air must go out the window if fresh air is to come in. If there were no vehicle for the disposal of dead matter, there would soon be no need for one—we would all be buried under a blanket of inert matter. Fungi, along with bacteria, are precisely that vehicle. They are nature's recyclers, the soil's replenishers. Plants deplete the soil by extracting minerals to manufacture their food. Animals, in turn, devour plants. In feeding on dead (or occasionally living) matter, fungi and bacteria reduce complex organic compounds to simpler building blocks, thereby enabling plants to re-use them. Thus, in a very profound sense, fungi are life-givers as well as destroyers. To associate them only with death and decay—as so many people do—is to do them, as well as our own ability to perceive, an injustice.

Fungi can be divided into three categories based on their relationship to their **substrate** (immediate environment). **Parasitic** fungi feed on living organisms. Most serious fungus pests (such as wheat rust) fall in this category, but relatively few mushrooms are parasitic. Their ranks include *Cordyceps* species (on insects, insect pupae, insect larvae, and other fungi); *Asterophora* and *Hypomyces* species (on other mushrooms); various polypores (on trees); and *Sparassis crispa* (on tree roots). Some, like the common honey mushroom, *Armillariella mellea,* are parasitic under certain conditions and saprophytic (see below) under others.

Saprophytic fungi subsist on dead or decaying matter (wood, humus, soil, grass, dung, and other debris). When there is an even distribution of nutrients in the environment, the mycelium of a terrestrial fungus may grow outward at the

6

A large fairy ring of *Marasmius oreades*.

same rate in all directions, periodically producing circles of mushrooms on its outer fringes. These circles or arcs are called **fairy rings**, presumably because people once thought fairies danced in them. Many mushrooms are capable of forming fairy rings, including the aptly named fairy ring mushroom *(Marasmius oreades)*, which grows on lawns. Each year the fairy ring gets larger as the mycelium grows outward, until something finally impedes its progress (usually a lack of food), and the mycelium dies or breaks up into arcs. By measuring the annual growth rate, it has been estimated that some fairy rings in the Midwest prairies are six hundred years old!

Mycorrhizal fungi comprise the third category. They form a symbiotic or mutually beneficial relationship with the rootlets of plants (mostly trees) called **mycorrhiza** (from *myco,* fungus, and *rhiza,* root). The mycelium forms a sheath of hyphae around the rootlets of the host and an exchange of nutrients takes place. The rootlets provide the fungus with moisture and organic compounds (such as carbohydrates), while the fungus aids the roots in the absorption of phosphorus, inorganic nitrogen, and other minerals, and apparently also provides added resistance to certain diseases. As a rule, mycorrhizal fungi cannot grow without their hosts, and studies have shown that trees deprived of their mycorrhizal partners do not compete successfully with those that have their normal complement. This is especially true in poor or exposed soils, where trees need all the help they can get, and mycorrhizal fungi have proved invaluable in reforestation projects.

Many mycorrhiza-formers are host-specific, i.e., they grow only with one kind of tree. For instance, *Suillus pungens* grows principally with Monterey pine, while *Amanita rubescens* is monogamous with live oak (in our area). A tree, however, may have several mycorrhizal associates whose relationships with the tree are qualitatively different (Alexander Smith reports finding over 50 species of mycorrhizal mushroms under an isolated Douglas-fir). In other words, each type of mushroom occupies a different **ecological niche.** Many factors are encompassed by the concept of niche, some of which we don't understand. A niche is not so much an organism's habitat as its "profession"—what it does for a living, or the "role" it plays in its biological community. Each kind of mushroom reproduces and germinates successfully within a certain humidity and temperature spectrum. Each extracts particular nutrients from its environment. Thus two species may occur in the same habitat, but occupy different niches. One may be taking the lignin from a log, another, the cellulose. Or one may be feeding on the heartwood, another on the sapwood. Or returning to the mycorrhizal fungi, one may be supplying phosphorus to the tree, and another, nitrogen. **Succession** also occurs—as one type of mushroom exhausts its nutrient supply, another takes its place. A living tree may harbor certain types of fungal growth. As soon as it dies, new species will appear. Eventually the wood is reduced to fragments or powder, at which point still other mushrooms take over, with growth habits better suited to the changed conditions.

NAMES AND CLASSIFICATION

NAMES, like automobiles, are largely vehicles of convenience. You can't claim to have a profound knowledge of human beings without knowing at least *some* of them on an individual basis. Recognition is a prerequisite to getting to know someone, and a name is helpful in associating that person with a unique set of identifying characteristics, whether it be his big nose and hairy face, or her long legs and swift smile. Rather than saying "the 6 ft. 7 in. acrobatic forward of the Philadelphia 76ers," we say "Julius Erving" or "Doctor J." Instead of "you know, that bright red mushroom with white spots that grows under pine," we say *Amanita muscaria,* or "fly agaric."

Names can also be descriptive. For instance, the red-headed woodpecker has a red head and pecks wood. What's more, names can reflect common bonds. Your last name identifies you as a member of a group with similar genes, and provides a clue as to your origins. Your first name defines you as an individual entity within that group.

Unfortunately, relatively few mushrooms have colloquial English names—a tribute, as pointed out previously, to our fungophobic roots. In this book I have used popular names where they exist, and in some cases have capriciously coined common names, but to do so in every case would only create confusion, as there is no assurance they would be accepted. Therefore, if you really want to get to know mushrooms, it is necessary to know their scientific names. People usually groan when they hear this, and to be sure, the long Latin names are intimidating. But so, at first, is a can opener—it's just a question of familiarity. In fact, you may already have mastered some Latin (scientific) names without realizing it— e.g., *Eucalyptus, Rhododendron, Hippopotamus.* Memorization is made easier by learning the meanings of the names. For instance, *Lactarius rubrilacteus* (*rubri*=red, *lacteus*=milk) exudes a red "milk" when cut. See "What It All Means" on p. 899 for more details. Don't get bogged down in pronounciation. It doesn't really matter *how* you say something as long as you *communicate* it. Even taxonomists don't agree on how some names should be pronounced!

As you begin to use scientific nomenclature, you'll discover its many advantages. Common names do *not* necessarily reflect natural affinities. Hedge nettle is not a nettle, and poison oak is by no means an oak. Likewise, the names meadow mushroom, honey mushroom, matsutake, and horse mushroom provide no clues as to which, if any, have common bonds or similar characteristics. Also, common names are *not* universal. For instance, *Boletus edulis* has dozens of regional names, and memorizing all of them would be almost as difficult as getting everyone to agree on one of them!

In contrast, scientific nomenclature transcends cultural and regional barriers. It is used by naturalists and biologists throughout the world, and it is designed to reflect natural relationships. It employs a **binomial system** in which each kind of organism has two names. The second name, the **species,** is the *kind* of organism; the first name, the **genus** (plural: **genera**) is a collection of species with very similar traits. The species epithet is meaningless without the genus name (or its abbreviation) attached (we never sign a check with only our first name!). The names are mostly Latin because that language was universally fashionable in learned circles when the binomial system was devised. The beauty of the binomial system is that it indicates commonality while simultaneously expressing singularity.* *Amanita calyptrata* (the coccora) and *Amanita phalloides* (the death cap) are different species belonging to the same genus, just as *Canis lupis* (wolf), *Canis latrans* (coyote), and *Canis familiaris* (dog) belong to the same genus. Yet the common names of these organisms don't indicate that they are closely related!

Some kinds of similarities are clearly more fundamental than others. Therefore, it was deemed necessary to erect a hierarchy of classification to indicate the *degree of commonality.* Genus and species are only two levels in that hierarchy. Just as species with common characteristics are grouped in a genus, several genera with common features are grouped in a **family** (e.g., the red fox, *Vulpes fulva,* belongs to the same family as the dog, but *not* to the same genus). Families are in turn grouped in an **order,** orders in a **class,** classes in a **subdivision** or **division** (or in the case of animals, in a **phylum**), and divisions in a **kingdom.** At the other end of the scale, slightly different populations of a single species are designated as **subspecies, varieties,** or **forms.**** Here is the complete classification scheme for three mushrooms—the blewit, shaggy mane, and inky cap.

Category	Blewit	Shaggy Mane	Inky Cap
Kingdom	Fungi	Fungi	Fungi
Division	Eumycota	Eumycota	Eumycota
Subdivision	Basidiomycotina	Basidiomycotina	Basidiomycotina
Class	Hymenomycetes	Hymenomycetes	Hymenomycetes
Order	Agaricales	Agaricales	Agaricales
Family	Tricholomataceae	Coprinaceae	Coprinaceae
Genus	*Clitocybe*	*Coprinus*	*Coprinus*
Species	*nuda*	*comatus*	*atramentarius*

All three of these organisms are fungi—they possess neither photosynthetic compounds nor seeds, and their vegetative phase is comprised of threadlike cells (hyphae). Since none are ameba-like, they are not slime molds and hence belong to the division Eumycota. All three produce their spores on microscopic cells called basidia, hence they belong to the subdivision Basidiomycotina. All forcibly discharge their spores from basidia which form on radiating blades (gills), therefore they belong to the class Hymenomycetes and the order Agaricales. However, the blewit has pinkish spores while the shaggy mane and inky cap have black spores—which is one reason they are placed in different families. Not only do the shaggy mane and inky cap belong to the same family (the Coprinaceae), but they also belong to the same genus, *Coprinus,* because

---

*The singularity of a *population* of organisms; our personal names, on the other hand, indicate the singularity of *individual* organisms.

**In this book the term "variety" is used in its ordinary English sense to mean a "type" of organism. Only when *one* species is being discussed and "variety" is abbreviated (var.) is it applied in its scientific sense. The same goes for "form" (f.).

both have gills that deliquesce (liquefy) at maturity. However, the shaggy mane has a tall, shaggy, cylindrical cap while the inky cap has a smoother, broader, oval or conical cap; consequently they are recognized as distinct species.

Note the suffixes at the levels above genus: —*ceae* denotes a family; —*ales* indicates an order; and —*cetes* denotes a class or larger category.

It must be remembered, however, that this elaborate classification scheme is contrived. *It is our attempt at boxing and categorizing nature.* There are common gene pools and definite lines of evolution, but no such clearcut categories exist. Thus, the definition and interpretation of species, genera, families, etc., *is largely a matter of opinion.* Disputes invariably arise, many of which have not been resolved. For instance, at the genus level and above, there is the problem of deciding which similarities among fungi are fundamental (indicators of common origin), and which are coincidental or superficial, or the result of convergent evolution.* The microscope has been a tremendous help in uncovering "hidden" similarities, but it has also exacerbated the confusion by introducing a vast new set of criteria on which to pass judgment. The result is a nomenclatural nightmare, from the upper echelons of the hierarchy right on down to the species level.

Mycological literature is as riddled with contradictions as a *Suilllus pungens* is with maggots. Anyone who has used more than one mushroom book can testify to the frustration of finding different names applied to the same fungus (synonyms), or one name applied to several different fungi (homonyms). For instance, *Clitocybe nuda* (the blewit) is better known as *Lepista nuda,* and was formerly known as *Tricholoma nudum.* It has been incorrectly called *Tricholoma personatum,* and in Europe is also known as *Rhodopaxillus nudus!* For an even more confusing example, see the list of synonyms for *Trametes occidentalis* on p. 550.

This confusion is partly the result of disagreement as to what exacly constitutes a "genus" or "species"—a difference in philosophy that is known in taxonomic circles as the battle between the "lumpers" and "splitters." The "lumpers" are conservative in their approach. They interpret genera or species broadly, allowing for a good deal of variation—in other words, they tend to stress similarities between mushrooms rather than differences. "Splitters," on the other hand, are forever describing new genera and species based on the most minute— but not necessarily insignificant—differences. Both approaches have their advantages and drawbacks, and both are self-defeating when carried to an extreme.**

The important thing to realize is that the system of classification used in this book is *by no means definitive.* It represents an amalgamation of various investigators' views of the fleshy fungi, plus the overriding consideration of *usefulness* to the amateur (since this book is designed for amateurs). Some of the names used will undoubtedly be invalidated in the near future. A few have not been validly published and are therefore placed in quotation marks. Synonyms have been provided and homonyms elucidated. But the inherent advantages of the binomial system of nomenclature will not be fully realized until stabilization is achieved and one name is agreed upon for each kind of mushroom. In exceptional cases like that of the blewit, it is perhaps easiest in the meantime to use the common English name—if there is one.

*The fins and torpedo-shaped bodies of sharks and killer whales are independent adaptations to a similar environment, *not* indicators of common origin. This phenomenon is called convergent evolution.

**One radical "lumper" I know recognizes only two kinds of mushrooms—the "pickers" and the "kickers" (those that deserve to be picked, and those that deserve to be kicked!).

COLLECTING MUSHROOMS

MUSHROOM hunting is not simply a matter of traipsing through the woods after it rains. It is an art, a skill, a meditation, and a process. If you proceed at a careful, deliberate rate, you'll enjoy much more success than if you rush around frantically picking whatever mushrooms you see, then stuff them in your basket, bring the whole mess home and dump it on your table. Mushrooms collected in this manner are likely to wind up in the garbage, unidentified and unappreciated.

Don't just collect, but *observe* the mushrooms—and their surroundings. In the process you'll discover many other clandestine wonders you were previously unaware of (see photographs on pp. 28-29). *Be selective*—pick only distinctive species in good condition. You enhance your chances of successful identification immeasurably by collecting *several specimens of each kind of mushroom.* This is absolutely essential, because you have no other means of assessing variation within a given species. Since mushrooms decay rapidly and identification can be a time-consuming process, don't pick every kind you see. It's better to fill your basket with many good examples of a few distinctive species (in which case your chances of identification are good) than with one or two specimens of many species (in which case your chances of identification are very poor).

Don't assume, however, that two mushrooms are the same species simply because they're growing together. Judge each on its own merits. If you're uncertain, *assume for safety's sake* that they're different, and treat them as such. As you become more adept at observation, your ability to identify fleshy fungi will "mushroom"—but only after a solid foundation or "mycelium" of experience is laid.

It is far better to learn a few species well than a large number superficially. The novelty of the "Easter egg" approach wears off quickly as its futility becomes apparent. If possible, choose a specific quarry for each hunt. Suppose it's January and you're going for a walk in a local live oak woodland. By using the chart on pp. 48-51 and consulting the description of the blewit *(Clitocybe nuda),* you find that the blewit is *often* abundant under oak in January. The next step is to familiarize yourself with its fieldmarks: bluish-purple color, stocky stature, absence of a veil, citrus odor, etc. Your mushroom hunt is thus transformed into a *blewit* hunt. Of course, nothing stops you from gathering other interesting fungi you encounter, but focusing your attention on the blewit insures that you'll learn something about blewits *even if you don't find any* (namely, that the locality and/or weather conditions were not conducive to its fruiting).

The desirability of this approach is underscored by the excitement it lends to the hunt. It is much more gratifying to find something you are specifically *trying* to find. And you'll be more likely to remember what it looks like and where it grows, so you can return to harvest more. By focusing on several species as the season unfolds, you will develop a quicker and keener appreciation of what grows where and what environmental factors they respond to. Many of the more distinctive species occur throughout the world, so the knowledge you accumulate will serve you elsewhere!

Always dig up unknown mushrooms so as not to miss the volva, if present. There's no room for carelessness, as shown here: *Agaricus arvensis,* at left, lacks a volva and is edible; *Amanita ocreata,* at right, is furnished with a volva and is deadly poisonous!

EQUIPMENT

Foraging for fleshy fungi requires little in the way of sophisticated paraphernalia. The bare essentials are:

A **rigid container** for carrying the mushrooms. There's no point in picking mushrooms unless you transport them home in decent condition. A broad, shallow basket is best, but a cardboard box or bucket will do unless it's raining. Paper bags sag and the mushrooms get crushed. DON'T USE PLASTIC BAGS! Mushrooms, like people, have to "breathe." Plastic bags trap moisture, making the mushrooms "sweat" and rot more rapidly.

Waxed paper is necessary when collecting mushrooms for identification. It provides support for mushrooms within the basket and also keeps them separated. *Never mix unknown species together.* Wrap each type separately and arrange them carefully in the basket with the heavier ones on the bottom. Tall specimens like *Agaricus* and *Amanita* species will bend unless placed upright (mushrooms exhibit negative geotropism: their caps turn away from gravity so as to orient the gills downward). Small paper bags are useful when harvesting familiar edible species, but again, don't use plastic bags!

A **knife** or **trowel** is a must for digging up mushrooms or detaching them from trees. A knife is also handy for cleaning edible species you are already familiar with, but *always* dig up unknown mushrooms so as to ascertain whether or not a volva (sack) or "tap root" is present. The telltale volva of the deadly Amanitas is usually buried in the ground!

A **pencil** and **small notebook** or **index cards** are useful for taking field notes and spore prints.

Bread, cheese, and **fruit** are essential if you're always hungry like I am. I make a practice of stocking my basket generously. Then each time I put a mushroom in my basket, I'm compelled to put something from the basket in my mouth! If you find some edible *Agaricus* buttons, put them in a sandwich! Not known for wasting opportunities, the French carry this tradition one step further—they bring wine, goblets, and a table cloth, and pause for a picnic every half hour. The advantage of this strategy is obvious—you needn't find any mushrooms to have a good time!

12

Binoculars are handy in open country (e.g., pastures). They enable you to distinguish at a distance giant puffballs *(Calvatia* species) and horse mushrooms *(Agaricus arvensis* and *A. osecanus)* from rocks and other assorted "pseudocarps." And of course, they allow you to watch birds and mammals as well.

A **three- or four-pronged rake or cultivator** and **a small hand cultivator** are necessary if you want to find truffles and false truffles, unless you have the services of a **truffle hound** or **muzzled pig**. The rakes will enable you to locate these elusive underground fungi by sifting through the forest duff and scraping the topsoil beneath it. (But remember: this practice can be unsightly as well as destructive, so don't do it on a widespread basis, be discreet, and *always* cover up the holes you dig.) See the chapter on truffles for more details.

Other optional equipment includes: a hand lens, compass, stick (for probing brambles and "mushrumps"), a field guide (for leisurely use on a sunny day), small jars or vials (for delicate specimens, such as Mycenas), a damp cloth or brush for cleaning edible species, rainboots and other rain gear, gloves for frigid winter mornings, and photographic equipment (usually too cumbersome except for special picture-taking expeditions).

FIELD NOTES

Is it growing on the ground or on a log? Is it near a tree? What kind(s)? Are familiar types of mushrooms growing nearby? Which ones? If it's growing on wood, is the wood coniferous or hardwood? Living, recently felled, or in an advanced stage of decay? If on the ground, is the humus layer deep? Is the ground disturbed? Is there a road, trail, parking lot, or laundromat nearby?

You should automatically ask yourself these questions every time you find a mushroom. *Observation begins in the field.* After all, in picking a mushroom you are leaving behind the vegetative portion of the fungus. It is folly to depart without some idea of the niche that fungus occupies in the larger scheme of things. Field (or mental) notes should include:

> **Date, weather conditions, abundance** (how many times you observed a particular species), **growth habit** (solitary, scattered, gregarious, clustered, in fairy rings, etc.), **substrate** (humus, soil, grass, moss, dung, wood, etc.), **vegetation** (the kinds of trees and shrubs within 50 feet).

> If growing on wood: **stage of decomposition, type of wood** (hardwood or conifer), **type of tree** (if discernible), **effects on the wood** (see chapter on polypores on p. 549).

> If growing on dung: **type of dung, stage of decomposition**

> If growing on ground: **type of ground** (disturbed, cultivated, hard-packed, sandy, charred, etc.)

Don't restrict your observations to specimens that you collect. After you've gathered a representative sampling of a particular species, continue to note its habitat each time you encounter it.

With terrestrial fungi, it is important to note all types of trees within 50 feet because mycorrhizal species grow in association with rootlets that may be quite a distance from the trunk of the host. Usually there are several kinds of trees in the vicinity, and you have no way of knowing which (if any) is the mycorrhizal associate. However, through repeated observation many possibilities can be eliminated. For instance, if you find *Suillus pseudobrevipes* growing under pine, you *cannot* conclude that there is a relationship between the two. But if you find that *Suillus pseudobrevipes* always grows with pine *and nowhere else,* then an intimate relationship of some kind can be inferred.

IDENTIFICATION AND TERMINOLOGY

IF YOU take the time to seek out mushrooms, it only makes sense to exercise the extra care and trouble necessary to get them home in beautiful condition. Handle them gingerly, don't leave them in stuffy places like cars, and don't shift them unnecessarily from box to box. *Always* conduct your studies on fresh material, preferably the day you pick them. *Coprinus* species digest themselves in a few hours, and many types quickly lose their original color or are devoured overnight by maggots. If you're pressed for time, at least sort them out and separate the worm-riddled specimens. Then refrigerate the ones you wish to save or spread them out in a cool, dry place where they can "breathe."

Now let's assume you've taken field notes, brought several species home, and are ready to study them. For the diligent and disciplined toadstool taxonomist, a detailed written description of each species is a must. For practically everyone else, compiling a written description is a tedious affair which tends to detract from the enjoyment and spontaneity of the hunt. However, it is an ideal tool for learning *how* to look at mushrooms critically, so try it at least a few times.

The basic terminology for identifying and describing gilled mushrooms is outlined here. Fruiting bodies with a radically different structure, such as puffballs, are illustrated and discussed in their respective chapters. Unfamiliar terms not illustrated or defined here can be looked up in the glossary. Remember to base your observations on *as many specimens of each species as possible*. The value of written descriptions is enhanced when accompanied by sketches, photographs, and spore prints of fresh material.

MACROSCOPIC CHARACTERISTICS

SIZE

Size is important for purposes of comparison. The terms **large, medium, small,** and **minute** cannot be given absolute measurements, but they communicate a characteristic *size range* that you will quickly learn to appreciate. The size of a fruiting body is dependent on three major factors: age, amount of moisture available, and genes. Since mushrooms grow very quickly, those that fruit during rainy weather are apt to be larger than those that fruit during a dry spell—subject, of course, to genetic constraints.

The measurements given in descriptions and keys represent *average* size ranges; those in parentheses indicate unusual dimensions, but do *not* take into account extremes due to extraordinary conditions. The metric system is used, but a conversion rule is provided on the back cover of this book. Also, it's easy to remember that 1 inch equals approximately 2½ centimeters, or 2 inches equals 5 centimeters (or 50 millimeters).

COLOR

Color is one of the most noticeable features of any fungus, but is also one of the most deceptive and variable. Unfortunately, beginners tend to attach undue importance to color at the expense of more critical characteristics. This inevitably leads to misidentification because many mushroom pigments are highly sensitive to environmental influence. Direct sunlight or prolonged rain can bleach a mushroom drastically, and I will never forget my experience following a prolonged rainy spell in which practically every mushroom growing under redwood had a reddish cap! (Apparently pigments from the redwood had dissolved in the drip-off and been absorbed by the mushrooms.) None of these mushrooms would have keyed out properly unless this phenomenon were taken into account! Color may also depend on age. An immature sea gull is brown, while an adult is black and white. Likewise, an immature *Hygrocybe conica* is red, orange, or yellow, but blackens as it ages.

Therefore, color should always be used *in conjunction with other characteristics.* The most striking feature of *Amanita muscaria* is its bright red cap, but a slew of other mushrooms are also bright red. Furthermore, *Amanita muscaria* has a yellow-capped and white-capped form, and even the red-capped variety will fade to orange or even whitish. It is the red cap *combined* with the presence of warts and a volva that render *A. muscaria* distinct.

The color of the stalk is not as variable because it is sheltered by the cap, but the gill color often changes as the spores mature. In fact, the disparity in gill color may be so great that young and old specimens can be mistaken for different species unless intervening stages are found.

Describing color is another problem. Color pictures are a great help, of course, but even they can't show the degree of variation within a given species. Standardized color charts are available, but they represent an extra expense and pose problems of their own. In this book, colors are described using familiar terms where possible, plus a few specific ones (e.g., vinaceous) which are defined in the glossary.

COLOR CHANGES

The tissue of many mushrooms oxidizes when exposed to the air—that is, it undergoes one or more color changes when bruised. This may occur instantaneously, like the blueing of the tubes and flesh in *Boletus erythropus*, or slowly (over several hours), like the reddening of the flesh in *Amanita rubescens*. Bruising reactions should be looked for on the surface of the cap, gills, and stalk, and on the cut flesh within the cap and stalk.

TEXTURE

The texture of the fruiting body is often significant—i.e., soft, watery, spongy, brittle, tough, leathery, corky, woody, etc. The dried fruiting bodies of some types (notably *Marasmius*) revive completely when moistened.

ODOR AND TASTE

It comes as a pleasant surprise to many people that mushrooms have distinctive tastes and odors. For instance, *Lactarius torminosus* is excruciatingly hot (peppery), while *Agaricus subrufescens* is sweet. *Marasmius copelandi* reeks of garlic and is often smelled before it is seen, *Clitocybe odora* smells like anise, *Russula fragrantissima* like maraschino cherries, and *Phyllotopsis nidulans* like sewer gas, while *Armillaria ponderosa* has a spicy odor—a provocative compromise between red hots and dirty socks.

The problem with odor and taste is in the terminology. They are essentially chemical tests, but each person's "laboratory" is different. What smells like sauerkraut to one person smells like sweet-and-sour sauce to another. Moreover, mushrooms with a "mild" odor may be very pungent when crushed, or may only develop an odor at maturity, and odoriferous species may be odorless if waterlogged. Consequently, odor and taste are not noted in this book except where useful in identification.

Incidentally, almost any mushroom can be safely tasted by chewing on a small piece of the cap *and then spitting it out.* However, it is *not* a good policy to sample unkown mushrooms, particularly Amanitas. Do so only when it is called for in the keys or descriptions.

CAP

The **cap** (or **pileus**) is the structure that supports the spore-producing surface (gills, pores, etc.). The skin of the cap is called the **cuticle** (or **pellicle,** if it peels easily). A **viscid** cap is sticky when moist, often **glutinous** (slimy) when wet, and sometimes assumes a glossy appearance as it dries. Debris stuck to the cap surface is a telltale sign that it was viscid when moist. Also, if you press your lip or wet finger against the surface, it will cling or feel slightly sticky. A **hygrophanous** cap appears watery (or even translucent) when moist and opaque when dry, and fades dramatically as it loses moisture (*Psathyrella candolleana*, a common lawn lover, is a good example). A **dry** cap is neither viscid nor hygrophanous. It often has a dull, unpolished appearance, but will naturally be moist or soggy in wet weather.

Other features to note are: **size, shape** (p. 17), **color and color changes, surface characteristics** (whether smooth, scaly, granulose, fibrillose, warty, etc.), and **margin** (inrolled, incurved, straight, or uplifted; and striate, translucent-striate, or not striate).

FLESH

Note **color, color changes, texture, thickness, odor when crushed,** and **taste** (don't swallow!).

GILLS

Gills (or **lamellae**) are the thin, radiating blades found on the underside of most mushrooms, including the commercially cultivated variety. Features to note include: **mode of attachment to the stalk** (p. 17), **spacing, thickness, depth, forking pattern** (if any), and **color** (in both immature and mature specimens). Shorter gills **(lamellulae)** are often interspersed with longer ones, but they don't usually have taxonomic significance.

STALK

The **stalk** is the stemlike structure on which the cap is mounted. Its function is to thrust the cap above the ground so the spores can be discharged into the air. Many wood-inhabiting forms lack a stalk because the wood serves the same purpose. The technical term for stalk is **stipe,** but I see no reason to clutter our vocabulary with yet another monosyllable starting with *st-*. Hereafter, the terms **stalk** and **stem** are used interchangeably.

Features to note include: **size, color and color changes, shape** (p. 17), **position** (p. 17), **texture** (fleshy or cartilaginous, hollow or solid or stuffed with a pith), **surface characteristics** (fibrillose, scaly, smooth, etc.), **viscidity,** and **presence or absence of an annulus and volva** (see below). Stalk width should be measured at the top unless otherwise indicated.

VEIL

A **veil** is a layer of specialized tissue that initially protects the developing mushroom and then breaks up or collapses so the spores can be released. Some mushrooms have more than one veil, others have only one veil, and many lack a veil altogether. A **persistent** veil leaves visible remnants on the stalk and/or the cap after breaking; an **evanescent** veil disappears and consequently can only be detected in the button stage. A **membranous** veil is skinlike or kleenexlike and usually persistent; a **fibrillose** veil is hairy (composed of fine fibers) and either disappears or forms a belt of collapsed hairs on the stalk; a **cortina** is a cobwebby fibrillose veil; a **glutinous** (slimy) veil is either evanescent or deposits a layer of slime on the stalk and/or cap.

There are two basic types of veils: a **partial veil** (or **inner veil**) extends from the margin of the cap to the stalk; it covers the gills (or pores) when young and frequently forms an **annulus** (collar or **ring**) on the stalk. A **universal veil** (or **outer veil**), on the other hand, surrounds most or all of the button mushroom. Sometimes it forms a **volva** (see below) after breaking; sometimes it adheres to the underside of the partial veil (as in *Agaricus arvensis*); in other instances it forms a "stocking" of scales on the stalk (as in *Lepiota clypeolaria*), and in still others it completely disappears. Needless to say, not all veils fit conveniently into one or the other of these two categories, and in this book the all-encompassing term **veil** is used for both universal and partial veils *except when the difference between the two is clearcut and of critical importance* (e.g., in *Amanita*).

ANNULUS

If an **annulus (ring)** is formed by the veil, note the color, texture (whether membranous or fibrillose), **shape** (collarlike, skirtlike, or sheathlike), and **position** on the stalk (superior or apical, median, inferior or basal).

SHAPE OF THE CAP

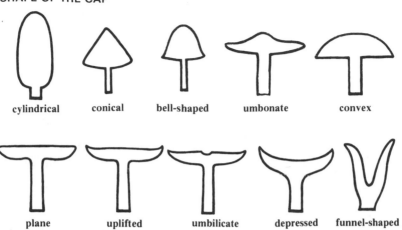

cylindrical   conical   bell-shaped   umbonate   convex

plane   uplifted   umbilicate   depressed   funnel-shaped

ATTACHMENT OF THE GILLS TO THE STALK (as seen in longitudinal section)

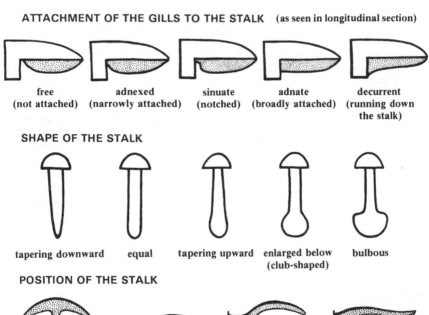

free   adnexed   sinuate   adnate   decurrent
(not attached) (narrowly attached) (notched) (broadly attached) (running down the stalk)

SHAPE OF THE STALK

tapering downward   equal   tapering upward   enlarged below (club-shaped)   bulbous

POSITION OF THE STALK

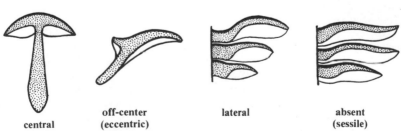

central   off-center (eccentric)   lateral   absent (sessile)

VOLVA

In some mushrooms the universal veil is developed to such an extent that, upon breaking, it leaves visible remains at the base of the stalk in the form of a sack, free collar, or series of concentric scales or rings. These remains are called the **volva**. The different types of volvas are illustrated under *Amanita* on p. 264. *Volvariella,* the stinkhorns, and certain stalked puffballs also have a volva. If small pieces of volval (universal veil) tissue adhere to the cap, they are called **warts;** a large piece of tissue is called a **volval patch.**

MYCELIUM

The mycelium is a loosely organized mass of threadlike cells (hyphae) which are invisible to the naked eye except when they bundle together to form thicker strands (called **rhizomorphs**). Close scrutiny of the forest humus will usually reveal the presence of numerous mycelial strands, but they are virtually indistinguishable from each other without fruiting bodies present. There are a few exceptions, like *Chlorosplenium aeruginascens,* an Ascomycete which stains its substrate (a log or piece of wood) bluish-green. As a rule, however, it is the fruiting body alone that provides clues to the identity of a fungus. This is one reason why the classification of mushrooms has lagged behind that of plants—we're handicapped at the outset because our studies are restricted to only one aspect of the organism, its "fruit."

SPORE COLOR AND SPORE PRINTS

Spore color is an extremely useful character, particularly in the gilled mushrooms. Unlike the color of the fruiting body, it is relatively constant for each species and *not* as susceptible to environmental influence. Though individual spores can't be seen with the naked eye, their color in mass can be ascertained by taking a **spore print**. Just cut off the cap of any mature mushroom and lay it gills down on a piece of *white* paper (covering it with a glass or bowl helps protect it from air currents). In 2-6 hours you'll *usually* have a spore print. Fruiting bodies which are too young or too old, or too soggy or too dry, will not give spore prints, and sterile specimens are sometimes encountered. Spore prints can also be obtained from boletes, teeth fungi, chanterelles, coral fungi, and sometimes polypores.

Always take a spore print to determine the spore color of an unknown mushroom, especially if you want to eat it. The color of the mature gills *may* indicate the spore color, but not often enough to be completely trusted. If you are frustrated by the delay, carry white index cards with you when you forage. Then each time you collect several individuals of the same species, you can decapitate one, place it on a card, wrap it in waxed paper, and put it gills

Black spore prints of *Panaeolus campanulatus.*

down in the basket in such a way that it won't be crushed. By the time you get home you'll have a spore print!

There are several short cuts for determining spore color, but I recommend them only for experienced collectors because they are not completely reliable. When mushrooms grow in a cluster, the lowermost caps will often be covered with spore dust from the upper ones (a wet finger will remove it). Spores borne aloft by air currents will often coat the cap of any mushroom, and falling spores are often trapped by the veil remnants (if present) or by the stalk, especially if it is sticky.

Spore color is traditionally broken down into several broad categories, and assigning your spore print to the correct category can be tricky at first. For instance, don't expect "pink" spores to be bright pink or "purple-brown" spores to be purple. "Pinkish" spores are really closer to flesh-color, while "purple-brown" describes spores which could just as well be called "oil-slick aubergine" (deep brown in mass and dark reddish-brown under the microscope). Only through practice and comparison will you learn to assess the color of a spore print correctly. The thickness of the spore print also affects the color (the heavier the deposit, the darker it will be), as does the moisture content.

Always take a spore print on white paper (a colored background distorts color perception). White spores will show up on white paper when viewed at an angle. Or you may wish to position the cap so that half of it is on white paper and half on black. Special cards can be designed for this purpose.

MICROSCOPIC CHARACTERISTICS

The microscope has had a great impact on the taxonomy of fungi, especially in the last fifty years. However, the vast majority of people do not have access to a microscope, so microscopic features are not stressed in this book. For those who *do* have a microscope, spore characteristics (shape, size, and ornamentation) have been listed for each species, as spores are the most easily seen microscopic cells. Spore size is measured in **microns** (or **micrometers**). A micron is one thousandth of a millimeter. For each family or order of mushrooms, some typical spores have been illustrated—not for the purpose of identification so much as to give those without a microscope an idea of what they look like.

Other microscopic criteria used in the taxonomy of gilled mushrooms may be even more

ARRANGEMENT OF THE GILL HYPHAE (as seen in cross-section)

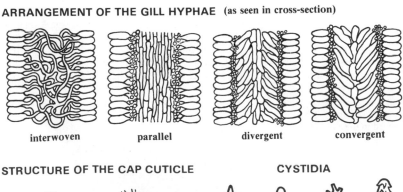

interwoven   parallel   divergent   convergent

STRUCTURE OF THE CAP CUTICLE   **CYSTIDIA**

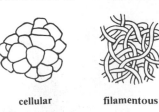

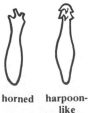

cellular   filamentous   lance-   flask-   horned   harpoon-
                                    shaped   shaped   like

significant than spore characters, but are not as easily observed. These criteria include: **orientation of the hyphae in the gill tissue, structure of the cap cuticle,** and **shape and size of cystidia** (specialized sterile cells) on the gills, stalk, and cap.

Spores can be examined by taking spore dust from a spore print and mounting it on a slide with a drop of water and a cover slip. Cystidia, basidia, and gill hyphae are best seen by making a *thin* cross-section of a gill with a razor blade and mounting it in a similar manner (it takes practice!). The cap cuticle, cap tissue, and stalk are also best observed in cross-section

CHEMICAL CHARACTERISTICS

The ways in which mushrooms react to different chemicals have also assumed great importance in their classification. For instance, the genus *Lyophyllum* was erected to embrace various white-spored agarics whose basidia contain granules that darken when heated in aceto-carmine. Since most people are not equipped to conduct tests of this sort, they are not stressed in this book. But there are two particularly useful chemicals that every serious mushroom hunter should try to get: a 5-10% aqueous solution of potassium hydroxide (KOH), and an iodine solution called Melzer's reagent (see the glossary for formula). Spores which stain bluish-gray to bluish-black in Melzer's reagent are said to be **amyloid.** Spores which turn brown or reddish-brown are **dextrinoid.** Many spores are neither amyloid nor dextrinoid. The test is most useful on spores that are not deeply colored to begin with. It is easily observed under the microscope, but can also be assessed by placing spore dust on a glass slide (paper itself may be amyloid), treating it with Melzer's reagent, then holding the slide over a piece of white paper in order to easily see the color change, if any (the reaction takes place within a few minutes). One of the ingredients in Melzer's reagent, chloral hydrate, is difficult to obtain, but other iodine solutions can be used in a pinch.

PRESERVING MUSHROOMS FOR FUTURE STUDY

Part of mushrooms' mystique is their ephemerality—they're literally here today and gone tomorrow, and there is no adequate means of preserving their beauty. For scientific purposes, however, they *can* be preserved by simply drying them out—they shrink and fade in the process, but their anatomical (microscopic) features remain intact. Of course, for dried material to have value it must be accompanied by a spore print and precise description of the fresh fruiting bodies, plus a photograph if possible. Drying is best accomplished on a rack or screen, using a light bulb or hot plate as a heat source. It is more important for the air to circulate freely (thus carrying off moisture) than for the temperature to be extremely hot. For this reason ovens are not suitable except in a pinch. Small or very fragile specimens may "cook" when heated, and should instead be air-dried with the help of silica gel.

For ornamental purposes, mushrooms can be freeze-dried or encased in cubes of plastic resin. Or they can be sliced and pressed in a book, like this *Chroogomphus.* Preserving mushrooms for consumption is discussed in the chapter on mushroom cookery (p. 888).

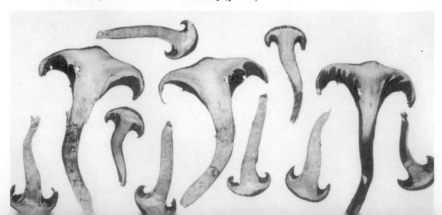

HOW TO USE THE KEYS

IF DESCRIPTIONS and photographs are the meat of this book, then **keys** are its skeletal structure. Keys are tools designed to aid in the identification process. With the exception of the two pictorial keys (pp. 52-55 and 61-62), the keys in this book are **dichotomous.** That is, they consist of a series of *contrasting paired statements* **(couplets).** You are asked to decide which statement in a given couplet is applicable to the mushroom in question. Having made your choice, you are then referred to the number of another couplet, where you again make a choice. This process is repeated until you are given the name of a mushroom (or group of mushrooms) and its appropriate page number (if a page number is not given, it means that particular mushroom is not treated or described beyond the key).

The dichotomous keys begin on p. 52 with "Key to the Major Groups of Fleshy Fungi." Under each major group of fungi you'll find a key to families; under each family, a key to genera; under each genus, a key to species. Keying will be a laborious process at first. However, as you gain experience you can take shortcuts. For instance, if you already know that your mushroom is an *Agaricus,* you needn't consult the keys to families and genera of gilled mushrooms. You can proceed directly to the *Agaricus* key in order to determine the species.

The *key* to the proper use of a dichotomous key is an understanding of its limitations. The following key to a banana, banana slug, hat, rabbit, and sea urchin will admirably illustrate the assets and pitfalls of the dichotomous key. It's designed for the express use of creatures from the planet Fazoog, but I'm sure they won't mind if we use it here.

    1. Object yellow 2
    1. Object not yellow 4

    2. Object more or less cylindrical 3
    2. Object not cylindrical .... **hat**

    3. Object peeling easily .... **banana**
    3. Object not peeling easily .... **banana slug**

    4. Object purple and spiny .... **sea urchin**
    4. Not as above 5

    5. Object moving of its own accord when poked, with four stumps or projections on its underside .... **rabbit**
    5. Not as above .... **hat**

Now let's pretend you're trying to identify a rabbit. You should *always begin with the first couplet*, which in this case asks you to decide whether or not the object is yellow. Presumably you'll choose "not yellow." Then, as indicated to the right of "object not yellow," you proceed to couplet #4 (thereby skipping #'s 2 and 3). Since the object is not purple *and* spiny, you move on to #5. At this juncture you are confronted with a more difficult decision. Does the object move of its own accord when poked *and* have four "stumps" (legs)? If so, it is a rabbit, at least according to the key.

But *keys can be worthless unless accompanied by detailed descriptions.* For instance, if you tried to identify a tiger using this key, you would arrive at "rabbit." Unless a description of a rabbit is provided, you have no way of knowing whether it is indeed a rabbit you have, or an entity *not included in the key*. Furthermore, if the word "edible" or "harmless" appeared next to "rabbit," you would find yourself in serious trouble. Imagine the consequences of attempting to eat a tiger under the mistaken impression that it is a rabbit (more likely the tiger would eat you!). Similarly, a good key to mushrooms does *not* contain information on edibility. Instead it refers you to a detailed description of the mushroom. *Always* compare the mushroom you are trying to identify with the appropriate description as indicated in the key (in a few cases, no page number is given, which means that the mushroom is not treated or described beyond the key). Resist the urge to "fit" the mushroom to the description or vice-versa. A major discrepancy between your mushroom and the description has *three possible explanations:*

(1) **IT IS THE MUSHROOMS' FAULT.** Mushrooms have not seen pictures or descriptions of themselves, and do not know what they are "supposed" to look like (even if they did know, why should they look the way they are *supposed* to—would you?). We can only summarize what a given species *usually* looks like. A good key will allow for a certain degree of variation. For instance, you will notice in the sample key that "hat" appears twice, because some hats are yellow and others, thank God, are not. The key does *not* account for other possibilities, however. Suppose you have a *dead rabbit*. Since it does not "move of its own accord when poked," it would be a *hat* according to the key. Or supposing you find a *peeled banana*, or a *blackened banana* (the kind my father likes to eat). It will not be yellow, and consequently will not key out properly. A better key would account for these possibilities, but *a key cannot account for every possible variation without becoming hopelessly unwieldy.* At first you'll have difficulty assessing just what is typical and what is not. The only solution to this problem is to *use several specimens of each type of mushroom*—at different stages of development if possible. You are then in a better position to decide what is typical (i.e., if you have ten rabbits and one of them is dead, you conclude there is something "wrong" with the dead one, and still use the key correctly.)

(2) **IT IS YOUR FAULT.** People often go astray because they misread a couplet or inadvertently go to the wrong number, or exercise poor judgment. There are several ways to minimize these mistakes. First and foremost, *always read both statements in a couplet* before deciding which one is true. Second, if you are unsure of your choice (as you undoubtedly will be at times), make a notation and choose the one that seems most likely. If it later proves to be wrong, you can go back and try the other choice. Remember the purpose of a key is not to *identify,* but to *eliminate.* There are more than 1000 possibilities as to what any unknown mushroom can be. If the keys eliminate 996 of these possibilities, they have done their job. It is then up to you to carefully compare the descriptions of the remaining three or four species, and decide which—if any—is the correct one. In this respect, keys are like mazes—if you arrive at a dead end, you can always turn around and go back!

Another pitfall to be aware of is language. Watch for qualifiers such as *and, if, usually, sometimes, generally, typically, when young,* and *at maturity;* i.e., "often" does *not* mean "always," and "typically" means "most often." Also watch out for the statement "not as above," as in the following excerpt from a dichotomous key to the poems of an obscure but brilliant Santa Cruz poet.

    1. Poem with a pointed social comment 2
    1. Not as above 77

In this case, "not as above" means "social comment absent, *or if present, then not pointed.*" However, if the first statement in the couplet said "social commment present, often pointed," then "not as above" would translate as "social comment absent."

(3) **IT IS THE KEY'S FAULT.** If you have followed the key correctly, and are certain your mushroom(s) is not aberrant, then a discrepancy between the mushroom and the description means either that the key is flawed (no key is perfect) *or* that you have a mushroom that is not included in the keys (no mushroom book is complete and none should pretend to be—our inventory of mushrooms is still in the preliminary stages!). You may then either consult another book in hopes of finding it, seek the counsel of a mushroom expert, discard it altogether, or give it an informal name of your own.

In addition to aiding in identification, keys can be effective teaching devices. In keying out a mushroom (instead of being told what it is or leafing through a bunch of pictures), you are forced to judge critically and develop an eye for detail. By the time you've keyed out a mushroom, you will have accumulated a considerable amount of information on what it is, *as well as what it is not.* You learn things about it that might otherwise go unnoticed. For instance, in keying out a banana slug using the sample key, *you must try to peel it*—something you ordinarily would *not* do. You would then learn that a banana slug does *not* peel easily. Another challenging activity is to try devising keys yourself—to common mushrooms, household objects, or people you know. You'll discover that it's not as easy as it looks!

QUESTIONS ABOUT MUSHROOMS

WHICH MUSHROOMS ARE GOOD TO EAT?

Fortunately, there is no easy answer to this question. If there was, everybody would pick mushrooms and there wouldn't be enough to go around!

The only way to determine the edibility of a mushroom is to eat it.

O.K.

But who wants to risk their life for the sake of one lousy (or marvelous) meal? Well, through just such a risky trial-and-error process we humans have painstakingly accumulated a reliable body of information on the edibility of many mushrooms. *By systematically learning the identifying characteristics of these mushrooms,* you can take advantage of other people's experiences—many of them unfortunate—by eating only the safest and most savory ones while avoiding the poisonous ones.

If possible, seek out the help of an experienced and *knowledgeable* mushroom hunter.* There's a subtle gap between the mushroom in the book and the mushroom in the bush, and she or he can help you bridge that gap. If you don't have a mushroom mentor, proceed with caution. Just because you identify a mushroom as an edible species doesn't mean you should *eat* it. It is better to collect it several times first, so that you become thoroughly familiar with it. When you misidentify a rock or lizard, it doesn't really matter, because you're not (hopefully) going to eat the thing. With mushrooms, it's different. The

---

*Mushroom hunters can be experienced without being knowledgeable. A *knowledgeable* hunter knows the names of the mushrooms she eats as well as those of all poisonous look-alikes.

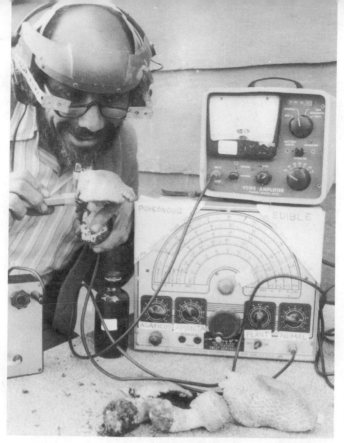

Even relatively sophisticated devices such as this "Toadstool Tester" are unreliable. There is no substitute for caution, experience, and a fundamental familiarity with the fleshy fungi.

cardinal rule is: *Don't eat any mushroom unless you are absolutely sure of its identity!* Or—WHEN IN DOUBT, THROW IT OUT! In other words, you're better off *not* eating an edible mushroom than eating a poisonous one, and it's more important *not* to eat a poisonous mushroom than to eat an edible one.

Empirical approaches such as the "silver coin" test are *without exception* pure poppycock. The lethal Amanitas do *not* blacken silver (fortunately, silver coins are a rarity nowadays, so this fallacious test is more difficult to carry out). Mushrooms partially eaten by mammals or insects are *not* necessarily fit for human consumption (one animal's meat is another's poison). Mushrooms that smell and taste good are *not* necessarily edible (again, the deadly Amanitas are said to be delicious.

Most insidious of all is the "intuitive" approach now in vogue. Intuition can play a part in finding mushrooms, but *not* in determining their edibility. It is a sad comment on our times that intuition is so often peddled as a *substitute* for critical observation. If our intuitions were infallible insofar as mushrooms are concerned, there would be no need for books on the subject. It takes commitment and a good deal of conscious effort to identify mushrooms correctly. The "intuitive" approach is appealing, I suspect, precisely because it promises a maximum amount of satisfaction for a minimum amount of effort.

WHAT'S THE DIFFERENCE BETWEEN A MUSHROOM AND A TOADSTOOL?

"Toadstool" usually carries the connotation of being poisonous, but since many people think that *all* wild mushrooms are poisonous, the two terms are virtually interchangeable in popular usage. The word "toadstool" is perhaps a testament to the old folk belief that toads gave warts to people who handled them, and made mushrooms poisonous by sitting on them.

WHEN AND WHERE DO MUSHROOMS GROW?

Learning to recognize a mushroom should not be an end in itself, but a means of getting to know the *fungus.* A field guide can teach you the physical features of the fruiting body (the *mushroom* aspect of the organism), but you must discover for yourself the traits of the *fungal* aspect—that is, its fruiting behavior and ecological idiosyncrasies. This fungal aspect can only be appreciated by spending a good deal of time in the woods and fields, and yet it is this aspect that is ultimately the most rewarding—getting to know organisms you can't see most of the time on an intimate basis.

Fungally speaking, we are smiled upon most favorably. Our mild coastal climate allows for a long mushrom season (late October through March), and at least a few species can be found any month of the year *(Agaricus augustus,* for instance, fruits prolifically during the dry summer months). However, the peak months for most mushrooms are the wettest ones—November through February. As you move north along the west coast, the season is progressively earlier (September through November or December) and more compressed; as you go south it is correspondingly later and more erratic. In the Rocky Mountains, on the other hand, the best time is usually August, while in eastern North America it is the summer and fall and in the Deep South it is the summer and winter or even year-round.

Just as some wildflowers bloom in March and others in July, so each kind of mushroom has its own biological clock. In our area the sulfur shelf *(Laetiporus sulphureus)* fruits on eucaylptus stumps in September and October (*before* the fall rains arrive), while the horn of plenty *(Craterellus cornucopioides)* seldom shows up before Christmas, morels (*Morchella* species) appear in the spring, and the blewit *(Clitocybe nuda)* fruits continuously throughout the mushroom season.

Since rainfall and temperature are major determining factors, no two mushroom seasons are quite alike. There are good mushroom years and bad mushroom years, but most typical is a year which favors the fruiting of some types at the expense of others. For instance, the warm winter rains of 1977-78 produced a bumper crop of *Agaricus* in our area, but *Chroogomphus* was practically absent. The previous year, however, there was a preponderance of *Chroogomphus* and only a smattering of *Agaricus.* It can be seen, then, that the terms "common" and "rare" can be misleading. "Common" means "often common," while "rare" actually means "rarely common," because some species may be absent for many years and then, at the beckoning of some mysterious signal, fruit suddenly in unprecedented quantity. As a result, you can search the same locality for years and still find new species!

To the beginner practically every mushroom is "rare." Experienced hunters usually find more mushrooms because they know exactly where to go. They stake out secret mycelial "patches" of their favorite fungi which they visit regularly and guard zealously. So unless it's the peak of the season, you're likely to find little in the way of collectable delectables unless you know exactly what you are looking for and where it is likely to grow. For instance, you *don't* go looking for manzanita boletes *(Leccinum manzanitae)* under pine. If you do, you'll come home empty-handed, although without realizing it you may pass up some delicious hedgehog mushrooms *(Dentinum repandum)*, especially if the area you search is overgrown with brambles and poison oak. The more you search, the more "patches" you discover and the more you develop an intuitive feel for when and where mushrooms grow. You learn to hunt as much with your nose and fingers and belly as with your eyes. Eventually you can walk through the woods when there are no mushrooms out, and successfully predict which kinds will come up when it's wetter.

And yet, a large part of mushroom hunting is timing, and a large part of timing is luck. So much depends on where you happen to be at what time. If you're too early, the mushrooms will be invisible, still under the mulch. If you're too late, they'll be "occupied" (riddled with maggots), already harvested, or too old to eat. Make as many generalizations and specifications as you like, and mushrooms will still defy them!

This element of uncertainty, though at times the cause of considerable frustration, is also the reason for much of the excitement. No matter how experienced and knowledgeable you are, you can never really be sure of what you'll find—you can only increase to some degree your chances of success. There is the acute despair of not finding what you hoped for, the frenzied delight and disbelief at finding more than you dreamed possible, the sure and comfortable satisfaction of finding exactly what you "knew" would be there.

CAN PEOPLE HARM MUSHROOMS BY PICKING THEM?

Since mushrooms are the "fruit" of a fungus, picking them is like picking apples or blackberries or figs—no harm is done providing you gather them carefully and do not unduly disturb the environment. Nor does selective picking interfere seriously with their ability to reproduce. Most mushrooms you find have already shed spores (unlike flowers, which have yet to form seeds), and will continue to shed them after they're picked. Also, the parent mycelium is usually perennial and will produce mushrooms periodically, and there is some evidence to suggest that removing a few mushrooms can *stimulate* the mycelium to produce more.

The last several years, however, have seen a tremendous increase in the amount of wild mushrooms being harvested *commercially*—particularly in northern California and the Pacific Northwest. These mushrooms are shipped to markets and restaurants all over the country as well as abroad. As commercial picking is only lucrative when large numbers of mushrooms are gathered, animosity is developing between those who pick for money and those who pick for pleasure.

The commercial pickers and their distributors claim that commercial harvesting provides valuable jobs in depressed areas (though the actual number is fairly small), and they augment this claim with the specious argument that they are providing a service to the public, i.e., making wild mushrooms available to a larger audience, or at least to more than just a select few. (The "select few", in this case, is almost any ambulatory person who wants to learn about mushrooms; the "larger audience" is anyone who has extra money with which to *buy* wild mushrooms, which are never cheap!)

The "select few," on the other hand, profess concern about over-picking and the possibility that certain edible species will be driven to extinction. However, there is no hard evidence to suggest that the latter is happening—or about to happen. I suspect what is *really* upsetting these mushroom hunters is that their habitual stomping grounds are being ravaged by commercial exploitation. This is a perfectly legitimate concern, if you ask me, for there *is* hard evidence that some—perhaps only a few—commercial pickers have not only "raided" but completely "cleaned out" areas where mushroom hunters with less time or mobility have been foraging for years. There *ought* to be enough mushrooms out there to support both commercial and personal picking, but mushroom hunting is one of many activities where a single selfish, inconsiderate boor can spoil it for everyone else. *So, you selfish inconsiderate boors—you know who you are!* (Or if you don't, we do!)

Returning, then, to the original question—mushrooms are a renewable resource, and the only sure way to eradicate a species is to eliminate its habitat (which, unfortunately, is becoming a common practice as wild areas succumb to pollution and the appetite of developers). So don't feel guilty about picking mushrooms, but don't, on the other hand, pick more than you can use **(COLOR PLATES 79, 81)**. Respect the environment and be considerate of those who follow in your footsteps—you may be one of them!

CAN PEOPLE HARM THEMSELVES BY PICKING MUSHROOMS?

No. Handling wild mushrooms is not dangerous. You may want to wash your

Poison oak **(left)** and poison ivy are the banes of many a would-be mushroom hunter. The leaves grow in threes, but are shed during the winter! If you're allergic to them, shower thoroughly after every hunt and put your clothes in the wash. Ticks **(right)** and boletivores (see p. 546) also cause problems.

Rattlesnakes are known to lurk near or under mushrooms, especially large boletes, and they do not always take kindly to being disturbed. (Maybe that's why boletivores carry long sticks!)

hands after prolonged contact with deadly Amanitas, but poisonous mushrooms can cause harm *only* if ingested. There *are,* however, some corollary dangers to mushroom hunting—see photos above and on previous page!

DO OTHER ANIMALS EAT MUSHROOMS?

Yes. As any mushroom hunter will tell you, many a marvelous meaty mushroom has been reduced in a matter of hours to a writhing mass of beatific maggots. These wiggly white "worms" with the black heads are the larvae of fungus gnats (Mycetophiladae). Eggs are usually laid at the base of the stalk, and the newly hatched maggots work their way up to the cap, gorging themselves

One of the unsung rewards of mushroom hunting is the discovery of a host of other fantastic clandestine creatures. Some of these, like slugs, love to munch on mushrooms. Others, like mantises and millipedes, use them for shelter or hiding places. **Left:** A banana slug grazing peacefully in a "fungal jungle" of oyster mushrooms. **Right:** A praying mantis, up close and personal.

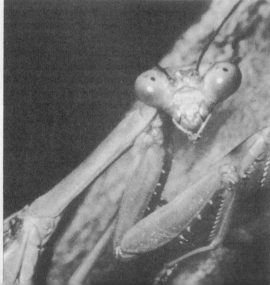

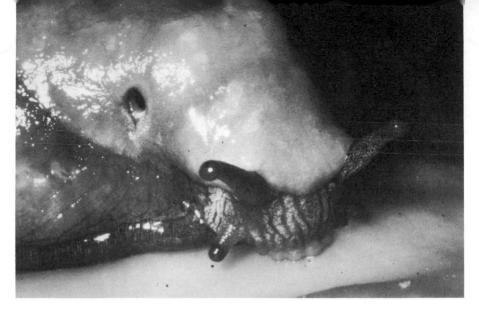

Close-up of a banana slug grazing on an oyster mushroom. Banana slugs can also be seen in Color Plates 57 & 152, and at the bottom of p. 28.

along the way. Many other insects feed on fungi, including springtails and various beetles, and flies like nothing better than to feast on stinkhorn slime.

Sowbugs seem to be very fond of *Agaricus augustus*. Slugs gormandize mushrooms whenever possible **(COLOR PLATES 57 & 152)** and have special "fungus-detectors" to aid them in the search. Tortoises have been known to interrupt their imperturbable peregrinations to munch on a mushroom or two. Rodents—particularly squirrels and chipmunks—ae confirmed fungophiles. They're fanatically fond of various boletes and agarics and those odoriferous underground fungi called truffles and false truffles. (A study by Joseph Hall of San Francisco State University revealed that the summertime diet of the Kaibab squirrel of Arizona consists almost *entirely* of fungi; during the summer captive Kaibab squirrels rejected other food when mushrooms were made available.)

Pigs have a passion for true truffles. In Europe they are used to hunt them down, but have to be muzzled so they don't devour them (see p. 842)! Even the ruminants—cattle, deer, etc.—occasionally indulge. I have seen cows grazing on giant puffballs, and Siberian reindeer are addicted to *Amanita muscaria* (like humans, they experience profound mental disturbances after eating it). They are also said to be extremely fond of human urine, and is it coincidence that the Siberians who ate *A. muscaria* also made a practice of drinking their own urine in order to recycle the intoxicants? R. G. Wasson, in *SOMA: The Divine Mushroom of Immortality,* quote Vladimir Jochelson, a Russian anthropologist:

> The reindeer have a keen sense of hearing and smell, but their sight is rather poor. A man stopping to urinate in the open attracts reindeer from afar, which, following the sense of smell, will run to the urine, hardly discerning the man, and paying no attention to him. The position of a man standing up in the open while urinating is rather critical when he becomes the object of attention from reindeer coming down on him from all sides at full speed.

Some mushrooms, incidentally, are carnivorous! *Cordyceps* species feed on insects, and it has recently been shown that several wood-rotting fungi (including the oyster mushroom) supplement their diet by digesting nematodes.

WHAT IS THE NUTRITIONAL VALUE OF MUSHROOMS?

Mushrooms are esteemed primarily for their flavor, but they can also be a healthy supplement to your diet. Each type, of course, has a different chemical composition, but in general their nutritive value compares favorably to that of most vegetables. They are rich in the B vitamins (including choline, which acts as a protective agent for your liver in case of mushroom poisoning), vitamin D, and vitamin K. Some are also high in vitamin A (e.g, the chanterelle, *Cantharellus cibarius*), and a few (e.g., *Fistulina hepatica*) contain vitamin C. Mushrooms are also rich in minerals such as iron and copper and various trace elements.

Like fruits, vegetables, and human beings, mushrooms are mostly water (85-95%). They have a low fat and carbohydrate content, and as a result, almost no calories—unless, of course, they are cooked in oil or butter. Some types are high in protein (especially *Agaricus, Lepiota,* and *Calvatia* species), and on a dry weight basis *Boletus edulis* contains more protein than any common vegetable except soybeans. However, some of this protein is indigestible, so mushrooms are *not* a viable substitute for meat or other high-protein foods. Cooking mushrooms maximizes their nutritive value by increasing their digestibility. Overcooking them, however, removes some of their vitamins and most of the flavor.

WHAT IS THE MEDICINAL VALUE OF MUSHROOMS?

Fungi contain many unique and powerful substances—penicillin, for instance, which was accidentally discovered in the bread mold *Penicillium*. Many mushrooms contain antibiotic substances, particularly those that don't decay quickly. The Chinese prize several polypores as much as ginseng (e.g., *Ganoderma lucidum, Grifola umbellata*) because of their alleged beneficial effects on health. There is some evidence to suggest they may strengthen or stimulate the immune system. The Japanese attribute similar properties to the shiitake *(Lentinus edodes).* More research should be conducted to determine what healing properties, if any, these and other fungi possess. It is a tribute to the general feeling of our society that mushrooms are harmful at worst and worthless at best that very little research of this kind is being done in this country.

CAN YOU GROW WILD MUSHROOMS?

Yes, but just *how* is beyond the scope of this book. Many wild mushrooms have been successfully cultivated on a commercial basis—particularly in Japan, a mushroom-loving country with relatively little wilderness in which to hunt them. Some of these mushrooms are now being grown in the United States, e.g., the shiitake *(Lentinus edodes)*, enokitake *(Flammulina velutipes)*, and oyster mushroom *(Pleurotus ostreatus).* In addition to the familiar cultivated mushroom *(Agaricus bisporus)*, so-called "magic mushrooms" such as *Psilocybe cubensis* are also grown, albeit illicitly. Several species are better suited to homescale than commercial production, including *Clitocybe nuda, Coprinus comatus, Agaricus subrufescens, Hericium* spp., and *Stropharia rugoso-annulata.* Still others, such as *Boletus edulis, Cantharellus cibarius,* and *Morchella* species, stubbornly resist attempts to raise them. As a rule, mycorrhizal fungi are difficult to raise because of their nutritional requirements (tree rootlets), while those that grow on wood, compost, dung, or disturbed ground are relatively easy. For more information on how to grow mushrooms, see the bibliography.

Human beings, incidentally, are not the only ones to cultivate fungi, and were

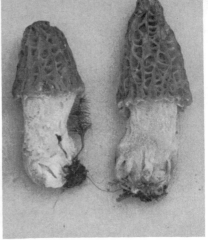

Left: The shiitake, *Lentinus edodes*, was probably the first mushroom cultivated by humans. A native of Japan, it is now grown in the United States and may establish itself in the wild. This cluster is fruiting from a shiitake kit (a "log" of compressed sawdust). **Right:** Morels are among the many delicious mushrooms that resist commercial cultivation. These are red morels (*Morchella elata* group, p. 790).

by no means the first. That honor belongs to certain ants and termites who raise fungi in their nests. On almost any walk through a tropical rain forest you can see long files of leaf-cutting ants carrying bits of leaves to their nests. These leaves are not food for the ants, but food for fungi *upon which the ants feed!*

HEY, MAAAAAAAAAN, DO ANY PSILOCYBIN MUSHROOMS GROW AROUND HERE?

I am asked this question more often than any other, and I must confess it irritates me—not because I object to the use of hallucinogenic drugs *per se*—but because of the attitude that usually—but not necessarily—accompanies their use. Most of the people who ask this question would rather change their way of looking at reality than face the difficult and discouraging task of transforming reality itself. Hence they see mushrooms as means to their own ends. They go out to the pastures and stuff their plastic bags with all the "LBM's" (Little Brown Mushrooms) they can find, then either pop the whole rotten mess into their mouths (a dangerous practice!), or expect someone like me—whose time is presumably less valuable than theirs—to sort them out. While clinging to the moronic belief that they constitute a "counterculture," they share our society's overriding urge for expediency. They make no attempt to learn about the organisms they eat and it has always struck me as ironic that people with such a low level of consciousness should be seeking "higher consciousness."

Excerpts from *Good Times* (January, 1971) epitomize their line of thinking: "Amanita Reality"; "us genetic revolutionaries the longhairs"; "circumcision of the second charka"; "magic mushroom generation"; "I could relate the black hole notion to the notion of total pollution in that either experience would seem kind of freaky"; and "*Amanita muscaria* is total revolt, and the revolution is just about won." From this mindless drivel it is clear why Marge Piercy laments:

> We grew up in Disneyland with ads for friends
> and believed we could be made new by taking a pill.
> We wanted instant revolution, where all we had to add
> was a little smoke.
> But there is no tribe who dance and then sit down
> and wait for the crops to harvest themselves
> and supper to roll over before the pot . . .

To answer the original question—yes, there are hallucinogenic mushrooms in California and the rest of North America, and most of them contain psilocybin and/or psilocin. But if you're just looking for a new high, you really ought to take up hang-gliding or bottle-throttling, or take your chances with what you get on the street. If, on the other hand, you have a genuine interest in our co-inhabitants on this planet, and you wish to explore altered states of consciousness, then the safest approach is to buy a *responsible* field guide to hallucinogenic mushrooms and use it in conjunction with this book. Possession of psilocybin and psilocin, incidentally, is prohibited by federal law (the Comprehensive Drug Abuse and Control Act).

LBM'S: LITTLE BROWN MUSHROOMS

THE CAP is brown, the stem a shade browner, the gills browner still. This can be said of nearly one half of all the mushrooms you find. On even the most casual jaunt through the woods, you'll find dozens and dozens of Little Brown Mushrooms (**COLOR PLATE 107**) sprouting at your feet, and very likely under them as well. The fact is, Little Brown Mushrooms ("LBM's") are so overwhelmingly abundant and uncompromisingly undistinguished that it is more than just futile for the beginner to attempt to identify them—it is downright foolish.

After spending a good 25% of my waking existence being downright foolish, I have come to the painful but inescapable conclusion that the only possible reason for there being more than one kind of Little Brown Mushroom is that their "creator" has an inexplicable fondness for prospective professionals in search of a profession, i.e., Little Brown Mushrooms provide an ever-expanding plethora of pleasant possibilities for lengthy treatises with intriguing and titillating titles such as, "A Preliminary Contribution toward a Partial Monograph of the Section Ignobiles of Subgenus Obfustucantes, Genus Immobilaria as it Occurs in Outer New Bunswick," or "More Useless and Uninteresting Agarics from Putrescent Point State Park."

I wouldn't begrudge this in the least were it not for the fact that this same "creator" is unequivocally cruel when it comes to rank amateurs such as you and I, who are not paid to peer down the narrow barrel of a microscope until we are bug-eyed or get scholar's thumb from flipping through all those worn-out stacks of abstruse Ph.D. theses—and whose curiosity must consequently be swallowed rather than satiated as we slam our big red mushroom book against the wall and decapitate that unpretentious "LBJ" (Little Brown Job) that didn't asked to be picked and "demystified" in the first place, but which we went ahead and picked anyway, and have ever since been attempting in vain to identify.

Though each new Little Brown Mushroom you find will bear a striking resemblance to the last Little Brown Mushroom you threw away, don't let this deceive you into assuming they are identical. Far from it! Somehow, each one finds a new and more minute way to be different, whether it be that the incrustation of the pileal epidermis is cheeriose rather than pretzeloid, or that the hymenial pleurocystidia contain mysterious particles with a refractive content when mounted in a 10% aqueous solution of Pepto-Bismol. It is almost as if they were deliberately challenging the taxonomist, who must tax his or her creative powers to their utmost in order to uncover the differences.

An "LBM" (Little Brown Mushroom), also known as an "LBJ" (Little Brown Job).,

Actually, thanks to the diligence and expertise of the professionals, we are slowly accumulating a large mass of knowledge on the Little Brown Mushrooms. Not a completely coherent mass, mind you (in some ways it still resembles a mess more than a mass), but six hours' painstaking perusal of the current literature will produce the name of at least one out of 50 of the featureless little fellows instead of one out of 51, as it used to be. Part of the problem, I suspect, is that new species are being designed and disseminated at approximately the same rate that old ones are being defined and differentiated, so that we will never attain the level of taxonomic command that we have of, say, the Little Brown Lizards or the Big Brown Toads.

Incidentally, to qualify as a bonafide Little Brown Mushroom, you *ought* to be little, and you most *definitely* have to be a mushroom, but you *don't* necessarily have to be brown, though it certainly simplifies things for you (by making it more difficult for us) if you are. That is to say, there are any number of boring buff Little Brown Mushrooms, wishy-washy white Little Brown Mushrooms, and gratuitous gray Little Brown Mushrooms (to say nothing of those whose color is so neutral or nebulous as to defy description), but there are no breathtakingly blue Little Brown Mushrooms, and by and large Little Brown Mushrooms are simply and unequivocally *brown*.

The point I'm trying to make, folks, is that it's sheer folly to embark upon the purchase of a field guide with the expectation of identifying each and every fleshy fungal fructification you find. Please, for your sake, don't expect the "pegs" to fit neatly into the "holes." *The "holes" are made by human beings and the "pegs" are not!* Some "pegs" will fit a number of holes, some won't fit anywhere and must wait for a hole to be gouged out. A few will fit snugly into one hole and one hole only, and these are the ones you should get to know. By concentrating on these larger and/or more distinctive types (see the chart on pp. 48-51), and proceeding at a measured, deliberate pace, you will experience a mounting satisfaction at your ability to demystify, identify, appreciate—and hopefully eat—some of the mushrooms you gather. Remember, what we know of mushrooms (or anything, for that matter) is substantially more than what we knew fifty years ago, but is still precious little compared to what we don't know.

The bulk of this book deals with the larger, more easily recognized fleshy fungi. However, in the interests of providing a broad overview, some of the more noteworthy (or rather, less unnoteworthy) "LBM's" have been included. Almost every genus boasts a few, but the majority belong to *Inocybe, Tubaria, Galerina, Collybia, Psathyrella, Marasmius, Mycena, Pholiota*, and *Cortinarius*.

An oak woodland, with clusters of honey mushrooms *(Armillariella mellea)* in the foreground. (Nancy Burnett)

HABITATS

DEVELOPING an awareness of biological communities is essential to any nature study. In this chapter, some of the more common or distinctive mushrooms are grouped according to habitat—the type of place where they're most likely to be found, or the type of tree with which they commonly grow. The environmental aspect of collecting is frequently neglected by mushroom hunters, who eagerly remove mushrooms from their place of growth without taking note of their surroundings. In many instances the place of growth will provide clues to the identity of the mushroom—and you'll soon discover that certain mushrooms always seem to grow together; that is, they are indicators of a certain habitat or biological community.

Most of the habitats discussed here are named according to dominant vegetation (oak, redwood, grass, etc.). Since only a few of our local trees are significant in mushroom identification, it makes a lot more sense to learn those trees *before* you learn the mushrooms, rather than trying to differentiate between several hundred kinds of mushrooms before learning the basic trees.

Always bear in mind that the different habitats discussed here overlap, so that in a given area there are usually several habitats present. For instance, a lawn with a Monterey pine will feature grass-loving fungi as well as species mycorrhizal with pine. And a road through the woods will offer many types of mycorrhizal or humus-inhabiting mushrooms plus those that grow in disturbed ground (the roadside). Be sure to check the list of cosmopolitan mushrooms ("Anywhere and Everywhere"), for these will turn up in almost any habitat. Also remember that mycorrhizal fungi do not necessarily grow *under* their host—they are associated with the tree's rootlets, which may be quite a distance away from the trunk. The habitats described in the following pages include several widespread ones (e.g., lawns or disturbed ground) and several specific to our area (e.g., cypress and redwood). Space does not permit an exhaustive discussion of forest trees outside our area, but a few are briefly dealt with under the heading "Other Trees."

34

PINE   (*Pinus* species)

Pine forests are favored by a melange of mycorrhizal mushrooms, plus multitudes of minute, saprophytic Mycenas. When the needle carpet is dry ("potato chip conditions"*), the larger mushrooms often hide underneath, manifesting themselves as low mounds or "mushrumps." Prized edible species include *Boletus edulis, Dentinum repandum, Clitocybe nuda, Sparassis crispa,* and *Tricholoma flavovirens. Suillus* and *Chroogomphus* species often fruit together in enormous quantities, along with the green and orange *Lactarius deliciosus* and the bright red or pink *Russula rosacea.* But the most spectacular fungal feature of our pine forests is unquestionably *Amanita muscaria,* with its fiery red cap studded with white "stars."

Of course, from region to region the different kinds of pines feature different mycorrhizal partners, and even in the same area there are usually some differences in the fungal associates of each pine species. For instance, *Suillus pseudobrevipes* and *Chroogomphus pseudovinicolor* grow with ponderosa pine in our area, *Suillus pungens* with Monterey and knobcone pine, and *Suillus fuscotomentosus* with knobcone and ponderosa pine, while *Hygrophorus gliocyclus* is abundant inland under digger pine, but also occurs along the coast with Monterey pine, and *Amanita caesarea* and *Lactarius indigo* occur with ponderosa pine in Arizona but are absent from California's ponderosa pine forests.

Agaricus subrutilescens
Albatrellus flettii
Amanita aspera
Amanita caesarea
Amanita muscaria
Amanita pachycolea
Amanita pantherina
Armillaria ponderosa
Boletus barrowsii
Boletus edulis
Boletus piperatus
Brauniellula nancyae
Callistosporium luteo-olivaceum
Chroogomphus species
Clavulina cristata group
Clitocybe deceptiva
Clitocybe nuda
Clitocybe sclerotoidea
Clitocybe species
Collybia butyracea
Collybia dryophila
Cortinarius cinnamomeus group
Cortinarius mucosus
Cortinarius obtusus group
Cortinarius phoeniceus

Cortinarius species
Cryptoporus volvatus
Dentinum species
Endoptychum depressum
Elaphomyces species
Gymnopilus sapineus
Gymnopilus spectabilis group
Hebeloma crustuliniforme
Hebeloma sinapizans group
Helvella lacunosa
Hydnellum species
Hygrophoropsis aurantiaca
Hygrophorus erubescens
Hygrophorus gliocyclus
Hygrophorus hypothejus
Hygrophorus purpurascens
Inocybe species
Inonotus tomentosus
Laccaria species
Lactarius deliciosus
Lactarius indigo
Lactarius rufus
Lactarius subflammeus
Lactarius vinaceorufescens
Lentinus ponderosus

Mycena species
Marasmius sp. (unidentified)
Mycena species
Naematoloma species
Phaeolus schweinitzii
Phellinus pini
Pholiota species
Pluteus atromarginatus
Rhizopogon species
Russula alutacea group
Russula brevipes
Russula emetica group
Russula rosacea
Russula sororia group
Sparassis crispa
Suillus species
Tricholoma flavovirens
Tricholoma imbricatum
Tricholoma pessundatum group
Tricholoma squarrulosum
Tricholoma terreum group
Tricholoma vaccinum
Tricholoma zelleri
Tricholomopsis rutilans

DOUGLAS-FIR   (*Pseudotsuga* species)

There are well over 1,000 kinds of mushrooms known to form mycorrhiza with Douglas-fir, and the great Douglas-fir forests of the Pacific Northwest are among the best fungal foraging grounds in the world. In our region, however, the Douglas-fir habitat is no better than average. The most prominent mycorrhizal associates are *Suillus, Gomphidius, Lactarius,* and *Russula* species. *Russula xerampelina* is among our few choice edibles that grow principally with Douglas-fir. *Laetiporus sulphureus* is occasionally found on logs and stumps, and *Cantharellus cibarius* and *C. subalbidus* are abundant under Douglas-fir farther north. *Tuber gibbosum,* when you can find it, is also excellent.

*Potato chip conditions: when it's so dry that it sounds like you're walking on a bed of potato chips; equally undesirable are "tidepool conditions," when the humus layer is so soggy it sounds like you're traipsing across a bed of sea anemones (squish squish squish).

The name "Douglas-fir" is hyphenated in this book because it is not really a fir. It is easily told from redwood (with which it often occurs locally) by its browner bark, spurred cones, and habit of dropping needles *without* twigs intact (redwood drops twigs with the reddish needles intact).

Amanita pantherina	*Lactarius vinaceorufescens*	*Russula placita* group
Cantharellus species	*Lepiota clypeolaria*	*Russula xerampelina*
Cortinarius species	*Mycena* species	*Strobilurus trullisatus*
Fomitopsis cajanderi	*Otidea* species	*Suillus caerulescens*
Fomitopsis pinicola	*Phaeolus schweinitzii*	*Suillus lakei*
Gomphidius species	*Phellinus pini*	*Suillus ponderosus*
Hygrophorus agathosmus	*Rhizopogon* species	*Tricholoma terreum* group
Inocybe species	*Russula gracilis* group	*Truncocolumella citrina*
Lactarius rubrilacteus	*Russula integra* group	*Tuber gibbosum*

REDWOOD   (*Sequoia sempervirens*)

Ironically, our largest tree supports a fungal phantasmagoria of dainty, fragile fungi, but only a smattering of the large, fleshy types. Relatively few redwood-lovers are wood-rotters, and even fewer (if any) are mycorrhizal—a tribute to the redwood's unique position among the conifers.

Caulorhiza (=*Collybia*) *umbonata*, with its conical or umbonate cap and long "tap root," is perhaps the most distinctive redwood-lover. Colorful waxy caps *(Hygrocybe* species) and blue-black to purple-black Leptonias abound, but pickings for the table are meager, at least south of San Francisco—*Agaricus augustus* (but usually near roads or trails), *Agaricus subrutilescens, Clitocybe deceptiva,* and sometimes *Cantharellus cibarius.* The largest inhabitant of our redwood forests is *Leucopaxillus albissimus,* which often forms impressive fairy rings, but unfortunately, is not a good edible.

Agaricus augustus	*Entoloma madidum*	*Mycena* species
Agaricus hondensis	*Geoglossum* species	*Nolanea* species
Agaricus praeclaresquamosus	*Helvella* species	*Polyporus hirtus*
Agaricus subrutilescens	*Hygrocybe* species	*Ramaria abietina*
Boletus zelleri	*Hygrophoropsis aurantiaca*	*Ramaria myceliosa*
Camarophyllus species	*Lepiota* species	*Ramariopsis kunzei*
Cantharellus cibarius	*Leptonia* species	*Stropharia ambigua*
Caulorhiza umbonata	*Leucopaxillus albissimus*	*Trichoglossum hirsutum*
Clavaria vermicularis	*Macrotyphula juncea*	*Tricholomopsis rutilans*
Clitocybe species	*Microglossum viride*	*Xeromphalina* species

CYPRESS   (*Cupressus* species)

Our cypress groves do not qualify as "woods" because nearly all of them were planted. They have few (if any) mycorrhizal associates, but there are a number of fleshy fungi that grow mainly under cypress. The harsh winds and salt spray to which coastal cypresses are subjected make for unpredictable hunting, but under favorable conditions there is a large burst of *Lepiota* and *Agaricus* species, many of them unclassified and/or endemic to California. From a gastronomic standpoint the best cypress-lovers are *Agaricus bisporus, A. lilaceps, A. bernardii, Lepiota rachodes,* and *Clitocybe nuda.* In southern Arizona, incidentally, there are large forests of Arizona cypress, and it would be very interesting to compare the fungal flora of those cypress tracts to that of coastal California.

Agaricus benesi	*Agaricus lilaceps*	*Clitocybe nuda*
Agaricus bernardii	*Agaricus pattersonae*	*Geastrum* species
Agaricus bisporus	*Agaricus perobscurus*	*Hygrocybe* species
Agaricus blandianus	*Agaricus* species	*Lepiota rachodes*
Agaricus californicus	*Agaricus xanthodermus*	*Lepiota* species
Agaricus fuscofibrillosus	*Battarrea phalloides*	*Tyromyces basilaris*
Agaricus fuscovelatus	*Camarophyllus* species	

OAK (*Quercus* species)

Oaks are endowed with a rich array of fleshy fungi that differ drastically from the conifer-lovers typical of northern California and the Pacific Northwest. In fact, oaks appear to have more mycorrhizal partners than any other angiosperms (hardwoods), and mushroom lovers are indeed fortunate that they are the dominant forest trees of the central California coast. Oaks also boast the longest mushroom season of any local forest type. The major fruiting, naturally, is in the fall and winter, but there is a characteristic spring crop highlighted by *Lactarius fragilis, L. rufulus, Clitocybe nuda, Tuber* species, and a trio of Amanitas—*A. ocreata, A. rubescens,* and *A. velosa.* Wood-inhabiting bracket fungi are prominent year-round, including *Trametes versicolor, Lenzites betulina, Stereum* species, and *Tremella mesenterica.* But to inveterate fungophiles, our oak woodlands are synonymous with chanterelles and blewits *(Cantharellus cibarius* and *Clitocybe nuda)*—the two not only grow together, they go together (in a variety of delicious dishes). Other choice edibles include *Boletus appendiculatus, B. regius, B. barrowsii, Craterellus cornucopioides, Lactarius fragilis, L. rufulus, Pleurotus ostreatus, Amanita velosa,* and *Tricholoma portentosum.* But beginners be careful—our most dangerous mushrooms, *Amanita phalloides* and *A. ocreata,* are also associated with oak!

The following list of oak-lovers applies mainly to live oak *(Q. agrifolia* and close relatives), the principal species in our area. Other kinds of oaks, mostly deciduous, occur inland and offer a more modest selection of fleshy fungi, perhaps because they receive less rainfall, or perhaps because they are deciduous.

Agaricus albolutescens
Agaricus californicus
Agaricus hondensis
Amanita baccata
Amanita constricta
Amanita magniverrucata
Amanita ocreata
Amanita pantherina
Amanita phalloides
Amanita rubescens
Amanita velosa
Armillariella mellea
Boletus appendiculatus
Boletus barrowsii
Boletus dryophilus
Boletus erythropus
Boletus flaviporus
Boletus "marshii"
Boletus regius
Boletus satanas
Bulgaria inquinans
Cantharellus cibarius
Chlorociboria species
Clavariadelphus pistillaris
Clitocybe nebularis
Clitocybe nuda
Collybia dryophila
Cortinarius glaucopus group

Cortinarius regalis
Cortinarius sodagnitus group
Cortinarius species
Craterellus cinereus
Craterellus cornucopioides
Crepidotus species
Daldinia species
Entoloma species
Exidia glandulosa
Genea species
Gyroporus castaneus
Hebeloma crustuliniforme
Hericium erinaceus
Hericium ramosum
Hygrophorus albicastaneus
Hygrophorus eburneus
Hygrophorus roseibrunneus
Hygrophorus sordidus
Hygrophorus species
Hymenogaster species
Hysterangium species
Inocybe sororia
Inocybe species
Lactarius alnicola
Lactarius argillaceifolius
Lactarius fragilis
Lactarius rufulus

Lenzites betulina
Mycena species
Omphalotus olivascens
Otidea species
Phellinus gilvus
Phyllotopsis nidulans
Pleurotus ostreatus
Pluteus cervinus
Pluteus lutescens
Polyporus decurrens
Psathyrella hydrophila
Psathyrella species
Russula cyanoxantha
Russula fragrantissima group
Russula maculata group
Russula subnigricans group
Schizophyllum commune
Stereum species
Trametes versicolor
Tremella foliacea
Tremella mesenterica
Tricholoma pessundatum group
Tricholoma portentosum
Tricholoma saponaceum
Tuber species
Tylopilus indecisus
Xylaria hypoxylon

TANOAK *(Lithocarpus densiflorus)*

Also known as tanbark oak, this close relative of the oaks forms dense stands with madrone at higher elevations in the coastal mountains, and also grows in sheltered basins with redwood. Among the prominent fungi of our tanoak woods are the lovely coral mushrooms *(Ramaria* species*)* and the diminutive garlic mushroom *(Marasmius* *copelandi),* which is often smelled before it is seen. The best edibles are *Agaricus silvicola, Armillaria ponderosa, Pleurotus ostreatus* (on decaying logs), *Boletus aereus, Boletus appendiculatus, Russula cyanoxantha,* and *Craterellus cornucopioides.*

Agaricus hondensis	*Cortinarius infractus*	*Macrotyphula juncea*
Agaricus silvicola group	*Cortinarius* species	*Marasmius copelandi*
Armillaria ponderosa	*Craterellus cornucopioides*	*Phellinus gilvus*
Boletus aereus	*Entoloma madidum*	*Pleurotus ostreatus*
Boletus appendiculatus	*Entoloma* species	*Polyporus hirtus*
Boletopsis subsquamosa	*Hydnellum caeruleum*	*Ramaria* species
Bulgaria inquinans	*Hydnum fuscoindicum*	*Russula albonigra*
Cantharellus subalbidus	*Hygrophorus russula*	*Russula cyanoxantha*
Clavariadelphus pistillaris	*Lactarius argillaceifolius*	*Tricholoma saponaceum*
Cortinarius collinitus group	*Lactarius subvillosus*	*Tricholoma zelleri*
Cortinarius cotoneus group		

CHINQUAPIN   *(Castanopsis chrysophylla)*

Also spelled "chinkapin," this cousin of the chestnuts has chesnut-like burrs that contain small edible nuts. Most of the local chinquapins are rather runty and ragged-looking but farther north they attain heights of 75-100 feet. Chinquapins are also closely allied to tanoak, and as might be expected, support a similar fungus flora, including *Ramaria* species and *Boletus aereus.* Among the wood-inhabitors, three in particular stand out: the sulfur tuft *(Naematoloma fasiculare)* and jack-o-lantern mushroom *(Omphalotus olivascens)* because of their brilliance, and the beefsteak fungus *(Fistulina hepatica)* because of its bizarreness. *Boletus aereus* is the best edible species.

Boletus aereus	*Marasmius copelandi*	*Ramaria* species
Entoloma species	*Naematoloma fasiculare*	*Stereum* species
Fistulina hepatica	*Omphalotus olivascens*	*Trametes versicolor*

MADRONE and MANZANITA   *(Arbutus menziesii* and *Arctostaphylos* species)

Madrone woods are my favorite foraging grounds. They boast their share of gastronomic delights (e.g., *Amanita calyptrata, Craterellus cornucopioides, Leccinum manzanitae),* but their chief attraction is their beauty. Madrones don't blot out the sun like redwoods, and they're more colorful (though not quite as venerable) as oaks. Every year they shed their reddish bark in sheets, revealing smooth, yellow, musclebound limbs beneath. In the summer the rags of stripped bark sizzle audibly in the sunlight as they curl up like pencil shavings. In the fall, the stripped bark combines with the large leaves and clusters of bittersweet orange-red berries to form a deep, incomparably rich, reddish-black humus. In such splendid company, it's hard to keep your mind on mushrooms!

Manzanitas are essentially miniature madrones. The most that can be said for foraging in manzanita thickets is that it's *different*—you spend the whole time crawling around on your hands and knees wishing you'd taken up stamp collecting or basket weaving or stayed home and watched the ball game.

It is interesting to note that a number of mushrooms normally associated with conifers cross over to madrone (they just can't resist!) and/or tanoak. These include *Amanita aspera, Amanita muscaria, Armillaria ponderosa, Tricholoma (=Armillaria) zelleri, Tricholoma aurantium, T. flavovirens, Hygrophorus chrysodon,* and *H. eburneus.*

Amanita aspera	*Boletus amygdalinus*	*Cortinarius* species
Amanita calyptrata	*Boletus flaviporus*	*Craterellus cornucopioides*
Armillaria ponderosa	*Cantharellus subalbidus*	*Entoloma* species
Boletus aereus	*Cortinarius balteatus*	*Hydnum fuscoindicum*

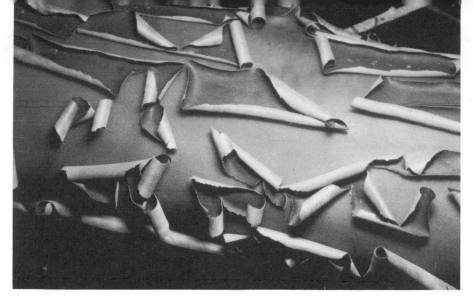

Watching bark peel can be as exciting as watching a bicycle rust, unless it's madrone bark, shown here.

Hygrophorus chrysodon
Hygrophorus eburneus
Leccinum manzanitae
Leccinum species
Lepiota atrodisca

Omphalotus olivascens
Phellinus igniarius
Russula cremoricolor
Russula emetica group
Tricholoma aurantium

Tricholoma flavovirens
Tricholoma manzanitae
Tricholoma saponaceum
Tylopilus humilus

EUCALYPTUS (*Eucalyptus* species)

This messy, aggressive intruder was originally brought to this country in the hope that its noxious fumes would combat malaria. It grows quickly, and its greedy, shallow roots inhibit the growth of many native plants and mushrooms *(Leucopaxillus albissimus, Coprinus plicatilis, Clitocybe nuda,* and various *Agaricus* species are among the few humus-inhabitors with the poor sense to tolerate it). However, eucalyptus makes fairly good firewood and if you chop them down you will not only prevent them from barging into someone's bedroom during the next windstorm, but you're also liable to get a nice crop of succulent sulfur shelves *(Laetiporus sulphureus)* sprouting from the cut stumps!

Agaricus californicus
Agaricus xanthodermus
Clitocybe nuda
Coprinus species
Hydnangium carneum

Hymenogaster albus
Hysterangium fuscum
Laccaria species
Laetiporus sulphureus
Lepiota rubrotincta

Leucopaxillus albissimus
Marasmiellus candidus
Marasmius plicatulus
Psathyrella species
Setchelliogaster tenuipes

RIPARIAN WOODLAND

Streams and rivers being indisputably damp, many people assume they constitute an ideal setting for mushrooms. The mixed hardwoods (alder, cottonwood, willow, maple, etc.) of our stream valleys and ravines do indeed support a characteristic fungus flora, but it is a surprisingly modest one that does not yield the bountiful harvests of our oak and pine forests. Log-rotters abound *(Pholiota, Psathyrella, Naematoloma, Armillariella, Pleurotus,* etc.), while the best edibles are *Armillariella mellea* and *Pleurotus ostreatus* in the fall and winter, *Verpa* and *Morchella* species in the spring.

Armillariella mellea
Clathrus archeri
Clavaria vermicularis
Coprinus micaceus
Disciotis venosa
Flammulina velutipes

Marasmiellus candidus
Morchella species
Mycena species
Naematoloma species
Pholiota species
Pleurotus ostreatus

Pluteus species
Psathyrella species
Sarcoscypha coccinea
Stropharia ambigua
Verpa bohemica
Verpa conica

OTHER TREES

Only a few of our native trees have been discussed so far, but the remaining types are not particularly significant from a mushroom identification standpoint. That is, the mushrooms they harbor (either beneath or on them) are likely to be cosmopolitan (see the "Anywhere and Everywhere" category). For instance, almost nothing fungal grows *under* bay laurel, but the *Ganoderma applanatum* group (especially *G. brownii*) is abundant *on* bay laurel as well as many other trees. Similarly, species of *Laccaria* (omnipresent mushroom "weeds") occur in droves under acacia, while the bush lupines along the coast harbor nice crops of *Flammulina velutipes* and a brown-capped form of *Pleurotus ostreatus* (both edible), and huckleberry is home to *Cortinarius balteatus* and the *C. cylindripes* group.

A few trees of special interest to mushroom hunters *that do not occur locally* are discussed briefly below, but space does not permit a detailed listing of their fungal associates.

LARCH *(Larix)*—Larches, or tamaracks as they are sometimes called, are deciduous northern conifers whose needles turn gold before dropping off. They feature a number of unique mycorrhizal fungi, particularly boletes, plus the usual array of wood-rotters and saprophytes. The following species *favor* larch over other conifers.

Gomphidius maculatus	*Fomitopsis officinalis*	*Suillus cavipes*
Hygrophorus speciosus	*Fuscoboletinus* species	*Suillus grevillei*

HEMLOCK *(Tsuga)*—Hemlocks are a common feature of the northern and montane forests of North America. In the West they usually mingle with other conifers, and so do their mycorrhizal mates. The following list includes only those species that seem overly fond of hemlock. Among them is *Boletus mirabilis,* a choice edible.

Boletus mirabilis	*Ganoderma tsugae*	*Ramaria* species
Chroogomphus tomentosus	*Hysterangium separabile*	*Rhizopogon rubescens*
Cortinarius subfoetidus	*Lactarius subpurpureus*	*Russula xerampelina*
Cortinarius species	*Pleurotus porrigens*	*Tylopilus* species
Elaphomyces granulatus		

SPRUCE *(Picea)*—The Sitka spruce stands of coastal northern California and the Pacific Northwest and the spruce forests of the North, Southwest, and Rocky Mountains are phenomenally rich in mycorrhizal mushrooms, particularly species of *Russula* and *Cortinarius* and the delectable *Boletus edulis.* A sizable number of spruce addicts have already been listed under pine; these species are excluded from the following list.

Agaricus smithii	*Clavariadelphus* species	*Lactarius scrobiculatus*
Albatrellus confluens	*Clitocybe odora*	*Leccinum* species
Albatrellus ovinus	*Cortinarius* species	*Macowanites americanus*
Armillaria albolanaripes	*Gomphidius glutinosus*	*Phaeocollybia* species
Armillaria caligata	*Gomphus* species	*Phellodon* species
Boletus pulcherrimus	*Hydnum* species	*Polyozellus multiplex*
Boletus rubripes	*Hygrophorus agathosmus*	*Rhizopogon parksii*
Calvatia fumosa	*Hygrophorus capreolarius*	*Russula* species
Calvatia subcretacea	*Hygrophorus olivaceoalbus*	*Russula xerampelina*
Cantharellus cibarius	*Hygrophorus pudorinus*	*Tricholoma saponaceum*
Catathelasma imperialis	*Lactarius alnicola*	*Tricholoma virgatum*
Catathelasma ventricosa	*Lactarius representaneus*	*Tylopilus pseudoscaber* group
Clavaria purpurea		

FIR *(Abies)*—Firs, like hemlocks, usually occur in association with other conifers. They feature many of the same fungi as their close relatives, the spruces. Some species, however, seem to *favor* fir over other conifers. These include:

Boletus calopus	*Echinodontium tinctorium*	*Nivatogastrium nubigenum*
Boletus regius	*Gastroboletus turbinatus*	*Oxyporus nobilissimus*
Cortinarius species	*Hericium abietis*	*Rozites caperata*

JUNIPER *(Juniperus)*—Junipers favor arid habitats (at least in the West) and have few if any mycorrhizal mates. Therefore, as might be expected, they offer a rather meager selection of fleshy fungi: a few saprophytes (e.g., *Geastrum* and *Tulostoma* species) plus some wood-rotters (e.g., *Daedalea juniperinus, Gloeophyllum striatum, Truncospora demidoffii*). Several of the species listed under cypress and "Sand and Deserts" also occur under juniper.

CEDAR and INCENSE CEDAR *(Thuja* and *Libocedrus)*—Cedars and incense cedars are not "true" cedars. The "true" or Old World cedars are members of the pine family, whereas American cedars have scales rather than needles and are more closely related to juniper and cypress. Our "cedars" harbor several unique fungi (e.g., *Tyromyces amarus,* which causes a heart rot of incense cedar), but they have few if any mycorrhizal mates. Like redwood and cypress, they are blessed mostly with puny saprophytes that grow best in a mushroom-poor environment (e.g., *Leptonia* and *Hygrocybe* species).

MAPLE *(Acer)*—See "Riparian Woodland."

BIRCH *(Betula)*—As any northerner will tell you, birch has dozens of mycorrhizal associates. In California it does not occur naturally, but planted trees in lawns and parks often feature *Lactarius torminosus, Lactarius pubescens,* and/or *Paxillus involutus.* Birch forests, on the other hand, are rich in other Lactarii as well as Russulas, Amanitas, and tasty Leccinums. Several of the more distinctive birch lovers are:

Amanita muscaria	*Lactarius pubescens*	*Leccinum* species
Amanita virosa group	*Lactarius thejogalus*	*Paxillus involutus*
Cortinarius armillatus	*Lactarius torminosus*	*Phellinus igniarius*
Cortinarius pholideus	*Leccinum scabrum*	*Piptoporus betulinus*
Fomes fomentarius		

Lactarius pubescens (see comments under *L. torminosus* on p. 73), left, and *Paxillus involutus* (see p. 477), right, commonly grow on lawns where birches have been planted. Both are poisonous, at least raw, and both have a strongly inrolled cap margin when young.

Polyporus badius (p. 562) is a small polypore that commonly grows on dead aspen. Note dark cap, dark stalk, and minute whitish pores.

ASPEN, POPLAR, and WILLOW *(Populus* and *Salix)*—In summer the aspen forests of northern and western North America are excellent foraging grounds for fungi, and in the fall they put on a dazzling display of leaf color **(COLOR PLATE 25).** Aspens are not close relatives of birches, but resemble them in several ways. For instance, both kinds of trees tend to have whitish bark and a northern distribution, and both feature many species of *Leccinum* (all edible) as well as the usual panoply of Russulas and Cortinarii. Cottonwoods (poplars) are in the same genus *(Populus)* as aspens, but usually occur near water. Among their mycorrhizal mates are the edible *Tricholoma populinum;* wood-rotters include *Pholiota destruens* and sometimes massive clumps of delectable oyster mushrooms *(Pleurotus ostreatus).* Willows *(Salix)* are closely related to cottonwoods and often mingle with them; consequently their fungal flora is similar. A few of the more distinctive and prominent species associated with aspen, poplar, and willow are:

Armillaria albolanaripes	Lactarius controversus	Pleurotus ostreatus
Armillaria species	Leccinum species	Polyporus badius
Armillaria straminea	Morchella species	Russula aeruginea
Clavicorona pyxidata	Phellinus igniarius	Stropharia kauffmanii
Cyptotrama chrysopeplum	Pholiota aurivella group	Tricholoma flavovirens
Flammulina velutipes	Pholiota destruens	Tricholoma leucophyllum
Helvella leucopus	Pholiota squarrosa	Tricholoma populinum
Hericium ramosum		

ALDER *(Alnus)*—Alders typically grow along streams (see "Riparian Woodland") or on logged-over or exposed areas with conifers. Characteristic alder addicts include:

Agaricus subrutilescens	Entoloma rhodopolium	Pleurotus ostreatus
Alpova diplophloeus	Pleurotus dryinus	Psathyrella species

DUNG AND MANURE

Cow patties—or almost any kind of dung—can be turned into miniature mushroom gardens by keeping them in a humid environment. They will yield a fascinating succession of dainty fungi, but few of them large enough to eat. The mycelium of a dung-inhabiting fungus has a shorter life span than that of other fungi due to the transient nature of the substrate. The spores are presumably ingested by cows (or other animals) with the grass they eat, but pass through the digestive system unscathed, and into the dung, where they germinate. In addition to the common dung addicts listed below, you'll find a multitude of minute Ascomycetes not treated in this book. Some of these are extremely specialized, living only on lizard dung, frog feces, oppossum excrement, etc.

Agaricus bisporus	Clitocybe tarda	Peziza fimeti
Agrocybe pediades group	Conocybe species	Peziza vesiculosa
Ascobolus species	Coprinus species	Psilocybe species
Bolbitius vitellinus	Coprobia species	Sphaerobolus stellatus
Cheilymenia species	Cyathus stercoreus	Stropharia semiglobata
Clitocybe nuda	Panaeolus species	Volvariella speciosa

Left: This "meadow muffin" (chunk of cow dung) shows extensive mycelial growth. The mushrooms belong to the *Agrocybe pediades* group (p. 468). **Right:** This is one clump of *Lentinus ponderosus* (p. 143) that won't be eaten!

PASTURES

Pastures and meadows offer a wide selection of easily identified edible mushrooms, notably *Agaricus campestris, A. cupreobrunneus, A. arvensis, A. crocodilinus,* and the *A. osecanus* group, plus *Lepiota naucina* and *Calvatia* species. Every fall there is a period of about one month when our grazed pastures are chock-full of mushrooms, and there is often another crop in the spring. The following list *excludes* mushrooms associated with trees (such as *Amanita velosa*), which frequently fruit at the edges of pastures, as well as those that grow on dung. It's interesting to note that several species which appear in pastures do *not* often grow on lawns (and vice versa).

Agaricus arvensis	*Astraeus* species	*Clitocybe tarda*
Agaricus campestris	*Bovista* species	*Lepiota naucina*
Agaricus crocodilinus	*Calvatia booniana*	*Marasmius oreades*
Agaricus cupreobrunneus	*Calvatia bovista*	*Melanoleuca melaleuca* group
Agaricus micromegathus	*Calvatia cyathiformis*	*Nolanea sericea* group
Agaricus osecanus group	*Calvatia gigantea* group	*Panaeolus campanulatus* group
Agaricus xanthodermus	*Clitocybe brunneocephala*	*Psilocybe semilanceata*
Agrocybe pediades group	*Clitocybe dealbata*	*Stropharia* species
Amanita sp. (unidentified)	*Clitocybe subconnexa* group	*Vascellum pratense*

LAWNS and GARDENS

To many people, mushroom hunting is synonymous with a day in the wild—plunging into the dripping depths of a forest, lunging through manzanita thickets, or meandering through puffball-pocked pastures. However, a mushroom hunt can and should begin in the mundane confines of your own yard (assuming you have one), and radiate outward gradually, like the mycelium of a fairy ring. After all, some of the most delicious *(Marasmius oreades, Agaricus campestris)* or bizarre *(Phallus impudicus)* mushrooms might be growing right outside your door!

Agaricus is the most prominent group of mushrooms in lawns and gardens, and more than one species of *Agaricus* often grow together. *A. campestris, A. bisporus,* and *A. bitorquis* are delicious, but care must be taken not to confuse them with *A. californicus* and *A. xanthodermus,* which are poisonous to some people. Other choice morsels include

43

Don't overlook your toolshed when looking for mushrooms! This hefty morel would have rotted away uneaten if my wife hadn't insisted that it was my turn to mow the lawn.

Lepiota rachodes, L. barssii, L. naucina, the omnipresent *Marasmius oreades,* and an occasional morel *(Morchella).*

Lawns and gardens are actually two distinctly different habitats, but are treated together here because they so frequently overlap. *Volvariella speciosa* and *Agaricus bisporus,* for instance, grow in gardens but not on lawns, whereas *Agaricus campestris* and *Marasmius oreades* grow on lawns but not in gardens. Mushrooms associated with trees are *excluded* from the following list though they often appear in people's yards. Most yards also qualify as "disturbed ground," and some may have dung or manure (in the form of compost or mulch) present.

Agaricus arvensis	*Clathrus ruber*	*Lysurus cruciatus*
Agaricus augustus	*Clitocybe brunneocephala*	*Lysurus mokusin*
Agaricus bernardii	*Clitocybe dealbata*	*Marasmius oreades*
Agaricus bisporus	*Clitocybe nuda*	*Melanoleuca* species
Agaricus californicus	*Clitocybe tarda*	*Morchella* species
Agaricus campestris	*Conocybe lactea*	*Naematoloma aurantiaca*
Agaricus comtulus	*Conocybe tenera* group	*Nolanea sericea*
Agaricus cupreobrunneus	*Coprinus atramentarius*	*Panaeolus foenisecii*
Agaricus micromegathus	*Coprinus comatus*	*Panaeolus subbalteatus* group
Agaricus osecanus group	*Coprinus micaceus*	*Phallus hadriani*
Agaricus perobscurus	*Coprinus plicatilis*	*Phallus impudicus*
Agaricus subrufescens	*Coprinus* species	*Pholiota* species
Agaricus xanthodermus	*Gymnopilus* species	*Pholiota terrestris*
Agrocybe pediades group	*Hohenbuehelia petaloides* group	*Pluteus petasatus*
Agrocybe praecox group	*Lepiota americana*	*Psathyrella candolleana*
Agrocybe sororia	*Lepiota barssii*	*Psathyrella* species
Bolbitius vitellinus	*Lepiota cepaestipes*	*Psilocybe* species
Bovista species	*Lepiota lutea*	*Scleroderma* species
Calocybe carnea	*Lepiota naucina*	*Stropharia* species
Calvatia species	*Lepiota rachodes*	*Vascellum pratense*
Chlorophyllum molybdites	*Lepiota* species	*Volvariella speciosa*

INDOORS

In some cases a mushroom hunt can begin *inside* your house. The brown cup fungus, *Peziza domiciliana,* is commonly found on walls, floors, carpets, tile, and plaster. *Lepiota lutea* consistently pops up in flower pots, though it sometimes finds its way outdoors.

Several fungi are pests of structural timber (e.g., *Serpula lacrymans*), and a mysterious *Coprinus* invaded the floor of my '67 Chevy Nova. However, the majority of fungi found indoors occur outside as well.

Coniophora puteana	*Lepiota cepaestipes*	*Pluteus* species
Conocybe species	*Lepiota lutea*	*Poria* species
Coprinus species	*Lepiota rachodes*	*Serpula lacrymans*
Galerina species	*Melanotus textilis*	*Thelephora* species
Gymnopilus species	*Mycena* species	*Volvariella* species
Hohenbuehelia petaloides group	*Peziza domiciliana*	

DISTURBED GROUND

"Disturbed" ground means roadsides, pathsides, gardens, vacant lots, building sites, and the perimeters of parking lots. Of course, it overlaps with lawns and gardens (or "cultivated ground"), but we do have a very distinct roadside fungus flora exemplified by the shaggy mane *(Coprinus comatus)*.

Just what makes certain mushrooms prefer (or even require) disturbed ground is a mystery. Obviously, the process of disrupting, overturning, or paving the soil must release certain otherwise inaccessible nutrients, and perhaps the moisture-absorbing properties and/or pH of the soil are changed. An unsurpassed edible mushroom like *Agaricus augustus* could be raised commercially if this puzzle were solved. Some mushrooms, such as the *Lyophyllum decastes* group, seem to *require* disturbed ground whereas other mushrooms only *prefer* it. Some show a liking for asphalt and will even poke up through it—notably *Coprinus comatus, Lyophyllum decastes, Agaricus bitorquis, Pisolithus tinctorius,* and *Scleroderma geaster*. Shaggy manes have even ruined tennis courts!

One asset of these fungi is that they're easily spotted from the road (or trail). Consequently, they are tailormade for people who have an extreme allergy to poison oak. And many are worth stopping for!—*Agaricus augustus, A. bitorquis, Coprinus comatus, Lepiota naucina, Lepiota rachodes,* and *Lyophyllum decastes*. The following list does *not* include species that commonly occur in undisturbed as well as disturbed ground.

Agaricus augustus	*Astraeus* species	*Lepiota rachodes*
Agaricus bernardii	*Calvatia gigantea* group	*Lyophyllum decastes* group
Agaricus bitorquis	*Clitocybe dilatata*	*Morchella* species
Agaricus californicus	*Coprinus atramentarius*	*Pholiota terrestris*
Agaricus praeclaresquamosus	*Coprinus comatus*	*Pisolithus tinctorius*
Agaricus species	*Coprinus micaceus*	*Scleroderma geaster*
Agaricus xanthodermus	*Lepiota naucina*	*Scleroderma* species
Aleuria aurantia		

BURNED AREAS

Many plants are adapted to growing in burned-over areas. The cones of the knobcone pine, for instance, release their seeds only when subjected to extreme heat. Similarly, certain mushrooms fruit only in burnt ground, and others show a definite fondness for it. After the eruptions of Mt. St. Helens, mushrooms were among the first organisms to "colonize" the devasted slopes of the volcano, and the most reliable way to cultivate morels is still to set fire to some woods and return the following spring! (This practice actually had to be outlawed in Europe.)

Anthracobia species	*Peziza praetervisa*	*Plicaria* species
Ascobolus carbonarius	*Peziza proteana*	*Psathyrella carbonicola*
Coltricia species	*Peziza violacea*	*Psathyrella* species
Coprinus species	*Pholiota brunnescens*	*Pulvinula* species
Geopyxis species	*Pholiota carbonaria*	*Pyronema omphalodes*
Lyophyllum species	*Pholiota fulvozonata*	*Rhizina undulata*
Morchella species	*Pholiota highlandensis*	*Tarzetta* species
Myxomphalia maura		

SAND and DESERTS

Odd as it may seem, a number of mushrooms grow in sand or sandy soil, including the delicious *Lepiota rachodes* (which I have found on Ano Nuevo Island) and various morels (*Morchella* species). The deserts of the West support a fascinating but exasperatingly elusive fungal flora in which the Gasteromycetes (puffballs and allies) figure prominently. Deserts do not necessarily have sandy soil, but there is considerable overlap in the two categories, so they are listed together here; those species which grow exclusively in the desert are marked with an asterisk.

Astraeus hygrometricus
Battarrea species
Calvatia species
Carbomyces species*
*Chlamydopus meyenianus**
*Coprinus asterophorus**
*Dictyocephalos attenuatus**

Disciseda species
Geastrum species
Geopora species
Longula texensis
*Montagnea arenarius**
Morchella species

Myriostoma coliforme
*Phellorina strobilina**
*Podaxis pistillaris**
*Schizostoma laceratum**
Scleroderma species
Tulostoma species

SNOWBANKS

The mountains of western North America feature a very characteristic collection of spring fungi. Some are popularly known as "snowbank mushrooms" because they fruit around the edges of melting snowbanks; others are more common shortly after the snow has disappeared. The "snowbank mushrooms" and other spring fungi are more dependent on snowmelt than rain for their moisture, but the latter can also have a stimulating effect. Obviously, the higher you go, the later the snow melts, so that in some areas the spring "snowbank" flora doesn't appear until early or even mid-summer.

This is one category in which Ascomycetes are more preponderant than Basidiomycetes. Morels *(Morchella* species) and false morels *(Gyromitra* and *Helvella* species) are what make the spring fungus flora famous, but dozens of other species occur also. Many of the larger agarics develop under the duff and show themselves only as low mounds or "mushrumps" in the humus. The following list includes some typical "snowbank mushrooms," plus many that fruit within a month after the snow disappears. The list does *not* include fungi with perennial fruiting bodies or species such as *Boletus edulis* that may occur in the spring but are more common in the summer and fall.

Agaricus silvicola group
Agrocybe praecox group
Amanita calyptrata
Amanita gemmata group
Armillaria albolanaripes
Armillaria olida
Auricularia auricula
Caloscypha fulgens
Calvatia species

Choiromyces alveolatus
Clitocybe albirhiza
Clitocybe species
Collybia albipilata
Cortinarius species
Cudonia monticola
Discina species
Disciotis venosa
Gautieria species

Geopora cooperi
Guepiniopsis alpinus
Gyromitra esculenta
Gyromitra gigas
Helvella leucomelaena
Hygrophorus goetzii
Hygrophorus marzuolus
Hygrophorus purpurascens
Hygrophorus subalpinus

Lyophyllum montanum (p. 175) usually grows nearly melting snow in the spring.

Lentinellus montanus
Lyophyllum montanum
Lyophyllum species
Melanoleuca species
Mitrula species
Morchella species
Mycena griseoviridis
Mycena overholtzii

Nivatogastrium nubigenum
Nolanea verna group
Peziza species
Phaeolus alboluteus
Pholiota species
Plectania nannfeldtii
Plectania species
Ramaria botrytis

Ramaria magnipes
Ramaria rasilispora
Sarcosoma species
Sarcosphaera crassa
Tyromyces leucospongia
Verpa bohemica
Verpa conica

SPECIALIZED HABITATS

Without specialization there would be far less diversity and little—if any—evolution. All fungi are to some extent specialized, but in some cases the specialization is more bizarre or extreme—or at least more apparent to us. For instance, it is said that certain poorly known Laboulbeniomycetes grow only on the left anterior appendage (left front foot) of their insect host!

Some kinds of fleshy fungi grow only in deep moss or in *Sphagnum* bogs. Others grow exclusively on other mushrooms: *Asterophora* species, *Claudopus parasiticus*, *Collybia tuberosa*, *Collybia racemosa*, *Collybia cookei*, *Collybia cirrhata*, *Volvariella surrecta*, *Boletus parasiticus*, *Psathyrella epimyces*, and *Hypomyces* species. *Cordyceps* species parasitize insects and certain truffles; and several *Omphalina* species are associated with algae. Many wood-rotters can grow on cones, but *Auriscalpium vulgare* (as well as *Baeospora myosura* and several *Strobilurus* species) are *restricted* to cones. The very specialized niche has helped *Auriscalpium* to survive, but if and when conifers disappear from the earth (as they are slowly doing), so will *Auriscalpium!*

ANYWHERE and EVERYWHERE

No habitat list is complete without mentioning some of our cosmopolitan fungi or "mushroom weeds." Many of these are restricted to wooded areas, others (those marked by an asterisk) are not. The best of the lot from an edibility standpoint are *Armillariella mellea*, *Clitocybe nuda*, and *Morchella* species.

*Agaricus californicus**
Agaricus hondensis
Agaricus praeclaresquamosus
Agaricus silvicola group
Agaricus subrutilescens
*Agaricus xanthodermus**
*Aleuria aurantia**
Amanita gemmata group
Amanita pachycolea
Amanita pantherina
Amanita vaginata
*Armillariella mellea**
Boletus chrysenteron
Boletus spadiceus
Boletus subtomentosus
Boletus truncatus
Boletus zelleri
Cantharellus cibarius
Cantharellus subalbidus
*Clavaria vermicularis**
Clitocybe inversa
Clitocybe nebularis
*Clitocybe nuda**
Clitocybe subconnexa group*
*Clitopilus prunulus**

Collybia dryophila
Collybia fuscopurpurea group*
Cortinarius glaucopus group
Cortinarius species
Elaphomyces species
Fomitopsis pinicola
Galerina species*
Ganoderma applanatum group*
Geastrum species*
*Helvella lacunosa**
Inocybe geophylla
Inocybe sororia
Inocybe species*
Laccaria species
Lactarius chrysorheus
Lactarius fragilis
Lactarius vinaceorufescens
*Laetiporus sulphureus**
*Lepiota rachodes**
Leucopaxillus albissimus
Leucopaxillus amarus
Lycoperdon species*
Melanoleuca melaleuca group*
Morchella species*
Mycena species*

*Naematoloma fasciculare**
Peziza species*
Phellinus pini
*Pholiota terrestris**
Phyllotopsis nidulans
*Pluteus cervinus**
Poria species*
*Psathyrella candolleana**
Psathyrella species*
Rhodocybe nuciolens
Russula albonigra
Russula brevipes
Russula sororia group
Russula species
Scleroderma cepa group*
Scleroderma citrinum
Scleroderma species*
Stereum species*
Thelephora species*
*Trametes versicolor**
Trichaptum species*
Tricholoma saponaceum
Tubaria species*
Tyromyces species
*Volvariella speciosa**

SEVENTY DISTINCTIVE MUSHROOMS
(A quick-reference check list)

COMMON NAME	SCIENTIFIC NAME	EDIBILITY
☐ **Delicious Milk Cap** (p. 68; plate 2)	*Lactarius deliciosus*	edible
☐ **Bleeding Milk Cap** (p. 68; plates 3, 6)	*Lactarius rubrilacteus*	edible
☐ **Indigo Milk Cap** (p. 69; plate 4)	*Lactarius indigo*	edible
☐ **Candy Caps** (pp.. 80-82; plate 10)	*Lactarius fragilis, L. rufulus, L. camphoratus*	edible
☐ **Short-Stemmed Russula** (p. 87)	*Russula brevipes*	edible
☐ **Emetic Russula** (p. 97)	*Russula emetica* group	poisonous
☐ **Rosy Russula** (p. 99; plate 13)	*Russula rosacea*	poisonous
☐ **Witch's Hat** (p. 116; plate 19)	*Hygrocybe conica*	doubtful
☐ **Ivory Waxy Cap** (p. 119)	*Hygrophorus eburneus*	edible
☐ **Oyster Mushroom** (p. 134; plates 27, 28)	*Pleurotus ostreatus*	edible
☐ **Jack-O-Lantern Mushrooms** (pp. 146-148; plates 40, 41)	*Omphalotus* species	poisonous
☐ **Blewit** (p. 153; plate 32)	*Clitocybe nuda*	edible
☐ **Man On Horseback** (p. 179; plate 33)	*Tricholoma flavovirens*	edible
☐ **White Matsutake** (p. 191; plate 37)	*Armillaria ponderosa*	edible
☐ **Honey Mushroom** (p. 196; plates 39, 42)	*Armillariella mellea* group	edible
☐ **Fairy Ring Mushroom** (p. 208; plates 38, 47)	*Marasmius oreades*	edible
☐ **Death Cap** (p. 269; plate 50)	*Amanita phalloides*	deadly poisonous
☐ **Destroying Angels** (pp. 271-273; plate 53)	*Amanita ocreata, A. verna, A. virosa, A. bisporigera*	deadly poisonous
☐ **Fly Agaric** (p. 282; plates 58, 59)	*Amanita muscaria*	poisonous

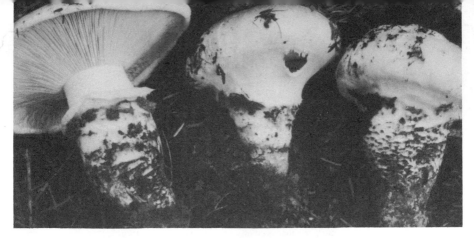

Armillaria ponderosa, better known as the matsutake or white matsutake, is a mushroom well worth knowing. Its odor is unique: something like a cross between "red hots" and dirty socks.

COMMON NAME	SCIENTIFIC NAME	EDIBILITY
☐ **Caesar's Amanita** (p. 284; plate 60)	*Amanita caesarea* group	edible
☐ **Coccora** (p. 284; plates 61-63)	*Amanita calyptrata*	edible with caution
☐ **Grisette** (p. 288)	*Amanita vaginata*	edible
☐ **Green-Spored Parasol** (p. 295)	*Chlorophyllum molybdites*	poisonous
☐ **Shaggy Parasol** (p. 297; plates 69)	*Lepiota rachodes*	edible with caution
☐ **Parasol Mushroom** (p. 298)	*Lepiota procera*	edible
☐ **Meadow Mushroom** (p. 318; plate 71)	*Agaricus campestris*	edible
☐ **Yellow-Staining Agaricus** (p. 329)	*Agaricus xanthodermus*	poisonous
☐ **Horse Mushrooms** (pp. 332-334; plate 78, 79, 81)	*Agaricus arvensis, A. osecanus* group	edible
☐ **The Prince** (p. 337; front cover, plate 77)	*Agaricus augustus*	edible
☐ **Shaggy Mane** (p. 345; plate 85)	*Coprinus comatus*	edible
☐ **Sulfur Tuft** (p. 382; plate 92)	*Naematoloma fasciculare*	poisonous
☐ **Scaly Pholiota** (p. 389; plates 96, 97)	*Pholiota squarrosa*	doubtful
☐ **Violet Cortinarius** (p. 446; plate 109)	*Cortinarius violaceus*	edible
☐ **Pine Spikes** (pp. 484-487; plate 113)	*Chroogomphus* species	edible

Boletus aereus is one of many delectable boletes. These young specimens are easily recognized by their white pore surface and dark cap that is covered with a fine whitish bloom.

COMMON NAME	SCIENTIFIC NAME	EDIBILITY
☐ **Slippery Jacks** (pp. 498-505; plates 114-121)	*Suillus* species	edible
☐ **Admirable Bolete** (p. 521; plates 127, 129)	*Boletus mirabilis*	edible
☐ **Butter Boletes** (pp. 525-526; plates 133, 134)	*Boletus appendiculatus, B. regius*	edible
☐ **Satan's Bolete** (p. 527; plate 137)	*Boletus satanas*	poisonous
☐ **Apple Bolete** (p. 528; plate 139)	*Boletus frostii*	edible
☐ **White King Bolete** (p. 529; plate 141)	*Boletus barrowsii*	edible
☐ **King Bolete** (p. 530; plates 142-144)	*Boletus edulis*	edible
☐ **Queen Bolete** (p. 531; plate 140)	*Boletus aereus*	edible
☐ **Manzanita Bolete** (p. 539; plate 145)	*Leccinum manzanitae*	edible
☐ **Aspen Boletes** (pp. 540-541; plate 147)	*Leccinum insigne* & close relatives	edible
☐ **Old Man Of The Woods** (p. 543)	*Strobilomyces floccopus*	edible
☐ **Beefsteak Fungus** (p. 553; plate 152)	*Fistulina hepatica*	edible
☐ **Hen Of The Woods** (pp. 564-565)	*Grifola frondosa, G. umbellata*	edible
☐ **Sulfur Shelf** (p. 572; plates 154, 155)	*Laetiporus sulphureus*	edible with caution
☐ **Artist's Conk** (p. 576)	*Ganoderma applanatum* group	not edible

COMMON NAME	SCIENTIFIC NAME	EDIBILITY
☐ **Varnished Conks** (pp. 577-578; plate 156)	*Ganoderma lucidum, G. tsugae, G. oregonense*	not edible
☐ **Turkey Tail** (p. 594; plate 158)	*Trametes versicolor*	not edible
☐ **Hericiums** (pp. 613-616; plates 163, 164)	*Hericium* species	edible
☐ **Hedgehog Mushrooms** (pp. 618-619; plates 161, 162)	*Dentinum repandum, D. umbilicatum*	edible
☐ **Pink-Tipped Coral Mushroom** (p. 656)	*Ramaria botrytis*	edible
☐ **Cauliflower Mushrooms** (p. 657; plate 172)	*Sparassis* species	edible
☐ **Scaly Chanterelle** (p. 661; plate 174)	*Gomphus floccosus* group	doubtful
☐ **White Chanterelle** (p. 662; plate 179)	*Cantharellus subalbidus*	edible
☐ **Chanterelle** (p. 662; plates 175, 177, 178)	*Cantharellus cibarius*	edible
☐ **Horn Of Plenty** (pp. 666-668; plate 182)	*Craterellus cornucopioides, C. fallax*	edible
☐ **Witch's Butter** (p. 673; plate 170)	*Tremella mesenterica*	edible
☐ **Giant Puffballs** (pp. 682-684; plates 184, 186)	*Calvatia gigantea* group, *C. booniana*	edible
☐ **Sierran Puffball** (p. 684)	*Calvatia sculpta*	edible
☐ **Dead Man's Foot** (p. 712; plates 185, 188, 189)	*Pisolithus tinctorius*	not edible
☐ **Stinkhorns** (pp. 768-770; plates 193, 194)	*Phallus impudicus* & close relatives	edible
☐ **Dog Stinkhorns** (p. 771; plates 195, 196)	*Mutinus caninus, M. elegans*	edible
☐ **Morel** (p. 787; plate 203)	*Morchella esculenta* & close relatives	edible
☐ **Black Morel** (p.790; plates 199, 202)	*Morchella elata* group	edible
☐ **False Morel** (p. 801; plate 206)	*Gyromitra esculenta*	edible/ poisonous
☐ **Fluted Black Elfin Saddle** (p. 815)	*Helvella lacunosa*	edible
☐ **Orange Peel Fungus** (p. 837; plate 208)	*Aleuria aurantia*	edible

KEY TO THE MAJOR GROUPS OF FLESHY FUNGI

1. Spores produced *on* mother cells called basidia; fruiting body variously shaped (see pp. 52-54) .... **Basidiomycotina**, p. 57
1. Spores produced *inside* mother cells called asci; fruiting body variously shaped (see p. 55) .... **Ascomycotina**, p. 782

BASIDIOMYCETES

Fruiting body with a cap and stalk, or just a cap; spores borne on gills (radiating blades) on underside of cap; spore print obtainable (if spores are being produced) .... **AGARICS** (GILLED MUSHROOMS), p. 58

close-up of the veined underside of a chanterelle

Fruiting body with a cap and stalk, usually vase-shaped or trumpetlike at maturity, *not* gelatinous; fertile underside of cap smooth, wrinkled, or with decurrent veins or folds .... **CHANTERELLES**, p. 658

longitudinal section showing the tube layer

Fruiting body with a cap and stalk, fleshy (not tough or woody); underside of cap with a spongy, often separable layer of tubes or pores; stalk more or less central; usually on ground but occasionally on wood .... **BOLETES**, p. 488

Fruiting body shelflike, bracketlike, crustlike, or with a cap and stalk; usually tough or woody but sometimes fleshy, usually on wood but sometimes terrestrial; spores produced in a layer of tubes or pores that usually line underside of cap. Stalk when present usually off-center or lateral (but sometimes central); tube layer *not* normally peeling away easily from cap . **POLYPORES & BRACKET FUNGI**, p. 549
(but if fruiting body is black and charcoal-like with concentrically zoned flesh, see **Flask Fungi**, p. 878)

Fruiting body variously shaped (with or without a cap), but always gelatinous or very rubbery; usually growing on wood .... **JELLY FUNGI**, p. 669

Fruiting body crustlike or bracketlike or occasionally with a cap and stalk, usually tough, *not* gelatinous; fertile surface smooth, wrinkled, veined, or warted but lacking pores, tubes, or spines; usually growing on wood .... **CRUST & PARCHMENT FUNGI**, p. 604
(also see **Morels, Elfin Saddles, & Cup Fungi**, p. 783)

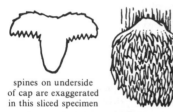

spines on underside of cap are exaggerated in this sliced specimen

Fruiting body bearing its spores on downward-pointing spines or "teeth"; spines either lining the underside of a cap or suspended like icicles from a cushion of tissue or a branched framework; stalk present or absent; on ground or wood .. .. **TEETH FUNGI**, p. 611

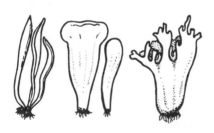

Fruiting body erect, unbranched (club-like) or profusely branched from a common base or "trunk" (coral-like); cap absent; spores borne on the smooth to slightly wrinkled surfaces of the upright clubs or branches ....
.... **CORAL & CLUB FUNGI**, p. 630
(also see **Earth Tongues**, p. 865)

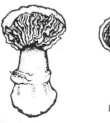

longitudinal section

Fruiting body reminiscent of an un-opened, aborted, or deformed agaric, consisting of a spore case or "cap" and stalk; stalk long or very short, usually percurrent (i.e., extending all the way to top of fruiting body); spore mass or fertile tissue usually composed of plates, branching cavities, or contorted "gills" which may or may not be exposed; spore print *not* obtainable; found mainly in deserts and mountains ....
.... **GASTROID AGARICS**, p. 724

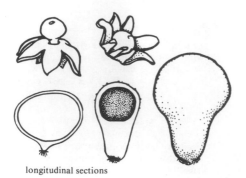

longitudinal sections

Fruiting body round to oval or pear-shaped, or the outer skin splitting into starlike rays; interior (spore mass) firm when young but powdery or dusty at maturity; spores borne inside a spore case or numerous lentil-like capsules; stalk absent or present only as a narrowed sterile base or rootlike fibers; usually growing above the ground ....
PUFFBALLS & EARTHSTARS, p. 677

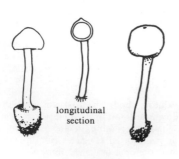

longitudinal section

Fruiting body with a puffball-like spore case mounted on a well-developed, clearly differentiated stalk; spore mass (interior of spore case) dusty or powdery at maturity; stalk terminating at base of spore case (i.e., *not* percurrent); often found in deserts, sand, or waste places ..
.... **STALKED PUFFBALLS,** p. 715

Fruiting body minute (often less than 1 cm broad), shaped like a miniature bird's nest with one or more small "eggs" inside (but top of "nest" often covered by a layer of tissue when young) ....
.... **BIRD'S NEST FUNGI,** p. 778

Fruiting body emerging from an "egg" whose skin forms a volva (sack) at base of mature fruiting body; fruiting body with a cap and stalk or just a stalk, or with arms, tentacles, or a branched or latticed framework; inside or outside surfaces of fresh fruiting body coated with a foul-smelling, greenish to brown to blackish spore slime ....
.... **STINKHORNS,** p. 764

interiors,
one with a columella
and one without

exterior

Fruiting body usually underground, round to oval or knobby; interior (spore mass) firm to spongy or gelatinous but not often powdery, usually composed of small chambers or cavities or occasionally plates; columella (internal stalk) present or absent .. **FALSE TRUFFLES,** p. 739

ASCOMYCETES

interiors,
one hollow and
one solid but marbled

exterior

Fruiting body usually underground, round to oval or knobby; interior hollow or with several large cavities or with canals or interior solid and marbled with sterile veins; texture of interior firm or sometimes powdery but *not* gelatinous; columella (internal stalk) only rarely present .... TRUFFLES, p. 841

Fruiting body growing on insects, spiders, or truffles *or* engulfing other mushrooms in a pimpled, powdery, or lumpy layer of tissue *or* growing on wood; if on wood then dark brown to black (or black beneath white spore powder) and tough, hard, or charcoal-like ....
.... FLASK FUNGI, p. 878

Fruiting body disc-shaped (flat) to cup-shaped, vaselike, or earlike (with or without a stalk) *or* fruiting body with a stalk and clearly differentiated cap; cap when present cup-shaped to saddle-shaped, irregularly lobed, brainlike, thimble-like, or pitted ....
MORELS, ELFIN SADDLES, & CUP FUNGI, p. 783
(also see **Earth Tongues**, p. 865)

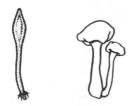

Fruiting body usually erect, simple, and unbranched (clublike), but usually with a differentiated (often flattened or swollen) fertile "head" or even a small, rounded or wrinkled cap; fruiting body usually small, if with a small cap then usually tough or gelatinous ....
.... EARTH TONGUES, p. 865
(also see **Morels, Elfin Saddles, & Cup Fungi**, p. 783)

The shaggy mane, *Coprinus comatus* (p. 345), is one of many Basidiomycetes. It is a familiar sight along roads and parking lots in cool wet weather.

Basidiomycetes

BASIDIOMYCOTINA

THE Basidiomycetes are a large group (subdivision) of fungi that bear spores on (but not inside) specialized cells called **basidia**. Two major classes of Basidiomycetes produce fleshy fruiting bodies or "mushrooms"—the Hymenomycetes and the Gasteromycetes, keyed below. Other classes of Basidiomycetes, such as the rusts and smuts, are not treated in this book.

Key to the Basidiomycetes

1. Basidia and spores borne externally (on the exposed surfaces of gills, tubes, spines, branches, lobes, etc.); spores forcibly discharged at maturity, i.e., a spore print often (but not always) obtainable; fruiting body with a cap and stalk, or clublike, or branched, or bracketlike, or crustlike (without a stalk or sometimes without a cap) or lobed or bloblike, etc. 2
1. Basidia and spores borne internally (inside the fruiting body *or* inside a spore case or small capsules); spores *not* forcibly discharged, thus a spore print unobtainable **Gasteromycetes,** p. 676

2. Fruiting body at first egglike with a gelatinous interior, the outer skin then breaking and an unbranched (often phallic), branched, or lattice-like fruiting body emerging; spores contained in a greenish to brown or black, foul-smelling slime that coats all or part of the fresh fruiting body (but slime dispersed by flies or rain); gills and tubes absent . . . **Gasteromycetes,** p. 676
2. Not as above . . . . **Hymenomycetes,** below

HYMENOMYCETES

IN this large divison of the Basidiomycetes the spore-bearing cells (basidia) form a layer or palisade **(hymenium)** on an exposed surface or surfaces of the fruiting body, and the spores are forcibly discharged at maturity. Terrestrial forms are usually furnished with a cap and stalk, but many of the wood-inhabiting types are shelflike, hooflike, bracketlike, crustlike, or bloblike. In the more primitive forms the hymenium is smooth to slightly wrinkled or warted, while in the more specialized (advanced?) types it takes the form of gills, tubes, spines, or upright branches.

The Hymenomycetes are traditionally divided into two orders: the Agaricales include the gilled mushrooms (agarics) and fleshy tube fungi (boletes), while the Aphyllophorales are an artificial group embracing the polypores, crust fungi, coral fungi, chanterelles, and teeth fungi. A third group, the jelly fungi (Tremellales, Dacrymycetales, and Auriculariales), are treated in this book as Hymenomycetes, although some taxonomists consider them to be only distantly related.

In the absence of a detailed fossil record, there is a great deal of speculation as to whether the Hymenomycetes evolved from the Gasteromycetes or vice-versa. It may actually be that evolution has flowed both ways. Spore dispersal is enhanced if the spores are actively (forcibly) discharged as in the Hymenomycetes, but the enclosed fruiting body of the Gasteromycetes is advantageous to spore development because it affords greater protection against the environment. Thus, many Hymenomycetes have adapted to inhospitable (cold, hot, or arid) environments by fruiting underground and/or never exposing their hymenium—thus becoming "gastroid" and losing their ability to discharge spores in the process. On the other hand, there may be lines of evolution leading from primitive Gasteromycetes to certain Hymenomycetes through the incremental development of a hymenium, stalk, and cap.

*Since most fungi decay quickly, they are far less apt to be fossilized than plants.

Key to the Hymenomycetes

1. Underside of cap with hundreds of pores (tube mouths) which may be obvious or very minute; tube layer visible when fruiting body is cut longitudinally (perpendicularly) 2
1. Not as above; pores and tubes absent 3

2. Fruiting body fleshy, rapidly decaying, usually but not always terrestrial; stalk typically central or nearly so; tubes usually—but not always—easily separable from cap (and often from each other); veil present or absent .... **Boletaceae** (a family in the **Agaricales**), p. 488
2. Fruiting body usually (but not always!) tough in age and usually (but not always!) growing on wood; stalk absent or lateral to off-center, *or* if central then tube layer not easily separable from cap; veil absent .... **Aphyllophorales**, p. 548

3. Underside of cap with radiating blades (gills) .... **Agaricales**, below
3. Gills absent (but spines, warts, folds, veins, or wrinkles may be present) 4

4. Fruiting body gelatinous (jellylike) or very rubbery; usually (but not always) growing on wood; basidia partitioned or forked (under the microscope) .... **Tremellales & Allies**, p. 669
4. Not as above .... **Aphyllophorales**, p. 548

Agarics (Gilled Mushrooms)

AGARICALES

THE agarics, or gilled mushrooms, bear their spores on radiating blades or plates called **gills.** They are by far the most familiar, numerous, and complex group of fleshy fungi, and are thought to have arisen from several different ancestors. All of the gilled mushrooms were originally grouped in a single massive genus, *Agaricus,* but in the 19th century a Swedish naturalist, Elias Fries (the "father" of mushroom taxonomy), divided *Agaricus* into a number of smaller genera based on easily ascertained features such as attachment of the gills, texture of the stalk, presence versus absence of a veil, and color of the spores.

In the ensuing years, agaricologists have tried to achieve a more "natural" classification (i.e., one that reflects common ancestry rather than superficial similarities), and in so doing have come to rely heavily on microscopic and chemical characters, such as the shape and arrangement of the cells in the gill tissue and cap cuticle and the shape, ornamentation, and amyloidity of the spores. Over one hundred genera of gilled mushrooms are now recognized, including most of Fries' original genera, albeit often in modified form (for example, *Agaricus* (not to be confused with the term "agaric"), in its modern sense, is restricted to the cultivated mushroom and its close relatives). The "Friesian" system, however, has remained popular among amateurs or "toadstool-testers" because of its simplicity.

Since modern agaric taxonomy is based on so many seemingly esoteric characters, and since most people have neither a microscope, nor the requisite chemical reagents, nor the necessary training to correctly use a microscope and reagents if they had them, a compromise between the modern and Friesian systems has been used in this book, to wit: most of the modern genera and families have been recognized, but the characters used to distinguish them are largely Friesian (macroscopic). The Agaricales have been divided into fourteen families, which are keyed out here. The boletes constitute a fifteenth family, but are treated separately. In order to use the key it is necessary to know the spore color—preferably by taking a spore print. It is also necessary to have some patience—or at least to realize that *subjectivity* (e.g., waxy versus non-waxy gills) and *fallibility* are the ineluctable byproducts of any key that relies on macroscopic (field) characters, when the critical distinctions between the different entities are in fact microscopic.

A pictorial chart of gilled mushroom genera has also been provided on pp. 61-62, thereby circumventing the family category. It is by no means infallible, but as already pointed out, neither is the dichotomous key, nor for that matter, are you and I.

Key to the Agaricales

1. Gills deformed or contorted or branched to form large cavities, sometimes never exposed; spores *not* forcibly discharged, hence spore print *not* obtainable; found mostly in deserts and mountains .... (see **Podaxales & Allies**, p. 724)
1. Not as above; spores forcibly discharged, hence a spore print obtainable *if spores are being produced*; gills exposed at maturity; common and widespread 2

2. Spore print white to buff, yellow, yellow-orange, or lilac-tinged 3
2. Spore print some other color (pinkish, salmon, yellow-brown, brown, rusty-orange, rusty-brown, chocolate-brown, purplish, greenish, black, etc.) 10

3. Universal veil enveloping young specimens and forming a volva at base of stalk when it ruptures *and/or* leaving numerous remnants (warts or flat patches) on cap .... **Amanitaceae**, p. 262
3. Neither volva nor warts present (but cap and stalk may have scales or fibrils) 4

4. Gills typically free *and* veil present; veil usually forming an annulus (ring) on stalk, or if not then stalk typically scaly or slimy below the veil 5
4. Not as above; veil absent, or if present then gills normally attached to stalk 6

5. Cap slimy or viscid when moist; stalk sometimes viscid also; gills tissue divergent (under the microscope); not common .... **Amanitaceae**, p. 262
5. Cap dry or only slightly viscid when moist; stalk dry; gill tissue not divergent; common .... **Lepiotaceae**, p. 293

6. Gills decurrent and foldlike (at least when young), i.e., gills thick, blunt, shallow, and usually forked or with cross-veins .... (see **Cantharellaceae**, p. 658)
6. Not as above; gills usually platelike or bladelike 7

7. Gills and/or flesh exuding a latex (milk or juice) when broken; stalk typically more than 3 mm thick; spores with amyloid warts or ridges .... **Russulaceae**, p. 63
7. Not as above 8

8. Fruiting body brittle and rather rigid, the stalk snapping open cleanly like a piece of chalk (i.e., without fibrous context); cap usually plane to depressed at maturity; stalk typically at least 3 mm thick; veil absent; usually but not always terrestrial; cap and stalk tissue typically containing nests of sphaerocysts; spores with amyloid warts or ridges ...... **Russulaceae**, p. 63
8. Not with above features 9

9. Gills soft and clean, with a waxy appearance or texture; cap often brightly colored; stalk more or less central; usually terrestrial; basidia long and narrow (at least six times as long as the spores); spores smooth .... **Hygrophoraceae**, p. 103
9. Not as above; gills not normally waxy; stalk central to lateral or absent; on ground or wood .... **Tricholomataceae**, p. 129

10. Spore print pinkish to flesh-colored, salmon, pinkish-cinnamon, or sordid reddish 11
10. Spore print some shade of orange, brown (including cinnamon-brown), green, purple, gray, or black 16

11. Universal veil present, usually forming a saclike volva at base of stalk .... **Pluteaceae**, p. 253
11. Not as above; volva typically absent 12

12. Gills typically free at maturity; growing on wood (but wood sometimes buried!) or sawdust or sometimes in lignin-rich humus; spores *not* angular .... **Pluteaceae**, p. 253
12. Not as above 13

13. Odor strong (fishy or like cucumber) and stalk velvety *or* arising from an underground, often hollow "tuber" *or* found on wood and fruiting body pinkish to yellow-orange with a reticulate (veined) cap surface; spores *not* angular; not common .... **Tricholomataceae**, p. 129
13. Not as above; common 14

14. Spore print pale pinkish, pinkish-buff, or pinkish-cream; gills sometimes purple or purple-tinged when fresh; spores *not* angular .... **Tricholomataceae**, p. 129
14. Spore print usually deeper in color (flesh-colored to sordid reddish, salmon, etc.); gills not often purple; spores usually angular in end and/or side view 15

15. Usually growing in clusters; flesh fragile; spores *not* angular; not common, at least in the western states .... **Coprinaceae**, p. 341
15. Not as above; spores typically angular in end *and/or* side view; very common and widely distributed .... **Entolomataceae**, p. 238

16. Spore print greenish to grayish-olive 17
16. Not as above 19

17. Gills adnate to decurrent .... **Gomphidiaceae**, p. 481
17. Gills free or only slightly attached 18

18. Gills blood-red to reddish to dark brown .... **Agaricaceae**, p. 310
18. Gills white to dull greenish or grayish .... **Lepiotaceae**, p. 293

19. Spore print purple-brown to purple-gray, purple-black, smoky-gray, black, chocolate-brown, or *deep* brown 20
19. Spore print rusty-orange to rusty-brown, cinnamon-brown, yellow-brown, dull brown, bright brown, cigar-brown, etc. (but may appear darker in heavy deposits!) 26

20. Gills usually decurrent (but sometimes adnate); spore print smoky-olive to smoky-gray to black; associated with conifers .... **Gomphidiaceae**, p. 481
20. Not as above; gills free to adnexed, adnate, or occasionally decurrent 21

21. Gills and/or cap auto-digesting (i.e., turning into an inky black mass) at maturity; spore print black .... **Coprinaceae**, p. 341
21. Not as above 22

22. Veil present, usually forming an annulus (ring) on stalk; gills free or nearly free at maturity, whitish to pinkish when young but becoming chocolate-brown or darker in age; cap *not* deeply striate; spore print chocolate-brown .... **Agaricaceae**, p. 310
22. Not as above 23

23. Cap brightly colored (yellow, green, orange, red, etc.) .... **Strophariaceae**, p. 367
23. Cap dull (some shade of brown, buff, gray, white, etc.) 24

24. Gills decurrent; cap margin inrolled when young; spore print reddish-brown or cocoa-brown .... **Paxillaceae**, p. 476
24. Not as above 25

25. Cap usually viscid when moist (but may dry out!); fruiting body fragile or not fragile; cap cuticle typically filamentous (under microscope) .... **Strophariaceae**, p. 367
25. Cap usually not viscid; fruiting body usually quite fragile; cap cuticle typically cellular (see illustrations on p. 19) .... **Coprinaceae**, p. 341

26. Growing on other mushrooms .... **Tricholomataceae**, p. 129
26. Not as above 27

27. Stalk absent or rudimentary; usually growing shelflike on wood 28
27. Stalk present; on wood or ground 30

28. Fruiting body very tough, leathery, or corky .... (see **Polyporaceae & Allies**, p. 549)
28. Not as above; fruiting body fleshy 29

29. Gills forked or with cross-veins or even with pores near base of cap ..... **Paxillaceae**, p. 476
29. Not as above .... **Cortinariaceae**, p. 396

30. Fruiting body small, fragile, often withering quickly; cap usually oval to conical or bell-shaped; usually growing in grass, dung, or gardens; cap cuticle typically cellular (see illustrations on p. 19) .... **Bolbitiaceae**, p. 466
30. Not with above features 31

31. Gills typically decurrent, often forked (or even forming pores near stalk) and often peeling easily from cap; margin of cap usually inrolled when young; veil absent; spore print yellowish to brown; stalk fleshy .... **Paxillaceae**, p. 476
31. Not as above; gills occasionally decurrent but usually not; veil present or absent; spore print variously colored (rusty-orange, rusty-brown, dull brown, etc.); very common ....
.... **Cortinariaceae**, p. 396

PICTORIAL KEY TO GILLED MUSHROOMS

Characteristics	Spore print white to buff, pale pinkish-buff, yellowish, yellow-orange, greenish, or tinged lilac		Spore print pinkish to reddish to flesh-color	Spore print brown to rusty-brown, yellow-brown, cinnamon-brown, or rusty-orange		Spore print black to gray, purple-gray, purple-brown, or deep brown
1. Stalk consistently off-center, lateral, or absent; usually growing shelflike on wood. (If stalk is usually central, see #2)	*Lentinus* & *Lentinellus* *Marasmius*	*Omphalotus* *Pleurotus* & Allies	*Claudopus* *Phyllotopsis* *Rhodotus*	**Crepidotus** **Paxillus**		
2. Volva present; annulus absent or present. (If volva absent, see #3)	*Amanita*		*Volvariella*	*Cortinarius*		*Agaricus* *Coprinus*
3. Veil present, usually forming an annulus or leaving remnants on stalk; gills free at maturity and close or crowded. (If not with above features, see #4)	*Amanita* *Chlorophyllum* *Lepiota*	*Limacella* *Melanophyllum*	*Chamaeota*	*Conocybe*		*Agaricus* *Coprinus* *Psathyrella*
4. Veil present, usually forming an annulus or leaving remnants on stalk; gills attached and stalk fleshy. (If not with above features, see #5)	*Amanita* *Armillaria* & Allies *Cystoderma* *Hygrophorus*	*Lentinus* *Limacella* *Squamanita* *Tricholoma*		**Agrocybe** **Cortinarius** **Gymnopilus** *Hebeloma*	*Inocybe* **Pholiota** *Rozites* & *Phaeolepiota*	**Chroogomphus** **Gomphidius** **Naematoloma** **Psilocybe** **Stropharia**
5. Veil absent or evanescent; gills decurrent, ranging to adnate; stalk fleshy. (If not with above features, see #6)	*Camarophyllus* *Cantharellus* *Clitocybe* & Allies *Craterellus* *Gomphus* *Hygrocybe* *Hygrophoropsis* *Hygrophorus* *Laccaria* *Lactarius*	*Lentinus* & *Lentinellus* *Leucopaxillus* *Lyophyllum* & Allies *Melanoleuca* *Omphalotus* *Pleurotus* & Allies *Russula* *Tricholomopsis*	**Clitocybe** **Clitopilus** & **Rhodocybe**	*Agrocybe* *Cortinarius* *Gymnopilus* *Hebeloma* *Inocybe*	**Paxillus** *Pholiota* *Phylloporus* *Ripartites*	**Chroogomphus** **Gomphidius**

Note: The more common genera are indicated in bold type

PICTORIAL KEY TO GILLED MUSHROOMS

Characteristics	Spore print white to buff, pale pinkish-buff, yellowish, yellow-orange, greenish, or tinged lilac	Spore print pinkish to reddish to flesh-color	Spore print brown to rusty-brown, yellow-brown, cinnamon-brown, or rusty-orange	Spore print black to gray to gray, purple-gray, purple-brown, or deep brown
6. Veil absent or evanescent; gills notched or adnexed, ranging to adnate (but not free); stalk fleshy. (If different than above, see #7.)	*Camarophyllus* *Collybia* & Allies *Hygrocybe* *Hygrophorus* *Laccaria* *Lactarius* *Lentinus* *Leucopaxillus* *Lyophyllum* *Melanoleuca* **Russula** *Tricholoma* *Tricholomopsis*	*Clitocybe* *Entoloma* **Rhodocybe**	*Agrocybe* *Cortinarius* *Gymnopilus* *Hebeloma* *Inocybe* *Phaeocollybia* *Pholiota* *Phylloporus*	*Naematoloma* *Panaeolus* *Psathyrella* *Psilocybe*
7. Veil absent; gills free at maturity and crowded or at least close. (If gills attached, see #8.)	*Hygrocybe* *Lepiota* *Lyophyllum* *Marasmius*, *Collybia* & Allies *Melanoleuca* *Mycena* **Russula** *Squamanita* *Tricholoma*	*Leptonia* *Nolanea* **Pluteus**	*Bolbitius* *Conocybe* *Galerina* *Inocybe*	*Agaricus* *Coprinus* *Panaeolus* *Psathyrella*
8. Veil present, usually forming an annulus or leaving remnants on stalk; stalk thin. (If veil is absent, see #9.)	*Cystoderma* *Hygrophorus*		*Conocybe* *Cortinarius* *Galerina, Tub-aria* & Allies *Pholiota*	*Coprinus* *Panaeolus* *Psathyrella* *Psilocybe* *Stropharia*
9. Cap conical to bell-shaped, at least when young (its margin usually straight); stalk thin. (If cap differently shaped, see #10.)	*Camarophyllus* *Hygrocybe* *Hygrophorus* *Marasmius* *Mycena* *Omphalina* & *Xeromphalina*	*Entoloma* *Macrocystidia* *Nolanea*	*Bolbitius* *Conocybe* *Cortinarius* *Galerina, Tub-aria* & Allies *Inocybe*	*Coprinus* *Naematoloma* *Panaeolus* *Psathyrella* *Psilocybe*
10. Cap convex to umbilicate or plane, the margin often incurved when young; stalk thin. (If not as above, go back to #1!)	*Asterophora* *Camarophyllus* *Cantharellus* *Clitocybe* & Allies *Craterellus* *Hygrocybe* *Hygrophoropsis* *Hygrophorus* *Laccaria* *Lentinellus* *Lyophyllum* & Allies *Marasmius*, *Collybia* & Allies *Melanoleuca* *Mycena* *Omphalina* & *Xeromphalina*	*Clitocybe* *Entoloma* *Leptoniaa* & Allies *Macrocystidia* *Nolanea* *Psathyrella* *Clitopilus* & *Rhodocybe*	*Agrocybe* *Bolbitius* *Cortinarius* *Galerina, Tub-aria* & Allies *Gymnopilus* *Inocybe* *Pholiota* *Riparites*	*Coprinus* *Naematoloma* *Panaeolus* *Psathyrella* *Psilocybe*

Note: The more common genera are indicated in bold type

spores

RUSSULACEAE

THIS family differs fundamentally from all other gilled mushrooms in the anatomy of the cap and stem tissue: nests or rosettes of large, roundish cells called **sphaerocysts** are interspersed with the usual filamentous hyphae, giving the mushroom a characteristic brittle, granular texture. In addition, the spores are conspicuously ornamented with strongly amyloid warts, spines, and/or ridges. These unique features suggest that the Russulaceae evolved independently from other agarics—perhaps from underground ancestors. Some toadstool taxonomists consequently place them in a separate order, the Russulales, rather than in the Agaricales.

There are two large genera in the Russulaceae: *Lactarius,* which exudes a milk or juice **(latex)** when broken, and *Russula,* which does not. Though they're often characterized as a "white-spored" family, their spore color ranges from white to yellow, buff, or ochraceous.

Lactarius and *Russula* are mycorrhizal and often fruit together in spectacular abundance. From an edibility standpoint they are good for beginners because few, if any, of the mild-tasting species are dangerous. Those species with an acrid or bitter taste are best avoided since many are poisonous or indigestible, at least raw. To determine whether or not a species is **acrid**, chew on a small piece of the cap and gills for a minute, then spit it out. If it is acrid, you'll experience a burning sensation on your tongue—in fact, you may have to wash out your mouth to get rid of the taste! If you're not sure whether it's acrid, then it's probably "mild." Be careful—some species really pack a wallop!

Key to the Russulaceae

1. Fresh fruiting body usually exuding a latex (milk or juice) when broken (the latex is best seen by cutting the gills near the stalk)* .... *Lactarius,* p. 64
1. Latex absent* .... *Russula,* p. 83

*In dry or very wet weather, *Lactarius* species with a scanty latex may exhibit none at all. The following characteristics will help distinguish such specimens from *Russula*:

If the fruiting body has greenish stains and colored flesh, it is probably a *Lactarius.*
If the cap is brightly colored and has white flesh, it is probably a *Russula.*
If the cap margin is striate, it is probably a *Russula.*
If the gills, stalk, or flesh are some color besides white or yellow when fresh, it may well be a
Lactarius. (Some Russulas, however, discolor reddish, gray, brown, or black in age.)
If the stalk is scrobiculate (pitted with darker spots), it is a *Lactarius.*
If the cap margin is bearded, it is a *Lactarius.*
If none of the above are applicable, try *Russula* first, then *Lactarius.*

The latex which distinguishes *Lactarius* from *Russula* is best seen by cutting the mushroom near the juncture of the gills and stalk. In many species, such as *L. argillaceifolius* var. *megacarpus,* shown here, the latex is milky white, but in some species it is brightly colored and in others practically absent.

LACTARIUS   (Milk Caps)

Typically terrestrial, mostly medium-sized woodland mushrooms. CAP usually depressed at
maturity, usually with an inrolled or incurved margin when young, sometimes concentrically zoned.
Flesh *crisp, brittle, usually exuding a latex when broken;* taste mild or acrid. GILLS attached
(usually adnate to decurrent), variously colored. STALK *typically rigid, brittle, snapping open
cleanly like chalk; central, often scrobiculate.* VEIL and VOLVA absent. SPORE PRINT *white to
buff or yellowish.* Spores with amyloid warts and/or ridges. Cap tissue nearly always with nests of
sphaerocysts.

LACTARIUS species are called milk caps because they exude a milky or juicy fluid
(**latex**) when broken (see Color Plate 6 and photo on p. 63). In dry or very wet weather
they may lack a distinct latex (especially if old), but their crisp or brittle flesh distinguishes
them from all but the Russulas (see footnote at bottom of p. 63 for a more detailed
comparison). A few Mycenas possess a latex, but they strike a radically different pose:
a conical to bell-shaped cap perched on a long, thin, fragile stalk.

The most important features to note on any *Lactarius* are the color of the latex, and
the color changes the latex undergoes or produces on surrounding tissue when exposed
to the air. The colors should *always* be noted on fresh material. The latex may react with
the air rapidly (within 45 seconds) as in *L. vinaceorufescens,* or slowly (within 5 hours)
as in *L. deliciosus,* or not at all, as in *L. fragilis.* The taste—whether mild or acrid (peppery)
—is also significant, as are microscopic features such as the structure of the cap cuticle
and pattern of ornamentation on the spores. It is also helpful to note whether or not the
stalk is **scrobiculate** (pitted with darker spots).

Lactarius contains a number of delectable collectables. The candy cap *(L. fragilis)* is the
tastiest of the local species, besides being one of the most abundant. The green-staining
milk caps *(L. deliciosus, L. indigo,* et al) are edible and popular, and in eastern North
America, *L. volemus, L. corrugis,* and *L. hygrophoroides* are *very* good. Many species,
on the other hand, are poisonous or indigestible. As a rule, the peppery-tasting ones
should be avoided as well as those with yellow- or purple-staining latex. In Europe, some of
the acrid species are eaten after pickling or parboiling, but it hardly seems worth the effort
or risk to do so.

Lactarius is a very large and complex genus, with over 200 species described from North
America. It is especially diverse in eastern—and more specifically southeastern—North
America. In fact, the milk caps are roughly to eastern hardwood forests what the
Russulas are to the coniferous forests of the West—so exceedingly diverse and ubiquitous
as to seem like "mushroom weeds." In California, however, there are fewer than 50 recorded species of *Lactarius,* and in our area there are only about a dozen. As a result,
they are much easier to identify than their bewildering brethren, the Russulas. Like the
Russulas, they are mycorrhizal, mostly terrestrial woodland fungi, but often appear at the
edges of woods or on tree-studded lawns. Many are specific to certain hosts—e.g., *L.
torminosus* grows with birch. In California the milk caps fruit from late summer through
the winter and sometimes rival the Russulas in abundance.

The publication in 1979 of *The North American Species of Lactarius* by L. R. Hesler and
Alexander Smith has resulted in a large number of name changes. For instance, the edible
species widely known as *L. sanguifluus* is now *L. rubrilacteus* and our *L. trivialis* has
become *L. argillaceifolius.* Those of you who find it hard enough to master *one* scientific
name for a mushroom without being responsible for *two* or *three* can salvage some solace
in the fact that while the *names* have changed, the *mushrooms* haven't. In other words,
so long as you know how to recognize a particular species, it hardly matters what you *call*
it—unless you are living with a taxonomist! Twenty-one species of *Lactarius* are described here and many others are keyed out and/or mentioned.

Key to Lactarius

1. Latex colored (red, orange, blue, brown, etc.), but often scanty or even absent; fruiting body often (but not always) staining greenish in age or where bruised (slowly, within 5 hours) . 2
1. Latex white to buff or clear (but may stain yellow or some other color when exposed!) 9

2. Latex blue; entire fruiting body blue to bluish-gray (or greenish-stained) ... *L. indigo,* p. 69
2. Fruiting body not blue (may be greenish-stained), or if partly blue then latex *not* blue 3

3. Latex carrot-orange to salmon-orange, yellow-orange, or rusty-orange (but may darken to red after exposure to air) 4
3. Not as above; latex red, purplish, brown, etc., or if orange in base of stalk then bluish in cap 5

4. Cap whitish when fresh; found in southeastern U.S. .. *L. salmoneus* (see *L. deliciosus,* p. 68)
4. Not as above; widespread and common .... *L. deliciosus* & others, p. 68

5. Latex dark red to purplish; fruiting body never bluish (but often greenish-stained) 6
5. Latex purple-brown to reddish-brown to brown or yellow-brown, or bluish in cap and red or orange in stalk; fruiting body sometimes with blue tints (especially young cap) 7

6. Associated with western conifers (mainly Douglas-fir and pine) 8
6. Associated with eastern conifers (mainly hemlock) *L. subpurpureus* (see *L. rubrilacteus,* p. 68)

7. Latex when present yellowish to brown or the color of grasshopper juice; found in eastern North America and the Southwest .... *L. chelidonium* (see *L. indigo,* p. 69)
7. Not as above .... *L. paradoxus* & others (see *L. indigo,* p. 69)

8. Cap whitish to pinkish-brown; found in Southwest .. *L. barrowsii* (see *L. rubrilacteus,* p. 68)
8. Not as above; cap reddish-brown to orangish or tan .... *L. rubrilacteus,* p. 68

9. Latex quickly becoming yellow upon exposure to air (usually within 45 seconds) 10
9. Not as above (but latex may *slowly* stain yellow or stain wounded tissue yellow) 15

10. Cap pinkish to pinkish-cinnamon to reddish or vinaceous *L. vinaceorufescens* & others, p. 74
10. Cap differently colored (white to yellowish, ochre, orange, etc.) 11

11. Cap white to yellowish to dark ochre; margin of cap bearded with woolly hairs (at least when young) or if not then associated with northern or mountain conifers 12
11. Not with above features 13

12. Cap pale yellow to dark ochre; stalk usually scrobiculate .... *L. scrobiculatus,* p. 73
12. Cap whitish when young (but often pale yellowish or yellow-stained in age); stalk not scrobi-culate or somewhat so only at maturity .... *L. resimus* (see *L. scrobiculatus,* p. 73)

13. Cap orangish; associated with conifers or birch in northern and western North America . 53
13. Not as above; either differently colored or associated with hardwoods such as oak 14

14. Cap orange to bright yellow-orange .... *L. croceus* (see *L. vinaceorufescens,* p. 74)
14. Not as above .... *L. chrysorheus* & others (see *L. vinaceorufescens,* p. 74)

15. Cap (and often stalk) with a velvety texture or appearance (use hand lens if unsure); cap gray to smoky-brown to dark brown or black, not viscid and not very large 16
15. Not as above (if velvety, then differently colored—white, orange, red-brown, etc.) 17

16. Gills close to crowded; associated with western conifers .... *L. fallax,* p. 77
16. Not with above features .... *L. lignyotus* & others (see *L. fallax,* p. 77)

17. Latex drying purplish or stained wounded areas purplish to dull lilac (often slowly); cap not whitish, or if whitish then also viscid or slimy when moist 18
17. Not as above 21

18. Cap distinctly hairy or fibrillose at or near the margin; cap yellow, ochre, buff, or dull orange .... *L. representaneus* & others, p. 75
18. Not as above; margin of cap naked 19

19. Cap pale yellow; cap *and* stalk viscid when moist *L. aspideoides* (see *L. representaneus,* p. 75)
19. Not with above features 20

20. Cap and stalk white and slimy, at least when young and moist; common under conifers in the Pacific Northwest .... *L. pallescens* (see *L. uvidus* group, p. 75)
20. Not as above; cap usually buff, gray, brown, purplish, etc. .. *L. uvidus* group & others, p. 75

21. Cap dark green to sordid greenish- or olive-brown (at least at center), sometimes honey-colored at margin; stalk similarly colored; latex white; taste acrid; cap staining magenta in KOH 22
21. Not with above features 23

22. Cap and stalk dark or dull greenish; associated with hardwoods (especially oak) in eastern North America .... *L. atroviridis* (see *L. olivaceoumbrinus,* p. 70)
22. Cap and stalk usually browner; found mainly with conifers *L. olivaceoumbrinus* & others, p. 70

23. Margin of cap distinctly bearded with woolly hairs, at least when young; cap sometimes whitish, but usually pale pinkish to pinkish-orange (at least at center); taste very acrid; associated with birch or sometimes aspen .... *L. torminosus* & others, p. 73
23. Not as above; if margin bearded then cap differently colored or habitat different 24

24. Gills well-spaced; cap whitish to pale grayish- or pinkish-tan or yellowish; latex and wounded areas staining rosy-salmon to rusty-orange (especially gills and stalk base); found under hardwoods in eastern North America (especially common in Southeast) *L. subplinthogalus*
24. Not as above 25

25. Cap pinkish-gray to purplish-gray to vinaceous-gray or vinaceous-buff (or brown tinged with these colors), *not* viscid; odor distinctly coconut-like; found mainly with birch (often in parks) or alder; northern .... *L. glyciosmus*
25. Not as above 26

26. Gills pinkish to pale pinkish; cap viscid when moist, white or with pinkish to lavender stains; taste acrid; associated with aspen, poplar, and willow .... *L. controversus,* p. 70
26. Not as above (cap if whitish usually *not* viscid) 27

27. Margin of cap with cottony tissue (at least when young); cap usually whitish when young, but may discolor yellowish, tan, etc. in age ..... *L. deceptivus* & others (see *L. piperatus,* p. 71)
27. Margin of cap naked or bearded with hairs but not as above 28

28. Cap predominantly whitish (but may discolor in age), not viscid; taste distinctly acrid (sometimes extremely so!); found mostly with hardwoods in eastern North America 29
28. Not as above (if cap whitish, then taste not acrid) 31

29. Latex staining wounded tissue pinkish; especially common in Southeast ..... *L. subvernalis*
29. Not as above 30

30. Cap and stalk minutely velvety; gills close to well-spaced *L. subvellereus* (see *L. piperatus,* p. 71)
30. Cap and stalk not velvety; gills crowded to *very* crowded .... *L. piperatus* & others, p. 71

31. Cap whitish to olive-buff (often with darker and paler areas); latex white, scanty, staining tissue reddish to pinkish-gray or pale yellow; taste *not* acrid; favoring pine *L. pallidiolivaceus*
31. Not as above 32

32. Latex white and very copious, staining wounded tissue brown (often slowly); taste mild; cap *and/or* stalk minutely velvety, never viscid; cap white to yellow, orange, red-brown, etc. (never grayish or olive); common under hardwoods in eastern North America 33
32. Not as above (but may have some of above features) 35

33. Cap and stalk whitish to buff or slightly yellowish .... *L. luteolus* (see *L. volemus,* p. 78)
33. Cap and stalk tawny to orange-brown, reddish-brown, or brown 34

34. Cap brown to reddish-brown or rusty-brown, often wrinkled; stalk typically 1-3 cm thick; odor usually mild .... *L. corrugis* (see *L. volemus,* p. 78)
34. Cap tawny, orange-brown, rusty-brown, or sometimes yellowish, not normally wrinkled; stalk typically less than 1.5 cm thick; odor often fishy in age .... *L. volemus,* p. 78

35. Cap with various shades of yellow, pale buff, ochre, copper, orange, or pinkish-orange, often (but not always) concentrically zoned; taste very acrid (peppery or burning) 36
35. Not with above features (but may have some of them) 40

36. Cap with various shades of yellow, ochre, dark ochre, buff, etc. 37
36. Cap oranger, pinker, or more coppery than above 38

37. Cap margin with fibrils which are brown to grayish in age *L. payettensis* (see *L. alnicola,* p. 71)
37. Not as above .... *L. alnicola* & others, p. 71

38. Margin of cap bearded when young *L. subvillosus* & *L. psammicola* (see *L. torminosus,* p. 73)
38. Not as above; margin of cap naked even when young 39

39. Associated with northern conifers .... *L. olympianus* (see *L. alnicola,* p. 71)
39. Associated with southern hardwoods (mainly oak) ..... *L. yazooensis* (see *L. alnicola,* p. 71)

40. Cap cinnamon to butterscotch-brown or buffy-brown; cap and stalk slimy or viscid when moist; taste acrid; common in eastern North America (rare or absent in West) .... *L. affinis*
40. Not with above features 41

41. Cap *and* stalk viscid or slimy when moist; cap dark brown to smoky-brown to charcoal-gray (without orangish, reddish, yellowish, or purplish shades), but may fade in age; associated with conifers 42
41. Not as above; cap viscid or dry; stalk not viscid and/or fruiting body differently colored . 43

42. Stalk more or less tan or paler .... *L. kauffmanii* (see *L. pseudomucidus,* p. 77)
42. Stalk darker or grayer (colored like cap or slightly paler) .. *L. pseudomucidus* & others, p. 77

43. Taste distinctly to excruciatingly acrid (sometimes latently so); cap pinkish-tan to reddish-tan to reddish-brown or brick-red (at least in age) or sometimes orangish 44
43. Not as above; cap differently colored and/or taste mild to only slightly acrid 48

44. Found under manzanita; cap small and viscid when moist *L. manzanitae* (see *L. rufus,* p. 79)
44. Not with above features 45

45. Associated with hardwoods in eastern North America 46
45. Not as above; usually but not always associated with conifers 47

46. Cap 6-20 cm or more broad, pinkish-tan to reddish-tan to brick-red (but whitish when young or where covered by humus), not zoned concentrically; gills white to buff .... *L. allardii*
46. Cap 5-15 cm broad, orange-brown to reddish-brown, often zoned concentrically; mature gills reddish-brown to dull purple-brown .... *L. peckii*

47. Cap dull or dark reddish to brick-red, vinaceous, or dark brown ....*L. rufus* & others, p. 79
47. Cap oranger or more brightly colored than above 48

48. Odor distinctly fragrant (sweet), at least when dried or cooked (and often when fresh) ... 49
48. Odor not fragrant or sweet, even when dried or cooked 52

49. Cap viscid, orange-brown to caramel-colored; odor coconut-like when dried; known only from California .... *L. cocosiolens* (see *L. subflammeus,* p. 79)
49. Not with above features 50

50. Latex clear (colorless); cap 4-15 cm broad; associated with northern conifers and especially common in boggy areas .... *L. helvus (=L. aquifluus)*
50. Latex white or watery white; cap 2-10 cm broad; widespread in many habitats 51

51. Stalk solid or stuffed (but sometimes with a central hollow in age), typically 0.5-3 cm thick; cap usually reddish but sometimes orangish; known only from California ..... *L. rufulus,* p. 82
51. Stalk often (not always!) hollow or partly hollow, especially in age, and usually slender (0.4-1 or occasionally 1.5 cm thick); cap usually burnt orange but sometimes redder, browner, or tawnier; widely distributed (including California) .... *L. fragilis* & others, p. 80

52. Cap small and olive-brown to dark grayish-brown or dark brown when young (but often tawny to cinnamon in age); common under alder in N. Calif. and Pacific Northwest *L. occidentalis*
52. Not with above features 53

53. Cap orange to reddish, pinkish, reddish-brown, or liver-colored 54
53. Cap grayish to violet-gray, dingy brown, olive brown, or dark brown 59

54. Cap viscid or slightly viscid when moist 55
54. Cap *not* viscid, often with a dull (matte) appearance 56

55. Cap scarlet to orange, orange-brown, or tawny .... *L. subflammeus* & others, p. 79
55. Cap pinkish to reddish to reddish-brown or liver-colored .... *L. subviscidus* & others, p. 80

56. Cap and stalk minutely velvety, orange to orange-brown; gills widely spaced, pallid; latex very copious; common under eastern hardwoods *L. hygrophoroides* (see *L. volemus,* p. 78)
56. Not as above 57

57. Associated with oak; mainly but not exclusively southern 58
57. Not as above; northern . *L. alpinus, L. thejogalus, L. oculatus* & others (see *L. rufulus,* p. 82)

58. Found in California .... *L. rufulus* & others, p. 82
58. Found in eastern North America .... *L. subseriiluus* & others (see *L. rufulus,* p. 82)

59. Cap medium-sized to large (4 cm or more broad), at least slightly viscid when moist 60
59. Not as above; cap small and violet-gray to gray or ochre-gray, usually with small scales at least in age; found in eastern North America, often on or near rotten wood .... *L. griseus*

60. Latex often discoloring injured tissue slowly; gills often dingy yellowish or yellow-brown in old age; found with hardwoods (mainly oak) *L. argillaceifolius var. megacarpus* & others, p. 76
60. Not as above .... *L. circellatus* & others (see *L. argillaceifolius* var. *megacarpus,* p. 76)

Lactarius deliciosus   (Delicious Milk Cap)   Color Plate 2

CAP 5-16 cm broad, broadly convex or with a depressed center and an inrolled margin when young, becoming depressed or shallowly funnel-shaped; surface viscid when moist but soon dry, smooth, often zoned; color variable: dull orange to carrot-orange or orange-brown, sometimes blotched with or entirely green; fading in age or dry weather to brownish, gray, dull greenish-gray, or even yellowish; margin inrolled when young. Flesh thick, brittle, orange to yellowish or greenish; taste mild or slightly bitter. **LATEX** very scanty, bright carrot-orange (but in some forms slowly staining dark red when exposed), eventually staining wounded or aged tissue greenish (within 5 hours). **GILLS** typically bright to dull orange, but varying to yellowish or orange-buff, greenish where wounded; adnate to decurrent, close. **STALK** 2-7 cm long, 1-2.5 cm thick, equal or narrowed at base, soon dry, sometimes scrobiculate; rigid, hollow in age, frequently maggot-riddled, colored like cap or paler. **SPORE PRINT** creamy yellowish-buff; spores 7-11 × 6-8 microns, broadly elliptical to nearly round, with amyloid ridges.

HABITAT: Scattered to gregarious or in troops under conifers (pine, spruce, etc.), common and widely distributed. It is abundant in our coastal pine forests in the late fall and winter, generally *after* the major crop of look-alike *L. rubrilacteus*.

EDIBILITY: Edible, but not necessarily delicious. Several varieties or forms occur and some are apparently better than others. Special treatment is required to overcome the grainy texture and latent bitterness. Its abundance and distinctiveness, however, make experimentation worthwhile. Some sources recommend slow cooking (e.g., baked in a casserole), others insist it should be cooked rapidly in a frying pan with *very* little butter. It is popular in Europe, and the Russians are fanatical about it—especially salted.

COMMENTS: This variable, cosmopolitan fungus strikes a discordant but colorful note with its unlikely combination of pistachio-green and carrot-orange—even more so when found in the company of the bright red *Russula rosacea*, as is so often the case in our area. The carrot-colored latex separates it from *L. rubrilacteus*, and confusion with other mushrooms is unlikely. The latex is so scanty as to often be non-existent, but the brittle flesh and greenish stains on the cap, stalk, and/or gills will identify it. The fruiting bodies persist for a long time and are often completely hollowed out by maggots. Several varieties of *L. deliciosus* have been described based on slight differences in color, staining reactions, and microscopic characteristics; one variety tends to have an areolate (scaly-cracked) cap in age. The Greeks were apparently fond of *L. deliciosus*, for it is depicted on a fresco from Herculaneum (buried in 79 A.D.). Other species: *L. thyinos*, of northern bogs, has carrot-orange latex but stains red, not green; in other respects it is quite similar. *L. salmoneus* is a smallish but distinctive southeastern species with a whitish cap, bright orange gills, flesh, and stalk (when fresh) and orange to salmon-orange latex. *L. pseudodeliciosus* is buff or dingy colored with yellow-orange to rusty-orange latex. All of these are edible and the latter two stain greenish in age or where wounded.

Lactarius rubrilacteus   (Bleeding Milk Cap)   Color Plates 3, 6

CAP 4-14 cm broad, broadly convex with a depressed center and inrolled margin when young, depressed or shallowly funnel-shaped in age; surface viscid when moist, smooth, reddish-brown to orange, orange-brown, or tan, or often concentrically zoned with these colors; duller and greenish-stained in age. Flesh thick, brittle, brownish to buff, reddish, etc.; taste mild or slightly bitter. **LATEX** scanty, dark red (but occasionally orange-red in old specimens), slowly staining wounded areas greenish. **GILLS** adnate to slightly decurrent, close, reddish or dull purplish-red, or tan with a dark reddish sheen; greenish where wounded. **STALK** 2-6 cm long, 1-2.5 cm thick, equal or narrowed below, firm, rigid, hollow, colored like cap or paler, sometimes scrobiculate. **SPORE PRINT** pale yellowish or buff; spores 7.5-10 × 6-8 microns, broadly elliptical to nearly round, with amyloid ridges.

HABITAT: Scattered or in large troops under conifers throughout the West; associated in our area with Douglas-fir and abundant in the fall and early winter. Where pines predominate it is largely supplanted by *L. deliciosus*, and where pine and Douglas-fir grow together, the two milk caps often mingle. In Europe, *L. sanguifluus* (which may be the same as *L. rubrilacteus*) is mycorrhizal with pine.

EDIBILITY: Edible, but not pleasing to everyone because of its granular texture; usually better, however, than *L. deliciosus* (see comments on edibility of that species).

COMMENTS: The dark red latex is the telltale trait of this handsome fungus, which for many years has been known to fungophiles as *L. sanguifluus*. In dry or very wet weather the latex may be absent, but in these conditions the greenish stains are usually quite pronounced. In fact, as with *L. deliciosus,* weathered fruiting bodies may be entirely green, and mature specimens often have tiny, aborted green "buttons" at their bases. Be sure not to confuse this species with the similarly colored *L. vinaceorufescens* and *L. chrysorheus*—they often grow with *L. rubrilacteus,* but have a white latex that quickly turns yellow. Other species: *L. barrowsii* has dark red latex, but its cap is much paler (whitish to pinkish-brown) and it occurs mainly with ponderosa pine. Another similar edible species, *L. subpurpureus* of eastern North America, has wine-red latex and a wine-red to silvery cap, and favors hemlock. See also *L. paradoxus* (under *L. indigo*).

Lactarius indigo   (Indigo Milk Cap)   **Color Plate 4**

CAP 4-15 cm broad, convex or centrally depressed with an inrolled margin when young, usually depressed in age; surface smooth, viscid when moist, then dry; indigo-blue when fresh but fading to grayish- or silvery-blue, sometimes with greenish stains; often zoned concentrically. Flesh pallid to bluish, brittle, slowly staining greenish; taste mild to slightly bitter-acrid. **LATEX** indigo-blue (bright dark blue), scanty, slowly staining wounded tissue greenish. **GILLS** adnate to slightly decurrent, close, indigo-blue becoming paler in age. **STALK** 2-6 cm long, 1-2.5 cm thick, indigo-blue to silvery- or grayish-blue, equal or narrowed at base, rigid, hollow in age, soon dry. **SPORE PRINT** creamy-yellowish; spores 7-9 × 5.5-7.5 microns, elliptical to nearly round, with amyloid warts and ridges.

HABITAT: Scattered to gregarious in summer and fall, mostly in oak and pine woods; found throughout southern and eastern North America, but most common along the Gulf Coast and in Mexico. I have found it in Arizona under ponderosa pine. If it should turn up in southern California it would certainly be a most welcome addition to our fungus flora.

EDIBILITY: Edible and very good—superior, at least, to the other greenish-staining milk caps, such as *L. deliciosus*. In Mexico it is sometimes sold in farmer's markets.

COMMENTS: The overall blue to blue-gray color and bright blue latex make this one of the safest and most memorable of all agarics. No other milk cap has blue latex, let alone a bluish fruiting body. Other species: *L. paradoxus,* common under pine in eastern North America (especially the South), often has a bluish-tinged cap when young, but has reddish-brown to purple-brown latex and gills. Another species, *L. chelidonium*, has a yellowish to dingy yellow-brown to bluish-gray cap and yellowish to brown latex "the color of grasshopper juice." It occurs commonly in the eastern United States and Southwest. *L. hemicyaneus* is a medium-sized to large species which usually has bluish flesh in the cap and orange to red-orange flesh in the base of the stalk. It may possibly occur in southern California, according to Greg Wright and Paul Harding. All of the above species are edible despite their tendency to stain green in age.

Lactarius olivaceoumbrinus   (Toadskin Milk Cap)

CAP 3-12 cm broad, convex with an inrolled margin becoming plane or shallowly depressed; surface viscid when moist, typically a mixture of dark olive, sordid olive-brown, and olive-buff, often spotted or zoned concentrically, but zones fading in age. Flesh thick, dingy olive, brittle, taste very acrid. **LATEX** copious, white, becoming greenish-gray (sometimes very slowly). **GILLS** adnate to decurrent, crowded, pallid becoming spotted or colored greenish or olive-gray. **STALK** 4-8 cm long, 1-3 cm thick; solid, becoming hollow in age; viscid when wet, then dry; rigid, colored more or less like cap; usually scrobiculate. **SPORE PRINT** pale buff; spores 7-10 × 6-9 microns, elliptical to nearly round, with amyloid warts and ridges. Cap surface staining magenta in KOH or ammonia.

HABITAT: Solitary or scattered or in small groups on ground under conifers in late summer and fall, Pacific Northwest and northern California, occasional.

EDIBILITY: Unknown. The sordid appearance and acrid taste are major deterrents.

COMMENTS: This species and its close relatives are among our most distinctive milk caps—easily recognized by their overall dingy greenish to murky olive-brown color and copious, acrid white latex. I know very few people who would call them *beautiful,* but that's precisely what they are—*beautiful*—in a grotesque sort of way that only a thoroughly jaded fancier of the fleshy fungi can appreciate. (For some reason, they remind me of toads —see Color Plate 1.) Other species: *L. sordidus* is a similar species with a somewhat browner cap and smaller spores; it also occurs in the Pacific Northwest as well as in eastern North America. *L. atroviridis* (**COLOR PLATE 1**) has a mottled dark greenish cap and stalk and grows under hardwoods (especially oak) in eastern North America. Neither is edible. The name *L. necator* (a European species) was previously applied to this group.

Lactarius controversus   (Poplar Milk Cap)

CAP (5) 7-20 cm broad, broadly convex becoming depressed; surface viscid when moist, then dry, white, or whitish with lavender to pinkish stains; margin at first inrolled. Flesh thick, firm, brittle, white; taste slowly acrid (burning). **LATEX** white, unchanging. **GILLS** adnate to slightly decurrent, crowded, narrow, pinkish to creamy-pink. **STALK** 2.5-7 cm long, 1-3 cm thick, equal or tapered downward, white, sometimes spotted; hollow in age; rigid. **SPORE PRINT** creamy to pale pinkish; spores 6-7.5 × 4.5-5 microns, elliptical, with amyloid warts and ridges.

Lactarius controversus has pinkish gills, white latex, acrid taste, and inrolled cap margin when young.

HABITAT: Scattered to gregarious under aspen, poplar, and willow; widely distributed. It can be found in the aspen forests of the Sierra Nevada in the late summer and fall. I have seen large fruitings in New Mexico in the late summer.

EDIBILITY: To be avoided due to the acrid taste.

COMMENTS: This handsome milk cap is one of several large, whitish species with white, unchanging latex and a very peppery taste. The viscid cap when moist and pinkish-tinted gills plus the association with aspen, poplar, and willow distinguish it from the other large white Lactarii (see *L. piperatus*), while the presence of a latex places it in *Lactarius*.

Lactarius piperatus   (Peppery White Milk Cap)

CAP (4) 6-16 cm broad, broadly convex and usually depressed centrally, becoming broadly vase-shaped in age; surface more or less smooth, dry, unzoned, white to creamy-white, but often developing dingy buff or tan stains in age; unpolished; margin naked. Flesh thick, crisp, brittle; odor mild, taste extremely acrid. **LATEX** white, copious, unchanging or drying yellowish (or in one variety staining wounded tissue dingy greenish). **GILLS** adnate to decurrent, narrow and very crowded, white to creamy, often forked. **STALK** 2-8 cm long, 1-3 cm thick, equal or tapered toward base, dry, white, smooth or with a whitish bloom, not scrobiculate. **SPORE PRINT** white; spores 4.5-7 × 5-5.5 microns, elliptical to nearly round with inconspicuous amyloid warts and ridges.

HABITAT: Solitary to widely scattered or gregarious on ground under hardwoods, common in the summer in eastern North America. It has also been reported from the west coast, but I have not seen it west of Minnesota.

EDIBILITY: Not recommended. In Russia and Scandinavia it is eaten, along with other acrid milk caps, after parboiling or pickling. However, it is rather difficult to digest and may even be poisonous if not properly prepared.

COMMENTS: The dull dry white cap, very crowded narrow gills, extremely peppery taste, and copious white "milk" form a distinctive set of characters. It has much the aspect of *Russula brevipes,* but of course that species lacks a latex. There are several similar, medium-sized to large whitish, acrid milk caps with white latex, including: *L. neuhoffii,* with larger spores and only moderately crowded gills that are sometimes pale pinkish-buff at maturity; *L. subvellereus,* with minutely velvety cap and stalk, and close to well-spaced gills; and *L. tomentoso-marginatus* and *L. deceptivus,* with a whitish cap that becomes dull tan or brownish in age and has conspicuous cottony tissue on the margin when young. All of these are common in eastern North America under hardwoods; the latter also grows with conifers. None have the slightly viscid cap and pinkish gills of *L. controversus.* Other species: *L. pseudodeceptivus* mimics *L. deceptivus,* but grows with northwestern conifers.

Lactarius alnicola   (Golden Milk Cap)   **Color Plate 5**

CAP 5-14 (20) cm broad, convex with a central depression and conspicuously inrolled margin when young, depressed or broadly funnel-shaped in age; surface viscid when moist, usually zoned concentrically with various shades of ochre and pale yellow, but sometimes nearly evenly colored; margin naked or slightly hairy but not bearded. Flesh thick, brittle, crisp, whitish; taste *very* acrid. **LATEX** white, unchanging or *very* slowly yellowing or staining wounded tissue yellowish. **GILLS** crowded, adnate to decurrent, whitish when young becoming buff or ochraceous-toned in age; wounded areas stained yellowish to yellow-brown. **STALK** 2-6 cm long, 1.5-3 cm thick, equal or with narrowed base, hard, usually hollow in age; pallid or tinged cap color, often (but not always) scrobiculate (with darker yellow to ochraceous spots). **SPORE PRINT** whitish to yellowish; spores 7.5-10 × 6-8 microns, elliptical, with amyloid warts and ridges.

Lactarius alnicola has a golden zoned cap and peppery white latex. This is our oak-loving variety, which has a *latently* acrid taste and bald cap margin. See color plate for conifer-loving form.

HABITAT: Scattered to gregarious under spruce and other conifers (despite the species epithet, which implies alder) in the Rocky Mountains and Pacific Northwest in the late summer and fall, and also abundant (a slightly different form—see comments) in our live oak woodlands in the fall and winter. Still another variant occurs in coastal sand dunes with willow and bush lupine.

EDIBILITY: Not edible. The excruciatingly peppery taste is a formidable deterrent.

COMMENTS: The overall yellowish-ochre color and zoned, non-bearded cap plus the frequently scrobiculate stalk, acrid taste, and white latex typify a very confusing group of milk caps which have traditionally passed under the European names of *L. insulsus* and *L. zonarius.* There is still doubt, however, as to just what those species are. The common member of this group in central and southern California differs from typical *L. alnicola* in its mycorrhizal mate (oak), somewhat smaller size and paler cap (yellowish-buff), frequently short, off-center stem, and *latently* acrid taste. Its latex usually stains surrounding tissue dingy yellow but does not turn yellow itself. Forms are encountered, however, in which the latex *does* yellow slowly, and forms in which it discolors tissue buff or grayish-buff, thus muddling the picture so thoroughly that I begin to yearn for the baseball season. (Ah, the simplicity, security, and symmetry of baseball, where each participant has a fixed name, number, and position, and a unique and indisputable set of statistics and characteristics.) Other species with a more or less zoned, non-bearded cap and acrid white latex include: *L. olympianus,* a common western conifer-lover, typically with an oranger cap and non-scrobiculate stalk; and two oak-loving easterners, *L. psammicola f. glaber,* with a yellowish cap and latex that stains wounded tissue buff to pinkish-cinnamon, and *L. yazooensis,* a southern species with an orangish cap. Similar species that are bearded are listed under *L. torminosus* and *L. scrobiculatus.* They include *L. subvillosus,* a common local species whose cap is often naked in age, but pinker or oranger than that of *L. alnicola.* Also worth mentioning is *L. payettensis,* a western conifer-lover with white scanty latex that yellows *slowly* and brownish to grayish fibrils on the cap margin. None of the above species should be eaten, at least until they are better known.

Lactarius torminosus   (Bearded Milk Cap)   Color Plate 7

CAP 4-12 cm broad, convex with a central depression and strongly inrolled margin when young, shallowly depressed in age; surface viscid when moist, yellowish-buff to pinkish-buff to whitish, the center usually pinkish to pinkish-orange when fresh, or sometimes pale pinkish to pinkish-orange throughout; margin bearded with a dense white mat of soft, woolly hairs (while inrolled) that may mimic a veil; hairs sparse or even absent in age. Flesh thick, firm, brittle, white or tinged pinkish; taste *very* acrid. LATEX often scanty, white and typically unchanging (but staining gills yellowish in var. *nordmanensis*). GILLS white to yellowish-tan or developing a pinkish tinge, crowded, narrow, adnate to slightly decurrent. STALK 2-7 cm long, 0.5-1.5 cm thick, rigid, equal or with a narrowed base, firm, dry, often hollow in age; colored like cap or paler, sometimes with dingy ochre spots. SPORE PRINT creamy-white; spores 7-10 × 6-8 microns, elliptical, with amyloid ridges.

HABITAT: Scattered to gregarious under or near birch or occasionally aspen; common and very widely distributed. In our area this species and *L. pubescens* (see comments) are common after the first fall rains on lawns with planted birch trees (often in the company of *Paxillus involutus*). The mycelium is probably imported on the roots of birch saplings.

EDIBILITY: Not recommended, as it is indigestible or even poisonous unless thoroughly cooked. However, in Russia and Scandinavia it is collected in large quantities and pickled, and in Norway it is roasted and added to coffee.

COMMENTS: The bearded, pinkish-tinged cap, white latex, acrid taste, and growth with birch are the trademarks of this attractive *Lactarius*. Buttons with an inrolled margin are reminiscent of *Paxillus involutus* (another birch-lover), but paler in color. *L. pubescens* is practically identical to *L. torminosus* but has smaller spores (6-8 microns long) and is more apt to have a whitish cap (see photo on p. 41). It is common with birch in parks and gardens as well as in forests, and is often mistaken for *L. torminosus*. One collection I made had greenish bands on the gills and stalk just like in *Russula brevipes* var. *acrior*. Other similar species: *L. subvillosus* is common in our tanoak-madrone woodlands. Its pinkish-orange to orange cap is bearded when young but often naked in age, leading to confusion with *L. alnicola*, which has a yellower cap. *L. psammicola* has an orangish, zoned, bearded cap but grows under eastern hardwoods. For bearded yellow-stainers, see *L. scrobiculatus*.

Lactarius scrobiculatus   (Scrobiculate Milk Cap)

CAP 4-15 (20) cm broad, convex to plane with the center usually depressed, in age often vase-shaped; surface viscid when moist, smooth or scaly in age, unzoned or faintly zoned, pale yellow to yellowish to bright ochre, the center sometimes darker; margin inrolled when young and typically bearded with hairs (but hairs absent or inconspicuous in var. *montanus* and var. *pubescens*). Flesh thick, firm, brittle, whitish but staining yellow when exposed; taste variable: strongly acrid to mild. LATEX copious or scanty, white, quickly staining yellow when exposed. GILLS adnate to slightly decurrent, whitish to pale or dull yellow, close. STALK 3-11 cm long, (1) 2-4 (5) cm thick, equal or narrowed at base, firm, dry, hollow in age; white to yellowish, usually scrobiculate (i.e., pitted with large, glazed, darker or brighter yellow to honey-colored spots). SPORE PRINT whitish to creamy or yellowish; spores 6-10 × 5-7.5 microns, broadly elliptical, with amyloid warts and ridges.

HABITAT: Solitary, scattered, or in groups under northern and mountain conifers; widely distributed. Its various varieties are common in the summer and fall in the Sierra Nevada, northern California, and other parts of the West, but absent in our area.

EDIBILITY: Not recommended; yellow-staining milk caps should be avoided.

COMMENTS: This robust but variable milk cap can have a mild to strongly peppery (acrid) taste and prominently bearded to virtually bald cap margin. However, it can

Lactarius vinaceorufescens (see description below) has white latex that quickly stains yellow when exposed. It is variable in other respects and intergrades with *L. chrysorheus,* another yellow-stainer.

be recognized by its overall pale yellow to dark ochre color, scrobiculate stalk, and white latex that *quickly* turns yellow or stains exposed tissue yellow. The latter feature distinguishes it from *L. alnicola,* which stains yellow slowly or not at all. Other species: *L. resimus* is a bearded, yellow-staining species whose stalk is scrobiculate only in age if at all. Its cap is white when young but often pale yellowish in age. It favors birch, aspen, alder, manzanita, and conifers. Also see *L. payettensis* (under *L. alnicola*).

Lactarius vinaceorufescens   (Yellow-Staining Milk Cap)

CAP 3-7 (9) cm broad, broadly convex with incurved margin, then plane or depressed; surface dry to slightly viscid, smooth, cinnamon-buff to pinkish, reddish-cinnamon, or dark reddish, often darkening to vinaceous-brown in old age; often faintly zoned or with darker watery spots; margin not bearded. Flesh brittle, staining yellow when cut; taste mild to somewhat bitter or acrid. **LATEX** white, quickly turning sulfur-yellow when exposed (within 45 seconds). **GILLS** adnate to slightly decurrent, close, pallid when young, then tinged cap color and eventually aging or staining dark reddish. **STALK** 3-7 cm long, 1-2.5 cm thick, equal, pallid or colored like cap but usually paler, often stained dark reddish in age; smooth, not scrobiculate; base often with hairs. **SPORE PRINT** white to yellowish; spores 6.5-9 × 6-7 microns, broadly elliptical to round, with amyloid warts and ridges.

HABITAT: Scattered or in groups or troops under both hardwoods and conifers; widely distributed. It is often abundant in our area from fall through early spring, particularly with manzanita, oak, Douglas-fir, and pine (often mingling with *L. fragilis, L. rubrilacteus, Russula emetica,* and *R. cremoricolor*). *L. chrysorheus* (see comments) is also common.

EDIBILITY: Reportedly poisonous—all yellow-staining milk caps should be avoided.

COMMENTS: This ubiquitous milk cap superficially resembles *L. rubrilacteus, L. rufus, L. subviscidus, L. fragilis,* and *L. rufulus,* but is easily distinguished by the prompt yellowing of the latex when exposed to the air. The rate at which it yellows depends on the moisture content and age of the mushroom, but it normally occurs in 5-30 seconds. *L. vinaceorufescens* has been confused in the past with *L. chrysorheus,* widespread and also common in our area, which has a somewhat paler (pallid to yellowish-cinnamon to pale pinkish), often zoned cap and gills and stalk that do not discolor as much in age. Other species with yellow-staining latex include: *L. maculatipes* and *L. croceus*, found under hardwoods in eastern North America, the first with a whitish to creamy-yellow, often spotted cap and slimy stalk when fresh, the latter with a bright saffron to yellow-orange or orange, viscid cap; and *L. xanthogalactus* of California, probably the same as *L. chrysorheus* or *L. vinaceorufescens*. For other yellow-staining species, see *L. scrobiculatus*.

Lactarius uvidus group   (Purple-Staining Milk Cap)

CAP 3-10 (12) cm broad, broadly convex to plane or shallowly depressed; surface smooth, viscid to slimy when moist, pallid becoming grayish, lilac-gray, or pale lavender-brown, sometimes obscurely zoned; margin naked. Flesh thick, white, staining lilac slowly when wounded; taste mild to slowly bitter or acrid. **LATEX** white or creamy, staining wounded areas lilac or dull purple. **GILLS** adnate to slightly decurrent, close, white to yellowish; wounded areas staining purplish, then eventually dingy tan. **STALK** 3-8 cm long, 1-2 cm thick, more or less equal, smooth, viscid when moist but soon dry, rigid, pallid or sometimes tinged cap color, often ochraceous-stained toward base. **SPORE PRINT** yellowish-white; spores 8-12 × 7-8 microns, elliptical, with amyloid warts and ridges. Cap surface staining green in KOH (potassium hydroxide).

HABITAT: Solitary, scattered, or in small groups in woods; widely distributed. In our area this species and its close relatives (see comments) are fairly common in mixed woods in the late fall and winter, but seldom fruit in quantity.

EDIBILITY: Said to be poisonous. All purple-staining milk caps should be avoided.

COMMENTS: The tendency of the latex to stain wounded tissue purple typifies this species (see photo on p. 897) and its close relatives. A stouter, conifer-loving version, *L. uvidus var. montanus,* has a darker, somewhat browner, nearly dry cap and resinous taste. Other purple-stainers include: *L. pallescens,* with a slimy or viscid whitish cap *and* stalk, common under conifers in northern California and the Pacific Northwest; *L. californiensis,* an acrid-tasting species whose purplish to brownish-gray cap often has yellowish tinges or stains; *L. cascadensis,* a large northwestern species with dull brownish gills in age, favoring swampy areas (under alder, etc.); *L. maculatus* of eastern North America, with a distinctly zoned cap and acrid taste; and *L. subpalustris,* larger, with a grayish to dingy brown, often spotted cap that does *not* stain green in KOH, plus a mild taste. For purple-staining species with a yellowish to buff cap, see *L. representaneus.*

Lactarius representaneus   (Purple-Staining Bearded Milk Cap)

CAP (4) 6-20 cm broad, broadly convex with a central depression and inrolled margin, becoming broadly depressed or vase-shaped in age; surface viscid when moist, sometimes zoned concentrically, pale yellow to golden-yellow to orange-buff, often developing rusty and sometimes purple stains; smooth at center but coarsely hairy (fibrillose) toward the margin, which is bearded with woolly yellowish hairs when young. Flesh thick, brittle, white, staining dull lilac or purplish slowly where cut or bruised; taste slowly bitter or acrid. **LATEX** copious, white or creamy, drying or slowly staining wounded tissue dull purple or lilac. **GILLS** adnate to decurrent, close, buff to dull ochre or orange-spotted, with lilac or purplish stains where bruised. **STALK** 4-12 cm long, 1-3 (4.5) cm thick, equal or thicker below, hard, whitish or more often colored like cap, usually prominently scrobiculate, sometimes lilac-stained; hollow or stuffed. **SPORE PRINT** whitish to yellowish; spores 8-12 × 6.5-9 microns, broadly elliptical, with amyloid warts and ridges.

HABITAT: Solitary to scattered or often gregarious on ground under northern conifers—especially spruce and fir; widely distributed but not found in our area. I have seen it under Engelmann spruce in Colorado and New Mexico in August, and under Sitka spruce in northern California in November. It is said to be abundant in Alaska.

EDIBILITY: Not recommended. The taste is not appealing and it may be poisonous.

COMMENTS: This often large, impressive milk cap is easily identified by its yellowish cap with the hairy margin, and purple-staining latex. It resembles *L. scrobiculatus* and *L. alnicola* in appearance, but those species have unchanging or yellow-staining latex. Other purple-staining species (see *L. uvidus*) are differently colored and do not have a bearded cap. Similar species include: *L. speciosus*, with a dull buff to tan cap and hairy margin, common under hardwoods in eastern North America (especially the South); and *L. aspideoides,* with a viscid, pale yellow cap that is not bearded. Neither should be eaten.

Lactarius argillaceifolius var. *megacarpus*   (Vulgar Milk Cap)

CAP 7-18 (27) cm broad, broadly convex-depressed with an inrolled margin when young, plane to depressed in age; surface viscid, drab grayish, violet-gray, or sometimes brownish tinted with violet, often fading in age to dingy tan or grayish-buff; sometimes obscurely zoned or with rusty-ochraceous spots. Flesh thick, pallid; taste mild or somewhat acrid. **LATEX** creamy-white, unchanging, but slowly staining wounded areas dingy brownish. **GILLS** pallid becoming dingy yellowish or buff, often darker (dingy yellow-brown) in old age, slowly staining brownish or grayish where bruised; adnate to slightly decurrent, close. **STALK** 6-15 cm long, 2-6 cm thick, firm, rigid, slightly viscid or dry; white to buff or tinted cap color, at times with ochre or yellow-brown stains. **SPORE PRINT** whitish to buff; spores 7-11 × 6-9 microns, elliptical to nearly round, with amyloid warts and ridges.

HABITAT: Solitary, scattered, or in small groups in humus under hardwoods (live oak, tanoak, madrone), fairly common in our area in the late fall and winter. It is known only from the Pacific Coast, but variety *argillaceifolius* (see comments) is widespread.

EDIBILITY: Possibly poisonous; to be avoided.

COMMENTS: The robust dimensions, dingy tan to grayish color, and white latex are the distinguishing features of this undistinguished mushroom. It is our largest and least attractive *Lactarius*. The taste is rather variable, but generally the latex is at least somewhat acrid while the flesh may be mild. The violet tints that sometimes pervade the cap are reminiscent of *L. uvidus* (with which it sometimes grows), but that species stains purple instead of dingy brown. *L. argillaceifolius* var. *argillaceifolius* is smaller and browner than var. *megacarpus,* and has a tendency to become dingy brown or yellow-brown overall in age. It is a characteristic summer mushroom in the hardwood forests of eastern North America. Other species: *L. caespitosus* is similar, but has a gray to olive-brown cap and grows under western conifers; *L. vietus* is also similar but favors birch and northern coni-

Lactarius argillaceifolius var. *megacarpus* can be told by its relatively large size, dingy color, and white latex that slowly stains the gills brownish. For a close-up of the latex, see photo on p. 63.

Lactarius pseudomucidus, a conifer loving milk cap with a dark cap and stem that are slimy when wet.

fers and often has a lilac- or vinaceous-tinged cap; *L. cinereus* has a gray to olive-gray cap, non-staining gills and stalk apex often tinged pale pinkish, and grows with beech; *L. circellatus* favors birch and northern conifers, but has non-staining gills and a brownish-gray to bluish-gray or lavender-tinged cap, and zebra-like amyloid stripes on the spores. For years *L. argillaceifolius* has been called *L. trivialis,* a similar slimy-stalked species whose gills discolor scarcely if at all; it occurs in the Pacific Northwest, usually under conifers.

Lactarius pseudomucidus   (Slimy Milk Cap)

CAP 3-8 (10) cm broad, broadly convex becoming plane or depressed; surface smooth, viscid (with a thick layer of slime when moist), evenly colored blackish-brown to dark gray; not zoned; margin naked. Flesh grayish, thin, fragile, taste slowly acrid. **LATEX** white, unchanging, but may stain gills tan to brownish, gray, or olive-gray. **GILLS** adnate to decurrent, white or tinged gray, sometimes discolored (brownish, etc.) in age. **STALK** 4-10 cm long, 0.5-1 cm thick, fragile, smooth, without spots, evenly colored like cap or paler (gray); very slimy when wet, usually thicker below. **SPORE PRINT** white; spores 7-9 × 6-7 microns, broadly elliptical, with amyloid ridges.

HABITAT: Scattered to gregarious under conifers in late summer and fall; known only from northern California and the Pacific Northwest, common.

EDIBILITY: Unknown. Only the most ardent slippery jack lover would fail to be deterred by the copious layer of slime coating the cap and stem.

COMMENTS: The dark grayish-brown cap and stalk under a thick layer of slime (when fresh) and the white latex are usually enough to distinguish this glutinous woodland mushroom. It is slimmer and slimier than either *L. argillaceifolius* or *L. trivialis,* and darker besides. It has been called *L. mucidus,* but that species, which is commoner in eastern North America, has a pale cap margin and only slightly viscid stalk. Another look-alike, *L. glutigriseus,* has a slightly viscid stalk that is paler at the apex and base. *L. kauffmanii,* also frequent in the Pacific Northwest and sometimes found with *L. pseudomucidus,* is similar but usually larger and stouter, with a distinctly tan or paler (not gray) stalk. Also see the species listed under *L. argillaceifolius* var. *megacarpus.*

Lactarius fallax   (Velvety Milk Cap)   **Color Plate 9**

CAP 2.5-7 (9) cm broad, convex to plane (often with an umbo) when young, often shallowly depressed in age; surface dry and more or less velvety, often wrinkled toward the center; evenly dark brown to nearly black, not zoned; margin often scalloped. Flesh rather thin, brittle, whitish, taste mild or slightly acrid. **LATEX** white, usually copious, slowly staining wounded tissue dull reddish. **GILLS** adnate to slightly decurrent, narrow, crowded, white to creamy-buff (but edges dark brown in one form). **STALK** 2.5-7 cm long, 0.5-1.5 cm thick, rather slender, more or less equal, colored like cap or somewhat paler; dry and unpolished or velvety. **SPORE PRINT** yellowish; spores 9-12 × 8-11 microns, more or less round, with amyloid warts and ridges.

HABITAT: Scattered or in small groups under conifers (especially fir) or occasionally on rotting wood; fairly common in the late summer and fall in the Pacific Northwest and northern California. I have yet to find it in our area.

EDIBILITY: Unknown.

COMMENTS: This species is one of several milk caps with a brown to nearly black, velvety cap. Though not good edibles, they are notable for their beauty, and therefore likely to attract the attention of even the casual collector. *L. fallax* is the most common member of the group in the West, and is easily distinguished from the similarly colored *L. pseudomucidus,* by its dry, velvety rather than smooth, slimy cap and stem. There are several closely related species found mainly in eastern North America, including: *L. lignyotus,* with broader, more widely-spaced gills; *L. gerardii,* with *very* well-spaced (distant) gills; *L. fuliginellus,* a hardwood-lover with close gills; and *L. fumosus*, with close gills and a paler (smoky-brown) cap. None of these are worth eating.

Lactarius volemus   (Weeping Milk Cap; Bradley)

CAP 5-10 (13) cm broad, convex becoming plane or depressed, or sometimes umbonate; surface dry, minutely velvety to nearly smooth, golden-tawny to orange-brown or rusty-orange (or pale yellowish in *var. flavus*). Flesh thick, firm but brittle, whitish; odor often becoming fishy in age; taste mild. **LATEX** white or creamy, very abundant, slowly staining brown or staining wounded tissue brown. **GILLS** close, white to creamy or sometimes pale tan, adnate to slightly decurrent. **STALK** 4-12 cm long, 0.8-1.2 (2) cm thick, equal or tapered downward, orange-brown to tawny (colored like cap or slightly paler), dry, unpolished, minutely velvety to nearly smooth; not scrobiculate. **SPORE PRINT** whitish; spores 7.5-10 × 7.5-9 microns, round or nearly round, with amyloid ridges (reticulate).

HABITAT: Widely scattered to gregarious on ground under hardwoods (especially oak) or in mixed woods; common in the summer and early fall in eastern North America.

EDIBILITY: Edible and delicious when properly cooked! Like most milk caps it has a slightly granular texture that displeases some people, but the flavor is excellent; slow cooking is best. *L. corrugis* and *L. hygrophoroides* (see comments below) are equally good if not better. All three species are usually free of maggots.

COMMENTS: This species is one of a trio of tasty milk caps (see below) that rank among the best edible mushrooms of eastern North America. To my knowledge they are not found in the West, but are to be looked for in southern Arizona, where several so-called "eastern" mushrooms occur (e.g., *Strobilomyces floccopus*). *L. volemus* is easily recognized by its dry, tawny to orange-brown cap and stalk which are often minutely velvety, its mild taste, and extremely copious white milk that slowly stains brown. The latex is so copious that the slightest nick of the gills will cause them to "weep," i.e., exude a stream of milky droplets. The fishy odor that often develops after it is picked or as it is cooked is also distinctive, but in no way reflects or affects its flavor. Also common under hardwoods in eastern North America are three similar species with copious white latex, mild taste, and minutely velvety cap and/or stalk: *L. hygrophoroides* is a beautiful, similarly colored or oranger species with well-spaced gills, a mild odor, and latex that does *not* stain brown; *L. corrugis* (**COLOR PLATE 8**) is larger and more robust, with a frequently wrinkled, dark brown to reddish-brown to rusty-brown cap, whitish to slightly yellowish or cinnamon-tinged gills, brown-staining latex, and a (typically) mild odor; *L. luteolus* has a white to buff or yellowish cap, brown-staining latex, and a frequently fishy odor. All of the above species are edible. See p. 79 for a photograph of *L. volemus.*

Left: *Lactarius volemus,* a delectable species with a dry cap and copious white latex. Note how stature is more slender than that of the closely related *L. corrugis* (Color Plate 8). **Right:** *Lactarius subflammeus* has a viscid orange cap and white latex. Note the rather long, slender stem.

Lactarius subflammeus (Orange Milk Cap)

CAP 2-6 (7) cm broad, convex becoming plane or shallowly depressed; surface smooth, viscid, scarlet when young soon fading to bright orange and eventually dull orange; not zoned concentrically; margin naked. Flesh thin, fragile; odor mild, taste slowly acrid. LATEX white, unchanging. GILLS adnate to decurrent, fairly close, whitish or colored like cap but paler. STALK 4-9 cm long, 0.5-1.5 cm thick, usually rather long and thicker toward the base; hollow, rigid but fragile, colored like cap; not viscid. SPORE PRINT whitish; spores 7.5-9 × 6.5-7.5 microns, elliptical, with amyloid ridges and warts.

HABITAT: Scattered to gregarious under pine, spruce, and other conifers; found in western North America in the late summer and fall and especially common in northern California and the Pacific Northwest. I have not seen it south of San Francisco.

EDIBILITY: Unknown; best avoided until better known.

COMMENTS: The bright orange viscid cap, stalk usually longer than the width of the cap, and white unchanging latex typify several similar species that have traditionally passed under the name *L. aurantiacus.* They can be told from the candy cap *(L. fragilis)* by their *viscid* cap and brighter color, plus their mild odor. *L. luculentus* is a very similar species with a mild to bitterish taste, an orange to tawny cap that is only slightly viscid, and gills that may stain brownish; it is also common in northern California and the Pacific Northwest. *L. cocosiolens,* discovered in California by Andrew Methven, has a viscid orange-brown to caramel-colored cap, mild taste, and copious white latex, and smells like coconut when it dries. *L. substriatus* is one of several look-alikes with yellow-staining latex.

Lactarius rufus (Red Hot Milk Cap)

CAP 4-12 cm broad, broadly convex becoming plane or depressed; surface usually not viscid; smooth, dark brick-red to bay-red or reddish-brown, not zoned; margin naked. Flesh dingy reddish, rather fragile; odor mild, taste strongly—but often latently—acrid. LATEX white, unchanging. GILLS crowded, adnate to slightly decurrent, whitish when young, flushed reddish in old age. STALK 4-11 cm long, 1-1.5 cm thick, equal, rigid but rather fragile, stuffed or hollow, dry, more or less colored like cap. SPORE PRINT pale yellowish; spores 7.5-11 × 5-7.5 microns, elliptical, with amyloid warts and ridges.

HABITAT: Scattered to gregarious or in troops under conifers; widely distributed. It is common in northern California and the Pacific Northwest from late summer through early winter, but I have not seen it south of San Francisco. It is also abundant in northern sphagnum bogs. Like *L. deliciosus,* its favorite mycorrhizal mates are pine and spruce.

79

EDIBILITY: Not recommended. Like several other peppery milk caps, it is harvested and canned commercially in Scandinavia. However, North American variants have not been thoroughly tested and it may be poisonous raw.

COMMENTS: This species occurs in droves under northern conifers. It can be told from other reddish milk caps by its red-hot taste, unchanging white latex, and fondness for conifers. Fresh young specimens are among the most acrid of all mushrooms. When you taste one, take only a *small* bite, and remember: the burning sensation can be delayed! Other species: *L. manzanitae* has a small (2-5 cm), viscid, orangish to coppery-red or brick-red cap, strongly acrid taste, and unchanging latex; it occurs under manzanita in California. *L. hysginus var. americanus* has a brown to dark vinaceous-brown, viscid cap and viscid stalk; it grows under conifers. For similar species with a milder taste, see *L. subviscidus, L. subflammeus,* and *L. rufulus.*

Lactarius subviscidus

CAP 1-4 (5) cm broad, shallowly depressed becoming broadly or more deeply depressed in age; surface smooth, viscid to thinly slimy when wet, dark reddish or reddish-brown to brick-red or paler (pinkish) in age, not normally zoned; margin sometimes faintly striate. Flesh thin, tinged cap color, fragile; odor mild, taste mild to slightly acrid. **LATEX** white, rather scanty, unchanging. **GILLS** close, adnate to decurrent, pinkish-buff to pinkish-cinnamon, or darker in age. **STALK** 2-5 cm long, 4-8 mm thick, more or less equal, usually hollow; smooth, not viscid, colored like cap or gills. **SPORE PRINT** white to yellowish; spores 8-10 × 7-8 microns, broadly elliptical, with amyloid ridges.

HABITAT: Scattered to gregarious on ground or rotten wood under conifers (spruce, pine, etc.) in the Pacific Northwest and California. It is fairly common in our area in the fall and early winter under pine, Douglas-fir, and manzanita.

EDIBILITY: Unknown.

COMMENTS: This small species looks like a candy cap *(L. fragilis* or *L. rufulus),* but the thinly viscid cap (often with debris stuck to it), mild odor, and white (rather than watery white) latex distinguish it. The latex does not stain yellow as in *L. vinaceorufescens,* but may slowly stain white paper yellow (overnight). The cap is not orange as in *L. subflammeus* and not as acrid (peppery) as *L. manzanitae* (see *L. rufus).* Similar species include: *L. riparius,* common under conifers, alder, and willow in seepage areas in the Sierra Nevada, with a brownish-red to brick-red cap that is slightly viscid when moist, mild taste, and copious unchanging white or watery-white latex; *L. atrobadius,* with a viscid liver-colored to blackish-red cap and non-yellowing latex; and *L. hepaticus,* with a viscid or dry cap and yellowing latex. For similar species with dry caps, see *L. rufulus.*

Lactarius fragilis   (Candy Cap)   Color Plate 10

CAP 2-7 cm broad, broadly convex or plane or sometimes with an umbo, becoming slightly to broadly depressed in age; surface dry (never viscid!), often uneven or somewhat wrinkled; usually colored burnt orange to cinnamon but sometimes reddish-brown; sometimes darker toward center but not zoned concentrically; margin often wavy or frilled. Flesh thin, fragile, tinged cap color; taste mild or slightly bitter; odor faintly fragrant or pungent becoming strongly aromatic (like maple syrup) upon drying. **LATEX** white or watery-white, unchanging, often scanty or absent. **GILLS** adnate becoming decurrent, close, pale pinkish-cinnamon or tinged cap color (or sometimes more yellowish), darkening to more or less cap color in age, or dusted whitish with spores. **STALK** (2)4-9 cm long, 0.4-1 (1.5) cm thick, usually rather slender; fragile, more or less equal, colored like

Lactarius fragilis. These mature specimens are rather regular in shape, but it is not uncommon for the caps to have frilled edges. Color ranges from burnt orange (see color plate) to cinnamon to reddish-brown and the stalk is usually hollow in age. Fragrant odor is also distinctive.

cap; often hollow in age, with hairs at base. **SPORE PRINT** white to pale yellowish; spores 6-9 microns, more or less round, with amyloid warts and ridges (reticulate). Cap and stalk tissue containing numerous nests of sphaerocysts.

HABITAT: Widely scattered to densely gregarious or in small clumps on ground in woods (often along trails and in other damp places, sometimes on wood). It is apparently confined to the Pacific Coast and Southeast, but the very similar *L. camphoratus* (see comments) occurs elsewhere. In our area it is normally abundant from the late fall through early spring, especially under oak but also with other trees (pine, Douglas-fir, etc.).

EDIBILITY: Edible and one of the very best of our late season mushrooms. Fresh specimens can be sauteed and used like any other mushroom; when chopped up and slowly dried, however, their flavor becomes sweet and they are great in pancakes, cookies, on cinnamon toast with sesame seeds, or even in ice cream! Fortunately, they are plentiful enough to merit collecting in spite of their small size, but be sure of your identification! A few poor souls, alas, are allergic to them.

COMMENTS: The most remarkable feature of candy caps is the sweet, persistent fragrance they develop when cooked or dried. If just a few candy caps are sauteed, their odor will permeate the entire house and linger for days. Stranger still, when eaten in quantity they imbue one's perspiration and bodily exudates with the unmistakable aroma of maple syrup—thereby provoking some puzzled stares! Their odor has also been likened to that of butterscotch, fenugreek, burnt sugar, and sweet clover. Fresh specimens may have only a slight odor, in which case the mild taste, watery-white unchanging latex, and overall burnt orange or "ferruginous" color separate them from most other mushrooms (but see below). In dry weather the latex is often absent, leading to confusion with *Laccaria laccata, Clitocybe inversa, Rhodocybe nuciolens, Collybia* and *Cortinarius* species, and various other similarly colored mushrooms. Fortunately, the odor is usually quite pronounced under those conditions. Be *sure* the latex does not stain yellow—*L. vinaceorufescens* often mingles with it and can be practically the same color.

The typical variety of *L. fragilis* occurs in the Southeast and is slightly tawnier. The western version, ***L. fragilis* var. *rubidus***, may actually be a distinct species. For years it has passed as ***L. camphoratus***. That species, however, is found in northern coniferous and hardwood forests and has a slightly redder, frequently nippled (umbonate) cap and non-reticulate spores. Still another candy cap, *L. rufulus,* occurs in California under oak, but tends to be larger, is often redder, usually has a thicker and firmer stem, and contains few if any sphaerocysts in its tissue (see description of *L. rufulus*). For similar species with a mild odor, see *L. subviscidus* and comments under *L. rufulus*.

Lactarius rufulus. This west coast native resembles *L. fragilis,* but differs microscopically and usually has a solid, thicker stem (as shown here) and larger cap. Color is variable but usually reddish.

Lactarius rufulus   (Rufous Candy Cap)

CAP 3-10 cm broad, broadly convex to plane or slightly umbonate, becoming wavy or depressed; surface often uneven or somewhat wrinkled, *not* viscid, color variable: usually dark red to reddish-brown or brick-colored, but sometimes oranger; evenly colored or with darker areas; margin incurved at first. Flesh firm, whitish to buff or tinged cap color, brittle; taste mild to faintly acrid, odor fragrant to mild or faintly pungent. **LATEX** white or watery white, unchanging, often scanty or even absent. **GILLS** adnate to decurrent, fairly close, pallid to cinnamon-buff, pinkish-buff, or tinged cap color (or yellower), often darker red or reddish-brown in age, at least on the margins. **STALK** 4-12 cm long, (0.5) 0.8-3 cm thick, equal or slightly tapered, colored more or less like cap, firm, usually solid or stuffed but sometimes partially hollow in age. **SPORE PRINT** creamy to yellowish; spores 7-9 microns, round or nearly round, with amyloid warts and ridges (at least partially reticulate). Nests of sphaerocysts scanty to completely absent in cap and stalk tissue.

HABITAT: Scattered to densely gregarious or tufted under oak, late fall through spring; known only from California. In my experience it is not as common as *L. fragilis* in regions of high rainfall (e.g., coastal central and northern California), but more common than *L. fragilis* in drier areas (e.g., inland and southern California).

EDIBILITY: Edible and good, but not quite as fragrant or tasty as *L. fragilis.*

COMMENTS: This oak-loving candy cap tends to be redder and more robust than *L. fragilis,* and often has a solid stem. The critical difference is microscope, however: the absence or near absence of swollen cells (sphaerocysts) in the cap and stem tissue is highly unusual for a *Lactarius!* *L. rufulus* can also be confused with *L. rufus,* but that species has an acrid taste and favors conifers, while *L. vinaceorufescens* has yellowing latex.

There is a whole slew of very similar, reddish-brown to dull rusty-orange milk caps with a mild to somewhat bitter taste and *little or no odor.* These are difficult to differentiate without a microscope, and far too numerous to ennumerate. As a group they have passed under the name *L. subdulcis,* but of course are not the "true" *L. subdulcis* of Europe (if they were, we would probably be calling them something else!). These species include: *L. sub-serifluus,* with clear latex and orange-brown cap, the oak-loving eastern counterpart of *L. rufulus* (it also lacks sphaerocysts in the cap tissue); *L. thiersii,* also lacking sphaerocysts but much smaller (cap 1-3 cm), known only from California; *L. oculatus,* with a red-brown to pinkish, often nippled, dry to slightly viscid cap, found in eastern *Sphagnum* bogs and under conifers; *L. alpinus var. mitis,* one of several species common under alder and northern conifers or in seepage areas; *L. hepaticus* (see comments under *L. subviscidus*); and *L. thejogalus,* with somewhat yellowing latex, occurring in swarms under birch. *L. herpeticus* should also be mentioned. For similar viscid-capped species, see *L. subviscidus.* None of these species should be eaten until better known.

RUSSULA (The Russulas)

Medium-sized to large, mostly terrestrial woodland mushrooms. CAP *plane or depressed at maturity,* viscid or dry, skin often peeling easily, margin often striate. *Flesh brittle, usually white;* taste mild or acrid. *Latex absent.* GILLS attached or free, brittle, *usually white to yellow, at least when young* (but may discolor in age). STALK central, *rigid, brittle, breaking open cleanly like chalk.* VEIL and VOLVA *absent.* SPORE PRINT *white to yellow, yellow-orange, or ochraceous.* Spores with amyloid warts and/or ridges. Cap tissue (and usually other tissue) containing nests or rosettes of sphaerocysts.

RUSSULA is distinct by virtue of its brittle (dry or granular) flesh, rigid fruiting body, white to yellow-orange spore print, and warty amyloid spores. "Brittle," of course, is a subjective term, but the crisper *Russula* species will snap audibly when broken, and in both the crisp and fragile types the stalk is fleshy and snaps open cleanly like a piece of chalk—i.e., there is no fibrous tissue present as in most fleshy-stemmed mushrooms. *Lactarius* species also snap open like chalk, but usually exude a latex when broken, while *Leucopaxillus* is somewhat brittle but has a tough, fibrous stem that does *not* break like chalk and often has white mycelium at the base. In addition to their brittle texture, the Russulas have a characteristic appearance that, though difficult to describe, makes them one of the easiest groups to recognize. The cap is plane to depressed at maturity and often broader than the length of the stem—with exceptions, of course.

Though *Russula* as such is a very clearcut group, mastering their identification is almost to be despaired of, for Russulas come in a bewildering panoply of reds, purples, pinks, yellows, oranges, browns, greens, and whites that is at once their most attractive and deceptive feature. The color pigments are unusually sensitive to environmental and genetic caprice, so that no two mushrooms look quite alike. Identification is made even more difficult by the fact that many species produce only a few fruiting bodies per mycelium—making it hard to gauge accurately the degree of color variation. The result is that many species are still poorly known or unclassified, while synonyms and homonyms abound. *Russula*-researchers have resorted to examining the spore ornamentation of different species with an electron microscope in an effort to establish fixed criteria for each species and dispel some of the confusion.

Russula xerampelina is one of the few Russulas worth gathering for the table. Note the shape, which is fairly typical of the genus *Russula,* and note how the stalk fractures like a piece of chalk.

The task of getting to know the Russulas is expedited by assigning them to groups. For instance, the section Compactae includes several large, hard-stemmed species with gills that usually alternate long and short (*R. albonigra* and *R. densifolia* are good examples). Another group, centered around *R. fragrantissima* and *R. sororia*, has a yellow-brown to brown, conspicuously striate cap and a strongly fragrant to unpleasant odor. A third group, mainly northern in distribution, has flesh which turns grayish when exposed (e.g, *R. claroflava*). The remaining Russulas can then be divided into four large, artificial groups based on spore color and taste: to wit, spores white and taste mild (e.g., *R. cyanoxantha*), spores white and taste acrid (e.g., *R. emetica*), spores yellow and taste acrid (e.g., *R. rosacea*), and spores yellow and taste mild (e.g., *R. xerampelina*).

Russulas are mycorrhizal with a broad range of hardwoods and conifers, and are almost always terrestrial. They grow not only in woods, but at the edges of pastures, in brushy areas, and on lawns near trees. They are among the most omnipresent mushrooms of our western coniferous forests, but are often concealed by leaves and needles and evident only as low mounds ("mushrumps") in the humus.

In eastern North America the Russulas, like the milk caps *(Lactarius)*, are often prominent during muggy weather when other mushrooms are relatively scarce, while in the Pacific Northwest and Rocky Mountains their season more closely coincides with that of other fungi. In our area Russulas run rampant throughout the mushroom season, subject, of course, to the whim of the weather. A metagrobolizing melange of species *(R. albonigra, R. xerampelina, R. cyanoxantha,* et al) usually erupts after the first fall rains, followed by another burst (spearheaded by *R. emetica*) in the winter.

Russulas are among the most maligned of all mushrooms. Even veteran mushroom hunters treat them mercilessly—throwing them over their shoulder or crushing them underfoot with disparaging remarks like "Oh, it's a JAR" ("Just Another Russula"). Their omnipresence, anonymity, and poor culinary reputation are partly responsible, I am sure. Also, their brittle flesh is irresistible to those who like to smash things. Furthermore, they have a habit of forming "mushrumps" in the humus which resemble those made by *Boletus edulis*. Frustrated boletivores can be uncompromisingly brutal (see p. 546), and their habitual hunting grounds are inevitably strewn with the broken bodies of Russulas. However, mushrooms were not created for the exclusive enjoyment of *Homo sapiens*, and it is wrong to judge them accordingly. Try to resist the sharp temptation to mash, maim, and mutilate them. Those who follow in your footsteps will appreciate your sensitivity and self-restraint.

Most of the mild-tasting Russulas are edible, but this should not be taken as a signal to sample them indiscriminately. Always identify what you intend to eat! Those which redden or blacken and/or have a peppery (acrid) taste should be avoided—at least some cause vomiting and diarrhea or worse. Thorough cooking may render them edible, but it hardly seems worth the effort or risk. Russulas are not widely revered as esculents, and it is true that their dry, granular flesh does not blend well with some dishes. However, *R. xerampelina* and *R. cyanoxantha* (to name just two of the local species) are marvelous delicacies in their own right, and *R. virescens* and *R. crustosa* of eastern North America are also excellent.

About 200 species of *Russula* are reported from North America, but the vast majority of them are "JAR's"—i.e., depressingly difficult to demystify. Only some of the more distinctive types are described here. You will doubtlessly encounter many, many more—some of them unclassified. If your "JAR" does not key out convincingly, you can send it to a specialist, or better yet, the nearest compost pile.

Key to Russula

1. Fruiting body medium-sized to very large, hard (especially the stalk); cap white to brown, black, or grayish (never brightly colored); gills typically adnate to decurrent and usually alternating long and short; cap cuticle (skin) *not* peeling easily, margin *not* striate 2
1. Not with above combination of characteristics 10

2. Fruiting body basically white (but often with brown to yellowish stains); flesh and stalk surface *not* staining when bruised or cut 3
2. Flesh and stalk surface changing color when bruised (within 10 minutes) and/or fruiting body not white 4

3. Cap 5-10 cm broad; taste very acrid; gills and stalk never blue-green .... **R. cascadensis** (see *R. brevipes*, p. 87)
3. Cap 7-20 cm or more broad; taste mild to slightly acrid, or if distinctly acrid then gills and/or stalk apex usually tinged blue-green or green .... **R. brevipes** & others, p. 87

4. Fruiting body turning dark gray to black *directly* when bruised (within 5-10 minutes), without an intermediate reddish phase; fruiting body blackening in age ... **R. albonigra** & others, p. 89
4. Not as above; if staining black, then with a preliminary reddish phase 5

5. Flesh and stalk surface distinctly staining reddish to orange when scratched or bruised (within 5 minutes), then eventually darkening to dark gray, dark brownish-gray, or black 6
5. Not as above (but may stain reddish *without* blackening afterward or may stain smoky-brown *directly*) 8

6. Gills thick and widely spaced .... **R. nigricans** (see *R. densifolia*, p. 90)
6. Not as above 7

7. Cap viscid when moist; taste usually acrid .... **R. densifolia,** p. 90
7. Cap not viscid; taste typically mild .... **R. dissimulans** (see *R. densifolia*, p. 90)

8. Gills usually with sordid reddish stains, often entirely reddish in old age; fruiting body *not* blackening; associated with oak in California and southern U.S. **R. subnigricans** group, p. 90
8. Not as above 9

9. Flesh typically staining faintly reddish when cut (within 20 minutes), and then eventually pale smoky-brown (or at times without reddish phase) **R. adusta** (see *R. subnigricans* group, p. 90)
9. Not as above 10

10. Fruiting body very firm and stalk hard; cap whitish becoming ochre, rusty-brown, or reddish-brown, margin *not* striate; all parts bruising or discoloring ochraceous to cinnamon-brown; odor often unpleasant in age, never sweet; spore print white; common in eastern North America .... **R. compacta** (see *R. subnigricans* group, p. 90)
10. Not as above 11

11. Odor heavy and sweet (like maraschino cherries or benzaldehyde), but in age often becoming fetid; cap medium-sized to large (5-15 cm broad), yellow-brown to straw-color, dull ochre, or yellow-orange .... **R. fragrantissima** group, p. 92
11. Not as above 12

12. Cap 5-12 cm broad, typically dark red with a nearly blackish center; taste usually acrid but sometimes mild; spore print whitish; common under hardwoods in eastern North America .... **R. atropurpurea (=R. krombholzii)**
12. Not as above 13

13. Flesh and stalk surface staining gray to ashy or black (sometimes slowly and sometimes with an intermediate reddish phase) 14
13. Not as above 17

14. Cap yellow to ochre 15
14. Cap orange to red, coppery, purple-red, etc. (sometimes mixed with yellow) 16

15. Taste mild .... **R. claroflava,** p. 92
15. Taste acrid .... **R. ochroleuca** (see *R. claroflava*, p. 92)

16. Cap red to orange or coppery-brown; flesh bruising gray directly . **R. decolorans** group, p. 91
16. Cap purplish, greenish, brownish, etc. (often a mixture of colors); flesh usually turning reddish *before* graying .... **R. occidentalis** & others (see *R. decolorans* group, p. 91)

17. Cap bright yellow to golden-yellow; mature gills yellowish   *R. lutea* (see *R. claroflava*, p. 92)
17. Not as above 18

18. Cap white or whitish to yellowish-white or pale yellow 19
18. Cap some other color (including yellow-brown) when fresh, but may fade to whitish in age 23

19. Cap cuticle (skin) thick and rubbery .... *R. crassotunicata* (see *R. cremoricolor*, p. 97)
19. Not as above 20

20. Spore print and gills white or whitish 21
20. Spore print and mature gills yellowish 22

21. Cap pale yellow to yellowish-white, viscid when moist, the margin usually striate in age; taste
     acrid; usually scattered to gregarious .... *R. cremoricolor* & others, p. 97
21. Cap dull white or tinged buff or pink at center, often dirty, viscid or dry; taste mild or acrid;
     usually occurring in small numbers .... *R. albidula* group, p. 96

22. Stalk typically staining yellow and then brown when scratched; odor fishy in age ....
     .... *R. sp.* **(unidentified)** (see *R. xerampelina*, p. 102)
22. Not as above .... *R. maculata* group, p. 100

23. Taste of flesh *and/or* gills distinctly acrid (taste them both!) 24
23. Taste mild or nearly so 32

24. Stalk white (but sometimes with stains or discolorations, especially at base) 25
24. Stalk red or flushed pink to rose 31

25. Cap yellow-brown to dull straw-color, hazel-brown, grayish-brown, or even darker at center;
     margin striate and with small bumps, at least in age; odor usually rather unpleasant; very
     common .... *R. sororia* group, p. 93
25. Not as above 26

26. Spore print and gills white or whitish 27
26. Spore print yellow to ochre; gills yellowish at least in age ....
     .... **Various JARs** (see *R. alutacea* group, p. 102 & *R. maculata* group, p. 100)

27. Stalk and flesh often slowly staining orange or pink when bruised (sometimes with an inter-
     mediate yellowish phase); cap typically multicolored (usually yellowish at center and splotched
     with yellow, orange, or red toward margin) .... *R. bicolor*
27. Not as above 28

28. Cap red suffused with gray or brown; associated with western conifers and especially common
     in the Rocky Mountains .... *R. montana*
28. Not as above (but cap may be bright red) 29

29. Cap bright red when fresh, but often fading to pink, orange, or sometimes even white ......
     .... *R. emetica* group, p. 97
29. Not as above; cap variously colored but not bright red 30

30. Many gills forked at quite a distance from the stalk; stalk firm, 1-3 cm thick, cap 5-15 cm broad
     .... *R. variata* (see *R. cyanoxantha*, p. 94)
30. Gills not forked, or forked only near stalk; stalk usually less than 1.5 cm thick, often quite fragile;
     cap up to 7 cm broad .... *R. fragilis* & others, p. 98

31. Stalk red to pink; cap dark red to bright red (but often fading in age) ..... *R. rosacea*, p. 99
31. Cap pink, olive-gray, etc., or a mixture of these colors; stalk usually tinged rose or grayish-rose
     .... *R. gracilis* group & others, p. 99

32. Cap surface dry and soon areolate (cracked into scales or plaques), greenish to blue-gray or with
     buff, yellowish, or even brownish tones; common under hardwoods in eastern U.S. 33
32. Not as above 34

33. Cap usually greenish or greenish-gray .... *R. virescens*, p. 95
33. Cap buff to yellowish or brownish, with green tones sometimes present also ....
     .... *R. crustosa* & others (see *R. virescens*, p. 95)

34. Cap predominantly greenish or grayish-green (sometimes mixed with other colors); stalk
     white or whitish (or sometimes brownish-stained, but never pink); odor *not* fishy in age; spore
     print white to creamy (pale yellow), *not* ochre or dark yellow 35
34. Not with above combination of characteristics; cap may have greenish tones, but typically the
     greenish tones not predominating 38

35. Spore print white; fruiting body often robust (stalk 1.5-3 cm thick or more); stalk *not* staining salmon in ferrous sulfate .... *R. cyanoxantha*, p. 94
35. Spore print pale cream; stalk typically 0.7-2 (2.5) cm thick, staining salmon in ferrous sulfate 36

36. Cap bluish-green to greenish, with a matt (dull) appearance .... *R. parazurea* (see *R. aeruginea*, p. 95)
36. Cap greenish to grayish, often with yellowish, rusty, or brownish areas, and viscid when moist 37

37. Cap usually greenish; found with aspen, birch, oak, and conifers .... *R. aeruginea*, p. 95
37. Cap often with other colors; found with oak and beech ... *R. grisea* (see *R. aeruginea*, p. 95)

38. Cap small to medium-sized (2-7 cm broad); surface dry, reddish to purple, but overlaid with a fine whitish powder or velvety bloom .... *R. mariae*, p. 96
38. Not as above; cap lacking whitish bloom 39

39. Spore print and gills white or whitish 40
39. Spore print yellow to ochre; gills yellowish in age 42

40. Cap typically with purplish, pinkish-lilac, bluish, and/or yellow tones; fruiting body sometimes large; gills often slightly greasy to the touch; stalk *not* staining salmon in ferrous sulfate .... *R. cyanoxantha*, p. 94
40. Cap usually browner, grayer, redder, or more flesh-colored than above; fruiting body usually medium-sized; stalk staining salmon in ferrous sulfate 41

41. Predominant color of cap typically flesh-color or reddish; cuticle (skin) often wrinkled; found mainly in eastern North America .... *R. vesca* (see *R. cyanoxantha*, p. 94)
41. Predominant color of cap typically some shade of brown, olive-brown, gray, or gray-brown; widespread .... *R. brunneola* (see *R. cyanoxantha*, p. 94)

42. Stalk typically staining yellow when scratched, then slowly becoming brownish; stalk usually rose-colored or at least with a blush of pink, but white in some forms; fruiting body medium-sized to very large; odor shrimpy or fishy in old age .... *R. xerampelina*, p. 102
42. Not as above 43

43. Cap red to pink, orange, or whitish, or a mixture of these colors; associated with hardwoods, especially oak; stalk white .... *R. maculata* group, p. 100
43. Not as above 44

44. Cap 2-6 cm broad, markedly fragile, variously colored but usually with at least some purple .... *R. placita* group & others, p. 100
44. Not as above; larger and/or not markedly fragile 45

45. Cap large, firm, sometimes massive (up to 35 cm broad!); stalk often tinged pinkish .... *R. olivacea* (see *R. alutacea* group, p. 102)
45. Not as above; medium-sized to fairly large, but not massive; stalk usually white 46

46. Spores and *mature* gills creamy; found with hardwoods .. *R. grisea* (see *R. aeruginea*, p. 95)
46. Spores and mature gills yellow to ochre; found with hardwoods or conifers 47

47. Cap medium-sized (3-12 cm broad), burgundy to livid red, purple, or rusty-red, or with a brownish center, or with reddish-buff tones; fruiting body quite firm *R. integra* group, p. 101
47. Not as above; medium-sized to fairly large (cap 5-20 cm broad) and often fragile in old age .... *R. alutacea* group & assorted JARs, p. 102

Russula brevipes   (Short-Stemmed Russula)

CAP 7-30 cm broad, broadly convex with a depressed center and inrolled margin, becoming broadly vase-shaped in age; surface dry, unpolished, minutely woolly or felty, white or whitish but often dirty and/or with yellowish to brown stains and discolorations; margin not striate. Flesh thick, crisp, brittle, white, not staining; taste mild or slowly acrid. **GILLS** thin, close or crowded, adnate to decurrent, usually alternating long and short, white or creamy (but tinged blue-green in one variety), often brownish-stained in age. **STALK** 2-7 (10) cm long, 2-5 cm thick, hard and rigid, often quite short and stout, equal or tapering downward, smooth, dry, dull white or brownish-stained, sometimes with blue-green tinge at apex. **SPORE PRINT** white to pale buff; spores 8-11 × 6.5-9 microns, elliptical, with amyloid warts and ridges.

"Better kicked than picked," *Russula brevipes* is one of the most common and mundane of all the Russulas. Note the centrally depressed or vaselike white cap. The stem can be slightly longer—or considerably shorter—than shown here and the cap is often incrusted with dirt.

HABITAT: Solitary or more often scattered or in groups or troops on ground in woods of all kinds; widely distributed and often abundant. It hugs the ground closely and may be visible only as a low mound or "mushrump" in the humus. In our area the major crop is typically in November or December, but it can be found most any time.

EDIBILITY: Edible, but better kicked than picked. In my experience it is insipid and granular, but some people are apparently fond of it. One authority recommends stewing it slowly in soups; another says to "throw it in with the sauerkraut." The flavor and texture are substantially improved when it is parasitized by *Hypomyces lactifluorum* (see p. 884).

COMMENTS: "So large, so mediocre" and "Better kicked than picked" typify the comments this tedious, vulgar, mundane mushroom received when overwhelmingly voted the "Most Boring Mushroom" of the 8th Annual Santa Cruz Fungus Fair. (Second place went to "the one behind the desk answering questions.") Actually, it is a rather attractive mushroom when first encountered, the pristine cleanness (clean pristineness?) of the gills contrasting prettily with the dirty hunks of humus hoisted aloft by the concave cap. However, as the above comments attest, one quickly tires of finding it—not only because of its ubiquitousness (it is by far the most common of the larger Russulas), but also due to its annoying habit of forming deceptively promising "mushrumps" in the humus (thwarted boletivores are not known for their ability to perceive beauty).

   R. brevipes is easily recognized by its dull white color, centrally depressed cap with an inrolled margin when young, hard stem, and crisp flesh that does not blacken or redden when bruised. It has the general aspect of a large *Lactarius* (e.g., *L. piperatus*), but lacks a latex, though the gills when young may be beaded with water droplets. *Var.* **acrior** is the form with an acrid taste and blue-green tints in the gills and stalk apex. Other species: ***R.*** ***romagnesiana*** is a rare but very similar eastern species with smaller spores; ***R. cascadensis*** also has smaller spores, but has an *intensely* acrid taste and smaller cap (4-12 cm broad) and occurs under conifers in the Pacific Northwest. ***R. delica*** is also close to *R. brevipes,* but is said to have thick, well-spaced gills. It was originally described from Europe and is apparently rare in North America.

Russula albonigra is a hard, robust species that is decidedly schizophrenic: white in youth, black in age (note the young whitish cap at the bottom). It also blackens when bruised.

Russula albonigra (Blackening Russula; Integrated Russula)

CAP 7-20 (25) cm broad, broadly convex or centrally depressed becoming broadly depressed in age; surface dry to slightly viscid, sometimes polished; smooth, at first white, soon becoming grayish or blackish-brown and finally entirely black; margin not striate. Flesh thick, crisp, brittle, white, bruising gray and then black; taste mild or slightly acrid. **GILLS** adnate to slightly decurrent, close to rather well-spaced, thick, brittle, usually alternating long and short; creamy-white staining gray or black, often entirely black in old age. **STALK** 3-13 cm long, 2-5 cm thick, very hard, stout, solid, rigid, smooth, equal or tapered downward, white becoming grayish or brownish-gray in age or where wounded, then black. **SPORE PRINT** white; spores 7-10 × 5.5-7.5 microns, broadly elliptical to nearly round, with amyloid warts and ridges.

HABITAT: Scattered or in groups or troops under both hardwoods and conifers; widely distributed. It is sporadically common in our oak-madrone woodlands but appears to be more abundant than it actually is because the fruiting bodies do not decay readily. In the Pacific Northwest it favors conifers.

EDIBILITY: Better punted than hunted. Although said to be edible if thoroughly cooked, it is hardly tempting and a closely related Oriental species is poisonous.

COMMENTS: This sometimes massive *Russula* is easily identified by its hard cap and stem and the blackening of all parts in age or when handled. It is decidedly schizophrenic in appearance—young specimens are reminiscent of *R. brevipes* and *Hygrophorus sordidus* in their whiteness, while elderly individuals are completely jet-black (hence the epithet *albonigra,* meaning "white-black"). Most dramatic, however, are those specimens in transition, i.e., with a black cap, black stem, and white gills (the gills generally take longer to blacken than the cap and stem). *R. sordida* is a passe synonym. *R. atrata* is a very similar west coast species that also blackens, but has a much thicker cap cuticle. Several allied species stain reddish *before* turning black (see *R. densifolia*), and *R. adusta* (see comments under the *R. subnigricans* group) darkens less dramatically.

Russula densifolia   (Reddening Russula)

CAP 5-15 cm broad, broadly convex to plane or depressed; surface viscid when moist, soon dry and often polished; smooth, whitish to pale buff becoming grayish or brownish or eventually blackish in old age; margin not striate. Flesh thick, crisp, white, slowly bruising reddish or orange-red, then eventually grayish-brown to black; taste usually acrid. GILLS adnate to slightly decurrent, close, brittle, whitish, developing sordid reddish to smoky stains in age, usually alternating long and short. STALK 4-10 cm long, 1-3 cm thick, hard, rigid, smooth, equal; whitish when fresh but staining like the flesh (reddish or orangish, then smoky-brown to black). SPORE PRINT white; spores 7-10 × 6-8 microns, broadly elliptical with amyloid warts and ridges.

HABITAT: Solitary, scattered, or in groups in woods; widely distributed. It is fairly common locally, along with *R. dissimulans* (see comments), but is not nearly as numerous as *R. albonigra, R. brevipes,* or *R. subnigricans.* The major crop is in the fall or early winter.

EDIBILITY: Better dribbled than nibbled. The sharp taste is said to disappear in cooking, but the end product is insipid at best and indigestible or even poisonous at worst.

COMMENTS: The slow staining of wounded tissue to reddish and then grayish-brown, sooty-gray, or black is the calling card of this coarse *Russula* and its close relatives (see below). The staining is a tactful compromise between *R. albonigra,* which blackens directly, and the *R. subnigricans* group, which stains reddish but not black. The full staining sequence may take half an hour and is best seen by scratching the stalk. Other coarse, closely related Russulas with the same staining reactions include: *R. dissimulans,* common and widely distributed, with a dry (not viscid) cap, mild taste, and slightly larger spores; and *R. nigricans,* fairly common (at least on the west coast), with very thick, well-spaced gills and smaller spores. Both of these species are "better beaten than eaten."

Russula subnigricans group   (Rank Russula)

CAP (5) 9-20 (30) cm broad, broadly convex to plane or depressed; surface slightly viscid when moist, smooth, whitish when young, soon becoming pale smoky-brown or dingy reddish-brown or dull ochraceous-stained, or often a sordid mixture of these shades; margin at first inrolled, not striate. Flesh thick, crisp, white, typically bruising *slowly* reddish (in 5-20 minutes), then usually browner or grayer within an hour, but sometimes staining grayish-brown directly (especially in stalk), or not at all; odor strong and unpleasant, at least in age or upon drying; taste mild to slightly bitter. GILLS thick, brittle, adnate to slightly decurrent, usually alternating long and short; pallid, soon stained dark or sordid reddish; close to fairly well-spaced. STALK (3) 7-13 cm long, (1.5) 3-7 cm thick, very hard, rigid, solid; equal or with tapered base, smooth, white, staining sordid reddish and/or smoky-brown in age or when handled. SPORE PRINT white; spores 6-10 microns, nearly round with low amyloid warts and ridges.

HABITAT: Solitary, scattered, or in groups under live oak; common in our area from late summer through early winter and also reported from Mississippi. In our area it is a good chanterelle-indicator (see comments).

EDIBILITY: Better trampled than sampled. The rank odor is hardly enticing and the Japanese version is extremely poisonous according to mycologist Tsuguo Hongo.*

COMMENTS: This vulgar giant of the genus is very close both macro- and micro-scopically to *R. subnigricans,* a species originally described from Japan. (Our version, upon which the above description is based, seems to differ only in its slightly larger size, closer gills, and rank odor at maturity.) The principal fieldmarks are its hard stem, brittle

---

*Hongo, coincidentally, is the Spanish word for "mushroom."

Russula subnigricans group. This large rank *Russula* develops dingy reddish stains on the gills, but never blackens. It seems to grow only with oak—usually in the company of chanterelles.

texture, large size, coarse appearance, and reddening of the gills in age or where wounded. The staining reactions elsewhere on the fruiting body are rather variable, but at no point does it blacken as in *R. albonigra, R. densifolia, R. dissimulans,* or *R. nigricans.* The cap color is rather amorphous—generally an unbecoming blend of murky gray, smoky-brown, and sordid reddish or ochraceous—but in one collection I made the fruiting body had bluish-green tinges as in *R. brevipes* var. *acrior.* Another interesting feature is its affinity for chanterelles, one or more of which will often grow right out of the base of its stem—or if not, then in the immediate vicinty! Other species: *R. adusta* is a somewhat similar, widespread species which has rather erratic staining reactions (the flesh normally bruises smoky-brown or grayish-black, but may show a slight reddish intermediate phase); its gills, however, do *not* redden in age. *R. compacta* of northern and eastern North America is also somewhat similar, but stains ochre or brown and has a pallid to rusty- or reddish-brown to ochre-brown cap. Both of these are "better stomped than chomped."

Russula decolorans group   (Graying Russula)

CAP 5-15 cm broad, nearly round becoming convex to plane or slightly depressed; surface smooth, viscid when moist, color variable: dull red to orange, red mixed with yellow, coppery-brown, or at times with cinnamon or tawny tones; margin striate in old age. Flesh thick, firm, white, becoming gray, ashy, or black when bruised or exposed (sometimes slowly); odor and taste mild. **GILLS** typically adnate to adnexed, fairly close, creamy becoming pale ochre or yellowish, sometimes grayish-stained in old age. **STALK** 4-12 cm long, 1-3 cm thick, equal or often thicker below, very firm when young (but often softer in age), white, staining and aging grayish. **SPORE PRINT** yellowish to pale ochre; spores 9-14 × 7-10 (12) microns, elliptical, with isolated amyloid warts.

HABITAT: Solitary to widely scattered or gregarious on ground in mixed woods and under conifers, associated principally with pine; widely distributed.

EDIBILITY: Better smashed than stashed—it is said to be edible, but there are several look-alikes of unknown edibility.

COMMENTS: The graying flesh, reddish to orange cap, and mild taste are the fallible fieldmarks of this firm, rather robust *Russula,* which is actually a close-knit "complex" of species. One striking form found in our coastal pine forests is *very* firm and has a somewhat shiny cap in dry weather. Other species: *R. paludosa* has a reddish cap and flesh that grays somewhat, but the taste is usually slightly bitter or acrid. *R. obscura* has a darker (purple-red to brownish) cap, mild taste, and flesh that stains pinkish or reddish before blackening. Both of the above species are quite firm and robust, have ochre spores, and grow with conifers. Another distinctive species, *R. occidentalis,* has a grayish to olive-brown to buff, brownish, or vinaceous cap (usually a mixture of these colors) and mild taste. Its stalk and flesh are white but usually stain reddish and then gray (or gray directly) when bruised. It is fairly common under mountain conifers in the Pacific Northwest.

Russula claroflava   (Yellow Russula)

CAP (3) 4-12 cm broad, nearly round becoming convex to plane or slightly depressed; surface slightly viscid, soon dry, yellow to golden-yellow, smooth; margin slightly striate in age. Flesh white, slowly staining grayish when rubbed; odor and taste mild. **GILLS** adnate to adnexed or free, close, creamy becoming pale ochre; sometimes grayish-stained in age. **STALK** 3-8 cm long, 1-2 cm thick, more or less equal, white to pale yellow, dry, smooth, aging or slowly bruising grayish. **SPORE PRINT** yellow-ochre; spores 8.5-10 × 7.5-8 microns, broadly elliptical with amyloid warts and ridges.

HABITAT: Scattered or in small groups in woods; mainly northern or montane in distribution and partial to birch, aspen, and various conifers. I have found it in New Mexico under aspen, but not in California. It fruits mainly in the summer and fall.

EDIBILITY: Edible.

COMMENTS: The pretty yellow cap and tendency to stain or age ashy-gray rescue this species from "JAR" status. *R. flava* is a synonym. *R. ochroleuca* is a similar species with a yellow-ochre cap, acrid taste, and flesh that grays *quickly.* Another yellow-capped species, *R. lutea,* is rather fragile, has a mild taste, ochre gills, and does *not* stain gray. All of these are fairly common in northern North America under conifers, birch, and aspen.

Russula fragrantissima group   Color Plate 11
(Fragrant Russula; Fetid Russula)

CAP 5-15 (20) cm broad, nearly round becoming convex, then plane or somewhat depressed; surface viscid or slimy when moist, smooth, yellowish to yellow-brown to bright ochre, straw-colored, orange-brown, or brown, often darker (or reddish-spotted) at the center; margin usually radially furrowed with small bumps (tuberculate-striate), at least in old age. Flesh thick, white or tinted cap color; odor heavy and penetrating—at first sweet (like maraschino cherries, almond extract, or benzaldehyde), but in age with a nauseating (fetid) component; taste nauseating; taste of gills acrid. **GILLS** creamy-white becoming yellowish or pale ochre, brownish-spotted in age, often beaded with water droplets when fresh; adnate to adnexed or free, fairly close. **STALK** 3-20 cm long, 1-4 cm thick, equal or somewhat tapered below, whitish to buff becoming yellowish or brownish-stained in age, especially near base; smooth or longitudinally lined, dry. **SPORE PRINT** pale orange-yellow; spores 6-9 × 6-8 microns, broadly elliptical to nearly round with amyloid warts and at least partially reticulate.

HABITAT: Solitary or scattered to gregarious on ground under both hardwoods and conifers, widely distributed. In our area it fruits mainly in the fall under oak, tanoak, and madrone; it can be found every year but is not nearly as numerous as the *R. sororia* group.

EDIBILITY: Unequivocally inedible and possibly poisonous—in other words, it is better punched than munched. The disgusting taste and strong odor should discourage most people from sampling it.

COMMENTS: Anyone under the mistaken impression that all mushrooms smell alike should get a whiff of this robust, odoriferous *Russula*. The penetrating odor is reminiscent of maraschino cherries, but with a fetid component that becomes more and more pronounced as the mushroom matures. The cap is often merely flat at maturity, rather than depressed as in most of the larger Russulas, and the stalk is often long in relation to the cap. The yellow-brown to ochre cap, fairly large size, and water droplets on the gills (when moist) are also distinctive. Other species in the "complex" include:: *R. laurocerasi* (possibly the one in the color plate), a slightly yellower, cleaner, slimmer, larger-spored version of *R. fragrantissima* whose odor is strongly fragrant through maturity, albeit with a subtle fetid component; it is also widely distributed and the two appear to intergrade. Both have passed under the name *R. foetens,* a European species with a nauseating odor even in youth and non-reticulate spores. Its North American counterpart, *R. subfoetens* (=*R. foetentula?*) is smaller, has a bright rusty-orange to rusty-yellow to brownish cap, and does not smell as strong. Finally, there is *R. ventricosipes,* which grows in sand under northern conifers and has a large, conspicuously swollen (ventricose) stalk.

Russula sororia group   (Comb Russula)

CAP 4-12 cm broad, convex becoming plane or centrally depressed; surface viscid when moist, smooth, dark grayish-brown to hazel-brown when young (rarely whitish), usually paler (yellow-brown to straw-colored or pale grayish-brown) toward margin or in age; margin usually furrowed with low bumps (tuberculate-striate) in age. Flesh brittle, firm, white, not bruising; odor unpleasant (rancid or "spermatic"); taste mild or slightly acrid; taste of gills slowly but distinctly acrid. **GILLS** white or creamy, but often brownish- or rusty-stained in age; close, adnate to adnexed or free. **STALK** 3-8 cm long, 1-2.5 cm thick, equal or tapered downward, whitish, but often rusty-stained near base; often with large cavities (usually 3) within. **SPORE PRINT** creamy to pale yellow; spores 6-9 × 5-7 microns, broadly elliptical with amyloid warts.

HABITAT: Scattered to densely gregarious in woods or at their edges, on lawns under trees, etc; common and widely distributed. In our area it is often abundant in the fall and winter, especially under pine, oak, and Douglas-fir.

Russula sororia is abundant under both hardwoods and conifers. The yellow-brown to grayish-brown cap has a striate margin in age.

EDIBILITY: Better stomped than chomped—it might be poisonous and doesn't taste good anyway.

COMMENTS: The brown to grayish-brown to dingy straw-colored cap, tuberculate-striate margin in age, peppery-tasting gills, unpleasant odor, and moderate size typify a group of Russulas that are at times exceedingly abundant. They differ from the *R. fragrantissima* group in their smaller size, grayer color, and different odor. They have passed under the names *R. pectinata*, a European species with a yellower cap, and *R. pectinatoides,* a North American species that is also yellower than *R. sororia,* and has a less acrid taste and darker (yellower) spores. Other species: *R. amoenolens* is the same as *R. sororia* (as defined here); *R. cerolens* is practically the same but has partially reticulate spores; *R. granulata* and *R. pulverulenta* have scurfy-granulose caps.

Russula cyanoxantha   (Variegated Russula)   Color Plate 12

CAP 5-18 cm or more broad, convex becoming plane or depressed; surface viscid when moist but soon dry, smooth, color variable, sometimes entirely but more often a mixture of: pinkish-lilac, dull purple, green, olive, yellow, blue-green, white, and/or brown; margin sometimes obscurely striate. Flesh thick, firm, white, crisp; odor mild or pleasant, taste mild. **GILLS** adnexed to adnate or slightly decurrent, many of them forked at least once; white or with a few brownish stains, slightly greasy to the touch. **STALK** 5-13 cm long, 1.5-5 cm thick, solid or stuffed, rather hard; more or less equal, smooth, white, but the flesh in base sometimes grayish. **SPORE PRINT** white; spores 7-10 × 6-8 microns, broadly elliptical to nearly round with isolated amyloid warts. Stalk *not* staining salmon in ferrous sulfate.

HABITAT: Solitary, scattered, or in groups or troops under hardwoods or sometimes conifers; widely distributed. It is often abundant shortly after the first fall rains in our live oak-tanoak-madrone forests, but fruits on into the winter. Like many Russulas, it is often concealed by fallen leaves.

EDIBILITY: Edible and choice; it is delectably sweet when young and rated highly in Europe. The hard, crisp buttons are hard to beat in omelets or with scalloped potatoes, but alas, they are also considered "edible and choice" by maggots!

COMMENTS: This variable species is more difficult to characterize than it is to recognize (see photo on p. 897). The telltale traits are its fairly large size, white gills and spores, firm white stalk, mild taste, and variegated cap (typically a mixture of pinkish-lilac, dull purple, green, yellow or even blue). In our form young specimens—especially those still covered by leaves—are often uniformly colored a beautiful pinkish-lilac-mauve. When exposed to light they gradually become spotted with yellow, green, etc., and in old age are often a washed-out whitish color at the center, with slight greenish or lilac tones showing on the margin. Another distinctive feature is its soft, somewhat greasy gills: they are not as brittle as those of most Russulas. *R. parazurea* (see comments under *R. aeruginea*) is easily mistaken harmlessly for greenish forms of *R. cyanoxantha,* but is usually slightly smaller, not so variegated, and has reticulate or partially reticulate spores. *R. variata (=R. cyanoxantha var. variata),* common in northern and eastern North America, differs in its more promiscuously forked gills, often acrid taste, and somewhat different cap color—tending more toward brownish-olive, greenish, dull purplish, or vinaceous-brown. Two other species, *R. vesca* and *R. brunneola,* also have a mild taste and whitish spores, but their stalks stain salmon or deep salmon in ferrous sulfate. The cap cuticle in *R. vesca* tends to be slightly redder or more flesh-colored (and is often wrinkled and/or recedes from the margin in age), while that of *R. brunneola* tends to have a brownish to yellowish-brown to olive-brown or grayish-olive cast. Both are widely distributed and easily confused with each other (as well as with *R. cyanoxantha*) when not "typically" colored. Both are edible, however, so it hardly matters. For similar species with creamy or yellow-tinged spores, see *R. parazurea* and *R. grisea* (under *R. aeruginea*).

Russula aeruginea   (Green Russula)

CAP 3-9 cm broad, convex becoming plane or slightly depressed; surface viscid when moist, dull green to dark green, sometimes with brown, gray, or yellowish tints or blotches; margin often striate. Flesh white, brittle; taste mild. **GILLS** adnate to adnexed or free, close, brittle, white becoming pale yellowish, often with brownish stains. **STALK** 4-8 cm long, 1-2 cm thick, equal or with tapered base, white or faintly yellow, base often with pale brown stains. **SPORE PRINT** creamy to pale yellow or pale orange-yellow; spores 6-9 × 5-7 microns, nearly round with amyloid warts and ridges.

HABITAT: Solitary, scattered, or in groups in woods; widely distributed. In California I have found it only a few times, but it is sometimes abundant under aspen in the Rocky Mountains and Southwest, and under oak in the South.

EDIBILITY: Edible. I have fried it.

COMMENTS: This moderate-sized *Russula* can be told by its smooth green cap, pallid or brownish-spotted gills, and pale yellow spore print. There is no red or purple in the cap as in some of the larger, yellow-spored Russulas (e.g., *R. olivacea* and *R. xerampelina*). There are several similar species, including: *R. parazurea,* a beautiful green-capped species with a matt appearance, firm white stem, and white spores, found under oak and tanoak in our area; and *R. grisea,* fairly common in our area under oak, with a slightly duller or grayer cap (green mixed with brown, gray, etc., or sometimes purplish or lilac) and yellowish spores. Also see *R. cyanoxantha* and the species listed under it.

Russula virescens   (Quilted Green Russula)

CAP 5-15 cm broad, nearly round at first becoming convex to plane or slightly depressed; surface dry and sometimes slightly velvety, markedly areolate at maturity (i.e., cracked into small flattened scales or patches); dull green to greenish-gray or sometimes dull bluish-green, at times with ochre, buff, or creamy discolorations; margin not striate or only very slightly so, often lobed. Flesh thick, white, brittle; odor and taste mild. **GILLS** white or creamy, fairly close, adnate to adnexed or even free. **STALK** 3-9 cm long, 1-3 (4) cm thick, more or less equal, dry, firm, brittle, white, more or less smooth. **SPORE PRINT** white or with a faint yellow tinge; spores 6-9 (10) × 5-7 microns, elliptical to nearly round, with amyloid warts and/or ridges.

HABITAT: Solitary to widely scattered or in groups under hardwoods, especially oak and beech; common (along with *R. crustosa*—see comments) during the summer months in eastern North America. It has also been recorded from Montana under birch and is to be expected in southern Arizona and New Mexico, where several "eastern" species occur.

EDIBILITY: Edible and choice—along with *R. cyanoxantha* and *R. xerampelina,* among the best of the Russulas for the table.

COMMENTS: One of the most distinctive of all the Russulas, this species is easily told by its greenish to greenish-gray, areolate (cracked or quilted) cap, white spores, and asso-ciation with hardwoods. It is one of the characteristic summertime mushrooms of eastern hardwood forests, where it sometimes fruits with the equally edible *Lactarius volemus* and/or *L. corrugis*. Other species: *R. crustosa* is a closely related edible species with an areolate cap that is slightly viscid when wet and more variable in color (usually browner or more ochre-buff, but sometimes with a greenish tinge); *R. maculosa* of the Southeast resembles *R. crustosa,* but has a paler cap and disagreeable odor in age.

Russula mariae   (Powdered Russula)

CAP 2-7 cm broad, convex becoming plane or slightly depressed; surface dry, purple to reddish-purple, amethyst, dark crimson, or maroon (sometimes toned with gray or olive), finely powdered or velvety from a whitish bloom; margin sometimes striate when old. Flesh thin, brittle, white; taste mild or slightly acrid. **GILLS** close, white becoming creamy or pale yellowish, typically adnate to adnexed. **STALK** 2.5-7.5 cm long, 1-2.5 cm thick, equal or tapering downward, firm, rigid, white or tinged reddish to purple. **SPORE PRINT** creamy-yellow; spores 7-10 × 6-8 microns, broadly elliptical to round, with amyloid warts and ridges.

HABITAT: Solitary or in groups on ground in woods; rather rare in California, where I have found it in the fall under oak and Douglas-fir, but a common mid-summer mushroom in the hardwood forests of eastern North America.

EDIBILITY: Edible, but hardly incredible.

COMMENTS: This exquisite, rather dainty *Russula* is one of the more easily recognized species in the genus. The dry, velvety or finely powdered reddish to purple cap, creamy spore print, and mild taste distinguish it. The "splitters" recognize two species, the purple-capped forms being called **R. alachuana.** The name *R. mariae* should not be confused with *R. mairei,* a member of the *R. emetica* group.

Russula albidula group   (Boring White Russula)

CAP 3-8 cm broad, convex becoming plane or shallowly depressed; surface smooth, slightly viscid when moist, then dry; white or tinged buff at center; margin not striate or very obscurely so. Flesh fragile, white; taste acrid. **GILLS** adnate to adnexed or free, brittle, close, white or creamy. **STALK** 2.5-7 cm long, 1-2.5 cm thick, equal or thicker below, dry, smooth, rigid, white; often spongy or hollow in age. **SPORE PRINT** white or creamy-white; spores 7-10 microns, nearly round with amyloid ridges.

HABITAT: Solitary or in small groups in woods or at their edges, widely distributed. It is fairly common in our area from late summer through the spring, usually with oak but never in large numbers (like lips and scissors, it tends to occur in pairs). It is partial to poor soil along trails or roadcuts.

EDIBILITY: Better whacked than sacked—the acrid taste is a deterrent.

COMMENTS: This plain, unprepossessing, profoundly forgettable *Russula* can be recognized by its white color, modest size, and acrid taste. It is smaller and more fragile than *R. brevipes* and does not have a strongly inrolled cap margin when young. There are several very similar, confusing, widespread, wishy-washy Russulas. They include **R. albella** and **R. albida,** with a mild to slightly bitter taste, and **R. anomala** and **R. subalbidula,** with an acrid taste, but distinguishing between them is a matter for the specialist. For a comparison of the *R. albidula* group to *R. cremoricolor,* see comments under the latter.

Russula albidula group. A dingy and often dirty whitish "JAR" of moderate size. *R. cremoricolor* (illustrated on next page) is usually much cleaner.

Russula cremoricolor resembles *R. emetica* except for its cap color, which is yellowish-white.

Russula cremoricolor   (Creamy Russula)

CAP 3-10 cm broad, convex to plane to centrally or broadly depressed; surface viscid when moist, smooth, white to pallid to very pale yellowish-white, the center often slightly darker (yellowish or buff); margin usually striate in age (but sometimes obscurely). Flesh brittle, white; odor mild or slightly fragrant; taste distinctly acrid. **GILLS** white, adnate to adnexed or ocasionally free, fairly close. **STALK** 3-10 cm long, 0.7-2.5 (3) cm thick, equal or slightly thicker below, dry, white or whitish, usually longitudinally lined or striate. **SPORE PRINT** white; spores 7.5-10 × 7.5-8 microns, round to broadly elliptical, with amyloid warts and ridges.

HABITAT: Solitary, widely scattered, or in groups or troops in woods; widely distributed. In our area it is often abundant in the late fall and winter in the same habitats as *R. emetica*. In the Pacific Northwest it favors conifers.

EDIBILITY: Better obliterated than refrigerated—the acrid taste is a deterrent.

COMMENTS: This species looks like an albino *R. emetica,* and save for the whitish cap, there is little difference. The cap, although whitish, is slightly yellower than that of *R. albidula*, and is usually cleaner and more markedly striate. Also, *R. cremoricolor* tends to fruit in larger numbers. Other species: *R. crenulata* and *R. raoultii* are two very similar species that are yellower (cap more or less pale yellow); the latter also has smaller spores. Another pale capped, bitter- or acrid-tasting species, *R. crassotunicata,* can be distinguished by its thick, rubbery or leathery cap cuticle that is often slightly scurfy and its white stalk that frequently develops brownish spots in age. It occurs under northern conifers. "Eating raoultii" (or any of the other species discussed here) is not recommended.

Russula emetica group   (Emetic Russula; The Sickener)

CAP 3-10 cm broad, rounded-convex becoming plane or broadly depressed; surface viscid when moist, smooth, bright red to scarlet, the center often darker; fading in age or wet weather to pink, orange, or blotched with white; margin eventually striate (but sometimes obscurely). Flesh white (but pink under the cuticle), brittle, odor mild, taste very acrid. **GILLS** white or creamy-white, brittle, close, adnate to adnexed or free. **STALK** 3-10 cm long, 0.7-2.5 cm thick, equal or thicker below, dry, white or whitish, usually with longitudinal lines. **SPORE PRINT** white; spores 7-11 × 6.5-9 microns, elliptical with amyloid warts and ridges.

97

Russula emetica group. The bright red cap (that often fades), white gills, white stem, and peppery taste make this an easy mushroom to identify.

HABITAT: Solitary, scattered, or in groups or troops in woods, widely distributed and common. Our member of the *R. emetica* group fruits late in the season (winter) in chaparral and mixed woods (oak, madrone, manzanita, and pine), often intermingled with *R. cremoricolor* and *Lactarius vinaceorufescens*. It also grows on very rotten wood—unorthodox behavior for a *Russula*.

EDIBILITY: Pleasing to the eyes of all, to the tongues of some, but to the stomachs of none. As its name implies, it is poisonous, at least raw, and can be used to induce vomiting. Parboiling may destroy the toxins, but it hardly seems worth the trouble or risk. It is collected by Himalayan villagers, presumably for use in curries (consult your local travel agent for details).

COMMENTS: This attractive mushroom is easily recognized by its red cap, white stem and gills, and red-hot taste. As in many Russulas, the skin peels rather easily from the cap. For years the name *R. emetica* has been applied to a confusing group of Russulas with the above features. I see no compelling reason to discontinue this trend, though the present consensus seems to be that the "real" *R. emetica* occurs principally with conifers and favors sphagnum bogs. Other species in the group include: ***R. silvicola***, an eastern species with a red to pinkish-yellow cap and slightly smaller spores, common with both hardwoods and conifers; and ***R. mairei***, colored like *R. emetica* but associated with hardwoods and with a firmer texture and smaller spores, 6-8.5 × 5-7 microns (it may well be our local species).

Russula fragilis   (Fragile Russula)

CAP 2-5 cm broad, convex becoming plane or depressed; surface smooth, viscid when moist, color variable: purplish to pinkish, olive-brown, greenish, or even yellow, or a mixture of these colors, but most often blackish at the center and pinkish or pinkish-yellow at the margin, with a grayish-olive zone in between; margin striate in age. Flesh thin, very fragile, white; odor variable; taste acrid. **GILLS** adnate to adnexed or even free, fairly close, white or creamy-white. **STALK** 2.5-7 cm long, 0.5-1.5 cm thick, equal or slightly thicker at either end, dry, whitish; fragile. **SPORE PRINT** whitish; spores 6-9 × 5-8 microns, broadly elliptical to nearly round, with amyloid warts and ridges.

HABITAT: Solitary, scattered, or in groups in woods of all kinds, but especially under conifers; widespread. I have not seen it in our area, but several similar species occur.

EDIBILITY: Not recommended—the acrid taste is a deterrent.

COMMENTS: This species is apt to befuddle the color-conscious beginner with its numerous color guises, but is rather distinctive when "typical," i.e., dark at the center, pinkish toward the margin and grayish-olive between. The other critical features are the fragile texture, peppery taste, white gills, and white spores. Other species: *R. aquosa* is a closely related but slightly larger species; its cap is viscid with a greasy feel and shiny appearance when wet, and is usually reddish to orange-red or pink with grayish, brownish, or yellow tints.

Russula gracilis group

CAP 2-8 cm broad, convex becoming plane or slightly depressed, sometimes retaining a slight umbo; surface smooth, viscid when moist, color variable: usually some shade of pink, but often with a dull greenish or grayish-olive to brownish-black center and pale pinkish margin, and sometimes a rather deep pink or violet-tinged; margin usually striate in old age. Flesh fragile, white; taste acrid. **GILLS** typically adnate to slightly decurrent, whitish becoming creamy and finally tinged yellow, close. **STALK** 3.5-8 cm long, 0.5-1.5 cm thick, equal or slightly thicker below, usually quite soft and fragile, whitish, but usually tinged or flushed pink to grayish-rose. **SPORE PRINT** creamy to pale ochre; spores 7-9 × 5-7 microns, elliptical, with isolated amyloid warts.

HABITAT: Widely scattered to gregarious on ground in woods; it is said to grow only with birch, but a very similar if not identical species is abundant in our area with Douglas-fir in the fall and winter.

EDIBILITY: Not edible due to the acrid taste.

COMMENTS: Also known as *R. gracillima,* this species can be told by its fragile pinkish-tinged stalk, yellowish spores, acrid (peppery) taste, and pinkish or pink-and-greenish cap. *R. fragilis* is quite similar in cap color but has whitish spores and a white stalk, while the *R. placita* group, which also occurs commonly with Douglas-fir in our area, has a mild taste, white stem, and darker cap. Other species: *R. pelargonia* also has a pinkish-tinged stalk and acrid taste and grows with conifers (especially Douglas-fir), but it has creamy spores, a more erratically colored cap (grayish, pinkish, brownish, reddish, etc.), and a distinct geranium-like odor (at least when the flesh is crushed).

Russula rosacea   (Rosy Russula)   Color Plate 13

CAP 3-12 cm broad, convex to plane or somewhat depressed; surface viscid when moist, smooth, dark red to bright red, fading in old age to pink or pink blotched with white. Flesh white, firm, brittle, odor mild, taste very acrid. **GILLS** creamy-white to pale yellow, adnate

Russula gracilis group. Our member of this species "complex" (shown here) is abundant under Douglas-fir. Note the slightly colored (pinkish-tinged) stalk.

to decurrent, close, brittle. **STALK** 4-10 cm long, 1-2.5 cm thick, equal or with a narrowed base, smooth, dry, pink or red, or at times only with a blush of red; hollow in age. **SPORE PRINT** pale yellow; spores 7-9 × 6-8 microns, nearly round with amyloid warts.

HABITAT: Scattered or in large troops under conifers, especially pine. Abundant on the Pacific Coast from fall through early spring, fruiting in our area mainly in coastal stands of Monterey pine, often in the company of *Lactarius deliciosus*.

EDIBILITY: To be avoided because of the acrid taste.

COMMENTS: This common, conspicuous, colorful inhabitant of our coastal forests is readily recognized by its red cap, red to rosy stem, creamy gills, and acrid (peppery) taste. It is one of our prettiest mushrooms—the clean, pale gills contrasting beautifully with the red cap and stem. At a distance it is sometimes mistaken for *Amanita muscaria,* which also grows commonly with pines but has a white stem with a ring (annulus) and volva. The pale yellow spores and red or pink stem distinguish *R. rosacea* from *R. emetica,* while the acrid taste separates it from *R. xerampelina* and the *R. alutacea* group. Other species: *R. sanguinea* is probably a synonym. A form with reddish-purple cap and stem—possibly distinct—is sometimes found in our area growing with *R. rosacea.*

Russula maculata group

CAP 4-10 (12) cm broad, broadly convex becoming plane or centrally depressed; surface viscid when moist, smooth, color variable: red to pink, reddish-orange, peachy-pink, yellowish-buff, or whitish, or often a mixture of these colors, or buffy-white with pinkish tints on the margin; margin obscurely striate in age. Flesh white, firm becoming fragile in age; taste mild or slightly peppery. **GILLS** creamy becoming pale ochre, close, adnexed, or free. **STALK** 4-10 (13) cm long, 1-3 cm thick, equal or thicker below, white, the base sometimes brownish-stained. **SPORE PRINT** yellow-ochre; spores 7-10 × 6-9 microns, nearly round, with amyloid warts and ridges.

HABITAT: Solitary, scattered, or in groups under hardwoods; common in our area in the fall and winter (sometimes spring) with live oak, but more widely distributed.

EDIBILITY: Unknown.

COMMENTS: The cap color in this species "complex" is quite variable, but within a well-defined range: red, pink, orange, yellowish-buff, and whitish. The association with oak, yellow spores, and rather fragile texture in age are also distinctive. The "true" *R. maculata* of Europe is said to have a mild to peppery taste and slightly fragrant odor. Our version has a consistently mild taste and little or no odor. Whether it is the "true" *R. maculata* or a closely related species is for the *Russula*-researchers to decide. There are a number of similar reddish-capped, mild-tasting species, such as *R. velenovskyi,* which do not fade as drastically and/ or differ microscopically.

Russula placita group   (Pleasing Russula)

CAP 2-7 (10) cm broad, convex to plane or somewhat depressed; surface viscid when moist, smooth, dark purple to wine-colored, reddish-violet, or brownish-purple, often darker at the center and paler toward margin, but sometimes brown to yellowish at center, and sometimes washed out in age; margin usually striate in age, at least obscurely. Flesh rather thin, fragile, white, odor and taste mild. **GILLS** close, white or creamy soon becoming yellow and finally dull ochre-yellow; adnate to adnexed or free. **STALK** 2-8 cm long, 0.5-2 cm thick, equal or slightly thicker at either end, white, dry, soft and spongy inside; fragile in age. **SPORE PRINT** yellow-ochre; spores 7-10 × 6-9 microns, broadly elliptical to nearly round, with amyloid warts and sometimes ridges.

Russula placita group. As defined here, this group includes many small, fragile Russulas with a mild taste, purplish cap, and yellow gills. They usually grow under conifers such as Douglas-fir.

HABITAT: Solitary to scattered or gregarious in woods, associated mainly with conifers; widely distributed. In our area this group is quite common in the fall and winter, especially with Douglas-fir (often accompanied by *R. xerampelina, R. integra,* and *Lactarius rubrilacteus*); elsewhere it often grows with spruce, fir, hemlock, and pine.

EDIBILITY: Edible but forgettable—thin-fleshed and virtually tasteless. Similar species are *probably*—but not definitely—harmless.

COMMENTS: A large number of small, fragile, mild-tasting, yellow-spored Russulas will fit the above description. Their caps vary considerably in color but usually have some purple or reddish-purple in them. Their fragility is such that it's difficult to transport them home without crushing them (the stalk, though fragile, breaks cleanly like chalk). They lack the shrimpy odor of *R. xerampelina* and do not stain brown, and are not as firm as *R. integra.* Their identification is best left to the specialist armed with monographs and a microscope. Some related species that bear mentioning are: **R. lilacea,** whose spores have isolated warts; **R. abietina,** cap often greenish or brownish; **R. puellaris,** with pale creamy-yellow spores; **R. caerulea,** with frequently umbonate cap; and **R. chamaeleontina,** with a small, thin, fragile, purplish to lilac to reddish to orange or yellow cap. The latter has proved invaluable to compulsive categorizers exasperated by the endless nuances of color in these Russulas. Aberrant forms with yellow spores, fragile texture, and a mild taste can simply be referred to *R. chamaeleontina* and conveniently forgotten about. *R. fragilis* has served the same purpose in the fragile, white-spored, acrid-tasting group.

Russula integra group

CAP 3-12 cm broad, convex becoming plane or slightly depressed; surface smooth, viscid when moist, color variable: usually burgundy to vinaceous-brown to livid red or rusty-red with the margin paler (reddish-buff), but in older specimens often buff or brownish in the central area and sometimes entirely reddish-buff; margin obscurely striate in old age. Flesh crisp, firm, white, not staining, odor and taste mild. **GILLS** adnate to adnexed or free, white becoming yellowish or pale ochre; close. **STALK** 3-9 cm long, 1-3 cm thick; usually equal or tapering upward, dry, white, sometimes with brownish or yellow-brown spots and stains. **SPORE PRINT** yellow-ochre; spores 8-11 × 6-9 microns, broadly elliptical with amyloid warts.

HABITAT: Scattered to gregarious under conifers, widely distributed. Common at times in our area in the fall and early winter. I have seen large fruitings under Douglas-fir.

EDIBILITY: Edible and good when young and crisp; however, there are many similar species of unknown edibility, so exercise caution.

COMMENTS: Also known as **R. polychroma,** this medium-sized species and its close relatives can be recognized by their mild odor and taste, yellow spores, white or brownish-stained stem, and reddish to wine-colored to brownish cap. They are larger than and not nearly so fragile as the *R. placita* group, and smaller than *R. xerampelina* and *R. alutacea.* As is commonly the case in *Russula,* this "complex" is in need of critical study.

Russula alutacea group

CAP 5-20 cm broad, convex to plane or somewhat depressed; surface viscid when moist but soon dry, smooth, color variable: dark red to red, purplish, purple-brown, or purple-red, or buff to straw-colored at the center or throughout, or with olive shades, or often a mixture of these colors; margin striate in age. Flesh brittle, white, firm at first, fragile in age; odor and taste mild. GILLS pale to dull ochre or straw-color; close, adnate to adnexed or free; brittle. STALK 3-10 cm long, 1.5-4 cm thick, more or less equal, smooth, dry, white or sometimes tinged pinkish. SPORE PRINT ochre-yellow; spores 8-11 × 6.5-9 microns, broadly elliptical with amyloid warts. Phenol solution staining stalk and flesh purple-red.

HABITAT: Solitary to scattered or gregarious in mixed woods and under conifers, widely distributed. In our area this group is especially abundant in old coastal pine forests in the fall and winter.

EDIBILITY: Not recommended. A large number of species will key out here—and we are not yet sure whether they're all edible.

COMMENTS: The above description embraces a number of medium-to-large, mild-tasting, yellow-spored Russulas. Distinguishing between them is a job for rabid *Russula* buffs. The most common variety in our area sometimes grows with *R. xerampelina,* but is easily distinguished by its mild odor and marked fragility in age; older caps often disintegrate in transit unless carefully collected and packed. Another local species is much firmer, with a reddish, often burnished cap (see comments under *R. decolorans*). *R. olivacea,* which occurs in Europe under hardwoods and in America under conifers, is a very firm, sometimes massive (cap to 35 cm broad!) species with a more frequently rose-tinted stem. It is closely related to the true *R. alutacea* by virtue of its purple-red phenol reaction. There are also a number of medium-sized to fairly large yellow-spored Russulas with an acrid taste. These include *R. veternosa* and *R. tenuiceps* of eastern North America (and of questionable occurrence in California); and an unidentified western conifer-lover that has a buffy-brown to pinkish-brown or flesh-colored cap, mild-tasting flesh, and acrid-tasting gills.

Russula xerampelina   (Shrimp Russula)   Color Plate 14

CAP 5-30 cm broad, convex becoming plane or centrally depressed; surface smooth, viscid when moist, color variable: typically red to dark red, purple, or brownish-olive, but often laced with (or sometimes entirely) green, brown, yellow-brown, purple-brown, etc.; margin usually striate in age (at least obscurely). Flesh thick, creamy-white, bruising yellowish and then brown or discolored brownish in old age, brittle; odor mild becoming shrimpy or crablike in old age; taste mild. GILLS close, adnate to adnexed, creamy-white becoming dull yellowish, staining like the flesh, drying brownish or grayish. STALK 3-12 cm long, 1-4 cm thick, equal or with slightly enlarged base, often longitudinally lined; dry, sometimes entirely rose-pink, but more often white with just a tinge of pink, or occasionally pure white; staining yellowish where scratched, then slowly brown; interior spongy and often discolored in age. SPORE PRINT yellowish; spores 8-11 × 6-8.5 microns, elliptical to nearly round, with amyloid warts. Flesh and stalk turning deep green in ferrous sulfate.

HABITAT: Solitary to gregarious under conifers; widely distributed and common, ranging as far north as the Arctic Circle. In our area it is associated almost exclusively with Douglas-fir and is often abundant in the fall, less so in the winter and early spring.

EDIBILITY: Edible and unforgettable—but one of the least appreciated of our edible fungi, perhaps due to the mediocrity of its brethren. The young, nutty caps are superb stuffed with grated cheese, chives, walnuts, and parsley and then broiled. In contrast to *R. cyanoxantha,* they are rarely riddled with maggots!

COMMENTS: This beautiful but extremely variable species can be recognized by the following *combination* of characteristics: (1) cap viscid when wet, usually with adhering debris when dry (2) stem usually—but not always—tinted pinkish and always staining yellowish and finally brown when handled or bruised (3) yellow spore print (4) mild (not peppery) taste (5) fishy odor *at maturity,* which is accentuated by cooking or drying (6) tendency of the gills to age or dry brownish to grayish. In addition, the young buttons are remarkably rotund and symmetrical—but this character can be appreciated only by religiously observing a large number of Russulas. Given the extreme variation in color and size, *R. xerampelina* may well be a composite species. At least two distinctive—but intergrading—varieties occur: one with a red to purple cap and lovely rosy stem, the other with a brownish-olive cap and white or nearly white stem (in addition to the color plate, there is a black-and-white photo on p. 83). Another distinctive variety occurs in our area with oak and manzanita; it differs in having a pale greenish-yellow to buff or whitish cap.

HYGROPHORACEAE   (Waxy Caps)

spores

Small to medium-sized, mostly terrestrial mushrooms. CAP typically smooth, often viscid and brightly colored. GILLS *soft, waxy,* usually thick and well-spaced, usually attached. STALK central; fleshy or hollow. VEIL absent or evanescent, rarely forming a slight ring (annulus). VOLVA absent. SPORE PRINT *white.* Spores round to elliptical, but often shaped like corn kernels when immature; typically smooth and not amyloid. Basidia long and narrow. Gill tissue interwoven, parallel, or divergent.

THESE are attractive, often colorful, white-spored mushrooms with soft, clean, waxy-looking gills. The flesh is not dry and brittle or chalky as in the Russulaceae, and the gills, though often thick, do not have the blunt edges typical of the chanterelles (Cantharella-ceae). However, there is little aside from the "waxy" gills to separate them from the numerous white-spored genera in the Tricholomataceae (particularly *Clitocybe, Mycena, Omphalina, Laccaria,* and *Marasmius).* The "waxiness" is admittedly an ambiguous character, but with a little experience is readily apparent—the gills are characteristically soft, clean, fleshy, and . . . they look (and often feel) *waxy.* (The cap sometimes also has a waxy appearance, hence the common name "waxy caps.") *Laccaria* species have somewhat waxy gills, but they have tough, fibrous stems and spiny spores. *Chroogomphus* and *Gomphidius* also have waxy gills, but their spores are smoky-black.

The waxy caps have traditionally been grouped together in a single large and diverse genus, *Hygrophorus,* but most mycologists now split them into two or more genera based primarily on the arrangement of the hyphae in the gill tissue (see illustrations, p. 19). As this character can only be determined with a microscope, the family is treated as one entity here, but three genera are recognized:

Camarophyllus (hyphae in gill tissue intricately interwoven): Fruiting body small to medium-sized, usually white or dull colored; cap usually not viscid or only slightly so; gills typically decurrent, thick, well-spaced, somewhat waxy; stalk dry; veil absent.

Hygrocybe (hyphae in gill tissue more or less parallel): Fruiting body small to medium-sized, usually brightly colored and quite waxy; cap dry to moist, viscid, or slimy; gills free to adnexed, adnate, or sometimes decurrent, obviously waxy; stalk dry or viscid or slimy, often slender and hollow; veil absent.

Hygrophorus in restricted sense (hyphae in gill tissue divergent from a central strand): Fruiting body medium-sized to fairly large, usually white or dull colored; cap viscid or slimy; gills typically adnate to decurrent, well-spaced or close, often only slightly waxy, usually white or pale colored; stalk slimy, viscid or dry, often fleshy (rarely hollow), apex often **punctate** (adorned with pointlike scales or granules); slimy or fibrillose veil sometimes present.

Hygrocybe species are the mushrooms most easily identified as "waxy caps," while *Hygrophorus* and *Camarophyllus* species are more likely to be confused with other genera, particularly *Clitocybe,* because their gills are not so obviously waxy. Some investigators still recognize only one genus, *Hygrophorus,* while others recognize only two by throwing *Camarophyllus* in with *Hygrocybe.* Several small, rare "satellite" genera also occur, but are not treated here.

The long, narrow protruding basidia that give the gills their "waxiness" are reminiscent of those in *Cantharellus,* leading some mycologists to infer that *Camarophyllus* (the most primitive genus of waxy caps) evolved from *Cantharellus*-like ancestors. (*Camarophyllus pratensis,* in fact, slightly resembles the chanterelle, *Cantharellus cibarius.*) A relationship with *Omphalina* and *Clitocybe* of the Tricholomataceae has also been suggested.

The waxy caps occur throughout the mushroom season but are partial to cold weather. In colder climates they may continue to fruit after the first frosts, while in our area they normally peak in the cold winter months (December-February). *Camarophyllus* and *Hygrocybe* species are mostly saprophytic. Like *Leptonia* species, they are particularly abundant in environments where other fleshy fungi are few. In coastal California this means under redwood and to a lesser extent cypress, but in other regions they are common in ungrazed fields, heaths, and boggy areas. *Hygrophorus* species, on the other hand, are largely mycorrhizal—hence in our area they occur with other mycorrhiza-formers under oak, madrone, pine, and Douglas-fir, rather than with redwood or cypress. *Hygrophorus* and *Hygrocybe* each contain about 100 species in North America, while *Camarophyllus* has about 40. Most waxy caps have wide distributions, and at least half of the North American species occur in California.

Waxy caps are without a doubt among our most colorful mushrooms, with virtually every hue in the rainbow represented: blue or green in *Hygrocybe psittacina,* pink in *Hygrocybe calyptraeformis,* bright yellow, orange, or red in a great many species, black in *Hygrocybe conica,* white in *Hygrophorus eburneus.* None are thought to be dangerously poisonous, and some species are rated highly as edibles. However, I have yet to find one to my liking. By and large they are too bland or too watery or too bland *and* too watery to be worth eating. (See comments on the edibility of *Hygrophorus sordidus* and *H. russula!*) Also, a few species (e.g., *Hygrocybe conica* and *H. punicea*) may cause illness. However, their beauty is reason enough for getting to know them, and they have the added attraction of being relatively easy to identify in the field! Consequently, a fairly extensive selection of species is offered here.

Key to the Hygrophoraceae

1. Fruiting body whitish except for yellow to golden-orange powder, flakes, or granules on the cap and/or stalk apex (but powder may smear in wet weather)   **Hygrophorus chrysodon**, p. 119
1. Not as above (if fruiting body white, then lacking yellow to golden-orange granules) 2

2. Fruiting body white to creamy, buff, or pale yellow when fresh 3
2. Fruiting body at least partially darker or differently colored (including bright lemon-yellow) 17

3. Stalk viscid or slimy when moist, at least over lower portion 4
3. Stalk not viscid 7

4. Cap tinged buff to yellowish or pale yellow, at least at center; especially common under pine (but not restricted to it) . . . . **Hygrophorus gliocyclus** & others, p. 120
4. Not as above 5

5. Stalk usually at least 1.5 cm thick; cap fairly large (5-15 cm broad) . . . .
. . . . **Hygrophorus ponderatus** (see *H. sordidus*, p. 122)
5. Stalk usually (but not always) less than 1.5 cm thick; cap medium-sized to fairly small 6

6. Cap conical when young, usually with a prominent umbo in age; gills *not* decurrent; found in eastern North America, rarely in California (mostly under cypress) . . . . . . **Hygrocybe pura**
6. Not as above; gills adnate to decurrent; widespread . . . . **Hygrophorus eburneus**, p. 119

7. Cap *not* conical; stalk 1-7 cm thick, or if thinner then cap viscid when moist and developing distinct yellow-brown to reddish-brown, peachy, salmon, or ochre tones in age 8
7. Not as above; stalk typically less than 1 cm thick and/or cap distinctly conical 10
8. Stalk often with an annulus (ring) or "volva" that is easily obliterated; typically found under mountain conifers .... *Hygrophorus subalpinus*, p. 121
8. Not as above; annulus or "volva" absent 9
9. Fruiting body white or discoloring slightly yellowish or buff; associated mainly with oak, occasionally with conifers .... *Hygrophorus sordidus* & others, p. 122
9. Either fruiting body discoloring more dramatically in age *or* else associated with conifers *and* possessing a densely scurfy stalk apex 74
10. Growing in grass; stalk tough and pliant; cap often umbonate (see *Marasmius oreades*, p. 208)
10. Not as above 11
11. Odor sharp and cedarlike or fragrant ...... *Camarophyllus russocoriaceus* & others, p. 109
11. Odor mild or not distinctive (or merely fungal) 12
12. Cap conical when young *and/or* gills typically notched or adnexed or at times adnate; gill hyphae parallel ..... *Hygrocybe subaustraliga & H. albinella* (see *Camarophyllus borealis*, p. 109)
12. Cap not normally conical; gills typically adnate to decurrent; gill hyphae not parallel 13
13. Cap viscid when moist 14
13. Cap not viscid 15
14. Cap thin and often translucent-striate when moist, usually less than 3 cm broad; found in many habitats; gill tissue interwoven ...... *Camarophyllus niveus* (see *C. borealis*, p. 109)
14. Not as above; found mainly under conifers; gill tissue divergent ....
    .... *Hygrophorus piceae* (see *H. eburneus*, p. 119)
15. Growing in groups or troops on pine needles; gills very widely spaced and usually with veins in between; cap broadly convex soon becoming plane or umbilicate (see photo at top of p. 207)
    .... (see *Marasmius* sp. (unidentified), p. 206)
15. Not as above (see top right photo on p. 109) 16
16. Fruiting body often developing a slightly yellowish tinge in age; cap dry ....
    .... *Camarophyllus virgineus* (see *C. borealis*, p. 109)
16. Not as above; cap sometimes slightly lubricous .... *Camarophyllus borealis* & others, p. 109
17. Cap small (up to 6 cm broad but usually less), dry, honey-colored or yellowish to orange or scarlet beneath a layer of small, gray to dark brown or blackish fibrillose scales 18
17. Not as above; either cap viscid or larger or lacking dark scales or differently colored 19
18. Background of cap scarlet to orange when fresh   *Hygrocybe turunda* (see *H. miniata*, p. 113)
18. Background of cap buff to dull yellowish or honey-colored; found in eastern North America
    .... *Hygrocybe caespitosa*
19. Fruiting body staining gray to black when handled (often slowly); cap conical, at least when young; common .... *Hygrocybe conica* & others, p. 116
19. Not as above; fruiting body not blackening when handled 20
20. Cap bright red, yellow, or orange, at least when fresh (but may fade to whitish) 21
20. Cap some other color (including pink, salmon, vinaceous, purple-red, pale dull orange, etc.) 42
21. Cap *and* lower portion of stalk viscid to slimy when moist; stalk typically 4 mm thick or more and *not* hollow; gills adnate to decurrent; gill tissue divergent 22
21. Not as above; if both cap and stalk viscid, then stalk typically slender, hollow, and fragile   23
22. Cap red to orange when young, at least at center .... *Hygrophorus speciosus*, p. 126
22. Cap at first dark brown to olive-brown at center, the brighter colors developing in age and the center often remaining brownish .... *Hygrophorus hypothejus*, p. 126
23. Cap dark to bright red when fresh (but may fade in age or as it dries to orange or yellow) . 24
23. Cap orange to yellow (sometimes fading to whitish), never bright red 31
24. Cap conical, up to 5 cm broad, bright red and not usually fading, viscid when moist; stalk also red and usually viscid; found in eastern North America .... *Hygrocybe ruber*
24. Not as above 25
25. Cap *and* stalk distinctly viscid to slimy when moist; cap usually small or minute ....
    .... *Hygrocybe reai* & others (see *H. miniata*, p. 113)
25. Not as above 26

26. Cap distinctly conical when young and usually retaining a pointed umbo in age; stalk yellow
.... *Hygrocybe cuspidata (see H. acutoconica*, p. 115)
26. Not as above 27

27. Gills usually distinctly decurrent; cap typically small or minute (less than 4 cm broad) ... 40
27. Not as above 28

28. Cap small (1-4 cm), *not* viscid, red to scarlet when moist but fading drastically as it dries to
orange or even yellow .... *Hygrocybe miniata* & others, p. 113
28. Not as above (if cap not viscid, then cap not fading drastically or cap larger) 29

29. Stalk white (base may be colored) .... *Hygrocybe laetissima* (see *H. punicea*, p. 114)
29. Stalk yellow to orange or red 30

30. Cap viscid when moist; stalk typically (0.5) 1-2 cm or more thick, usually yellow or reddish fading
to orange-yellow; base of stalk often whitish .... *Hygrocybe punicea* & others, p. 114
30. Cap not viscid or only slightly viscid; stalk typically 3-8 mm thick, usually red to reddish-orange
with a yellow to orange base .... *Hygrocybe coccinea* & others, p. 114

31. Cap *and* stalk slimy or very viscid when moist; cap generally 4 cm broad or less 32
31. Not as above (cap may be slimy, but stalk dry to only slightly viscid) 35

32. Fruiting body showing pink, flesh-colored, or greenish (rarely bluish) tones somewhere on
fruiting body; gills not normally decurrent .... *Hygrocybe psittacina*, p. 118
32. Not as above; fruiting body yellow to orange (but may show some white or have a lilac tinge) 33

33. Growing on rotting conifers .... (see *Mycena lilacifolia*, p. 236)
33. Not as above; usually found on the ground 34

34. Gills typically decurrent *and* cap usually depressed centrally ....
.... *Hygrocybe nitida* (see *H. flavescens*, p. 115)
34. Not as above .... *Hygrocybe chlorophana* & others (see *H. flavescens*, p. 115)

35. Cap distinctly viscid or slimy when moist, 2-7 cm broad or more when expanded; gills not
decurrent 36
35. Cap not viscid, or if slightly viscid then less than 3 cm broad or gills decurrent 37

36. Cap distinctly conical, at least when young .... *Hygrocybe acutoconica*, p. 116
36. Cap broadly convex to plane, not conical .... *Hygrocybe flavescens* & others, p. 115

37. Gills brilliant orange to yellow-orange, retaining their color even after the cap fades; common
in eastern North America, occasional in the Pacific Northwest *Hygrocybe marginata*, p. 112
37. Not as above; common and widespread 38

38. Gills usually not decurrent; cap red to orange-red when fresh and moist (but fading) ....
.... *Hygrocybe miniata* & others, p. 113
38. Not as above; gills usually decurrent 39

39. Found in humus; cap yellow to orange-yellow (never scarlet, red, or orange) ....
.... *Hygrocybe parvula* & others (see *H. flavescens*, p. 115)
39. Not as above 40

40. Stalk usually quite long (4-10 cm); found mainly in eastern North America on ground, in moss,
or on rotten wood .... *Hygrocybe cantharellus* (see *H. miniata*, p. 113)
40. Not as above; found on ground, or if in lichen, moss, or wood, then stalk not so long 41

41. Typically found in humus or soil; cap red or scarlet when fresh ....
.... *Hygrocybe subminiata* (see *H. miniata*, p. 113)
41. Not as above; often found on wood, lichens, moss   (see *Omphalina* & *Xeromphalina*, p. 221)

42. Cap sharply conical when young, usually retaining a pointed umbo in age, coral-pink to pink,
pinkish-orange, or salmon; gills pink .... *Hygrocybe calyptraeformis*, p. 117
42. Not as above 43

43. Entire stalk or at least the lower part distinctly viscid to slimy *when moist*; cap also viscid   44
43. Stalk dry to slightly lubricous but not viscid; cap dry or viscid 54

44. Cap small; stalk slender (usually less than 6 mm thick), hollow, *not* white; veil absent 45
44. Cap usually medium-sized to fairly large; stalk typically at least 5 mm thick, sometimes white;
veil present or absent 47

45. Cap blackish to dark brown or grayish ..... *Hygrocybe unguinosa* (see *H. psittacina*, p. 118)
45. Not as above; cap brighter or lighter in color 46

46. Cap sometimes green or greenish-tinged; gills typically adnate . *Hygrocybe psittacina,* p. 118
46. Green shades absent; gills often decurrent ...... *Hygrocybe laeta* (see *H. psittacina,* p. 118)

47. Cap dull yellowish becoming browner or grayer in age; stalk at least 1 cm thick; gills white or developing greenish stains, but not yellow; found mainly under hardwoods in eastern North America .... *Hygrophorus paludosus*
47. Not as above 48

48. Cap (or center of cap) chestnut-brown to cinnamon-brown, reddish-brown, pinkish-tan, salmon, or yellow-brown 49
48. Cap olive-brown to grayish-brown, gray, blackish, etc. 51

49. Cap and gills rufous to rusty-orange, apricot, or salmon-buff; growing under oaks in eastern North America .... *Hygrophorus subsalmonius*
49. Not as above 50

50. Odor fragrant, like almond extract ..... *Hygrophorus variicolor* (see *H. bakerensis,* p. 126)
50. Odor more or less mild .... *Hygrophorus laurae* & others (see *H. roseibrunneus,* p. 125)

51. Lower portion of stalk sheathed with gray to brown, olive-brown, or blackish fibrils, scales, or granules; cap, gills, or stalk *not* yellow ..... *Hygrophorus olivaceoalbus* & others, p. 127
51. Not as above 52

52. Gills creamy to yellow or orange; cap and stalk often exhibitng some yellow or orange tones .... *Hygrophorus hypothejus,* p. 126
52. Yellow and orange tones absent 53

53. Cap evenly blackish to deep olive-brown when young; found mainly in eastern North America .... *Hygrophorus fuligineus* (see *H. hypothejus,* p. 126)
53. Cap paler when young or with a paler margin; widespread .... *Hygrophorus fuscoalbus* & others (see *H. hypothejus,* p. 126)

54. Gills persistently bright orange to yellow-orange, even after cap fades; cap 1-5 cm broad, stalk less than 6 mm thick .... *Hygrocybe marginata,* p. 112
54. Not as above 55

55. Cap green (citrine-green), at least when mature .... *Hygrocybe virescens,* p. 118
55. Not as above 56

56. Cap gray to grayish-brown, dark grayish-brown, olive, deep olive-brown, blackish, or umber; gills and cap lacking violet or blue tones 57
56. Cap differently colored (tan, reddish-brown, vinaceous, etc.), or if colored as above then cap or gills with a violet, lilac, or bluish tinge when fresh 63

57. Cap conical or with a pointed umbo, dull or dark gray .... *Hygrocybe acuta* (see *Camarophyllus recurvatus,* p. 112)
57. Not as above 58

58. Odor fragrant (like almond extract), but sometimes faint .... *Hygrophorus agathosmus* & others, p. 128
58. Odor mild or not as above 59

59. Base of stalk usually pale pinkish to pinkish-orange (especially interior) .... (see *Tricholoma saponaceum,* p. 184)
59. Not as above 60

60. Stalk sheathed by grayish to dark grayish-brown fibrils, scales, or granules .... *Hygrophorus inocybiformis* & others (see *H. olivaceoalbus,* p. 127)
60. Not as above 61

61. Fruiting body rather small (cap typically less than 5 cm broad, stalk typically less than 1 cm thick and sometimes hollow) .... *Camarophyllus recurvatus* & others, p. 112
61. Not as above; fruiting body medium-sized; stalk usually 1 cm thick or more 62

62. Cap viscid to slimy when moist; gills white or pink .... *Hygrophorus calophyllus,* p. 129
62. Cap viscid or dry, often appearing streaked; gills white or gray .... *Hygrophorus camarophyllus* & others (see *H. calophyllus,* p. 129)

63. Stalk rooting deeply or with a "tap root" encased by grayish-brown volva-like material; cap white becoming pinkish to vinaceous-red; gills white; known only from southern California under oak; solitary or clustered .... *Hygrophorus marianae*
63. Not as above; widespread and common 64

64. Gills (or entire fruiting body) soon streaked or stained coral-red to vinaceous, pinkish-red, or purple-red (especially in age), sometimes also with yellow stains 65
64. Not as above (but gills may be naturally pinkish) 68

65. Associated with hardwoods (mainly oak and tanoak) . . . . *Hygrophorus russula,* p. 123
65. Associated with conifers (mainly pine and spruce) 66

66. Cap reddish at first but staining yellow and becoming pale to bright yellow in age; taste bitter; especially common in the Rockies . . . . . *Hygrophorus amarus* (see *H. purpurascens,* p. 124)
66. Not as above (but fruiting body may yellow somewhat); widespread (including Rockies) . 67

67. Fibrillose veil present, at least when young . . . . *Hygrophorus purpurascens,* p. 124
67. Veil absent . . . . . . *Hygrophorus erubescens & H. capreolarius* (see *H. purpurascens,* p. 124)

68. Gills lilac or violet or pinkish to dingy flesh-colored, attached and sometimes slightly decurrent but not *deeply* decurrent; stalk tough and fibrous, pliant, often fibrillose; cap often minutely scaly or scurfy, *not* viscid; spores usually spiny . . . . (see *Laccaria,* p. 171)
68. Not as above; spores smooth 69

69. Gills and/or cap with a violet, purple, or bluish tinge when fresh . . . .
. . . . *Camarophyllus subviolaceus* & others, p. 111
69. Not as above 70

70. Gills brown to vinaceous-brown, adnate to decurrent; cap some shade of brown when moist, fading as it dries; found with hardwoods in eastern North America *Hygrophorus kauffmanii*
70. Not as above; gills paler 71

71. Growing in grass; cap buff to tan, often umbonate; gills whitish; stalk thin, tough, pliant . . . .
. . . . (see *Marasmius oreades,* p. 208)
71. Not as above 72

72. Cap 4.5 cm broad or less, translucent-striate when moist 73
72. Not as above; cap typically opaque 74

73. Stalk 4-7 mm thick; cap vinaceous-brown to pinkish-gray or buff, slightly viscid when moist; gills pinkish-gray or paler, distinctly decurrent; terrestrial . . . .
. . . . *Camarophyllus colemannianus* (see *C. subviolaceus,* p. 111)
73. Not as above; growing on ground or wood . . . . (see **Tricholomataceae,** p. 129)

74. Stalk 1-3 cm thick, the apex distinctly scurfy (punctate); cap 5-20 cm broad, pale tan to pinkish, flesh-colored, pinkish-orange, or occasionally whitish; gills usually tinged pink (but sometimes whitish); cap viscid when moist, the margin inrolled until maturity; odor *not* like almond extract; associated with conifers (especially spruce) . . . . . . *Hygrophorus pudorinus,* p. 124
74. Not with above combination of characteristics 75

75. Gills distinctly paler than center of cap (white to creamy or pale yellow) 76
75. Gills colored like the cap or slightly paler or darker (pinkish to salmon, etc.) 79

76. Odor faintly fragrant (like almond extract) to strongly aromatic 77
76. Not as above; odor usually mild or somewhat potato-like 78

77. Odor like almond extract . . . . *Hygrophorus bakerensis* & others, p. 126
77. Odor strongly aromatic (but not as above) *Hygrophorus pacificus* (see *H. bakerensis,* p. 126)

78. Cap pink or rose (but may fade!), stalk *lacking* sordid yellowish patches below; found under mountain conifers, often near melting snow *Hygrophorus goetzii* (see *H. pudorinus,* p. 124)
78. Not with above features . . . . *Hygrophorus roseibrunneus* & many others, p. 125

79. Odor sickeningly sweet; fruiting body pinkish-buff to pinkish-cinnamon; cap not viscid . . . . .
. . . . *Camarophyllus graveolens*
79. Not as above 80

80. Cap whitish to buff, sometimes with darker spots or zones; odor often faintly fruity (if kept in a closed space); gills pinkish-cinnamon to salmon or ochre-salmon; associated with conifers (but not redwood); gill hyphae divergent . *Hygrophorus saxatilis* (see *H. pudorinus,* p. 124)
80. Not with above features 81

81. Cap whitish when young *Hygrophorus albicastaneus* & others (see *H. roseibrunneus,* p. 125)
81. Cap not white when young (but may be quite pale in age or after fading) 82

82. Gills adnexed to adnate or slightly decurrent; stalk 1-4 cm thick, cap 5-13 cm broad; odor mild or radish- to cucumberlike; gill hyphae divergent; usually associated with oak . . . .
. . . . *Hygrophorus nemoreus* (see *Camarophyllus pratensis,* p. 110)
82. Not as above; gills usually decurrent . . . . *Camarophyllus pratensis,* p. 110

Left: *Hygrophorus subalpinus* is a robust whitish conifer-loving waxy cap (see description on p. 121). Note the slight annulus (ring) on specimen at left. **Right:** *Camarophyllus borealis* and its close relatives are white to yellowish with a slender build, decurrent gills, and little or no odor.

Camarophyllus borealis   (Snowy Waxy Cap)

CAP 1-5 cm broad, convex or obtusely umbonate, often expanding in age to plane or depressed; surface smooth, moist or lubricous but not viscid, watery white to dull white. Flesh thin, soft, white, odor mild. **GILLS** usually decurrent, well-spaced, thick, soft, somewhat waxy; white. **STALK** 2-9 cm long, 2-5 (8) mm thick, equal or tapering downward, smooth, firm, dry, often curved or sinuous, white. **VEIL** absent. **SPORE PRINT** white; spores 7-9 (12) × 4.5-6.5 microns, elliptical, smooth. Gill tissue interwoven.

HABITAT: Scattered or in groups on ground and humus in woods or at their edges, widely distributed. In our area this species and its relatives *C. virgineus* and *C. niveus* (see comments) are fairly common throughout the mushroom season, but seldom occur in the large numbers typical of *Hygrophorus eburneus*. The best fruitings usually occur in the winter, often in relatively dry weather.

EDIBILITY: Edible but fleshless, flavorless, and savorless.

COMMENTS: Also known as *Hygrophorus borealis,* this is one of several rather small, whitish waxy caps with a dry to slightly viscid (not slimy) cap, slender non-viscid stalk, and interwoven gill tissue. Others include: *C. niveus,* common, cap slightly viscid and slightly striate when moist; *C. virgineus,* especially common in California, cap dry and usually tinged yellow in age or dry weather, and spores 8-12 microns long; *C. angustifolius,* dull white with spores 5-8 microns long; *C. cremicolor*, with pale yellow gills when young; and three species with parallel gill hyphae and non-decurrent gills: *Hygrocybe subaustraliga,* small, cap scarcely viscid, gills notched or adnexed, fairly common; *H. albinella*, cap white and conical at first, stalk dry, rare; and *H. fornicata*, with a grayish-tinged viscid cap. All of these have been placed in *Hygrophorus*. None have the slimy cap and stalk of *Hygrophorus eburneus*. *Alboleptonia sericella* and *Inocybe geophylla* can be similar but have pinkish and brown spores respectively. See also the undescribed *Marasmius* on p. 206.

Camarophyllus russocoriaceus   (Cedar Waxy Cap)

CAP 1-3 (5) cm broad, convex to plane or slightly umbonate to slightly depressed; surface smooth, not viscid, white or often tinged pale tan to yellowish, especially at the center. Flesh whitish, thin; odor fragrant when fresh (like arborvitae or cedar). **GILLS** well-spaced, usually decurrent, thick, slightly waxy, white or whitish. **STALK** 3-10 cm long, 2-5 (8) mm thick, equal or tapered downward, often long in relation to cap; dry, smooth, colored more or less like cap, often curved or sinuous. **VEIL** absent. **SPORE PRINT** white; spores 6.5-9 × 4-6 microns, elliptical, smooth. Gill tissue interwoven.

Camarophyllus russocoriaceus is a slim waxy cap that smells like cedar or Russian leather. Note off-white to buff-colored cap and long, slender stem.

HABITAT: Widely scattered or in small groups in woods or at their edges, known only from the west coast (in addition to Europe). Common in our area in the fall and winter, especially at higher elevations in the coastal mountains, but rarely in large numbers.

EDIBILITY: Not recommended—it has a slightly medicinal flavor, at least raw.

COMMENTS: The piquant cedarlike odor distinguishes this plain-looking waxy cap from *C. borealis* and other look-alikes. Other species: *Hygrophorus pusillus* is a somewhat similar species with a viscid cap that is usually tinged cream or pale brownish-flesh-color. It has a slight fragrant odor, divergent to nearly parallel gill tissue, and occurs in groups or troops under conifers in northern California and the Pacific Northwest.

Camarophyllus pratensis   (Meadow Waxy Cap)

CAP 2-9 (10) cm broad, obtuse to broadly convex becoming broadly umbonate or plane to depressed in age; surface smooth or sometimes cracking in age (especially at center), not viscid; rufous to dull orange or pale dull orange, fading to salmon-buff, pinkish-tan, buff, tawny, or even whitish; margin often wavy. Flesh white or tinged cap color; odor mild. **GILLS** typically decurrent, same color as cap or paler (usually pale dull orange or salmon-buff to nearly white); well-spaced, thick, broad, somewhat waxy. **STALK** 3-10 cm long, 0.5-2 cm thick, whitish or tinged cap color, equal or tapering downward, dry, smooth or fibrillose. **VEIL** absent. **SPORE PRINT** white; spores 5.5-8 × 3.5-5 microns, elliptical to nearly round. Gill tissue interwoven.

HABITAT: Solitary to scattered or gregarious in damp places—common and very widely distributed—from Europe to Iceland to South America. In our area it fruits mainly in the winter under redwood, but as its name implies (*pratensis*=fields), it can also occur in open or grassy situations.

EDIBILITY: Edible and rated highly in Europe (one author even puts in on a par with morels!). However, my own experience with it gave me no great cause for enthusiasm. Unlike the similarly-colored chanterelle, it is readily attacked by maggots.

COMMENTS: Better known as *Hygrophorus pratensis,* this common waxy cap is extremely variable in size and shape, but can generally be recognized by its pale, dull orange

110

Camarophyllus pratensis is variable in both shape and color, but the gills are nearly always decurrent and widely spaced, and the cap is never slimy.

color and decurrent gills. The cap and stem are not viscid, but the gills are thick and rather waxy. Robust individuals are slightly reminiscent of chanterelles *(Cantharellus cibarius),* but differ in their duller color and broad, well-developed gills with acute rather than blunt edges. A similarly colored species, **Hygrophorus nemoreus**, is fairly common in our area with oak in the fall and winter. It is larger and fleshier than *C. pratensis* (cap 5-13 cm broad, stalk 1.5-4 cm thick) and has a slightly viscid to dry, orangish or cinnamon-orange cap, adnexed to adnate or slightly decurrent gills, a mild to radish- or cucumberlike odor, and divergent gill tissue. It is large enough and common enough to be worth sampling, but I can find no information on its edibility.

Camarophyllus subviolaceus   (Violet-Gray Waxy Cap)

CAP 2-6 cm broad, broadly convex to plane or slightly depressed with an uplifted margin; surface slightly viscid when moist, hygrophanous: violet-gray to violet-brown at least at the margin when moist (center often paler), fading as it dries to gray or pallid; margin translucent striate when fresh. Flesh rather thin, colored like cap or paler; taste mild to bitter or slightly acrid. **GILLS** decurrent, whitish soon becoming smoky-violet or gray with a violet tinge; well-spaced, thick, soft, rather waxy. **STALK** 3-7 cm long, 0.4-1 cm thick, equal or tapered downward, dry, smooth, white or tinged cap color, often curved. **VEIL** absent. **SPORE PRINT** white; spores 6-8 × 4-6 microns, elliptical, smooth. Gill tissue interwoven.

Camarophyllus subviolaceus has widely spaced decurrent gills that are violet-gray when fresh.

HABITAT: Scattered or in small groups in woods and swamps throughout northern North America, late summer through early winter. I have found it several times in the Pacific Northwest but it is not particularly common.

EDIBILITY: Unknown.

COMMENTS: Better known as *Hygrophorus subviolaceus*, this beautiful species has been included because its violet-tinged gills are unusual for a waxy cap. There are several similarly-colored species, including: *C. pallidus*, with smaller, round or nearly round spores and a viscid, violet-gray cap; *C. angelesianus*, with amyloid spores and a viscid cap that does not fade appreciably, usually found at high elevations in the spring and summer; *C. rainierensis*, with a strong green corn odor and slightly viscid cap; and *C. cinereus*, with a non-viscid cap, mild taste, and larger spores. Other species include: *C. colemannianus*, closely related, but with a viscid, brown to pinkish-gray to buff cap and pinkish-gray to pale buff colors; *Hygrocybe caerulescens*, a bluish-tinged species with parallel gill tissue; and *Hygrocybe purpureofolia*, with lavender to dull purplish gills and a brown cap that fades to yellow-orange, found in eastern North America.

Camarophyllus recurvatus   (Little Brown Waxy Cap)

CAP 1-3 cm broad, convex becoming plane or slightly depressed, sometimes with a small pointed umbo; surface smooth or cracking into small scales, dry to slightly viscid, dark to pale olive-brown or at times pallid; margin often wavy or pleated. Flesh thin, olive-brown, odor mild. **GILLS** decurrent, well-spaced, broad, thick, soft and rather waxy, white or grayish-white. **STALK** 2-5 cm long, 3-6 mm thick, equal or tapering downward, smooth, not viscid, whitish or colored like cap. **VEIL** absent. **SPORE PRINT** white; spores 7-10 × 4-6 microns, elliptical, smooth. Gill tissue interwoven.

HABITAT: Scattered to gregarious under conifers and in open, grassy areas; widely distributed. I've seen large fruitings in northern California but have yet to find it in our area.

EDIBILITY: Unknown.

COMMENTS: Also known as *Hygrophorus recurvatus*, this is one of a number of undistinguished, brown to grayish waxy caps that qualify as "LBM's." As they are unlikely to interest the average collector, only a few others are worth mentioning: *Hygrocybe ovina*, with deeply notched to adnate gills and a convex to plane, grayish-brown cap that often has a paler or yellower margin; *Hygrocybe nitrata*, with a nitrous odor; *Camarophyllus paupertinus*, with a small (1-2 cm) cap and exceedingly strong, disagreeable odor (known only from northern California under redwoods); *Hygrocybe acuta*, with a sharply conical, dull gray to gray-brown cap, growing under conifers on the west coast; and *Hygrocybe atro-olivacea*, with a minutely scaly cap that is dark at the center or throughout.

Hygrocybe marginata   (Orange-Gilled Waxy Cap)

CAP 1-5 cm broad, conical or convex becoming umbonate, plane, or with uplifted margin; surface smooth, hygrophanous but not viscid: deep yellow to yellow-orange or orange when moist, sometimes with an olive tinge, fading to pale yellow or whitish as it dries. Flesh thin, fragile, waxy. **GILLS** slightly decurrent to adnate, adnexed, or even free; well-spaced, broad, thick, soft, waxy; brilliant orange or at times orange-yellow, not fading. **STALK** 4-10 cm long, 3-6 mm thick, more or less equal, smooth, fragile, not viscid; pale orange to buff or pale yellow-orange, often curved; hollow. **VEIL** absent. **SPORE PRINT** white; spores 7-10 × 4-6 microns, elliptical, smooth. Gill tissue parallel to somewhat interwoven.

HABITAT: Solitary to scattered or in small groups in humus and on very rotten wood, under both hardwoods and conifers; common in eastern North America in summer and fall, known also from the Pacific Northwest and to be expected in northern California. I have not seen it in our area.

EDIBILITY: Edible but of negligible substance and flavor.

COMMENTS: The gills' ability to retain their bright orange color long after the cap has faded is the outstanding attribute of this modest waxy cap, also known as **_Hygrophorus marginatus_**. The cap is at times slightly greasy or tacky to the touch, but never truly viscid. Several color forms occur, including one with an olive cap (fairly common in Washington) and another with yellow gills (in eastern North America).

Hygrocybe miniata   (Miniature Waxy Cap)

CAP 1-4 cm broad, convex to plane or slightly depressed; surface smooth or minutely scaly, hygrophanous but not viscid: bright red to scarlet when moist, quickly fading as it dries to orange and finally yellow. Flesh very thin, waxy, colored more or less like cap. **GILLS** adnexed to adnate or very slightly decurrent, soft, waxy, thick, broad, red to orange, yellow, or peachy (fading like cap). **STALK** 2-5 (8) cm long, 2-4 mm thick, equal, smooth, dry, red to orange or yellow (fading like the cap but more slowly). **VEIL** absent. **SPORE PRINT** white; spores 6-10 × 4-6 microns, elliptical, smooth. Gill tissue parallel.

HABITAT: Solitary, scattered, or in groups or small tufts on ground, rotting logs, or in moss; widely distributed and common. In our area it fruits in the late fall and winter in a variety of habitats, but usually in deep shade. The largest fruitings I've seen were in a dense stand of pole-size redwood saplings and in a plot of cypresses mixed with oaks.

EDIBILITY: Edible but of negligible substance and according to most sources, bland. However, Captain Charles McIlvaine says: "The gunner for partridges will not shoot rabbits; the knowing toadstool-seeker will pass all others where _H. miniatus_ abounds."

COMMENTS: The small size, waxy gills, and convex to plane, non-viscid cap that fades markedly from red to yellow or orange are the far-from-infallible fieldmarks of this dainty fungus. It is better known as **_Hygrophorus miniatus_**. The cap is neither viscid nor conical, and is often minutely scaly (squamulose). Its frequent occurrence on rotten wood is unorthodox behavior for a waxy cap. Other species: **_H. squamulosa_** has thicker, firmer flesh but is otherwise similar. **_H. cantharellus_** is very similar but has a longer stem and decurrent gills; **_H. moseri_** has a less scaly cap and decurrent gills. **_H. subminiata_** has a minute (up to 1.5 cm broad), scarlet to orange, slightly viscid cap plus decurrent, whitish to pale yellowish gills and a _very_ thin (about 2 mm) stalk; it is fairly common in our area, especially under redwood. **_H. turunda_** is a widespread miniature northern species with

Hygrocybe miniata has a bright red cap that fades to orange or yellow as it loses moisture. Note small size (the largest is 2 cm broad) and convex (not conical) cap.

minute brown to grayish squamules (scales) on a red to yellow background. There are also several minute species with a slimy-viscid cap *and* stalk when fresh, including: *H. reai,* with a bitter-tasting cap; *H. minutula,* with a stalk that fades to yellowish in age; and *H. subminutula,* whose stalk scarcely fades from red. All of the above species have a red to scarlet cap when moist, and all were originally placed in *Hygrophorus.* None are worth eating.

Hygrocybe coccinea   (Righteous Red Waxy Cap)   Color Plate 20

CAP 1.5-4 (6) cm broad, obtusely conical or convex becoming plane or slightly umbonate; surface smooth, dry or tacky, deep red, blood-red, or bright red when fresh, fading somewhat in age or developing paler streaks or splotches. Flesh thin, reddish to orange, waxy. **GILLS** adnate to adnexed or free, reddish to orange or peachy, or red with yellow edges; thick, broad, soft, waxy. **STALK** 3-7 (10) cm long, 3-8 mm thick, equal, smooth, not viscid, hollow; usually red to reddish-orange with a yellow base (but base sometimes appearing whitish from mycelium); typically *not* fibrillose-striate. **VEIL** absent. **SPORE PRINT** white; spores 7-10.5 × 4-5 microns, elliptical, smooth. Gill tissue parallel.

HABITAT: Solitary, scattered, or in small groups in woods and other wet places; widely distributed, but not as common in North America as *H. punicea.* Like *H. punicea,* it is frequent in our area in the winter under redwoods or in mixed woods. Curiously, in some regions (such as England) it grows mainly in open fields!

EDIBILITY: Said to be edible, but easily confused with *H. punicea.* I haven't tried it.

COMMENTS: Also known as *Hygrophorus coccineus,* this exquisite bright red waxy cap ranks among our most beautiful mushrooms and is definitely worth seeking out. It is redder than *H. miniata* and does not fade nearly as drastically. It is often confused with *H. punicea,* but that species is usually more robust, has a distinctly viscid cap (when moist), and most often has a yellow to orange, fibrillose-striate stem with a whitish base rather than a red stem with a yellow base (see color plates 20 and 21). Other species: *H. marchii* is a similar species with a viscid red cap that soon dries out and fades to orange or yellow-orange. Like *H. coccinea,* it is a rather small, slender species (stalk 3-6 mm thick) that calls to mind *H. flavescens* but for its color. It is widely distributed but I have not seen it locally.

Hygrocybe punicea   (Scarlet Waxy Cap)   Color Plate 21

CAP 2.5-12 (14) cm broad, obtusely conical or convex when young, then plane to broadly umbonate or with uplifted margin in age; surface smooth, viscid or lubricous when moist, deep red to bright red, fading (often in streaks or splotches) to reddish-orange and finally orange. Flesh rather thin, waxy, watery reddish-orange to yellow-orange. **GILLS** adnate to adnexed or free, reddish-orange to yellow or peachy, well-spaced, thick, broad, soft, waxy. **STALK** 3-12 cm long, 0.5-2 cm thick, equal or narrowed at base, smooth, not viscid, often fibrillose-striate; yellow or sometimes red soon fading to orange or yellow; base usually whitish but sometimes yellow. **VEIL** absent. **SPORE PRINT** white; spores 7-12 × 4-6 microns, elliptical to oblong, smooth. Gill tissue parallel.

HABITAT: Scattered to gregarious in cool, damp places (usually in woods); widely distributed and fairly common in our area in the winter and early spring, especially under redwood. It is usually the last waxy cap to appear (mid-January or later).

EDIBILITY: Listed as edible in most books, but poisonous at least to some people! I know two toadstool-testers who had very unpleasant experiences with it. It is an efficient concentrator of the malleable metallic element cadmium, which is decidedly deleterious when consumed on a regular basis.

COMMENTS: Better known as *Hygrophorus puniceus*, this beautiful mushroom is as easy to recognize as it is difficult to overlook. The bright red sticky cap and indisputably waxy gills make it stand out vividly in the dim, damp milieu where it thrives. It is just as common in our wintertime woods as *H. coccinea*, but much more conspicuous because of its larger size. For a comparison of the two, see comments under *H. coccinea*. Other species in the *H. punicea* "complex" include: *H. laetissima*, with an even brighter red cap plus a white stalk when young; *H. splendidissima*, stalk white becoming reddish-striate in age; and *H. aurantiosplendens*, whose cap fades more markedly. All of these favor redwoods in California and could just as well be treated as variations of *H. punicea*.

Hygrocybe flavescens   (Golden Waxy Cap)   Color Plate 22

CAP 2-7 cm broad, broadly convex becoming plane or with margin slightly uplifted; surface smooth, viscid when moist, bright lemon-yellow to golden-yellow (or sometimes orange toward the center). Flesh thin, yellow, waxy. **GILLS** typically adnexed or free, soft, thick, waxy, yellow or pale yellow. **STALK** 4-9 cm long, 3-10 mm thick, equal, smooth, sometimes tacky or slightly viscid but not slimy; easily splitting, often grooved, yellow to yellow-orange with a whitish base. **VEIL** absent. **SPORE PRINT** white; spores 7-9 × 4-5 microns, elliptical, smooth. Gill tissue parallel.

HABITAT: Solitary to widely scattered or in small groups on ground, usually in woods, widely distributed. Common in our area in the winter, especially under redwood but also with oak, madrone, etc. It usually peaks well after *H. acutoconica* has finished fruiting, sometimes when the bright red waxy caps *(H. punicea* and *H. coccinea)* appear.

EDIBILITY: Edible, but far from incredible: it is watery and has little substance or taste. One author says it "would make a colorful and novel addition to a salad." So would a banana slug.

COMMENTS: The bright yellow to lemon-yellow color, convex to plane cap, and waxy gills identify this beautiful mushroom, better known as *Hygrophorus flavescens*. The cap is never red as in *H. punicea*, nor is it conical as in *H. acutoconica*. In rainy weather the surface of the cap can be quite viscid or even slippery—giving it the appearance of a just-licked lollipop—but in dry weather it is often somewhat silky or shiny. Other widely distributed yellow to orange species with non-conical caps include: *H. chlorophana*, often confused with *H. flavescens*, but slightly smaller with a distinctly slimy stalk when moist; *H. ceracea*, with a slightly viscid cap and stalk and adnate to decurrent gills; *H. citrinopallida*, with a slimy-viscid cap and stalk that fade from yellow to whitish and gills that retain their yellow color; *H. flavifolia*, with a slimy-viscid cap and white, slimy-viscid stalk; *H. parvula*, with a non-viscid stalk and decurrent gills; and *H. nitida*, always yellow when fresh, with a centrally depressed cap (even when young), gills that are soon deeply decurrent, and a frequently viscid stalk. The latter species is widely distributed but especially common in mossy or boggy areas in eastern North America. In both *H. nitida* and *H. flavifolia* the cap often fades to creamy or whitish as it loses moisture or ages.

Hygrocybe acutoconica   (A Cute Conic Waxy Cap)

CAP 2-7 (10) cm broad, bluntly or acutely (sharply) conical when young, sometimes expanding in age but usually retaining pointed umbo; surface smooth, viscid or slimy when moist, bright yellow to yellow-orange or orange (the orange usually toward center). Flesh soft, thin, waxy, yellow. **GILLS** adnexed to free, thick, soft, broad, waxy, yellow, never blackening. **STALK** 5-8 (12) cm long, 3-6 (10) mm thick, equal or thicker below, usually longitudinally striate and/or twisted, easily splitting; smooth, moist or slightly viscid, yellow to yellow-orange, but usually white at base; not blackening when handled, but base may bruise or age grayish to nearly black. **VEIL** absent. **SPORE PRINT** white; spores 9-15 × 5-9 microns, elliptical, smooth. Gill tissue parallel.

Hygrocybe acutoconica. This common yellow to orange waxy cap does *not* blacken when handled. The cap is sharply conical when young, but eventually expands as shown at left.

HABITAT: Scattered to gregarious on ground in woods or under trees, widely distributed. In our area it fruits in the fall and early winter, mostly under redwood and oak. It is nearly as common as its blackening counterpart, *H. conica,* but has a shorter season.

EDIBILITY: Harmless, fleshless, flavorless (see comments on edibility of *H. flavescens*).

COMMENTS: Better known as *Hygrophorus acutoconicus*, this cute, conic waxy cap differs from *H. conica* in its yellowish gills and "failure" to blacken when handled. Scientists may scoff at our tendency to anthropomorphize (attribute human qualities to inhuman beings or inanimate objects), yet our language leaves us little choice. For instance, the word "failure" (with its attendant implications of inadequacy, its connotation of "attempting to, but not succeeding") is often utilized by taxonomists to denote "absence." Before I'm accused of lending credence to this trend, let me earnestly put this question to you philosophical few among the myopic many (or you fungophilic few among the mycophobic many): Is the "failure" in this case an innate inability to succeed? An admirable example of genetic (or aesthetic) restraint? A coincidental and pointless byproduct of circumstance? Or none of the above? . . . Other species: *H. langei* and *H. aurantiolutescens* are very similar if not the same; *H. cuspidata* (called *H. persistens* by some) is also similar but has a conical red cap when fresh that fades with age, plus an orange to yellow stem. It is widely distributed, but I have yet to find it south of San Francisco.

Hygrocybe conica   (Witch's Hat; Conical Waxy Cap)   Color Plate 19

CAP 1-5(12)) cm broad, bluntly to sharply conical, sometimes expanding in age but usually retaining a pointed umbo; surface smooth, dry to moist or slightly viscid, sometimes red but more often orange, yellow, or olive-yellow; blackening in age or upon handling; margin sometimes uplifted in old age. Flesh thin, waxy, blackening in age. **GILLS** adnexed or free, thick, broad, soft, waxy, usually whitish but sometimes tinged yellow, soon becoming grayish and finally black. **STALK** (2) 4-20 cm long, 0.3-1 (1.5) cm thick, equal, not viscid or only slightly so, usually striate and/or twisted, hollow in age and easily splitting; pallid to yellow, olive-yellow, orange, or red, with a whitish to gray base, but turning gray or black when handled or in age. **VEIL** absent. **SPORE PRINT** white; spores 8-14 × 5-7 microns, elliptical, smooth. Gill tissue parallel.

HABITAT: Solitary to scattered or gregarious on ground in damp places (usually in woods); widely distributed and very common, but rarely fruiting in large numbers. It is known from a wide variety of habitats in North America, but in our area shows a pronounced preference for redwood and cypress. It fruits in the fall and winter and, along with

116

Hygrocybe conica (better known as *Hygrophorus conicus)* is one of our most common waxy caps. Note conical cap (which may expand somewhat in age) and tendency to blacken.

H. acutoconica, is usually the first of the brightly colored waxy caps to appear. Bob Winter says it often grows on lawns in Fresno, California.

EDIBILITY: Not recommended. It was once considered poisonous, perhaps due to its blackening qualities, but also because four deaths in China were (mistakenly?) attributed to it. Now it's generally regarded as harmless, but Larry Stickney, *chef extraordinaire,* meandering mainstay of the Mycological Society of San Francisco, and an inimitable mushroom-loving marvel of a man, says it "elicited an odd sensation of lightheadedness and numbness" when he tried it. At any rate, it hardly seems worth experimenting with such a thin-fleshed, watery, tasteless morsel.

COMMENTS: The conical cap and tendency of all parts to blacken when handled immediately identify this cosmopolitan mushroom. It is likely to be among the first waxy caps encountered by beginners and is quite beautiful and waxy-looking when growing in the woods. By time it is brought home, however, it is often barely recognizable because of the gray and black stains that develop. Old, withered, completely black speimens are sometimes found growing alongside vividly colored fresh ones. When growing in deep humus (e.g., redwood duff) the stalk is often quite long, but when growing in bare soil or grass or shallow humus it is apt to be shorter and/or stouter; *Hygrophorus conicus* is a synonym. The "splitters" recognize several closely related blackening species, including: *Hygrocybe nigrescens*, with a bluntly conical, red to scarlet cap, often found under oak; *H. singeri,* cap and stalk distinctly viscid; and *H. olivaceoniger*, small, thin, and olive-green.

Hygrocybe calyptraeformis (Salmon Waxy Cap)   **Color Plate 23**

CAP 2.5-7 cm broad, acutely (sharply) conical at first, then expanding somewhat but retaining a pointed umbo; surface dry or slightly viscid, smooth, coral-pink to pink to salmon when young (rarely tinged lavender), paler (or pinker or oranger, sometimes with whitish areas) in age; margin often splitting at maturity and sometimes uplifted. Flesh thin, watery pinkish, waxy. **GILLS** adnate to adnexed or even free, thick, soft, waxy, pink to pale pink. **STALK** 3-16 cm long, 3-8 (10) mm thick, usually rather long and slender, equal, smooth, often longitudinally striate and/or twisted; fragile, hollow, and easily splitting, white or tinged pinkish (or rarely lavender), not truly viscid. **VEIL** absent. **SPORE PRINT** white; spores 6-9 × 4.5-6 microns, elliptical, smooth. Gill tissue parallel.

HABITAT: Solitary to widely scattered or in small groups on ground in woods or at their edges (often rooted deeply in the humus); widely distributed but not common. I have found it only twice in our area, under redwoods, in the winter.

EDIBILITY: Edible.

COMMENTS: In my fickle fungal opinion, this striking species is one of the most beautiful and elegant of all the waxy caps, its rarity serving to accentuate its beauty. In a genus noted for its bright colors, this species still manages to stand out—its pointed pink to salmon cap and pink gills render it distinct. *H. psittacina* and *H. laeta* can be pinkish, but are smaller, usually slimy, and are never as sharply conical. *Nolanea salmonea* of eastern North America is also similar, but gives a pinkish spore print and does not have waxy gills. *Hygrophorus calyptraeformis* is a synonym.

Hygrocybe virescens   (Lime-Green Waxy Cap)

CAP 2-5 cm broad, obtusely conical or convex becoming plane, umbonate, or with uplifted margin in age; surface smooth, moist or tacky but not truly viscid; color variable when young (honey-yellow or green mixed with dull orange, etc.), but soon becoming citrine-green to yellow-green ("lime-green") overall. Flesh thin, fragile, greenish, waxy. **GILLS** adnexed to free, whitish or with citrine-green tints, thick, soft, waxy. **STALK** 3-7 cm long, 3-8 mm thick, more or less equal, smooth, hollow in age, not viscid; lime-green with a whitish base. **VEIL** absent. **SPORE PRINT** white; spores 7-10 × 5-6.5 microns, elliptical, smooth. Gill tissue parallel.

HABITAT: In scattered groups or tufts under redwood, northern California, apparently quite rare. I have seen only one substantial fruiting—in December of 1971 at Van Damm State Park in Mendocino County. Subsequent expeditions to the same area failed to turn it up.

EDIBILITY: Unknown.

COMMENTS: The striking "lime-green" color of the cap and stalk make this one of the easiest of all waxy caps to identify. Unfortunately, it is one of the most difficult of all waxy caps to find! The above description is adapted from that of Alexander Smith, who originally found it near Trinidad, California. Whereas many waxy caps are cosmopolitan (particularly Hygrocybes), this one appears to have a sharply limited habitat and geographical distribution. The only other green waxy cap, *H. psittacina,* is smaller, slimmer, and slimier, and is green when young rather than in age.

Hygrocybe psittacina   (Parrot Waxy Cap)   Color Plate 18

CAP 1-3 cm broad, bell-shaped or convex to broadly umbonate or plane in age; surface smooth, slimy or viscid when moist, usually shiny when dry; color extremely variable: at first dark green to bright green or olive-green, but soon fading to some shade of yellow, pink, orange, rufous, vinaceous, ochre-buff, tawny, etc.; margin translucent-striate when moist. Flesh thin, soft, waxy, odor mild. **GILLS** adnate to very slightly decurrent, but sometimes seceding; well-spaced, soft, thick, waxy; at first greenish, then fading like the cap (but often yellower or redder), and often retaining slight greenish tints. **STALK** 2-6(8) cm long, 2-5 mm thick, equal or tapering upward, hollow, smooth, very slimy or viscid when moist, greenish when young but soon fading to yellow or cap color (pink, orange, etc.). **VEIL** absent. **SPORE PRINT** white; spores 6.5-10 × 4-6 microns, elliptical, smooth. Gill tissue parallel.

HABITAT: Solitary to scattered or in small groups in damp soil, moss, humus, etc.; widely distributed. Fairly common in our area in late fall and winter but easily overlooked. Like most of our Hygrocybes it favors redwood, but also occurs under other trees as well as on mossy roadbanks or in grass. The largest fruiting I've seen was in a swampy hardwood forest in Pennsylvania.

EDIBILITY: Edible, but slimy and insubstantial. Raw specimens make a colorful but slippery supplement to salads, sliding down your throat before you can savor them.

COMMENTS: Also known as *Hygrophorus psittacinus,* this slippery little fungus is a cinch to recognize when young and fresh—there is no other small green agaric with a glutinous cap and stem! As it dries out, however, it changes color drastically, and faded specimens have been known to fool the most seasoned *Hygrocybe*-hound. The rate, extent, and nature of the color changes seem to vary according to environmental conditions, but close inspection will often reveal a faint olive tinge somewhere on the fruiting body. Even when there is no green present, the small size and slimy-viscid stalk will distinguish it from all but *H. laeta* (see below). The stalk can be so slimy that it is difficult to pluck. Other species: *H. psittacina var. californica* is a rare fungus with a beautiful blue rather than gorgeous green cap in youth. *H. laeta* is a widely distributed species of variable color (violet-gray to pinkish, orange, vinaceous, etc.), but is never green when young and usually has decurrent gills. It, too, has a slimy-viscid cap and stalk. *H. unguinosa* has a slimy-viscid, gray to dark brown to nearly black cap and stem, and is also widely distributed. For viscid-stalked Hygrocybes with bright yellow to orange or red caps, see comments under *H. flavescens* and *H. miniata.*

Hygrophorus chrysodon   (Flaky Waxy Cap)   Color Plate 17

CAP 2.5-8 (10) cm broad, convex to plane or broadly umbonate; surface smooth, viscid when moist, white except for delicate yellow to golden-orange flakes or granules on the margin (or sometimes tinted yellow throughout); margin at first inrolled. Flesh thick, soft, white. **GILLS** typically decurrent, well-spaced, white, soft, rather waxy. **STALK** 3-10 cm long, 0.5-2 cm thick, equal, viscid when moist, white except for a ring of yellow to golden-orange granules at apex (or in rainy weather, the granules dispersed throughout). **VEIL** evanescent, leaving slime on stalk. **SPORE PRINT** white; spores 7-10 × 3.5-5 microns, elliptical, smooth. Gill tissue divergent.

HABITAT: Solitary to gregarious in woods; widely distributed, but not very common in our area. It is said to favor conifers, but in coastal California it shows a definite preference for madrone and tanoak, fruiting sporadically during the late fall and winter.

EDIBILITY: Edible, but slimy and bland (for more details see *H. sordidus*).

COMMENTS: The exquisite flakes of gold on the cap margin and / or stalk apex (see color plate!) are unique to this beautiful species. However, rain may obliterate the granules or disperse their yellow pigment over the entire cap and stem surfaces, leading to confusion with *H. gliocyclus, H. eburneus,* and other white or yellowish waxy caps.

Hygrophorus eburneus   (Ivory Waxy Cap)

CAP 2-7 (10) cm broad, obtuse to convex becoming broadly umbonate to plane or with uplifted margin; surface smooth, extremely slimy or viscid when wet, pure white or sometimes slightly yellowish in old age or occasionally with small pinkish spots; margin at first inrolled. Flesh soft, white, odor mild. **GILLS** adnate to decurrent, well-spaced, thick, soft, waxy, pure white to very slightly yellowish in old age. **STALK** 4-15 (18) cm long, 0.3-1 (2) cm thick, equal or tapered toward base, usually rather slender; smooth or with punctate apex, slimy-viscid when moist, pure white, but sometimes discoloring slightly yellowish or pinkish in age. **VEIL** evanescent, depositing slime on stalk. **SPORE PRINT** white; spores 6-9 × 3.5-5 microns, elliptical, smooth. Gill tissue divergent.

Hygrophorus eburneus. Sometimes called "Cowboy's Handkerchief," this common pure white waxy cap is easily told by its slimy cap and stem. In wet weather the cap is coated with such a thick layer of slime that it looks as if someone blew their nose on it.

HABITAT: Scattered to gregarious or tufted on ground in woods, very widely distributed but most common on the west coast. In our area it is mycorrhizal with hardwoods and is the most common fall and winter *Hygrophorus* of our oak-madrone woodlands. In other regions it may grow with conifers, e.g., in Oregon it is often abundant in oak-pine forests and in New Mexico it fruits in the late summer and fall under pinyon pine.

EDIBILITY: Edible and collected by some people in spite of its sliminess (see comments on edibility of *H. sordidus*).

COMMENTS: The pure white slimy cap *and* slimy stem plus the soft, waxy white gills typify this cosmopolitan *Hygrophorus,* which is the type species of the genus. Sometimes the layer of slime is so thick that it's difficult to pick up the mushroom. At other times the slime dries out, in which case there is likely to be debris glued irrevocably to the cap—a sure sign that *H. eburneus* ranks with *Limacella illinita* and *H. gliocyclus* (among others) as the "slipperiest and slimiest gilled fungus among us." (*L. illinita* is also white, but has free gills, while *H. gliocyclus* is squatter, slightly yellower, and is monogamous with pine). *H. piceae* is a very similar species with a pure white viscid cap and *non-viscid* stalk. It is common under conifers, especially spruce, in the Pacific Northwest and northern California. For similar species with a buff- or ochre-tinged cap, see comments under *H. gliocyclus.*

Hygrophorus gliocyclus (Glutinous Waxy Cap)

CAP (2) 4-10 (15) cm broad, convex or obtuse becoming broadly umbonate, plane, or even shallowly depressed; surface smooth, very slimy or viscid when moist, white to pale cream, usually more yellowish or buff toward the center; margin at first inrolled. Flesh white, fairly firm. **GILLS** adnate to decurrent, fairly well-spaced, thick, soft, somewhat waxy; whitish to pale or dingy yellow, or sometimes tinged faintly pinkish. **STALK** 2-6 cm long, (0.6) 1-2.5 cm thick, typically rather squat, equal to ventricose (swollen in the middle) or tapered at base, smooth, very slimy when moist except at apex; dingy white to creamy, solid. **VEIL** evanescent, leaving slime on stalk and sometimes an obscure ring (annulus). **SPORE PRINT** white; spores 8-11 × 4.5-6 microns, elliptical, smooth. Gill tissue divergent.

HABITAT: Scattered to gregarious in needle duff under pines, widely distributed. Fairly common in our area in fall, winter, and early spring, sometimes growing with *H. hypothejus.* I have found it along the coast with Monterey and bishop pines, and in great numbers inland with Coulter, digger, and ponderosa pines.

120

Hygrophorus gliocyclus, mature specimens. This species is just as slimy as *H. eburneus,* but is stockier and slightly yellower. It is common under pine.

EDIBILITY: Edible and "choice," according to some, but like *H. eburneus,* disagreeable to collect because of the copious slime (see comments on edibility of *H. sordidus*). One book describes how to remove the slime from the *mushrooms,* but says nothing about removing the slime from your *hands.*

COMMENTS: The creamy-white to yellowish cap, stocky stature, thick layer of slime on the cap and stem (when moist), somewhat waxy gills, and association with pine are the hallmarks of this humdrum *Hygrophorus.* It is thicker, squatter, and yellower than *H. eburneus,* and the stalk is indisputably viscid, in contrast to *H. sordidus.* A closely-related, equally slimy pine-lover, **H. flavodiscus,** is also widespread but has pinkish gills in youth; **H. glutinosus** of eastern North America and the Pacific Northwest is the same color as *H. gliocyclus* but favors hardwoods and shows reddish-brown spots on the stalk apex as it dries; **H. whiteii** of northern California has the color of *H. gliocyclus* but the stature of *H. eburneus*; **H. cossus** is a slimy white species that develops ochraceous tones on the cap as it ages and has a distinctive odor (like "goat moth larvae"), but is more common in Europe than North America; **H. chrysaspis** also discolors in age but has no odor and gills that dry dark brown. *H. ponderatus* (see *H. sordidus*) is large and white.

Hygrophorus subalpinus   (Subalpine Waxy Cap)

CAP 4-15 (25) cm broad, broadly convex becoming plane or slightly depressed; surface viscid when moist but soon dry, smooth, pure white or developing a slight yellowish tinge in age; margin sometimes with veil remnants. Flesh thick, firm at first but soft in age, white; odor and taste mild. **GILLS** white when young, often creamy or tinged dingy yellowish in age; typically adnate to decurrent, narrow, close, soft and/or waxy. **STALK** 3-10 cm long, 1-5 (7) cm thick at apex, usually thick and stout, often with a rounded basal bulb when young (but often more or less equal in age); firm, dry, solid, white. **VEIL** somewhat membranous, disappearing or forming a narrow, flaring or flange-like, median to inferior ring on the stalk (just above the bulb). **SPORE PRINT** white; spores 8-10 × 4.5-6 microns, elliptical, smooth. Gill tissue divergent.

HABITAT: Solitary to gregarious on ground under conifers; known only from the mountains of western North America. It is a common "snowbank" species in the Sierra Nevada, Cascades, and Rocky Mountains, but also appears in the summer and late fall. I have yet to find it on the coast.

EDIBILITY: Deer consider it a delicacy; I don't. Although tempting because of its robustness and penchant for growing in the spring when there are few edible agarics out and about, it decays quickly and does not have the greatest texture and flavor (see comments on edibility of *H. sordidus*).

COMMENTS: This hefty white *Hygrophorus* is easily told from other waxy caps (*H. sordidus,* et al) by its thick dry stalk and whitish color, the presence of a veil that often forms an annulus (ring), and association with mountain conifers (see photograph at top of p. 109). The annulus, when present, can mimic the volva of an *Amanita* because it tends to sit so low on the stalk. Amanitas, however, do not have decurrent gills and are rarely as robust. The presence of an annulus can also lead to confusion with *Armillaria ponderosa* and *A. olida* (another springtime conifer-lover), but both of those mushrooms have strong odors. *Russula brevipes* is also somewhat similar, but has brittle flesh and lacks a veil.

Hygrophorus sordidus   (Sordid Waxy Cap)

CAP 5-20 cm broad or more, convex becoming plane or with uplifted margin; surface smooth, viscid when moist but not often slimy and soon dry; white or sometimes tinged yellowish-buff at center; margin at first inrolled. Flesh thick, firm, white. **GILLS** adnate to decurrent, soft but only slightly waxy; white, sometimes dingy yellowish in age. **STALK** 6-10 cm long, 1.5-4 cm thick, equal or narrowed at base, solid, smooth, firm, white, *not* viscid. **VEIL** absent. **SPORE PRINT** white; spores 6-8 × 4-5.5 microns, elliptical, smooth. Gill tissue divergent.

HABITAT: Solitary to widely scattered or gregarious on ground in woods, associated principally (if not exclusively) with oaks, widespread but primarily southern. In our area it fruits with live oak in the fall and winter. It is quite abundant some years, but almost completely absent others.

EDIBILITY: Edible but *not* choice. I ruined an otherwise superb curry in my sole attempt to make it palatable. The sliced caps neither absorbed the surrounding spices nor contributed any special flavor of their own. The turmeric did turn them yellow, however, making them look for all the world like undercooked and overfed banana slugs—gummy, amorphous masses of slime that coagulated around or completely engulfed the peas and savory chunks of potato, rendering the entire dish inedible (though unforgettable). Since its attractive appearance belies its "slugulose" qualities, I can only conclude that the scientific soul who named it had a similar and equally sordid experience with it.

COMMENTS: Our heftiest *Hygrophorus,* its size alone distinguishes it from most other white members of the genus. The stalk is not viscid as in *H. gliocyclus* and *H. eburneus,* nor is there a veil as in *H. subalpinus.* It is likely to be mistaken for a *Clitocybe* or

Hygrophorus sordidus. A robust white waxy cap with a viscid cap, dry stem, and white gills that are only slightly waxy. *H. ponderatus* (not illustrated) is an equally robust, viscid-stalked white species.

Russula because of its fleshiness, but the gills are soft, and at least to the experienced *Hygrophorus*-hunter, *waxy*. *H. penarius* has a slightly darker cap but is otherwise similar. *H. perfumus* of the Sierra Nevada has a fragrant odor. *H. ponderatus* is a similar but slightly more flavorful, widespread species whose stalk is viscid or lubricous (at least over the lower portion); in California it usually grows with conifers.

Hygrophorus russula   (Russula-Like Waxy Cap)

CAP 5-13 cm broad, convex to plane or with uplifted margin in age; surface viscid when wet but soon dry; coral-pink to vinaceous-red, usually streaked with purple-red or vinaceous fibrils; smooth or minutely scaly, occasionally staining yellowish when rubbed or in age; margin often paler or whitish and incurved when young. Flesh thick, white or tinged pink; odor and taste typically mild. **GILLS** usually adnate but sometimes adnexed or slightly decurrent, close to crowded (120-150 reach the stalk), soft, slightly waxy; white at first but soon flushed pink and developing purple-red to vinaceous stains in age. **STALK** 3-10 cm long, 1.5-3.5 cm thick, usually rather stout; solid, dry, smooth, equal or tapered below; white, soon stained or streaked pink to reddish or vinaceous. **VEIL** absent. **SPORE PRINT** white; spores 6-8 × 3-5 microns, elliptical, smooth. Gill tissue slightly divergent.

HABITAT: Scattered to gregarious or in rings in mixed woods and under hardwoods, associated mainly with oaks; widely distributed and common in eastern North America, but infrequent on the west coast. In our area I find it occasionally in the late fall and early winter in tanoak-madrone woods at higher elevations in the coastal mountains.

EDIBILITY: Edible and choice, according to some, but my lingering distaste for fleshy waxy caps (see comments on edibility of *H. sordidus*) has discouraged me from sampling it. One authority calls it "the best of the family for the table"—a classic example of damning with faint praise, if you ask me. On the other hand, local *Hygrophorus*-hound Luen Miller, who suffers from an acute case of "McIlvaine-mania,"* says of *H. russula:* "Not as good as *H. sordidus,* but as an edible species it is not to be despised. It has a noble waxy texture and makes a toothsome meal. Although it lacks the supreme succulence of *H. sordidus,* it largely makes up for it in possessing a copious supply of wax, which coats the mouth and throat for hours after eating it, the way good ice cream does. All authorities pronounce it excellent. It is delightful scalloped or stewed, or better yet, made into catchup and poured over food. The Tartars, I am told, know it as 'poor man's candle' since the dried 'carps offer a weak flickering flame when lit."

*See footnote on p. 419!

Hygrophorus russula, mature specimens. The entire fruiting body develops dark red streaks and stains; the gills are close together and only slightly waxy.

COMMENTS: A beautiful, robust *Hygrophorus,* best recognized by its coral-pink to vinaceous (wine-colored) or vinaceous-red color, absence of a veil, and growth with hardwoods. As the name implies, it has somewhat the stature of a *Russula,* but the stalk does not snap open cleanly like chalk. It was originally placed in *Tricholoma,* but the gills are usually adnate rather than notched, and quite soft and waxy. For similar waxy caps associated with conifers, see *H. purpurascens.*

Hygrophorus purpurascens   (Purple-Red Waxy Cap)

CAP (3) 6-12 (20) cm broad, convex becoming broadly convex or plane; surface slightly viscid when wet, otherwise dry; whitish to coral-pink with darker (vinaceous-red to purple-brown) splashes, streaks, and/or fibrils; margin incurved at first and usually paler. Flesh thick, firm, white; odor mild, taste mild or bitter. **GILLS** adnate to decurrent, white at first but soon flushed pink, then spotted or stained purple-red to vinaceous; fairly close, soft, slightly waxy. **STALK** (3) 5-12 (15) cm long, 1-2.5 cm thick, equal or tapered below, solid, firm, not viscid; colored or stained more or less like the cap. **VEIL** fibrillose, white, forming a slight superior or apical ring or hairy zone on stalk, or disappearing entirely. **SPORE PRINT** white; spores 5.5-8 × 3-4.5 microns, elliptical, smooth. Gill tissue divergent.

HABITAT: Solitary to scattered, gregarious, or in troops under conifers, especially spruce and pine; widely distributed in northern and western North America. It does not occur in our area, but is fairly common in the Sierra Nevada and Cascades from spring through fall, and fruits prolifically in late summer in the spruce forests of the Rocky Mountains.

EDIBILITY: Edible, but some variants are unpalatable (bitter) even when cooked—as I can personally attest.

COMMENTS: This waxy cap and its close relatives (see below) are very common at times in our western mountains. They resemble *H. russula* in their reddish-pink to vinaceous-red color, but favor conifers rather than hardwoods. *H. purpurascens* is distinct by virtue of its fibrillose veil, which is often evident only in young, unexpanded specimens. Its close relatives include: *H. erubescens,* quite similar but lacking a veil and with a frequent tendency to slowly stain yellow when bruised or left overnight in the refrigerator. It is normally a fairly robust mushroom, but a slender form occurs rarely in our area with pine. *H. capreolarius* is a similar but smaller, more slender (stalk about 1 cm thick), spruce-loving species with well-spaced gills and fruiting body evenly colored dark vinaceous-red in age. I have collected it several times under Sitka spruce in northern coastal California, but its distribution parallels that of *H. purpurascens. H. amarus* is also similar, but is very bitter-tasting and tends to develop pronounced pale yellow to bright yellow tones on the cap (or whole fruiting body) in age. It may or may not have a very slight fibrillose veil (check young specimens!); it's fairly common under spruce and fir in the Rocky Mountains.

Hygrophorus pudorinus   (Spruce Waxy Cap)   **Color Plate 16**

CAP 5-15 (20) cm broad, obtuse or convex with an inrolled margin, becoming broadly convex or plane in age; surface smooth, viscid when moist, pale tan to pinkish or flesh-color, or pinkish-orange toward the center and pinker or paler at the margin. Flesh thick, firm, white or tinged cap color; odor mild to faintly fragrant. **GILLS** adnate to decurrent, fairly close, soft, waxy, sometimes whitish but more often pinkish to pale flesh-color; *not* developing reddish stains. **STALK** 4-20 cm long, 1-3 cm thick, equal or narrowed below, solid, dry to tacky but not truly viscid, whitish to buff or pinkish or colored like cap; lower portion fibrillose, upper portion conspicuously punctate (with tiny whitish scurfy scales or tufts that darken to reddish-brown when dried or in age, and turn yellow-orange in KOH). **VEIL** absent. **SPORE PRINT** white; spores 6.5-9.5 × 4-5.5 microns, elliptical, smooth. Gill tissue divergent.

HABITAT: Scattered or in groups on ground under conifers, particularly spruce; widely distributed but erratic in its fruiting habits. It does not occur in our area (probably due to the absence of spruce), but I have seen large fruitings in December with Sitka spruce in northern California, and in August with Engelmann spruce in the Southwest.

EDIBILITY: Edible but mediocre. Some variants are said to have a "turpentine-like taste," but I cannot vouch for the accuracy of this comparison, since I have never tasted turpentine. (Have you?)

COMMENTS: This attractive, rather robust waxy cap is difficult to characterize but relatively easy to recognize (see color plate). The viscid pinkish to pinkish-salmon or pale tan cap, punctate stalk apex, and waxy gills are good fieldmarks. The viscid cap distinguishes it from the similarly colored *Camarophyllus pratensis.* Several varieties and color forms have been described, including var. *fragrans,* large and tall, with a yellow-orange stalk base and tendency to stain yellow or orange when bruised. Another beautiful waxy cap, *H. saxatilis,* occurs with conifers in the Pacific Northwest, especially on rocky hillsides. It has a viscid, whitish to pale buff cap (sometimes with darker watery spots) and lovely pale apricot to pinkish, decurrent gills. Still another distinctive species, *H. goetzii,* is rather small and slender, with a rosy or pinkish cap and creamy gills often tinged with the cap color. It is sometimes common in the late spring and summer in the Sierra Nevada and Cascades, often near or even *in* melting snow.

Hygrophorus roseibrunneus   (Rosy-Brown Waxy Cap)

CAP 2-10 cm broad, convex becoming broadly umbonate or plane, or with an uplifted margin in age; surface smooth, viscid when moist, reddish-brown to pinkish-cinnamon to rosy-brown or brown at least at the center and often overall; margin often paler or whitish. Flesh white, soft, odor mild. **GILLS** usually adnate, but sometimes decurrent or adnexed; close, soft, somewhat waxy, white. **STALK** 3-12 cm long, 0.5-1 (1.5) cm thick, equal or tapered below, often curved near base; smooth, white, not viscid; apex often prominently punctate. **VEIL** absent. **SPORE PRINT** white; spores 6-9 × 3.5-5 microns, elliptical, smooth. Gill tissue divergent.

HABITAT: Solitary to widely scattered or in groups, associated with oak in our area and sometimes common in the late fall and winter (along with *H. brunneus*). It was first collected in Palo Alto, California, but also occurs in eastern North America.

EDIBILITY: Edible. I have tried it.

COMMENTS: The reddish- to rosy-brown color of the cap (or center of the cap), non-viscid stalk, waxy white gills, and mild odor are the fallible fieldmarks of this species. The punctate stem apex is characteristic of many waxy caps, but in this one it is especially attractive. *Leucopaxillus amarus* is similarly colored but has a dry cap, non-waxy gills, bitter taste, and white mycelium at the base of the stem. Just as common in our area is *H. brunneus*, very similar but for its yellow-brown to tawny-brown cap (or cap center). A species endemic to California, *H. albicastaneus,* is also similar, but ranges from pure white (when young) to yellow-brown to peach- or ochre-tinged (in age). It has a non-viscid stalk and usually grows with oak, but stains rusty or fulvous in potassium hydroxide. *H. subpungens* is a small, slender-stemmed species with a pinkish-brown to yellow-brown to salmon-centered cap and very faint fruity odor; it occurs locally under alder, oak, and Douglas-fir. *H. tennesseensis,* widespread, is also similar but favors conifers and has a bitter taste and potato-like odor. There are also several similar species with a viscid stalk, including: *H. laurae,* with a whitish cap margin, favoring hardwoods; *H. discoideus,* slightly darker, favoring conifers; and *H. vernalis,* a western springtime "snowbank" species with a pinkish-buff to pale vinaceous cap and sordid yellowish patches on the lower half of the stalk.

Hygrophorus bakerensis   (Brown Almond Waxy Cap)

CAP 4-15 cm broad, obtuse to convex becoming broadly convex or plane; surface smooth, slimy or viscid when moist, cinnamon-brown to yellow-brown or tawny, the margin usually paler or whitish. Flesh thick, white; odor sweet but sometimes faint, like almond extract or crushed peach pits. GILLS white to creamy or pinkish-buff, usually decurrent but varying to adnate; soft, somewhat waxy. STALK 4-15 cm long, 0.8-2.5 cm thick, equal or tapered downward, smooth, solid, dry; white to pinkish-buff. VEIL absent. SPORE PRINT white; spores 7.5-10 × 4.5-6 microns, elliptical, smooth. Gill tissue divergent.

HABITAT: Widely scattered to gregarious under conifers, known only from the Pacific Northwest and northern California; very common at times in the fall and early winter.

EDIBILITY: Edible but bland (it doesn't taste like it smells).

COMMENTS: This species is one of the commonest and most characteristic waxy caps of the Pacific Northwest. The reddish-brown to yellow-brown cap, waxy decurrent gills, non-viscid stalk, and almondy odor set it apart. The gills and stalk are sometimes beaded with droplets in moist weather, but there is no latex as in *Lactarius.* The gills are usually decurrent, in contrast to *Collybia oregonensis,* a similarly colored, strongly fragrant mushroom with adnexed or notched, non-waxy gills. *H. agathosmus* has the same odor, but its cap is gray, not brown. Other fragrant waxy caps from the Pacific Northwest and California include: *H. monticola* and *H. vinicolor,* similar but with larger spores (the latter with an unpleasant taste); *H. variicolor,* also similar but with lower half of stalk distinctly viscid or glutinous; and *H. pacificus,* a strongly aromatic (but not almondy) species with a russet to tawny to pinkish-buff, often lobed or wavy cap, pale yellowish gills, and larger spores.

Hygrophorus speciosus   (Larch Waxy Cap)   Color Plate 15

CAP 2-5 (8) cm broad, convex to broadly umbonate or expanding to plane or even centrally depressed; surface slimy or viscid when moist, smooth, bright orange-red to orange, often fading to orange-yellow; margin often paler. Flesh white or tinted yellow, soft. GILLS adnate to decurrent, white to pale yellow, the edges usually yellow; well-spaced, thick, soft, waxy. STALK 3-10 cm long, 0.4-1 (1.5) cm thick, equal or thicker below, white or with yellow to orange stains over lower portion; viscid to slimy when moist, at least below. VEIL single (var. *speciosus*) or double-layered (var. *kauffmanii*); outer layer evanescent, leaving slime on stalk; inner layer when present fibrillose, sometimes forming a slight ring on stalk. SPORE PRINT white; spores 8-10 × 4.5-6 microns, elliptical, smooth. Gill tissue divergent.

HABITAT: Scattered to densely gregarious in woods and bogs under conifers, especially larch but also pine; common in the late summer and fall throughout the range of larch, but also frequent with ponderosa pine in the Southwest.

EDIBILITY: Edible; I haven't tried it.

COMMENTS: This is one of the few brightly colored waxy caps belonging to the genus *Hygrophorus* in its strictest sense (i.e., waxy caps with divergent gill tissue). It is easily told from the numerous red to orange Hygrocybes by its habitat, viscid stalk, and white to pale yellow *decurrent* gills. The cap is never olive-brown as in *H. hypothejus. H. pyrophilus* is similar; it has a red cap, less viscid stalk, and was found in burnt ground near Mt. Shasta.

Hygrophorus hypothejus   (Olive-Brown Waxy Cap)

CAP 2-8 cm broad, convex to broadly umbonate, plane, or depressed; surface smooth, viscid or slimy when wet, color variable: typically dark brown to olive-brown at the center and greenish-yellow to yellow at the margin when young (but sometimes entirely olive-brown), often developing yellow-orange to reddish-orange tones in age, especially near the margin. Flesh thin, yellowish to whitish; odor mild. GILLS decurrent or occasionally

Hygrophorus hypothejus commonly grows with pine, often in the company of *H. gliocyclus*. Gills are pale yellow to orange, decurrent, and fairly waxy, and the cap is slimy when wet.

adnate, well-spaced, thick, soft, waxy, at first pallid but soon becoming pale yellow, and in age sometimes brightly colored (like margin of cap). **STALK** (3) 5-15 cm long, 0.5-1.5 (2) cm thick, equal or tapered downward; yellow at apex, otherwise pallid or variously colored (like cap) and viscid or slimy when moist. **VEIL** evanescent, leaving slime on stalk and sometimes an obscure fibrillose ring. **SPORE PRINT** white; spores 7-9 × 4-5 microns, elliptical, smooth. Gill tissue divergent.

HABITAT: Scattered to gregarious or in troops under conifers, particularly pine—very widely distributed and often abundant in cool weather. It is common from late fall through early spring in our coastal pine forests.

EDIBILITY: Bountiful, but bland. See *H. sordidus* for details.

COMMENTS: This waxy cap is best recognized by its olive-brown to olive-yellow cap when young which is sticky or slimy and convex to depressed in age, the decurrent gills, and association with pine. The colors are extremely variable, especially in age, and young specimens bear little resemblance to old ones until you find stages in between. Both slim and relatively robust specimens can be found. Other species with an olive-brown to grayish cap and viscid stalk include: *H. fuligineus,* very similar but with a darker (olive to blackish) cap, common in cool weather under conifers (mainly in eastern North America); *H. fusco-albus,* with larger spores (9-13 microns long) and *H. limacinus,* with even larger (10-17 microns) spores; and two species which lack an inner (fibrillose) veil: *H. occidentalis,* small-spored, cap brown to grayish with a pallid margin, often under oak; and *H. mega-sporus*, with spores 12-18 microns long. All of these species favor conifers and have white to grayish gills; none develop the bright colors typical of mature *H. hypothejus.*

Hygrophorus olivaceoalbus   (Sheathed Waxy Cap)

CAP 3-12 cm broad, convex or broadly umbonate to more or less plane; surface slimy or viscid when moist, dark brown to nearly black at the center, usually paler (grayish to olive-brown) toward the margin, and usually with a streaked appearance from darker fibrils. Flesh thick, white, soft, odor mild. **GILLS** adnate to decurrent, thick, soft, waxy, pure white or pallid, sometimes becoming grayish. **STALK** (3) 5-12 (15) cm long, 1-3 cm thick, equal or thicker below; smooth and white above the ring, viscid to slimy (when moist) below and sheathed with blackish to grayish-brown, olive-brown, or gray fibrils which break up into patches or scaly rings. **VEIL** double-layered, the outer layer evanescent and depositing slime on stalk, the inner layer fibrillose and sometimes forming an obscure ring at top of fibrillose sheath. **SPORE PRINT** white; spores 9-12 × 5-6 microns, elliptical, smooth. Gill tissue divergent.

HABITAT: Solitary, scattered, or in groups on ground under conifers (especially spruce) in western and northern North America, fruiting from late summer through early winter. I have seen it in northern California under Sitka spruce and in the southern Rocky Mountains under Engelmann spruce and blue spruce.

EDIBILITY: Said to be edible, but bland and slimy.

COMMENTS: This handsome *Hygrophorus* can be distinguished from similarly colored waxy caps by the gray to dark brown fibrillose sheath on the stem. The contrasting colors are very striking. The coastal variety (also called **H. persoonii**) is usually quite dark and slimy; the Rocky Mountain version is often paler (grayer) and only slightly viscid if at all. Other species: **H. inocybiformis** has a grayish-brown sheath and veil, but its stalk is *not* viscid, and its dark gray cap is only 3-7 cm broad; it occurs in the Pacific Northwest under conifers. **H. tephroleucus, H. pustulatus,** and **H. morrisii** are even smaller and have small, pointlike grayish scales on the stalk. See also the species listed under *H. hypothejus.*

Hygrophorus agathosmus   (Gray Almond Waxy Cap)

CAP 3-10 cm broad, convex to plane or with margin uplifted; surface smooth, viscid when moist, dull gray to ashy-gray, brownish-gray, or at times grayish-olive; margin incurved at first. Flesh soft, whitish; odor sweet, like almond extract (but sometimes faint). **GILLS** adnate to slightly decurrent, close or well-spaced, soft, waxy, white or sometimes grayish in age. **STALK** 4-10 (16) cm long, 0.5-1.5 (2) cm thick, equal or narrowed below, smooth, not viscid; white or tinged gray. **VEIL** absent. **SPORE PRINT** white; spores 7-10.5 × 4.5-5.5 microns, elliptical, smooth. Gill tissue divergent.

HABITAT: Scattered to gregarious under conifers, widely distributed and fairly common in cool weather, but rather infrequent in our area. It is partial to spruce, but is associated locally with Douglas-fir (probably because there isn't any spruce).

EDIBILITY: Edible, but bland. It is unfortunate that it doesn't taste like it smells.

COMMENTS: The grayish cap, waxy gills, dry stalk, almondy fragrance, and association with conifers are the telltale traits of this fine fungus. The odor is sometimes faint, but if several are placed together in a closed container it usually becomes evident. *H. calophyllus* and *H. camarophyllus* are somewhat similar but do *not* have an almondy odor. Other species: **H. odoratus** is a small, slender version of *H. agathosmus* with an almondy odor and larger spores (11-14 microns long); it occurs in the Pacific Northwest under conifers. **H. occidentalis** has a viscid stalk, weaker odor, and grows under oak or conifers.

Hygrophorus agathosmus, mature specimens. Note the Douglas-fir needles stuck to the cap, indicating that it was viscid. Cap is grayish, gills white and waxy, odor almondy. It also grows with spruce.

Hygrophorus calophyllus   (Gray-Brown Waxy Cap)

CAP 4-11 cm broad, convex to broadly umbonate or plane; surface smooth, viscid or slimy when moist, evenly colored deep olive-brown to dark gray-brown to umber, the margin sometimes slightly paler. Flesh soft, white, thick; odor mild or faintly fragrant. **GILLS** white or flushed a delicate pink, well-spaced, thick, soft, waxy, usually decurrent. **STALK** 5-12 cm long, 1-1.5 cm thick, equal or narrowed below, smooth, not viscid; apex white, otherwise colored like the cap but usually slightly paler. **VEIL** absent. **SPORE PRINT** white; spores 5.5-8 × 4-5 microns, elliptical, smooth. Gill tissue divergent.

HABITAT: Solitary or in small groups on ground under conifers, western North America, not common. I have found it only once in our area, with Douglas-fir, in December.

EDIBILITY: Edible.

COMMENTS: Whether white or pink, the gills contrast sharply with the dark cap and stem, making this a most attractive mushroom. Often the gills will develop their pinkish tint only as they mature, leading one to wrongly assume that the developing spores are pink! The stalk is not viscid as in *H. hypothejus* and *H. olivaceoalbus,* and the soft, clean, waxy gills distinguish it from *Clitocybe.* Very similar but more common is ***H. camarophyllus,*** with a dry to only slightly viscid (never slimy), streaked, grayish-brown cap and white to grayish-tinted (never pink) gills. It also fruits under conifers, sometimes near melting snow, and is easily mistaken for a *Clitocybe.* ***H. marzuolus*** grows almost exclusively in the spring, often near snow. It has well-spaced gills and is entirely pallid when young, but soon becomes grayish overall (including the gills).

spores

TRICHOLOMATACEAE

THIS is by far the largest and most diverse family of pale-spored agarics. The spore print is usually white, but ranges to buff, yellowish, pale lilac, or pinkish. A stem is normally present, but in several of the shelflike, wood-inhabiting species it is rudimentary or even absent. The gills are typically attached to the stem, but in some of the smaller forms they are free. Though most of the species are terrestrial, many grow on wood—whereas other pale-spored families are almost exclusively terrestrial. They are largely woodland fungi, but a few, such as *Marasmius oreades,* grow in grass.

The best way to recognize the family is to eliminate the other pale-spored families. The gills are not normally soft, thick, and waxy as in the Hygrophoraceae, nor shallow, blunt, and foldlike as in the Cantharellaceae; the fleshy forms do not have noticeably dry, brittle flesh as in the Russulaceae (and their tissue lacks sphaerocysts—a microscopic feature), nor do they have a volva as in the Amanitaceae, and the few species with a veil do not have the free gills typical of the Lepiotaceae. (There are some exceptions to the above, but they are discussed under individual species or genera.)

The size and complexity of the Tricholomataceae are such that no generalizations can be made regarding edibility—other than that some are poisonous and some are not. Many have not been adequately tested and others are too small to be of value. The safest approach is to learn the distinguishing characteristics of each edible species—and there *are* some plentiful collectable delectables that are well worth getting to know! The most notable are the oyster mushroom *(Pleurotus ostreatus),* man-on-horseback *(Tricholoma flavovirens),* honey mushroom *(Armillariella mellea),* matsutake *(Armillaria ponderosa),* fairy ring mushroom *(Marasmius oreades),* and blewit *(Clitocybe nuda).* Several lesser-known species (e.g., *Lentinus ponderosus)* are also excellent.

The Tricholomataceae embraces more genera than any other family of gilled mushrooms, and its taxonomy is still in a state of flux. Many of the genera are defined by esoteric chemical and anatomical (microscopic) characteristics, which presents obvious problems when attempting to construct (or use) a key based solely on field characters. It helps to break down the genera into three large groups: wood-inhabiting, shelflike forms (typified by *Pleurotus*); fleshy-stalked, mostly terrestrial forms (typified by *Tricholoma* and *Clitocybe),* and thin, fragile- or cartilaginous-stalked, terrestrial or wood-inhabiting types (typified by *Mycena, Collybia,* and *Marasmius*).

In the following key, only the more distinctive genera are included; the more difficult or obscure ones are then keyed out under the distinctives ones, sometimes on a species-by-species basis.

Key to the Tricholomataceae

1. Growing on other mushrooms; gills thick and widely spaced or poorly formed to practically absent .... *Asterophora,* p. 200
1. Not growing on other mushrooms, or if so then gills well-developed, thin, close 2

2. Fruiting body pinkish to salmon, orange, or yellow-orange; cap surface conspicuously reticulate (netted or veined and pitted); cap 2-5 cm broad; stalk central or off-center, tough; spore print pinkish; found on dead eastern hardwoods (e.g., maple); infrequent .. *Rhodotus palmatus*
2. Not as above 3

3. Stalk absent, or if present then typically off-center to lateral; usually growing on wood (or woody material such as coffee bean waste or wood chip mulch) or *on* moss 4
3. Stalk present, well-developed, more or less central; growing on ground or wood 6

4. Fruiting body tough and leathery or corky; gill edges longitudinally split and cap densely hairy *or* cap concentrically zoned or grooved (and often velvety) and gills often mazelike (forming elongated pockets) or wavy; found on hardwoods ..... (see **Polyporaceae & Allies,** p. 549)
4. Fruiting body fleshy, or if tough then not as above 5

5. Edges of gills conspicuously serrated (toothed) or eroded, even when fresh ....
      .... *Lentinus & Lentinellus,* p. 141
5. Not as above; gill edges usually entire (but sometimes wavy) ..... *Pleurotus* & **Allies,** p. 132

6. Stalk arising from an underground "tuber" (the "tuber" cylindrical or bulbous, often hollow); cap usually scaly, fibrillose, or granulose; not common .... *Squamanita,* p. 197
6. Not as above (but stalk may have a tapered underground "tap root") 7

7. Cap granulose (covered with a layer of mealy or powdery granules that are sometimes washed off by rain); veil present (check young specimens!), sometimes forming an annulus (ring) on stalk; stalk granulose below the veil .... *Cystoderma,* p. 198
7. Not as above; veil absent, or if present then cap and stalk not granulose 8

8. Veil present, usually forming a distinct annulus (ring) on stalk ... *Armillaria* & **Allies,** p. 189
8. Veil absent or rudimentary and evanescent, not forming an annulus 9

9. Gills and stalk bruising dark gray to black (sometimes slowly) or developing such stains in age; basidia with siderophilous granules; not common in our area . *Lyophyllum* & **Allies,** p. 173
9. Not as above 10

10. Stalk fleshy, usually at least 5 mm thick 11
10. Stalk usually thin and hollow or stuffed and either fragile or cartilaginous (tough), typically 5 mm thick or less (occasionally thicker but then with a tough cartilaginous outer rind) . 23

11. Fruiting body partially or completely purple, violet, or lilac when fresh (at least the gills); odor *not* radishlike; usually on ground or compost 12
11. Not as above 13

12. Gills thickish and fairly well-spaced; stalk fibrous and/or fibrillose; spore print white or lilac-tinged .... *Laccaria,* p. 171
12. Gills close, not thick; spore print dull or dingy pinkish .... *Clitocybe* & **Allies,** p. 148

13. Copious white mycelial mat usually present at base of stalk or in substrate; stalk and gills white or yellowish but not gray; cap and stalk dry, dull, unpolished, often tough; stalk *not* hollow; spore print white; spores with amyloid warts; found in woods or under trees *Leucopaxillus,* p. 166
13. Not with above features; spore print variously colored (white, buff, pinkish, etc.) 14

14. Spore print white, yellowish, or buff (or in one case brownish) 15
14. Spore print pinkish to pinkish-buff 27
15. Typically growing in dense clusters in disturbed soil (along roads, paths, etc., but sometimes also in woods); gills whitish to gray; stalk at least 1 cm thick; caps typically at least 3 cm broad; spore print white (*not* buff); basidia with siderophilous granules ... ***Lyophyllum* & Allies,** p. 173
15. Not with above features 16
16. Gills pinkish, flesh-colored, cinnamon, or somewhat vinaceous, thickish and fairly well-spaced; cap up to 6 cm broad; stalk rather tough and fibrous, *not* white; spores spiny ***Laccaria,*** p. 171
16. Not with above features 17
17. Gills typically adnate to decurrent 18
17. Gills typically notched, adnexed, or even free (occasionally adnate but not decurrent) ... 21
18. Gills *and* flesh olive-yellow to yellow to orange; cap *not* viscid; gills *not* repeatedly forked; growing on or near wood (but wood often buried or not visible) or from roots 19
18. Not with above features 20
19. Fruiting body orange to yellow-orange or with olive tones; cap and stalk smooth, without scales; associated with hardwoods .... ***Omphalotus,*** p. 146
19. Fruiting body pale yellow to yellow, often with differently colored scales or fibrils on cap or stalk; associated mainly with conifers .... ***Tricholomopsis,*** p. 144
20. Cap viscid or slimy when moist *and/or* gills thick, widely spaced, and clean or waxy-looking .... (see **Hygrophoraceae,** p. 103)
20. Not as above .... ***Clitocybe* & Allies,** p. 148
21. Growing on or near wood (sometimes very rotten or buried); flesh and gills often (but not always!) yellow to pale yellow .... ***Tricholomopsis,*** p. 144
21. Not as above 22
22. Cap smooth, without fibrils or scales, usually white to gray to grayish-brown or dark brown when fresh and moist; gills crowded; found in many habitats but especially in grassy or land-scaped areas *or* in mountains soon after snow melts; spores amyloid .. ***Melanoleuca,*** p. 169
22. Not as above; cap variously colored (yellow, greenish, brown, grayish, white, reddish-brown, etc.); found in woods or with trees; spores not amyloid .... ***Tricholoma,*** p. 176
23. Spore print pinkish to ochre-brown; cap bell-shaped to conical when young, reddish-brown to dark brown to blackish (the margin often paler); stalk similarly colored, minutely velvety; odor usually strong and fishy or reminiscent of cucumber; giant cystidia present on gills; not common in our area (but widespread) .... ***Macrocystidia cucumis***
23. Not as above 24
24. Cap conical or bell-shaped when young (but may expand in age), often translucent-striate when moist, margin *not* usually incurved when young; stalk *not* polished or tough ***Mycena,*** p. 224
24. Not as above 25
25. Gills purple to pinkish, flesh-colored, or dingy cinnamon, rather thick and well-spaced; cap convex to plane or uplifted, *not* conical or bell-shaped; stalk tough, fibrous, *not* white; growing on ground; spores usually spiny .... ***Laccaria,*** p. 171
25. Not as above 26
26. Cap small (up to 2.5 cm broad), cap and stalk golden-yellow to pale yellow and covered with scurfy or mealy particles; gills creamy to yellow, usually adnate to decurrent; growing on logs and sticks of hardwoods in eastern North America and on aspen in the Southwest and Rocky Mountains .... ***Cyptotrama chrysopeplum (=Xerulina chrysopepla)***
26. Not as above 27
27. Gills typically adnate to decurrent 28
27. Gills notched to adnexed or free, or sometimes adnate ***Marasmius, Collybia,* & Allies,** p. 201
28. Gills *and/or* cap pale yellowish to yellow, orange, pinkish, or greenish when fresh (may fade!); fruiting body small or minute (cap often less than 2.5 cm broad); stalk usually 1-3 mm thick; cap usually depressed centrally in age; often found on logs or with grass, moss, or lichen .... ***Omphalina* & *Xeromphalina,*** p. 221
28. Not as above 29
29. Stalk thin, tough, and pliant or if thick then with a tough outer cartilaginous rind; gills usually adnate, if decurrent then usually widely spaced ..... ***Marasmius, Collybia,* & Allies,** p. 201
29. Not as above; gills often decurrent, usually close or crowded ..... ***Clitocybe* & Allies,** p. 148

PLEUROTUS & Allies

Small to large *wood-inhabiting mushrooms usually growing shelflike.* CAP smooth or hairy, dry or viscid. Flesh soft, rubbery, pliant, or tough. GILLS adnate to decurrent, edges typically not serrated. STALK *absent or if present usually lateral or off-center,* occasionally central. VEIL absent (except in *Pleurotus dryinus*). VOLVA absent. SPORE PRINT *pale (white to yellowish, pale lilac, or pinkish).* Spores smooth, not amyloid except for *Panellus.*

THIS is an artificial grouping of pale-spored, wood-inhabiting agarics with a consistently off-center to lateral or absent stalk. The most common genus, *Pleurotus,* can be recognized by its rather soft, fleshy fruiting body. *Hohenbuehelia* includes a number of species once placed in *Pleurotus* but now segregated because of their semi-gelatinous to rubbery-pliant flesh and large, thick-walled sterile cells (cystidia) on the gills. *Panus* and *Panellus* incorporate those forms with a tough, often hairy fruiting body (the latter with amyloid spores), while *Phyllotopsis* includes species with pinkish, sausage-shaped spores.

All of the above genera are small but widely distributed. The oyster mushroom *(Pleurotus ostreatus)* and its relatives are among our best edible mushrooms, but most of the other species are either too small, too tough, or too rare to be of value.

Key to Pleurotus & Allies

1. Veil present when young but often disappearing in age; flesh very thick; cap medium-sized to large, typically with grayish fibrils, hairs; or scales (but sometimes whitish or ochre); gill edges normally entire (not serrated); growing on hardwoods (often living) *Pleurotus dryinus,* p. 136
1. Not with above features 2

2. Veil present, at least when young, sometimes forming an annulus (ring) on stalk 3
2. Veil absent in all stages 4

3. Fruiting body small, tough, brownish; cap less than 2.5 cm broad; veil membranous; spores amyloid; growing on dead hardwoods .... *Tectella patellaris (=Panus operculatus)*
3. Not as above; larger .... (see *Lentinus & Lentinellus,* p. 141)

4. Cap minute (2-8 mm broad), bluish-gray to bluish-black or grayish-black; fruiting body shaped like an inverted cup; gills widely spaced; stalk absent; growing on hardwoods, shrubs, vines, etc., but not on mosses .... *Resupinatus applicatus*
4. Not as above 5

5. Cap small (less than 2.5 cm) and hairy, white to brownish; gills forking or veinlike and very wavy (crisped); on hardwoods in eastern North America *Trogia crispa* (see *Schizophyllum,* p. 590)
5. Not as above 6

6. Cap small (less than 2.5 cm), often rubbery or gelatinous, white to grayish; gills often veinlike or even absent; usually (but not always) growing *on* mosses; mostly northern *Leptoglossum*
6. Not as above 7

7. Cap small or minute (up to 2.5 cm broad), white to creamy or tinged purplish or pinkish .. 8
7. Not as above (larger and/or differently colored) 11

8. Gills very widely spaced and often with veins between them; stalk usually present and darkening in age from base upward; common along west coast ... (see *Marasmiellus candidus,* p. 206)
8. Not as above 9

9. Fruiting body white or whitish; gills fairly well-spaced to widely-spaced; fruiting body not tough; spores not amyloid .... *Pleurotus (=Cheimonophyllum) candidissimus*
9. Not as above; spores amyloid 10

10. Gills pinkish-gray; cap usually with purplish or vinaceous tints; growing in groups on dead hardwoods (especially birch) .... *Panellus ringens*
10. Gills pallid to pale pinkish-gray; cap more or less same color and often with a gelatinous layer when young; found on dead conifers (especially larch) .... *Panellus mitis*

11. Gills usually veined (especially near stalk or base of cap) *or* repeatedly forked; gills orange to yellowish or olive-yellow, or if not then lower stalk brown and velvety (see **Paxillaceae,** p. 476)
11. Not with above features 12

12. Gills orange to yellow-orange, olive-yellow, or yellowish 13
12. Gills some other color (white, gray, brownish, violet, etc., but may age or discolor yellowish) 15

Pleurotus ostreatus, gill detail. The white gills, white or lilac-tinged spores, off-center stalk, and fleshy texture typify *Pleurotus.* (Dan Harper)

13. Cap densely hairy or fuzzy; stalk absent or rudimentary ...... ***Phyllotopsis nidulans,*** p. 140
13. Not as above; cap not hairy, or if hairy then stalk present 14

14. Cap hairy and taste bitter-acrid *or* cap viscid when moist or old and flesh white 26
14. Cap not hairy; flesh not white .... (see ***Omphalotus,*** p. 146)

15. Cap fairly large to very large (10-50 cm broad when mature); either growing on hardwoods and cap coarsely hairy *or* found on or near conifers and stalk usually well-developed 16
15. Cap smaller, or if large then not as above 17

16. Cap coarsely hairy; growing on hardwoods .... ***Panus strigosus,*** p. 140
16. Stalk well-developed; growing on or near conifers .... (see ***Lentinus ponderosus,*** p. 143)

17. Cap with dense brown hairs and fibrils, funnel-shaped or with a deeply depressed center; gills decurrent, very crowded, pallid; stalk well-developed; common in the tropics and also along the Gulf of Mexico .... ***Panus (=Lentinellus) crinitis***
17. Not as above 18

18. Cap only 2-4 cm broad, viscid when moist, pallid to pale orange becoming pinkish to caramel-brown; gills close to fairly well-spaced (not crowded); spores amyloid . ***Panellus longinquus***
18. Not with above features; common 19

19. Gills pallid (white to creamy, yellowish, or gray) 20
19. Gills darker (tan to brown, reddish-brown, violet-tinted, etc.) 26

20. Gills narrow (shallow) and crowded; fruiting body small to medium-sized 21
20. Gills broad or fairly broad (deep), well-spaced to close, but not crowded; fruiting body medium-sized to large 23

21. Cap pure white when fresh, but may become creamy in age ..... ***Pleurotus porrigens,*** p. 135
21. Not as above 22

22. Cap dark grayish-brown to bluish-black; usually found in the wild ....
   .... ***Hohenbuehelia atrocaerulea*** (see *H. petaloides* group, p. 136)
22. Cap some shade of brown or paler; found in wild or often in gardens, flower pots, etc. ......
   .... ***Hohenbuehelia petaloides*** group & others, p. 136

23. Stalk absent or rudimentary, not well-developed; cap with ridges and/or spines; taste unpleasant .... ***Hohenbuehelia mastrucatus*** (see *H. petaloides* group, p. 136)
23. Not as above 24

24. Cap white to tan or pinkish-tinged, often breaking up into scales in age; stalk present; found on living hardwoods (elm, etc.), often high up in the tree; widespread but not yet reported from California; edible but rather tough .... ***Hypsizygus tessulatus (=Pleurotus ulmarius)***
24. Not as above 25

25. Stalk well-developed and often long (4-22 cm); gills usually adnexed or notched; cap creamy or tinged pinkish, often with watery spots; usually on living hardwoods . ***Pleurotus elongatipes***
25. Stalk absent, or if well-developed then gills typically decurrent; cap white, gray, brown, greenish, etc.; very common, especially on hardwoods .... ***Pleurotus ostreatus*** & others, p. 134

26. Cap small (3 cm broad or less); gills ochre-buff to brownish to pale cinnamon or tawny-olive in old age; taste usually acrid or bitter .... ***Panellus stipticus***, p. 138
26. Not as above; usually larger 27
27. Cap densely hairy .... ***Panus rudis***, p. 139
27. Not as above; cap more or less smooth (or breaking up into scales) 28
28. Cap viscid when moist or in age, yellow-green to olive-green, olive-brown, or with violet tones; stalk absent or lateral; growing shelflike .... ***Panellus serotinus***, p. 137
28. Cap not viscid, violet to reddish-brown, tan, etc., but never greenish or yellow; stalk often well-developed, lateral to off-center or central .... ***Panus conchatus***, p. 138

Pleurotus ostreatus   (Oyster Mushroom)   Color Plates 27, 28

CAP 4-15 cm broad or more, oyster- or fan-shaped, convex becoming plane or sometimes funnel-shaped; surface smooth, slightly lubricous when moist but not viscid; color variable: white to gray, grayish-brown, tan, or dark brown (sometimes yellowish in old age); margin inrolled when young, often wavy or lobed. Flesh thick, white, firm but soft, tougher near the stalk; odor and taste mild. **GILLS** fairly close, broad, decurrent (if stalk is present), white or tinged gray but often discoloring yellowish in old age. **STALK** absent or if present usually short, stout, and off-center or lateral (but sometimes central); 0.5-4 cm long and thick; solid, firm, dry, usually hairy or downy at least at base. **VEIL** absent. **SPORE PRINT** white to pale lilac or lilac-gray; spores 7-9 × 3-4 microns, oblong to elliptical, smooth, not amyloid.

HABITAT: Occasionally solitary but usually in shelving masses or overlapping rows or columns on hardwood logs and stumps; sometimes also on standing trees, rarely on conifers; common throughout most of the northern hemisphere. Its preferred hosts include elm, cottonwood, alder, and sycamore, but in our area it favors oak and tanoak, producing large crops after the first fall rains and smaller crops thereafter through the spring. I have also seen stupendous fruitings (several hundred pounds!) growing in clusters in a treeless field where crushed coffee beans were dumped. It is easily cultivated on a wide variety of substrates, including compressed sawdust, shredded Time magazines, and presumably coffee grounds. If an "oyster log" is dragged home from the wild and kept moist, it will produce crops regularly.

EDIBILITY: Edible and delicious—breaded and fried it is superb and remarkably reminiscent of seafood. Be sure to check for small beetles between the gills (these can be removed by dousing the mushroom briefly in water), and of course, for maggots. The tough stem or basal stump of tissue should be removed. Large specimens can be pounded like abalone to make them tender. *P. ostreatus* and its Asian counterpart, *P. sajor-cajou,* are now cultivated commercially and sold fresh under the name "tree oysters."

COMMENTS: Pure, pale, and graceful, the oyster mushroom is easily distinguished by its white gills, tender flesh, smooth cap, and shelflike growth habit on wood. The cap color and position of the stem depend to some extent on the location of the fruiting body. When growing out of the side of a log, the stem is lateral or absent, since there is no need to elevate the cap. When growing from the top of a log, however, the stem can be central, leading to confusion with *Clitocybe*. The cap is generally darker in sunlight and correspondingly paler in dim surroundings, but distinct color forms also seem to occur, including a brown-capped form that grows on bush lupine along the ocean and a giant thick-fleshed form that is common on cottonwood in inland valleys. In fact, *P. ostreatus* has long been recognized as a "collective" species, i.e., a group of closely related but distinct forms. Fortunately, they all appear to be edible, so that their exact taxonomy needn't concern you (at least, it doesn't concern me!). Other species: *P. sapidus* is now regarded as a synonym for *P. ostreatus; P. columbinus* is a rare species with a bluish- or greenish-tinted cap, but is otherwise very similar; *P. cornucopiae* grows in dense, upright clusters on woody debris in

So you think oyster mushrooms or "tree oysters" *(Pleurotus ostreatus)* always grow on trees? **Left:** Clusters growing in a treeless field on decomposing coffee beans. **Right:** A cluster growing out of a kitchen chair. (The owner of the chair claims to be a sloppy eater who unknowingly helped incubate the developing mycelium by sitting in the chair. A fortuitous leak in his roof elicited this crop of oyster mushrooms. He now waters the chair regularly!)

the Rocky Mountains and probably elsewhere. It has a lined or ridged, nearly central stalk and depressed or funnel-shaped cap that is open (incised) on one side. It is edible when young but often develops a bitter or unpleasant taste in age. For a smaller, thinner, white species growing on conifers, see *P. porrigens.*

Pleurotus porrigens (Angel Wings)

CAP 4-8 (10) × 2-5 cm, fan-shaped to tongue- or petal-shaped; surface smooth, not viscid, pure white to milky-white, but sometimes creamy in old age; margin at first incurved, often lobed or wavy. Flesh very thin, pliant, white, odor and taste mild. **GILLS** crowded, thin, narrow, white or yellowish, decurrent if a stalk is present. **STALK** absent or present only as a narrowed, stubby white base. **VEIL** absent. **SPORE PRINT** white; spores 6-7 × 5-6 microns, nearly round, smooth, not amyloid.

HABITAT: In shelving groups or overlapping clusters on old rotting conifers, especially hemlock; widely distributed. It is very common in the fall in the Pacific Northwest and northern California, but I have yet to find it south of San Francisco.

EDIBILITY: Edible, but in my humble fungal opinion, bland and insubstantial. However, some people proclaim it superior to *P. ostreatus.*

Left: *Pleurotus porrigens,* a thin white cousin of the oyster mushroom that grows on dead conifers. **Right:** *Pleurotus dryinus* (see p. 136) resembles other oyster mushrooms, but boasts a veil, remnants of which can be seen clinging to the margin of the cap.

COMMENTS: Also known as *Pleurocybella porrigens* and *Pleurotellus porrigens*, this species can be distinguished from other types of *Pleurotus* by its thin, pliant, white fruiting body and narrow, crowded gills. It is to rotting conifers what *P. ostreatus* is to rotting hardwoods—i.e., common and cosmopolitan. The shining white fruiting bodies stand out vividly in the forest gloom, looking so exquisitely pure and unsullied that it is easy to see how they acquired the nickname "Angel Wings." Other species: *P. lignatilis (=Clitocybe lignatilis)* is a similar whitish species with narrow, crowded gills and a more prominent stem. It frequently has a farinaceous odor and prefers hardwoods rather than conifers.

Pleurotus dryinus   (Veiled Oyster Mushroom)

CAP 4-20 cm broad, broadly convex sometimes becoming plane or slightly depressed in age; surface dry, with soft grayish fibrils or scales, but sometimes whitish or in age yellowish; margin at first inrolled. Flesh very thick, white, firm; odor mild to pungent or fragrant. **GILLS** decurrent, fairly close, often veined or forking on the stalk; white, but sometimes discoloring yellowish in age. **STALK** 3-10 cm long, 1-3 cm thick, usually off-center but sometimes central; rather tough, often short, equal or tapered downward; solid, whitish. **VEIL** membranous, white to grayish, forming a slight ring on stalk or leaving remnants on cap margin or disappearing entirely. **SPORE PRINT** white; spores 9-12 (17) × 3.5-5 microns, elliptical, smooth, not amyloid.

HABITAT: Solitary or in small groups on hardwoods (usually living); widely distributed but not common. Alder is a favorite host; it is also reported on oak, and I've found it growing locally from the wound of a living madrone, in December.

EDIBILITY: Edible but rather tough.

COMMENTS: The thick firm flesh and soft hairs or scales on the cap are good fieldmarks; so is the veil when it is visible (see photograph at bottom of p. 135). It might be confused with *Lentinus lepideus*, which has serrated gills, or *Panus strigosus*, which is differently colored. *P. corticatus, Panus dryinus,* and *Armillaria dryina* are synonyms.

Hohenbuehelia petaloides group   (Shoehorn Oyster Mushroom)

CAP 2-7 × 3-7 (10) cm, spatula- to funnel-shaped or shoehorn-like when upright (i.e, split or open on one side), fan- or petal-shaped when shelflike; tapering to a stemlike base; surface smooth or with a whitish bloom when young and often downy toward the base, moist to somewhat rubbery-gelatinous but not viscid except when very wet; some shade of brown, tan, or grayish-brown; margin often lobed or wavy, at first incurved or inrolled. Flesh pliant, usually white (sometimes watery tan). **GILLS** narrow, thin, crowded, deeply decurrent, white or tinged gray, often becoming yellowish or creamy in age and often becoming crisped (wavy) in dry weather. **STALK** lateral or off-center, continuous with cap, often short (1-4 cm long), equal or tapered downward, up to 2.5 cm thick, white or grayish; fuzzy, downy, or minutely hairy. **VEIL** absent. **SPORE PRINT** white; spores 7-9 × 4-5 microns, elliptical, smooth, not amyloid. Gills with large, thick-walled cystidia.

Hohenbuehelia petaloides group. Note the crowded gills and shoehorn-shaped fruiting body. The white specks on the caps at left are an abnormality (probably a fungal parasite).

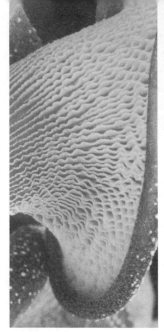

Hohenbuehelia petaloides group. **Left:** Close-up of gills. The edges often become wavy as the gills dry out, but are not serrated (toothed). **Right:** These young specimens remind me of penguins. They came up in a potting mix composed largely of wood chips.

HABITAT: Usually in groups or small clusters on rotting or buried wood, sawdust mulch, etc.; widely distributed. This species "complex" is common year-round in our area in nurseries, flower pots, landscaped areas where wood chip mulch has been used, etc. It occurs less commonly in the wild, usually on rotting conifers such as hemlock. I have seen it in Yosemite National Park in the spring and fall.

EDIBILITY: Edible, but not choice (according to most sources); I haven't tried it.

COMMENTS: Also known as *Pleurotus petaloides,* this species and its close relatives are best distinguished by their crowded gills, brown cap, white spores, and lateral stem or stemlike base. Terrestrial fruiting bodies are often reminiscent of upright, rolled-up leaves or shoehorns, while those that grow shelflike on wood are usually fan-shaped as in other oyster mushrooms. The gelatinized layer of tissue beneath the surface of the cap is seldom evident unless the specimens are waterlogged. Under the microscope, however, it is often discernible. *H. geogenia* is a very similar species with a hazel-brown to yellow-brown cap that differs microscopically. Like *H. petaloides,* it is apt to be looked for in *Clitocybe,* but the cap is open or split on one side and the stalk is usually off-center. Other species: *H. atrocaerulea* has a brown to bluish-black, mussel-shaped cap which is felty at least toward the base; it grows on wood, including yucca. *H. angustatus* has a pale (pinkish-buff) cap and round spores. *H. mastrucatus* has a grayish cap, unpleasant taste, and broad gills that are fairly well-spaced. All of these species were formerly included in *Pleurotus.*

Panellus serotinus (Late Oyster Mushroom)

CAP 2.5-10 (15) cm broad, kidney- or fan-shaped; surface viscid when moist or in age, color variable: olive-green to yellow-green, ochre, greenish-brown, or with violet tones; margin incurved, often lobed or wavy. Flesh thick, firm, white, with a gelatinous layer under the cuticle. **GILLS** adnate to decurrent, close, pale orange to ochraceous to pale yellow, often fading in age. **STALK** absent or if present, laterally attached, short, and stubby (0.5-2.5 cm long); yellow to brownish or colored like cap, but hairy or velvety. **VEIL** absent. **SPORE PRINT** yellowish; spores 4-6 × 1-2 microns, sausage-shaped, smooth, typically amyloid (at least in dried specimens).

HABITAT: Scattered or in shelving groups on dead hardwood logs and branches (especially wild cherry), sometimes also on conifers; widely distributed. Fairly common in the Pacific Northwest in the fall and winter, but I have yet to find it in our area. Like *Flammulina velutipes,* it is a cold-weather fungus, and its appearance is usually a sign that the mushroom season is almost over.

EDIBILITY: Edible but mediocre; it sometimes develops a bitter taste as it ages.

COMMENTS: The viscid, greenish to yellowish or violet-tinted cap and pale yellow to orange gills, plus the short, stubby stem and growth on wood make this an easy mushroom to recognize. According to Alexander Smith, the spores of some forms do not display the amyloid reaction until dried out or stored in a herbarium.

Panellus stipticus

CAP 0.5-3 cm broad, spatula-, kidney-, or fan-shaped, convex to plane or depressed near the stalk; surface dry, minutely hairy or scurfy, buff to ochre-buff, tan, brownish, or cinnamon-brown, sometimes concentrically zoned. Flesh thin, tough, white or pale yellowish; taste usually acrid or astringent. **GILLS** close, narrow, often forked, brownish to pale cinnamon or ochre-buff; adnate to decurrent, often luminescent. **STALK** 0.5-2 cm long, 3-8 mm thick, off-center to lateral, usually narrowed at base, often somewhat flattened; same color as cap or paler (to nearly whitish). **VEIL** absent. **SPORE PRINT** white; spores 3-5 × 1.5-3 microns, elliptical to oblong or sausage-shaped, smooth, amyloid.

HABITAT: Usually gregarious or in clusters or overlapping tiers on dead hardwoods; widely distributed, more common in eastern North America than in the West. It occurs in California but I have yet to find it in our area. It usually fruits in the fall, but the fruiting bodies do not rot quickly and consequently can be found practically year-round.

EDIBILITY: Inedible due to its small size, tough texture, and bitter taste.

COMMENTS: But for its luminescent gills, this listless little wood-rotter wouldn't attract enough attention to merit mention. Because of the brownish gills it can be mistaken for a *Crepidotus* or small *Paxillus,* but the spore print is white and the texture much tougher. The species epithet refers to its use as a styptic (blood-clotter). For other listless *Panellus* species, see the key to *Pleurotus* and Allies.

Panus conchatus   (Smooth Panus; Conch Panus)

CAP 4-17 cm broad, broadly convex becoming plane or broadly depressed in age; surface dry, smooth or minutely downy, often cracked into small scales in age; vinaceous-brown or violet-tinted when young and moist, fading to brownish, reddish-brown, or tan in age or as it dries; margin often wavy, at first inrolled. Flesh rather tough, firm, white; taste mild. **GILLS** decurrent, fairly close, narrow, often forking near stalk, tan to buff, or when moist often violet-tinted. **STALK** 2-5 cm long, 0.5-3 cm thick; off-center to lateral or sometimes central, usually tapered downward; solid, tough, colored more or less like cap; covered with fine hairs at least when young. **VEIL** absent. **SPORE PRINT** white; spores 5-7 × 2.5-3.5 microns, elliptical, smooth, not amyloid.

HABITAT: Solitary or in small groups or clusters on hardwood logs, stumps, and fallen branches; widely distributed but rather infrequent in our area, where it fruits in the winter.

EDIBILITY: Tough but apparently harmless.

COMMENTS: Also known as *P. torulosus,* this mushroom is almost entirely violet when young and moist, but fades in age to tan or reddish-tan. The color, plus the growth on wood, decurrent gills, and non-hairy cap are good fieldmarks. The stem is sometimes

Panus conchatus (=*P. torulosus*). These specimens are brownish, but were distinctly violet or vina-ceous when younger (just like *P.rudis*). Note smooth (not hairy!) cap and growth on wood.

central, leading to confusion with *Clitocybe,* and it is consequently keyed out under that genus. The cap is not hairy as in *P. rudis,* nor do the gills have serrated edges as in *Lentinus.*

Panus rudis   (Hairy Panus)

CAP 2.5-10 cm broad, fan-shaped or wedge-shaped to somewhat irregular in outline, convex becoming plane or depressed; surface dry, covered with dense, coarse, stiff, velvety hairs, reddish-brown to tan, but often violet when fresh and wet; margin incurved, often lobed. Flesh tough, thin, white; taste slightly bitter. **GILLS** decurrent, close, narrow, edges entire; white, creamy, or colored like cap. **STALK** a short, stout plug of tissue up to 2 cm long, off-center to lateral or sometimes central; tough, solid, hairy like the cap and more or less same color. **VEIL** absent. **SPORE PRINT** white or yellowish; spores 5-7 × 2-3 microns, elliptical, smooth, not amyloid.

HABITAT: Usually in groups on rotting hardwood stumps and logs, widely distributed. I have seen it several times in the fall on tanoak, but it is rather rare in our area.

EDIBILITY: Edible but very hairy. You'd do better to brush your teeth with it than eat it.

Panus rudis grows on hardwood stumps and logs. The cap is hairy, the stalk off-center to lateral. Young specimen at left has a violet cap, mature individuals at right are tan.

COMMENTS: The hairy cap, tough texture, white spores, and short lateral to off-center stem set this singular fungus apart. As in *P. conchatus*, fresh wet caps are a gorgeous deep violet, but soon fade to reddish- or pinkish-brown.

Panus strigosus   (Giant Panus)

CAP 10-40 cm broad or more, fan-shaped to broadly convex, plane, or slightly depressed; surface dry, with coarse hairs, white to buff or creamy, discoloring yellowish in old age or when dried. Flesh thick, rather tough, white or yellowish, taste mild. **GILLS** broad, white to buff or even tinged lilac or brownish, becoming yellowish in old age; usually decurrent, edges typically entire. **STALK** 2-15 cm long, 1-4 cm thick, usually off-center or lateral; solid, tough, white to buff or aging yellowish; equal or thicker below, coarsely hairy especially toward base. **SPORE PRINT** white; spores 10-13 × 3-5 microns, oblong, smooth, not amyloid.

HABITAT: Solitary or clustered (but rarely more than four together), usually in wounds of living hardwoods; widely distributed but rare, at least in the West. It favors maple and birch, but occurs in Arizona on walnut. Three gigantic local specimens were brought to me, but the collector didn't note the host. Each cap was more than two feet in diameter!

EDIBILITY: Reportedly edible, but too rare and tough to be of consequence.

COMMENTS: But for the hairy cap this humongus fungus might be mistaken for a giant oyster mushroom *(Pleurotus ostreatus). Lentinus ponderosus* rivals it in size but grows on conifers and is not as hairy. The species epithet means "strigose," which means hairy.

Phyllotopsis nidulans

CAP 2-8 cm broad, more or less fan-shaped to scallop-shaped in outline, broadly convex to plane; surface dry, often covered at first with a white chamois-like, cottony pubescence, otherwise pale orange to orange-buff, yellow-orange, or fading to buff, and densely hairy or fuzzy; margin at first inrolled. Flesh colored like cap or paler; odor typically strong and disagreeable (like sewer gas or rotten eggs), but sometimes mild. **GILLS** close, narrow, orange-buff to orange-yellow or pale orange. **STALK** absent or rudimentary. **VEIL** absent. **SPORE PRINT** pale pinkish to apricot-pink to pinkish-brown; spores 5-8 × 2-4 microns, sausage-shaped, smooth, not amyloid.

HABITAT: In groups or shelving masses on rotting logs and stumps (of both hardwoods and conifers); widely distributed. In our area it is common on dead oaks in the fall and

Phyllotopsis nidulans. A common wood-inhabitor, easily told by its hairy or fuzzy cap, orangish to yellow-orange gills, and obnoxious odor. Note the absence of a stalk.

winter, and in the Sierra Nevada and Rocky Mountains I have seen it on aspen.

EDIBILITY: Unknown. The odor is so disgusting that only a zealot with the iron constitution of Charles McIlvaine would consider eating it.

COMMENTS: Formerly known as *Claudopus nidulans* and *Panellus nidulans,* this rather attractive pale orange shelving mushroom is easily recognized by its peach-fuzzlike cap and obnoxious odor (the latter feature, however, is lacking in some collections). *Paxillus panuoides* is somewhat similar, but has yellowish-buff spores and veined or forked gills. *Crepidotus* species have brown spores, *Panellus* species have white to yellowish spores, while *Claudopus* species have pinkish spores but do not have orange gills.

LENTINUS & LENTINELLUS

Small to medium-sized or very large fungi *usually growing on wood.* CAP often hairy or scaly. Flesh firm or tough. GILLS usually adnate to decurrent, *edges usually toothed, serrated, or ragged.* STALK absent to lateral, off-center, or central. VEIL absent or sometimes present and forming a slight annulus (ring) on stalk. VOLVA absent. SPORE PRINT *white to yellowish or buff.* Spores smooth or rough, amyloid *(Lentinellus)* or not amyloid *(Lentinus).*

THESE pale-spored, wood-inhabiting agarics can usually be recognized by their ragged or serrated gill edges (see photo on next page). Other white-spored wood-inhabitors do not normally have serrated gills unless they are very old or weathered. In *Lentinellus* the fruiting body is small to medium-sized and the spores are amyloid, while in *Lentinus* the fruiting body is sometimes gigantic and the spores are not amyloid. In both genera the stalk can be central, leading to confusion with *Tricholomopsis, Clitocybe, Armillariella,* and other wood-rotters, but more often than not it is off-center to lateral or even absent.

Both genera are small and neither is particularly common, at least in our area. Some of the fleshier *Lentinus* species are edible, but by and large this group is not one to tempt the "toadstool-tester." A notable exception is *Lentinus (=Tricholomopsis) edodes,* the renowned shiitake or "black mushroom" of Oriental cuisine. For centuries it has been grown on log "teepees" in Japan, and more recently on sawdust mixtures. It is now being grown commercially in North America and is sold fresh or dried in many markets and specialty shops. Shiitake "logs" (cultivation kits) are even available for those who want to grow their own (see photo on p. 31). In North America the shiitake has not yet been found in the wild, but it may very well escape cultivation and establish itself on native oaks (or other hardwoods). For this reason it is keyed out below, along with several other species.

Key to Lentinus & Lentinellus

1. Stalk typically well-developed and central to off-center or sometimes lateral 2
1. Stalk typically absent or present only as a stubby lateral point of attachment to wood, not well-developed .... *Lentinellus ursinus* & others, p. 144

2. Veil present when young, often leaving remnants on stalk and/or cap in age 3
2. Veil absent (check young specimens if possible) 6

3. Found on hardwoods along Gulf of Mexico and in tropics; fruiting body *Collybia*-like but tough; gills usually staining reddish-brown when bruised .... *Lentinus detonsus*
3. Not as above 4

4. Usually growing on conifers (including fence posts and railroad ties); cap whitish to buff or yellow, often with darker scales .... *Lentinus lepideus,* p. 142
4. Not as above; growing on hardwoods or cultivated commercially 5

5. Cap with a dense coating of dark brown to black hairs or small scales which become sparser in age, revealing the whitish to buff background; veil leaving a slight ring on stalk or remaining intact (covering the gills and never breaking) .... *Lentinus tigrinus*
5. Cap brown to dark brown, often with whitish veil remnants (especially near margin); native to Asia, cultivated in U.S. but not yet naturalized (see photo on p. 31) ...... *Lentinus edodes*

Gill detail in *Lentinus ponderosus.* The serrated edges are characteristic of *Lentinus* and *Lentinellus.*

6. Cap more than 8 cm broad when mature; stalk over 1 cm thick; fruiting body tough or hard; found on or near conifers (but wood sometimes buried!) ..... ***Lentinus ponderosus,*** p. 143
6. Not as above; smaller 7
7. Growing in clusters on hardwoods, stalks often fused; caps some shade of brown, often irregular or misshapen and deeply depressed in age; not uncommon in eastern North America, much rarer in the West .... ***Lentinellus cochleatus***
7. Not as above 8
8. Stalk usually grooved and stuffed (inside) with soft whitish tissue; cap and stalk reddish-brown to pinkish-brown, smooth; taste usually acrid; found on ground or woody debris; spores amyloid .... ***Lentinellus omphalodes***
8. Not with above features 9
9. Cap 3-8 cm broad, pinkish-tan to tan; found on conifers (e.g., Sitka spruce) along the Pacific Coast .... ***Lentinus kauffmanii***
9. Cap up to 4 cm broad, brown to orange-brown or cinnamon; found on hardwoods; widely distributed .... ***Lentinus sulcatus***

Lentinus lepideus   (Train-Wrecker)

CAP 5-15 (20) cm broad, convex to plane; surface dry or slightly viscid, whitish to buff or pale yellow, but usually with darker (brownish) scales; margin sometimes beaded with droplets when young. Flesh thick, tough, white, but often aging or bruising yellow; not decaying readily, odor usually distinctive (pungent or fragrant). **GILLS** usually decurrent but sometimes notched or adnate, whitish to buff or in one form yellow, often bruising brownish and/ or yellowish in age; edges entire when young but often serrated in age. **STALK** 3-5 cm long, 1-3 cm thick, central to somewhat off-center, tapered at base; solid, tough and hard, colored more or less like cap, usually with brownish to reddish-brown scales or fibrils below ring. **VEIL** membranous, forming a pallid, superior to apical ring on stalk which may be slight or disappear in age. **SPORE PRINT** whitish; spores 9-12 × 4-5 microns, almost cylindrical, smooth, not amyloid.

HABITAT: Solitary, scattered, or in small groups on conifer logs, stumps, fence posts, and other lumber, sometimes also on oak; widely distributed and fairly common in cool weather in the coniferous forests of the West, but rare in our area and less frequent than *L. ponderosus* in the Sierra Nevada. It used to be common on railroad ties, resulting in derailments and the common name, "train-wrecker." It causes a brown rot in its host.

EDIBILITY: Edible and quite good, but the tough flesh requires thorough cooking. Use only young caps—older specimens may have an unpleasant taste.

COMMENTS: The scaly cap, serrated gills (at least in age), membranous veil, white spores, and growth on wood (sometimes buried!) set apart this species. Like *L. ponderosus,* it decays slowly, and old or weathered specimens can be difficult to recognize. Several variants occur, including a yellow one.

142

Lentinus ponderosus. **Left:** Mature specimens growing from a log buried by a landslide. Exposed caps are apt to be scalier than sheltered ones. (Bob Winter) **Right:** Young specimens. Note decurrent gills. See p. 142 for close-up of the serrated gills, and p. 43 for a picture of a clump growing in a lake!

Lentinus ponderosus (Ponderous Lentinus)

CAP 10-50 cm broad or more, convex to plane or somewhat depressed; surface dry or slightly tacky, at first smooth but in most cases soon breaking up into large scales, revealing the white flesh beneath; color variable depending on age and exposure: white to tan, yellowish, brownish, or pinkish-brown, usually discoloring yellowish or orangish in age. Flesh thick, tough, not readily decaying, white (but may age or bruise yellowish); odor often fragrant. **GILLS** typically decurrent, fairly close, white to yellowish, but often developing orangish to rusty-brown stains in age; edges serrated or torn, at least at maturity. **STALK** 5-20 cm long, 2-8 cm thick, central or off-center, usually with a narrowed, rooting base; solid, hard, tough, whitish aging yellowish to brown or rusty-orange, often with brownish patches or scales. **VEIL** absent. **SPORE PRINT** white; spores 8-12 × 3-5.5 microns, elliptical, smooth, not amyloid.

HABITAT: Solitary or in groups or clusters on or near dead conifers (but often appearing terrestrial) in the late spring, summer, and early fall; known only from western North America. Although not common in most areas, it is very conspicuous because of its large size. I have never seen it at low elevations, but it is often abundant in the Sierra Nevada in the summer on lodgepole pine. I have also collected it on ponderosa pine in the Southwest. It produces a brown rot in its host.

EDIBILITY: Edible and choice, but thorough cooking or parboiling is required because of its toughness. Biologist Bob Winter of Fresno says that it is avidly sought by Japanese-Americans as a shiitake- and matsutake-substitute, perhaps because of its chewy texture.

COMMENTS: This large, tough mushroom rivals *Catathelasma imperialis* and *Clitocybe gigantea* for the title of "Most Humongus Gilled Fungus Among Us." The former, however, has a veil and the latter has a fragile cap, and both are terrestrial, whereas *L. ponderosus* lacks a veil and grows on or near wood. *Panus strigosus* can also be very large, but grows on hardwoods and has a hairy cap and non-serrated gills. The fragrant odor of *L. ponderosus* is sometimes reminiscent of the matsutake *(Armillaria ponderosa),* but that species is terrestrial and has a prominent veil. Smaller specimens can be confused with *L. lepideus,* which is rather similar in overall aspect but also has a veil. The cap color and degree of scaliness vary considerably from specimen to specimen (depending on age, temperature, and exposure to direct sunlight), but the size, toughness, and decurrent gills with serrated edges are distinctive.

Lentinellus ursinus. Note hairy cap, ragged or serrated gill edges, and near absence of stalk.

Lentinellus ursinus

CAP 3-10 × 2-5 cm, kidney- to fan-shaped in outline, broadly convex becoming plane; surface dry, dark brown to brown, yellow-brown, or reddish-brown, with sparse to dense, brown to dark brown pubescence (fine hairs), at least toward the stalk; margin usually smooth, often paler and lobed, at first incurved. Flesh thin; taste slowly acrid or bitter. **GILLS** decurrent (if stalk present), close, broad, dingy white to pinkish-brown with ragged or coarsely toothed edges. **STALK** absent or rudimentary. **VEIL** absent. **SPORE PRINT** white; spores 2.5-5 × 2-3.5 microns, nearly round, with minute amyloid spines.

HABITAT: On rotting logs and stumps, usually in groups or shelving clusters; widely distributed. It grows on both hardwoods and conifers but is not common in our area. I have found it in the late fall and winter on Douglas-fir and live oak.

EDIBILITY: Inedible due to the bitter or acrid taste.

COMMENTS: This flaccid, fleshless, featureless fungus could carelessly be mistaken for for a decrepit oyster mushroom *(Pleurotus ostreatus)* were it not for the ragged gills and hairy cap. *L. flabelliformis* is a similar but slightly smaller species with whitish pubesence on the cap. Another widespread species, *L. vulpinus,* also has whitish pubescence (at least at the base of the cap), but is often ribbed or reticulate and sometimes has a stalk; it favors hardwoods. Still another species, *L. montanus,* can be told by its well-spaced gills and tendency to fruit on dead conifers, usually at higher elevations after the snow melts.

TRICHOLOMOPSIS

Medium-sized mushrooms *usually found on or near rotting wood.* CAP smooth or scaly, not viscid. *Flesh often yellow.* GILLS *attached, usually yellow.* STALK *typically central, fleshy.* VEIL absent or evanescent. VOLVA absent. SPORE PRINT *white.* Spores smooth, not amyloid. Cystidia abundant on the edges of the gills.

THIS is a small genus of wood-inhabiting agarics formerly distributed among *Tricholoma, Clitocybe,* and *Collybia.* In most species the fruiting body is largely yellow, but in *T. platyphylla* it is white to grayish-brown. *Tricholomopsis* may occasionally appear

144

terrestrial, but the yellow gills and yellow flesh, absence of a veil, and central, fleshy stalk are distinctive. None of its members are particularly good eating. Of the species keyed below, only *T. rutilans* is common in our area.

Key to Tricholomopsis

1. Lower portion of stalk dark rusty-brown to blackish-brown and velvety from a coating of minute hairs; usually growing in tufts or clusters .... (see *Flammulina velutipes*, p. 220)
1. Not as above 2
2. Cap and stalk yellow, or yellow beneath a layer of colored fibrils or scales; flesh and gills pale yellow to yellow 3
2. Not as above .... *T. platyphylla* & others, p. 146
3. Cap grayish to black at the center or with small grayish-brown to olive-brown to blackish scales .... *T. decora* (see *T. rutilans*, below)
3. Not as above 4
4. Cap and/or stalk with reddish to purple-red scales or fibrils . *Tricholomopsis rutilans*, below
4. Cap basically yellow, without differently colored scales or fibrils, at least when young (but may have brownish fibrils or streaks in age) *T. sulfureoides* & others (see *T. rutilans*, below)

Tricholomopsis rutilans   (Plums and Custard)

CAP 3-12 cm broad, convex becoming plane; surface dry, yellow, but covered with dark red to purple-red scales or fibrils which become sparser in age or toward the margin. Flesh thick, firm, pale yellow, odor mild. **GILLS** adnate or notched, close, yellow to pale yellow. **STALK** 5-10 (18) cm long, 1-2.5 cm thick, equal or slightly thicker below, dry, yellow with reddish or reddish-purple scales like those on the cap (but usually sparser and sometimes entirely yellow in old age. **VEIL** absent. **SPORE PRINT** white; spores 5-7 × 3-5 microns, elliptical, smooth. Cystidia on gill edges numerous, club-shaped.

HABITAT: Solitary, tufted, or in small groups on or near rotting conifers, wood chips, and humus rich in lignin; widely distributed. Fairly common in our area in cool weather (late fall, winter), but rarely fruiting in large numbers. I find it most often with redwood and pine.

EDIBILITY: Edible, but according to *Chroogomphus*-connoisseur Ciro Milazzo— who was born and raised in Brooklyn—"it tastes like rotting wood." Since I've never tasted rotting wood, I cannot attest to the validity of this statement. I wasn't born and raised in Brooklyn either.

Tricholomopsis rutilans is a beautiful wood-loving agaric with dark red scales on the cap and stalk. The scales are not always as dense or prominent as those on the young specimens pictured here. Also, the cap tends to broaden with age. *T. decora* (not illustrated) has olive-brown to blackish scales that are much sparser than those of *T. rutilans*.

COMMENTS: A real beauty when fresh, this mushroom is easily recognized by the dark red or purple-red fibrillose scales on a yellow background (hence its British name, "plums and custard"). This color combination is practically unique among fleshy-stemmed, white-spored agarics. If you find what looks to be a small *T. rutilans* with an entirely yellow stem, you probably have *T. flammula,* a questionably distinct species. *T. decora* is also quite similar, but has gray to brownish or black cap center and/or scales. I find it occasionally on rotting redwood, but it is more common farther north. Finally, there are several entirely yellow species also found on rotting conifers, including: *T. flavissima,* with a fibrillose, fringed cap margin; and *T. sulfureoides,* partial to hemlock, with an evanescent veil, and a cap that develops small brownish scales or streaks in age. *T. rutilans* was originally placed in *Tricholoma, T. decora* in *Clitocybe,* and *T. sulfureoides* in *Pleurotus.*

Tricholomopsis platyphylla   (Broad-Gill)

CAP 4-12 cm broad, convex to plane or centrally depressed; surface smooth, not viscid, often streaked; dark brown to grayish-brown, or sometimes pallid with a darker center and/or fibrils. Flesh pallid, thin. GILLS adnate or more often notched, well-spaced, very broad (deep), often splitting or with eroded edges in age, white or grayish. STALK 6-12 cm long, 1-3 cm thick, equal or thicker below, white or flushed cap color, hollow in age with a tough outer rind; base usually with white mycelial cords attached. VEIL absent. SPORE PRINT white; spores 7-9 × 4-7 microns, elliptical, smooth. Cystidia abundant on gill edges.

HABITAT: Solitary or in small groups on or near rotting logs and stumps, especially of hardwoods, widely distributed. In eastern North America it is common in the spring and early summer when few other mushrooms are out and about, and it is also said to be quite common in Arizona. In our area, however, it is rather rare and fruits in the fall and winter.

EDIBILITY: Not recommended. Some people are adversely affected by it and the flavor is poor. Also, it is not particularly easy to identify.

COMMENTS: This mushroom has few obvious relatives and has consequently been placed in several different genera, including *Collybia* and *Oudemansiella.* It is somewhat reminiscent of *Pluteus cervinus,* but does not have pinkish spores. If not clearly growing on wood it can be mistaken for a *Tricholoma* or robust *Collybia,* but the very broad, frequently eroded gills and white mycelial cords (rhizomorphs) are distinctive. The latter may only be evident if the mushroom is dug out carefully and completely, and even then are sometimes absent. *T. fallax* is a closely related species with yellowish-tinted gills and stalk, found on conifers in the Rocky Mountains.

OMPHALOTUS

Golden-yellow to olive-yellow to bright orange mushrooms growing from hardwood trees, stumps, and roots; often clustered. CAP smooth. GILLS well-developed, with acute edges, *typically decurrent, often luminescent when fresh.* STALK central or off-center, fleshy. VEIL and VOLVA absent. SPORE PRINT *white or tinged yellow.* Spores smooth, not amyloid.

POPULARLY known as "jack-o-lantern mushrooms," these are brightly colored agarics with a fleshy stem and decurrent gills that often glow in the dark. Their luminescence is best seen by sitting alone in a dark closet with the mushroom while eating a grilled cheese sandwhich. Unless you are a voracious eater, this method helps combat boredom while allowing your eyes to adjust to the darkness. After a few minutes an eerie silvery-green glow will become visible, growing gradually brighter with each bite (of the cheese sandwhich) until each gill is clearly outlined. Fresh, actively-sporulating specimens glow the brightest, but even they will not always cooperate.

Jack-o-lantern mushrooms are strictly wood-inhabitors, but frequently appear terrestrial because they like to grow on roots or old, buried stumps. Two to four species occur in North America, but only one on the west coast. They contain muscarine and are poisonous.

Key to Omphalotus

1. Fruiting body pumpkin-colored (bright yellow-orange to orange); common in eastern North America, Mexico, possibly the Southwest .... *O. olearius* (see *O. olivascens*, below)
1. Fruiting body golden-yellow to yellow-orange, but usually toned with olive (sometimes other colors also present); restricted to the west coast .... *O. olivascens*, below

Omphalotus olivascens   Color Plates 40, 41
(Jack-O-Lantern Mushroom; Western Jack-O-Lantern Mushroom)

CAP 4-16 (25) cm broad, broadly convex becoming plane or depressed; surface smooth, not viscid, color variable: bright golden-yellow to orange with olive tones often present also, varying to dull orange, brownish-orange, olive, or slightly reddish. Flesh rather thin, pliant, colored more or less like cap; odor mild. **GILLS** olive to bright yellow-orange (often yellow with olive tints), decurrent, close, usually luminescent when fresh. **STALK** 4-20 cm long, 1-4 (8) cm thick, central to off-center, equal or tapered downward, solid, dry, more or less colored like cap or gills, or dingier olive. **SPORE PRINT** white to yellowish; spores 6-8 × 5.5-7 microns, elliptical to nearly round, smooth.

HABITAT: In tufts or clusters or occasionally solitary on or around hardwood trunks, stumps, and buried wood; known only from the west coast, but replaced elsewhere by *O. olearius* (see comments). Common in our area from fall through early spring, especially on oak, manzanita, madrone, and chinquapin.

EDIBILITY: Poisonous! Profuse sweating and gastrointestinal distress are typical symptoms; muscarine is one of the toxins (see p. 894). It is sometimes eaten under the mistaken impression that it is a chanterelle.

COMMENTS: The bright yellow-orange to olive color, decurrent gills, pale spores, and tendency to grow in clusters distinguish this handsome mushroom. The chanterelle *(Cantharellus cibarius)* is somewhat similar but has thick, shallow, blunt, foldlike gills and

Omphalotus olivascens often—but not always—grows in clusters. The gills are well-developed, decurrent, and have thin edges. Entire fruiting body is golden-yellow to orange or olive (see color plates), *including the flesh.*

white flesh; the false chanterelle *(Hygrophoropsis aurantiacus)* is smaller and has oranger, repeatedly forked gills, while *Gymnopilus* species have dark orange to rusty-brown spores. All lack the olive tones characteristic of mature *O. olivascens.*

The common jack-o-lantern mushroom of eastern North America is essentially the same as *O. olivascens* except that it is pumpkin colored (bright orange to yellow-orange, without any olive tones). Formerly called *Clitocybe illudens,* it is now called *O. olearius* (or *O. illudens* by those who consider it distinct from the *O. olearius* of Europe). According to one report, its luminescence is sometimes bright enough to read a newspaper by. And then there's the tale of the shipwrecked sailor on an uninhabited island, who wrote a last message by the light of a jack-o-lantern mushroom, using the ink from a shaggy mane and the stalk of an *Agaricus* as a pen. Unfortunately, he starved to death because he was afraid to eat any of the mushrooms he found!

CLITOCYBE & Allies

Fairly small to large mushrooms found mostly on the ground, sometimes on rotten wood. CAP convex to plane or often depressed to funnel-shaped at maturity, rarely viscid. GILLS *usually adnate to decurrent* and usually white to gray or buff. STALK central, *usually fleshy,* but often slender. VEIL and VOLVA *absent.* SPORE PRINT *white to buff, yellowish, or dull pinkish.* Spores smooth or roughened but not ridged or angular; usually not amyloid.

THIS is a large and complex group of soft, fleshy, pale-spored mushrooms with no veil and a central, usually fleshy stem. The spore color is typically white, buff, or yellowish, but in some species—at one time honored with their own genus, *Lepista*—it is dull or pale pinkish. How, then, can you separate *Clitocybe* from other pale-spored mushrooms? Mostly by a process of elimination: the gills are not soft and waxy as in the Hygrophoraceae, nor orange as in *Hygrophoropsis* and *Omphalotus,* nor thick, blunt, shallow, and foldlike as in *Cantharellus;* the flesh is not granular and brittle as in *Russula,* there is no latex as in *Lactarius,* and the white-spored species do not have the notched gills characteristic of *Tricholoma*—though this is a somewhat capricious character, since the attachment depends to some extent on the age and shape of the cap; the "Lepistas" are easily confused with the Entolomataceae (particularly the genus *Entoloma*), but the latter have deeper, more vividly colored spores which are angular or longitudinally ridged under the microscope; the wood-inhabiting Clitocybes generally have a central stalk, thus eliminating *Pleurotus, Panus,* and other shelflike types; the smaller Clitocybes with slender stems are apt to be mistaken for *Omphalina,* in which the fruiting body is *very* small, and *Collybia,* which has a cartilaginous stem of a different texture from the cap, and adnexed to adnate but not decurrent gills; finally, there are a number of small but common genera (e.g., *Laccaria, Leucopaxillus, Lyophyllum*) that are best distinguished by learning the individual species. Whew!!

Except for the blewit *(C. nuda)* and its close relatives, *Clitocybe* is a lackluster group whose aesthetic and gustatory value is practically nil. Confirmed *Clitocybe* experts will be the first to admit that the anonymous throngs of white to grayish Clitocybes that litter our wintertime woods are exceedingly difficult to differentiate. Clitocybes are most prevalent in coniferous forests, but also occur under hardwoods and in grass or manure. Like the Tricholomas, they are largely cold weather fungi, most abundant in our area from December through February.

Though the blewit is a safe and popular edible mushroom, several Clitocybes are poisonous, including the small grass-inhabiting species, *C. dealbata.* The larger forms are slightly easier to identify but may be just as difficult to digest. Several have a disagreeable odor, notably *C. nebularis* and *C. robusta*. On the other hand, the anise-scented types (e.g., *C. deceptiva* and *C. odora*) are edible and quite good.

Over 200 species of *Clitocybe* occur in North America. About 50 are listed for California, but a diligent effort to catalog our Clitocybes would probably double that number. Only some of the more easily identified species are described here, including a few smallish types with amyloid spores that are now placed in the genera *Myxomphalia, Clitocybula,* and *Cantharellula*. If your *"Clitocybe"* does not key out satisfactorily, check the Hygrophoraceae (if the spores are white) or the Entolomataceae (if the spores are pinkish).

Key to Clitocybe & Allies

1. Odor distinctly licorice- or aniselike 2
1. Not as above (but odor may be sweet, e.g., like root beer) 3

2. Fruiting body entirely or partially blue-green to greenish or grayish-green .. *C. odora,* p. 161
2. Fruiting body lacking blue or green tints .... *C. deceptiva* & others, p. 162

3. Spore print pink or pinkish; fruiting body often (but not always) purple 4
3. Spore print white to yellowish or buff, or tinged lilac or brownish; fruiting body not commonly purple, and if purple then growing on wood 13

4. Stalk 3-7 mm thick; growing in grass, manure, straw, etc. .... *C. tarda,* p. 152
4. Stalk thicker, or growing in woods 5

5. Spore print bright pink; cap and stalk dingy cinnamon to vinaceous-brown; gills decurrent; growing in woods (especially pine), mainly in eastern North America ...... *C. martiorum*
5. Not as above; spore print dull pinkish to pinkish-buff or flesh-colored 6

6. Gills (and usually rest of fruiting body) *distinctly* bluish-purple to purple or pale purple (lilac) when fresh (but often fading in age); very common .... *C. nuda,* p. 153
6. Not as above; gills not distinctly purple 7

7. Stalk distinctly purple to lilac-tinged when fresh .... *C. saeva* (see *C. nuda,* p. 153)
7. Not as above 8

8. Fruiting body with subtle vinaceous or lilac tints when fresh ....
   .... *C. glaucocana* & *C. graveolens* (see *C. tarda,* p. 152)
8. Not as above 9

9. Fruiting body white when fresh, soon developing rusty to reddish stains ....
   .... (see *Collybia maculata,* p. 217)
9. Not as above 10

10. Cap pinkish to orangish to orange-brown or reddish-brown; spores angular in end view ....
    .... (see **Entolomataceae,** p. 238)
10. Not as above; cap white to watery brown, tan, hazel, buff, pinkish-buff, gray, etc. 11

11. Typically growing in clusters .... *C. subconnexa* group & others, p. 155
11. Typically growing scattered to gregarious or in rings 12

12. Cap watery brown to tan, hazel, buff, or whitish; growing in pastures, lawns, or sometimes in woods (especially oak) .... *C. brunneocephala* & others, p. 154
12. Cap white becoming pinkish-buff to dingy buff or tan in age; growing in woods; widespread .... *C. irina* (see *C. brunneocephala,* p. 154)

13. Growing on burnt ground or debris; cap small (less than 4 cm), blackish to grayish-brown to gray or olive-brown .... **Myxomphalia maura** & others, p. 165
13. Not growing in burned areas, or if so, then very differently colored 14

14. Gills forked repeatedly and usually orange *or* odor very fragrant (somewhat like root beer) .... (see **Paxillaceae,** p. 476)
14. Not as above 15

15. Growing on wood (sometimes very rotten or buried) or on coffee grounds 16
15. Growing on ground 24

16. Gills bright yellow to orange; fruiting body small . (see *Omphalina* & *Xeromphalina,* p. 221)
16. Not as above 17

17. Fruiting body entirely or partially vinaceous, purplish, or reddish-tinged when fresh and moist
    .... (see *Pleurotus* & Allies, p. 132)
17. Not as above 18

18. Cap open or incised on one side; gills very crowded, narrow . (see *Pleurotus* & Allies, p. 132)
18. Not as above 19

19. Gill edges serrated or eroded *and/or* fruiting body tough, hard, fairly large to very large (cap
    10-50 cm broad when mature; stalk 2-5 cm thick) ..... (see *Lentinus* & *Lentinellus,* p. 141)
19. Not as above 20

20. Cap ochre-brown to pinkish-brown or cinnamon when fresh, *not* typically growing in clumps or
    dense clusters .... *C. americana* & others (see *C. inversa,* p. 156)
20. Not as above; cap differently colored (brownish to gray or white), or growing in clumps . 21

21. Stalk white, usually tough and short, at least 1 cm thick; cap without hairs or scales, often pale;
    spore print frequently tinged lilac .... (see *Pleurotus ostreatus,* p. 134)
21. Not as above 22

22. Typically growing in clumps or dense clusters on wood; cap *not* white (or if watery whitish, then
    stalk usually rather long and less than 5 mm thick) 23
22. Typically growing solitary, scattered, or in small groups but not clumps; cap variously colored
    24

23. Stalk typically less than 5 mm thick; growing on rotting conifers ....
    .... (see *Clitocybula familia* & *C. abundans* under *Collybia acervata,* p. 215)
23. Stalk generally thicker than above; found in southern United States, usually on hardwoods .
    .... (see *Armillariella tabescens* under *A. mellea,* p. 196)

24. Spore print brownish; fruiting body small and white or grayish; cap often somewhat hairy, espe-
    cially toward margin; not common *Ripartites* (*R. tricholoma* is the most widespread species)
24. Not as above 25

25. Stalk thin (usually 1-3 mm thick); fruiting body small; cap white or yellow, pinkish, vinaceous,
    or tinged faintly gray 26
25. Not as above; stalk thicker or fruiting body differently colored 27

26. Growing on ground in groups or troops under pine or other conifers *or* growing on logs, sticks,
    berry canes, etc.; gills *very* widely spaced .... (see *Marasmius, Collybia,* & Allies, p. 201
26. Not as above .... (see *Omphalina* & *Xeromphalina,* p. 221)

27. Cap orange-buff to orange-brown to reddish-brown, cinnamon, purplish-brown, pinkish-
    brown, pinkish-tan, or tan when fresh 28
27. Cap white to buff, grayish, olive-gray, olive-brown, greenish, brown, or darker 34

28. Gills widely spaced .... (see *Camarophyllus pratensis,* p. 110)
28. Gills fairly close to crowded 29

29. Fruiting body medium-sized to large; stalk 1-4 cm thick; cap usually 7 cm or more broad when
    mature and at that stage usually depressed or funnel-shaped .... *C. maxima,* p. 157
29. Fruiting body smaller, thinner, or differently shaped; not as above 30

30. Usually growing in grassy areas or lawns; stalk typically 5 mm thick or less; cap not usually
    depressed .... (see *Calocybe carnea,* p. 176)
30. Not as above; growing in woods 31

31. Cap vinaceous-red to purplish-red, gills yellow-ochre; growing at high altitudes under conifers
    .... (see *Calocybe onychina* under *C. carnea,* p. 176)
31. Not as above 32

32. Gills white to pale buff; cap tan to pinkish-tan, flesh-colored, etc. 33
32. Gills often pale pinkish-cinnamon or colored like cap in age; cap orange to orange-brown,
    cinnamon, reddish-brown, reddish-tan, etc. .... *C. inversa* & others, p. 156

33. Stalk white, buff, or tinged only slightly with the cap color .... *C. gibba,* p. 157
33. Stalk colored like cap or darker .... *C. squamulosa* (see *C. gibba,* p. 157)

34. Spore print pale yellowish to buff; odor rancid; stalk at least 1 cm thick 35
34. Spore print white, or if yellowish to buff, then not as above 36

35. Cap white .... *C. robusta* (see *C. nebularis,* p. 159)
35. Cap grayish to buff or brownish .... *C. nebularis,* p. 159

36. Growing in grass, straw, or compost, often in groups or rings but *not* in massive clusters; cap usually less than 5 cm broad, white to grayish, buff, or tinged pinkish 37
36. Not as above; growing in woods or under trees, or differently colored 38
37. Cap broadly convex to plane or umbonate; gills usually notched but sometimes adnate or very slightly decurrent; spores amyloid; not often growing in rings .... (see *Melanoleuca*, p. 169)
37. Cap convex becoming plane or depressed; gills adnate to decurrent, grayish-white to buff or pinkish-buff; spores not amyloid; often growing in rings ...... *C. dealbata* & others, p. 163
38. Cap white or whitish 39
38. Not as above 41
39. Typically growing in large clusters along roads and paths in the Pacific Northwest and Rocky Mountains .... *C. dilatata*, p. 159
39. Not as above 40
40. Cap medium-sized to very large (8 cm broad or more) .. *C. candida* (see *C. gigantea*, p. 158)
40. Cap smaller, generally less than 8 cm broad .. *C. variabilis* & others (see *C. albirhiza*, p. 161)
41. Cap and/or gills greenish to bluish-green, more than 2 cm broad; growing in woods in eastern U.S. or under western mountain conifers .... *C. aeruginosa* & others (see *C. odora*, p. 161)
41. Not as above 42
42. Growing in tight clumps (occasionally solitary) from a fleshy mass of tissue which is often buried .... *C. sclerotoidea*, p. 164
42. Not growing in clumps from a fleshy mass of tissue 43
43. Base of stalk or surrounding humus with conspicuous white mycelial threads; common under mountain conifers, especially as or just after the snow melts .... *C. albirhiza*, p. 161
43. Not as above 44
44. Gills gray; cap with a hoary bloom when young; growing under mountain conifers, usually near melting snow .... (see *Lyophyllum montanum*, p. 175)
44. Not as above 45
45. Base of stalk (or flesh in base) pinkish to pale orange .. (see *Tricholoma saponaceum*, p. 184)
45. Not as above 46
46. Gills forked repeatedly; cap grayish to grayish-brown; gills sometimes reddish-stained; usually growing in moss in northern and eastern North America .... *Cantharellula umbonata* (see *Clitocybe cyathiformis*, p. 164)
46. Not as above 47
47. Stalk very thick (2-5 cm); gills whitish to buff to dingy tan or yellowish, but not gray 48
47. Not as above 50
48. Cap 10-40 cm broad, white becoming buff or dingy brownish in age, thin and easily broken at maturity, margin often obscurely ribbed .... *C. gigantea*, p. 158
48. Not as above 49
49. Odor strongly unpleasant; cap grayish to dingy tan *C. septentrionalis* (see *C. gigantea*, p. 158)
49. Not as above; cap more or less grayish-brown .... *C. crassa* (see *C. nebularis*, p. 159)
50. Stalk less than 4 mm thick; cap small (up to 5 cm broad but usually less than 3 cm), dark greenish to olive-brown, sooty-brown, ashy-gray, or blackish 51
50. Not as above; usually larger 53
51. Cap minutely scaly or scurfy at center .... *C. epichysium* (see *Myxomphalia maura*, p. 165)
51. Not as above 52
52. Cap and gills yellow-green to green, dark green, olive, etc. .... *C. atroviridis* & others (see *C. odora*, p. 161)
52. Not as above; not greenish . *Omphaliaster* & *Fayodia* spp. (see *Myxomphalia maura*, p. 165)
53. Stalk clothed with dark scurfy scales; gills well-spaced; cap dark brown to blackish; growing on rotten wood or in rich humus ..... *Clitocybula atrialba* (see *Clitocybe cyathiformis*, p. 164)
53. Not as above; stalk not clothed with dark scurfy scales 54
54. Cap typically incised or open on one side; gills crowded and narrow (shallow) .... (see *Hohenbuehelia petaloides* group, p. 136)
54. Not as above 55

55.   Typically growing in clusters; stalk usually 1 cm thick or more 56
55.   Not as above 57
56.   Cap tan to brown or purple-brown; spore print off-white to pale buff; known from Alaska ..
      .... *C. polygonarum* (see *C. subconnexa* group, p. 155)
56.   Cap gray, brown, tan, etc.; spore print white; widespread   . (see *Lyophyllum* & Allies, p. 173)
57.   Gills distinctly grayish to grayish-brown, at least at maturity 58
57.   Gills white to buff or yellowish 59
58.   Stalk 1 cm thick or more at apex; cap 6-15 cm broad or more *C. harperi* (see *C. nebularis*, p. 159)
58.   Stalk usually less than 1 cm thick; cap 2-5 (7) cm broad .... *C. cyathiformis*, p. 164
59.   Stalk 1-3 cm thick at apex; usually growing on or near rotting wood ....
      .... *C. avellaneialba* (see *C. clavipes*, p. 160)
59.   Not as above; stalk up to 1.2 cm thick at apex; growing on ground 60
60.   Cap brown to grayish-brown or olive-brown 61
60.   Cap paler .... *C. coniferophila* & others (see *C. albirhiza*, p. 161)
61.   Gills distinctly decurrent; cap usually depressed at maturity .... *C. clavipes* & others, p. 160
61.   Gills only very slightly decurrent if at all; cap broadly convex to plane or slightly umbonate
      .... (see *Melanoleuca*, p. 169)

Clitocybe tarda

CAP 1-6 (9) cm broad, convex with an incurved margin, then plane to broadly funnel-shaped or at times umbonate; surface smooth, not viscid, flesh-colored to brownish or grayish with a faint lilac or vinaceous tinge when moist, fading as it dries. Flesh thin, odor mild or slightly fragrant. **GILLS** adnate to slightly decurrent or at times notched, close, grayish to brownish-buff or pinkish-buff, often with a lilac tint when fresh. **STALK** 2-6 cm long, 3-8 mm thick, usually slender, equal or slightly thicker below, colored more or less like cap or paler, fibrillose. **SPORE PRINT** dingy pale pinkish; spores 6-8 × 3-5 microns, elliptical, finely roughened.

HABITAT: Scattered to gregarious or clustered or sometimes in rings in grass, dung, manure, straw heaps, old fields, compost piles, etc. Widespread and not uncommon in our area after heavy rains, late fall through spring.

EDIBILITY: Edible but thin-fleshed and not particularly easy to identify. I haven't tried it.

COMMENTS: Also known as *Lepista tarda* and *Tricholoma sordidum,* this is a smaller, slimmer version of the blewit. It isn't purple, but when fresh and moist it often has a slight lilac or vinaceous hue—especially the gills. It can easily be mistaken for a *Melanoleuca,* but the spore print is pale pinkish. The cap color is difficult to characterize but is generally some shade of buff, brown, or even gray. Two other pinkish-spored "Lepistas" with subtle purple tints are: *C. graveolens,* with a strong, disagreeable odor (like "moldy hay"), and *C. glaucocana.* Both of these are much larger and more robust than *C. tarda,* and in fact they closely resemble the blewit *(C. nuda)* but for their subtler color (older specimens can scarcely be distinguished). Both species are rare and fortunately, not poisonous.

Clitocybe tarda is a slim, trim version of the blewit, but has only a *slight* violet tinge (if any), and is thus easily confused with other grayish or brownish Clitocybes. The spore print, however, is pinkish.

Clitocybe (=Lepista) nuda, the blewit, has a characteristic shape that is hard to describe but easy to recognize. Note stocky build and inrolled margin when young. The monstrosity on the left is a stem which kept growing after the cap was cut off.

Clitocybe nuda   (Blewit)   Color Plate 32

CAP 4-14 (18) cm broad, convex with an inrolled margin when young, becoming broadly umbonate to plane, or with an uplifted, often wavy margin in age; surface smooth, lubricous when moist but not viscid, often somewhat lustrous when dry; purple, or purple shaded with brown or gray when fresh, soon fading to brownish, flesh-color, tan, etc., but the margin often retaining purple tones well into maturity. Flesh thick, rather soft, purplish to lilac-buff; odor faintly fragrant when fresh (like frozen orange juice), taste pleasant to slightly bitter. **GILLS** close, adnate to adnexed or notched, or sometimes decurrent; purple or pale purple to bluish-purple or grayish-purple when fresh, fading to buff, pinkish-buff, or brownish in age. **STALK** 2.5-7 (10) cm long, 1-2.5 (3) cm thick at apex, equal or more often with an enlarged base; dry, fibrillose, purple to pale purple or colored like the gills; base often covered with downy purple mycelium. **SPORE PRINT** dull pinkish to pinkish-buff; spores 5.5-8 × 3.5-5 microns, elliptical, roughened.

HABITAT: Scattered to gregarious, often in rings or arcs—in woods, brush, gardens, compost piles, i.e., wherever there is organic debris; widely distributed. It seems to favor cool weather but is common in our area throughout the mushroom season. A favorite abode is in brambles under live oak, often in the company of chanterelles; it is also common under pine and cypress, and I have found it on Ano Nuevo Island in beach grass and elephant seal dung. A single mycelium will produce several crops a year, so check your patches regularly. I know of one fairy ring sixty feet in diameter that produces about 200 blewits each time it fruits! Known as a "trash inhabitor" because of its fondness for virtually any type of decomposing organic matter, it can be grown on a wide range of substrates, including shredded newspapers and compost.

EDIBILITY: Edible and very popular—a favorite with beginners and gourmets alike, and one of the most plentiful edible wild mushrooms in our area. It has the dubious distinction of being one of the few purple foods that actually tastes good. It is even popular in fungophobic England and Scotland, where it is sometimes sold in markets.

COMMENTS: The ubiquitous blewit is the quintessential embodiment of spunk and persistence—cut one down and two will grow back! Decapitated stems will often continue to grow as if nothing had happened—a new cap will not form, and a grotesque (but edible) cancerous-looking pale purple growth will take its place. The blewit's trademarks are its beautiful purple to bluish-purple color with inrolled cap margin when young, stout stature, absence of a veil, faintly fruity fragrance, and dull pinkish spores. The cap has a characteristic lubricous feel when moist, but may look quite different—polished and silvery-violet— when dry. The amount of purple present varies considerably depending on the age and moisture content of the mushroom, and possibly the habitat or geographical area (some forms, such as the one commonly found under cypress, tend to be quite pale, with only a slight violet tinge). Old faded blewits are barely recognizable, but by that stage are usually bug-bitten anyway.

Other purple mushrooms include: *Inocybe lilacina,* with brown gills (when mature), brown spores, and a small umbonate cap; many *Cortinarius* species, with a cobwebby veil when young and rusty-brown spores; *Mycena pura,* small and slender with white spores; the *Laccaria amethystina* group, with white or lilac-tinged spores and a long, tough, fibrous stem. Of these, only the *Inocybe* and possibly the *Mycena* are poisonous. There are also several bluish *Entoloma* and *Leptonia* species, but they are not nearly as purple.

Synonyms for the blewit are almost as numerous as the blewit itself. They include: **Tricholoma nudum, Rhodopxillus nudus, Lepista nuda,** and incorrectly, *Tricholoma personatum.* "Blewit," incidentally, is a corruption of "blue hat"—though the blewit is more purple than blue. In Europe the blewit is often called the "wood blewit," to distinguish it from the "field blewit" or "blue-leg," *C. saeva (=Lepista saeva, Tricholoma personatum).* The latter is very similar to *C. nuda* in shape and stature and is equally delicious, but shows purple only on the stem—the cap and gills being grayish to pinkish-buff to watery tan (or the gills tinged vinaceous). Also, it tends to grow in pastures or grass rather than in the woods. It is infrequent in North America but has been reported from California. See also *C. tarda,* and the species discussed under it.

Clitocybe brunneocephala

CAP 4-13 cm broad, convex with an inrolled margin becoming broadly umbonate to plane or uplifted; surface moist or lubricous but not viscid, smooth, watery brown to tan, hazel, buff, or even whitish (usually darker when young). Flesh thick, pallid, odor mild or pleasant. **GILLS** usually notched but often adnate or slightly decurrent, close, buff to grayish-buff or pale brown, then dusted pinkish with spores. **STALK** 2-5 (10) cm long, 1-3 (4) cm thick (usually about 2); equal or enlarged below, often stout and relatively short, solid, dry, smooth, buff or colored like the cap (but usually paler). **SPORE PRINT** rosy-buff or dull pinkish; spores 5-8 × 3-4 microns, elliptical, minutely roughened.

HABITAT: Scattered to gregarious, often forming fairy rings, late fall through early spring, mainly in lawns and pastures, but also at the edges of woods or under trees (cypress, oak, etc.); known only from California. It was very abundant in our area during the warm and wet winter of 1977-78, but has been rather rare since.

EDIBILITY: Edible and quite good—I have tried it. Be sure not to confuse it with poisonous Entolomas, however, or *C. olesonii* (see comments below).

COMMENTS: Listed in the first edition as "*Lepista* sp. (unidentified)," this interesting relative of the blewit has recently been rechristened *C. brunneocephala* by clitocybiologist Howard Bigelow. The shape, stature, and characteristically lubricous feel of the cap when moist are very reminiscent of the blewit and blue-leg, but there is no purple anywhere on the fruiting body. The gills are usually notched, but as in most "Lepistas" their attachment varies considerably. I nearly always find it growing in grass, but a very similar species with a pungent or unpleasant odor, *C. olesonii,* is common under oak in southern California

Clitocybe brunneocephala is built like a blewit but is never purple. It is sometimes common locally in lawns and pastures as well as under oaks.

and the Sierra Nevada foothills. Both *C. brunneocephala* and *C. olesonii* (whose edibility I haven't determined) are larger and stouter than *C. tarda,* and do not grow in clusters like *C. subconnexa* and *C. densifolia.* They can be separated from most Entolomas by their duller spore color, non-angular spores, and in the case of *C. brunneocephala,* by the grassland milieu. Other pinkish-spored species: *C. praemagna,* a western prairie and sagebrush species, is quite similar but has a white cap when young that becomes tan to dull brownish in old age; *C. (=Lepista) irina* is a widespread woodland species that often has a fragrant blewit-like odor. Its cap is white to pinkish-buff, dingy buff, or pale tan, and its spores are slightly larger than those of *C. brunneocephala.*

Clitocybe subconnexa group

CAP 3-10 cm broad, convex with an incurved margin becoming plane or with uplifted margin; surface smooth, dry, satiny white at first, often discolored or spotted slightly in age. Flesh thick, whitish, rather brittle; odor usually mild or faintly pleasant. **GILLS** adnate to decurrent, crowded, narrow, pallid soon becoming buff, then dull pinkish as the spores ripen. **STALK** 3-10 cm long, 0.5-2 (3) cm thick, equal or thicker below, smooth, dry, dull grayish to buff with a whitish silky-fibrilose coating. **SPORE PRINT** pinkish-buff or flesh-colored; spores 4.5-6 × 3-4 microns, elliptical, minutely roughened.

HABITAT: In groups on ground, usually tufted or clustered, widely distributed. In our area both this species and *C. densifolia* (see comments) are fairly common from fall through early spring in woods or at their edges, brushy areas, and open places.

EDIBILITY: Edible, but not recommended. It is good when fresh according to some sources, but can develop an unpleasant astringent taste in age. There is also the possibility of confusing it with *C. dilatata* or a poisonous *Entoloma.*

COMMENTS: This common species can be identified by its whitish color, adnate to decurrent gills, dull pinkish spores, and tendency to grow in clusters—though solitary fruiting bodies can be found. A very similar and equally common species, *C. densifolia,* also grows in clusters, but has smaller spores, narrower gills, and a whitish cap that becomes dingy buff to grayish in age. Both species, like the blewit, seem to grow almost anywhere. They are larger and fleshier than *Clitopilus prunulus* and do not smell as nice, nor have they longitudinally ridged spores. They differ from *Clitocybe brunneocephala* in their more typically decurrent gills and clustered growth habit, and from *C. dilatata* in their pinkish spores. Other species: *C. fasciculata (=Tricholoma panaeolum var. caespitosus)* is

155

Clitocybe subconnexa group gives a dull pinkish spore print, has adnate to decurrent gills, and grows in clusters in a wide variety of habitats.

also very similar, but has a rancid-farinaceous odor and taste and tends to grow in clusters along roads; *C. subalpina* is a brown to dark brown clustered species known from the Pacific Northwest; *C. polygonarum* has a tan to brown or purple-brown cap, but has off-white to pale buff spores; it also grows in clusters and is common in Alaska.

Clitocybe inversa

CAP 2-10 cm broad, broadly convex or centrally depressed with an incurved margin, becoming broadly depressed or even funnel-shaped in age; surface dry, dull orange to pale orange-brown, orange-tan, tan, reddish-tan, ochre-buff, or cinnamon-brown; margin often paler. Flesh thin; odor mild or sharp. **GILLS** distinctly decurrent, close, buff to pale pinkish-cinnamon or colored like cap but paler. **STALK** 3-10 cm long, 4-8 mm thick, equal or thickened below, typically rather slender and often curved, colored like cap or paler, smooth or with whitish hairs at base. **SPORE PRINT** white to creamy-yellowish; spores 4-5 × 3.5-4 microns, nearly round, minutely prickly (or appearing smooth).

HABITAT: Scattered to gregarious or tufted on ground in woods; widespread but particularly common on the west coast, from Alaska to southern California. In our area it is fairly common in the late fall and winter in mixed woods and under oak or pine, sometimes in large fairy rings.

EDIBILITY: Not recommended

Clitocybe inversa is a cheerful orange-brown to cinnamon color. **Left:** Several typical examples. **Right:** Close-up of the decurrent gills.

COMMENTS: Also known as *Lepista inversa,* this species can be told by its cheerful color, depressed cap, decurrent gills, and white or pale spores. *C. gibba* is rather similar but differently colored, while *Rhodocybe nuciolens* has pinkish spores. The common form of *C. inversa* in our area has a slightly yellowish spore print and sharp, spicy or pepperlike odor. It may actually be *C. flaccida,* a very similar species. In eastern North America *C. gilva* is quite common; it differs in having a plane to only slightly depressed, yellowish to dull pinkish cap. Other similarly- colored Clitocybes include: *C. sinopica,* in humus or on burnt ground, with an orange-brown to rusty-cinnamon cap, yellowish spores, and farinaceous odor and taste; *C. ectypoides,* northern, growing on rotting conifers, with a minutely scaly ochre-brown cap; and *C. americana,* the most common and widespread wood-inhabiting *Clitocybe,* with a watery brown to pinkish-brown or cinnamon cap that fades to whitish as it dries, growing on hardwood stumps and logs.

Clitocybe gibba   (Funnel Cap)   **Color Plate 34**

CAP 3-8 cm broad, plane or with a central depression, soon becoming funnel-shaped; surface smooth, not viscid, tan to pinkish-tan, flesh-colored, or pinkish-cinnamon, but fading in age; margin often wavy. Flesh thin, whitish; odor mild or faintly sweet (like cyanide). **GILLS** deeply decurrent, crowded, white or pale buff. **STALK** 3-8 cm long, 0.4-1 cm thick, usually rather slender, equal or thicker below, smooth or faintly fibrillose; whitish to buff, the base often with white down. **SPORE PRINT** white; spores 5-8 × 3.5-5 microns, elliptical, smooth.

HABITAT: Solitary to scattered or in small groups on ground in woods, especially under oak but also with conifers; widely distributed. In coastal California it fruits in the winter and early spring but is not common; in New Mexico it is common in August along with *C. squamulosa* (see comments).

EDIBILITY: Edible and excellent, but not a good mushroom for beginners—too many species of unknown edibility resemble it. Many mushroom books rate it as mediocre, but it's probably a case of each author taking another's word for it.

COMMENTS: The pale pinkish-tan cap which is funnel-shaped at maturity plus the crowded, whitish, decurrent gills and pallid slender stem render this species distinct. *C. infundibuliformis* is a lengthy synonym. It is paler than *C. inversa,* especially in its gill color, and more slender than *C. maxima. C. squamulosa* is a very similar species in which the stalk is the same color as the cap or darker. It is a prominent feature of our western coniferous forests, particularly at higher elevations in the spring and fall. It is interesting to note that *C. gibba* releases hydrogen cyanide gas into the atmosphere. So does the common fairy ring mushroom, *Marasmius oreades,* as well as several other mushrooms and plants such as lupine and almond. The gas is not released in sufficient quantity to harm humans, however.

Clitocybe maxima   (Large Funnel Cap)

CAP (4) 7-20 (30) cm broad, at first broadly convex or plane with an inrolled margin, becoming depressed or broadly funnel-shaped; surface dry, usually smooth, pale pinkish-tan or flesh-colored varying to reddish-tan, sometimes spotted; margin sometimes paler and sometimes lobed or radially ribbed. Flesh thin, white, firm becoming flaccid; odor mild or rather unpleasant. **GILLS** soon deeply decurrent, close, white to pale buff or tinged pinkish-buff. **STALK** 3-10 (15) cm long (1) 1.5-3 (4) cm thick, equal or with an enlarged base; solid, fairly firm, whitish to buff or becoming colored like cap, fibrillose-striate, base usually with white down. **SPORE PRINT** white; spores 6-9 × 4-6 microns, elliptical, smooth.

Clitocybe maxima is essentially an overgrown version of *C. gibba* (see color plate). Along with its close relative *C. geotropa,* it is fairly common under conifers in the Rocky Mountains and Pacific Northwest. Note deeply decurrent gills and depressed or vase-shaped cap.

HABITAT: Solitary or more often in groups or rings on ground in woods or clearings; widely distributed and fairly common in the late summer and fall in the mountains of western North America. I've see large fruitings near Mr. Rainier in Washington and in the Sacramento Mountains of southern New Mexico, but I have not found it in California.

EDIBILITY: Said to be edible; I haven't tried it.

COMMENTS: This large, handsome *Clitocybe* is easily told by its thick stem, decurrent whitish gills, and pinkish-tan, broadly funnel-shaped cap. It is essentially an overgrown *C. gibba,* and some consider it a variety of that species. *C. geotropa* is a very similar edible species with more or less round spores. It tends to be longer-stemmed and more leather-colored when young, and its cap often has a broad central umbo, even when funnel-shaped. It is also widely distributed, but rare; I have seen it only in the southern Rockies.

Clitocybe gigantea   (Giant Clitocybe)

CAP (5) 8-35 (45) cm broad, convex becoming plane and then broadly funnel-shaped; surface smooth, not viscid; white becoming buff or pale tan, especially at the center; easily broken when mature; margin at first inrolled, often obscurely ribbed in age. Flesh white, in age becoming fragile; odor and taste pleasant to slightly disagreeable. **GILLS** close, pallid soon becoming pale buff or creamy-buff, often pale dingy tan in old age; crowded, decurrent, at least some forked. **STALK** 4-10 cm long, 2-4 (6) cm thick, often rather stout, equal or enlarged slightly at either end; white, or in age pallid with darker fibrils. **SPORE PRINT** white; spores 6-8 × 3-4.5 microns, elliptical, smooth, weakly amyloid.

HABITAT: Solitary to scattered or gregarious, often in arcs or huge rings, in open woods and grassy (but usually wooded) areas; widely distributed, but most common in the Pacific Northwest and Rocky Mountains. I have seen it and the very similar *C. candida* (see comments) in the mountains of northern New Mexico and southern Arizona. Neither species has been reported from California, but they may well occur.

EDIBILITY: Temptingly large and handsome, but poor in flavor and difficult to digest.

158

COMMENTS: This mammoth mushroom is often so wide in proportion to its thickness that the cap breaks or crumbles if handled carelessly. Its size and fragility, plus the buff-colored decurrent gills, pale cap, and white spore print are diagnostic. Young, compact specimens can be mistaken for *C. irina, C. praemagna,* or *C. robusta,* but have white spores. Older individuals resemble *Leucopaxillus,* but are not nearly so tough, and lack the white mycelial mat at the stem base. *C. candida* is a very similar but more common species with a slightly "smaller" cap (6-30 cm) that remains white until old age. Another large species, *C. septentrionalis,* has a dull grayish to dingy tan cap, strong unpleasant odor, and short thick stalk. I have seen it in groups or clumps under ponderosa pine in both Oregon and New Mexico. Along with *C. gigantea* and *C. candida* it is placed in the genus *Leucopaxillus* by some mycologists because of its weakly amyloid spores. There are also numerous large grayish Clitocybes with white, non-amyloid spores. The most common in our area is *C. harperi* (see comments under *C. nebularis*).

Clitocybe dilatata

CAP 2-15 cm broad, convex to plane or often somewhat misshapen; surface dry, smooth, gray when young, but soon white or chalky-white, sometimes with buff areas; margin at first incurved, often wavy in age. Flesh white to grayish, firm; odor mild, taste typically somewhat sour or disagreeable. GILLS adnate to decurrent, whitish to buff, close. STALK 5-12 cm long, 0.5-3 cm thick, equal or enlarged below, whitish, fibrillose, fibrous, often curved. SPORE PRINT white; spores 4.5-6 × 3-3.5 microns, elliptical, smooth. Basidia lacking siderophilous granules.

HABITAT: In densely-packed groups or clusters in sandy or gravelly soil along roads, trails, etc.; very common in the fall in the Pacific Northwest and Yukon. I have not seen it in California, but it may occur in the northern part of the state, and Chuck Barrows reports it from New Mexico.

EDIBILITY: Probably poisonous—it is thought to contain muscarine.

COMMENTS: The white spores and growth habit in clusters (often quite large) in disturbed ground is unusual for a *Clitocybe* and serves to distinguish this species from its brethren. It is most likely to be confused with the *Lyophyllum decastes* group (especially *L. connatum*), which also grows along roads, and with the *C. subconnexa* group, which has pinkish spores. *C. cerussata var. difformis* is a synonym.

Clitocybe nebularis   (Cloudy Clitocybe)

CAP 6-25 cm broad or more, convex becoming plane or depressed; surface dry, finely fibrillose or often with a hoary bloom, gray to grayish-brown to buff, often darker at the center, sometimes with watery spots or appearing streaked; margin at first incurved, often wavy or lobed in age. Flesh thick, white; odor rancid and disagreeable. GILLS adnate to decurrent, close, whitish becoming dingy yellowish or buff. STALK 6-15 cm long, 1.5-4 cm thick, base often enlarged and covered with white down; whitish, or with dingy brownish fibrils, firm but easily broken. SPORE PRINT pale buff to yellowish; spores 5.5-8.5 × 3.5-4.5 microns, elliptical, smooth.

HABITAT: Scattered to gregarious, often in large rings, under both hardwoods and conifers; widely distributed but especially common on the west coast. It is fond of cold weather and in our area seldom appears before December.

EDIBILITY: Edible but far from incredible. It is indigestible unless thoroughly cooked, its flavor is said to be poor, and its rank odor doesn't exactly make you want to rush home and throw it in the frying pan.

Clitocybe nebularis is a common large cold-weather *Clitocybe.* Note decurrent gills and grayish cap which is broadly convex at first but becomes depressed in age (the ones on left look darker than they are because of shadow). Cap margin is often frilled or lobed and the skunklike odor can't be missed.

COMMENTS: The outstanding feature of this large, drab, cold-weather *Clitocybe* is the unpleasant odor, which has been likened to that rancid flour, rotting cucumbers, skunk cabbage, mice cages, and beer barf. In size and stature it is reminiscent of *Leucopaxillus albissimus,* but is grayer, decays much more rapidly, and has pale yellowish spores. *C. robusta (=C. alba)* is a closely related species with the same stature, spore color, and rancid odor. However, it has a pure white, sometimes lustrous cap. I have found it several times in our area in mixed woods but it is more common in eastern North America. There are also a number of large, undistinguished grayish Clitocybes that have white spores and a more or less mild odor. These include: *C. harperi,* with darker (grayish) gills, fairly common in our area and throughout the West under conifers; and *C. crassa,* a springtime Rocky Mountain species with a thick, massive stalk.

Clitocybe clavipes   (Club-Foot)

CAP 2-10 cm broad, broadly convex or plane becoming depressed or funnel-shaped; surface more or less smooth, not viscid, brownish to grayish-brown or olive-brown, often paler toward the margin. Flesh pallid, odor mild or often somewhat fragrant or fruity. **GILLS** decurrent, fairly close, at first white, then yellowish-buff in age. **STALK** 2-7 cm long, 0.5-1.2 cm thick at apex, typically club-shaped (enlarged below), pallid with grayish or sordid olive-buff fibrils; base often spongy and covered with white down. **SPORE PRINT** white; spores 6-8.5 × 3-5 microns, elliptical, smooth.

HABITAT: Solitary, scattered, or gregarious under conifers and in mixed woods, late fall and winter; widely distributed. It is particularly abundant in pine plantations in eastern North America, rather infrequent in California (although several similar species occur). I have not seen it in our area.

EDIBILITY: Edible, but not recommended. Not only is there the danger of confusing it with poisonous species, but it has reportedly caused coprine-like poisoning when consumed with alcohol (see p. 896).

COMMENTS: I must often strain to find something interesting to say about Clitocybes, and this species—the type of the genus—is no exception. The drab grayish-brown cap, decurrent gills, white spores, and club-shaped stem are characteristic, but there are many

Clitocybe clavipes has a swollen stem base, decurrent gills, and grayish-brown cap, but there are dozens of similar, difficult-to-differentiate Clitocybes.

more or less similar species. One of these, *C. avellaneialba,* is fairly common in the Pacific Northwest and northern California in humus or near decayed logs. It is somewhat larger, with a darker cap, white gills and stem, and elongated spores (8-10 microns long). Another western species, *C. leopardina,* is also similar but has a viscid, watery-spotted cap and more or less equal stalk. It has been found by Greg Wright in southern California.

Clitocybe albirhiza   (Snowmelt Clitocybe)

CAP 2-10 cm broad, convex becoming plane to broadly umbonate, then depressed; surface smooth or with whitish down, or sometimes with riverlike lines; watery brown to pale buff to cinnamon-, pinkish-, or grayish-buff. Flesh thin, colored like cap; odor unpleasant or mild. **GILLS** adnate to decurrent, close, pale buff or colored more or less like cap. **STALK** 3-8 cm long, 0.4-2 cm thick, equal or tapered at either end, often hollow in age, sometimes fluted; colored more or less like cap, the base with a dense mass of white mycelial threads. **SPORE PRINT** white; spores 4.5-6 × 2.5-3.5 microns, elliptical, smooth, not amyloid.

HABITAT: Scattered to densely gregarious, often in small clusters or rings, on ground under conifers; common in the mountains of western North America, especially in the spring shortly after the snow melts (sometimes found with *Lyophyllum montanum*).

EDIBILITY: Unknown.

COMMENTS: This species is merely one of a metagrobolizing myriad of mundane white to buffy-brown or grayish Clitocybes that are exceedingly difficult to distinguish. In other words, they are better neglected than collected! It can be separated from its numerous look-alikes, however, by the dense mass of white mycelial threads (rhizomorphs) at the base of the stem and in the surrounding humus. Its growth in the spring is also distinctive. Other smallish Clitocybes include: *C. cerussata,* with whitish cap; *C. variabilis,* with pure white cap when fresh and spores 5.5-8 microns long, found under conifers; *C. candicans,* also with a white cap, but favoring hardwoods; and *C. coniferophila,* one of several conifer-loving species with a dingy buff cap. All of these lack the rhizomorphs characteristic of *C. albirhiza,* and in California are more likely to fruit in the fall or winter than the spring.

Clitocybe odora   (Blue-Green Anise Mushroom)

CAP 2.5-10 cm broad, convex to plane or somewhat depressed; surface smooth, not viscid, bluish-green to dingy greenish, gray, or grayish-brown with a slight blue-green tinge, to nearly whitish in dry weather. Flesh whitish or tinged cap color; odor strongly fragrant, anise-like, at least when fresh. **GILLS** adnate to decurrent, close, blue-green to greenish or dark green in one form, whitish to buff or pinkish-buff in another. **STALK** 2-6 (9) cm long, 0.5-1.5 (3) cm thick, equal or thicker at either end; smooth, white to buff or cap-colored. **SPORE PRINT** pinkish-cream or buff; spores 6-8 × 3-5 microns, elliptical, smooth.

Anise mushrooms. **Left:** *Clitocybe deceptiva* (formerly *C. suaveolens*), a young specimen. **Right:** Mature specimens of *Clitocybe deceptiva,* a whitish to buff species, and *Clitocybe odora,* a partially or completely greenish or bluish-green species (far right). Both smell like anise when fresh.

HABITAT: Scattered or in groups in woods; widely distributed. It favors oak in eastern North America, but in the Rocky Mountains and Pacific Northwest it is often abundant under conifers. In our area it is rare—I have seen it only twice, under oak in the winter.

EDIBILITY: Edible, but best used as a flavoring agent because of the strong taste.

COMMENTS: The blue-green to dull greenish color combined with the anise or licorice odor immediately identify this mushroom. The form with blue-green tints in the gills and stalk is *var.* **pacificus,** while the typical variety shows the color mainly in the cap. The other anise-scented Clitocybes (*C. deceptiva,* etc.) are never blue-green. *C. aeruginosa* of eastern North America has a greenish cap, but has white spores and does not smell like anise; a very similar but unidentified greenish species with a distinctive (but not anise-like) odor occurs in the Sierra Nevada under conifers. There are also several small (cap 1-2.5 cm broad, stalk 1-4 mm thick) greenish species that connect *Clitocybe* to *Omphalina.* These include: *C. atroviridis (=Omphalina chlorocyanea?),* with a green to blackish-green to olive-gray cap and paler gills, found in grass, moss, or on lichens; *Omphalina grossula,* greenish-yellow to olive-yellow, favoring lichens and moss; and *O. wynniae,* with a greenish-yellow to olive-brown cap and yellow to greenish-yellow gills, usually found on rotting conifers.

Clitocybe deceptiva   (Anise Mushroom)

CAP 1-6 cm broad, broadly convex becoming plane or depressed; surface smooth, not viscid, color variable: pale brown to grayish-brown fading to buff or watery whitish, sometimes with a darker marginal band; margin at first incurved. Flesh thin, pallid, odor distinctly fragrant and anise-like when fresh. **GILLS** adnate to slightly decurrent, close, white or cap-colored or slightly darker. **STALK** 2-7 cm long, 1.5-5 (7) mm thick, equal or thicker at base, smooth, more or less colored like cap, usually rather slender. **SPORE PRINT** pale pinkish-buff; spores 6.5-9 × 4-4.5 microns, elliptical, smooth.

HABITAT: Scattered to gregarious in damp places under conifers; western North America. In our area this species and its close relatives (see comments) are often common in the fall, winter, and early spring under redwood and pine, or occasionally oak.

EDIBILITY: Edible, but best used as a flavoring agent (in cakes, breads, cookies, etc.), because of its strong, anise-like (but not sweet) flavor. Be *sure* each and every specimen smells like anise—there are numerous look-alikes without the anise odor, some of which grow in the same habitats and are poisonous!

COMMENTS: Better known as *C. suaveolens,* the distinct anise odor of this species and its close relatives separate them from the dozens of other small, nondescript white to grayish or buff-colored Clitocybes (see *C. albirhiza*). Similar anise-smelling species include: *C. fragrans,* practically identical, but with a paler cap and white spores, and growing

162

under hardwoods as well as conifers; *C. oramophila,* with a dull pinkish or flesh-colored cap when moist and pale creamy spores; and *C. obsoleta,* with a strongly fragrant but not aniselike odor.

Clitocybe dealbata   (Sweat-Producing Clitocybe)

CAP 1-4 (5) cm broad, convex becoming plane or depressed; surface smooth, not viscid, white to dingy white, grayish, or buff, sometimes with a pinkish tinge in wet weather, sometimes with watery spots. Flesh thin, grayish to white; odor mild. **GILLS** grayish-white to buff or pinkish-buff, close, adnate to decurrent. **STALK** 1-4 cm long, 2-7 mm thick, equal or slightly thicker at either end, colored like cap, smooth, rather tough. **SPORE PRINT** white or rarely creamy; spores 4-5.5 × 2-4 microns, elliptical, smooth.

HABITAT: Scattered to gregarious or in rings, in pastures and other grassy places; widely distributed and common locally in the fall and winter. In the Pacific Northwest it frequently mixes company with *Marasmius oreades,* but in our area is more often found with *Agaricus campestris* and *A. cupreobrunneus.*

EDIBILITY: Poisonous—and potentially fatal to small children in the "grazing" stage! It produces profuse sweating, salivation, diarrhea, etc. Muscarine is the main toxin (see p. 894 for details).

COMMENTS: The dingy grayish-white color, close decurrent gills, white spores, small size, and growth in grass distinguish this drab, undistinguished mushroom, also known as *C. sudorifica.* In our area it frequents pastures while the edible and somewhat similar fairy ring mushroom *(Marasmius oreades)* grows mainly on lawns; in other regions, however, they may grow together. However, the fairy ring mushroom has a browner, frequently umbonate cap and well-spaced gills that are never decurrent. The edible *Clitopilus prunulus* is also similar, but has pinkish spores. There are numerous woodland look-alikes in the genus *Clitocybe,* and all should be strictly avoided. Closely related, poisonous grass-inhabiting species include: *C. augeana,* very similar but with a farinaceous odor, sometimes rivulose in age (developing riverlike lines or cracks on the cap), and growing in manure and compost as well as in grass; *C. rivulosa,* with cap often rivulose and tinged flesh-color, growing in grass; and *C. morbifera,* with a grayish-brown cap, in grass.

Clitocybe dealbata is a small poisonous grass-inhabiting mushroom with closely spaced, adnate to decurrent gills. Overall color is dingy gray to buff, leading to confusion with the edible fairy ring mushroom *(Marasmius oreades).*

Clitocybe sclerotoidea. Note growth habit in small, tight clumps. Each clump arises from a mass of tissue thought to be aborted *Helvella lacunosa.*

Clitocybe sclerotoidea   (Parasitic Clitocybe)

CAP 0.5-4 (5) cm broad, convex to plane or slightly depressed centrally; surface dry, unpolished, with a fine whitish fibrillose coating which rubs off; pallid to sordid buff to brownish or grayish, sometimes with darker watery spots. Flesh whitish, odor mild. **GILLS** adnate or slightly notched becoming decurrent, pale buff or pinkish-buff darkening to gray, olive-gray, or grayish-brown. **STALK** 2-4 (8) cm long, 3-10 mm thick, equal or thicker at either end, solid, colored like cap or paler from soft, matted, downy white hairs; arising from a fleshy mass of tissue which is often at least partially buried. **SPORE PRINT** white; spores 8-11 × 3-4 microns, subfusiform (elongated), smooth.

HABITAT: Typically in small, tight clumps on ground under pine, known only from the Pacific Coast; common at times in our area in the winter and spring. The fluted black elfin saddle *(Helvella lacunosa)* is often found nearby.

EDIBILITY: Unknown, and like myself, likely to remain so.

COMMENTS: I can find hidden virtues in almost any fungus, but this drab *Clitocybe* defies me—its mediocrity is downright stupefying. It serves as a compelling reminder that organisms do not exist for our enjoyment alone, and should not be judged accordingly. It can be recognized by its small size, dingy color (or lack of color), and habit of growing in compact clumps from a fleshy mass of tissue. Small, aborted individuals are usually present in each clump, and occasionally large, solitary specimens are encountered. The tissue mass from which they arise ("sclerotium") is actually composed of hyphae from both *C. sclerotoidea* and *Helvella lacunosa*—suggesting that the *Clitocybe* is parasitic on the *Helvella.* It is interesting to note, however, that the *Helvella* occurs with both hardwoods and conifers, while *C. sclerotoidea* is confined to pine, at least in my experience.

Clitocybe cyathiformis

CAP 2.5-8 cm broad, centrally depressed with an inrolled margin, becoming funnel-shaped in age; surface smooth, not viscid, dark brown to dark gray-brown, but fading in age to grayish or paler brown. Flesh thin, pallid; odor mild. **GILLS** at first adnate but soon deeply decurrent, pallid becoming grayish or grayish-brown, close. **STALK** 5-12 cm long, 0.4-1 cm thick, equal or thicker below, often rather long, colored like cap or paler, fibrillose, often with whitish down at base. **SPORE PRINT** white; spores 7-11 × 5-6 microns, elliptical, smooth, amyloid.

HABITAT: Solitary or in small groups in humus or on rotten logs, in woods and at their edges; widely distributed. Not uncommon in our area in the fall and winter, especially under redwood.

EDIBILITY: Not recommended. It is said to be edible, but there are many very similar species of unknown edibility.

COMMENTS: This species is one of numerous small to medium-sized, grayish, white-spored agarics with decurrent gills and a depressed to funnel-shaped cap. Because of its amyloid spores it has been placed in several different genera, including *Cantharellula* and *Pseudoclitocybe*. Similar species with amyloid spores include: *Cantharellula umbonata*, common in moss beds in northern and eastern North America, with crowded, narrow, whitish, forked gills that stain reddish in age; and *Clitocybula atrialba*, cap 2-10 cm broad and blackish-brown, gills well-spaced and white to grayish, stalk clothed with dark scurfy scales, growing in rich humus or on rotting hardwoods in the Pacific Northwest. The first of these were formerly placed in *Cantharellus*, the second in *Clitocybe*. Also see *Myxomphalia maura* and *Clitocybe albirhiza*.

Myxomphalia maura

CAP 1-3.5 (5) cm broad, convex or centrally depressed with an incurved margin, becoming plane or centrally depressed; surface viscid when moist but soon dry and often shiny, smooth, dark grayish-brown or olive-brown to blackish-brown, fading to gray or paler as it dries. Flesh thin, white to grayish. **GILLS** adnate to slightly decurrent, close, white to pale grayish (usually paler than cap). **STALK** 2-6 cm long, 2-5 (6) mm thick, more or less equal, smooth, colored more or less like cap or slightly paler, but not fading as quickly. **SPORE PRINT** white; spores 4.5-6.5× 3.5-4.5 microns, broadly elliptical to nearly round, smooth or very minutely ornamented, amyloid.

HABITAT: Solitary, scattered, or in groups on burnt soil and debris, especially under conifers; widely distributed and fairly common in our area in the appropriate habitat, fall through spring. I often find it with *Pholiota brunnescens* and *Psathyrella carbonicola*.

EDIBILITY: Unknown, but too puny to be of value.

Left: *Clitocybe cyathiformis*, mature specimens. **Right:** *Myxomphalia maura*, a small mushroom that favors recently burned areas. Note the adnate to decurrent gills.

COMMENTS: The habitat on burnt soil distinguishes this undistinguished mushroom from a host of similar grayish-brown look-alikes. *Lyophyllum atratum (=Collybia atrata)* is another species that grows in burned-over areas; it is quite similar but has darker (gray) gills and non-amyloid spores (the two are sometimes found growing together). *Fayodia anthracobia* is a minute charcoal-lover (cap up to 1 cm broad) with round, warted, amyloid spores. *Omphaliaster asterosporus* and *O. borealis* have warted or starlike spores, but are larger (cap 2-5 cm broad), have grayish to grayish-brown gills, and prefer moss (under conifers) to burnt soil. Another small widespread species, *Clitocybe (=Omphalina) epichysium*, has a dry cap about 2 cm broad that is dark sooty-brown to olive-brown or ashy and minutely scaly or scurfy at the center. It is fairly common in our area in mixed woods and under conifers, but not in burnt soil. See also *C. cyathiformis.*

LEUCOPAXILLUS

Medium-sized to large terrestrial mushrooms. CAP convex to plane or depressed, margin usually inrolled when young; *surface dry, unpolished. Flesh rather tough and dry.* GILLS attached, close, *usually white or yellowish.* STALK typically central, fleshy, *tough, often with a conspicuous white mycelial mat at base.* VEIL and VOLVA absent. SPORE PRINT *white.* Spores amyloid, rough.

THIS is a small but very conspicuous group of robust mushrooms with a dry, unpolished cap, white to pale yellow gills, a tough, fleshy stem, and white, amyloid spores. In some species the gills peel rather easily from the cap (as in the brown-spored genus *Paxillus).* The flesh is rather dry and brittle as in *Russula,* but the stature of the fruiting body is quite different and the fibrous stalk does not snap open cleanly like chalk. Also, most *Leuco-paxillus* species have a copious white mycelium that permeates the surrounding duff and frequently adheres to the base of the mushroom when it is plucked—a character not found in *Russula.*

Confusion with *Clitocybe* and *Tricholoma* is also likely (*Leucopaxillus* species were originally placed in those genera), but they are not so tough and rot more readily. In contrast, *Leucopaxillus* species are remarkably resistant to bacterial decay—antibiotic substances have been isolated in some of them and their fruiting bodies, like those of *Laccaria,* persist for weeks and thus appear to be more common than they actually are.

Leucopaxillus albissimus, mature and dried-up specimens. Note the tremendous variation in size, shape, and stature. Some of these are quite old and shrivelled.

Leucopaxillus albissimus. Young white specimens. (Ralph Buchsbaum)

Less than ten species of *Leucopaxillus* are known from North America. The two common ones in our area fruit prolifically and are likely to be among the first agarics you encounter. They are strictly woodland fungi and may be mycorrhizal. Though alluringly large and firm, they are difficult to digest because of their toughness. However, they are not known to be poisonous, so intrepid toadstool-testers may opt to experiment with the mild-tasting forms. They have one compelling advantage—a single large specimen can fill a basket!

Key to Leucopaxillus

1. Gills pale clear yellow when young; cap 9-30 cm broad, buff to dingy tan or yellow-brown; found under hardwoods in eastern North America and southern Arizona . . . . *L. tricolor*
1. Not as above; gills not yellow when young (but may be dingy yellowish in old age) 2

2. Cap brown to reddish-brown, medium-sized; taste bitter . . . . *L. amarus,* p. 168
2. Cap white, buff, pale tan, yellowish, etc; medium-sized to very large; taste mild or bitter . . . .
. . . . *L. albissimus* & others, below

Leucopaxillus albissimus   (Large White Leucopaxillus)

CAP 5-20 (40) cm broad or more, convex with an inrolled margin when young, becoming plane or depressed in age; surface dry, dull, unpolished, white becoming buff, yellowish, or even tan (at least toward center) in age, often cracked or splitting in age or dry weather; margin sometimes obscurely ribbed. Flesh thick, tough, white; odor unpleasant or fragrant, taste also variable (mild or bitter). **GILLS** white to slightly yellowish in old age, attached (typically decurrent at least by lines, but ranging to adnate or even adnexed); close. **STALK** 5-20 cm long, 1-5 cm thick, tough, solid, equal or enlarged at the middle or below (often with a narrowed base); white, discoloring slightly in age; base usually imbedded in a white mycelial mat; smooth or often scaly (especially in mid-portion). **SPORE PRINT** white; spores 5.5-8.5 × 4-6 microns, elliptical, warty, amyloid.

HABITAT: Solitary, scattered, or gregarious (often in large rings) in woods, mainly under conifers; widely distributed. It is common in our area in the cool winter months, particularly under redwood, but also with other trees, including eucalyptus. It is slow to grow and even slower to decay.

EDIBILITY: A tempting specimen, but I don't recommend it—even the non-bitter forms are coarse and difficult to digest. Thorough cooking would be necessary.

COMMENTS: Several varieties of this widespread species have been described based on differences in cap color, taste, and spore characters. The above description is based primarily on the variety common in our area, var. ***paradoxus (=L. paradoxus)***. It has a rather strong but not unpleasant odor and a distinctive but not usually bitter taste. It is white when young, but usually becomes dingy yellowish, buff, or pale tan as it ages or dries out. A white mycelial mat can usually be seen permeating the duff around the mushroom or adhering to its base. However, its most outstanding attributes are the large size, tenacious toughness, and resolute resistance to decay, as evidenced by the following two "tried and true" methods for distinguishing it from *Clitocybe nebularis* and other large look-alikes:

1) Choose a firm (not waterlogged) specimen and throw it against a wall, housemate's head, or other dense, hard, thick object. If it (the mushroom) remains more or less intact, it is probably *L. albissimus;* if it shatters, chances are it *was* a *Clitocybe.*

2) Choose a firm (not waterlogged) specimen and leave it out on the porch, or hide it in your housemate's closet. If after one week it (the mushroom) shows no visible signs of decay, it is probably *L. albissimus;* if unchanged after one month it is *definitely L. albissimus.* If, on the other hand, it shows visible or smellable signs of decay (i.e., has begun to rot or is reduced to a heap of writhing maggots), chances are, once again, that it *was* a *Clitocybe.* (Note: *L. albissimus* can behave like a *Clitocybe* if too old or maggot-riddled.)

Other species: ***L. albissimus* var. *lentus*** is quite similar to var. *paradoxus,* but is usually smaller (stalk only 0.8-1.5 cm thick and smooth); *var.* ***piceinus*** is similar in color, but has a bitter taste; *var.* ***typicus*** also has a bitter taste but is pure white; ***L. laterarius*** is a hardwood-loving species with a very bitter taste, pinkish-buff-tinged cap, and round spores; it is particularly common in eastern North America.

Leucopaxillus amarus   (Bitter Brown Leucopaxillus)   Color Plate 31

CAP 5-12 (15) cm broad, broadly convex with an inrolled margin when young, becoming plane or slightly depressed; surface dry, unpolished, smooth, dark brown to brown, pecan-brown, or reddish-brown, evenly colored or paler at the margin, sometimes cracked or faded in age; margin often obscurely ribbed. Flesh thick, firm, dry, white; odor mild or pungent; taste very bitter. **GILLS** typically adnate but ranging from notched to slightly decurrent by lines; close, white. **STALK** 4-10 cm long, 0.5-2 (4) cm thick, equal or enlarged below, smooth, dry, solid, white or sometimes discolored brownish below; base imbedded in a copious white mycelial mat. **SPORE PRINT** white; spores 4-6 × 3-5 microns, nearly round, warted, amyloid. Cystidia numerous on gill edges.

HABITAT: Scattered to gregarious (often in rings) under conifers and oaks; common and widely distributed. In our area it is often abundant in the fall and winter.

EDIBILITY: Unequivocally inedible—it smells like "creepy crawlers" and tastes like a mildewed army tent. If you're lost in the woods and have nothing to eat, you'd do better to follow the example of Charlie Chaplin and stew your boots before venturing to make a meal of this mushroom.

COMMENTS: Also known as ***L. gentianeus,*** this mundane mushroom can usually be recognized by its dull, boring brown to reddish-brown cap, white gills, bitter taste, absence of a veil, and moldy-looking white mycelium that usually permeates the surrounding humus and frequently adheres to the stem when it is plucked. The latter feature helps to distinguish it from brown-capped Tricholomas (e.g., *T. imbricatum*). *Hygrophorus roseibrunneus* can also be similar in color, but has soft and waxy gills. As with *L. albissimus,* several varieties and forms have been described. In addition to the color plate, *L. amarus* is shown on p. 918.

MELANOLEUCA

Small to medium-sized terrestrial mushrooms. CAP *usually smooth,* broadly convex to plane or often umbonate; often hygrophanous. GILLS *close or crowded, attached (usually adnate or notched), usually white.* STALK central, typically *rather stiff, straight, semi-cartilaginous;* often slender. VEIL and VOLVA absent. SPORE PRINT *white or creamy.* Spores minutely warted or roughened, amyloid. Cystidia often present on gill edges.

THESE rather unimposing whitish to gray or brownish mushrooms have been aptly characterized as "overgrown Collybias," for the stature of the fruiting body is intermediate between that of *Tricholoma* and *Collybia.* The straight, semi-cartilaginous stem, close to crowded (and often notched) gills, and smooth, dull-colored, frequently umbonate cap are the principal fieldmarks.

Melanoleucas can grow almost anywhere, but in our area are found most often on lawns or shaded areas in pastures and parks. They are also quite common under conifers in the Sierra Nevada and other mountain ranges. I can find no mention of poisonings attributed to *Melanoleuca,* but the North American species are not well known. Two widespread species are described here, and several others are keyed out.

Key to Melanoleuca

1. Gills tan to ochre, creamy-ochre, or pale pinkish-cinnamon, at least in age; cap 5-13 cm broad; stalk usually quite tall .... *M. cognata,* p. 170
1. Not as above; gills typically whitish, sometimes with a slight yellow or pinkish tinge 2

2. Cap viscid when moist, 4-8 cm broad, white or in age often tinged or stained yellow; stalk white, usually scurfy, 0.7-1.5 cm thick; fairly common in central and southern California in a variety of habitats .... *M. lewisii*
2. Not as above 3

3. Cap yellowish-brown fading to yellowish-white; fairly common under hardwoods in eastern North America .... *M. alboflavida*
3. Not as above; cap some shade of brown or gray when moist (but may fade in age) 4

4. Growing under mountain conifers, usually shortly after the snow melts 5
4. Growing in lawns, pastures, etc., sometimes also in woods, but not typically as above .... *M. melaleuca* group, below

5. Stalk fleshy, usually more than 1 cm thick; cap medium-sized to fairly large .... *M. evenosa* group, p. 171
5. Stalk usually less than 1 cm thick; cap medium-sized to fairly small .... *M. graminicola* & others (see *M. melaleuca* group, below)

Melanoleuca melaleuca group

CAP 2-7 (10) cm broad, broadly convex to plane to shallowly depressed, often with a low broad umbo; surface smooth, hygrophanous but not viscid; dark brown to gray to grayish-brown when moist, often fading to buffy-tan, gray, or even paler in sunlight (or as it dries). Flesh thin, whitish, odor mild. GILLS narrow, close or crowded, white, usually notched but varying to adnate or even very slightly decurrent. STALK 2-8 cm long, 3-6 (12) mm thick, equal or with a slightly swollen base, rather slender, stiff, whitish or with darker (brown) fibrils, apex sometimes minutely scaly or scurfy. SPORE PRINT white; spores 6-8 × 4-5.5 microns, elliptical, minutely warted, amyloid. Harpoonlike cystidia typically present on gill edges.

HABITAT: Scattered to gregarious on ground in open places (lawns, pastures, etc.), under trees, along roads and trails, in straw and wood chips, and also in the woods; common and widely distributed, but usually fruiting at lower elevations. Common in our area throughout the mushroom season but most numerous in the winter and early spring.

Melanoleuca melaleuca group is common in grassy areas, but also grows under trees and in the woods. Cap is usually dark when young and moist (see specimen on right), but may fade considerably in age or sunlight. Note crowded white gills.

EDIBILITY: Edible. The caps are delicious fried in butter, but since it is not an easy mushroom to recognize, I hesitate to recommend it. Also, the edibility of closely related species has not been adequately ascertained.

COMMENTS: The above description will actually fit a number of closely related Melanoleucas whose exact identities are best left to *Melanoleuca*-masters. As a group they are characterized by a smooth, hygrophanous, dark brown to grayish (or paler) cap, close white gills, straight stalk, and strongly amyloid, warted spores. Specimens growing in the open are generally much paler than their counterparts growing in the shade. Related species include: *M. brevipes,* common on lawns, with a short brown to whitish stalk and abundant cystidia on the gills; and *M. polioleuca,* with a hoary bloom on the cap when fresh. There are also several species that fruit prolifically in the spring and early summer in the coniferous forests and alpine meadows of western and northern North America. The most common of these, *M. graminicola,* is similar to *M. melaleuca* in color and stature, but lacks cystidia on the gills. Another, *M. evenosa,* is larger and more robust (see description). Although probably harmless, none of these species should be eaten until better known. *M. vulgaris* is a synonym for the tongue-twisting *M. melaleuca.*

Melanoleuca cognata

CAP (5) 7-13 cm broad, broadly convex to plane, usually with a broad umbo, sometimes becoming slightly depressed in age; surface smooth and sometimes shiny, dry or slightly viscid, brown to ochre-brown, fading to pale tan in age, the center sometimes darker. Flesh whitish; odor often slightly sweet, rancid, or "peculiar" (Smith). **GILLS** pallid becoming tan, creamy-ochre, deep ochre, or pale pinkish-cinnamon; crowded, attached (usually notched). **STALK** 6-12 cm long, 1-2 cm thick, equal or with a swollen base, longitudinally lined or twisted-striate, straight, colored more or less like cap or paler, the base sometimes brownish-stained. **SPORE PRINT** creamy or yellowish; spores 7-10 × 4.5-6 microns, elliptical, minutely warted, amyloid. Cystidia abundant on edges and faces of gills.

HABITAT: Solitary, scattered, or in small groups on ground in mixed woods and under conifers in the spring, summer, and early fall; widely distributed but not common. It does not occur in our area but is to be looked for in the Sierra Nevada. It is fairly frequent in the spruce-fir-aspen forests of the southern Rocky Mountains and Southwest.

EDIBILITY: Edible. I haven't tried it.

COMMENTS: In contrast to many of its kin, this *Melanoleuca* is fairly easy to recognize. The broad, frequently umbonate brown to ochre-tan cap, tall straight stem, and brown to tan or ochre mature gills are good fieldmarks.

170

Melanoleuca evenosa group   (Robust Melanoleuca)

CAP 5-19 cm broad, broadly convex to plane or wavy; surface smooth, dry to slightly viscid when moist, pallid to brown to grayish-brown or in age sometimes slightly ochre. Flesh firm, thick, white; odor often rather pleasant. **GILLS** crowded, white (or in age sometimes brownish- or ochre-stained); usually notched or adnexed. **STALK** 3-7 cm long, (1) 1.5-3 (3.5) cm thick, equal or slightly thicker below, very firm, stiff, solid; white with brownish or cap-colored fibrils; apex or upper portion usually dandruffy or minutely scaly (or sometimes dandruffy throughout). **SPORE PRINT** white; spores 8-11 × 4-5 microns, elliptical, minutely warted, amyloid.

HABITAT: Solitary, widely scattered, gregarious, or even clustered in duff under conifers or in grassy clearings, at the edges of forests, etc.; fairly common in the mountains of western North America during the spring and early summer.

EDIBILITY: Unknown. Several Melanoleucas are edible and this one is fleshy enough to warrant *cautious* experimentation, but be sure of your identification!

COMMENTS: Anyone who regularly hunts the Cascades and Sierra Nevada during the springtime will come across this species or species "complex," which also goes under the name *M. subalpina*. Because of its robust stature (see photo on p. 897) it is likely to be mistaken for a *Tricholoma*. However, the crowded white gills, scurfy stalk, smooth cap, amyloid spores, and springtime growth under conifers should distinguish it from most Tricholomas. Several other Melanoleucas grow under mountain conifers, but are more slender (see comments under the *M. melaleuca* group). *M. cognata* can be fairly large, but is usually taller and slimmer with tan to ochre gills in age.

LACCARIA

Small to medium-sized terrestrial mushrooms. CAP convex to plane, centrally depressed, or uplifted; not viscid. GILLS attached, *rather thick, slightly waxy, pinkish to flesh-colored, cinnamon, purple, or lilac*. STALK *tough and fibrous*, elastic, often slender, more or less central. VEIL and VOLVA absent. SPORE PRINT *white to pale lilac*. Spores usually spiny, not amyloid.

THIS small but common genus can be distinguished in the field by its thick, purple to pinkish or flesh-colored gills and tough, fibrous stem. The gills may be somewhat waxy-looking as in the waxy caps (Hygrophoraceae), but the tough stalk is distinctive and the spores are usually spiny under the microscope. In small individuals the stalk may be slender but is not noticeably fragile, and the cap is not conical or bell-shaped as in *Mycena*.

Laccarias form mycorrhizae with many kinds of trees and shrubs, and are among the few mycorrhizal species that are easily raised to fruition in the laboratory. Their omnipresence in coniferous forests has caused them to be called "mushroom weeds." However, they also grow with hardwoods. They seem particularly fond of sandy, boggy, or other poor soils, and are a conspicuous fungal feature of our coastal pine forests. They persist for weeks without decaying, and thus appear to be more plentiful than they actually are.

About 16 species of *Laccaria* occur in North America. Most are quite variable in shape and color and thus difficult to distinguish in the field, but they split nicely into two groups —the ones that are purple and the ones that aren't. All are thought to be edible and most are fairly good, or at least better than some authorities would have us believe. Two representative species are described here.

Key to Laccaria

1. Downy mycelium at base of stalk *and/ or* gills purple to lilac when fresh (may fade in age!) 2
1. Violet tones completely absent, even when fresh (but may be vinaceous-tinged) . . . .
. . . . *L. laccata* & others, p. 172

2. Growing in sand (often buried); spores smooth *L. trullisata* (see *L. amethystina* group, below)
2. Not as above (but may grow in sandy soil); spores spiny 3
3. Fruiting body medium-sized to large (cap 5-20 cm broad; stalk 1-3 cm thick); common under hardwoods (especially oaks) in eastern North America . . . . *L. ochropurpurea*
3. Not as above; small to medium-sized (cap usually less than 7 cm broad; stalk usually less than 1 cm thick, but at times up to 1.5 cm); widespread in many habitats 4
4. Gills distinctly purple when fresh . . . . *L. amethystina* group, below
4. Gills with only a tinge of violet (if any) when fresh *L. bicolor* (see *L. amethystina* group, below)

Laccaria laccata   (Lackluster Laccaria)   Color Plate 29

CAP 1.5-6 cm broad, convex becoming plane to centrally depressed or sometimes even with a hole in the center, the margin often uplifted in age; surface not viscid, often minutely scaly; color variable: flesh-colored to orangish, brownish-cinnamon, reddish-tan, or pinkish-brown when moist, much paler as it dries; margin often wavy or irregularly lobed in age. Flesh thin, tinged cap color, odor mild or sometimes radishlike. **GILLS** thick, well-spaced, somewhat waxy, pale pinkish to flesh-colored or reddish-tan, dusted white by spores at maturity; attachment variable but typically adnate to slightly decurrent. **STALK** 2-10 cm long, 3-10 mm thick, more or less equal, tough and fibrous, elastic, often fibrillose, often somewhat twisted or compressed; same color as moist cap or darker (reddish-brown), usually rather slender; downy mycelium at base white (if present). **SPORE PRINT** white; spores 7-10 × 6-9 microns, round or nearly round, spiny.

HABITAT: Scattered or in groups or troops in woods or near trees, especially in poor or sandy soil or in boggy areas; very common and widely distributed. In our area it can be found almost anywhere at anytime, but favors cool weather and pine. *L. proxima* and *L. altaica* (see comments) are also common locally.

EDIBILITY: Edible and fairly good, especially if seasoned; the tough stems should be discarded. Be certain of your identification, however—there are many similarly-colored mushrooms!

COMMENTS: This cosmopolitan mushroom has a knack of turning up in droves when one is seeking more exotic or unusual species, and is thus spurned and scorned as a mushroom "weed." To be sure, it is as unostentatious as it is ubiquitous, without the lovely color of the *L. amethystina* group, yet in its own unexciting way it is quite beautiful. Like the honey mushroom, it is so vexingly variable in size, color, and shape that even veteran collectors have trouble recognizing some forms! The telltale traits are the overall color (pinkish to orangish or dull cinnamon), thick well-spaced gills, white spores, and tough or fibrous stem. *L. proxima* is a very similar, widespread species. It tends to be slightly larger with a more fibrillose stalk and scalier cap, but can only be told with certainty by its broadly elliptical (rather than round) spores. There are also several similarly-colored but smaller species (cap 0.5-4 cm broad) that usually grow in wet or boggy areas and have striate caps when moist. They include: *L. ohiensis* and *L. altaica* (both of which have been called *L. striatula*), with larger spores (8-13 microns), the latter especially common in northern latitudes; and *L. tortilis,* a very small species with only a few (10-15) gills and even larger spores. See also *L. bicolor* (under *L. amethystina* group), which has violet mycelium.

Laccaria amethystina group   (Amethyst Laccaria)   Color Plate 30

CAP (1) 2-4 (7) cm broad, convex becoming plane, centrally depressed (sometimes with a hole in the middle), or with an uplifted margin in age; surface often minutely scurfy or scaly, not viscid; purple to brownish-purple when fresh and moist, fading to brown, gray, buff, or whitish (see comments!) as it loses moisture; often cracked in dry weather; margin often wavy or lobed in age. Flesh thin, violet-tinged; odor mild. **GILLS** well-spaced, thick, somewhat waxy, deep or bright amethyst-purple when fresh, gradually fading to dull purple or grayish-purple and eventually dusted white by spores; attachment variable but usually adnate to slightly decurrent. **STALK** 5-12 cm long, 0.3-1 (1.5) cm thick, more or less equal, typically rather long and slender, often curved or twisted; tough and fibrous, elastic, usually conspicuously fibrillose, colored more or less like moist cap or browner, fading as it dries; base usually covered with downy purple or white mycelium. **SPORE PRINT** white or tinged lilac; spores 7-11 microns, round, spiny (but see comments!).

HABITAT: Scattered to densely gregarious on ground in forests and at their edges; widespread and common. In our area this species and its look-alikes (see comments) are particularly abundant under coastal pines. They can fruit most any time but are partial to cold weather and are often abundant when other collectable delectables are scarce.

EDIBILITY: Edible, and a good choice for beginners because of its distinctive color and convenient availability. It has a nice texture but not much flavor, so try mixing it with potatoes and seasoning it with garlic. The tough, hairy stems should be discarded.

COMMENTS: The lovely amethyst-purple gills when fresh combine with the rather long and tough, fibrous-hairy stem to set apart this cosmopolitan mushroom and its close relatives. The color fades rather dramatically as the mushroom loses moisture, but the gills usually retain their purple tint well into maturity. The common "Amethyst Laccaria" along the west coast has recently been given a new name, *L. amethysteo-occidentalis,* based on its broadly elliptical (rather than round) spores and the tendency of its cap to fade to brown rather than buff or gray as in the "true" *L. amethystina* (or *L. amethystea*). Our local variety, however, sometimes fades to buff or even white, so that microscopic examination is often necessary to distinguish it from *L. amethystina* (the latter may occur in our area, but if so is uncommon). The specimens in the color plate are probably *L. amethysteo-occidentalis,* but are labeled *L. amethystina* group because their spores were not examined. To muddle matters even more, *L. bicolor* is also common in our area, especially under pine. It has broadly elliptical spores, but has only a slight violet or vinaceous tinge (if any) when fresh, except for the violet downy mycelium at the base of the stalk (which may, however, fade to white in age, leading to confusion with *L. laccata*). Other purple-gilled species: *L. trullisata* has long (16-22 microns) smooth spores and grows only in sand or sand dunes (often buried!). *L. ochropurpurea* is a robust eastern species (see key to *Laccaria*).

LYOPHYLLUM & Allies

Small to medium-small mushrooms, or if larger then *usually growing in dense clusters.* CAP *typically some shade of gray or brown,* but sometimes white; not viscid. GILLS attached or sometimes free, *usually white or gray.* STALK fleshy or thin, central, usually white, gray, or brown. VEIL and VOLVA absent. SPORE PRINT *white.* Spores smooth or spiny, not amyloid. Cystidia on gills typically inconspicuous or absent. Basidia with siderophilous (carminophilous) granules.

LYOPHYLLUM is a nondescript amalgamation of white-spored mushrooms at one time dispersed among *Tricholoma, Collybia,* and *Clitocybe.* They are puzzling to the amateur because the various species appear to have little in common aside from their drab color—some are slender and *Collybia-* or *Mycena*-like; others are robust and *Tricholoma-* or *Clitocybe*-like. The unifying feature is rather esoteric: the basidia contain particles which darken dramatically when heated in acetocarmine! (Wouldn't you?)

Lyophyllum is a fairly sizable genus but only the *L. decastes* group, which fruits in large, dense clusters along roads and trails, is conspicuous. The rest qualify as "LBM's" and are treated here only perfunctorily (they are differentiated largely on microscopic characteristics anyway). Also included here is one species of *Calocybe*—a small, rare genus which, like *Lyophyllum,* has siderophilous granules in the basidia, but typically has a more brightly colored cap.

Key to Lyophyllum & Allies

1. Cap pinkish-tan to reddish (sometimes fading to tan); stalk usually 5 mm thick or less ......
    .... *Calocybe carnea* & others, p. 176
1. Not as above; cap grayish to brown, black, or white 2
2. Gills and/or stalk staining gray or black where bruised (sometimes slowly), or developing
    grayish to black spots in age ....:..... *L. semitale* & others (see *L. montanum,* p. 175)
2. Not as above 3
3. Growing on burnt ground or debris; stalk thin (up to 3 mm); cap small, dark brownish to black
    .... *L. atratum* (see *Myxomphalia maura,* p. 165)
3. Not growing on burnt ground and/or stalk thicker 4
4. Stalk fairly thick (1-2.5 cm thick), usually white; typically growing in dense clusters (rarely
    solitary) along roads and paths or less commonly in undisturbed woods 5
4. Not as above; stalk usually slender 6
5. Caps white, often somewhat misshapen; taste usually sour or disagreeable; known only from
    western North America .... (see *Clitocybe dilatata,* p. 159)
5. Not as above; caps not white, or if white then taste more or less mild *L. decastes* group, below
6. Fruiting in spring in western mountains, usually near melting snow; fruiting body grayish
    when fresh; gills gray; cap usually with a hoary bloom at first .... *L. montanum,* p. 175
6. Not as above 7
7. Found in sphagnum bogs; cap usually with an umbo; stalk thin, fragile, often long *L palustre*
7. Similar to above but not growing in bogs .... *L. rancidum* & others

Lyophyllum decastes group   (Fried Chicken Mushroom)

CAP 3-12 cm broad, convex to plane or with slightly uplifted margin in age; surface smooth, often with a soapy (lubricous) feel when moist, but not viscid; color variable: dark brown to grayish-brown to yellowish-brown, watery tan, or paler; margin often lobed. Flesh firm, white; odor mild. **GILLS** adnate to slightly decurrent or often notched, close, white or pallid but sometimes becoming straw-colored in age. **STALK** 3-10 cm long, 1-2.5 cm thick, solid, equal or tapering downward, often curved, smooth, dry, white, sometimes discolored brownish in age, especially at base. **SPORE PRINT** white; spores 4-6 microns, round or nearly round, smooth. Basidia with siderophilous granules.

HABITAT: Gregarious on the ground, usually in large compact clumps and often half-hidden by leaves and grass; fruiting mainly in disturbed areas—along roads and beaten paths (sometimes even forcing its way up through asphalt), in waste places, around old sawdust piles, or sometimes in the woods; widely distributed. It is common along the west coast from Alaska to southern California and fruits in our area from early fall through spring or even summer. It was the single most abundant mushroom on the University of California, Santa Cruz campus (a very "disturbed" place!) during the 1976-77 drought. A couple hundred pounds could easily have been harvested.

EDIBILITY: Edible and quite popular in some regions, but it hardly tastes like chicken as its common name implies. I've found it to be crunchy, but bland. Its abundance and habit of fruiting when other good edibles are scarce make experimentation worthwhile. Be careful, however—this is a species "complex," and some mild cases of poisoning have been attributed to it. These may simply have been "allergic" reactions, or may have resulted from confusion with poisonous species (e.g., *Clitocybe dilatata*).

COMMENTS: Also known as *Clitocybe multiceps* and *Tricholoma aggregatum,* this exceedingly common species "complex" is best recognized by its fondness for disturbed places and its growth in dense clusters which may contain a hundred or more individuals and weigh as much as 15 pounds! Within the *L. decastes* group at least three species have

Lyophyllum decastes group is especially common along roads and paths. Note clustered growth habit, white stalk, and absence of a veil. Another clump is shown on p. 898.

been recognized on the basis of cap color: *L. connatum* has a white or whitish cap; *L. loricatum* has a blackish-brown to dark brown cap when young, often with a hoary sheen or metallic luster, and a thick cartilaginous cap cuticle; as it ages, however, it fades to paler brown or tan, and is then indistinguishable from "typical" *L. decastes,* which has a brown to grayish-brown to tan cap. *L. loricatum* is the most common of the three in our area, but *L. decastes* also occurs. A fourth, unidentified species is fairly common in our live oak woodlands. It has a grayish cap and grayish gills and its edibility is unknown. Still another species grows under mountain conifers soon after the snow melts.

Though all of these species tend to grow in clusters, solitary individuals occasionally occur, and these are apt to baffle the beginner. Since the cap color and gill attachment are so variable, it is best only to eat those growing in large clumps in disturbed ground. Make sure the spore deposits on the lower caps of mature clusters are white. Poisonous Entolomas sometimes grow in clusters, but have deep pinkish spores. The *Clitocybe subconnexa* group also grows in clusters and has pinkish spores, while *C. dilatata* is white-spored with a whitish cap and disagreeable taste. Most other fleshy, clustered terrestrial mushrooms have darker spores and/or a veil.

Lyophyllum montanum   (Snowbank Lyophyllum)

CAP 2-6.5 cm broad, convex to broadly umbonate or plane; surface not viscid, gray to lead-colored beneath a hoary (whitish) bloom which may wear off; often becoming somewhat yellower (to dingy yellow-brown) in old age. Flesh thin, brownish to whitish, odor mild. **GILLS** dark gray to gray, close, adnate to nearly free (usually adnexed). **STALK** 3-7 cm long, 0.5-1.5 cm thick, equal or thicker below, hoary and colored more or less like cap when fresh, or slightly browner; base often with white mycelial down. **SPORE PRINT** white; spores 6.5-8 × 3.5-4 microns, elliptical to oblong, smooth. Basidia with siderophilous granules.

HABITAT: Solitary, scattered, or in small groups on ground under conifers (particularly spruce and fir), usually near melting snow; fairly common in the spring and early summer in the mountains of western North America.

EDIBILITY: Unknown.

175

COMMENTS: There are a number of nondescript grayish Lyophyllums that are very difficult to identify. This one, however, can be told by its snowbank milieu and hoary cap surface when young (see photo on p. 46). Other species: *L. semitale* is one of more than 50 difficult-to-differentiate Lyophyllums that stain gray, black, or brown when bruised (often slowly). Its cap and gills are grayish and it is widespread under conifers or less commonly hardwoods. *L. infumatum* also blackens, but has whitish gills when young.

Calocybe carnea

CAP 1.5-4 cm broad, convex to plane or slightly umbonate; surface dry, more or less smooth, usually pinkish to pinkish-brown, but varying to dark reddish or fading to pale tan. Flesh thin, whitish. **GILLS** crowded, narrow, white, adnate to slightly decurrent or notched. **STALK** 1.5-4 cm long, 2-5 mm thick, equal, colored more or less like cap, smooth or finely fibrillose. **SPORE PRINT** white; spores 4-6 × 2-3 microns, elliptical, smooth. Basidia with siderophilous granules.

HABITAT: Scattered or in groups in lawns, grassy clearings in woods, and other open places; widely distributed but not common. I have found it only once, on a lawn in Santa Cruz, California, during the seventh inning stretch of the fourth game of the 1978 World Series (I don't remember the date).

EDIBILITY: Not recommended. It is said to be edible but is easily confused with poisonous species.

COMMENTS: Formerly known as *Clitocybe socialis* and *Lyophyllum carneum,* this pretty little lawn-lover can be recognized by its pinkish cap, crowded white gills, and white spores. It appears to have a wide distribution but nowhere does it seem to be common. A similar but slightly larger species with a purple-red cap and yellow-ochre gills, *C. onychina,* occurs in the spruce-aspen forests of the Rocky Mountains in the summer.

TRICHOLOMA

Medium-sized to large *terrestrial*, mostly woodland fungi. CAP viscid or dry, smooth, fibrillose, or scaly. GILLS *typically notched or adnexed,* occasionally adnate. STALK central, *fleshy.* VEIL *absent* (except in *T. zelleri* and a few others). VOLVA absent. SPORE PRINT *white.* Spores smooth, not amyloid. Gills only rarely with cystidia.

TRICHOLOMA is a large and prominent, well-defined group of terrestrial white-spored mushrooms with notched gills and a central, fleshy stem. It corresponds in stature to *Entoloma* (deep pinkish spores) and *Hebeloma* (brown spores), but has little else in common with those fungi. Among the white-spored genera, *Collybia* differs in its cartilaginous stalk, *Tricholomopsis* grows on or near wood, *Hygrophorus* has gills that are soft and waxy, *Russula* has dry, brittle flesh, and *Clitocybe* has adnate to decurrent gills and/or pinkish spores. *Melanoleuca, Leucopaxillus,* and *Lyophyllum* differ microscopically and are keyed out in *Tricholoma,* while *Armillaria* and *Catathelasma* have a veil.

Tricholoma contains several excellent edible species as well as some poisonous ones. The man on horseback, *T. flavovirens,* is the only species I consider safe for beginners. The brown-capped and gray-capped species should be strictly avoided until you are intimately familiar with each and every one of them. Then, and only then, should species like *T. portentosum* become part of your culinary repertoire. Tricholomas, incidentally, form the bulk of the wild mushrooms served by the Czarnecki family at their fabulous wild mushroom restaurant in Reading, Pennsylvania.

Tricholomas are almost exclusively woodland fungi. Most species are mycorrhizal—especially with pine and oak, but also with spruce, aspen, fir, and other trees. They are partial to cold weather and in our area reach their peak in December or January. In colder regions they may continue to fruit after there is snow on the ground!

Though *Tricholoma* is an easy genus to recognize, its species are perplexing even to the professional. Several kinds are known only from a single locality and microscopic characteristics are not as helpful as in, say, *Mycena*. Over 100 species occur in North America and perhaps 50 in California. Fifteen are described here.

Key to Tricholoma

1. Cap yellow to greenish-yellow, at least at margin, or yellow overlaid with purple-red fibrils   2
1. Cap not yellow at margin 8

2. Stalk sheathed with cottony or shaggy scales .... (see *Armillaria* & Allies, p. 189)
2. Not as above 3

3. Cap *and/or* stalk with reddish-purple fibrils or scales; flesh pale yellow to yellow ....
   .... (see *Tricholomopsis rutilans*, p. 145)
3. Not as above 4

4. Flesh yellow; odor typically unpleasant and pungent (like coal-tar gas); gills yellow, widely spaced; cap *not* viscid .... *T. sulphureum*, p. 179
4. Not as above 5

5. Cap with blackish to dark brown to purple-gray or grayish center and/ or radiating fibrils . 6
5. Cap lacking dark radiating fibrils, entirely yellow or yellow at the margin and reddish-brown to brown toward the center 7

6. Cap yellow to greenish-yellow with blackish to dark brown center and radiating fibrils .....
   .... *T. sejunctum* & others, p. 180
6. Not as above; cap streaked with grayish to purple-gray fibrils .... *T. portentosum*, p. 180

7. Gills yellow .... *T. flavovirens*, p. 179
7. Gills white .... *T. leucophyllum* (see *T. flavovirens*, p. 179)

8. Veil present, usually forming a distinct annulus (ring) on stalk 9
8. Veil absent or evanescent, not forming an annulus 11

9. Cap small (up to 5 cm broad), gray to brownish-gray or bluish-gray; associated mainly with willow; odor usually farinaceous .... *T. cingulatum*
9. Not as above (but may have farinaceous odor) 10

10. Stalk rather slender and fragile; veil fibrillose, usually forming only a slight ring on stalk ....
    .... (see *Limacella glioderma*, p. 291)
10. Stalk usually at least 1 cm thick; veil membranous, forming a persistent ring   *T. zelleri,* p. 188

11. Stalk belted with rusty-orange scales or scurf up to a well-defined line near apex, sometimes also beaded with orange droplets; cap viscid when moist, yellow-orange to tawny, orange, rusty-brown, or greenish .... *T. aurantium* & others, p. 187
11. Not as above 12

12. Base of stalk (or interior at base) usually pale pinkish to pinkish-orange; cap not viscid, its color variable but usually greenish to olive-gray, sometimes shaded with brown, or grayish at center and pallid at margin; surface smooth or cracking, but *without* differently colored fibrils or hairs .... *T. saponaceum*, p. 184
12. Not as above 13

13. Stalk with a tapered "tap root" extending deep into humus; stalk usually with a tough or cartilaginous outer rind .... (see *Caulorhiza umbonata* & others, p. 218)
13. Not as above 14

14. Odor pungent and very unpleasant (like coal-tar gas); cap pallid; gills widely spaced; associated with conifers .... *T. platyphyllum* (see *T. sulphureum*, p. 179)
14. Not as above 15

15. Cap white to creamy when fresh (but may discolor or age yellow, reddish, tan, brown, etc.); stalk *lacking* small yellow dandruffy scales or granules at apex 16
15. Cap distinctly colored or at least with colored scales or fibrils, even when young, *or* if cap white then stalk with yellow scales or granules at apex or growing under mountain conifers soon after the snow melts 21

16.   Fruiting body often huge (15-75 cm broad!); known from Florida .... *T. titans*
16.   Not as above 17

17.   Fruiting body soon developing reddish or rusty stains; taste often bitter; stalk usually longer
      than width of cap .... (see *Collybia maculata,* p. 217)
17.   Not as above 18

18.   Cap with at least a few scattered grayish scales or a grayish center ..... *T. pardinum,* p. 183
18.   Not as above 19

19.   Cap viscid when moist, white or in age often tinged or stained yellow, 4-8 cm broad; stalk usually
      scurfy-scaly; spores amyloid .... (see *Melanoleuca,* p. 169)
19.   Not as above; spores not amyloid 20

20.   Stalk usually hollow and as long or longer than width of cap, often slender; usually growing
      under redwood (in California) .... (see **Hygrophoraceae,** p. 103)
20.   Not as above; stalk not hollow; occurring with various trees but not redwood ....
      .... *T. resplendens* & others, p. 183

21.   Found under mountain conifers in spring; odor strongly cucumber-like or fishy ....
      .... (see *Armillaria* & **Allies,** p. 189)
21.   Not as above 22

22.   Cap black to gray, grayish-brown, or purplish-gray, or whitish with gray to blackish scales or
      fibrils 23
22.   Not as above 32

23.   Cap viscid when moist (but may dry out), often streaked ... *T. portentosum* & others, p. 180
23.   Cap not viscid 24

24.   Cap whitish with scattered pale gray to dark gray scales, at least at center; stalk thick (1.5-3 cm)
      and fleshy .... *T. pardinum,* p. 183
24.   Not as above; cap usually darker 25

25.   Taste acrid .... *T. acre* (see *T. virgatum,* p. 181)
25.   Taste not acrid 26

26.   Cap densely hairy, scaly, or fibrillose-scaly (scales sometimes sparse in age); cap *not* usually
      umbonate at maturity; veil sometimes present when very young 27
26.   Cap smooth to radially fibrillose or streaked, sometimes conical or umbonate at maturity; veil
      absent 29

27.   Fruiting body fairly small (cap usually less than 8 cm broad; stalk less than 1.5 cm thick) . 28
27.   Fruiting body fairly robust (cap 5-18 cm broad; stalk typically at least 1 cm thick) ....
      .... *T. atroviolaceum* & others (see *T. virgatum,* p. 181)

28.   Stalk with small grayish to blackish scales .. *T. squarrulosum* (see *T. terreum* group, p. 182)
28.   Stalk lacking grayish to black scales .... *T. terreum* group & others, p. 182

29.   Growing on or near wood, or if not then stalk usually attached to white mycelial cords that
      arise from wood; gills broad (deep) .... (see *Tricholomopsis platyphylla,* p. 146)
29.   Not as above; typically terrestrial 30

30.   Cap streaked with radiating fibrils, conical when young, often with a pointed umbo in age ..
      .... *T. virgatum,* p. 181
30.   Not as above; cap without fibrils, broadly convex to plane or with a low umbo 31

31.   Cap with a hoary bloom when young; gills grayish; growing under mountain conifers, usually
      near melting snow; spores *not* amyloid .... (see *Lyophyllum montanum,* p. 175)
31.   Not with above features (but may grow near snow); gills usually white or whitish 32

32.   Gills crowded, pale; spores amyloid, or if not then odor often heavy and sweet ....
      .... (see *Marasmius, Collybia,* & **Allies,** p. 201)
32.   Not as above 33

33.   Fibrillose veil present when young, sometimes forming a slight ring on stalk 34
33.   Veil absent 35

34.   Cap at least slightly viscid when moist; growing with both hardwoods and conifers ....
      .... (see *Limacella glioderma,* p. 291)
34.   Cap dry, not viscid; associated with conifers (mainly pine and spruce) ... *T. vaccinum,* p. 186

35.   Stalk apex with small yellow scales or granules ....
      .... *T. acerbum* & others (see *T. pessundatum* group, p. 185)
35.   Not as above 36

36. Cap viscid when moist, dark to medium reddish-brown to reddish-tan, the margin often paler; gills often developing reddish-brown spots or stains in age 37
36. Cap not viscid (but may be colored as above) 39
37. Associated with poplar or cottonwood, usually in sandy soil . . . . *T. populinum,* p. 185
37. Not as above; associated with other trees 38
38. Odor distinctly farinaceous or cucumberlike . . . . *T. pessundatum group* & others, p. 185
38. Odor more or less mild . . . . *T. ustale* (see *T. pessundatum* group, p. 185)
39. Spore print creamy to yellowish; gills tan to ochre or pale pinkish-cinnamon in age; stalk straight, usually long; cap smooth, without scales; spores amyloid (see *Melanoleuca,* p. 169)
39. Not as above; spore print white; spores not amyloid 40
40. Typically growing in disturbed soil (along roads, paths, etc.); cap smooth, grayish-brown to dark brown to brown or tan, etc. (but never reddish-brown) . . . . . (see *Lyophyllum,* p. 173)
40. Not as above; growing in woods 41
41. Cap grayish-olive-brown . . . . *T. sp.* (unidentified) (see *T. virgatum,* p. 181)
41. Cap brown to reddish-brown, cinnamon-brown, pinkish-brown, flesh-colored, etc. 42
42. Stalk 1-3 cm thick; cap often smooth when young and scaly in age, especially toward the margin; margin naked . . . . *T imbricatum,* p. 186
42. Stalk usually less than 1.5 cm thick; cap scaly or fibrillose-scaly even when young; margin usually with hairy or woolly veil remnants when young . . . . *T. vaccinum,* p. 186

Tricholoma sulphureum   (The Stinker)

CAP 2-8 cm broad, convex to umbonate or sometimes plane; surface dry, smooth, yellow or sometimes olive-yellow, or tinged brownish to grayish-brown at the center. Flesh rather thin, yellow; odor usually strong and repulsive, like coal tar gas. **GILLS** typically adnexed or notched, broad, thick, well-spaced, yellow. **STALK** 4-10 cm long, 0.5-1 cm thick, often rather long in relation to cap; more or less equal, smooth, dry, yellow to olive-yellow or with darker fibrils. **VEIL** absent. **SPORE PRINT** white; spores 8-12 × 5-6 microns, elliptical, smooth.

HABITAT: Scattered to gregarious on ground under both hardwoods and conifers, widely distributed. It is fairly common in the Pacific Northwest under conifers and also occurs in northern California in the late fall and winter, but I have yet to find it in our area.

EDIBILITY: Indisputably inedible because of the obnoxious odor; possibly poisonous.

COMMENTS: While several Tricholomas have a distinctly disagreeable odor, this species has a downright disgusting one. By this trait alone it can be distinguished from the edible man on horseback, *T. flavovirens,* which it rather resembles in color, and from *T. sejunctum* as well. Other species: Another mushroom with the same odor is *T. platyphyllum (=T. inamoenum?),* a white to creamy species also found under conifers in the Pacific Northwest.

Tricholoma flavovirens   (Man On Horseback)   Color Plate 33

CAP 4-15 (20) cm broad, convex becoming plane or with margin uplifted in age; surface viscid when moist, smooth, entirely yellow or brown to reddish-brown toward the center and yellow at the margin, or sometimes olive-yellow; margin at first inrolled. Flesh thick, firm, white; odor farinaceous or sometimes like coconut. **GILLS** close, broad, notched or adnexed, yellow. **STALK** 3-10 cm long, 1-3 (4) cm thick, equal or enlarged slightly at either end, dry, smooth, solid; white to pale yellow or sometimes with darker stains at base. **VEIL** absent. **SPORE PRINT** white; spores 6-8 × 4-5 microns, elliptical, smooth.

HABITAT: Scattered to densely gregarious under pines (rarely other conifers), often partially buried or visible only as "mushrumps" in the duff; widely distributed. Common in our area in the late fall and winter, usually in grassy, sandy, or shrubby areas with pine present; however, a bright yellow form is found occasionally with madrone, and in the Southwest it is quite common under aspen.

EDIBILITY: Edible and excellent—one of the least appreciated and most flavorful of our fleshy fungi, though a few people are adversely affected by it. The viscid cap should be brushed clean in the field or the skin peeled off, and the sand removed (if it is present).

COMMENTS: Formerly known as *T. equestre,* this delectable mushroom is as dependably yellow as the blewit is purple. It lacks the blackish radial fibrils at the center of the cap characteristic of *T. sejunctum,* and it also lacks the veil characteristic of *Armillaria albolanaripes* and various yellow *Cortinarius* species. *T. sulphureum* is similarly colored but has a dry rather than viscid cap and usually smells awful. The yellow color plus the sticky cap (when moist), absence of a veil, white spores, and habit of hiding under pine needles combine to make it one of the safest—as well as tastiest—of gilled mushrooms. None of which explains the misnomer "Man On Horseback"—it doesn't look anything like a horse, and most of the horseback riders I see are women . . .
    One local patch of *T. flavovirens* produces crops that *sometimes* have a very strong coconut odor and taste, and at other times have the usual mealy (farinaceous) odor. A very similar edible species, *T. leucophyllum,* is common under aspen in the Rocky Mountains and Southwest, and may actually mingle with *T. flavovirens.* It differs only in having white rather than yellow gills.

Tricholoma sejunctum

CAP 3-8 (10) cm broad, convex to plane or broadly umbonate; surface slightly viscid or tacky when moist, smooth, yellow or greenish-yellow with dark innate (flattened) fibrils or streaks radiating from the blackish to brown center; sometimes with small scales in age. Flesh white or tinged yellow; odor farinaceous, taste often bitter or nauseating, but in some forms mild. **GILLS** fairly close, typically notched; at first whitish or creamy-white, but often becoming yellow near the margin of the cap or occasionally yellowish throughout. **STALK** 5-8 (12) cm long, 1-2 (3) cm thick, more or less equal or somewhat swollen below, firm, smooth, whitish, but often developing yellowish tints. **VEIL** absent. **SPORE PRINT** white; spores 5-7 × 4-5.5 microns, broadly elliptical or elliptical, smooth.

HABITAT: Scattered or in groups under both hardwoods and conifers, fruiting mainly in the fall; widely distributed. It is fairly common in the Pacific Northwest, and I have seen luxuriant fruitings in California in mixed woods and under manzanita.

EDIBILITY: Not recommended—it is insipid at best and poisonous at worst.

COMMENTS: This species is likely to be mistaken for the edible *Tricholoma flavovirens* because of its yellowish cap color, but the radiating blackish or dark brown fibrils or streaks at the center of the cap and the tendency of the gills to show yellow only near the cap margin should distinguish it. Other species: *T. cheilolamnium* is very similar but has a dry cap; it is fairly common in the Pacific Northwest under conifers.

Tricholoma portentosum   (Streaked Tricholoma)

CAP 4-12 cm broad, convex to obtusely umbonate or plane, or with uplifted margin in age; surface viscid when moist, smooth but with a streaked or radially fibrillose appearance, pale gray to dark gray, brownish-gray, or purplish-gray, the center sometimes nearly black and margin often paler; yellow tints occasionally present also, especially in age. Flesh white or tinged gray, fairly thick; odor and taste mild to farinaceous. **GILLS** adnexed or notched, at first white, becoming grayish or sometimes pale yellow in age;

Tricholoma portentosum has a dark (purple-gray to blackish) streaked viscid cap, whitish gills, white stalk, and white spores. Note how the gills are notched.

fairly close. **STALK** 5-10 cm long, 1-2.5 cm thick, more or less equal, firm, smooth, dry, white or sometimes tinged yellow. **VEIL** absent. **SPORE PRINT** white; spores 5-7 × 3-5 microns, elliptical, smooth.

HABITAT: Scattered to gregarious on ground in woods, widely distributed. In most regions it occurs with conifers, particularly pine, but in our area it favors live oak and tanoak. It fruits in the late fall and winter and is one of the last Tricholomas to appear.

EDIBILITY: Edible and excellent, with a strong hearty flavor—but be sure of your identification before eating it!

COMMENTS: The viscid cap separates this species from a multitude of grayish, dry-capped Tricholomas, some of which are poisonous (see *T. pardinum* and *T. virgatum*). The notched gills, fleshy white stalk, white spore print, absence of a veil, and gray to purple-gray streaked cap are also important fieldmarks. *T. sejunctum* is somewhat similar, but much yellower as a rule. *Entoloma madidum* is also similar, but has pinkish spores (and mature gills) and usualy has a darker (cap-colored) stalk. *T. niveipes* is a very similar edible pine-loving easterner; it never develops yellow tones and has narrower spores.

Tricholoma virgatum

CAP 3-8 (10) cm broad, conical to broadly conical to nearly plane with a pointed umbo; surface dry, grayish to grayish-brown or grayish-purple (the center often darker, margin paler), streaked with radiating fibrils or fibrillose scales. Flesh thin, white becoming grayish; odor mild or earthy, taste usually sharp or acrid. **GILLS** adnexed or notched, white to grayish, close. **STALK** 6-12 (15) cm long, (0.5) 1-2 cm thick, more or less equal, solid, smooth or fibrillose, white or tinged gray. **VEIL** absent. **SPORE PRINT** white; spores 6-7.5 × 5-6 microns, elliptical, smooth.

HABITAT: Solitary or scattered to gregarious in mixed woods and under conifers, widely distributed; occasionally found in our area in the winter.

EDIBILITY: Not recommended—it may be poisonous, and it resembles *T. pardinum,* which is *definitely* poisonous.

COMMENTS: The dry, fibrillose-streaked grayish cap which is conical when young affords a good means of recognizing this species, which is also known as *T. subacutum.* The cap is never viscid as in *T. portentosum,* and the fibrils and/or scales are radially arranged, in contrast to the *T. terreum* group. Other species: *T. atroviolaceum* is a large (cap 5-18 cm), robust species with a convex to plane (not conical), densely fibrillose-scaly, blackish to dark grayish-brown or violet-tinted cap. It has grayish or pinkish-tinged gills, a thick, brownish, sometimes bulbous stalk, and a farinaceous odor; it is not uncommon in California and the Pacific Northwest under conifers. A similar robust, unidentified species is common in our area under hardwoods; it has a slightly paler (olive, dark gray, or brown) cap and paler stalk. *T. acre*, fairly common under hardwoods in eastern North America and the Rocky Mountains, is also similar but slightly smaller (cap 3-9 cm), with a grayish fibrillose cap, white to grayish gills, and a distinctly acrid taste.

Tricholoma terreum group   (Mouse Tricholoma)

CAP 2-5 (7) cm broad, conical or convex-umbonate becoming more or less plane in age; surface dry, hairy or felty from a dense layer of mouse-colored (gray to black) scales; scales often sparser and color often paler (silvery-gray or even white) in age and/or toward margin. Flesh thin, fragile, white to gray; odor and taste mild to slightly farinaceous. **GILLS** adnate to adnexed or notched, fairly close, white to gray. **STALK** 2-5 (8) cm long, 0.5-1 cm thick, usually rather slender, equal or slightly thicker below, white or tinged gray, smooth (but see comments), dry. **VEIL** absent or cobwebby and evanescent. **SPORE PRINT** white; spores 6-8 × 3.5-5.5 microns, elliptical, smooth.

HABITAT: Scattered to densely gregarious on ground under conifers; widely distributed. Fairly common in our area in the late fall and winter under pine and Douglas-fir. The largest fruiting I've seen was on a lawn under a huge pine.

EDIBILITY: Not recommended. The flavor of our local variety is good, but this is a species "complex" and is best avoided until better known. There is also the danger of confusing larger specimens with *T. pardinum* or *T. virgatum*.

COMMENTS: The small size (for a *Tricholoma*), fragile flesh, and dry, scaly or "furry" mouse-colored cap characterize a complex of confusing, closely related species best left to the specialists. It is probable that they are all edible, but we can't be *sure* until they are better known. The most common North American member of the club is said to be *T. myomyces*, which has an evanescent cortina (cobwebby veil) and smaller spores (the "true" *T. terreum* lacks a cortina). Other species in the group include: *T. squarrulosum,* also very common in our area, but with mouse-colored fibrils or scales on the *stalk* as well as the cap (see photo below), and several slightly larger species, e.g., *T. orirubens,* with gills that redden in age; *T. argyraceum (=T. scalpturatum?),* which favors hardwoods and has gills that yellow in age, and *T. acre* (see comments under *T. virgatum*).

Two small, common, mouse-colored conifer-lovers. **Left:** *Tricholoma squarrulosum* (see comments above) has a fibrillose-scaly cap *and* stalk. **Right:** Close-up of the smooth white stalk of *T. terreum* (left hand specimen) and the scaly stalk of *T. squarrulosum* (on right).

Tricholoma pardinum   (Tiger Tricholoma)

CAP 5-16 (25) cm broad, convex to plane; surface dry, whitish with small pale gray to dark gray fibrillose or spotlike scales, at least at center. Flesh thick, firm, white; odor farinaceous. GILLS notched or adnexed, close, white (rarely flushed pinkish), not stained or spotted gray. STALK 4-15 cm long, 1.5-3 cm thick, equal or enlarged below, white or sometimes tinged gray, smooth, firm, solid. VEIL absent. SPORE PRINT white; spores 7-10 × 5-6.5 microns, elliptical, smooth.

HABITAT: Solitary to scattered or gregarious on ground in woods, widely distributed in northern North America. It is sometimes abundant under conifers in the Pacific Northwest and Rocky Mountains, but in our area it grows with tanoak and madrone in the winter. It is sporadically common, i.e., abundant once every several years but otherwise infrequent to rare.

EDIBILITY: Poisonous! It causes severe and persistent gastroenteritis that may require hospitilization.

COMMENTS: Anyone tempted to eat Tricholomas should learn to recognize this poisonous species. It is larger and fleshier than other grayish Tricholomas (*T. terreum, T. virgatum*, etc.), and often paler. At times it is nearly white with a few very pale grayish scales, and could then be mistaken for one of the white Tricholomas (see *T. resplendens*). The stalk is not pinkish-orange in the base as in *T. saponaceum*.

Tricholoma resplendens   (White Tricholoma)

CAP 4-10 cm broad, convex becoming plane or with margin uplifted; surface viscid when moist, often shiny when dry; smooth, white or with yellowish center, often discoloring slightly brownish in age, especially toward center. Flesh firm, white, odor mild. GILLS notched or adnexed to occasionally adnate, white, close. STALK 3-8 cm long, 1-2.5 cm thick, equal or narrowed at base, white, smooth, dry, solid. VEIL absent. SPORE PRINT white; spores 5-7.5 × 3.5-4.5 microns, elliptical, smooth.

HABITAT: Solitary or scattered or in small groups in woods, especially with oaks; of questionable occurrence in California, but I have encountered a very similar species under live oak in the winter. In eastern North America this species or one very similar is sometimes common.

EDIBILITY: Not recommended—it is said to be edible, but is easily confused with poorly known or poisonous species (see comments below).

COMMENTS: This species is one of several white or whitish Tricholomas, and can be distinguished by the viscid cap when moist (however, see *Melanoleuca lewisii* in the key to that genus, for it also has a viscid cap when moist, but tends to age yellowish and has amyloid spores). Tricholomas with a non-viscid, whitish cap include: *T. sulphurescens,* which stains or discolors yellow quite readily; *T. venenata,* poisonous and common under hardwoods in eastern North America, with cap, gills, and stalk that discolor brownish-buff in age or where injured, and a bitter taste; *T. album,* taste also bitter or sharp, but not discoloring so much; *T. columbetta,* edible, cap often spotted with blue, green, or pink in age, gills crowded, and odor mild; and *T. gambosum (=T. georgii, Calocybe gambosa),* edible, large and fleshy, with a white to creamy-buff cap, crowded gills, and strongly farinaceous odor, favoring grassy areas and open woods. Whether any of these species occur in California is uncertain, but they are to be expected. They might be confused with certain *Hygrocybe* species (e.g., *H. subaustralis, H. fornicata),* but the latter have hollow or stuffed stems and slightly waxy gills, and are most common under redwood in our area. Also see *T. manzanitae* (under the *T. pessundatum* group), which can be whitish in youth.

Tricholoma saponaceum is an extremely variable species. Cap at top is greenish-gray, while the small one at the bottom is brown. In all specimens, however, the flesh in the stalk base is pinkish to orange.

Tricholoma saponaceum (Soapy Tricholoma)

CAP 4-12 (18) cm broad, convex to plane or with uplifted, often wavy margin; surface dry or moist but not viscid, smooth or cracking into scales in dry weather; color variable: olive to greenish-gray, gray, yellowish-olive, brownish-olive, grayish-brown, coppery, or with rusty tints, or sometimes dingy gray at center and pallid toward the margin. Flesh thick, white, but may stain slowly yellowish or pinkish when bruised; odor and taste mild to farinaceous, soapy, or "of wash rooms." **GILLS** adnate to adnexed or notched, well-spaced, rather thick, white or tinged olive or yellowish, sometimes stained reddish. **STALK** 5-12 (20) cm long, 1-3 cm thick, shape variable but often thickest in the middle and tapered below to a somewhat rooting base; solid, smooth or with small scales, white or tinted variously with the cap color; base usually with pale pinkish to pinkish-orange interior. **VEIL** absent. **SPORE PRINT** white; spores 5-6 × 3-4 microns, elliptical, smooth.

HABITAT: Solitary, scattered, tufted, or in groups or troops under both hardwoods and conifers, widely distributed. Common under conifers (especially spruce) throughout much of the West, but in our area found under live oak, tanoak, and madrone in the late fall and winter.

EDIBILITY: Inedible—it has an insipid or soapy taste and may actually be poisonous.

COMMENTS: As evidenced by the lengthy description, this is a vexingly variable species and is rather difficult to recognize when greenish shades are not evident on the cap. One fairly infallible (or less unreliable) feature, however, is the pinkish-orange color of the flesh at or near the base of the stalk. This feature is normally visible in the majority of specimens from any group, and serves to distinguish them from other Tricholomas. The gills are rather soft and well-spaced, leading to confusion with *Hygrophorus,* but the cap is not truly viscid. The cap is most often some shade of grayish-olive or yellowish-green, but may develop brown or coppery tones, especially in dry weather. It may crack into scales, but lacks the fibrillose scales of *T. virgatum, T. pardinum,* and others. In the Pacific Northwest and Rocky Mountains *T. saponaceum* is at times so overwhelmingly abundant that it has been called a mushroom "weed." In our area it is not quite so prevalent and its beauty is more readily appreciated. One especially attractive form has a bluish-green cap.

184

Tricholoma pessundatum group   **Color Plate 35**

CAP 5-14 (18) cm broad, convex then plane or with slightly uplifted margin; surface viscid when moist, entirely reddish-brown to reddish-tan, or often with a paler (or even whitish) margin; smooth or sometimes finely scaly in age; margin often lobed, at first inrolled, sometimes faintly ribbed. Flesh thick, firm, white; odor strongly farinaceous or like linseed oil. **GILLS** white, but often developing sordid reddish or reddish-brown spots and stains; typically notched or adnexed, but at times adnate or even free; close. **STALK** 4-10 (14) cm long, 1-3 cm thick, equal or swollen or tapered below, solid, firm; whitish or developing sordid reddish or brownish stains or fibrils, especially over lower portion. **VEIL** absent. **SPORE PRINT** white; spores 4-6 × 2.5-4 microns, elliptical, smooth.

HABITAT: Scattered to densely gregarious in forests or under trees, widely distributed. It is partial to conifers in most areas (including the Pacific Northwest), but is locally common in the late fall and winter with live oak, and less commonly pine.

EDIBILITY: To be avoided, as evidenced by the following severely censored excerpt from an erstwhile colleague's memoirs: "This common, viscid, red-brown *Tricholoma* is delicious when stewed slowly with zucchini and served steaming hot on rice with chicken chow mein and white wine. Suffering from an acute attack of overconfidence, M. Henis and C. Cole tried it in this manner one winter evening in order to determine its edibility. They subsequently staggered thrugh an all-night ordeal of nausea, vomiting, and diarrhea, in which not only the mushroom, but everything else, was expelled . . ."

It is probable, then, that this common, viscid, red-brown *Tricholoma* is poisonous, although a violent allergic reaction on the part of M. Henis and C. Cole cannot be ruled out entirely. It is suggested that those foolish enough to try it (or any other mushroom of unknown edibility) should do so in extremely small amounts—*without* the rice, chicken chow mein, and white wine, and by all means, regardless of one's nutritional needs, culinary quirks, or dietary deficiencies, *without* the zucchini.

COMMENTS: Several robust, viscid, reddish-brown Tricholomas with reddish-spotted gills will more or less fit the above description, and I leave it to licensed tricholomatologists to decide whether or not ours is the "true" *T. pessundatum.* Closely related species include: *T. albobrunneum,* said to have a weaker odor and cap finely streaked with darker lines; *T. ustaloides,* with a transient cortina (hairy veil) and sharply defined white zone at the stalk apex; *T. flavobrunneum (=T. fulvum),* with pale yellow gills when young and yellow-tinted flesh in the stem; and *T. ustale,* which lacks a farinaceous odor. (See also *T. populinum.*) All of the above have viscid caps when moist, and none have the belted scales on the stalk or the veil characteristic of *T. aurantium* and *T. zelleri* respectively. Other species: *T. manzanitae* is a manzanita- and madrone-loving Californian with pale yellow granules or dandruffy scales at the stalk apex and a viscid cap that ranges from white (when young) to pinkish, orangish, or brown (often with reddish stains). *T. acerbum* is also said to have a yellow-dandruffy stalk, but has a strongly inrolled, ribbed cap margin (at least until maturity) and a fragrant odor and/or sharp taste. None of the above should be eaten.

Tricholoma populinum   (The Sandy; Poplar Tricholoma)

CAP 5-16 cm broad, convex with an inrolled margin becoming plane or with uplifted margin; surface viscid when moist, then dry, often radially streaked or with watery spots; smooth, dull reddish-cinnamon to pale dingy reddish-brown, the margin usually paler or whitish. Flesh firm, white, thick; odor and taste strongly farinaceous. **GILLS** typically adnexed or notched, close, white, developing reddish-brown spots and stains, especially on edges. **STALK** 2.5-7.5 cm long, 1-3 cm thick, equal or enlarged below, solid, firm, dull whitish, developing dingy reddish-brown stains in age or after handling. **VEIL** absent. **SPORE PRINT** white; spores 5-6 × 3.5-4 microns, elliptical, smooth.

HABITAT: Scattered to densely gregarious, frequently fruiting in large rings or dense masses in sandy soil or along rivers, apparently always in association with poplar or cottonwood. It fruits in cool weather and is widely distributed in western North America. I have found it only once in our area, in December, but it is commoner inland. It is said to be abundant in the John Day country of eastern Oregon in the fall and early winter, and I have seen very old specimens (perhaps from the previous fall?) in the spring near Pecos, New Mexico, while looking for morels and wild asparagus.

EDIBILITY: Edible and popular in the Pacific Northwest, but be *absolutely sure* it is associated with cottonwood—the similar *T. pessundatum* group is poisonous!

COMMENTS: The association with cottonwood and somewhat paler cap distinguish this species from other viscid, red-brown Tricholomas (see the *T. pessundatum* group). Its penchant for growing in densely-packed masses or long arcs is also distinctive.

Tricholoma vaccinum

CAP 4-7 (10) cm broad, broadly conical to convex, becoming umbonate or plane; surface dry, covered with dark reddish-brown to rusty-cinnamon-brown to pale pinkish-brown, tan, or flesh-colored fibrils or scales on a buff background; often darker at center; margin with hairy veil remnants at least when young, often splitting in age. Flesh white or pallid; odor usually farinaceous but sometimes mild. **GILLS** adnate becoming notched, close, whitish or buff when young, but usually tinged flesh-color to pale cinnamon in age; sometimes also with darker stains. **STALK** 3-8 cm long, 0.8-1.5 cm thick, equal or thicker at either end, dry, smooth or with brownish to reddish-brown fibrils or small scales, usually hollow at least in age. **VEIL** woolly-fibrillose, not forming an annulus (ring) on stalk, but usually leaving traces on cap margin. **SPORE PRINT** white; spores (4) 6-7.5 × 4-5 microns, elliptical, smooth.

HABITAT: Scattered or in small tufts, groups, or large troops under conifers, especially pine and spruce; common and very widely distributed, fruiting from late summer through early winter. I have seen enormous fruitings under spruce in the Rocky Mountains and under pine on the northern California coast.

EDIBILITY: Listed as mildly poisonous by some authors. Like cheap coffee and frozen French fries, it is best avoided.

COMMENTS: One of the commonest Tricholomas of the coniferous forests of North America, this species often fruits with *T. imbricatum,* but is apparently replaced by that species in our local coastal pine forests. The two are quite similar, but *T. vaccinum* has a scalier cap, frequently hollow stalk, and woolly veil which normally leaves hairs on the cap margin. Several color forms occur, ranging from dark reddish-brown to pale pinkish-brown, and the size is also variable. Usually it is smaller and more slender than *T. imbricatum,* but in northern California and Oregon a fairly robust, reddish-brown form occurs.

Tricholoma imbricatum

CAP 4-12 (20) cm broad, convex with an inrolled margin, becoming convex-umbonate to plane or uplifted; surface dry, dark brown to brown or cinnamon-brown, with flattened fibrils that may break up into scales in age, especially toward margin (which may be obscurely ribbed). Flesh thick, firm, white; odor mild or faintly farinaceous. **GILLS** adnexed, notched, or even adnate; close, white or tinged flesh-color, often discoloring brown in age, especially on the edges. **STALK** 4-12 cm long, 1-3 cm thick, solid, firm, dry, equal or swollen below with a tapered, sometimes rooting base; white or buff becoming brownish in age, especially over lower portion (apex usually pallid); fibrillose or minutely scaly in age. **VEIL** absent. **SPORE PRINT** white; spores 5-7 × 3.5-5 microns, elliptical, smooth.

Tricholoma imbricatum, young specimens. This very common conifer-lover has a dull brown dry cap. In age the cap often flattens out or becomes wavy. For close-up of gills, see photo on next page.

HABITAT: Solitary to scattered or densely gregarious under conifers, particularly pine and spruce (often hidden by needles); widely distributed. It is a prominent fungal feature of our coastal pine forests in the winter and early spring.

EDIBILITY: Reportedly edible, but not recommended because it is easily confused with members of the poisonous *T. pessundatum* group. It is rather tough anyway.

COMMENTS: This common *Tricholoma* can be told from other members of the genus by its dry, dull brown cap, solid stem, absence of a veil, and generally coarse, robust appearance, though slender individuals also occur. The species epithet, which means "shingled," is somewhat misleading, since the cap is usually quite smooth in youth and only somewhat scaly in age (it is rarely truly "shingled" in the way that the tooth fungus *Hydnum imbricatum* is). Its closest relative, *T. vaccinum,* has a scalier cap, evanescent veil, and hollower stem, while the *T. pessundatum* group has a viscid cap (at least when moist). It looks somewhat like a *Russula,* but has a tough, fibrous stem. *Leucopaxillus amarus* also has a brown cap, but its gills remain white and it has a bitter taste and amyloid spores.

Tricholoma aurantium

CAP 4-10 cm broad, convex becoming obtusely umbonate or plane; surface viscid when moist, smooth or breaking into small scales (especially at center); color variable: yellow-orange to tawny, bright rusty-orange, orange-brown, orange-tan, or even orange-red, sometimes splashed with olive-green or in one form entirely deep olive-green when young; margin at first inrolled, sometimes beaded with orange droplets when moist. Flesh thick, white; odor and taste strongly farinaceous and disagreeable (like rancid oil or cucumber). **GILLS** adnate to adnexed or notched, close, whitish, often developing rusty-brown or reddish-brown spots and stains. **STALK** 3-8 cm long, 0.8-2 cm thick, equal or thicker at either end, solid, firm, belted with rusty-orange scales or scurfy flakes up to a well-defined line near apex, pallid above the line; in wet weather sometimes beaded with orange droplets near the line. **VEIL** absent or very rudimentary. **SPORE PRINT** white; spores 4-6 × 3-5 microns, elliptical to nearly round, smooth.

HABITAT: Solitary or scattered to gregarious on ground in woods, widely distributed. Throughout most of the West it is common under conifers or sometimes aspen; in our area it can be found under madrone in the winter, but is fairly rare.

EDIBILITY: Indisputably unpalatable due to the obnoxious odor and taste.

COMMENTS: This species strongly resembles *T. zelleri* but lacks a membranous veil. Its sharply defined line near the stalk apex, however, is suggestive of a veil and probably represents a rudimentary one. The viscid, orange to orange-brown or olive-splashed cap, rusty-spotted gills, and strong odor help distinguish it. Our local form (var. *olivascens*?)

Left: *Tricholoma imbricatum,* close-up of gills. **Right:** *Tricholoma aurantium,* showing characteristic belts of scales or granules on stalk.

is frequently a deep olive-green when young, and when beaded with orange droplets is quite striking. *T. aurantio-olivaceum* is similar in many respects, but is smaller and odorless, and often has olive-stains on the gills in age; it occurs in the Pacific Northwest.

Tricholoma zelleri

CAP 4-15 cm broad, convex becoming plane or broadly umbonate; surface viscid when moist, bright orange to yellow-orange, or orange-brown, or sometimes splashed with olive-green; margin at first hung with veil remnants. Flesh thick, white, slowly bruising orange-brown; odor and taste strongly rancid-farinaceous. **GILLS** white, developing rusty-orange-brown stains, close, adnate or notched. **STALK** 4-13 cm long, 1-3 cm thick, usually tapered downward, solid, dry, pallid above the ring, usually somewhat scaly or with orange or brown stains below. **VEIL** white, membranous, forming a flaring or ragged, median to superior ring on stalk which frequently collapses in age. **SPORE PRINT** white; spores 4-5.5 × 3-4 microns, elliptical, smooth, not amyloid.

HABITAT: Scattered to gregarious on ground in woods, northern North America.

Tricholoma (=Armillaria) zelleri has the stature of a typical *Tricholoma,* but possesses a well-developed membranous veil that usually forms an annulus (ring) on the stalk.

Extremely abundant under conifers in the Pacific Northwest (often in the same areas as the matsutake, *Armillaria ponderosa*), but rather rare in our region and fruiting mainly in tanoak-madrone woods at higher elevations in the coastal mountains (like *A. ponderosa*), in the late fall and early winter. I have also seen it fruiting in large numbers with *Armillaria caligata* under spruce in the southern Rockies.

EDIBILITY: Not edible because of the unpleasant taste and smell; however, it is a good matsutake-indicator!

COMMENTS: This is essentially a veiled version of *T. aurantium*—same color, odor, taste, and habitat. It is better known as *Armillaria zelleri,* but because of the obvious affinity with *T. aurantium,* it is now placed in *Tricholoma*. The sticky yellow-orange to orange, brown, or greenish-splashed cap, plus the attached gills, presence of a membranous veil, and white spores are diagnostic. It might possibly be confused with *Limacella glioderma,* which has a redder cap, fibrillose veil, and fragile stem. Other species: *T. robustum* and *T. focale* are very similar if not the same (they are said to be reddish-brown in color).

ARMILLARIA & Allies

Medium-sized to very large, fleshy mushrooms found on ground or wood. CAP convex to plane. GILLS *attached.* STALK central, *fleshy.* VEIL *typically present, well-developed, usually forming an annulus (ring) on stalk.* VOLVA absent. SPORE PRINT *white or tinged yellow.* Spores smooth, amyloid *(Catathelasma),* to weakly amyloid or not amyloid *(Armillaria* and *Armillariella).*

GROUPED here are three small genera of fleshy, white-spored mushrooms with a cottony or membranous veil that usually forms a distinct ring on the stem. There are no warts on the cap nor is there a volva on the stalk as in *Amanita,* the gills are not free as in *Lepiota* and *Limacella,* nor soft and waxy as in *Hygrophorus,* and the cap and stalk are not covered with mealy granules as in *Cystoderma. Tricholoma* intergrades somewhat with *Armillaria,* but as defined here does not usually have a veil (but see *T. zelleri!*).

The principal genus, *Armillaria,* is comprised of terrestrial forest mushrooms. The honey mushrooms, *Armillariella,* somewhat resemble *Armillaria* but grow on wood, often in large clusters. They were originally placed in *Armillaria* and some mycologists retain them in that genus while transferring the Armillarias of this book to *Tricholoma* and a separate genus, *Floccularia.* The third genus, *Catathelasma,* is terrestrial like *Armillaria,* but has a double-layered veil, decurrent gills, amyloid spores, and a hard, often massive fruiting body. It seems to be restricted to coniferous forests and is rather rare.

Several mushrooms in this group are prized edibles. The matsutake of Japan *(Armillaria matsutake)* and its magnificent North American counterpart *(A. ponderosa)* are highly esteemed by Japanese- and Korean-Americans. The honey mushroom *(Armillariella mellea)* is well known and popular among fungophiles, while being well known and singularly *unpopular* among gardeners and farmers. It grows wherever there are trees and shrubs (even grape vines) and is almost as common in towns as in the woods. It has been called the most serious plant disease in California gardens, because once it has infected a bush or tree, there is no cure. One can only hope that it will co-exist with—rather than kill—its host, and make the best of a sad situation by harvesting the mushroom bounty when it appears! Six species of *Armillaria, Armillariella,* and *Catathelasma* are depicted here.

Key to Armillaria & Allies

1.   Growing on wood (may be buried!) or at the bases of trees, sometimes in large clusters 2
1.   Growing widely scattered to gregarious on ground (not normally in large clusters) 5
2.   Cap with rusty-brown scales; stalk sheathed with similarly colored fibrils below the veil; gills
        typically *not* decurrent; found on hardwoods in eastern North America; rare .. *A. decorosa*
2.   Not as above; common 3

3. Cap some shade of yellow, tan, or brown; taste usually bitter (but not detectable by everyone); stalk fibrous, with a stringy white pith inside, often long 18
3. Not as above; cap paler in color or the stalk solid and hard 4

4. Cap often scaly; nearly always on or near conifers ..... (see *Lentinus* & *Lentinellus,* p. 141)
4. Not as above; usually found on hardwoods .... (see *Pleurotus* & Allies, p. 132)

5. Gills typically decurrent; fruiting body hard and thick-fleshed, often large (cap 7-40 cm broad!); odor variable but not spicy-fragrant; spores amyloid; found with northern conifers 6
5. Gills not decurrent, or if decurrent then not as above (not hard and thick-fleshed, etc.) 7

6. Cap dull white to grayish .... *Catathelasma ventricosa* (see *C. imperialis,* p. 195)
6. Cap dingy yellowish to olive-brown to dark brown . *Catathelasma imperialis* & others, p. 195

7. Odor pleasingly spicy-fragrant (somewhat like cinnamon); fruiting body whitish when young, but often developing cinnamon-brown or yellowish stains in age .....*A. ponderosa,* p. 191
7. Not as above; not spicy-fragrant, or if so then fruiting body darker when young 8

8. Lower portion of stalk shaggy or conspicuously scaly *and* fruiting body showing at least some yellow 9
8. Not with above combination of characteristics 10

9. Cap smooth or with flattened fibrils .... *A. albolanaripes* & others, p. 194
9. Fresh cap with yellow scales, at least near margin *A. straminea* (see *A. albolanaripes,* p. 194)

10. Cap and stalk covered with cinnamon-brown to chestnut-brown or vinaceous-brown threads (fibrils) which may break up into scales; veil membranous, usually forming a ring (annulus) on stalk; odor *sometimes* spicy-fragrant; typically found in summer or fall *A. caligata,* p. 192
10. Not as above; differently colored or veil not membranous or found in spring and early summer (shortly after snow melts); odor not spicy-fragrant 11

11. Odor typically farinaceous, cucumberlike, fishy, or like raw peanuts, or unpleasantly pungent (crush flesh in the cap if unsure) 12
11. Odor typically mild, not distinctive, or merely fungal 15

12. Found under mountain conifers shortly after the snow melts; cap becoming ochre-buff to grayish, brownish, violet-gray, etc. (occasionally whitish) .... *A. olida,* p. 193
12. Not as above (habitat or season usually different) 13

13. Cap bright yellow-orange to orange-brown, reddish-brown, rusty, pinkish-brown, brick-red or splashed with green; odor farinaceous or rancid 14
13. Not as above 20

14. Veil fibrillose, often disappearing; stalk up to 1.5 cm thick, often fragile (see *Limacella,* p. 291)
14. Veil membranous, forming a ring; stalk not fragile .... (see *Tricholoma zelleri,* p. 188)

15. Fruiting body white and gills usually decurrent *or* fruiting body developing reddish to vinaceous-red stains or streaks and veil fibrillose, evanescent; found with conifers, especially at higher elevations .... (see **Hygrophoraceae,** p. 103)
15. Not as above 16

16. Stalk with cottony or shaggy scales below the veil; fruiting body grayish or paler; found under mountain conifers, especially in Rockies .. *A. fusca* & others (see *A. albolanaripes,* p. 194)
16. Not as above 17

17. Stalk tough and fibrous, usually with a stringy white pith inside; cap usually with small dark hairs or scales, especially toward center; taste usually latently bitter (but not detectable by everyone); gills usually decurrent (but sometimes adnate); on wood or ground; common   18
17. Not as above; on ground 20

18. Veil absent .... *Armillariella tabescens* (see *A. mellea,* p. 196)
18. Veil present, at least when young 19

19. Stalk usually bulbous or thicker at base; veil cottony, not typically forming a prominent ring; often on ground, not in *large* clusters .. *Armillariella bulbosa* (see *A. mellea* group, p. 196)
19. Not as above; often with a prominent ring, often clustered *Armillariella mellea* group, p. 196

20. Stalk viscid; odor alkaline; fruiting body whitish; rare .... *A. viscidipes*
20. Not as above; stalk not viscid; fruiting body white or variously colored; common 21

21. Cap 2-4 (6) cm broad, not white; stalk 3-6 mm thick; odor usually farinaceous; usually found with willow; spores *not* amyloid; not common .... (see *Tricholoma,* p. 176)
21. Not with above features; common .... (see *Amanita,* p. 263)

Armillaria ponderosa, showing the veil that sheathes the stalk. Also see the color plate and the photo on p. 49. (The latter photo shows a young specimen with an unbroken veil.)

Armillaria ponderosa (White Matsutake; Matsutake) **Color Plate 37**

CAP 5-20 (35) cm broad, convex to plane; surface dry or slightly viscid when moist, at first white, but in age developing pale cinnamon to pinkish-brown or yellow-brown stains or with fibrils that become these colors; margin at first inrolled and cottony. Flesh thick, very firm, white; odor distinctly spicy-aromatic (like cinnamon). **GILLS** white, discoloring or spotted rusty-brownish to cinnamon in age; crowded, adnate to adnexed or notched. **STALK** 4-15 cm long, 1-5 cm thick, solid, tough, hard, equal or with a narrowed base; white above the ring, usually scaly or fibrillose below and colored more or less like cap. **VEIL** thick, membranous, sheathing the stalk, at first white; forming a prominent cottony ring which flares outward at first, then collapses against the stalk in age. **SPORE PRINT** white; spores 5-7 × 4.5-5.5 microns, broadly elliptical to nearly round, not amyloid.

HABITAT: Widely scattered to gregarious on ground in forests, thickets, and pine barrens; found throughout northern North America, but particularly abundant in the Pacific Northwest, where it is harvested commercially. In the mountains of Idaho, Washington, and Oregon it is common under mixed conifers and second-growth Douglas-fir, while on the coast it favors sandy pine forests. It also likes to lurk in thickets of ericaceous shrubs (e.g., rhododendron, huckleberry, manzanita), which makes for very difficult collecting. In coastal California, however, it prefers tanoak-madrone stands to conifers, though it also fruits in sandy soil under manzanita with a pine canopy. In our area "patches" are hard to find but fairly reliable, producing one crop each year, generally in November or December.

EDIBILITY: Edible and highly prized by Asian-Americans. In San Francisco and San Jose it sells fresh for as much as $25 a piece! However, its tough, chewy texture does not appeal to everyone. Special techniques are required to render it tender while highlighting its unique flavor. (If you bring some to me, I will give you a free demonstration!)

COMMENTS: The unique spicy odor—a provocative compromise between "red hots" and dirty socks—is the hallmark of this magnificent mushroom. Its robust stature, whitish color (at least when young), and prominent veil are also distinctive. It might be mistaken for a "JAR" (Just Another Russula) or an *Amanita,* but the veil and odor distinguish it. Old cinnamon-stained specimens may be rather unattractive, but the young, firm, white, cottony buttons are undeniably gorgeous. The only other mushrooms with the same odor are the fragrant form of *A. caligata,* the matsutake of Japan, plus some specimens of *Lentinus ponderosus* (which grows on wood), and *Inocybe pyriodora* (a small poisonous species with brown spores). *Hygrophorus subalpinus* and *Catathelasma* species are somewhat similar, but lack the odor. The white matsutake has recently been transferred to *Tricholoma* and given a new name, ***Tricholoma magnivelare.***

Armillaria caligata, mature specimens. Note the dark fibrils on the cap and stalk, and the prominent annulus (veil).

Armillaria caligata

CAP 4-12 cm broad, broadly convex becoming plane or with uplifted margin; surface dry, covered with flattened cinnamon-brown to chestnut or vinaceous-brown fibrils which typically separate and cluster in age to form small scales or patches, revealing the whitish to pinkish flesh beneath. Flesh thick, white; odor variable: distinctly spicy-fragrant to fruity, mild, or unpleasant; taste mild to nutty, bitter, or disagreeable. **GILLS** close, adnexed to adnate (rarely slightly decurrent); white, the edges developing brownish stains. **STALK** 4-9 cm long, 1-3 cm thick, more or less equal, solid, firm, white or pallid above the ring, fibrillose or scaly below and colored like cap. **VEIL** membranous, sheathing the stalk, forming a distinct flaring ring which collapses in age; underside colored like cap, upper surface white. **SPORE PRINT** white; spores 5-8 × 4-5.5 microns, broadly elliptical, smooth, not amyloid.

HABITAT: Solitary to scattered or in groups on ground in woods, widely distributed. In the West, it fruits mainly under mountain conifers in the summer and fall, but isn't common. In eastern North America it occurs under oaks and ericaceous shrubs. The largest fruiting I've seen was under spruce in the Rocky Mountains. I haven't found it in our area.

EDIBILITY: Edible—the forms which do not have a disagreeable taste and/ or odor are said to be as good as *A. ponderosa!*

COMMENTS: Also known as *Tricholoma caligatum,* this species is easily separated from its cousin *A. ponderosa* by the cinnamon-brown to purple-brown fibrils on the cap and stalk. It might be confused with *Hygrophorus purpurascens,* but the latter has a fibrillose, evanescent veil and a redder (not as brown) cap, plus slightly waxy gills. The above description encompasses several varieties and forms of *A. caligata.* The western version typically has dark fibrils and the spicy-cinnamon odor of *A. ponderosa.* Eastern material, on the other hand, is apt to be more cinnamon-colored with a mild to fruity to pungent or downright disgusting odor. Other species: The matsutake of Japan, *A. (=Tricholoma) matsutake*, is very close to the fragrant western variety of *A. caligata* and may actually be the same species. It is edible, of course, and highly prized.

192

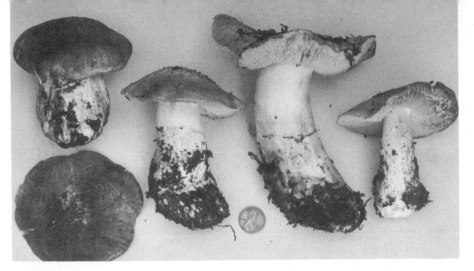

Armillaria olida is a prominent "snowbank" mushroom of the Sierra Nevada and Cascades. These specimens are fairly typical, except that the cap is sometimes paler. Note slight annulus formed by the veil. The odor is also distinctive (see description).

Armillaria olida   (Cucumber Armillaria)

CAP 6-15 cm broad, convex becoming plane to broadly umbonate, or in age often depressed or with an uplifted margin; surface dry to somewhat viscid, color variable: whitish (when still under the duff) to gray to bluish-gray, purplish-gray, brown, or developing olive, buff, or ochre tones; often overlaid with white cottony or fibrillose veil remnants (these sometimes scattered but more often merged to form a central patch). Flesh thick, firm, white; odor very distinctive: usually like cucumber, watermelon rind, rotting potatoes, or freshly mowed grass, but sometimes like old fish. **GILLS** close, white or tinged gray, typically adnexed or notched but sometimes adnate when very young and free in old age. **STALK** 6-14 cm long, 1.8-4 cm thick, equal or swollen below (or even with a bulb), solid, very firm, dry, white or at times pale buff, usually sheathed below the ring by veil remnants which are often ochre- or cinnamon-tinged. **VEIL** cottony or fibrillose, usually forming a slight median to superior ring on the stalk, but sometimes disappearing. **SPORE PRINT** white; spores 9-12.5 × 4.5-6.5 microns, elliptical, smooth, not amyloid.

HABITAT: Solitary to gregarious or in small clumps of 2-3 individuals in duff under mountain conifers, often partially buried or forming "mushrumps"; known only from the West, fruiting during the spring or shortly after the snow melts. It is quite common in the Sierra Nevada and is sometimes a good morel-indicator. I have seen large fruitings in Yosemite National Park in April.

EDIBILITY: Unknown—or at least not commonly eaten. The taste is said to resemble the odor, which is not particularly pleasing.

COMMENTS: When the veil remnants are not obvious, this robust springtime *Armillaria* is likely to be mistaken for a *Tricholoma*. In fact, as the definition of *Armillaria* is narrowed and that of *Tricholoma* is broadened to include more veiled species, it will probably be transferred to that genus. Its distinctive cucumber or fishy odor, grayish cap, and growth under mountain conifers in the spring plus the frequent presence of veil remnants on the cap and/or stalk form a distinctive combination of characteristics. The cap color can be reminiscent of *T. portentosum* and *T. virgatum,* but those species lack a veil and are not so strongly scented.

193

Armillaria albolanaripes, mature specimens. The golden to yellow-brown color (see color plate) and the shaggy stalk are distinctive. *A. straminea* (not illustrated) is similar, but has a scaly cap.

Armillaria albolanaripes (Sheathed Armillaria)   **Color Plate 43**

CAP 5-12 cm broad, convex or slightly umbonate to plane; surface moist or slightly viscid, yellow to golden-yellow or more often brown at the center and yellow at the margin; with flattened fibrils or scales which darken in age. Flesh white or tinged yellow; odor mild. **GILLS** adnexed or notched, white to pale yellow, close. **STALK** 2-8 cm long, 1-2.5 cm thick, equal or thicker below, dry, white above the ring, sheathed with soft cottony scales below, the scales white at first, yellow- or brown-tipped in age. **VEIL** white, cottony, fragile, leaving a ragged superior ring on stalk and/or remnants on cap margin. **SPORE PRINT** white; spores 5-8 × 3-5 microns, elliptical, smooth, weakly amyloid.

HABITAT: Solitary or in scattered groups in woods and along paths; widely distributed. It is common under conifers in the mountains of western North America in spring, summer, and early fall, but I find it only rarely in our area, usually under oak.

EDIBILITY: Edible but insipid—I have fried it. The closely related *A. straminea* (see comments) is said to be a popular edible mushroom in Colorado.

COMMENTS: This handsome mushroom is easily identified by its yellow-brown cap, creamy to pale yellow gills, and soft cottony scales on the stem. The veil can be seen in young specimens but does not always form a distinct ring. The scaly stem might be mistaken for a volva, but there are no warts on the cap. *Amanita aspera* is similarly colored but has white gills and a warted cap. In the montane aspen-conifer forests of the West, *A. albolanaripes* has several close relatives, including: *A. pitkinensis,* with a grayer cap and stalk and only slight yellow tints (but gills yellowish in old age); *A. fusca,* a grayish version with no yellow at all; and *A. straminea (=A. luteovirens),* more widely distributed and colored like *A. albolanaripes,* but with conspicuous bright yellow scales on the cap (or at least the cap margin) as well as on the stalk. In some regions, whitish or "albino" forms of *A. straminea* and *A. albolanaripes* can also be found. All of the above species have amyloid spores and are given their own genus, *Floccularia,* by some taxonomists (the same ones who retain the honey mushrooms in *Armillaria* and transfer the matsutakes to the genus *Tricholoma*).

194

Cathathelasma imperialis. Note the decurrent gills, scaly or fibrillose cap, and prominent annulus (ring) on stalk. *C. ventricosa* (not illustrated) is similar but has a paler cap.

Catathelasma imperialis   (Imperial Mushroom)

CAP 10-40 cm broad, convex to plane; surface slightly viscid when moist but soon dry, smooth, fibrillose-scaly, or cracked into scales or plaques(areolate); dark brown to brown, dingy yellow-brown, or olive-brown. Flesh very thick (up to 15 cm!) and hard, white; odor and taste sharply farinaceous. **GILLS** decurrent, pallid or buff to yellowish or pale grayish-olive (in age), close, many forked. **STALK** 12-18 cm long, 3-8 cm thick, tapered below to a bluntly pointed base; dry, dingy brownish to pinkish-buff below the ring; solid, hard. **VEIL** membranous, double-layered, its lower surface often areolate while still covering the gills; typically forming a double ring, the upper one thick, striate above, and often flaring; the lower one sheathing the stalk as a thin membrane or gelatinous zone, or indistinct. **SPORE PRINT** white; spores 10-15 × 4-5.5 microns, cylindrical, smooth, amyloid.

HABITAT: Solitary, scattered, or in groups on ground under conifers (mainly spruce and fir), late summer and fall, northern North America. It is more common in the Rocky Mountains than on the west coast. It does not occur in our area, but *C. ventricosa* (see comments) is fairly common under Sitka spruce in northern California.

EDIBILITY: Edible and very tempting because of its size—but tough. Buttons are as large as baseballs, and just as hard!

COMMENTS: This mountain of a mushroom may well qualify as the "Most Humongus Gilled Fungus Among Us." Its often gargantuan size, hard flesh, dingy brownish cap, growth on the ground, double veil, and white spores set it apart. The cap may be smooth, areolate, or fibrillose-scaly. The buttons are smaller, of course, but still distinct by virtue of their hardness. *Lentinus ponderosus* is also gigantic, but grows from stumps and lacks a veil. A slightly "smaller" sister species, *C. ventricosa,* also occurs under northern conifers. Its stalk may be quite long but its non-viscid cap averages "only" 7-35 cm across and is dingy whitish to grayish. It looks something like the white matsutake *(Armillaria ponderosa)* but lacks the spicy odor of that species. It is said to be a good edible in spite of its hard, unpleasant-tasting raw flesh. Other species: *C. singeri* of the Rocky Mountains looks like a *Hygrophorus* with its dingy yellowish viscid cap, but has amyloid spores; *C. macrospora* has broad spores. All of these species grow and decay slowly.

Armillariella mellea group   (Honey Mushroom)   **Color Plates 39, 42**

CAP 3-15 cm broad or more, convex becoming plane or sometimes broadly umbonate or in age uplifted; surface viscid or dry, usually with scattered minute dark brown to blackish fibrillose scales or erect hairs, especially toward the center; color variable: yellow, yellow-brown, tawny, tan, pinkish-brown, reddish-brown, etc. Flesh thick and white when young, sometimes discolored in age; odor mild, taste usually latently bitter. **GILLS** adnate to slightly decurrent or sometimes notched; white to yellowish or sordid flesh-color, often spotted darker in age. **STALK** 5-20 cm long, 0.5-3 (5) cm thick, tough and fibrous with a stringy pith inside; usually tapered below if growing in large clusters, or enlarged below if unclustered and on the ground; dry, whitish above the ring, soon yellowish to reddish-brown below and often cottony-scaly when very young. **VEIL** cottony-membranous, white to yellowish, forming a superior ring on stalk or occasionally disappearing. **SPORE PRINT** white; spores 6-10 × 5-6 microns, elliptical, smooth, not amyloid.

HABITAT: In small or massive clusters on stumps, logs, and living trees, or scattered to gregarious (occasionally solitary) on ground—but growing from roots or buried wood; common on a wide variety of trees and shrubs, and practically worldwide in distribution. In our area it occurs year-round, but is most common in the fall and early winter. I have seen truly stupendous fruitings on oak as well as walnut (in an orchard) and other trees. It is a virulent parasite of timber, fruit, and garden trees, but can also be a harmless saprophyte on dead trees or on the dead wood (heartwood) of living trees. It is called "oak root fungus" in California because of its insatiable appetite for oaks, and "shoe string root rot" because of the stringy black mycelial strands (rhizomorphs) by which the mycelium spreads. These "runners" may extend up the host's trunk or infect neighboring trees by traversing great distances through the soil. On oak trees the mycelium can frequently be seen as whitish fanlike growths between the bark and wood. It generally feeds on the roots and lower trunk of its host, reducing it to a pathetic white, spongy pulp. The mycelium is also thought to be the culprit responsible for the "aborted" fruiting bodies of *Entoloma abortivum*. Actively growing mycelium may phosphoresce at night, giving the wood an eerie luminous aura called "foxfire." Inhabitants of subarctic regions are said to mark their trails with bits of glowing wood infected by *A. mellea.*

EDIBILITY: Eminently edible. Use only firm caps and discard the tough stalks. It is an abundant food source, crunchy in texture, and a very passable substitute for the shiitake *(Lentinus edodes)* in stir-fried dishes. The bitter taste cooks out, but some forms are better than others, and some (e.g., those that grow on buckeye or hemlock) can cause digestive upsets. The common name, incidentally, is a reference to its color (which, like honey, is extremely variable), not its taste (which isn't the least bit sweet).

COMMENTS: There is very little that can or cannot be said about the honey mushroom. Also known as *Armillaria mellea,* it is among the most variable and cosmopolitan of the fleshy fungi, and in its innumerable guises will confound you time and time again. Especially variable are its color, shape, viscidity, and manner of growth, but there are several key, *relatively* constant features that distinguish it: (1) the presence of a veil (2) the tough, fibrous stalk (3) frequent presence of small dark hairs on cap (4) the bitter taste when raw (some people, however, are unable to detect it) (5) the growth on wood (though it may be buried) (6) the white or faintly yellowish spores (in any mature cluster the lower caps will be covered with white spore dust).

There are at least two distinct, widespread variants (one study recognized 14 different "species" in the *A. mellea* complex). One has a yellow to yellow-brown cap that is viscid or dry but becomes slimy in wet weather. It also has a yellow-tinged veil and tapered stalk, and usually grows in clusters. The second variety, on the other hand, has a hairier

This form of the honey mushroom *(Armillariella mellea)* usually grows in small tufts on the ground rather than in large clusters on wood (as shown in color plates). Also, the stalk is usually swollen at the base and the veil is fragile and cottony. It approaches the European form now called *A. bulbosa.*

pinkish-brown to reddish-brown or dingy brown cap with a white cottony veil and frequently enlarged stem base. It grows scattered or in small tufts, often on the ground. This form, which is close to *A. bulbosa* (a European species), is especially confounding to beginners. Intermediate forms abound also. In view of the extreme variability, beginners should eat *only those clearly growing in clusters on wood,* and be certain that the **spores are whitish.** The poisonous *Galerina autumnalis* grows on wood and has a ring on the stalk, but is smaller and more fragile, with a smoother cap and *brown* spores. *Pholiota* species also have brown spores, while *Gymnopilus* has rusty-orange spores. In the eastern and southern United States you may encounter *A. tabescens,* a very similar, clustered, wood-inhabiting, white-spored mushroom that *lacks* a veil (and annulus) and has a dry cap. It is also edible.

SQUAMANITA

Fairly small to medium-sized terrestrial, mainly woodland mushrooms. CAP *usually scaly, fibrillose-scaly, or granulose.* GILLS *usually attached.* STALK typically central, *arising from a conspicuous, cylindrical to bulbous, often hollow, underground "tuber."* VEIL typically present, sometimes forming a slight annulus (ring) on stalk. VOLVA absent or present as a collar or scaly rings above the "tuber." SPORE PRINT *white or pink.* Spores smooth, thin- or thick-walled; amyloid, dextrinoid, or neither. Hyphae in gill tissue typically parallel or nearly so.

THIS small, rare, oddball genus is distinct by virtue of the underground bulb or "tuber" from which the stem arises. Many mycologists place it in the Agaricaceae (along with *Lepiota, Agaricus,* and *Cystoderma*) rather than in the Tricholomataceae, but its affinities are unclear. Since some *Amanita* species have a swollen, rooting stem base that could be mistaken for a "tuber," *Squamanita* has been keyed out under that genus. Amanitas, however, differ fundamentally in their divergent rather than parallel gill tissue (a microscopic feature, see p. 19).

Squamanita is unlikely to be encountered by the average mushroom hunter. One odoriferous species is described here and two others are keyed out.

Key to Squamanita

1. Cap and stalk grayish to purple-gray, lilac-gray, or darker, but covered with an ochre-brown granulose coating, at least when fresh .... *S. paradoxum* (see *S. odorata*, below)
1. Not as above; cap and stalk often scaly, but granulose layer absent 2
2. Fruiting body purple-gray to purple-brown except for the yellowish to buff tuber; odor distinctly fruity (somewhat like grape soda) .... *S. odorata*, below
2. Purplish tones absent; cap ochre to ochre-brown to buff, with a whitish or grayish tuber (or clusters of tubers); found in eastern North America .... *S. umbonata*

Squamanita odorata

CAP 1-4.5 cm broad, obtusely bell-shaped or convex, expanding somewhat in age but usually retaining a broad umbo; surface dry, densely and coarsely scaly or fibrillose-scaly, the scales often erect; usually more fibrillose toward margin; brownish-purple to purplish-gray or lilac-gray, often darker in age. Flesh colored like cap; odor strongly and persistently fruity-fragrant (like grape soda or grape juice). GILLS adnate or notched, fairly well-spaced, colored more or less like cap. STALK 1-3.5 cm long, (2)3-10(15) mm thick, arising from a swollen, sometimes hollow, juglike underground "tuber" 1-2.5 cm high and up to 2 cm thick; colored more or less like cap and covered with conspicuous scales like those on the cap, except for the smooth, sometimes silky apex and yellowish to buff-colored "tuber"; hollow or partially hollow in age. VEIL not forming a distinct ring on stalk. SPORE PRINT pinkish; spores 6.5-9 × 4-6 microns, elliptical, smooth, not amyloid.

HABITAT: Usually in groups or clumps on ground in woods; widely distributed but apparently very rare. I have examined specimens collected under conifers in Washington.

EDIBILITY: Unknown. Although too rare to be of value, the odor is certainly intriguing.

COMMENTS: Formerly known as *Coolia odorata*, this little mushroom is as bizarre as it is rare. The coarsely scaly purplish cap and stalk, similarly colored gills, yellowish-buff "tuber," and strong grapelike odor make a most distinctive set of features. Other species: *S. paradoxum* (=*Dissoderma paradoxum*) is gray to lilac- or purplish-tinted in age beneath an ochre-brown granulose coating. It also occurs in the Pacific Northwest, but is rare.

CYSTODERMA

Small to medium-sized, terrestrial or wood-inhabiting mushrooms. CAP dry, *with a coating of mealy or powdery granules*, at least when fresh. GILLS typically whitish or pallid, *usually attached.* STALK central, *lower portion sheathed with mealy granules or scales.* VEIL *present, often forming an annulus (ring) on stalk.* VOLVA absent. SPORE PRINT *white.* Spores smooth, sometimes amyloid but not dextrinoid.

THE outstanding feature of this small genus is the layer of mealy granules that coats the cap and lower stem. Rain may wash the granules off the cap, but the stem normally retains them. A veil is always present and in several species it forms a prominent ring. *Armillaria, Armillariella,* and *Catathelasma* have a veil and attached gills, but are larger and lack the granulose coating. Most Cystodermas were originally placed in *Lepiota* and some mycologists retain them in the same family. Lepiotas, however, typically have free gills, while in *Cystoderma* the gills are usually attached to the stalk.

Cystodermas are common in northern coniferous forests, especially in beds of moss. About 20 species occur in North America. Several are very attractive but little is known of their edibility. Two species are described here; both are rare in our area.

Key to Cystoderma

1. Stalk generally 8 mm thick or more; fruiting body medium-sized 2
1. Stalk generally less than 8 mm thick; fruiting body rather small or sometimes medium-sized 5

Cystoderma fallax, mature specimen. Note the umbonate cap, prominent annulus (ring), and coating of granules on the stalk and cap. It grows singly as well as in small groups or clusters.

2. Spore print pale yellow-brown to orange-buff; large ... (see *Rozites* & *Phaeolepiota,* p. 411)
2. Spore print white or whitish; medium-sized 3
3. Cap white when young (but often pale cinnamon or buff in age); rare .... *C. ambrosii*
3. Not as above; cap not white 4
4. Cap more or less orange; veil forming a persistent, well-developed annulus (ring) on stalk; found on rotting hardwoods, mainly in eastern North America .... *C. granosum*
4. Cap cinnabar-red to rusty-orange, etc.; veil evanescent, not usually forming a well-developed annulus; widely distributed .... *C. cinnabarinum* (see *C. amianthinum,* p. 200)
5. Veil typically forming a distinct, well-developed annulus (ring) on stalk 6
5. Veil evanescent or merely forming a ragged zone at top of granular sheath on stalk 7
6. Cap white or tinged pinkish or lilac; rare .... *C. carcharias*
6. Cap rusty-brown to tawny-brown; widely distributed and common .... *C. fallax,* below
7. Growing on wood; spores amyloid .... *C. gruberianum* (see *C. amianthinum,* p. 200)
7. Growing on ground or in moss; spores amyloid or not amyloid 8
8. Cap often (but not always) radially wrinkled, tawny to ochraceous to brown or rarely white; spores amyloid .... *C. amianthinum,* p. 200
8. Cap dark reddish-brown to brick-colored to tawny or paler (rarely white), but not wrinkled; spores not amyloid .... *C. granulosum* (see *C. amianthinum,* p. 200)

Cystoderma fallax

CAP 2-5 cm broad, convex to plane or frequently with an umbo; surface dry, with conspicuous mealy granules which are erect at first but flattened and more powdery in age (or often wear away completely); cinnamon-brown to rusty-orange to tawny-ochre; margin often hung with remnants from the veil. Flesh thin, whitish or tinged cap color. **GILLS** adnexed to adnate, close, white to pale pinkish-buff or tinged yellow. **STALK** 3-7 cm long, 3-5 (7) mm thick, equal or enlarged below, smooth and pallid above the ring, sheathed with cinnamon-brown to rusty-ochre granules or flaky scales below. **VEIL** forming a large, delicate but persistent, often flaring ring on the stalk; ring median to superior, smooth and pallid on upper side, colored like the cap underneath. **SPORE PRINT** white; spores 3.5-5.5 × 3-4 microns, broadly elliptical to nearly round, smooth, amyloid.

HABITAT: Solitary, scattered, or in small groups or tufts on ground under conifers or in mixed woods, sometimes also on rotting wood; widely distributed and common in the summer and fall in the Pacific Northwest and Rocky Mountains. Fruiting in the fall and early winter in our area, but rather rare.

EDIBILITY: Unknown.

COMMENTS: One of the most attractive and delicately adorned of our woodland fungi, this *Cystoderma* is easily identified by its rusty-orange to cinnamon color, prominent ring on the stalk, whitish gills which are attached to the stem, and granulose coating on the cap and stem (rain may wash the granules off the cap, but not the stem). The illustration does not do it justice, but since when is justice usually done?

Cystoderma amianthinum

CAP 2-5 cm broad, bell-shaped or somewhat conical becoming convex or umbonate to nearly plane; surface dry, prominently wrinkled (radially) in one form; covered with mealy or powdery granules which may wear off in age, tawny-ochre to ochre-brown, ochre-buff, or yellowish; margin often hung with veil remnants. Flesh thin, odor mild or strongly pungent. **GILLS** adnexed to adnate, crowded, white or creamy or tinged yellow-orange. **STALK** 2.5-7 cm long, 3-8 mm thick, equal or slightly enlarged below, smooth and whitish above the veil, sheathed with granules or granulose scales below and colored like the cap. **VEIL** fragile, forming a slight ring on stalk or often disappearing. **SPORE PRINT** white; spores 4-7 × 3-4 microns, elliptical, smooth, amyloid. Cap cuticle staining rusty-brown to reddish-brown in KOH (potassium hydroxide).

HABITAT: Solitary, scattered, or in groups under or near conifers, especially in moss; widely distributed in northern regions and probably the most common member of the genus. I have seen it in late summer, fall, and early winter in northern California and the Pacific Northwest, but it does not seem to occur south of San Francisco.

EDIBILITY: Not recommended. Some sources list it as edible, but it doesn't have much substance and can be confused with poisonous species (e.g., *Lepiota castanea*).

COMMENTS: This petite mushroom is quite attractive when growing amongst colorful lichens or in beds of bright green moss. It is best recognized by its granulose cap and stalk, ochre color, and fragile veil which disappears or forms only a slight ring on the stalk (rather than a prominent one, as in *C. fallax*). It is easily mistaken for a small *Lepiota,* but the gills are usually attached to the stem rather than free. In one variety the cap is conspicuously wrinkled, in another it is not. Other species: *C. granulosum* is similar, but has a reddish-brown to tawny, non-wrinkled cap and non-amyloid spores; *C. gruberianum* is a small species that grows on rotten wood; *C. cinnabarinum* is a larger, farflung species with a rusty-orange to beautiful cinnabar-red or vermillion cap and stalk. Whitish-capped forms of *C. amianthinum* and *C. granulosum* also occur, but are rare.

ASTEROPHORA

Small mushrooms *parasitic on other mushrooms.* CAP often powdery. GILLS *thick and well-spaced or poorly formed to practically absent.* STALK present. VEIL and VOLVA absent. SPORE PRINT *white to brownish when obtainable.* Spores mostly produced asexually, smooth or spiny.

THIS small genus contains a staggering total of two species. Both are outlandish oddballs that grow exclusively on other agarics, particularly *Russula* and *Lactarius* species. They differ from *Collybia tuberosa* and other mushroom-inhabiting mushrooms in having thick and well-spaced or poorly formed gills. They are also unique in that they produce very few spores on basidia. Instead the hyphae block off to form asexual spores called **chlamydospores.** *Asterophora* is listed in some books as *Nyctalis.*

Key to Asterophora

1. Cap more or less round and puffball-like, white becoming brownish and powdery as spores mature; gills often malformed or practically absent . . . . *A. lycoperdoides,* below
1. Not as above; cap not powdery; gills thick, well-spaced, usually decurrent, eventually disintegrating into powdery spores . . . . *A. parasitica* (see *A. lycoperdoides,* below)

Asterophora lycoperdoides

CAP 0.5-2 cm broad, nearly round; surface dry, whitish becoming brown and powdery from spores. Flesh thin, odor farinaceous. **GILLS** often malformed or barely present; well-spaced, thick, whitish. **STALK** 1-3 cm long, 3-8 mm thick, more or less equal, white becoming brownish. **SPORE PRINT** white when obtainable; spores 5-6 × 3.5-4 microns, elliptical, smooth. Chlamydospores 12-18 microns, round, bumpy or spiny, thick-walled, brownish.

HABITAT: In colonies on old mushrooms, particularly species in the *Russula densifolia* group. Widely distributed but not common; very rare in our area.

EDIBILTY: Unknown.

COMMENTS: This oddball might be mistaken for a puffball because of its poorly formed gills and powdered round cap. However, no puffballs are known to be parasitic on gilled mushrooms! *A. parasitica* is also widely distributed, but even rarer than *A. lycoperdoides.* It has thick, well-spaced, decurrent gills and a white to grayish, brownish, or lilac-tinged cap, plus smooth and elliptical chlamydospores.

MARASMIUS, COLLYBIA, & Allies

Minute to medium-sized mushrooms, some of which shrivel up in dry weather and then revive when moistened, others of which do not. CAP *usually convex to plane,* but sometimes bell-shaped; not viscid in most cases; *margin usually incurved when young.* GILLS *usually free, adnexed, or notched,* but sometimes adnate (or in *Marasmius,* even decurrent). STALK *usually thin and pliant, tough, cartilaginous, or wiry;* usually central. VEIL and VOLVA absent. SPORE PRINT *white to buff or rarely tinged pinkish.* Spores smooth, usually not amyloid. Cells in the upper layer of the cap cuticle usually forming a palisade *(Marasmius),* or not forming a palisade *(Collybia).*

THESE minute to medium-sized mushrooms typically have adnexed to free gills and a cartilaginous or wiry stem. The cap is typically convex to plane or if conical then with an incurved margin when young (rather than straight as in *Mycena).* Two large genera (*Marasmius* and *Collybia*) plus several smaller ones are treated together here because they are difficult to separate in the field. The traditional trademark of *Marasmius* is its astonishing reviving ability. If dried-up specimens are placed in a bowl of water they will quickly swell up, magically reassuming their original shape and dimensions. In the wild, species of *Marasmius* often seem to spring up in droves right after or *during* a rain, when in fact they were already there, shrivelled up and inconspicuous. In addition, they can often be told by their tough texture and wiry or hairlike stem. (*Xeromphalina* is somewhat similar, but usually has decurrent and/ or more brightly colored gills.)

   Collybia has traditionally been separated from *Marasmius* on the basis of its slightly fleshier, non-reviving fruiting body. However, some species have been shuttled back and forth between *Marasmius* and *Collybia* because they revive *somewhat* when moistened. Recognizing the arbitrary nature of this character, taxonomists now differentiate *Collybia* from *Marasmius* primarily on microscopic features such as the structure of the cap cuticle. As a result, *Collybia,* as currently defined, includes a few species which *do* revive, while *Marasmius* includes some that do not. The gills in *Collybia* are usually adnexed or even free; in some cases they are more broadly attached, leading to confusion with *Clitocybe.* The stalk is usually thin and pliant; if thick, it has a cartilaginous outer rind that helps distinguish it from *Tricholoma*.

Both *Collybia* and *Marasmius* are "troubled" taxonomically. They have been fertile fodder for the "splitters" (see p. 10), who have recently erected a number of "satellite" genera. Some of these are easily distinguished in the field (e.g., *Caulorhiza* and *Oudemansiella* usually have a "tap root"; *Flammulina* has a viscid cap and velvety stem; *Crinipellis* has dextrinoid hairs on the cap; *Strobilurus* usually grows on cones; *Callistosporium* is olive-brown and yellow); others differ microscopically (e.g., *Micromphale* and *Marasmiellus*); still others are not recognized here (e.g., *Rhodocollybia* and *Microcollybia*).

Marasmius is a very large genus centered in the tropics. As might be expected, it is more diverse in the humid deciduous forests of eastern North America than in the West. Most species are saprophytic on sticks and leaves and many are exquisitely constructed. Collybias, on the other hand, are by and large a listless lot. They are also saprophytic on humus and wood and are among our most common woodland agarics; a few may be mycorrhizal.

Most of the genera treated here are difficult from a taxonomic standpoint and have little to offer the mushroom-eater. Two exceptional exceptions are the fairy ring mushroom or "Scotch Bonnet," *Marasmius oreades* (forgive my promiscuous use of superlatives, but it is an exceptionally flavorful fungus!), and the garlic mushrooms (*M. copelandi* and allies). A representative sampling of "marasmioid" and "collybioid" fungi is presented here and several additional species are keyed out but not described.

Key to Marasmius, Collybia, & Allies

1. Typically growing on fallen cones (sometimes buried!) or magnolia pods; rarely found on rotten wood, and if so then stalk thin, more or less rooting, and hairy over lower portion 2
1. Typically growing on ground, wood, or other mushrooms 5
2. Cap conical or bell-shaped when young, often reddish- or vinaceous-tinged; gills frequently with reddish to dark purple edges .... (see *Mycena*, p. 224)
2. Not as above 3
3. Found on fallen magnolia "cones" ..... *Strobilurus conigenoides* (see *S. trullisatus*, p. 211)
3. Found on cones of conifers 4
4. Gills *very* crowded; spores amyloid .. *Baeospora myosura* (see *Strobilurus trullisatus*, p. 211)
4. Not as above .... *Strobilurus trullisatus* & others, p. 211
5. Stalk with numerous side-branches, at least on lower portion .... *C. racemosa*, p. 213
5. Stalk lacking side-branches 6
6. Cap minute (typically 1 cm broad or less) and pale; stalk whitish, very thin; usually (but not always) colonizing the blackened remains of other mushrooms *C. tuberosa* & others, p. 212
6. Not growing on other mushrooms; if small, then not as above 7
7. Fruiting body minute (cap less than 1 cm broad), stalk short, gills adnate to decurrent; growing on bases of madrones *Micromphale arbuticola* (see *Marasmius androsaceus* group, p. 208)
7. Not as above 8
8. Gills violet or lilac when fresh (but they may fade!) 9
8. Gills not violet or lilac 10
9. Gills crowded; stalk not white; odor mild; spores amyloid; usually found on rotten wood; widely distributed in northern latitudes .... *Baeospora myriadophylla*
9. Gills well-spaced; stalk whitish to pale gray; odor usually unpleasant; spores not amyloid; usually terrestrial; restricted to eastern North America (?) ... *C. iocephala (=M. iocephalus)*
10. Odor distinctly garlic- or onionlike, at least when flesh is crushed; cap small or minute (usually less than 2.5 cm broad); often gregarious but not normally clustered 11
10. Not as above; odor may be fetid or otherwise distinctive, but if garliclike then cap typically 2 cm broad or more or fruiting bodies often clustered 12
11. Stalk smooth (hairless); habitat variable .... *M. scorodonius* (see *M. copelandi*, p. 207)
11. Stalk minutely hairy (use hand lens); found on leaves .... *M. copelandi* & others, p. 207
12. Cap small or minute, with coarse tawny to brown hairs; stalk thin (less than 2 mm thick), also hairy (minutely so), wiry-tough; hairs on cap dextrinoid . *Crinipellis piceae* & others, p. 210
12. Not as above; cap usually without hairs 13

13. Odor fetid; stalk velvety; gills yellowish to brown or tinged reddish; cap and stalk brown to red-brown; found on sticks or bark in eastern U.S. **Micromphale foetidum (=Marasmius foetidus)**
13. Not as above 14
14. Odor sweet and heavy (like benzaldehyde); cap brown to reddish-brown, vinaceous-brown, or dark brown, at least toward the center; stalk usually at least 5 mm thick 15
14. Not as above 16
15. Stalk white (but may develop vinaceous or brownish stains below) ... **C. oregonensis,** p. 218
15. Stalk brown to dark brown or vinaceous-brown **C. subsulcatipes** (see *C. oregonensis,* p. 218)
16. Growing in grass, often in arcs or rings; stalk tough; gills fairly well-spaced (*not* crowded or close); cap white to tan, buff, or brownish but not gray or vinaceous, usually less than 6 cm broad; spore print white; very common and widespread ....**M. oreades,** p. 208
16. Not as above 17
17. Stalk with a tapered underground "tap root" (dig up carefully!); spore print white; fruiting body without reddish or rusty stains; cap opaque (*not* normally translucent-striate when moist) 18
17. "Tap root" lacking or not well-developed, or if present then cap translucent-striate when moist *or* fruiting body often reddish-stained *and* spore print pinkish-buff 19
18. Cap blackish, dark brown, grayish, whitish, or yellowish-brown (but if the latter then usually viscid when moist); found from the Rockies eastward **Oudemansiella radicata** & others, p. 219
18. Not as above; cap chestnut-brown to warm tan or yellow-brown, not viscid ....**Caulorhiza umbonata** & others, p. 218
19. Spore print pinkish; growing in grass, straw, or manure ....(see *Clitocybe tarda,* p. 152)
19. Not with above combination of features 20
20. Gills yellow; cap and stalk olive to olive-brown or yellowish (but may develop dark reddish-brown tones as it dries); usually on rotten wood .. **Callistosporium luteo-olivaceum,** p. 211
20. Not as above 21
21. Stalk dark, stiff, bristle-like, less than 1 mm thick; cap typically less than 1 cm broad (rarely 2 cm); cap *not* whitish when fresh (but may fade!); substrate (twigs, needles, leaves) usually with black horsehair-like rhizomorphs (mycelial threads) .... **M. androsaceus** group & others, p. 208
21. Not with above features (but may have some of them) 22
22. Gills adnate to decurrent and fairly well-spaced; cap 2-4 cm broad and predominantly whitish (may be slightly darker at center), usually wrinkled; stalk becoming brownish from the base upward; terrestrial in the forests of the Pacific Northwest .... **M. umbilicatus**
22. Not as above 23
23. Gills adnate to decurrent, white, very widely spaced; cap white or tinged gray to olive-gray or even slightly yellowish, translucent-striate when moist; stalk whitish; found under pine in coastal California, often in large numbers ....**M. sp. (unidentified),** p. 206
23. Not as above 24
24. Fruiting body small or minute (cap usually 2 cm broad or less); cap white or whitish or tinged pale yellowish (the center may be tinged brown), but may develop reddish or pinkish stains in age; stalk less than 3 mm thick 25
24. Not as above; either differently colored or larger 29
25. Gills free or nearly free; margin of cap usually with veil remnants *and/or* the cap and stalk minutely powdered; stalk whitish; growing on ground ..... (see *Lepiota seminuda,* p. 307)
25. Not as above 26
26. Stalk black beneath a coating of minute white hairs; cap 1-2 cm broad; found on leaves or twigs in eastern North America ....**Marasmiellus nigripes** (see *M. candidus,* p. 206)
26. Not as above 27
27. Growing on fallen leaves .... **M. delectans** & others (see **Marasmiellus candidus,** p. 206)
27. Growing on sticks, berry canes, wood, etc. 28
28. Gills very widely spaced; stalk relatively short (less than 3 cm long); abundant on the west coast, infrequent elsewhere ....**Marasmiellus candidus,** p. 206
28. Not as above; abundant in eastern North America, rare or absent elsewhere on continent ... .**M. rotula** (see *Marasmiellus candidus,* p. 206)
29. Growing in compact bundles on rotting conifers (the wood sometimes buried or very decomposed—see Color Plate 49); margin of cap incurved when young 30
29. Not as above (but may grow tufted on rotting conifers or in dense clusters on ground) ... 31

30. Stalk white to grayish .... *Clitocybula familia* & *C. abundans* (see *Collybia acervata*, p. 215)
30. Stalk vinaceous-brown to reddish-brown (or somewhat paler when dry), at least at apex . 55

31. Stalk solid or firmly stuffed (not hollow), straight and equal except for very base (which may be slightly swollen), sometimes scurfy or dandruffy at apex or throughout, or longitudinally lined (but without hairs); gills *crowded*, white (except for one large species with tan to pinkish-cinnamon gills); cap typically rather flat (broadly convex to plane, sometimes with a blunt umbo); surface of cap usually smooth and dark brown to grayish, sometimes ochre-brown, yellowish, or whitish (but not reddish-brown or vinaceous-brown); *spores amyloid*, usually roughened; found in many habitats, but especially in grass or landscaped ground or under mountain conifers soon after the snow melts .... (see *Melanoleuca*, p. 169)
31. Not as above; spores typically neither amyloid nor roughened; stalk sometimes hollow, sometimes clothed with minute hairs but not often scurfy; gills crowded to widely spaced; usually found in woods or near trees (but not always) 32

32. Cap with gray to black hairs or fibrillose scales ..... (see *Tricholoma terreum* group, p. 182)
32. Not as above 33

33. Stalk very thin (less than 1.5 mm); cap flesh-colored to light brown, often wrinkled, up to 12 mm broad; stalk very minutely hairy (pubescent), *not* shiny; found on needles and twigs of redwood, spruce, fir *Micromphale sequoiae* & others (see *Marasmius androsaceus* group, p. 208)
33. Not as above 34

34. Cap grayish to dark brown, olive-brown, or black; growing in moss, *Sphagnum* bogs, or on burnt ground, or sometimes simply associated with conifers 59
34. Not as above; cap differently colored or habitat different 35

35. Stalk smooth (hairless) or finely powdered, or with hairs only at the base 36
35. Stalk pubescent or velvety (covered with minute hairs) over at least the lower half by maturity (use hand lens if unsure) 51

36. Gills reddish-brown to dark brown or blackish-brown; cap and stalk similarly colored (but cap may fade); flesh staining green in KOH; fairly common in eastern North America and the Pacific Northwest .... *C. alkalivirens*
36. Not as above; gills typically paler 37

37. Base of stalk with a litter-binding mycelial pad; cap yellowish-brown to reddish-brown; stalk 1-3 mm thick, shining; growing in groups or dense clusters on hardwoods leaves and debris in eastern North America .... *M. cohaerens* & others
37. Not as above 38

38. Growing in grass; fruiting body small (cap usually less than 2.5 cm broad) and vinaceous- or reddish-tinged; stalk *not* tough and polished; cap *not* pleated ....
.... *Mycena* sp. (unidentified) (see *Marasmius oreades*, p. 208)
38. Not as above 39

39. Cap and stalk pale or whitish (cap may have grayish-brown center); gills fairly well-spaced; known from California and South America .... *M. albogriseus* (see *M. oreades*, p. 208)
39. Gills close or crowded, or if well-spaced then cap and stalk differently colored 40

40. Gills widely spaced; stalk usually polished; cap often (but not always) pleated 41
40. Not as above; gills typically fairly close or crowded 42

41. Cap bay-brown to reddish-brown, brown, or wine-red; stalk 5-13 cm long; common on west coast .... *M. plicatulus*, p. 209
41. Not as above; either cap differently colored or stalk shorter or found elsewhere ....
.... *M. siccus* & others (see *M. plicatulus*, p. 209)

42. Cap striate when moist and often translucent; cap typically conical or bell-shaped when young
.... (see *Mycena*, p. 224)
42. Not as above 43

43. Stalk tough, grooved or twisted, often with a rooting base, tan to brown (not white!); growing on hardwoods; rare (not positively known from North America) .... *C. fusipes*
43. Not as above; common 44

44. Stalk 0.5-2.5 cm thick, often with a rooting base, white or yellowish (but may develop reddish stains below); cap usually over 4 cm broad and whitish, but often becoming reddish, pinkish, or vinaceous-brown at the center and sometimes entirely those colors from the beginning; found under conifers, usually on decayed wood or lignin-rich humus .... *C. maculata*, p. 217
44. Not as above; either differently colored, smaller, or with a different habitat 45

45. Stalk dark red except at apex; cap tan to buff, often plane at maturity; growing in tufts or clusters in humus and under trees; not common .... *C. marasmioides* (see *C. acervata*, p. 215)

45. Not as above (if tufted or clustered then stalk differently colored, including reddish-brown) 46

46. Gill edges *coarsely* ragged or toothed, even when young; spore print white; spores amyloid; widespread (but not reported from California) ...... (see *Lentinus & Lentinellus*, p. 141)

46. Gill edges entire or finely serrated (or in age sometimes coarsely serrated); spore print white or slightly colored; spores rarely amyloid; very common in California and elsewhere 47

47. Cap yellowish to light brown, often fading to whitish; stalk whitish or tinged yellow; found under eastern hardwoods (or mixed woods) *M. strictipes & M. nigrodiscus* (see *M. oreades*, p. 208)

47. Not as above 48

48. Usually growing in grass and cap typically pinkish *or* growing under mountain conifers and cap vinaceous-red to purplish-red with ochre-yellow gills .... (see *Lyophyllum* & Allies, p. 173)

48. Not as above 49

49. Spore print white to pale cream; cap averaging 1-5 cm broad (occasionally larger); cap color variable but often tawny; gills white or pale yellow, their edges often entire; spores not dextrinoid; common under hardwoods and conifers .... *C. dryophila* & others, p. 215

49. Spore print cream to buff or pinkish-buff; cap averaging 3-8 cm broad (sometimes larger), various shades of brown but not tawny; gills white or with reddish stains, the edges often finely scalloped at maturity; at least some of the spores dextrinoid; found mainly (but not exclusively) under conifers 50

50. Cap vinaceous- to reddish-brown, not fading appreciably; gills sometimes reddish-stained .. .. *C. extuberans* & others (see *C. butyracea*, p. 216)

50. Cap reddish-brown, brown, tan, or even grayish; gills not reddish-stained *C. butyracea*, p. 216

51. Cap velvety, more or less orange-brown; gills adnate to decurrent; growing on hardwoods in eastern North America .... (see *Omphalina & Xeromphalina*, p. 221)

51. Not as above; stalk may be velvety but cap not velvety and gills not decurrent 52

52. Cap usually viscid when moist (but may dry out!); lower portion of stalk rusty-brown to blackish-brown and velvety *when mature* (usually smooth and pallid when young); found on wood (sometimes buried!) .... *Flammulina velutipes*, p. 220

52. Not as above; cap not normally viscid 53

53. Odor garlicky *or* taste distinctly acrid (burning) .... *C. polyphylla & C. peronata* (see *C. confluens*, p. 213)

53. Not as above (but taste may be somewhat bitter) 54

54. Gills usually crowded (sometimes merely close), white or tinged flesh-color; cap not prominently wrinkled; stalk pubescent (covered with minute white hairs) at least over the lower half; usually growing in tufts or clusters 55

54. Not as above; gills darker or more widely spaced or hairs on stalk brown to gray or tawny or cap distinctly wrinkled, etc. 57

55. Stalk reddish to reddish-brown or vinaceous-brown beneath the pubescence (may fade slightly in age), 2-6 mm thick; found on ground or wood but not normally on lawns 56

55. Stalk buff or whitish (or pale brown toward base), (2) 5-10 mm thick; growing on rotten wood, wood chips, or lawns; not common .... *C. luxurians* (see *M. oreades*, p. 208)

56. Growing in compact bundles on rotting conifers, the wood often buried or decomposed (see Color Plate 49) .... *C. acervata*, p. 215

56. Found on ground under both hardwoods and conifers, often clustered but not in compact bundles; base of stalk often with a litter-binding mycelial mat .... *C. confluens*, p. 213

57. Stalk with an enlarged, spongy base; cap and stalk reddish-brown to tan; restricted to eastern North America .... *C. spongiosa*

57. Not as above 58

58. Gills white or pallid .... *C.* **spp. (unidentified)** (see *C. confluens*, p. 213)

58. Gills soon darker (but may be dusted white by spores) ...... *C. fuscopurpurea* group, p. 214

59. Gills usually adnate to decurrent; odor usually mild; on burnt ground, moss, etc., but not normally in *Sphagnum* .... (see *Myxomphalia maura* & others, p. 165)

59. Gills usually adnexed to adnate; often in *Sphagnum* bogs, or if not then odor often rank or rancid .... (see *Lyophyllum* & Allies, p. 173)

Marasmiellus candidus   Color Plate 36

CAP 0.6-2.5 cm broad, convex to plane or with slightly depressed center; surface dry, shining white or translucent white, but often stained deep pinkish or reddish in old age; often striate or grooved at maturity. Flesh very thin, pliant, soft, odor mild. GILLS few and far between, usually interspersed with smaller gills or veins; adnexed or adnate to slightly decurrent, white like the cap but often pinkish- or reddish-stained in old age. STALK 0.5-3 cm long, 1-2 mm thick, equal or slightly tapered at either end, often rather short, central or off-center but not lateral, tough, smooth, often curved, white or with gray to pinkish-gray base, darkening gradually to brownish-black from the base upward as it ages. SPORE PRINT white; spores 10-15 × 3.5-6 microns, spindle-shaped to elongated tear-shaped, smooth, not amyloid.

HABITAT: In groups or rows on dead sticks, branches, berry canes, etc.; widely distributed but most abundant along the Pacific Coast. It fruits in wet weather, mainly in the fall and early winter in our area. It is especially abundant along creeks overgrown with brambles, and on rotting oak, eucalyptus, bay laurel, cedar, etc.

EDIBILITY: Utterly inconsequential.

COMMENTS: Also known as *Marasmius candidus, M. magnisporus,* and *Marasmiellus albuscorticis,* this dainty mushroom is reminiscent of a small shell. Although small, its shining white cap stands out vividly in the forest gloom. The exceedingly well-spaced (distant) gills are its outstanding feature. The stem, which may be off-center, is tough and darkens at maturity. The entire fruiting body may develop pinkish or sordid vinaceous tones as it ages, leading one to falsely (but reasonably) assume that the spores are pink. In eastern North America it is largely replaced by the equally beautiful "pinwheel *Marasmius,*" *Marasmius rotula.* This little gem has a longer (1.5-8 cm), central black stem that is 1-2 mm thick and a white cap (1-2 cm broad) with darker center. It grows on decaying hardwoods, usually in large groups. Other whitish-capped eastern species include: *Marasmius delectans,* growing on leaves, stalk pallid to yellowish above and dark brown below; and *Marasmiellus nigripes*, stalk black beneath a coating of minute white hairs and spores triangular or jack-shaped. Another whitish-capped easterner, *Marasmius epiphyllus*, grows on sticks and leaves and has widely spaced, veined gills and a relatively long, hairlike stem; it also occurs in the West, as does its oak-leaf-inhabiting look-alike, *M. querco-phyllus*. All of the above species are too small to be of culinary value.

Marasmius sp. (unidentified)   (Pine Needle Pinwheel)

CAP 1.5-4.5 cm broad, broadly convex to plane or umbilicate; surface not viscid, smooth or wrinkled, translucent-striate when moist, pure white or with a grayish to olive-gray tinge, sometimes becoming slightly yellowish in age; margin often wavy. Flesh very thin, fragile, soft; odor mild. GILLS adnate to slightly decurrent, widely spaced, usually with veins in between, white. STALK 3-7 cm long, 2-5 mm thick, equal or tapered below or with a swollen base; often flattened, smooth, colored like cap or slightly yellower, the base often with brownish stains; hollow, usually with hairs at the base. SPORE PRINT white; spores 10-12 × 4-6 microns, elliptical, smooth, not amyloid. Cap cuticle cellular.

HABITAT: Scattered to densely gregarious or in troops on pine needles, often abundant in our coastal pine forests in the late fall, winter, and spring.

EDIBILITY: Unknown, but much too miniscule to be of value.

COMMENTS: The widely spaced, adnate to decurrent gills, pale color, small size, frequently umbilicate cap, white spores, and growth under pine distinguish this pretty little mushroom. It looks like a pint-sized *Clitocybe* or an *Omphalina* or *Camarophyllus*, and is keyed out under those genera. Actually, the term "unidentified" is not quite appropriate

This dainty undescribed *Marasmius* is a common feature of our coastal pine forests. Note how widely spaced the gills are.

because it has been "identified" by Dennis Desjardin, Rolf Singer, Howard Bigelow, and other mycologists as an undescribed species. It is not as tough as most *Marasmius* species and does not revive when moistened, but is assigned here to *Marasmius* for lack of a better alternative. According to Desjardin, it may be the prototype for a "new" genus.

Marasmius copelandi   (Garlic Mushroom)

CAP 0.5-2 (2.5) cm broad, convex to plane or centrally depressed; surface smooth or often wrinkled or striate, light brown to buff or flesh-colored, sometimes fading to whitish as it dries; not viscid. Flesh thin, whitish; odor distinctly garlic- or onionlike; taste garlicky to slightly acrid. **GILLS** pallid to flesh-colored or colored like the cap, attached (usually adnate or notched, sometimes seceding), often somewhat crisped. **STALK** 2-7 cm long, 1-3 mm thick, equal or slightly thicker at either end, tough, hollow, minutely hairy; dark purple-brown to reddish-brown with a paler apex, the base often blackish-brown (but the hairs may appear whitish when dry). **SPORE PRINT** white; spores 12.5-16 × 3-4 microns, more or less narrowly pip-shaped, smooth, not amyloid.

HABITAT: Scattered to gregarious on fallen leaves; common along the west coast. In our area it is often abundant on tanoak leaves and chinquapin burrs in the fall and winter. A large-spored variant occurs farther north on the leaves of salal and other shrubs.

EDIBILITY: Edible. It can be used as a seasoning or garlic substitute, but should be cooked only slightly if at all. It makes up for its small size by fruiting in large numbers.

COMMENTS: True to its name, this species and its close relatives smell and taste like garlic. In fact, they are often smelled before they are seen. Aside from the odor,

Marasmius copelandi is one of several small brownish species with a strong garlic- or onionlike odor. It has passed under several names, but marasmiologist D. Desjardin says that *M. copelandi* is correct.

there is little else to separate them from other "LBM's." The mycelium must also smell like garlic, because in wet weather our tanoak humus will often have a distinct garlic odor— even when no fruiting bodies are present! There are several very similar "garlic mush-rooms," including: *M. olidus,* found on oak leaves in eastern North America, with slightly shorter spores; *M. prasiosmus,* a European species with whitish gills and even smaller spores (of uncertain occurrence in North America); *M. scorodonius,* widespread, with a reddish-brown to pallid cap and *smooth* (hairless) stem, found on needles, twigs, grass stems, etc.; *M. alliaceus,* a European species with a long black, minutely hairy stalk; and *M. thujinus,* with a minute (1-3 mm broad) cap and garlic odor if crushed, found under northern conifers. Several of these names have been applied to our local garlic mushroom, but marasmiologist Dennis Desjardin says that *M. copelandi* is the correct name.

Marasmius androsaceus   (Horsehair Fungus)

CAP 2-10 (20) mm broad, convex to plane, the center often depressed; surface dry, soon radially wrinkled or striate, reddish-brown to pale brown or flesh-colored, fading in age. Flesh very thin, pliant, reviving when moistened; odor mild. GILLS narrow, well-spaced, pallid becoming flesh-colored or brownish, usually adnexed to adnate. STALK 2-7 cm long, less than 1 mm thick, equal, hairlike, tough, stiff, entirely black or black with a brown to reddish-brown apex; black horsehair-like rhizomorphs usually emanating from base or visible in surroundings. SPORE PRINT white; spores 6-9 × 2.5-4.5 microns, elliptical or pip-shaped, smooth, not amyloid. Cystidia present on gill edges.

HABITAT: Scattered or in troops on needles, twigs, or leaves; widespread. In our area it is fairly common, along with similar species (see below) in wet weather.

EDIBILITY: Unknown—hardly worth the trouble to find out.

COMMENTS: Several dainty marasmioid fungi will more or less fit the above description. They are barely visible when shrivelled up, but rain revives them. They differ from the more numerous Mycenas in their dark, hairlike stem and tougher texture. *M. androsaceus* and *M. pallidocephalus* (very similar, but partial to conifer needles and lacking cystidia on the gill edges) usually possess black rhizomorphs, and sometimes form stems with no caps. *Micromphale sequoiae* is similar but grows only on redwood needles, has a brown stalk, and lacks thick rhizomorphs; likewise *Micromphale perforans*, which grows on spruce and fir needles. *Micromphale arbuticola,* found in swarms at the bases of madrones, is a minute brownish bark-inhabitor with a slight garlic-onion odor; *Marasmius capillaris*, partial to dead leaves in eastern North America, has a black stalk and minute brownish cap with a white center. See also *M. epiphyllus* and *M. quercophyllus* under *Marasmiellus candidus*.

Marasmius oreades   (Fairy Ring Mushroom)   Color Plates 38, 47

CAP 1-5 (6) cm broad but usually 2-4 cm; at first bell-shaped or umbonate with an incurved margin, then convex or plane but often retaining an obtuse umbo, the margin often uplifted in old age; surface smooth, dry, color variable: reddish-tan to light brown, tan, buff, or even white; margin faintly striate when moist. Flesh tough, pliant, pallid, reviving when moistened; odor agreeable. GILLS adnate, adnexed, or free, fairly well-spaced, broad, white to pale tan, sometimes discoloring brownish in old age. STALK 2-8 cm long, 1.5-6 mm thick, equal or tapering downward, tough and pliant, smooth, colored like the cap or paler (whitish). SPORE PRINT white; spores 7-10 × 4-6 microns, elliptical to somewhat irregular, apiculate, smooth, not amyloid.

HABITAT: Gregarious in grass, usually in arcs or rings; widely distributed and very common in lawns, parks, cemeteries, etc.; also common in pastures in the Pacific North-west, but rarely straying far from suburbia in our area. Found year-round except during cold spells, but most abundant in California in late spring, summer (on watered lawns), and fall. Several crops are produced each year, but its presence can be detected even when it isn't fruiting—just look for "fairy rings" (patches of brown grass rimmed by a lush zone of darker green grass). The living mycelium on the periphery of the ring stimulates the grass to grow, while the dried-up mycelial matter within the circle inhibits growth.

EDIBILITY: Delectably delicious—one of the few "LBM's" worth learning. What it lacks in substance it makes up for in abundance. Discard the tough stems and use the caps whole. They're superb is just about anything—omelets, soups, sauces, stir-fried dishes, even cookies. Or simply saute in butter and serve on toast! What's more, they dry easily, don't decay quickly, and are usually free of maggots. Don't pass up shrivelled, sun-dried specimens—they are easily resurrected or can be stored in an airtight jar for later use.

COMMENTS: At first glance this seems like yet another Boring Ubiquitous Mushroom ("BUM"). However, a closer look reveals that it is really quite attractive, with a lean, clean, subtle symmetry all its own. Many "BUM's" and "LBM's" grow on lawns, but the fairy ring mushroom can be distinguished by the following features: (1) cap obtusely umbonate in many specimens (2) white spores (the grass beneath mature caps is often dusted with white spore powder) (3) broad (deep) white to buff gills which are fairly well-spaced and *not* decurrent (4) thin, *tough* stem (5) growth in grass (6) the ability of dried specimens to revive dramatically when moistened. Be especially careful not to confuse it with the poisonous *Clitocybe dealbata,* which is white-spored and grows in grass (often with *M. oreades*), but has thin, crowded, adnate to decurrent gills and a convex to plane (*not* umbonate) cap. The growth in rings is *not* a good means of distinguishing it, as many mushrooms are capable of growing in circles, including *C. dealbata.* Other species: *M. albogriseus* is fairly common in central and southern California under trees, shrubs, and chaparral or even in grass. It tends to have a grayer or browner or yellower cap than *M. oreades,* at least at the center, and a hollow stem (the stalk of *M. oreades* is usually stuffed with a white pith), but is otherwise quite similar. *M. (=Collybia) strictipes* of eastern North America is also somewhat similar but grows in the woods and has closer gills and a yellowish cap. Another woodland easterner, *M. nigrodiscus,* is larger (cap up to 11 cm broad) and often has a striate stalk, but is similar in color to *M. albogriseus. Collybia luxurians* sometimes grows in grass but has close gills and a reddish-brown cap. Finally, there is an unidentified *Mycena* that often grows in grass. However, it is smaller than *M. oreades,* its stalk is not as tough, it is usually reddish- or vinaceous-tinged, and its cap is convex to plane (not umbonate).

Marasmius plicatulus   (Pleated Marasmius)   **Color Plate 45**

CAP 1-4 (5) cm broad, obtusely conical to bell-shaped, often expanding to convex or plane or with uplifted margin in age; surface dry, with a velvety or frosted appearance when fresh, furrowed or wrinkled in age or upon drying; bay-brown to reddish-brown, brown, wine-red, or maroon. Flesh thin, pliant; odor mild. **GILLS** adnate to nearly free, well-spaced, broad, white to buff to pinkish or tinged cap color. **STALK** 5-13 cm long, 1.5-3 mm thick, equal, tough but brittle, usually long and thin, smooth, polished, reddish-black to deep chestnut below, often paler (pinkish or sometimes pallid) above; base often with whitish mycelium. **SPORE PRINT** white; spores 11-15 × 5-6.5 microns, elliptical, smooth, not amyloid.

HABITAT: Widely scattered to gregarious in humus under trees and shrubs; apparently endemic to the west coast. Common in our area in the late fall and winter (at least one month after the rainy season begins) under eucalyptus, oak, conifers, in brambles, etc.

EDIBILITY: Like myself, too tough and thin to be edible.

COMMENTS: One of the most exquisite of all mushrooms—the frosted wine-red to brown cap, widely spaced (distant) pallid gills, and long, shining reddish-black stalk are

Left: *Marasmius plicatulus* is one of our most beautiful mushrooms. See color plate for close-up of gills. **Right:** *Marasmius haematocephalus*, a gorgeous tropical species (see comments below). *M. siccus* of eastern North America (not illustrated) closely resembles it. (Michael Fogden)

distinctive. The stalk is so brittle that the mushroom must be dug up (rather than plucked) to keep it intact. The conical cap may lead to confusion with *Mycena,* but the tough, polished stem is characteristic of *Marasmius.* There are several similar and equally exquisite species with distant gills, including: *M. bellipes,* with smaller spores and a small cap (up to 15 mm broad) and short stalk; *M. borealis*, whose cap is not pleated or striate; *M. siccus* and *M. fulvoferrugineus* of eastern North America, with even more distant gills, a shorter (2-7 cm) stalk, an orange-brown to ochre-tawny to rusty-brown, deeply ribbed or pleated cap (like a miniature umbrella) and spores 15-21 microns long (I have seen the oranger of the two, *M. siccus,* preserved nicely in plastic cubes); and *M. haematocephalus* (see photo above), a gorgeous tropical leaf-inhabitor with a dark red cap.

Crinipellis piceae

CAP 3-7 (10) mm broad, convex to broadly convex or nearly plane, the center sometimes slightly depressed; surface whitish to buff or tinged tawny except for the dark (tawny-brown to brown or blackish) center, which is often surrounded by a dark circle; covered with coarse tawny to brownish hairs and sometimes minute scales, not viscid; margin often ciliate (fringed with projecting hairs). Flesh very thin, white. **GILLS** white, close, adnexed or free. **STALK** 2-6 cm or more long, up to 1 mm thick, more or less equal, very thin and tough, brown to blackish-brown beneath a coating of minute hairs. **SPORE PRINT** white; spores 7-10 × 3-4.5 microns, cylindrical, smooth, not amyloid. Hairs on cap dextrinoid.

HABITAT: Solitary, widely scattered, or in groups on twigs, needles, and debris of conifers, especially spruce; known only from the west coast (and Asia). It fruits practically year-round in damp weather and is, according to Dennis Desjardin, the most numerous "marasmioid" fungus of the coastal forests of northern California. It does not seem to occur in our area, but neither does spruce.

EDIBILITY: Much too miniscule to merit attention.

COMMENTS: This minute *Marasmius*-like mushroom is easily told by the coarse dextrinoid hairs on the cap, frequently ciliate cap margin, and thin, dark stalk. Other species: *C. campanella* is a slightly larger northern species with a rusty-orange to chestnut-brown hairy cap and stem and a tendency to grow on conifer twigs (especially cedar) that are still on the tree. *C. zonata* of eastern North America is a "large" (cap 1-2.5 cm) species with coarse tawny hairs; it grows on dead wood. *C. stipitaria* has a minute central nipple on the cap. None of these are worth eating.

Callistosporium luteo-olivaceum

CAP 1.5-6.5 cm broad, convex or slightly umbonate becoming plane or shallowly depressed; surface not viscid, often minutely scurfy at first but becoming smooth; dark olive to olive-brown or olive-yellow, often becoming yellower (yellow-brown to honey-colored) in age and developing dark reddish-brown tones when dried. Flesh thin, pallid or yellow or tinged cap color; odor mild to pungent or slightly fruity; taste mild or slightly bitter. **GILLS** yellow to golden-yellow, tending to redden when dried; close, notched or adnexed or at times adnate. **STALK** 2.5-7 cm long, 0.3-1 cm thick, equal or slightly thicker at either end, often flattened, smooth to fibrillose or scurfy (especially over the lower portion), sometimes streaked in age; colored like cap or slightly darker, tending to turn deep reddish-brown from the base upward as it dries. **SPORE PRINT** white; spores 4.5-6.5 × 3-4.5 microns, elliptical to nearly round, apiculate, smooth, not amyloid but many of them staining vinaceous in KOH (potassium hydroxide).

HABITAT: Solitary, scattered, or in small groups or tufts on rotten wood (often buried!) under conifers; widely distributed but not common. I have found it in our coastal pine forests in the winter. It is said to occur on hardwoods also, particularly in the tropics.

EDIBILITY: Unknown.

COMMENTS: This distinctive mushroom has the stature of a *Collybia* and was originally placed in that genus. However, the olive and yellow coloration plus the tendency to grow on rotten wood (sometimes very decomposed!) distinguish it. *C. graminicolor* is a similar but smaller (cap up to 2 cm broad) northwestern species with larger spores.

Strobilurus trullisatus

CAP 0.5-1.5 cm broad, convex to plane or slightly depressed; surface dry, often striate or wrinkled, minutely granular, white to pinkish-buff or brownish. Flesh very thin. **GILLS** typically adnate to adnexed, close, white or tinged pinkish-buff. **STALK** 2-5 cm long, 1-1.5 mm thick, equal, dry, minutely granular; apex white, lower portion yellowish to brownish or tawny; base with yellow to tawny-orange hairs and mycelial threads. **SPORE PRINT** white; spores 3-6 × 1.5-3 microns, elliptical, smooth, not amyloid.

HABITAT: In colonies (usually 4-10) on old Douglas-fir cones or rarely cones of other conifers; common throughout the range of Douglas-fir. In our area it usually fruits after the first fall rains.

EDIBILITY: Who knows? Who cares?

COMMENTS: Also known as *S. kemptonae* and *Collybia trullisata,* this little mushroom is one of several species that grow *only* on rotting cones. (Some wood-inhabiting mushrooms, such as *Mycena purpureofusca,* may grow on cones but are not restricted to them.) Other *Strobilurus* species include: *S. conigenoides,* a whitish species that grows only on

Strobilurus trullisatus is a nondescript mushroom that grows exclusively on cones. *Baeospora myosura* (not illustrated) is rather similar but has very crowded gills.

the fallen seed pods ("cones") of magnolias in eastern North America; *S. occidentalis,* which occurs on the cones of Sitka spruce; *S. albipilatus (=Collybia albipilata),* with a pinkish-buff to brown or dark brown cap, common in the Sierra Nevada and other western mountains on pine cones and other coniferous debris; and *S. lignitilis,* which tends grow on buried decaying wood and has a grayish-brown cap. Finally, there is *Baeospora myosura (=Collybia conigena),* which is slightly larger than *S. trullisatus* and has a smooth, buff or tan to pinkish-brown cap, *very* crowded gills, a coarsely hairy stem base, and amyloid spores. It grows on the cones of various conifers but favors Douglas-fir, at least in California.

Collybia tuberosa

CAP 3-10 mm broad, convex to plane or centrally depressed; surface smooth, dry, whitish to buff, sometimes with a darker (yellowish to brownish or pinkish-buff) center. Flesh very thin, white. **GILLS** white or rarely tinged pinkish, adnate to adnexed, close or crowded. **STALK** 1-3 cm long, up to 1 mm thick, equal, dry, minutely downy, white or tinged brown, often arising from a small orange-brown to reddish-brown to blackish, appleseed-like body or "tuber" (sclerotium). **SPORE PRINT** white; spores 3-6 × 2-3 microns, elliptical, smooth, not amyloid.

HABITAT: In colonies on the blackened remains of old mushrooms, particularly larger *Russula* and *Lactarius* species (e.g., *R. albonigra*), occasionally in humus; widely distributed. I have found it only once in our area, in December, but the very similar *C. cookei* and *C. cirrhata* (see comments) are fairly common.

EDIBILITY: Unequivocally inconsequential.

COMMENTS: This dainty little *Collybia* is one of four widespread species that colonize decayed mushrooms. The host may be so deteriorated, however, that it is not recognizable as a mushroom. The other three species are: *C. cookei,* practically identical but with rounder, more prominent, tan to yellow or yellow-orange sclerotia (and occasionally found on rotten wood or in humus); *C. cirrhata,* also very similar but with white mycelial threads instead of sclerotia, found in humus as well as on old mushrooms (and particularly common in the Sierra Nevada); and *C. racemosa,* which has stubby lateral branches on the stem (see description). All of these are placed in *Microcollybia* by some taxonomists.

Left: *Collybia tuberosa* and its close relatives grow in colonies on decaying mushrooms and other debris. Sclerotia (beadlike bodies from which the mushrooms arise) are not visible in this picture. **Right:** *Collybia racemosa* is easily told by its small size and unique branching stalk.

Collybia racemosa   (Branched Collybia)

CAP 3-10 mm broad, bluntly conical to convex becoming umbonate to nearly plane; surface smooth, not viscid, dark gray to gray or brownish-gray, the margin often paler. Flesh very thin. GILLS typically adnexed, close, gray or brownish-gray. STALK 3-8 cm long, 0.5-3 mm thick, with numerous short lateral side-branches, especially over the lower half or two-thirds; gray to brownish-gray, often entirely buried and sometimes originating from a small blackish beadlike body (sclerotium). SPORE PRINT white; spores 4-5.5 × 2-3 microns, oblong to elliptical, smooth, not amyloid. Asexual spores (conidia) often produced on the swollen tips of the side-branches.

HABITAT: In small groups or colonies on old decayed or blackened mushrooms (e.g., *Russula albonigra*) or occasionally in coniferous duff; widely distributed but seldom encountered, perhaps because it is so easily overlooked. I have found it only once in our area, in December, but it is said to be fairly common in the Sierra Nevada.

EDIBILITY: Unknown, but much too puny and rare to be of value.

COMMENTS: This curious *Collybia* is the only one with side-branches on the stem (see photo on p. 212). The stalk may meander somewhat through its substrate, branching along the way, and the sclerotium from which it originates is not always evident. Stems without caps are sometimes found, implying that the formation of asexual spores on the side-branches can be enough to "satisfy" the mycelium's reproductive urges.

Collybia confluens   (Tufted Collybia)

CAP 2-5 cm broad, convex to plane or slightly umbonate, the margin sometimes uplifted or wavy in age but incurved at first; surface smooth, hygrophanous: reddish-brown to pinkish-cinnamon or flesh-colored when moist (often darker at center and pallid at margin), fading to pinkish-buff, grayish-pink, or whitish as it dries. Flesh thin, white. GILLS crowded, narrow, adnate soon becoming adnexed or even free, whitish to flesh-colored. STALK 3-10 cm long, 2-5 mm thick, equal, hollow, pliant, sometimes flattened or grooved, tough, usually darker than cap (reddish-brown), but covered with a minute white pubescence (downy hairs); base often with white mycelial mat attached. SPORE PRINT white or tinged yellow; spores 7-9 × 3-4 microns, narrowly elliptical, smooth, not amyloid.

HABITAT: Gregarious, often in tufts or clusters, on ground in woods; widely distributed. In the Pacific Northwest it is quite common under conifers, but in eastern North America it favors hardwoods. I have not seen it in our area, but similar species (see comments) occur.

EDIBILITY: Edible with caution; it is tough and similar species have not been tested.

COMMENTS: The fine white pubescence on the stalk (use hand lens!) and crowded gills help to separate this species from *C. dryophila* and *C. butyracea*. It shrivels up in dry weather and revives somewhat when moistened, and as a result was originally placed in *Marasmius*. A similar, unidentified species whose gills are adnate to adnexed and not so crowded and whose stalk lacks the litter-binding mycelial mat is quite common in our area in the fall and winter; it sometimes rivals *C. dryophila* for abundance but appears later in the season and, like *C. confluens,* it has a minutely downy stem. Another local, downy-stemmed species appears to be undescribed. It resembles the *C. fuscopurpurea* group, but has whitish gills and a pallid to brownish-tan cap. It differs from *C. confluens* in having attached gills which are fairly well-spaced, and it usually grows in tufts or clusters on decaying wood, wood chips, or in lignin-rich humus. A third species, *C. polyphylla,* has a garlicky to slightly unpleasant odor and taste, while a fourth, *C. peronata (=Marasmius urens)* has an acrid (burning) taste. The latter two species are widely distributed but do not seem to occur in California.

Collybia fuscopurpurea group. These ubiquitous "LBM's" have a tough texture and usually grow in groups or clusters.

Collybia fuscopurpurea group

CAP 1-4 cm broad or slightly larger, convex becoming plane, slightly depressed, or with the margin uplifted in age; surface dark reddish-brown to brown, purple-brown, or chocolate-brown when fresh, usually paler (near tan) when dry, usually radially wrinkled or finely striate. Flesh whitish or colored like cap, thin, reviving somewhat when moistened; odor mild. **GILLS** attached (usually notched or adnexed), fairly well-spaced, pallid or pale pinkish-tan becoming more or less cap color, then dusted whitish by spores. **STALK** 2-10 cm long, 1-4 mm thick, equal or tapered downward, dry, rather tough and pliant, brown with the apex usually paler; clothed with minute grayish to brownish hairs (unless very wet) throughout or over the lower two-thirds; often curved near the litter-binding base. **SPORE PRINT** white; spores 6-8 × 3-4 microns, elliptical, smooth, not amyloid.

HABITAT: Densely gregarious (often in clusters) among leaf litter, humus, and woody debris in woods, under trees (especially oak), or in wood chips or landscaped areas; widely distributed. Common year-round in our area but particularly abundant in the early fall.

EDIBILITY: Unknown.

COMMENTS: Formerly known as *Marasmius fuscopurpureus,* this common species "complex" often forms dense swarms in leaf litter and woody debris, especially where the ground has been recently disturbed. Like most "LBM's," it blends in well with its surroundings. It can be recognized by the reddish-brown to dark brown fruiting body, convex to plane cap (never bell-shaped as in *Mycena*!), white spores, pubescent (finely hairy) stem, and gregarious disposition. However, separating the numerous species within the "complex" is a job for specialists. One such specialist, Dennis Desjardin, says that several of California's representatives in this "complex" (including the most common one) appear to be undescribed. There are a number of other difficult-to-identify *Collybia* and *Marasmius* species with a pubescent stalk and litter-binding mycelium. Several of these (e.g., *M. cohaerens*) are described in the key to *Marasmius* and *Collybia;* others are discussed under *C. confluens.*

Collybia acervata   (Clustered Collybia)   **Color Plate 49**

CAP 1-4 (5) cm broad, convex with an incurved margin, becoming broadly convex in age; surface smooth, hygrophanous, not viscid; dark reddish-brown when fresh and moist, fading to pale reddish-brown, pinkish-buff, or paler (sometimes with darker and lighter zones) as it dries. Flesh thin, pallid. **GILLS** close or crowded, narrow, typically adnexed or notched or free, white to dingy pinkish or vinaceous-buff. **STALK** 4-12 cm long, 2-6 mm thick, more or less equal, dry, hollow, pliant but brittle, smooth above, with fine whitish hairs over lower half or at base; reddish-brown to vinaceous-brown or sometimes paler in age. **SPORE PRINT** white; spores 5-7 × 2-3 microns, elliptical, smooth, not amyloid.

HABITAT: In compact bundles or clusters on rotting conifers (but often appearing terrestrial); widely distributed. It is common in the summer and fall in the Sierra Nevada and Rocky Mountains and in the Pacific Northwest, but apparently absent in coastal California south of San Francisco.

EDIBILITY: Inedible. It is said to have a bitter taste when cooked and is apparently slightly poisonous to some people.

COMMENTS: The bundled growth habit (see color plate!), white spores, and reddish-brown stem make this an easy mushroom to recognize. It always grows on wood, but may appear terrestrial if its host is buried or in a very advanced stage of decay. The cap varies in color according to the amount of moisture present, but is never translucent-striate as in *Mycena*. Other species: *C. marasmioides (=C. erythropus, C. bresadolae)* tends to grow in tufts or clusters on the ground, but has a pale tan to creamy-buff cap and beautiful dark red (paler at apex) stem. A mushroom meeting this description occurs in our area but is rare. *Clitocybula (=Collybia) familia* has the aspect of *C. acervata* (densely clustered growth habit on rotting conifers—see photo on p. 898), but it has a watery white to smoky-gray to somewhat brownish or tan (never reddish-brown) cap, white to grayish stem, and round, amyloid spores. It is edible and widely distributed, but in California it seems to be restricted to the Sierra Nevada. *Clitocybula (=Collybia) abundans* resembles *C. familia,* but has a smaller fibrillose cap and elliptical, amyloid spores.

Collybia dryophila   (Common Collybia; Oak-Loving Collybia)

CAP 1-5 (7) cm broad, broadly convex with an incurved margin, becoming plane or with with an uplifted, often wavy margin in age, sometimes also slightly umbonate; surface smooth, hygrophanous: chestnut-brown to reddish-brown, yellow-brown, tawny, or ochre when young and moist, but fading to tan, pinkish-tan, yellowish-tan, or buff as it dries. Flesh thin, white. **GILLS** crowded, usually notched or adnexed, white to pale yellow. **STALK** 2-8 cm long, 2-6 mm thick, equal or with a swollen base, slender, smooth, hollow, rather tough and cartilaginous; pale cream or colored like cap (but often paler); white mycelium often visible at base or in surrounding humus. **SPORE PRINT** white or pale cream; spores 5-7 × 2-3.5 microns, elliptical, smooth, neither amyloid nor dextrinoid.

HABITAT: Scattered to gregarious or in small tufts in woods or near trees, often forming arcs or rings; widely distributed. It is abundant in our area shortly after the first fall rains, but is less common thereafter. As its name implies (*dryophila*=oak-loving), it is fond of oak, but is also common under pine and other conifers, as well as around the edges of old sawdust piles. At higher elevations it often fruits in the spring as well as in the summer and fall.

EDIBILITY: Edible, but some people are apparently sensitive to it. Only the caps are tender enough to eat. It is a proficient concentrator of mercury, but occasional consumption should pose no threat as mercury is a cumulative poison.

Collybia dryophila is a cosmopolitan "LBM" with white or yellowish gills that are usually notched, adnexed, or even free. Note the relatively slender equal stalk and small size.

COMMENTS: The appearance of this species in large fairy rings under oaks is a sure sign that the coastal California mushroom season is under way. The hygrophanous reddish-brown to tawny, rusty, or tan cap, more or less adnexed gills, growth on the ground, and smooth (hairless) stalk distinguish it from all but *C. butyracea,* which has a greasier (buttery) cap, buff-colored spores, and ragged gill edges. Both species are reminiscent of the fairy ring mushroom *(Marasmius oreades),* but grow under trees and have crowded gills; neither grows in the tight bundles characteristic of *C. acervata,* and neither has downy hairs on the stem like *C. confluens.* In some regions *C. dryophila* is frequently covered by lumps or masses of somewhat jelly- or tumorlike tissue, caused by the fungus **Christiansenia mycetophila.** However, I have not observed this phenomenon in California. The yellow-gilled form of *C. dryophila* is sometimes listed as a separate species, **C. subsulphurea.**

Collybia butyracea (Buttery Collybia)

CAP 3-8 (12) cm broad, convex becoming plane or uplifted, often with a broad umbo; surface smooth, greasy or slippery when moist but not truly viscid; dark reddish-brown to chesnut-brown to dull brown (or grayish-brown in one form), fading as it dries or ages to tan, grayish-tan, reddish-tan, or ochre-buff; margin at first incurved. Flesh thin, soft, whitish or watery. **GILLS** close or crowded, free or adnexed, white, the edges usually uneven, eroded, or finely scalloped at maturity. **STALK** 2-10 cm long, 4-10 mm thick, equal or more often thicker below and/or pinched at the base; often longitudinally striate or twisted; rather tough and cartilaginous, hollow at least in age, smooth; colored like cap in age but often buff when young; base usually with white mycelial down. **SPORE PRINT** creamy to yellowish, buff, or pinkish-buff; spores 6-8 × 3-3.5 microns, elliptical, smooth, many of them dextrinoid.

HABITAT: Scattered to gregarious or tufted in humus under conifers (especially pine) or occasionally hardwoods; widely distributed. It is sometimes common in our coastal pine forests in the late fall, winter, and early spring.

EDIBILITY: Edible; about like *C. dryophila.* Care must be taken to identify it correctly!

COMMENTS: Also known as ***Rhodocollybia butyracea,*** this handsome and widespread species is often confused with *C. dryophila.* However, its cap is apt to be greasier when moist, the gills are more apt to be finely scalloped or eroded, and the spores are slightly more colored in deposit and at least somewhat dextrinoid. A grayish-brown version of the species *(form* **asema)** also occurs in North America, especially on the west coast. Other species: *C. badiialba* and *C. distorta* are two closely related species with round or nearly round spores, a reddish-brown to vinaceous-brown cap, and white gills that may or may

216

Collybia butyracea is often confused with *C. dryophila,* but is slightly different in color and slightly larger, and often has a club-shaped stalk (thicker at base).

not develop reddish stains in age. Both favor conifers and grow in humus or on rotten wood. The former occurs in coastal California and the Pacific Northwest while the latter is more common in eastern North America. Another species with a vinaceous-brown cap, *C. extuberans,* has elliptical spores and is widely distributed.

Collybia maculata   (Spotted Collybia)

CAP (2.5) 4-12 cm broad, convex with an incurved margin at first, becoming broadly umbonate to plane or undulating; surface dry or lubricous but not truly viscid, smooth, typically white to buff, pale tan, or pinkish-tinged, often darker or redder in age, especially at the center and often developing rusty or reddish spots and stains. Flesh thick, white; odor mild or somewhat fragrant; taste usually bitter (but mild in *var. occidentalis*). **GILLS** white to pale pinkish-buff, often developing rusty or reddish stains (like the cap) in age; adnate to adnexed or notched to nearly free, crowded. **STALK** 4-15 cm long, 0.5-2 (3) cm thick, often tapered below to form a more or less rooting base; solid (at least when young), firm, smooth or fibrillose, often striate or grooved and easily splitting longitudinally; white to pale buff, usually developing rusty or reddish stains below. **SPORE PRINT** creamy to pale peach or pinkish-buff; spores 5-11 × 4-6 microns, round or nearly round in the typical variety, elliptical in *var. occidentalis*; smooth, at least a few dextrinoid.

HABITAT: Solitary, scattered, or in groups or tufts on decaying wood or lignin-rich humus under conifers; widely distributed. I have not seen it in our area, but it is quite common in the Pacific Northwest and Sierra Nevada in the late spring, summer, and fall.

EDIBILITY: Inedible because of the frequently bitter taste.

COMMENTS: Also known as *Rhodocollybia maculata,* this rather large species is closely related to *C. oregonensis* and has the same stature, but typically has a paler cap and mild odor. The pinkish to rusty-reddish spots and stains and tendency to grow on rotten wood are good fieldmarks. It could conceivably be mistaken for a *Tricholoma,* but the stalk is longer than in most Tricholomas and it splits more easily. It is a variable species, and a number of varieties have been designated based on differences in color, taste, and spore shape. The above description embraces several such varieties. Others include: *var. scorzonerea,* a large version with yellowish gills and a frequently yellowish-tinged stalk and/or cap; and two small varieties from the Pacific Northwest with a vinaceous-brown to blackish cap.

217

Left: *Collybia oregonensis* is an amazingly fragrant mushroom. **Right:** *Oudemansiella radicata* (p. 219), a common mushroom in eastern North America, is rather variable in color but always has broad gills (as shown here) and a deeply rooting stalk (not visible).

Collybia oregonensis   (Fragrant Collybia)

CAP 4-10 cm broad, broadly convex to plane; surface smooth, slightly viscid when moist but soon dry; deep chestnut-brown to vinaceous-brown or reddish-brown at the center, often paler or pinker or redder toward the margin, and gradually fading overall in age. Flesh thin, white or reddish-stained; odor heavy and sweet (like benzaldehyde), taste somewhat bitter. **GILLS** crowded, creamy or pale yellow to buff, often reddish-stained in age; adnexed, notched, or seceding; edges sometimes overlapping each other and eroded in age. **STALK** 6-20 (30) cm long, (0.5) 0.8-2 cm thick, usually quite long and deeply rooted; equal above or swollen slightly near ground level, tapered gradually below to a point; dry, whitish, but usually developing reddish to reddish-brown stains especially over the lower portion; solid or hollow in age. **SPORE PRINT** whitish to buff; spores 6-8 × 3.5-5 microns, elliptical, smooth, at least some of them dextrinoid.

HABITAT: Solitary or in small groups around old stumps and in lignin-rich humus; apparently endemic to the Pacific Coast, not common. I've seen it in the Sierra Nevada, and in one local spot (in mixed woods) where it fruits every year, usually in the fall.

EDIBILITY: Unknown. Despite its sweet odor it has a somewhat bitter taste.

COMMENTS: The strong, heavy benzaldehyde or almond extract odor and vinaceous-brown cap are distinctive. The fragrance is reminiscent of *Russula fragrantissima,* but without the fetid component of that species. The stalk often roots deeply but does not form a true pseudorhiza ("tap root") as in *Caulorhiza umbonata.* The gills are not waxy as in the almond-smelling *Hygrophorus* species (e.g., *H. bakerensis*) and the odor is usually thicker and more pronounced. *C. subsulcatipes,* known only from Washington, has a similar but weaker odor, a brown to dark vinaceous-brown stalk with a "tap root," a brown to vinaceous-buff cap, and dull vinaceous gills in age. Both of these species are placed in a separate genus, *Rhodocollybia,* by some collybiologists.

Caulorhiza umbonata   (Redwood Rooter)

CAP 3-15 cm broad, at first conical or umbonate, then expanding but usually retaining the umbo; surface smooth, dry, chestnut-brown to warm tan or yellow-brown; margin at first incurved. Flesh thin, pallid; odor mild. **GILLS** notched or adnexed to nearly free, close, broad, white to yellowish or tinged cap color. **STALK** 6-50 cm long or more, 0.4-1.5

Caulorhiza (=Collybia) umbonata is a flagrant fungal feature of our redwood forests. Note the conical or umbonate cap and long "tap root" which may extend as much as three feet into the humus.

(2) cm thick, most of it underground in the form of a long, tapered "tap root"; smooth, twisted-striate, buff to tan or colored like cap but paler; rather tough and cartilaginous. **SPORE PRINT** white; spores 5-8 × 3-5 microns, elliptical, smooth, amyloid.

HABITAT: Solitary, scattered, or in groups near or under redwood (rarely elsewhere); restricted to the range of coastal redwood *(Sequoia sempervirens)*. It is very common in our area in the fall and winter.

EDIBILITY: Like myself, not firmly established. However, if it were poisonous we would probably know by now. I've sampled small quantities without ill effects.

COMMENTS: Anchored by its long "tap root," this distinctive species, like a dandelion, resists being uprooted. The "tap root," pallid gills, umbonate cap, and white spore print, plus the association with redwood distinguish it. The "tap root" is brittle, however, and will break off and stay behind in the ground unless it is dug up very carefully. *Collybia umbonata* is a synonym. *Collybia subsulcatipes* (see comments under *C. oregonensis*) is somewhat similar but has a sweet odor. *Caulorhiza hygrophoroides* is a closely related hardwood-lover with a smaller (up to 5 cm) cap. None of these have the grayish-brown colors and exceptionally broad gills of *Oudemansiella radicata* and its relatives.

Oudemansiella radicata (Beech Rooter)

CAP 2.5-12 cm broad, bell-shaped becoming broadly convex to plane or with a broad umbo, the margin sometimes uplifted in old age; surface smooth or radially wrinkled, often viscid or tacky when moist, usually dark brown to grayish-brown, but in some variants grayish and in others whitish, blackish, or olive- to yellowish-brown; margin at first incurved. Flesh thin, whitish; taste mild. **GILLS** broad, fairly well-spaced, thickish, usually adnexed or notched; white. **STALK** 5-25 cm long, 0.3-1 (1.5) cm thick, usually thickest at or near ground level, with a tapered underground portion or "tap root"; whitish at apex or in upper portion, colored more or less like the cap below, dry, rather stiff and brittle, longitudinally lined or twisted-striate; smooth or in one form scurfy. **SPORE PRINT** white; spores 12-18 × 9-12 microns, broadly elliptical, smooth, not amyloid.

HABITAT: Solitary, scattered, or in small groups on or near hardwood stumps and roots (often appearing terrestrial) in woods, grassy clearings, etc.; fairly common in eastern North America, spring through fall (especially on beech), apparently absent in the West.

EDIBILITY: Edible but not choice—or so I am told. Only the caps are worth eating.

COMMENTS: Formerly known as *Collybia radicata,* this common eastern mushroom can be told by its broad white, well-separated gills (see photo on p. 218) and rooting stalk. The latter snaps easily, however, so that if it is not dug up carefully the telltale "tap root" will remain behind in the ground. The color of the cap is quite variable, but is usually in the dark brown to grayish-brown range. The viscidity of the cap is not always evident. *Tricholomopsis platyphylla* is somewhat similar but does not normally have a "tap root," while *Melanoleuca* species are terrestrial and have narrower, crowded gills. Other species: *O. longipes* is a similarly colored but slightly smaller species with a dry cap and much smaller spores; it occurs in the Rocky Mountains on hardwoods such as aspen.

Flammulina velutipes   (Velvet Foot; Velvet Stem)   **Color Plate 26**

CAP 1-5 (7) cm broad, convex to plane or broadly umbonate; surface smooth, slimy or viscid when moist, reddish-brown to yellow-brown, yellow-orange, orange-brown, or tawny, the margin often paler (yellower); fading in dry weather; margin at first incurved. Flesh thin, white or yellowish. **GILLS** adnate to adnexed or notched, white to pale yellow, close. **STALK** 2-11 cm long, 3-5 (12) mm thick, equal or thicker below, slender, tough, often curved, sometimes slightly off-center; smooth and pallid to yellowish to orange-brown when young, but developing a rusty-brown to blackish-brown velvety pubescence (tiny hairs) from the base upward as it matures. **SPORE PRINT** white; spores 6.5-9 × 3-5 microns, elliptical to pip-shaped, smooth, not amyloid. Gills with cystidia.

HABITAT: In tufts or clusters on or near stumps, logs, and roots of hardwoods (but sometimes appearing terrestrial); very widely distributed. It is fairly common in our area from fall through spring on poplar and willow, and in coastal sand dunes on bush lupine. In the Rocky Mountains it is abundant in the late summer on aspen; in eastern North America it fruits from the late fall through early spring on elm and other hardwoods.

EDIBILITY: Edible, but the sticky skin should be removed before cooking. In colder regions it is an important edible and is called the 'winter mushroom" because it fruits very late in the season (even during winter thaws) when other fungi are not available. In our balmy climate, however, its season coincides with that of other, more flavorful mushrooms, so it is not often gathered. A cultivated form of it called the "snow-puff mushroom" or enokitake can be bought in many markets. It looks something like a pure white bean sprout with its long, smooth (not velvety!) stem and negligible cap.

COMMENTS: The smooth, sticky, yellow-orange to brownish cap and stalk which is dark brown and velvety (at least below) at maturity plus the white spores, absence of a veil, and clustered growth habit on dead hardwoods typify this hardy mushroom. It was originally placed in *Collybia,* but is now given a genus of its own due to its viscid cap and the prominent sterile cells (cystidia) on the gills. In arid climates, however, the cap may appear dry rather than viscid. The fruiting bodies can apparently survive freezing, and after thawing out will continue to produce spores!

Flammulina velutipes is easily told by velvety brown stalk, sticky cap, cheerful color (see color plate), and growth on wood.

OMPHALINA & XEROMPHALINA

Small to minute, often brightly colored, saprophytic mushrooms. CAP *usually plane to depressed or umbilicate at maturity, not conical;* margin often incurved when young. GILLS often yellow, orange, or pinkish, *typically adnate to decurrent.* STALK usually central, *thin, cartilaginous, often pliant or rather tough,* hollow. VEIL and VOLVA absent. SPORE PRINT *white to pale yellow.* Spores smooth, amyloid *(Xeromphalina)* or not amyloid *(Omphalina).*

THESE are dainty, brightly colored mushrooms with decurrent gills and a cartilaginous stem. They were originally grouped together in the obsolete genus *Omphalia,* but *Xeromphalina* was created for the species with a dark stem, tougher texture, ability to revive somewhat when moistened, and amyloid spores.

Omphalina is interesting from a botanical standpoint because several species grow only with lichens. It is now thought that the mushrooms are actually the fruiting bodies of the fungal component of the lichen (a lichen is a symbiotic relationship between an alga and a fungus). Other Omphalinas grow in grass, moss, or on soggy logs. They are easily confused with Mycenas (which usually have a conical or bell-shaped cap), while even experts cannot agree on the distinction between *Omphalina* and some of the small, slender Clitocybes.

Xeromphalinas occur on logs and in humus and are more common than Omphalinas, at least in California. Their tough, usually dark stem and ability to revive after being dried are reminiscent of *Marasmius,* but their yellow to orange gills and amyloid spores are distinctive. In addition to *Xeromphalina* and *Omphalina,* the genus *Gerronema* is recognized by some mycologists; it embraces several Omphalinas not specifically associated with lichens. None of these mushrooms are large enough to eat. If your mushroom does not key out convincingly below, check *Mycena, Marasmius,* and the Hygrophoraceae.

Key to Omphalina & Xeromphalina

1. Stalk tough, wiry, or horny, at least the lower portion dark brown to orange-brown or reddish-brown when mature 2
1. Not as above 5
2. Typically growing on wood (sometimes buried), often in clusters or dense groups 3
2. Typically growing on ground or humus, usually scattered to gregarious but not clustered . 4
3. Cap less than 2.5 cm broad; growing on conifers .... *X. campanella,* p. 222
3. Cap typically at least 2 cm broad; cap velvety, more or less orange-brown; growing on hardwoods in eastern North America and the tropics .... *X. tenuipes*
4. Taste bitter; gills adnate .... *X. fulvipes* (see *X. cauticinalis,* p. 222)
4. Taste more or less mild; gills adnate to decurrent .... *X. cauticinalis,* p. 222
5. Cap *and* stalk slimy-viscid when moist, more or less yellow .. (see *Mycena lilacifolia,* p. 236)
5. Not as above 6
6. Fruiting body orange (or at times pinkish-orange or fading to yellow or orange-buff); gills well-spaced and waxy-looking; found in groups or clusters on rotting conifers, especially in the Pacific Northwest and California .... *O. luteicolor* (see *O. ericetorum,* p. 223)
6. Not as above 7
7. Cap and/or gills greenish-yellow to green, olive, olive-brown, or dark green; growing on rotting conifers or on ground, moss, or lichens . *O. wynniae* & others (see *Clitocybe odora,* p. 161)
7. Not as above; green shades absent 8
8. Cap 0.5-4 cm broad, ochre to grayish-ochre; gills yellow to orange-yellow; growing on rotting conifers; spore print often yellow- or salmon-tinged .... *O. chrysophylla*
8. Not as above; cap usually less than 2.5 cm broad 9
9. Cap minute, up to 1.5 cm broad, yellow-orange to orange or orange-red; stalk orange; gills pallid or yellowish; growing in moss .... *O. (=Rickenella* or *Mycena) fibula*
9. Not as above; differently colored or larger or growing elsewhere 10
10. Cap grayish to gray-brown; gills often yellowish (especially in age); found on rotten wood in eastern North America .... *O. strombodes*
10. Not as above .... *O. ericetorum* & others, p. 223

Xeromphalina fulvipes (see comments under *X. cauticinalis*) forms carpets of miniature fruiting bodies under redwood and other conifers. Gills are typically adnate and taste is bitter.

Xeromphalina cauticinalis

CAP 0.5-2.5 cm broad, convex becoming plane or with a small central depression; surface smooth, not viscid, reddish-brown to tawny or ochraceous-tawny, fading to yellowish. Flesh very thin, pliant; taste mild. **GILLS** adnate to decurrent, yellow, with veins between. **STALK** 2-8 cm long, 1-2.5 mm thick, equal or more often with a small bulb at the base; pliant, tough, tawny or yellowish above, dark brown below; base with tawny mycelium. **SPORE PRINT** white; spores 5.5-7 × 3-4 microns, elliptical, smooth, amyloid.

HABITAT: Scattered to densely gregarious on conifer needles and debris; common in western North America. In our area it appears (along with *X. fulvipes*—see comments) in the late fall and winter under redwood and other conifers.

EDIBILITY: Unknown.

COMMENTS: This is one of several terrestrial Xeromphalinas. The decurrent yellow gills distinguish it from *Mycena,* and the stalk is much tougher than in *Hygrocybe.* Other species: *X. fulvipes* is similar but has a bitter taste, adnate gills, and minute orange hairs on the stem. I have found it under redwood several times, but it is not as common as *X. cauticinalis; X. picta* is a minute terrestrial species with a greatly swollen stalk apex.

Xeromphalina campanella   Color Plate 44

CAP 0.3-2.5 cm broad, convex becoming broadly convex with a depressed center; surface smooth, not viscid, yellow-brown to tawny to orange-brown or cinnamon-brown; margin striate when moist. Flesh thin, pliant, yellowish. **GILLS** yellowish to dull orange, fairly well-spaced, decurrent with veins in between. **STALK** 1-5 cm long, 0.5-3 mm thick, equal or enlarged at base, pliant, tough, smooth, horny, often polished; yellowish above, brown to reddish-brown below; usually curved; base with bright tawny hairs. **SPORE PRINT** white to pale buff; spores 5-8 × 3-4 microns, elliptical, smooth, amyloid.

HABITAT: In groups or dense clusters on rotting conifers; widely distributed. It is common throughout the West, fruiting in our area in the fall, winter, and early spring.

EDIBILTY: A miniscule morsel that is hardly worth eating.

222

COMMENTS: The yellow to orange, decurrent gills plus the thin polished stem and small size typify this dainty mushroom. Its occurrence on conifers is so dependable that its presence can be taken as "proof" that its host is some type of conifer. It might be mistaken for a *Mycena,* but the shape of the mature cap is quite different. *Galerina autumnalis* is also somewhat similar, but has a veil and brown spores. The cluster pictured in the color plate is darker (browner) than normal, but shows the shape and growth habit quite well. Other species: *X. kauffmanii* is very similar but occurs with hardwoods in eastern North America; *X. orickiana* has a dark reddish-brown cap, grayish to brownish gills, and grows grows on redwood.

Omphalina ericetorum   (Lichen Agaric)

CAP 0.5-2.5 (3.5) cm broad, at first plane with an incurved margin, becoming deeply depressed or funnel-shaped in age; surface smooth, not viscid, dull cinnamon to brownish, fading to yellowish or straw color or paler as it ages; margin striate, often wavy. Flesh very thin, pliant. **GILLS** decurrent, well-spaced, sometimes veined, pale yellowish. **STALK** 1-3 cm long, 1-3 mm thick, equal or enlarged at base, often curved, smooth, pliant, pale reddish-brown above, yellow-brown to pale brown below; often pale yellowish in age. **SPORE PRINT** white to yellow; spores 7-9 × 4-6 microns, elliptical, smooth, not amyloid.

HABITAT: Usually in groups on old lichen-laden conifer logs or scum-covered soil, always associated with the lichen *Botrydina vulgaris;* widely distributed in the cool temperate zone, but rare in our area. It is one of the commonest mushrooms of the Arctic.

EDIBILITY: Much too puny to be of value.

COMMENTS: The small size, depressed cap, thin stem, decurrent gills, and distinctive habitat are the principal fieldmarks. The gills may appear somewhat waxy as in *Hygrocybe* and *Camarophyllus,* but the latter don't normally grow with lichens. *O. umbellifera* is a synonym. There are several similar species, including: *O. hudsoniana,* similarly colored but associated with the lichen *Coriscium viride; O. postii,* with a pinkish-red to orange cap, found in moss; *O. pyxidata* (see photo below), fairly common in our area in grassy or open places, with pinkish- or vinaceous-tinged gills that fade to yellowish or creamy in age; and *O. luteicolor* (**COLOR PLATE 24**), a beautiful waxy-gilled orange to salmon-colored species that fades with age and is common in the Pacific Northwest, northern California, and the Sierra Nevada on rotting conifers, usually in groups or clusters.

Two small Omphalinas with different habitats (both are discussed above). **Left:** *Omphalina ericetorum* is associated with lichens (which, however, are often very inconspicuous). **Right:** *Omphalina pyxidata* grows in moss, grass, and soil. Note how both species have decurrent gills. Similar greenish species are discussed under *Clitocybe odora* (p. 162).

MYCENA

Small to minute saprophytic mushrooms that do not revive when moistened (after being dried). CAP *typically conical to bell-shaped when young,* but often expanding in age; often translucent-striate; *margin usually straight when young, rarely incurved.* GILLS usually attached. STALK central, *thin, usually hollow; fragile or cartilaginous.* VEIL and VOLVA absent. SPORE PRINT *white.* Spores smooth, amyloid or not amyloid. Cystidia usually present on gills.

THIS is a very large group of very small mushrooms with a thin, fragile or cartilaginous stem and bell-shaped to conical cap (at least when young). The gills are not waxy as in the Hygrophoraceae; the dried fruiting body does not revive when moistened as in *Marasmius,* and is not particularly tough; *Collybia* and *Omphalina* are similar but usually have a convex to plane or umbilicate cap. Other small mushrooms with conical to bell-shaped caps *(Coprinus, Conocybe, Nolanea,* etc.) do not have white spores.

In sheer numbers Mycenas are more abundant than any other mushrooms, but because of their diminutive dimensions, most people are oblivious to their presence. In the wake of heavy rains they fruit in untold quantity in the woods, especially on needle beds under conifers, where they form thick carpets of delicate domes. They are strictly saprophytic—on logs, stumps, sticks, leaves, soil, and humus. They turn up occasionally in lawns, gardens, and flower pots, but do *not* grow in dung like *Coprinus* and *Bolbitius.*

The most common forms are gray or brown, but a few, such as *M. acicula,* are brightly colored. Alexander Smith's monumental monograph lists 218 North American species, but dozens more occur. The overwhelming majority cannot be identified positively without a microscope (some are so small they can barely be seen with the naked eye!). The size and shape of the sterile cells (cystidia) on the gills plays a particularly important role in their identification. However, *Mycena* can be divided into manageable groups based on gross features such as size, color, odor, habitat, and viscidity. In addition, a few species exude a juice or latex when the base of the stem is broken and squeezed. *Lactarius* species also possess a latex, but are much larger and fleshier with thicker stems.

In every mixed bag of individuals I take mushroom hunting, there's always one or two with a keen eye for detail. In the normal course of events, I begin with a brief spiel about the marvels of the mushroom world before dispatching the group to rush about madly in search of the elusive 25-lb. *Boletus edulis* (trampling myriad Mycenas with every step), while I make a sly beeline for my secret *Boletus* patch. These one or two keen-eyed individuals, however, are content to remain where they are, meticulously examining every leaf, twig, and cone, and uncovering in the process not only a multitude of marvelously minute Mycenas, but an astonishing assortment of other clandestine creatures—slugs, centipedes, spiders, snakes, salamanders, etc.

As I am rather small myself, I harbor a profound respect for these exceptional individuals. In a society where we are taught from birth to think big, it is encouraging to find some who are still able to make the distinction between quality and quantity, who appreciate the fact that size alone is not a measure of intrinsic worth.

My lust satiated and my basket laden with *Boletus,* I return later to find these patient and perceptive souls sprawled out on their bellies in the exact spot where I left them—for they consider the day well spent if they discover fifty Mycenas whose combined mass is no bigger than their thumb! Finding myself in good company, I bring out the cheese and bread —and if I'm lucky they bring out the wine—and we proceed to have an impromptu picnic, sharing our discoveries, then savoring the silence around us while awaiting the riotous return of the rest of the group.

For the benefit of these discerning individuals I have included nineteen species of *Mycena* in this chapter, which is nineteen more species than the average individual cares to know. Of course, they constitute only a fraction of the "YAM's" ("Yet Another Mycena") that can be found. They are much too small to eat, and some may actually be poisonous, but they deserve to be better known. They are exquisite in their daintiness and are among the most attractive of the fleshy (fleshless?) fungi. If your *"Mycena"* does not key out convincingly, check *Marasmius & Collybia, Omphalina & Xeromphalina,* and the waxy caps (Hygrophoraceae).

Key to Mycena

1. Base of stalk exuding a red to blackish-red juice when cut or squeezed 2
1. Not as above 4
2. Growing on wood .... *M. haematopus,* p. 231
2. Growing on ground or humus 3
3. Flesh in cap exuding a reddish juice when cut and squeezed ...... *M. sanguinolenta,* p. 232
3. Flesh in cap exuding a watery orange-yellow juice when cut and squeezed ....
.... *M. subsanguinolenta* (see *M. sanguinolenta,* p. 232)
4. Fruiting body bright orange to yellow; gills yellow with orange margins; found on hardwoods in eastern North America (usually clustered) .... *M. leaiana* (see *M. lilacifolia,* p. 236)
4. Not as above 5
5. Stalk (and often the cap) viscid or slimy *when fresh and moist* 6
5. Stalk not viscid or slimy; cap not normally viscid either 10
6. Stalk yellow, orange, greenish-gray, or lilac-tinged when fresh, at least at the apex 7
6. Stalk white, gray, or brown (not brightly colored) 9
7. Gills usually decurrent; found on rotting conifers .... *M. lilacifolia,* p. 236
7. Not as above 8
8. Cap olive-brown to blackish; stalk discoloring brown from the base upward; common under mountain conifers when or shortly after the snow melts ....
.... *M. griseoviridis* (see *M. epipterygia,* p. 237)
8. Not as above .... *M. epipterygia* & others, p. 237
9. Cap dry (not viscid) .... *M. rorida,* p. 237
9. Cap viscid or slimy, at least when moist .... *M. vulgaris* & others (see *M. rorida,* p. 237)
10. Fruiting body minute (cap typically less than 1 cm broad and stalk less than 1 (2) mm thick), or if slightly larger then growing on bark (often mossy) of living trees 11
10. Fruiting body small but not minute (stalk generally at least 1 mm thick; cap at least 5 mm broad and often larger); growing in a variety of habitats but not usually on bark of living trees . 14
11. Cap brightly colored (red, pink, orange, or yellow) .... *M. acicula* & others, p. 228
11. Not as above (cap white, gray, brown, vinaceous, etc.) 12
12. Typically growing on the bark of trees 13
12. Typically growing on leaves, twigs, humus, stems, or occasionally on bark ....
.... *M. capillaris* & others, p. 227
13. Cap brownish to vinaceous-brown or vinaceous-buff ....
.... *M. corticola* & *M. madronicola* (see *M. clavularis* group, p 227)
13. Cap white to grayish or grayish-brown .... *M. clavularis* group & others, p. 227
14. Gills marginate (the edges of at least the larger ones significantly darker and differently colored than the faces—use hand lens if unsure!) 15
14. Gills not marginate (edges same color or paler than the faces) 22
15. Gill edges scarlet to bright orange 16
15. Gill edges pink, dark reddish, purple, yellowish, etc., but not scarlet to bright orange 17
16. Gill edges scarlet; hairs at base of stalk orange .... *M. strobilinoides,* p. 228
16. Gill edges orange; hairs at base of stalk yellow-orange ....
.... *M. aurantiomarginata* (see *M. strobilinoides,* p. 228)
17. Gill edges greenish-yellow to yellow to yellow-brown ....
.... *M. citrinomarginata* & others (see *M. capillaripes,* p. 229)
17. Gill edges pink to reddish, purplish-red, reddish-brown, etc. 18
18. Edges of at least the larger gills pink to rosy; growing on ground .... *M. capillaripes,* p. 229
18. Gill edges darker than above (but faces may be pink or rosy); growing on ground or wood 19

19. Gills pink to rosy with sordid reddish edges; growing on ground under conifers . . . .
     . . . . **M. rosella** (see M. capillaripes, p. 229)
19. Not as above 20

20. Odor radishlike; typically growing on ground . **M. pelianthina** & others (see *M. pura,* p. 230)
20. Odor not radishlike; growing on wood, cones, wood chips, or in lignin-rich humus 21

21. Gill edges very dark purple or grayish-purple . . . . **M. purpureofusca,** p. 229
21. Gill edges rosy to vinaceous-brown . . **M. elegantula** & others (see *M. purpureofusca,* p. 229)

22. Lilac, purplish, blue, or bluish-green tints usually present somewhere on fruiting body . . . 23
22. Not as above 24

23. Lilac or bluish tints often present; odor radishlike . . . . **M. pura,** p. 230
23. Blue or bluish-green tints usually present when fresh; odor more or less mild . . . .
     . . . . **M. amicta** & **M. subcaerulea** (see *M. pura,* p. 230)

24. Cap brightly colored (bright coral-pink to orange or yellow) when fresh (may fade in age) 25
24. Cap typically some shade of gray, white, brown, dull vinaceous, etc. (not brightly colored) 26

25. Growing on walnut or hickory nut shells . . . . **M. luteopallens** (see *M. strobilinoides,* p. 228)
25. Growing on ground or in humus . . . **M. amabilissima** & others (see *M. strobilinoides,* p. 228)

26. Typically growing on ground, needles, twigs, leaves, etc. 27
26. Typically growing on logs, stumps, branches, etc. 35

27. Found in grass or disturbed soil; cap brick-red to vinaceous-tinged, soon convex or plane, less
     than 2 cm broad . . . . **Mycena sp. (unidentified)** (see *Marasmius oreades,* p. 208)
27. Not as above 28

28. Found in grass; stalk tough; cap white to brown, never grayish (see **Marasmius oreades,** p. 208)
28. Not as above 29

29. Extreme base of stalk exuding a droplet of milky fluid when squeezed (fresh specimens only!)
     . . . . **M. galopus,** p. 232
29. Not as above 30

30. Odor of crushed flesh distinctly alkaline, bleachlike, antiseptic, or at least sharp . . . .
     . . . . **M. leptocephala** & others (see *M. alcalina,* p. 234)
30. Odor mild or farinaceous, but not as above 31

31. Stalk with whitish fibrils or particles; cap olive-brown when fresh . . . . . **M. scabripes,** p. 233
31. Not as above 32

32. Cap margin not often incurved at first; cap often translucent-striate when moist, up to 4 (5) cm
     broad, grayish to buffy-brown; stalk usually less than 4 (6) mm thick; spores often amyloid 33
32. Not as above; margin of cap usually incurved when young *and/or* cap opaque or differently
     colored; spores typically *not* amyloid . . . . (see **Marasmius, Collybia,** & Allies, p. 201)

33. Stalk often with a tapered underground rooting portion; cap sordid tan to buffy-brown; usually
     growing near hardwood stumps . . . . **M. galericulata** & others, p. 235
33. Not as above 34

34. Stalk base typically exuding a droplet of clear liquid when squeezed; gills often reddish-stained
     in age . . . . **M. atroalboides** (see *M. galopus,* p. 232)
34. Not as above . . . **M. murina** group & sundry assorted YAMS (see *M. murina* group, p. 234)

35. Gills (and often cap) developing sordid pink to reddish spots or stains in age . . . .
     . . . . **M. maculata** & others, p. 235
35. Not as above (but gills may be tinged evenly pinkish in age) 36

36. Growing solitary or in rows, pairs, or groups, but not normally clustered **M. subcana,** p. 233
36. Not as above 37

37. Usually growing on hardwood stumps; stalk often rooting in substrate **M. galericulata,** p. 235
37. Not as above; usually on conifers 38

38. Fruiting body relatively large (cap 2-5 cm broad), grayish; usually growing near melting snow
     (or shortly after snow melts) in mountains . . . . **M. overholtsii** (see *M. subcana,* p. 233)
38. Not as above 39

39. Cap translucent-striate at maturity (when moist), gray to brownish-gray or paler; flesh very
     thin . . . . **M. occidentalis** & many others (see *M. subcana,* p. 233)
39. Not as above . . . . (see **Marasmius, Collybia,** & Allies, p. 201)

Mycena capillaris   (Miniscule Mycena)

CAP 2-6 (8) mm broad, bluntly conical or bell-shaped to convex, sometimes plane in age; surface smooth, gray, soon fading to white, translucent-striate when moist. Flesh very thin, odor mild. GILLS adnate to nearly free, well-spaced, grayish then white. STALK 3-7 cm long, about 1 mm thick, threadlike, equal, very fragile, smooth, dark gray when young becoming whitish in age. SPORE PRINT white; spores 8-10 × 4-5 microns, elliptical, smooth, not amyloid.

HABITAT: Solitary or in small groups on fallen oak (and beech) leaves; widely distributed. Fairly common in our area in wet weather, but difficult to distinguish from similar species.

EDIBILITY: Unequivocally inconsequential.

COMMENTS: The mycelium of this miniscule whitish *Mycena* is usually confined to a single leaf, so that you can pick up the leaf and carry it home, mushroom and mycelium intact. There are multitudes of other minute white to grayish or brownish Mycenas. They are especially abundant after long periods of wet weather, but are extremely difficult to separate (let alone see!) without a microscope. Their ranks include: *M. tenerrima* and *M. osmundicola,* even smaller, with well-spaced, nearly free gills, growing on bark, twigs, and woody debris (the former with a very short, usually curved stem); *M. delicatella,* with close gills, gregarious on needles and twigs under conifers; *M. albidula,* with well-spaced, decurrent gills and a convex cap, on leaves and bark of hardwoods; *M. stylobates,* with a flat circular disc at base of stalk; *M. paucilamellata,* cap only 1-2 mm broad and gills very few and foldlike to practically absent; *M. ignobilis,* growing in mud; and *M. juncicola,* vinaceous-tinged, growing on the culms (stems) of sedges. These Mycenas are among the tiniest of all gilled mushrooms. *Marasmius* species can also be very tiny but tend to have a convex to plane cap and dark, wiry or hairlike stem. See also the *M. clavularis* group.

Mycena clavularis group   (Bark Mycena)

CAP 4-7 mm broad, rounded to convex or somewhat bell-shaped; surface smooth, striate and often grooved or fluted toward margin; white or translucent, the center and striations frequently grayish or grayish-brown. Flesh very thin, pallid. GILLS few and widely spaced, white or tinged gray, attached to the stalk or separating to form a collar around it. STALK 0.5-2.5 cm long, 0.5-2 mm thick, equal except for a small basal bulb or disc; white or translucent, smooth, fragile; base often with long white hairs, inserted in bark. SPORE PRINT white; spores 8-10.5 microns, round or nearly round to broadly elliptical, smooth, amyloid.

HABITAT: Scattered to gregarious on mossy bark of trees (both living and dead); widely distributed but not often noticed. In our area it is common in wet weather, especially on the lower trunks of living oak and madrone trees.

EDIBILITY: Academic—a tremendous number would be needed for a meal!

COMMENTS: The above description will apply to a number of small Mycenas that typically grow in groups or troops (but not clusters) on the bark of trees (see photograph on p. 229). Whether the common form in our area is the "true" *M. clavularis* or a very similar species is for the *Mycena*-experts (of which there are few) to decide. There are several species with a similar growth habit, including: *M. tenerrima,* a minute (cap up to 4 mm broad) whitish species; *M. madronicola,* a brownish to vinaceous-buff or grayish species with adnate to slightly decurrent gills; and *M. corticola,* which has a dark purplish to vinaceous-brown cap that finally fades to pale grayish-brown, and grows abundantly on the bark of both hardwoods and conifers in eastern North America, often by the hundreds.

Mycena acicula

CAP 3-7 (10) mm broad, convex or bell-shaped, but sometimes expanding in age; surface not viscid, coral-red when young, soon fading (often from margin inward) to bright orange-yellow or yellow. Flesh very thin, yellow, odor mild. **GILLS** attached (usually adnate), pale orange to yellow or whitish. **STALK** 1-7 cm long, less than 1 mm thick, threadlike, equal, brittle, orange-yellow to yellow, smooth except for hairy base. **SPORE PRINT** white; spores 9-11 × 3.5-4.5 microns, elongated-elliptical, smooth, not amyloid.

HABITAT: Solitary, scattered, or in small groups on leaves and debris in woods, especially along streams and in other wet places; widely distributed. Common throughout the mushroom season in our area, but easily overlooked. I have seen it in large numbers in a thicket of oak saplings and blackberries along a stream.

EDIBILITY: Unknown, but a great many would be needed for a mouthful!

COMMENTS: This minute *Mycena* is a delight to behold, its coral-red to yellow cap contrasting vividly with the dim backdrop of decaying leaves and humus. One usually has to get down on hands and knees to find it! *M. oregonensis* is a similar miniature species whose cap is yellow to yellow-orange from infancy; it grows on needle carpets as well as oak leaves. For larger brightly colored Mycenas, see *M. strobilinoides.*

Mycena strobilinoides   (Flame Mycena)

CAP 1-2 cm broad, conical to bell-shaped; surface smooth, not viscid, striate when moist; red, soon fading to scarlet, then slowly to orange and finally yellow or even whitish; margin often scalloped. Flesh thin, yellowish; odor mild. **GILLS** adnate to slightly decurrent, well-spaced, yellow to pale pinkish-orange, the edges scarlet (at least when young). **STALK** 3-6 cm long, 1-2 mm thick, equal, thin, fragile, orange to yellow; base with orange hairs. **SPORE PRINT** white; spores 7-9 × 4-5 microns, elliptical, smooth, amyloid.

HABITAT: Scattered to densely gregarious in needle beds under conifers (especially pine); widely distributed, but most abundant in the mountains of the Pacific Northwest and northern California. It fruits in mild, moist weather. I have not seen it in our area.

EDIBILITY: Unknown.

COMMENTS: This little mushroom is likely to attract attention because of its beautiful color. The scarlet to yellow cap and stalk, scarlet-edged gills, and orange hairs at the stem base make a most distinctive combination of characters such that even beginners should have no trouble identifying it. It is one of the more prominent Mycenas in the Pacific Northwest, sometimes forming colorful "carpets" so thick that it is difficult to examine them closely without stepping on some. *M. aurantiomarginata* is a closely related species with an olive-brown to orange-tinted cap, brilliant orange gill edges, and yellow-orange hairs at the stalk base. It is commoner at lower elevations in the Pacific Northwest and I find it fairly frequently in our area as well. Brilliantly colored Mycenas *without* differently colored gills edges include: *M. monticola,* cap coral-red to pink, stalk coral-pink when fresh; *M. amabilissima,* cap bright pinkish-red fading to whitish and stalk soon fading to whitish; and *M. adonis,* with a smaller (5-12 mm broad), scarlet cap that becomes orange or yellow-orange and then fades to orange-buff, plus a pale yellow to whitish stalk. All three of these occur mainly on needle carpets under conifers. *M. luteopallens,* on the other hand, grows on old walnuts and hickory nuts in eastern North America. It is bright orange to yellow before fading. All of the above have stalks which are typically at least 1 mm thick. For minute, brightly colored species with thinner or threadlike stalks, see *M. acicula.*

Left: *Mycena clavularis* group (p. 227) typically grows on bark of standing trees. **Right:** *Mycena capillaripes* is one of many species that form dense carpets of delicate domes on duff under conifers.

Mycena capillaripes

CAP 0.5-2.5 cm broad, oval becoming bell-shaped or conical, often with a blunt umbo, sometimes nearly plane in age; surface smooth, not viscid, usually some shade of vinaceous-gray or gray, the center often browner; translucent-striate when moist, usually wrinkled or furrowed when dry. Flesh thin; odor typically nitrous or bleachlike (especially when crushed), but sometimes radishlike. **GILLS** attached (usually adnate), well-spaced, typically pallid or vinaceous-gray, at least the longer ones with pale pink or rosy edges (use hand lens!). **STALK** 3-10 cm long, 1-3 mm thick, equal or slightly thicker at either end, hollow, smooth, fragile, colored more or less like cap or paler above, sometimes slightly yellower in age; base with coarse white hairs. **SPORE PRINT** white; spores 7-12 × 4-6 microns, elliptical, smooth, amyloid.

HABITAT: Scattered to densely gregarious on conifer duff (but occasionally also under hardwoods); widely distributed. In our area it often occurs in droves under Monterey pine in the fall and winter.

EDIBILITY: Inconsequential.

COMMENTS: This is one of several Mycenas that carpet our pine forests with a miniature "fungal jungle" of petite parasols. The faint reddish or vinaceous tints on the cap plus the pink-margined gills are distinctive. *M. rosella* is a widely distributed pink to grayish-rose species with pale to bright rosy gills that have darker (pale to dark reddish) edges. It is uncommon in our area but often abundant under pine and other conifers farther north. Other terrestrial species with colored gill margins include: *M. citrinomarginata,* with pale yellow to yellow-brown gill edges and spores 8-11 microns long; *M. olivaceobrunnea,* with smaller spores, a smaller cap, and yellowish gill edges; and *M. elegans,* with pale greenish-yellow gill edges. In these species the cap is typically some shade or mixture of olive-brown, gray, and/or yellow.

Mycena purpureofusca

CAP 0.5-2.5 cm broad or occasionally larger, obtusely conical with a slightly incurved margin, becoming broadly conical or bell-shaped or sometimes expanding to nearly plane in old age; surface smooth, not viscid, dark purplish with a paler (lilac) margin when young, fading to purplish-gray or vinaceous-lavender; translucent-striate when moist and mature. Flesh thin, pliant, odor mild. **GILLS** attached (usually adnate), fairly close, pallid to

229

Mycena purpureofusca commonly grows on pine cones and other woody debris.

grayish with dark grayish-purple edges. **STALK** 3-10 cm long, 1-3 mm thick, equal, hollow, cartilaginous, colored more or less like cap or paler above; base with white hairs and sometimes rooting. **SPORE PRINT** white; spores 8-14 × 6-8.5 microns, broadly elliptical, smooth, not amyloid.

HABITAT: Solitary or in small groups or tufts on conifer wood and debris; widely distributed and frequent in our area in the fall and winter. I find it often on old pine cones.

EDIBILITY: Unknown, and like myself, likely to remain so.

COMMENTS: The purplish conical cap, dark purplish gill edges, and growth on wood or cones typify this species. The stalk does not exude a red latex like *M. haematopus.* The color and stature are reminiscent of *Marasmius plicatulus,* which has a very tough, brittle stalk and pallid gill edges and grows on the ground. *Collybia fuscopurpurea* has a similar name but has a convex to plane cap and is also terrestrial. Other species: *Mycena elegantula* is closely related, but has a vinaceous-brown to pinkish cap and pale pink to vinaceous gill edges; it also grows on coniferous wood and debris, including cones. *M. rubromarginata* grows on spruce and fir and has a browner cap and reddish-brown gill edges.

Mycena pura   (Lilac Mycena)

CAP 2-5 (6.5) cm broad, obtusely umbonate or convex becoming broadly convex or plane, sometimes with an uplifted margin; surface smooth, hygrophanous, color variable: various shades or mixtures of lilac, pink, gray, and blue-gray, but varying also to blue-green with a yellowish center, or sometimes even whitish tinged purple or blue at the center; translucent-striate when moist, otherwise opaque. Flesh thin, soft; odor and taste radishlike. **GILLS** adnate or adnexed, usually tinged cap color (lilac, etc.), but sometimes grayish or white; edges pallid. **STALK** 3-7 (10) cm long, 2-7 mm thick, equal or thicker below, hollow, smooth, pallid or colored like the cap or paler. **SPORE PRINT** white; spores 5-9 × 3-4 microns, elliptical to cylindrical, smooth, amyloid.

HABITAT: Solitary or in groups or small tufts on ground in woods, widely distributed. Common in our area from fall through early spring, but rarely in large numbers. I find it most often with oak, Douglas-fir, and pine.

EDIBILITY: Edible according to some sources, but one study revealed traces of the toxin muscarine. Definitely not recommended—it is too small to be of value anyway.

Mycena pura is a terrestrial species that sometimes resembles a miniature blewit. Note how the cap becomes convex or plane in age.

COMMENTS: This is one of the larger Mycenas as well as one of the more beautiful. The color is exceedingly variable, but there is usually a trace of lilac somewhere on the fruiting body, especially the stem. The convex rather than conical cap is atypical for a *Mycena*, and for this reason it is keyed out under *Collybia*. The radishlike odor is another important fieldmark. The gill edges are not purple or pink as in *M. capillaripes* and *M. purpureofusca*, and it is strictly terrestrial. It never grows in huge clusters and does not normally fruit in dense troops either. I have seen fairly robust specimens which could have been mistaken for blewits *(Clitocybe nuda)*. *Inocybe lilacina* is similarly colored, but has brown spores. *M. pelianthina* and *M. rutilantiformis* are two very similar species with a radishlike odor. They differ in having dark reddish to purple gill edges, and the latter typically shows yellow at the apex of the stalk. They are not as highly colored as *M. pura* and both grow on the ground under hardwoods. *M. amicta* is a western species whose cap and stalk are blue to blue-green or tinged those colors. It grows on coniferous debris, while *M. subcaerulea* is its eastern, deciduous-woodland counterpart.

Mycena haematopus (Bleeding Mycena) Color Plate 46

CAP 1-3.5 (5) cm broad, oval to bell-shaped when young, with the margin often extending beyond the gills; in age sometimes convex or plane with an umbo and uplifted margin; surface smooth, not viscid, reddish to vinaceous-brown to reddish-brown or pinkish-brown, margin often paler (vinaceous-gray); striate at maturity when moist; margin often scalloped in age. Flesh thin, exuding a dark red juice when cut. **GILLS** attached (usually adnate or adnexed), fairly close, pallid, often developing reddish stains; edges white or in one form reddish. **STALK** 3-8 (14) cm long, 1-3 mm thick, equal, dull reddish or reddish-brown, or sometimes pallid; smooth, fragile; base with coarse hairs, exuding a dark red juice when cut or squeezed. **SPORE PRINT** white; spores 7-11 × 5-7 microns, elliptical, smooth, amyloid.

HABITAT: Solitary or more commonly tufted or in groups on decaying logs and stumps (mostly hardwoods). Very widely distributed and common, fruiting in our area from fall through spring.

EDIBILITY: Edible according to some sources, but too small to be of value.

COMMENTS: One of the most distinctive and common Mycenas, this species resembles several others in color, but if the base of the stalk is squeezed a dark red bloodlike fluid will emerge. Its growth on wood separates it from other "bleeding" Mycenas (see *M. sanguinolenta*). Also distinctive is the frequent presence of a sterile band of tissue around the cap margin of young specimens. It is suggestive of a veil but breaks up in age or forms a scalloped rim around the cap.

Left: *Mycena sanguinolenta* is a common terrestrial species that "bleeds" when broken. **Right:** *Mycena galopus* looks like numerous other grayish or brownish Mycenas, but its base exudes a milky droplet when squeezed.

Mycena sanguinolenta   (Terrestrial Bleeding Mycena)

CAP 0.3-1.5 (2.5) cm broad, conical to bell-shaped or sometimes convex in age; surface smooth, not viscid, color variable: some shade of pale reddish-brown to bright reddish-brown to orange-brown, etc; margin often vinaceous and furrowed at maturity. Flesh thin, reddish, exuding a dark red juice when cut; odor mild. **GILLS** attached (typically adnate), well-spaced, pallid or tinged flesh-color or reddish, edges dark reddish-brown. **STALK** 2-7.5 cm long, 1-2 mm thick, equal, hollow, fragile, colored more or less like cap, smooth; base with white hairs and exuding a dark red juice when broken or squeezed. **SPORE PRINT** white; spores 8-11 × 4-6 microns, elliptical, smooth, weakly amyloid. Cystidia present on gill faces and edges.

HABITAT: Solitary or widely scattered to gregarious or tufted on leaf mold and needles in woods or at their edges; widely distributed. Common in our area in fall and winter, usually scattered under oak. In northern California and the Pacific Northwest it sometimes fruits in large numbers under conifers, especially spruce.

EDIBILITY: Unknown, and like most of us, destined to remain so.

COMMENTS: True to their names, this species and its close cousin, *M. haematopus,* "bleed" when broken. As such they are the easiest Mycenas to recognize and among the most common as well. *M. sanguinolenta* is easily distinguished from *M. haematopus* by its growth on the ground, absence of a sterile band of marginal tissue on the cap, and somewhat different color. The striate or furrowed cap is reminiscent of *Marasmius plicatulus,* but that species has a *tough,* polished stem and does not "bleed." Other species: *M. subsanguinolenta* is closely related but has dark red to orange juice, is slightly yellower overall, and lacks cystidia on the gill faces. It often fruits in large numbers under conifers in northern California.

Mycena galopus   (Milky Mycena)

CAP 0.5-2.5 cm broad, conical to bell-shaped or umbonate, the margin sometimes curling up in age; surface smooth, not viscid, at first grayish-black (at least at center), but soon fading to gray or grayish-brown, then pale gray or even whitish; translucent-striate when moist. Flesh thin, soft, odor mild. **GILLS** well-spaced, attached (usually adnate), white to grayish. **STALK** 4-8 (12) cm long, 1-2 mm thick, equal, fragile, hollow, smooth, dark grayish-brown to gray below, paler above; base with white hairs, exuding a droplet of milky juice when squeezed. **SPORE PRINT** white; spores 9-13 × 5-6.5 microns, elliptical, smooth, not amyloid or sometimes weakly amyloid.

HABITAT: Scattered to gregarious on ground and humus in woods, widely distributed. Common in our area during wet weather, especially under redwood and pine, but tending to fruit widely scattered over a large area rather than in dense troops or "carpets" like some of our other pine-loving species (e.g., *M. murina, M. capillaripes*).

EDIBILITY: Who knows?

COMMENTS: This ubiquitous *Mycena* will be encountered by anyone who takes their "LBM's" seriously. At first glance there is little to distinguish it from the myriad of other anonymous brown or grayish Mycenas. However, if you break the very base of the stalk and squeeze gently, a drop of milky or cloudy fluid will emerge. This is best observed in the field, while the fruiting body is fresh. A similar species, *M. atroalboides*, exudes liquid that is clear rather than milky, and often has reddish spots on the gills in old age. It is not common in our area but is often abundant farther north under conifers.

Mycena scabripes   (Rough-Stemmed Mycena)

CAP 1.5-3 (5) cm broad, obtusely conical becoming convex, umbonate, or plane; surface not viscid, blackish-brown or dark brown at the center, olive-brown toward the margin, but fading somewhat in age to olive-gray or gray; not distinctly striate. Flesh thin, brownish-gray, odor mild. GILLS adnate to adnexed or free in age, pallid or tinged cap color, sometimes spotted reddish-brown in age. STALK 3-9 cm long, 2-4 mm thick, equal, hollow, grayish or colored like cap, with a thin but distinct coating of silky white to grayish fibrils and particles. SPORE PRINT white; spores 7-9 × 4-5 microns, elliptical, smooth, amyloid.

HABITAT: Scattered to gregarious or in small tufts in humus and debris in woods; widely distributed but not common, or at least seldom noticed. I have found it in our area in the fall and winter under redwood and oak.

EDIBILITY: Who knows? Who cares?

COMMENTS: The distinguishing feature of this undistinguished *Mycena* is the presence of silky white fibrils and particles on the stem. The olive-brown, more or less umbonate cap is also distinctive.

Mycena subcana   (Neutral Gray Mycena)

CAP 0.5-2.5 (3) cm broad, oval or obtusely conical to bell-shaped, becoming convex in age; surface smooth or with a hoary sheen, grayish to pale gray, usually darker at the center,

Left: *Mycena scabripes* is a nondescript species with a minutely scaly or dandruffy stalk. Right: *Mycena subcana* typically grows singly or in pairs or groups (but not clusters) on sticks and branches.

sometimes fading to whitish; not viscid, translucent-striate nearly to center when moist. Flesh thin, grayish, odor mild. **GILLS** attached (usually adnate), well-spaced, whitish to pale gray. **STALK** 1.5-6 cm long, 1.5-3 mm thick, equal or enlarged at base, smooth, same color as cap or paler, with downy white hairs at base. **SPORE PRINT** white; spores 8-10 × 5-6 microns, elliptical, smooth, amyloid.

HABITAT: Solitary or in small groups (not large clusters!) on dead sticks and branches, sometimes on living trees; fairly common on the west coast. In our area it fruits in the fall and winter on conifers (especially redwood); I have also found it on tanoak.

EDIBILITY: Who knows? Who cares? I don't.

COMMENTS: This grayish species is representative of a multitude of indifferent grayish, wood-inhabiting "YAM's." The salient macroscopic features of *M. subcana* are its color, its translucent-striate cap, and its preference for dead sticks and branches rather than stumps or logs (a completely different niche!). Unlike many of its kin, it does not grow in clusters, but instead fruits alone or in attractive rows of solitary or sometimes paired specimens. Typifying those that grow in *clusters* on stumps or logs is *M. occidentalis,* a grayish species found throughout the West on conifers. Also clustered on conifers is *M. laevigata,* a whitish species with grayish-black to bluish-gray tints on cap and stalk when very young and yellowish discolorations in age; and *M. overholtsii,* a large grayish species found in clumps near melting snow in the spring. Another species, *M. parabolica,* has a growth habit in between that of *M. subcana* and *M. occidentalis*—it grows in groups or *small* clusters on rotting wood and has a sooty black cap that fades to gray in age.

Mycena alcalina   (Alkaline Mycena)

CAP 1-4 cm broad, conical to bell-shaped or convex-umbonate, expanding somewhat in age; surface smooth, dark brownish-black to grayish-black when very young, soon fading to gray, grayish-brown, or yellow-brown; striate when moist. Flesh thin, fragile, odor faintly to strongly alkaline (like bleach). **GILLS** rather close, adnate to slightly decurrent, whitish to grayish, sometimes stained reddish-brown in age. **STALK** 3-10 cm long, 1.5-3 mm thick, equal, pallid to grayish or often sordid yellowish-brown in age; fragile, hollow. **SPORE PRINT** white; spores 8-11 × 5-7 microns, elliptical, smooth, amyloid.

HABITAT: In groups or tufts on decaying conifer logs, or sometimes densely gregarious on needles under conifers; common and widespread throughout North America. In some regions it is most abundant in the spring, but in our area it fruits in the fall and winter.

EDIBILITY: Who knows? Who cares? I don't. Do you?

COMMENTS: The sharp, bleachlike odor is usually quite distinct in this species, and helps distinguish it from the throngs of other grayish Mycenas that grow on wood. The odor is best detected by crushing the cap. Another common species with an alkaline odor, *M. leptocephala,* is terrestrial, growing on sticks, needles, or even in grass. It is only rarely umbonate, and slightly smaller than *M. alcalina.* Other species: *M. metata* is terrestrial, with a slightly browner cap and faint but sharp odor; *M. iodiolens* has an antiseptic odor. For similar species without a distinctive odor, see *M. subcana* and the *M. murina* group.

Mycena murina group   (Yet Another Mycena)   **Color Plate 48**

CAP 0.7-3 cm broad, conical to bell-shaped, striate nearly to the center when moist; surface smooth, not viscid, dark gray to gray when moist, paler (gray to whitish) when faded, darker when young. Flesh thin, odor mild. **GILLS** grayish with pallid edges, attached, fairly well-spaced. **STALK** 3-8 cm long, 1-4 mm thick, equal, fragile, smooth, pale to dark gray with pallid apex; base with white hairs. **SPORE PRINT** white; spores 8-11 × 5-7 microns, elliptical, smooth, amyloid.

HABITAT: Scattered to densely gregarious by the hundreds on needle duff under conifers, widely distributed. This species and its look-alikes are often abundant in our coastal pine forests in the late fall and early winter. They also appear under hardwoods, in grass, even in planter boxes and flower pots—i.e., wherever there is organic matter to nourish them.

EDIBILITY: Who knows? Who cares? I don't. Do you? Do you care if I do?

COMMENTS: This species, which has passed under the name *M. stannea*, is one of a slew of nondescript, odorless, gray to brownish, terrestrial Mycenas that are very prevalent under conifers. They are differentiated from each other largely on the basis of microscopic characteristics such as the size and shape of spores and sterile cells (cystidia). The one pictured in the color photograph is not quite as gray as typical *M. murina* and may well be a distinct species, but to the average mushroom hunter it is a "YAM" ("Yet Another Mycena"), and thus hardly worth the time and trouble to distinguish. Other "YAM's" include *M. latifolia, M. filopes, M. abramsii, M. pseudotenax, M. atroalboides, ad infinitum*. For those species with a sharp or bleachlike odor, see comments under *M. alcalina,* and for those with colored gill margins, see *M. capillaripes.*

Mycena maculata   (Reddish-Spotted Mycena)

CAP 1.5-5 cm broad, conical to bell-shaped expanding to broadly convex or plane with a broad umbo, the margin sometimes uplifted in age; surface smooth, blackish-brown soon fading to brown or brownish-gray, but usually spotted with reddish-brown and sometimes entirely watery gray; striate when moist. Flesh thin, firm, grayish, slowly bruising reddish-brown. GILLS adnate to adnexed, whitish to pale gray, soon stained with reddish spots, sometimes entirely sordid reddish in old age; edges pallid. STALK 4-8 (12) cm long, 2-5 mm thick, nearly equal, smooth, hollow, cartilaginous, pallid above, colored like cap or paler below; base with dense white hairs and soon stained reddish-brown, sometimes rooting. SPORE PRINT white; spores 7-9 × 4-5 microns, elliptical, smooth, amyloid.

HABITAT: In small tufts or large clusters on logs and stumps, especially of conifers; widely distributed. It is common in our area in cool, wet weather on redwood, Douglas-fir, etc., occasionally also on hardwoods. "The most abundant *Mycena* on conifer wood in the Pacific Northwest," says Alexander Smith, author of a monograph on the genus.

EDIBILITY: Who knows? Who cares? I don't. Do you? Do you care if I do? I won't if you don't.

COMMENTS: This rather unattractive species can be recognized by the sordid reddish stains that develop in age and its growth in clusters on rotten wood. The stalk sometimes has a "tap root" as in *M. galericulata.* It doesn't "bleed" like *M. haematopus,* and the gill edges aren't purple as in *M. purpureofusca. M. occidentalis* (see comments under *M. subcana)* is also common in clusters on conifers, but does not develop reddish stains. Another reddish-staining species, *M. rugulosiceps,* has larger spores and grows on hardwoods (alder, rhododendron, etc.) as well as on conifers. *M. inclinata* is a closely related species that is especially common in eastern North America. It often develops reddish or pinkish tones as it ages but favors hardwoods and has fibrillose flecks on the lower half of the stalk. See top of next page for a photograph of *M. maculata.*

Mycena galericulata

CAP 2-5 (7) cm broad, conical when young becoming broadly bell-shaped to umbonate to plane or with an uplifted margin in age; surface moist but not viscid, often radially wrinkled nearly to center; buffy-brown to grayish, dingy tan, cinnamon-brown, or grayish-brown. Flesh watery gray; odor faintly to distinctly farinaceous or radishlike. GILLS adnate to adnexed, often with cross-veins between, white to grayish or in age flushed pale

Left: *Mycena maculata* (see p. 235) is one of many Mycenas that grow in clumps on rotten wood. **Center:** *Mycena galericulata,* young specimens with broadly conical caps; note how stalk roots deeply. **Right:** *Mycena galericulata,* mature specimens which have opened up so as to resemble Collybias.

pinkish, but not spotted or stained. **STALK** 5-10 (14) cm long, 2-4 (6) mm thick, equal, cartilaginous, hollow, smooth, often rooting; grayish-white or colored more or less like cap, often darker below. **SPORE PRINT** white; spores 8-11 × 5.5-7 microns, elliptical, smooth, amyloid.

HABITAT: Solitary to scattered, gregarious, or in small clusters on decaying hardwood stumps, logs, and debris; widely distributed and not uncommon in our area in the fall and winter. One form—possibly a distinct species—grows on manzanita burls.

EDIBILITY: Edible, but not recommended—it is one of the few Mycenas large enough to eat, but many of its difficult-to-distinguish relatives have not been tested.

COMMENTS: This widespread species can be mistaken for a *Collybia* because the cap is convex to plane in age. The telltale traits are its pale color, frequently long, rooting stem, and penchant for growing scattered or in small, loose clusters. The gills are often tinged pinkish in age, but do not become reddish-spotted as in *M. maculata.* Other species: *M. vitilis* has a long stem that often has a rooting base that is covered with white hairs below the ground. It grows in humus under hardwoods, especially alder.

Mycena lilacifolia

CAP 0.8-2.5 cm broad, convex to helmet-shaped, the center slightly depressed in age; surface viscid to slimy and translucent-striate when moist, smooth, lilac when very young, but soon bright to pale yellow. Flesh thin; odor mild. **GILLS** well-spaced, usually decurrent, pale lilac becoming whitish to pale yellow or pinkish. **STALK** 1-4 cm long, 1-2 mm thick, equal or enlarged at the base, smooth, slimy-viscid when moist, same color as gills but soon yellow; base often with lilac mycelium. **SPORE PRINT** white; spores 6-7 × 3-3.5 microns, elliptical, smooth, not amyloid.

HABITAT: Solitary, scattered, or in groups on rotting conifers, widely distributed. It usually fruits in cool weather and is fairly common in our coastal pine forests.

EDIBILITY: Unknown, but too slippery to be of value.

COMMENTS: This species is best distinguished from other viscid Mycenas by its yellow color and growth on conifers. The gills will sometimes retain their lilac tint longer than the cap and stalk, but I have found specimens which showed absolutely no trace of their original lilac color. It might be looked for in *Omphalina* because of the decurrent gills (see photo on p. 237), but the slimy stem and cap distinguish it. Viscid, brightly colored wood-inhabiting species with *non-decurrent* gills include: *M. leaiana,* bright orange to yellow, growing in clusters on hardwoods in eastern North America, and certain varieties of *M. epipterygia* (see description on next page), which grow on conifers.

236

Mycena lilacifolia (p. 236) is one of several Mycenas with a slimy cap and stem. Note decurrent gills.

Mycena epipterygia (Yellow-Stemmed Mycena)

CAP 0.8-2 (2.5) cm broad, oval becoming umbonate, bell-shaped, or convex; surface smooth, slimy or viscid when moist, color variable: mostly yellow with olive or grayish tones, sometimes fading in age. Flesh thin, yellowish; odor mild or faintly fragrant. **GILLS** attached, whitish or tinged yellow, fairly well-spaced. **STALK** 5-9 cm long, 1-2 mm thick, equal, fragile, hollow, smooth, slimy or viscid when moist, yellow, but sometimes fading to whitish in old age. **SPORE PRINT** white; spores 8-10 × 5-6 microns, elliptical, smooth, amyloid.

HABITAT: Scattered to gregarious in needle duff under conifers, widely distributed. It is partial to cold weather, fruiting in northern regions in the late fall until frosts set in, and in the winter in our coastal pine forests. One variety with a rank odor occurs on decaying conifers. The related *M. griseoviridis* (see comments) frequently fruits nearly melting snow in the spring and summer.

EDIBILITY: Unknown.

COMMENTS: This is one of several attractive Mycenas with a viscid cap and viscid, yellow to greenish-gray stem. The viscid layers may dry out in age, but the colors are distinctive. Similar terrestrial species include: *M. viscosa,* with a strong rancid odor and taste; *M. epipterygioides,* with a dark olive-gray cap that does not fade to white; and *M. griseoviridis,* with a deep olive to olive-brown to blackish cap, a yellow stem that becomes brown or deep vinaceous-brown from the base upward, a cucumberlike odor, and thorny cystidia on the gill edges. The latter species is often common under mountain conifers soon after the snow melts, but is not restricted to that habitat. Any of these species can be mistaken for a waxy cap (e.g., *Hygrocybe psittacina*), but their gills are not noticeably waxy.

Mycena rorida (Slippery Mycena)

CAP (2) 5-10 (15) mm broad, at first rounded or bell-shaped, then expanding to plane or nearly so; surface dry, pale brown or brownish-gray to tan, fading to white or yellowish-white. Flesh thin, odor mild. **GILLS** adnate to decurrent, well-spaced, white. **STALK** 2-3 (5) cm long, about 1 mm thick, equal, whitish, covered with a sheath of slime when fresh and moist. **SPORE PRINT** white; spores 8-12 × 4-6 microns, elliptical, smooth, amyloid.

HABITAT: Solitary or scattered to gregarious on ground and debris under conifers and in mixed woods; widely distributed, but not common in our area. I have found it under redwood in November.

EDIBILITY: Too small and slippery to be of value.

COMMENTS: This little mushroom is easily identified by its pale color, dry cap, and sheath of slime on the stem. Plucking it can be a difficult proposition if you try to grasp it by the stalk. Other Mycenas with a pallid to grayish, viscid stem include: *M. tenax,* with a gray-brown to pale gray cap and strong disagreeable odor; *M. vulgaris,* similar but with an extremely viscid cap and slight odor; and *M. clavicularis,* with a dry grayish cap and mild odor. All of these occur in groups or troops under conifers.

spores

ENTOLOMATACEAE

THIS is the largest family of pinkish-spored mushrooms and also the most difficult. It is unusual among agarics in having spores which are distinctly angular (several-sided) in side and/ or end view. The gills are typically attached to the stalk (in contrast to the other major pinkish-spored family, the Pluteaceae), and the stalk is not cleanly separable from the cap. The spore color, though characterized as "pinkish," actually ranges from deep flesh-color to salmon- or rosy-pink to dull reddish or sordid pinkish-cinnamon, whereas the pinkish-spored members of the Tricholomataceae (e.g., *Clitocybe*) have pale or dull pinkish, non-angular spores.

The Entolomataceae have a very confusing nomenclatural history. Four principal genera have traditionally been recognized—*Clitopilus, Entoloma, Nolanea,* and *Leptonia* —but the presence of intermediate forms has led some mycologists to consolidate all but *Clitopilus* in a single giant genus, *Entoloma,* while recognizing a few smaller genera (e.g., *Rhodocybe*) that differ microscopically. To muddle matters more, the name *Rhodophyllus* is often used instead of *Entoloma,* and Rhodophyllaceae instead of Entolomataceae. The result is that each species has a surfeit of superfluous synonyms—for instance, *Leptonia sericella, Alboleptonia sericella, Entoloma sericellum,* and *Rhodophyllus sericellus* are all names for the same fungus! (Human beings are certainly a confusing—and confused—lot!)

The mushrooms do little to ameliorate matters, for they are exasperatingly difficult to demystify. Some *Entoloma*-experts have resorted to such abstruse activities as measuring the urea concentration of the fruiting body or determining the type of symmetry exhibited by the spores in their efforts to delineate the various genera and species!

In North America the Entolomataceae are largely woodland fungi, but some species grow in open or grassy areas. The vast majority are terrestrial (in contrast to the wood-inhabiting genus *Pluteus*), but a few occur on rotting wood. Several species are poisonous, and in view of the difficulty in identifying them, beginners should avoid the entire group. Particularly dangerous are some of the large fleshy Entolomas—a fact which clearly invalidates the old adage that "any mushroom with pink gills is edible" (a reference, no doubt, to *Agaricus*).

From two to nine genera are recognized, depending on whether a "lumper" or "splitter" is doing the recognizing. Several of the genera are small and rare (e.g., *Claudopus* and *Pouzarella*); these are included in the following key but not treated beyond it.

Key to the Entolomataceae

1. Growing on dung or on other mushrooms; widely distributed but rare 21
1. Growing on wood, humus, or ground 2

2. Stalk lateral to absent; fruiting body small, usually on wood (rarely on ground) 3
2. Stalk present, usually central or slightly off-center; on ground or sometimes on wood 7

3. Gills pale orange, orange, or yellow *or* gill edges split lengthwise (see **Tricholomataceae**, p. 129)
3. Not as above 4

4. Odor strongly unpleasant (skunklike) .... *Claudopus graveolens*
4. Not as above 5

5. Cap felty or hairy, white, 4-20 mm broad; spores *not* angular ...... (see **Crepidotus**, p. 405)
5. Not as above; spores angular under the microscope 6

6. Cap brownish-orange to grayish-brown beneath a dense layer of whitish fibrils; odor usually farinaceous; stalk silky-hairy, base often with white mycelial threads . *Claudopus byssisedus*
6. Cap whitish to gray or pinkish-gray, not as hairy as above, 3-7 mm broad; odor mild ......
.... *Claudopus depluens*

7. Gills purple or purple-tinged; spores *not* angular .... (see **Clitocybe** & Allies, p. 148)
7. Not with above features; gills rarely purple, and if so then spores angular 8

8. Gills often decurrent (but sometimes adnate or even notched); cap and stalk never bluish, violet, or black; spores warted or longitudinally lined in side view, angular only in end view .... *Clitopilus & Rhodocybe,* below
8. Not as above; spores usually angular in side view or not angular at all 9
9. Cap *and/or* stalk blue, blue-black, blue-gray, violet, steel-gray, or black, at least when young and fresh 10
9. Not as above 12
10. Stalk usually 5 mm thick or more; usually growing on ground 11
10. Stalk slender, usually hollow and fragile or cartilaginous, typically 1.5-5 mm thick at apex, or if thicker then growing on wood .... *Leptonia & Allies,* 248
11. Stalk fibrillose-scaly, often with a metallic luster or iridescence; cap usually fibrillose-scaly also .... *Leptonia & Allies,* p. 248
11. Stalk smooth to finely fibrillose, or if fibrillose-scaly then cap smooth ..... *Entoloma,* p. 242
12. Fruiting body small and white (or in age tinged yellowish or pinkish) *Leptonia & Allies,* p. 248
12. Not as above 13
13. Cap hairy and/or scaly and base of stalk with coarse, stiff hairs; not common ... *Pouzarella*
13. Not as above 14
14. Stalk fleshy, 4 mm thick or more 15
14. Stalk slender, usually hollow and fragile or cartilaginous, usually less than 5 mm thick ... 17
15. Spore print pale to dull pinkish; spores *not* angular .... (see **Tricholomataceae,** p. 129)
15. Spore print deep flesh-color to salmon-pink to pinkish-cinnamon or darker; spores angular under the microscope 16
16. Cap conical to bell-shaped or with a pointed umbo (but may expand with age) *Nolanea,* p. 245
16. Not as above (but cap may be bluntly conical when young) .... *Entoloma,* p. 242
17. Growing on or near wood; spores smooth or at least not angular 18
17. Usually terrestrial (but occasionally on wood); spores angular under microscope 19
18. Gills usually free (or nearly free) and close together .... (see **Pluteus,** p. 254)
18. Not as above .... (see **Tricholomataceae,** p. 129)
19. Stalk yellow-green *or* gills pale with pink edges .... *Leptonia & Allies,* p. 248
19. Not as above 20
20. Cap conical to bell-shaped or umbonate, or if convex to plane then smooth or silky; cap *not* umbilicate .... *Nolanea,* p. 245
20. Cap convex to plane or umbilicate and usually finely fibrillose, scurfy, scaly, or fibrillose-scaly, at least at the center (and particularly in age) .... *Leptonia & Allies,* p. 248
21. Parasitic on other mushrooms (mostly chanterelles and polypores); fruiting body minute, grayish to white; small stalk usually present .... *Claudopus parasiticus*
21. Not as above; growing on dung .... *Clitopilus & Rhodocybe,* p. 239

CLITOPILUS & RHODOCYBE

Small to medium-sized, mostly terrestrial mushrooms. CAP convex to plane or depressed. GILLS *usually adnate to decurrent.* STALK thick or slender, fleshy or rather fragile, central or off-center. VEIL and VOLVA absent. SPORE PRINT *pinkish to salmon, flesh-colored, or pinkish-cinnamon.* Spores angular in end view only; in side view longitudinally ridged *(Clitopilus)* or bumpy to warty *(Rhodocybe).*

THESE two small genera of pinkish-spored mushrooms typically have adnate to decurrent gills and spores which are angular only in end view. They are a rather lackluster lot, most likely to be confused with species of *Entoloma,* which usually have notched gills, and *Clitocybe,* which have paler, non-angular spores. Several small species of *Leptonia* and *Eccilia* have decurrent gills, but their spores are angular in both side and end view. There are also several Rhodocybes with notched gills; these are likely to be confused with *Collybia* or *Entoloma* unless their spores are examined under the microscope.

Both *Clitopilus* and *Rhodocybe* are small genera, but some of their species are quite common. Two are described here (one from each genus) and several others are keyed out.

Key to Clitopilus & Rhodocybe

1. Odor like a candy store or bubble gum (break open cap if unsure) ..... (see *Nolanea,* p. 245)
1. Not as above 2

2. Cap reddish-brown to orange-brown, cinnamon, pinkish, yellowish-ochre, or tan 3
2. Cap white or grayish to grayish-brown, olive-brown, etc. 4

3. Cap 2-12 cm broad, opaque; stalk usually at least 5 mm thick; fairly common in California, but more widely distributed .... *R. nuciolens* & others, p. 241
3. Cap usually less than 5 cm broad and translucent-striate when moist; stalk 1-7 mm thick; widespread .... *R. nitellina* (see *R. nuciolens,* p. 241)

4. Found on dung or compost; cap small, stalk often poorly developed ..... *C. passeckerianus*
4. Not as above 5

5. Cap gray to grayish-brown; "aborted" fruiting bodies often found that are somewhat puffball-like: whitish to pinkish-tan and bumpy on the outside, marbled within; common in eastern North America; spores angular in side view .... (see *Entoloma,* p. 242)
5. Not as above 6

6. Gills decurrent; cap white to grayish and typically 3 cm or more broad 7
6. Not as above; cap smaller and/or dark brown to olive-brown, etc. 9

7. Taste bitter; cap often cracked concentrically .... *R. mundula* (see *R. nuciolens,* p. 241)
7. Not as above 8

8. Stalk often somewhat off-center and fairly slender (4-12 mm thick); odor pleasant; typically *not* growing in clusters; spores longitudinally lined under microscope .. *C. prunulus,* below
8. Stalk often quite thick; sometimes clustered; spores *not* lined   (see *Clitocybe* & Allies, p. 148)

9. Spores finely warted in side view, angular only in end view *R. caelata* (see *R. nuciolens,* p. 241)
9. Spores angular in side and end views .... (see *Leptonia* & Allies, p. 248)

Clitopilus prunulus   (Sweetbread Mushroom)

CAP 3-10 cm broad, convex becoming plane or centrally depressed; surface dry and slightly felty or in one form slightly viscid when moist, smooth, white to gray; margin often lobed or wavy. Flesh rather thin, white; odor strongly but pleasantly farinaceous (like sweetbread). **GILLS** decurrent or occasionally adnate, fairly close, narrow, whitish to gray, then dusted pinkish by spores. **STALK** 2-8 cm long, 0.4-1.2 cm thick, central or off-center, solid, equal or tapering downward, white or grayish. **SPORE PRINT** flesh-colored or salmon; spores 9-12 × 5-7 microns, elliptical, longitudinally ridged, angular only in end view.

Clitopilus prunulus looks like a *Clitocybe* with its decurrent gills and white to gray color, but its spores are pinkish and show ridges under the microscope. The tree frog shown here is very tiny!

HABITAT: Solitary to scattered or in small groups on ground in open woods and grassy places near trees; widely distributed. It is fairly common in our area in the fall and winter, usually under oak or pine.

EDIBILITY: Edible and choice, but not recommended. It is rated highly in Europe but is easily confused with poisonous mushrooms such as *Clitocybe dealbata*.

COMMENTS: The pinkish spores, decurrent gills, white to grayish cap, and strong odor of meal are the fallible fieldmarks of this unremarkable fungus. The poisonous *Clitocybe dealbata* closely mimics it but is somewhat smaller and has *white* spores. *Clitocybe sub-connexa* and its close relatives have pinkish spores, but grow in clusters and are more robust. The name *Clitopilus orcellus* has been given to forms of *C. prunulus* with a white, slightly sticky cap, but it is no longer regarded as a distinct species by most mycologists.

Rhodocybe nuciolens

CAP 2-7 (11) cm broad, broadly convex becoming plane or slightly depressed or some-times slightly umbonate; surface smooth, not viscid, dull pinkish to pinkish-brown, pinkish-tan, or sometimes cinnamon-brown to brick-colored, fading as it dries (especially toward margin), sometimes with watery spots or bands. Flesh white or pale ochre; odor usually rather spicy. **GILLS** whitish becoming pinkish as spores mature; close, adnate or notched or sometimes slightly decurrent. **STALK** 2-8 (15) cm long, 0.4-1.5 (4) cm thick, equal or occasionally thicker at base, whitish or colored like cap (but usually paler). **SPORE PRINT** rosy-buff; spores 5.5-7.5 × 3.5-4.5 microns, elliptical and warty in side view, angular in end view.

HABITAT: Solitary, scattered, or in groups or tufts in humus or on rotten wood; common in central and southern California in the fall and winter, but more widely distributed. It occurs under both hardwoods and conifers, I find it most often with redwood.

EDIBILITY: Uncertain. Greg Wright of Los Angeles says it has an excellent flavor, but that one collection caused sweating, chills, and other symptoms of muscarine poisoning.

COMMENTS: The dull pinkish to brick-colored cap and pinkish spores are the distin-guishing features of this eminently undistinguished fungus. The gill attachment is quite variable, leading to confusion with *Clitocybe inversa* (which has white or buff spores), candy caps *(Lactarius fragilis* and *L. rufulus)*, and various Collybias, but the pinkish spores and pinkish gills at maturity are distinctive. There are several other Rhodocybes that are difficult to key to genus without examining the spores. These include: *R. nitellina,* cap up to 5 cm broad and orange-brown to reddish-brown to pinkish-cinnamon or ochre with a translucent-striate margin when moist; *R. roseiavellanea,* one of several uncommon wood-inhabiting species with a more or less tan cap; *R. mundula,* cap 3-7 cm broad, grayish, and bitter-tasting; *R. caelata,* with a small (1-3 cm) grayish cap and decurrent gills; and *R. aureicystidiata*, also small, but with a darker cap that bruises dark or dingy reddish.

Rhodocybe nuciolens is variable in size, shape, and color, but always has pinkish spores. These are rather small specimens.

ENTOLOMA

Medium-sized to fairly large *terrestrial* mushrooms. CAP typically convex to plane or uplifted, *usually not scaly.*GILLS *typically attached (usually notched),* fairly close to well-spaced, usually pinkish or flesh-colored in old age. STALK *fleshy,* smooth or slightly scaly, central. VEIL and VOLVA absent. SPORE PRINT *pinkish to flesh-colored or cinnamon-pinkish.* Spores angular.

ENTOLOMA is easily the most conspicuous genus in the Entolomataceae. As defined here it is comprised of fleshy, pinkish-spored, terrestrial mushrooms that correspond in shape and stature to the white-spored genus *Tricholoma.* The cap and stalk are smooth to only slightly scaly; the gills are usually notched but vary in their attachment (however, they are not truly free as in *Pluteus,* and only rarely decurrent as in *Clitopilus*). In contrast to *Nolanea* and *Leptonia,* the stalk is typically fleshy, but there are numerous nondescript forms with a "somewhat fleshy" stalk, giving the "lumpers" (see p. 10) a good excuse to broaden the definition of *Entoloma* to include both *Nolanea* and *Leptonia.*

Because of their larger size Entolomas are much easier to see than Nolaneas and Leptonias—but they are just as difficult to differentiate, if not more so. Most species are a forgettable shade of brown, gray, dingy olive, or dull yellow, but a few feature the blue and violet hues so typical of *Leptonia.*

Entolomas are almost exclusively terrestrial—a good means of distinguishing them from deer mushrooms *(Pluteus)*—and grow mostly in the woods. Some are probably mycorrhizal (in Europe several are associated exclusively with trees and shrubs of the rose family—an unusual habitat for mushrooms). In our area they are particularly common under hardwoods (oak, tanoak, madrone) in the late fall and winter, and are often most abundant when other mushrooms are rather scarce.

Entolomas should *not* be eaten. Several species are quite poisonous—e.g., *E. lividum* and *E. rhodopolium*—and the few that are definitely known to be edible are not easy to recognize, with the possible exception of *E. abortivum,* an eastern species that frequently produces aborted or misshapen fruiting bodies.

Entoloma is a fairly large genus even in its strictest sense. Three representative species or species "complexes" are described here, but according to *Entoloma*-expert David Largent, several of our local species are endemic and undescribed.

Key to Entoloma

1. Cap red-brown to orange-brown, cinnamon, or pinkish   (see *Clitopilus & Rhodocybe,* p. 239)
1. Not as above; cap blue, gray, black, grayish-brown, grayish-yellow, olive-brown, whitish, etc.
   (but may be dusted pinkish by spores) 2
2. Cap gray to grayish-brown and gills usually decurrent; "aborted" fruiting bodies often found
   (infected by mycelium of *Armillariella mellea*) that are whitish to pinkish-tan and bumpy on
   the outside, marbled within; common in eastern North America .... *E. abortivum*
2. Not as above; gills typically *not* decurrent; "aborted" fruiting bodies absent 3
3. Cap *and/or* stalk bluish-gray to steel-gray to dark blue, violet-tinged, indigo-blue, or black
   when fresh 4
3. Not as above 6
4. Only the stalk bluish or violet; cap brown to dark brown and smooth; stalk rather slender; found
   on wood or ground   *E. trachysporum var. purpureoviolaceum* (see *Leptonia carnea,* p. 250)
4. Not as above 5
5. Cap at least slightly viscid when moist; stalk (0.5) 1-3 cm thick; cap blue-gray to steel-gray to
   violet-gray to black; common .... *E. madidum,* p. 243
5. Cap and/or stalk deep indigo-blue; cap not viscid, stalk usually more slender than above; not
   common .... *E. nitidum* (see *Leptonia carnea,* p. 250)
6. Odor nitrous or bleachlike when fresh (crush flesh if unsure)   ... *E. nidorosum* group, p. 244
6. Odor mild or at least not as above 7

7. Cap white or whitish when fresh 8
7. Cap not white 9
8. Spores *not* angular; rare .... (see *Hebeloma*, p. 463)
8. Spores angular under microscope; not uncommon ....
.... *E. prunuloides* & *E. speculum* (see *E. rhodopolium* group, below)
9. Gills tinged yellow when young; stalk 1-3 cm thick; mainly found in eastern North America
.... *E. lividum* (see *E. rhodopolium* group, below)
9. Not as above; widely distributed .... *E. rhodopolium* group & others, below

Entoloma madidum   (Midnight Blue Entoloma)   Color Plate 51

CAP (3.5) 5-15 (20) cm broad, obtuse or convex becoming plane or remaining broadly umbonate, the margin sometimes uplifted in age; surface viscid to slightly lubricous when moist, smooth or wrinkled, often with a fibrillose or streaked appearance; blue-gray to nearly black to violet-gray or steel-gray, sometimes fading in age to fuscous or occasionally with ochraceous stains or paler areas. Flesh thick, firm, white; odor usually farinaceous. **GILLS** notched or adnexed or at times appearing free, close, white or tinged cap color, becoming pinkish or flesh-colored as the spores ripen. **STALK** 4-13 cm long, (0.5) 1-3 cm thick, solid, firm, fleshy, fibrillose, equal or with a tapered base or often thicker in middle; upper portion blue-gray or colored like cap, base whitish to slightly yellowish (but occasionally white throughout). **SPORE PRINT** flesh-color to pinkish-cinnamon; spores 6.5-8.5 (10) × 6-8 microns, nearly round to broadly elliptical but angular (5-6 sided).

HABITAT: Solitary, scattered, or gregarious on ground in woods; widely distributed but especially common along the west coast in the fall and winter. In our area two slightly different forms occur—one with conifers (especially redwood), the other with hardwoods such as tanoak and madrone. The European version is said to grow in grass!

EDIBILITY: Edible, with meaty flesh and a good flavor. However, many fleshy Entolomas are toxic, so be certain of your identification!

COMMENTS: One of the few Entolomas that can be recognized easily in the field—the black to bluish-gray or violet-tinged cap, pinkish spores, notched or adnexed gills that are whitish when young, and firm, fleshy stem render it distinct. It has the aspect of a *Russula*, but is not as brittle and does not have white or yellow spores. It is much larger and fleshier than the small bluish-black Leptonias *(L. parva*, etc.) and is not nearly as purple as the blewit *(Clitocybe nuda)*. The viscid or lubricous cap (when moist) separates it from *E. nitidum* and *Leptonia carnea*, which are also more vividly colored. *Tricholoma portentosum* is quite similar, but has white spores and a white stem.

Entoloma rhodopolium group

CAP 3-12 (15) cm broad, convex or broadly umbonate becoming plane or slightly depressed; surface smooth, not viscid or only slightly so, gray to grayish-brown to yellowish-gray, usually paler as it dries; margin often wavy or splitting in age. Flesh firm, white; odor mild to slightly farinaceous, but not bleachlike. **GILLS** adnate or notched, whitish or tinged cap color, becoming pinkish as the spores ripen; fairly close to rather well-spaced. **STALK** 4-13 cm long, 0.5-1.5 (2) cm thick, fleshy, equal or tapered below, dry, white or tinged gray. **SPORE PRINT** pinkish-salmon to deep flesh-color; spores 8-11 × 7-9 microns, nearly round but angular.

HABITAT: Solitary, scattered, or in groups or clusters on ground in woods, mainly under hardwoods; widely distributed. In California it is quite common in the fall and winter under alder, but has many look-alikes that occur with oak, tanoak, and madrone.

EDIBILITY: Poisonous—causing vomiting, diarrhea, and abdominal cramps that may require hospitalization! All fleshy Entolomas should be strictly avoided, although some are said to be edible—they are difficult to differentiate and not well known.

COMMENTS: The above description will fit a number of medium-sized to large, showy woodland Entolomas with a smooth, grayish to yellowish or tan cap, fleshy stem, attached (usually notched) gills, and pinkish spores. They are the most common and conspicuous of our Entolomas (along with the *E. nidorosum* group) as well as the most dangerous. As a group they can be recognized by their color and *Tricholoma*-like stature (i.e., they are never as slender as the similarly-colored Nolaneas). Distinguishing between the various Entolomas within the group (as loosely defined here) is strictly a matter for specialists. The largest and most infamous of the lot, *E. lividum,* (also called *E. sinuatum*), has a tan to grayish or pallid cap that is 6-20 cm broad or more, pale yellowish gills when young, and a thick (1-3 cm) stalk; it is common in eastern North America, but has not yet been reported from California. Other species include: *E. grayanum,* a common fleshy eastern species; *E. clypeatum,* whose cap is grayish-brown to olive-buff and streaked with darker fibrils; *E. prunuloides,* with a large white viscid cap; and *E. speculum,* slender, with a dry white or whitish, often umbonate cap. The latter species occurs in California, but the others may not. (According to *Entoloma*-expert David Largent, many of California's large, fleshy Entolomas are undescribed and probably endemic.)

Entoloma nidorosum group   (Nitrous Entoloma)

CAP 3-12 cm broad, convex becoming plane or depressed or occasionally broadly umbonate; surface smooth, not viscid or slightly so when moist; olive-brown to brownish to grayish-brown or pale yellowish-brown when moist (but see comments), paler (grayish to grayish-yellow or even whitish) as it dries; margin often wavy in age. Flesh pallid; odor distinctly nitrous (bleachlike) when fresh, but fading after sitting awhile; taste rather rancid-oily. **GILLS** white or pallid, becoming pinkish as the spores mature; usually adnexed or notched, sometimes adnate; fairly well-spaced to fairly close. **STALK** 4-12 cm long, 0.3-1.5 cm thick, more or less equal or narrowed at base, whitish or faintly yellow, fibrillose-striate, firm, fleshy, solid. **SPORE PRINT** pinkish-salmon to deep flesh-color; spores 7-9 × 6-8 microns, nearly round but angular.

Entoloma nidorosum group. The larger Entolomas are notoriously difficult to identify, but this species "complex" has a distinctive bleachlike odor. Some are stouter than the specimens shown here, others are more slender. The gills are usually adnexed or notched.

HABITAT: Solitary to scattered or gregarious (sometimes tufted) on ground in woods, widely distributed. Common in our area under hardwoods in the fall and winter.

EDIBILITY: To be avoided. The very similar *E. ferruginans* (see comments) is apparently edible, as it has been mistaken inexplicably (but harmlessly) for the blewit *(Clitocybe nuda)*, in spite of its unpleasant taste and unpurple color. Several similar Entolomas are quite poisonous, however (see the *E. rhodopolium* group).

COMMENTS: The bleachlike odor of fresh specimens is usually quite pronounced, and serves to separate this species complex from other gray to brown, fleshy Entolomas. The stature of *E. nidorosum* is typically rather erect and slender, but the stalk is distinctly fleshy, in contrast to species of *Nolanea*. Other Entolomas with a bleachlike odor include: *E. pernitrosum,* with a *persistent* odor (i.e., one that doesn't fade); and *E. ferruginans,* which is one of the commonest gilled mushrooms in the live oak woodlands of southern California, and the most common *Entoloma* there (it also occurs in our area and probably throughout the state). It differs from *E. nidorosum* in its more variable cap color (hazel to buff, pale yellow, olive-brown, dark brown, or nearly black), thicker stalk (up to 2.5 or even 3 cm thick) which is often tinged with the cap color, and slightly larger spores. Whether it is indeed a distinct species endemic to California, or merely a robust version of *E. nidorosum,* is for the *Entoloma*-experts to decide.

NOLANEA

Smallish, mostly saprophytic, terrestrial mushrooms. CAP usually smooth, often silky when dry, *typically conical, bell-shaped, or distinctly umbonate,* at least when young, but sometimes convex. GILLS *typically attached* (adnate to adnexed or sometimes decurrent), but sometimes appearing free. STALK *typically slender and fragile or cartilaginous,* usually hollow and easily splitting; central. VEIL and VOLVA absent. SPORE PRINT *pinkish to flesh-colored to salmon- or cinnamon-pinkish.* Spores angular (sometimes shaped like jacks or ice cubes!), not amyloid.

THESE are small, drab, gray to brownish mushrooms with a thin, cartilaginous or fragile stem and a conical to bell-shaped or distinctly umbonate ("nippled") cap. The shape and stature of the fruiting body are reminiscent of the white-spored genus *Mycena,* but the spore color, of course, is pinkish. The cap is convex to plane in a few species, but never (or only rarely) umbilicate as in *Leptonia. Entoloma* differs by its fleshy stem, but this is a somewhat ambiguous character, and the two genera intergrade.

Nolanea species are differentiated largely on microscopic characteristics such as the presence or absence of cystidia (sterile cells) on the edges of the gills. They are especially common under conifers, but also occur with hardwoods and in lawns and pastures.

Most are of unknown edibility but some, such as *N. verna,* are said to be poisonous. Three widely distributed species are described here. They are similar to one another in appearance but differ in habitat and season. Two colorful and distinctive eastern species are also included in the key, plus the infrequently encountered genus *Eccilia.*

Key to Nolanea

1. Odor sweet, like a candy store or bubble gum (break open the cap if unsure) 2
1. Odor mild or distinctive, but not as above 3
2. Stalk (and usually the cap) yellowish .... *N. icterina* (see *N. verna* group, p. 247)
2. Cap and stalk more or less grayish-brown .... *N. fructufragrans* (see *N. verna* group, p. 247)
3. Cap pink to yellow or orange and conical or bell-shaped (or if expanding in age then retaining a distinct umbo); found in eastern North America 4
3. Not as above; cap not usually brightly colored 5
4. Cap yellow to yellow-orange .... *N. murraii*
4. Cap more or less salmon-colored .... *N. salmonea*

5. Stalk minutely velvety; cap reddish-brown to blackish (at least at center); odor usually strong, like cucumber or fish .... (see **Tricholomataceae,** p. 129)
5. Not as above 6

6. Fruiting mainly in the spring, especially in areas with snowfall ...... *N. verna group,* p. 247
6. Not as above 7

7. Usually growing in grass or open areas; odor of crushed flesh farinaceous (like raw meal or cucumber) .... *N. sericea* & others, below
7. Usually growing in woods or under trees; odor mild or various, sometimes farinaceous ... 8

8. Cap conical to bell-shaped, or if expanding somewhat then usually retaining a distinct umbo .... *N. staurospora* & others (see *N. sericea,* below & *N. verna* group, p. 247)
8. Cap broadly conical to convex or plane 9

9. Gills decurrent .... *Eccilia*
9. Gills not decurrent .... *N. stricta* & others, below

Nolanea stricta   (Strict Nolanea)

CAP 2.5-7 cm broad, conical or bell-shaped expanding to broadly convex or even plane, but usually with a distinct, sometimes pointed umbo; surface smooth, hygrophanous, watery cinnamon to grayish-brown when moist, paler and often somewhat streaked when dry. Flesh thin, fragile, odor mild. **GILLS** broad, adnate to deeply notched, adnexed, or free; pallid becoming flesh-colored in age, fairly close. **STALK** 5-15 cm long, 2-5 (10) mm thick, typically long and straight (strict), hollow, fragile, splitting easily, equal or tapered slightly upward, longitudinally striate; whitish or colored like cap (but usually paler); base whitish and often downy. **SPORE PRINT** salmon-pink or pinkish-cinnamon; spores 9-13 × 6-9 microns, elliptical but angular. Cystidia absent on gill edges.

HABITAT: Solitary, scattered, or in small groups on ground or very rotten wood, under both hardwoods and conifers; widely distributed. I have found it—or a very similar species—in our area in the late fall, winter, and spring.

EDIBILITY: Unknown—to be avoided.

COMMENTS: Also known as *Rhodophyllus strictior* and *Entoloma strictius,* this is one of a number of attractive but difficult-to-identify, pinkish-spored, woodland mushrooms with a long, straight ("strict"), slender stalk that splits or breaks easily, and a gray to brown cap. Similar species include: *N. hirtipes,* with a grayish-brown cap, farinaceous odor, and cystidia on the gill edges; and *Leptonia jubata,* with a convex to plane or uplifted, minutely hairy or velvety cap. For smaller species with a conical or "nippled" cap, see comments under *N. verna.*

Nolanea sericea   (Silky Nolanea)

CAP 2-4 (6.5) cm broad, convex becoming plane or with a slight umbo; surface smooth but not viscid, hygrophanous: dark brown (umber) to dark olive-brown to grayish-brown or hazel-brown when moist, paler or grayer with a silky, often streaked appearance when dry. Flesh thin, watery; odor usually farinaceous, at least when crushed. **GILLS** adnate to adnexed, notched, or nearly free, pallid to grayish or tinged cap color, then dusted pinkish with spores. **STALK** 2.5-5 (10) cm long, (2) 3-8 mm thick, equal or with a slightly enlarged base, often short, fragile, longitudinally fibrillose-striate and silky when dry; grayish-brown or colored like cap, the base whitish. **SPORE PRINT** deep salmon-pinkish or cinnamon-pinkish; spores 8-13 × 6-9 microns, elliptical but angular-nodulose. Cystidia absent on gill edges.

HABITAT: Scattered to gregarious in pastures, on lawns or hillsides, in waste places, sometimes also in woods or under willows; widely distributed and common in our area from the fall through early spring. Unlike many mushrooms, it tends to fruit *during* rainy spells rather than following them.

Nolanea sericea is common in pastures and other grassy places and occasionally turns up in the woods as well. The cap usually has a silky sheen in dry weather.

EDIBILITY: Unknown.

COMMENTS: Only when this species is found in grassy places can the beginner hope to distinguish it from the throngs of other brown to grayish Nolaneas, Entolomas, and Leptonias. Be sure to look for the characteristic silky sheen that develops in dry weather or as the cap loses moisture. The cap is not normally conical or bell-shaped as in many Nolaneas (see *N. verna*), but may have a slight umbo (the very similar *N. hirtipes* has a more prominent umbo or "nipple" and features cystidia on the gill edges). The pinkish spores and gray-brown to brown color help separate *N. sericea* from our other common grass-inhabiting mushrooms. Other species: *N. edulis* is a very similar but smaller species that is common in open places in southern California; *N. staurospora* grows in grassy as well as shaded or wooded areas and has a conical to bell-shaped cap that may become convex or umbonate as it matures, and ranges in color from dark brown to date-brown, grayish-cinnamon, or grayish-brown, but fades as it dries to tan or paler. Furthermore, it has "cruciate-nodulose" spores (shaped like stars or jacks). I mention this feature because it is impossible to appreciate the relief they (the spores) provide unless you've spent many long, tedious hours scrutinizing the round to elliptical spores sported by the overwhelming majority of fleshy fungi. Apart from the spores, however, it is just another "LBM" that is easily confused with dozens of other "LBM's." Enough said.

Nolanea verna group   (Springtime Nolanea)

CAP 2-5 cm broad, conical or bell-shaped, often with a pointed umbo or "nipple," or expanding sometimes to convex or plane (but often retaining the umbo); surface dry, usually smooth, finely fibrillose and often silky or glossy; color variable: dark to light brown, yellow-brown, olive-brown, tan, grayish, or yellowish. Flesh thin, odor mild. **GILLS** fairly close, adnate to adnexed but often seceding, pale grayish or buff or colored like cap, becoming salmon or reddish as spores mature. **STALK** 2.5-10 cm long, 3-7 (10) mm thick, equal, straight, often flattened or twisted-striate, hollow, fragile and easily splitting; pallid or more often colored like the cap (or paler). **SPORE PRINT** rosy-salmon to flesh-colored; spores 8-11 × 7-8 microns, elliptical but angular.

HABITAT: Widely scattered or in groups on ground in woods, mainly under conifers; widely distributed and very common in the spring and early summer shortly after the snow melts (about the same time as morels!). I have seen large fruitings in the Cascades and Sierra Nevada, but have not yet found it in our area.

EDIBILITY: Said to be poisonous, causing moderate to severe gastrointestinal distress.

Nolanea verna group is common during the spring, especially under mountain conifers (often when the morels are out). Note the conical cap when young.

COMMENTS: Also known as *Entoloma vernum,* this species or species "complex" has many look-alikes in the genus *Nolanea;* but fruits almost exclusively in the spring. The pinkish spores distinguish it from *Mycena,* and the slender, cartilaginous stem separates it from *Entoloma.* The yellowish form that is common in the Sierra Nevada may actually be *N. cuneata* (according to David Largent). Similar Nolaneas include: *N. holoconiota,* also common in western mountains, with a brown to yellow-brown cap and pale yellow to pale orangish stalk that is minutely powdered above; the *N. mammosa-N. papillata* group, with a small (1-4 cm) nippled, bell-shaped to conical cap and long thin stalk, usually fruiting in the late summer, fall, or winter; and *N. staurospora,* with star-shaped spores (see comments under *N. sericea*). Also worth mentioning are two small species that smell like a candy store or "tooty-fruity gum" (Largent): *N. fructufragrans,* with a more or less grayish-brown cap and stalk; and *N. icterina,* with a yellowish to olive-yellow cap and stalk and adnexed to slightly decurrent gills. None of these are worth eating.

LEPTONIA & Allies

Small to medium-sized, saprophytic mushrooms growing on ground or sometimes on wood. CAP *typically convex to plane, depressed, or umbilicate; often fibrillose or with small scales,* and often blue, black, or violet. GILLS *typically attached* (adnexed to adnate or slightly decurrent), but sometimes appearing free. STALK *typically slender and fragile or cartilaginous,* but sometimes fleshy; smooth or scaly; often blue to gray, black, or violet. SPORE PRINT *pinkish to rosy-pinkish, flesh-colored, pinkish-cinnamon, etc.* Spores angular, not amyloid.

IN stature *Leptonia* corresponds roughly to the white-spored genera *Collybia* and *Omphalina.* The cap is typically convex to plane or umbilicate, rather than conical as in *Nolanea,* and the stalk—with a few exceptions—is thin and fragile rather than fleshy as in *Entoloma.* Leptonias can sometimes be recognized by their color alone: many species are breathtakingly blue, vividly violet, or beautifully black, and some, such as *L. carnea,* have an iridescent quality that makes them among the most striking of all mushrooms. Others, however, are gray or brownish like umpteen *Entoloma* and *Nolanea* species.

Leptonias fruit throughout the mushroom season, but tend to be most abundant when or where other mushrooms are scarce. Fairly few fleshy fungi, for instance, are fond of fruiting under fern fronds in redwood duff, but that is precisely the niche in which *Leptonia* thrives in California. In the Pacific Northwest, on the other hand, it is frequent under alder and western red cedar (also rather unfruitful foraging grounds for fleshy fungi), and in England and Scotland it is said to favor heaths. Most of the species are terrestrial; a few grow on rotten wood, but can be separated from small species of *Pluteus* by their attached, more widely spaced gills.

Though Leptonias are among our most beautiful mushrooms, little is known of their edibility. David Largent recognizes over 100 species in his monograph on *Leptonia,* but even the distinctively colored ones are rather difficult to identify. A mere four species are described here, plus one species of *Alboleptonia,* a small genus of whitish mushrooms often included in *Leptonia.*

Key to Leptonia & Allies

1.   Cap *and/or* stalk blue, blue-black, blue-gray, steel-gray, violet, or black, at least when young
      and fresh 2
1.   Not as above 12

2.   Cap smooth (not scaly!) and brown to dark brown; stalk blue or violet-blue and often iridescent,
      usually at least 4-5 mm thick .... (see *Entoloma,* p. 242)
2.   Not with above features 3

3.   Growing on wood; cap usually fibrillose-scaly; rare *L. cyanea* & others (see *L. carnea,* p. 250)
3.   Growing on ground and/or cap smooth 4

4.   Stalk slender, hollow, and either fragile or cartilaginous, usually 1-5 mm thick at apex 5
4.   Stalk usually 5 mm thick or more 11

5.   Fresh stalk fibrillose-scaly and iridescent or metallic .. *L. occidentalis* (see *L. carnea,* p. 250)
5.   Not as above 6

6.   Gills pallid or grayish with distinctly darker (blue-gray to black) edges ....
      .... *L. serrulata* (see *L. parva,* p. 251)
6.   Not as above (but gills may be entirely blue-gray) 7

7.   Cap and/or stalk with distinct violet or purple tints when fresh 8
7.   Not as above 9

8.   Gills pale yellow when young .... *L. zanthophylla* (see *L. nigroviolacea,* p. 250)
8.   Not as above .... *L. nigroviolacea* & others, p. 250

9.   Cap bluish-black to black when fresh (but may fade to brownish or gray in age) 10
9.   Cap yellow-brown to reddish-brown to grayish-brown when fresh *L. gracilipes* & others, p. 252

10. Gills blue-gray to blue .... *L. nigra* (see *L. parva,* p. 251)
10. Gills whitish or tinged gray .... *L. parva* & others, p. 251

11. Cap and stalk deep indigo-blue (or stalk apex sometimes with violet tints); known only from
      the west coast .... *L: carnea,* p. 250
11. Cap grayish-brown to deep violet; stalk usually paler; found mainly in eastern North America
      .... *L. porphyrophaea* (=*Entoloma violaceum*)

12. Fruiting body entirely white when fresh (but may develop yellowish, ochraceous, or pinkish
      tones in age) .... *Alboleptonia sericella* & others, p. 252
12. Not as above 13

13. Gills whitish with brown to gray-brown edges *L. fuligineomarginata* (see *L. gracilipes,* p. 252)
13. Not as above 14

14. Edges of gills pink; cap rose-tinted when fresh; rare .... *L. rosea*
14. Not as above 15

15. Stalk yellow-green, usually staining blue-green when bruised; cap greenish-yellow to yellow-
      brown to olive-brown; rare .... *L. incana*
15. Not as above 16

16. Stalk white; cap umbilicate in age .... *L. exalbida* (see *L. gracilipes,* p. 252)
16. Not as above 17

17. Gills decurrent .... *Eccilia*
17. Gills adnexed to adnate or with a very slight decurrent tooth 18

18. Cap striate and usually umbilicate in age .... *L. undulatella* (see *L. gracilipes,* p. 252)
18. Not as above; cap not umbilicate .... (see *Nolanea,* p. 245)

Leptonia carnea   (Indigo Leptonia)   **Color Plate 52**

CAP 2.5-7 cm broad, convex becoming broadly umbonate or plane, the margin sometimes uplifted in old age; surface dry, densely fibrillose or fibrillose-scaly, deep indigo-blue, at times almost iridescent, scarcely fading. Flesh yellowish or whitish, but tinged violet under the cuticle; odor typically farinaceous. GILLS adnate to adnexed or notched, pallid or bluish-gray, slowly becoming pinkish as spores mature. STALK 5-10 cm long, 0.5-1 (1.5) cm thick, equal or thicker below, dry, densely fibrillose or minutely scaly, deep indigo-blue, the apex sometimes violet-tinted; base often whitish, or in age yellow-orange within. SPORE PRINT pinkish; spores 9-13 × 6.5-10.5 microns, elliptical but angular.

HABITAT: Solitary or in small groups in woods; known only from California, and apparently rare. I have found it several times near or under redwood in late fall and winter, during otherwise uneventful mushroom hunts. Its deep color camouflages it well.

EDIBILITY: Undetermined, but too rare to be of value.

COMMENTS: The stunning indigo-blue color and fibrillose-scaly cap and stem render this breathtakingly beautiful mushroom distinct (the color plate does not do it justice). Its rareness makes it all the more pleasurable to find. Its only rival for intensity of color is *Cortinarius violaceus,* which has rust-brown spores and is deep violet. The stem is thicker than that of the numerous blue-black or violet-black Leptonias (see *L. parva* and *L. nigroviolacea*), but thinner than that of *Entoloma madidum.* The gills are not purple as in the blewit *(Clitocybe nuda)* and the stalk and cap are much scalier. There are several similarly colored species, including: *Entoloma nitidum,* with a smoother (less scaly) cap and stalk and smaller spores (7-9 microns long), not common; *Entoloma trachysporum var. purpureoviolaceum,* with a smooth dark brownish cap and metallic blue stalk; *L. occidentalis var. metallica,* smaller, with an iridescent or metallic blue, densely scaly-fibrillose stalk (up to 5 mm thick at apex), mild to slightly fragrant odor, and white to blue-gray flesh, also not common; and three rare wood-inhabiting species with fibrillose-scaly stalks: *L. cyaneonita,* bluish-black, with an iridescent stalk; and *L. cyanea* and *L. violaceonigra,* both violet-tinted, the former with pallid gills and the latter with bluish- or violet-tinged gills when young.

Leptonia nigroviolacea   (Violet-Black Leptonia)

CAP 1-3.5 cm broad, convex or obscurely umbonate becoming plane to slightly umbilicate; surface dry, fibrillose, usually with small scales at least at the center; entirely violet-black or deep violet, or sometimes bluish-black or fading to violet-gray. Flesh thin, whitish or tinged violet. GILLS fairly well-spaced, white or pale buff or sometimes tinged gray, becoming pinkish in age as spores mature; attachment variable but usually adnexed to adnate. STALK 3-8 cm long, 2-5 mm thick, equal or tapered slightly, smooth or fibrillose, often with a few small scales, violet-black to violet-gray, the extreme apex usually pallid; hollow; usually with white mycelial down at base. SPORE PRINT rosy-pinkish; spores 9.5-11.5 (13) × 7-8.5 microns, elliptical but angular. Cystidia absent.

HABITAT: Solitary, scattered, or in groups or tufts on ground in woods; widely distributed. In our area it is fairly common in the late fall and winter under redwood, but is not restricted to that habitat.

EDIBILITY: Unknown.

COMMENTS: This is one of several dainty, violet-tinged Leptonias. They are not as numerous as the blue-black models (see *L. parva*), but are just as attractive. *L. occidentalis* var. *metallica* (see *L. carnea*) is similar but has a more iridescent, densely fibrillose-scaly stalk. Other species which characteristically exhibit violet tints include: *L. convexa,* whose cap has erect scales on a bluish-gray, reddish-brown, or grayish-brown background; *L. zanthophylla,* with pale yellow gills when young; and *L. diversa,* whose cap is frequently umbilicate in age and has purple-blue scales on a paler background, and whose gills are often tinged bluish-gray. For violet-tinted, wood-inhabiting species, see comments under *L. carnea*

Leptonia parva   (Blue-Black Leptonia)

CAP 0.7-4 cm broad, broadly convex becoming plane to shallowly depressed or umbilicate; surface dry, finely fibrillose when young, often breaking up into small scales in age; blue-black to black when fresh, often somewhat paler or grayer in age; margin often finely striate or fibrillose-striate. Flesh thin, white or tinged cap color. **GILLS** very slightly decurrent to adnate or adnexed, but sometimes seceding; whitish or with a slight tinge of blue-gray, becoming pinkish in old age from spores. **STALK** 1.5-8 cm long, 2-4 (7) mm thick, more or less equal, smooth or in one form with minute bluish-black scales over upper half; blackish-blue to bluish-gray, often becoming paler gray in age; base often with whitish mycelial down. **SPORE PRINT** rosy-pinkish; spores 8-12 × 6-8 microns, elliptical but angular. Cystidia typically absent on gill edges.

HABITAT: Scattered or in small groups in forest humus, widely distributed. This species and its numerous look-alikes (see below) are especially common in coastal California in redwood duff (often under ferns), and in the Pacific Northwest under western red cedar. They occur throughout the mushroom season in our area, but are easily overlooked because of their dark color.

EDIBILITY: Unknown, but too small to be of value.

COMMENTS: There are a number of petite blue to black or violet-tinted Leptonias that will more or less fit the above description (David Largent recognizes over 100 *Leptonia* species in his monograph). They are quite difficult to distinguish in the field, but you needn't know their exact identities to appreciate their beauty. *Entoloma madidum, E. nitidum,* and *Leptonia carnea* are similar in color but are larger and have thicker, fleshier stems. *L. parva* is characterized by its whitish gills in youth and nearly smooth cap that becomes minutely scaly (especially toward the center) in age. Some of its more common or distinctive cousins include: *L. decolorans,* very similar and common, gills often bluish-gray with cystidia on their edges; *L. corvina,* also similar and common, but with a cap that fades to brown in age, except at the center; *L. rectangula,* with a black cap that is nearly

Left: *Leptonia nigroviolacea,* a dainty violet-tinged species. **Right:** *Leptonia parva,* a common black or steel-gray species.

Left: *Leptonia parva,* or a very similar species. **Right:** Gill detail in *Leptonia serrulata,* showing the finely scalloped black edges characteristic of that species.

rectangular in profile when young and deeply depressed (but not umbilicate) in age, and gray gills; *L. nigra,* a beautiful species with evenly blue-gray gills and minutely hairy blue-black cap that is convex to plane but never umbilicate, fairly common; *L. serrulata,* gills white to gray with dark blue to blue-black, minutely serrated edges (see above photograph), common and widely distributed; and *L. cupressa,* cap blackish but gills and stalk brownish-gray, found mainly under cypress, not common. All of these are typically terrestrial. For similar wood-inhabiting species or those with iridescent, fibrillose-scaly stalks, see comments under *L. carnea,* and for small terrestrial species with a distinct violet tinge, see *L. nigroviolacea.*

Leptonia gracilipes

CAP 1.5-5 cm broad, convex becoming plane to shallowly depressed or umbilicate; surface dry, minutely scaly toward the center, the margin finely fibrillose or smooth; dark yellow-brown to grayish-brown or dark brown, often fading in age to paler brown or even orange-brown, the center usually darker. Flesh thin, fragile, grayish- or olive-tinted. **GILLS** adnexed to adnate or with a slight decurrent tooth, white or pallid when young, becoming pinkish in age from maturing spores, or sometimes brownish. **STALK** 2-8 cm long, 1-5 mm thick, equal, often flattened or grooved, smooth, dark blue-gray fading to blue-gray or gray in age. **SPORE PRINT** rosy-pinkish; spores 8-12 × 6.5-8 microns, elliptical-angular.

HABITAT: Scattered or in small groups in humus under ferns, alder, and conifers; widely distributed. I have found it several times in our area under redwood, in the fall and winter, when few other mushrooms were out and about.

EDIBILITY: Unknown.

COMMENTS: The slender blue-gray stalk and yellow-brown to grayish-brown cap characterize this rather forgettable species. Its stature is similar to that of *L. parva,* but the cap is never bluish-black. Other species: *L. asprella* is similar, but has grayer gills and a grayer cap; *L. vinaceobrunnea* has a blue-gray stalk and reddish-brown to vinaceous-brown cap; *L. fuligineo-marginata* has a violet-brown to dark reddish-brown to grayish-brown cap and stalk, but its gills are white with dark reddish-brown to grayish-brown edges; *L. undulatella* is brown to yellow-brown with a striate cap; and *L. exalbida* has an umbilicate, dark yellow-brown cap and white stalk.

Alboleptonia sericella   (Little White Leptonia)

CAP 1-3 (5) cm broad, convex becoming plane or centrally depressed; surface smooth, dry, often silky or finely fibrillose; pure white to translucent white or tinged yellow (especially in age), or sometimes even pinkish. Flesh very thin, whitish, fragile. **GILLS** adnate

Alboleptonia sericella is a nondescript fragile whitish mushroom with pinkish gills at maturity.

to slightly decurrent or notched, well-spaced, white, becoming rosy-pinkish in age. **STALK** 1.5-5 cm long, 1-4 mm thick, equal, smooth, fragile, white or discoloring like the cap. **SPORE PRINT** bright flesh-color; spores 9-13 × 6-8 microns, elliptical but angular (nodulose).

HABITAT: Scattered to gregarious in damp soil in woods, thickets, along trails, etc.; widely distributed. Fairly common in our area in the fall and winter.

EDIBILITY: Unknown.

COMMENTS: Also known as *Leptonia sericella* and *Rhodophyllus sericellus,* this is the only common small white mushroom with pinkish spores, a thin stem, and attached gills. Its fragility makes it difficult to get specimens home in one piece. *Camarophyllus* species and *Inocybe geophylla* are somewhat similar, but have white and brown spores, respectively. Other species: *A. adnatifolia* is pure white with slightly smaller spores; *A. ochracea* ages or bruises ochraceous.

PLUTEACEAE

spores

FORMERLY called the Volvariaceae, these are small to medium-sized mushrooms with a central stem, free gills at maturity, and smooth, pinkish to sordid reddish spores. In the other major pinkish-spored family, the Entolomataceae, the gills are usually attached to the stem and the spores are angular or longitudinally ridged under the microscope.

There are two principal genera in the Pluteaceae: *Volvariella* has a volva; *Pluteus* does not. A third genus, *Chamaeota,* is extremely rare and not treated in this book. It lacks a volva, but has a partial veil that usually forms an annulus (ring) on the stalk.

Because of the free gills and volva in *Volvariella,* the Pluteaceae are thought to be related to the white-spored Amanitaceae. Besides the disparity in spore color, however, there is a fundamental anatomical difference: the gill tissue is convergent in the Pluteaceae, divergent in the Amanitas (see p. 19).

Key to the Pluteaceae

1. Universal veil and volva absent 2
1. Universal veil present, typically forming a volva (sack) at base of stalk 4

2. Partial veil present, usually forming an annulus (ring) on the stalk 3
2. Veil and annulus absent .... **Pluteus,** p. 254

3. Spore print white or with only a slight pinkish tinge .... (see **Lepiotaceae,** p. 293)
3. Spore print distinctly pinkish or deep flesh-color; mainly tropical; rare .... *Chamaeota*

253

4. Spore print pale, with only a slight pinkish tinge; volva whitish ...... (see *Amanita,* p. 263)
4. Spore print distinctly pinkish to reddish-cinnamon; volva variously colored *Volvariella,* p. 258

PLUTEUS

Small to medium-sized, *wood-inhabiting* mushrooms. CAP convex to plane. Flesh usually soft. GILLS *typically close or crowded and free (at least at maturity),* usually pinkish in old age. STALK typically central and cleanly separable from the cap. VEIL and VOLVA *absent.* SPORE PRINT *flesh-colored to deep pinkish, pinkish-cinnamon, or sordid reddish.* Spores smooth, usually elliptical. Gill tissue convergent.

THESE pinkish-spored mushrooms have a central stalk and free, close gills and grow almost exclusively on wood. The wood, however, may be buried or decomposed, making the mushrooms appear terrestrial. They are most often confused with the pinkish, angular-spored Entolomataceae, which are usually terrestrial with gills attached to the stem.

Pluteus is frequently encountered but rarely abundant. Most species have soft flesh and decay rapidly. However, the larger types such as *P. petasatus* and *P. cervinus* are good edibles when fresh and firm. *P. salicinus* and *P. cyanopus* may contain traces of psilocybin and/ or psilocin; the edibility of many of the smaller species is unknown.

It is a fairly sizable genus, with over 50 species in North America and more than 20 in California. As they are segregated primarily on microscopic features such as the structure of the cap cuticle and the shape of sterile cells (cystidia) on the gills, only five species are described here.

Key to Pluteus

1. Cap red to orange, fading to yellow .... *P. aurantiorugosus* (see *P. lutescens,* p. 257)
1. Not as above (cap may be yellow but never red or orange) 2
2. Stalk yellow 3
2. Stalk not yellow 5
3. Cap velvety or granulose; stalk white or pinkish when young, yellowish only in age ....
      .... *P. flavofuligineus,* p. 258
3. Not as above; stalk yellow even when young 4
4. Cap brown or olive-brown .... *P. lutescens,* p. 257
4. Cap yellow or ochre .... *P. admirabilis* (see *P. lutescens,* p. 257)
5. Cap white .... *P. pellitus* (see *P. cervinus,* p. 255)
5. Cap not white (but may have colored fibrils or scales on a whitish background) 6
6. Gill edges dark brown to nearly black; usually found on conifers ....
      .... *P. atromarginatus* (see *P. cervinus,* p. 255)
6. Not as above; gill edges same color as gills or paler (whitish, pinkish, etc.) 7
7. Base of stalk blue- or greenish-stained in age or when bruised ....
      .... *P. salicinus* & *P. cyanopus* (see *P. longistriatus,* p. 257)
7. Not as above 8
8. Cap plushlike, velvety, or granulose but not conspicuously striate. ....
      .... *P. flavofuligineus* & others, p. 258
8. Cap not velvety or granulose, or if so, then also conspicuously striate 9
9. Cap 3-12 cm broad or more, not striate; stalk 4 mm thick or more 10
9. Cap 5 cm broad or less, sometimes striate; stalk 5 mm thick or less 11
10. Cap whitish or pallid with brownish fibrils or scales; often growing in clusters (and often appearing terrestrial) .... *P. petasatus,* p. 255
10. Cap dark brown to pale brown, grayish-brown, etc., sometimes with a fibrillose-streaked appearance .... *P. cervinus* & others, p. 255
11. Cap yellow or yellowish .... *P. leoninus* (see *P. lutescens,* p. 257)
11. Cap not yellow .... *P. longistriatus* & many others, p. 257

Pluteus petasatus is a robust species with small scales or fibrils on the cap. It often grows in clusters, and like other species of *Pluteus,* has free crowded gills.

Pluteus petasatus

CAP 4-15 (20) cm broad, convex, or becoming broadly umbonate to plane in age; surface usually not viscid, whitish with a brownish to grayish center or with darker (brown to grayish-brown) fibrils or scales; margin usually whitish. Flesh fairly thick and firm, white; odor mild or radishlike. **GILLS** crowded, broad, free at least in age, whitish for a long time, then eventually pinkish. **STALK** 4-10 cm long, 0.7-3 cm thick, equal or swollen above or in the middle; firm, whitish or discoloring below in age; sometimes streaked with fibrils. **SPORE PRINT** pinkish to deep flesh-color; spores 6-10 × 4-6 microns, elliptical, smooth. Cystidia on faces of gills with long necks and "horns."

HABITAT: Gregarious or clustered on sawdust or wood chips in gardens, along roads, etc. (often appearing terrestrial); widely distributed. It is fairly common in our area during the mushroom season and sometimes even fruits in the summer.

EDIBILITY: Edible and very good—the best of the genus for the table. It is much firmer and meatier than the better known *P. cervinus.*

COMMENTS: The robust stature, pale cap with darker fibrils or scales (or a darker center), and tendency to grow in clusters distinguish this species from *P. cervinus.* Because the gills remain white for so long it is liable to be mistaken for a white-spored mushroom, but the free crowded gills and absence of a veil point to *Pluteus. P. magnus* (see *P. cervinus)* can also be robust, but has a dark brown to nearly black cap.

Pluteus cervinus   (Deer Mushroom; Fawn Mushroom)

CAP 3-12 (15) cm broad, obtuse or convex becoming broadly convex to broadly umbonate or plane; surface smooth or radially streaked with fibrils, slightly viscid when moist and often somewhat wrinkled when young; dark brown to pale brown to grayish-brown or dingy fawn, the margin sometimes paler. Flesh soft, white; odor usually radishlike. **GILLS** close or crowded, broad, soft, white becoming pinkish, finally dingy reddish or flesh-colored; free at maturity. **STALK** 5-13 cm long, 0.5-2 (2.5) cm thick, equal or thicker at base, dry, white or with grayish to brownish longitudinal fibrils. **SPORE PRINT** flesh-colored to pinkish-brown; spores 5-8 × 4-6 microns, elliptical, smooth. Cystidia on faces of gills with long necks and 2-4 "horns."

Pluteus cervinus, often called the deer mushroom, is a common species with a brown cap and free crowded gills that turn pinkish in old age.

HABITAT: Solitary or in groups on decaying wood, debris, sawdust piles, or humus rich in lignin; widely distributed and common. In our area it usually appears with *Pleurotus ostreatus* after the very first fall rains, then continues to fruit sporadically through the remainder of the mushroom season. It is partial to (but not restricted to) hardwoods, but *P. atromarginatus* (see comments) is common in our area on conifers.

EDIBILITY: Edible and quite good when fresh and firm, but the flaccid or waterlogged specimens one usually finds are apt to be insipid.

COMMENTS: Also known as *P. atricapillus,* the deer mushroom is our most common and conspicuous *Pluteus.* It is rather nondescript, but can be safely identified by its brown cap, pinkish spores, close free gills, absence of a veil, and growth on wood. The cap color is quite variable, but typically some shade of brown. Special care should be taken not to confuse it with poisonous *Entoloma* species, which typically have attached (often notched) gills and grow on the ground. *P. cervinus,* however, may appear terrestrial and have gills which are slightly attached when young, so it's a good precaution to eat only those that are *clearly* growing on wood. There are several similar *Pluteus* species, all apparently edible, including: *P. magnus,* more robust (stalk 1-3 cm thick) and with a dark, frequently wrinkled cap, often growing in clusters on sawdust piles, etc.; *P. atromarginatus,* common throughout much of the West on wood and debris of conifers, with a brown to dark brown cap and dark brown to black gill edges (see photograph!); and *P. pellitus,* which has a white cap (sometimes brown at center) and grows on dead hardwoods. The latter looks like a destroying angel (*Amanita ocreata, A. virosa,* etc.) from a distance, but lacks a volva and has pinkish gills in age (and pinkish spores). It occurs in our area, but is rare.

Gill detail in *Pluteus atromarginatus.* This species resembles *P. cervinus,* but has dark-edged gills, as shown here.

Left: *Pluteus lutescens*, a yellow-stemmed species. **Right:** *Pluteus longistriatus*, one of several small and nondescript, infrequently-encountered species.

Pluteus lutescens   (Yellow-Stemmed Pluteus)

CAP 1.5-5 cm broad, convex becoming broadly umbonate or plane; surface sometimes wrinkled at the center, not viscid, dark brown to olive-brown to yellowish-brown or yellowish-olive. Flesh thin, white or pale yellow; odor mild. **GILLS** free at maturity, fairly close, broad, whitish to pale yellow, but finally pinkish from ripening spores. **STALK** 2-7 cm long, 2-6 mm thick, more or less equal, straight or curved, fragile; pale yellow, the base usually brighter yellow. **SPORE PRINT** pinkish to deep flesh-color; spores 6-7 × 5-6 microns, nearly round, smooth. Cystidia on faces of gills club-shaped to flask-shaped.

HABITAT: Solitary or in small groups on rotting hardwood logs, sticks, and debris; widely distributed. Occasional in our area in the late fall and winter, especially on dead oak.

EDIBILITY: Said to be edible, but too small and infrequent to bother with.

COMMENTS: A fragile but beautiful *Pluteus,* easily identified by its yellow stem and brownish to olive cap plus the free gills and pinkish spores. *P. nanus var. lutescens* is a passe pseudonym. Other colorful species include: *P. admirabilis,* very similar but with a yellow to ochre cap and yellow stalk, especially common in eastern North America; *P. leoninus,* similar to *P. admirabilis,* but with a white stalk when young; and *P. aurantiorugosus (=P. coccineus),* a beautiful but rare species with a bright red to orange-red cap that fades to bright yellow as it ages, plus a white to yellow or orange-yellow stalk.

Pluteus longistriatus   (Pleated Pluteus)

CAP 1.5-5 cm broad, convex to plane or with a slightly depressed center; surface sometimes minutely scaly or granulose, at least at the center, not viscid; gray to brownish-gray, conspicuously striate in age. Flesh very thin, soft, pallid; odor mild. **GILLS** close, soft, whitish becoming pinkish in age; free at maturity. **STALK** 2-8 cm long, 1.5-3 mm thick, finely fibrillose-striate, pallid or tinged cap color, equal or with an enlarged base, straight or curved, hollow. **SPORE PRINT** pinkish to deep flesh-color; spores 6-7.5 × 5-5.5 microns, nearly round, smooth. Cystidia on gills club-shaped or flask-shaped.

HABITAT: Solitary, scattered, or in small groups on decaying branches and sticks of hardwoods; widely distributed but not common. I have found it several times in our area on oak debris, and also on wet wood in a bathroom!

EDIBILITY: Unknown.

COMMENTS: This is one of many small, nondescript *Pluteus* species that are unlikely to catch the eye of the average mushroom hunter except when they are sharing a log with more spectacular fare (such as *Pleurotus ostreatus*). The conspicuously striate cap is the principal fieldmark of this species. Others include: *P. californicus,* cap hazel to hazel-brown or greenish-gray, with a striate margin when moist and a reddish-brown spore print; *P. seticeps*, with a brown striate cap; *P. cyanopus,* with a chestnut- to cinnamon-brown, faintly striate cap and hazel to grayish-olive stalk; and *P. salicinus,* with a grayish-brown to greenish- or bluish-gray, non-striate cap and whitish stalk that stains blue at the base. None of these are worth eating. *P. chrysophaeus* should also be mentioned.

Pluteus flavofuligineus

CAP 2-7 cm broad, convex becoming broadly umbonate or broadly convex to plane; surface not viscid, appearing velvety to minutely granulose; dark brown or olive-brown when young, developing yellow tones in age from the margin inward (eventually often entirely yellow or ochre-yellow or yellowish with a brownish center). Flesh thin, pallid; odor mild. **GILLS** free but sometimes appearing adnexed; close, whitish or tinged yellow when young, becoming pinkish as spores mature. **STALK** 4-11 cm long, 4-6 (8) mm thick, equal or tapering upward, smooth; whitish or pinkish when young, usually becoming yellowish in old age. **SPORE PRINT** pinkish to deep flesh-color; spores 6-7 × 4.5-6 microns, nearly round, smooth. Cystidia on gills flask-shaped.

HABITAT: Solitary or in small groups on rotting hardwood logs, branches, etc.; widely distributed, but not common in the West. I find it every winter in our oak woodlands.

EDIBILITY: Unknown.

COMMENTS: The velvety to granulose, brown or yellow cap combined with the free gills, pinkish spores, and growth on dead wood are good fieldmarks. The stalk is often rather tall for a *Pluteus*—as much as three times as long as the diameter of the cap! *P. lutescens* is somewhat similar, but has a clear yellow stalk even when young, and does not have a velvety cap. Other species: *P. granularis* is also similar, but has a velvety brown cap *and* stem.

VOLVARIELLA

Small to medium-large, saprophytic or parasitic mushrooms found on wood, soil, humus, or other mushrooms. CAP oval to convex or plane, sometimes viscid. GILLS *free at maturity,* close, pallid becoming *pinkish to flesh-colored or sordid reddish* at maturity. STALK central, usually hollow, cleanly separable from cap. VEIL universal, membranous, *forming a saclike volva at base of stalk.* SPORE PRINT *pinkish to deep flesh-color or sordid reddish.* Spores smooth, usually elliptical. Gill tissue convergent.

THE pinkish to reddish spores and presence of a volva separate *Volvariella* (formerly *Volvaria*) from all other mushrooms. The volva is always saclike and there is no partial veil or annulus. But for the spore color this genus resembles *Amanita*, and young specimens with pallid gills are often mistaken for that genus. Confusion with *Pluteus* is also possible if the volva is overlooked or destroyed.

Several Volvariellas are edible. In fact, the paddy straw mushroom, *V. volvacea,* is to the tropics what *Agaricus bisporus* is to the temperate zone—the principal mushroom of commerce. It is cultivated widely in southern Asia—usually on straw in rice paddies—and shipped abroad so you can pay exhorbitant prices for it in exotic food stores. I was presented a can of it ten years ago, on some unforgettable occasion I can no longer remember. I am sorry to say it has languished in the back of my cupboard ever since—there are always so many fresh mushrooms in my refrigerator!

Volvariella is principally a tropical genus, and fewer than a dozen species occur in temperate North America. They frequent forests, cultivated fields, gardens, straw heaps, and greenhouses; a few grow on wood and one is parasitic on other mushrooms. Only one species, *V. speciosa,* is truly common; several others are encountered infrequently.

Key to Volvariella

1. Growing on other mushrooms .... *V. surrecta* (see *V. bombycina,* p. 261)
1. Growing on ground, wood, compost, in greenhouses, etc. 2

2. Cap typically 5-20 cm broad when mature 3
2. Cap typically 2-5 (6) cm broad when mature 5

3. Growing on wood; cap covered with silky fibrils; rare .... *V. bombycina,* p. 261
3. Not as above 4

4. Volva brown or grayish-brown (at least the upper portion or edge); cap not normally viscid; found in tropics, subtropics, and warm environments *V. volvacea* (see *V. bombycina,* p.261)
4. Volva white to pale gray; cap viscid when moist; very common in temperate zone in cultivated fields, gardens, roadsides, manure, etc. .... *V. speciosa,* below

5. Volva white, with long hairs; cap gray to grayish-brown *V. villosavolva* (see *V. smithii,* p. 261)
5. Not as above; volva without long hairs 6

6. Cap gray to pinkish-gray; volva brown to grayish .... *V. taylori* (see *V. smithii,* p. 261)
6. Cap whitish when fresh, but may discolor overall in age and center often tinged another color 7

7. Volva ochre-stained to brownish .... *V. smithii,* p. 261
7. Volva white to grayish 8

8. Stalk minutely hairy (pubescent); cap margin not striate .... *V. hypopithys,* p. 260
8. Stalk not pubescent; cap margin often striate in age ... *V. pusilla* (see *V. hypopithys,* p. 260)

Volvariella speciosa   (Common Volvariella)   Color Plate 68

CAP 5-15 cm broad, at first oval, then convex to plane or broadly umbonate; surface smooth or occasionally with patches of universal veil tissue, viscid when moist; dull white to gray or grayish-brown (rarely fulvous), or often whitish with a darker center; often with a metallic luster when dry; margin sometimes striate. Flesh soft, white. **GILLS** crowded, broad, free, white becoming flesh-colored and finally sordid reddish. **STALK** 5-20 cm long, 1-2.5 cm thick, equal or thicker at base, dry, whitish, more or less smooth. **UNIVERSAL VEIL** membranous, forming a white to pale grayish saclike volva at base of stalk; volva often buried in soil, sometimes inconspicuous. **SPORE PRINT** deep flesh-color to pinkish-brown; spores 11.5-21 × 7-12 microns, elliptical, smooth.

HABITAT: Solitary to scattered or gregarious in cultivated soil—gardens, vacant lots, roadsides, manure, compost, straw heaps, fallow and planted fields, etc.—occasionally also in the woods; widely distributed. It occurs year-round in our area, but is especially common in the spring; I've seen enormous fruitings in Brussels sprouts fields. In the San Joaquin Valley I've seen fallow fields littered with thousands of Volvariellas.

EDIBILITY: Edible, but mediocre. It was once thought to be poisonous, perhaps due to confusion with *Amanita.* Should you try it, be sure to take a spore print—some Amanitas can have pinkish gills in old age!

COMMENTS: This mushroom can cause quite a stir when it appears, bold and uninvited, in the middle of your cabbage patch. It is our only common *Volvariella,* and is recognized as such by its saclike volva, absence of a ring (annulus), and deep pinkish spores. It is quite attractive when it first emerges, but quickly becomes flaccid and waterlogged. *V. speciosa var.* **gloiocephala** is said to differ from the typical variety in its darker (pearl-gray to fulvous) pileus (cap) with a striate margin, and smaller spores. One dismal day of illegal trespassing

Left: A young specimen of *Volvariella speciosa* (see color plate for older ones). **Right:** *Volvariella hypopithys* is a slender white species with silky hairs on the cap.

in a smelly old Brussels sprouts field yielded the following: twenty soggy specimens with pale pileus and non-striate margin; six soggy specimens with dark pileus and striate margin; five soggy specimens with dark pileus and non-striate margin; and eleven soggy specimens with pale pileus and striate margin. If you can detect a meaningful mycelial thread running through all of this, then you are a better mycologist than I . . . *

Volvariella hypopithys   (Petite White Volvariella)

CAP 2-6 cm broad, at first somewhat bell-shaped, then convex to nearly plane; surface dry, fibrillose (with fine silky hairs that sometimes form small scales), pure white or tinged yellowish at the center, but sometimes discoloring overall in old age; margin usually fringed with long, silky white hairs and not striate or only faintly so. Flesh thin, soft, white. **GILLS** fairly close, free (at least at maturity), white becoming dingy pinkish or flesh-colored. **STALK** 2-9 cm long, 2-5 mm thick, equal or slightly thicker below, fragile, white or pallid; densely pubescent (clothed with short or long silky hairs) at least over the middle and upper portion. **UNIVERSAL VEIL** membranous, forming a small, often inconspicuous or barely visible volva at base of stalk; volva white, saclike, usually lobed. **SPORE PRINT** pinkish or flesh-colored; spores 6-8.5 (10) × 3.5-6 microns, broadly elliptical, smooth.

HABITAT: Solitary or in small groups in woodland humus; widely distributed but rare. I have found it only once in our area, under live oak, in February.

EDIBILITY: Said to be edible, but much too small to be worth eating.

COMMENTS: This rare and beautiful *Volvariella* is easily recognized by its petite dimensions, white color, silky-fibrillose cap, and white pubescent stalk. The inconspicuous white volva and slender stature (see photograph) are also distinctive, and help to separate it from *V. smithii,* which is slightly stockier and has a more prominent, brownish-stained volva. Another widely distributed but rare white species, *V. pusilla (= V. parvula),* is even more petite than *V. hypopithys.* It has a slightly larger white or grayish volva, a non-pubescent stalk, and a striate cap margin in age, and is more frequent in gardens, greenhouses, lawns, and roadsides than in the woods. See also the species under *V. smithii.*

*Just be thankful, as I am, that I didn't measure the spores!

260

Volvariella smithii is a small rare species with a prominent volva. Note size in relation to cypress cone at upper right.

Volvariella smithii   (Smith's Volvariella)

CAP 2-5 cm broad, broadly conical to convex or plane; surface not viscid, usually somewhat fibrillose at the margin; white, but often tinged buff to pinkish-buff at center. Flesh rather soft, white. **GILLS** white becoming pinkish, broad, close, free (at least in age). **STALK** 3-5 cm long, 3-7 mm thick, equal or enlarged at base, white, covered with fine white hairs (pubescent). **UNIVERSAL VEIL** membranous, forming a large, prominent, saclike volva at base of stalk; volva brownish to ochre or ochre-stained, usually lobed. **SPORE PRINT** pinkish to sordid flesh-color; spores 4.5-7 × 3-4 microns, elliptical, smooth.

HABITAT: Solitary or in small groups in soil or humus in woods; known only from the west coast, rare. The specimens in the photograph were growing under pine at New Brighton Beach State Park near Aptos, California, in February.

EDIBILITY: Unknown, but too small and too rare for anyone to care.

COMMENTS: This is one of several small, uncommon, white to grayish Volvariellas. It is also one of several mushrooms to be named after Alexander Smith, the foremost authority on North American mushrooms. The brownish to ochraceous-stained volva, whitish cap, and small size distinguish it. Other small species include: *V. taylori,* widely distributed but rare, with a gray to pinkish-gray cap and a brown to grayish volva; and *V. villosavolva* of eastern North America, with a gray to grayish-brown cap and a white volva that has long mycelial hairs. For small, pure white Volvariellas, see *V. hypopithys.*

Volvariella bombycina   (Silky Volvariella)

CAP 5-20 cm broad, oval becoming bell-shaped or convex and in old age sometimes nearly plane; surface dry and covered with long silky fibrils (usually more coarsely fibrillose at margin), white to yellowish (often palest at margin). Flesh thin and rather soft or flaccid, white. **GILLS** crowded, free, broad, white when young becoming flesh-colored or pinkish as the spores mature. **STALK** 6-20 cm long, 1-3 cm thick, usually tapered upward or enlarged below, often curved, smooth, white, firm. **UNIVERSAL VEIL** membranous, often areolate or scaly, forming a thick, long (deep), saclike volva which sheathes the base of the stalk; volva white to yellowish or dingy brown, often lobed. **SPORE PRINT** pinkish to deep flesh-color; spores 6.5-10 × 4.5-6.5 microns, elliptical, smooth.

HABITAT: Solitary or in small groups on dead hardwoods or fruiting from wounds in living trees; rare but conspicuous, found mostly in warm weather. It is widely distributed east of the Mississippi and has also been found in riparian woodlands in New Mexico and southern California. Elm and maple are among its favorite hosts; it also grows on beech, oak, magnolia, and various other hardwoods. Our rainy season is apparently too cold for it.

EDIBILITY: Edible, and according to reports, choice. I haven't tried it.

COMMENTS: It is easier to identify this striking mushroom than to find it. The silky cap, pale color, deep saclike volva, and growth on wood make it unmistakable. The latter feature plus the pinkish spores (and mature gills) separate it from *Amanita,* which also has a volva. A yellow-capped variety has been found in Florida. Other species: *V. volvacea,* the edible "Paddy Straw Mushroom," sometimes turns up in greenhouses or straw and compost in warmer climates, and is fairly common outdoors along the Gulf Coast. It has a brown or partly brown volva and a gray to dark brown or blackish, often streaked cap. *V. surrecta* is a rare but distinctive northern species that grows parasitically on other mushrooms, particularly *Clitocybe nebularis.* It is small to medium-sized, with a white to grayish cap and an ample volva that is usually lobed. Since its favorite host is common in California, it may occur there rarely.

spores

AMANITACEAE

MEMBERS of this family have white spores, white to pale-colored gills, a universal veil that envelops the young mushrooms, and in most cases, a volva. Many are also furnished with a partial veil and the stalk is cleanly separable from the cap.

There are two genera: In *Amanita* the universal veil is membranous, warty, powdery, or cottony; in *Limacella* it is glutinous, manifesting itself as a layer of slime on the cap and sometimes the stalk.

The Amanitaceae are most likely to be confused with the white-spored Lepiotaceae, which have neither volva nor viscid cap. Microscopically the two families are distinct by virtue of the amyloid or non-amyloid (but not dextrinoid) spores and divergent gill tissue of the Amanitaceae, as opposed to the typically dextrinoid spores and parallel to interwoven gill tissue of the Lepiotaceae (see p. 19).

Amanita is of paramount importance to toadstool-testers because it contains the deadliest of all mushrooms, as well as some of the most delicious and beautiful. *Limacella* is too rare to be of culinary value.

Amanita "eggs" resemble puffballs while still enveloped by the universal veil. However, when sliced open lengthwise (perpendicular to the ground), they reveal the embryonic outline of cap, gills, and stalk (left and center), while a puffball (right) is solid within. Note difference in shape between the deadly poisonous *A. phalloides* "egg" (center) and the edible *A. calyptrata* "egg" (left).

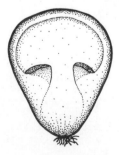
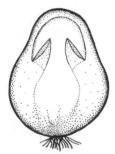
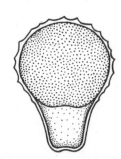

Key to the Amanitaceae

1.   Volva present at base of stalk as a sack, free collar, or series of concentric rings *Amanita,* below
1.   Volva absent or indistinct 2
2.   Cap usually viscid or slimy when moist, without warts or veil material; stalk sometimes slimy
     also .... *Limacella,* p. 291
2.   Cap usually with warts or universal veil material, often dry or slightly viscid but not usually
     slimy; stalk never slimy .... *Amanita,* below

AMANITA   (The Amanitas)

Medium-sized to large *terrestrial* fungi found mostly in woods. CAP *smooth (bald) or with warts or
a cottony patch or other veil tissue.* GILLS *typically white, creamy, yellow, or pale gray, close,*
attached or free. STALK central, usually hollow or stuffed in age and cleanly separable from cap.
PARTIAL VEIL *often present and in most species forming a membranous ring on stalk.*
UNIVERSAL VEIL *present, usually forming a volva at base of stalk in the form of a sack, rim,
collar, or concentric scales.* SPORE PRINT *white.* Spores smooth, amyloid or not amyloid. Gill
tissue divergent, at least when young.

LEARNING to recognize this genus should be an overriding priority for all mushroom
hunters, since Amanitas are responsible for 90% of mushroom-induced fatalities. The
outstanding attributes of any *Amanita*—what makes even a rotten *Amanita* not just
another rotten mushroom, but a rotten *AMANITA* (and therefore worthy of your
attention and respect)—are the white spores, pallid gills, and presence of a universal veil.
The universal veil completely envelops the young mushroom, but breaks as the stalk
elongates, usually forming a **volva** (sack or collar or scales) at the base of the stalk and often
depositing remnants on the cap in the form of a single large piece of tissue (**volval patch**) or
many smaller pieces (**warts).** Obviously, it is important to carefully dig up any unfamiliar
mushroom so as not to miss the volva, if it is present. It is also a good idea to examine the
surrounding soil to be doubly sure pieces of the volva aren't left behind.

   Most Amanitas—including the most dangerous ones—are also furnished with a partial
veil which, upon breaking, often forms a skirtlike ring (annulus) near the top of the stalk.
At one time those species without a partial veil were placed in a separate genus, *Amanitopsis.* A feature emphasized by most mushroom books is that the gills in *Amanita* are free.
This is *not* necessarily the case, however, and this feature is not stressed here.

   The only other genus of agarics consistently equipped with a volva is *Volvariella,* which
has pinkish spores. *Agaricus, Coprinus,* and *Cortinarius* occasionally form a volva, but
have brown to black spores, while *Lepiota* superficially resembles *Amanita* but typically
lacks a volva and nearly always has free gills.

   Amanita is divided into two large groups (subgenera) based on whether or not the spores
are amyloid. Half of the species described here (including the most dangerous ones!) have
amyloid spores. It should be emphasized, however, that some species with non-amyloid
spores are also poisonous. These two large groups are in turn subdivided according to the
type of volva (see p. 264). If the universal veil is **membranous** (skinlike), a loose sack or cup
is formed at the base of the stalk and the cap is usually bald or adorned with a volval patch.
If the universal veil is **friable** (easily crumbling), it manifests itself as a series of concentric
scales or rings around the base of the stem. If the universal veil is semi-friable and
interwoven with the base of the stalk, it will form a collar or free rim (as in *A. pantherina*),
but not a true sack. When the universal veil is friable or semi-friable, numerous pieces of
tissue (warts) are usually deposited on the cap. These warts are typically white, gray, or
yellow and with a few exceptions (e.g., *A. magniverrucata*) are readily removable—unlike
the colored scales of an *Agaricus* or *Lepiota.* In fact, they are often washed off by rain.

   In some Amanitas, such as *A. rubescens* and *A. silvicola,* a distinct volva is not formed.

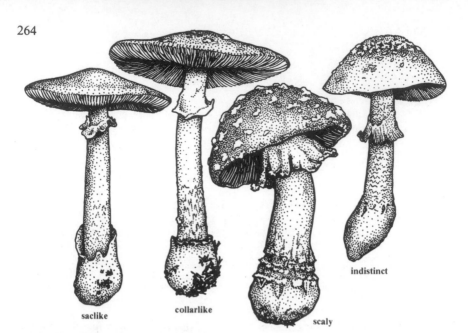

saclike   collarlike   scaly   indistinct

Different types of volvas in *Amanita.* Left to right: *A. phalloides, A. pantherina, A. muscaria, A. rubescens.*

However, vestiges of the universal veil on the cap (in the form of warts or cottony tissue) signify *Amanita*. In the rare instances in which neither volva nor universal veil remnants are visible, Amanitas can still be recognized by their uncanny "*Amanita* aura." They are so unequivocally elegant and graceful that you quickly learn to tell an *Amanita* without having to dig it up!

Amanita is a study in antithesis. At one extreme are the most poisonous of all mushrooms—the death cap and destroying angels *(A. phalloides, A. ocreata, A. virosa,* etc.). Every fungophile should learn the telltale signs of these deadly fungi (for more details, see pp. 892-893). At the other end of the spectrum are three of the most exquisitely flavored of the fleshy fungi—*A. caesarea, A. calyptrata,* and *A. velosa.* The rest of the Amanitas fall somewhere between these extremes: several are hallucinogenic and/or poisonous but not normally fatal (*A. muscaria* and *A. pantherina*); others are edible but scarcely incredible (*A. pachycolea* and *A. rubescens*); still others are of unknown edibility (*A. aspera* and *A. magniverrucata*).

I for one do *not* subscribe to the wholesale philosophy (as expounded by many mushroom mentors) that Amanitas should not be eaten under any circumstances. In my humble fungal opinion, it is just as easy to carelessly overlook the volva and mistake a deadly *Amanita* for an edible mushroom of another genus as to mistake a deadly *Amanita* for the coccora *(A. calyptrata)* or grisette *(A. vaginata).* True, *it is sheer stupidity to risk your life for the sake of a single meal, however delectable it may be.* But the key word here is *risk*—and in the case of a few species such as *A. calyptrata, A. caesarea,* and *A. vaginata,* I don't consider it a risk for *discriminating* amateurs to eat them, provided they become thoroughly familiar with their characteristics and those of their lethal counterparts. Simplistic slogans or catchwords such as "Do not eat-a the *Amanita*" often accomplish the precise opposite of what they intend. Rather than encouraging people to use their eyes and nose and the gray mass between their ears, to approach each and every mushroom with discrimination, intelligence, and respect, such adages reinforce people's desire for expediency by fostering an unhealthy, mindless reliance on shortcuts and glib generalizations. Those who need simple rules should learn how to play dominoes or Scrabble rather than eat wild mushrooms. Adages such as the above can even be misconstrued to read: "If a mushroom isn't an *Amanita* it won't kill you"—a dangerous assumption!

Too many people eat and enjoy edible Amanitas for me not to recommend them. But at the risk of being redundant, let me reiterate some rules of the trade. Unless you are ABSOLUTELY, INDISPUTABLY, and IRREFUTABLY sure of your Amanita's identity, don't eat it! (The one adage with which I wholeheartedly concur is: "When in doubt, throw it out!"). If possible, have an experienced collector verify your identification, and collect the species several times before venturing to eat it. Above all, *don't rely on a single characteristic* (such as striate vs. non-striate margin—see photo on p. 286) to distinguish between edible and deadly poisonous species. Each individual mushroom is subject to a different set of environmental and genetic factors—therefore each will be slightly different. Only by using a *combination* of critical characteristics can you rest assured that you have a savory coccora or grisette instead of a death cap or destroying angel. Furthermore, *don't* assume that two or more Amanitas growing together are the same species. Judge each and every mushroom on its own merits. Finally, always keep in mind the possibility of encountering a species not described in this book—or any other book!

My reason for lecturing on the Amanitas at such length is that they never fail to attract attention and admiration. You certainly needn't eat them to enjoy them, for they are among the most beautiful and graceful of all fungi, the epitome of impeccability and elegance.

The fly agaric *(A. muscaria),* with its fiery red cap and white "stars" is the most spectacular example, of course. Down through the ages it has been compared to bull testicles and male genitalia and worshipped as the earthly incarnation of infinity, divinity, and virility (more for its appearance, I suspect, than for its properties). It is one of the commonest mushrooms of our pine forests, yet one never tires of finding it. The variation in color, size, shape, and "constellations" is such that each and every one presents a new and deliciously different feast for the eye. It's hard to resist taking one or two home to show off to impressionable neighbors or friends, but never is the ephemerality of life so emphatically underscored as when they come over the next day to pay their respects, only to find a writhing mass of beatific maggots where your blazing incarnation of the cosmos had been!

In contrast to the flamboyant splendor of the fly agaric is the subdued and radiant warmth of the coccora *(A. calyptrata)*. I say "warmth" because the huge eggs are so soft and cottony that they look positively warm inside. Finding a family of them in rich red madrone humus is like stumbling onto the nest of a rare and secretive woodland bird. And watching a coccora "hatch"—the round, orange head emerging from its cottony cocoon— is like watching the sun rise from a blanket of clouds, a quietly inspiring reaffirmation of life best experienced alone.

Most Amanitas are mycorrhizal. As a result, they are most common in the woods or near trees. Some, however, grow in grass or open ground. *Amanita* attains its greatest diversity in the warm temperate zone. In the southeastern United States, for instance, there is a very diverse and bewildering *Amanita* flora that is beyond the scope of this book. The west coast has fewer species but several are endemic. In our region two distinct floras can be recognized. The first, comprised of northern species, occurs primarily with madrone and conifers in the late fall and winter. *A. muscaria, A. calyptrata,* and *A. silvicola* are prominent examples. The second group has a more southerly distribution and is associated with oak. Some members of this group fruit in the fall (e.g., *A. phalloides*), others in the late winter and spring (e.g., *A. velosa, A. rubescens, A. ocreata*—see photo on next page).

Amanita is a far more diverse genus than once thought. Despite a wealth of studies conducted on the genus, no all-encompassing monograph has been published. (A volume by David Jenkins is in press.) Particularly perplexing are the "Lepidellas": those species with amyloid spores and a friable universal veil that leaves remnants on the cap margin. Several hundred species of *Amanita* occur in North America, including many poorly known or unnamed ones; California has over 25 species. Twenty Amanitas are described here and many others are keyed out. Their elegance and individuality make them a fascinating and rewarding group to study, even for those not armed with a microscope. If your *"Amanita"* doesn't key out convincingly, try *Cystoderma, Armillaria,* and *Lepiota.*

Three species of *Amanita* that commonly fruit in the spring in association with live oak: *Amanita velosa* (two at left), a small specimen of *A. ocreata* (top), and *A. rubescens* (three at right).

Key to Amanita

1. Volva saclike (i.e., forming a true sack that sheathes base of stalk as shown on p. 264); cap usually bald or with a cottony or membranous patch of universal veil tissue or occasionally with several patches or non-friable warts 2
1. Volva collarlike (i.e., intergrown with base of stalk but with a free rim), scaly, warty, powdery, or indistinct but *not* saclike (see p. 264); cap often with many small pieces of universal veil tissue (warts), powder, etc., occasionally with larger pieces 15

2. Volva *tough*, thick, large; cap and/ or stalk often shaggy, fibrillose, or with cottony patches of veil tissue; cap white or tinged brown (especially at center), often bruising brown or reddish, margin often striate in age; stalk similarly colored; annulus (ring) absent; spores *oblong or elliptical, amyloid;* fairly common in eastern North America, rare in West   *A. volvata* & close relatives
2. Not with above features 3

3. Margin of cap distinctly striate (at least when mature); spores *not* amyloid 4
3. Margin of cap not striate or only faintly so (occasionally striate in age); spores amyloid   .. 12

4. Partial veil present when young, usually (but not always!) forming an annulus (ring) on stalk 5
4. Partial veil and annulus absent or rudimentary (but stalk sometimes scaly) 8

5. Gills and stalk yellow to yellow-orange; cap bright red to orange (but may fade to yellow or paler in age or sunlight) .... *A. caesarea group* & others, p. 284
5. Not as above; gills and stalk typically white to creamy or very pale yellow 6

6. Volva often small and inconspicuous; cap brown to gray or sometimes nearly white; growing in mixed woods and under hardwoods in eastern North America .... *A. spreta*
6. Not as above; known only from western North America 7

7. Cap brown to yellow or whitish, but if brown then usually with a yellow margin; partial veil typically (but not always!) forming a prominent skirtlike annulus on stalk   *A. calyptrata,* p. 284
7. Cap variously colored, but the margin not yellow; annulus (ring) usually pressed closely to the stalk or poorly defined 48

8. Cap dark brown to gray or grayish-brown 9
8. Cap white to pale tan, beige, orangish, pinkish, orange-brown, or reddish-brown 10

9. Fruiting body medium-sized to large; cap dark gray to dark brown when young, often paler in age and often developing a darker band near inner edge of striations; gill edges usually brown; known only from the West .... *A. pachycolea,* p. 290
9. Fruiting body medium-sized to rather small and slender; cap usually gray, but sometimes grayish-brown or brown; gill edges not brown; widely distributed 49

10. Cap white with *long* striations; widespread, but rare in West   *A. alba* (see *A. vaginata,* p. 288)
10. Not as above; if cap white then with shorter striations and usually found near western oaks in winter and spring (or even summer); common 11

11. Cap pinkish-tan to orangish, beige, or paler .... *A. velosa* & others, p. 286
11. Cap orange-brown to reddish-brown, tawny, etc.; common in eastern North America, infrequent in West .... *A. fulva,* p. 287

12. Cap greenish to yellow-green, brownish-olive, grayish-olive, or nearly white when young, often duller (dingy tan, etc.) or with a metallic luster in age .... *A. phalloides,* p. 269
12. Not as above; cap usually white or whitish when fresh (but may discolor by maturity) ... 13

13. Cap white, but discoloring pinkish, brownish, or yellowish (at least centrally) in age; associated with oak; found in California and the Southwest and Texas ...... *A. ocreata,* p. 271
13. Cap usually remaining white or found elsewhere 14

14. Cap white; partial veil usually forming a distinct annulus (ring) on stalk (which may disappear in age!); very common in eastern North America, also found in the Pacific Northwest .... *A. virosa* & others (see *A. ocreata,* p. 271)
14. Not as above; partial veil absent or evanescent; found in eastern North America (do not eat!) .... *A. peckiana* & others

15. Universal veil remnants yellow to grayish-yellow (check cap for warts and base of stalk for volva); cap *not* whitish 16
15. Universal veil remnants not yellow; cap may or may not be whitish 22

16. Partial veil absent; gills yellow; found in eastern North America .... *A. parcivolvata* (see *A. caesarea,* p. 284)
16. Not as above 17

17. Cap bright red to orange-red (but may fade in age); stalk white .... *A. muscaria,* p. 282
17. Cap orange to yellow, yellow-brown, or dark brown; stalk white or yellow 18

18. Lower stalk sheathed with shaggy scales .... (see *Armillaria* & Allies, p. 189)
18. Not as above 19

19. Cap salmon to salmon-pink when fresh; found in mountains of eastern U.S. ..... *A. wellsii*
19. Not as above 20

20. Base of stalk staining reddish in age or where bruised; found in eastern North America .... *A. flavorubescens* (see *A. aspera,* p. 278)
20. Not with above features (if staining as above, then found in West) 21

21. Cap yellow to yellow-brown to dark brown; stalk usually white; common along the Pacific Coast .... *A. aspera,* p. 278
21. Cap yellow-orange to yellow; stalk often colored similarly; fruiting body rather small; common in eastern North America, rare in the West .... *A. flavoconia,* p. 278

22. Cap pale yellow-green to pale yellow to nearly whitish with thin grayish, whitish-buff, or pinkish to lavender-gray warts (which may wash off); cap margin *not* striate; stalk with an abrupt, soft, rounded bulb at base; spores amyloid; common in eastern North America, especially under hardwoods but also with conifers .... *A. citrina* (see *A. porphyria,* p. 279)
22. Not as above 23

23. Cap brown to olive-brown or paler; stalk lacking grayish patches, terminating in an abrupt basal bulb that is usually split or chiseled longitudinally; flesh usually staining reddish-brown; spores amyloid; common in eastern North America (especially under hardwoods); also reported from the Pacific Northwest (but rare) .. *A. brunnescens* (see *A. porphyria,* p. 279)
23. Not as above 24

24. Cap brightly colored (red, orange, or yellow); partial veil present, usually forming an annulus (ring) on stalk 25
24. Not as above (cap may be sordid reddish or reddish-brown, pinkish-tan, etc.) 26

25. Volva usually a series of concentric rings at apex of bulbous stalk base, but sometimes only a single ring or collar; cap medium-sized to large and bright red to orange, apricot, yellow-orange, or yellow (yellow form rare in coastal California, but common in the Sierra Nevada and most of eastern North America) .... *A. muscaria,* p. 282
25. Volva usually a single collar at top of basal bulb or often indistinct, but sometimes consisting of several rings; cap small to medium-sized (occasionally large), usually pale yellow (but sometimes brighter), at times completely covered by veil material; widespread *A. gemmata,* p. 281

26. Partial veil absent (check young specimens if possible); cap margin distinctly striate; spores not amyloid 27
26. Partial veil present, or if absent, then cap margin *not* striate; spores amyloid or not 31

27. Cap gray to grayish-brown, brown, or darker 28
27. Cap pale yellow, orange-buff, salmon, pinkish, tan, or nearly white 30

28. Cap powdery-mealy; volva if present also mealy *A. farinosa* (see *A.* sp. (unidentified), p. 275)
28. Not as above 29

29. Cap gray or sometimes grayish-brown, with or without warts; upper limb of volva usually well-developed (but falls off easily); common in California . . . . *A. constricta,* p. 289
29. Cap gray to brown to dark brown to nearly black, usually with warts; upper limb of volva not well-developed; widespread, but rare in California . . . *A. inaurata* (see *A. constricta,* p. 289)

30. Cap yellow to creamy to whitish; volva usually collarlike (with free rim) . *A. gemmata,* p. 281
30. Cap orange-buff to pale pinkish-orange to pinkish-tan, beige, or sometimes whitish; volva not typically collarlike; often growing in the open (but near trees) . . . . *A. velosa,* p. 286

31. Cap entirely brown when young, breaking up into large brown scales in age; flesh in stalk usually staining orange or saffron (and eventually reddish) when cut   (see *Lepiota rachodes,* p. 297)
31. Not as above 32

32. Cap with erect, often pyramidal brown warts which usually come off easily; stalk and/or underside of veil with similar warts; spores usually dextrinoid, *not* amyloid   . (see *Lepiota,* p. 293)
32. Not with above features (but may have some of them); common 33

33. Some part of fruiting body usually with sordid reddish stains (especially the stalk); flesh *slowly* staining dingy reddish when bruised or cut; maggot tunnels also reddish *A. rubescens,* p. 276
33. Not as above 34

34. Fresh fruiting body white or whitish (but may age buff, yellowish, brownish, pinkish, etc.) 35
34. Not as above; cap distinctly colored or with colored veil material even when young 42

35. Stalk terminating in a fairly conspicuous bulb, *not* typically with a rooting portion below the bulb 36
35. Stalk without bulb, or if with a bulb, then also with a tapered rooting base below the bulb   40

36. Volva typically present as a distinct free rim (collar) or series of concentric rings at apex of basal bulb; spores *not* amyloid 37
36. Not as above; spores amyloid 38

37. Volva typically consisting of a single tight-fitting collar around bulb apex; cap often tinged yellowish or brownish at center; often rather slender *A. cothurnata* (see *A. pantherina,* p. 280)
37. Volva usually a series of concentric rings; cap white to grayish-white or tinged buff; not unusually slender . . . . *A. muscaria,* p. 282

38. Cap surface with rather soft and cottony universal veil tissue, lacking conspicuous warts; known only from western North America . . . . *A. silvicola,* p. 273
38. Not as above (if cap cottony, then found elsewhere) 39

39. Cap usually with brown warts; stalk often rather stout (up to 8 cm long); known only from California, associated with live oak . . . . *A. sp. (unidentified)* (see *A. rubescens,* p. 276)
39. Not as above 40

40. Cap without warts or warts obscure; growing in sand . . . . *A. baccata,* p. 273
40. Not as above; cap usually with distinct, well-developed warts 41

41. Cap covered with large, exaggerated warts; fairly common with oak and pine, known only from California . . . . *A. magniverrucata,* p. 274
41. Found elsewhere, or if found in California then warts smaller and often concentrated at center of cap . . . . *A. cokeri* & many others (the "Lepidellas") (see *A. magniverrucata,* p. 274)

42. Stalk arising from a well-developed cylindrical to jug-shaped, sometimes hollow, underground "tuber"; rare . . . . (see *Squamanita,* p. 197)
42. Not as above (but stalk may root deeply or have a bulbous base) 43

43. Cap yellow to creamy, the margin usually striate or tuberculate-striate . . *A. gemmata,* p. 281
43. Not with above features 44

44. Volva indistinct, powdery, or scaly; cap margin not normally striate 45
44. Volva usually present as a free rim or collar on basal bulb *and/or* margin of cap striate in age 47

45. Found in California 46
45. Found in eastern North America (especially common in Southeast) 50

46. Usually found in open ground (pastures, etc.); partial veil often disappearing; cap grayish-white to gray to brownish-gray, small . . . . *A. sp. (unidentified),* p. 275
46. Associated with oak; partial veil usually forming a persistent, prominent annulus; cap white to brownish . . . . *A. sp. (unidentified)* (see *A. rubescens,* p. 276)

47. Warts gray *and/or* stalk gray or with grayish patches; spores amyloid . . *A. porphyria,* p. 279
47. Stalk white; warts usually white or pallid; spores not amyloid . . . . *A. pantherina,* p. 280

48. Cap brown .... *A. calyptratoides* (see *A. calyptrata*, p. 284)
48. Cap orangish to salmon, pinkish-tan, buff, or even whitish .... *A. velosa,* p. 286
49. Lower part of volva tightly constricted around stalk, the upper part flaring outward (see photo on p. 289) .... *A. constricta,* p. 289
49. Not as above; volva saclike .... *A. vaginata,* p. 288
50. Cap pallid beneath a thin coating of pinkish-tan to brownish-orange universal veil material and often with several large chunky warts at center; found in woods in Southeast . *A. roseitincta*
50. Not as above 51
51. Cap distinctly pinkish when fresh; usually found in open; southern .... *A. salmonea*
51. Not as above 52
52. Cap with concentrically arranged brown to grayish-brown universal veil remnants (scales); partial veil evanescent; known only from the Southeast .... *A. hesleri*
52. Not as above; cap usually grayish or with grayish warts 53
53. Cap pale brown to dark grayish-brown, with large grayish warts or patches; partial veil usually forming a distinct annulus (ring) on stalk .... *A. spissa* (see *A. pantherina*, p. 280)
53. Not as above; cap grayish .... *A. onusta & A. cinereoconia* (see *A.* sp. (unidentified), p. 275)

Amanita phalloides   (Death Cap)   Color Plate 50

CAP 4-16 cm broad, at first nearly oval, then convex to plane; surface smooth, viscid or tacky when moist, often shiny when dry or with a metallic luster; color variable: green to brownish-olive, yellow-green, yellowish, or sometimes white, often darker toward center and paler at margin, and often fading in age (to grayish-green, light brown, olive-buff, dull yellowish, etc.); sometimes with one or more patches of thin, silky, white universal veil tissue; margin typically *not* striate. Flesh white; odor at first mild, but later quite pungent or nauseating (like raw potatoes or chlorine). GILLS adnate to adnexed or free, close, white or tinged faintly greenish. STALK 5-18 cm long, 1-3 cm thick, tapering upward or equal with an enlarged base; white or tinged cap color, smooth or with minute scales and fibrils; solid or hollow. PARTIAL VEIL membranous, white or tinged yellow-green, forming a persistent but fragile, superior, skirtlike ring which may disappear in age. UNIVERSAL VEIL membranous, white, forming a saclike volva that sheathes base of stalk; volva thin and rather fragile, usually buried in ground and sometimes disintegrating. SPORE PRINT white; spores 7-12 × 6-9 microns, broadly elliptical to nearly round, smooth, amyloid.

Amanita phalloides, showing the classical features of the deadly Amanitas: presence of a sack (volva) at the base of the stalk and a skirt or ring (annulus) near the top of the stalk, plus white or whitish gills. (Don't expect every specimen to be so "classical," however!)

Fruiting body development in the deadly *Amanita phalloides*. At first it is enclosed by the universal veil, which ruptures to form a volva at the base of the stem as the stalk elongates.

HABITAT: Solitary, scattered, or in groups or troops in woods or on lawns near trees; widely distributed. In our area it is very common under live oak in the fall and early winter and may even turn up in the summer if moisture is sufficient. In the late winter and spring, however, it is largely supplanted by *A. ocreata*. Perhaps an adventitious (but hardly advantageous!) introduction from Europe, it has taken a fancy to our native oaks and spread like the plague, so that it is now the most abundant *Amanita* of our live oak woodlands. Heavy fall rains often elicit a stupendous crop—I have counted as many as one hundred specimens under a single oak! I've never seen it growing without live oak in the vicinity, but in southern Oregon it grows with other oaks and in eastern North America it has turned up in numerous localities under conifers as well as hardwoods. Apparently it can form mycorrhiza with a wide range of hosts!

EDIBILITY: DEADLY POISONOUS! Learn to recognize this species before eating any mushroom with gills! It is particularly dangerous because the symptoms are delayed, not appearing for from 6 to 24 hours after ingestion, by which time there is relatively little modern medicine can do except to treat the victim symptomatically. In the last decade there have been one or two deaths in California every time a bumper crop has appeared. However, none of the victims were knowledgeable fungophiles, let alone "mushroom experts" (as they are often called by the press). The flavor, incidentally, is described by survivors as excellent—despite the awful odor which develops in old specimens. For an account of symptoms and treatment, see pp. 892-893.

COMMENTS: There is no *rational* reason why anyone should mistake the death cap for an edible mushroom—but since when were human beings completely rational? The telltale signs are: (1) white gills (2) white spores (3) partial veil covering the gills, then breaking to form a skirtlike ring or annulus near the top of the stalk (4) membranous white sack (volva) at the base of the stalk (5) margin of the cap not striate (however, I have seen mature specimens with small striations on the margin where the cap tissue had apparently

270

Maturing specimens of *Amanita phalloides*. Note how cap opens out and the partial veil breaks to form an annulus (ring) high up on the stalk. Cap color is variable: green, yellow, brownish, even white. The cap is usually bald, but sometimes has a thin white patch of universal veil tissue (as shown here).

collapsed against the gills!). The ring is sometimes obliterated and the volva can be carelessly overlooked, so just to be safe, don't eat mushrooms with any two of these characteristics unless you are *absolutely sure* what they are. Your life is at stake! The cap color—usually greenish, but extremely variable—and the pungent odor in age (it reeks of death) plus the association with live oak (in our area) are good secondary field-marks. In eastern North America, *A. phalloides* can be confused with *A. citrina* and *A. brunnescens* (see the key to *Amanita* and comments under *A. porphyria*).

Amanita ocreata   (Destroying Angel; Death Angel)

CAP 4-15 cm broad, nearly round or oval becoming convex and finally plane; surface viscid when moist but soon dry, smooth, white when young but in age often discoloring pinkish, buff, yellowish, or brownish, especially toward the center; sometimes with a very thin white patch of universal veil tissue; margin usually (but not always!) *not* striate. Flesh thick, firm when young, white; odor mild, becoming disagreeable in old age. **GILLS** at first adnate or adnexed, sometimes free in age, close, white. **STALK** 6-20 cm long, 1-3 cm thick, tapered upward or equal with an enlarged base, white, often finely powdered or scaly at apex or occasionally throughout; hollow or solid. **PARTIAL VEIL** membranous, white, forming a very fragile superior or apical skirtlike ring or shredding into pieces or disappearing entirely. **UNIVERSAL VEIL** membranous, white, forming a saclike volva that sheathes the base of the stalk; volva often large and ample, but sometimes thin. **SPORE PRINT** white; spores 9-14 × 7-10 microns, broadly elliptical to nearly round, smooth, amyloid. Flesh turning bright yellow in KOH (potassium hydroxide).

HABITAT: Solitary to widely scattered or in small groups on ground under oaks, known from Marin County in California east to the Sierra Nevada foothills and south to Arizona and Texas. In our area it is associated with live oak and is common in the winter and spring,

Amanita ocreata. This deadly poisonous *Amanita* is pure white when young but usually discolors (pinkish, brownish, ochre, etc.) on the cap as it ages. Note the fragile partial veil, voluminous volva, and white gills. See Color Plate 53 for its eastern counterpart.

usually after *A. phalloides* has finished fruiting. It is quite numerous, but doesn't normally produce the huge crops typical of the latter species. Related "destroying angels" are abundant under hardwoods and conifers in eastern North America, and to a lesser extent, the Pacific Northwest (see comments).

EDIBILITY: DEADLY POISONOUS! It doesn't enjoy the same notoriety as *A. phalloides,* but is just as dangerous, and in some regions (such as southern California), it is the more common of the two. It recently caused several deaths near San Diego. The victims were apparently starving illegal aliens who ate the mushrooms out of desperation.

COMMENTS: This elegant, pristine-pure, lethal-looking *Amanita* is our only white mushroom with both an annulus (ring), a true sack (volva) at the base of the stalk, and a non-striate (or only rarely striate) cap margin. The species epithet *ocreata* means "sheathed," a reference to the voluminous volva which may extend as much as halfway up

Amanita ocreata varies greatly in stature, as shown here. **Left:** A robust specimen that could easily be mistaken for the pale form of *A. calyptrata* (and is just as big). **Right:** A slender, graceful mature specimen that could be mistaken for a washed-out *A. velosa.*

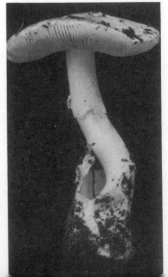

the stalk, though it is often much smaller. The partial veil is quite fragile, however, and often turns to shreds rather than forming a distinct ring (annulus). The sinister name "destroying angel" also embraces three closely related, deadly poisonous, pristine-white "veiled threats"—*A. virosa* (COLOR PLATE 53), *A. vernu,* and *A. bisporigera.* All three are common in eastern North America — especially under hardwoods — and have been reported from the Pacific Northwest, but not from California. They closely resemble *A. ocreata,* but are often more slender and do not discolor as much in age; also, *A. verna* supposedly doesn't yellow in KOH. *A. virosa* is distinguished by its round spores and scalier stalk, while *A. verna* has elliptical spores and *A. bisporigera* has 2-spored basidia and round spores. *A. alba* and the pale (white to creamy) form of *A. calyptrata* are easily confused with these species, but always have a striate cap margin and non-amyloid spores. Be *very* careful—*A. ocreata* sometimes has a striate cap margin (see photo on p. 286)! Other species: *A. mutabilis* is an eastern species with a white or pinkish-tinged cap. It has a saclike volva and amyloid spores, stains pinkish when bruised, and often smells like anise.

A manita silvicola   (Western Woodland Amanita)

CAP 5-12 cm broad, convex to plane; surface dry or slightly viscid when moist, white (but occasionally discolored in age), covered with flattened cottony or fluffy-powdery universal veil tissue; margin often hung with veil remnants and extending beyond the gills, not striate. Flesh white; odor mild or slightly soapy. GILLS white, close, adnate to adnexed or free, edges finely powdered or cottony. STALK 5-12 cm long, 1.5-2.5 cm thick, usually rather stout, terminating in a basal bulb which is up to 5 cm broad; white (or sometimes brownish-stained), usually powdery or with cottony scales. PARTIAL VEIL white, delicate, forming a slight ring or fibrillose zone or disappearing entirely. UNIVERSAL VEIL cottony, white, forming a scaly or indistinct volva consisting of cottony white zones or patches at base of stalk which often disintegrate or remain in the ground. SPORE PRINT white; spores 8-12 × 4.5-6 microns, elliptical, smooth, amyloid.

HABITAT: Solitary or in small groups in mixed woods and under conifers (e.g., Douglas-fir); known only from the Pacific Northwest (where it is fairly common in the fall, particularly in campgrounds) and California. I have found it only once in our area, in December.

EDIBILITY: Unknown. Do not experiment!

COMMENTS: This attractive *Amanita* can be told at a glance by its cottony white cap, rather stout stature (for an *Amanita*), and enlarged stem base (see photo on next page). The cap lacks the prominent warts characteristic of *A. magniverrucata, A. cokeri,* etc., and the volva is not saclike as in the destroying angels *(A. ocreata, A. virosa,* et al). Also, the stalk lacks the tapered, pointed, rooting base so characteristic of *A. baccata.* The attached gills and indistinct volva suggest *Armillaria,* but it has that ineffable *Amanita* "aura." Also see *A. smithiana* (under *A. magniverrucata*), a larger version of *A. silvicola.*

A manita baccata   (Sand Amanita)   Color Plate 54

CAP 4-10 (12) cm broad, convex becoming plane; surface dry or slightly tacky, white (but often dirty and sometimes discolored buff in age), with obscure mealy to powdery warts or flattened universal veil remnants which may disappear in age; margin often hung with veil remnants, not striate. Flesh white, soft, fragile; odor mild or slightly pungent. GILLS close or crowded, white, becoming dingy yellowish in age, usually adnate or adnexed, but sometimes free. STALK 5-18 cm long, 0.5-2.5 cm thick, with a long, tapered, often pointed rooting base below a slight to distinct bulb; dry, white, sometimes with yellowish or buff stains; usually somewhat fibrillose or with delicate ragged or powdery scales; apex often striate. PARTIAL VEIL white, cottony, very fragile; disappearing or forming a thin, poorly defined superior ring on stalk. UNIVERSAL VEIL friable, white, forming

Left: *Amanita silvicola* lacks the saclike volva of *A. ocreata* and the tapered rooting base of *A. baccata*. **Right:** *Amanita baccata* grows in sand; note pointed rooting base (see color plate for view of cap).

an indistinct volva or a scaly zone just above the bulb. **SPORE PRINT** white; spores 10-15 × 4-6.5 microns, more or less cylindrical, smooth, amyloid.

HABITAT: Solitary or in small groups in sand or sandy soil (often buried), associated with oak and/or pine; distribution spotty—reported from southern Europe, northern Africa, and Michigan—frequent in our area in the fall and winter, but never in large numbers.

EDIBILITY: Unknown.

COMMENTS: Also known as *A. boudieri,* this is an inelegant but oddly charming *Amanita.* The tapered, rooting stem base, soft flesh, absence of prominent warts on the cap, cylindrical spores, and habitat in sand form a most distinctive combination of features. The cap may scarcely poke above the ground and along with the rooting base, is usually covered with dirt or sand (see photo). The fruiting body decays rapidly and is hard to keep in one piece. For these reasons and many more, it is one of my favorite Amanitas.

Amanita magniverrucata   (Pine Cone Amanita)   Color Plate 55

CAP 7-20 (30) cm broad, nearly round becoming broadly convex or plane; surface dry or slightly viscid when moist, covered with large (up to 2 cm broad and 1 cm high!), persistent, strongly-attached warts, the warts pyramidal at first becoming truncated or more flattened in age; color white to creamy-white becoming yellowish-buff or tan in old age (sometimes with darker stains); margin usually with cottony veil remnants and extending beyond the gills, not striate. Flesh thick, firm, white; odor unpleasant in age (like chlorine or dirty socks). **GILLS** adnate to adnexed or free, close, white or creamy, delicately powdered. **STALK** 7-12 (20) cm long, 1-4 cm thick at apex, rooting deeply in ground; equal above or tapering upward from a thicker base or bulb (up to 6 cm broad but sometimes inconspicuous), the rooting portion below the bulb tapered downward; white throughout or with brownish to yellowish-buff stains; firm, rather tough. **PARTIAL VEIL** white, membranous, usually forming a fragile, superior, skirtlike ring on stalk. **UNIVERSAL VEIL** warty and friable, forming a scaly volva consisting of concentrically arranged rows of warts or scales at apex of bulb, these sometimes disappearing in age. **SPORE PRINT** white; spores 8.5-12.5 × 5.5-8.5 microns, elliptical to nearly round, smooth, amyloid.

HABITAT: Solitary to gregarious under live oak and pine; known only from California, but related species (see comments) are widespread. In our area it is not uncommon in the fall, winter, and spring. It develops slowly and persists for weeks without decaying.

Left: *Amanita magniverrucata,* top view of cap. Note the exaggerated warts. **Right:** *Amanita cokeri* is a somewhat similar whitish species with smaller warts on the cap.

EDIBILITY: Unknown—do not experiment! It belongs to the "Lepidellas," a subgroup of *Amanita* that contains both poisonous and harmless species.

COMMENTS: One of our most spectacular mushrooms—the large, erect warts on the cap set it apart, making it look like a white pine cone or a glob of meringue (buttons resemble the Sierran puffball, *Calvatia sculpta*). It is one of several large, white, warty Amanitas that for years have passed under the names *A. strobiliformis* and *A. solitaria.* However, our species is quite distinct because of its exaggerated warts, and has recently been rewarded with a name of its own. It has several large counterparts in eastern North America with an unpleasant, often chlorine-like odor and more fragile warts, including: *A. chlorinosma,* *A. polypyramis* (especially large), and *A. rhopalopus* (with a large, deeply rooting base); *A. ravenelii,* with brownish imbricate (shingled) warts or scales, and *A. daucipes,* which often has a pinkish- or orange-tinged stalk. *A. cokeri* is a common eastern oak- and pine-loving species that also occurs in California. It has smaller (to 5 mm broad) but firmly-attached warts mainly at the center of the cap, a mild odor, a rooting bulb on the stalk, and an annulus (ring) and volva (see photo above). *A. smithiana* grows under conifers in the Pacific Northwest. It looks something like a matsutake or large *A. silvicola* with its poorly developed cottony warts, unpleasant odor, and ragged or shaggy stalk with an enlarged but scarcely pointed base. There are many other closely related white or pallid species (the "Lepidellas"), particularly in southeastern North America (e.g., *A. abrupta,* with a broad, abrupt basal bulb; *A. longipes,* usually odorless with a pointed, often flattened, rooting bulb; *A. thiersii,* with a shaggy stalk; *A. atkinsoniana* and *A. cinereopannosa,* with a grayish veil; and *A. praegraveolens,* with a whitish to pinkish-tan, scaly cap and non-bulbous stalk). "Lepidellas" grow on lawns as well as in the woods, and are often difficult to identify without a microscope. They should not be eaten.

Amanita sp. (unidentified) (Anonymous Amanita)

CAP 2.5-5 (8) cm broad, convex becoming plane; surface dry, whitish to gray, covered with darker (gray to brownish-gray) mealy or powdery warts which are easily obliterated; margin not striate. Flesh white, odor rather pungent in age. **GILLS** usually adnate or adnexed (rarely free), close, creamy-white becoming dingy yellowish or yellowish-orange in old age. **STALK** 2-4 (7.5) cm long, 0.4-1 cm thick, equal or tapering downward, sometimes with a short, rooting base or swollen slightly above the base; white and striate above the ring, dingy whitish or tinged cap color below and somewhat scaly. **PARTIAL VEIL** membranous but very fragile, disappearing or forming a superior, median, or even basal ring on stalk. **UNIVERSAL VEIL** friable, mealy-powdery, forming an indistinct or scaly volva in the form of obscure grayish scales or powdery-mealy warts over lower portion of stalk. **SPORE PRINT** white; spores 7-11 × 5-8.5 microns, elliptical, smooth, amyloid.

Amanita sp. (unidentified). An odd, unimposing species that is locally abundant in pastures and open ground. Note the small size, stocky stature, and grayish warts on the cap. The fragile partial veil and volva are easily obliterated.

HABITAT: Scattered to gregarious in pastures, open fields, hard-packed ground, under trees, etc. Fairly common in our area in the fall and winter and sometimes very abundant. It usually grows in the open, often mingling with *Agaricus campestris* and *Agaricus cupreobrunneus,* which it superficially resembles. Apparently it is not mycorrhizal.

EDIBILITY: Unknown. Do not experiment!

COMMENTS: This anomalous, anonymous *Amanita* is quite unamanitalike with its short stem, compact mealy-warty, grayish cap, and predilection for growing in meadows. The volva is so fragile that it may disappear, but the warts on the cap signify *Amanita*. *A. farinosa* is a similar species (cap small, powdery-mealy, gray to brownish-gray, etc.), but has non-amyloid spores, a striate cap, no partial veil, and grows in forests or at their edges. It is widespread but rare in the West. Two similar grayish species, *A. cinereoconia* and *A. onusta,* are larger and apparently restricted to eastern North America; the latter usually grows in sandy soil.

Amanita rubescens   (Blushing Amanita; The Blusher)

CAP 4-12 (20) cm broad, convex becoming plane or shallowly depressed; surface slightly viscid or dry, at first covered with white, pinkish, brownish, or grayish warts, the background white at first, then flushed sordid reddish, pinkish, reddish-brown, brown, etc.; margin usually not striate. Flesh firm, white, *slowly* reddening when bruised; odor mild, taste mild or latently bitter. **GILLS** adnate to adnexed when young, sometimes free in age; close, white or pallid, sometimes stained reddish. **STALK** 5-14 (20) cm long, 1-3.5 cm thick, equal or enlarged downward to a swollen base (bulb); at first white, soon stained sordid reddish, reddish-brown, or pinkish below the ring and often somewhat scaly; white or tinged pinkish above. **PARTIAL VEIL** membranous, white or tinged reddish, forming a fragile, superior, skirtlike ring on stalk. **UNIVERSAL VEIL** friable, forming an indistinct or scaly volva (i.e., disappearing or leaving sordid reddish scaly zones at base of stalk). **SPORE PRINT** white; spores 7.5-10.5 × 5-7 microns, elliptical, smooth, amyloid.

HABITAT: Solitary to scattered or in groups in woods and under trees, partial to oak but also found with conifers; common and widely distributed. In our area it is monogamous with live oak and usually fruits twice—a small flush after the first fall rains, and a larger crop in the late winter and spring (February-April) when *A. velosa* and *A. ocreata* "bloom." It can also turn up in the summer—in fact, I've found it every month of the year!

EDIBILITY: Not recommended. It is edible when cooked, but indigestible or even poisonous raw, and it is easily confused with poisonous species. It is highly esteemed in Europe (chiefly France), but our local version does not have a good flavor.

276

Amanita rubescens is extremely variable in size, stature, and color, but always stains dingy reddish (often slowly). These whitish-capped specimens are small; more can be seen below and on p. 266.

COMMENTS: The "blushing" of the cap, stem, and flesh is the one infallible fieldmark of this fickle fungus. The blushing process is slow to manifest itself and is best seen on the lower stalk or around the edges of maggot tunnels (see photograph below). In other respects it is an exasperatingly variable *Amanita.* Young specimens may be pure white (in our form), while older individuals usually develop strong reddish or brownish tones. The warts may be evenly disposed over the entire cap surface, or concentrated at the center, or more prevalent toward the margin, or completely absent (especially in rainy weather). Mature specimens are sometimes mistaken for *A. pantherina* (if they are brownish), or even *A. muscaria* (if they are reddish), but the indistinct volva, reddish stains, and amyloid spores separate it. Pure white buttons, on the other hand, resemble the *A. strobiliformis* group (see *A. magniverrucata*), but lack the tapered, rooting base of those species. They can also be confused with the "false blusher" (see below). The attached gills and absence of a volva can lead to confusion with *Armillaria,* but once again the reddish stains distinguish it.

The "true" *A. rubescens* that is so common in eastern North America differs from our form in several respects. Its universal veil is grayish to dirty pinkish and its cap is soon reddish-brown to flesh-color to brown or olive-tinged, while ours is often white. Also, it is a much larger, taller, and more stately fungus, and is often parasitized by a pallid to pinkish mold (see photo on p. 884), whereas our form is not. In fact, about all that our version has in common with the "true" *A. rubescens* is its blushing behavior—and it may eventually prove to be a distinct variety or species. (Perhaps we should appoint an *ad hoc* committee to investigate the problem!)

Also common under live oak in our area is the "false blusher"—an unidentified and probably unnamed species that closely mimics our *A. rubescens* in every respect save one: it does not "blush." Its bulb is often brownish-stained, however, and like *A. rubescens* it is whitish when young, browner in age, and has warts on the cap. Its edibility is unknown —another good reason not to eat *A. rubescens!*

Amanita rubescens. **Left:** Mature brown-capped specimens; note warts on cap and indistinct volva. **Right:** Even the maggot tunnels stain reddish, as shown in this close-up.

Amanita aspera   (Yellow-Veiled Amanita)   **Color Plate 56**

CAP 4-12 (15) cm broad, nearly round to convex, then plane; surface viscid when moist, dark brown to grayish-brown, yellow-brown, or bright yellow, covered with yellow mealy or powdery warts which become flattened and grayish to dingy buff in age or occasionally disappear; margin not striate or only faintly so. Flesh white or tinged yellow, soft. **GILLS** white or creamy-yellow, close, adnate to adnexed or free. **STALK** 5-15 (20) cm long, 0.7-2.5 cm thick, equal or tapering upward, the base often enlarged; white to pale yellow above the ring, white to yellow, buff, or grayish-tinged below; base often with orangish, reddish, or brown stains (on exterior or interior). **PARTIAL VEIL** membranous, white or pale yellow above, bright yellow to grayish-yellow on underside; forming a superior, skirt-like ring on stalk. **UNIVERSAL VEIL** powdery, friable, forming a scaly volva (scaly zones and/or yellow to grayish-yellow powdery scales) at base of stalk, but volva easily obliterated. **SPORE PRINT** white; spores 8-12 × 6-8 microns, elliptical, smooth, amyloid.

HABITAT: Solitary to scattered or in small groups under both hardwoods and conifers; common in our area and throughout the Pacific Northwest, but seldom in large numbers. It often grows with *A. muscaria* in our coastal pine forests and the two make a colorful pair; it is also frequent in our oak-madrone woodlands. Rather than fruiting in one large, spectacular burst like *A. calyptrata* and *A. muscaria,* it usually keeps a lower profile, fruiting rather sporadically throughout the mushroom season.

EDIBILITY: To be avoided. Chemical analysis has failed to reveal the presence of amanita-toxins, but this does *not* mean it is edible. True, the European form is edible (according to one source), and if our form were poisonous we would probably know by now —but why tempt fate when there are so many other safe, savory mushrooms available?

COMMENTS: This elegant *Amanita* is easily recognized by its powdery yellow universal veil and yellow partial veil. Fresh specimens are among our most lovely mushrooms. The cap color ranges from yellow through yellow-brown to dark brown and is apt to bewilder the color-conscious beginner. *A. pantherina* and *A. gemmata* are often confused with *A. aspera,* but have non-amyloid spores and white veils plus a collarlike volva. *A. flavorubescens (=A. flavorubens)* is a very similar if not identical species with a yellow-orange to yellow or yellow-brown cap. It is common in eastern North America, especially under oak. *A. flavoconia* is also similar, but is smaller and never brown and is rare in the West.

Amanita flavoconia

CAP 2.5-7.5 (10) cm broad, nearly oval becoming convex or plane; surface viscid when moist, bright orange to yellow-orange or yellow, or orange at the center and yellow at the margin; adorned with scattered yellow warts which may disappear in age or rainy weather; margin not striate or only faintly so. Flesh rather thin, white. **GILLS** adnate to adnexed or free, close, white, sometimes with yellow edges. **STALK** 4-10 cm long, 0.5-1.5 cm thick, tapering upward or equal with an enlarged base; smooth or somewhat scaly, yellow or sometimes white. **PARTIAL VEIL** membranous, forming a superior, skirtlike ring which usually has a yellow underside. **UNIVERSAL VEIL** friable, forming a scaly volva consisting of powdery yellow scales and patches at base of stalk which wear off easily or disappear in age. **SPORE PRINT** white; spores 7-11 × 3.5-5 microns, elliptical, smooth, amyloid.

HABITAT: Solitary to scattered or gregarious in woods (mainly under hardwoods, but also with conifers); often abundant in summer and early fall in eastern North America, but rare in the West. (It occurs in Arizona and has been found in northern California.)

EDIBILITY: Unknown. Do not experiment!

COMMENTS: This beautiful *Amanita* looks like a miniature *A. muscaria,* and is often mistaken for the yellow-orange form of that species. However, it does not have concentric rings at the base of the stalk and is much smaller. The yellow veil remnants are reminiscent of *A. aspera,* but the cap is orange to yellow rather than yellow to brown, and the stalk is often yellow as well. Other species: *A. frostiana* of eastern North America is similar, but has whitish to buff warts, a more distinct volva, and non-amyloid spores.

Amanita porphyria   (Booted Amanita)

CAP 3-12 cm broad, convex to plane or broadly umbonate; surface slightly viscid when moist, gray to grayish-brown, often with a subtle purplish cast; usually adorned with scattered grayish (or sometimes whitish) warts, these sometimes merging to form a patch or often wearing off or washing away; margin not striate or only faintly so. Flesh white; odor often turniplike in age. **GILLS** adnate to adnexed or free, close, white or sometimes aging or bruising grayish. **STALK** 5-18 cm long, 1-1.5 cm thick at apex, equal above with a large, soft, abrupt, rounded or flattened bulb at the base; bulb often cleft; white or pale gray above the ring, gray or with grayish to grayish-brown or purple-gray patches below. **PARTIAL VEIL** membranous, forming a superior, skirtlike ring which often collapses or disappears in age; ring gray or grayish-yellow. **UNIVERSAL VEIL** somewhat membranous but also friable; forming a collarlike volva (free rim) around the apex of bulb or leaving scattered grayish patches or disappearing. **SPORE PRINT** white; spores 7-10 microns, round, smooth, amyloid.

HABITAT: Solitary, scattered, or in small groups under conifers; widely distributed in northern North America. It is fairly common in the Pacific Northwest but rather rare in California. I have yet to find it in our area.

EDIBILITY: Poisonous?? Do not experiment!

COMMENTS: The abrupt, soft, rounded, basal bulb and grayish color distinguish this conifer-loving *Amanita.* The volva may have a free margin but is not truly saclike. In eastern North America there are two related species with a prominent basal bulb: *A. citrina (=A. mappa)* is very common (mainly under hardwoods) and rather slim, with a pale yellow to yellow-green to whitish cap and buffy-white to pinkish or lavender-gray warts; *A. brunnescens,* also fond of hardwoods, has a brown to olive-brown or paler cap, a stalk that stains reddish-brown, and a longitudinally split or chiselled basal bulb; it has also been reported from the Pacific Northwest. Neither of these Amanitas should be eaten.

Amanita porphyria. Note large basal bulb and grayish patches on stalk. Cap is purple-gray to brown.

Amanita pantherina, young button at right, mature specimen at left, intermediate stages between. Note the collarlike volva (i.e., the free rim of the basal bulb).

Amanita pantherina   (Panther Amanita)

CAP 5-15 (25) cm broad, nearly round or convex becoming plane to slightly depressed; surface viscid when moist, color variable: dark brown to light brown, tan, dull yellowish, or paler (see comments), often darker at center and paler toward margin; adorned with many white to pale buff universal veil remnants (warts), but these often washed off by rain; margin usually striate. Flesh firm, white. **GILLS** adnate to adnexed or free, close, white or pallid. **STALK** 5-15 (20) cm long, 1-3 cm thick, tapering upward or equal with a basal bulb; dry, white or aging buff; usually smooth above the ring, often scaly below. **PARTIAL VEIL** membranous, white, forming a superior or median skirtlike ring whose margin is often ragged or toothed. **UNIVERSAL VEIL** friable, white, usually forming a collarlike volva (i.e., adhering to the bulb except for a free rim at apex of bulb), sometimes also with scaly or ragged zones above the free rim, or sometimes indistinct (no free rim). **SPORE PRINT** white; spores 9-13 × 6.5-9 microns, elliptical, smooth, not amyloid.

HABITAT: Solitary to scattered or gregarious on ground in woods; widely distributed, but especially common under conifers from the Rocky Mountains westward. In our area it fruits from fall through spring and is quite common under pine, oak, and Douglas-fir, though rarely in the large numbers characteristic of *A. muscaria.* In the Pacific Northwest and Sierra Nevada it is perhaps the most omnipresent of all the Amanitas.

EDIBILITY: Poisonous! It contains the same toxins as *A. muscaria,* but apparently in higher concentrations—large doses can be fatal! It is said to be one of the most common causes of mushroom poisoning in the Pacific Northwest.

COMMENTS: The dark brown to tan or dull yellowish cap with whitish warts and the free rim or collar at the top of the basal bulb are the fieldmarks of this ubiquitous species. As is often the case in *Amanita,* there is considerable variation in cap color as well as in size. Yellowish forms are difficult to distinguish from *A. gemmata*—generally they are duller and often somewhat browner at the center, but the two species appear to intergrade (see comments under *A gemmata*). The cap is never red or bright orange-yellow as in *A. muscaria,* and the stalk lacks the grayish patches characteristic of *A. porphyria. A. spissa* of eastern North America is somewhat similar, but has amyloid spores and a gray universal veil that leaves a powdery-scaly rather than collarlike volva. *A. cothurnata* is a slender eastern species with a whitish cap that is often tinged yellowish or brown at the center. A mushroom meeting this description occurs in our area, but is rare. Some amanitologists consider *A. cothurnata* to be a variety of *A. pantherina.*

Amanita gemmata   (Gemmed Amanita; Jonquil Amanita)

CAP 4-10 (14) cm broad, rounded becoming convex to plane; surface slightly viscid when moist, creamy to pale yellow, golden-yellow, or buff (but see comments), often slightly darker at the center; covered with whitish universal veil remnants (warts) which may merge to form a patch or may disappear in age; margin striate or not. Flesh white, fairly thick. **GILLS** adnate to adnexed or free, close, white. **STALK** 5-13 cm long, 0.5-2 cm thick, tapered upward or equal with an enlarged base; dry, white or tinged yellowish, usually smooth above the ring and sometimes scaly below. **PARTIAL VEIL** typically present (but absent in *var. exannulata*), membranous, white; forming a fragile, superior to median, skirtlike ring on stalk, or disappearing. **UNIVERSAL VEIL** friable to somewhat cottony, white, usually forming a collarlike volva (free rim) at apex of basal bulb, sometimes also with scaly zones above rim, or sometimes forming a thin sheath instead of a rim, or sometimes the volva more or less indistinct. **SPORE PRINT** white; spores 8-13 × 6-9 microns, elliptical, smooth, not amyloid.

HABITAT: Solitary, scattered, or in groups in woods or along forested paths and roads; widely distributed. In our area it occurs in mixed woods and under live oak shortly after the first fall rains and less commonly thereafter. In the Sierra Nevada and Pacific Northwest it is common under conifers in the spring as well as the summer and fall. The mountain form differs slightly from the typical form and may eventually be classified as a separate species.

EDIBILITY: Poisonous! Some books list it as edible, but other sources say it contains the same toxins as *A. pantherina* and *A. muscaria.*

COMMENTS: Also known as *A. junquillea,* this attractive *Amanita* can be distinguished from typical *A. pantherina* by its yellow to creamy cap and more modest size. (*A. pantherina* ranges from small to very large.) However, a confusing series of "hybrids" exist whose cap color and degree of toxicity are intermediate between the two species. *A. aspera* and *A. flavoconia* differ by having yellow veil remnants, while the yellow-capped form of *A. muscaria* is larger, with a volva composed of concentric rings. Sometimes the universal veil of *A. gemmata* forms a continuous layer over the entire cap, but does not peel off easily like the volval patch of *A. calyptrata.* The latter also differs in its large, thick, saclike volva. Other species: *A. russuloides* of eastern North America is a similar, slimmer version of *A. gemmata* and may just be a regional variation. *A. breckonii* has a double, nearly basal annulus (ring) and a short, tapered, rooting base beneath the bulb. It occurs in our area under pine and perhaps oak, but is rare. A small species with an orange cap (sometimes yellow toward margin) and no partial veil also occurs rarely in our area. It may be a form of var. *exannulata* or it may be an undescribed species.

Amanita gemmata. **Left:** Typical specimens; note collarlike volva. **Right:** A vernal mountain form that may actually be a distinct species. It also has a yellow cap, but is often stockier. It usually appears when the morels are out.

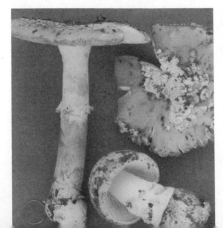 

Amanita muscaria   (Fly Agaric; Fly Amanita)   **Color Plates 58, 59**

CAP 5-30 (40) cm broad, round becoming convex and finally plane or slightly depressed; surface viscid when moist, color variable: bright red to blood-red, scarlet-red, or orange-red when fresh (*var. muscaria* and *var. flavivolvata*), but often fading to orange, yellow-orange, or paler; bright yellow-orange to yellow, then fading (*var. formosa*); yellow with a peachy center (*var. persicina*); or white to buff to silvery-grayish-white (*var. alba*); covered at first with a dense coating of universal veil fragments (warts) which are usually white (but are yellow in *var. flavivolvata* and often buff or tan in *var. alba*); warts flattened in age, often wearing away or washed off by rain; margin usually at least somewhat striate. Flesh firm when young, soft in age; thick, white. **GILLS** adnate to adnexed or free, close, broad, white. **STALK** 5-20 (30) cm long, 1-3 (4) cm thick at apex, tapering upward or equal with a basal bulb up to 6 cm broad; white or whitish, or somewhat discolored in age; smooth or somewhat ragged-scaly below the ring; often fragile in old age. **PARTIAL VEIL** membranous, usually forming a thin, persistent, median to superior, skirtlike ring on stalk which may collapse in age; ring white or with yellow patches, margin often torn or toothed. **UNIVERSAL VEIL** friable, forming a scaly volva at apex of bulb consisting of one or more (usually 2-4) concentric rings. **SPORE PRINT** white; spores 9-13 × 6.5-9 microns, broadly elliptical, smooth, not amyloid.

HABITAT: Solitary or scattered to densely gregarious or in large rings in forests and at their edges, also with planted trees. Common throughout most of the northern hemisphere—its favorite mycorrhizal mates include pine, spruce, fir, birch, and aspen. The bright red form with white warts (var. *muscaria*) is the common one in western North America as well as Europe and Asia. In our area it fruits along with the yellow-warted variety from fall to early spring and is often abundant in our coastal pine forests and along freeways where pines have been planted. It also turns up occasionally under oak and madrone. In the Rocky Mountains it is common in late summer with spruce and in the Pacific Northwest in the fall with various conifers. The red-capped form with yellow warts (var. *flavivolvata*) has a southern distribution but is quite common in our area. The form with a bright yellow to yellow-orange cap (var. *formosa*) is the dominant one in eastern North America. In California it is common in the Sierra Nevada but rather rare in the Coast

Amanita muscaria. **Left:** A button completely covered with white warts (the universal veil). **Right:** "Bloody mirror of the galaxy." This large button was frisbee-sized when fully expanded. (Joel Leivick)

Amanita muscaria var. *alba* closely resembles the more common red- and yellow-capped varieties, but has a white to grayish-buff cap.

ranges. The grayish-white form (var. *alba*) is more common in northern regions. It has been found in northern California, but is rare. The peach-centered variety (var. *persicina*) is most common in the Southeast. It is interesting to note that wherever *A. muscaria* grows it is often accompanied by *Boletus edulis,* and can serve as a "red flag" indicator for that species. I've even seen a very large ring of *A. muscaria* locally that numbered over 100 fruiting bodies, with three bulky *B. edulis* growing right in the middle of it!

EDIBILITY: Poisonous and hallucinogenic—esteemed by both maggots and mystics. Fatalities are extremely rare, but it is undoubtedly dangerous in large or even moderate amounts. The effects vary from person to person, mushroom to mushroom, and from region to region and season to season, so that there is no way to determine in advance what one's reaction will be. Too many people have had unpleasant experiences for me to recommend it (for an account of its effects, see p. 894). Tales of "getting off" by nibbling on a piece of the veil or licking the "stars" off the cap are frivolous. However, the skin apparently contains greater concentrations of the toxins than the rest of the fruiting body. The name "fly agaric," incidentally, is derived from the ancient practice of using the mushroom (often mixed with milk) to stupefy flies. It is also notable for its ability to concentrate vanadium—a rare, malleable, ductile metal used to add tensile strength to steel—from the soil.

COMMENTS: The brilliant red form of this mushroom needs no introduction—it is known to every myopic middle-class "mystic" in America. Its caricature appears on key chains, incense holders, posters, candles, curtains, calling cards, and calendars. Large plastic reproductions can be found on lawns and in display windows, amusement parks, and fantasy decor. The irony of it is that few people realize that such a mushroom actually exists! The color, of course, is its outstanding fieldmark, plus the numerous warts on the cap, which look like curds of cottage cheese, or to the more cosmically inclined, like "stars." (The warts may be washed off by rain, however.) The white gills, presence of an annulus (ring), and scaly volva separate it from other bright red mushrooms (notably the Russulas), and, unless you're color blind, it is difficult to confuse it with any of our other Amanitas. The cap may fade drastically as it ages, especially in direct sunlight or after a soaking rain, but will usually retain at least some vestige of its original splendor. The other color forms, particularly the white one, are not quite so distinctive, and are best recognized by their volva, which usually consists of a series of scaly concentric rings above the basal bulb. In our area the fly agaric grows bigger than any other *Amanita*—I have seen "monsters" nearly two feet broad! In such specimens the stalk has difficulty supporting the weight of the cap and will break at the slightest provocation, or even topple over of its own accord.

Amanita caesarea group   (Caesar's Amanita)   Color Plate 60

CAP 7-20 (25) cm broad, nearly oval to convex, becoming broadly convex or plane; surface slightly viscid when moist, smooth, or sometimes with one or two pieces of thick white universal veil tissue; bright red to orange-red or orange, often fading to yellowish or even paler in sunlight or heavy rain; margin conspicuously striate. Flesh thick, yellow under the cuticle, otherwise white. GILLS close, broad, adnate to adnexed or free, bright chrome-yellow or at times egg-yellow, or pale yellow with darker yellow edges. STALK 5-15 (20) cm long, 1.5-3 cm thick, equal or slightly thicker below (but may appear bulbous from the thick volva); smooth or with small scales, same color as gills or slightly paler; stuffed with a pith or jelly, eventually hollow. PARTIAL VEIL membranous, forming a persistent, superior and often striate, skirtlike ring which is colored like the gills. UNIVERSAL VEIL thick, membranous, white, forming a large, rather tough, lobed, saclike volva which is attached only to the base of stalk and may have a flange within. SPORE PRINT white or faintly yellow; spores 8-12 × 6-8 microns, elliptical, smooth, not amyloid.

HABITAT: Scattered to gregarious or sometimes in large rings, in pine and oak woods in southern Europe and much of the warm temperate zone. Its North American distribution parallels that of *Lactarius indigo:* it is common in the summer under ponderosa pine in Arizona and New Mexico, as well as with various pines in Mexico and Central America, and a slightly different variety (see comments) occurs in the southeastern United States north to Quebec. It has not been recorded from California but is to be looked for in the warmer parts of the state (it is apparently supplanted by *A. calyptrata*). I have seen enormous fruitings near the Grand Canyon, in August and September.

EDIBILITY: Edible and highly prized in Europe, where it is considered among the very best of all mushrooms (it is said to have been a favorite of the Caesars). In America, however, it is not so highly regarded. Its flavor is very similar to that of *A. calyptrata,* which is to say, not to everyone's liking. Because of its brilliant color it is by far the safest of the Amanitas for the table (but not necessarily the best—see *A. velosa!*).

COMMENTS: This resplendent prince of the Amanitas is easily distinguished by its bright yellow gills and stalk, brilliant red to orange cap with striate margin, and thick, volumi-nous, white saclike volva. It is a robust fungus that calls to mind *A. calyptrata.* In fact, it may mimic the latter in having one or more thick patches of veil tissue on the cap, especially in dry climates. The variety in eastern North America has a slimmer stalk (sometimes with reddish fibrils) plus a usually bald, sometimes umbonate cap that is often yellow toward the margin in age. It is now considered a distinct species, *A. hemibapha* (formerly *A. umbonata*). Another common eastern species, *A. parcivolvata*, is similar in color (red to yellow-orange striate cap and yellow gills), but lacks a partial veil and does not have a saclike volva; it is harmless anyway. The poisonous *A. muscaria* is easily dis-tinguished by its warted cap, *white* gills and stalk, and *scaly* rather than saclike volva.

Amanita calyptrata   (Coccora; Coccoli)   Color Plates 61, 62, 63

CAP 7-25 cm or more broad, rounded becoming convex and finally plane; surface viscid when wet, usually with a large, thick, central, cottony whitish patch of universal veil tissue; otherwise typically dull orange to orange-brown, bronze, yellow-brown, or dark brown, usually with a paler (yellow) margin, *or* pale yellow to creamy or whitish (the pale form) *or* occasionally greenish; margin usually clearly striate. Flesh firm, thick, white or creamy; odor mild becoming slightly unpleasant in age. GILLS broad, close, adnate becoming adnexed or free, white to pale creamy-yellow. STALK 7-25 cm long, 1-3.5 cm thick, equal or tapering upward, smooth, creamy to yellowish (typical form) or creamy to white (pale form) or white to greenish (greenish form); hollow, the cavity often filled with a cottony or jellylike substance. PARTIAL VEIL membranous, colored like stalk or slightly darker,

Amanita calyptrata (= A. calyptroderma), mature specimens. Note striate cap margin. (Joel Leivick)

forming a large but fragile, superior, skirtlike ring on stalk which is easily obliterated. **UNIVERSAL VEIL** membranous, white, forming a large saclike volva that sheathes the stalk base; volva thick, ample, felty or cottony, often lobed, often with a collar or flange within. **SPORE PRINT** white; spores 8-11 × 5-6 microns, elliptical, smooth, not amyloid.

HABITAT: Solitary to widely scattered or gregarious on ground in woods; known only from the west coast. The distribution of the typical (darker) form parallels that of its favorite mycorrhizal mate, madrone: it is common in northern California and southern Oregon, where madrone is also common, less frequent in Washington, where madrone is likewise less numerous, then common again on Vancouver Island in British Columbia, where madrone is abundant. In our area it usually fruits after the first fall rains, sometimes in tremendous quantity. The pale form, on the other hand, is a late winter and spring mushroom. In our area it favors oak, but never fruits in the large quantities characteristic of the fall (typical) form. In the Sierra Nevada the pale form is prevalent under pine and other conifers, often forming "mushrumps" (especially in the spring).

EDIBILITY: Edible and popular, but with a rather strong, fishy flavor that doesn't appeal to everyone. Too many people eat and enjoy the typical form for me not to recommend it. However, be *absolutely sure* of your identification, review the comments on pp. 264-265, and avoid the pale and greenish forms, which are easily confused with poisonous species (see photo on next page!). The caps are superb stuffed and then broiled. The hollow stems can be sliced crosswise (to make rings) and marinated. The flesh does not keep well, so use what you pick as soon as possible. Italian-Americans stalk it with a passion, undoubtedly because of its resemblance—in both appearance and flavor—to their beloved *A. caesarea* of Italy. "Coccora," "coccoli," and "cocconi" (another nickname) are presumably derived from an Italian word for cocoon—a very apt description of the large, soft, cottony "eggs."

COMMENTS: For many years this species has been called *A. calyptroderma,* but the name *A. calyptrata,* first applied to the greenish form, may be the correct one according to the International Code of Botanical Nomenclature (whose purpose it is to dispel confusion) because it was published first. The typical (darker) form and the pale form are quite different in color, as evidenced by the color plates. In my opinion they merit *at least* varietal status (just like the varieties of *A. muscaria*) because they do *not* appear to intergrade and are ecologically distinct (see comments above). The rare greenish form, on the other hand, may be environmentally-induced (mycologist Harry Thiers suggests it is caused by cold temperatures). The typical form of this magnificent mushroom is distinguished by its (1) large size (2) orange to brown or yellow-brown cap with a yellow margin (3) cottony

285

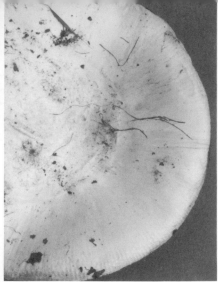

Left: The pale winter-spring form of *Amanita calyptrata*. **Right:** Close-up of cap in *Amanita ocreata* (p. 271-272). The pale form of *A. calyptrata* is easily confused with the deadly *A. ocreata* (see photos on p. 272). *A. ocreata* usually lacks striations (fine lines) on the margin of the cap, but as evidenced here, striate or partially striate specimens do occur—some of them with volval patches!

white volval patch or "skullcap" on the cap (4) striate cap margin (5) creamy or pale yellow tints to the stalk and veil (6) thick, voluminous, saclike volva (7) non-amyloid spores. Even beginners have little trouble recognizing the typical (darker) form once they've seen it several times. In the deadly poisonous *A. phalloides* the cap is usually (but not always) greenish to greenish-yellow or paler, there is no volval patch or if one is present it is very thin, the cap margin is only very rarely striate, the spores are amyloid, and an unpleasant pungent odor often develops in age. Veteran toadstool-testers can differentiate them at a glance by their color, but beginners should *not* place undue emphasis on such a capricious character. The veil tissue on the cap of *A. gemmata* and *A. pantherina* can mimic a "patch," but does not peel off easily, and the volva in those species is collarlike or indistinct, not saclike. The pale form of *A. calyptrata* is easily confused with poisonous Amanitas such as *A. ocreata* and *A. gemmata,* and I cannot recommend it to any but the most seasoned and intrepid toadstool-tester. The same goes for the greenish form. Another variety, originally called *A. calyptratoides,* may be a distinct species. I have found it only once in our area, but it is quite common in southern California under oaks. It sometimes approaches *A. velosa* in size and stature, and has a very poorly formed, appressed or evanescent annulus (ring), a brown cap without a yellow margin, and a modest volval patch. It is edible, but care must be taken in identification!

Amanita velosa (Springtime Amanita) Color Plate 64

CAP 3-12 (15) cm broad, at first nearly oval, soon convex and finally plane; surface viscid when moist, smooth, pinkish-tan to orange-buff to salmon, beige, "the color of a brown hen's egg," or sometimes much paler (even white), often fading; usually with a white patch of universal veil tissue or several large, thick pieces; margin distinctly striate. Flesh fairly thick, white, odor rather pungent in age. **GILLS** adnate to adnexed or free, close, white, sometimes dull pinkish in old age. **STALK** 5-12 (15) cm long, 0.5-1.5 (2.5) cm thick, equal or tapering upward; white or tinged cap color, apex often powdery or striate, lower portion smooth or broken into rings or scales. **PARTIAL VEIL** usually absent, but a rudimentary ring (annulus) sometimes present as a cottony or fibrillose-scaly zone. **UNIVERSAL VEIL** membranous, forming a saclike volva which sheathes the stalk base; volva white or tinged cap color, usually buried, ample but sometimes disintegrating or obscure. **SPORE PRINT** white or tinged very slightly pinkish; spores 8.5-12.5 × 6-10 microns, nearly round to elliptical, smooth, not amyloid.

HABITAT: Widely scattered to gregarious, common in winter and spring, associated in our area with live oak but often growing out in the open (in fields and around their edges, brushy areas, lawns, etc.), up to 40 feet away from its mycorrhizal host (which is often stunted by poor soil conditions). It also occurs in the aspen-conifer forests of the Sierra Nevada and with various oaks in the Sierra Nevada foothills and southern Oregon. It is apparently endemic to the west coast and more common in our area than anywhere else. With *A. rubescens* and *A. ocreata* it forms a striking triumvirate of springtime Amanitas that are monogamous locally with live oak. It is interesting to note that when all three form mycorrhiza with the same tree, they frequently occupy distinctly different zones or niches: *A. velosa* typically grows in the open at the outer fringes of the host's roots; *A. ocreata* grows in deep shade near the trunk; and *A. rubescens* occupies the intermediate zone or "shade border" near the perimeter of the branches.

EDIBILITY: Edible and incredible! Most "objective" fungophiles rate it far superior to the better-known *A. calyptrata* and *A. caesarea*. At any rate, it is far sweeter. As a bonus, it fruits in the spring when there is a paucity of other collectable delectables. However, it is *not* a good choice for beginners. Faded specimens should be avoided, as they are easily confused with *A. ocreata* and other poisonous Amanitas.

COMMENTS: The pinkish-tan to orange-buff, beige, or paler cap with striate margin and volval patch or large warts, absence of a partial veil, saclike volva and habit of fruiting in the open (unusual behavior for an *Amanita!*) are the fallible fieldmarks of this handsome but variable fungus. In our area it is the last of its clan to appear—often not showing up until February—but as if to make up for its tardiness, it lingers on through April or even May. *A. calyptrata* and *A. calyptratoides* both have volval patches but are differently colored, tend to grow in the woods, and have a partial veil. However, *A. velosa* often has a rudimentary ring and one specimen I found had a full-fledged membranous partial veil that covered the gills when young and broke to form an annulus (ring). This lends credence to the modern trend toward de-emphasizing gross structural ("Friesian") features and paying more attention to chemical and anatomical (microscopic) similarities. The obsolete genus *Amanitopsis* (Amanitas without a partial veil) was incorporated into *Amanita* on just such a pretext. The trend may bode ill for the multitudes without microscopes, but it better reflects natural relationships. *A. crocea*, an eastern and Rocky Mountain species, is similar and equally delicious, but has an oranger, usually bald cap and orangish scales on the stalk.

Amanita fulva   (Tawny Grisette)

CAP 4-10 cm broad, oval becoming convex to plane or umbonate; surface slightly viscid when moist, smooth or rarely with large whitish fragments of universal veil tissue; orange-brown to reddish-tan, tawny, or tan; margin deeply grooved (striate). Flesh thin, white, odor mild. GILLS close, adnexed or free, white or creamy. STALK 7-16 cm long, 0.4-1.5 cm thick, more or less equal, usually rather long, white or tinged cap color; fibrillose or with a few scales, hollow in age. PARTIAL VEIL absent. UNIVERSAL VEIL membranous, forming a large, loose, persistent, lobed, saclike volva which sheaths the stalk base; volva or "bag" white to pale tan or rusty-stained. SPORE PRINT white; spores 8-10 microns, round, smooth, not amyloid.

HABITAT: Solitary to scattered or gregarious under both hardwoods and conifers, often in boggy areas; widely distributed but not yet reported from California. I have seen large fruitings in the aspen-conifer forests of the southern Rocky Mountains in late summer.

EDIBILITY: Edible, but be sure of your identification!

COMMENTS: This species is essentially a "fulvous" version of the grisette, *A. vaginata,* and for many years was regarded as a mere color form of that species. The orange-brown to

Amanita fulva. Note conspicuously striate (grooved) cap and absence of a partial veil (annulus). Cap color ranges from orange-brown to reddish-tan to tan.

reddish-tan, markedly grooved cap, absence of a partial veil, large "bag" at the base of the stem, and rather tall, slender stature distinguish it. It is quite handsome when fresh. Other species: *A. crocea* is similar but has an oranger cap and pale orange scales on the stalk.

Amanita vaginata   (Grisette)

CAP 3-10 cm broad, at first oval, then convex and finally plane or with a slight umbo; surface slightly viscid when moist, gray to grayish-brown, smooth or sometimes with a white patch or patches of universal veil tissue; margin grooved (deeply striate). Flesh soft, white to grayish, thin. **GILLS** white or tinged gray, adnate to adnexed or free, close. **STALK** 7-15 (20) cm long, 0.5-2 cm thick, usually rather long and slender; equal or tapering upward, smooth and white or often covered with delicate grayish to grayish-brown scales. **PARTIAL VEIL** absent. **UNIVERSAL VEIL** membranous, forming a saclike volva that sheathes the stalk but is attached only at the very base; volva white or tinged gray (occasionally rusty-stained), loose, lobed. **SPORE PRINT** white; spores 8-12 microns, round or nearly round, smooth, not amyloid.

HABITAT: Solitary to scattered or in small groups in woods or under trees; common and widely distributed. In our area it occurs under both hardwoods and conifers, but is largely supplanted by *A. pachycolea* and *A. constricta.*

EDIBILITY: Edible when cooked, and fairly good. Though prized in France, it tends to be flaccid, thin-fleshed, and does not refrigerate well. It is one of the safest Amanitas for the table, but see comments on pp. 264-265 before eating it!

COMMENTS: The combination of deeply striate gray to gray-brown cap, white gills, absence of a ring, and membranous sack or "bag" at the base of the stem typifies a group of Amanitas collectively called *A. vaginata.* The group is especially complex in the Southwest, where a number of forms with white to gray or brown caps are common in ponderosa and pinyon pine forests after summer thundershowers. Several color forms of *A. vaginata* have been described, some of which are now considered distinct species. One is *A. fulva* (see description); another is *A. alba,* which has a pure white, grooved (striate) cap, a saclike volva, and no partial veil. Although edible, *A. alba* should not be eaten because of its resemblance to the destroying angels *(A. ocreata, A. virosa,* etc.). I have found it only twice in our area, but it is said to be fairly common in some regions. *A. vaginata* has also been called *Amanitopsis vaginata* and *Vaginata plumbea.*

Left: *Amanita vaginata,* mature specimens. The cap does not necessarily feature the small patch of universal veil tissue shown here; note saclike volva. **Right:** Close-up of the constricted, flaring volva of *Amanita constricta* (which otherwise resembles *A. vaginata*).

Amanita constricta   (Constricted Grisette)

CAP 4-13 cm broad, oval becoming convex or plane to slightly umbonate; surface slightly viscid when moist, smooth or covered with a large patch of white to buff or grayish universal veil tissue which often separates later into several pieces; color beneath the veil tissue (if present) gray, or sometimes brownish-gray. Flesh fragile, rather thin, white to grayish. **GILLS** adnate to adnexed or free, close, white or grayish. **STALK** 10-16 cm long, 1-2 cm thick, equal or tapering upward, stuffed or hollow in age; smooth or more often belted with numerous delicate grayish scales. **PARTIAL VEIL** absent. **UNIVERSAL VEIL** membranous, gray to white or buff, often bruising reddish when wet; forming a constricted, flaring volva (i.e., lower portion of volva constricted around the stalk, upper margin flaring outward) which may disintegrate in age. **SPORE PRINT** white; spores 9.5-13 × 8-10.5 microns, elliptical to nearly round, smooth, not amyloid.

HABITAT: Solitary to scattered or in groups in woods, associated mainly with oaks; apparently endemic to California or at least the west coast. Common in our area throughout the mushroom season but most abundant in the winter.

EDIBILITY: Edible, but see comments on pp. 264-265 before eating it!

COMMENTS: The gray, conspicuously striate cap, absence of a partial veil, and peculiar constricted volva are the hallmarks of this *Amanita.* It is the most common grisette in our area. It is easily distinguished from both *A. pachycolea* and *A. vaginata* by its constricted rather than saclike volva, and frequent presence of universal veil tissue on the cap. *A. inaurata (=A. strangulata, A. ceciliae)* is a closely related species "complex" whose volva is also "strangled," usually forming a belt of grayish tissue around the stalk base. It has a gray to grayish-brown to blackish cap decorated with gray to charcoal-gray warts (which may wear off in age) and it has round spores (10-14 microns). It is fairly common in the Pacific Northwest as well as in eastern North America, but I have yet to find it in our area.

289

Amanita pachycolea is easily told by its large size and dark brown to grayish-brown striate cap. The cap is often darkest at the inner edge of the striations (a feature also shown in the color plates).

Amanita pachycolea (Western Grisette) Color Plates 65, 66

CAP 7-20 (25) cm broad, at first nearly oval, then convex or somewhat bell-shaped, finally plane or with uplifted margin and often a low, broad umbo; surface smooth, viscid when moist, dark brown when young, brown to grayish-brown or paler in age (usually darker at the center and paler toward margin or with a darker brown band at inner edge of the striations); sometimes completely washed out in old age, occasionally with a thick patch or patches of universal veil tissue; margin grooved (deeply striate) for 1.5-3 cm. Flesh white, rather soft; odor sometimes unpleasant in age. **GILLS** close, broad, adnate becoming adnexed or free; white with dark brown edges when fresh, sometimes discoloring dingy orange or yellow-orange in old age. **STALK** 12-30 cm long or more, 1-3 cm thick, equal or tapering upward gradually, usually covered with fine, delicate, grayish-brown to brown particles or fibrillose scales on a pallid background; stuffed or hollow in age. **PARTIAL VEIL** absent. **UNIVERSAL VEIL** membranous, forming a very large, loose, lobed, saclike volva which sheathes lower portion of stalk for up to 12 cm but is attached only at the very base; volva ample, thick, white or rusty-stained. **SPORE PRINT** white; spores 11-14.5 × 10-12.5 microns, broadly elliptical to round, smooth, not amyloid.

HABITAT: Solitary, scattered, or in small groups in mixed woods and under conifers; apparently restricted to the Pacific Coast. It is fairly common in our area in the fall and winter with pine and oak, but rarely fruits in large numbers.

EDIBILITY: Edible, but not choice. Though meatier than *A. vaginata*, it develops a rather strong fishy taste in age. See comments on pp. 264-265 if you plan to eat it.

COMMENTS: This lofty *Amanita* is easily recognized by its brown, deeply striate cap, absence of a ring (partial veil), and huge sheathing "bag" at the base of the stem. It is one of the most strikingly beautiful of all mushrooms, differing from its close relatives in the *A. vaginata* group by its larger size, brown rather than gray cap, and frequent presence of rusty stains on the huge volva. The volva is so voluminous that even washed out specimens are easily recognized. The stalk in mature individuals is so long and fragile that it is difficult to transport them home in one piece. Other species: *A. **umbrinolutea*** is similar, but is said to have white gills, smaller spores, and a wide distribution.

LIMACELLA

Medium-sized, mostly terrestrial fungi. CAP *usually smooth and viscid or slimy.* GILLS *free or nearly free, close, white or pallid.* STALK typically central, hollow or stuffed, dry or viscid. PARTIAL VEIL *present,* sometimes forming a ring, sometimes evanescent. UNIVERSAL VEIL *slimy, not forming a volva.* SPORE PRINT *white.* Spores smooth, not amyloid (but rarely dextrinoid). Gill tissue divergent, at least when young.

THE viscid to slimy cap, presence of a veil, typically free gills, and absence of a volva typify this small, rather rare genus. Many Limacellas used to be placed in *Lepiota,* but the divergent gill tissue suggests a closer relationship to *Amanita.* The stature of the fruiting body is somewhat like *Amanita,* but the universal veil takes the form of a layer of slime that coats the cap and often the stem, and does not form a volva.

About a dozen species are known from North America. They are woodland fungi with whimsical fruiting habits. Generally they are rare, but every so often there is a large, localized fruiting. Little is known of their edibility. Five species are keyed here and two are described.

Key to Limacella

1. Stalk distinctly viscid or slimy 2
1. Stalk not viscid (but patches of slime from cap may drip onto it) 4
2. Partial veil forming a distinct membranous annulus (ring) *L. roseicremea* (see *L. illinita,* p. 292)
2. Veil slimy, not forming a membranous annulus 3
3. Cap (and slime) entirely brown to bright reddish-brown to golden-brown 6
3. Cap white to creamy, or brownish at center and white toward margin . . . . . *L. illinita,* p. 292
4. Partial veil fibrillose, disappearing or forming a slight ring or ragged zone on stalk; cap reddish-brown, chestnut-brown, pinkish-brown, or orange-brown . . . . *L. glioderma,* below
4. Not as above; partial veil forming a distinct membranous (flaring or skirtlike) ring 5
5. Cap white or pale buff . . . . *L. solidipes* (see *L. illinita,* p. 292)
5. Cap dull ochre to pale tan to creamy-pink; gills developing olive-gray stains in age or upon drying in one variety, not developing them in another . . . . *L. guttata (=L. lenticularis)*
6. Cap and stalk bright reddish-brown . . . . *L. glischra* (see *L. glioderma,* below)
6. Cap and stalk more or less golden-brown . . . . *L. kauffmanii* (see *L. glioderma,* below)

Limacella glioderma

CAP 3-8 cm broad, convex to plane or broadly umbonate; surface viscid to nearly dry, bright to dull brick-red, reddish-brown, cinnamon, chestnut-brown, or at times orange-brown, sometimes fading in age to pinkish-tan; cuticle often breaking up into small fibrillose scales or pulling away from the margin, revealing the pinkish flesh underneath; margin often hung with veil remnants. Flesh thin, tinged pinkish; odor distinctly but pleasantly farinaceous. GILLS adnexed or notched, or in age becoming free; close, whitish or tinged cap color. STALK 4-12 cm long, 0.5-1 (1.5) cm thick, often rather slender and fragile, equal or slightly thicker at either end; dry or occasionally with a few patches of slime from the cap; whitish or pallid above the ring, with cap-colored scales, patches, and/ or fibrils below. PARTIAL VEIL fibrillose, whitish, forming a slight superior ring or ragged zone on stalk or disappearing entirely. UNIVERSAL VEIL evanescent (usually not visible), not forming a volva. SPORE PRINT white; spores 3-5 microns, round, smooth, not amyloid. Gill tissue divergent.

HABITAT: Solitary, scattered, or in groups on ground in woods; widely distributed, and the most common *Limacella* in our area. I find it regularly in the late fall and winter with oak, madrone, and conifers.

EDIBILITY: Unknown, and like myself, likely to remain so.

Limacella glioderma. Note ragged or patchy stalk and evanescent partial veil.

COMMENTS: The viscid reddish-brown cap, evanescent veil, and dry fragile stem distinguish this species from other Limacellas. Because the gills may be slightly attached to the stalk, this fungus has been placed in *Tricholoma* and *Armillaria.* However, the divergent gill hyphae (a microscopic feature) indicate *Limacella.* The prevalent form in our area has only a slightly viscid cap, but fairly glutinous specimens are also encountered. *L. glischra* is a widespread but rare species with a bright reddish-brown cap and stalk, *both* of which are very slimy when moist. *L. kauffmanii* is a southern species with a more or less golden-brown, viscid-slimy cap and stalk. Neither of these is worth eating.

Limacella illinita   (White Limacella)   Color Plate 67

CAP 2-8 cm broad, at first rounded or oval, then convex, finally plane or broadly umbonate; surface smooth, very slimy or at least viscid, white to creamy-white in typical variety, brownish at the center and white at the margin in *var. argillacea*; margin often hung with slimy veil remnants. Flesh thin, soft, white. **GILLS** notched or free, white, close, **STALK** 5-9 (13) cm long, 0.5-1 cm thick, equal or tapering upward, slimy or viscid, white in typical variety, but tinged tan or buff in var. *argillacea.* **PARTIAL VEIL** fibrillose beneath a layer of slime, evanescent, not forming a distinct ring. **UNIVERSAL VEIL** slimy, coating stalk and cap but not forming a volva. **SPORE PRINT** white; spores 4.5-6.5 × 4-6 microns, broadly elliptical to round, smooth, not amyloid. Gill tissue divergent.

HABITAT: Scattered or in groups in woods, widely distributed; probably the most common *Limacella* in North America. In our area the typical variety turns up occasionally after heavy rains in the fall and winter in mixed woods and under Douglas-fir. I have seen var. *argillacea* only once—a large fruiting under tanoak. The typical variety is also said to occur in sand dunes. In the South it is said to be common on lawns as well as in woods.

EDIBILITY: Unknown. The slime is a formidable deterrent.

COMMENTS: The slimy white or brown-tinged cap and stem combine with the free gills to set this slippery mushroom apart. *Hygrophorus eburneus* is just as glutinous, but has adnate to decurrent gills and is much more common. *L. solidipes* is a widely distributed but rare species with a viscid white cap, large flaring to skirtlike ring, and dry white stalk. It also occurs in mixed woods in our area and could be mistaken for *Lepiota naucina*, but the viscid cap and woodland habitat distinguish it. Other species: *L. roseicremea*, known only from Washington, is similar but has a creamy or rose-tinged cap plus a distinctly membranous partial veil and annulus (ring).

292

 spores   LEPIOTACEAE   (Parasol Mushrooms)

Saprophytic, mostly terrestrial mushrooms of variable size. CAP typically dry or only slightly viscid, often scaly with a smooth center when mature. GILLS *typically free, white to pallid or yellow,* close. STALK cleanly separable from cap, central, base often enlarged. VEIL *present, usually forming a ring (annulus) on stalk, or if not then stalk usually scaly below the veil.* VOLVA *typically absent.* SPORE PRINT *white to pale buff or dull greenish.* Spores smooth, usually dextrinoid, but not amyloid. Gill tissue parallel or interwoven.

THIS family is defined here to include one very large and diverse genus, *Lepiota,* plus a single green-spored mushroom, *Chlorophyllum molybdites,* which could just as well be considered an aberrant species of *Lepiota.* The "splitters" have erected several additional genera for Lepiotas with thick-walled spores (e.g., *Leucocoprinus, Leucoagaricus,* and *Macrolepiota*), but until one system of classification is firmly agreed upon, it seems best for the purposes of this book to retain them in *Lepiota.*

Among the white-spored genera, *Lepiota* is most likely to be confused with *Cystoderma* and *Limacella*—both of which were originally included in *Lepiota. Cystoderma,* however, has attached rather than free gills plus a granulose cap and stalk, while *Limacella* has a sticky or slimy cap. *Lepiota* also bears a superficial resemblance to *Amanita,* but lacks a volva.* There are also fundamental microscopic differences: the gill tissue is parallel to interwoven in *Lepiota,* divergent in *Amanita* (see p. 19), and the spores are typically dextrinoid in *Lepiota,* while in *Amanita* they are frequently amyloid but never dextrinoid.

Aside from spore color, *Lepiota* closely resembles the common genus *Agaricus* (which has chocolate-brown spores) in having free gills, a veil, and no volva. (Some taxonomists even consider the Lepiotas to be a subfamily of the Agaricaceae.) There are ecological similarities between *Lepiota* and *Agaricus,* as well as morphological ones. Both are large genera centered in the warm temperate zone and tropics. (*Lepiota* is one of the largest and most bewildering genera of tropical agarics.) And both are at their best in moist, relatively warm weather. Furthermore, both are largely if not exclusively saprophytic—in contrast to the Amanitas, which are mostly mycorrhizal. The larger Lepiotas or "parasol mushrooms" frequent roadsides, waste places, lawns, gardens, pastures, and open woods. Many of the smaller species occur in both heavily forested and cultivated areas. In our area, cypresses are unusually rich in *Lepiota* (and *Agaricus*) species—some of them unclassified.

The large, white-spored parasols are among our most delicious edible mushrooms. However, allergic reactions are reported for virtually every edible *Lepiota,* and extreme care must be taken not to confuse them with poisonous species. *L. procera* of eastern North America is the best of the best, *L. rachodes* and *L. barssii* are the best in the West, while *L. naucina* is the best of the rest. The dozens of smaller Lepiotas should be strictly avoided. Not only are they devilishly difficult to differentiate, but several (e.g., *L. josserandii, L. helveola,* and *L. castanea*) are said to contain potentially fatal amanita-toxins! The green-spored parasol, *Chlorophyllum molybdites,* can cause *severe* gastrointestinal distress, and the powdery yellow greenhouse species, *L. lutea,* is also said to be poisonous.

At least 200 species of *Lepiota* are thought to occur in North America. Many have very limited or erratic distributions. Only some of the larger or more distinctive species are presented here. The diligent *Lepiota*-lover will doubtlessly uncover many species that cannot be identified, particularly in California and the southern United States.

Key to the Lepiotaceae (Lepiota & Allies)

1. Fruiting body medium-sized to large; stalk usually (but not always) at least 6 mm thick; veil usually forming a distinct collarlike or sleevelike annulus (ring) on stalk which may or may not be thick, ragged, and double-edged 2
1. Fruiting body small to medium-sized; stalk usually less than 6 mm thick; veil disappearing, or if forming an annulus then the annulus not typically thick and double-edged 10

---

* L. rachodes, however, may have a volva-like rim on its basal bulb.

2. Both scales and background of cap white or whitish, the center often tinged yellow or yellow-brown; margin distinctly striate at maturity; common in Gulf Coast region in many habitats; spore print white; (may be very poisonous!) .... *L. humei*
2. Not as above 3

3. Cap (and often stalk or underside of veil) covered with pointed, often pyramidal warts (at least when young) that usually rub off easily 11
3. Not as above; scales on cap absent, or if present then caused by the breaking up of the cap cuticle and not rubbing off easily 4

4. Spore print greenish or grayish-green; usually growing in grass or gardens; fruiting in warm or hot weather .... *Chlorophyllum molybdites,* p. 295
4. Spore print white or pallid, not greenish or grayish 5

5. Fruiting body tall (12-40 cm), cap 7-25 cm broad; stalk brown or breaking up into delicate brown scales or granules below the ring (which may wear off); found in woods, old pastures, etc., in eastern North America (also in southern California and the Southwest?)   *L. procera,* p. 298
5. Not as above 6

6. Cap smooth when young, but soon breaking up into brown to reddish scales or fibrils 7
6. Not as above; cap white to gray, grayish-brown, or buff 9

7. Cap red to reddish-brown, pink, pinkish-orange, or cinnamon-buff, at least at center; fruiting body *not* staining when bruised or in age; stalk typically up to 1 cm thick *L. rubrotincta,* p. 305
7. Not as above; stalk interior usually staining yellow-orange to reddish when cut or bruised . 8

8. Fruiting body aging or drying dark reddish-brown to vinaceous; stalk often spindle-shaped (i.e., swollen above the base) .... *L. americana,* p. 301
8. Not as above; stalk terminating in a swollen base or bulb .... *L. rachodes,* p. 297

9. Cap covered with flattened gray to grayish-brown fibrils when young, sometimes scaly in age .... *L. barssii,* p. 303
9. Cap white to gray or buff, usually smooth (but sometimes breaking up into small scales in age and typically without fibrils when young) .... *L. naucina,* p. 299

10. Cap (and sometimes stalk) covered with small, erect, pointed or pyramidal scales which rub off easily; cap *not* striate 11
10. Not as above 13

11. Stalk with pointed scales below the veil 12
11. Stalk with few if any scales (but underside of veil often has them); mainly found in eastern North America and the Southwest .... *L. acutesquamosa* (see *L. eriophora,* p. 303)

12. Veil usually forming a distinct annulus (ring) on stalk; stalk 1-2 cm thick ....
.... *L. asperula* (see *L. eriophora,* p. 303)
12. Veil disappearing in age; stalk 4-8 mm thick .... *L. eriophora,* p. 303

13. Cap conspicuously striate, at least in age, the surface usually powdery or mealy or with small scales when young; flesh thin and fragile; veil usually (but not always) forming a distinctive sleevelike or collarlike annulus (ring) on stalk (examine several specimens) 14
13. Not with above features; cap not normally striate 16

14. Fruiting body yellow or mostly yellow, at least when fresh .... *L. lutea* & others, p. 302
14. Fruiting body not yellow (but may stain yellow) 15

15. Cap and base of stalk purplish to pinkish-brown or purple-brown when young, the cap surface breaking up into purplish scales; found in nurseries and greenhouses ....
.... *L. lilacinogranulosa* (=*Leucocoprinus lilacinogranulosus)*
15. Not as above .... *L. cepaestipes* & others, p. 301

16. Stalk shaggy or cottony below the veil; cap margin often shaggy also; growing in woods; spores 12-20 microns long .... *L. clypeolaria* & others, p. 309
16. Not with above features 17

17. Fruiting body whitish or with a brownish cap, the stalk typically staining reddish; common in grazed pastures (but not on dung) in Florida .... *L. sp. (unidentified)**
17. Not as above 18

18. Veil usually forming a distinct collarlike annulus (ring) on stalk (examine several specimens) 19
18. Veil usually disappearing or forming only a fibrillose or cottony zone or obscure annulus   29

*Popularly known as "Peele's Lepiota" because it was discovered by Stephen Peele of Florida, this species is said to be hallucinogenic; it should not be eaten until better known, because several similar species are poisonous!

19. Cap, stalk, *and/or* flesh staining bright orange to red (or staining yellow and then orange to reddish) when bruised or rubbed, then often discoloring to dark brown or purplish 20
19. Not as above 24

20. Fruiting body aging or drying dark reddish to vinaceous; stalk usually spindle-shaped (i.e., swollen above or at the base); cap often with coarse scales; growing in compost, sawdust, around old stumps, etc., usually in groups or clusters . . . . ***L. americana*** & others, p. 301
20. Not as above 21

21. Tropical or found along Gulf of Mexico; staining yellow before reddening or darkening . 22
21. Not as above; not staining yellow 23

22. Gills tinged pale yellow; center of cap and scales on cap blackish-brown . . . . . . ***L. sanguiflua***
22. Not as above . . . . ***L. tinctoria*** & others (see *L. americana*, p. 301)

23. Gills bruising pinkish or red-orange . . . . ***L. roseifolia*** & others (see *L. flammeatincta*, p. 304)
23. Gills not staining when bruised . . . . ***L. flammeatincta*** & others, p. 304

24. Cap black or dark gray at center, usually with scattered gray to greenish-gray scales . . . .
. . . . ***L. atrodisca,*** p. 304
24. Not as above; cap scales not gray or black 25

25. Cap white to yellowish or pale tan 26
25. Cap purple, red, pink, reddish-brown, or brown, at least at the center 27

26. Taste strongly acidic; cap typically with small scales in age; usually growing with alder . . . . . .
. . . . ***L. pulcherrima***
26. Not as above; cap usually smooth, yellowish to tan or grayish at center and paler toward margin
. . . . ***L. sequoiarum*** & others, p. 307

27. Cap or fibrils on cap purplish . ***L. roseilivida*** (see *L. rubrotincta*, p. 305 & *L. cristata*, p. 306)
27. Not as above; cap (or scales) some shade of red, pink, orange, brown, etc. 28

28. Cap 3-8 cm broad, typically with radiating fibrils . . . . ***L. rubrotincta*** & others, p. 305
28. Cap usually 5 cm broad or less, with small scattered or concentrically arranged scales . . . 32

29. Gills yellow . . . . ***L. luteophylla*** (see *L. lutea*, p. 302)
29. Gills not yellow 30

30. Cap and stalk tinged lavender to lilac, without scales; odor unpleasant; growing in woods; rare . . . . ***L. bucknallii***
30. Not as above 31

31. Cap small (usually 2.5 cm broad or less), white or tinged pinkish to pinkish-cinnamon; smooth or minutely powdery but without contrasting colored scales . . . . ***L. seminuda,*** p. 307
31. Not as above 32

32. Stalk typically without scales; cap small (usually 1-5 cm broad); annulus (ring) often present on stalk . . . . ***L. cristata*** & others, p. 306
32. Stalk typically with scales below the veil or cap larger; distinct annulus usually lacking . . 33

33. Stalk and often the cap developing strong ochraceous to rusty-orange tones in age or after handling; fruiting body small (stalk typically 1-3 mm thick) . . . . ***L. castanea,*** p. 307
33. Not as above (but reddish tones may develop); fruiting body small to medium-sized 34

34. Scales on cap and stalk blackish to dark brown . . . . ***L. felina*** (see *L. josserandii*, p. 308)
34. Not as above; scales brown to reddish . . . . ***L. josserandii*** & others, p. 308

Chlorophyllum molybdites   (Green-Spored Parasol)

CAP (5) 10-30 (40) cm broad, oval or nearly round, then convex to broadly cone-shaped, plane, or umbonate; surface dry, at first smooth but soon breaking up into light brown to brown or pinkish-brown scales on a white background; scales flat or curled, usually rather few at maturity and concentrated toward the center. Flesh thick, white, soft in age; not staining when bruised or bruising slowly reddish or occasionally bruising orange in the stalk. **GILLS** free, broad, close, white to dingy yellowish, then slowly becoming grayish to greenish in old age. **STALK** 5-25 cm long, 1-2.5 cm thick at apex, equal or thicker at base; smooth, firm, white or brownish-stained. **VEIL** membranous, white, forming a persistent, superior, double-edged ring on stalk; ring becoming brownish on underside and usually

Chlorophyllum molybdites is best distinguished from other parasol mushrooms by its greenish spore print. Note how the gills are still white in the fully expanded specimen in background.

movable in age. **SPORE PRINT** grayish-olive to greenish; spores 8-13 × 6.5-9 microns, elliptical, smooth, thick-walled, with an apical pore, dextrinoid. Cap cuticle composed of narrow, interwoven, mostly repent hyphae (but often upright at center of cap).

HABITAT: Solitary to scattered or in groups or large rings on lawns and other grassy places, also in gardens; fruiting in summer or during warm weather and widely distributed in the tropics and warm temperate zone. It is common in most of eastern and southern North America as well as inland northern California. I have seen enormous fruitings on lawns in Fresno, Los Angeles, Palo Alto, and San Diego, but have never seen it on the central California coast, perhaps because of the cool summers.

EDIBILITY: Poisonous! Some people eat it without ill effect but many suffer *severe* gastrointestinal distress. It is probably the most common cause of mushroom poisoning in the United States, a tribute to its growth on lawns, tempting size (it borders on being irresis-

Chlorophyllum molybdites growing on a lawn. Note broadly cone-shaped cap of mature specimen at top right, and drumstick shape of buttons in foreground. Gills at far right have started to darken.

tible) and resemblance to *Lepiota rachodes, Coprinus comatus, Agaricus* species, etc.

COMMENTS: This distinctive summer mushroom always attracts attention because of its large size and handsome appearance (frisbee-sized specimens are commonplace). In much of inland and southern California and the Southwest, it is one of the commonest urban lawn mushrooms or "toadstools." It so closely resembles other large Lepiotas that some lepiotologists merely consider it an aberrant species with greenish spores (i.e., *Lepiota molybdites*). It is most likely to be mistaken for *Lepiota rachodes,* especially in the button stage. It differs, however, in its habitat (usually grass), fondness for hot weather, less pronounced staining reactions (though this character is variable), and less bulbous stem. The only completely reliable feature, however, is the spore color. But beware—the gills may remain whitish well into maturity, so that making a spore print is the only sure way to determine the spore color! Young buttons look like drumsticks when they first emerge from the ground. They may not show any white on the cap, whereas adults may be almost entirely white. *Lepiota morgani* is an older name for it.

Lepiota rachodes   (Shaggy Parasol)   **Color Plate 69**

CAP 5-20 cm broad, oval to convex or marshmallow-shaped, becoming plane in age or slightly umbonate; surface dry, at first pale brown to reddish-brown, cinnamon-brown, or brown, soon breaking up into large, coarse or shaggy scales as the cap expands, revealing the white to dingy buff background, the center usually remaining smooth and brown; margin usually fringed or shaggy. Flesh thick, white, fairly firm, typically bruising yellow-orange to orange and then reddish or brown when cut. **GILLS** broad, free, white, close, sometimes dingy brownish in old age or when handled. **STALK** 5-18 cm long, 1-3 cm thick at apex, enlarged below (*var.* **rachodes**), or with a large, abrupt basal bulb which may have a raised rim (*var.* **hortensis**); white when young, developing brownish stains in age or upon handling; smooth, dry; interior usually staining bright orange or saffron and then reddish to dark brown when cut. **VEIL** membranous, white or with a brown, ragged margin; forming a large, thick, collarlike, double-edged, superior ring on stalk which is usually movable in age. **SPORE PRINT** white; spores 6-13 × 5-9 microns, elliptical, smooth, with a large apical pore, thick-walled, dextrinoid. Cap cuticle composed of enlarged, erect cells.

HABITAT: Usually in groups or rings on ground under trees (particularly conifers) and bushes, in gardens and compost piles, near stables, on ant hills, along roads and other disturbed places, even in basements and greenhouses, sometimes also in open fields or in

Lepiota rachodes var. *hortensis* differs from the typical variety (shown in color plate) by having a more abrupt and pronounced, even rimmed (volva-like) basal bulb. Note how cap cuticle is smooth at first, then cracks into scales. Specimen at far right has been sliced to show the staining of the flesh.

the woods; widely distributed, but most common in western North America. It fruits whenever conditions are favorable, i.e. moist and mild, and often produces several crops a year. "Patches" are not particularly numerous, but can be *very* prolific—I've seen several hundred specimens under a single tree! Variety *hortensis* is quite common in our area, especially under planted conifers (e.g., cypress) and in sandy soil along the coast (I have found it on Ano Nuevo Island in sand and elephant seal dung, and in the splash zone at Pebble Beach!). The typical variety is more common in the Pacific Northwest.

EDIBILITY: Edible and excellent, with caution. A number of people have had severe "allergic" reactions to it, particularly on the west coast, and it is easily confused with *Chlorophyllum molybdites*. It has an exceptionally strong, nutty flavor but the water content is high. For the best and safest results, fry it on high heat in an open pan. Delicious!

COMMENTS: The outstanding features of this outstanding mushroom are the large brown scales on the cap, free white gills, prominent collarlike ring (not skirtlike as in *Amanita*!), basal bulb or thickened stem base, and brusing of the flesh to orange and then reddish. (The latter feature is best seen by cutting the stem.) The brightness and duration of the color changes vary according to the moisture content and age of the mushroom. In variety *hortensis* (the common one in our area), the basal bulb may have a raised rim which can be mistaken for a volva. However, the brown cap scales and collarlike ring point to *Lepiota*. The poisonous *Chlorophyllum molybdites* looks *very* similar, but has dull greenish spores. *L. rachodes* (sometimes spelled *L. rhacodes* and also known as *Macrolepiota rachodes* and *Leucoagaricus rachodes*) can also be mistaken for an *Agaricus* (especially *A. augustus*), but the white spores and gills distinguish it.

Lepiota procera   (Parasol Mushroom)

CAP 7-25 cm broad or more, at first oval, then expanding to convex, plane, or umbonate; surface dry, at first smooth and brown, soon breaking up into brown scales and patches except for the smooth, dark central umbo; flesh between the scales at first white but soon weathering to buff, grayish, or brownish and usually with a shaggy or torn-up appearance. Flesh white or tinged reddish, but not staining orange or red when cut; soft in age. **GILLS** free, broad, close, white when young, but sometimes discoloring to pinkish, tan, or dingy brownish in old age. **STALK** 12-40 cm long, 0.8-1.5 cm thick, typically very long and relatively slender with an enlarged base; pallid but the surface below the ring covered with numerous small brown scales or flakes which often separate into belts or wear away in age. **VEIL** membranous, forming a thick, superior, brown and white, double-edged ring on stalk; ring collarlike and movable. **SPORE PRINT** white; spores 12-18 × 8-12 microns, broadly elliptical, thick-walled, with an apical pore, smooth, dextrinoid.

HABITAT: Solitary to widely scattered or in small groups in open woods and at their edges, in old pastures, along trails, etc.; fairly common in the summer and fall in eastern North America (especially New England and the South) and Mexico. It (or something similar) also occurs in southern California, and it may very well grow in southern Arizona and New Mexico, where a number of so-called "eastern" species (e.g., *Lactarius indigo, Strobilomyces floccopus*) occur.

EDIBILITY: Edible and one of the very best of all agarics! Prized by connoisseurs for its strong, meaty-nutty flavor, it is now being grown commercially in Europe. It does not rot readily and is usually free of maggots. The leathery, sun-dried specimens frequently found in dry weather can be just as good as fresh ones if treated properly.

COMMENTS: Also known as *Macrolepiota procera* and *Leucoagaricus procerus,* this lofty, imposing mushroom is the most distinctive of all the Lepiotas. It is also the safest and the tastiest, although "allergic" reactions have been reported. It is much taller than *L.*

The parasol mushroom, *Lepiota procera*. **Left:** Side view of mature specimen. **Right:** Top view of two mature specimens. The fruiting bodies look like long drumsticks before the cap opens out.

rachodes, and doesn't stain red or orange when cut. The spore print is white (not greenish as in *Chlorophyllum molybdites*), and the brown stalk that breaks up into delicate branlike scales (which may wear away) is also distinctive. There is no volva at the base of the stalk as in *Amanita,* and the solid brown "nipple" or umbo at the center of the cap is also distinctive. The stalk is so slender in relation to its height that it is not uncommon for large specimens to topple over in old age from the weight of the cap.

Lepiota naucina   (Smooth Parasol; Woman On Motorcycle)

CAP 4-10 (15) cm broad, oval to nearly round when young, then broadly convex to plane; surface dry, typically dull white but at times gray, buff, or creamy, the center sometimes tinged pinkish-buff; usually smooth, but sometimes with numerous small branlike particles, or breaking up into scales in age; one form staining yellow when handled, another brown. Flesh thick, white, not bruising; odor mild, or in one form unpleasant. **GILLS** free, close, white, but often becoming buff, pinkish, or grayish-pink in old age, and finally brownish. **STALK** 5-15 cm long, 0.5-1.5 cm thick, equal or with an enlarged base; dry, white, without scales, sometimes staining yellow when bruised, and usually discoloring brownish in age or upon handling. **VEIL** membranous, white, forming a distinct, persistent, superior, double-edged, collarlike or sleevelike ring on stalk which is usually movable in age (and may fall off). **SPORE PRINT** white or very faintly pinkish; spores 7-9 × 5-6 microns, broadly elliptical, thick-walled, smooth, with an apical pore, dextrinoid.

HABITAT: Solitary to scattered or gregarious in grassy areas (lawns, pastures, etc.), sometimes also along roads, freeways, and in other disturbed places, and sometimes under trees or even in the woods; widely distributed and common. In our area it is often abundant in the fall and early winter, and occasionally encountered in the spring and summer as well.

EDIBILITY: Edible and very good, but not recommended. Either some persons are sensitive to it, or certain variants (perhaps the yellow-staining or ill-smelling ones) are toxic. According to one source, it is one of the most frequent causes of mushroom poisoning in the Pacific Northwest, yet it is listed as "edible" in every mushroom book! One former friend of mine who had eaten it previously was made quite ill by a cream-of-*L. naucina* soup which I had painstakingly prepared. There is also the more serious danger of carelessly confusing it with a deadly *Amanita*!

Lepiota naucina in its favorite milieu: grass. Note free gills and annulus. (Nancy Burnett)

COMMENTS: The white to grayish cap, membranous annulus (ring), free gills, white spores, and fondness for grass are the hallmarks of this cosmopolitan mushroom. Its appearance on lawns and in cemeteries marks the beginning of our fall mushroom season. A good way to become acquainted with it is to bicycle around town: there it will be—tall, white, stately, in graceful groups on lawns. Later it appears in pastures. It is a beautiful mushroom when young, the smooth unexpanded cap being reminiscent of a motorcycle helmet. The shape, though difficult to describe, is very distinctive (see photographs). Though the stalk base is often enlarged, there is never a sack (volva) as in the deadly Amanitas, and the ring is not skirtlike, nor is the cap viscid as in *Limacella*. The staining reactions are also an important fieldmark. In some specimens the cap and stem stain yellow quite dramatically in the tradition of *Agaricus xanthodermus*. Usually, however, they discolor yellow-brown to brown after handling and in age—especially on the stem. The frequent darkening of the old gills to pink or brown leads to confusion with *Agaricus*, and once again points out the folly of equating gill color with spore color. I have also seen buttons of *Agaricus californicus* interspersed with *L. naucina* that were virtually indistinguishable. A final hint: If your *"L. naucina"* fails to turn brown when cooked, double-check your identification, because you may have something else! ***Lepiota leucothites, L. naucinoides,*** and ***Leucoagaricus naucinus*** are synonyms.

Lepiota naucina, sometimes called the "Woman on Motorcycle" because of the helmet-shaped young caps.

Lepiota americana, a distinctive reddening species with a spindle-shaped stalk.

Lepiota americana   (American Parasol)

CAP 3-15 cm broad, oval becoming convex, plane, or broadly umbonate; surface dry, smooth at first, soon breaking into coarse vinaceous- to reddish-brown or pinkish-buff scales, the center usually remaining smooth; background white, but often reddening with age. Flesh white, bruising yellow to orange when young and fresh (especially in stalk), but aging or drying reddish to vinaceous. GILLS free, close, white, but may stain or age like the flesh. STALK 7-14 cm long, 0.5-2 cm thick, enlarged at base or often spindle-shaped (swollen at or below the middle, with a narrowed base); smooth, at first white, but aging or drying reddish to dark vinaceous. VEIL membranous, forming a white, double-edged, superior ring on stalk which may disappear in age. SPORE PRINT white; spores 8-14 × 5-10 microns, elliptical, smooth, with an apical pore, thick-walled, dextrinoid.

HABITAT: In groups or clusters in sawdust and compost piles, around old stumps, in waste places, rich soil, etc. Widely distributed and fairly common in eastern North America, but rare in California. I have found it only once in our area, growing with *Agaricus subrufescens,* in the summer and early fall.

EDIBILITY: Edible, but be sure of your identification! I haven't tried it.

COMMENTS: This species is easily recognized by its tendency to darken to reddish or burgundy as it ages or dries, and the yellow to orange staining of fresh specimens (best seen by cutting the stalk). In addition, the shape of the stalk is quite unusual—often very slender at the apex and blatantly bulbous at or toward the base (see photo). *L. badhamii* of Europe is very similar, but is said to be poisonous. There are also several closely related species in the southeastern U.S., including one unnamed variant with pale green spores and several species (e.g., *L. tinctoria*) with smaller, paler, and more numerous cap scales.

Lepiota cepaestipes   (Onion-Stalk Parasol)

CAP 2-8 cm broad when expanded, oval becoming broadly conical or bell-shaped to nearly plane or umbonate, eventually drooping; surface dry, powdery or mealy becoming somewhat scaly or fibrillose-scaly in age, white to pale pinkish (but may be darker when young and yellowish to brownish in age); margin clearly striate at maturity. Flesh thin, white, sometimes bruising yellowish. GILLS white, crowded, free. STALK 4-14 cm long, 3-6 mm thick, equal or swollen in places, with an enlarged base; smooth or powdery, slender, white, but may discolor yellowish when handled. **VEIL** white, forming a persistent, superior but easily detachable ring on stalk. **SPORE PRINT** white; spores 6-10 × 5-8 microns, elliptical, with an apical pore, smooth, thick-walled, weakly dextrinoid.

HABITAT: In groups or clusters in rich soil, wood chips, around old stumps, straw piles, gardens—in other words, in decomposing organic matter of almost any kind; widely distributed, but much more common in eastern North America than in the West. In our area it fruits in the summer or during warm, moist weather.

EDIBILITY: Not recommended. Though traditionally listed as edible, it has adverse effects on some people and is very thin-fleshed besides.

COMMENTS: Also known as *Leucocoprinus cepaestipes,* this species is easily recognized by its mealy, whitish, striate cap. It is essentially a whitish version of *L. lutea,* and the two used to be considered color forms of the same species. There are many closely related tropical and southern species, including: *L. breviramus,* similar but with soft warts on the cap (including the center) and usually growing scattered or tufted; *L. longistriatus,* with a tan to pale tan fibrillose cap; and *L. brebissonii,* with a small (2-3 cm) white cap with a dark gray to brownish center. I have found the latter on lawns in Berkeley, California, in the summer. Because of their mealy, striate caps, all of these species are placed in their own genus, *Leucocoprinus,* by many mycologists.

Lepiota lutea   (Yellow Parasol; Flower Pot Parasol)   **Color Plate 70**

CAP 2.5-6 cm broad when expanded, oval becoming broadly conical or bell-shaped, then eventually umbonate or plane and finally drooping; surface dry, powdery, mealy, and/or minutely scaly, usually scalier and less powdery in age; bright yellow to greenish-yellow or pale yellow, the center sometimes brown or buff; fading quickly after it matures or becoming browner; margin conspicuously striate nearly to center at maturity. Flesh very thin, yellow. **GILLS** free, crowded, yellow or pale yellow. **STALK** 3-10 cm long, 1.5-5 mm thick, usually slender and enlarged somewhat at or toward base base; dry, smooth or powdery like the cap, yellow. **VEIL** yellow, forming a small, superior, collarlike ring on stalk which may disappear. **SPORE PRINT** white; spores 8-13 × 5.5-8 microns, elliptical, with an apical pore, thick-walled, smooth, dextrinoid.

HABITAT: Solitary, tufted, or in groups in flower pots, greenhouses, and planter boxes, or if it is warm enough, outdoors (in lawns, gardens, etc.); widely distributed, fruiting indoors most anytime, outdoors mainly in the summer. The specimens in the color plate were growing in a planter box in front of the Bank of America in downtown Los Angeles.

EDIBILITY: Poisonous to some people, according to some sources.

COMMENTS: This brilliant yellow greenhouse mushroom boasts a plethora of pseudonyms, including *Leucocoprinus luteus, L. birnbaumii,* and *L. cepaestipes var. luteus.* It can be the object of considerable consternation to plant lovers when it pokes up in one of their flower pots. However, it won't hurt the plant (or you, unless you eat it). If one should appear, consider yourself lucky and take advantage of the situation—sprinkle it lightly as you would any other houseplant, and watch new individuals develop from tiny "pinheads" within a few days. The bright yellow color and striate cap set apart *L. lutea* from other Lepiotas. It has a *Coprinus*-like stature, but the spores are white and it doesn't deliquesce (though like many tropical fungi, it withers quickly). *Bolbitius vitellinus* is also yellow, but has a slimy-viscid cap. Other species: *L. flavescens* is another yellow greenhouse species with smaller, nearly round spores (5-7 microns broad); *L. fragilissimus* has a very thin, fragile cap and pale yellow to white gills and is a tropical and subtropical woodland species. *Lepiota luteophylla* has a brownish, non-striate cap and bright yellow gills. It was originally found in California but has also turned up in Michigan.

Lepiota eriophora   (Sharp-Scaled Parasol)

CAP 2-7.5 cm broad, oval or convex when young, broadly convex to umbonate or plane at maturity; surface dry, white to buff, covered with small, erect, pointed brown scales which rub off easily. Flesh white, rather thin, not staining. **GILLS** free, white to buff or tinged pinkish, close. **STALK** 2-10 cm long, 3-7 (10) mm thick, equal or enlarged slightly at base, dry, smooth at apex, covered with brown scales (like cap) below, but these often wearing away in age. **VEIL** evanescent, *not* forming a membranous ring on stalk, but sometimes leaving a fibrillose zone. **SPORE PRINT** white; spores 3-6 × 2-3 microns, oblong, smooth, dextrinoid.

HABITAT: Solitary, scattered, or in small groups in woods and in rich soil, widely distributed. Occasional in our area in the fall and winter, especially under cypress.

EDIBILITY: Unknown.

COMMENTS: This is one of a number of confusing, *Amanita*-like Lepiotas with erect, pointed scales on the cap. It was described in the first edition under the name *L. hispida*, a possible synonym. Closely related species include: *L. asperula*, with an annulus (ring) and stalk 1-2 cm thick; *L. scabrivelata*, a small subtropical species with yellow warts on the cap and stalk; and *L. acutesquamosa*, medium-sized to large (cap 5-15 cm) with large brown warts on the cap but few if any on the stalk, and veil often with brown scales on its underside but not necessarily forming an annulus (ring). The latter, which is said to be edible, is fairly common in eastern North America and also occurs in the Southwest.

Lepiota barssii   (Gray Parasol)

CAP 4-15 cm broad or more, nearly round to convex, becoming broadly convex to plane or umbonate; surface dry, covered with fine, flattened gray to brownish gray radiating fibrils, the center often darker; sometimes becoming scaly in age or direct sunlight. Flesh fairly thick, white to grayish, not staining when bruised. **GILLS** close, free, white or staining dingy yellowish to brownish. **STALK** 5-13 cm long, 0.5-2.5 cm thick, equal or more often swollen below, but with a tapered base (no bulb!); smooth; white or stained brownish. **VEIL** membranous, white, forming a superior, collarlike ring on stalk which may be movable or drop off at maturity. **SPORE PRINT** white; spores 7-11 × 5-6 microns, elliptical, smooth, dextrinoid.

HABITAT: Scattered to gregarious in cultivated or composted soil, lawns, gardens, pastures, and plowed fields; known only from the west coast. It is fairly common in Washington and Oregon, but rather infrequent in our area, where it fruits mainly in the summer and fall.

EDIBILITY: Edible and choice—but be sure of your identification!

COMMENTS: The grayish fibrillose cap, membranous ring, free white gills, absence of a basal bulb on the stalk, and habitat in cultivated ground are the telltale traits of this fine fungus. The stalk is often buried deep in the soil, especially if it is growing in mulch. Specimens growing on lawns or hard soil tend to be smaller, with a shorter stem. The flesh does not stain noticeably when cut as in *L. rachodes*. The veil is persistent but may be obliterated, in which case the free gills are an important fieldmark. It is much stouter and paler than *L. atrodisca,* and does not grow in the woods. It is darker and more fibrillose than *L. naucina,* and it can also be confused with an unidentified *Amanita* (p. 275) which has warts on the cap, amyloid spores, and yellowish gills in old age. In the Willamette Valley in Oregon *L. barssii* is quite common and frequently attains a large size; most of the specimens I've seen in California, however, have been smaller, averaging 4-10 cm broad. Other species: *L. excoriata* is vaguely similar, but has a cuticle that recedes from the margin as the cap expands.

Lepiota atrodisca has black to greenish-gray cap scales and a smooth stem. Note sleevelike annulus.

Lepiota atrodisca   (Black-Eyed Parasol)

CAP 1-5 (6.5) cm broad, oval or convex becoming broadly umbonate or plane, the margin sometimes uplifted in age; surface dry, white with flattened black, gray, or greenish-gray scales or fibrils, the center usually darker (blackish). Flesh thin, white, not bruising. **GILLS** free, white or creamy, close. **STALK** 2.5-10 cm long, 3-7 mm thick, usually slender but sometimes rather stout, enlarged somewhat at the base; smooth, dry, white or discoloring slightly upon handling. **VEIL** membranous, forming a fragile, sleevelike, white or black-edged ring at or above middle of stalk, or sometimes disappearing. **SPORE PRINT** white; spores 6-8 × 3-5 microns, elliptical, smooth, dextrinoid.

HABITAT: Solitary, scattered, or in small groups on ground or rotting wood under both hardwoods and conifers, fall and winter, apparently endemic to the west coast. During cold, dry weather I have seen enormous numbers in woodland thickets (oak-manzanita-hazel nut)—an unusual abode for a *Lepiota*—but it is not restricted to that habitat.

EDIBILITY: Unknown—do not experiment! It is too small to be of importance anyway.

COMMENTS: This attractive little woodland *Lepiota* is easily recognized by its unusual grayish-black scales. *L. felina* (see comments under *L. josserandii*) has dark brown to blackish scales on both the cap and stem, and an evanescent veil; it is poisonous.

Lepiota flammeatincta   (Flaming Parasol)

CAP 1.5-5 (7.5) cm broad, convex when young, then plane or with uplifted margin or sometimes broadly umbonate; surface dry, with nearly black to dark purple-brown to brown or reddish-brown fibrils or scales (sometimes very sparse) on a whitish background; surface quickly staining scarlet to scarlet-orange when bruised, then slowly turning dark brown or dark purple-brown. Flesh white, staining slightly pinkish, reddish, or orange when bruised, then fading. **GILLS** free, close, white, not staining when bruised. **STALK** 3-10 (15) cm long, 2-6 mm thick, equal or slightly thicker below, slender, white above the ring, fibrillose (like cap) below and quickly bruising scarlet to scarlet-orange like the cap surface, then discoloring dark brown. **VEIL** membranous, white, forming a median to superior, sleevelike ring on stalk, or disappearing. **SPORE PRINT** white; spores 6-8.5 × 4-5 microns, elliptical, smooth, dextrinoid.

HABITAT: Solitary or in small groups in woods and under trees, along trails, etc.; not uncommon along the Pacific Coast. In our area it fruits in the fall and early winter, but seldom in large numbers. *L. roseifolia* (see comments) is also fairly common.

EDIBILITY: Unknown.

COMMENTS: This is one of our most striking Lepiotas because of the spectacular scarlet-

304

Lepiota flammeatincta. Surface of cap and stalk stain bright reddish-orange when rubbed, but gills do not. A very similar species, *L. roseifolia* (not illustrated) has pinkish-staining gills.

staining ("flaming") of the cap and stem. The dark brown color that wounded areas subsequently assume is also characteristic. **L. roseifolia** is a very similar species whose gills turn pinkish or red-orange when bruised (within five minutes); it favors cypress in our area. An unidentified local species with a lovely purple to vinaceous-pink cap when young (browner in age) also occurs, usually under oak. Its cap and especially the stalk *sometimes* stain orange or orange-red when rubbed. **L. brunnescens** and **L. roseatincta** are two eastern scarlet-staining species; the latter has a pinkish-red cap.

Lepiota rubrotincta   (Red-Eyed Parasol)

CAP 3-8 cm broad, oval or rounded becoming convex, finally plane or broadly umbonate or with an uplifted margin; surface dry, at first uniformly pinkish-brown or reddish, then breaking up into flat, radially arranged fibrils or scales which vary in color from cinnamon-buff to coral-pink, reddish, or pinkish-orange; background whitish and the center remaining smooth and usually darker (deep red to chestnut); margin often splitting in age. Flesh

Lepiota rubrotincta. Cap has fibrils and a smooth dark center; common in the fall. (Joel Leivick)

thin, white, not bruising. **GILLS** free, white, close, not bruising. **STALK** 4-16 cm long, 0.4-1 cm thick, usually rather slender and equal or thicker below, often extending fairly deep into the humus; smooth, white, discoloring somewhat in age and becoming hollow. **VEIL** membranous, white, forming a thin, fragile but persistent ring on stalk; ring median to superior, typically sleevelike above and flaring below. **SPORE PRINT** white; spores 6-10 × 4-6 microns, elliptical, smooth, dextrinoid(?).

HABITAT: Solitary to scattered or in small groups in humus, usually in woods; widely distributed and sometimes common in our area, especially after the first fall rains.

EDIBILITY: Not firmly established. Katy Caldwell of Santa Cruz claims to have eaten a small quantity without ill effect, but it is easily confused with other Lepiotas, some of which are poisonous.

COMMENTS: This beautiful mushroom is one of our more common and easily identified woodland Lepiotas. It is larger and taller than *L. cristata,* and has a more persistent ring. The color of the cap fibrils varies considerably, but usually has a reddish tone. The smooth, dark center is suggestive of a human breast. Other species: ***L. roseilivida*** is somewhat similar but smaller, has a purplish cap, and occurs under various western conifers; ***L. glatfelteri*** has a vinaceous-brown fibrillose cap and has been found in rich soil and under cypress in Santa Barbara. Also similar is the beautiful unidentified species mentioned under *L. flammeatincta*; its cap is purple or pink at first and its stalk often bruises orange.

Lepiota cristata   (Brown-Eyed Parasol)

CAP 1-5 (7) cm broad, convex becoming broadly convex to plane or broadly umbonate; surface dry, soon breaking up into small tawny to brown or red-brown scales (often concentrically arranged) on a white background, the center usually remaining smooth and darker (brown to red-brown). Flesh thin, white, not bruising; odor mild or sweet and fruity or pungent. **GILLS** free, white to buff, close. **STALK** 2-8 cm long, 2-5 mm thick, equal or thicker below, slender, white to pinkish-buff and often darker toward base; more or less smooth, fragile. **VEIL** white, sometimes disappearing, at other times forming a thin, fragile, median to superior ring on stalk. **SPORE PRINT** white to pale buff; spores 5-8 × 3-5 microns, bicornute (wedge-shaped and spurred at one end), smooth, dextrinoid.

HABITAT: Scattered or in groups on ground in woods, under trees (especially redwood and cypress) or shrubs, on lawns, etc. Widespread; common in our area in fall and winter.

EDIBILITY: To be avoided—perhaps poisonous.

Left: *Lepiota cristata.* Note absence of prominent scales on stalk. **Right:** *Lepiota seminuda,* our smallest *Lepiota.* Note the free whitish gills and fragile veil that leaves remnants on cap margin.

COMMENTS: The chestnut to dark brown cap center, fragile ring, small size, and smooth stem characterize a number of Lepiotas which can only be differentiated microscopically. *L. castaneidisca,* for instance, has elliptical rather than bicornute spores but is otherwise identical. In my experience it is just as common in our area as *L. cristata.* There are many other similar species too numerous to mention. They include: *L. decorata,* with reddish to pink scales; *L. roseilivida,* with a purplish or purplish-pink, fibrillose cap; and *L. "tomentodisca,"* with pale to dark reddish-brown scales. In all three of these species the center of the cap is minutely hairy (tomentose), a character best seen with a hand lens.

Lepiota sequoiarum   (Boring Lepiota)

CAP 1.5-4 cm broad, oval becoming convex or finally plane; surface dry, smooth, without distinct scales, yellowish to tan at the center, pallid toward the margin. Flesh thin, white, not bruising. GILLS white, free, close. STALK 2-7 cm long, 2-5 mm thick, equal or thicker below, slender, white, smooth. VEIL membranous, white, forming a superior ring on stalk which often collapses in age. SPORE PRINT white; spores 7-9 × 3.5-4 microns, elliptical, smooth, dextrinoid.

HABITAT: Scattered or in small groups in woods and planted areas, infrequent. I have found it several times in the fall and winter under redwood. It is one of several small Lepiotas apparently endemic to the west coast.

EDIBILITY: Unknown—do not experiment!

COMMENTS: The smooth cap, small size, membranous ring, free white gills, and white spores are the decisive features of this lackluster *Lepiota.* The less said about it the better. A somewhat similar unknown species with a grayer cap when young occurs on wood chips.

Lepiota seminuda   (Lilliputian Lepiota)

CAP 1-2 (4) cm broad, conical or convex becoming plane; surface dry, white or tinged pinkish (especially at center), smooth or minutely powdery-mealy; margin often with veil fragments. Flesh very thin, white. GILLS free, white to pale pinkish, close. STALK 2-5 cm long, 1.5-3 mm thick, equal or slightly thicker below, white to dingy pinkish or tinged cinnamon toward base, thin and fragile, smooth or minutely mealy like the cap. VEIL evanescent, leaving remnants on the cap margin, but not a distinct ring on stalk. SPORE PRINT white; spores 3-5 × 2-3 microns, elliptical, smooth.

HABITAT: Solitary or scattered in humus under hardwoods or conifers; widespread. Occasional in our area in the fall and winter (e.g., under redwood) but easily overlooked.

EDIBILITY: Unknown. It is hardly worth troubling with such a trifle.

COMMENTS: Also known as *L. sistrata,* this is our smallest and most delicate *Lepiota.* Its lilliputian dimensions might lead to confusion with *Mycena* or *Collybia,* but it has free gills and a veil when young. It is smaller than *L. sequoiarum,* and the veil does not normally form an annulus (ring) on the stalk (see photo at bottom of p. 306).

Lepiota castanea   (Petite Parasol)

CAP 1-3 (4) cm broad, convex-umbonate to nearly plane; surface dry, covered with rusty-ochraceous to chestnut-brown to cinnamon-brown scales on a pallid to yellow-brown or ochraceous background (scales often densest at center). Flesh very thin, yellowish-buff; odor sometimes faintly sweet. GILLS close, usually free, white to buff or rusty-stained. STALK 3-8 cm long, 1-3 mm thick, very fragile and slender, equal or enlarged slightly at base; smooth above the veil, covered with rusty-ochraceous to dark chestnut-brown scales (like those on cap) below, often staining orange or yellow-orange when handled or in age.

Two deadly poisonous Lepiotas. **Left:** *Lepiota castanea.* Note scales on lower stalk. **Right:** *Lepiota josserandii* is easily confused with *L. cristata* (p. 306), but usually lacks an annulus (ring) on stalk.

VEIL fibrillose, evanescent, *not* forming a distinct ring on stalk, but sometimes leaving remnants on cap margin. **SPORE PRINT** white; spores 9-13 × 3.5-5 microns, bullet-shaped, smooth, dextrinoid.

HABITAT: Solitary to widely scattered or in small groups in rich humus in woods, especially under conifers. It has a wide distribution and is not uncommon in our area in the fall and winter, but never seems to fruit in large numbers and is likely to be overlooked.

EDIBILITY: POISONOUS! Like *L. josserandii,* it contains deadly amanita-toxins.

COMMENTS: The rusty-orange to chestnut-colored scales on cap and stem, free or nearly free gills, and lack of a distinct annulus (ring) typify this petite *Lepiota*. It might be mistaken for a *Cystoderma*, but the gills are usually free and the spores are dextrinoid. It is quite fragile and difficult to transport home in one piece.

Lepiota josserandii   (Deadly Parasol)

CAP 2-5 (7) cm broad, convex to plane or sometimes umbonate; surface dry, with cinnamon-brown to pinkish- or reddish-brown scales, the center darker and the background whitish to ochraceous, but reddish or rosy tints often developing in age or upon drying; margin often fringed with veil remnants. Flesh thin, pallid; odor faintly sweetish or musty, especially if several are left in a closed container. **GILLS** free or adnexed, close, white to creamy-yellow, not bruising. **STALK** 3-7 cm long, 0.3-1 cm thick, equal or enlarged downward, pallid at apex, fibrillose-scaly to near smooth and cap-colored (or slightly pinker) below. **VEIL** fibrillose, evanescent, not forming a distinct ring on stalk, but sometimes leaving a hairy zone. **SPORE PRINT** whitish; spores 6-8 × 2.5-4.5 microns, elliptical, smooth, dextrinoid.

HABITAT: Solitary, scattered, or in groups in cultivated ground, under bushes and trees (including cypress), on lawns, etc.; widely distributed and not uncommon in California. I have found it several times in our area in the summer and fall.

EDIBILITY: POISONOUS! Deadly amanita-toxins have been isolated in this species and several relatives. Fortunately, they are unlikely to be eaten because of their small size and infrequent occurrence. However, *L. josserandii* has been implicated in at least one fatality (in Albany, New York).

COMMENTS: This undistinguished little *Lepiota* (see photo at top of p. 308) is described here because it poses a threat to those who sample mushrooms wantonly, or think that only the Amanitas are deadly poisonous. It belongs to a large group of small, poorly-known, difficult-to-identify Lepiotas with no annulus (ring) or only a slight one, and a more or less fibrillose-scaly or cottony stem. Until they are better known, *none* should be eaten. Several occur in our area under cypress, including a beautiful pinkish-hued species that *might* be *L. subincarnata*, another poisonous species. Other deadly poisonous species include: *L. helveola,* very similar to *L. josserandii,* but with larger spores (7-10 microns long) and a slightly more membranous veil; and *L. felina,* with blackish to dark brown scales on both the cap and stalk. See also *L. castanea.* Other species: *L. cortinarius* is a medium-sized species (cap 3-10 cm broad; stalk 0.6-2 cm thick) that grows under northern conifers such as spruce. Its cap cuticle soon breaks up into numerous concentrically-arranged small brown to reddish-brown scales, its veil is evanescent, and its gills are remote from the stalk or attached to a collar. It does not age or stain reddish or pinkish; its edibility is unknown.

Lepiota clypeolaria   (Shaggy-Stalked Parasol)

CAP 2-8 cm broad, oval or bell-shaped becoming convex to nearly plane with a low, broad umbo; surface dry, soon breaking up into yellow-brown or brown scales except for the smooth, darker center; often yellower toward margin, which is soft and ragged from cottony veil remnants. Flesh white, not staining appreciably; odor sometimes pungent. **GILLS** free, close, white or creamy. **STALK** 4-12 (18) cm long, 3-7 (10) mm thick, usually slender and about equal, fragile; sheathed with soft, shaggy or cottony scales below the veil, usually yellow or colored like cap. **VEIL** cottony, leaving remnants on cap margin or a slight cottony-fibrillose ring on stalk. **SPORE PRINT** white; spores 12-20 × 4-6 microns, fusiform (elongated), smooth, dextrinoid.

HABITAT: Solitary, scattered, or in small groups in woods, widely distributed. In our area it favors conifers such as Douglas-fir, and is fairly common in the fall and winter. A related species or variant (see comments) occurs under oak in the winter and spring.

Lepiota clypeolaria, mature specimens. The shaggy stalk plus the smooth dark cap center typify this beautiful woodland mushroom. The cap is oval or convex before it expands.

Lepiota clypeolaria (form *"nabiscodisca"*). This form is not as shaggy as the one shown on the previous page, and does not show as much yellow. In our area it favors oak.

EDIBILITY: Said to be poisonous—all slender woodland Lepiotas are best avoided.

COMMENTS: The ragged or shaggy appearance of the cap margin and stalk, smooth "eye" at the center of the cap, absence of a distinct annulus (ring), free whitish gills, and long narrow spores are diagnostic. In prime condition it is one of our most beautiful and exquisitely adorned mushrooms. In age, however, it assumes a rather decrepit appearance —the stalk is weak and collapses easily. In North America *L. clypeolaria* is probably a "collective" species—that is, there are several varieties which may be distinct species. In addition to the form described above, we have in our area an oak-loving variety with fusiform spores, a more evenly colored and less ragged cap, and a whitish, only slightly shaggy stalk. I call it var. *"nabiscodisca"* because of its cookie-colored cap cuticle, but it may prove to be the "true" *L. clypeolaria* of Europe (and the *L. clypeolaria* described above may be *L. ventriosospora*—but such issues are best left to licensed lepiotologists). Other species: *L. clypeolarioides* of the Pacific Northwest lacks yellow tones, is not as shaggy, and has elliptical spores (6-9 microns long).

AGARICACEAE (Agaricus)

spores

Mostly medium-sized to large, terrestrial, saprophytic mushrooms. CAP smooth or scaly, typically neither brightly colored nor viscid. Flesh usually white. GILLS close, *free or nearly free* (at least at maturity), pallid or pinkish when young, *chocolate-brown to blackish-brown in age.* STALK central, fleshy, cleanly separable from cap. VEIL *present, membranous or cottony, usually forming an annulus (ring) on stalk.* VOLVA *typically absent* (except in *A. bitorquis* and relatives). SPORE PRINT *chocolate-brown.* Spores smooth, mostly elliptical or almond-shaped.

FROM both an economic and gastronomic standpoint this is unquestionably the most important group of gilled mushrooms, for it includes the familiar cultivated or "grocery store" mushroom as well as a large number of other delectable collectables. As defined here, the Agaricaceae include only one common genus, *Agaricus* (sometimes listed in older books as *Psalliota*). It is a remarkably clearcut, "natural" genus, and as such is a cinch to recognize: the stalk is typically furnished with an annulus (ring) but lacks a volva, the gills are pinkish or pallid when young but become chocolate-brown and are free at maturity, and the spore print is *always* chocolate-brown (the color of dark chocolate, *not* milk chocolate). *Stropharia* is somewhat similar to *Agaricus,* but usually has a viscid cap and attached gills that are never pink, while *Amanita* has white spores and white or pallid gills plus (usually) a volva. *Lepiota* and *Chlorophyllum* are very similar in general appearance, but have white and greenish spores respectively. (The Lepiotas and several miscellaneous smaller genera are included in the Agaricaceae by many mycologists.)

These photographs show the key fieldmarks of any *Agaricus:* presence of a veil that usually forms an annulus on the stalk plus gills that are free and chocolate-brown at maturity. **Left:** Close-up of a mature *Agaricus augustus* (p. 337). **Right:** Mature specimens of *A. arorae* (an unusual species—see p. 325). Note how cap separates easily from stalk.

Because so many of its species are edible, one often hears *Agaricus* characterized as a "safe" genus. This is hardly the case, however.True, no *Agaricus* is *deadly* poisonous, but *some* species produce mild to severe vomiting and diarrhea in *most* people, and *most* of the edible species (including the cultivated mushroom, *A. bisporus*) produce mild to severe vomiting and diarrhea in *some* people. In fact, *Agaricus* species are the most frequent cause of mushroom poisoning in our area, if not in California—not a very good track record for a "safe" genus! It therefore behooves you *Agaricus*-eaters to sample each and every species cautiously to determine your reaction to it, and to be *absolutely sure* of your identification!

Unfortunately, being "absolutely sure of your identification" is easier said than done, because *Agaricus* species are perplexingly polymorphic (variable in size, shape, color, etc.), and as a result, devilishly difficult to differentiate from each other. In view of the frequency with which *Agaricus* species are eaten, and the fact that there is no single "Simple Simon" rule for distinguishing the edible ones from the poisonous, it seems prudent to detail some of the more important characters used in separating them:

1) **Staining reactions** should be noted *when fresh* on the surfaces of the cap (near the margin) and stalk, the flesh in both the cap and stalk, and the flesh in the *extreme* base of the stalk. Some species do not stain at all; others are **rufescent,** (i.e., they stain red to orange or vinaceous when cut) or **lutescent** (i.e., they stain yellow to yellow-orange when bruised or rubbed repeatedly); still others are **latently lutescent** (i.e., they stain yellow only when a drop of 10% potassium hydroxide (KOH) or sodium hydroxide (NaOH) is applied to the surface of the cap near the margin).* Actually, potassium and sodium hydroxide dramatize the yellow-staining of *all* the naturally lutescent species, but in our area are especially useful for distinguishing the edible, non-staining *A. campestris* from the inedible or poisonous, latently lutescent *A. californicus.*

2) **Odor** should be noted by gently crushing the cap tissue as well as the flesh in the *very* base of the stalk. Three odors are especially prevalent in *Agaricus:* **mild** (or "fungal") to faintly fruity, as in the cultivated mushroom, *A. bisporus;* **sweet** (like anise or almond extract) as in *A. augustus;* and **phenolic** (an unpleasant chemical odor reminiscent of phenol, carbolic acid, creosote, library paste, ink, tar, or bleach), as in *A. xanthodermus.* However, these odors, when present, are not always obvious and it sometimes takes an experienced nose to detect them. It is interesting to note that there is often a correlation between the degree of lutescence and strength of odor—as a rule, the more quickly and brightly an *Agaricus* stains yellow, the stronger it will smell!

*Many household cleaning agents contain potassium or sodium hydroxide and can be substituted successfully. Drano (one teaspoon per ¼ cup water) and Lysol both work, but don't eat the tissue tested with these chemicals!

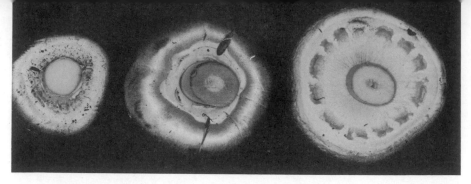

Veil detail in *Agaricus*. Left to right: *A. campestris*, with a thin cottony veil; *A. californicus*, with a thicker membranous veil and inrolled, lobed cap margin; and *A. arvensis*, whose membranous veil shows a cogwheel pattern of patches on its underside. (Ralph Buchsbaum)

3) **Veil characteristics:** The veil in *Agaricus* is actualy composed of two layers of tissue. In some species only one layer is clearly visible, while in others the lower layer (universal veil) can be seen as distinct patches of tissue on the underside of the upper layer (partial veil), as shown in the photograph above. The type of annulus(ring) formed by the ruptured veil is also significant: whether **skirtlike (pendant), sheathlike (peronate),** or **intermediate** between the two (see illustrations on this page).

4) **Spore size** is extremely useful in delimiting species, as are reactions to other chemicals besides potassium and sodium hydroxide. However, these features are not stressed here since most people will be unable to ascertain them. The best spores for measuring purposes, incidentally, are those that are deposited naturally (in the wild) on the annulus or stem.

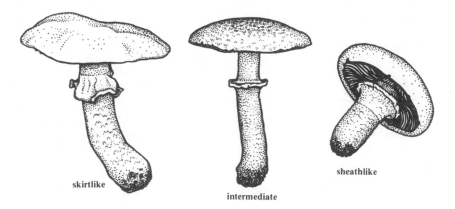

skirtlike

intermediate

sheathlike

Different types of rings in *Agaricus*. Left to right: *A. silvicola, A. praeclaresquamosus, A. bitorquis.*

What makes identification of *Agaricus* species so difficult is that all of the above characters with the exception of spore size are easily influenced or altered by the environment, as are grosser features such as the size, shape, color, and degree of scaliness of the fruiting body. Specimens growing in the open, for instance, can be differently colored or squatter or more scaly than specimens of the same species growing in deep shade. Dry or old specimens are apt to manifest their staining reactions and odors much more slowly or subtly than young, fresh individuals, and soggy specimens may not display their normal odor and staining reactions at all! It is obvious, then, that habitat and environmental conditions are among the first things to be taken into account when you are attempting to identify an unfamiliar *Agaricus*.

312

One practical way for beginners to overcome some of the vagaries and difficulties discussed is to approach the genus *Agaricus* at the level of section (subgroup) rather than species. The overwhelming majority of North American species will fit into one of seven groups or "sections." A few, such as *A. subrutilescens* and *A. arorae,* do not fit conveniently into any of them, and may well represent "bridges" between them. The seven sections are:

Section **Agaricus:** Not staining; annulus thin and slight or even absent (when present usually intermediate); KOH-negative; edible.

Section **Hortenses:** Not staining *or* rufescent; annulus skirtlike or intermediate; stature usually rather squat; KOH-negative; edible.

Section **Bitorques:** Not staining *or* rufescent; annulus sheathlike; flesh hard; stalk solid; KOH-negative; edible or sometimes too tough to eat.

Section **Sanguinolenti:** Rufescent; annulus skirtlike; stature usually erect; KOH-negative; edible.

Section **Xanthodermati:** Latently or strongly but fleetingly lutescent (i.e., staining yellow, then eventually discoloring brownish or vinaceous); annulus skirtlike or intermediate; odor phenolic (sometimes faint!); KOH-positive; poisonous!

Section **Arvenses:** Latently or strongly but *persistently* lutescent; annulus skirtlike; odor usually sweet; veil typically with two distinct layers; KOH-positive; edible.

Section **Minores:** Very much like section Arvenses and sometimes grouped with it, but with a small, usually slender and fragile fruiting body, and veil often with only one distinct layer.

Learning to recognize these "sections" has several advantages. The most obvious is that it hones critical skills while dealing with only seven entities instead of dozens. Moreover, you learn to group species according to their *chemistry,* which is precisely what edibility is all about! For instance, it appears that the phenol-smelling species (section Xanthodermati) poison a majority of people who eat them. The phenol odor is not always evident in the field, but usually becomes quite pronounced if the mushrooms are cooked. (The taste of these species is rather astringent or metallic, yet there are people who eat them not only with impunity, but with gusto!) It is also interesting to note that people who have an allergy to an edible species are likely to have allergies to other members of the same section, but not necessarily to species in other sections . For instance, someone allergic to *A. augustus* is likely to be adversely affected by *A. arvensis* also. Similarly, someone who can eat the cultivated mushroom *(A. bisporus)* will probably not have trouble with other species in the section Hortenses. Thus it clearly pays for prospective *Agaricus*-eaters to learn something of the chemistry and interrelationships of the common species, in addition to their critical fieldmarks.

The best known—and perhaps the most mediocre—of the edible *Agaricus* species is undoubtedly the cultivated mushroom, *A. bisporus.* The meadow mushroom or "champignon," *A. campestris,* is also very popular, but the best species are somewhat lesser known—*A. augustus, A. subrufescens, A. arvensis, A lilaceps,* and *A. bernardii* are choice mushrooms if there are any, and *A. bitorquis* has few peers. *Agaricus* species are also notable for their beauty. Some rival the Amanitas for stateliness and elegance, while others are as plump and meaty as boletes.

Agaricus species are saprophytic and are among the most conspicuous urban and suburban mushrooms. Not only is the cultivated or "button" mushroom, *A. bisporus,* to be found in practically every grocery store and produce stand, but its untamed (and in some cases unnamed) cousins fruit prolifically on lawns (where they are often kicked over like other "toadstools"), in cemeteries, under hedges and shrubbery, along roads and sidewalks, in gardens, and on compost piles. Many species are rural as well, fruiting by the bushel in pastures and fields, and some also favor the woods, but comprise a much smaller percentage of the fleshy fungi found there. Few, if any, are mycorrhizal. It is interesting to note that in our area cypresses—particularly old ones—harbor a remarkable wealth of species, including both cosmopolitan and rare or endemic ones.

Agaricus species fruit whenever conditions are favorable—that is, moist and mild. In our area the greatest variety can be found shortly after the first fall rains, but another crop appears in the spring, and if it is mild enough, in the winter. Some species also fruit on watered lawns in the summer, and *A. augustus* is a common summertime mushroom in the coastal fog belt.

Agaricus is a large genus centered in the tropics and subtropics, hence one would expect to find more species in the southern United States than in the north. About 200 species are estimated for North America, but no all-encompassing critical study has been made. In California the total number of known species is about 60, most of which occur in our area. Many of the 24 species described here have wide distributions, but some are presently known only from California.* Readers outside "our area" will undoubtedly enounter several species they can't identify, but with a good nose and a keen eye they should be able to assign most of them to one of the seven "sections" already described.

Key to the Agaricaceae (Agaricus)

Note: As already pointed out, many of the characters used in this key are subject to environmental influence—particularly the intensity (or even presence) of staining reactions and odor. It is therefore imperative to have several specimens of each species in hand, preferably collected in conditions neither exceptionally wet nor unusually dry.

1. Very base of stalk giving off an unpleasant odor (like phenol or library paste) when crushed; base sometimes also staining yellow when cut or crushed 2
1. Not as above; odor not phenolic; base of stalk typically not staining yellow (but may stain orange or yellow-orange), or if staining yellow then odor sweet 7
2. Extreme base of stalk staining *bright* yellow when cut or nicked; other surfaces often staining yellow also 3
2. Base of stalk not staining bright yellow (but may stain faintly yellow) 4
3. Cap with inky-gray to grayish-brown to brown or dark brown fibrils or fibrillose scales; surface of cap not bruising yellow or only sometimes bruising yellow 6
3. Cap white or discoloring grayish to buff or tan; surface usually bruising bright yellow when rubbed repeatedly, especially near margin .... *A. xanthodermus,* p. 329
4. Cap with pale pinkish-brown to fawn-colored, tan, or reddish-brown fibrils (or at times nearly white); annulus (ring) on stalk thick and feltlike; stalk 1-3 cm thick at apex, the base enlarged; cap 6-15 cm or more broad; growing in woods or under trees .... *A. hondensis,* p. 326
4. Not with above features 5
5. Cap 3-9 cm broad, entirely white or whitish with a brown center, or sometimes brownish throughout; stalk 0.5-1 (1.5) cm thick; found in many habitats on west coast, but especially common in urban and suburban areas, in grass, under cypress and oak, etc. 52
5. Not as above; if found in West then cap (4) 5-25 cm broad, with inky gray to grayish-brown, brown, or dark brown fibrils or fibrillose scales (never entirely white) and stalk 1-3 cm thick, and usually growing in woods or along roads and paths through the woods 6
6. Found in eastern North America; cap typically finely fibrillose ....
.... *A. placomyces & A. pocillator* (see *A. praeclaresquamosus,* p. 329)
6. Widespread but most common in West; cap fibrillose to scaly *A. praeclaresquamosus,* p. 329
7. Some part of fruiting body (especially cap surface) staining yellow when bruised *and/or* the flesh smelling sweet when crushed (like almond extract or anise); cap surface typically yellowing in KOH 8
7. Not as above (odor rarely slightly almondy, but if so then flesh reddening when cut) 25

*Several of the species depicted or mentioned in this chapter have been named and described by Rick Kerrigan in an article which, at the time of this writing, is about to be published. The provisional names he has given them are used here with his permission. They are: *A. fuscovelatus, A. arorae, A. sequoiae, A. vinaceovirens, A. blandianus, A. smithii, A. summensis, A. perobscurus,* and *A. rubronanus.* Another provisional name, *A. pinyonensis,* is used here with the permission of Bill Isaacs and Chuck Barrows.

8. Cap white or whitish when young and fresh (but may discolor yellow to buff or amber in age) 9
8. Cap not white (i.e., with distinctly colored cuticle, fibrils, or scales, at least at center) 16
9. Growing in grass (lawns, pastures, etc.) *or* in pinyon-juniper forests 10
9. Not typically growing in above habitats 12
10. Cap small (2-5 cm broad), often with a brownish to buff-tinged center; stalk less than 1 cm thick;
    spores 5.5 microns long or less .... *A. comtulus* (see *A. micromegathus,* p. 340)
10. Not as above; usually larger and spores larger 11
11. Stalk stuffed or sometimes hollow; fruiting body medium-sized, or if large than not particularly
    squat; cap surface usually smooth (or with a few fissures); widely distributed, but favoring
    lawns over pastures in California; spores 7-8.5 microns long .... *A. arvensis,* p. 332
11. Not as above; either growing in pinyon-juniper forests *or* if growing in grass, then stalk usually
    solid, fruiting body often large (cap 7-50 cm broad), the cap smooth to conspicuously scaly or
    warty, stature often robust or squat; especially common in pastures and prairies but also on
    lawns; spores either smaller or larger than above 41
12. Fruiting body often robust, quickly staining amber when bruised (especially the cap surface)
    and often aging amber overall; cap often fibrillose; odor *strongly* almondy or anise-like; im-
    mature gills often staining yellow; known only from the West ..... *A. albolutescens,* p. 335
12. Not as above 13
13. Fruiting body robust (stalk 2 cm thick or more); cap typically with slightly colored (yellowish)
    fibrils; rare .... *A. summensis & A. augustus* (white form) (see *A. augustus,* p. 337)
13. Not as above; common 14
14. Cap surface typically staining distinctly yellow when rubbed repeatedly, at least at the margin;
    common and widespread .... *A. silvicola group,* p. 334
14. Not as above; yellow-staining weak or erratic; found in eastern North America *or* in western
    mountains 15
15. Cap white, 4-8 (15) cm broad; odor pungent or faintly sweet; spores only 4-5 microns long; fairly
    common in woods of eastern North America .... *A. cretacellus*
15. Cap white or silvery; gills remaining pink for a long time; found in western mountains ......
    .... *A. chionodermus* (see *A. silvicola* group, p. 334)
16. Fruiting body fairly small; cap 1-5 (7) cm broad; stalk usually less than 1 cm thick 17
16. Fruiting body medium-sized to very large; not as above 19
17. Growing in grass; odor usually anise-like .... *A. micromegathus* & others, p. 340
17. Growing in woods or under trees; odor mild or aniselike 18
18. Odor distinctly anise-like; cap fibrils brown to reddish-brown ....
    .... *A. semotus* (see *A. micromegathus,* p. 340)
18. Odor faint or absent; cap fibrils pink, purple, purple-gray, or vinaceous when fresh ....
    .... *A. diminutivus group,* p. 340
19. Growing in pastures or grass; fruiting body robust, often squat; cap often with large warts or
    scales, but sometimes smooth, whitish becoming yellowish or pale brownish ....
    .... *A. crocodilinus* (see *A. osecanus* group, p. 333)
19. Not growing in grass, or if growing in grass then not as above 20
20. Growing in compost or rich soil; cap with pallid to pale brown to faintly grayish, pale pinkish-
    brown, fawn-colored (or sometimes brown or tawny) fibrils, sometimes slightly ruddy or
    yellowish in age; stalk *not* scaly or shaggy (or only obscurely so) below the veil; stalk often with
    a swollen base; odor *strongly* sweet; spores 5.5-7 microns long ..... *A. subrufescens,* p. 336
20. Not as above; often but not always growing in woods 21
21. Cap with pink or purple tints; not common .... *A. lilaceps,* p. 323
21. Not as above 22
22. Stalk typically rather slender (1-1.5 (2) cm thick at apex), but usually with a bulb at base; fibrils on
    cap ochraceous-orange to tawny; found in the coastal forests of northern California and the
    Pacific Northwest, especially under Sitka spruce ...... *A. smithii* (see *A. augustus,* p. 337)
22. Not as above 23
23. Fibrils on cap yellow when young, ochraceous or tawny in age; rare ....
    .... *A. summensis* (see *A. augustus,* p. 337)
23. Not as above; fibrils brown to gray to tawny, but not yellow 24

24. Fibrils on cap distinctly gray to grayish-brown to nearly black (at center) when young, but often paler in age; scales on stalk below the veil often scanty; gills usually passing through a pinkish phase as they mature; known only from California .... *A. perobscurus,* p. 339

24. Not as above; fibrils on cap brown to tawny; stalk usually conspicuously shaggy or scaly below the veil, at least when young; gills rarely pink; widely distributed ...... *A. augustus,* p. 337

25. Flesh normally staining red to orange or vinaceous when cut or rubbed repeatedly 26

25. Not as above (but cap and flesh may have reddish stains in old age or wet weather) 38

26. Flesh in base of stalk staining orange to yellow-orange when cut; flesh elsewhere reddening at least somewhat; cap with broad, flattened, chocolate-brown scales, often depressed centrally at maturity; growing in woods, very rare (reported from the Pacific Northwest) .. *A. lanipes*

26. Not as above 27

27. Unbroken veil (i.e., underside of broken veil) soon brown to grayish-brown, purple-brown, or chocolate-brown to nearly black 28

27. Not as above 30

28. Stalk averaging 2.5-4 cm thick at apex, often bulbous   *A. pattersonae* (see *A. lilaceps,* p. 323)

28. Stalk usually more slender, or if thick then *without* a bulb at base 29

29. Underside of unbroken veil soon purple-gray to brown or dark brown; flesh staining only slightly (often orangish) when cut; stalk typically less than 2 cm thick ...... *A. fuscovelatus,* p. 324

29. Not as above 30

30. Veil forming a sheathlike annulus (ring) on stalk; cap whitish to buff or sometimes dingy brownish, sometimes cracked into scales; stalk solid; fruiting body very firm or hard; growing in disturbed soil, mud flats, on lawns near ocean, etc.; coastal in distribution *A. bernardii,* p. 322

30. Not as above; veil forming a skirtlike to intermediate annulus 31

31. Growing in compost or manured soil .... *A. bisporus,* p. 319

31. Not as above 32

32. Fruiting body large and dense (cap 8-25 cm broad when mature, stalk 2-7 cm thick; cap brown or sometimes with pinkish, purplish, and/ or ochre or yellow-orange tones; veil often yellowtinged before breaking, forming a skirtlike annulus (ring) on stalk; found mainly under cypress and other planted trees, but not in manure .... *A. lilaceps,* p. 323

32. Not as above 33

33. Margin of cap inrolled when young; stature of fruiting body usually rather stout or squat; annulus (ring) on stalk intermediate or skirtlike 54

33. Not as above; annulus usually skirtlike 34

34. Flesh staining distinctly red 35

34. Flesh staining dingy vinaceous or vinaeous-brown or dingy reddish (usually weakly) 38

35. Cap white or whitish, sometimes with faintly grayish or brownish fibrils ....
.... *A. benesi* (see *A. fuscofibrillosus,* p. 325)

35. Not as above; cap darker 36

36. Common under mountain conifers (especially spruce and fir) in southern Rocky Mountains and Southwest .... *A. amicosus* (see *A. arorae,* p.325)

36. Not as above 37

37. Stalk usually less than 1 cm thick; cap 3-7 (10) cm broad; cap surface yellowing in KOH; known only from California, usually growing in mixed woods and under oak .... *A. arorae,* p. 325

37. Not as above; stalk usually thicker *and/ or* growing under cypress; cap surface *not* yellowing in KOH; widespread .... *A. fuscofibrillosus* & others, p. 325

38. Veil forming a sheathlike or even volva-like annulus on stalk *or* a large collarlike band (with both upper and lower edges free from stalk); found in hard-packed or disturbed soil or sometimes in grass or under cypress; texture very firm; stalk solid .. *A. bitorquis* & others, p. 321

38. Not as above; veil not sheathlike or bandlike and/ or stalk with a central hollow in age ... 39

39. Cap white or whitish when fresh (but may discolor tan, yellowish, or buff in age) 40

39. Cap colored at least at the center, or with colored fibrils or scales 42

40. Cap medium-sized to very large (10-50 cm broad when mature); either growing in grass *or* in pinyon-juniper forests 41

40. Not as above; smaller and/ or habitat different 42

41. Found with pinyon and juniper in the Southwest .... *A. pinyonensis,* p. 331
41. Found in grassy places   *A. osecanus* group & *A. crocodilinus* (see *A. osecanus* group, p. 333)

42. Cap 1-4 cm broad when expanded; stalk 2-6 mm thick; fruiting body fragile 43
42. Not as above; typically larger (cap 3 cm broad or more; stalk 5 mm thick or more) 44

43. Cap grayish to grayish-brown, granular or powdery; margin usually hung with veil remnants; gills deep pink to blood-red, then darker; spore print reddish or tinged greenish when moist, drying darker brown; growing in greenhouses, leaf litter, rich soil, bogs, etc.; widely distributed but not common .... *Melanophyllum echinatum*
43. Not as above; cap with reddish-brown to pinkish or purplish fibrils, *not* granular or mealy; growing in woods .... *A. diminutivus* group, p. 340

44. Cap silvery-white or white; fruiting body often rather erect and slender; growing in mountain forests of the West .... *A. chionodermus* (see *A. silvicola* group, p. 334)
44. Not as above 45

45. Cap covered with dark brown to purple-brown or wine-colored fibrils or fibrillose scales; stalk shaggy or sheathed by cottony white scales below the veil, at least when young; common in woods .... *A. subrutilescens,* p. 326
45. Not as above 46

46. Growing in the open (usually in grass, occasionally in hard-packed soil); veil typically thin and somewhat cottony, forming only a slight annulus (ring) on stalk or disappearing (annulus usually intermediate, *not* skirtlike); gills usually pink or brownish in button stage; cap surface *not* yellowing in KOH 47
46. Not as above; habitat different *and/or* veil typically membranous and forming a distinct intermediate to skirtlike annulus; gills pinkish, brown, *or* white in button stage 49

47. Cap white or sometimes with pale brown to grayish fibrils or scattered scales; very common and widespread .... *A. campestris* & others, p. 318
47. Cap covered with brown to grayish-brown or reddish-brown fibrils, even when young ... 48

48. Cap often ruddy or reddish in age or after handling; growing in lawns or hard-packed soil along roads; not common .... *A. rutilescens* (see *A. cupreobrunneus,* p. 319)
48. Not as above; growing in lawns or pastures; common .. *A. cupreobrunneus* & others, p. 319

49. Stature erect (stalk more than 5 cm long); cap white or pallid (pale buff) 50
49. Not as above; cap darker *and/or* stature rather robust and stocky 52

50. Stalk 8-30 cm long; found in redwood and mixed forests of coastal northern California, usually near rivers or in bottomlands .... *A. sequoiae* (see *A. pinyonensis,* p. 331)
50. Not as above; stalk generally up to 9 cm long; habitat usually different 51

51. Veil usually disappearing or leaving scaly zones on mid- or lower stalk rather than forming a distinct annulus (ring); found under trees or on coastal bluffs; rare .... *A. altipes*
51. Not as above; common 52

52. Cap white or pallid; veil ample, usually with a cogwheel pattern of patches on underside; stalk rather stout (3-6 cm long); growing in pastures, sometimes in rings; known only from the central California coast; not common .... *A.* **sp. (unidentified)** (see *A. campestris,* p. 318)
52. Not as above; very common 53

53. Cap covered with purplish to purple-gray or purple-brown fibrils, the margin usually white; found in woods and under trees in southeastern North America .... *A. rhoadsii*
53. Not as above 54

54. Gills pinkish to brownish in button stage; fruiting body usually stout, stalk usually 1-4 cm thick; found in manure, compost, rich soil, hard-packed ground, and under cypress (or rarely in the woods); odor never phenolic; cap surface *not* yellowing in KOH or rarely yellowing slightly 55
54. Gills pallid or whitish in button stage (but often pink after veil breaks); stalk 0.5-1 (1.5) cm thick; found almost anywhere (including lawns, gardens, woods, under cypress), but only rarely in manure or compost piles; odor sometimes phenolic; cap surface yellowing in KOH .... *A. californicus,* p. 327

55. Growing in hard-packed ground, often in towns or cities (often developing underground, then pushing through); veil thick, often forming a double ring; cap surface usually breaking up in age to form large brown to reddish-brown scales .... *A. vaporarius*
55. Not as above; found in compost, manure, under cypress, or occasionally in other habitats; cap white or with brown fibrils or fibrillose scales; common ...... *A. bisporus* & others, p. 319

Agaricus campestris   (Meadow Mushroom)   Color Plate 71

CAP 4-11 (15) cm broad, convex or dome-shaped for a long time, then often becoming plane; surface dry, smooth or silky-fibrillose, pure white, or sometimes with a few grayish to brown or cinnamon-buff fibrils or fibrillose scales; margin extending beyond the gills, often hung with veil remnants. Flesh thick, white, not staining when bruised but sometimes discoloring brownish or reddish in age or wet weather (especially just above the gills); odor mild. GILLS close, free at maturity, pale pink in button stage, then bright pink, becoming purple-brown to chocolate-brown and finally blackish-brown. STALK 2-6 (10) cm long, 1-2.5 cm thick, usually with a tapered base; firm, white, smooth above the veil, often with a few fibrils below; stuffed or hollow (but see comments). VEIL thin, somewhat cottony, white, forming a thin ring on stalk or leaving remnants on cap margin or disappearing entirely; ring rarely well-formed, intermediate (sometimes flaring) or rarely skirtlike, median to superior. SPORE PRINT chocolate-brown; spores 6.5-8.5 × 4-5.5 microns, elliptical, smooth. Cap surface *not* yellowing in KOH. Basidia mostly 4-spored.

HABITAT: As its name implies (*campestre*=field), this is a grassland species, occurring throughout the world from sea level to above timberline. It usually grows in groups or rings in lawns, pastures and meadows, cemeteries, golf courses, baseball fields, etc., but can also be solitary. It fruits practically year-round in our area, but is most abundant in the fall and early winter, when stupendous crops are sometimes produced in our pastures. Thousands of pounds, in fact, go unpicked every year. Though they can often be seen from the road, fungophobic Americans drive right by them, some undoubtedly on the way to the store to buy mushrooms!

EDIBILITY: Edible and excellent, both raw and cooked. It is the most widely picked mushroom in English-speaking countries (where it is also known as the "pink-bottom"), and is the popular *champignon* of France. Those who equate it with the cultivated mushroom *(A. bisporus)* do it an injustice. Its flavor is *far* superior, though the texture is softer. When the gills are bright pink and the cap pure white, it is as beautiful as it is delicious.

COMMENTS: The silky white cap, bright pink gills when young, poorly defined or evanescent annulus (ring), tapered stem, absence of a volva, stocky stature, chocolate-brown spores, and growth in grass are the principal fieldmarks of this universal favorite. It is the type species of *Agaricus* as well as the entire order Agaricales. The gills are pinkish even in the button stage, unlike *A. californicus*. The cap is white or has scattered brownish scales, but is not brown and fuzzy in youth as in *A. cupreobrunneus*. The cap remains dome-shaped for a long time, and its surface does not bruise yellow, though I have encountered a solid-stemmed variety in New Mexico whose cap sometimes ages faintly yellowish. This variety may be *A. **solidipes**,* an edible southern species with slightly larger spores than *A. campestris* (but identical in most other respects, including habitat).

In most regions the meadow mushroom is a perfectly safe mushroom for beginners. Unfortunately, in California it is frequently confused with the mildly poisonous phenol-smelling *Agaricus* species. Compare it especially carefully with *A. californicus* (see comments under that species), which has a more persistent, membranous veil and whitish gills in the button stage. A small, anonymous *Amanita* (p. 275) sometimes grows with *A. campestris* in pastures and looks very similar from the top. However, its gills are white to to yellow-orange, never pink or chocolate-brown. The poisonous destroying angels *(Amanita ocreata, A. virosa,* etc.) can also be distinguished by their white gills, plus the presence of a volva (sack) at the base of the stem. Other species: An unidentified, probably unnamed *Agaricus* occurs occasionally in our pastures. It resembles *A. campestris* in size, shape, and color, but has an ample veil with a cogwheel pattern of patches on its underside. Its veil, in fact, is reminiscent of *A. arvensis*, but the fruiting body is smaller and does not have a sweet odor. It stains yellow in potassium hydroxide (KOH), but does not normally yellow naturally, and its edibility has not been determined (it might be a member of the *A. xanthodermus-A. californicus* group, and therefore unsavory or toxic).

Agaricus cupreobrunneus (Brown Field Mushroom) **Color Plate 73**

CAP 2-7 cm broad, convex becoming broadly convex to plane or slightly uplifted; surface dry, more or less tomentose (hairy) from a layer of small brown to grayish- or reddish-brown fibrils or fibrillose scales; margin usually extending beyond gills. Flesh white, rather soft and fragile, not staining when bruised but often discoloring reddish to brownish in old age or wet weather, especially just above the gills; odor mild. GILLS free at maturity, close, pale dingy pinkish in button stage, becoming pinkish-brown to purple-brown, then chocolate-brown to blackish-brown. STALK 2-4 cm long, 0.7-1.2 (2) cm thick, usually equal but sometimes thicker at either end; white, smooth above the veil, often somewhat scurfy or scaly below; hollow or stuffed. VEIL thin, white, somewhat cottony, forming a thin ring on stalk; ring more or less median, intermediate (sometimes flaring), well-formed or disappearing, sometimes with a slight second ring below it. SPORE PRINT chocolate-brown; spores 7-9 × 4-6.5 microns, elliptical, smooth. Cap surface *not* yellowing in KOH. Basidia mostly 4-spored.

HABITAT: Solitary to scattered or gregarious (often in rings) in pastures, lawns, and other grassy places, but showing a special affinity for poor soil; very common in California and probably widespread. In our area it fruits prolifically in the fall and early winter, often growing with and outnumbering *A. campestris.*

EDIBILITY: Edible. The flavor is comparable to that of *A. campestris,* but the texture is much softer and more fragile. Also, the buttons tend to develop underground, making it difficult to find specimens with pink gills.

COMMENTS: For many years this species has passed as a small, brown, fuzzy form of *A. campestris.* It shares with that species certain fundamental characteristics: e.g., a poorly defined or even absent annulus (ring), stocky stature, and growth in grass. However, it has a softer texture than *A. campestris,* and is smaller and brown-capped from the button stage on. Other species: *A. porphyrocephalus,* widely distributed, is very similar but has dull reddish-brown to purple-brown fibrils or scales on the cap, firmer flesh, and smaller spores. *A. rutilescens* is also similar but has a ruddier complexion and a tendency—at least in some forms—to stain reddish. It grows on lawns or in hard-packed soil along roads, but is not common. *A. argenteus* is a grass-loving species with a silky or silvery cap with brown to grayish-brown fibrils, larger spores, and a better-defined annulus. It is common in the southern United States. All three of the above species are edible.

Agaricus bisporus   (Cultivated Mushroom; Button Mushroom)

CAP 3-16 cm broad, convex when young, often plane or even slightly depressed in age; surface dry, in one form entirely white, but more often with flattened pale brown to brown fibrils which in age or dry weather often break up into fibrillose scales; margin inrolled when young, often extending beyond the gills. Flesh thick, very firm, white, usually (but not always) discoloring somewhat (brown to reddish or pinkish-orange) when cut and rubbed repeatedly; odor mild or faintly fruity. GILLS free at maturity, close, pinkish or pale brown when young, purple-brown to chocolate-brown in age, and finally blackish-brown. STALK 2-8 cm long, 1-3 (4) cm thick, usually stout, very firm, equal or enlarged at base; white or turning dingy brownish with age, smooth or slightly cottony-scaly below ring. VEIL membranous, cottony, white, two-layered, typically forming a delicate, median to superior ring on stalk which may collapse in age; ring intermediate or sometimes skirtlike, its upper surface often striate. SPORE PRINT chocolate-brown; spores 5.5-8.5 × 4-6.5 microns, elliptical, smooth. Cap surface *not* yellowing in KOH. Basidia mostly 2-spored.

HABITAT: Scattered to densely gregarious or in clumps in compost and manure, rich soil, along paths and in gardens, and also very common under cypress, but only rarely found in woods or on lawns; very widely distributed. In our area it "fruits" year-round in markets. The major fruiting in the wild is typically in the fall or winter.

Agaricus bisporus (= A. brunnescens) looks like the cultivated mushroom, which it is. Specimens at far right are mature; next to them is a button. Cap usually has brown fibrils.

EDIBILITY: Edible—and if popular demand is any indication—choice (over a half billion pounds are cultivated *annually* in the United States!). However, as with any mass-produced agricultural product, flavor has been sacrificed for appearance, keeping quality, yield, and disease resistance. Pesticides are used, of course, and the result is, in Valentina Wasson's felicitous phrase, "a sickly simulacrum" of what a mushroom should be. The "wild" form is slightly better—especially the young, hard buttons.

COMMENTS: Also known as **A. brunnescens** and **A. hortensis** (the latter name is applied only to the white form), the cultivated mushroom mimics the meadow mushroom, *A. campestris,* but has 2-spored basidia, a well-developed ring, a browner cap, slightly reddening flesh, and does not normally grow in grass. If you use mushroom compost (from a mushroom farm) in your garden or put old cultivated mushrooms in your compost pile, you are likely to get some sooner or later. The largest fruitings I've seen have been around old compost piles and under cypress, where, curiously, it is one of our most common mushrooms. Care should be taken not to confuse it with *A. californicus,* which is also common in gardens and under cypress. The latter is more slender and usually smells slightly of phenol, at least when cooked. For a more detailed comparison, consult couplet #54 of the key to *Agaricus.* A very similar edible species with more rapidly reddening flesh, skirtlike ring, and 4-spored basidia, **A. blandianus,** also occurs quite commonly under cypress; **A. subfloccosus** is another similar edible species with 4-spored basidia, but it has a whitish cap with small, cottony scales (squamules) near the margin, and frequently has a strong odor as well. It occurs under conifers, including cypress, and is widely distributed, but not common. See also *A. spissicaulis* (under *A. arorae*).

The cultivated or "button" mushroom, *Agaricus bisporus.* Large specimens from a mushroom farm.

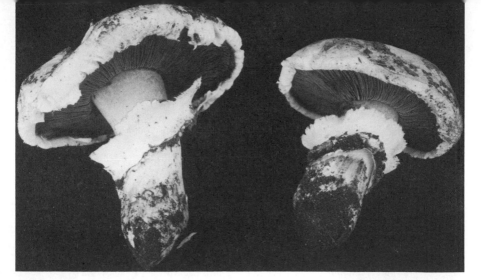

Agaricus bitorquis (=A. rodmani), mature specimens. Note prominent, flaring bandlike annulus. The flesh is exceptionally firm, especially in the stalk. For a view of it in its natural habitat, see color plate.

Agaricus bitorquis (Banded Agaricus; Urban Agaricus) Color Plate 80

CAP 4-18 cm broad or more, broadly convex becoming plane or with a slight central depression; surface dry, smooth or occasionally cracking into scales; white or whitish but often dirty; not bruising yellow but sometimes discoloring sordid yellowish to tan in old age; margin inrolled when young, often extending beyond gills. Flesh thick, very firm, white, not staining when bruised (but may discolor slightly); odor mild. GILLS close, free or nearly free; pallid, soon becoming grayish-pink, then deep reddish-brown to chocolate-brown, finally blackish-brown. STALK 2-10 (18) cm long, 1-3 (4) cm thick, very firm, solid, equal or slightly thicker below, but often with a narrowed or pointed base; white, without scales. VEIL membranous, white, thick, forming a prominent, persistent, more or less median (or even basal) ring on stalk; ring bandlike (upper edge free or flaring and lower edge free) or sheathlike (like a volva, with only the upper edge free). SPORE PRINT chocolate-brown; spores 5-7 × 4-5.5 microns, broadly elliptical, smooth. Cap surface *not* yellowing in KOH.

HABITAT: Solitary, scattered, or in groups or rows on road shoulders, in hard-packed soil, along sidewalks, around playgrounds, and in other disturbed areas; often fruiting underground (looking for them can be like digging for clams!). Widely distributed but not common on the coast, where it is largely replaced by *A. bernardii*. I have heard of gigantic fruiting bodies (10 inch buttons!) of either this species or *A. bernardii* buried deep in alluvial soil in riverbeds in the Livermore-Pleasanton area of California. In colder regions *A. bitorquis* often appears along roads where salt is sprayed in the winter. In the Southwest it is often abundant along highways, but is hard to spot because of its subterranean habit. However, shaggy manes *(Coprinus comatus)*, which are hard *not* to spot, are often an indicator of its presence. In our area I have found it in the fall, winter, and spring.

EDIBILITY: Edible, and in my fickle fungal opinion, the best of all *Agaricus* species. In Europe it is cultivated commercially because of its immunity to the virus disease that plagues *A. bisporus*. It is larger, meatier, and much firmer than either *A. bisporus* or *A. campestris*. One devotee goes so far as to say: "Life is like an *Agaricus bitorquis*—all good except for a few gills." Life may not be so consistently good, but *A. bitorquis* certainly is!

COMMENTS: The white cap, bandlike or sheathing ring (see photographs), and unchanging flesh plus the solid stem, firm texture, and fondness for hardpacked soil distinguish this hardy, handsome fungus from other species of *Agaricus*. There is no anise or phenol odor and it does not stain yellow or red. When it fruits underground you must search

321

Agaricus bitorquis is common inland in disturbed or hard-packed soil. Note how the large bandlike ring can mimic a volva (specimens at top and far right). *A. bernardii* (not illustrated) is a similar species that is common along the coast. It has reddening flesh and often has a scaly cap.

very carefully for the telltale cracks in the soil that mark its presence (see color plate!). When the annulus (ring) is sheathlike it sometimes resembles a volva, but the chocolate-brown spores and gills prevent confusion with *Amanita*. For years it has been called *Agaricus rodmani; A. edulis* is another name for it. *A. chlamydopus* is a similar species with a cottony white cap, sheathing veil, stout stature, and larger spores. It is particularly common in the southern United States on lawns and along roads, and is edible. *A. vinaceo-virens* is a cypress-loving species with a repulsive briny odor and a slightly scaly or scurfy whitish to light brown cap. It often develops vinaceous tints in age and sometimes stains greenish as well. Its edibility is unknown.

Agaricus bernardii (Salt-Loving Agaricus)

CAP 5-15 cm or more broad, convex to plane or somewhat depressed centrally; surface dry, smooth or often breaking up to form scales or warts; white or buff, but may discolor dingy brownish in age; margin inrolled when young. Flesh thick, very firm, white, staining reddish-orange to reddish, vinaceous, or reddish-brown when cut (sometimes slowly); odor mild to pungent or briny. **GILLS** free at maturity, close, soon grayish-pink or pinkish, then reddish-brown and finally deep chocolate-brown or blackish. **STALK** 4-10 (13) cm long, 2-4 (8) cm thick, solid, very firm, equal or tapered below (rarely thicker below), smooth, white. **VEIL** membranous, thick, somewhat rubbery, white, forming a more or less median, sheathlike (or occasionally bandlike) ring on stalk (the upper edge often flaring). **SPORE PRINT** chocolate-brown; spores 6-7.5 × 5-6 microns, broadly elliptical, smooth. Cap surface *not* yellowing in KOH.

HABITAT: Solitary or scattered to densely gregarious in sand, sandy soil, disturbed areas, on lawns, and in various saline habitats; occurring along the Atlantic and Pacific coasts and in Europe. In our area it is quite common in the fall and winter, usually appearing after the first soaking rains. In Santa Monica, California, I have seen it in the summer on lawns. It apparently has an even greater tolerance (or liking) for salt than does *A. bitorquis,* and like that species, sometimes fruits underground.

EDIBILITY: Edible and choice—almost the equal of *A. bitorquis,* but a little chewier and sometimes with a slightly salty or briny taste.

COMMENTS: Also known as *A. halophilus* and *A. maritimus,* this is a characteristic coastal species. It has the general appearance of its close relative, *A. bitorquis,* (i.e., hard flesh, sheathing or even bandlike veil, and whitish color), but differs in staining reddish when cut and is more apt to have a warty or scaly cap. In dry weather, however, the staining may be slow and/or slight.

Agaricus lilaceps is one of the largest and firmest of its tribe. Note thick stalk, skirtlike annulus (ring), and tendency of the flesh to darken (redden) when cut. The veil is often tinged yellow before it breaks.

Agaricus lilaceps (Giant Cypress Agaricus)

CAP 7-25 cm broad or more, broadly convex when young, usually plane in age; surface dry, smooth or covered with flattened brown to pale brown or cinnamon-brown fibrils which only rarely break up into scales, but sometimes developing strong pink, lilac, purplish, ochre, or tawny-orange tones (see comments). Flesh very thick and firm, white, usually staining vinaceous or dark reddish (often slowly) when cut; odor mild to faintly sweet or fruity. **GILLS** free at maturity, close, pallid or pale pinkish, soon becoming reddish- or purplish-brown, then chocolate-brown to blackish-brown. **STALK** 5-22 cm long, (2) 3-6 (9) cm thick, equal or enlarged below, very firm and thick, white or often brownish-stained (but sometimes developing pink to purplish hues as on the cap), usually buried deeply in the humus; smooth or fibrillose (especially below), base often with yellow or ochre stains. **VEIL** membranous, fairly thick, underside sometimes white but very often yellow at first, with white to brownish patches that soon wear away; rupturing to form a skirtlike or sometimes flaring, superior ring on stalk. **SPORE PRINT** chocolate-brown; spores 5-7 × 4-5 microns, broadly elliptical, smooth. Cap surface negative in KOH or staining very faintly yellow.

HABITAT: Scattered to densely gregarious or clustered on ground under old cypresses (or occasionally other trees), or where cypresses have been cut down; known only from California. It is locally common in the fall and winter or whenever it is damp enough; I have seen large fruitings in Monterey and San Mateo counties.

EDIBILITY: Edible and choice! It is one of the meatiest of all edible mushrooms, but like *A. bitorquis* and *A. bernardii,* it requires thorough cooking to render it tender.

COMMENTS: This massive *Agaricus* has a rather interesting history. It was originally described as a lilac- or pinkish-hued species on the basis of specimens collected in Pacific Grove, California, in the 1930's. However, careful study by Rick Kerrigan suggests that it is *typically* a brown-capped species with reddening flesh, and only develops the brighter colors in certain localities (e.g., the Monterey Peninsula) and/or under certain conditions (perhaps cold weather or refrigeration). In any event, it is easily recognized by its impressive size, thick hard stem, slowly reddening flesh, and frequently yellow-tinged veil (when unbroken). It is our heaviest *Agaricus*—not as broad as *A. crocodilinus* nor as tall as *A. augustus,* but much denser. As in many other species of *Agaricus,* soggy or weathered specimens will not necessarily display the "correct" staining reactions, but are usually

323

recognizable by their size, stature, and color. The common brown-capped form was provisionally called *A. "luteovelatus"* by Rick Kerrigan until it became evident that it was actually *A. lilaceps.* Another large edible cypress-lover with reddening to barely reddening flesh, *A. pattersonae,* has a scalier, darker brown cap. Its veil is brownish on the underside rather than yellow, and frequently separates into two distinct rings. It is slightly smaller than *A. lilaceps* (stalk 2.5-4 cm thick) and sometimes grows with other trees (e.g., redwood).

Agaricus fuscovelatus   (Purple-Veiled Agaricus)

CAP 3-9 cm broad, obtusely bell-shaped to convex when young, plane in age or with a slightly uplifted margin; surface dry, covered with brown to cinnamon-brown or dark reddish-brown fibrils or scales on a white to dingy, gray, or violet-gray background. Flesh thick, firm, white, staining slightly or slowly yellowish-orange to pinkish-orange or reddish when cut (especially in stalk); odor mild or faintly fruity. **GILLS** pallid or grayish-pink becoming pale brown, then chocolate-brown or darker; close, free or nearly free. **STALK** 4-12 cm long, 0.7-2 (2.5) cm thick, equal or enlarged slightly below, smooth or slightly scaly near base; white, discoloring somewhat in age; stuffed or hollow. **VEIL** thin, membranous; underside at first with a layer of white felty patches, soon becoming purplish-gray to purple-brown or chocolate-brown to nearly black and scaly; veil often not breaking free from the cap until the cap has completely expanded, then forming an apical or superior, fragile, skirtlike, striate ring on stalk. **SPORE PRINT** chocolate-brown; spores 6.5-8 × 5-6.5 microns, broadly elliptical, smooth. Cap surface *not* yellowing in KOH.

HABITAT: Scattered to densely gregarious or clustered under cypress or sometimes other trees such as cedar and *Cryptomeria* (an exotic bushy conifer); known only from coastal California. It is sometimes common in the fall and winter, less so in the spring, but does not seem to fruit every year. Under a single cypress tree in a cemetery I have seen more than 300 fruiting bodies mingled with dozens of *A. bisporus!*

EDIBILITY: Edible, but rather tough (especially the stem).

COMMENTS: The most peculiar feature of this most peculiar *Agaricus* is the chocolate-brown to purplish-gray or "fuscous" color of the unbroken veil. Its overall appearance is also distinctive, but hard to describe: the stalk tends to be straight and cylindrical and the veil, which is close to the top of the stem, remains intact for an unusually long time before rupturing (it sometimes stretches over the gills like a piece of tissue paper even after the cap has fully expanded and spores are being produced!). The staining reaction is best seen by cutting the stalk in half crosswise. *A. lilaceps* and *A. pattersonae* (see comments under *A. lilaceps*) are somewhat similar but much larger.

Agaricus fuscovelatus. Note the dark color of the unbroken veil and the more or less equal (cylindrical) stalk. The flesh turns orange or reddish when cut.

Agaricus fuscofibrillosus (Bleeding Agaricus)   **Color Plate 74**

CAP 5-13 cm broad, convex to plane; surface dry, smooth, with flattened brown to reddish-brown fibrils which only rarely break up into scales. Flesh thick, firm, white, quickly staining red when cut or bruised; odor mild or faintly fruity. GILLS close, free at maturity, pinkish becoming reddish-brown, finally chocolate-brown or darker. STALK 4-10(14) cm long, 0.8-2 (3) cm thick, equal or thicker below, hollow or stuffed; smooth and white above the ring, whitish below or with brownish to vinaceous fibrils or scales, or in one variant sheathed by a "boot" of brownish to vinaceous fibrils; staining red quickly when bruised. VEIL membranous, white or becoming brownish in age; forming a superior, skirtlike ring on stalk. SPORE PRINT chocolate-brown; spores 5-6 (7) × 4-4.5 microns, elliptical, smooth. Cap surface *not* yellowing in KOH.

HABITAT: Solitary to scattered or gregarious on ground under cypress or rarely other trees; sporadically common in our area in the late fall and winter, but probably more widely distributed (in habitats other than cypress).

EDIBILITY: Edible and "rich" according to Shalom Compost, who put it in a soup. Reddish-staining Inocybes (e.g., *I. pudica*) are poisonous but smaller, with paler spores.

COMMENTS: This *Agaricus* is easily recognized by its "bleeding" flesh and brown to reddish-brown fibrillose cap. It is one of several "bleeders" that have traditionally been lumped under the names *A. haemorrhoidarius* and *A. silvaticus.* (The "true" *A. haemorrhoidarius* has distinct fibrillose scales on the cap and is said to be fairly common in the forests of eastern North America, but has not yet been found in California; *A. silvaticus* has been reported from the Pacific Northwest.) *A. fuscofibrillosus,* on the other hand, has an innately fibrillose cap that does not normally become scaly, and grows almost exclusively with cypress (in our area). There are several other red-staining species, including: *A. benesi(=A. albosanguineus),* cap pure white or tinged grayish to pale cinnamon, widely distributed but rare except in our area, where it is fairly frequent under cypress; *A. pattersonae,* a larger, thicker-stemmed species (see comments under *A. lilaceps*); *A. rubronanus,* a very small species (cap 2-3 cm broad, stalk 3-6 mm thick); and a large, stately, delicious but unidentified species with dark reddish-brown fibrils or fibrillose scales and minute spores (4-5 microns long). I have found the latter several times in mixed woods, but it is not common. *A. arorae* also "bleeds," but usually grows with oak and stains yellow in KOH.

Agaricus arorae

CAP 3-7 (10) cm broad, convex to plane; surface dry, with brownish to reddish-brown fibrils or fibrillose scales, at least at the center, on a white to reddish background; staining reddish when rubbed. Flesh white, reddening when cut (usually quickly, but sometimes slowly); odor mild or faintly fruity. GILLS free at maturity, close, pinkish becoming purple-brown, then chocolate-brown or darker. STALK 5-14 cm long, 0.5-1.5 (2) cm thick, equal or enlarged at base, hollow in age; white, sometimes with scaly zones, staining reddish when bruised and often aging reddish-brown or vinaceous. VEIL membranous, thin, white (or dingy in age), forming a median to superior, skirtlike, fragile ring on stalk. SPORE PRINT chocolate-brown; spores 4-5.5 × 3-4 microns, broadly elliptical, smooth. Cap surface staining yellow in KOH.

HABITAT: Solitary to widely scattered or in small groups on ground in mixed woods and under oaks, fruiting mainly in October and November, sometimes common. It is known only from Santa Cruz County, California, but probably has a wider distribution.

EDIBILITY: Unknown.

COMMENTS: This odd species appears to be a compromise or "bridge" between the red-staining and yellow-staining (rufescent and lutescent) sections of *Agaricus.* The flesh when fresh bruises red quite dramatically, but the surface of the cap turns yellow in potassium hydroxide! In the field it can be separated from other "bleeding" species such as *A. fusco-fibrillosus* by its slimmer build (see photo on p. 311) and fondness for oak. *A. amicosus* also reddens when cut and yellows in KOH; it is common under mountain conifers in the Southwest and southern Rockies, and is edible. *A. spissicaulis* has a stocky stature like *A. bisporus* (and is sometimes mistaken for that species) and slowly reddening flesh, yellows somewhat in KOH, has a slight almond odor, and is widely distributed but rare.

Agaricus subrutilescens (Wine-Colored Agaricus)   Color Plate 75

CAP 5-15 (20) cm broad, convex or with a somewhat flattened top, becoming broadly umbonate to plane or with an uplifted margin; surface dry, covered with brown to purple-brown or wine-colored fibrils or fibrillose scales (sometimes only at the center) on a whitish to dingy background. Flesh white, firm, not staining when bruised; odor mild or slightly fruity. **GILLS** close, free at maturity; at first whitish, then pinkish, then darkening slowly to pinkish-brown and finally chocolate- or blackish-brown; when pinkish usually bruising brighter rosy-pink when cut. **STALK** 5-20 cm long, 0.5-1.5 (4)) cm thick, equal or thicker below, smooth and white or reddish above the ring, sheathed with soft cottony white scales below which break up into fibrillose patches in age or wear away; stuffed or hollow, often fragile in age. **VEIL** membranous, white, forming a thin, usually superior, skirtlike ring on stalk. **SPORE PRINT** chocolate-brown; spores 4.5-6 × 3-4 microns, elliptical, smooth. Cap surface usually staining greenish-olive in KOH (sometimes slowly).

HABITAT: Solitary, scattered, or in small groups in woods, usually under conifers; common in the fall and winter along the west coast, also reported from Japan. In our area it favors redwood, but occurs in Oregon under alder, in northern California under Sitka spruce, in southern California under oak, and in the Sierra foothills in mixed woods.

EDIBILITY: Edible and choice, but has been known to cause rather severe gastric upsets in some people. It is not as meaty as many *Agaricus* species, but it *is* delicious.

COMMENTS: The dark purple-brown cap and shaggy white stem distinguish this handsome mushroom from other *Agaricus* species. Along with *A. hondensis*, it is our most common *Agaricus* of deep, undisturbed woods. *A. hondensis*, however, has a paler cap and smooth (not shaggy) stalk. Because of the greenish KOH reaction and a number of other chemical peculiarities, *A. subrutilescens* is something of an anomaly. It does not fit well into any of the existing "sections" of *Agaricus*, but the rosy-staining gills suggest a possible relationship to the red-staining species (section Sanguinolenti). It is usually a rather tall and slender mushroom, but robust individuals can be found, particularly under pine. A small Texan species, *A. vinaceo-umbrinus*, also stains greenish in KOH.

Agaricus hondensis   (Felt-Ringed Agaricus)

CAP 6-15 (20) cm broad, convex becoming plane; surface dry, smooth, whitish or with pale pinkish-brown to pinkish-gray to fawn-colored flattened fibrils or fine fibrillose scales (at least at center), the fibrils often darkening in age to brown, reddish-brown, or reddish-gray, but in one northern form darker brown from the beginning. Flesh thick, white, unchanging or staining pale yellowish when bruised, then often slowly discoloring pinkish; odor of crushed flesh mild or faintly phenolic, but usually *distinctly* phenolic in base of stalk. **GILLS** pale pinkish to pinkish-gray becoming brown, then chocolate-brown or darker; free at maturity, close. **STALK** 7-20 cm long, 1-2.5 cm thick but with a thicker or bulbous base; firm, smooth, without scales, white or discoloring dingy pinkish or brownish in age or after handling; flesh in extreme base usually staining pale yellowish when bruised. **VEIL** membranous, white, forming a thick, feltlike, superior ring on stalk; ring skirtlike

Agaricus hondensis. A common woodland species with a large, thick, felty ring, smooth stem, and pale brown to reddish-brown fibrils on the cap. Tempting, but not edible!

but often flaring outward instead of collapsing against stalk. **SPORE PRINT** chocolate-brown; spores 4.5-6 × 3-4 microns, elliptical, smooth. Cap surface staining yellow in KOH.

HABITAT: Solitary or in groups, troops, or rings in woods, particularly where there are thick accumulations of fallen twigs and other debris. Very common in our area in the late fall and winter under both hardwoods and conifers; like *A. subrutilescens,* which grows in similar habitats, it is apparently restricted to the west coast. The main fruiting typically follows close on the heels of *A. praeclaresquamosus.* It was originally described from La Honda, California.

EDIBILITY: Poisonous to many people, causing stomach distress, vomiting, etc. It's difficult to imagine a more delicious-looking mushroom, but it has an unpleasant, astringent-metallic taste even when cooked.

COMMENTS: This handsome, alluring woodland *Agaricus* is distinguished by the pale cap fibrils which often darken in age, the thick felty annulus(ring), smooth naked stalk, and chocolate-brown spores. The phenol odor is often absent in the cap but usually quite pronounced when cooked *or* if the base of the stalk is broken open and crushed. It is frequently mistaken for edible species, particularly *A. subrutilescens* and *A. augustus,* both of which have cottony scales on the stem when young. It has also been confused with *A. silvaticus,* a species with reddening flesh (see *A. fuscofibrillosus*).

Agaricus californicus　　(California Agaricus)　　Color Plate 72

CAP 3-9 (12) cm broad, at first convex or marshmallow-shaped, the margin inrolled and often somewhat lobed, then expanding to broadly convex or plane or broadly umbonate; surface dry, smooth or with fibrils or small scales; color variable: most often white to silvery gray with a brown center, but sometimes entirely white and at other times with brown to grayish-brown fibrils or scales throughout; usually darker in age and often with a metallic luster; often pinkish- or reddish-stained in wet weather; margin often extending beyond gills. Flesh thick, white, unchanging or staining slightly yellowish when crushed, then eventually sordid reddish or brown; odor of crushed flesh phenolic but often faint. **GILLS** close, free at maturity, usually pallid until the veil breaks, then bright pink to pinkish-brown, finally chocolate-brown to blackish-brown. **STALK** 3-8 (12) cm long, 0.5-1 (1.5) cm thick, equal or slightly enlarged at base, stuffed or hollow, smooth, without scales; white, but often discoloring pinkish to dingy brown in age or after handling. **VEIL** membranous, thick, white, often with felty patches on underside; forming a persistent,

327

Agaricus californicus is often confused with *A. campestris* and *A. bisporus*. Note how the gills are whitish in the button stage, and how the veil forms a thick persistent annulus (ring) on the stalk. Also compare the shape of the young caps to that of *A. campestris* and *A. bisporus*.

superior to median ring on stalk or occasionally clinging to cap margin; ring skirtlike or intermediate. **SPORE PRINT** chocolate-brown; spores 5-7 × 4-5 microns, elliptical, smooth. Cap surface staining yellow in KOH.

HABITAT: Solitary to scattered or densely gregarious (but not often in rings) on ground, growing anywhere and everywhere, but especially abundant in suburbia—on lawns, in gardens, parks, under trees (eucalyptus, acacia, oak, cypress, etc.), also in the woods, but only rarely in pastures (except under trees). Found year-round in our area, but most abundant in the fall, when it often mingles with *Lepiota naucina* and other *Agaricus* species. It is known only from the west coast, but may very well have a wider distribution.

EDIBILITY: Mildly poisonous to some people (causing stomach upsets), and frequently mistaken for the common meadow mushom *(A. campestris)*, with which it often grows.

COMMENTS: This ubiquitous mushroom gives my students more trouble than any other, not only because it closely mimics edible species, but also because of the endless variation it exhibits. "Typical" forms—if they can be said to exist—do not stain bright yellow like *A. xanthodermus*, but yellow slightly in cooking, in addition to giving off a phenol odor. The persistent membranous ring, modest size, rather long stalk, slight phenol odor when fresh (often not evident to those unfamiliar with the odor) and whitish (not pinkish!) gills in the button stage (i.e., *before* the veil breaks) help distinguish it from *A. campestris, A. cupreobrunneus,* and *A. bisporus.* The cap varies tremendously in color and scaliness, but is usually brownish at least at the center, and often somewhat shiny. As a rule, specimens growing in sheltered situations (such as under oak) are paler or even pure white, while those growing in the open can be quite dark. Specimens on lawns are usually rather firmly rooted in the ground and often have to be dug up to avoid breaking the stem. For a comparison with *A. campestris,* see that species and the key to *Agaricus.* Since it is the most ubiquitous *Agaricus* in coastal California (one quickly tires of finding it), the name *A. californicus* is certainly appropriate. It has also been called the "Fool's Agaric," but I don't care for this name because fools aren't the only ones to be fooled by it!

328

Agaricus praeclaresquamosus   (Flat-Top Agaricus)   **Color Plate 76**

CAP 5-25 cm broad, at first convex or somewhat marshmallow-shaped, then broadly convex or plane; surface dry, covered with flattened inky-gray to grayish-brown or brown fibrils or fibrillose scales (at least at center) on a whitish background, but often developing reddish or pinkish stains in wet weather; in one form bruising yellow. Flesh thick, white, unchanging or staining slightly yellow when bruised and then slowly discoloring brownish or vinaceous; odor of crushed flesh phenolic (especially in base of stalk). **GILLS** close, free at maturity, at first pallid, then grayish or light pink, becoming reddish-brown to chocolate-brown and finally blackish-brown. **STALK** 7-18 cm long, 1-3 (4) cm thick, equal or enlarged below or sometimes tapering to a point if growing in clusters; stuffed or hollow, smooth, without scales; white, but often discoloring reddish-brown to dingy brown in age or upon handling; flesh in *extreme* base usually (but not always) staining bright yellow when cut. **VEIL** membranous, white, thick, feltlike, somewhat rubbery, often splitting at the margin; rupturing to form a persistent, superior, skirtlike or intermediate ring on stalk. **SPORE PRINT** chocolate-brown; spores 4-6.5 × 3-4.5 microns, elliptical, smooth. Cap surface staining yellow in KOH.

HABITAT: Solitary or in groups or clusters in woods or under trees, especially along roads and paths; widely distributed, but especially common along the west coast. It is frequent in our area in the fall and winter, and occasional in the early spring. The main crop is usually in November or December, but I have seen gorgeous fruitings of the large form (see comments) under redwood in Humboldt County, California, in September.

EDIBILITY: Poisonous to many, causing vomiting and diarrhea, but some people eat it with impunity. Like *A. hondensis,* it is tempting, but has an unpleasant metallic taste.

COMMENTS: Better known as *A. **meleagris,*** this beautiful omnipresent *Agaricus* can be told by the inky to grayish-brown fibrils on the cap, plus the thick veil, smooth stem, and phenolic odor of the crushed flesh. In addition, the base of the stalk often bruises bright yellow when nicked, as in *A. xanthodermus.* The above description embraces two forms: one medium-sized (but generally larger than *A. californicus*) and sometimes clustered, the other very large and strikingly beautiful (about the size of *A. augustus*) and often gregarious but rarely clustered. The latter is especially tempting for the table (see color plate!), but can be distinguished from *A. augustus* by the smooth stem, grayer cap, and different odor. The larger form is usually found under redwood, whereas the medium-sized form grows under various trees, including redwood and oak. Other species: *A. **placomyces,*** a common woodland mushroom in eastern North America, is very similar to the medium-sized form of *A. praeclaresquamosus,* but has a more fibrillose (less scaly) cap, a less rubbery veil, and often shows yellowish to brown droplets on the underside of the veil or on the stalk. Another eastern species, *A. **pocillator,*** closely resembles *A. placomyces,* but lacks veil droplets and often has a double ring, is usually smaller and slimmer, and sometimes has a rimmed (cuplike) basal bulb. It is especially common in the South. Both of these species are eaten by some people, but are best avoided.

Agaricus xanthodermus   (Yellow-Staining Agaricus)

CAP 6-15 (20) cm broad, round to somewhat marshmallow-shaped or convex, then expanding to broadly convex or plane; surface dry, smooth, pure white to gray or grayish-buff, or often whitish at the margin and buff to tan toward the center; often discoloring brownish in old age and sometimes breaking up into scales; typically staining bright yellow quickly when rubbed repeatedly, especially on the margin, but then slowly discoloring brownish or vinaceous; margin inrolled somewhat when young and often lobed. Flesh thick, firm, white, turning yellowish when crushed; odor phenolic (unpleasant). **GILLS**

Agaricus xanthodermus growing with *Alyssum* (the flowers) along a road. Note how the sliced specimen at far right is stained yellow at base of stalk. Cap color ranges from white to grayish to buff.

close, free at maturity, at first white, then pinkish or grayish-pink, finally chocolate-brown to blackish-brown. **STALK** 5-12 (18) cm long, 1-2 (3) cm thick, equal or with an enlarged base, smooth, stuffed or hollow, without scales; white, usually bruising yellow, then brownish; in age often discolored brownish; flesh in *very* base turning bright yellow when cut. **VEIL** membranous, white or yellow-stained, usually with patches on underside; forming a large, thick, feltlike, median to superior ring on stalk; ring skirtlike or intermediate, often flaring at first. **SPORE PRINT** chocolate-brown; spores 4.5-6 × 3-4.5 microns, broadly elliptical, smooth. Cap surface staining yellow in KOH.

HABITAT: Scattered to densely gregarious under trees and hedges, in yards, on lawns, along roads and paths, also in woods, pastures, and under cypress; widely distributed. Abundant in our area from fall through spring and even showing up in the summer. I have seen stupendous fruitings in an old olive orchard, under acacia and eucalyptus, and in numerous cypress plantations as well as with oak.

EDIBILITY: Poisonous to many people—causing headaches, nausea, vomiting, and diarrhea. It is a very tempting, meaty mushroom but the unpleasant odor becomes unbearably obnoxious when the mushroom is cooked, and the taste is awful as well. Yet there is no accounting for taste. I know one person who gathers it—and nothing else!

COMMENTS: The tendency of all parts to stain bright yellow, the phenolic odor, and the white to grayish or tan cap plus the presence of a veil and chocolate-brown spores are the trademarks of this cosmopolitan and variable species. The yellow-staining is usually most dramatic on the margin of the cap, but is most *reliable* in the *extreme* base of the stalk, which turns bright chrome-yellow when nicked. The latter is perhaps the best fieldmark, since

Agaricus xanthodermus. Note yellow stains on cap and the prominent annulus in mature specimens.

These handsome white examples of *Agaricus xanthodermus* might easily be mistaken for *Agaricus arvensis* or another edible species. Their phenolic odor, however, is usually quite pronounced and the very base of the stalk turns bright yellow when cut.

only the equally poisonous *A. praeclaresquamosus* stains so brilliantly in the base of the stalk. Several edible *Agaricus* species, such as *A. augustus* and *A. albolutescens*, stain yellow on the cap, but do not subsequently discolor brownish. The edible horse mushroom *(A. arvensis)* and giant horse mushroom *(A. osecanus* group*)* superficially resemble *A. xanthodermus*, but do *not* stain yellow at the base of the stalk and do *not* smell like phenol.

*Agaricus pinyonensis**   (Pinyon Agaricus)

CAP 6-20 cm broad, convex or shaped like an inverted bowl, expanding slightly in age but not normally becoming plane; surface smooth or occasionally breaking up to form a few small scales; entirely white or with a tan- to buff-tinged center, often discoloring buff to very pale tan overall in old age; sometimes yellowing when bruised. Flesh *very* thick (at least 2 cm) and firm, white, typically not staining; odor mild or merely mushroomy. **GILLS** pinkish when young, then darkening to purple- or chocolate-brown and finally blackish-brown; close, free at maturity; narrow in relation to the flesh. **STALK** 5-20 cm long, 2-5 cm thick or more, often swollen below (just above the base); white and smooth or nearly so, sometimes discoloring like the cap in old age; firm; solid, stuffed, or with a hollow center. **VEIL** thick, membranous, white, usually with patches on underside when young; typically forming a median to superior skirtlike ring on stalk. **SPORE PRINT** chocolate-brown; spores 6.5-8 × 4-5 (?) microns, broadly elliptical, smooth. Cap surface yellowing in KOH.

HABITAT: Solitary or in groups on ground under pinyon and juniper; known only from New Mexico, but probably occurring throughout the Southwest in the appropriate habitat. It fruits mainly in August and September, but does not appear every year.

EDIBILITY: Delectably delicious—a favorite of fungus-fanciers in the Southwest. Its discoverers and namers, Bill Isaacs and Chuck Barrows of Santa Fe, New Mexico, rate it among the best in the genus ("firm like *A. bitorquis* and milder than *A. campestris*").

COMMENTS: This magnificent white mushroom is remarkable for its thick flesh, firm texture, and fine flavor. It is somewhat reminiscent of *A. bitorquis,* but its veil is quite different and it apparently grows only with pinyon and juniper. Its name (which has yet to be officially published and may actually turn out to be *A. "barrowsii"*) is therefore apt, but its affinities within the genus *Agaricus* are rather unclear. It may be related to *A. arven-*

*Provisional name by Isaacs & Barrows

Left: *Agaricus pinyonensis* is a robust whitish species of the Southwest. Note how thick the flesh is! **Right:** *Agaricus arvensis* is a sweet edible species with a distinct anise odor and a tendency to discolor yellow in age. For close-up of unbroken veil in this species, see photo on p. 312.

sis, but does not normally smell sweet and does not *necessarily* stain yellow when bruised. Other species: A medium-sized to fairly large (cap 4-14 cm) whitish species, *A. sequoiae,* grows solitary to gregarious or in large clusters in the redwood and mixed forests of coastal northern California, usually near rivers or at least in bottomlands. It has a longer (8-30 cm) and more slender stem that is usually equal, and it does not stain yellow when bruised or in KOH, but sometimes has a yellow-tinged veil. It has a mild odor and is edible according to Rick Kerrigan, who named it.

Agaricus arvensis (Horse Mushroom)

CAP (4) 7-20 cm broad, oval or convex becoming broadly convex or plane; surface dry, smooth or sometimes cracking into small scales, especially at the center; white to creamy, buff, or yellowish (especially toward center), usually bruising yellow if rubbed (especially when young); margin sometimes hung with veil remnants. Flesh thick, firm, white, unchanging or yellowing slightly when crushed; odor sweet (like anise or almond extract) when young, often somewhat musty in age. **GILLS** close, free at maturity, pallid becoming grayish (rarely pinkish), then chocolate-brown or darker. **STALK** 5-12(17) cm long, 1-3 cm thick, equal or slightly enlarged below, stuffed or hollow, smooth or with small cottony scales below the ring; white, sometimes bruising or aging yellowish, but extreme base *not* bruising yellow when cut. **VEIL** membranous, white or tinged yellow, with cottony patches on underside that often split to form a starlike or cogwheel pattern; forming a fragile, superior, skirtlike ring on stalk. **SPORE PRINT** chocolate-brown; spores 7-8.5 × 5-6 microns, elliptical, smooth. Cap surface staining yellow in KOH.

HABITAT: Solitary, scattered, or in groups in grassy areas—lawns, pastures, etc.; common and widely distributed. It is frequently encountered in our area on lawns and in cemeteries in the spring, summer, and fall, but seems to be replaced in our pastures by the giant horse mushroom *(A. osecanus* group*)*.

EDIBILITY: Edible (for most people) and excellent. Buttons have a slightly sweetish flavor intermediate between that of *A. campestris* and *A. augustus.* Be sure not to confuse it with the deadly white Amanitas, which have white spores, white gills, and a volva.

COMMENTS: The white to pale yellowish cap, sweet odor when young, well-developed membranous veil which often has a cogwheel pattern of patches on its underside (see photo on p. 312), chocolate-brown spores, and growth in grass typify this beautiful, cosmopolitan mushroom. The cap will often age or stain yellowish—more so than that of *A. croco-*

dilinus and *A. osecanus.* However, the base of the stalk does *not* stain bright yellow when cut—an easy way to distinguish it from the poisonous, yellow-staining *A. xanthodermus.* The stature of *A. arvensis* is generally more robust than that of *A. silvicola,* its smaller-spored sylvan look-alike, but it is not nearly as ponderous as that of the giant horse mushroom *(A. osecanus* group*)* or *A. crocodilinus.* The growth in grass is quite characteristic, though a swamp-inhabiting form (*var. palustris*) has been described. There are numerous variants around *A. arvensis* that have not been critically evaluated, including a small, slender form (cap only 3-9 cm broad) that occasionally occurs in our area on lawns. *A. fissuratus* is a very similar coastal species with a frequently cracked (fissured), white to yellowish cap and slightly larger spores. It is common in the Puget Sound area of Washington, but probably has a wider distribution, and is edible.

Agaricus osecanus group   Color Plates 78, 79, 81
(Giant Horse Mushroom)

CAP 7-40 cm broad, rounded or convex, then broadly convex to plane; surface dry, often very smooth (like kid leather), but frequently with minutely cottony scales in youth and sometimes cracking into scales or warts in age or dry weather; white, or discoloring buff or yellowish in age (especially toward center); surface bruising slightly yellowish when young and fresh (but usually not staining in age). Flesh thick, very firm when young but quite soft in age, not usually bruising yellow; odor slightly sweet (like almond extract or or anise) when young, but often unpleasant (musty or like wet straw) at maturity. **GILLS** close, free in age, white becoming grayish or grayish-pink, then reddish-brown, finally chocolate-brown to blackish-brown. **STALK** 4-16 cm long, (1.5) 2.5-5 (8) cm thick, often spindle-shaped (swollen at or below the middle) but sometimes equal; firm, usually solid; smooth above the ring, often with small pointed but easily-obliterated scales below; white, sometimes aging yellowish or rusty-yellow; flesh in base *not* staining bright yellow when nicked. **VEIL** membranous, white, with patches on underside which form a cogwheel pattern or adhere to stalk as downward-pointing scales; rupturing to form a median to superior, skirtlike or collapsed ring on stalk. **SPORE PRINT** chocolate-brown; spores (5) 6-7 × 4.5-5.5 microns, elliptical, smooth. Cap surface yellowing in KOH.

HABITAT: Solitary to scattered or in groups or rings in pastures, lawns, and other grassy areas; distribution uncertain, but extremely abundant at times in our coastal pastures (when they are "popping" it is not unusual to pick 100 pounds from a single pasture—along with some giant puffballs!) There is usually one crop in the fall and another, larger one in the spring, but it is also to be expected during warm spells in the winter. Its range extends at least into northern California and Oregon, where the very similar *A. crocodilinus* (see comments) is also common.

EDIBILTY: Edible (for most people), and very popular with my group of acquaintances. The young, hard buttons are best—mature specimens have very soft flesh and a musty odor and do better in the compost pile than the frying pan.

COMMENTS: This fine and forthright fungus can be spotted from afar, shining like a light bulb when its spotless white skin catches the sun. It is essentially a larger, squatter, robuster version of the horse mushroom *(A. arvensis),* but differs in having a *solid,* often swollen stem and smaller spores. Also, the sweet odor and yellow bruising reaction of young specimens are weaker than in typical *A. arvensis* (and may actually be non-existent), while the musty odor that develops in age is stronger. The appearance of the cap—whether warty or smooth—seems to depend largely on age, exposure, and weather conditions (when growing in the open or exposed to the wind it is more likely to be warty or cracked). Be careful not to confuse it with the poisonous *A. xanthodermus,* which may also be large and white, but has a phenol odor, yellows more rapidly when bruised, and stains *bright* yellow in the base of the stem.

Agaricus crocodilinus, mature specimens. It is common for the caps to be warty, but they can also be smooth. Note the squat stature. Each of these caps is about one foot across. (Bill Everson)

yellow in the base of the stem. An equally edible but larger-spored species, *A. crocodilinus* ("Crocodile Agaricus"—see photo above) is common in parts of northern California and the Pacific Northwest, and has also been reported from the prairies of eastern New Mexico and Colorado. It is just as large or even larger than the giant horse mushroom, and looks virtually the same. True, it is more apt to be warty or scaly and is often slightly browner in age, but it can also be perfectly smooth and pure white, so that in regions where both occur the two species can only be separated with certainty by measuring the spores (in *A. crocodilinus* they are 8-11 (14) microns long). Our giant horse mushroom, incidentally, may very well be an unnamed, endemic species. It belongs to a group or "complex" that is very confused taxonomically, and is listed here as a member of the *A. osecanus* group largely because of its spore size. However, it differs somewhat from the "true" *A. osecanus* of Europe, and could just as well be called *A. nivescens*—another European species with a solid stem and a shade smaller spores. Until a critical study of this group is completed, it is perhaps best to call it "giant horse mushroom," because its common name cannot be subsequently invalidated or proved "incorrect!" Since all of these species are equally and unequivocally edible, the exact (or inexact) differences between them needn't concern you. At least, they don't concern me! Other species: *A. macrosporus (=A. villaticus)* of Europe is a large-spored species very similar to, if not identical with, *A. crocodilinus;* the suggestive moniker *A. urinescens* has been given to still another large-spored variant—a fitting tribute to the indiscreet odor that frequently develops in old age.

Agaricus silvicola group   (Woodland Agaricus)

CAP 5-12 (18) cm broad, convex becoming plane; surface dry, smooth or silky-fibrillose, sometimes obscurely fibrillose-scaly in age; white, usually aging yellowish, especially at the center, and staining at least slightly yellow when bruised, particularly on margin. Flesh firm, white, unchanging or yellowing slightly when crushed; odor sweet (like anise or almond extract), at least when young. **GILLS** close, free at maturity, white becoming gray or pinkish-gray, then brown and finally chocolate-brown or darker. **STALK** 5-14 (20) cm long, 1-2 (2.5) cm thick, usually enlarged below, stuffed or hollow, smooth or with small cottony scales below ring; white or pinkish at apex, white below, but often aging or bruising yellowish; base *not* staining bright yellow when cut. **VEIL** membranous, white or stained

Left: This stately member of the *Agaricus silvicola* group is fairly common in our area under tanoak. Cap is white and stains yellow. **Right:** *Agaricus albolutescens*, also common, resembles *A. silvicola* but is usually more robust, has a stronger odor, and stains more dramatically.

yellow, with patches on underside that sometimes form a cogwheel pattern; forming a prominent, superior, skirtlike ring on stalk. **SPORE PRINT** chocolate-brown; spores 5-6.5 × 3.5-4.5 microns, elliptical, smooth. Cap surface staining yellow in KOH.

HABITAT: Solitary, scattered, or in small groups in woods; widely distributed and common, but rarely fruiting in large numbers. In our area it is fairly common in the fall and winter under oak, tanoak, and conifers; at higher elevations it fruits under conifers.

EDIBILITY: Edible (for most people) and choice, with caution. Make sure there is no volva and that the mature gills are not white—I have seen it mix company with deadly Amanitas that were *very* similar in size, shape, and color!

COMMENTS: Also spelled *A. sylvicola,* this species or species "complex" is recognized by its white cap, tendency to stain or age yellow, anise odor (sometimes faint, but usually detectable when the young flesh is crushed), skirtlike annulus (ring), chocolate-brown spores, and woodland milieu. It rather closely resembles *A. arvensis,* but is usually more erect and less robust, has smaller spores, and does not grow in grass. There are several variants in need of critical study. The "typical" *A. silvicola* grows mainly under conifers, at least in the West, but a large, very stately and striking variety (see photo) is fairly common in our area under tanoak, while a form with an abruptly bulbous stem *(=A. abruptibulbus)* occurs in California as well as in eastern North America. None of these forms stain as dramatically as *A. albolutescens,* and they do not smell as strong nor are they as robust as that species. Other species: *A. chionodermus* looks similar, but has bright pink to reddish gills for a long time, a mild odor, and a white to silvery-white, often fibrillose cap. It grows mainly under conifers and is edible and apparently widely distributed—I have seen it in New Mexico. See also *A. sequoiae* (under *A. pinyonensis*), *A. summensis* (under *A. augustus*), and *A. albolutescens.*

Agaricus albolutescens   (Amber-Staining Agaricus)

CAP 7-18 cm broad, convex to broadly convex or plane; surface usually dry, smooth or fibrillose, at first white but quickly staining amber to yellow-orange when bruised and often entirely yellowish to yellow-orange to amber or ochraceous in age; margin often hung with veil remnants. Flesh thick, white, usually bruising yellowish if crushed; odor strongly sweet and aniselike or almondy. **GILLS** free at maturity, close, pallid becoming grayish or grayish-pink, then eventually chocolate-brown or darker; often bruising yellow when immature. **STALK** 5-14 cm long, 1.5-3 cm thick, usually enlarged below (up to 5 cm

This squat fragrant *Agaricus* is common in the Sierra Nevada during the spring and early summer. It appears to be a form of *A. albolutescens.*

thick), firm, white or discoloring yellowish, smooth above the ring, smooth or slightly cottony-scaly below; flesh in base *not* usually bruising *bright* yellow, but exterior of base may. **VEIL** membranous, white or yellow-stained, with patches on underside that sometimes form a cogwheel pattern; rupturing to form an ample, superior, skirtlike ring on stalk. **SPORE PRINT** chocolate-brown; spores 5-7 × 3.5-4.5 microns, elliptical, smooth. Cap surface staining yellow in KOH.

HABITAT: Solitary, scattered, or gregarious on ground in woods; known only from the West. In our area it is quite common from late fall through early spring under oak (often in the company of *A. hondensis*). In the Sierra Nevada a vernal variant occurs.

EDIBILITY: Delectably delicious, with a strong sweet flavor. As with most *Agaricus* species, some people are adversely affected by it.

COMMENTS: This woodland species is reminiscent of *A. silvicola,* but is easily distinguished by its squatter or more robust stature, much stronger odor, and amber-staining cap. It also tends to occur in larger numbers than *A. silvicola,* at least in my experience. (I have seen fairy rings containing nearly 100 specimens!) The rapid yellow-staining of the cap can lead to confusion with the poisonous *A. xanthodermus,* which eventually discolors brownish after staining yellow, smells like phenol, and stains bright yellow in the *base* of the stalk. *A. arvensis* is also somewhat similar but usually grows in open, grassy places. In the Sierra Nevada and other mountain ranges, what appears to be a variant of *A. albolutescens* is often common (especially in the spring). It has a fibrillose cap and its stem is tougher and more cylindrical (less bulbous) than the coastal version, but it is otherwise quite similar (see photograph above).

Agaricus subrufescens (Almond Mushroom) Color Plate 82

CAP (6) 8-25 cm broad, round or marshmallow-shaped becoming convex to plane; surface dry, smooth but with fine flattened fibrils which may break up into minute scales except at the center (but in dry weather sometimes cracking into large warts); fibrils pallid to buff to pale brown, pale pinkish-brown, or fawn-colored, often becoming browner or ruddier with age; background white to pinkish-buff, often becoming yellow in age or when bruised. Flesh thick, firm, white, not bruising yellow or only very slightly; odor strongly sweet (like almond extract), especially when young. **GILLS** free at maturity, close, whitish becoming grayish or pinkish, then reddish-brown and finally chocolate-brown or darker. **STALK** 6-15 cm long, 1.5-4 cm thick, equal or with an enlarged base, smooth or with a few fibrils or scales below the ring; white, but often staining or aging yellow; base often staining yellow to yellow-orange but flesh in base *not* staining bright yellow when cut; base sometimes with white mycelial threads attached. **VEIL** thick, membranous, white, usually with patches (often obscure) on underside; forming a superior, skirtlike ring on stalk. **SPORE PRINT** chocolate-brown; spores 5.5-7 × 4-5 microns, elliptical, smooth. Cap surface staining yellow in KOH.

336

Agaricus subrufescens is a rare but remarkable species. The odor and flavor are strongly almondy, but the stalk is not nearly as shaggy as that of *A. augustus,* and the cap is paler.

HABITAT: Scattered to densely gregarious or clustered in compost, manure, and rich soil; originally described from eastern North America and apparently widely distributed, but rare. I have seen it fruit by the hundreds in manure-filled trenches on a berry farm near Watsonville, California, in the late spring, summer, and early fall (sometimes accompanied by *Lepiota lutea* and *L. americana*). It is easily grown at home in a vegetable garden (if you use compost) and mycelium from the just-described "patch" is now being marketed for this purpose. It has the advantage of liking it *hot,* meaning you can enjoy it at the same time as your other summer vegetables (*A. bisporus* is also grown easily in gardens, but likes it much cooler).

EDIBILITY: Edible (for most people) and delicious! The almondy odor and taste is much stronger than that of *A. augustus*—so strong, in fact, that some people don't care for it. I put several specimens in a cream-of-mushroom soup and those who tasted it swore I had dumped in a bottle of almond extract! It was apparently cultivated around the turn of the century, but lost out to *A. bisporus*—which raises an interesting prospect: if it had become the "cultivated mushroom" instead of *A. bisporus,* would the word "mushroomy" have a completely different (sweet and almondy) connotation?

COMMENTS: The strong almond extract odor and taste plus the robust stature, presence of an annulus, chocolate-brown spores, and light brownish cap make this a fairly easy mushroom to recognize. The cap is quite similar in color to that of *A. hondensis,* but the habitat and odor are completely different, while the paler color and smoother stem distinguish it from *A. augustus.* Other species: In eastern North America, an edible *Agaricus* with tawny scales on the cap (like *A. augustus*), a shaggy stalk, and weaker almond odor has also gone under the name *A. subrufescens,* apparently because of its similarly-sized spores. Whether it is a distinct species or merely an extreme form of *A. subrufescens* is for licensed *Agaricus*-experts to decide.

Agaricus augustus   (The Prince)   Front Cover, Color Plate 77

CAP 7-30 (40) cm broad, usually marshmallow-shaped but sometimes convex, slowly expanding to plane or with an uplifted margin; surface dry, covered with numerous dark brown to warm brown or tawny-brown fibrils or fibrillose scales on a white background that usually becomes yellowish, buff, or ochre in age, giving an overall golden tone to many mature specimens; center often darker; surface sometimes breaking up into warts in dry weather and bruising yellow when rubbed, at least when young. Flesh thick, white, firm; odor sweet (like almond extract), especially when young. **GILLS** close, free at maturity, remaining pallid for a long time, finally turning grayish-brown (or very briefly pinkish), eventually chocolate-brown to blackish-brown. **STALK** 8-35 cm long, 1-4 (6) cm thick, equal or enlarged slightly below, the base usually buried deep in ground; white, but often aging or bruising yellowish; smooth above the ring, sheathed with white or brown-tipped

Agaricus augustus. Note the marshmallow-shaped caps of these unexpanded buttons. In this stage they are delicious! See also Color Plate 79, photo on front cover, and photo on p. 311.

scales below (but these often wearing away in age); rather tough and fibrous. **VEIL** membranous, with white to brown cottony patches on underside (but these sometimes disappearing); forming a large, ample, superior, skirtlike ring on stalk. **SPORE PRINT** chocolate-brown; spores 7.5-10 × 5-6 microns, elliptical, smooth. Cap surface staining yellow in KOH.

HABITAT: Solitary or in groups or clumps on ground in the woods (especially under redwood), but usually near roads and paths, in clearings, and other places where the soil has been disturbed; sometimes also in flower beds, composted areas, under trees in towns, in arboretums, etc.; widely distributed, but frequent only along the Pacific Coast and perhaps more common in our area than anywhere else. It shows a definite preference for warm weather, fruiting in the spring, summer, and fall—or in the winter if it's mild enough. Several crops are produced each year, so visit your patches regularly. It is curious that such a large mushroom requires so little moisture to fruit—a little fogdrip is all it needs!

EDIBILITY: Edible (for most people) and one of the very best! It's especially significant because it fruits in the spring and summer, when edible fungi are scarce in our area. It's like getting two mushrooms in one: delectably sweet and almondy when young, strong and mushroomy at maturity. Unfortunately maggots, slugs, sowbugs, and centipedes are fond of it too!

COMMENTS: This prince of a mushroom is distinguished by its large size—caps of "LP" dimensions (one foot) are not uncommon—almond extract odor, yellow-staining cap with brown fibrils or scales, prominent annulus (ring), chocolate-brown spores, and shaggy stem (when young) which is usually buried in the ground. The veil is large and exquisitely constructed, and the marshmallow shape of the young caps (i.e., somewhat flattened on top) is also characteristic, though several other mushrooms (e.g., *A. praeclaresquamosus, Lepiota rachodes*) hae the same shape. The gills remain whitish for a long time—falsely suggesting that the spores are white—but eventually they will darken to brown and finally chocolate-brown. The poisonous *A. praeclaresquamosus* is often common in the same habitats, but has grayer cap fibrils, a smoother stem, and smells like phenol. *A. hondensis* also has a smooth stem, while the edible *A. perobscurus* has darker cap fibrils when young, a smoother stem, and subtler odor. Other species: *A. **smithii*** (also called *A. **perrarus***) is a

338

similar but somewhat smaller, more slender version with an ochraceous to ochraceous-brown cap and almondy odor. It is common in northern California and the Pacific Northwest under coastal conifers, particularly Sitka spruce. A pale form of *A. augustus* also occurs, but is rare. It resembles the typical form but has pale yellow cap fibrils which give the cap an overall creamy-white appearance. Somewhat similar to this pale form is *A. summensis*—a large, robust species whose cap is whitish with yellowish to ochraceous-tawny fibrils. It has a thick, often bulbous stem (2.5 cm thick or more) and is rare. I have seen it in our area under forest trees along roads. Finally, *A. augustus* has a smaller-spored look-alike in eastern North America (see comments under *A. subrufescens*). All of the above species are edible and delicious.

Agaricus perobscurus   (The Princess)

CAP 7-20 cm broad, convex or marshmallow-shaped becoming broadly convex to plane or with margin slightly uplifted; surface dry, at first covered with dark (grayish-brown to dark brown or blackish) fibrils which tend to break up in age to form scattered, sparse grayish-brown to dingy olive-brown fibrillose scales on a whitish to ochre-buff background (but center usually remaining dark); bruising yellow only slightly if at all. Flesh thick, white, not usually bruising yellow; odor typically of almond extract but often faint or even absent. **GILLS** close, free at maturity, pallid becoming pinkish or pinkish-brown, then brown and finally chocolate-brown or blackish-brown. **STALK** 7-15 cm long, 1-3 cm thick, equal or more often enlarged at base; white or discoloring slightly in age, smooth above the ring, slightly scaly or fibrillose below, but often appearing smooth, especially in age. **VEIL** membranous, with cottony patches on underside (but these often disappearing), forming an ample, superior, skirtlike ring on stalk. **SPORE PRINT** chocolate-brown; spores 6.5-7.5 × 4-5 microns, elliptical, smooth. Cap surface staining yellow in KOH.

HABITAT: Solitary or in groups under or near trees (usually planted) or shrubs, sometimes also at the edges of woods; known only from coastal California. It is fairly frequent in the San Francisco Bay region from late fall through spring or even summer, but uncommon to rare elsewhere (not nearly as numerous as *A. augustus*). I usually find it under pine, cypress, and acacia, but have never seen it in the redwood forests where *A. augustus* thrives.

EDIBILITY: Edible and choice—about as good as *A. augustus*.

COMMENTS: This "sister" to *A. augustus* is likely to be mistaken for that species, but has a subtler almond odor and less pronounced tendency to stain or age yellow, a smoother

Left: *Agaricus perobscurus* is a smoother-stalked version of *A. augustus*. Note how cap is quite dark in these young specimens. Right: *Agaricus semotus* (see comments under *A. micromegathus* on p. 340) is a small woodland species with a distinct anise odor and pinkish to reddish to brownish cap fibrils.

(less shaggy) stalk, and consistently smaller spores. In addition, the gills pass through a more marked pinkish phase before turning brown, and the fibrils on the cap are much darker when young (the fibrils may be quite sparse in age, however, giving the cap an overall paler appearance). It is also likely to be mistaken for the poisonous *A. praeclaresquamosus,* but that species has a totally smooth stalk, a phenol odor, often stains bright yellow in the base of the stem, and is more common in habitats characteristic of *A. augustus* than those favored by *A. perobscurus.*

Agaricus micromegathus   (Anise Agaricus)

CAP 2-5 (8) cm broad, convex to plane; surface dry, with silky fibrils or fibrillose scales (sometimes very sparse toward margin); fibrils at first pinkish in one form, yellow-brown in another, but in both forms becoming brownish to grayish-brown in age on a whitish to buff background; yellow to orange stains often developing in age or upon handling. Flesh thin, fragile, white, unchanging or bruising yellowish; odor distinctly sweet (aniselike or almondy). **GILLS** free at maturity, close, pallid or grayish, then pinkish, finally chocolate-brown or darker. **STALK** 2-6 cm long, 3-10 mm thick, equal or slightly enlarged at base, white or staining yellow to orange upon handling (especially below); stuffed or hollow, fragile at maturity. **VEIL** membranous, thin; forming a fragile, superior to median ring on stalk, or disppearing entirely; ring skirtlike or intermediate. **SPORE PRINT** chocolate-brown; spores 4.5-5.5 × 3.5-4 microns, broadly elliptical, smooth. Cap surface staining yellow in KOH.

HABITAT: Solitary to scattered or gregarious in lawns, pastures, fields, and other grassy or open places; widely distributed. Fairly common in our area in mild weather, especially in the fall, often in the company of *A. campestris* and *A. cupreobrunneus.*

EDIBILITY: Edible, but thin-fleshed and fragile. Its fragrance suggests that the flavor is good, but it doesn't often occur in enough quantity to invite collecting. Chuck Barrows of Santa Fe, New Mexico, attained a small degree of notoriety by smothering an ice cream cone with raw *A. semotus* (a very similar species—see comments).

COMMENTS: This is one of several puny, poorly-known species with an anise odor, slight annulus (ring), and fragile flesh. They have a tendency to become yellow or rusty-yellow in age, and can be distinguished from most other *Agaricus* species by their size alone. Similar species with a distinct anise odor include: *A. semotus* (see photo on p. 339), with brown to reddish fibrils on the cap and slightly longer stalk, found in woods and under trees (especially oak); and *A. comtulus* (also spelled *A. comptulus*), with a white to yellowish cap that is usually tinged brown or buff at the center, found in grass. Both of these are edible and common, but rarely collected because of their small size. Also see *A. diminutivus.*

Agaricus diminutivus group   (Diminutive Agaricus)

CAP 1-4 cm broad, oval or convex becoming plane or slightly umbonate; surface dry, with flattened pink to purplish-pink to amethyst-gray to reddish-brown fibrils, at least at the center; margin often paler. Flesh thin, white, not staining; odor mild or faintly fragrant (like anise). **GILLS** free at maturity, close, pallid or pink becoming reddish-brown, then chocolate-brown or darker. **STALK** 2-7 cm long, 2-6 mm thick, equal or with a small basal bulb, white or pallid, but in age often stained yellowish or orange below the ring; stuffed or hollow, fragile. **VEIL** membranous, thin, white; forming a fragile superior to median, skirtlike ring on stalk which often disappears in age. **SPORE PRINT** chocolate-brown; spores 4.5-6 × 3.5-4.5 microns, broadly elliptical, smooth. Cap surface yellowing in KOH.

HABITAT: Solitary, widely scattered, or in small groups in humus in woods, widely distributed. The most common variant in our area fruits in the late fall and winter in mixed woods and under live oak, but farther north this group is common under conifers.

Agaricus diminutivus group. Note small size, slender stature, and fragile veil that may or may not form a distinct ring on the stalk.

EDIBILITY: Presumably edible, but too small to be of value and sometimes confused species of *Inocybe* (e.g., *I. pudica*), most of which are poisonous.

COMMENTS: Members of this species "complex" are easily distinguished from other *Agaricus* species by their petite size, reddish-pink to purplish cap color, and woodland habitat. The form illustrated is the common one in our area. It has a beautiful amethyst-tinged cap and seems to fit descriptions of *A. purpurellus* (a member of the *A. diminutivus* group) quite well. However, there are probably more names (e.g., *A. dulcidulus, A. amethystina*) in this group than there are species, so a critical study is necessary before any positive identifications can be made.

spores

COPRINACEAE

THIS is a large family of fragile mushrooms with deep brown to black spores and a cartilaginous or fragile stem. The gills are not decurrent as in *Gomphidius* and *Chroogomphus,* and those species with dark brown spores do not normally have free gills and an annulus (ring) on the stalk, as in *Agaricus.* The cap cuticle is cellular (composed of round or pear-shaped cells) as opposed to filamentous as in the Strophariaceae, but since this distinction is microscopic, it is better to learn the characteristics of each genus than attempt to distinguish the two families *in toto* (for instance, *Coprinus* has gills that digest themselves, and *Panaeolus* usually grows on dung or manure). As a rule, however, the fragile white to brown species with a dry or only slightly tacky cap belong to the Coprinaceae, while those with a viscid *and/or* brightly colored cap belong to the Strophariaceae. Both families should be checked when in doubt. There are three common genera in the Coprinaceae, keyed below.

Key to the Coprinaceae

1. Mature gills (and often the cap) digesting themselves, i.e., either turning into an inky black fluid or withering away .... *Coprinus,* p. 342
1. Gills not digesting themselves 2
2. Spore print black; sides (faces) of gills often mottled; growing in grass, dung, or manure ....
.... *Panaeolus,* p. 353
2. Not as above; growing in humus, wood chips, on wood, etc., or if in grass then spore print dark brown or purple-brown; not normally growing in dung .... *Psathyrella,* p. 361

COPRINUS   (Inky Caps)

Minute to medium-sized or sometimes large, ephemeral, saprophytic mushrooms. CAP usually oval, cylindrical, or conical when young, *usually striate at maturity in the smaller species and often translucent or deliquescing.* GILLS free or attached, gray to black and *autodigesting at maturity.* STALK usually white and hollow, *thin and fragile, or if thick, then cartilaginous.* VEIL present or absent, sometimes forming an annulus (ring) on stalk. VOLVA typically absent (but a rudimentary one present in some species). SPORE PRINT *typically black* (but deep brown in a few species). Spores mostly elliptical, smooth or roughened, with a germ pore; discoloring in concentrated sulfuric acid. Cap cuticle usually cellular.

MEMBERS of this genus are called inky caps because the gills and often the cap digest themselves at maturity, turning into an inky black fluid that drips to the ground. The **auto-digestion** or **deliquescing** of the gills plus the black spore print are the main diagnostic features of *Coprinus*. In *Bolbitius* the gills sometimes liquefy but the spore print is rusty-brown, while in *Psathyrella* and *Panaeolus* the spore print may be black but the gills do not deliquesce.

The autodigestion process is a unique method of spore dispersal that should not be confused with the normal process of decay that occurs in most mushrooms. Rather than maturing at an even rate, the spores near the margin of the cap ripen first. Enzymes are simultaneously released which dissolve the surrounding tissue, causing the edge of the cap to spread out and curl back. This pulls the gills apart, enabling the spores to be discharged into the air. If you look at the gills of a shaggy mane *(Coprinus comatus)*, you'll see that they are crowded together like the pages of a book. If autodigestion did not occur, the spores would be discharged onto adjoining gills and their dispersal would be greatly impeded. Of course, many spores are trapped in the inky liquid that drips to the ground, but millions more are successfully discharged into the air.

In the larger inky caps, both the cap and gills dissolve, leaving nothing but a few rags of tissue stuck to the top of the stalk. This phenomenon was undoubtedly the inspiration for Shelley's memorable lines:

> Their mass rotted off them flake by flake
> Til the thick stalk stuck like a murderer's stake,
> Where rags of loose flesh yet tremble on high
> Infecting the winds that wander by.

In the smaller species, such as *C. plicatilis,* an inky fluid is not necessarily formed. There is so little substance to the gills that they wither instead of liquefying, and the cap may be so thin that it is **translucent-striate** (the gills can be seen through it as radiating lines).

Inky caps frequent areas inhabited by livestock and human beings. They fruit whenever it is moist and mild enough, and are among our most common urban and suburban mushrooms. The smaller types are especially prevalent on dung and manure, while the larger ones crop up in gardens, cellars, along roads, on or around stumps, in disturbed soil, and humus. A few grow in the woods. As we alter the environment and create new "niches," we can expect new species to evolve. One winter, following three weeks of rain, an unidentified *Coprinus* sprouted from the orlon carpet of my leaky '67 Rolls Canardly (it rolls down one hill and canardly make it up the next). My friends naturally assumed it was the result of carrying around so many mushrooms in my car, yet it was a species I had never seen before!

Several of the larger inky caps are good edibles, and the shaggy mane, *C. comatus,* is a popular and unmistakable favorite. Most species, however, are much too thin and insubstantial to warrant collecting. A few are said to be poisonous, and the common inky cap, *C. atramentarius,* contains a compound which reacts with alcohol in the body to produce a very peculiar set of symptoms (see p. 896). *Coprinus* species should be eaten before they

Autodigestion in *Coprinus comatus*. **Left:** The gills have just begun to blacken and liquefy at the cap margin. **Right:** Dry weather has caused the autodigestion process to cease with a few rags of tissue left. In wet weather it continues until only the stalk remains. (Rick Kerrigan)

deliquesce, and consequently should not be kept overnight. They are among the most easily digestible of all fungi, but their high water content makes them unsuitable for some dishes. As they frequently occur in large numbers, you may wish to preserve them for later use. This is best effected by marinating, drying, or sauteeing and then freezing. The "ink" from the fleshier species, incidentally, makes a very passable ink if diluted with a little water.

Coprinus is a large genus, with over 200 species known. It is a difficult group to study because the fruiting bodies are so ephemeral (sometimes lasting only a few hours) and because many species fruit where you don't normally look for mushrooms. Consequently, the North American species are not well known. As their favored habitats (dung, asphalt, etc.) are ubiquitous, most inky caps are widely distributed and likely to occur in California. The smaller ones are especially difficult to identify—even with the aid of a microscope. A few distinctive and/or common types are presented here.

Key to Coprinus

1. Growing on dung, manure, straw, or compost 2
1. Growing on ground, wood chips, wood, or indoors 9
2. Partial veil typically forming a distinct annulus (ring) on stalk 3
2. Annulus absent or rudimentary 5
3. Cap minute (usually less than 1 cm broad); annulus disclike or saucerlike, sometimes as large as the cap; common on horse dung . . . . **C. ephemeroides,** p.352
3. Not as above; fruiting body larger 4
4. Cap 2-5 cm high when unexpanded, at first entirely white or white with brownish scales . . . . .
. . . . **C. sterquilinus** & others (see *C. comatus*, p. 345)
4. Not as above 9
5. Cap at first covered with white to buff or gray universal veil remnants (hairs, fibrils, or powdery or flaky scales) 6
5. Universal veil absent; cap hairless or with only a few minute hairs 8
6. Cap white or pallid with white to pale gray powdery or mealy scales . . . .
. . . . *C. niveus* & others (see *C. radiatus* group, p. 351)
6. Not as above (but cap may be covered at first with white fibrils or hairs) 7

343

7.   Cap minute (usually less than 8 mm high before expansion) ...... *C. radiatus group,* p. 351
7.   Cap usually larger than above .... *C. fimetarius* & others (see *C. lagopus* group, p. 350)

8.   Cap minute (1-5 mm high before expansion) .... *C. miser* (see *C. plicatilis,* p. 352)
8.   Cap small, but typically larger than above .... *C. ephemerus* (see *C. plicatilis,* p. 352)

9.   Cap cylindrical before expansion and 4-25 cm or more tall, usually at least somewhat shaggy,
     entirely white or with a brown center and/or brownish scales; partial veil present when young,
     often forming a movable annulus (ring) that may drop off; cosmopolitan *C. comatus,* p. 345
9.   Not as above 10

10. Stalk tough and woody .... (see **Podaxales & Allies,** p. 724)
10. Not as above 11

11. Growing in deserts or arid regions, typically terrestrial (*not* on dung); cap usually with one or
     more thick, feltlike, persistent patches of universal veil tissue; stalk sometimes with a volva
     at the base 12
11. Not as above 14

12. Volval patch (veil tissue) on cap usually single and stellate (with starlike arms or points) . 13
12. Volval patch or patches irregular, not stellate; base of stalk bulbous with a volva-like rim ...
     .... *C. xerophilus*

13. Stalk typically with a volva-like basal bulb .... *C. asterophora*
13. Basal bulb and volva lacking or rudimentary, at least at maturity .... *C. asterophoroides*

14. Gills with yellow edges; stalk with yellow hairs .... *C. sulphureus*
14. Not as above 15

15. Base of stalk or immediate vicinity with cinnamon to yellow-brown or yellow-orange mycelial
     threads or a woolly mycelial mat (oozonium); growing on wood (often indoors) ....
     .... *C. radians* (see *C. micaceus,* p. 348)
15. Not as above 16

16. Cap typically 5 cm broad or more when expanded, dark brown (or becoming black) beneath a
     layer of whitish universal veil tissue which breaks up into discrete patches or flakes or "spots";
     stalk fairly thick (usually 0.5-1.5 cm); terrestrial .... *C. picaceus,* p. 346
16. Not as above (cap may have a coating of whitish hairs, but if so, then differently colored or stalk
     thinner) 17

17. Cap at first clothed with white to grayish hairs (see Color Plate 86), gray to dark brown or black
     beneath the hairs; stalk typically long, slender, fragile ... *C. lagopus group* & others, p. 350
17. Not as above 18

18. Cap with large flaky patches or flat scales which may wear off in age; growing in clusters (occa-
     sionally scattered) on dead hardwoods (but wood may be buried) in eastern North America
     . *C. quadrifidus, C. variegatus,* & *C. americanus* (not good edibles—some variants are bitter)
18. Not as above 19

19. Cap whitish except at center; partial veil usually forming a membranous ring on stalk; growing
     on rotting wood .... *C. alnivorus* (see *C. comatus,* p. 345)
19. Not as above 20

20. Cap at first with a dense coating of white universal veil fibrils, whitish to pale tawny beneath the
     fibrils or if darker then growing in ashes; stalk 3-10 mm thick *C. domesticus* & others, p. 349
20. Not as above; cap bald or with whitish particles, or if with whitish fibrils then usually darker
     or brighter than above (beneath the fibrils) and not found in ashes; stalk thin or thick .. 21

21. Stalk typically 6 mm thick or more; cap 2-8 cm broad and/or high, brown to grayish ....
     .... *C. atramentarius,* p. 347
21. Stalk typically 1-6 mm thick; cap up to 5 cm broad or high, but if larger than 3 cm then typically
     buff to yellow-brown or reddish-brown when fresh 22

22. Cap conspicuously pleated in age (see Color Plate 87) and often expanding to plane; usually
     growing scattered or at least not in clumps .... *C. plicatilis,* p. 352
22. Not as above; often in clumps or clusters 23

23. Stalk typically 3-8 cm long and 2-6 mm thick .... *C. micaceus* & others, p. 348
23. Stalk typically 1-4 cm long and 1-2 mm thick 24

24. Gills crowded .... *C. impatiens* (see *C. disseminatus,* p. 352)
24. Gills well-spaced or fairly close but not crowded .... *C. disseminatus,* p. 352

Shaggy manes in egg batter and bread crumbs, anyone? This beautiful cluster of *Coprinus comatus* is at the perfect stage for eating. How do they taste? See photo on p. 890 for the answer.

Coprinus comatus   (Shaggy Mane)   Color Plate 85

CAP 4-15 (25) cm tall, cylindrical or columnar, expanding somewhat as margin curls up until it is more or less bell-shaped, then deliquescing from the bottom up; surface not viscid, white with a brown to pale cinnamon-brown or buff center, soon breaking up into shaggy white to brown scales (universal veil remnants) which often recurve in age; margin striate in age and often tattered. Flesh soft, white. **GILLS** very crowded (like pages in a book), free or nearly so, at first white, then passing through delicate shades of pink, pinkish-red, or vinaceous; finally black and inky (deliquescing). **STALK** 5-20 (40) cm long, 1-2 cm thick, tapering upward or with an enlarged, more or less pointed base; smooth, white, hollow or stuffed with a pith, cleanly separable from cap. **PARTIAL VEIL** membranous, forming a small, white, movable, inferior ring which often drops to the base of the stalk or falls off. **SPORE PRINT** black; spores 10-16 (18) × 7-9 microns, elliptical, smooth, with a germ pore.

HABITAT: Sometimes solitary but more often scattered to densely gregarious or in loose clumps on hard ground and grassy areas, rich or disturbed soil, etc. Common throughout the northern hemisphere, especially along roads and trails. In our area the major crop is in the fall or early winter, but it can be found most any time. In the Southwest I have seen troops of them crowding highway shoulders for miles—both in the spring and fall. For some reason "shags" seem to gravitate toward asphalt and often burst up through it. They've been known to ruin tennis courts, and one is reported to have lifted a 10 pound slab of concrete in a heroic attempt to proliferate its species. Alexander Smith describes a fruiting of over 1000 specimens on a baseball field, "extending in a line from first base to short left field."

EDIBILITY: Edible and delicious—one of the best known and safest of all wild mushrooms. The flavor is very delicate but the texture is marvelous—not slimy as in okra, but

345

succulent as in octopus. For a delicious snack, slice them in half and dip them in egg batter and bread crumbs (or flour), then saute them briefly and serve hot! Use only young caps—darkened ones are mostly water—and don't pick more than you can eat in two meals unless you plan to preserve them, for they will deliquesce quickly. In rare instances they may react with alcohol in the body to produce effects similar to those of *C. atramentarius*. However, I have cooked them in wine many times with no ill effects.

COMMENTS: Shaggy manes are the soldiers among mushrooms, the sentinels of the roads. Their tall, shaggy cylindrical heads are distinctive even in silhouette, and when inky individuals are found in the vicinity of young ones, there can be no doubt as to their identity. The poisonous *Chlorophyllum molybdites* can be similar when young, but expands in age and doesn't deliquesce. *Podaxis pistillaris* is quite similar in shape and color, but lacks gills. "Shags" are as pleasing to the eye as to the palate, especially in the delicate reddish tints the gills assume as they mature. "Shags" also possess a special spontaneity—seemingly popping up overnight after a rain, while most mushrooms fruit several days after. Extreme aridity may arrest development so that the spores never mature and the gills remain white, or the gills blacken and then shrivel up rather than deliquescing. Several forms and varieties of the shaggy mane have been described, including a small, oval one only 5-6 cm high. The height of the larger forms seems to depend partially on the depth of the humus they must transcend—I have found specimens in redwood duff by a road that were two feet tall—but the "normal" height is 6-20 cm. Other species: *C. colosseus* is a giant version of the shaggy mane with stalk 35-50 cm long and spores 17-20 microns long; it has been described from Washington by Fred Van De Bogart, but is rare. Several small (cap less than 5 cm high) versions of the shaggy mane also occur, including: *C. palmeranus,* terrestrial; *C. alnivorus,* wood-inhabiting; *C. sterquilinus* and *C. umbrinus,* on dung or compost piles, the former with large spores and the latter with a brownish stem base; and *C. spadiceisporus,* on the dung of wild animals (rabbit, deer, etc.). All of these have the ring (annulus) characteristic of the shaggy mane and are too small and rare to be worth eating.

Coprinus picaceus   (Magpie Mushroom)

CAP (3) 5-8 cm broad, oval or cylindrical when young, broadly bell-shaped in age; surface at first covered with whitish universal veil material which soon breaks up into discrete patches, flakes, or "spots" on a dark brown to eventually black background; finely striate nearly to the center in age, and deliquescing. Flesh thin, fragile, soft. **GILLS** white, soon becoming reddish- or vinaceous-tinted, finally black and inky (deliquescing); crowded, more or less free. **STALK** 7-15 (25) cm long, 0.5-1.5 cm thick, base usually enlarged and with woolly hairs; hollow or stuffed, white, smooth, fairly fragile. **PARTIAL VEIL** absent. **SPORE PRINT** black; spores (13) 14-20 × 10-13 microns, elliptical, smooth.

HABITAT: Solitary or in small groups on ground in woods and along paths, etc.; widely distributed, but not common. I have found it only once in our area, but it is not uncommon in southern California when it is damp enough. In Europe it grows under beech and is said to be partial to alkaline soil.

EDIBILITY: Poisonous, at least to some people.

COMMENTS: The mottled brown-and-white or black-and-white cap makes this one of the easiest of all inky caps to recognize. Large specimens are reminiscent of the shaggy mane in shape—but the resemblance ends there. Greg Wright of Los Angeles, who has had more experience with this species than I, says the odor is also distinctive—like that of "mothballs, a hot grass pile, or burnt hair, or rank." Other species: In my own pathetic excuse for a garden I have found a somewhat smaller, mottled (black-and-white), unidentified *Coprinus* with a distinct annulus (ring) on the stalk.

Coprinus atramentarius, the common inky cap. Note hollow stalk in the beautiful slice at bottom.

Coprinus atramentarius   (Inky Cap; Tippler's Bane)

CAP 2-8 cm high and/ or broad, round or oval and often somewhat lobed when young, becoming conical to bell-shaped, or in old age convex; deliquescing from the margin toward the center; surface dry, smooth or with silky whitish fibrils (universal veil tissue) when young, grayish-brown to grayish-tan when young (the center often browner or with small brown scales and the margin sometimes pallid), becoming lead-gray to inky-gray in age; margin striate or grooved, usually tattered or splitting in age. Flesh thin, soft, pallid or grayish. **GILLS** free or nearly free, crowded, at first white, soon gray or with pinkish to vinaceous tints, finally black and deliquescing (becoming inky). **STALK** 4-15 (20) cm long, 0.6-1.5 cm thick, equal or with a narrowed or enlarged base; white or with grayish to brownish fibrils below; hollow. **PARTIAL VEIL** fibrillose and evanescent or leaving a median to basal ring or ridged zone on stalk, or sometimes even a rudimentary volva near base. **SPORE PRINT** black; spores 7-12 × 4-6 microns, elliptical, smooth.

Coprinus atramentarius usually grows in groups, tufts, or clusters. **Left:** These specimens have rounded smooth caps. **Right:** A large scaly-capped cluster.

HABITAT: Scattered to densely gregarious or in massive clumps in cultivated areas, lawns, gardens, roadsides, around or on old stumps, etc., sometimes also in the woods; widely distributed and very common, fruiting in our area practically year-round. It used to be an unwanted intruder in cultivated mushroom beds, and is one of our characteristic "suburban" mushrooms. The largest fruiting I've seen in the wild was under aspen in New Mexico. Several crops are generally produced each year.

EDIBILITY: Edible when young and fairly good—but sometimes reacting with alcohol in the body to produce a peculiar type of poisoning (see p. 896 for details). The grayish-brown caps look rather unappetizing—and precisely for this reason were the very first wild mushroom I ventured to eat. Inky cap-and-salami sandwhiches became a staple item in my teen-age diet until I discovered finer and more flavorful fungal foods, such as "Sparassis Sole" and "Agaricus Elegante."

COMMENTS: The lead-gray to brownish, bell-shaped caps of this species are a familiar sight in vacant lots and gardens. Several varieties occur, including one with a pointed or umbonate cap (*var. acuminatus*) and one with a copious universal veil. The typical variety, however, has a non-umbonate cap and only slight universal veil remnants on the cap (if any). The young unexpanded buttons provide an unexpected visual treat when sliced open lengthwise, but true to their name, they turn into an unsightly inky black mess as they mature, suitable for writing but not biting. The inky cap sometimes rivals the shaggy mane in size and abundance and may even mix company. I have seen gigantic 10-lb. clusters fruiting from an old cut stump in a field, but it does not necessarily grow in clumps. Other species. *C. insignis (=C. alopecia)* of eastern North America is a similar species with rough spores. It grows on hardwood stumps, especially maple, and is poisonous to some people.

Coprinus micaceus (Mica Cap; Glistening Inky Cap)

CAP 2-4 cm high when young, 1.5-5 cm broad when expanded; at first oval, soon bell-shaped, then expanding to convex; surface sprinkled at first with minute glistening whitish particles (universal veil remnants) which often disappear in age, otherwise tan to yellow-brown to ochre, buff, fulvous, or cinnamon-brown (margin often paler), becoming grayer in age, especially toward margin; striate at least half way to center, the margin usually tattered or split at maturity. Flesh thin, pallid or white, soft. **GILLS** crowded, adnate to adnexed or free, palllid soon becoming gray or brownish, finally black; deliquescing (sometimes partially, sometimes completely). **STALK** 3-8 (12) cm long, 2-6 mm thick, more or less equal, fragile, hollow, smooth, white or discoloring buff. **PARTIAL VEIL** absent or rudimentary, but stalk sometimes with a slight basal ring (presumably formed by the universal veil). **SPORE PRINT** dark brown to black; spores 7-11 × 4-6 microns, elliptical but often compressed (flattened) somewhat, smooth.

Coprinus micaceus. Note white stalk, striate cap, and tendency to grow in clusters. Tattered specimens at right have autodigested.

A clump of *Coprinus micaceus* glistening in the rain. Note how the cap is striate nearly to the center. The fine mica-like particles (veil remnants) often present on the cap have been washed off by the rain.

HABITAT: Gregarious, usually in clusters, on wood or woody debris, around stumps, or on roots or buried wood (thus appearing terrestrial); widely distributed and very common. It fruits practically year-round in our area. Like *C. atramentarius,* it is a common feature of suburbia but also occurs in the woods (on cottonwood, oak, alder, etc.).

EDIBILITY: Edible, but thin-fleshed and watery. The flavor is said to be good, however, and since it often occurs in large numbers, it is easy to gather enough for a meal. It does not always digest itself completely, another point in its favor.

COMMENTS: The cap color of this ubiquitous *Coprinus* varies considerably, but typically falls somewhere between tan and rusty-yellow. It is striate, but not translucent as in the much smaller *C. disseminatus.* It might be mistaken for a *Psathyrella* if the gills do not liquefy, but old caps become very tattered as the margin gets eaten away. The mica-like particles responsible for its name are not always evident because they wear away quickly. Other species: *C. silvaticus* is a very similar species that lacks the particles when young and has larger spores. *C. radians* is also similar, but can usually be distinguished from *C. micaceus* by its deliquescing gills and the presence of a cinnamon-brown to yellow-brown or yellow-orange woolly mycelial mat or network of mycelial strands (oozonium) in the surrounding substrate. It often grows indoors (on wet wood in basements, bathrooms, etc.), and bountiful crops sprout periodically from the woodwork of a popular cafe in Santa Cruz, California. Its employees are divided on how to respond: whether to advertise its presence as a unique organic addition to the existing decor, or whether to hush it up by organizing daily "search and squish" patrols, for fear of alienating the more squeamish members of their clientele.

Coprinus domesticus   (Domestic Inky Cap)

CAP 1-5 cm high, oval to bell-shaped, expanding slightly in age; surface buff, pallid, or yellowish with a tawny to ochre or fulvous center, at first covered with silky white to buff fibrils or patches (universal veil remnants) which often disperse or disappear in age; conspicuously striate nearly to the center; margin often splitting in age. Flesh thin, pallid. **GILLS** crowded, free or adnexed, white becoming gray to reddish-gray, finally black and inky (deliquescing). **STALK** 3-8 cm long, 3-10 mm thick, equal or with swollen base, white,

Left: *Coprinus domesticus,* immature specimens. Cap is grayish-buff to whitish and conspicuously striate. **Right:** An unidentified charcoal-loving *Coprinus* (see comments under *C. domesticus*). Note striations and prominent universal veil remnants on cap.

hollow. **PARTIAL VEIL** evanescent or sometimes forming a ragged ring or ridged zone on stalk, or even a rudimentary volva at its base. **SPORE PRINT** black; spores 7-10 × 3.5-4.5 microns, elliptical or bean-shaped, smooth.

HABITAT: Gregarious in small, scattered clumps on wood chips, compost, and vegetable debris in a large garden in Saratoga, California; probably widely distributed. Crops were produced in the above locality more or less continuously except in very cold weather.

EDIBILITY:"Of very good flavor when cooked," says Greg Wright of Los Angeles— but I don't recommend it because of possible confusion with other species.

COMMENTS: The name *C. domesticus* has apparently been applied to more than one species, so its use here must be regarded as tentative. However, our specimens compare favorably to the original (European) description of the species. It resembles *C. micaceus*, but has a somewhat paler cap with silky white fibrils (instead of particles) when young, and deliquesces to a greater degree. A rather similar but unidentified species with a darker (ochre to tawny) cap and a white fibrillose universal veil (see photo above) occurs in our area on ashes.

Coprinus lagopus group (Woolly Inky Cap) Color Plate 86

CAP 1.5-5 (7) cm high when unexpanded, 2-6 cm broad when expanded; at first oval or cylindrical to acorn-shaped or conical, then expanding to nearly plane with the margin splitting and often recurving (curling up and back); deliquescing in wet weather; surface dry, grayish to brownish-gray becoming blackish, but at first clothed with delicate white to grayish hairs or fibrils (universal veil tissue) which break up into patches and often wear away in age; striate nearly to the center beneath the hairs, at least at maturity. Flesh very thin, soft. **GILLS** more or less free, narrow, fairly close, pallid soon becoming grayish, then black and inky (but in dry weather merely withering). **STALK** 4-15 (20) cm long, 2-5 mm thick, equal or tapering upward, very fragile, hollow, white, at first clothed with minute white hairs (like cap); base with long woolly white hairs. **PARTIAL VEIL** absent or evanescent. **SPORE PRINT** blackish; spores 10-13 × 6-7 microns, elliptical, smooth.

HABITAT: Solitary, scattered, or in dense troops in leaf litter, woody debris, burned areas, etc. (usually in the woods); widely distributed. Along with closely related species (see comments), it is fairly common in our area in the fall, less so in the winter and spring. The fragile fruiting bodies last only a few hours before consuming themselves.

350

EDIBILITY: Probably harmless—but also fleshless and flavorless.

COMMENTS: The above description encompasses a "complex" of closely related inky caps that are one big headache for compulsive *Coprinus*-categorizers. In addition to the "true" *C. lagopus,* there is the very similar *C. lagopides,* which has smaller spores (7-10 microns long) and appears to be fairly common in our area. There are also several look-alikes that are somewhat smaller and grow on dung, manure, or wet straw. These include: *C. fimetarius* and *C. macrorhizus* (the latter with a rooting stem base); *C. macrocephalus,* with large spores (11-16 microns long); and *C. tectisporus,* which grows in greenhouse soil and only partially deliquesces. To make matters more confusing, the name *C. cinereus* is often used, sometimes for a woodland species like *C. lagopides* but with a taller cap, and sometimes as a synonym for *C. fimetarius* and *C. macrorhizus.* You needn't know the exact names of these inky caps, however, to appreciate their beauty (see color plate!), and they are easily recognized as a group by their slender, fragile, white stalk and striate grayish cap that is clothed exquisitely with white hairs when young. The dung- and straw-inhabiting forms, incidentally, are often abundant in our area and are *extremely* ephemeral —the caps deliquesce within a few hours, leaving behind only the long, slender, fragile, bean-sprout-like stalks. (For a miniature dung-inhabiting version, see the *C. radiatus* group.) Other species: *C. arenatus* is also similar to *C. lagopus,* but grows in sand; *C. narcoticus* grows in manure and rich soil and has a strong, unpleasant odor. Finally, there is *C. "chevicola,"* which fruited one season on the floor of my ramshackle '67 Chevy Nova Supersport. It superficially resembled *C. lagopus,* but was smaller and differed in several other respects, not the least of which was its choice of abode!

Coprinus radiatus group   (Miniature Woolly Inky Cap)

CAP minute, 2-8 mm high, 1-4 mm broad before expanding, at first oval or cylindrical, then expanding to plane or nearly plane (with the margin usually splitting and curling up), then deliquescing; surface gray or with a brownish center, but at first covered with long white to grayish hairs and/or scales (universal veil tissue); striate nearly to the center beneath hairs. Flesh very thin, soft. **GILLS** narrow, well-spaced in age, free or nearly free, pallid soon becoming gray to black, deliquescing with the cap. **STALK** 1-5 cm long, about 1 mm thick, more or less equal, hollow, very fragile, white, with rooting, hairy base. **PARTIAL VEIL** absent. **SPORE PRINT** black; spores 11-14 × 6-8 microns, elliptical, smooth.

HABITAT: Solitary or in groups on dung, compost, and manure; widely distributed and fruiting whenever it is damp enough. It is one of our commonest dung fungi, especially on piles of horse manure mixed with straw.

EDIBILITY: Who knows? Who cares?

COMMENTS: The small size and dense white to grayish hairs on the cap distinguish this diminutive dung-lover. It is essentially a miniature version of the equally common *C. cinereus-C. fimetarius-C. macrocephalus* complex (see comments under the *C. lagopus* group), and is itself a "complex" of difficult-to-distinguish forms. Because of its small size and brief life span it is best observed in a "dung garden" (dung or manure placed in a moist chamber). Other dung lovers include: *C. pseudoradiatus,* very similar but with smaller spores and a non-rooting stalk base; *C. niveus,* larger, with a mealy snow-white cap (at least when fresh); *C. semilanatus,* like *C. niveus,* but with a slightly darker cap (beneath the veil remnants); and a host of small species which cannot be reliably identified without a microscope. Also see *C. ephemeroides,* which has a large annulus (ring), *C. ephemerus* and *C. miser* (under *C. plicatilis*), which have a bald or nearly bald cap, and comments under the *C. lagopus* group.

Coprinus ephemeroides   (Ringed Dung Inky Cap)

CAP minute, 1-5 mm broad, convex or in age nearly plane; surface with minute granules or particles (but may appear smooth), grayish or tinged yellowish to ochre at the center; practically transparent; striate. Flesh very thin (almost non-existent). GILLS usually adnexed or free, becoming grayish or black, narrow, soon shrivelling or deliquescing. STALK 0.5-6 cm long, about 1 mm thick or less, equal or with a small bulb at base, very fragile, whitish, smooth; base sometimes with a rudimentary volva (in one form). PARTIAL VEIL membranous, forming a large, persistent, more or less median ring on stalk; ring disclike or saucerlike. SPORE PRINT black; spores 6-9 × 3-6 microns, angular (appearing almost triangular), smooth.

HABITAT: Solitary or in groups on dung and manure (especially of horses) and straw piles; widely distributed and common. Fruiting in our area whenever it is moist enough.

EDIBILITY: Academic—a tremendous number would be needed for one mouthful!

COMMENTS: One of our tiniest gilled mushrooms, this inky cap is easily recognized by its growth on dung and the large saucerlike annulus (ring) on the stalk. The ring is sometimes as broad as the cap, giving the illusion of a stalk with two caps! The name *C. ephemeroides* should not be confused with *C. ephemerus,* a slightly larger dung-lover that lacks an annulus (see comments under *C. plicatilis*). *C. bulbilosus* is a synonym.

Coprinus plicatilis   (Pleated Inky Cap)   Color Plate 87

CAP 0.5-1.5 cm high when young and 1-3 cm broad when expanded; oval or cylindrical to conical when young, broadly convex or plane in age; surface buff to yellow-brown, usually with a darker (cinnamon-brown or fulvous) center, in age becoming grayish except for center; deeply grooved (pleated) nearly to center; margin sometimes recurved in age. Flesh *very* thin, fragile. GILLS free (but attached to a collar around stalk apex), well-spaced, narrow, soon gray and eventually black, but tending to wither rather than liquefy. STALK 3-7.5 cm long, 1-3 mm thick, more or less equal, very fragile, thin, smooth, white or buff, hollow. PARTIAL VEIL absent. SPORE PRINT black; spores 10-13 × 6.5-10 microns, broadly elliptical, smooth.

HABITAT: Solitary, scattered, or in small groups in lawns and other grassy areas, under trees, in woods, along paths and roads, etc; widely distributed and fairly common in our area, especially in the fall and early winter. I have seen it fruit prolifically with the similarly named *Marasmius plicatulus* in grass under eucalyptus.

EDIBILITY: Unequivocally inconsequential due to its thin flesh.

COMMENTS: The exquisitely pleated cap, small size, and terrestrial habit distinguish this petite, parasol-like inky cap (see color plate!). The free collar to which the gills are attached is also distinctive, but is not evident in one form (perhaps a distinct species). It is larger than *C. disseminatus* and occurs in smaller numbers, and its cap lacks the white patches or hairs (universal veil remnants) typical of *C. lagopus* and most other inky caps. Other species: *C. ephemerus* also lacks hairs on its cap, but is slightly smaller and grows in dung, straw, and compost; *C. miser* is a minute species (cap 1-5 mm high, reddish-brown to orange-brown when fresh, but fading) that it also common on dung.

Coprinus disseminatus   (Little Helmet; Fairy Bonnet)

CAP 5-10 mm broad, oval soon becoming bell-shaped, then sometimes convex; surface minutely scurfy when young, then smooth; pallid or buff with a cinnamon-brown to honey-brown center, becoming grayish toward the margin in age; deeply striate or pleated to the

Coprinus (=Pseudocoprinus) disseminatus. This minute attractive mushroom usually grows in swarms and does not deliquesce. Note the bell-shaped striate cap and widely spaced gills.

center and translucent at maturity. Flesh *very* thin, soft. **GILLS** adnate to adnexed or appearing free, fairly well-spaced, at first white but soon gray, finally black or slightly paler; not deliquescing. **STALK** 1.5-4 cm long, 1-2 mm thick, equal, white or buff, hollow, fragile, smooth, often curved. **PARTIAL VEIL** absent. **SPORE PRINT** dark brown to black; spores 7-10 × 4-5 microns, elliptical, smooth, with a large apical germ pore.

HABITAT: Densely gregarious (sometimes hundreds) on or near decayed wood and debris, or on buried wood; usually found in woods or grassy areas, widely distributed. It fruits throughout the mushroom season in our area, but is not particularly common. In eastern North America it is often abundant.

EDIBILITY: Too small to be of any value.

COMMENTS: This dainty little fungus grows in troops that do indeed resemble "little helmets" or "fairy bonnets." Because the gills do not deliquesce, the "splitters" have rewarded it with the name *Pseudocoprinus disseminatus.* It has the general aspect of *C. micaceus,* but is much smaller. The translucent, pleated cap separates it from *Psathyrella,* (to which it has also been assigned); it is also reminiscent of *Mycena,* but the spore print is black or nearly black, not white. Other species: *C. impatiens* is a similar but slightly larger species with crowded, at least somewhat deliquescent gills; it usually occurs in smaller groups of up to one dozen individuals.

PANAEOLUS

Small to medium-sized, fragile, *dung- or grass-inhabiting* mushrooms. CAP usually *bell-shaped or conical when young (but sometimes convex).* GILLS typically attached, gray to deep brown to black when mature; *sides (faces) often mottled.* STALK typically thin, *brittle or fragile.* VEIL present or absent, but not usually forming an annulus (ring). VOLVA absent. SPORE PRINT *typically black* (but dark brown in *P. foenisecii*). Spores mostly elliptical, with a germ pore; retaining their color in concentrated sulfuric acid. Cap cuticle cellular.

THIS is a small genus of little brown mushrooms ("LBM's") with a bell-shaped to conical cap and thin, brittle stalk. The sides (faces) of the gills often have a mottled appearance (see photograph on p. 356) due to uneven maturation of the spores, but they do not deliquesce as in *Coprinus.* Psathyrellas are similar but do not typically grow in dung, and those that grow in grass tend to have a convex cap and/or dark brown spores. *Psilocybe* and *Conocybe* are common in dung, but do not have black spores.

Panaeolus is abundant in pastures, lawns, dung, and manure heaps, fruiting whenever it's moist. It often mixes company with other nondescript "LBM's" (*Conocybe, Agrocybe, Stropharia,* etc.), and would rapidly be relegated to the ranks of fungal forgetability were it not for the fact that some of its members contain traces of psilocybin and other pupil-dilating (the term "mind-expanding" being open to debate) compounds. However, in their search for a more promising and less painful reality, people will stoop to anything, even if it grows on cow patties and is only two inches tall. Thus, after every rain, our pastures are marred by hordes of "magic mushroom" hunters, inevitable plastic bags in hand.

Actually, there is considerable confusion as to which species are pupil-dilating. Traces of psilocybin have been isolated in virtually every species, but its presence and concentration is contingent upon a number of genetic, geographic, and environmental factors. My own experience with the *P. campanulatus-P. sphinctrinus* complex indicates that a sizable amount *may* produce mild hilarity, or even a transitory state of pseudoerotic effervescence. More often, however, it will produce a queasy stomach and if gulped down too eagerly, hiccups. It hardly seems worth the effort to harvest and digest the necessary number of fungal fructifications (30-50), but some people will do anything to "alter" reality, and it can be argued, I suppose, that doing anything is better than doing nothing.

Indiscriminate sampling of "LBM's" is foolish, of course, since some are poisonous, so care must be taken to identify your *Panaeolus* correctly. Six species are described here.

Key to Panaeolus

1. Cap *and/or* stalk staining blue to blue-green when bruised or in age 2
1. Not as above 3

2. Fruiting body staining blue to blue-green only at base of stalk (or mycelium) and only faintly; common in temperate regions . . . . *P. subbalteatus* group, p. 358
2. Fruiting body staining blue to blue-green in most parts (cap, stalk, flesh); largely tropical and subtropical . . . . *P. cyanescens* & others, p. 358

3. Cap striate or pleated *and/or* very tiny (less than 5 mm broad or high) . (see *Coprinus,* p. 342)
3. Not as above 4

4. Partial veil present (covering the gills when young), either forming an annulus (ring) or fibrillose zone on stalk or leaving toothlike remnants on cap margin 5
4. Veil absent (check several specimens in different stages if possible) 7

5. Veil usually forming a distinct annulus on stalk; cap pale (white to buff or pale tan), often viscid when moist . . . . *P. semiovatus,* p. 355
5. Veil typically leaving remnants on cap margin (but stalk sometimes with a black ring of spore dust); cap usually darker than above, not viscid 6

6. Cap reticulate (netted) or coarsely wrinkled (see photo at bottom of p. 357) . . . . *P. retirugis* (see *P. campanulatus* group, p. 356)
6. Not as above . . . . *P. campanulatus* group, p. 356

7. Mature gills and spore print dark brown; growing in grass but not on dung *P. foenisecii,* p. 360
7. Mature gills and spore print black (gills may be brown when young); in dung, grass, etc. . . 8

8. Cap whitish to buff or dingy yellowish when fresh, 4-10 cm broad; stalk solid, usually at least 4 mm thick . . . . *P. solidipes,* p. 355
8. Not as above (but cap may fade to whitish as it dries out) 9

9. Fruiting body very small; stalk whitish or nearly translucent; on dung (see *Psathyrella,* p. 361)
9. Not as above; common 10

10. Cap typically bell-shaped or conical, even at maturity . . . . *P. acuminatus* & others (see *P. campanulatus* group, p. 356)
10. Cap typically convex to plane, at least in age 11

11. Gills brown before becoming black; cap often with a darker marginal band as it begins to dry out; often but not always growing in small clusters; common   *P. subbalteatus* group, p. 358
11. Not as above; not common . . . . *P. fimicola* (see *P. campanulatus* group, p. 356)

Panaeolus solidipes   (Solid-Stemmed Panaeolus)   **Color Plate 83**

CAP 4-10 cm broad, rounded to convex or broadly bell-shaped; surface smooth or wrinkled, not viscid or only slightly so, buff to whitish, or dingy yellowish in age (or gray from spore dust); sometimes breaking up into small scales as it matures. Flesh fairly thick, whitish. **GILLS** adnate to adnexed, close; edges whitish; faces pallid becoming mottled with gray and black, and finally entirely black. **STALK** (4) 8-20 cm long, (3) 5-15 mm thick, equal or thicker below, solid, rather tough, smooth or longitudinally twisted-striate; often beaded with droplets when young and moist, especially at apex; white to buff or tinged gray from spores. **VEIL** absent. **SPORE PRINT** black; spores 14-22 × 9-14 microns, elliptical, smooth. Chrysocystidia present on gills.

HABITAT: Scattered to gregarious on dung (especially of horses), manure, and straw; widely distributed. In my experience it is not particularly common, but often prolific when it fruits. I have seen large numbers near Saratoga, California, in January, and near Flagstaff, Arizona, in September. O.K. Miller says it is common during the summer in Alaska and the Yukon.

EDIBILITY: Edible. I haven't tried it, but McIlvaine says, "it is one of the best of the toadstools." The relatively large size makes it the only non-hallucinogenic *Panaeolus* worth eating.

COMMENTS: The large size and solid stem are remarkable for a *Panaeolus*. Along with the black spores, pale cap, absence of a veil, and growth on dung, these features make it easy to recognize. *P. phalaenarum* and *P. sepulchralis* are synonyms, and it has also been placed in the genus *Anellaria* (along with *P. semiovatus,* to which it is closely related). The latter species, however, has a veil which usually forms a ring (annulus) on the stalk.

Panaeolus semiovatus   (Ringed Panaeolus)

CAP 2-6 (9) cm broad and 2-6 cm high, oval at first, then bluntly conical or parabolic; surface viscid when moist, often shiny when dry, smooth or slightly wrinkled, pale tan becoming buff or whitish in age, or tinged gray from spores; margin sometimes hung with veil remnants. Flesh soft, pallid. **GILLS** adnate to adnexed or seceding; edges whitish, faces pallid becoming brown or grayish, then mottled with black, and finally entirely black. **STALK** (5) 8-15 (18) cm long, (3) 5-10 (12) mm thick, equal or with enlarged base, stuffed or hollow, whitish to buff, apex often striate. **VEIL** membranous, usually forming a superior to median ring on stalk which is blackened by falling spores, but sometimes merely leaving a fibrillose zone. **SPORE PRINT** black; spores 15-22 × 8-12 microns, elliptical to somewhat pip-shaped, smooth. Chrysocystidia present on gills.

HABITAT: Solitary or in groups on dung and manure, especially of horses; widely distributed and fairly common in mild, wet weather. It is not uncommon in our area, but the largest fruitings I've seen were in New Mexico, in the summer.

EDIBILITY: Edible, according to most sources. However, there is one dubious report of psilocybin-containing specimens from Colorado.

COMMENTS: Also known as *Panaeolus separatus* and *Anellaria separata,* this species is easily recognized by its white to buff cap, presence of an annulus (ring), and growth in horse dung (see photograph at bottom of p. 357). Since the cap is slightly viscid, it might be mistaken for a *Psilocybe* or *Stropharia*. The mottled gills, however, signify *Panaeolus*. Specimens without a distinct annulus (ring) resemble *P. solidipes,* but that species has a thicker solid stem and lacks a veil even in the button stage. The paler cap color helps distinguish *P. semiovatus* from the *P. campanulatus* group, which can have a "ring" of black spore dust on the stem.

Panaeolus campanulatus group. This close-up clearly shows the mottled gills and tendency of the veil to break up into toothlike remnants that cling to the margin of the cap.

Panaeolus campanulatus group   (Bell-Shaped Panaeolus)

CAP 1-4 cm broad and/or high, bluntly conical or bell-shaped, scarcely expanding in age; surface not viscid, often shiny when dry, smooth or finely wrinkled or often cracking to form scales (especially in sunlight); some shade of brown, gray, or olive-gray when fresh (or in one form reddish-brown), paler (olive, tan, or buff) when faded, or dusted black by spores; margin hung with small, white, toothlike veil remnants, at least when young. Flesh thin, fragile. **GILLS** adnate or adnexed but often seceding, fairly close; edges whitish, faces gray becoming mottled with black; entirely black in age. **STALK** 6-15 cm long, 1-3 (5) mm thick, equal or thicker at apex, often quite long and thin, very brittle or fragile, brown to grayish, minutely powdered. **VEIL** evanescent, usually seen in young specimens as flaps of sterile tissue on margin of cap, and sometimes in older specimens as a thin band of black spore dust on the stalk. **SPORE PRINT** black; spores 13-18 × 7.5-12 microns, elliptical, smooth.

HABITAT: Solitary or in "families" on or near dung or in grass where cattle have grazed, sometimes also in compost; widely distributed and very common. It is often abundant in our coastal pastures in the fall, winter, and spring.

EDIBILITY: Some strains of this species "complex" (see comments) have mild hallucinogenic effects when eaten raw in large quantities (see p. 354). More often, however, there is no effect. Traces of psilocybin as well as serotonin have been isolated.

COMMENTS: The gray to brown, more or less bell-shaped cap, long thin brittle stalk, mottled gray to black gills, and toothlike veil remnants on the margin of the cap when young are characteristic of a closely-knit group of dung-lovers: *P. campanulatus,*

356

Panaeolus campanulatus group. Conical to bell-shaped cap, black gills, long brittle stalk, and growth on or near dung typify this common species.

P. sphinctrinus, and **P. papilionaceus.** Since even the experts cannot agree on the exact differences (if any) between them, it would be presumptuous to attempt to differentiate them here. At any rate, they are among our most common cow patty ("meadow muffin") mushrooms, along with *Psilocybe coprophila* and *Stropharia semiglobata.* The stalk is so fragile that it is difficult to bring home a specimen in one piece. Similar species include: **P. retirugis,** a smaller species with an often conspicuously veined or reticulate cap (see photograph), also found in our area, but not nearly as common and seemingly partial to horse dung; the **P. acuminatus-P. rickenii group,** with a conical to bell-shaped cap and no veil (check young specimens!); and **P. fimicola,** which also lacks a veil, but has a more convex or hemispherical (domed) cap. None of these are worth eating.

Two common inhabitants of horse dung. **Left:** *Panaeolus retirugis* (see comments above) is easily distinguished by its wrinkled or netted (reticulate) cap. **Right:** *Panaeolus semiovatus (= P. separatus)* has a smooth, pale cap and a veil that usually forms a distinct annulus (ring) on the stalk (see description on p. 355).

Panaeolus cyanescens   (Blue-Staining Panaeolus)

CAP 1.5-4 cm broad, bell-shaped to convex; surface smooth or sometimes cracked, not viscid, brown when moist, fading to grayish or whitish as it dries; margin often wavy or split in age. Flesh thin, bruising blue or bluish-green. **GILLS** adnate to adnexed or seceding, gray to black, the faces usually mottled. **STALK** 6-12 cm long, 2-4 mm thick, equal or with a slight bulb at base, usually long and slender, smooth, pallid to yellowish, grayish, or pinkish, the base brownish or tinged flesh-color; bruising bluish at least somewhat when handled. **VEIL** absent. **SPORE PRINT** black; spores 12-14 × 8.5-11 microns, elliptical, smooth.

HABITAT: Solitary to widely scattered or in groups on or near dung in pastures; widely distributed in the tropics and subtropics. It is fairly common along the Gulf Coast of the United States and also occurs in Mexico and Hawaii. Two similar species (see comments) have been reported from California.

EDIBILITY: Hallucinogenic—as might be expected of a blue-staining *Panaeolus*. *P. tropicalis* (see comments) is also potent, and both species are gathered by "magic mushroom" hunters in Hawaii.

COMMENTS: Blue-staining *Panaeolus* species are largely tropical and subtropical in distribution. They can be distinguished from the blue-staining Psilocybes by their black spore print, non-viscid cap (unless *very* wet), and growth on dung. Other blue-staining species include: *P. tropicalis,* with smaller spores (10-12 microns long) and no pinkish tinge to the stalk; and *P. cambodginensis,* with an olive to ochre-brown cap when moist, and smaller spores. All of these are placed by some in a separate genus, *Copelandia*. The latter two have been found in southern California on well-fertilized lawns.

Panaeolus subbalteatus group   (Belted Panaeolus)

CAP 2-6 cm broad, convex or bluntly conical becoming broadly convex to broadly umbonate to plane or with uplifted margin; surface smooth or wrinkled, in age sometimes breaking into scales (fissured), not viscid; color variable: brown to reddish-brown or cinnamon-brown when moist, fading as it dries to tan, buff, or even whitish (or grayish from spores), often with a darker (reddish-brown to brown or dark gray) marginal zone when partially dry. Flesh thin, brownish. **GILLS** adnate to adnexed or seceding, close,

Panaeolus subbalteatus group growing on a lawn fertilized with horse manure. The cap is convex to plane or wavy rather than conical, and usually develops a dark marginal band (as shown at left) when it begins to lose moisture.

Panaeolus subbalteatus group. Note clustered growth habit, relatively thick stem (for a *Panaeolus*), and convex cap. These were growing in a garden fertilized with compost from a mushroom farm.

broad, at first pale watery brown or reddish-brown, darkening gradually to black; edges whitish, faces usually mottled in age. **STALK** 4-10 cm long, (1) 3-6 (10) mm thick, equal or tapered at either end, hollow but not fragile, brown to reddish-brown, but often appearing whitish from a fine powder, or dusted gray by spores; apex often paler; usually longitudinally striate throughout; base (and mycelium) occasionally staining faintly bluish when bruised. **VEIL** absent. **SPORE PRINT** black; spores 10-14 × 7-9 microns, elliptical, smooth.

HABITAT: Scattered to densely gregarious—often in small clumps—in manure, compost, and fertilized lawns; widely distributed. In our area it fruits practically year-round but is most common in warm weather. I've seen several hundred specimens in a garden overgrown with vetch and mulched with compost from a mushroom farm. (It is one of the major mushroom "weeds" in cultivated mushroom beds.)

EDIBILITY: Hallucinogenic—the psilocybin content varies from moderate to low, perhaps due to differences in the nitrogen concentration of the substrate. A middle-aged Santa Cruz woman who mistook them for cultivated mushrooms *(Agaricus bisporus)* was hospitalized with hallucinations (a frightening experience if you're unprepared for them!), but recovered shortly. It is one of the more popular "recreational" species on the west coast, and is easily cultivated.

COMMENTS: The above description encompasses what may be a single variable species or a complex of closely related forms. The key features are the convex to broadly umbonate to plane cap (not bell-shaped!), brown gills when young, rather firm stalk, absence of a veil, black spore print, and frequent presence of a darker marginal band on the cap as it dries out. The tendency to grow in small clusters is also unusual for a *Panaeolus*. The black spores and non-viscid cap separate it from *Psilocybe,* and the larger size and black gills in age distinguish it from the common, lawn-inhabiting *Panaeolus foenisecii,* which may also have a zoned cap. *P. subbalteatus* is the most common "psilocybin" mushroom in California, but does *not* normally stain blue. However, the base of the stalk may occasionally stain blue *very* slowly, and it is sometimes coated with cottony, bluish-tinged mycelium. One variant (also hallucinogenic) that is common on fertilized lawns in our area has a slimmer stalk (1-3 mm thick) and smaller cap. Whether it is a "new" species or merely a diminutive form of *P. subbalteatus* is for the *Panaeolus*-pundits to decide.

Panaeolus foenisecii often grows on lawns with *Marasmius oreades* and *Conocybe lactea.* Cap is dark brown when moist, paler when dry, and often features a dark marginal band (shown at left) in intermediate stages. Gills and spores are brown, never black, and there is no veil.

Panaeolus foenisecii (Haymaker's Panaeolus)

CAP 1-3 (4) cm broad, bluntly conical to bell-shaped, expanding to convex, broadly umbonate, or nearly plane; surface smooth or cracking into scales in dry weather, hygrophanous but not viscid; chestnut-brown to dark brown or cinnamon-brown when moist, fading as it dries to dingy buff or tan, often with a darker marginal band when partially dry. Flesh thin, fragile. **GILLS** adnate to adnexed or seceding, fairly close, brown becoming deep brown, deep grayish-brown, or chocolate-brown, the faces often somewhat mottled and the edges paler or whitish. **STALK** 4-8 cm long, 1.5-4 mm thick, equal or with an enlarged base, fragile, more or less smooth, white to dingy brownish (often becoming brown from the base upward). **VEIL** absent. **SPORE PRINT** deep brown or purple-brown; spores 12-17 × 6-9 microns, elliptical, roughened.

HABITAT: Scattered or in groups on lawns and other grassy places; very common in warm wet weather (or on watered lawns) throughout the northern hemisphere. It is one of our characteristic spring, summer, and early fall lawn mushrooms, often mingling with *Conocybe lactea, Marasmius oreades, Psathyrella candolleana,* and the *Agrocybe pediades* group. In contrast to most *Panaeolus* species, it doesn't grow on dung, and in our area it doesn't even seem to grow in pastures.

EDIBILITY: Harmless in small quantities, but potentially poisonous to toddlers in the "grazing" stage, since chemical analysis has revealed traces of psilocybin in some collections. However, western material is apparently "inactive."

COMMENTS: This pixieish *Panaeolus* is one of our very common lawn-inhabiting "LBM's." Its small size, thin fragile stem, absence of a veil, growth in grass (not on dung!), and dark brown spores and gills distinguish it. Because the gills and spore print are not black as in other *Panaeolus* species and the spores are not smooth, the "splitters" (see p. 10) have erected a special genus for it, *Panaeolina.* Alexander Smith, on the other hand, dumps it into *Psathyrella* in his voluminous testament to that genus (his arguments for placing it in that genus are actually quite persuasive, but he has the weight of tradition against him in this case). Other species: *P. castaneifolius* is a similar diminutive lawn lover with a slightly thicker (3-6 mm) stem, purple-black gills, and purple-black, slightly roughened spores; I have not seen it in our area but it may occur. *P. subbalteatus* is also similar, but has black gills at maturity and black spores and is usually larger. *P. acuminatus* (see comments under *P. campanulatus* group) also grows on lawns, but has smooth black spores and lacks the marginal band on the cap.

PSATHYRELLA

Small to medium-sized mushrooms *found mostly on wood or in humus.* CAP conical to convex or plane, often hygrophanous, not usually viscid, *typically some shade of brown, buff, or gray.* Flesh *markedly fragile.* GILLS typically dark brown to black at maturity, usually attached but not decurrent. STALK *usually slender, fragile, and white or pallid.* VEIL absent or present, usually not forming an annulus (ring) on stalk. VOLVA absent. SPORE PRINT *deep brown to purple-brown or blackish, or rarely reddish.* Spores smooth or rough, with a germ pore; discoloring in concentrated sulfuric acid. Cap cuticle cellular.

FEW fleshy fungi have less to offer the average mushroom hunter—not to mention the average human being—than the Psathyrellas. They constitute an immense, monotonous, and metagrobolizing multitude of dull whitish, buff, grayish, or brownish mushrooms with a fragile stem, fragile flesh, and purple-brown to blackish spores. Psathyrellas are so nondescript and unassuming that it's much easier to define what they are *not* than what they *are:* their gills do not deliquesce as in *Coprinus,* their flesh never bruises blue as in *Psilocybe,* their cap is not colorful as in *Stropharia* and *Naematoloma,* and only rarely is it viscid; and they don't grow in dung like *Panaeolus.* Some Psathyrellas are quite attractive, however, and all have their indispensable "roles" to fulfill.

Psathyrellas are largely wood inhabitors, but often appear terrestrial because they feed on wood in the final stages of decay, after all the other wood-lovers have had their fill. Some species, such as *P. candolleana,* are ubiquitous, but most favor damp, shady situations, especially along trails or streambeds. In California, a very good *place* to look for them is in stands of willow and alder, and a very good *time* to look for them is in the late fall or early winter, but I can't think of a very good *reason* to look for them.

Psathyrellas are listed in older books under several different genera, including *Psathyra* and *Hypholoma.* Several are edible, but most have not been tested and few are fleshy or distinctive enough to warrant collecting. Alexander Smith has authored an abstruse monograph in which he describes more than 400 North American species. Most of them can only be identified if one has a microscope plus a special fondness for esoteric undertakings. My advice is to leave the Psathyrellas to professional psathyrellologists, who are paid to wrestle with such matters. A mere seven species are described here.

Key to Psathyrella

1. Growing either in sand dunes or on burnt ground or burnt debris or on other mushrooms . 2
1. Not as above; growing in grass, humus, wood chips, on wood, dung, etc. 5

2. Growing on shaggy manes *(Coprinus comatus)*; northern in distribution . . . . . . **P. epimyces**
2. Not as above 3

3. Growing in sand dunes or sand, often barely poking above ground . **P. ammophila** & others
3. Growing on burnt ground or debris 4

4. Cap with whitish fibrils when young (but these soon wearing away) . . . **P. carbonicola**, p. 366
4. Not as above 5

5. Stalk usually with a membranous, well-defined ring (annulus) after veil breaks . . . .
. . . . **P. longistriata** & others, p. 362
5. Not as above; ring absent or fibrillose and evanescent 6

6. Cap (2) 5-10 cm broad and distinctly fibrillose or fibrillose-scaly, at least when young, yellow-brown to rusty-brown or sometimes dark brown to blackish-brown; stalk typically at least 5 mm thick . . . . **P. velutina** & others, p. 366
6. Not as above 7

7. Cap distinctly striate or pleated nearly to the center; usually growing in groups or clusters . . .
. . . . (see *Coprinus*, p. 342)
7. Not as above; if growing in clusters then not striate or striate only near margin when moist 8

8. Clustered on ground or wood chips, each cluster arising from a buried "tap root"; stalks usually at least twice as long as widths of caps . . . . *P. multipedata* (see *P. hydrophila,* p. 364)
8. Not as above 9

9. Typically growing in clusters on or near wood, stumps, etc. 10
9. Not as above 13
10. Spore print distinctly reddish-tinted *P. sublateritia* & *P. conissans* (see *P. hydrophila*, p. 364)
10. Spore print some shade of gray, brown, or black 11
11. Veil present when young, but often disappearing in age 12
11. Veil absent .... *P. spadicea* (see *P. hydrophila*, p. 364)
12. Cap yellowish to honey-colored when moist, fading to buff or whitish as it dries; especially
    common on lawns and in towns .... *P. candolleana* & others, p. 363
12. Cap rusty-brown to cinnamon-brown when fresh, fading to tan as it dries; found mainly in the
    woods .... *P. hydrophila*, p. 364
13. Stalk 0.8-1.5 cm thick, white; cap 4-10 cm broad, rounded becoming convex or even plane, at first
    pallid from silky fibrils, but dark gray to grayish-brown in age; found under aspen in the Rocky
    Mountains, oak in California (see photo at bottom of p. 363) .... *P. uliginicola*
13. Not as above; stalk generally less than 1 cm thick 14
14. Growing in grass, dung, or manure 15
14. Not as above (may occasionally grow in grassy places, but not associated with grass) 17
15. Growing in dung; not common .... *P. stercoraria* & others
15. Growing in grass; common 16
16. Cap yellowish or honey-colored when moist, fading to buff or whitish as it dries; veil present
    when young but often disappearing in age .... *P. candolleana* & others, p. 363
16. Not as above .... (see *Panaeolus*, p. 353)
17. Odor typically fruity or pungent (like root beer, cat urine, etc.); cap often radially wrinkled
    and striate; cap and/ or stalk often vinaceous- or purple-tinged .... *P. bipellis*
17. Not as above 18
18. Veil present when young (covering the gills when *very* young), usually leaving remnants on
    or near the cap margin 19
18. Veil absent or rudimentary and quickly disappearing 20
19. Cap soon becoming broadly convex to plane, honey-colored or paler when moist, fading to
    buff or whitish as it dries out .... *P. candolleana* & others, p. 363
19. Not as above; cap remaining broadly conical to bell-shaped or becoming convex only in old
    age *and/ or* differently colored .... *P. longipes* group & others, p. 364
20. Cap white or whitish .... *P. sp.* (unidentified) (see *P. carbonicola*, p. 366)
20. Cap not white; stalk usually at least twice as long as cap width *P. gracilis* group & others, p. 365

Psathyrella longistriata   (Ringed Psathyrella)

CAP 2.5-8 (10) cm broad, conical to convex becoming broadly convex to plane or umbonate in age; surface smooth or slightly wrinkled, with whitish veil fibrils when very young, not viscid; some shade of brown, but sometimes fading in age. Flesh thin, fragile. **GILLS** adnate to adnexed or seceding, close, pallid becoming dark brown or purple-brown. **STALK** 4-10 cm long, 0.4-1 cm thick, equal, pallid or white, fragile, hollow, with scattered whitish scales (veil remnants) below ring. **VEIL** membranous, white, forming a persistent, superior, usually striate ring on stalk which is eventually darkened by spores. **SPORE PRINT** deep brown to nearly black; spores 7-9 × 4-5 microns, elliptical, smooth.

HABITAT: Solitary to scattered or in small groups in woods, particularly under conifers; fairly common in the Pacific Northwest, less so in California—I have found it but a few times in our area, in the winter.

EDIBILITY: Unknown.

COMMENTS: The persistent ring on the stalk is unusual for a *Psathyrella*, and helps to distinguish this western species. *Panaeolus semiovatus* and *Stropharia semiglobata* are somewhat similar but grow in dung or grass, not in the woods. Another *Psathyrella* with a prominent ring, *P. kauffmanii*, occurs under hardwoods in eastern North America, and is quite common in the aspen forests of the Rocky Mountains and Southwest.

Psathyrella candolleana is a common suburban species with a convex to plane or wavy cap and very fragile flesh. The cap is usually honey-brown to yellowish when fresh but fades as it dries. Note the dark spore dust on cap at upper right and the slight veil remnants on margins of caps at left.

Psathyrella candolleana (Suburban Psathyrella)

CAP 2-7 (10) cm broad, bluntly conical or convex when young, becoming broadly convex to plane or broadly umbonate in age; surface smooth or with a few scattered patches of whitish fibrils (veil remnants) when young; hygrophanous but not viscid; brown, honey-colored, or yellowish when moist, fading quickly as it dries to buff or whitish (the center often remaining darker); margin often hung with veil remnants, at least when young. Flesh very thin, fragile. **GILLS** close, adnate but sometimes seceding, at first whitish, soon grayish or grayish-purple, finally dark brown. **STALK** 4-10 cm long, 2-7 (10) mm thick, equal, hollow, fragile, white or whitish, often silky or scurfy. **VEIL** white, usually disappearing, but sometimes forming a fibrillose ring on stalk. **SPORE PRINT** deep brown; spores 7-10 × 4-5 microns, elliptical, smooth.

HABITAT: Scattered to gregarious or tufted in lawns, gardens, on or about old hardwood stumps, on buried roots or debris, etc.; widely distributed and very common in urban and suburban settings, sometimes also in the woods. It fruits year-round in our area but is most common in the late spring, summer, and early fall.

EDIBILITY: Edible. Some authors describe it as delicious, but it is very fragile, insubstantial, and not particularly easy to identify. Its main asset is its easy availability.

COMMENTS: The fragile, usually convex cap that is more or less honey-colored when fresh and pallid in age, plus the fragile white stalk, deep brown spores, and fondness for suburbia mark this ubiquitous but boring mushroom. The cap color is so variable that it is more of a hindrance to identification than a help. The situation is complicated by the presence of some very similar suburbanites, including *P. hymenocephala,* with a more cinnamon-colored cap when young, and *P. incerta* (formerly *Hypholoma incertum* and often listed as a synonym for *P. candolleana*), with smaller spores and a pale yellowish cap. All of these are so fragile that getting them home in one piece is difficult!

Left: *Psathyrella uliginicola* or a similar species (see couplet #13 on p. 362) occurs under oak in California; it is unusually robust for a *Psathyrella*, with a convex to plane cap that is striate in old age. **Right:** A cluster of *Psathyrella candolleana* with clearly visible universal veil remnants. (Greg Wright)

Psathyrella hydrophila commonly grows in clusters on wood. **Left:** A large clump growing from an old (buried) stump. **Right:** Close-up showing the white stalk, dark gills, and fragile veil remnants on margin of cap.

Psathyrella hydrophila   (Clustered Psathyrella)

CAP 2-5 (7) cm broad, bluntly conical or more often convex when young, expanding to nearly plane in age; surface smooth, hygrophanous but not viscid, dark reddish-brown to orange-brown or rusty-brown when moist, fading as it dries to pale tan or sometimes grayish; margin often darker brown and/or hung with veil remnants. Flesh thin, fragile. **GILLS** crowded, buff to pale brown, becoming chocolate-brown or dark brown in age; adnate to adnexed. **STALK** 3-7 (10) cm long, 2-6 (10) mm thick, equal, hollow, fragile, smooth, white to faintly grayish or sometimes brownish in old age, especially below. **VEIL** fibrillose, evanescent, leaving remnants on cap margin and occasionally an obscure zone of hairs on stalk. **SPORE PRINT** deep brown; spores 4-6 × 3-4 microns, elliptical, smooth.

HABITAT: Gregarious, usually in tufts or large, dense clusters on hardwood stumps, logs, and buried wood; widely distributed. Common in our area from fall through early spring.

EDIBILITY: Not recommended. Reportedly harmless, but some collections are bitter and closely related species haven't been tested.

COMMENTS: This is one of several Psathyrellas that grow in attractive, often large clumps on decaying wood. The principal fieldmarks are the white or pallid stalk, dark brown spores, fragile texture, and smooth, hygrophanous, usually convex cap. It is likely to be mistaken for a *Naematoloma,* but is much more fragile and not as brightly colored, or for a *Galerina* or *Pholiota,* which have paler brown spores. Other clustered Psathyrellas include: *P. spadicea,* similar, but with a thicker (4-10 mm) stalk and no veil, growing mainly on *Populus* (poplar, cottonwood, aspen); *P. fuscofolia,* also lacking a veil, but with stalk only 2-4 mm thick, on decaying wood; *P. circellatipes,* also slender-stemmed, with spores 12-15 microns long, favoring aspen; *P. multipedata,* with clusters originating from a common, deeply rooted base or "pseudorhiza"; *P. sublateritia,* with a reddish-hued spore print (rather unusual for a *Psathyrella*); and *P. conissans,* with a pinkish-red spore print (highly unusual for a *Psathyrella*).

Psathyrella longipes group

CAP 2-7 cm broad, broadly conical or bell-shaped, expanding slightly in age (rarely plane); surface smooth, hygrophanous, brown to cinnamon-brown or yellow-brown when moist, fading to tan, buff, or whitish as it dries; margin often adorned with widely spaced, tooth-like veil remnants. Flesh thin, fragile. **GILLS** close, adnate but often seceding, at first whitish, soon darkening to brown, finally dark brown to blackish. **STALK** 6-16 cm long, 2-8 mm thick, equal or enlarged slightly at base, fragile, hollow, white. **VEIL** white,

Left: *Psathyrella longipes* group, mature specimens. Note relatively large size (for a *Psathyrella*) and widely spaced veil fragments on margin of cap. Right: *Psathyrella gracilis,* or one of its numerous look-alikes. Note hygrophanous bell-shaped cap and long, slender, fragile stem.

somewhat membranous, but soon disappearing except for remnants on margin of cap. **SPORE PRINT** deep brown to blackish; spores 10-15 × 6.5-9 microns, elliptical, smooth.

HABITAT: Solitary, scattered, or in small groups on ground or debris in woods; found mostly in the West. Fairly common in our area from fall through early spring, under both hardwoods and conifers.

EDIBILITY: Unknown.

COMMENTS: The broadly conical to bell-shaped cap, long white fragile stalk, small veil remnants on the cap margin, relatively large size (for a *Psathyrella*), and dark brown to almost black spores are characteristic, but there is a multiplicity of *Psathyrella* look-alikes that are best identified with a microscope. In the "true" *P. longipes,* the spores are slightly flattened in end view; in the very similar but more common *P. elwhaensis,* they are not. Other species include: *P. atrofolia,* widely distributed and common, cap honey-brown fading to buff, with veil remnants on the margin (but not widely spaced or toothlike) and spores 8-10 microns long; *P. subnuda,* slightly larger, with a more cinnamon-colored convex cap when moist and practically no veil; and *P. conopilea,* with a reddish-brown to orange-brown conical cap that fades to buff, and spores 14-18 microns long, especially common in southern California under oaks and in gardens, but widely distributed. All of the above species tend to grow scattered to gregarious on the ground or in lignin-rich humus, rather than in clusters on wood. They do not grow in dung or manure like *Panaeolus.* See also *P. gracilis,* which lacks a veil.

Psathyrella gracilis group   (Graceful Psathyrella)

CAP 1.5-4 (5) cm broad, conical to bell-shaped becoming convex or nearly plane; surface smooth, not viscid, brown to dull yellowish-brown when moist, paler (or with a pinkish tinge) as it dries, and sometimes grayish from spore dust; translucent-striate when moist. Flesh very thin, fragile. **GILLS** close, broad, adnate but often seceding, pallid becoming grayish to brown and then dark brown. **STALK** 6-12 cm long, 1-3 mm thick, more or less equal, straight, fragile, thin, white or stained darker by spores. **VEIL** absent or rudimentary (and quickly disappearing). **SPORE PRINT** dark purple-brown to nearly black; spores 10-15 × 6-8 microns, elliptical, smooth.

HABITAT: Scattered or in groups or troops on ground, debris, and wood chips in woods, parks, under trees, etc.; common and widely distributed. I have seen colossal fruitings on wood chip mulch in a park at regular intervals throughout the mushroom season.

EDIBILITY: Too thin and fragile to be worthwhile.

COMMENTS: This species lacks the veil remnants on the cap margin characteristic of the *P. longipes* group. The long, thin, straight, whitish stalk, brown to grayish cap that is so thin as to be translucent-striate when moist, absence of a veil, dark spores, and terrestrial growth habit are the main fieldmarks. There are many similar Psathyrellas which can only be differentiated microscopically.

Psathyrella carbonicola   (Charcoal Psathyrella)

CAP 1.5-6 cm broad, bluntly conical to convex; surface dry, at first covered with a dense coating of whitish fibrils, these eventually wearing away to reveal the chocolate-brown to brown background; fading in age. Flesh rather thin. **GILLS** close, adnate but sometimes seceding, pale brown becoming dark brown in age. **STALK** 3-7 cm long, 2-6 mm thick, more or less equal, rather fragile, sheathed with white fibrils or scales below the veil, at least when young; base often brownish in age. **VEIL** white, evanescent or forming a slight fibrillose superior ring on stalk. **SPORE PRINT** dark brown; spores 6-8 × 3-4 microns, elliptical, smooth.

HABITAT: Scattered to densely gregarious or clustered on charred soil and wood, common after forest fires in northern and western North America. In our area I have seen large fruitings in the winters following controlled burns. It is often accompanied by *Pholiota highlandensis, P. brunnescens,* and *Myxomphalia maura.*

EDIBILITY: Unknown.

COMMENTS: The whitish fibrillose coating on the cap and stalk when young plus the growth on burnt ground make this one of the easiest of all Psathyrellas to identify. Other species: *P. canoceps* is one of several small species with a coating of silky white fibrils on the cap, but it does not grow in burned areas; a small, unidentified, whitish-capped species with black spores is common in our area in woods and under trees.

Psathyrella velutina

CAP (2) 5-10 cm broad, obtuse to convex or broadly umbonate, becoming nearly plane; surface dry, densely fibrillose or fibrillose-scaly but sometimes nearly smooth in age; dull yellow-brown to tawny to rusty-brown or sometimes darker brown; margin often paler, splitting in age, and hung with veil remnants. Flesh rather thick, brownish to ochre. **GILLS** crowded, adnate to notched or seceding, pale yellowish becoming light brown to rusty-brown, finally deep brown as spores mature; faces usually mottled in age, edges white and sometimes beaded with droplets. **STALK** 5-15 cm long, (0.3) 0.5-1.5 (2) cm thick, equal or swollen at base, fibrillose or scaly, dry, whitish above, light brown to dingy tawny or ochre below. **VEIL** fibrillose-cottony, usually forming an obscure superior hairy or cottony ring or zone on stalk which is darkened by spores. **SPORE PRINT** blackish-brown; spores 8-12 × 5-8 microns, elliptical, minutely roughened, with a prominent snoutlike germ pore.

HABITAT: Solitary or in groups or small clusters in grassy places, roadsides, around sawdust and compost piles, on gravelly ground, or sometimes in the woods; widely distributed, but infrequent in our area. I have found it in the fall and spring.

EDIBILITY: Edible, but not recommended.

COMMENTS: Also known as *Lacrymaria velutina,* this species is unusually large and sturdy for a *Psathyrella,* and is likely to be looked for in another genus. The fibrillose to fibrillose-scaly cap and stalk, obscure hairy annulus (ring), and blackish-brown spores are distinctive. *P. lacrymabunda* is often listed as a synonym, but has smooth spores according to Smith. Both species are sometimes placed in a separate genus, *Lacrymaria.* Other species: *P. maculata* is somewhat smaller, has blackish- to grayish-brown fibrillose patches on the cap, and grows in clumps on alder in the Pacific Northwest. Also see *P. uliginicola* (couplet #13 of the key) and *P. carbonicola.*

spores

STROPHARIACEAE

THIS is a fairly common family of saprophytic mushrooms with brown to purple-brown to purple-black spores and attached gills. A veil is usually present, but does not necessarily form an annulus (ring) on the stalk. The gills are not normally decurrent as in *Gomphidius* and *Chroogomphus,* nor are they usually free as in *Agaricus,* nor do they deliquesce as in *Coprinus.* The small, fragile species resemble *Psathyrella* and *Panaeolus,* but tend to have a viscid and/or brightly colored cap. (For a comparison of the brown-spored species with genera in the Cortinariaceae, see comments under the genus *Pholiota.*) The mushrooms in this family also share several anatomical (microscopic) characteristics: the cap cuticle is usually filamentous rather than cellular (see p. 19), the spores are smooth and often have a germ pore, and the gills frequently feature special sterile cells **(chrysocystidia)** which have a highly refractive golden content when mounted in potassium hydroxide (KOH).

Four genera are recognized here, all of which intergrade to some extent: *Pholiota* has dull brown to rusty-brown spores and is consequently placed in the Cortinariaceae by some mycologists; *Stropharia, Psilocybe,* and *Naematoloma* have deep brown to purplish or black spores, and are sometimes lumped together (by the "lumpers," of course) in a single giant genus, *Psilocybe.* The latter three genera are differentiated largely on microscopic characteristics such as the presence or absence of chrysocystidia, but can usually be told in the field by the combination of characteristics outlined in the key.

This is not an important family from a gastronomic standpoint. However, it is *the* most significant group for "magic mushroom" hunters, because *Psilocybe* is the principal genus of hallucinogenic or "pupil-dilating" mushrooms. The active principles are psilocybin and psilocin (see p. 895 for details, and also read comments on pp. 31-32).

Key to the Strophariaceae

1. Spore print dull brown to cinnamon-brown or rusty-brown .... *Pholiota,* p. 384
1. Spore print purple-brown to purple-gray, purple-black, or black 2

2. Lower portion of stalk and/or other parts of fruiting body staining blue or green when handled (sometimes slowly) .... *Psilocybe,* p. 368
2. Not as above (but cap may be blue or blue-green to begin with) 3

3. Growing on dung or manure 4
3. Not as above (but may grow in grass) 6

4. Cap white to yellow, yellow-brown, or pale tan 5
4. Cap darker (orange-brown to reddish-brown, grayish-brown, dark brown) *Psilocybe,* p. 368

5. Spore print black; cap white to buff or very pale tan .... (see *Panaeolus,* p. 353)
5. Spore print purple-brown to purple-black; cap usually yellowish or darker *Stropharia,* p. 374

6. Veil membranous or cottony-membranous, usually forming a distinct ring (annulus) on stalk .... *Stropharia,* p. 374
6. Veil absent, or if present then fibrillose and disappearing or merely forming a fibrillose zone on stalk which may subsequently be darkened by falling spores 7

7. Cap small (usually less than 4 cm broad), viscid when moist, some shade of brown, gray, dull olive, buff, or if whitish then usually narrowly conical or bell-shaped .... *Psilocybe,* p. 368
7. Not as above; cap white or brightly colored (yellow, red, green, etc.), or if dull-colored then not viscid, even when wet; cap usually 2 cm broad or more 8

8. Cap viscid or slimy when moist; veil present at least when young; growing solitary to scattered or gregarious, but not usually clustered .... *Stropharia,* p. 374
8. Cap usually not viscid; veil absent or present; often (but not always!) growing in tufts or clusters .... *Naematoloma,* p. 381

PSILOCYBE

Small to medium-sized, saprophytic mushrooms found in a variety of habitats. CAP smooth, *with a viscid (when moist), often separable pellicle (skin); usually some shade of brown, gray, yellow-brown, or buff.* GILLS *typically attached,* dark at maturity. STALK usually slender, *often turning blue or green when handled.* VEIL *often present, but not usually forming a distinct annulus (ring) on stalk* (except *P. cubensis*). VOLVA absent. SPORE PRINT *purple-gray to purple-brown to nearly black.* Spores mostly elliptical, smooth, with a germ pore. Chrysocystidia typically absent on the gills. Cap cuticle filamentous.

PSILOCYBE enjoys a notoriety grossly disproportionate to its visibility, for it embraces some of the most exalted and sought-after of all mushrooms—as well as some of the most mundane. The exalted ones are the hallucinogenic ("pupil-dilating") species popularly known as "magic mushrooms." They contain psilocybin and/or psilocin, and cause startling changes in one's perceptions and sensations if consumed in sufficient quantity. The changes are similar to those provoked by LSD (see p. 895). The mundane ones are those that *don't* contain psilocybin or psilocin—and since most *Psilocybe* species don't contain those compounds, it follows that most Psilocybes are mundane.

The Psilocybes as a group are difficult to characterize: the majority are listless little brown mushroom ("LBM's") with a viscid cap (when moist) and dark (purplish to nearly black) spores. The hallucinogenic species usually turn blue or greenish when bruised, especially on the stem, but almost any "LBM" can be mistaken carelessly for a *Psilocybe*—with potentially disastrous results! A good spore print is crucial, as it will eliminate the brown-spored genera *(Galerina, Inocybe, Conocybe, etc.),* which contain many poisonous species. Among the dark-spored genera, *Coprinus* has deliquescing gills, *Psathyrella* typically has a non-viscid cap and never stains blue; *Panaeolus* species with a viscid cap grow on dung and have black spores; and *Naematoloma* and *Stropharia* species are usually brightly colored, while the cap color in *Psilocybe* (with the notable exception of *P. cubensis*) is typically some shade of brown, gray, or buff.

Contrary to popular belief, Psilocybes do not grow exclusively on that brown stuff that sounds like a bell. Rather, they occur in a wide variety of habitats: in grass, on wood chips and mulch in landscaped areas, on decaying wood, and in humus or beds of moss in forests and bogs. The hallucinogenic species are particularly abundant in two disparate locales: the Pacific Northwest and southern Mexico. In our area, alas, they are like solar eclipses— seemingly rare, though actually more common than any one person's experience would indicate. In other words, they are not something you can really *look* for. It is more a matter of geography—being in the right place at the right time. (This situation may change, however, as introduced species like *P. cyanescens* spread.)

Since it was discovered that Native Americans near Oaxaca, Mexico, used certain mushrooms to induce altered states of consciousness, Psilocybes have received an inordinate amount of attention in the North American press. Underground newspapers and magazines are full of frivolous articles on "getting off," there is a glut of "magic mushroom" field guides and cultivation manuals available, and, as so often happens, it has become difficult to sort fact from fiction. Those wishing to pursue the subject should read the comments on pp. 31-32 and then invest in a *responsible* book on "psilocybin" mushrooms (e.g., *Psilocybe Mushrooms and Their Allies* by Paul Stamets), and *use it in conjunction* with a general field guide.

Whatever you do, *don't* rely on shortcuts and *don't* sample Psilocybes indiscriminately! True, most of the hallucinogenic species exhibit a blueing reaction when bruised, but the "optimum" dosage varies greatly from species to species, and as with any hallucinogenic drug, there is no way to predict the specific effects it will have on *you.* Furthermore, I have seen people mistake the staining reactions of other mushrooms (for instance, the blackening of the flesh in *Hygrocybe conica*) for the blueing reaction in *Psilocybe*. All the more reason to develop a systematic knowledge of mushrooms' habits and characteristics *before* venturing into the realm of the "LBM's."

Psilocybe is a fairly large and difficult genus. Only a few species are "pupil-dilating," and those that aren't are too small or too rare to be of food value. Five species are described here and several others are keyed out.

Key to Psilocybe

1. Some part of fruiting body aging or bruising blue to green (check cap, veil, stalk base, flesh) 2
1. Not aging or bruising blue or green 8
2. Fruiting in warm, muggy weather; mainly tropical and subtropical 3
2. Fruiting in cool or cold weather; mainly temperate 4
3. Growing on or near dung or manure; cap whitish to yellowish or yellow-brown . . . .
   . . . . *P. cubensis*, p. 373
3. Not as above . . . . *P. caerulescens* & others (see *P. cubensis,* p. 373)
4. Cap narrowly conical to bell-shaped and not expanding much, usually less than 2.5 cm broad; veil absent or rudimentary 5
4. Not as above; cap usually expanding, at least in age; veil present, at least when young 6
5. Growing mostly in grass . . . . *P. semilanceata*, p. 370
5. Growing mostly under conifers . . . . *P. pelliculosa* & others (see *P. semilanceata*, p. 370)
6. Growing on hardwood logs and debris in eastern North America; blueing only very slowly when bruised . . . . *P. caerulipes* (see *P. stuntzii*, p. 372)
6. Not as above; found mainly on west coast 7
7. Most parts of fruiting body staining distinctly blue to blue-green when bruised; veil usually disappearing or forming only a very slight ring on stalk . . . . . *P. cyanescens* & others, p. 371
7. Fruiting body blueing only slightly if at all; veil often forming a distinct (but small and fragile) ring on stalk that is sometimes greenish- or bluish-tinged . . . . *P. stuntzii*, p. 372
8. Growing on dung or manure 9
8. Not as above (but may grow in grass) 10
9. Veil absent or rudimentary . . . . *P. coprophila* & others, p. 370
9. Veil present, often forming a slight fibrillose ring on stalk *P. merdaria* (see *P. coprophila*, p. 370)
10. Cap convex to plane, 1.5-4 cm broad, dark reddish-brown to brown (but often paler toward margin or fading overall in age to pale tan); stalk 2-4 mm thick, usually with white mycelium at base; typically growing in dense groups or clusters on lawns or in other disturbed areas . .
    . . . . *P. castanella (=P. californica)*
10. Not as above; cap differently shaped and/or habitat or color different 11
11. Veil present, usually forming a small annulus (ring) on stalk . . . . *P. stuntzii*, p. 372
11. Not as above; veil absent or if present, usually disappearing 12
12. Cap typically conical or bell-shaped and scarcely expanding; usually growing in grassy areas
    . . . . *P. semilanceata*, p. 370
12. Cap differently shaped and/or growing in moss or wet places 13
13. Growing in woods or grass in eastern North America (especially the South); cap dark brown to rusty-brown, fading to ochre or yellow-brown, bell-shaped to convex . . . .
    . . . . (see *Naematoloma ericaeum* under *N. dispersum*, p. 384)
13. Not as above 14
14. Cap tawny to dull orange-brown; base of stalk with hairs; found in wet areas, especially common in the Pacific Northwest . . . . *P. (=Galerina) corneipes*
14. Not as above 15
15. Growing in swamps or bogs; stalk very long (5 cm or more) and thin; cap reddish-brown to blackish-brown . . . . *P. atrobrunnea*
15. Growing in moss; stalk usually less than 6 cm long; cap dark reddish-brown to tawny or yellowbrown . . . . *P. montana*

Psilocybe coprophila is the most common and widespread member of its genus. Note small size and growth on dung. Cap is brown to red- or orange-brown and viscid when moist.

Psilocybe coprophila   (Meadow Muffin Mushroom)

CAP 0.5-2 (3) cm broad, hemispherical to convex or broadly bell-shaped, or at times nearly plane in age; surface viscid when moist, then dry, smooth or with minute white particles or patches at margin (when young), dark reddish-brown to cinnamon-brown, brown, or orange-brown, fading to tan or sometimes grayish-brown as it dries (or darker from spore dust); margin striate when moist. Flesh thin, brownish. **GILLS** adnate to slightly decurrent, fairly well-spaced, grayish-brown becoming deep purple-brown or black. **STALK** 1-4 cm long, 1-4 mm thick, more or less equal, fibrillose, pallid to yellowish or brown, darkening in age but not bruising blue. **VEIL** absent or rudimentary and evanescent. **SPORE PRINT** purplish-brown to nearly black; spores 11-14 × 6.5-8.5 microns, elliptical, smooth. Chrysocystidia absent on gills.

HABITAT: Solitary or in small colonies on dung and manure (especially cow patties); widely distributed. It is one of our most common dung fungi, fruiting whenever it is damp enough. It can be "raised" by bringing home a "meadow muffin" and keeping it moist.

EDIBILITY: Generally regarded as harmless, but some strains apparently contain enough psilocybin to be rewarded with the euphemistic label "active." A large number would be needed to produce any noticeable effects, however.

COMMENTS: The small size, viscid brownish cap, purple-brown to dark brown spore print, and absence of an annulus (ring) distinguish this diminutive dung addict. Also found on dung is *P. (=Stropharia) merdaria,* which is similar, but has a dingy yellow to orange-brown or cinnamon-brown cap, and a veil which often forms a fibrillose annulus (ring) on the stalk. It is common also, and widely distributed. Other species: *P. angustispora* is a minute conical brown inhabitant of the Pacific Northwest; it grows on the dung of elk, sheep, etc.

Psilocybe semilanceata   (Liberty Cap)

CAP 0.5-2.5 cm broad and high, narrowly conical to bell-shaped with a pointed umbo, scarcely expanding in age; surface smooth, chestnut-brown to brown or olive-brown and at least slightly viscid when moist, fading to tan, olive-buff, or even yellowish as it dries; margin sometimes with bluish or olive stains. Flesh very thin, pallid. **GILLS** adnate to adnexed or seceding, pallid, soon becoming gray, then finally dark purple-brown or chocolate-brown; edges whitish. **STALK** 3-10 cm long, 1-2 (3) mm thick, equal, often curved or sinuous, pliant, whitish or with brownish base, sometimes with a bluish or blue-green tinge in age, especially at base. VEIL absent or rudimentary. SPORE PRINT purple-brown; spores 11-14 × 7-9 microns, elliptical, smooth. Chrysocystidia absent on gills.

HABITAT: Widely scattered to gregarious in pastures, tall grass, etc., but not on dung; widely distributed. It is especially common west of the Cascades from northern California to British Columbia. It fruits from late summer through early winter or sometimes in the spring. I have not found it in our area.

EDIBILITY: Hallucinogenic (see p. 895), and often gathered for recreational use despite its small size. It is not as potent as *P. cyanescens,* but is much stronger than *P. pelliculosa.*

COMMENTS: The liberty cap is one of the most distinctive "pupil-dilating" Psilocybes. The sharply conical or "peaked" cap, dark spore print, small size, and growth in grass (often tall) make it distinct. The flesh and stalk bruise blue only slightly, if at all, but may age olive or slightly bluish. Another "liberty cap," *P. pelliculosa,* is often mistaken for *P. semi-lanceata.* It is common under conifers in the Pacific Northwest and northern California. In addition to its different habitat, it is not quite as conical and has a more pronounced tendency to bruise or age blue-green. It is only mildly hallucinogenic, with 20-40 caps constituting an "average" dose. A third species, *P. silvatica,* has spores less than 10 microns long; it also grows under conifers and occurs across the northern half of the continent. There are also a number of small to minute, more or less conical, brown or gray Psilocybes (e.g., *P. montana*) that do not contain psilocybin. They grow mostly in bogs or in beds of moss and are very difficult to distinguish (see key to *Psilocybe*). All of the species discussed above, including the liberty cap, resemble *Panaeolus* species, but have a viscid pellicle (skin) that peels easily from the cap, slightly paler spores, and do not grow in dung.

Psilocybe cyanescens   (Potent Psilocybe)   Color Plate 88

CAP 1.5-4 (5) cm broad, soon convex to broadly convex, then plane or with an uplifted, often wavy margin; surface smooth, viscid when moist, dark brown or reddish-brown becoming caramel-brown, then fading as it dries to tan, yellowish-brown, or paler; sometimes with blue or blue-green stains, especially near margin. Flesh thin, bruising blue or blue-green. GILLS typically adnate but sometimes seceding, fairly close, brown or cinnamon-brown becoming dark smoky-brown or sometimes bluish-stained; edges whitish. STALK 3-8 cm long, 2-6 (8) mm thick, equal or with an enlarged base, sometimes curved; dry, whitish, but staining blue to bluish-green when handled or bruised. VEIL fibrillose or cobwebby, copious but disappearing or at most forming a very slight ring or hairy zone on stalk. SPORE PRINT purple-brown to purple-gray or purple-black; spores 9-12 × 5-9 microns, elliptical, smooth. Chrysocystidia absent on gills.

HABITAT: Widely scattered to densely gregarious on wood chips, sawdust, mulch, and humus, and on lawns rich in lignin; partial to coniferous debris, but also fond of alder and eucalyptus. It is fairly common in the San Francisco Bay area in cold weather (December-February), especially in landscaped areas and mulched flower beds, and is also fairly common in Oregon, Washington, and British Columbia. It is easily cultivated, and its aggressive mycelium responds readily to transplanting—given the proper conditions.

EDIBILITY: Hallucinogenic (see p. 895) and extremely potent, especially raw. Along with *P. baeocystis* and *P. strictipes* (see comments), it is the most powerful known hallucino-genic mushroom in the north temperate zone. Only one or two caps are needed to induce marked changes in perception and sensation. Psilocin is primarily responsible, with psilocybin and possibly other compounds contributing to the effects. Together they are said to constitute 0.6% of the mushroom on a dry weight basis—substantially more than the better known *P. cubensis.* A six-year-old Washington boy died after ingesting an unknown quantity of *P. baeocystis* along with other unidentified mushrooms. However, effects on adults other than those typical of psilocybin- and psilocin-ingestion have not been reported.

COMMENTS: The dark spores, viscid caramel-brown cap that fades as it dries, evanescent veil, and blueing of the stalk and flesh are the fallible fieldmarks of this potent *Psilocybe.* There are two very similar, closely-related, equally potent species on the west coast: *P. baeocystis,* with a less copious veil and more conical cap that is usually olive-brown when young; and *P. strictipes,* with a long, slender stalk (10-13 cm long, 2-3 mm thick). They may well occur in California, but earlier reports of *P. baeocystis* from San Francisco were apparently based on *P. cyanescens.* Spore prints should be taken to distinguish all three of these species from deadly Galerinas and other "LBM's."

Psilocybe stuntzii   (Stuntz's Blue Legs)

CAP 1-4 (5) cm broad, bluntly conical becoming convex to broadly umbonate, plane, or with an uplifted margin in age; surface smooth, viscid when moist, color variable: deep olive-brown to chestnut-brown when young, but often fading as it ages or dries to dingy yellow-brown or yellowish-buff; margin striate when moist and often tinged greenish. Flesh thin, pallid to brownish. **GILLS** adnate or adnexed, pallid soon becoming grayish or brownish; close or fairly well-spaced. **STALK** 2-6 cm long, 1.5-4 mm thick, equal or thicker at either end, often curved, yellowish to brown or sometimes with darker or bluish stains, especially below; not viscid, often with mycelial threads at base. **VEIL** membranous but thin; forming a fragile ring or fibrillose zone on stalk which is often blue or bluish-green but eventually may be darkened by falling spores or may disappear. **SPORE PRINT** dark purple-brown; spores 8-12 × 6-8 microns, elliptical, smooth. Chrysocystidia absent on gills.

HABITAT: Scattered to densely gregarious or clustered on wood chips, mulch, etc., in lawns, gardens, and landscaped areas; also under conifers and in fields. It is known only from the west coast and is especially common in the Puget Sound region of Washington in the fall, early winter, and spring. (The appearance of large numbers in wood chip mulch on

Psilocybe stuntzii growing in a flower pot mulched with fir bark. Note the annulus (ring) on stalk and the viscid cap (at least when moist). The white globules are fertilizer pellets.

the University of Washington campus in Seattle more than a decade ago helped spur the "magic mushroom" craze that subsequently swept the Pacific Northwest.) In our area I have found it in mulched flower pots.

EDIBILITY: Weakly hallucinogenic (see p. 895), but popular with "magic mushroom" hunters because it often fruits in large numbers. Be sure not to confuse it with deadly *Galerina* species, which can look quite similar, but have rusty-brown spores—they will even grow intermixed with *P. stuntzii!*

COMMENTS: Like the liberty cap *(P. semilanceata)*, this "LBM" contains psilocybin, but bruises blue only weakly if at all. However, the dark spore print and ring on the stalk plus the viscid cap when moist are distinctive. It is named after the late, great Dr. Daniel Stuntz of the University of Washington, who was the first person to collect it—or more precisely, the first to collect specimens for the herbarium rather than for consumption! The common name, which was coined by Gary Lincoff, is apparently a reference to the tendency of the stalk or "leg" to stain bluish, though it often won't stain. Other species: *P. caerulipes* of eastern North America also blues only slightly if at all; it has a more or less evanescent veil and grows on hardwood logs and debris. Like *P. stuntzii,* it is weakly hallucinogenic.

Psilocybe cubensis   (Magic Mushroom)

CAP 1.5-8 (10) cm broad, broadly conical or oval or bell-shaped (often with an umbo) when young, gradually expanding to convex, broadly umbonate, or plane; surface smooth or with small whitish veil remnants when young, viscid when moist, soon dry, color variable: whitish with a brown to yellowish center, or entirely yellow to yellowish-buff to yellow-brown, or sometimes cinnamon-brown when young and sometimes dingy olive in old age; bruising and aging bluish; margin sometimes hung with veil remnants. Flesh firm, white, staining blue or blue-green when bruised. **GILLS** close, adnate to adnexed or seceding to free; pallid, soon becoming gray, then deep purple-gray to nearly black; edges whitish. **STALK** 4-15 cm long, 0.4-1.5 cm thick, equal or more often thicker below, dry, white or sometimes yellowish to yellow-brown, aging or bruising blue or blue-green; smooth. **VEIL** membranous, white or bluish-stained, usually forming a thin, fragile, superior ring on stalk which is blackened by falling spores. **SPORE PRINT** dark purple-brown to blackish; spores 11-17 × 7-12 microns, elliptical, smooth, thick-walled, with a large apical germ pore. Cystidia present on faces of gills, but chrysocystidia absent.

HABITAT: Solitary or in groups on dung and manure, especially in cattle pastures; widely distributed in the tropics and subtropics—Colombia, Central America, Mexico, etc.—and in the Gulf Coast region of the United States. Since it is being cultivated on a widespread basis, it may eventually turn up in the warmer parts of California—as it did in Santa Cruz during muggy summer weather (on compost from a "magic mushroom" farm). It can be cultivated on a variety of simple grain or compost mediums, but strict temperature control and sterile conditions are necessary to induce growth and prevent contamination.

EDIBILITY: Hallucinogenic (see p. 895). It is not as powerful on a dry weight basis as *P. cyanescens,* but is larger. At one time the demand for "magic mushrooms" far out-stripped the supply, with the result that many of the "psilocybin" mushrooms sold on the street were actually grocery store mushrooms *(Agaricus bisporus)* laced with LSD or other substances (one sample I examined proved to be a soggy chanterelle soaked in ammonia!). Now that cultivation procedures have been refined, you're much more likely to get the real thing. It is often sold dried, though drying decreases the potency. The mycelium is also "active." Ironically, natives of Oaxaca, Mexico, consider this species inferior to others, such as *P. mexicana* (perhaps because it was unknown to them until the Spaniards introduced cattle). Instead of using it themselves, they sell it to eager gringos!

Psilocybe cubensis is best told by its shape, whitish to golden-brown cap color, and tendency to stain blue when handled, especially on the stalk. It is common on dung in tropical and subtropical regions.

COMMENTS: This is the largest, handsomest, and best known of all the "psilocybin" mushrooms. The brownish to yellow or pallid cap, membranous ring on the stem, and tendency to bruise blue are the main fieldmarks—plus its growth in fields, usually on cow patties. It resembles the yellow-capped Stropharias (e.g., *S. semiglobata*) so closely that it is called *Stropharia cubensis* by many mycologists. The Stropharias, however, do not stain blue. In shape it resembles *Agrocybe praecox,* but the spore print is darker, and it has a decidedly different "aura"—or so I'm told by at least one certified fruitcake! Other species. The "Landslide Mushroom," *P. caerulescens,* is another common, blue-staining, hallucinogenic species that occurs along the Gulf Coast as well as in Mexico. It has an *evanescent* veil, bitter taste, and an olive-black cap when young that may become redder (e.g., reddish-brown) in age. As its name implies, it grows in recent landslides, as well as on sugar cane mulch and other debris. *P. tampanensis*, discovered in Florida, is a slender-stemmed hallucinogenic species that forms 1-2" underground "tubers."

STROPHARIA

Small to medium-sized, saprophytic mushrooms. CAP *usually viscid when moist and often brightly colored;* typically convex to plane or umbonate. GILLS *typically attached,* dark brown to gray, purple-gray, or black at maturity. STALK often fleshy, but sometimes slender; central. VEIL *present, usually forming an annulus (ring) on stalk.* VOLVA absent. SPORE PRINT *deep brown to purple-brown, purple-black, or black.* Spores typically elliptical and smooth, with a germ pore. Chrysocystidia usually present on gills. Cap cuticle filamentous. Acanthocytes usually present in mycelium.

THIS is a medium-sized genus of brightly colored mushrooms with a well-developed, persistent veil, attached gills, and dark (purple-brown to black) spores. The cap is usually viscid and some shade of yellow, yellow-brown, orange, red, green, blue, or white. In most cases the veil is membranous and forms a distinct annulus (ring) on the stalk, but in some species, such as *S. semiglobata,* it forms only a fibrillose zone (as in *Naematoloma* and *Psilocybe),* and in others, e.g., *S. ambigua,* it leaves copious remnants on the cap margin. An interesting and apparently unique microscopic feature of *Stropharia* is the presence of **acanthocytes** (needlelike calcium oxalate crystals) in the mycelium of many —if not all—species.

Stropharia is most apt to be mistaken for *Agaricus,* which has chocolate-brown spores and free gills which are frequently pink when young, and for *Agrocybe,* which has a browner (never purple-brown) spore print and dry to only slightly viscid cap. *Stropharia* intergrades to some extent with *Psilocybe,* but in the latter (as defined here) either the stalk bruises bluish-green or the veil does not form an annulus or the cap is brownish to buff. Most *Stropharia* species are found in humus, grass, or dung, but a few grow on decayed wood or wood chips. Woodland species can be mistaken for *Naematoloma,* but as a rule they rarely grow in the clusters typical of that genus, and their veil is more persistent.

Many Stropharias are attractive, but only the large and distinctive *S. rugoso-annulata* is commonly eaten. Some species may actually be poisonous, though there are conflicting opinions on this point. Seven representatives of the genus are described here.

Key to Stropharia

1. Cap brick-red to reddish to orange *and* small (typically less than 6 cm broad) ....
   .... (see *Naematoloma aurantiaca* & others, p. 382)
1. Differently colored and/or larger 2
2. Cap and/or stalk blue to green when fresh (or partially so), but often fading or developing
   yellowish tones in age .... *S. aeruginosa* & others, p. 380
2. Not as above 3
3. Growing in dung, manure, grass, compost, or mulched or landscaped areas 4
3. Not as above; usually growing in woods 9
4. Veil membranous, usually forming an annulus (ring) that is often striate or grooved (on upper
   surface); stalk generally 1-2 cm thick *or* if thinner then usually rather short 5
4. Not as above; stalk usually at least 5 cm long *and* slender 7
5. Cap 4-15 cm or more broad, wine-red to reddish-brown to tan, yellow-brown, or even grayish-
   brown; stalk at least 1 cm thick .... *S. rugoso-annulata,* p. 378
5. Not as above; usually smaller and more slender; cap white to yellowish or yellow-brown .. 6
6. Cap white or tinged yellowish to ochre at center .. *S. melanosperma* (see *S. coronilla,* p. 377)
6. Cap golden-brown to yellowish or creamy .... *S. coronilla,* p. 377
7. Stalk viscid or slimy below the veil, at least when moist ..... *S. semiglobata* & others, p. 376
7. Stalk not viscid 8
8. Cap smooth (without scales) and white to yellowish when fresh, and usually less than 6 cm broad;
   stalk neither cottony nor scaly ..... *S. umbonatescens* & others (see *S. semiglobata,* p. 376)
8. Not as above 9
9. Cap scaly or fibrillose and *not* viscid; stalk thick (at least 1 cm) and scaly below the ring
   (annulus) .... *S. kauffmanii,* p. 380
9. Not as above 10
10. Cap small (less than 5 cm broad), without scales, chestnut-brown to olive-brown or olive-gray
    when moist, sometimes fading to yellow-brown or buff as it dries; stalk only 2-4 mm thick;
    annulus (ring) often greenish- or bluish-tinged .... (see *Psilocybe,* p. 368)
10. Not as above 11
11. Stalk slender, brownish, with small scales below the annulus (ring); cap tawny-orange to yellow-
    ish to brown or olive-brown, small (less than 8 cm broad), often umbonate and with small,
    often concentrically arranged scales (but these sometimes washed off); growing on rotten
    wood, sawdust, debris, etc.; especially common in the Pacific Northwest ....
    .... *S. squamosa* (see *S. kauffmanii,* p. 380)
11. Not as above 12
12. Cap typically some shade of yellow or cream; veil typically shredding, most of it remaining on
    cap margin or forming only a slight annulus (ring) on stalk 13
12. Not as above; veil usually forming a large annulus on stalk 14
13. Stalk usually at least 0.6 cm thick, often cottony or shaggy; found in woods *S. ambigua,* p. 377
13. Stalk more slender (less than 1 cm thick), not shaggy or only slightly so; growing in a wide
    variety of habitats .... *S. riparia* (see *S. ambigua,* p. 377)
14. Cap yellowish, often with darker (brownish) spots; found under hardwoods and in lawns and
    other open places in eastern U.S. (especially the South) .. *S. hardii* (see *S. coronilla,* p. 377)
14. Cap dull brown or shaded with gray, purple, etc.; found in northern North America, usually
    under conifers .... *S. hornemannii,* p. 379

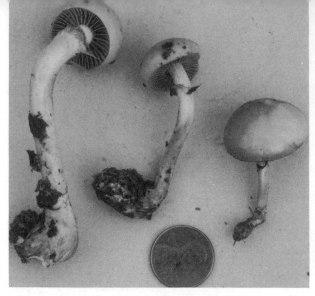

Stropharia semiglobata. **Left:** Small specimens. **Right:** A larger one. It resembles *Psilocybe cubensis,* but does not stain blue. The veil does not always form a distinct ring on stalk. Note long slender stem.

Stropharia semiglobata (Hemispherical Stropharia)

CAP 1-5 (6) cm broad, hemispherical (rounded) or broadly bell-shaped, becoming convex or rarely plane; surface smooth, viscid or slimy when moist, pale yellow to straw colored, yellowish-buff, or yellow-brown; margin often paler, sometimes hung with whitish veil remnants. Flesh pale or watery yellowish, thin. **GILLS** typically adnate but sometimes seceding, at first grayish, then dark purple-brown to black. **STALK** 5-8 (13) cm long, 2-6 mm thick, typically long and slender, equal or slightly enlarged at base; somewhat fibrillose above the veil, viscid or slimy below (when moist); white to yellowish. **VEIL** slimy, delicate, forming a fragile, superior, fibrillose ring or zone on stalk which is soon blackened by falling spores, or sometimes disappearing entirely. **SPORE PRINT** dark purple-brown to black; spores 15-19 × 7.5-10 microns, elliptical, smooth. Chrysocystidia present on gills.

HABITAT: Solitary or in small groups on dung, manure, rich soil, straw, and grazed or fertilized grass; widely distributed, and one of our most common dung fungi. In our area it fruits whenever it is damp, often in the company of *Psilocybe coprophila* and the *Panaeolus campanulatus* group.

EDIBILITY: Edible, but slimy and mediocre according to most sources. As usual, however, Captain Charles McIlvaine disagrees, stating that, "the caps are equal to any mushroom—tender, good, and harmless."

COMMENTS: Our most common and uninteresting *Stropharia,* this species is sometimes mistaken for the considerably more interesting *Psilocybe cubensis,* which has a more prominent ring and stains blue or green when bruised. The viscid to slimy yellowish cap, slender viscid or slimy stem, and dark gills set *S. semiglobata* apart. The veil may form a slight ring, but is not as membranous as that of *S. coronilla.* There are several closely related species or variants which are difficult to distinguish, including: *S. stercoraria,* with cap usually plane at maturity, stalk not viscid or only slightly so, and spores larger; *S. siccipes,* with a dry, sometimes rooting stalk and smaller spores; *S. (=Psilocybe) umbonatescens,* with a conical to distinctly umbonate cap, dry stalk, and similarly-sized spores; and *Psilocybe (=Stropharia) merdaria* (see comments under *P. coprophila*). All of these are partial to dung, manured ground, and grass, while *S. albonitens,* another similar species, grows on the ground in a variety of habitats. It has a whitish or yellow-tinged, often umbonate cap and a membranous but thin or evanescent veil. *Agrocybe* species are also similar in color and shape, but have browner gills and a brighter or browner spore print. See also *S. riparia* (under *S. ambigua*).

Stropharia coronilla, various stages of development. This grass-inhabiting species looks like a small *Agaricus,* but has attached gills and a striate or grooved annulus (ring). Note relatively short stalk.

Stropharia coronilla   (Garland Stropharia)

CAP 2-6 cm broad, convex to plane or slightly uplifted in age; surface usually smooth (but in one form with small orangeish scales), slightly viscid when moist, golden-brown to yellowish, yellowish-buff, or creamy. Flesh soft, white. **GILLS** adnate or at times adnexed in age, close, pallid becoming grayish, then purplish or purple-gray to purple-black. **STALK** 2-5 cm long, 3-6 (10) mm thick, usually rather short, more or less equal, not viscid; whitish, minutely scaly or cottony above the ring, fibrillose to smooth below; base often with white mycelial threads. **VEIL** membranous, white, forming a persistent, median to superior ring on stalk; ring usually striate or grooved on upper surface and soon darkened by spores (but in one form not striate). **SPORE PRINT** dark purple-brown to blackish; spores 7-11 × 4-5.5 microns, elliptical, smooth. Chrysocystidia present on gills.

HABITAT: Scattered to gregarious on lawns and baseball fields, also in pastures and other grassy areas; widely distributed but not particularly common, at least in our area. It fruits most any time but is most frequent in the fall.

EDIBILITY: Dubious—poisonous according to some; hardly worth experimenting with.

COMMENTS: This attractive little *Stropharia* is often mistaken for an *Agaricus,* but has attached rather than free gills and a frequently grooved or lined (striate) ring. It is shorter than *S. semiglobata,* the ring is more prominent, and the stalk is not viscid. Forms with a deeply grooved ring have been called *S. bilamellata.* A similar, widespread species with a paler (whitish or tinged creamy to ochre at center) cap, *S. melanosperma,* grows on dung or in pastures. *S. hardii* is a slightly larger (cap up to 10 cm broad) eastern species that grows in grass as well as in woods. Its veil is more apt to disappear in age and it has smaller spores. Like *S. coronilla,* it is easily confused with *Agaricus,* but has attached rather than free gills.

Stropharia ambigua   (Questionable Stropharia)   **Color Plate 89**

CAP 3-15 cm broad, obtuse to convex, becoming plane or even uplifted in age; surface smooth, viscid or slimy when moist, yellow to yellowish-brown to yellowish-buff, tawny, or sometimes nearly white; margin hung with cottony white veil remnants. Flesh white, thick, soft. **GILLS** pale gray, gradually darkening to purplish-gray or purplish-black; close, typically adnate but sometimes seceding. **STALK** 6-18 cm long, 0.5-2 cm thick, more or less equal, often long; stuffed or hollow; silky and white above the veil, clothed with soft, dry,

377

delicate, cottony white scales below (but these sometimes wearing off); often yellowish toward base in age; base often with white mycelial threads attached. **VEIL** soft, white, cottony, leaving shreds or strands on the cap margin and sometimes a superior ring or ragged zone on the stalk. **SPORE PRINT** dark purplish to nearly black; spores 11-14 × 6-7.5 microns, elliptical, smooth. Chrysocystidia present on gills.

HABITAT: Solitary to scattered or in groups in rich humus, usually under conifers, but also with alder and other hardwoods; known only from the Pacific Coast. It is fairly common in our area from late fall through early spring, especially in dank, cold places (in rain forests, along woodland streams and gullies, etc.).

EDIBILITY: Edible ? According to one authority, it tastes "like old leaves." As I do not make a habit of chewing on old leaves, I cannot attest to the validity of this comparison.

COMMENTS: There is nothing ambiguous or questionable about this elegant, stately fungus. It is our most common woodland *Stropharia* and at its best is one of the most exquisitely beautiful of all mushrooms—well worth seeking out. The soft, delicate white scales that sheathe the stem and the strands of veil tissue on the cap margin are very striking. *Amanita gemmata* is somewhat similar in color but has white gills; *S. hornemannii* has a duller cap and more prominent ring. Other species: *S. riparia (=S. magnivelaris?)* is a somewhat similar but slightly smaller and slimmer (stalk less than 1 cm thick) species that is fairly common in the West in a variety of habitats (under aspen and alder, along streams, under mountain conifers, etc.). Its veil is thinner than that of *S. ambigua* and more kleenex-like, and often leaves remnants on the cap near the margin rather than dangling from the margin itself. Also, the stalk is not nearly so shaggy as that of *S. ambigua,* and the cap, although similar in color, is more apt to be slightly umbonate.

Stropharia rugoso-annulata   (Wine-Red Stropharia)

CAP 4-15 (20) cm broad, obtusely bell-shaped or convex becoming broadly umbonate to plane; surface smooth, slightly viscid or dry, color variable: wine-red to purple-brown to reddish-brown when fresh, but fading to tan, straw-color, or even grayish as it ages. Flesh thick, fairly firm, white. **GILLS** adnate or notched (but sometimes becoming free in age), crowded, at first whitish but soon gray and finally purple-gray to purple-black with whitish edges. **STALK** 7-12 (25) cm long, 1-3 (7) cm thick, often enlarged at base, white or discoloring yellowish to brownish in age; base often with white mycelial threads attached.

Left: Close-up of the shaggy veil of *Stropharia ambigua.* **Right:** *Stropharia rugoso-annulata.* Cap color ranges from wine-red to tan. The cultivated version is often much more robust.

VEIL membranous, white; forming a thick, persistent, superior ring on stalk that is soon blackened by falling spores; ring grooved or lined (striate) on upper surface and often split radially into segments. **SPORE PRINT** deep purple-brown to black; spores 10-15 × 6-9 microns, elliptical, smooth. Chrysocystidia present on gills.

HABITAT: Scattered to gregarious in mulch, wood chips, straw, lawns, gardens, and other cultivated areas; widely distributed. It is quite common in New England and has also turned up in Washington. I have yet to find it in our area, but it is bound to turn up sooner or later (probably sooner).

EDIBILITY: Edible, and the best of the genus for the table. It is easily grown at home and widely cultivated in Europe (particularly eastern Europe). The flavor is fairly good—reminiscent of "undercooked potatoes soaked in burgundy," according to Rick Kerrigan.

COMMENTS: This large, handsome *Stropharia* is easily recognized by its wine-red to tan cap, purple-black spores, and grooved, often segmented or clawlike annulus (ring). Its growth in planted areas suggests that it is an "alien," but its origin is unknown. The gills are never pink as in *Agaricus,* and they are attached to the stalk—at least when young—and the spores are purple-black rather than chocolate-brown.

Stropharia hornemannii

CAP 4-12 (15) cm broad, obtuse to convex, becoming broadly umbonate or plane; surface viscid or slimy when moist, smooth or with a few whitish scales (veil remnants) near margin; dull brown to dingy purple-brown, grayish-brown, grayish-purple, or smoky reddish-brown, often fading in age to yellow-brown or grayish-tan. Flesh thick, soft, white; taste rather disagreeable. **GILLS** typically adnate but sometimes seceding, broad, close, pale gray becoming purple-gray to dull purple-brown to nearly black. **STALK** 5-15 cm long, 0.5-2.5 cm thick, more or less equal, silky-smooth above the ring, sheathed with soft, dry, delicate, cottony white scales below, at least when young; base often with white mycelial threads. **VEIL** membranous, white, forming a persistent, flaring or skirtlike, superior ring on stalk which is darkened by falling spores. **SPORE PRINT** purple-brown to purple-black; spores 10-14 × 5.5-7 microns, elliptical, smooth. Chrysocystidia present on gills.

Stropharia hornemannii has a shaggy stalk like that of *S. ambigua,* but its veil is much more membranous and forms a prominent annulus (ring) on the stalk.

HABITAT: Solitary, scattered, or in small groups on ground or rotting wood under conifers; widely distributed in northern North America. Like *S. ambigua,* it is quite common in the late summer and fall in the Pacific Northwest and northern California, but does not seem to range as far south as that species (I have yet to find it in our area). In Mt. Rainier National Park in Washington the two species can often be found growing together on the Longmire Trail.

EDIBILITY: Unknown, but not worth experimenting with because of its poor taste.

COMMENTS: This beautiful mushroom resembles *S. ambigua,* but has a duller (browner or grayer) cap, and its veil forms a prominent, well-developed annulus (ring) rather than leaving copious strands on the cap margin. *S. depilata* is apparently a synonym. Other species: *Pholiota albivelata* is somewhat similar, but smaller (cap 4-8 cm broad, pinkish-brown to vinaceous-brown), with a white stalk that is scurfy or scaly below the membranous, persistent, striate annulus. It is placed in *Pholiota* because it has cinnamon-brown to dark yellow-brown spores; however, it has acanthocytes on the mycelium, suggesting that, like *P. subcaerulea,* it belongs in *Stropharia.* It occurs under conifers in the Pacific Northwest and northern California, but rarely fruits in large numbers.

Stropharia aeruginosa   (Blue-Green Stropharia)   **Color Plate 91**

CAP 2-6 (8) cm broad, broadly bell-shaped to convex becoming broadly umbonate or nearly plane; surface viscid when moist, smooth or with a few whitish scales (veil remnants) near margin; bright green to blue-green when fresh, developing yellow tones in age. Flesh soft, white or tinged blue. **GILLS** typically more or less adnate (but may secede), fairly close, at first pallid but soon grayish, finally purple-brown or chocolate-brown. **STALK** 3-8 cm long, 3-8 (12) mm thick, more or less equal, smooth and pallid above the ring, colored like the cap or paler below, and usually with small cottony scales; often slightly viscid when moist. **VEIL** membranous, white, soft, forming a fragile, superior ring on stalk which often disappears in age. **SPORE PRINT** dark purple-brown to purple-black; spores 6-10 × 4-5 microns, elliptical, smooth. Chrysocystidia present on gills.

HABITAT: Solitary or in small groups in rich soil, humus, woody debris, or sometimes even in grass; widely distributed. I have not found it in our area, but it is quite common in southern California, especially in the winter under oak. In the Southwest I've collected it under aspen in the summer, and in the Pacific Northwest I've seen it under conifers.

EDIBILITY: Not recommended. It is deemed poisonous by many authors, but is supposedly eaten in Europe.

COMMENTS: This is one of the most distinctive of all the agarics, easily identified by its viscid, greenish to blue-green or yellow-green cap and gray to chocolate-brown or blackish-brown gills. Two similar species occur under conifers in the Pacific Northwest: *S. albocyanea,* with a paler (whitish to yellowish) cap and stalk tinged variously with blue or blue-green; and *Pholiota subcaerulea,* with a bluish cap (when fresh) and paler (cinnamon-brown) spores and mature gills.

Stropharia kauffmanii   **Color Plate 90**

CAP 5-15 cm broad, convex to plane; surface *not* viscid; covered with brown to yellow-brown or grayish-brown scales on a dull yellowish or tan background; margin sometimes hung with veil remnants. Flesh thick, white. **GILLS** adnate or notched, narrow, thin, close or crowded, pallid becoming gray, then purple-gray to purple-black; edges often eroded. **STALK** 6-10 cm long, 1.5-3 cm thick, equal or slightly swollen at base; whitish or creamy, with erect or recurved fibrillose scales, especially below the ring; base often with white mycelial threads. **VEIL** membranous, forming a fragile, white, superior ring on stalk which is soon darkened by falling spores, or often disappears in age. **SPORE PRINT** dark purple-brown to purple-black; spores 6-8 × 4-4.5 microns, elliptical, smooth, apical germ pore absent or minute. Chrysocystidia present on gills.

HABITAT: Solitary, scattered, or in groups in rich humus, around brush piles and decayed woody debris, etc.—usually under hardwoods such as alder, cottonwood, and aspen, spring through fall. It occurs in northern California and the Pacific Northwest, but in my experience is more common in the aspen forests of the Southwest and southern Rockies than anywhere else. There it fruits, like most other mushrooms, in the summer.

EDIBILITY: A tempting mushroom, but I can find no information on it.

COMMENTS: The dry (non-viscid) scaly cap is unusual for a *Stropharia,* and combines with the scaly stem and violet-gray gills to distinguish this beautiful species. Its closest relative is probably **Pholiota fulvosquamosa,** which differs in having brown gills and brown spores. Both species are in some respects more typical of *Agaricus* than of the Strophariaceae, but have gills attached to the stem (at least when young). Another species with a scaly stalk, *S. (=Psilocybe) squamosa,* has a smaller, viscid cap (see key to *Stropharia*). It is widely distributed, but particularly common in the Pacific Northwest.

NAEMATOLOMA

Small to medium-sized mushrooms *found mostly on wood.* CAP smooth, *often brightly colored,* usually not viscid. GILLS attached, dark at maturity. STALK *usually slender,* more or less central. VEIL *typically present but evanescent or forming only a slight annulus (ring) on stalk.* VOLVA absent. SPORE PRINT *usually deep brown to purple-brown,* rarely cinnamon-brown. Spores typically elliptical, with a germ pore. Chrysocystidia present on gills. Cap cuticle filamentous.

THIS is a small group of dark-spored mushrooms with a brick-red to cinnamon-brown, yellow, greenish-yellow, or orange-red cap. A veil is usually present in young specimens, but does not normally form a ring, and the cap is not usually viscid, thereby helping to distinguish it from *Stropharia. Psathyrella* is often confused with *Naematoloma,* for good reason—the two genera were once grouped together in the genus *Hypholoma* (a name now used interchangeably with *Naematoloma*). However, Psathyrellas can be told in the field by their fragile flesh, white or pallid stalk, and brown to buff or whitish cap, and in the laboratory by their cellular cap cuticle. *Pholiota* can also be confused with *Naematoloma,* but as defined here has a brown to cinnamon-brown spore print.

Naematolomas are partial to cold weather, but occur throughout the mushroom season. The common species in our area fruit in tufts or large clusters on decaying wood, but may appear terrestrial if the wood is buried. Some species, however, fruit in a scattered to gregarious pattern on the ground, usually in cold northern bogs or under conifers in lignin-rich humus. Naematolomas may also occasionally turn up on lawns, but none grow on dung. They are not choice edibles, and at least one, *N. fasciculare,* is poisonous. Four species are described here.

Key to Naematoloma

1. Cap more or less brick-red, the margin usually pallid; typically growing in clumps on dead hardwoods in eastern North America .... *N. sublateritium* (see *N. aurantiaca,* p. 382)
1. Not as above 2

2. Cap orange to red or sometimes reddish-brown to brick-red; growing scattered to gregarious or tufted on ground, in wood chips, mulch, etc. .... *N. aurantiaca* & others, p. 382
2. Not as above (but cap may be orange or tawny at center and yellow at margin) 3

3. Stalk white; fruiting body fragile; cap cuticle cellular .... (see *Psathyrella,* p. 361)
3. Not as above 4

4. Typically growing in tufts or clusters on wood (sometimes buried) or roots, or occasionally densely gregarious on sawdust or wood chips 5
4. Typically growing widely scattered to gregarious (but not normally clustered) on ground or in bogs or moss, mainly under conifers .... *N. dispersum* & others, p. 384
5. Gills yellow to greenish when young .... *N. fasciculare,* below
5. Gills pallid to grayish or grayish-brown when young .... *N. capnoides* & others, p. 383

Naematoloma aurantiaca   (Orange Naematoloma) **Color Plate 93**

CAP 1.5-5.5 cm broad, convex becoming broadly umbonate or plane; surface slightly viscid or dry, smooth, bright scarlet to red-orange or orange, or at times brick-red to rusty-reddish to reddish-brown; margin often hung with whitish veil remnants. Flesh pallid, not bruising blue. **GILLS** close, pallid or yellowish when young, then grayish-brown or grayish-olive and finally purple-brown to purple-black in old age; adnate or notched, sometimes seceding. **STALK** (2) 3-7 (10) cm long, 2-6 (10) mm thick, equal or with the base slightly swollen or narrowed; white or tinged yellow above, developing bright orange to reddish-orange stains over the lower half; base sometimes with white to yellow mycelial threads. **VEIL** membranous but very thin, whitish, soon disappearing or forming a slight, easily-obliterated ring on stalk. **SPORE PRINT** dark purple-brown; spores 10-14 × 6-9 microns, elliptical, smooth. Chrysocystidia present on faces of gills.

HABITAT: Scattered to gregarious on wood chips, sawdust, and humus rich in lignin, but often appearing on lawns, in gardens, etc.; distribution uncertain and erratic, but fairly common in California. It is common in the parks of the San Francisco Bay area. It also occurs in Los Angeles, and I have found it in Santa Cruz growing from fallen eucalyptus seed pods. It likes the same habitats as *Psilocybe cyanescens,* but has a longer season, fruiting from fall through spring, or even in the summer if it is wet enough.

EDIBILITY: Unknown.

COMMENTS: Also known as *Stropharia aurantiaca,* this bright, attractive little mushroom is easily recognized by its reddish-orange cap, dark spore print, and tendency to grow in parks or gardens. It is probably an introduced species, but its origin is unknown. Other species: *Stropharia thrausta (=S. squamosa var. thrausta)* is a similarly colored species with a frequently umbonate cap and reddish to orange scaly stem. It is widely distributed (I have seen it in New Mexico), but does not occur in our area. *N. sublateritium* of eastern North America has a brick-red cap with a paler margin and much smaller spores. It fruits in clusters on dead hardwoods in the fall and early winter, and is edible and quite common.

Naematoloma fasciculare   (Sulfur Tuft)   **Color Plate 92**

CAP (1) 2-5 (9) cm broad, at first broadly conical or bell-shaped, soon becoming convex, then broadly umbonate to plane; surface smooth, not viscid, bright sulfur-yellow to greenish-yellow, or at times yellow-orange (especially when young), the center sometimes darker (orange-tan to orange-brown); margin often hung with small veil remnants. Flesh thin, yellow; taste very bitter (rarely mild). **GILLS** close, typically adnate but sometimes seceding, at first sulfur-yellow, becoming greenish-yellow or olive, then finally dusted purple-brown to nearly black with spores. **STALK** 5-12 cm long, 3-10 (15) mm thick, equal or tapering downward, yellow to tawny, but often developing rusty or brownish stains from base upward; often curved or sinuous, dry, firm. **VEIL** thin, pale yellow, evanescent, or leaving slight vestiges on cap margin or an obscure fibrillose zone on stalk which is subsequently blackened by falling spores. **SPORE PRINT** purple-brown to deep purple-gray; spores 6-8 × 3.5-5 microns, elliptical, smooth. Chrysocystidia present on faces of gills.

HABITAT: Gregarious, usually in tufts or dense clusters on decaying wood of both hardwoods and conifers, but sometimes growing from buried wood or roots and thus appearing terrestrial; widely distributed. It is abundant in our area in the fall and winter, less so in the spring, and is one of the first woodland mushrooms you're liable to encounter.

EDIBILITY: Poisonous. In Europe and Asia it has caused several deaths; in America only gastrointestinal upsets have been reported. Fortunately, the bitter taste is a deterrent.

COMMENTS: The yellow to greenish-yellow gills, dark spores, bitter taste, and clustered growth habit are the principal fieldmarks of this attractive, cosmopolitan mushroom *Hypholoma fasciculare* is a synonym. In age the clustered caps often assume a grayish or purple-brown tinge from a coating of spore dust. *N. capnoides* is quite similar in cap color, but its gills are gray to purple-brown (never yellow), and it has a mild taste and grows only on conifers. Several similar Naematolomas grow on the ground (see *N. dispersum*). Other species: *N. subviride* is a small southern version of *N. fasciculare; N. dispersum var. idahoense* has a tawny to cinnamon-brown cap and bitter taste, and grows in clusters on conifers. For similar brown-spored species, see comments under *Pholiota malicola* group.

Naematoloma capnoides   (Conifer Tuft)

CAP 2-7 cm broad, convex or slightly umbonate to plane; surface smooth, not viscid, yellow to tawny, orange-brown, rusty-brown, or cinnamon, the margin often yellower and hung with veil remnants. Flesh thin, pallid; taste mild. **GILLS** close, usually adnate but often seceding, at first pallid, then grayish, finally dark gray to purple-brown. **STALK** 5-10 cm long, 3-8 (10) mm thick, equal or tapered downward, dry, slender, pallid or yellowish above, often rusty-brown to tan or brownish below. **VEIL** evanescent or leaving small patches of tissue on cap margin and sometimes an obscure fibrillose zone on stalk. **SPORE PRINT** purple-brown to deep purple-gray; spores 6-7.5 × 3.5-4.5 microns, elliptical, smooth. Chrysocystidia present on faces of gills.

HABITAT: Gregarious, usually in clusters, on rotting conifers; widely distributed, but especially common in the Pacific Northwest. I have seen it in our area in the fall and winter on Douglas-fir, but it is not common.

EDIBILITY: Edible, but thin-fleshed and not particularly tasty.

COMMENTS: Also known as *Hypholoma capnoides,* this nondescript *Naematoloma* resembles its ubiquitous relative, *N. fasciculare,* in cap color, but has a mild taste and grayish rather than yellow or greenish-yellow immature gills. The stalk is often quite long in relation to the cap. Several Pholiotas look similar (see *Pholiota malicola* group), but have brown or cinnamon spores. Other species: A pale-capped version occurs under mountain conifers in the spring; it may or may not be a distinct species; *N. radicosum* of eastern North America is similar in color but has a deeply rooting base and bitter taste.

Naematoloma capnoides. This slender-stemmed conifer-lover grows in tufts or clusters. It resembles *N. fasciculare* (see color plate) but does not have yellow or greenish gills.

Naematoloma dispersum   (Dispersed Naematoloma)

CAP 1-4 cm broad, bell-shaped, but sometimes expanding to convex or even plane with an umbo; surface smooth, not viscid, tawny to tawny-orange, fading to yellowish; margin often hung with veil remnants. Flesh thin; taste typically somewhat bitter. **GILLS** usually adnate, but sometimes seceding, close, pallid, becoming dingy olive or olive-gray, then finally purple-brown with paler edges. **STALK** 6-12 cm long, 2-5 mm thick, equal, usually long and slender, rather tough and pliant but sometimes also brittle; yellowish above and brown to dark reddish-brown below. **VEIL** fibrillose or cobwebby, evanescent or leaving a fibrillose zone on upper stalk. **SPORE PRINT** purple-brown; spores 7-10 × 4-5 microns, elliptical, smooth. Chrysocystidia present on faces of gills.

HABITAT: Widely scattered to gregarious in humus and debris under conifers; widespread, but particularly common in logged-over areas of the Pacific Northwest from late summer through early winter. It occurs in northern California, but not in our area.

EDIBILITY: Unknown.

COMMENTS: Also known as *Hypholoma dispersum,* this is one of several *Naematoloma* species that characteristically grow in a scattered to gregarious fashion rather than in clusters. The tendency of the gills to develop olive tints in age is one of its distinctive features, as is its tall slender stalk and purple-brown spore print. It may grow in small tufts, but not in the large clumps typical of *N. fasciculare,* and the stem is not white as in *Psathyrella.* There are a number of similar Naematolomas that are difficult to distinguish, but of little interest to the average collector since none are good edibles. Their ranks include: *N. olivaceotinctum,* with a greenish-tinted cap (and often gills) in age; *N. squalidellum* and *N. polytrichi,* with yellow gills that become greenish-yellow in age; *N. ericaeum,* a southern species with a slightly viscid brown to ochre cap; *N. udum,* with a browner cap and larger spores (14-18 microns long), "growing on muck in bogs, especially along rabbit runways" (A.H. Smith); *N. elongatum,* with dull cinnamon spores; and *N. myosotis,* growing in bogs, with a viscid cap and dull cinnamon spores (the latter two species are also placed in *Pholiota*). The above species are not as fragile as the Psathyrellas and do not have white stems. All tend to grow in groups on troops on the ground rather than in clumps on wood.

PHOLIOTA

Mostly medium-sized mushrooms *found on wood or woody debris* or sometimes on ground. CAP *usually viscid and/or scaly.* GILLS typically adnexed to adnate to slightly decurrent. STALK *central to somewhat off-center but not lateral, usually fleshy* but frequently slender; *often scaly below the veil.* VEIL *present;* membranous, slimy, or fibrillose, disappearing *or often forming an annulus (ring) on stalk.* VOLVA absent. SPORE PRINT *dull brown to cinnamon-brown or sometimes rusty-brown.* Spores mostly elliptical, smooth, usually with a germ pore. Chrysocystidia often present on faces (sides) of gills. Cap cuticle typically filamentous.

PHOLIOTA is the largest genus of brown-spored, wood-inhabiting agarics. The fruiting body is usually larger and fleshier than in *Galerina* and *Tubaria,* a veil is always present in young specimens, and the stalk is well-developed and central to somewhat off-center. The veil can be fibrillose, cottony, slimy, or membranous, and in most cases leaves visible remains on the stem in the form of an annulus (ring) or a coating of scales over the lower portion of the stalk.

Among the brown-spored wood-rotters, *Crepidotus* lacks both a veil and stem, *Paxillus* has distinctly decurrent gills and lacks a veil, while *Gymnopilus* may have a veil but has a dry cap and brighter (rustier or oranger) spores. *Naematoloma* and *Stropharia* intergrade somewhat with *Pholiota,* but are traditionally separated by their darker (purple-brown to blackish) spore color, as is *Psathyrella.* A few Pholiotas are terrestrial, and these are more difficult to distinguish. In the field they can usually be told from *Cortinarius* and *Hebeloma* by their viscid or slimy cap, presence of a veil, frequently slender build (only rarely is the stalk bulbous), and sometimes clustered growth habit; microscopically they are distinct by virtue of their smooth spores.

Many Pholiotas qualify as "LBM's" and are not likely to interest the average mushroom hunter. However, some are quite striking (e.g., *P. squarrosa* and the *P. aurivella* group). As few are worth eating and several are mildly poisonous, the genus is best avoided by beginners. (Some species listed as edible in older books have since been shuffled to genera such as *Rozites* and *Agrocybe*.) In their monograph on North American Pholiotas, Alexander Smith and L.R. Hesler recognize over 200 species. Since microscopic examination is required to determine the shapes or types of sterile cells (cystidia) on the gills, many of these species cannot be identified in the field. The coniferous forests of the West are uniquely rich in *Pholiota* species, but for some reason our area is poorly represented.

Key to Pholiota

1. Growing on recently charred soil or wood (ashes) 2
1. Not typically growing in ashes 3
2. Stalk typically 5-10 mm thick .... ***P. brunnescens,*** p. 393
2. Stalk thinner (2-6 mm) ***P. highlandensis, P. carbonaria,*** & others (see *P. brunnescens,* p. 393)
3. Cap blue to blue-green when fresh (but may discolor yellowish or tan in age); found in the West, usually under conifers .... ***P. subcaerulea*** (see *Stropharia aeruginosa,* p. 380)
3. Not as above (if blue or greenish-blue when young, then found in eastern North America) . 4
4. Growing on wood (occasionally buried) or in wood chip mulch 5
4. Growing on the ground 15
5. Cap more or less bright orange-pink when fresh and lacking scales; taste bitter ....
    .... ***P. astragalina,*** p. 387
5. Not as above 6
6. Growing on dead wood of mountain conifers in the spring (when or shortly after the snow melts); cap and veil viscid to slimy; cap and stalk lacking prominent scales; stalk typically at least 6 mm thick; cap yellowish to rusty-tawny but not brightly colored; known only from the West
    .... ***P. sp. (unidentified)*** ("Snowbank Pholiota"—see photo below)
6. Not growing when the snow is melting, or if so then not as above 7

Left: These young specimens of *Pholiota squarrosoides* (see comments under *P. squarrosa*) are flagrantly scaly, but older caps are viscid and not as scaly (see color plate). **Right** "Snowbank Pholiota." This unidentified (and probably unnamed) species is common in the northern Sierra Nevada and Cascades on dead conifers soon after the snow melts. The yellowish-buff to tawny or rusty cap is viscid or slimy and has few if any scales; the veil is also viscid.

7. Growing on poplar, cottonwood, aspen, or willow; stalk 1-4 cm thick at apex, usually thicker at base, lacking large erect scales (but may have patches of tissue); cap white to creamy or buff when fresh (but may darken to ochre or brownish in age) .... *P. destruens,* p. 395
7. Not as above; color or habitat different or stalk thinner and/or scalier 8

8. Both cap and stalk yellowish-brown to dark rusty-brown and covered with powdery or granulose scales when fresh (those on cap may wash off or disintegrate); cap 1-4 cm broad and dry (*not* viscid!); stalk slender (1.5-4 mm thick); taste usually bitter or metallic; odor not distinctive .... *Phaeomarasmius erinaceellus (=Pholiota erinaceella)*
8. Not with above features 9

9. Cap bright yellow to orange, bright tawny, golden, or bright rusty-brown when fresh (but sometimes darker rusty-brown when very young); cap usually decorated with large scales, spots, patches, or "straps" of tissue (which may be darker and may wash off) 10
9. Not as above; if brightly colored then cap bald or with a few scattered fibrillose veil remnants (one species tawny to whitish with erect or recurved scales) 11

10. Scales on cap and stalk bright yellow; cap dry to slightly viscid .... *P. flammans,* p. 391
10. Scales usually darker than background color; cap distinctly viscid to slimy when moist, often shiny in dry weather .... *P. aurivella group* & others, p. 390

11. Cap and lower stalk becoming reddish-brown to dark vinaceous-brown at maturity (but often orange-brown when young) and decorated with brown fibrillose scales; cap viscid; usually solitary or in twos or threes, mainly in eastern North America and the Southwest ....
.... *P. albocrenulata,* p. 392
11. Not with above features 12

12. Cap *and* stalk with prominent erect or recurved scales (which may be obliterated or flattened somewhat in age); stalk typically less than 1.5 cm thick 13
12. Not as above (but cap *or* stalk may have recurved scales and both may have flattened scales 15

13. Cap never viscid; gills often (but not always) greenish-tinged in age; odor mild or garlicky; growing on hardwoods (especially aspen) and conifers .... *P. squarrosa,* p. 389
13. Cap often dry at first but a viscid layer beneath the scales usually evident in age; gills not greenish-tinged; odor mild or slightly fruity; on wood or ground 14

14. Typically found on hardwood stumps or logs; cap and stalk very scaly when young, the scales tawny or paler .... *P. squarrosoides* (see *P. squarrosa,* p. 389)
14. Usually growing on the ground, less commonly on wood; cap and stalk very scaly to only slightly so, the scales some shade of brown (darker than in above species) ..... *P. terrestris,* p. 389

15. Growing in deep moss, bogs, or mucky places; stalk equal and usually hollow, long and thin (length usually at least 40 times the width); cap small, yellow or olive; mostly northern .. 34
15. Not as above 16

16. Lower stalk (i.e., below the veil) with distinct scales (at least in most specimens), the scales sometimes recurved but often small 17
16. Not as above 19

17. Cap smooth (without scales) and hygrophanous: rusty-brown to reddish-brown or orange-brown and striate when moist, fading to yellowish, ochre, or buff as it loses moisture (often two-toned in intermediate stages); often found in large clusters .... *P. mutabilis,* p. 395
17. Not as above 18

18. Growing on ground or wood chips (often along roads or trails), usually in tufts or clusters; cap with scales at first (which may wash off or wear away), dark brown to light brown, grayish-brown, yellow-brown, or tawny; common .... *P. terrestris,* p. 389
18. Not as above 19

19. Cap yellow to bright ochre or orange when fresh 20
19. Cap some other color 23

20. Cap distinctly viscid when moist 21
20. Cap not viscid 22

21. Lower portion of stalk (i.e., below the veil) with distinct scales 23
21. Not as above .... *P. malicola group* & others, p. 388

22. Fruiting body staining orange-brown when bruised *P. multifolia* (see *P. malicola* group, p. 388)
22. Not as above .... *P. malicola group* & others, p. 388

23. Cap viscid or slimy when moist, often with adhering debris when dry, not hygrophanous . 24
23. Cap dry or hygrophanous or sometimes slightly viscid (but if so, then soon drying out) .. 30

24. Veil membranous, typically forming a well-developed annulus (ring) on stalk . . . .
. . . . *P. albivelata* (see *Stropharia hornemannii*, p. 379)
24. Not as above; veil typically disappearing or merely forming a slight ring or fibrillose zone   25
25. Cap whitish to pinkish-buff, pinkish-gray, or even grayish *P. lenta* (see *P. lubrica* group, p. 392)
25. Not as above 26
26. Cap shaded variously with green, olive, purple, brown, and/ or gray when young, often yellow to
orange in age; found mainly in eastern North America (especially the South) . *P. polychroa*
26. Not as above 27
27. Stalk lacking scales, even when young; cap usually olive-brown to brown or ochre with a yellow
to dingy greenish-yellow margin . . . . *P. spumosa*, p. 394
27. Not as above; stalk cottony or with small scales below the veil (at least when young and fresh)
and/ or differently colored 28
28. Cap usually with rows of concentrically-arranged veil remnants when young and often ap-
pearing fibrillose or streaked in age . . . . *P. decorata* (see *P. lubrica* group, p. 392)
28. Not as above 29
29. Veil slimy or viscid in wet weather; cap bright to dark reddish-brown, vinaceous-brown, or
brown and completely smooth (bald) . . . . *P. velaglutinosa* (see *P. lubrica* group, p. 392)
29. Veil not slimy or viscid; cap brown to rusty-brown, orangish, etc., the margin often paler or
even whitish . . . . *P. lubrica* group & others, p. 392
30. Veil usually forming a ring (annulus) on stalk or leaving skinlike fragments on cap margin   31
30. Veil disappearing or leaving only a slight hairy zone on stalk (check several specimens!) . . 32
31. Stalk thick or slender, lacking obvious scales or cottony material below the veil; cap smooth
(bald), usually at least 4 cm broad when mature; white rhizomorphs (mycelial threads) often
present at base of stalk or in surrounding humus; cap cuticle cellular (see *Agrocybe*, p. 467)
31. Not as above; fruiting body usually small and stalk usually slender; cap cuticle not cellular
. . . . (see *Galerina, Tubaria,* & Allies, p. 399)
32. Cap and stalk pale yellow to pale cinnamon or pinkish-cinnamon; cap *not* translucent-striate
when moist; cap convex to plane or slightly umbonate . . *P. scamba* (see *P. spumosa*, p. 394)
32. Not as above 33
33. Stalk with scattered patches of veil remnants over lower portion; cap convex to plane at maturity,
with an incurved margin when young; gills *not* decurrent; very common in the spring and early
summer under mountain conifers (on rotten wood or in lignin-rich humus), but also occurring
in other habitats . . . . *P. vernalis* (see *P. mutabilis*, p. 395)
33. Not as above . . . . (see *Galerina, Tubaria,* & Allies, p. 399)
34. Cap viscid when moist, olive or olive-tinged . . . .
. . . . *P. myosotis (=Naematoloma myosotis)* (see *N. dispersum*, p. 384)
34. Cap not viscid, yellow to olive-yellow . . . .
. . . . *P. elongatipes (=Naematoloma elongatum)* (see *N. dispersum*, p. 384)

Pholiota astragalina   (Pinkish-Orange Pholiota)

CAP 2-5 cm broad, bell-shaped or obtuse becoming convex, umbonate, plane, or with an
uplifted, often wavy margin in old age; surface smooth, viscid or slimy when wet but soon
dry; bright reddish-orange to pinkish-orange, the margin sometimes paler, fading somewhat in age and often developing blackish discolorations; margin often hung with veil
remnants when young. Flesh thin, orange to yellow; taste bitter. **GILLS** typically adnexed
or notched or even free, close, bright yellow or yellow-orange, discoloring where bruised or
in age. **STALK** 5-12 cm long, 4-7 mm thick, equal or tapered toward base, fibrillose, without scales, hollow, sometimes sinuous, pale yellow, the base often oranger or discoloring
brownish. **VEIL** yellowish, leaving remnants on cap margin or disappearing. **SPORE
PRINT** brown; spores 5-7 × 3.5-4.5 microns, elliptical, smooth. Chrysocystidia present.

HABITAT: Scattered to gregarious or in small clusters on rotting conifers; widely distributed. It is especially common in the Pacific Northwest and northern California in
the late summer and fall; I have not seen it in our area.

EDIBILITY: Inedible because of the bitter taste.

COMMENTS: The brilliant pinkish-orange cap plus the yellow gills, brown spores, and growth on conifers make this one of our most distinctive as well as beautiful Pholiotas. It is reminiscent of *Naematoloma,* but has paler (browner) spores and a viscid cap when wet.

Pholiota malicola group   (Forgettable Pholiota)

CAP 3-8 (15) cm broad, convex becoming plane or slightly umbonate; surface viscid or dry, smooth or with a few veil remnants at margin, yellow to ochraceous-tawny, orange, or orange-buff, fading somewhat in age; margin often wavy. Flesh pallid or yellowish; odor mild or faintly fragrant; taste mild. **GILLS** close, adnexed or notched, yellowish becoming rusty-brown or cinnamon-brown. **STALK** 4-15 cm long, 0.4-1 (2.5) cm thick, equal or tapered downward or enlarged at base; solid, dry, fibrillose but not scaly; pallid or yellowish above, darker (tawny or colored like cap) below, becoming rusty-brown in age from falling spores. **VEIL** fibrillose, disappearing or forming a slight ring or zone of fibrils near top of stalk. **SPORE PRINT** rusty-brown; spores 7.5-11 × 4.5-5.5 microns, elliptical, smooth, with an apical germ pore. Chrysocystidia absent.

HABITAT: In groups or clusters on rotting logs and stumps, wood chips, etc.; widely distributed. It is fairly common in our area in the fall and winter.

EDIBILITY: Unknown.

COMMENTS: This forgettable *Pholiota* is a member of the so-called "*P. alnicola* complex"—a group of difficult-to-distinguish, scale-less (but not veil-less), yellowish to rusty or tawny, wood-inhabiting species. They are reminiscent of Naematolomas in their color and clustered growth habit, but differ in having rusty-brown or cinnamon-brown spores. They also approach *Gymnopilus,* but the species in that genus have oranger spores. Clumps growing in wood chip mulch might be confused with *P. terrestris,* but that species has small scales on the stalk below the veil. Closely related and/ or superficially similar Pholiotas include: *P. flavida,* with slightly smaller spores and a frequently fragrant or at least distinctive odor; the "true" *P. alnicola,* with a mild to bitter taste and yellowish or olive-tinged cap; *P. spinulifera* and *P. fibrillosipes,* with smaller spores, a viscid cap and conspicuous cystidia on the gill faces, found in groups or clusters in soil, mulch, and sawdust (veil remnants whitish in former, yellow to orange in latter); *P. subochracea,* with a viscid, pale yellow to ochre cap, long (5 cm or more) stem, chrysocystidia, and small spores (5-6 microns long), found in the Pacific Northwest; *P. prolixa,* like *P. subochracea* but larger-spored, common in eastern North America, often in large clusters in low hardwood forests or bottomlands; and *P. multifolia,* also eastern, with a dry yellow cap and tendency to bruise rusty-brown or orange-brown. None of these are worth eating.

Pholiota malicola group commonly grows in tufts or clusters on dead wood and in wood chip mulch. These rather small specimens have a yellow cap and practically smooth (scale-less) stalk.

Pholiota terrestris is common on lawns and along roads and trails. It usually grows in tufts or clusters on the ground, rarely on wood. Note scaly stalk (the scales can be large or small); spore print is brown. (Joel Leivick)

Pholiota terrestris (Terrestrial Pholiota)

CAP (1) 2-8 (10) cm broad, obtusely conical or convex becoming plane or somewhat umbonate; surface usually with dry fibrillose scales, but viscid or slimy in wet weather beneath the scales, which sometimes wear off; color variable: dark brown to light brown, grayish- or yellow-brown, or tawny, the scales darker; margin sometimes streaked, often hung with veil remnants. Flesh white to watery yellow or brown, thin. **GILLS** attached (usually adnate), close, at first pallid to grayish, then dull brown to dull cinnamon-brown. **STALK** 3-10 (13) cm long, (2) 4-10 mm thick, equal or narrowed below, slender, solid or becoming hollow, dry, pallid to buff, or brownish toward base; covered with brown scales or patches below the veil. **VEIL** fibrillose, whitish, forming a slight superior, fibrillose ring or zone on stalk, or disappearing. **SPORE PRINT** brown; spores 4.5-7 × 3.5-4.5 microns, elliptical, smooth. Chrysocystidia present on gills.

HABITAT: In groups or clusters on the ground, especially along roads, paths, and in other disturbed areas; also on lawns, lignin-rich debris, rarely on or around old stumps. It is widely distributed and especially common along the west coast (including our area) in the fall and winter. Single individuals are occasionally found, but tufts or clusters are the rule.

EDIBILITY: Edible but thin-fleshed, insipid, and usually wormy to boot.

COMMENTS: Our most common *Pholiota,* this is the only veiled, brown-spored mushroom that *habitually* grows in clusters on the ground (several dark-spored Psathyrellas also do). The scaly stem and evanescent veil are good secondary fieldmarks, but the color, viscidity, and degree of scaliness exhibited by the cap vary considerably according to age and weather conditions. The veil does not form a prominent ring as in the white-spored honey mushroom *(Armillariella mellea),* but after collapsing it traps falling spores and turns brown as in *Cortinarius. P. squarrosoides* (see comments under *P. squarrosa*) is very closely related to *P. terrestris,* but has paler scales and is not terrestrial.

Pholiota squarrosa (Scaly Pholiota)   **Color Plates 96, 97**

CAP 3-10 (15) cm broad, obtuse or convex becoming broadly bell-shaped to slightly umbonate or plane; surface dry, pale tan to straw color, buff, or pale yellow-brown, or in age darker yellow-brown or sometimes greenish-yellow toward the margin; covered with a dense layer of upright or recurved, often darker (brown) scales; margin incurved at first

389

and often fringed copiously with veil remnants. Flesh pale yellowish, rather pliant; odor mild in some forms, distinctly garlic- or onionlike in others; taste mild or rancid. **GILLS** crowded, adnexed to adnate to slightly decurrent, pale yellowish to buff or tinged gray, then often developing a greenish tinge before finally becoming brown or dull rusty-brown. **STALK** 4-12 cm long, 0.5-1.5 cm thick, equal or tapered downward, solid, smooth above the veil, covered with erect or recurved scales below (like on cap); colored like cap or becoming darker brown or reddish-brown below. **VEIL** membranous-fibrillose, forming a fragile, often torn, superior ring on stalk or only leaving shreds on cap margin. **SPORE PRINT** dull rusty-brown; spores 5.5-9 × 3.5-5 microns, elliptical, smooth, with a germ pore. Chrysocystidia present on gills.

HABITAT: In tufts or dense clusters on wood, usually at the bases of trees (both hardwoods and conifers); widely distributed. It is very common on aspen and spruce in the Rocky Mountains and Southwest during the summer. It is also common on aspen in the Sierra Nevada in the summer and fall, but I have never seen it on the coast.

EDIBILITY: Not recommended. Some people eat it regularly but others have suffered severe stomach upsets and old specimens are often rancid-tasting.

COMMENTS: A beautiful and memorable mushroom in its prime, the erect or recurved scales on the cap and stalk plus the tan to pale tan color and brown spore print set this species apart. The cap is not viscid as in most Pholiotas and the garlic odor, when present, is also distinctive. It is usually found on wood, whereas *P. terrestris,* which is closely related, typically grows on the ground. Other species: *P. squarrosoides* **(COLOR PLATE 98)** is a very similar and edible scaly species with a somewhat paler (whitish to pale tawny or yellowish or light brown) cap and mild odor. Its gills are never greenish and in age or wet weather the cap becomes viscid from a gelatinous layer beneath the scales. It occurs on hardwoods such as maple and alder, and is much more common than *P. squarrosa* in northern California and the Pacific Northwest. In addition to the color plate, a young cluster is shown on p. 385. Another similar but unidentified species with a *very* distinctive citrus fragrance occurs in New Mexico.

Pholiota aurivella group   (Golden Pholiota)   Color Plate 95

CAP (3) 5-16 cm broad, broadly bell-shaped or convex becoming broadly umbonate or plane; surface very sticky-gelatinous or slimy when moist (but may dry out), pale to dark yellow, tawny, golden-orange, or rusty-orange (or sometimes rusty-brown when young), decoratd with darker scales (the scales large and flattened to slightly recurved and triangular to strap-shaped or spotlike) that sometimes wear away or wash off in age; margin often hung with veil remnants when young. Flesh pallid to yellowish, soft in age. **GILLS** close, adnate or notched, pallid to yellow becoming brown to rusty-brown or even brownish-orange in age. **STALK** 4-15 cm long, 0.4-2.5 cm thick, equal or tapered in either direction, central or off-center, dry and more or less smooth above the veil, scaly below (the scales usually not viscid); yellow to pale yellow-brown or colored like the cap (but often paler). **VEIL** fibrillose, whitish or yellowish, forming a slight ring or fibrillose zone on upper stalk or disappearing. **SPORE PRINT** brown; spores 8.5-10 × 5-6.5 microns (but see comments!), elliptical, smooth, with a germ pore. Chrysocystidia often present.

HABITAT: Gregarious (often tufted or clustered) on living or dead hardwoods and conifers; widely distributed. In our area this species "complex" occurs rarely on hardwoods in the fall and winter, but it is a very prominent fungal feature of the coniferous forests of the Rocky Mountains, Southwest, Sierra Nevada, and Pacific Northwest, especially in the summer and fall. In the Southwest I have also seen large fruitings on aspen.

Pholiota aurivella group. Note extremely slimy cap with large scattered scales. Cap color ranges from yellow to bright tawny- or rusty-orange. These specimens, which have bright rusty caps, were growing on a dead oak; they are probably *P. limonella* or *P. squarroso-adiposa* (see comments below).

EDIBILITY: To be avoided. Some books list it as edible, but many people have suffered gastric upsets after eating members of this species "complex." The texture is rather soft and gelatinous anyway, and it is said to taste "like marshmallows without the sugar."

COMMENTS: This striking mushroom and its look-alikes (see below) are easily recognized by their scaly stalk, brown spores, and yellow to orange, viscid cap with large spotlike darker scales. The honey mushroom *(Armillariella mellea)* is somewhat similar, but is not as brightly colored and has white spores. Distinguishing species within the *P. aurivella* "complex," however, is not so easy. Many variants have been described (e.g., *P. connata,* with a thinly viscid stalk and *P. abietis,* with pale brown immature gills), but recent cultural studies have revealed the presence of three widespread, non-interbreeding species which differ principally in spore size: *P. aurivella,* with spores 8.5-10 × 5-6.5 microns, *P. limonella (=P. squarroso-adiposa),* with spores 6.5-9.5 × 3.5-5.5 microns and apparently the most common of the three in North America; and *P. adiposa,* a hardwood-lover with even smaller spores (5-6 × 3-4 microns) and often viscid scales on the stalk. To complicate matters, there are some other closely related species, including: *P. aurivelloides,* with even larger spores than *P. aurivella; P. hiemalis,* found on northern conifers usually late in the fall, with viscid scales on the stalk and pallid, yellow-edged gills when young; and *P. filamentosa,* with a lemon-yellow to greenish-yellow cap and a thick, persistent annulus (ring) on the stalk, also found with conifers. Whew!

Pholiota flammans   (Flaming Pholiota; Yellow Pholiota)

CAP 3-8 (10) cm broad, convex or obtuse becoming broadly umbonate to nearly plane; surface brilliant yellow or at times dark yellow or tawny at the center, covered with bright yellow scales which may wear off in age; dry or in wet weather sometimes viscid beneath the scales; margin usually fringed with veil remnants. Flesh fairly firm, yellow. **GILLS** usually adnexed or notched, bright yellow becoming rustier in age, close. **STALK** (3)5-10 cm long, 4-10 mm thick, equal or slightly thicker at base, bright yellow or the base slightly darker; smooth above the veil, sheathed with a dense layer of recurved yellow scales below; not viscid. **VEIL** bright yellow, disappearing or forming a slight superior ring or fibrillose-cottony zone on stalk. **SPORE PRINT** brown; spores 3-5 × 2-3 microns, oblong to elliptical, smooth. Chrysocystidia present on gills.

HABITAT: Solitary or tufted on conifer logs and stumps; widely distributed in northern North America, not common. In California it fruits in the fall and winter, but is rare.

EDIBILITY: Edible, but not choice (see comments on edibility of *P. aurivella*).

COMMENTS: The brilliant yellow color sets apart this beautiful mushroom. It is most likely to be confused with the *P. aurivella* group, but the cap is dry to only slightly viscid and the scales on the stalk and cap are yellow rather than rusty, cinnamon, or brown. *P. adiposa* (see comments under the *P. aurivella* group) is somewhat similar, but favors hardwoods.

Pholiota albocrenulata

CAP 3-10 (15) cm broad, broadly conical or convex becoming broadly umbonate to nearly plane; surface viscid or slimy when moist, orange-brown to dark rusty-brown or reddish-brown, becoming dark vinaceous-brown in age, and adorned with scattered brown fibrillose scales (veil remnants); margin often fringed with veil remnants. Flesh thick, whitish; taste mild or bitter. **GILLS** close, notched or adnate to slightly decurrent, whitish becoming grayish and finally brown, the edges finely scalloped and white or beaded with tiny white droplets. **STALK** 3-10 (15) cm long, 0.5-1.5 cm thick, more or less equal, rather fibrous, stuffed or hollow; pallid or grayish above, brown to reddish-brown (like cap) below, with scattered brown scales below the veil; often curved. **VEIL** fibrillose-cottony, forming a slight superior fibrillose ring or zone on stalk, or disappearing. **SPORE PRINT** brown; spores 10-15 (18) × 5-8 microns, elliptical, smooth.

HABITAT: Solitary or in small groups (twos and threes) on stumps, logs, and living trees, usually of hardwoods (especially maple and elm); widely distributed, but seldom found in quantity and apparently absent on the west coast. I have collected it twice in New Mexico in August—once on a ponderosa pine and once on an aspen.

EDIBILITY: Said to be harmless, but seldom eaten.

COMMENTS: Though not often encountered, this is a striking mushroom by virtue of its reddish-brown to dark brown, scaly, viscid cap and white-edged gills. Microscopically it is close to *Stropharia,* and is placed in that genus by some mycologists (along with several other Pholiotas, e.g., *P. subcaerulea* and *P. albivelata*).

Pholiota lubrica group   (Lubricous Pholiota)

CAP 3-10 cm broad, convex becoming plane or sometimes with an uplifted, wavy margin; surface viscid or very slimy when moist, smooth (but see comments), color variable: dark reddish-brown to rusty-brown, rusty-orange, or ochraceous-tawny at the center, often paler (yellowish or even whitish) toward the margin, which is often hung with veil remnants when young. Flesh fairly thick, whitish to watery yellow or greenish-yellow. **GILLS** adnate to adnexed, close; whitish, yellow, or sometimes greenish-yellow, becoming brown or dull cinnamon as the spores mature. **STALK** 5-10 cm long, (0.4) 0.8-1.5 cm thick, more or less equal, smooth above the veil, usually with small scales below which may wear off; white or yellow, the scales often darker; sometimes brownish-stained in age. **VEIL** fibrillose to somewhat membranous, forming a slight superior ring or fibrillose zone on stalk. **SPORE PRINT** brown; spores 5.5-7 × 3-4.5 microns, elliptical, smooth. Chrysocystidia absent.

HABITAT: Solitary, scattered, or in groups in humus and on woody debris in woods; widely distributed. Fairly common in our area in the late fall, winter, and early spring, especially under pine.

EDIBILITY: Unknown.

Left: *Pholiota velaglutinosa* has a vinaceous-brown cap and slimy veil. **Right:** *Pholiota ferrugineo-lutescens* has a rusty-orange cap, dry veil, and scaly stalk. Both species are discussed under the *Pholiota lubrica* group.

COMMENTS: The above description will more or less fit a large number of Pholiotas with a viscid-slimy cap and scaly stalk that grow solitary to gregarious (but not often clustered) in humus or on rotting wood. They are particularly prevalent under conifers but also occur with hardwoods, and their identification is best left to pholiotologists. The "true" *P. lubrica* is said to have a dark brown to reddish-brown cap with a paler or whitish margin. Some of the other commoner species or variants are: *P. ferruginea,* with an oranger cap; *P. ferrugineo-lutescens,* occasional in our area, with a slightly oranger cap and a thicker white stem that stains yellow; *P. sublubrica,* with brownish veil remnants on the cap and a thicker (1-1.5 cm) stem; *P. velaglutinosa,* with a bright to dark reddish-brown or brown cap that lacks veil remnants and a viscid or glutinous veil, sometimes abundant in our coastal pine forests (see photograph); *P. lenta,* with a pale cap (whitish to buff or sometimes tinged ochre or gray or pinkish) that features whitish veil remnants; and *P. decorata,* with a fibrillose-streaked cap that is dark vinaceous-brown to reddish-brown with a paler margin and has rows of concentrically-arranged veil remnants when young, common in the Pacific Northwest and Rocky Mountains. Also see *P. spumosa,* which lacks scales on stalk, and *P. albivelata* (under *Stropharia hornemannii*), which has a membranous ring.

Pholiota brunnescens   (Charcoal Pholiota)

CAP 2-7 cm broad, convex to plane or with an uplifted, often wavy margin; surface viscid or very slimy when moist, smooth or with small scattered whitish veil remnants; chestnut-brown to dark reddish-brown to orange-brown, tawny, or dark yellow-brown, sometimes fading in age to dull orange; margin often paler. Flesh rather soft, dingy brownish; odor usually mild. **GILLS** adnate to adnexed, crowded, narrow, whitish or grayish or pale yellowish, becoming dull cinnamon-brown to brown in age. **STALK** 4-6(9) cm long, (4)7-10 mm thick, equal, whitish to pale yellow, often darkening somewhat below in age or staining tawny when handled; covered with numerous small yellowish, fibrillose scales (usually arranged in concentric belts) below the veil. **VEIL** fibrillose, lemon-yellow, usually disappearing or forming only a slight superior ring on stalk. **SPORE PRINT** brown; spores 6-7 × 4-4.5 microns, elliptical, smooth. Chrysocystidia absent.

HABITAT: Scattered to densely gregarious or clustered on wood or soil in recently burned areas; locally common in its favored habitat throughout the mushroom season. It is known only from the West, but the similar *P. highlandensis* (see comments) is widespread.

Pholiota brunnescens is one of several charcoal-loving Pholiotas. Note slimy cap and the scales on the stalk. The scales are yellow when young but turn brown as they trap discharged spores.

EDIBILITY: Unknown.

COMMENTS: This is one of several closely-related Pholiotas that fruit only on charred soil or wood. It is distinguished by its sticky-slimy, tawny to orangish to dark reddish-brown cap, scaly stem, brown spores, and yellow veil. The other common ash-lovers have slimmer stems (2-6 mm thick). They include: *P. highlandensis (=P. carbonaria* of Europe), widely distributed, with a whitish veil and slightly smaller cap; *P. carbonaria,* with a rusty-red to reddish veil; *P. fulvozonata,* with an orange-brown or russet-colored veil; and *P. subangularis,* with a small cap that is only slightly viscid (if at all) and a pallid veil.

Pholiota spumosa   (Slender Pholiota)

CAP 2-6 (8) cm broad, obtusely conical or convex becoming umbonate or plane; surface smooth or appearing fibrillose or streaked, viscid or slimy when moist; color variable but usually olive-brown when young becoming brown to tawny or tawny-ochre at the center and yellow to dingy greenish-yellow toward the margin. Flesh yellow or greenish-yellow, soft, thin; odor mild. **GILLS** close, adnate or notched, yellow to pale greenish-yellow, becoming grayish or tawny and finally brown. **STALK** 3-7 (10) cm long, 4-6 (8) mm thick, more or less equal, fibrillose but not scaly, yellow to pale greenish-yellow above, becoming sordid brownish below or from the base upward. **VEIL** pale yellowish or whitish, delicate, fibrillose, disappearing or leaving slight remnants on cap margin and stalk. **SPORE PRINT** brown or dull rusty-brown; spores 6-9 × 4-4.5 microns, elliptical, smooth. Chrysocystidia absent.

HABITAT: Solitary to widely scattered to gregarious or tufted on ground and debris under conifers, widely distributed. Fairly common in our area in the fall and winter under pine, but rarely fruiting in large numbers.

EDIBILITY: Unknown.

COMMENTS: This species belongs to a complex group of Pholiotas with yellow to greenish-yellow flesh and gills, a viscid to slimy cap, and an evanescent veil. The veil is not slimy as in *P. velaglutinosa,* nor is the stem scaly below the veil as in the *P. lubrica* group. Other species: *P. graveolens* is similar but has a strong odor. In our area I have also found *P. scamba,* a small species (cap 1.5-3 cm broad) with a whitish to pale yellow or cinnamon-tinged cap and thin (1-3 mm) fibrillose or woolly stalk. It grows in groups or small clusters on dead conifer logs, sticks, etc. Also see comments under the *P. lubrica* group.

Pholiota destruens   (Destructive Pholiota)

CAP 5-20 cm broad, convex becoming broadly convex or rarely plane; surface slightly viscid when moist, white to creamy, buff, or at times ochre to brownish, covered with soft or cottony, whitish to buff scales or patches which may be come matted or washed off in age; margin often shaggy from veil remnants. Flesh thick, white, firm. GILLS adnate or notched, close, white becoming dull brown to deep rusty-cinnamon in age. STALK (3)5-15 cm long, 1-3 cm thick, equal or enlarged below, central or off-center, solid, hard, white, but often developing brownish stains below in age; smooth above the veil, at first clothed with whitish to buff scales and patches below. VEIL cottony, white, forming a slight superior ring on stalk, or disappearing. SPORE PRINT cinnamon-brown; spores 7-9.5 × 4-5.5 microns, elliptical, smooth, with a germ pore. Chrysocystidia absent.

HABITAT: Solitary or in groups or clusters on dead cottonwood and poplar or sometimes aspen or willow, especially on the cut ends of logs; widespread. It is particularly common in the valleys and bottomlands of the West, where Lombardy poplar and cottonwood are so prevalent. I have seen it in Oregon, New Mexico, and the Sacramento Valley in California, but not on the coast. It usually fruits late in the season.

EDIBILITY: Edible, but rather tough and poorly-flavored.

COMMENTS: This large *Pholiota* is distinct by virtue of its pale overall color, soft whitish veil remnants (scales or patches) on the cap, thick hard stalk, brown spore print, and occurrence on poplar (or sometimes willow). Its name refers to the fact that it rapidly destroys the wood on which it feeds.

Pholiota mutabilis   (Changeable Pholiota)

CAP 1.5-6 cm broad, obtuse becoming convex, broadly umbonate, or even plane; surface smooth, lubricous or slightly viscid when wet, hygrophanous: rusty-brown to orange-brown, reddish-brown, or tawny when moist, fading from the center outward to yellowish-brown, ochre, or yellowish-buff as it dries (often two-toned: yellowish at center and browner toward margin); margin translucent-striate only when moist. Flesh thin, white or tinged brown. GILLS adnate to slightly decurrent, close, pallid soon becoming brown or dull cinnamon. STALK 3-10 cm long, 2-10 (12) mm thick, equal or tapered toward base, stuffed or hollow, smooth and whitish above the ring, becoming brownish below and covered with numerous small, often recurved scales (at least when fresh); base sometimes blackish-brown in age. VEIL whitish, forming a small membranous or fibrillose superior ring on stalk, or sometimes disappearing. SPORE PRINT cinnamon-brown; spores 5.5-7.5 × 3.5-5 microns, elliptical, smooth, with a germ pore. Chrysocystidia absent.

HABITAT: Typically in clusters—often large—on logs, stumps, or occasionally buried wood; widely distributed and very common, late summer through early winter, but I have not seen it in our area. Although it is said to prefer hardwoods, I usually find it on conifers in northern California and the Pacific Northwest. The fruitings are sometimes so massive that the substrate (log or stump) is hidden from view.

EDIBILITY: Edible, but not recommended. Experienced collectors sometimes harvest the large clusters, but it is easily confused with the poisonous *Galerina autumnalis* and numerous other "LBM's" of unknown edibility.

COMMENTS: Also known as *Kuehneromyces mutabilis* and *Galerina mutabilis,* this brown-spored "LBM" is best recognized by it smooth, hygrophanous, often two-toned cap, scaly stalk with a ring, and penchant for growing in dense clusters. The poisonous *Galerina autumnalis* is quite similar, but lacks scales on the stem, has roughened spores, and does not usually grow in large clusters. Several *Naematoloma* and *Psathyrella* species

Pholiota mutabilis is a nondescript "LBM" that usually fruits in clusters. Note how the cap is striate when moist and becomes two-toned as it loses moisture. Also note how stalk is scaly below the ring.

are also similar but have darker spores. Other species: **P. vernalis** is closely related and similar, but lacks scales on the stem (or has only a few patches) and does not grow in such large clusters. It is common in the mountains of western North America shortly after the snow melts in the spring, but occurs elsewhere also.

CORTINARIACEAE

spores

ALMOST every brown-spored, terrestrial, woodland mushroom you find will belong to the Cortinariaceae, and many wood-inhabiting ones will also. Like their white-spored counterparts, the Tricholomataceae, they are a vast, diverse, and baffling group—with even more species, but fewer genera. The gills are not deeply decurrent and/or poroid as in the Paxillaceae, and the cap cuticle is typically filamentous and the spores pore-less, in contrast to the Bolbitiaceae. The latter characters are microscopic, but with a little practice the two can be distinguished in the field. The Cortinariaceae are primarily woodland fungi, a dominant fungal feature of cool temperate forests, whereas the Bolbitiaceae are warm-weather fungi found mostly in grass, gardens, and dung. Also, with the exception of *Galerina* and *Tubaria,* the Cortinariaceae tend to be larger, fleshier, and/or less fragile than the Bolbitiaceae.

As applied to the Cortinariaceae, the term "brown-spored" is somewhat misleading. The spore color actually ranges from orange or bright rusty-orange (*Gymnopilus*) to rusty-brown *(Cortinarius)* to dull brown, ochre-brown, or yellow-brown. Mushrooms with a filamentous cap cuticle and purple-brown to purple-black spores are traditionally placed in the Strophariaceae. However, the difference between "brown" and "purple-brown" is not always clearcut, and the two families seem to intergrade via *Pholiota* (a brown-spored member of the Strophariaceae that is keyed out here in the Cortinariaceae) and *Galerina*. Another family, the Crepidotaceae, is recognized by some authorities; it is largely tropical and composed almost entirely of "LBM's" included here under the Cortinariaceae.

There are very few esteemed edibles in the Cortinariaceae. The gypsy mushroom *(Rozites caperata)* is perhaps the best known and most widely collected, but there are many more species that are definitely known to be poisonous (particularly in *Galerina, Cortinarius, Hebeloma,* and *Inocybe*), and hundreds of others have yet to be tested. Identification is very difficult in this family, and having access to a microscope does little to expedite the tedious and labyrinthine identification process, because so little has been published on the family and so many species are poorly known or still unnamed. Several representatives from each common genus are included in this book.

Key to the Cortinariaceae

1. Typically growing on wood (may be buried!), wood chips, bark mulch in nurseries, sawdust, fabrics, or ashes (on burnt ground or wood); neither the flesh nor the gills lilac or violet .. 2
1. Typically growing on the ground (but occasionally on very rotten wood); gills and/ or flesh violet or lilac *in some cases* 8

2. Stalk poorly developed (lateral to off-center) or absent at maturity; on wood or various fabrics ....**Crepidotus,** p. 405
2. Stalk well-developed, central to somewhat off-center but not lateral; *not* found on fabrics . 3

3. Spore print rusty-orange to bright rusty-brown; cap *not* normally viscid or slimy; gills usually yellow to orange or rusty-orange; stalk fleshy though sometimes slender **Gymnopilus,** p. 407
3. Not as above; spore print different colored (including duller rusty-brown or ochre-brown) 4

4. Fresh fruiting body small and dark red to wine-colored .. **Galerina, Tubaria, & Allies,** p. 399
4. Not as above 5

5. Veil absent 6
5. Veil present (check young specimens because it may disappear) 7

6. Gills free at maturity; spore print pinkish-cinnamon .... (see **Pluteaceae,** p. 253)
6. Not as above 25

7. Stalk fleshy, or if thin then not particularly fragile and sometimes scaly below the veil; cap usually larger than 5 cm broad (but sometimes smaller), sometimes dry but more often viscid and usually *not* translucent-striate when moist; cap sometimes brightly colored, *sometimes* growing in large clusters .... (see **Strophariaceae,** p. 367)
7. Stalk thin (less than 5 mm), usually hollow and fragile and without scales; cap small (usually less than 4 cm) and dry, or if viscid then typically also hygrophanous and translucent-striate when moist; cap usually dull-colored, *not* typically growing in large clusters 25

8. Spore print reddish or with a greenish or olive tinge when moist; gills reddish when young; cap powdery and small (up to 5 cm broad) and dull grayish to brownish; not common .... (see **Agaricaceae,** p. 310)
8. Not as above; very common 9

9. Membranous veil present when young, usually forming a distinct annulus (ring) on stalk . 10
9. Veil absent, or if present then cobwebby, silky, hairy, or slimy but not membranous 13

10. Stalk thick and fleshy (at least 8 mm thick, usually more) *and/ or* gills purplish when young 11
10. Not as above 12

11. Veil covering the gills well into maturity or tending to shred radially rather than break away from the cap; fruiting body tending to develop underground (occasionally surfacing in wet weather); associated with mountain conifers .... **Cortinarius,** p. 417
11. Not as above; developing above the ground .... **Rozites & Phaeolepiota,** p. 411

12. Cap distinctly viscid or slimy when fresh, sometimes with scales; stalk often scaly or cottony or shaggy below the veil .... (see **Strophariaceae,** p. 367)
12. Cap not viscid or only slightly so and not scaly; stalk not normally cottony, shaggy, or scaly (but may be fibrillose) .... **Galerina, Tubaria, & Allies,** p. 399

13. Stalk tapered below to form a "tap root" that extends deep into humus; veil *absent*; spore print typically rusty-brown to cinnamon-brown .... **Phaeocollybia,** p. 413
13. Not as above; "tap root" absent, or if present then a veil also present (when young) and/ or spores differently colored 14

14. Gills brightly colored (yellow, orange, red, blue, green, purple), at least when young; cap *not* typically translucent-striate when moist 15
14. Gills not brightly colored (but may be dingy yellow, tawny, etc.) and/or cap translucent-striate when moist 17

15. Cobwebby, silky, or hairy veil present when young; cap viscid or dry; spore print rusty-brown to cinnamon (but not rusty-orange); gills variously colored; common .. ***Cortinarius,*** p. 417
15. Veil absent; cap not typically viscid; spore print orange to rusty-orange *or* dull brown to yellow-brown or olive-brown; gills yellow to orange or rusty-orange; not very common 16

16. Spore print orange to rusty-orange or bright rusty-brown .... ***Gymnopilus,*** p. 407
16. Spore print brown to olive-brown .... (see **Paxillaceae,** p. 476)

17. Volva present at base of stalk; spore print reddish or pinkish-cinnamon (see **Pluteaceae,** p. 253)
17. Not as above; volva absent or spore color browner or rustier 18

18. Cap typically viscid or slimy when moist, usually smooth (bald) or with a few scattered scales or veil remnants 19
18. Cap typically *not* viscid, but sometimes slightly tacky in wet weather; cap surface smooth or silky to fibrillose, woolly, scaly, or powdery 24

19. Stalk distinctly viscid or slimy from the remains of a slimy veil; spore print typically rusty-brown to cinnamon .... ***Cortinarius,*** p. 417
19. Not as above 20

20. Cap usually small (up to 4 cm broad) and striate when moist; stalk thin, i.e., 1-3 (5) mm thick; fruiting body fragile; often found in moss .... ***Galerina, Tubaria,* & Allies,** p. 399
20. Not as above 21

21. Veil present, at least when young (check several specimens if unsure), usually but not always leaving traces on the stalk in the form of hairs, scales, or an annulus (ring) 22
21. Veil absent in all stages .... **Hebeloma,** p. 463

22. Spore print typically rusty-brown to cinnamon-brown; veil cobwebby or silky; stalk usually lacking scales below the veil, usually (but not always) at least 1 cm thick; spores roughened .... ***Cortinarius,*** p. 417
22. Not as above 23

23. Odor often radishlike or spermatic and/or taste bitter; gill edges often whitish; stalk apex often powdery, scurfy, or with small flakes; cap viscid but not often slimy; spore print dull brown (not cinnamon- or rusty-brown); *not* typically growing in large clusters from a common base .... **Hebeloma,** p. 463
23. Not as above; sometimes growing in clusters .... (see **Strophariaceae,** p. 367)

24. Spore print yellow-brown to tan to dull brown, *or* if rusty-brown or cinnamon then fruiting body often tawny or yellowish, translucent-striate when moist, and often growing in moss; spores smooth or roughened 25
24. Spore print typically rusty-brown to cinnamon-brown; cobwebby or silky veil present when young, often leaving hairs on stalk; stalk thick or thin but not usually fragile; spores roughened .... ***Cortinarius,*** p. 417

25. Odor spermatic or like green corn (crush the cap if unsure!); spore print dull brown or dull yellow-brown .... **Inocybe,** p. 455
25. Not as above 26

26. Spore print dull brown or dull yellow-brown; cap usually opaque and small or medium-sized, often umbonate and/or with easily-splitting margin (especially in age); surface of cap usually silky, hairy, or scaly; gill edges often whitish; gills typically *not* decurrent; very apex of stalk often minutely powdered, flaky, or scurfy; often growing in groups but not usually in clusters .... **Inocybe,** p. 455
26. Not as above 27

27. Veil present at least when very young; stalk with small scales (often somewhat concentrically arranged) below the veil .... (see **Strophariaceae,** p. 367)
27. Not as above; veil absent or present; stalk not normally scaly, but may have fibrils 28

28. Stalk typically 4 mm or more thick; cap typically 4-10 cm broad 29
28. Not as above (usually smaller) .... ***Galerina, Tubaria,* & Allies,** p. 399

29. Spore print reddish- or pinkish-cinnamon .... (see **Entolomataceae,** p. 238)
29. Not as above .... (see **Bolbitiaceae,** p. 466)

GALERINA, TUBARIA, & Allies

Small to minute, mostly brown mushrooms found on wood, moss, or ground. CAP typically smooth or nearly so, often translucent-striate toward margin when moist. GILLS *brown to rusty or tawny at maturity, typically adnexed to adnate (Galerina)* or *slightly decurrent (Tubaria).* STALK *thin (generally less than 5 mm thick), fragile or cartilaginous,* usually hollow and central. VEIL absent or present, sometimes forming an annulus (ring) on stalk. VOLVA absent. SPORE PRINT *ochre-brown to cinnamon-brown or brown.* Spores smooth or roughened, usually lacking a germ pore. Cystidia usually present on gill edges. Cap cuticle usually filamentous.

THIS is an artificial grouping of little brown mushrooms ("LBM's") with brownish spores and a thin, usually fragile stem. *Galerina* is a large genus (over 200 species) formerly divided among *Pholiota* and the obsolete genus *Galera.* It can be recognized in the field by its small size, yellowish to brown color, and conical to convex cap, plus the thin stem and brown spores. *Pholiota* intergrades with *Galerina,* while *Conocybe* species also have brown spores but are usually more conical and have a different type of cell structure in the cap cuticle (cellular rather than filamentous). Also, Conocybes usually grow in grass, dung, or cultivated soil, whereas Galerinas are partial to wood, humus, and moss in forests and bogs (however, some Galerinas grow in grass and some Conocybes in moss).

Tubaria is a small but ubiquitous genus with slightly decurrent gills and a convex to plane or depressed cap. It is most likely to be confused with *Agrocybe,* which does not usually have decurrent gills, and with *Pholiota,* with which it intergrades. Several other small genera are also treated or mentioned here (e.g., *Naucoria, Alnicola,* and *Simocybe*), but they are differentiated largely on microscopic characters and are of little interest to the average mushroom hunter.

Galerina and *Tubaria* are especially common during dry spells when other mushrooms are scarce. Several Galerinas are known to contain deadly amanita-toxins. Many others have not been tested and all are difficult to identify—a compelling reason to avoid eating *all* Galerinas, Tubarias, and other "LBM's."

Two Galerinas and two Tubarias are described here, plus one rare but distinctive mushroom of uncertain disposition, *Naucoria vinicolor.* If your "LBM" does not key out convincingly, throw it away! For more comments on "LBM's," see p. 32.

Key to Galerina, Tubaria, & Allies

1. Fruiting body dark red to vinaceous (wine-colored) when fresh .. *Naucoria vinicolor,* p. 404
1. Not as above 2

2. Veil present when young, usually forming a small or large annulus (ring) on the stalk (check several specimens if possible) 3
2. Veil absent, or if present then not forming an annulus 11

3. Annulus prominent, distinctly membranous and often flaring and/or movable; cap dry and *not* translucent-striate; cap conical or bell-shaped when young (but may expand), 0.5-2.5 cm broad; spores with a germ pore, cap cuticle cellular .... (see *Conocybe,* p. 470)
3. Not as above; cap usually striate when moist (at least at margin) and/or differently shaped 4

4. Growing on wood or wood chips 5
4. Growing on ground (in humus, moss, etc.) 9

5. Cap thinly to thickly viscid when moist 6
5. Cap not viscid 7

6. Cap thinly viscid and translucent-striate when moist; stalk lacking obvious scales below the annulus (ring); spores roughened .... *Galerina autumnalis,* p. 401
6. Not as above; spores smooth .... (see *Pholiota,* p. 384)

7. Cap often with hoary white fibrils or particles when young; gills usually adnate to decurrent .... *Tubaria confragosa,* p. 403
7. Not as above 8

8. Stalk with small or prominent scales or patches of veil remnants below the veil; spores smooth
    .... (see *Pholiota,* p. 384)
8. Not as above; spores usually roughened ..... *Galerina marginata* (see *G. autumnalis,* p. 401)

9. Cap reddish-cinnamon when moist; often growing on lawns; known only from the Pacific
    Northwest .... *Galerina venenata* (see *G. autumnalis,* p. 401)
9. Not as above; usually growing in moss, grass, or swampy ground and/or differently colored 10

10. Cap translucent-striate when moist .... *Galerina paludosa* (see *G. autumnalis,* p. 401)
10. Not as above .... (see *Agrocybe,* p. 467)

11. Gills usually slightly decurrent; cap convex to plane or slightly depressed (*not* conical and
    not typically tawny or yellow) 12
11. Not as above 13

12. Cap and stalk white to grayish; cap often somewhat hairy, especially toward margin; found
    in woods; widespread but not common .... (see *Clitocybe* & Allies, p. 148)
12. Cap brown when fresh but often fading to buff or whitish as it dries; found in many habitats
    .... *Tubaria furfuracea* & others, p. 402

13. Cap more or less olive-brown; growing on wood; spores smooth .... *Simocybe centunculus*
13. Not with above features 14

14. Stalk or base of stalk blue to blue-green in age; growing in moss or grass (see *Conocybe,* p. 470)
14. Not as above 15

15. Cap bright yellow when fresh (but often fading) *and* conspicuously striate and viscid; entire
    fruiting body withering or dissolving quickly .... (see *Bolbitius,* p. 473)
15. Not as above 16

16. Spore print ochre-brown to pinkish; odor usually strong and fishy or cucumberlike; cap bell-
    shaped to conical when young, reddish-brown to dark brown or blackish (margin often paler);
    stalk also dark, minutely velvety; gills with giant cystidia ... (see Tricholomataceae, p. 129)
16. Not as above 17

17. Growing on decaying wood; cap hairy or scaly or silky and/or odor spermatic; cap *not* viscid
    .... (see *Inocybe,* p. 455)
17. Not as above 18

18. Cap convex to plane or broadly umbonate 19
18. Cap conical to bell-shaped or acutely umbonate 24

19. Lower stalk usually with small scales or veil remnants; veil present at least in young specimens;
    cap often viscid when moist; spores smooth; usually found on wood, sometimes on ground, but
    not usually in moss .... (see *Pholiota,* p. 384)
19. Not as above; spores smooth or roughened; growing in many habitats, often in moss 20

20. Cap viscid when moist and conspicuously striate; flesh *very* thin; stalk white or yellow and very
    fragile (2-3 mm thick); entire fruiting body withering or decaying quickly; cap cuticle cellular
    .... (see *Bolbitius,* p. 473)
20. Not as above 21

21. Gills adnate; cap brown but often fading to tan, buff, or even whitish as it loses moisture; spore
    print ochre to dull brown or tan; spore walls nearly colorless and thin, easily breaking; common
    .... *Tubaria furfuracea* & others, p. 402
21. Not as above 22

22. Spore print brown; found mostly under alder and willow 23
22. Spore print usually rusty-brown to cinnamon-brown (brighter than above); found in many
    habitats 24

23. Cap dark reddish-brown .... *Alnicola scolecina*
23. Cap brown to tan or yellow-brown .... *Alnicola melinides, A. escharoides,* & others

24. Cap usually yellowish to yellow-brown or tawny or sometimes rusty-brown; veil present or
    absent; usually growing in moss or grass or on wood; cap usually translucent-striate when
    moist; stalk fragile; spores roughened or smooth ... *Galerina heterocystis* & others, p. 402
24. Not as above; sometimes growing in moss, but usually on ground and not normally in grass or
    on wood; veil always present (at least when very young); stalk not often fragile; spores
    roughened .... (see *Cortinarius,* p. 417)

Galerina autumnalis is a deadly poisonous "LBM." Growth on wood, brown spores, and scale-less stalk with a small annulus (ring) are the main features.

Galerina autumnalis   (Deadly Galerina)

CAP 1-4 (6.5) cm broad, convex to nearly plane or slightly umbonate; surface smooth, viscid when moist, dark brown to yellow-brown or tawny, fading to tan or yellowish as it dries; margin translucent-striate when moist. Flesh thin, watery brown; odor mild or slightly farinaceous. **GILLS** attached (slightly decurrent to adnexed) but often seceding, close, yellowish to pale brown becoming rusty-brown or brown. **STALK** 2-10 cm long, 3-6 (10) mm thick, equal or thicker below, dry, hollow, pallid to brownish, often darker below in age, fibrillose below the veil; base often with white mycelial strands. **VEIL** fibrillose or somewhat membranous, usually forming a thin, superior, white ring on stalk which is subsequently darkened by falling spores or often disappears in age. **SPORE PRINT** rusty-brown; spores 8-11 × 5-6.5 microns, elliptical, roughened and/or wrinkled.

HABITAT: Scattered to gregarious or tufted on rotting wood and debris of both hardwoods and conifers; widely distributed. Fairly common in our area from fall through early spring, especially during relatively dry years.

EDIBILITY: DEADLY POISONOUS—it contains amanita-toxins! Fortunately, it is rarely eaten because of its diminutive dimensions and mundane appearance.

COMMENTS: Since this drab "LBM" is deadly poisonous, it is important to learn its distinguishing characteristics: (1) the rusty-brown spores (2) the small size and thin stem (but in areas of high rainfall overgrown specimens often occur) (3) the veil which usually (but not always!) forms a thin whitish superior annulus (ring) (4) the growth on wood (sometimes buried or very decayed). The annulus may turn brown in age or even disappear, so it is best to avoid any mushroom that is remotely similar, including the edible *Pholiota mutabilis,* which typically grows in large clusters and has small scales on the stem. There are several very similar deadly poisonous Galerinas with a thin annulus, including: *G. marginata,* with a moist but not viscid cap, found mainly on decaying conifers; and *G. venenata,* with a reddish-cinnamon cap that fades to pinkish-buff or whitish in age, growing on lawns or buried wood and known only from the west coast. Other species: *G. paludosa* also has a white superior annulus, but it has a long thin stem and grows in bogs; its edibility is unknown.

Galerina heterocystis

CAP 5-20 mm broad, bluntly conical becoming bell-shaped or convex or sometimes umbonate; surface smooth, hygrophanous: pale yellow to pale cinnamon to tawny and translucent-striate when moist, paler (more or less buff) when dry. Flesh very thin, fragile. GILLS close, usually attached but not decurrent, pale yellowish becoming pale cinnamon-brown. STALK 1-8 cm long, 1-3 mm thick, more or less equal, tubular, fragile, pallid to pale yellowish darkening to brown or cinnamon; lower portion faintly fibrillose. VEIL absent or rudimentary. SPORE PRINT pale cinnamon-brown; spores 11-17 × 6.5-8.5 microns, more or less elliptical, roughened to nearly smooth.

HABITAT: Scattered to gregarious in damp mossy or grassy places (usually in or near woods); widely distributed. In our area this species and its numerous look-alikes are especially common during relatively dry weather when other mushrooms are scarce.

EDIBILITY: Unknown. Do not experiment!

COMMENTS: There are dozens of Galerinas that will more or less fit the above description, and they can only be differentiated microscopically. Unlike *G. autumnalis,* the veil is evanescent or even absent, but like that species the cap is usually translucent-striate when fresh. Some are found in the woods or at their edges, others frequent seepage areas, still others grow on logs or in bogs; most are moss-inhabiting. *Conocybe* species may also key out here, but are generally more sharply conical, not as translucent, and have a cellular cap cuticle; they favor lawns, dung, or cultivated ground, but a few grow in moss. *Psilocybe* and *Psathyrella* species are also similar, but have darker spores. Other Galerinas with an evanescent or absent veil include: *G. semilanceata,* with a fibrillose veil, usually found on the ground; *G. cedretorum,* larger (cap 1-3 cm), with a more or less convex cap and no veil, found in humus or debris under conifers; *G. hypnorum,* minute, with a rudimentary veil, found on mossy logs; *G. triscopa,* also minute, but with a sharply conical cap when young, growing on logs; and *G. tibicystis,* growing only in *Sphagnum* bogs. None of these should be eaten.

Tubaria furfuracea   (Totally Tedious Tubaria)

CAP 1-3 (4) cm broad, convex becoming plane or slightly depressed; surface smooth to finely fibrillose or often with minute whitish flecks and patches (veil remnants); hygrophanous but not viscid, brown to reddish-brown, cinnamon-brown, or tan when moist, fading to buff, pinkish-buff, or whitish as it dries (often fading in center first); margin striate when moist. Flesh thin, brownish. GILLS close, adnate to slightly decurrent, pale tawny to cinnamon or brown. STALK 2-6 cm long, 1-4 mm thick, equal or slightly thicker below, colored more or less like cap or paler, sometimes with whitish flecks, fibrillose, fragile; base usually with whitish mycelium. VEIL whitish, fibrillose, evanescent. SPORE PRINT ochre-brown to pale ochre; spores 6-9 × 4-6 microns, elliptical, smooth.

HABITAT: Scattered to gregarious on ground, sticks, and woody debris in wet places—

Tubaria furfuracea, the "Totally Tedious Tubaria." Note adnate to slightly decurrent gills and tendency of cap to fade as it loses moisture. A similar viscid-capped species is also common locally.

Tubaria furfuracea is a quintessential "LBM." These specimens are taller than most and their gills are rather widely spaced. Could they be a distinct species? Does anyone know? Does anyone care?

woods, vacant lots, landscaped areas, along trails, etc.; common and widely distributed. It seems to be most abundant when and where other mushrooms are scarce—perhaps because it is only likely to be *noticed* when and where other mushrooms are scarce. I have seen enormous fruitings in wood chip mulch at Golden Gate Park in San Francisco.

EDIBILITY: Unknown.

COMMENTS: This is it, folks—your quintessential "LBM" (see p. 32)—as boring as it is ubiquitous and as innocuous as it is inconspicuous. The brownish spores, adnate to decurrent gills, thin stalk, and whitish-flecked, hygrophanous cap are the most distinctive (or least undistinctive) fieldmarks. To say more about it would do the more interesting mushrooms in this book an injustice. Other species: The "Totally Tedious Tubaria" is easily mistaken for the "Truly Trivial Tubaria," *T. pellucida,* which, however, is slightly smaller (cap 0.5-1.5 cm broad) and has slightly smaller spores. *T. tenuis*, otherwise known as the "Truly Trivial AND Totally Tedious Tubaria," is also smaller, but has a completely glabrous (bald) cap and widely spaced gills. A small unidentified viscid-capped species also occurs in our area. There are various other "LBM's" belonging to obscure genera such as *Simocybe* and *Alnicola*. You can find more information on these in the key to *Galerina, Tubaria,* & Allies, but don't you have something more exciting to do?

Tubaria confragosa   (Not So Tedious Tubaria)

CAP 1-5 cm broad, broadly convex to more or less plane or slightly uplifted; surface moist or dry but not viscid; hygrophanous, brown to vinaceous-brown or reddish-brown to reddish-cinnamon when moist, markedly paler (buff to cinnamon-buff) as it dries out; smooth or often appearing hoary at first from a thin layer of whitish fibrils or minute scales (especially toward the margin, which is striate when moist). Flesh thin, fragile, colored like cap. **GILLS** cinnamon to rusty-brown to reddish-cinnamon to brown, adnate to slightly decurrent, close. **STALK** 2-8 cm long, 1.5-6 mm thick, equal or thicker below, soon hollow, colored like cap or paler, usually with fibrils or a few small scales below the ring; base typically with white mycelial mat. **VEIL** usually forming a membranous, often flaring superior ring on stalk, but sometimes disappearing or leaving only a fibrillose zone. **SPORE PRINT** brown to dark reddish-cinnamon; spores 6.5-9 × 4-6 microns, broadly elliptical, smooth.

HABITAT: Gregarious (often clustered) on rotting logs, sawdust, etc.; widely distributed. I have seen large fruitings in the fall and winter in wood chip mulch, sometimes accompanied by *T. furfuracea.*

Tubaria confragosa, the "Not So Tedious Tubaria," is reminiscent of the more common "Totally Tedious Tubaria" *(T. furfuracea),* but usually has an annulus (ring) on the stalk.

EDIBILITY: Unknown.

COMMENTS: The presence of a well-defined annulus (ring)—at least in *many* specimens —rescues this "LBM" from the obscurity it so richly deserves. It is slightly larger than *T. furfuracea,* and is more likely to be cespitose (clustered). It is also known as ***Phaeomarasmius confragosus*** and ***Pholiota confragosa,*** but the hygrophanous, non-viscid cap and smooth to only slightly scaly stalk separate it from the Pholiotas of this book.

Naucoria vinicolor   Color Plate 94

CAP 1-4 cm broad, convex to plane or obtusely umbonate; surface smooth or finely fibrillose or occasionally fibrillose-scaly, not viscid, dark red to wine-red (vinaceous), the center often darker. Flesh thin, tinged cap color; odor mild. **GILLS** slightly decurrent to adnate, adnexed, or notched, close; dark red or vinaceous when young, soon becoming cinnamon or rusty-brown as spores mature. **STALK** 1-7 cm long, 2-6 mm thick, equal or slightly thicker below, colored more or less like cap but often overlaid with a fine whitish silky-fibrillose coating. **VEIL** fibrillose, vinaceous, disappearing or forming a slight hairy zone on stalk which turns cinnamon from falling spores. **SPORE PRINT** cinnamon-brown; spores 6.5-8 × 4-5 microns, elliptical or bean-shaped, smooth.

HABITAT: In small tufts or clusters (or sometimes solitary) on dead wood; known only from California, but perhaps more widely distributed. I have found it several times on oak in the fall and winter, but it is rare.

EDIBILITY: Too rare to be of value. Greg Wright, who has found it in Los Angeles County, reports that it is harmless, with a "mealy, moderately mushy" texture and a flavor "that suggests bland beef." (I just can't wait to try it! Can you?)

COMMENTS: This interesting mushroom was originally described in 1909. It does not belong to the genus *Naucoria* in its modern sense, but has not officially been transferred to another genus—perhaps because of its rarity, but perhaps also because its "correct" genus is in doubt! The fibrillose veil and cinnamon-brown spores give it the aspect of a small *Cortinarius* (e.g., *C. sanguineus*), but it always seems to grow on wood, usually in small clusters. Its wine-red color and small size plus the spore color and growth on wood are the distinguishing fieldmarks. *Lactarius fragilis* sometimes grows on wood but has a latex and/or a fragrant odor, lacks a veil, and has white or yellowish spores. *Naematoloma aurantiaca* is similarly colored but has darker spores and is usually terrestrial.

CREPIDOTUS

Small to medium-sized mushrooms *typically growing shelflike on wood.* CAP usually round to kidney-shaped in outline, surface smooth or hairy. Flesh soft, thin. GILLS usually brownish at maturity. STALK *usually absent or rudimentary,* or if present then lateral or off-center. VEIL and VOLVA absent. SPORE PRINT *dull brown to yellow-brown, cinnamon-brown, or pinkish-brown.* Spores smooth or roughened, lacking an apical germ pore. Cystidia often present on gills. Cap cuticle filamentous.

THIS is a fairly common but lackluster group of wood-inhabiting mushrooms with little or no stem. They superficially resemble the oyster mushrooms *(Pleurotus)* as well as *Phyllotopsis* and *Claudopus,* but have brown spores. Other brown-spored, wood-inhabiting mushrooms (*Pholiota, Galerina,* etc.) have well-developed stems. *Crepidotus* is given its own family, the Crepidotaceae, by some mycologists.

    Crepidotus species are worthless as food—they are flaccid and decay rapidly. They favor decaying hardwood logs, branches, and twigs (especially oak), but a few species occur on conifers or even in soil—in which case they may have a lateral to off-center (but not central) stem. Two widespread representatives are depicted here.

Key to Crepidotus

1. Fruiting body tough and leathery or woody . . . . (see **Polyporaceae & Allies,** p. 549)
1. Not as above; fruiting body fleshy 2

2. Cap white or whitish 3
2. Cap tawny to pale ochre, brownish, or red (but may sometimes fade to whitish in age) 4

3. Cap 1-4 cm broad and nearly smooth (bald)   *C. applanatus* & others (see *C. herbarum,* below)
3. Cap smaller (up to 2 cm broad) and distinctly hairy or downy . *C. herbarum* & others, below

4. Cap bright red (scarlet to cinnabar-red), small (up to 15 mm broad); gill edges also scarlet to red; found on hardwoods in eastern North America; rare . . . . *C. cinnabarinus*
4. Not as above; cap not red 5

5. Found on fabrics or old carpets, seat covers of abandoned cars, mattresses, "rotting blue jeans," or on wood; fruiting body yellow-brown to brown or cinnamon; stalk present, often darker, usually off-center, curved, and slender . . . . *Melanotus textilis*
5. Not as above; found on wood or lignin-rich humus 6

6. Gills often forked or veined, especially near base of cap . . . . (see **Paxillaceae,** p. 476)
6. Not as above 7

7. Gills yellow to orange or ochre-orange . . . . *C. crocophyllus* (see *C. mollis,* p. 406)
7. Gills whitish to grayish, brownish, or dull cinnamon . . . . *C. mollis,* p. 406

Crepidotus herbarum   (Little White Crep)

CAP 0.5-2 cm broad, kidney-shaped to nearly round in outline; surface hairy or downy, white, not viscid. Flesh very thin, white. GILLS fairly well-spaced, white becoming pale ochre or brownish; radiating from point of attachment to substrate. STALK absent or rudimentary. SPORE PRINT pale yellow-brown; spores 6-8 × 3-4 microns, pip-shaped to lance-shaped or somewhat elliptical, smooth.

Crepidotus variabilis (see comments under *C. herbarum*) is one of several small whitish shelving species with brown or pinkish-brown spores.

HABITAT: Scattered or in groups or troops on fallen branches, twigs, herbaceous stems, and debris(usually of hardwoods); widely distributed. Common in our area throughout the mushroom season, but often overlooked.

EDIBILITY: Unknown, and like most of us, destined to remain so.

COMMENTS: This is one of several small whitish *Crepidotus* species that typically grow on twigs and branches, sometimes completely covering their host. It is likely to be mistaken for a *Claudopus* or small *Pleurotus,* but has brownish spores. Similar species include: *C. versutus,* with dull brown, elliptical spores 9-10 microns long; *C. variabilis,* with pale brown to pinkish-brown, minutely warty, elliptical spores 5-7 microns long; *C. fusisporus,* with pinkish-buff, fusiform (elongated) spores; *C. applanatus,* very common and widely distributed, with round, minutely spiny spores and a larger(1-4 cm), nearly smooth, white cap that often becomes brownish in age; and *C. maculans,* also larger and smooth, but usually blackish-spotted in age.

Crepidotus mollis   (Jelly Crep; Flabby Crepidotus)

CAP 1-5 (8) cm broad, fan- or kidney-shaped to nearly round in outline, convex to plane; surface gelatinous in wet weather beneath a dense to rather sparse coating of fulvous to rusty-ochre to brown fibrils (hairs) or small fibrillose scales; in age often smooth or with very few fibrils and varying in color from tawny to pale ochre to brown, or fading to whitish. Flesh soft, thin, pallid, soon flaccid. **GILLS** close, whitish becoming brown or dull cinnamon; radiating from base of cap. **STALK** absent or rudimentary. **SPORE PRINT** dull brown to yellowish-brown; spores 7-11 × 4.5-6.5 microns, elliptical, smooth.

HABITAT: Usually in groups or overlapping tiers on the bark of dead hardwoods (or rarely conifers); very widely distributed and common. In our area it is frequent throughout the mushroom season, especially on live oak.

EDIBILITY: Unknown.

COMMENTS: Also known as *C. fulvotomentosus* and *C. calolepis,* this is our most common and conspicuous *Crepidotus,* often fruiting in attractive masses on dead oaks. The relatively large size(for a *Crepidotus),* gelatinous texture when wet, brown spores and mature gills, and fibrillose scales on the cap when young are the main fieldmarks. In age it becomes quite flabby and is likely to attract your attention for this reason if for no other. *C. crocophyllus* is a similar, widely distributed species with yellow to ochre-orange immature gills. I have found it several times in our area.

Crepidotus mollis is a common shelving mushrooms with a flabby cap and brown spores. Young caps shown here are covered with brown fibrils or scales, but older or rain-battered caps can be bald.

GYMNOPILUS

Medium-sized to large mushrooms *found mostly on wood.* CAP smooth or scaly, dry. GILLS notched to slightly decurrent, *usually yellow to rusty-orange,* STALK *more or less central, fleshy.* VEIL *usually (but not always) present,* sometimes forming an annulus (ring). VOLVA absent. SPORE PRINT *orange to rusty-orange to bright rusty-brown.* Spores typically elliptical and roughened, without an apical pore. Cystidia present, at least on gill edges. Cap cuticle filamentous.

THIS is a clearcut group of rusty-orange-spored mushrooms formerly divided among *Pholiota* and the defunct genus *Flammula.* The fruiting body is typically reddish-brown to rusty-orange to yellow, and a veil is often present. The vast majority of species grow on wood but at times may appear terrestrial. *Pholiota* and *Cortinarius* are the genera most often confused with *Gymnopilus. Pholiota,* however, usually has a viscid cap and duller (brown to cinnamon-brown) spores, while *Cortinarius* grows on the ground.

    Gymnopilus species are found primarily in the woods, but sometimes turn up on cut stumps, wood chip mulch, and in nursery flats and flower pots. They are quite common, but rarely fruit in the large numbers typical of, say, *Hebeloma, Cortinarius,* and *Inocybe.*

    About 75 species of *Gymnopilus* are known from North America and over 25 from California. Several are quite striking but none are good edibles. A few—specifically *G. spectabilis, G. validipes,* and *G. aeruginosus*—are "pupil-dilating" (hallucinogenic). Four species are depicted here.

Key to Gymnopilus

1. Cap usually tinged or variegated with blue or blue-green when young (sometimes also with other colors) or staining bluish when bruised 2
1. Not as above; blue or blue-green shades absent 3

2. Cap with fibrils or scales; veil present, at least when young . . . . *G. aeruginosus,* p. 409
2. Cap nearly smooth; veil absent . . . . *G. punctifolius* (see *G. aeruginosus,* p. 409)

3. Cap with fibrils or scales of a radically different color than the background (i.e., reddish, pinkish, brownish, etc.); not common 4
3. Cap smooth or with fibrils or scales that are roughly the same color as background (i.e., some shade of yellow, orange, rusty-orange, or reddish-brown); common 6

4. Cap scales or fibrils red to purple-red, reddish-brown, or pink 5
4. Cap scales or fibrils tawny to dark brown to blackish-brown . . . .
. . . . *G. fulvosquamulosus* & *G. parvisquamulosus* (see *G. luteofolius,* p. 409)

5. Fruiting body (especially cap) sometimes with bluish or blue-green stains (examine several specimens!); veil evanescent, *not* typically forming a distinct annulus (ring) . . . .
. . . . *G. aeruginosus* & others, p. 409
5. Not as above; veil often forming a slight annulus . . . . *G. luteofolius* & others, p. 409

6. Cap medium-sized to gigantic (up to 40 cm broad or more!); stalk 1-7 cm thick; veil membranous or fibrillose, often forming an annulus (rng) on stalk; usually growing in clusters (but occasionally solitary) . . . . *G. spectabilis* group, p. 410
6. Not as above; smaller and veil absent or evanescent; not often clustered 7

7. Veil present when young (but often disappearing in age) 8
7. Veil absent (check young specimens!) 10

8. Veil whitish . . . . *G. penetrans* & *G. flavidellus* (see *G. sapineus,* p. 408)
8. Veil pale yellow or yellowish 9

9. Cap golden-yellow to tawny-orange, often with minute scales or scattered fibrillose patches
. . . . *G. sapineus,* p. 408
9. Cap darker (tawny to cinnamon, russet, or reddish-brown), smooth . . . .
. . . . *G. luteocarneus* (see *G. sapineus,* p. 408)

10. Usually growing on ground; taste mild . . . . *G. terrestris* (see *G. sapineus,* p. 408)
10. Typically found on wood; taste usually bitter *G. liquiritae* & *G. bellulus* (see *G. sapineus,* p. 408)

Gymnopilus luteocarneus (see comments under *G. sapineus,* below) is one of several small orangish species with an evanescent veil or no veil at all. Mature caps can be somewhat larger than the ones shown here, but never approach the size of the *G. spectabilis* group. **Left:** A young specimen on wood and an older one on a cone. **Right:** Mature specimen; note how stalk base is darkened by falling spores.

Gymnopilus sapineus (Common and Boring Gymnopilus)

CAP (1) 2-5 (9) cm broad, convex to nearly plane or sometimes obscurely umbonate; surface dry, usually with minute scales or scattered patches of fibrils, often cracking in age; golden-yellow to tawny-orange, the margin sometimes paler. Flesh yellowish, firm; taste usually bitter. **GILLS** attached (usually adnate), close, yellow becoming rusty-yellow to rusty-cinnamon in age. **STALK** 2.5-7 cm long, 3-7 (10) mm thick, equal or tapered slightly below; yellowish-buff to yellow, becoming brownish-yellow in age or when handled; fibrillose. **VEIL** yellowish, fibrillose, disappearing or leaving a few hairs near top of stalk which turn rusty from falling spores. **SPORE PRINT** amber- to rusty-orange to bright rusty-brown; spores 7-10 × 4-5.5 microns, elliptical, minutely roughened.

HABITAT: Solitary to scattered or in small tufts or groups on rotting logs, cones, and humus rich in lignin; widely distributed, but rarely fruiting in large numbers. In our area it is common under conifers (especially pine) throughout the mushroom season. Similar species turn up occasionally in nursery flats and flower pots.

EDIBILITY: Unknown; the small size and bitter taste are deterrents.

COMMENTS: There are several small, eminently undistinguished and evidently indistinguishable *Gymnopilus* species that will more or less fit the above description. As a group they can be recognized by their yellow to orange-brown color, modest size, growth on wood, dry cap, and rusty-orange spores. Some have an evanescent fibrillose veil, others have no veil at all. *G. sapineus* is distinct by virtue of the small scales or fibrillose patches on the cap. The others are hardly worth differentiating, but some of them are: *G. luteocarneus,* common in California on conifers, with a smooth, darker (tawny-cinnamon to russet) cap and pale yellowish, evanescent veil; *G. penetrans* and *G. flavidellus*, with a smooth yellow to yellow-orange cap and whitish evanescent veil, the latter with a stalk that stains orange-brown when handled; and the following species with *no* veil and a smooth, tawny to orange, rusty, or cinnamon-brown cap: *G. liquiritiae,* growing on wood, taste bitter; *G. bellulus,* on wood, taste bitter, but cap only 1-2.5 cm broad and smooth to minutely scurfy; and *G. terrestris,* usually terrestrial under conifers, with a mild taste.

Left: A close relative of *G. aeruginosus* (perhaps the same?), this unidentified *Gymnopilus* has dark red fibrils on the cap and stalk when young and often exhibits bluish stains. Note absence of annulus (ring) on stalk (see comments below). **Right:** Close-up of *G. luteofolius* showing cap fibrils and annulus.

Gymnopilus aeruginosus

CAP (2) 5-15 (23) cm broad, convex to nearly plane; surface dry, fibrillose-scaly (or cracked in age); color variable: at first dull bluish-green or variegated with green, yellow, salmon, red, or vinaceous, sometimes fading to buff or pinkish-buff in age; scales tawny to reddish to dark brown. Flesh whitish or tinged blue or green; taste bitter. **GILLS** adnexed to adnate or slightly decurrent, often seceding; buff to yellow-orange or ochre, close. **STALK** (3) 5-12 cm long, (0.4) 1-1.5 (4) cm thick, more or less equal, colored more or less like the cap, smooth or fibrillose, dry. **VEIL** yellowish, fibrillose, often scanty, leaving an evanescent zone of hairs near top of stalk. **SPORE PRINT** rusty to rusty-orange or rusty-cinnamon; spores 6-9 × 3.5-4.5 microns, elliptical, roughened.

HABITAT: In groups or clusters on stumps, logs, and sawdust of both hardwoods and conifers; widely distributed, but most common in the Pacific Northwest. I have seen large fruitings in the fall, winter, and spring in wood chip mulch.

EDIBILITY: Hallucinogenic. The blue-green tones are indicative of psilocybin and/or psilocin, as in *Psilocybe*.

COMMENTS: The tendency of the cap to exhibit a blue-green tinge when young (often variegated with pink, vinaceous-red, and/or other colors) plus the scaly cap, yellowish gills, rusty-orange spores, and growth on wood make this species quite distinctive. *G. harmoge* may be a synonym. A similar unidentified species (perhaps the same?) with vinaceous-red fibrils on the cap and stalk when young (see photo) is common on wood chips. Its cap often shows bluish or blue-green stains and it is hallucinogenic. *G. punctifolius* is colored like *G. aeruginosus*, but has a nearly smooth (bald) cap and no veil.

Gymnopilus luteofolius

CAP 2-12 cm broad, convex or obtuse becoming nearly plane; surface dry, at first covered with dense, dark red to purple-red or reddish-brown, fibrillose scales, these fading slowly to pinkish-red or yellowish-red; surface finally yellowish in old age as scales disperse; margin inrolled at first. Flesh thick, reddish to lavender, then fading to yellowish; taste bitter. **GILLS** notched to adnate or slightly decurrent, fairly close, yellow becoming bright rusty-orange or rust-colored as spores mature. **STALK** 3-10 cm long, 0.3-2 cm thick, fleshy, equal or enlarged below (or tapered downward if clustered); solid, dry, fibrillose, more or less colored like cap, becoming yellowish or rusty-stained. **VEIL** fibrillose to somewhat membranous, yellowish; forming a hairy, superior ring on stalk which may disappear or trap falling spores. **SPORE PRINT** bright rusty-orange; spores 5.5-8.5 × 3.5-4.5 microns, elliptical, roughened.

HABITAT: In groups or clusters on decaying coniferous wood, sawdust, and humus rich in lignin (rarely on hardwoods); widespread, but not common. I have seen one fantastic local fruiting, under pine at New Brighton Beach State Park, with *Pluteus cervinus* and large clusters of *Pholiota terrestris* and *Naematoloma fasciculare* (all certified wood-lovers) also present.

EDIBILITY: Unknown, but the closely related *G. aeruginosus* is said to be hallucinogenic.

COMMENTS: This distinctive mushroom can be told by the dark red fibrils or scales on the cap (and often the stem), plus the yellow gills, rusty-orange spores, and relatively persistent veil. The much more common *Tricholomopsis rutilans* features the same attractive color combination, but has white spores and lacks a veil, while *G. aeruginosus* and relatives have a transient veil and often have bluish stains. *G. pulchrifolius* is a smaller species with a pink to pale pink cap. Others with distinctively colored cap scales include: *G. fulvosquamulosus,* with tawny to brown scales on a yellow background; and *G. parvisquamulosus,* with dark brown to blackish-brown scales on an ochraceous-tawny background. Both have a mild taste and fibrillose to slightly membranous veil. I have found them in our area, but they are rare.

Gymnopilus spectabilis group   Color Plate 99
(Big Laughing Mushroom; Giant Gymnopilus)

CAP 5-40 cm or more broad, convex becoming broadly convex or nearly plane; surface dry, smooth to silky or fibrillose, often breaking up to form small scales; bright yellow-orange to yellowish-buff when young (or at times nearly whitish from overlaid fibrils), often somewhat darker in age (rusty-orange to golden-tawny to orange-brown or reddish-brown); margin at first incurved, sometimes wavy, sometimes with veil remnants. Flesh thick, firm, yellowish; taste bitter. **GILLS** notched to adnate or slightly decurrent, close, ochre-buff or pale yellow becoming more or less rusty-orange to rusty-brown in age. **STALK** (3) 5-25 cm long, 1-6 (10) cm thick, usually swollen in the middle or below, the base often narrowed; solid, firm, dry, rusty-orange to rusty-yellow or paler, fibrillose below the ring. **VEIL** pallid to pale yellowish or rusty-stained, membranous or fibrillose, sometimes disappearing, but usually forming a superior ring on stalk; ring soon stained with rusty-orange spores, often collapsing or disappearing in age. **SPORE PRINT** bright rusty-orange; spores 7-10.5 × 4.5-6 microns, elliptical, roughened or wrinkled.

HABITAT: Usually in clusters (occasionally solitary) on or around stumps and trees; widely distributed. This species "complex" favors conifers on the west coast, hardwoods in eastern North America. In our area it is quite common on old pine stumps in the fall, winter, and early spring, and occasionally turns up on eucalyptus also.

EDIBILITY: Inedible due to the bitter taste. Forms in Asia and eastern North America apparently contain psilocybin and/ or psilocin and are hallucinogenic (hence the Japanese name, "Big Laughing Mushroom"). On the west coast, however, it is apparently "inactive." An Ohio woman had an unforgettable experience after inadvertently nibbling on one. She found herself in an alien world of fantastic shapes and glorious colors, and while concerned friends were rushing her to the hospital, she was heard to mutter, "If this is the way you die from mushroom poisoning, then I'm all for it . . ."

COMMENTS: This species "complex" is easily recognized by its overall yellow-orange to rusty-orange color, rusty-orange spores, robust size, bitter taste, and frequent presence of a ring formed by the veil. Several *Pholiota* species are similar, but have duller spores and viscid caps; the honey mushroom *(Armillariella mellea)* is also somewhat similar, but has white spores and whitish or flesh-colored gills, while the jack-o-lantern mushrooms

Gymnopilus spectabilis group is easily told by its yellow to rusty-orange color (see color plate), membranous veil, and rusty-orange spores. Note how small the penny is! This large western form is considered to be a separate species, *G. ventricosus,* by some authorities.

(Omphalotus) have white or yellow-tinted spores, lack a veil, and grow only on hardwoods. "Typical" *G. spectabilis* grows on both hardwoods and conifers and has a somewhat fibrillose veil. It was originally called ***Pholiota spectabilis*** and is also known as *G. junonius* (probably its "correct" name). The large, non-hallucinogenic, conifer-loving western form with a swollen stalk and membranous veil is called *G. **ventricosus*** (if truly distinct). It is one of our largest and most spectacular mushrooms, commonly attaining pizza size (more than one foot in diameter) with clusters weighing 10 pounds or more. Young specimens present an entirely different appearance: squat and compact with hard, very thick, yellow flesh and very narrow (shallow) gills. Other species in the *G. spectabilis* "complex" include: *G. **subspectabilis,*** with larger spores; and *G. **validipes*** of eastern North America, a hallucinogenic species with a mild to only slightly unpleasant taste and a fibrillose veil.

ROZITES & PHAEOLEPIOTA

Medium-sized to large, *terrestrial,* woodland mushrooms. CAP *dry;* often wrinkled *(Rozites)* or *powdery-granulose (Phaeolepiota).* GILLS attached, tawny to orange-buff to brownish at maturity. STALK central, *fleshy, smooth to fibrillose (Rozites)* or *powdery-granulose (Phaeolepiota).* VEIL *present, membranous, usually forming a distinct annulus (ring) on stalk.* SPORE PRINT *rusty-brown to tawny or orange-buff.* Spores elliptical, roughened to nearly smooth.

THOUGH not closely related, these two small genera are treated together here because they have a number of features in common: both grow on the ground, both have yellow-brown to rusty-brown spores and a well-developed membranous veil that forms a distinct annulus on the stalk, and both were originally placed in the genus *Pholiota.*

    Rozites has somewhat wrinkled or roughened spores and has been aptly characterized as "a *Cortinarius* with a membranous veil." *Phaeolepiota,* on the other hand, has a granulose or powdery cap and stem and has consequently been called "a brown-spored *Cystoderma*." (Some mycologists go so far as to place it alongside *Cystoderma* in the Agaricaceae; others retain it in *Pholiota* of the Strophariaceae.)

    Each genus includes a single well-known species. Both are edible and quite good, but some people are apparently "allergic" to *Phaeolepiota.*

411

Key to Rozites & Phaeolepiota

1. Cap *and* stalk covered with a granulose or powdery layer (but the granules sometimes wearing or washing off, especially those on the cap); stalk 1.5 cm thick or more; found mainly in the Pacific Northwest .... *Phaeolepiota aurea,* below
1. Cap and stalk not powdery or granulose 2

2. Gills purplish when young or some part of fruiting body purple-tinged (see *Cortinarius,* p. 417)
2. Not as above; purple or violet shades absent 3

3. Cap distinctly viscid or slimy when moist 4
3. Cap not viscid (or only very slightly so and soon drying out) 5

4. Gills covered by a cobwebby partial veil when young; annulus (ring) on stalk formed by the outer (universal) veil; spore print rusty- to cinnamon-brown .... (see *Cortinarius,* p. 417)
4. Not as above .... (see *Pholiota,* p. 384)

5. Cap warm tan to yellow-brown or orange-brown, often wrinkled radially (especially in age) and sometimes with a thin silky or hoary bloom when young; spore print rusty-brown; found in forests .... *Rozites caperata,* below
5. Not as above; cap dark brown to olive-brown to tan, creamy, etc.; spore print dull to dark brown but not rusty-brown; found in forests as well as suburban habitats .. (see *Agrocybe,* p. 467)

Rozites caperata   (Gypsy Mushroom)   Color Plate 101

CAP 5-15 cm broad, oval becoming somewhat bell-shaped to broadly convex, plane, or obscurely umbonate; surface dry, usually distinctly wrinkled or corrugated radially, at first covered with a thin white to grayish coating of silky fibrils (especially at center); warm tan to yellow-brown or orange-brown, margin often paler. Flesh thick, white, firm. **GILLS** adnate to adnexed or notched, close, at first pallid, soon dull tawny or brown, sometimes transversely banded with darker and lighter zones. **STALK** 5-13 cm long, 1-2.5 cm thick, equal or slightly enlarged at base, solid, firm, white to pale tan or pale ochre; apex often striate or scurfy, base sometimes with an obscure volvalike zone. **VEIL** white, membranous, forming a more or less median ring on stalk. **SPORE PRINT** rusty-brown; spores 11-15 × 7-10 microns, elliptical, roughened or warty. Some cystidia present on gill edges.

HABITAT: Scattered or in groups on ground in woods; widely distributed in northern regions. It favors mossy, old-growth coniferous forests but also grows under hardwoods, especially when huckleberry is in the vicinity. I have seen large fruitings in Washington and Idaho in the late summer and fall. It also occurs in northern California and the Sierra Nevada, but I have yet to find it in our area.

EDIBILITY: Edible, and in my humble fungal opinion, the best of the Cortinariaceae. It is especially good with rice after a long, hard day of backpacking. The tough stems should be discarded.

COMMENTS: Originally called *Pholiota caperata,* the gypsy mushroom is easily told by its warm brown or yellowish, wrinkled cap that has a hoary sheen when young, plus the membranous veil which forms an annulus (ring) and the rusty-brown spores. *Agrocybe praecox* is somewhat similar, but does not have a wrinkled cap and usually grows in cultivated ground; *Phaeolepiota aurea* differs in its powdery-granulose cap and stem, while similarly-colored *Cortinarius* species do not have a membranous veil.

Phaeolepiota aurea

CAP 6-20 (30) cm broad, obtuse to convex becoming broadly convex, plane, or broadly umbonate; surface dry, granular to somewhat powdery (the granules sometimes wearing away in age), orange to orange-tan, tawny-yellow, or golden-brown, often somewhat paler in age; margin usually hung with veil remnants. Flesh thick, pallid or yellowish. **GILLS** adnate to notched or free, close, pallid or pale yellowish becoming tawny to orange-brown. **STALK** 5-15 (25) cm long, (1) 2-4 (6) cm thick, thicker toward base, orange to buff or colored like cap and granulose or powdery below the ring. **VEIL** membranous, colored like cap, sheathing the stalk and breaking to form a superior flaring or funnel-like ring which eventually collapses or becomes skirtlike; ring smooth on upper surface and granulose on underside. **SPORE PRINT** pale yellow-brown to orange-buff; spores 10-14 × 5-6 microns, elliptical, smooth to minutely roughened.

HABITAT: In groups or clusters in rich humus and soil under both hardwoods and conifers; known from the Pacific Northwest and Alaska (also Europe), fruiting in the late summer and fall. It is rather rare, but when it fruits it often does so in large quantities. An ideal place to look for it is under alder along roads and trails.

EDIBILITY: Edible for most people, but mildly poisonous to some. It is a tempting specimen, but try it cautiously if you must.

COMMENTS: Also known as *Pholiota aurea* and *Togaria aurea,* this large, beautiful mushroom is as distinctive as it is rare. No other large, brownish-spored mushroom is golden-brown to pale orange with a granulose coating on both the cap and stem. *Cystoderma* species are granulose but much smaller and have white spores; *Agaricus augustus* is similarly colored but has chocolate-brown spores and an almondy odor, while *Pholiota* species, *Rozites caperata,* and *Agrocybe praecox* lack the granulose coating.

PHAEOCOLLYBIA

Medium-sized, terrestrial, woodland mushrooms. CAP *usually viscid or slimy when moist.* GILLS adnexed to free, *usually rusty-brown or cinnamon at maturity.* STALK *with a long, tapered, deeply rooting base or "tap root."* VEIL and VOLVA absent. SPORE PRINT *cinnamon- to rusty brown.* Spores elliptical, roughened, without an apical germ pore. Cystidia present, at least on gill edges.

THE presence of a "tap root" and absence of a veil distinguish this small, well-marked genus from its closest relative, *Cortinarius.* The "tap root" (**pseudorhiza**) is merely an extension of the stem that roots deeply in the humus. It gradually tapers to a point, becoming so thin and fragile that it can't be traced to its origin (probably tree roots). Most other agarics with "tap roots" (e.g., *Caulorhiza* and *Oudemansiella*) have white spores.

Phaeocollybias are quite rare in most regions (including ours), but where they do occur they often fruit in large numbers. The Pacific Northwest is an exception, for it is there that the genus is most common and diverse. Many types seem to occur only with Sitka spruce, while several favor western hemlock, redwood, or Douglas-fir; few, if any, occur with hardwoods. About two dozen species are known from North America, many of them restricted to the Pacific Northwest. Little is known of their edibility. Three representatives are described here and several others are keyed out.

Key to Phaeocollybia

1. Cap dark green to olive, at least when fresh and moist (but may fade as it loses moisture) .. 2
1. Cap never greenish or olive 4
2. Gills violet or lilac when young .... *P. fallax* (see *P. olivacea,* p. 414)
2. Gills whitish to pale brown when young, or at least not violet 3
3. Stalk typically slender (less than 1 cm thick at apex); cap 1-6 cm broad ....
.... *P. festiva & P. pseudofestiva* (see *P. olivacea,* p. 414)
3. Stalk typically about 1 cm thick (or more) at apex; cap 3-11 cm broad .... *P. olivacea,* p. 414

4.   Gills violet or lilac when young .... *P. lilacifolia* (see *P. californica*, p. 415)
4.   Gills pallid to yellowish, tan, brownish, or cinnamon when young 5
5.   Cap typically with grayish tones (grayish-brown to grayish-buff) at maturity or when faded 6
5.   Not as above; cap dark brown to amber-brown, cinnamon, reddish, or orangish 7
6.   Stalk typically 8 mm thick or more at apex .... *P. gregaria* (see *P. olivacea*, p. 414)
6.   Stalk typically 3-7 mm thick at apex .... *P. scatesiae* (see *P. olivacea*, p. 414)
7.   Stalk slender (up to 7 mm thick at apex) and cap rather small (up to 5 cm broad) 8
7.   Stalk thicker than above (usually at least 5 mm at apex) and cap 3-7 cm broad or larger .. 10
8.   Stalk 1.5-3 mm thick at apex; cap not viscid or only slightly so ....
     .... *P. similis* (see *P. californica*, p. 415)
8.   Not as above; stalk usually at least 3 mm thick at apex, or if thinner than cap viscid or slimy 9
9.   Cap amber-brown to yellowish to salmon or salmon-ochre; spores at least 8 microns long ...
     .... *P. attenuata* & *P. laterarius* (see *P. californica*, p. 415)
9.   Cap cinnamon-brown to reddish or orangish   *P. radicata* & others (see *P. californica*, p. 415)
10. Cap (4) 6-25 cm broad, cinnamon to dark reddish-brown to liver-colored, dark brown, or
     umber; surface viscid or slimy; stalk 1-4 cm thick at apex 11
10. Not as above; either smaller, slimmer, or oranger than above, or cap not viscid 12
11. Cap dark brown to umber or sometimes dark pinkish-brown; cap conical when young ......
     .... *P. spadicea* (see *P. kauffmanii*, p. 416)
11. Cap usually redder or more liver-colored than above and/ or not conical when young ....
     .... *P. kauffmanii* & others, p. 416
12. Cap slimy or viscid when moist .... *P. californica* & others, p. 415
12. Cap not viscid .... *P. deceptiva* (see *P. californica*, p. 415)

Phaeocollybia olivacea   (Olive Phaeocollybia)

CAP 3-11 cm broad, obtusely conical or convex becoming plane or uplifted, but usually retaining an umbo; surface smooth, viscid or slimy when moist, hygrophanous: deep green to olive-green when moist, fading as it dries (to olive-buff, etc.). Flesh thin, olive; odor cucumberlike or radishlike when fresh. **GILLS** pallid becoming pale brown, then rusty-brown; adnexed or free, close, edges often eroded or wavy. **STALK** 10-22 cm long or more, (0.5) 1-2.5 cm thick at apex, equal above or swollen at ground level, then tapering below the ground to form a "tap root"; cartilaginous, smooth, stuffed with a pith; watery olive to yellowish above, rusty-orange to reddish-brown below or occasionally overall. **SPORE PRINT** rusty-brown; spores 8-11 × 5-6 microns, elliptical with snoutlike apex, roughened.

HABITAT: Scattered to densely gregarious, often in large rings, on ground in mixed woods and under conifers; fruiting in the fall and winter, known only from the west coast. It is said to be common in southern Oregon and northern California, but is rare in our area. Budding boletivore Craig Mitchell has located one very prolific fairy ring near Big Basin State Park.

EDIBILITY: Unknown.

COMMENTS: This striking mushroom is easily recognized by its slimy olive-green cap, rather thick rooting stem, rusty-brown spores, and absence of a veil. It is somewhat unusual among the Phaeocollybias in that it often fruits in mixed woods (i.e., it may or may not be associated with conifers). There are three other Phaeocollybias with an olive cap when fresh: *P. festiva, P. pseudofestiva,* and *P. fallax.* All are considerably slimmer, smaller, and more conical (cap 1-6 cm broad, stalk less than 1 cm thick at apex) than *P. olivacea,* and all have slightly smaller spores. *P. fallax* is easily told by its violet or lilac gills when young, but the other two can only be differentiated microscopically. Also worth mentioning are two species which have grayish-toned caps: *P. scatesiae,* with a slim (3-7 mm) stalk, dingy yellowish gills when young, and a dull cinnamon cap that fades

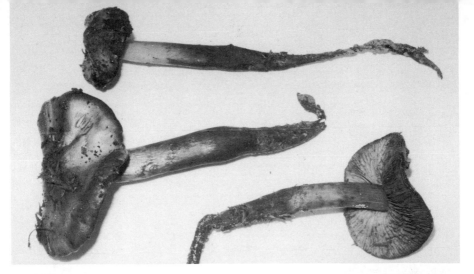

Phaeocollybia olivacea. Greenish to olive-gray cap, rusty-brown spores, and "tap root" are distinctive.

to grayish-brown, often fruiting in massive clusters of up to several hundred fruiting bodies; and *P. gregaria,* with a thicker (8-15 mm) stalk and a yellowish-gray to brownish-gray or grayish-buff cap. All of the above species occur with northern conifers and are very rare or absent in our area.

Phaeocollybia californica   Color Plate 100

CAP (2) 3-7 (10) cm broad, conical with an inrolled margin when young, often expanding in age to plane but retaining a distinct umbo; surface smooth, viscid when moist, hygrophanous: amber-brown to cinnamon-brown or orange-brown when moist, fading as it dries, but in age often rusty-spotted or darkening to reddish-brown. Flesh thin; odor pungent (somewhat radishlike). **GILLS** adnate to adnexed or even free, pallid or tinged cap color, darkening to brown or rusty-brown in age; close. **STALK** 10-20 cm long or more, 4-10 mm thick, equal above, gradually tapering below the ground to form a "tap root"; cartilaginous, hollow, smooth, colored more or less like cap or paler, or more often dark reddish or reddish-brown below and paler (apricot-buff) above. **SPORE PRINT** rusty-brown; spores 8-11 × 5-6 microns, elliptical with a small "beak" at apex, roughened.

HABITAT: Densely gregarious or in large loose clusters on ground under conifers, often forming arcs or fairy rings; known only from the west coast, not common. In our area it is quite rare, but one large fairy ring under redwood and Douglas-fir fruits in colossal quantities every winter.

EDIBILITY: Unknown.

COMMENTS: The presence of a "tap root," absence of a veil, rusty-brown spores, and amber-brown to orangish or reddish-brown color are the distinctive features of this species and its look-alikes (see below). *Caulorhiza umbonata* also features a "tap root," but is much more common and has pallid gills, a non-viscid cap, and white spores. Other dark cinnamon to orangish Phaeocollybias include: *P. piceae,* with a bitter taste and more or less mild odor, favoring Sitka spruce; *P. dissiliens,* also fond of Sitka spruce, but with distinctly smaller spores and a thicker stalk that easily splits lengthwise; *P. lilacifolia,* with violet gills when young; *P. attenuata,* with a thinner stem (3-6 mm thick at apex) and a viscid, amber-brown to yellowish-buff cap; *P. laterarius,* also slender but with a salmon to pale salmon-ochre cap when moist (known from Michigan); *P. similis,* with a *very* thin stalk (1.5-3 mm) and a non-viscid or only slightly viscid cap; and *P. deceptiva,* with a cinnamon to yellowish-brown, non-viscid cap and a thicker (8-15 mm) stalk that does

Phaeocollybia californica. Note conical cap, "tap root," and absence of veil. Also see color plate.

not markedly redden with age. Finally, there is a group of species (e.g., *P. radicata, P. jennyae, P. christianae, P. sipei*) with a slender (3-6 mm) stem, small cap, and smaller spores (less than 8 microns long) that are best differentiated from each other microscopically. Nearly all of the above occur in the Pacific Northwest; a few are more widely distributed or have counterparts (e.g., *P. rufipes*) in eastern North America. None are as large as *P. kauffmanii* and none show the olive or grayish tones of *P. olivacea* and the species listed under it.

Phaeocollybia kauffmanii   (Giant Phaeocollybia)

CAP 7-25 cm broad, obtuse or convex when young becoming broadly umbonate to nearly plane in age; surface smooth, viscid to very slimy; cinnamon to pale reddish-cinnamon or reddish-brown becoming more or less liver-colored in age, but fading to reddish-orange, rusty-reddish, or apricot as it loses moisture; margin remaining inrolled for a long time, not striate. Flesh thick, firm, colored like cap or paler; odor usually farinaceous, at least when crushed. **GILLS** close or crowded, free or adnexed, buff becoming brown to rusty-brown in age. **STALK** (15) 20-40 cm long, 1.5-4 cm thick at apex, tapered downward to form a long "tap root" below the ground; pinkish-brown to pinkish-buff above, reddish-brown to dark purplish-brown below and darkening with age (sometimes nearly black); smooth; stuffed with a pith, the outer rind tough and cartilaginous. **SPORE PRINT** cinnamon-brown; spores 8-11 × 4.5-7 microns, elliptical with an apical beak, roughened.

HABITAT: Solitary, scattered, or in groups under conifers (particularly Sitka spruce) in the late summer and fall; known only from the west coast. It is fairly common during some seasons, particularly in Oregon and northern California, rare during others. I have not seen it in our area.

EDIBILITY: Unknown. Its large size is tempting but the glutinous cap is not.

COMMENTS: This giant of the genus can usually be recognized by its size alone. The colors are quite variable depending on age and moisture, but are usually darker and redder than those of *P. californica* and the species discussed under it. No other *Phaeocollybia* has a stalk so consistently thick nor a cap so broad. The obtuse rather than conical shape of young caps is also distinctive. Other species: *P. oregonensis* is a very similar but slightly smaller species with smaller spores; *P. spadicea* is a fairly large species (cap 4-12 cm broad, stalk 1-2 cm thick) with a dark brown to umber cap when moist. Both occur in the Pacific Northwest.

416

CORTINARIUS

Woodland fungi nearly always growing on the ground. CAP viscid or dry, smooth to fibrillose or scaly. GILLS typically attached but not often decurrent; variously and often brightly colored when young but *usually rusty-brown to cinnamon-brown in age.* STALK thick or thin but usually fleshy, central, the apex typically *not* powdery or dandruffy. VEIL *present as a cobwebby or silky cortina which often leaves hairs on stalk;* fibrillose or slimy universal veil often present also. VOLVA typically absent (but rimmed bulb sometimes present). SPORE PRINT *rusty-brown to cinnamon-brown* or at times ochraceous-tawny. Spores round to elliptical, roughened, lacking an apical pore. Cystidia usually absent on faces (sides) of gills but sometimes present on the edges. Cap cuticle typically filamentous.

CORTINARIUS is the largest genus of gilled mushrooms, with an estimated 1000 species—many of them still unclassified. Cortinarii can be dry or slick, thin or thick, large or small, chunky or tall, but as a group they are easy to recognize, for there are three fungamental features that unite all 1000+ species: rusty-brown to cinnamon-brown spores, the presence of a cortina, and a terrestrial woodland growth habit.

The **cortina,** which is the hallmark of *Cortinarius,* is an exquisitely constructed veil of silky or cobwebby fibers. It shrouds the gills of the young mushroom **(COLOR PLATE 105),** but collapses as the cap opens, often leaving hairs on the stem which are subsequently stained rusty-brown by spores **(COLOR PLATE 103).** In some species the cortina may completely disappear, but the combination of rusty-brown spores (and often, rusty-brown mature gills) and growth on the ground is a fairly infallible indicator of *Cortinarius.* In many species the cortina is augmented by a universal veil that leaves scaly belts, fibrils, and/or a sheath of slime on the stem (and sometimes the cap).

Among the brown-spored mushrooms, *Cortinarius* is most likely to be confused with *Inocybe* and *Hebeloma,* which are terrestrial and may have a cortina, but which have duller brown spores and often have whitish-edged gills and a powdery or dandruffy stalk apex; with *Phaeocollybia,* which has rusty-brown spores but lacks a cortina; and with *Gymnopilus* and *Pholiota,* both of which are primarily wood-inhabiting. (Cortinarii can occasionally be found on wood in an advanced stage of decay, but you will never find them on the mossy crotch of a live oak or the cut end of a felled fir.)

Within these parameters—presence of a cortina and often a universal veil, rusty-brown to cinnamon-brown or "ochraceous-tawny" (whatever that may mean) spores, and growth on the ground near or under trees—Cortinarii vary tremendously in shape, size, color, and habit. Many are Boring Ubiquitous Mushrooms ("BUM's"), but a sizable number are big, bold, and blatantly beautiful. In fact, the Cortinarii, along with the waxy caps (Hygrophoraceae), are the most colorful of all the agarics: blues, violets, yellows, oranges, browns, and olives predominate, but reds, bright greens, and whites are not rare. To make any headway in identification it is *essential* to know the color of *fresh* individuals—particularly the color of the *immature* gills, since it changes as soon as the spores begin to form. As Cortinarii grow older, the bright colors and individuality of their youth are drowned in brownness, until the enormity of their uniformity makes identification of species hopeless (on the other hand, the sameness of their mundane-ness helps to signify that they are Cortinarii).*

You should also note whether the cap and stalk are viscid *when they are fresh and moist.* Scent is another important variable—many species have indifferent odors, but some are enticingly aromatic (e.g., *C. percomis*), others are downright disgusting (e.g., *C. camphoratus*), while still others are aromatic *and* disgusting (e.g., *C. subfoetidus*).

Even if you take meticulous notes on *fresh* specimens in *all* stages of development, you will *still* find Cortinarii difficult to identify. Perhaps this is why professional mycologists

---

*Note: These lines were composed at 3:00 A.M. under the influence of severe boredom.

seldom bother to name the multitudes of Cortinarii they encounter, and why Elias Fries, the "father" of mushroom taxonomy, said in 1838: "No genus is more natural nor more sharply distinguished from others . . . but the species are so intimately related among themselves that to distinguish the separate ones is almost to be despaired of." *Cortinarius* species deserve to be better known, however, so don't let your inability to *name* them prevent you from learning to *recognize* them. Naming them, after all, is not an end in itself, but a means of expediting the recognition process. If you really *must* have labels, you can give them descriptive nicknames of your own.

The labyrinthine process of identifying a *Cortinarius* can be shortened to some extent by learning to place each species in its proper subgenus. Five (or sometimes seven) large subgenera have traditionally been recognized, and even raised to the rank of genus by some mycologists:

Subgenus **Myxacium:** both cap *and* stalk viscid or slimy.

Subgenus **Bulbopodium:** only the cap viscid; stalk with an abrupt, often rimmed basal bulb (see photograph on p. 439 and Color Plate 105).

Subgenus **Phlegmacium:** only the cap viscid; stalk equal to club-shaped or swollen, but lacking a rimmed bulb.

Subgenus **Telamonia:** neither the cap nor the stalk viscid; cap hygrophanous and smooth or occasionally fibrillose.

Subgenus **Cortinarius** (now divided into four subgenera—see below): neither the cap nor the stalk viscid; cap often fibrillose or scaly but not typically hygrophanous.

The three subgenera with viscid caps are the easiest to recognize. Myxacium is especially distinctive by virtue of the slimy universal veil that coats the cap and stalk. Bulbopodium and Phlegmacium feature many large and/or colorful species but intergrade somewhat (the bulb can be *somewhat* abrupt or *slightly* rimmed), causing some investigators to recognize only the latter.

The hygrophanous-capped subgenus Telamonia is the largest and most difficult group of Cortinarii, with at least 400 species—most of them brownish or cinnamon-brown and many of them small and nondescript (in other words, typical "LBM's" in the grand, bland tradition of *Inocybe, Galerina,* and *Tubaria*).

Members of the dry-capped subgenus Cortinarius are now divided up by *Cortinarius*-categorizers into *four* subgenera, based largely on the types of pigments they contain:

Subgenus **Cortinarius** (in its most restricted sense): contains a single striking *deep* violet, fibrillose-scaly species *(C. violaceus).*

Subgenus **Sericeocybe:** has violet to pale violet or lilac-white colors.

Subgenus (or genus) **Dermocybe:** composed mainly of small, slender-stemmed, brightly colored, conifer-loving species with yellow, olive, orange, or red pigments called anthraquinones (which are water-soluble and hence make wonderful dyeing agents).

Subgenus **Leprocybe:** a large and diverse group whose color ranges from olive to yellow, yellow-brown, rusty-orange, or brown, but whose pigments differ from those found in Dermocybe.

Cortinarius is one of the most prominent features of our woodland fungal flora, but its fruiting behavior is notoriously erratic. Sometimes there is one sudden spectacular, *Russula*-like eruption of Cortinarii; at other times they fruit in a rather uninspired, desultory fashion throughout the rainy season. A few species, such as *C. glaucopus,* are common every year, but many types will be exceedingly abundant one season, then rare or absent for the next two or ten (or even twenty!) years.

Most if not all Cortinarii are mycorrhizal. They attain their maximum numbers and diversity in the cool coniferous forests of the north temperate zone, but are also common under hardwoods (particularly "Bulbopodiums" and "Phlegmaciums"). In our area each

Cortinarius crocolitus (see comments under *C. collinitus*, p. 431) is one of hundreds of Cortinarii. Its stalk is marked by yellow-brown or ochre belts formed by the universal veil.

forest type offers its own characteristic clique of Cortinarii: e.g., *C. infractus, C. cotoneus,* and *C. collinitus* with tanoak; *C. glaucopus, C. regalis,* and *C. sodagnitus* with live oak; *C. cinnamomeus, C. phoeniceus* var. *occidentalis,* and *C. obtusus* with pine; and *C. balteatus, C. cylindripes,* and *C. vibratilis* with ericaceous trees and shrubs (huckleberry, manzanita, and madrone). It is in northern latitudes or higher altitudes (i.e., with *northern* conifers, especially spruce), however, that the Cortinarii come into their own, and their astonishing diversity becomes overwhelming.

Some of the older popular mushroom manuals list *Cortinarius* as a "safe" (albeit mediocre*) genus—*but nothing could be farther from the truth. Cortinarius,* in fact, contains *several deadly poisonous species* (particularly in the subgenus Leprocybe—see *C. gentilis!*) whose effects on the human body are greatly delayed (for up to three weeks!). Therefore it is best not to eat ANY *Cortinarius,* especially in view of the large number of poorly known or difficult-to-identify species. (Even some of those proclaimed to be edible may not be safe, since there is no assurance that the eater identified them correctly.) *Cortinarius* is a tempting *potential* food source, however, so if you are adamant about experimenting, then stick to the ones with *viscid* caps (none of which, *as yet,* are known to be deadly poisonous), and wait *at least two weeks*—rather than the usual two days—after your first "test" before indulging in more.

Because Cortinarii are so prominent and colorful, the selection of species in this book is fairly sizable, at least in comparison to other giant—but more mundane—genera such as *Inocybe* and *Psathyrella.* Over 130 species of *Cortinarius* are mentioned (not all of them native to California) and 34 are fully described. Remember, however, that these 130+ species represent only a fraction of the total number that occur. The fieldmarks listed are often insufficient for positive identification, and many of the descriptions are meant to serve as models or "archetypes" for a large number of species (e.g., brown "Telamonias" with a club-shaped stalk and medium-sized cap are typified by *C. laniger*). Thus most of the names in this chapter should be seen as *suggestions*—rather than statements—as to what your mysterious *Cortinarius* might be. Those wishing to verify their identifications should consult a more definitive source, such as Alexander Smith's keys to North American Cortinarii (which are not available to the public), or Daniel Stuntz's key to species of the Pacific Northwest and Meinhard Moser's keys to those of Europe, which are (see Suggested Readings and Primary References).

*Even Captain Charles McIlvaine, that intrepid turn-of-the-century toadstool-tester who sung the praises of such flaccid, featureless, fleshless fungi as *Coprinus ephemerus* ("choice as a flavoring") and *Psathyrella candolleana* ("tender: one of the best"), was non-plussed by the Cortinarii, for on more than one occasion he characterized their flavor as "more or less that of rotten wood." However, one colleague of mine has a dissenting interpretation of McIlvaine's comparison: he reasons that McIlvaine's tastes (which included a fondness for stinkhorns) were so bizarre and eccentric that "rotten wood" was undoubtedly high up on his list of "best edibles," and consequently, so were the Cortinarii!

Key to Cortinarius

1. Fruiting body typically developing underground (but cap may surface in wet weather), the veil membranous or somewhat membranous and persistently covering the gills (i.e., either remaining intact through maturity or shredding radially but not rupturing); found mostly under mountain conifers 2

1. Not as above; if developing underground then veil not persistent and membranous 4

2. Cap purplish or lavender-tinged .... *C. velatus* (see *C. magnivelatus,* p. 442)

2. Cap white to yellow, ochre, or yellow-brown (not purplish) 3

3. Cap whitish when fresh; veil tough, often not rupturing .... *C. magnivelatus,* p. 442

3. Cap yellow to yellow-brown or tan (or rusty-stained); veil not as tough or thick as above, often shredding radially .... *C. verrucisporus* & others (see *C. magnivelatus,* p. 442)

4. Cap viscid or slimy when moist (it may dry out, but look for adhering debris) 5

4. Cap not normally viscid (in wet weather it may be slightly viscid or tacky but never slimy)   61

5. Cap radially corrugated or coarsely wrinkled, yellow-brown to ochre, tawny, or rusty-brown; gills purplish at first but soon brown to rusty-cinnamon; stalk with a basal bulb when young, but bulb often obscure in age; found under hardwoods in eastern North America, often in large numbers .... *C. corrugatus*

5. Not as above 6

6. Fruiting body massive; stalk 3-7 cm thick and lacking a basal bulb; flesh also very thick; cap brownish to russet at the center (often with small flattened scales) and yellowish at margin (lacking pronounced violet tones) .... *C. ponderosus* & others, p. 432

6. Not as above; either smaller or with violet on cap or stalk with a basal bulb 7

7. Stalk (at least the lower part) viscid or slimy when moist, sometimes enlarged below but typically *without* an abrupt, rimmed basal bulb (in very wet or very dry weather the viscidity may not be evident) 8

7. Stalk typically *not* viscid (but if cap is slimy, some of the slime may drip off onto stalk base); stalk with or without an abrupt, rimmed basal bulb 21

8. Fruiting body small (cap 2-5 cm broad) and gray, the cap and young gills with a slight lilac tinge that often disappears in age; stalk whitish with a bluish-tinged apex when young; found under conifers; rather rare .... *C. sterilis*

8. Not as above 9

9. Fruiting body purple, violet, or bluish in some part, at least when young and fresh 10

9. Violet or bluish shades completely absent 17

10. Stalk typically club-shaped (thicker at base) 11

10. Stalk equal, tapered downward, or spindle-shaped (swollen slightly in the middle and tapered below) 13

11. Cap violet or lavender well into maturity   *C. iodes* & others (see *C. cylindripes* group, p. 430)

11. Not as above 12

12. Center of cap rusty or tawny becoming chestnut-brown in age, the margin paler; gills violet-tinged when very young .... *C. castaneicolor* (see *C. cylindripes* group, p. 430)

12. Not as above .... *C. griseoluridus* & others (see *C. cylindripes* group, p. 430)

13. Lower portion of stalk with conspicuous transverse or concentric bands, plaques, or scaly belts formed by the universal veil; cap yellow-brown to rusty-orange or rusty-brown at maturity (margin may be bluish or violet when young) .... *C. collinitus* group, p. 431

13. Not as above (stalk occasionally with some scaly belts, but if so then cap darker or differently colored) 14

14. Cap purple or lilac, at least when young .... *C. cylindripes* group & others, p. 430

14. Not as above 15

15. Cap blackish to deep chestnut-brown to dark olive-brown, often fading to cinnamon-brown or yellowish-brown in age, the margin often becoming radially wrinkled or grooved; stalk long, usually rooting deeply in soil; common under northern conifers 16

15. Not as above; cap differently colored, at least when young (i.e., paler or brighter) ....
.... *C. pseudosalor* & *C. delibutus* (see *C. cylindripes* group, p. 430)

16. Cap blackish to deep chestnut-brown when young, fading to cinnamon in age; gills *not* violet
     or bluish when young .... *C. vanduzerensis,* p. 432
16. Cap colored as above or ranging to olive- or yellow-brown; gills violet or bluish-tinged when
     young .... *C. elatior* (see *C. collinitus* group, p. 431)
17. Stalk usually club-shaped or thickened below, at least when young; fruiting body sometimes
     rather small 18
17. Stalk more or less equal or tapered downward; fruiting body medium-sized or larger 20
18. Fruiting body medium-sized (cap usually 5 cm broad or more); stalk usually at least 1 cm thick
     at apex, typically with ochre-brown patches of veil tissue below; odor and taste not distinctive
     .... *C. pallidifolius* (see *C. collinitus* group, p. 431)
18. Not as above; fruiting body rather small (cap usually less than 5 cm broad); odor typically
     fragrant *or* taste of cap surface bitter 19
19. Odor fragrant; flesh, young gills, and stalk yellow .. *C. citrinifolius* (see *C. percomis,* p. 436)
19. Odor typically mild but taste very bitter (at least of the cap surface); flesh and stalk whitish
     .... *C. vibratilis,* p. 429
20. Lower portion of stalk typically with transverse or concentric bands, plaques, or scaly belts
     of universal veil tissue; found under both hardwoods and conifers *C. collinitus group,* p. 431
20. Not as above; associated principally with conifers .... *C. mucosus,* p. 429
21. Immature gills sooty olive to olive, greenish, or yellow-green, or soon becoming these colors 22
21. Immature gills differently colored (including yellow) 28
22. Stalk lacking an abrupt, rimmed basal bulb; cap sooty-olive when young (may be browner or
     tawnier in age), the skin often bitter-tasting; gills usually sooty-olive also ....
     .... *C. infractus* & others, p. 435
22. Not as above; stalk usually with an abrupt, often rimmed basal bulb, *at least when young* 23
23. Immature gills with violet faces *or* entire gills with a fleeting violet phase at first 24
23. Gills never violet 25
24. Gills at first with violet faces and olive edges *C. montanus* & others (see *C. cedretorum,* p. 439)
24. Gills violet-tinged when very young but soon entirely olive to greenish or greenish-yellow ...
     .... *C. scaurus group,* p. 440
25. Cap reddish to reddish-brown or orange-brown when fresh, at least at the center ....
     .... *C. orichalceus* & *C. rufo-olivaceus* (see *C. scaurus* group, p. 440)
25. Not as above 26
26. Cap olive; gills olive to olive-yellow; stalk yellowish or olive-tinged; found mainly under conifers
     in the Pacific Northwest .... *C. prasinus* (see *C. scaurus* group, p. 440)
26. Not as above; usually more brightly colored, at least in part; widespread 27
27. Gills green; stalk pale bluish when fresh; found in eastern North America under hardwoods
     .... *C. virentophyllus* (see *C. scaurus* group, p. 440)
27. Not as above; especially common in the West .... *C. scaurus group,* p. 440
28. Some part of fruiting body lilac, violet, purple, or bluish when fresh and young (check cap sur-
     face, stalk, *immature* gills, and flesh) 29
28. Fruiting body completely lacking violet, lilac, or bluish tones 48
29. Flesh (at least in base of stalk) staining wine-red when bruised 72
29. Not as above (but flesh may stain purple or dull lilac) 30
30. Gills *and/or* flesh staining purple or dull lilac when bruised (if already purple, then staining
     darker purple) .... *C. mutabilis group* & others, p. 437
30. Not as above 31
31. Immature gills yellow *or* purplish with yellow or olive edges 32
31. Not as above 33
32. Flesh partly violet or lilac and partly yellowish or whitish when fruiting body is sliced open
     lengthwise (see Color Plate 105); stalk apex not bluish ..... *C. cedretorum* & others, p. 439
32. Not as above; stalk apex often bluish; gills entirely olive or with olive edges ....
     .... *C. montanus* (see *C. cedretorum,* p. 439)
33. Whitish fibrillose or felty veil remnants usually present on cap (often in the form of a patch)
     .... *C. calyptratus* & *C. calyptrodermus* (see *C. regalis,* p. 443)
33. Not as above 34

34.   Stalk with an abrupt, rimmed basal bulb, *at least when young* 35
34.   Stalk merely equal or club-shaped, lacking a sharply defined, rimmed basal bulb 43

35.   Basal bulb with a whitish volva-like sheath; cap grayish to bluish-gray; growing mainly under
        spruce .... *C. volvatus* (see *C. regalis,* p. 443)
35.   Not as above 36

36.   Cap bright yellow to yellow-brown or ochre (the center sometimes oranger), even when young
        and fresh .... *C. calochrous* & others (see *C. fulmineus,* p. 441)
36.   Cap differently colored (but may develop ochre or yellowish shades in age) 37

37.   Cap dark reddish-brown to chestnut, becoming paler reddish-brown in age; gills pale lilac at
        first, darkening to reddish-brown; found mainly under conifers, especially in southern Oregon
        and northern California .... *C. subpurpureophyllus*
37.   Not as above 38

38.   Cap dark olive or greenish to steel-gray or brown, often streaked or mixed with ochre, blue, or
        gray, and often developing rusty to fulvous tones when older (especially toward center);
        immature gills gray to bluish-gray to bluish-violet (or occasionally violet) ....
        .... *C. glaucopus* group & others, p. 437
38.   Not as above; immature gills lilac, violet, or lavender, or if bluer then cap typically bluish to
        bluish-gray or at least differently colored from above 39

39.   Associated with conifers in the Pacific Northwest; fruiting body modest in size (stalk typically
        less than 1.5 cm thick at apex); gills lilac to pinkish-lilac when young . *C. olympianus,* p. 439
39.   Not as above; especially common under hardwoods 40

40.   Fruiting body (or at least the immature gills and stalk apex) with a bluish tinge when fresh
        .... *C. caesiocyaneus* (see *C. olympianus,* p. 439)
40.   Not as above (gills and stalk apex more violet or purple than blue) 41

41.   Mature fruiting body with a small but sharply defined basal bulb; taste usually bitter (at least
        of the cap cuticle); cap with violet or lilac tones at least at the margin; common under oak in
        California .... *C. sodagnitus* group, p. 438
41.   Not as above; found mainly under hardwoods in eastern North America 42

42.   Cap and basal bulb discoloring ochre or buffy-tan fairly quickly ....
        .... *C. velicopia* (see *C. olympianus,* p. 439)
42.   Not as above .... *C. michiganensis, C. aggregatus,* & others (see *C. olympianus,* p. 439)

43.   Cap usually lilac or lavender at least at the margin; stalk usually very thick (2-5 cm) *or* if not then
        odor usually sweetish (like overripe pears) 44
43.   Not as above 45

44.   Stalk very thick (2-5 cm); flesh thick and firm .... *C. balteatus* & others, p. 433
44.   Stalk typically 2 cm thick or less; odor usually sweetish .... *C. subfoetidus,* p. 434

45.   Stalk with pale tawny patches or fibrils, thinly viscid at first; cap tawny to chestnut-brown with
        a paler (buff-colored) margin .... *C. castaneicolor* (see *C. cylindripes* group, p. 430)
45.   Not as above 46

46.   Cap yellow to yellow-brown, tawny, or fulvous *C. varius* & others (see *C. multiformis,* p. 442)
46.   Cap differently colored 47

47.   Cap reddish-brown to liver-brown or dull purplish; gills usually violet or lilac when young ..
        .... *C. variicolor* & others (see *C. multiformis,* p. 442)
47.   Not as above; gills usually bluer or grayer .... *C. glaucopus* group, p. 437

48.   Odor distinctly fragrant and sweet (break open the flesh if there is any doubt) 49
48.   Not as above (but odor may be distinctive) 51

49.   Odor aniselike; stalk with a basal bulb, at least when young *C. odorifer* (see *C. percomis,* p.436)
49.   Not as above 50

50.   Stalk and flesh yellow or yellowish .... *C. percomis,* p. 436
50.   Stalk and flesh white or whitish .... *C. luteoarmillatus* (?) (see *C. percomis,* p. 436)

51.   Odor strongly pungent (like green corn or corn silk); gills and cap yellowish when young, the cap
        becoming browner or redder in age; stalk usually with darker fibrillose patches or zones; found
        under conifers in the Pacific Northwest .... *C. superbus*
51.   Not as above 52

52. Gills yellow to yellow-orange or yellow-brown when young *and* stalk with an abrupt, usually rimmed basal bulb (at least when young) 53
52. Not as above 54

53. Cap bright green to citrine-green when fresh (sometimes brownish at the center); common under oak in California .... *C.* **sp. (unidentified)** (see *C. scaurus* group, p. 440)
53. Cap not green .... *C. fulmineus* & others, p. 441

54. Cap brownish to dingy-flesh colored, but usually with a patch or patches of white fibrillose or felty veil material, only slightly viscid; stalk with a basal bulb; found under oak in California .... *C. regalis,* p. 443
54. Not as above 55

55. Cap white or whitish (or at times ranging to buff or *pale* tan) 56
55. Cap darker or brighter than above 57

56. Stalk with a basal bulb; found under hardwoods in eastern North America ...... *C. albidus*
56. Not as above; stalk equal or club-shaped (thicker below) but lacking a prominent basal bulb; often found under conifers 138

57. Lower portion of stalk with concentric belts or transverse bands of colored (usually ochre) veil tissue .... *C. crocolitus* (see *C. collinitus* group, p. 431)
57. Not as above 58

58. Cap radially corrugated or wrinkled, cinnamon to pinkish-cinnamon to cinnamon-brown; found in the Pacific Northwest under conifers .... *C. corrugis*
58. Cap not radially corrugated 59

59. Veil copious, usually leaving remnants on cap margin and/ or a fibrillose white sheath or patches on the stalk (and often a slight ring or annulus) .... *C. turmalis* (see *C. multiformis*, p. 442)
59. Not as above; veil not so copious (but may leave a few remnants) 60

60. Stalk 1-4 cm thick and very firm, usually stout, more or less equal or narrowed slightly at the base .... *C. crassus* (see *C. balteatus*, p. 433)
60. Stalk with a basal bulb (at least when young) *or* not as thick and stout as above .... *C. multiformis* & others, p. 442

61. *Fresh* fruiting body at least partially purplish, bluish, lilac, lavender-gray, lavender-white, etc. (check immature gills, stalk apex, flesh, and cap) 62
61. Fresh fruiting body completely lacking violet, bluish, or pale lavender shades 87

62. Lower part of stalk with reddish bands, patches, or "bracelets" formed by the universal veil *or* stalk with a fiery orange to red base 63
62. Not as above (but upper stalk may have a ring of rusty hairs from the cortina) 64

63. Stalk 3-7 mm thick at apex, with reddish to vinaceous patches, fibrils, or "bracelets"; cap 2-5 cm broad .... *C. boulderensis* (see *C. armillatus*, p. 448)
63. Usually larger than above; stalk with a bright orange or red base .... *C. rubripes,* p. 449

64. Stalk typically with a persistent, somewhat membranous annulus (ring) formed by the veil, usually club-shaped or bulbous; gills broad, well-spaced, deep purplish becoming dark cinnamon-brown; cap deep purplish to brownish to copper-brown, often fibrillose or hoary; found mainly under hardwoods in eastern North America *C. torvus* (see *C. evernius*, p. 450)
64. Not as above 65

65. Cap hygrophanous (fading markedly as it loses moisture) and smooth (bald) or at most with a few whitish veil remnants near the margin 66
65. Cap smooth, silky, fibrillose, hairy, or scaly, but if smooth then not markedly hygrophanous 72

66. Fruiting body small (cap 1-3 cm broad; stalk less than 5 mm thick) .... *C. pulchellus* & *C. subflexipes* (see *C. obtusus* group, p. 452)
66. Fruiting body medium-sized (larger than above) 67

67. Fresh fruiting body violet-tinged only at apex of stalk ... *C. brunneus* (see *C. laniger*, p. 451)
67. Fresh fruiting body showing more violet than above (but violet tones may fade) 68

68. Stalk with conspicuous remnants from the universal veil 69
68. Stalk with inconspicuous veil remnants (or lacking them) .... *C. impennis* & *C. adustus* (see *C. evernius*, p. 450)

Cortinarius pholideus occurs mostly with birch. Note prominent scales on cap and stalk.

69.   Universal veil remnants on stalk yellowish to buff   . **C. subpurpureus** (see *C. evernius,* p. 450)
69.   Not as above 70

70.   Stalk more or less club-shaped (thicker at base) .... **C. lucorum** (see *C. evernius,* p. 450)
70.   Stalk equal or tapering downward 71

71.   Stalk often with a narrow, flaring annulus (ring); cap pale when fresh ....
      .... **C. urbicus** (see *C. evernius,* p. 450)
71.   Not as above .... **C. evernius** & others, p. 450

72.   Flesh bluish or violet but staining wine-red to wine-brown when cut or bruised, *at least in base of
      stalk;* cap bluish or bluish-gray to grayish-lavender becoming deep purplish-brown in age;
      cap surface usually fibrillose or with flattened scales; gills dark bluish or bluish-violet when
      young; stalk with a basal bulb which often disappears in age; not common ..... **C. cyanites**
72.   Not as above; flesh not staining wine-red when bruised; common 73

73.   Cap distinctly hairy, scaly, or fibrillose-scaly 74
73.   Cap smooth or silky or very slightly scaly (but may have universal veil remnants on it) ... 78

74.   Stalk with only a tinge of violet or bluish-gray and soon fading to whitish; cap brownish to buff
      beneath a white to grayish-white fibrillose-hairy layer (cap lacking violet or bluish shades);
      found under conifers .... **C. plumiger** (see *C. evernius,* p. 450)
74.   Not as above 75

75.   Entire fruiting body *deep* violet (sometimes nearly black) when fresh (the gills may be rusty-
      dusted in age) .... **C. violaceus,** p. 446
75.   Not as above; fruiting body paler, browner, or with only a tinge of violet 76

76.   Cap *and* stalk with brown to cinnamon-brown scales; stalk *not* radically bulbous (see above
      photo); fairly common under hardwoods (especially birch) in eastern North America, rarely
      in the Pacific Northwest .... **C. pholideus** (see *C. squamulosus,* p. 445)
76.   Not with above features; stalk not usually conspicuously scaly 77

77.   Stalk radically bulbous; cap with brown to dark chocolate-brown scales; found under hard-
      woods in eastern North America, rare (or absent) in West .... **C. squamulosus,** p. 445
77.   Not as above; stalk not radically bulbous and/or habitat different 78

78.   Flesh in stalk rusty-brown to cinnamon-, tawny-, or yellow-brown (often marbled); associated
      with conifers .... **C. traganus,** p. 447
78.   Not as above (flesh in stalk usually white to buff or tinged violet) 79

Cortinarius bolaris favors beech and other hardwoods in eastern North America. Note distinctive red fibrils or scales on the cap and stalk.

79. Stalk with buff to tan, ochre, or brownish-yellow patches, zones, or "bracelets" formed by the universal veil .... *C. anomalus* & *C. caninus* (see *C. alboviolaceus,* p. 447)
79. Not as above; universal veil remnants differently colored (if present) 80

80. Odor distinctive: either sweetish (like overripe pears), spermatic, or very unpleasant 81
80. Not as above; odor mild, radishlike, or merely fungal 83

81. Odor sweetish, resembling overripe pears .... *C. fragrans* (see *C. traganus,* p. 447)
81. Odor unpleasant or spermatic 82

82. Odor usually spermatic; fruiting body rather small, the cap often umbonate; spore print dull brown .... (see *Inocybe,* p. 455)
82. Odor unpleasant but not spermatic; fruiting body medium-sized; spore print rusty brown to cinnamon-brown .... *C. camphoratus* & others (see *C. traganus,* p. 447)

83. Stalk very thick (2-5 cm thick at apex) and firm; cap viscid when moist (but soon dry and often shiny) .... *C. balteatus,* p. 433
83. Stalk typically 2 cm thick or less at apex and/or cap not viscid when young 84

84. Cap and lower stalk viscid when very young, often with brown or olive-brown patches or spots (of dried slime) when older; found under conifers; rare .... *C. griseoviolaceus* (see *C. cylindripes* group, p. 430)
84. Cap and stalk not viscid (or cap only very slightly so in wet weather) and not spotted as above (but may develop ochre, tan, buff, or brownish discolorations in age); found under both hardwoods and conifers; common 85

85. Cap predominantly pale violet to lilac-white or silvery-violet or pale bluish-violet (but buff to ochre stains may develop in age) 86
85. Cap differently colored (either grayish-buff or soon becoming brown to ochre throughout) .... *C. subpulchrifolius* & *C. malachius* (see *C. alboviolaceus,* p. 447)

86. Lower portion of stalk sheathed with conspicuous white universal veil tissue (fibrils); stalk often rather long and club-shaped (thicker below but not abruptly bulbous) *C. alboviolaceus,* p. 447
86. Stalk often thick and stout with a prominent basal bulb and/or lacking conspicuous white universal veil tissue .... *C. argentatus* & *C. subargentatus* (see *C. alboviolaceus,* p. 447)

87. Growing in sand (but associated with pine); gills rusty-orange to brownish-orange when young .... *C. aureifolius* (see *C. cinnamomeus* group, p. 453)
87. Not with above combination of characteristics 88

88. Cap olive-brown to olive-yellow to yellow-brown with small darker (often blackish) hairs or fibrillose scales, especially toward center .... *C. cotoneus* group, p. 445
88. Not as above 89

89. Cap and stalk with reddish scales, fibrils, and/or streaks on a paler (whitish to yellowish) background; found under hardwoods, especially beech, in eastern North America (see above photo) .... *C. bolaris* (see *C. squamulosus,* p. 445)
89. Not as above (but cap and stalk can be entirely reddish or the stalk can have reddish universal veil remnants) 90

425

90. Stalk white to brownish with 1-4 reddish to reddish-brown bands or belts of universal veil tissue over the lower portion *or* stalk with a bright red to orange base or a sheath of vermillion fibrils (universal veil remnants) 91

90. Not as above (but stalk may be entirely red or orange or the stalk may have a single zone of rusty hairs formed by the cortina, usually near apex) 94

91. Cap 1-3 cm broad; stalk wholly or partially sheathed by vermillion fibrils ....
.... *C. miniatopus* (see *C. rubripes*, p. 449)

91. Not as above; cap usually larger 92

92. Stalk base red to bright orange; associated with hardwoods, especially oak *C. rubripes*, p. 449

92. Stalk with 1-4 reddish bands or "bracelets"; associated with conifers or birch 93

93. Typically associated with birch .... *C. armillatus*, p. 448

93. Typically associated with conifers (especially in the Pacific Northwest) or with hardwoods other than birch .... *C. haematochelis* & *C. subtestaceus* (see *C. armillatus*, p. 448)

94. Stalk 6-18 cm long and 0.5-2 cm thick at apex, usually whitish and club-shaped (thicker at base); gills pale yellow when young; cap creamy to pale buff to yellowish or tawny-brown (often darkest at center), silky or minutely scaly; found under hardwoods (especially beech) in eastern North America .... *C. flavifolius*

94. Not as above 95

95. Growing on lawns or buried wood; stalk typically 3-5 mm thick and often with a small annulus (ring); cap typically less than 4 cm broad, reddish-brown to cinnamon, fading to more or less buff as it loses moisture; found in the Pacific Northwest ....
.... (see *Galerina, Tubaria,* & Allies, p. 399)

95. Not as above 96

96. Stalk yellow to bright yellow but the gills whitish to brown (i.e., *not* brightly colored) when young; cap smooth and hygrophanous, fading to yellowish as it loses moisture 97

96. Not as above; stalk not yellow (but may have yellowish veil remnants), or if stalk yellow then immature gills usually brightly colored also or cap minutely scaly 98

97. Gills whitish when young; cap brown to tan when moist .... *C. renidens*

97. Gills brownish or cinnamon when young; cap bronzy-brown to deep olive-brown when moist
.... *C. angulosus*

98. Gills colorful (yellow, greenish, orange, red, dark red, or bright rusty-orange) when young; stalk often colorful also; cap usually *not* hygrophanous 99

98. Gills dull (whitish, grayish, brownish, or dull cinnamon) when young; stalk usually dull also (brownish, gray, white, etc.); cap usually hygrophanous 110

99. Immature gills orange-red to rusty-orange; cap *hygrophanous* (dark rust-red but fading as it loses moisture); stalk usually paler than cap and gills; found only on the west coast in mixed woods and under conifers ...... *C. californicus* (see *C. phoeniceus* var. *occidentalis*, p. 454)

99. Not as above (if colored as above, then cap not hygrophanous or found elsewhere) 100

100. Immature gills red to dark red 101

100. Immature gills some shade of yellow, olive, orange, or bright cinnamon 104

101. *Entire* fruiting body orange-red to red, dark red, or rust-red 102

101. At least the stalk (and sometimes cap) yellowish or ochre to yellow-brown 103

102. Fruiting body cinnabar-red to rusty-red or dark red; found mainly in eastern North America, especially under hardwoods .... *C. cinnabarinus* & others (see *C. sanguineus*, p. 454)

102. Fruiting body blood-red to dark red; found mainly under conifers in northern and western North America .... *C. sanguineus*, p. 454

103. Cap red to dark red or reddish-brown .... *C. phoeniceus* var. *occidentalis*, p. 454

103. Cap ochre to brown or yellowish ....
.... *C. semisanguineus* (see *C. phoeniceus* var. *occidentalis*, p. 454)

104. Cap orange-red to bright rusty-red to coppery-red (or occasionally orange-brown); immature gills bright yellow to greenish-yellow or sometimes orangish; associated mainly with conifers and willow; not common .... *C. uliginosus*

104. Not as above; cap not reddish 105

105. Cap tawny to rusty-orange and covered with numerous minute, erect scales ....
.... *C. rainierensis* (see *C. gentilis*, p. 444)

105. Not as above; cap smooth to fibrillose or with a few scales only 106

106.   Cap smooth and hygrophanous, orange-brown to yellow-brown or ochre when moist, fading as it loses moisture; gills ochre-yellow to cinnamon-brown or brown (not particularly bright); stalk deep orange-brown or cinnamon-colored, often with a few yellow veil remnants when young; fruiting body small (stalk less than 5 mm thick) .... *C. gentilis*, p. 444
106.   Not as above; cap smooth to fibrillose or slightly scaly, *not* hygrophanous; fruiting body variously colored but often brighter or more olive than above (especially the stalk) 107

107.   Immature gills yellow to olive-yellow or greenish 108
107.   Immature gills saffron (orange-yellow) to orange or rusty-orange 109

108.   Stalk club-shaped and at least 2.5 cm thick at base; fruiting body bright yellow becoming oranger or browner in age .... *C. callisteus* (see *C. cinnamomeus* group, p. 453)
108.   Not as above; stalk more slender and/or equal; fruiting body sometimes with olive tones (but sometimes without) .... *C. cinnamomeus* group & others, p. 453

109.   Cap 3-9 cm broad; stalk usually 0.5-1.5 cm thick; gills ochre to tawny or orangish-cinnamon when young; occurrence in North America questionable ....
        .... *C. orellanus* & *C. speciosissimus* (see *C. gentilis*, p. 444)
109.   Not as above; usually smaller or more slender; gills sometimes brightly colored (saffron, orange, etc.); common in North America ....
        .... *C. croceofolius* & others (see *C. cinnamomeus* group, p. 453)

110.   Cap and stalk whitish to grayish or pale tan, but the flesh staining yellow when bruised and finally turning (or aging) reddish; widely distributed, but especially common along the west coast under conifers .... *C. rubicundulus (=C. pseudobolaris)*
110.   Not as above 111

111.   Fruiting body small and rather fragile; cap sometimes umbonate, averaging 1-4 (5) cm broad; stalk averaging 2-6 mm thick (examine several specimens if unsure!) 112
111.   Fruiting body medium-sized or larger; cap not umbonate or only very broadly so, averaging 4-15 cm broad; stalk averaging 0.5-3 cm thick (check several specimens if unsure) 121

112.   Stalk usually with yellow or yellowish veil remnants, at least when young (check several specimens); gills ochre-yellow to cinnamon-brown when young .... *C. gentilis*, p. 444
112.   Not as above 113

113.   Gills staining black when bruised ...... *C. washingtonensis* (see *C. obtusus* group, p. 452)
113.   Not as above 114

114.   Cap with a prominent black umbo (otherwise slightly paler) ....
        .... *C. nigrocuspidatus* (see *C. obtusus* group, p. 452)
114.   Not as above (but cap may be entirely blackish) 115

115.   Cap very dark when moist (deep vinaceous-brown to blackish-brown or black), but often paler as it loses moisture 116
115.   Cap paler than above when moist 119

116.   Stalk with conspicuous white universal veil material, at least when young and fresh ... 117
116.   Stalk with only slight universal veil remnants (if any) 118

117.   Odor geranium-like (at least when flesh is crushed); found in bogs or other wet places .....
        .... *C. paleaceus* (see *C. obtusus* group, p. 452)
117.   Not as above .... *C. stemmatus* (see *C. obtusus* group, p. 452)

118.   Cap blackish-brown when moist; gills and stalk also very dark; found under conifers ......
        .... *C. uraceus*
118.   Cap dark vinaceous-brown when moist; gills tawny-yellowish to pale brownish when young
        .... *C. decipiens* (see *C. obtusus* group, p. 452)

119.   Cap (and sometimes stalk) minutely scaly ....
        .... *C. psammocephalus* & *C. incisus* (see *C. obtusus* group, p. 452)
119.   Cap more or less smooth 120

120.   Cap typically with an acute (sharp) umbo   *C. acutus* & others (see *C. obtusus* group, p. 452)
120.   Cap typically with an obtuse (blunt) umbo or with none at all *C. obtusus* group & others, p. 452

121.   Fruiting body (including stalk) rusty-orange to yellowish or yellow-brown; cap minutely scaly and *not* hygrophanous; not definitely known to occur in North America ....
        .... *C. orellanus* & *C. speciosissimus* (see *C. gentilis*, p. 444)
121.   Not as above; very common in North America 122

122. Cap brownish to dingy flesh-colored, but often with a whitish patch of universal veil tissue; stalk thick (1.5-3.5 cm at apex), with a large, abrupt, often rimmed basal bulb; fairly common in California under oak .... *C. regalis*, p. 443
122. Not as above 123

123. Cap covered with cinnamon-brown to chocolate-brown scales, not markedly hygrophanous; either stalk also with scales or stalk radically bulbous; common under hardwoods in eastern North America, rare in the West 124
123. Not as above 125

124. Stalk radically bulbous and smooth to fibrillose or only slightly scaly *C. squamulosus*, p. 445
124. Stalk equal or slightly thicker below, with brown scales like those on cap .... *C. pholideus* (see *C. squamulosus*, p. 445)

125. Cap broadly cone-shaped and finely silky-scaly from the universal veil when young; stalk equal, sheathed with white universal veil remnants, soon becoming hollow; cap dark to pale cinnamon .... *C. hemitrichus*
125. Not as above (but may be similar in some respects, such as color) 126

126. Cap blackish-brown when moist; gills and stalk also very dark; found under conifers .... *C. uraceus*
126. Not as above; usually paler or brighter in color 127

127. Stalk typically tapered downward or spindle-shaped (swollen near ground and tapered below), the base often rooting 128
127. Stalk typically equal to club-shaped or bulbous 130

128. Cap cinnamon-brown when moist .... *C. duracinus* (see *C. laniger*, p. 451)
128. Cap vinaceous-brown to cocoa-colored when moist 129

129. Stalk with conspicuous white universal veil remnants *C. damascenus* (see *C. laniger*, p. 451)
129. Universal veil remnants on stalk absent or evanescent .... *C. privignus* & *C. cacao-color* (see *C. laniger*, p. 451)

130. Cap silky to fibrillose or fibrillose-scaly and white to silvery-gray or pale brownish-gray *or* cap brown beneath a whitish to grayish fibrillose layer 131
130. Cap darker, at least when moist, and usually more or less smooth or minutely scurfy (but cap may fade to buff or paler as it loses moisture) 133

131. Cap brown to buff beneath a fibrillose-scaly layer .... *C. plumiger* (see *C. evernius*, p. 450)
131. Cap paler than above and silky to fibrillose (but sometimes becoming pale brownish-gray in age) 132

132. Cap silky-fibrillose, white to silvery, sometimes becoming pale brownish in age .... *C. pinetorum* (see *C. laniger*, p. 451)
132. Cap whitish or silvery, usually with a faint hint of lavender or violet (examine several specimens if unsure), silky; flesh and stalk apex often with a violet tinge also .... *C. alboviolaceus* & others, p. 447

133. Cap yellow-brown when moist, fading to pale yellow-orange or orange-buff as it loses moisture .... *C. armeniacus* (see *C. laniger*, p. 451)
133. Cap redder or darker than above, at least when moist 134

134. Gills widely spaced; cap smooth or scurfy .... *C. distans* (see *C. laniger*, p. 451)
134. Not as above 135

135. Stalk more or less equal .... *C. biformis* & *C. dilutus* (see *C. laniger*, p. 451)
135. Stalk usually club-shaped or swollen below, at least when young (but sometimes equal, so examine several specimens if unsure) 136

136. Stalk with conspicuous universal veil remnants, at least when young 137
136. Stalk with only slight traces of the universal veil, if any (but may have remnants from the cortina) .... *C. triformis* & others (see *C. laniger*, p. 451)

137. Cap dark or dull brown when moist; stalk similarly colored beneath whitish to brownish veil remnants, the apex sometimes vinaceous- or violet-tinged *C. brunneus* (see *C. laniger*, p. 451)
137. Cap cinnamon to reddish-brown or chestnut-brown when moist; stalk similarly colored beneath whitish veil remnants, *not* vinaceous- or violet-tinged at apex *C. laniger* & others, p. 451

138. Cap surface or cuticle bitter-tasting .... *C. crystallinus* (see *C. vibratilis*, p. 429)
138. Not as above .... *C. lustratus* & others

Cortinarius vibratilis. Note small size, slimy cap, and viscid stalk that is swollen at base.

Cortinarius vibratilis   (Bitter Cortinarius)

CAP 1.5-5 cm broad, broadly bell-shaped to convex to nearly plane; surface smooth, viscid or slimy when moist, yellowish to tawny, fulvous, orange-brown, or yellow-orange (or in one form brown) when fresh, fading as it ages or dries (sometimes to pale tan). Flesh thin, whitish; odor mild, taste bitter (especially cap surface). **GILLS** close, notched to adnate or slightly decurrent, at first whitish, soon becoming dull ochre or tawny and finally cinnamon-brown. **STALK** 3-7 cm long, 3-10 mm thick at apex, usually clearly thickened below (club-shaped) but sometimes equal; viscid or slimy when moist, but in age sometimes viscid only at the base; white, smooth or nearly so. **UNIVERSAL VEIL** slimy, colorless. **CORTINA** scanty, sometimes leaving a few hairs on stalk which trap falling spores. **SPORE PRINT** rusty-brown; spores 7-9 × 4-5 microns, elliptical, roughened.

HABITAT: Solitary to widely scattered or in small groups in forest humus; widely distributed and fairly common, but rarely fruiting in large numbers. It usually occurs with conifers, but in our area I have collected it under manzanita, madrone, and huckleberry, in December.

EDIBILITY: Inedible due to the bitter taste. (If *Russula brevipes* is "better kicked than picked," then *C. vibratilis* is "better kicked than licked!")

COMMENTS: The small size, viscid tawny cap, and viscid white stalk plus the bitter taste —which can usually be detected by pressing your tongue to the surface of the cap—are the hallmarks of this petite *Cortinarius*. It typifies a group of species within the subgenus Myxacium which have a club-shaped (thickened) stalk, at least when young. Other species: *C. crystallinus* is similar in size, shape, and bitter taste but is whitish to buff or pale tan overall and has a non-viscid stalk. See also *C. citrinifolius* (under *C. percomis*).

Cortinarius mucosus   (Slimy Cortinarius)

CAP 4-10 cm broad, convex to plane or broadly umbonate, or even uplifted in age; surface smooth and very slimy when moist, chestnut- to reddish-brown or bright orange-brown, often paler (tawny or ochre) in age. Flesh whitish; odor and taste mild. **GILLS** close, usually more or less adnexed, whitish to grayish when very young, becoming pale ochra-

429

ceous or tawny, then cinnamon-brown from ripening spores. **STALK** 4-15 cm long, 1-2 (2.5) cm thick, more or less equal, slimy or viscid when moist, white or whitish but sometimes rusty-stained from spores; *not* breaking up into conspicuous scaly belts. **UNIVERSAL VEIL** whitish, slimy, disappearing or leaving indistinct remains on stalk. **CORTINA** usually forming a superior, hairy-fibrillose zone on stalk which is stained rusty-brown by spores. **SPORE PRINT** rusty-brown; spores 11-18 × 5-7.5 microns, elongated-elliptical, roughened.

HABITAT: Widely scattered to gregarious in mixed woods and under conifers (particularly pine and spruce); widely distributed, but especially common in the West in the late summer and fall, and in the southern United States. I have seen large fruitings in Washington, New Mexico, Arizona, and northern California, but not in our area. *Cortinarius*-classifier Joe Ammirati says it has been found under birch and willow in the Brooks Range in northern Alaska; it also occurs with hemlock.

EDIBILITY: Better eschewed than stewed. It is too slippery to be worthwhile, and its edibility is unknown.

COMMENTS: The smooth, viscid, tawny to orange-brown or reddish-brown cap and viscid, whitish, equal stalk are the distinguishing features of this "slimy sucker." The cap color is reminiscent of *C. collinitus,* but the stalk lacks the scaly belts characteristic of that species, and the taste is not bitter nor the stalk club-shaped as in *C. vibratilis. Phaeocollybia* species are somewhat similar, but lack a cortina and have a long, rooting stalk. *C. turmalis* (see comments under *C. multiformis*) is also similar, but has a dry stalk.

Cortinarius cylindripes group   (Slimy Purple Cortinarius)

CAP 3-12 cm broad, convex or broadly bell-shaped becoming broadly umbonate to plane or with an uplifted margin; surface smooth or in age sometimes wrinkled, viscid to extremely slimy when moist, at first lavender or violet, but fading quickly or slowly (often from center outward) to yellowish, brownish-ochre, tawny, brown, or paler. Flesh rather thin, violet-tinged becoming whitish; taste mild. **GILLS** at first lavender or grayish-lavender, soon grayish or grayish-cinnamon with whitish, finely scalloped edges, finally rusty-cinnamon; adnate to adnexed or notched, close. **STALK** 5-15 cm long, 1-2.5 (3) cm thick, equal or at times spindle-shaped when young (swollen slightly in the middle); viscid or slimy throughout or at least over lower portion; violet to silvery-violet or lavender in part or throughout, but often fading somewhat (like cap) in age; smooth or with whitish to violet scales above. **UNIVERSAL VEIL** slimy, violet, forming a slime sheath on stalk. **CORTINA** obscured by slime, evanescent or leaving a few hairs on stalk. **SPORE PRINT** dark cinnamon-brown; spores 12-17 × 5.5-8 microns, elliptical, roughened.

HABITAT: Scattered to densely gregarious in woods; widely distributed and locally abundant (along with similar species) in the late fall and winter. Like *C. balteatus,* it tends to fruit with ericaceous shrubs and trees, e.g., manzanita, madrone, and/ or huckleberry.

EDIBILITY: Better eyed than fried; the snotlike universal veil will deter most.

COMMENTS: This species may not be the most distinctive or desirable of our local Cortinarii, but it unquestionably qualifies as the slimiest. The yellowish to brownish or lavender cap, slimy universal veil, and viscid, more or less equal, violet or lavender stalk are the principal diagnostic characters of *C. cylindripes,* but there are numerous look-alikes, including: *C. stillatitius,* with only slightly roughened spores; *C. salor,* with smaller, round spores; *C. pseudosalor,* similarly colored in age, but without a purplish cap when young; *C. splendidus,* with a small (3-5 cm) liver-brown to tawny cap, also found with conifers; *C. iodes (=C. heliotropicus)* and *C. iodioides,* with a purple cap that doesn't fade much plus a club-shaped to spindle-shaped stalk when young (the latter also has bitter-tasting slime);

Cortinarius cylindripes group includes a number of purple or partially purple species with a slimy cap and stalk. The stalk is equal or slightly spindle-shaped and not conspicuously banded.

C. griseoviolaceus, a small violet-tinged conifer-lover whose cap and stalk are only thinly viscid (if at all); and *C. griseoluridus,* with a club-shaped stalk when young and a violet to olive or ochre cap, found under conifers in the Rocky Mountains. All of these have viscid, violet-tinged stalks that lack the distinctive belts characteristic of *C. collinitus,* and none have caps as dark or as wrinkled as those of *C. vanduzerensis* and *C. elatior.* Other violet-tinged Myxaciums include: *C. castaneicolor,* found under northern conifers, with a whitish to tawny, club-shaped stalk, pale violet gills when very young, and a buff to tawny to chestnut-brown (darker at the center) cap; and *C. delibutus* and *C. sphaerosporus* (two very similar if not identical species), with a bright yellow to ochre or straw-colored cap, violet-tinged gills at first, and an equal to club-shaped whitish stalk that often has yellowish veil patches (or slime) below and often has a violet-tinged apex.

Cortinarius collinitus group   Color Plate 104
(Belted Slimy Cortinarius)

CAP 3-10 cm broad, bell-shaped or convex becoming broadly umbonate to plane; surface smooth, slimy or viscid, color variable: yellowish to orange-yellow, orange-brown, tawny, ochre, fulvous, or even reddish-brown, the margin often paler or yellower (or in one form bluish-violet) and sometimes striate. Flesh firm, whitish to yellowish-buff. **GILLS** close, adnate to adnexed or notched, pallid or pale grayish (or violet-tinged in one form) when young, becoming brown and finally rusty-brown in age. **STALK** 5-15 cm long, 0.5-2 cm thick, equal or tapering downward, usually rooting somewhat in the humus; viscid or slimy when moist, color variable but usually whitish above, and lower portion usually breaking up into irregular whitish, yellowish, ochre, and/or rusty-brown (or in one form violet or bluish-violet) bands, patches, or scaly rings (but these sometimes obscure). **UNIVERSAL VEIL** fibrillose beneath a layer of slime, leaving bands or patches on stalk. **CORTINA** pallid, usually forming a ring of hairs near top of stalk which turns rusty-brown from falling spores. **SPORE PRINT** rusty-brown; spores 10-15 × 6-8.5 microns, elliptical, roughened.

HABITAT: Solitary, scattered, or in groups under both hardwoods and conifers; widespread in the northern hemisphere and fairly common in our area in the late fall and winter under tanoak and madrone at higher elevations in the coastal mountains (often mingling with *C. cotoneus* and *C. infractus*). It is also abundant under aspen in the southern Rocky Mountains and the Southwest and with conifers in the Pacific Northwest.

EDIBILITY: Better neglected than collected. In spite of the slime it is said to be edible, but it is a species "complex," and some of the variants are poorly known or untested.

COMMENTS: This species "complex" is relatively easy to recognize by virtue of the slimy-viscid cap and belted or scaly, viscid stalk. It is a very confused group taxonomically, however, and names like *C. trivialis* and *C. mucifluus* have been used to describe some of

431

the variants which differ in color and spore size. Our local version, which *might* be *C. trivialis,* does *not* show violet on the stem, and neither does the common aspen-loving form. *C. elatior* is a similar species that is not so strikingly banded. It has a dark brown to olive-brown or yellow-brown cap that is often radially wrinkled or grooved at the margin in age, violet or bluish-tinged gills when young, and a tapered, rooting stalk that is also bluish- or violet-toned (and slimy). It favors conifers and is fairly common in northern California and the Pacific Northwest. Other species: *C. pallidifolius* has pallid gills when young and a viscid club-shaped stalk with ochre-brown patches or zones; it is also found under conifers in the Pacific Northwest, especially fir. *C. crocolitus (=C. triumphans?)* and close relatives have a yellowish to brown viscid cap and a *dry* stalk marked with yellow-brown to ochre belts or patches (see photo on p. 419); they favor hardwoods.

Cortinarius vanduzerensis

CAP 4-10 cm broad, somewhat conical becoming broadly conical or convex; surface smooth, very slimy when moist, deep chestnut-brown to nearly black when young, becoming paler chestnut and finally cinnamon-brown in age, often wrinkled radially or corrugated at maturity, especially toward margin. Flesh pallid to cinnamon-buff; odor mild. **GILLS** close, adnate or adnexed, pallid or buff becoming pale brown and finally dull cinnamon-brown. **STALK** 8-20 cm long, 1-2 cm thick, equal or often tapered slightly toward the base, often deeply rooted, viscid to very slimy, bluish-violet to violet to dark lavender above, paler below, often fading in age. **UNIVERSAL VEIL** slimy, sheathing at least the lower half of the stalk and occasionally breaking up into concentric zones. **CORTINA** forming a ring of hairs near top of stalk or disappearing. **SPORE PRINT** rusty-brown; spores 11-15 × 7-9 microns, elliptical, roughened.

HABITAT: Solitary, scattered, or in groups under conifers; known only from northern California and the Pacific Northwest, where it is fairly common in the fall and early winter. I have found it several times under Sitka spruce.

EDIBILITY: Unknown, but much too slippery to be of value.

COMMENTS: The dark slimy cap, slimy violet or bluish-violet stalk, and pallid to brown gills when young form a distinctive combination of features. It closely resembles *C. elatior* (see comments under the *C. collinitus* group) in shape, size, and the wrinkled cap at maturity, but does not have the bluish-violet immature gills of that species. The stalk often roots deeply in the ground, but the presence of a veil distinguishes it from *Phaeocollybia.* The stalk occasionally shows concentric ring-zones, but not to the extent of *C. collinitus.* (Violet-stalked forms of the latter also have a brighter or paler cap.) For other similar "Myxaciums" with paler caps, see *C. cylindripes* group. Also see photo on p. 433.

Cortinarius ponderosus   (Ponderous Cortinarius)

CAP 10-35 cm broad, convex becoming nearly plane in old age; surface viscid to slimy when moist, brown to cinnamon or russet at the center, with small, flattened, spotlike, russet to rusty-brown or brown scales; margin usually yellow or ochre-yellow to yellow-olive or tawny, remaining inrolled for a long time. Flesh *very* thick and firm, whitish or buff; odor mild to somewhat spermatic (*Inocybe*-like). **GILLS** close, slightly decurrent to adnate or sometimes adnexed, lilac-tinged when young (at least near margin), becoming dingy ochre and finally rusty-brown as spores mature; often with rusty or russet spots or stains in age. **STALK** 8-20 cm long, (3) 4-7 cm thick, equal or slightly enlarged below *or* tapered at the base; solid, very firm and hard; viscid below when moist; apex pallid or colored like gills, lower portion with yellow-ochre to rusty-brown or russet stains and/ or fibrils or fibrillose patches. **UNIVERSAL VEIL** slimy, yellow, disappearing or leaving

Left: *Cortinarius vanduzerensis* has a dark slimy cap and long, violet-tinged stalk. **Right:** *Cortinarius ponderosus* (or something very similar—see comments below). This species may be the largest of all the Cortinarii, but this is a rather small specimen.

slime on stalk. **CORTINA** often leaving a hairy zone on upper stalk. **SPORE PRINT** rusty-brown; spores 8.5-11 × 5-6 microns, elliptical, roughened.

HABITAT: Solitary to scattered or gregarious or in arcs on ground in mixed woods (oak, ponderosa pine, etc.) in northern California and southern Oregon in late fall and winter; not common, but often prolific when it fruits. A very similar if not identical species occurs under oak and madrone (see comments).

EDIBILITY: Better chucked than plucked. It is certainly large enough to be tempting, but its edibility has not been determined, and several Cortinarii are very poisonous.

COMMENTS: This massive *Cortinarius* can almost be identified by its size alone. The stalk is exceptionally hard and thick and the fruiting bodies develop gradually over a considerable length of time. An equally ponderous but possibly distinct *Cortinarius* occurs in our area. It differs in several minor respects: the stalk is dry or only slightly viscid (tacky), the intermediate color phase of the gills is grayish rather than ochre, the cap is viscid but not usually slimy (and the slime on the cap, when present, is not yellow), and the upper half of the stalk is slightly different in color. In size, shape, and overall aspect, however, it is remarkably similar (see photograph). Whether it is just a variation of *C. ponderosus* or a distinct species is uncertain, but collectors should have no trouble recognizing it. For other large species without a pronounced basal bulb, see *C. balteatus.*

Cortinarius balteatus

CAP (3) 6-20 (25) cm broad, broadly convex with an inrolled margin becoming plane in age; surface viscid when moist but soon dry and often shiny, smooth, without scales but sometimes appearing streaked or blotched; dull violet or lavender to lavender-gray when young (but sometimes whitish while still covered by humus), soon becoming tawny to tawny-olive or brown from the center outward, the margin usually remaining lavender until old age. Flesh thick, firm, whitish to grayish to buff or slightly violet-tinted under the cuticle; odor mild. **GILLS** adnate to adnexed or notched, close, whitish or tinged violet to violet-gray when young, gradually becoming brown as spores mature. **STALK** 5-12 cm long, (1) 2-4 (5) cm thick, equal above, often slightly narrowed at base (but sometimes thicker); firm, solid, dry, pallid or lavender becoming brownish-stained in age, especially below (or staining yellow in one form). **CORTINA** disappearing or leaving hairs on stalk which trap falling spores. **SPORE PRINT** dull cinnamon-brown; spores 9-11 × 5-6 microns, elliptical, roughened.

Cortinarius balteatus is a large firm purple-tinged species. Note how the stalk lacks a basal bulb.

HABITAT: Scattered to densely gregarious on ground under hardwoods and sometimes conifers, usually with members of the Ericaceae (huckleberry, manzanita, madrone, etc.) also present; widely distributed. In our area it is sometimes common in the fall and early winter under madrone. Usually only one crop is produced but the fruiting bodies develop and decay slowly, and so may be found over a period of several weeks. I have also seen it in the Oregon Cascades (near Crater Lake) in both the spring and fall.

EDIBILITY: Unknown. As Alexander Smith points out, it is firm and meaty enough to warrant experimentation—but in light of the recent rash of poisonings caused by species of *Cortinarius,* who wants to be the one to try it?

COMMENTS: This cumbersome *Cortinarius* can be recognized by its violet-tinged gills when young, violet-tinted cap margin, and hard thick stem which is *not* abruptly bulbous. The cap is only thinly viscid and may not seem so at all in dry weather. In wet weather our form is often full of tiny springtail beetles (columbalids). Other cumbersome, viscid-capped Cortinarii without an abrupt basal bulb include: *C. largus,* very similar but with violet-tinged flesh and an ochre to reddish-brown cap in age; *C. balteato-cumatilis,* also quite similar but with an odor "like overripe pears" (Stuntz); and *C. crassus,* with a yellow-brown to russet or reddish-brown cap, short thick stalk, and pallid to buff gills which gradually darken to cinnamon. The latter is fairly common under western conifers, especially at higher elevations (I have seen it in the Rockies).

Cortinarius subfoetidus

CAP 2.5-10 cm broad, broadly umbonate to convex or plane in age; surface viscid or slimy when moist, smooth or appearing fibrillose, bright lavender or violet or at times bluish-lavender, in age sometimes fading at the center to buff or paler. Flesh tinged cap color, paler in age; odor distinctly fragrant, but with a fetid (nauseating) component. **GILLS** close, adnexed or notched to adnate, at first lilac or violet, but becoming pale to dull brown and finally rusty-brown in age. **STALK** 5-8 cm long, 0.7-2 cm thick, more or less equal, colored like cap or paler (often fading), the apex often whitish; not viscid. **UNIVERSAL VEIL** fibrillose, forming a pale lavender to violet sheath over lower portion of stalk. **CORTINA** transient, sometimes leaving a few hairs at stalk apex which trap falling spores. **SPORE PRINT** rusty-brown; spores 7-10 × 5-5.5 microns, elliptical, roughened.

HABITAT: Solitary to widely scattered or in small groups in duff and moss under conifers; known only from the Pacific Northwest, fruiting primarily in the fall, especially with hemlock and/or fir. I have seen it in northern Idaho, where, according to Alexander Smith, it is more common than anywhere else.

EDIBILITY: Better wasted than tasted.

COMMENTS: The sickeningly sweet odor (which has been likened to that of overripe pears) and bright lavender to violet color combine to make this one of the easiest of all Cortinarii to recognize. Alas, the casual hunter may not get a chance to recognize it because it is not a very common species. The viscid cap distinguishes it from *C. traganus, C. camphoratus* and other "Sericeocybes," while the stalk is neither viscid as in the *C. cylindripes* group nor bulbous as in *C. olympianus* and other purplish "Bulbopodiums." Although uncommon, it attracts more than its share of attention because of its beautiful color. *C. balteato-cumatilis* (see comments under *C. balteatus*) in similar in odor and color but is much more robust (stalk typically *at least* 2 cm thick).

Cortinarius infractus   (Sooty-Olive Cortinarius)

CAP 4-13 cm broad, obtuse or convex becoming broadly umbonate or plane; surface viscid when moist, dark olive to sooty-olive becoming slightly browner (dingy yellow-brown) in age or sometimes mottled with tawny shades; margin sometimes wavy and/or faintly zoned. Flesh thick, firm, whitish or tinged violet or ochre-buff; taste bitter (at least of the skin). **GILLS** notched to adnexed or adnate, close, dark olive to sooty-olive or olive-brown, or in some forms tinged violet at first, becoming dark cinnamon-brown in old age. **STALK** 3-13 cm long, 0.5-2.5 cm thick, equal or more often enlarged slightly below; solid, fibrillose, dry, whitish or tinged cap color (or in some forms tinged violet at apex), and often dingy olive-brown toward the base. **CORTINA** pale grayish, usually forming a hairy median or superior zone on stalk which turns rusty-brown from falling spores. **SPORE PRINT** rusty-brown; spores 7-9 × 5-7 microns, elliptical to nearly round, roughened.

HABITAT: Scattered to gregarious on ground under both hardwoods and conifers; widely distributed and common. In our area it fruits in the fall and winter, usually under tanoak and madrone (often accompanied by *C. collinitus, C. cotoneus,* and *Entoloma* species).

EDIBILITY: Better overlooked than cooked—it is probably bitter-tasting.

Cortinarius infractus is a common sooty-olive species with a viscid cap when moist. Note collapsed hairs (cortina) on stalk. These are fairly mature specimens; young ones have smaller caps.

COMMENTS: A distinctive *Cortinarius* by virtue of the sooty-olive color of its cap and gills. The cap is viscid when moist, but there is no abrupt bulb at the base of the stalk as in the *C. scaurus* group. A number of varieties have been recognized, some of which show violet in the stalk apex and/or immature gills. The common version in our area, however, is violet-less. *C. immixtus* is a similar species with brighter (yellowish-olive-green) immature gills, a mild taste, and larger spores; it has been reported from the Pacific Northwest.

Cortinarius percomis   (Fragrant Cortinarius)

CAP 4-8 cm broad, convex to broadly convex with an inrolled margin when young, expanding in age; surface smooth, viscid when moist, bright to dull yellow to bright ochre or at times tinged fulvous or greenish-yellow. Flesh yellow; odor strongly aromatic and sweet (see comments); taste mild to slightly unpleasant. **GILLS** close, adnexed or notched to adnate, pale yellow to sulfur-yellow when young, becoming browner (dingy olive-brown to rusty-cinnamon) as spores mature. **STALK** 4-8 cm long, 0.7-1.5 (2.5) cm thick at apex, equal or more often club-shaped (thicker below), but lacking a sharply-defined bulb; bright yellow to yellowish-white, dry, fibrillose, solid, firm. **UNIVERSAL VEIL** sulfur-yellow, fibrillose, often leaving remnants (fibrils) on stalk. **CORTINA** usually collapsing to form a ring of hairs on upper stalk which turn rusty-brown from falling spores. **SPORE PRINT** rusty-brown; spores 10-13 × 5-6.5 microns, elliptical, roughened.

HABITAT: Solitary, scattered, or in small groups on ground under conifers; widely distributed. It is fairly frequent in western North America in the late summer and fall, but seldom in quantity. In our area a similar fragrant species (see comments) occurs with oak.

EDIBILITY: Unknown. I have cautiously sampled our local variety (see comments), and found its flavor to be a distinct disappointment (it didn't live up to the odor).

COMMENTS: The strong, remarkably sweet odor (best detected by breaking open the flesh) and yellow color are the outstanding characteristics of this attractive *Cortinarius*. The odor is difficult to describe but is reminiscent—to different people—of citrinella, toilet bowl cleaner, or papaya. In our area *C. luteoarmillatus* (?), with a similar odor, is common in the fall and winter under live oak. It has a pale ochre to pale tan, buff, or pale yellowish cap, white flesh, and a white stalk with or without yellowish universal veil zones. *C. citrinifolius* looks and smells like *C. percomis,* but has a viscid stalk; it occurs in the Pacific Northwest. *C. odorifer* has an aniselike odor and a rimmed basal bulb (at least when young) and a greenish-yellow to cinnamon to vinaceous-brown cap. *C. osmophorus* is a fragrant European species that closely mimics our local variety in color, but it also has a rimmed bulb. None of these are worth eating.

Cortinarius luteoarmillatus (or a very similar species—see above comments) is common in California under oak. Note how stalk is club-shaped (thicker below) but not bulbous. Sweet odor is distinctive.

Cortinarius mutabilis group   (Purple-Staining Cortinarius)

CAP 4-10 cm broad, convex becoming plane or undulating; surface smooth, viscid when moist, at first violet or grayish-violet, but usually suffused in age with buff, ochre, or brown, especially toward the center (margin often remaining violet). Flesh thick, pallid or pale violet becoming dark purple when bruised or cut and then rubbed (sometimes staining slowly); odor mild. **GILLS** adnate or more often adnexed or notched, close, dull violet or grayish-lilac when young, slowly becoming pale brownish as spores mature; turning deep lilac or purple when bruised. **STALK** 4-10 cm ong, 1-2 cm thick at apex, thicker or swollen at base but without an abrupt bulb; pallid to pale violet or colored like cap, staining deep lilac or purple when bruised. **STALK** 4-10 cm long, 1-2 cm thick at apex, thicker or swollen at base but without an abrupt bulb; pallid to pale violet or colored like cap, staining which traps falling spores. **SPORE PRINT** rusty-brown; spores 7-9 × 4.5-5.5 microns, elliptical, roughened.

HABITAT: Scattered to gregarious or in small clumps on ground or very rotten wood under conifers, western North America; especially common at higher elevations in the late summer and fall. I have seen it in the Rocky Mountains and Cascades, but not on the coast.

EDIBILITY: Unknown, like most of us.

COMMENTS: This species "complex" is variable in a number of respects, but can be recognized by the purple-staining gills and stalk surface. The flesh also stains purple, but the reaction seems to vary somewhat in brightness and duration. Other species: *C. purpurascens* is a purple-staining species with a purple-brown to reddish-brown, often variegated cap and more sharply defined basal bulb on the stem; *C. occidentalis* stains purple, but has a bluish cap that becomes grayish in age; *C. subpurpurascens* has purple-staining gills, but the flesh does not stain. All of these grow mainly with conifers.

Cortinarius glaucopus group

CAP 4-12 (17) cm broad, convex becoming plane or with margin slightly uplifted; surface smooth, viscid or tacky when moist, color extremely variable: rich greenish-brown to olive-black, greenish-gray, steel-gray, bluish-gray, etc., often streaked or mottled with yellowish or ochre fibrils and often becoming fulvous, cinnamon, or rusty-colored in age (typically from the center outward); margin sometimes yellowish-olive and often wavy. Flesh firm, thick, pallid to grayish or tinged violet-blue, becoming ochraceous in age; also ochraceous in base of stalk. **GILLS** close, adnate to adnexed or notched, bluish-violet to bluish-gray or sometimes gray when young, darkening to rusty-brown as spores mature. **STALK** 4-10 cm long, 1-3 cm thick at apex, more or less equal above, but with a bulb at base, the bulb abrupt and rimmed when young but often oblique or poorly defined in age; color variable but usually violet, blue, or green at apex and variously colored (often paler) below, often browner in age; dry, solid, firm. **CORTINA** pallid or pale bluish-violet, usually leaving hairs on stalk which are stained rusty-brown by falling spores. **SPORE PRINT** rusty-brown; spores 6-10 × 4-5.5 microns, elliptical, slightly roughened.

HABITAT: Scattered to densely gregarious or in small clumps or sometimes in troops, associated with both hardwoods and conifers; widely distributed but particularly common in western North America. Our members of this "complex" fruit prolifically in the fall and winter and are the most common "Bulbopodiums" of our live oak woodlands. In the Rocky Mountains and Pacific Northwest, however, it favors conifers, especially spruce.

EDIBILITY: Better drowned and beaten than browned and eaten. Many members of this group have not been adequately tested and some are apparently bitter.

COMMENTS: This species "complex" is extremely variable in color, but can be told in the button stage by its conspicuous bulb and blue-gray to bluish-violet gills. In age it becomes

Cortinarius glaucopus group is common under both hardwoods and conifers. In the button stage (not shown here) the basal bulb is usually quite large and abrupt, but becomes less pronounced as the stalk elongates.

cinnamon- or rusty-colored and is easily confused with dozens of other Cortinarii. Younger specimens can resemble the blewit *(Clitocybe nuda)*, but have a cortina and rusty-brown spores. *C. pseudoarquatus* is a similar species with a more buff or brownish cap and significantly larger spores. For other bluish-violet "Bulbopodiums" (most of which are more violet than blue and/or have differently colored caps), see comments under the *C. sodagnitus* group and *C. olympianus.*

Cortinarius sodagnitus group

CAP 3-7 (10) cm broad, convex to plane; surface smooth, viscid when moist, at first bright violet or lilac overall, but soon becoming brown to ochraceous or buff from the center outward (or becoming somewhat silvery in dry weather), the margin retaining violet tones well into maturity. Flesh thick, firm, whitish with a violet tinge at top of stalk and ochre tint in basal bulb; not staining appreciably; taste bitter (at least of the skin). **GILLS** violet when young, becoming duller and suffused with brown or cinnamon-brown, but the edges usually remaining violet until old age; close, adnate to adnexed or notched. **STALK** 2.5-8 cm long, 0.5-2 cm thick, with a small but sharply defined, rimmed bulb at the base (1-3 cm broad); violet like the cap or pallid when young, discoloring brownish or rusty from the base upward, but usually retaining distinct violet zone at apex even in old age; solid, not viscid. **CORTINA** violet-tinged, leaving a median to superior zone of hairs on stalk, the hairs often rusty-stained by spores. **SPORE PRINT** rusty-brown; spores 10-12 × 5.5-6.5 microns, elliptical, roughened. Cap surface staining red or wine-red in KOH or NaOH.

HABITAT: Solitary, scattered, or gregarious on ground under hardwoods; distribution uncertain. This species "complex" is quite common in our live oak woodlands in the fall and winter. It was originally described from Europe, where it favors beech.

EDIBILITY: Better tossed than sauced—it is probably bitter.

COMMENTS: This beautiful member of the subgenus Bulbopodium can be told by its rather small but well-defined, abrupt, rimmed bulb at the base of the stem (see photo at top of p. 439), by its violet colors when young, and by the tendency of the cap margin and stem apex to remain violet even in old age. Fresh specimens mimic the blewit *(Clitocybe nuda)* in size, shape, and color, but have a cortina and rusty-brown spores. Closely related species include: *C. dibaphus,* with a slightly pinker or more lilac cap and gills which turn brown more quickly (but otherwise very similar, if not the same); and *C. caesiostramineus,* with a slightly grayer cap. There are dozens of other purple or violet "Bulbopodiums"— see comments under *C. olympianus* for some of them.

438

Cortinarius sodagnitus group. Note small but very prominent basal bulb. Common in California under oak.

Cortinarius olympianus

CAP 3-7 (10) cm broad, convex becoming plane, the margin inrolled at first; surface smooth, slimy or viscid when moist, pale violet or lilac, sometimes fading to lilac-white, or sometimes tinged yellowish at center. Flesh fairly thick, white to grayish. **GILLS** pale lilac to pinkish-lilac when young, becoming browner in age but usually retaining a lilac tinge for a long time; notched or adnexed to adnate, close, not staining when bruised. **STALK** 3-7 cm long, 0.7-1.2 cm thick at apex, solid, firm, not viscid, with a distinct basal bulb that is rimmed, at least when young; lilac or pale violet above, often brownish-stained at base. **CORTINA** scanty, lilac-white, often leaving a few hairs on stalk which trap falling spores. **SPORE PRINT** rusty-brown; spores 8-10 × 5-6 microns, elliptical, slightly roughened.

HABITAT: Scattered to gregarious on ground under conifers, fairly common in the Pacific Northwest in the late summer and fall, especially where there is fir, spruce, and/or hemlock. I have seen large fruitings in Mt. Rainier National Park in late September.

EDIBILITY: Unknown.

COMMENTS: This is one of several viscid-capped Cortinarii with a rimmed bulb and overall lilac to violet-white color. There are several similarly colored species that favor hardwoods and are especially prominent in eastern North America. These include: *C. michiganensis* and *C. caesiocyaneus,* both larger with conspicuous bulbs and pale violet to pale bluish-violet overall color (when fresh), the latter with a whitish universal veil; *C. velicopia,* similar to the previous two species but discoloring ochre to buff or tan on the cap and basal bulb with age (often rather quickly); *C. aggregatus,* with darker violet gills when young and only a small, oblique bulb at the stem base; *C. caerulescens,* with a darker blue or violet-blue cap when young; and *C. cyanites* (see couplet #72 of the key to *Cortinarius*). There are also several species that are very similar in size, shape, and color but do *not* have viscid caps, e.g., *C. argentatus* and *C. subargentatus* (see comments under *C. alboviolaceus* and photo on p. 448).

Cortinarius cedretorum   Color Plate 105

CAP 5-15 cm broad, convex with an inrolled margin when young, becoming broadly convex or plane; surface smooth, viscid or slimy when moist, at first yellow but darkening slowly to cinnamon-brown, orange-brown, or reddish-brown from the center outward, the margin often remaining yellowish. Flesh thick, firm, pallid to pale yellow with distinct lavender or violet areas near cap cuticle and in stalk; often ochraceous-yellow in base of stalk; odor mild. **GILLS** close, adnate to adnexed, yellow or sometimes greenish-yellow when young, dull cinnamon-brown in age, but often with an intermediate lavender to dull purple phase. **STALK** 4-12 cm long, 1.5-3 cm thick at apex, with a conspicuous, abrupt basal bulb which is rimmed, at least when young and is 3-7 cm broad; solid, not viscid, color variable: pale lavender or yellow at apex or throughout, becoming dingy or often rusty-stained in age, the bulb usually yellow. **CORTINA** pale yellow to greenish-yellow, usually leaving hairs on the stalk which trap falling spores. **SPORE PRINT** rusty-brown; spores 11-14 × 6.5-8 microns, elliptical, roughened.

HABITAT: Solitary to scattered or gregarious on ground in woods, mainly in western North America. It is sometimes common in our area under oak in the late fall and winter; in the Pacific Northwest it favors conifers and in the Southwest I've seen it in mixed woods of aspen, spruce, and fir.

EDIBILITY: Unknown.

COMMENTS: The yellow gills and viscid yellow cap when young plus the beautiful violet-tinged flesh (see color plate!) and bulbous stalk form a distinctive set of characteristics. In our local form the purplish phase of the gills is often transient and difficult to detect. *C. aureofulvus* is a similar oak-loving species which may show violet in the flesh but not in the gills. *C. atkinsonianus* shows the same transient violet gill phase as *C. cedretorum*, but has an olive-yellow to yellowish to deep reddish-brown cap and grows with hardwoods in eastern North America. *C. montanus* has violet-tinged gills with olive or yellowish-olive edges when young, and a rusty-brown to olive-brown, yellow-brown, or variegated/streaked cap. It is often common under old-growth conifers in western mountains, particularly the Cascades.

Cortinarius scaurus group

CAP 3-10 cm broad, convex to broadly convex with the margin inrolled somewhat at first, expanding in age; surface smooth, viscid when moist, color variable: usually olive-green to dark olive-brown, but in some varieties brownish at the center and greenish to greenish-yellow at the margin, and in others bright grass- or citrine-green, often with a browner center or brownish spots and fibrils. Flesh pallid to yellowish, thick, firm; odor and taste usually mild. **GILLS** close, notched or adnexed to adnate, typically olive to greenish-yellow (in some forms with a fleeting initial lavender or bluish-violet stage), but in some forms yellow when young, all forms becoming browner or rustier as spores mature. **STALK** 4-10 cm long, 0.5-1.5 (2) cm thick at apex, equal above, the base with a more or less rimmed, abrupt bulb when young that may become oblique or obscure in age; usually greenish like the cap or yellower or even pallid (in age), the apex sometimes bluish, the bulb often sulfur-yellow; solid, not viscid. **CORTINA** olive to yellow, usually leaving hairs on stalk which turn rusty-brown from falling spores. **SPORE PRINT** rusty-brown; spores 9-13 × 6-7.5 microns, elliptical, roughened.

HABITAT: Solitary to scattered or gregarious on ground in woods; widely distributed. Some members of this group (see comments) favor conifers and are fairly common in the Pacific Northwest; others grow under oak and other hardwoods, including one unidentified variety that is common in our area in the fall and winter with live oak (see comments).

EDIBILITY: Due to difficulty in identification, this entire group should be avoided. The group as a whole is *probably* edible, but I wouldn't bet my life on it—and that's just what you do when testing a *Cortinarius!*

COMMENTS: The above description has been broadened to include a number of bright green to greenish-yellow species with a viscid cap and well-defined bulb (at least when young). Mycologists prefer to refer to them collectively as the *C. scaurus* group (when they refer to them at all, which is seldom), for they are a confusing lot, both in the field and in the literature. They are readily distinguished as a group, however, by their color. The "true" *C. scaurus* of Europe is said to have a spotted cap, slender stalk (less than 1 cm thick), and grow with conifers. Some of the other species in the group include: *C. herpeticus,* very similar but with a thicker stalk and often with violet-tinged immature gills and flesh and a whitish basal bulb; *C. prasinus,* with an olive-colored cap, olive to olive-yellow gills, and yellowish or olive-tinged stalk, fairly common in the Pacific Northwest under conifers; *C. virentophyllus,* with a deep green cap that fades to yellow as it ages, green gills, and a pale bluish stem, found under hardwoods in eastern North America; *C. flavovirens*, with a farinaceous odor, greenish-yellow gills, and yellow stalk, reported from southern California; *C. orichalceus* and *C. rufo-olivaceus*, with a reddish to reddish-brown cap (at least at center), and greenish to yellowish stalk (the latter species is quite large and robust, the former is smaller and can have an oranger cap); and finally, a distinctive but unidentified species which is sometimes common under live oak in California. It has a bright citrine-green cap (sometimes shaded with brown at the center), yellow gills when young, and a yellowish stalk with a bulb at the base. Whether it is one of several similar European species, or is unnamed and endemic to California is unclear, but it *is* definitely a member of the *C. scaurus* group, which should satisfy all but the most hardcore *Cortinarius* -categorizers. See also *C. montanus* (under *C. cedretorum*).

Cortinarius fulmineus

CAP 5-15 cm broad, convex becoming plane, the margin at first incurved; surface viscid when moist, entirely yellow or with only the margin yellow and the center fulvous to ochre to orange-brown, sometimes with small spotlike scales. Flesh thick, firm, yellow to yellowish-white or buff; odor and taste mild or radishlike. GILLS close, adnate or notched, yellow to yellow-brown or ochre, becoming ochre-cinnamon and finally rusty-brown as spores mature. STALK 3-7 cm long, 1-3 cm thick at apex, often rather short and stout, with a large rimmed bulb at the base; solid, firm, not viscid; yellowish-white or yellow, sometimes becoming ochraceous in age. CORTINA pallid or yellowish, often leaving hairs on stalk which turn rusty-brown from falling spores. SPORE PRINT rusty-brown; spores 8-10 × 4.5-6 microns, elliptical, roughened.

HABITAT: Solitary to widely scattered or gregarious on ground under hardwoods; widely distributed. It is not uncommon in our area in the fall and winter in mixed woods and under live oak.

EDIBILITY: Unknown.

COMMENTS: This species typifies a number of yellowish to rusty-orange Cortinarii with a viscid cap, yellowish to orange gills, and a distinctly rimmed bulb when young. The rimmed bulb places it in the colorful subgenus Bulbopodium, and the yellowish or pallid flesh separates it from another "Bulbopodium," *C. cedretorum.* Other species: *C. fulgens* is similar but brighter orange in color; *C. elegantior* is also similar but has larger spores (12-16 microns long), while *C. elegantioides* has even larger spores and a decidedly bitter taste. There are also several beautiful "Bulbopodiums" with a bright yellow cap (or yellow-margined with a redder, browner, or oranger center) and lilac to violet gills at first. These include: *C. metarius,* common under spruce and fir in western North America, with a fairly broad basal bulb; *C. calochrous,* widespread under both hardwoods and conifers and very similar to *C. metarius,* but with a smaller bulb and slightly shorter spores; *C. citrinipedes,* cap yellowish with a darker brown margin; and *C. cyanopus (=C. amoenelens)*, widespread, with an ochre cap, bitter-tasting cap cuticle, and deep violet gills when young.

Cortinarius multiformis

CAP 4-12 cm broad, convex becoming plane or broadly umbonate, the margin at first inrolled; surface smooth, viscid when moist, ochre to ochre-buff to tawny or yellow-brown or occasionally more reddish or fulvous and sometimes becoming rustier in age, often with a whitish silky bloom when young. Flesh thick, firm, pallid; odor mild or slightly pungent. **GILLS** adnate to adnexed or notched, close, at first whitish, then tan or watery brown, finally rusty-cinnamon. **STALK** 4-10 cm long, 1-2 (2.5) cm thick at apex, usually with a bulb at the base that is abrupt when young but often obscure or even absent in age; dry, solid, firm, white or discoloring tan or ochre. **UNIVERSAL VEIL** disappearing or leaving a thin whitish film on cap. **CORTINA** scanty, white, disappearing or leaving a few hairs on stalk which trap falling spores. **SPORE PRINT** cinnamon-brown; spores 7-11 × 4-6 microns, elliptical, minutely roughened.

HABITAT: Solitary to scattered or gregarious on ground in woods (mainly under conifers); widely distributed. It is common throughout much of the West, but infrequent or absent in our area. I have seen large fruitings under spruce near Santa Fe, New Mexico.

EDIBILITY: Unknown. It is eaten in Europe, but I can find no information on the North American version, and there are too many similar species for me to recommend it.

COMMENTS: The viscid, ochre to tan cap and pallid to tan immature gills serve to distinguish this species from most other Cortinarii. The cap and gill color suggest a *Hebeloma*, but the spores are cinnamon-brown and the odor is not radishlike. The hoary white film on young caps is reminiscent of the gypsy mushroom *(Rozites caperata)*, but the veil is not membranous. There are many similar Cortinarii with non-viscid caps that sometimes mingle with *C. multiformis* (see *C. laniger*). Other viscid-capped species with an equal to club-shaped (but not abruptly bulbous) stalk include: *C. turmalis,* with a tawny to fulvous or yellow-brown to darker brown cap and a well-developed white veil that usually leaves a distinct fibrillose sheath and/or ring on the stalk; *C. cliduchus* and *C. latus,* without such a copious veil but otherwise similar to *C. turmalis* (the latter with a paler cap); *C. crassus*, a stout, thick-stemmed species (see comments under *C. balteatus*); *C. varius,* a widespread species with a fulvous to bright yellow-brown or ochre cap and *violet* gills when young plus a white stalk; *C. variicolor*, with a purplish to brownish cap, lilac gills when young, and a fibrillose, whitish or lilac-tinged stalk; and *C. claricolor*, with bluish-gray to brownish gills, a yellower cap, and more copious universal veil. All but the last species favor conifers and occur fairly commonly in the Pacific Northwest and/or Rocky Mountains.

Cortinarius magnivelatus

CAP 3-10 cm broad, convex becoming plane or irregularly undulating; surface moist or dry but not viscid, smooth or fibrillose, entirely white or white at margin and ochre toward the center, often stained buff or ochraceous in age or in bruised areas; margin incurved. Flesh thick, firm, white to buff; odor mild. **GILLS** close, slightly decurrent to adnate or adnexed, whitish when very young, becoming buff and then tawny-brown to brown; often forked near the stalk. **STALK** 4-6 cm long, 1-3 cm thick at apex, often (but not always) enlarged below, colored like cap or whiter, solid. **VEIL** thick, tough, membranous, elastic, whitish, typically covering the gills through maturity or shredding radially but not normally detaching from the cap and stalk. **SPORE PRINT** rusty-brown; spores 9-14 × 6-8 microns, elliptical, roughened.

HABITAT: Solitary to gregarious under mountain conifers (especially fir and pine), usually buried in the duff; fairly common throughout the higher mountains of California, particularly in the late spring and summer, but probably more widespread.

EDIBILITY: Unknown.

Two conifer-loving Cortinarii with persistent veils. *Cortinarius magnivelatus* (two at right) has a tough, whitish veil that is reminiscent of a rubber band. *Cortinarius verrucisporus* (three at left) is yellower and has a short stalk and thinner veil that tends to tear radially (see comments below).

COMMENTS: The whitish color, underground growth habit, and tough, thick, persistent veil distinguish this *Cortinarius* from most others. Presumably the latter features are adaptations to the pendulum-like changes of weather that occur in our western mountains. Many Cortinarii (particularly "Bulbopodiums") tend to fruit under the duff in dry weather, but they do not have the persistent membranous veil of this species. That the veil does not break suggests that this species is on its way to becoming "gastroid." The spores, however, are forcibly discharged (hence a spore print is obtainable), unlike the truly gastroid genus *Thaxterogaster* (see p. 734). The membranous nature of the veil can lead to confusion with other brown-spored genera, but the overall aspect is clearly "cortinarioid." Several other semi-gastroid Cortinarii occur in western North America. Like *C. magnivelatus,* they have a *persistent* veil and tend to grow under the duff, although the mature caps may surface if conditions are wet enough. These species include: *C. wiebeae,* discovered in the Cascades, similarly colored but larger with thin, fragile gills that are brown or rusty (not whitish) when young; *C. verrucisporus,* fairly common in the Sierra Nevada, with a yellow to yellow-brown or rusty-stained cap and a thinner yellowish veil that stretches from the stalk or stalk base to the cap margin and usually tears radially (see above photo), plus broad gills, a frequently underdeveloped stalk, and very coarsely warted spores; *C. bigelowii,* fairly common in Idaho, with a pale yellow-brown to yellowish cap, whitish veil, and a short, fat bulbous stalk; and *C. velatus* of the Sierra Nevada, easily distinguished by its purplish to lavender-tinged cap and somewhat thinner, semi-cobwebby (but persistent) veil. Other semi-gastroid species undoubtedly occur in the West but have yet to be described.

Cortinarius regalis

CAP (4) 6-15 cm broad, convex with an inrolled margin, becoming broadly umbonate to plane or with margin slightly uplifted in age; surface dry or very slightly tacky, usually with a large patch or patches of white fibrillose universal veil tissue at first; background brownish to dingy flesh-colored to dull brown with a vinaceous tinge, often appearing somewhat streaked. Flesh thick, pallid or tinged vinaceous or cap color; odor mild to somewhat musty or sometimes fruity. **GILLS** usually adnexed or notched, pallid in button stage, becoming grayish-brown to dull watery-brown and eventually dark brown. **STALK** 6-16 cm long, (1) 2-3.5 cm thick at apex, with a large, abrupt and/ or rimmed bulb at base (3-6 cm thick); dry, solid, firm pallid or tinged cap color, fibrillose. **UNIVERSAL VEIL**

whitish, felty-fibrillose, often leaving remnants on cap and fibrils or hairs on stalk which become rusty-stained. **CORTINA** white, disappearing or also leaving hairs on stalk. **SPORE PRINT** dull rusty-brown; spores 7-11 × 5-6 microns, elliptical, finely roughened.

HABITAT: Solitary, scattered, or in groups on ground in woods; known only from California, where it usually grows with oak. I find it regularly in the fall and winter.

EDIBILITY: Unknown.

COMMENTS: The fairly large size, thick bulbous stalk, absence of distinct violet hues, and whitish veil material on the cap combine to distinguish this *Cortinarius*. The bulb is typical of the subgenus Bulbopodium, but the cap is not truly viscid, making this species something of an anomaly. Viscid-capped species with conspicuous whitish veil remnants on the cap *and* a prominent bulb at the base of the stalk include: *C. calyptratus,* medium-sized, with violet gills when young, occurring under conifers in northern California; and *C. calyptrodermus,* larger, with deep violet gills when young, occurring under hardwoods in eastern North America. Another viscid-capped species with purplish (or purplish-blue) gills when young, *C. volvatus,* has a universal veil which forms a whitish sheath or "volva" on the basal bulb, but does not normally leave conspicuous remnants on the cap.

Cortinarius gentilis   (Deadly Cortinarius)

CAP 1-5 cm broad, conical or bell-shaped at first, expanding somewhat in age but usually retaining an umbo; surface smooth, *not* viscid but somewhat hygrophanous, tawny to ochre to orange-brown or rusty-yellow, fading in age or as it dries. Flesh thin, yellowish. **GILLS** fairly well-spaced, adnexed to adnate, ochre-yellow to cinnamon-brown becoming browner in age. **STALK** 3-10 cm long, 3-5 (7) mm thick, more or less equal, colored like the cap or more cinnamon-colored; often (but not always) with traces of the yellow veil when young; not viscid. **UNIVERSAL VEIL** yellow, disappearing or leaving a few patches or hairs on stalk. **CORTINA** also yellow, disappearing or leaving hairs on stalk. **SPORE PRINT** rusty-brown; spores 7-9 × 5.5-7 microns, elliptical, minutely roughened.

HABITAT: Scattered to gregarious or in troops in moss or duff under conifers; widely distributed. It is sometimes abundant in the summer and fall in the Rocky Mountains and Pacific Northwest, but is not likely to be collected by the average mushroom hunter.

EDIBILITY: DEADLY POISONOUS! Fortunately, it is seldom eaten because of its small size. Along with several other Cortinarii in the subgenus Leprocybe (notably *C. orellanus* and *C. speciosissimus*), it contains toxins which destroy the liver. Symptoms are greatly delayed (it may be two weeks before they appear!), making the culprit difficult to pinpoint.

COMMENTS: This species would be just another "LBM" were it not for the fact that it is deadly poisonous. The yellow-brown to dull orange-brown overall color, somewhat hygrophanous cap, small size, yellow veil, and rusty-brown spores form a distinctly undistinguished set of distinguishing features. It is most likely to be confused with the *C. cinnamomeus* group, which is more brightly colored, does not have a hygrophanous cap, and contains different pigments (anthraquinones). Related species thought to be deadly poisonous include *C. rainierensis, C. orellanus,* and *C. speciosissimus.* The latter two species have caused a rash of deaths in Europe. They are somewhat larger (cap 3-9 cm broad and tawny-brown to orange-brown, fulvous, or cinnamon-colored) than *C. gentilis; C. speciosissimus* has an umbonate cap, *C. orellanus* has an un-umbonate cap and a yellowish stem. They may well occur in North America, but their presence has not yet been definitely established. *C. rainierensis,* on the other hand, was originally described from the Pacific Northwest (under conifers). It is rusty-orange to tawny or brownish-orange, and its cap is covered with small, more or less erect or protruding scales.

Cortinarius cotoneus group   (Scaly Cortinarius)

CAP 3-10 cm broad, obtuse to convex becoming plane; surface dry, covered with small, dark, fibrillose scales and appearing blackish to olive-brown at the center where the scales are densest, and greenish to olive-brown, olive-yellow, or yellow-brown toward the margin (sometimes paler overall in age). Flesh brownish to pale olive-yellow; odor often somewhat radishlike. **GILLS** adnate to adnexed or notched, rusty-yellow to olive-yellow to dull yellowish or yellow-brown, becoming dull orange to dark cinnamon-brown in age. **STALK** 4-15 cm long, 0.8-3 cm thick, equal or thicker at base; dry, solid, pale olive-yellow or yellowish above, tawny to olive or colored like cap below, often dingier or browner throughout in age. **UNIVERSAL VEIL** yellow or olive-yellow, often leaving hairy fibrils or patches on lower stalk. **CORTINA** disappearing or leaving a zone of hairs on upper stalk which trap falling spores. **SPORE PRINT** rusty-brown; spores 7-9 × 5-7.5 microns, elliptical to nearly round, roughened.

HABITAT: Solitary to scattered or gregarious on ground in woods, widely distributed. Our representative of this group is fairly common in tanoak-madrone woods in the late fall and winter, often mingling with *C. collinitus* and *C. infractus*. In the Pacific Northwest, however, it is common under conifers.

EDIBILITY: Unknown, but possibly very dangerous! It belongs to the same group (subgenus Leprocybe) as the deadly poisonous *C. gentilis, C. orellanus,* and *C. speciosissimus.*

COMMENTS: The dry cap with small blackish scales on an olive-yellow to yellow-brown background distinguishes this *Cortinarius* "complex" from most others. It is one of several distinctive Cortinarii with a hairy and/or scaly cap. Others include: *C. clandestinus,* practically identical but with pallid gills when very young; the deep violet *C. violaceus* (see description); and a striking trio of eastern and northern hardwood-lovers: *C. bolaris, C. pholideus,* and *C. squamulosus* (see the latter species for more details).

Cortinarius squamulosus   Color Plate 108
(Bulbous Scaly Cortinarius)

CAP 4-10 cm broad, convex to broadly umbonate or plane; surface dry, fibrillose, soon breaking up into dense fibrillose scales, brown tinged with purple at first, usually more or less chocolate-brown in age, the background paler (yellower or more cinnamon). Flesh rather thick, whitish to grayish or pinkish-tinged; odor often somewhat spicy, especially in age. **GILLS** sometimes adnate at first but usually deeply notched by maturity, close, dark purplish or purplish-brown soon becoming brown to cinnamon-brown or chocolate-brown. **STALK** 6-15 cm long, 1-2 cm thick at apex, with a conspicuous basal bulb 3-6 cm broad; smooth to fibrillose or slightly scaly, at first purplish-tinged, becoming more or less cap-colored, often with a median bandlike ring (from universal veil?). **CORTINA** whitish to brownish, disappearing or leaving a few hairs on upper stalk. **SPORE PRINT** dark rusty-brown; spores 6-8 × 5-7 microns, elliptical to nearly round, roughened.

HABITAT: Solitary to gregarious under hardwoods (especially oak), often in low wet woods; not uncommon in the late summer and fall in northeasern North America. I have seen impressive fruitings in Minnesota and Wisconsin, but have yet to find it on the west coast. One source reports it from California but does not mention the habitat.

EDIBILITY: "Consistency very pleasant and flavor fairly good," says McIlvaine, who ate almost anything. However, several relatives are either poisonous or haven't been tested.

COMMENTS: Like *C. violaceus,* this species is one of the few truly distinctive Cortinarii (i.e., one that beginners can recognize in the field), but like that species it is not particularly common. The dry, purplish-brown to brown, densely scaly cap plus the radically bulbous stem are its main features. Another scaly-capped species, *C. pholideus,* is common under birch in the Northeast, often on or near rotten logs, and also occurs rarely in the Pacific Northwest. Its stalk is equal or only slightly thicker at the base and is decorated with distinctive brown to cinnamon-brown scales like those on the cap (see photo on p. 424) and the fruiting body is slightly more cinnamon-colored overall than *C. squamulosus.* It might be mistaken for a *Pholiota* because of the brown scales on the stem and tendency to grow on or near rotting logs, but the *dry* scaly cap, violet-tinged stalk apex (when young), and roughened, cinnamon-brown spores distinguish it. Still another distinctive hardwood-loving easterner, *C. bolaris,* is quite different: its cap and stalk feature reddish scales, fibrils, and/or streaks on a whitish to yellowish background (see photo on p. 425 and couplet #89 of the key to *Cortinarius*).

Cortinarius violaceus   (Violet Cortinarius)   Color Plate 109

CAP 3.5-12 (15) cm broad, convex becoming broadly convex, broadly umbonate, or plane; surface dry, densely covered with minute erect, tufted hairs or small scales, giving it a rough, somewhat velvety appearance; deep violet to nearly black, often with a metallic luster in age; margin often somewhat paler and fringed or ragged. Flesh thick, deep violet becoming grayish-violet; odor mild or cedarlike. **GILLS** adnate becoming adnexed or notched, fairly well-spaced, deep violet or colored like cap, then dusted with cinnamon-brown spores. **STALK** 6-18 cm long, 1-2.5 cm thick at apex, equal or more often thicker below, dry, fibrillose or woolly, deep violet, solid, firm. **CORTINA** violet, soon disappearing or leaving a few indistinct hairs near top of stalk which may catch falling spores. **SPORE PRINT** rusty-brown; spores 13-17 × 7-10 microns, broadly elliptical to oblong, roughened. Cystidia present on both the faces and edges of the gills.

HABITAT: Solitary or in twos and threes under conifers, sometimes next to rotting logs; widely distributed but quite rare except in certain localities, such as the old-growth coniferous forests of Mt. Rainier and Olympic national parks in Washington. In our area it shows up occasionally in the late fall and winter in mixed woods; in northern California it is sometimes quite common under Sitka spruce; in Europe it is said to favor hardwoods.

EDIBILITY: Edible, but not choice. Its principal appeal is its beauty.

COMMENTS: There are dozens upon dozens of violet Cortinarii, but none are as deeply colored as this one. Its hue alone makes it as unique as it is unforgettable—the color photograph hardly does it justice. The cap is dry and rough due to the presence of many small scales or tufted fibrils, another distinctive feature. The color is at times so deep that it borders on black, making it difficult to pick out in the shade of the forest. Its only rivals for color are some of the deep blue or violet Leptonias (see *L. carnea* and *L. nigroviolacea*), but they lack a cortina and have pinkish spores. Several colorful Cortinarii make wonderful dyes (e.g., *C. sanguineus, C. semisanguineus*), but *C. violaceus,* despite its intense color, is not one of them.

It is ironic to note that *C. violaceus,* the species chosen to typify the immense genus *Cortinarius,* is perhaps the *least* typical of the 1000+ species! Besides its unique color it is practically the only species to possess sterile cells (cystidia) on both the edges and faces of the gills. This is the kind of difference to which the "splitters" (see p. 10) owe a living, and *C. violaceus* might indeed have its own little genus by now, were it not that under the existing rules of the International Code of Botanical Nomenclature, the "new" genus would have to be *Cortinarius* (since *C. violaceus* is the type species or "model" of *Cortinarius*), and all of the other 1000+ Cortinarii would have to be placed in a new genus (or genera), thus compounding the confusion that already exists. Not even the most ardent "splitter" wants to be responsible for a mess like that! Perhaps as compensation, *C. violaceus* is itself split by some "splitters" into two species: *C. violaceus,* with oblong-elliptical spores, and *C. hercynicus,* with rounder spores.

Cortinarius traganus   (Lilac Conifer Cortinarius)   **Color Plate 106**

CAP 4-13 cm broad, obtuse or convex becoming plane or broadly umbonate; surface smooth, dry, finely silky or fibrillose, violet to lilac, sometimes with white wedge-shaped sectors, occasionally also with rusty or ochraceous stains; margin often hung with veil remnants. Flesh rusty-brown to tawny- or yellow-brown (often marbled) in the stalk, usually paler in the cap and yellower in the stalk base; odor often faintly pungent or sweet. **GILLS** pale cinnamon or ochre-buff becoming cinnamon- or rusty-brown in age; fairly well-spaced, adnexed to adnate. **STALK** 5-12 cm long, 1-3 (5) cm thick, usually enlarged below, solid, dry, finely fibrillose, lilac or purplish, sometimes also with white areas. **CORTINA** pale lilac, leaving hairs on upper stalk which turn rusty-brown from falling spores. **SPORE PRINT** rusty-brown; spores 7-10 × 5-6 microns, elliptical, roughened.

HABITAT: Solitary, scattered, or gregarious (sometimes in clumps) under conifers; widely distributed. It is a common late summer and fall mushroom in the mossy, old-growth forests of the Pacific Northwest and northern California, but does not seem to extend south into our area. I have often found it growing with the gypsy mushroom, *Rozites caperata.*

EDIBILITY: Not firmly established (like so many of us!), but poisonous according to one source and merely "indigestible" according to another.

COMMENTS: This characteristic inhabitant of northern coniferous forests is easily recognized by its beautiful lilac or lavender color, rusty-brown gills and spores, and rusty to tawny-brown flesh in the stalk. The latter feature distinguishes it from most other members of the subgenus Sericeocybe of *Cortinarius* (see *C. alboviolaceus*), and the sweetish odor ("like overripe pears") that is frequently present is also distinctive. The cap is not viscid as in *C. olympianus, C. sodagnitus,* and others, and the presence of a cortina separates it from the edible blewit *(Clitocybe nuda).* Some authorities consider the American variant to be a distinct species, *C. pyriodorus,* while reserving the name *C. traganus* for the European version, which is said to smell "like goats." Other species: *C. fragrans* also smells like overripe pears, but has a slightly paler cap and whitish or lavender-tinged flesh in the stalk. *C. caesiifolius* has a buff to brownish cap, bluish gills when young, and a fleeting, disagreeable odor. *C. camphoratus* has a lavender to lilac-white cap (fading to buff), violet or bluish-lilac gills and stalk when young, and a powerful, remarkably raunchy odor of rotting meat or vegetables. The stench alone distinguishes it from other lilac Cortinarii, although a more deeply colored species, *C. amethystinus,* is said to smell "like burnt hair." *C. camphoratus* grows almost exclusively with conifers and is widely distributed. I have found it in northern California under Sitka spruce, in November.

Cortinarius alboviolaceus   (Silvery-Violet Cortinarius)

CAP 3-8 cm broad, obtusely bell-shaped becoming convex or broadly umbonate to nearly plane; surface dry, silky-shining, pale violet soon becoming pale silvery-violet, lilac-white, or even whitish. Flesh pallid to pale violet; odor mild. **GILLS** adnate or adnexed or notched, fairly close, pale violet to purple-gray, then eventually cinnamon-brown as spores ripen. **STALK** 4-12 cm long, 0.5-1 (1.5) cm thick at apex, usually club-shaped or enlarged at base, dry, silky, violet or pale violet above, clothed with whitish silky fibrils below (but pale violet underneath the fibrils). **UNIVERSAL VEIL** white, silky, usually forming a thin, soft, silky sheath over lower half of stalk. **CORTINA** white, evanescent or leaving hairs at top of stalk which trap falling spores. **SPORE PRINT** rusty-brown; spores 7-10 × 4-6 microns, elliptical, minutely roughened.

HABITAT: Scattered to gregarious or in small clumps in forest humus, associated mainly with hardwoods; widely distributed. It turns up occasionally in our area in the late fall and winter under tanoak and other trees, but I have never seen it in large numbers.

EDIBILITY: Unknown.

Left: *Cortinarius alboviolaceus.* Note the rather long stalk sheathed with copious universal veil remnants. **Right:** *Cortinarius argentatus* (see comments under *C. alboviolaceus*) is also pale violet or bluish-violet but has a thicker, stouter stalk with a bulb at the base.

COMMENTS: The beautiful silvery-violet to lilac-white overall color plus the dry silky cap, mild odor, and whitish universal veil sheath on the stem distinguish this attractive mushroom from its close relatives in the subgenus Sericeocybe (a closely-knit group of dry-capped Cortinarii with violet to pale lilac pigments). The flesh is never rusty-brown or tawny as in *C. traganus,* and the gills are violet-tinged when young. *Inocybe lilacina* is somewhat similar but smaller and more umbonate, with dull brown spores and a spermatic odor. Similar species include: *C. argentatus,* lilac-blue to violet-gray but with a thicker, stouter, prominently bulbous stalk (see above photo) that lacks a silky white sheath and with a radishlike odor and tendency to turn ochre in age, found mostly under hardwoods (especially eastern); *C. subargentatus,* favoring hardwoods, very similar to *C. argentatus,* but lacking the odor; *C. subpulchrifolius,* with a grayish-buff cap that often develops ochraceous or rusty stains in age, dull purplish immature gills, and a dull purplish, equal or club-shaped stalk that is sheathed by veil remnants, found under hardwoods in eastern North America; *C. malachius,* favoring conifers, violet-tinged at first but its cap soon becoming ochre or brownish; and *C. caninus* and *C. anomalus,* growing under both hardwoods and conifers, the former with a violet-tinged cap that becomes reddish-brown and buff to tan universal veil zones on the stalk, the latter with a violet-tinged to gray or brown cap, distinctly violet flesh and violet stalk apex, and ochre veil zones on the stalk. For more odoriferous and/ or brightly colored "Sericeocybes," see *C. traganus,* and for violet or violet-tinged hygrophanous Cortinarii, see *C. evernius.*

Cortinarius armillatus   (Bracelet Cortinarius)

CAP 5-13 cm broad, obtuse to broadly bell-shaped or convex, becoming nearly plane or broadly umbonate; surface smooth or sometimes with small scales in age, not viscid and only slightly hygrophanous (if at all); dull tawny or yellow-brown to rusty-brown, orange-brown, or reddish-brown; margin sometimes hung with veil remnants. Flesh thick, pallid or brownish; odor usually radishlike. **GILLS** fairly well-spaced, broad, adnate to adnexed or notched, pale or dull cinnamon becoming rusty-brown as spores mature. **STALK** 7-15 cm long, 1-2.5 cm thick at apex, club-shaped (thicker below), dry, whitish to brownish with one or more (usually 2-3) dull red bands or "bracelets" below the cortina. **UNIVERSAL VEIL** fibrillose, forming reddish bands on stalk. **CORTINA** whitish and copious, often leaving hairs on upper stalk that turn rusty-brown from falling spores. **SPORE PRINT** rusty-brown; spores (7) 9-13 × 5.5-7.5 microns, elliptical, roughened.

Left: *Cortinarius armillatus,* a birch-lover with reddish bracelets on stalk. **Right:** *Cortinarius boulderensis* (see comments below) has reddish or vinaceous bracelets but is smaller and purple-tinged.

HABITAT: Solitary or more often scattered or in groups on ground in woods; associated primarily if not exclusively with birch and found throughout the range of birch (therefore absent in California). It is especially common in northeastern North America in the late summer and fall, but I have also seen it in northern Idaho and British Columbia.

EDIBILITY: A good edible—as I can personally attest. It is one of the most distinctive of the Cortinarii, but you should be *very* cautious nevertheless!

COMMENTS: The hallmarks of this common eastern species are the reddish bracelets on the stalk, yellow-brown to cinnamon-brown cap, and association with birch. In the Pacific Northwest a slightly smaller conifer-lover, *C. haematochelis,* occurs. It also has reddish bracelet(s) on the stem, but its spores are much smaller and nearly round. A third species with reddish (or vinaceous) fibrillose patches or "bracelets," *C. boulderensis,* is even smaller (cap 2-5 cm broad, stalk 3-7 mm thick at apex), with a violet-brown to reddish-brown or brownish cap and violet-tinged immature gills and stalk apex. It also grows under conifers in the Pacific Northwest, but I have found something very similar—if not the same—in our area in mixed woods (see photo above). *C. subtestaceus* is a similar but smaller easterner with reddish-brown to brown "bracelets." Also see *C. anomalus* (under *C. alboviolaceus*) and *C. crocolitus* (under the *C. collinitus* group), both of which have more ochraceous-colored bands or patches on the stalk.

Cortinarius rubripes (Red-Footed Cortinarius)

CAP 4-12 cm broad, obtuse to convex becoming plane or broadly umbonate; surface hygrophanous but not viscid, watery cinnamon to reddish-brown fading to tawny, ochraceous, or ochre-buff, sometimes tinged with pink or sometimes with concentric zones; finely silky-fibrillose in age, the margin sometimes tinged violet when young. Flesh thin, reddish-brown or pallid; odor mild. **GILLS** fairly well-spaced, adnate to adnexed or notched and seceding; often tinged violet in button stage, but soon tawny to cinnamon, and finally cinnamon-brown. **STALK** 4-9 cm long, 0.5-1.5 cm thick at apex, enlarged below, dry, pallid or brownish above, the base and mycelium fiery orange to red-orange or peach-colored. **UNIVERSAL VEIL** red-orange, leaving remnants at base of stalk. **CORTINA** disappearing or leaving a few hairs on upper stalk. **SPORE PRINT** rusty-brown; spores 7-10 × 5-6 microns, elliptical, roughened.

HABITAT: Solitary to scattered or gregarious under hardwoods; originally described from Michigan, but it—or something very much like it—also grows in our oak-tanoak-madrone woods in December and January. In eastern North America it is sometimes abundant after late summer rains.

EDIBILITY: Unknown.

COMMENTS: The bright orange to fiery red stalk base and overall brown to reddish-brown color combine to distinguish this *Cortinarius* from most others. A European species, *C. builliardi,* is probably the same, in which case that would be the "correct" name for our species. *C. armillatus* and *C. haematochelis* are somewhat similar, but normally show one to four dull reddish bands on the stalk rather than having a bright red-orange base. *C. miniatopus* is a small species (cap 1-3 cm broad) with orangish gills when young and a slender stalk that is wholly or partially sheathed by vermillion fibrils (densest at the base); it is not uncommon under conifers in the Pacific Northwest.

Cortinarius evernius

CAP 3-10 cm broad, conical to bell-shaped at first (and scarcely wider than the stalk), then becoming convex or obtusely umbonate to plane; surface smooth, markedly hygrophanous: violet or brown with a purple tinge when moist, quickly fading to vinaceous or reddish-brown or paler as it loses moisture; margin at first with whitish silkiness. Flesh thin, at first violet or violet-tinged, but fading as it dries or ages. **GILLS** at first violet with whitish edges, but quickly fading to brown and then darkening to cinnamon-brown; adnate or becoming notched, well-spaced. **STALK** 7-15 (20) cm long, 0.8-2 cm thick, equal or narrowed toward base, usually rather long and often extending deep into the humus or moss; not viscid; pale to deep violet when fresh (darker below), but often covered with veil material at first, and fading as it dries. **UNIVERSAL VEIL** whitish or violet-tinged, forming fibrillose patches or zones on stalk which may disappear in age. **CORTINA** whitish, often disappearing. **SPORE PRINT** rusty-brown; spores 8-10 × 5-6 microns, elliptical, slightly roughened.

HABITAT: Solitary, scattered, or in groups under conifers, often in moss; widely distributed but mainly northern. It is not a common species, but several difficult-to-distinguish look-alikes (see comments) are quite ubiquitous, including at least two species in our area.

EDIBILITY: Unknown. Do not experiment!

COMMENTS: This species has been included to represent a large number of medium-sized, hygrophanous Cortinarii in the subgenus Telamonia that are violet or violet-tinged when fresh and moist. They are very difficult to identify and some are so markedly hygrophanous that they lose their original color in an hour or two—or in the time it takes to bring them home! Other "Telamonias" showing violet shades when fresh and moist include: *C. lucorum,* a fairly common conifer-lover that differs from *C. evernius* only in having a more or less club-shaped stem; *C. saturninus,* with a brown cap and more or less equal stalk that is violet above and whitish below; *C. brunneus,* which sometimes shows violet at the stalk apex (see comments under *C. laniger*) and has an equal to club-shaped stalk; *C. subpurpureus,* with yellow or yellowish-buff veil patches on the lower portion of the stalk; *C. adustus* and *C. impennis,* with little or no veil remnants on the stalk; and *C. plumiger,* with a grayish-violet to watery violet or grayish-blue-tinged stalk that quickly fades to whitish and a cap that is brownish to buff beneath a white to grayish-white, fibrillose-hairy layer. All of these are partial to conifers, but another species, *C. torvus,* favors hardwoods and is the most distinctive of the lot. It has broad, widely-spaced gills that are deep purplish when young, a brownish to deep purplish cap, sweetish odor, and club-shaped or bulbous stalk that often has a well-formed annulus (ring)—an unusual feature for a *Cortinarius*! It is fairly common in eastern North America but I can find no mention of it being found in the West. Another species that tends to form an annulus, *C. urbicus,* does occur in the West, but it has a much paler cap than *C. torvus.*

Cortinarius laniger   (Brown Cortinarius)

CAP 3-12 cm broad, broadly bell-shaped or convex becoming broadly convex or sometimes plane; surface smooth, hygrophanous but not viscid, dark reddish-brown or chestnut-brown to reddish-brown or cinnamon-brown when moist, fading as it loses moisture to pale reddish-brown or tan; often covered with a white silky coating when very young, especially at margin. Flesh whitish or colored like cap. **GILLS** fairly close or rather well-spaced, adnate to adnexed or notched, pale rusty-brown becoming a beautiful cinnamon-brown. **STALK** 3-10 cm long, 1-3 cm thick, enlarged at base (at least when young), dry, solid, firm; cinnamon to brown (like cap) beneath a coating or patches of fibrillose whitish veil remnants. **UNIVERSAL VEIL** fibrillose or slightly cottony, white, usually leaving a whitish silky coating or patches and zones of white fibrils on stalk and sometimes even a slight annulus (ring). **CORTINA** disappearing or leaving a few hairs on upper stalk. **SPORE PRINT** rusty-brown; spores 7-10 × 5-6 microns, elliptical, roughened.

HABITAT: Scattered to gregarious on ground under conifers; widely distributed. I have seen large fruitings under spruce in the Rocky Mountains, and have also found it in the Pacific Northwest. In our area similar species (see comments) are common under oak.

EDIBILITY: Unknown. Do not experiment!

COMMENTS: This medium-sized brown *Cortinarius* with the hygrophanous cap is one of a multitude of medium-sized brown Cortinarii with hygrophanous caps that belong to the baffling subgenus Telamonia. *C. laniger* is best recognized by its cinnamon-brown color and club-shaped stalk that is covered by whitish veil remnants, and by the complete absence of violet shades on the fruiting body. Similar "BUM's" (Boring Ubiquitous Mushrooms) are legion. Some have an equal to club-shaped or bulbous stalk like that of *C. laniger*. These include: *C. bulbosus*, with a paler or duller brown, swollen stalk and slightly smaller spores, found under both hardwoods and conifers; *C. bivelus,* especially common in the southern Rockies, with a cinnamon-brown to tawny-brown cap and larger spores; *C. triformis,* with a strongly hygrophanous cap (dark reddish-brown when moist, yellow-brown or tan as it dries out), but with a transient veil that leaves only slight traces (if any) on the stalk, found under both hardwoods and conifers but favoring oak in California; *C. distans,* with minute branlike scales on the cap and widely spaced gills, found under oak, ponderosa pine, etc.; *C. armeniacus,* with a swollen or bulbous stalk and deep yellow-brown cap that fades to orange-buff or pale yellow-orange as it loses moisture; *C. brunneus,* a widespread nondescript conifer-lover with an equal to club-shaped, dull brown stalk whose apex sometimes has a slight violet or vinaceous tinge; *C. dilutus* and *C. biformis,* both smaller with a more or less equal stalk and deep reddish-brown (moist) to buff (dry) cap, favoring northern conifers (the former with nearly round spores); and *C. pinetorum,* a fungal feature of western coniferous forests with a fibrillose, white to silvery-gray to pale brownish-gray cap that may be slightly tacky when wet, plus brownish gills and a whitish, more or less club-shaped stalk. Other species have a stalk which is spindle-shaped and/or tapered below (or occasionally equal). These include: *C. privignus* and *C. cacaocolor,* with a dark or dull vinaceous-brown to cocoa-colored cap when moist (the former developing a silky whitish sheen as it loses moisture) and stalk which lacks persistent veil remnants; *C. damascenus,* similar to the previous two species but with an equal or tapered stalk that is covered with a persistent coating of white veil fibrils; and *C. duracinus,* which has a whitish stalk and a cinnamon-brown cap that fades to buff or pale tan as it dries out. The above species are but a few of the hundreds of brownish "Telamonias." As already pointed out, recognition of these fungi is very difficult and none should be eaten. For medium-sized "Telamonias" with a violet tinge when moist, see *C. evernius,* and for smaller species, see the *C. obtusus* group.

Cortinarius obtusus group   (Little Brown Cortinarius)

CAP 1-4 cm broad, bell-shaped to conical at first, expanding in age but usually retaining a blunt umbo; surface smooth, hygrophanous but not viscid, brown to reddish-brown to tawny-brown and faintly striate when moist, fading to tan or paler as it loses moisture. Flesh thin, brownish; odor usually radish- or iodine-like (but often faint). **GILLS** light to dull brown becoming cinnamon-brown in age, close, adnate to adnexed. **STALK** 4-8 cm long, 2-6 mm thick, equal or often slightly thicker in the middle and tapered below; dry, hollow in age, fragile; colored like the cap or often yellower or paler, whitish or pallid as it loses moisture. **UNIVERSAL VEIL** and **CORTINA** scanty, pallid or whitish, disappearing or leaving a few fibrils on stalk. **SPORE PRINT** brown to rusty-brown; spores 8-10 × 4-5.5 microns, elliptical, roughened.

HABITAT: Scattered to densely gregarious on ground under conifers, especially pine and spruce; widely distributed. This species and its numerous look-alikes (see comments) are common in most regions, including ours.

EDIBILITY: Unknown. Do not experiment!

COMMENTS: The above description encompasses several small boring, brownish Cortinarii ("LBM's") sometimes referred to as the *C. obtusus-C. acutus* complex. (*C. acutus* is slightly slimmer, smaller, and paler than *C. obtusus,* with a more acute umbo; another species, *C. scandens,* is bluntly umbonate like *C. obtusus,* but has smaller spores.) The "true" *C. obtusus* and *C. acutus* both have rather pale stems (usually paler than the cap or even whitish). Because of their small size, they might be confused with Inocybes or Galerinas. The former, however, have duller brown spores while the latter usually have a yellower fruiting body and typically grow in moss or on wood. Other small "Telamonias" are legion, especially under northern or mountain conifers. Even experts have difficulty differentiating them from each other and there is absolutely no reason for the amateur to even try. Nevertheless, I will mention a few of them for the categorically-inclined: *C. subcuspidatus* and *C. fasciatus* resemble *C. acutus,* but the first has yellowish veil material on the stalk, or if not, can be told by its dark cinnamon-brown gills when young, and the latter has a darker stalk and tends to grow in clusters. *C. stemmatus* is a slightly larger species with a blackish-brown cap that fades to reddish-brown or reddish-cinnamon as it loses moisture plus a mild odor and whitish veil material on the stalk (see photograph); it is common under conifers (especially pine in our area). *C. nigrocuspidatus* can usually be told by its dark cap with a prominent black umbo, and *C. incisus* by its minutely scaly, brownish to reddish-brown cap and white veil remnants on the stalk. *C. psammocephalus* has tawny-brown fibrillose scales or patches on both the cap *and* stalk. *C. paleaceus* has a blackish-brown cap that fades to dingy brown in age, whitish veil remnants on the cap

Cortinarius stemmatus (see comments under *C. obtusus*) is one of dozens of small, brownish, difficult-to-demystify Cortinarii. The cap is blackish-brown when moist but turns reddish-brown as it loses moisture. Note umbonate cap.

margin and stalk (at least when young), and a geranium-like odor; it grows in mossy bogs or other wet places under spruce. *C. decipiens* also has a dark cap (vinaceous-brown or darker when moist), but it lacks white veil remnants on the stalk, while *C. washingtonensis* can easily be told by its black-staining gills. All of the above species are especially common under northern conifers, but some may also occur with hardwoods. Finally, there are a number of small "Telamonias" that are distinctly violet or violet-tinged when fresh and moist, including: *C. pulchellus,* violet overall when young and especially fond of alder; and *C. subflexipes,* which has a rusty-brown to ochraceous cap but shows violet in the gills and stalk apex when young. For larger "Telamonias," see *C. laniger* and *C. evernius.*

Cortinarius cinnamomeus group

CAP 1.5-6 cm broad, obtusely conical becoming convex to more or less plane or often with an umbo; surface dry, finely fibrillose, yellowish to olive-yellow to tawny, ochre-buff, yellow-brown, or olive-brown, becoming more cinnamon-colored in age. Flesh thin, yellowish or olive-yellow. **GILLS** adnate to adnexed or notched, close, yellow to olive-yellow becoming tawny, then rustier in age as spores ripen. **STALK** 2.5-10 cm long, 3-5 (10) mm thick, more or less equal, often curved, dry, finely fibrillose or with small fibrillose scales, yellow to yellow-brown or ochraceous, often browner toward base and duller in age. **CORTINA** yellow or yellow-olive, disappearing or leaving inconspicuous hairs on upper stalk. **SPORE PRINT** rusty-brown; spores 6-9.5 × 4-5 microns, elliptical, minutely roughened.

HABITAT: Widely scattered to gregarious on ground under conifers, especially pine; widely distributed, and common in our area from late fall through early spring.

EDIBILITY: To be avoided! It is said to be edible, but some very similar species (see *C. gentilis)* are deadly poisonous!

COMMENTS: Also known as *Dermocybe cinnamomea,* this contrary little *Cortinarius* and its numerous look-alikes are not cinnamon-colored at all. Rather, the dominant colors are yellow, yellow-olive, and ochre. The color plus the slender stem, dry cap, and evanescent fibrillose veil (cortina) are characteristic of a large complex of species in the subgenus Dermocybe which are best left to the specialist. It may actually be that the "true" *C. cinnamomeus* does not occur here, but a number of very similar species certainly do, including: *C. cinnamomeo-luteus,* said to be very common in the Pacific Northwest; *C. humboldtensis* (from northern California) and *C. olivaceopictus,* the latter with reddish to orange fibrils on the stalk; *C. raphanoides,* olive-yellow, with a radish odor; and *C. thiersii,* with larger spores, especially common under mountain conifers. Another group of

Cortinarius cinnamomeus group is common under conifers, but identifying the various species in the "complex" is very difficult. Some forms have sharply conical caps **(right),** others do not **(left).**

species, typified by *C. croceofolius,* have saffron to rusty-yellow or orange gills when young and a browner or more cinnamon-colored cap. They are just as common as the *C. cinnamomeus* group, and their ranks include: *C. aurantiobasis,* with larger spores; *C. zakii,* with *deep* orange gills when young and stalk coated with brown to vinaceous-brown fibrils; and *C. aureifolius,* a very distinctive species that usually grows in sand (often partially buried!) under pines, and has orange to brownish-orange (or in one form, yellowish) gills, a cinnamon-brown to dark reddish-brown cap, and larger spores. Finally, there is *C. callisteus,* a bright yellow to orange-ochre species with a thick club-shaped stem (at least 1 cm thick at apex and 2.5 cm or more thick at base). None of these species should be eaten. (Note: some authorities have a concept of *C. cinnamomeus* which more closely fits the *C. croceofolius* of this book, and vice-versa, at least insofar as gill color is concerned.)

Cortinarius phoeniceus var. occidentalis

CAP 2.5-8 cm broad, convex becoming plane or broadly umbonate; surface dry, finely fibrillose or silky, maroon-red or dark red to reddish-brown; margin sometimes lobed. Flesh thin, reddish near cuticle, olive-brownish near gills; odor mild or slightly radishlike. GILLS deep red, vinaceous-red, or blood-red, becoming rustier as spores ripen; adnate to adnexed or notched. STALK 3-10 cm long, 0.4-1.2 cm thick, more or less equal, dry, fibrillose, yellowish or ochre or sometimes brownish in old age. CORTINA scanty, yellowish, disappearing or leaving a few hairs on upper stalk. SPORE PRINT rusty-brown; spores 6-8 × 4-5.5 microns, elliptical, minutely roughened.

HABITAT: Solitary to scattered or in groups under conifers in the Pacific Northwest and California; sometimes common in the fall and early winter. In our area it also grows with madrone and/or huckleberry, but never in large numbers.

EDIBILITY: Not recommended, but an excellent choice for dyeing (not dying!).

COMMENTS: The beautiful deep red gills and maroon-red to reddish-brown cap are set off nicely by the yellowish stem, making this *Cortinarius* (or *Dermocybe*) relatively easy to identify. Other species: *C. semisanguineus* also has red gills, but has a slender yellow stem *and* yellowish to ochre or olive-brown cap; it occurs in California, but is more common in eastern North America and the Pacific Northwest. Another species, *C. californicus,* has a smooth, *hygrophanous* cap that is dark rusty-red when moist but fades as it loses moisture. It has orange-red to rusty-orange or "burnt sienna" gills, and a rather long (7-15 cm) stalk that is paler than the cap (dull orange or even paler, often with orange fibrils from the cortina). It is known only from west coast and occurs in a variety of habitats, but seldom in large numbers. For completely red species, see *C. sanguineus.*

Cortinarius sanguineus   (Blood-Red Cortinarius)

CAP (1) 2-5 cm broad, obtusely bell-shaped becoming convex to plane or broadly umbonate; surface dry, silky or finely fibrillose, colored evenly deep carmine-red to blood-red or deep red. Flesh thin, blood-red to reddish-purple; odor mild or faintly pleasant. GILLS close, adnate to adnexed, blood-red to dark blood-red, but in age dusted with cinnamon-colored spores. STALK 4-10 cm long, 3-5 (8) mm thick, equal or slightly thicker below, same color as cap or slightly darker, dry, base with yellowish to yellow-orange or sometimes pinkish downy hairs. CORTINA red or at least tinged red, often scanty and disappearing or leaving a few hairs on upper stalk. SPORE PRINT rusty- to reddish-brown; spores 6-9 × 4-6 microns, elliptical, minutely roughened.

HABITAT: Solitary to widely scattered or in small groups, mainly under conifers and often in beds of moss; widely distributed but generally rare. In favorable years it fruits in the fall under spruce and fir along the northern California coast, but never in large numbers. I have found it as far south as Mendocino County in California.

EDIBILITY: Unknown. Do not experiment! Like many species in the subgenus Dermocybe, it can be used to dye wool or yarn, yielding a brilliant array of pinks, reds, and purples.

COMMENTS: Also known as *Dermocybe sanguinea,* this beautiful mushroom is easily recognized by its blood-red to dark red color, small size, and evanescent cortina. It might be mistaken for a brightly colored *Hygrocybe,* but the spores are not white and the gills are not waxy. There is no yellow in the cap and/ or stalk as in *C. phoeniceus* var. *occidentalis* and *C. semisanguineus,* and the cap is not hygrophanous as in *C. californicus* (see comments under *C. phoeniceus* var. *occidentalis*). Other completely red Dermocybes include: *C. puniceus,* said to differ by its ochraceous to golden-brown cortina, dark red to purple-red color, and preference for hardwoods; and *C. cinnabarinus,* also favoring hardwoods (but not limited to them), with a slightly larger, cinnabar-red to rusty-red or dark red fruiting body and larger spores. The latter species has often been reported from eastern North America, but according to *Cortinarius*-experts Alexander Smith and Joseph Ammirati, American material differs slightly from the European version. (At least one common hardwood-lover has been recognized as a "new" species, *C. marylandensis.*)

INOCYBE

Small to medium-sized, *typically terrestrial* mushrooms. CAP *often umbonate or conical; surface typically dry and silky, woolly, scaly, or radially fibrillose;* margin often splitting at maturity; *odor of flesh often distinctive* (usually unpleasant). GILLS adnate to adnexed or free, usually brown at maturity (but often with whitish edges). STALK central, often rather slender; apex often powdery or with small flakes. VEIL absent, or if present then fibrillose or cobwebby and usually evanescent. VOLVA absent. SPORE PRINT *brown.* Spores smooth, warty, angular, or nodulose, lacking an apical germ pore. Gills with cystidia, at least on edges. Cap cuticle filamentous.

INOCYBE is a large, listless, and lackluster assemblage of mundane, malodorous brown mushrooms that are of little interest to the average mushroom hunter except that many are poisonous. The best means of recognizing an *Inocybe* is by its characteristically silky, fibrillose, minutely scaly, and/ or woolly cap which is often umbonate and seldom viscid. The spore color is some shade of brown, and is generally duller than that of *Cortinarius.* In addition, most Inocybes have a noticeable odor—occasionally sweet or fruity as in *I. pyriodora,* but more often unpleasant (pungent, spermatic, fishy, or like fresh green corn but not often radishlike as in *Hebeloma*).

Like *Cortinarius,* Inocybes are largely terrestrial and mycorrhizal and are a major fungal facet of temperate forests. Unlike *Cortinarius,* they are not the least bit colorful. They come in an endless, senseless procession of boring browns, yucky yellows, gratuitous grays, and wishy-washy whites, with only *I. lilacina* (among the common species) deviating from the norm. Almost without exception they not only qualify as "LBM's" (p. 32), but as "BUM's" ("Boring Ubiquitous Mushrooms"), and it is necessary to know the size and shape of the spores and cystidia before an accurate identification can be made. Even then, unraveling them is a trying and tedious task whose futility is only exceeded by its pointlessness, and underscored by the sad fact that most Inocybes are poisonous. I would venture to guess, in fact, that *Inocybe* contains a higher percentage of poisonous species than any other major mushroom genus, including *Amanita*! The toxin is muscarine, which can be fatal in large amounts (see p. 894). It is therefore advisable to avoid *all* Inocybes, even those rumored to be edible.

Inocybe is a very large genus—almost as immense, in fact, as *Cortinarius.* (The late Daniel Stuntz listed over 400 North American species in section Inocybium alone, i.e., those with smooth spores and cystidia on the faces of the gills!) A mere nine Inocybes are described here, which for most intents and purposes are eight too many! Several other species are keyed out.

Key to Inocybe

1. Odor fragrant or fruity (though sometimes sickening, as in *rotten* fruit) 2
1. Odor spermatic, pungent, fishy, like green corn, mild, etc., but not fragrant or fruity 5

2. Cap olive to olive-brown at center .... *I. corydalina* (see *I. pyriodora,* p. 459)
2. Not as above 3

3. Either cap conical with reddish-brown to wine-red radiating fibrils *or* cap with a whitish
   patch of universal veil tissue and odor with a green corn component 7
3. Not as above 4

4. Odor of sweet peas .... *I. suaveolens* (see *I. pyriodora,* p. 459)
4. Odor spicy or fruity or otherwise (but not of sweet peas) ...... *I. pyriodora* & others, p. 459

5. Stalk base or lower portion dull bluish-green, olive, or dull greenish .. *I. calamistrata,* p. 462
5. Not as above 6

6. Cap grayish-lilac to grayish-brown or pinkish-brown with a smooth white or creamy, umbonate
   center (*or* in one form cap entirely whitish and stalk pinkish); stalk typically pruinose (granular-
   dandruffy) most of its length, with a bulb at base ..... *I. albodisca* (see *I. maculata,* p. 458)
6. Not as above (but may have some of above features) 7

7. Cap small (2-5 cm broad) and yellowish to brown, but with a patch of whitish universal veil
   tissue at the center (or sometimes overall), at least in most specimens; stalk white when young
   and *not* shaggy or woolly .... *I. lanatodisca* & others (see *I. maculata,* p. 458)
7. Not as above 8

8. Cap 3-8 cm broad, at first covered with a whitish universal veil that leaves patches of tissue near
   margin; stalk 5-10 mm thick, also sheathed with woolly remnants of the veil, dingy brownish
   beneath the remnants; widespread, but especially common in the Rockies ...*I. leucoblema*
8. Not as above 9

9. Stalk smooth or fibrillose or occasionally with a few small fibrillose patches or scales 10
9. Stalk distinctly scaly (i.e., with numerous small scales similar in color to those on cap) ... 25

10. Some part of fruiting body lilac to violet when fresh (but may fade or discolor in age) ....
    .... *I. lilacina* & others, p. 461
10. Not as above (but stalk or cap may be pink, carmine, vinaceous, etc.) 11

11. Cap white, smooth to very finely silky and *not* discoloring appreciably in age or when bruised;
    stalk *not* pruinose (granular-dandruffy) or pruinose only at apex ...... *I. geophylla,* p. 460
11. Not as above (but cap may be white initially and then discolor in age, or cap may have white veil
    remnants) 12

12. Cap and stalk whitish at first, but usually developing pink, red, or orange tones as it ages; very
    common, especially under conifers .... *I. pudica,* p. 460
12. Not as above 13

13. Cap conical when young, often with a pointed umbo in age, the surface radially fibrillose (and
    often splitting in age) but not woolly or scaly, creamy to straw-colored, yellow-brown or brown
    (but not reddish); gills *sometimes* with an olive or greenish tinge 14
13. Not as above 15

14. Odor resembling that of fresh green corn; cap creamy to straw-colored .... *I. sororia,* p. 457
14. Odor spermatic; cap usually somewhat darker (yellow-brown to brown), at least at the center
    .... *I. fastigiata* & others (see *I. sororia,* p. 457)

15. Cap brownish with a *very* prominent gray to dark brown or black "nipple" (umbo) at center;
    widely distributed ....*I. fuscodisca*
15. Not as above 16

16. Cap with reddish-brown to carmine or wine-red fibrils, more or less conical when young; stalk
    whitish at least at apex; found mainly with oak (in California) .... *I. jurana,* p. 458
16. Not as above 17

17. Stalk pale pink to salmon-pink; cap reddish-brown, at least at center ....
    .... *I. laetior* & *I. oblectabilis* (see *I. jurana,* p. 458)
17. Not as above 18

18. Stalk entirely pruinose (minutely granular-scurfy-dandruffy) *and* discoloring dark gray to dark brown or vinaceous-brown from the base upward in age .... *I. picrosma, I. leucomelaena,* & others
18. Not as above 19
19. Fruiting body fairly small (cap typically up to 4 cm broad; stalk less than 5 mm thick) ... 20
19. Fruiting body medium-sized (cap usually at least 3 cm broad and stalk at least 5 mm thick 23
20. Cap with loosely-arranged or torn-up fibrils that give it a shaggy or ragged appearance, especially at margin; color more or less dark brown .... *I. lacera* (see *I. lanuginosa,* p. 462)
20. Not as above; cap usually woolly or scaly or differently colored 21
21. Cap conical to bell-shaped, the surface often splitting radially but *not* scaly; stalk with a basal bulb .... *I. mixtilis* (see *I. sororia,* below)
21. Not as above; cap usually scaly or woolly 22
22. Cap dark brown to blackish, with small erect scales; often growing on *rotten* wood; spores warty .... *I. lanuginosa,* p. 462
22. Not as above; usually terrestrial ..... *I. flocculosa* & many others (see *I. lanuginosa,* p. 462)
23. Cap yellow-brown to ochre or dull cinnamon (sometimes darker at center), fibrillose or in age becoming somewhat scaly; stalk similarly colored at first but becoming brown to smoky-brown from the base upward; common under conifers in the Pacific Northwest ..... *I. olympiana*
23. Not as above; cap usually brown to dark brown or chestnut-brown (but sometimes overlaid with whitish veil remnants) 24
24. Stalk with a large turniplike bulb at base; cap usually silky-fibrillose or smooth; spores angular .... *I. napipes* (see *I. sororia,* below)
24. Stalk equal or with small bulb; cap sometimes scaly; spores smooth *I. maculata* & others, p. 458
25. Cap relatively large (5-12 cm broad); stalk fairly thick (at least 5 mm); both cap and stalk with dark brown scales (those on stalk arranged in more or less concentric rings) .... *I.* sp. (unidentified) (see *I. calamistrata,* p. 462)
25. Not as above; smaller 26
26. Cap and stalk yellowish to yellow-brown to cinnamon .... *I. terrigena* & *I. caesariata* (see *I. calamistrata,* p. 462)
26. Cap and stalk with darker brown scales .... *I. hystrix* (see *I. calamistrata,* p. 462)

Inocybe sororia   (Corn Silk Inocybe)

CAP 2-8 (10) cm broad when expanded, sharply conical or bell-shaped when young, often expanding in age but retaining a prominent umbo, the margin usually uplifted or splitting in old age; surface dry, silky, radially fibrillose, creamy to pale yellowish, honey-yellow, or straw-colored, often somewhat browner at the center or in age. Flesh thin; odor usually strongly pungent, like freshly husked or green corn. **GILLS** close or crowded, pallid becoming yellowish, then olive-yellow and finally brownish-gold or brown, the edges usually paler; adnate to adnexed, often seceding to free. **STALK** 3-14 cm long, 2-5 (10) mm thick, equal or with a slightly enlarged base, white or tinged cap color, fibrillose and often scurfy. **VEIL** absent. **SPORE PRINT** brown; spores 10-13 (17) × 5-8 microns, elliptical or bean-shaped, smooth. Cystidia on gills thin-walled.

HABITAT: Solitary, scattered, or in small groups on ground in woods; widely distributed. In our area it is common in the fall and winter under live oak and pine; farther north it favors conifers. I have seen large fruitings locally as well as in Oregon.

EDIBILITY: Poisonous! It contains high concentrations of muscarine.

COMMENTS: A common and prominent *Inocybe*, easily recognized by its sharply conical to umbonate, creamy to yellowish, fibrillose cap and odor of fresh green corn. The stem is often quite long and slender. *I. fastigiata* is a very similar, widespread species with a strongly spermatic odor and a slightly darker (yellow-brown to brownish-ochre) cap, at least at the center. Other species: *I. cookei* has a silky yellowish cap, small but prominent bulb at the base of the stem, slight odor, and smaller spores; *I. napipes* has a blunter

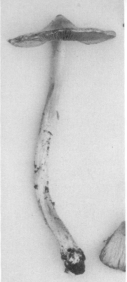

Inocybe sororia. **Left:** Young, relatively robust specimens. **Right:** A slender mature specimen. Note the conical or umbonate cap. Green corn odor is also distinctive.

brownish cap, a cortina (cobwebby veil) when young, a turniplike bulb at the base of the stem, and warty (nodulose) spores; *I. **mixtilis*** also has warty spores, but is smaller than *I. napipes* and lacks a cortina. The latter two species are fairly common under conifers in the Pacific Northwest, but I have not seen them locally. All of the above are poisonous.

Inocybe jurana   (Reddish Inocybe)

CAP 2-8 cm broad when expanded, conical or bell-shaped, expanding in age but usually retaining an umbo; surface dry, radially fibrillose, buff or brownish with darker brown fibrils, but soon flushed carmine, reddish-brown, or wine-red (vinaceous); center sometimes with small scales, margin often splitting in age. Flesh tinged pinkish or vinaceous, thin; odor mild to somewhat fruity or unpleasant. **GILLS** pallid becoming grayish-brown or dull brown, close, adnate to adnexed or even free. **STALK** 2-8 cm long, 4-10 (15) mm thick, equal or slightly enlarged at base, dry, white or flushed cap color below. **VEIL** absent. **SPORE PRINT** brown; spores 9-15 × 5-8 microns, elliptical or bean-shaped, smooth. Cystidia on gills thin-walled.

HABITAT: Solitary, scattered, or in small groups on ground in woods and at their edges; widely distributed. In our area I have found it twice under live oak in the winter.

EDIBILITY: Not recommended. According to some sources it is edible, but why tempt fate? Several similar species are poisonous.

COMMENTS: The deep reddish or vinaceous-toned cap is the hallmark of this illustrious *Inocybe.* In other respects it rather resembles *I. sororia,* but does not smell like fresh green corn. Other species: *I. **laetior*** has a bright salmon-pink stem and dark reddish-brown centered cap with a yellowish margin; it occurs under conifers in the Pacific Northwest and is one of numerous Inocybes "discovered" and named by Daniel Stuntz. *I. **oblectabilis*** has a pale pink stalk and reddish-brown cap.

Inocybe maculata   (Brown Inocybe)

CAP 2-8 cm broad, conical or bell-shaped, becoming convex or plane with an obtuse umbo; surface dry, radially fibrillose, dark brown to chestnut-brown, at first with whitish down (at least at center) which may break up into small scales or wear away; margin often lobed, splitting in age. Flesh thin, whitish; odor mild or slightly aromatic. **GILLS** pale grayish becoming grayish-brown to olive-brown or dull brown, close, adnate to adnexed or seceding. **STALK** 3-10 cm long, 0.5-1.2 cm thick, equal or with an enlarged base, fibrillose-striate, white or flushed cap color (brownish). **VEIL** absent except for whitish down on cap. **SPORE PRINT** brown; spores 9-12 × 4.5-6.5 microns, elliptical or bean-shaped, smooth. Cystidia on gills thin-walled.

HABITAT: Solitary or in small groups in woods and under trees; widely distributed. Occasional in our area in the fall and winter.

EDIBILITY: Poisonous; it contains muscarine.

COMMENTS: Any medium-sized to large *Inocybe* with a dark brown fibrillose or hairy cap and smooth spores can be referred here and conveniently forgotten about. I have encountered forms with an obnoxious odor, others which were mild and still others with a slightly aromatic or truffle-like smell. Undoubtedly more than one species is involved. Another fairly large species, *I. serotina,* has a brown cap with a whitish to yellowish-buff center and a bulb at the base of the stem. *I. napipes* (see comments under *I. sororia*) also has a basal bulb and can be quite large. There are also several smaller species that feature white universal veil tissue at the center of the cap. These include: *I. lanatodisca,* cap brown to yellow-brown with a moldy-looking patch of whitish fibrils at the center which may disperse in age, a whitish stalk (at least when young), and cystidia only on the edges of the gills plus a pungent to sweetish odor, fairly common in our area under oak but also occurring with conifers; and *I. sindonia,* similar to *I. lanatodisca,* but with cystidia on the faces of the gills as well as the edges. Another species, *I. albodisca,* has a lilac-gray to pinkish-brown (or sometimes whitish) cap with a smooth whitish umbonate center and a basal bulb on the stalk; it is common in northern latitudes under both hardwoods and conifers. None of the above species should be eaten.

Inocybe pyriodora   (Fragrant Inocybe)

CAP 2-7 cm broad, bell-shaped or obtusely conical becoming broadly umbonate; surface dry, at first silky-smooth but often fibrillose or fibrillose-scaly (especially at center) in age (or with torn-up appearance); white, the fibrils becoming dingy ochre to yellow-brown or brownish, sometimes pinkish- or reddish-stained in age. Flesh thin, white, slowly staining pinkish or reddish when cut; odor fragrant, usually spicy (cinnamon- or matsutake-like) but sometimes like overripe pears or unpleasant in old age. **GILLS** adnate to adnexed or notched, close, whitish becoming dull cinnamon, brown, or reddish-tinged. **STALK** 4-10 cm long, 3-15 mm thick, more or less equal, white throughout or colored like cap except for very apex, sometimes developing pinkish stains; smooth or finely fibrillose. **VEIL** fibrillose or cobwebby, evanescent. **SPORE PRINT** brown; spores 7-10 (12) × 4.5-7.5 microns, elliptical to bean-shaped, smooth. Cystidia on gills with slightly thickened walls.

HABITAT: Solitary, scattered, or in small groups in woods or at their edges, under brush, along trails, etc., favoring hardwoods but also occurring with conifers; widely distributed. It is not uncommon in our area in the fall and winter in mixed woods and under oak.

EDIBILITY: To be avoided—despite the enticing odor it is probably poisonous.

COMMENTS: This is one of several nondescript Inocybes with a spicy or fruity fragrance. The odor is often reminiscent of the matsutake *(Armillaria ponderosa),* but is sometimes fruity. Other fragrant species include: *I. bongardii,* cap scaly, odor fruity, spores larger; *I. corydalina,* cap whitish with brownish fibrils and an olive to olive-brown center and a strongly pungent or sickeningly sweet odor (like rotting fruit); *I. godeyi,* a reddening species with a slight fruity odor and abrupt bulb at the base of the stem; *I. cookei,* which has a slight fruity odor and abrupt basal bulb but does not redden; *I. hirtella,* with an almond extract or cyanide odor when fresh; and *I. suaveolens,* which smells like sweet peas and has nodulose-angular spores and is fairly common in the Pacific Northwest.

Inocybe pudica is a whitish species that develops reddish to salmon-orange tones as it ages. It is especially common under conifers.

Inocybe pudica   (Blushing Inocybe)

CAP 2-6 (8) cm broad, conical or bell-shaped when young, becoming convex to nearly plane at maturity, but usually retaining an umbo; surface dry or tacky, silky-fibrillose to nearly smooth, white at first but developing pinkish to reddish or orange stains as it ages. Flesh thin, white; odor unpleasant or spermatic. **GILLS** pallid or flushed pinkish or orange, becoming grayish-brown or dull brown as spores mature, close, adnate to adnexed, notched, or free. **STALK** 4-8 cm long, 0.4-1 cm thick, equal or enlarged at base, smooth or silky-fibrillose, firm; white, but discoloring like the cap. **VEIL** fibrillose or cobwebby, whitish, evanescent. **SPORE PRINT** brown; spores 7-10 × 4-6 microns, elliptical or bean-shaped, smooth. Gills typically with both thick- and thin-walled cystidia.

HABITAT: Scattered or in groups or troops on ground under conifers; widespread, but especially common on the west coast. In our area it is often abundant in the late fall, winter, and early spring under pine and Douglas-fir. It seems to favor brushy areas or immature second-growth stands.

EDIBILITY: Poisonous; like most Inocybes, it contains muscarine.

COMMENTS: This is one of the most common Inocybes of the western United States as well as one of the most easily recognized. The tendency of the white fruiting body to "blush" pink, red, or orange plus the growth under conifers and relatively smooth cap are the principal fieldmarks. The cap may be slightly tacky or slippery in wet weather, but is not truly viscid. *I. pyriodora* is somewhat similar but has a fragrant or spicy odor.

Inocybe geophylla   (Little White Inocybe)

CAP 1-3 (4) cm broad, conical to bell-shaped, expanding in age to plane but often retaining an umbo; surface dry, white or sometimes with a slight yellow tinge in age; smooth to silky or finely fibrillose; margin sometimes uplifted or split in age. Flesh thin, white; odor disagreeable (spermatic). **GILLS** adnate to adnexed or notched, close, at first pallid, then grayish, finally dull brown. **STALK** 2-6 cm long, 2-5 mm thick, equal or slightly thicker at base, firm, white or grayish-white, finely fibrillose. **VEIL** fibrillose, whitish, evanescent or leaving a slight hairy zone on stalk. **SPORE PRINT** brown; spores 8-10 × 5-6 microns, elliptical, smooth. Cystidia on gills thick-walled.

HABITAT: Scattered to gregarious on ground (or occasionally very rotten wood) in woods; widely distributed and common. It is abundant in our area throughout the mushroom season, especially under live oak, pine, and Douglas-fir.

EDIBILITY: Poisonous; it contains muscarine.

Inocybe geophylla is small but common. Note silky white cap and brown gills; odor is spermatic.

COMMENTS: The small size, white umbonate cap, and dull brown gills at maturity characterize this ubiquitous little *Inocybe*. It might be mistaken at first glance for a small waxy cap *(Camarophyllus)*, *Mycena*, or *Alboleptonia*, but the spores and mature gills are brown. It also resembles *I. pudica*, but does not "blush."

Inocybe lilacina   (Lilac Inocybe)

CAP 1.5-4 (5) cm broad, obtusely conical or bell-shaped, expanding in age but often retaining an umbo; surface dry, silky-fibrillose to nearly smooth, pale to deep lilac, but often with pinkish, gray, or brownish tones (or ochre at center), and often paler or whiter in age. Flesh thin, white or tinged lilac; odor disagreeable (spermatic). **GILLS** adnate to adnexed or notched, close, pallid or tinged lilac, then grayish to dull brown, the edges often whitish. **STALK** 2.5-6 cm long, 0.3-1 cm thick, equal or slightly thicker at base, finely fibrillose, firm, colored more or less like cap or with white areas; base whitish. **VEIL** fibrillose, evanescent or leaving a very slight hairy zone on stalk. **SPORE PRINT** brown; spores 7-9 × 4-5 microns, elliptical, smooth. Cystidia on gills thick-walled.

HABITAT: Solitary, scattered, or in groups on ground in woods; widely distributed. In our area it can be found in the late fall and winter under pine and Douglas-fir, but is not as numerous as its close relative *I. geophylla*.

EDIBILITY: Poisonous; like most Inocybes it contains muscarine.

COMMENTS: Also known as *I. geophylla* var. *lilacina*, this species is about the only common *Inocybe* that is the least bit colorful, although faded caps can be so pale that they resemble *I. geophylla*. The umbonate cap, relatively small size, and brown spores distinguish it from the blewit *(Clitocybe nuda)* and most other purple or lilac mushrooms. Several species of *Cortinarius* resemble it, but are generally larger and have brighter (rusty-brown) spores. There are also several other Inocybes which are violet or violet-tinged when fresh. *I. obscurioides,* for instance, is one of several species with a violet stalk or stalk apex and a brownish cap that typically shows violet only at the margin (if at all).

Inocybe lilacina is lilac-tinged when fresh. Note umbonate cap of most specimens.

Inocybe lanuginosa   (Woolly Inocybe)

CAP 1.5-3 (4) cm broad, convex to bell-shaped becoming umbonate, plane, or with an uplifted margin; surface dry, densely woolly or scaly, dark brown to brownish, the scales small and often erect. Flesh thin, watery brownish; odor mild to slightly unpleasant (spermatic). GILLS adnate, adnexed, or notched, fairly close, pallid becoming grayish, then dull brown or cinnamon-brown. STALK 2-5 (8) cm long, 2-4 (7) mm thick, more or less equal, fibrillose or fibrillose-scaly, colored more or less like cap except for paler apex. VEIL cobwebby, pallid, evanescent. SPORE PRINT brown; spores 8-10.5 × 5-7 microns, elliptical with blunt warts (nodulose). Cystidia on gills typically thin-walled.

HABITAT: Solitary or in small groups on very rotten wood or in forest humus; widely distributed. It favors conifers but also grows with hardwoods; I have not seen it in our area, but there are dozens of similar species.

EDIBILITY: Unknown, but do not experiment—it may very well contain muscarine!

COMMENTS: There are innumerable little brown Inocybes with hairy or minutely scaly caps. They are differentiated largely on microscopic characters such as the shape and size of their cystidia and spores. Most have an unpleasant or spermatic odor, and none should be eaten. This one is distinctive because it often grows on very rotten wood—highly unusual behavior for an *Inocybe!* Other brownish species—all typically terrestrial—include: *I. lacera,* widespread in a variety of habitats but especially common under aspen, with a torn-up or raggedly scaly cap and long, nearly cylindrical, sometimes warty spores; *I. flocculosa,* one of dozens of species with a hairy cap, smooth (bald) stalk, and smooth spores; and *I. agardhii,* one of several species with a woolly-scaly cap that hang out with alder and willow. For still other species, see the key to *Inocybe.*

Inocybe calamistrata   (Scaly Inocybe)

CAP 1-4 cm broad, bell-shaped to convex, expanding only slightly in age; surface dry, breaking up into densely-arranged scales, dark brown to coffee-brown; margin *not* typically splitting. Flesh thin; odor spermatic or fishy. GILLS adnate to adnexed or free, close, brown to dull cinnamon-brown or colored like cap, the edges whitish. STALK 4-10 cm long, 3-5 mm thick, equal or tapering upward, firm, covered with recurved scales (like cap) which may be obliterated or wear away in age; brown to dark brown (like cap), the base or lower portion dingy greenish-blue to olive-green (both inside and out). VEIL disappearing (not forming a distinct ring on stalk). SPORE PRINT brown; spores 9-13 × 4.5-6.5 microns, elliptical-oblong, smooth.

HABITAT: Solitary, scattered, or in small groups on ground under conifers or sometimes hardwoods; widely distributed, but not common. I have seen it in several localities, but have yet to find it in our area.

EDIBILITY: Unknown. A recent study revealed the presence of psilocybin.

COMMENTS: The blue-green to olive-green stem base and scaly convex cap which does *not* split radially in age are the main features of this uninteresting agaric. It is included here as a representative of those Inocybes with scales on both the cap *and* the stem. Others include: *I. hystrix,* with small, brown to dark brown scales on both the cap and stalk; *I. terrigena,* larger and fleshier (in fact, somewhat reminiscent of a *Pholiota*), with golden-brown to cinnamon scales on the cap and stalk and sometimes a slight annulus (ring), widespread but especially common under aspen in the southern Rocky Mountains; *I. caesariata,* somewhat similar in color to *I. terrigena* but with smaller scales, fairly common in eastern North America; and finally, a very robust, unidentified local species with a dark brown scaly cap and several zones of concentrically-arranged dark brown scales on the stem (cap 5-12 cm broad, stalk 0.5-1.5 cm thick!). None of these have the blue-green or olive-green stem base of *I. calamistrata.*

HEBELOMA

Small to medium-sized, *terrestrial* mushrooms. CAP *typically viscid when wet and more or less smooth; white to buff, tan, or some shade of brown.* Flesh *often with a radishlike odor.* GILLS adnate to adnexed or notched, *usually brown at maturity;* often with whitish edges. STALK fleshy, central; apex often powdery or with small flakes. VEIL *usually absent or if present then cobwebby-fibrillose and evanescent.* VOLVA absent. SPORE PRINT *brown* (or rarely reddish-brown). Spores typically elliptical, smooth or roughened, lacking a germ pore. Cystidia typically present on gill edges, often conspicuous. Cap cuticle filamentous.

THIS is yet another faceless and featureless collection of brownish mushrooms. Those that are too large to qualify as "LBM's" most certainly fall into the category of "BUM's" ("Boring Ubiquitous Mushrooms"). The stalk is fleshy and the gills attached but not decurrent, giving the fruiting body the stature of a *Tricholoma* or *Entoloma.* The odor is usually mild or radishlike, not spermatic as in *Inocybe,* and the cap is typically smooth and viscid rather than dry and fibrillose or scaly. *Hebeloma* is closely related to *Cortinarius,* but that genus has rusty-brown spores and a cobwebby veil (cortina). Some Hebelomas have a cortina, but can be distinguished by their duller spore color and/or the presence of sterile cells (cystidia) on the edges of the gills, plus the frequently powdered or flaky stem apex. *Agrocybe* is also similar, but has a cellular cap cuticle and usually grows in grass, dung, wood or wood chips, or cultivated soil. Hebelomas, in contrast, are largely mycorrhizal. As a result, they are found in forests or at their edges or on tree-studded lawns and cemeteries. (However, *H. syriense* feeds on decaying corpses and is said to have led to the discovery of at least one crime!)

About 200 species of Hebeloma occur in North America, but none are exceptionally distinctive or colorful. Whites, browns, tans, and buffs predominate, making identification on the basis of color hopeless. More minute measures of individuality must be taken into account, such as the size and shape of the cystidia and spores. Many mushroom-hunters, however, prefer to recognize only two varieties—the big brown ones and the not-so-big brown ones! Hebeloma should not be eaten, as some are definitely poisonous and the group as a whole is poorly known. Three widespread species are described here.

Key to Hebeloma

1. Spore print reddish to pinkish-brown or pinkish-cinnamon 2
1. Not as above; spore print dull brown or at times rusty-brown 3
2. Cap whitish when fresh, but often aging ochre, gray, or brownish; spores elliptical (*not* angular) .... *H. sarcophyllum* (see *H. crustuliniforme,* p. 464)
2. Not as above; spores more or less angular .... (see *Entoloma,* p. 242)
3. Odor spermatic; cap often (but not always) umbonate and rather small . (see *Inocybe,* p. 455)
3. Not as above 4
4. Veil present when young, disappearing or leaving remnants on the cap margin and/or a slight ring (annulus) or fibrillose sheath on the stalk 5
4. Veil absent or rudimentary 6
5. Lower portion of stalk sheathed with conspicuous veil remnants; cap often with veil remnants also, especially near margin .... *H. strophosum* (see *H. mesophaeum* group, p. 465)
5. Stalk not conspicuously sheathed with veil remnants .... *H. mesophaeum* group, p. 465
6. Cap small (typically less than 4 cm broad) and white to grayish, often hairy at margin; gills adnate to decurrent and peeling easily from cap .... (see *Clitocybe* & Allies, p. 148)
6. Not as above 7
7. Growing on or near corpses, mainly in eastern North America; cap brown to reddish-brown . ... *H. syriense*
7. Growing in woods, near trees, etc., but not on corpses; cap variously colored (including brown or reddish-brown); common and widespread 8

8. Cap typically brown to cinnamon-brown or dark reddish-brown (but sometimes paler brown and often shaded with gray or overlaid with a pallid sheen); stalk usually with small protruding scales .... *H. sinapizans* group & others, p. 465
8. Cap white to buff, pale brown, tan, crust-colored, or yellow-brown; stalk often with a dandruffy or scurfy apex but not normally with scales 9
9. Odor sweet .... *H. sacchariolens* (see *H. crustuliniforme,* below)
9. Odor typically mild or radishlike, etc., but not sweet 10
10. Cap viscid or slimy when wet; stalk apex usually dandruffy (with small white flakes); gills often with white edges; odor usually (but not always) radishlike; found in woods or on lawns near trees .... *H. crustuliniforme* & others, below
10. Not as above; cap only slightly viscid; found in wood chips, gardens, etc. (see *Agrocybe,* p. 467)

Hebeloma crustuliniforme   (Poison Pie)

CAP 3-11 cm broad, convex or broadly convex with an inrolled margin, becoming plane to obtusely umbonate or with an uplifted margin in age; surface viscid when moist, smooth, whitish to buff, pale tan, or crust-brown (usually darker toward center and paler at margin); margin naked. Flesh thick, white; odor distinctly radishlike. **GILLS** crowded (often appearing to slightly overlap one another), adnate or notched, pallid when young, becoming watery brown and finally dull brown; edges white and minutely scalloped, often beaded with water droplets in wet weather and brown-spotted when dry. **STALK** 4-13 cm long, 0.5-1.5 (2) cm thick, usually equal except for an enlarged base; solid, fibrillose; white or tinged cap color; apex powdered or with flakes or granules; base sometimes with white mycelial threads. **VEIL** absent. **SPORE PRINT** brown; spores 9-13 × 5.5-7.5 microns, elliptical or almond-shaped, smooth or minutely roughened. Cystidia abundant on edges of gills.

HABITAT: Solitary, scattered, or in groups or troops on ground in woods or at their edges and on lawns or cemeteries near trees; widely distributed and common. In our area it is usually abundant from late fall through early spring under pine, and to a lesser extent, oak. It is by far our most common *Hebeloma.*

EDIBILITY: Poisonous—it causes mild to severe gastrointestinal distress.

COMMENTS: In spite of my general disdain for Hebelomas, I must admit this is a most attractive mushroom when fresh. The smooth, viscid, pallid to pie-colored cap, attached

Hebeloma crustuliniforme is a common poisonous species with a whitish to pale tan cap, brown gills (at least at maturity), and radishlike odor. For photo of young buttons, see p. 466.

1. A "toadstool," *Lactarius atroviridis* (see comments under
L. olivaceoumbrinus, p. 70).

2. *Lactarius deliciosus*, p. 68; edible but not necessarily delicious.

3. Close-up of the "bleeding" latex of *Lactarius rubrilacteus,* p. 68.

4. *Lactarius indigo* (Indigo Milk Cap), p. 69.

5. *Lactarius alnicola* (Golden Milk Cap), p. 71, has a very peppery taste.

6. *Lactarius rubrilacteus* (also known as *L. sanguifluus* or the Bleeding Milk Cap), p. 68.

7. *Lactarius torminosus* (Bearded Milk Cap), p. 73; common under birch.

8. *Lactarius corrugis,* an edible eastern milk cap (see comments under *L. volemus,* p. 78).

9. *Lactarius fallax* (Velvety Milk Cap), p. 77.

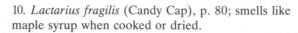

10. *Lactarius fragilis* (Candy Cap), p. 80; smells like maple syrup when cooked or dried.

11. *Russula fragrantissima* group, p. 92; can be fragrant or foul-smelling.

12. *Russula cyanoxantha*, p. 94; like many Russulas, its color is extremely variable.

13. *Russula rosacea* (Rosy Russula), p. 99.

14. *Russula xerampelina* (Shrimp Russula), p. 102; as beautiful as it is delicious!

15. *Hygrophorus speciosus*, p. 126; a beautiful larch- and pine-loving waxy cap.

16. *Hygrophorus pudorinus*, p. 124, favors spruce.

17. Close-up of the golden-flaked stalk of *Hygrophorus chrysodon*, p. 119.

18. *Hygrocybe psittacina (= Hygrophorus psittacinus)*, p. 118, is dark green or bright green when fresh but soon fades.

Ray Gipson

19. *Hygrocybe conica (= Hygrophorus conicus* or Witch's Hat), p. 116, blackens when bruised or handled.

20. *Hygrocybe coccinea* (= *Hygrophorus coccineus*), p. 114.

21. *Hygrocybe punicea* (= *Hygrophorus puniceus*), p. 114.

22. *Hygrocybe (=Hygrophorus) flavescens* (Golden Waxy Cap), p. 115.

24. *Omphalina luteicolor* (see comments under *O. ericetorum*, p. 223) grows in groups on rotting conifers.

23. *Hygrocybe (=Hygrophorus) calyptraeformis* (Salmon Waxy Cap), p. 117; pointed pink or salmon-colored cap is distinctive.

25. Aspens and cottonwoods (p. 42) supply *Pleurotus* and *Flammulina* in the fall and winter, morels in the spring, *Leccinum* in the summer.

26. *Flammulina velutipes* (Velvet Foot), p. 220.

27. *Pleurotus ostreatus* (Oyster Mushroom), p. 134, growing on bush lupine by the ocean.

28. *Pleurotus ostreatus* (Oyster Mushroom), p. 134.

29. *Laccaria laccata* (Lackluster Laccaria), p. 172.

30. *Laccaria amethystina* group (Amethyst Laccaria), p. 172.

31. *Leucopaxillus amarus* (Bitter Brown Leucopaxillus), p. 168.

32. *Clitocybe (= Lepista) nuda* (Blewit), p. 153. Some forms have only a slight lilac tinge; others are perfectly purple.

33. *Tricholoma flavovirens* (Man on Horseback), p. 179.

34. *Clitocybe gibba,* p. 157.

35. *Tricholoma pessundatum* group, p. 185.

36. *Marasmiellus candidus,*
p. 206.

37. *Armillaria ponderosa* (= *Tricholoma magnivelare* or White Matsutake), p. 191.

38. *Marasmius oreades* (Fairy Ring Mushroom), p. 208, a common suburban lawn mushroom.

40. *Omphalotus olivascens*
(Western Jack-O-Lantern
Mushroom), p. 147.

39. Close-up of a cluster of
Armillariella (= Armillaria) mellea
group (Honey Mushroom), p. 196.

41. *Omphalotus olivascens* (Western Jack-O-Lantern Mushroom), p. 147.

42. *Armillariella (=Armillaria) mellea* group (Honey Mushroom), p. 196.

43. *Armillaria albolanaripes*, p. 194.

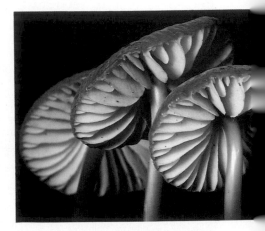

44. *Xeromphalina campanella*, p. 222; these specimens are somewhat browner than normal.

45. *Marasmius plicatulus*, p. 209; one of our most beautiful mushrooms.

46. *Mycena haematopus* (Bleeding Mycena), p. 231.

47. *Marasmius oreades* (Fairy Ring Mushroom), p. 208.

48. *Mycena murina* group (Yet Another Mycena), p. 234, often fruits by the hundreds under pines.

49. *Collybia acervata,* p. 215, grows in bundles under conifers.

50. *Amanita phalloides* (Death Cap), p. 269; don't expect this deadly species to be as green and pristine as these perfect specimens!

51. *Entoloma madidum*, p. 243.

52. *Leptonia carnea*, p. 250, is dark blue and often iridescent.

54. *Amanita baccata* (Sand
Amanita), p. 273.

53. *Amanita virosa* (Destroying Angel)
(see comments under *A. ocreata*,
p. 271).

55. *Amanita magniverrucata*, p. 274, has exaggerated warts on the cap.

56. *Amanita aspera* (also called *A. francheti*), p. 278; cap color ranges from bright yellow to dark brown.

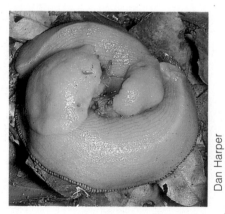

57. Banana slugs (see p. 29) prefer mushrooms to almost anything but other banana slugs. In this stage they are commonly called "Poor Man's Peaches."

Dan Harper

58. The yellow-capped variety of *Amanita muscaria* (Fly Agaric), p. 282.

59. *Amanita muscaria* (Fly Agaric), p. 282; the red-capped variety.

60. *Amanita caesarea* group (Caesar's Amanita),
p. 284, has bright yellow to yellow-orange gills.

61. *Amanita calyptrata* (= *A. calptroderma* or Coccoli, Coccora), p. 284.

62. *Amanita calyptrata,*
p. 284; the dark-capped
(fall) form.

63. *Amanita calyptrata,*
p. 284; the winter-spring
form with a pale yellow to
whitish cap.

64. *Amanita velosa* (Springtime Amanita), p. 286, often grows
at the edges of meadows when spring wildflowers are in bloom.

65. *Amanita pachycolea* (Western Grisette), p. 290; young specimens with dark brown caps.

66. *Amanita pachycolea* (Western Grisette), p. 290; mature specimen with a paler cap.

67. Close-up of the glutinous veil in *Limacella illinita*, p. 292.

68. *Volvariella speciosa*, p. 259, is abundant in gardens and cultivated fields.

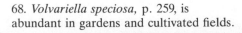

69. *Lepiota rachodes* (Shaggy Parasol), p. 297.

70. *Lepiota lutea* (= *Leucocoprinus luteus*), p. 302, grows in lawns, gardens, flower pots.

71. *Agaricus campestris* (Meadow Mushroom or Field Mushroom), p. 318.

72. *Agaricus californicus*, p. 327, a mildly poisonous mushroom often mistaken for the meadow mushroom.

73. *Agaricus cupreobrunneus* (Brown Field Mushroom), p. 319.

74. *Agaricus fuscofibrillosus,* p. 325, is one of several species that "bleed" (stain red) when cut.

75. *Agaricus subrutilescens* (Wine-Colored Agaricus), p. 326; can be slender (as shown here) or robust.

76. *Agaricus praeclaresquamosus* (= *A. meleagris*), p. 329; poisonous to many people.

77. *Agaricus augustus* (The Prince), p. 337; it is also shown on the front cover.

78. *Agaricus osecanus* group (Giant Horse Mushroom), p. 333, differs microsopically from the better known *A. arvensis.*

79. A Greedy Person (see p. 27) laden with giant horse mushrooms (*Agaricus osecanus* group, p. 333). Turn to next page to see what happens to Greedy People.

80. Crack in the soil caused by *Agaricus bitorquis* (= *A. rodmani*), p. 321, a delicious edible mushroom that often grows underground. This picture is dedicated to those purists who insist that mushrooms *must* be photographed exactly as they occur in nature.

81. What happens to Greedy People (see p. 27 and Color Plate 79).

82. *Agaricus subrufescens* (Almond Mushroom), p. 336; smells and tastes like almond extract.

83. *Panaeolus solidipes* (= P. phalaenarum), p. 355.

84. A long-stemmed form of *Endoptychum depressum*, p. 730; it looks like an *Agaricus* but lacks gills.

85. *Coprinus comatus* (Shaggy Mane), p. 345.

86. *Coprinus lagopus* group, p. 350.

87. *Coprinus plicatilis*, p. 352; the cap is oval when younger.

88. *Psilocybe cyanescens*, p. 371, stains blue when bruised.

89. *Stropharia ambigua* (Questionable Stropharia), p. 377.

90. *Stropharia kauffmanii,*
p. 380.

91. *Stropharia aeruginosa*
(Blue-Green Stropharia),
p. 380.

92. *Naematoloma fasciculare* (Sulfur Tuft), p. 382.

93. *Naematoloma aurantiaca*, p. 382.

94. *Naucoria vinicolor,*
p. 404, a rare mushroom.

95. *Pholiota aurivella* group (Golden
Pholiota), p. 390; the caps are very
slimy when wet.

96. *Pholiota squarrosa* (Scaly Pholiota), p. 389.

97. *Pholiota squarrosa* (Scaly Pholiota), p. 389.

98. *Pholiota squarrosoides* (see comments under *P. squarrosa*, p. 389); youngsters are much scalier than these mature specimens.

99. *Gymnopilus spectabilis* group, p. 410. The bark has been
stripped away to show the base of the cluster; it is not
normal for the caps to be so misshapen.

100. *Phaeocollybia californica,* p. 415, grows in troops or loose bundles.

101. *Rozites caperata* (Gypsy Mushroom), p. 412; note wrinkled cap.

102. *Agrocybe praecox* group, p. 469.

103. Close-up of the collapsed cortina (cobwebby veil) of a *Cortinarius* (p. 417); the rusty-brown color is caused by a coating of spores.

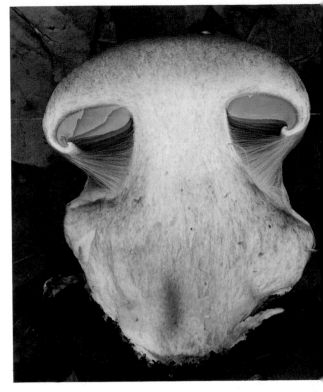

104. *Cortinarius collinitus* group, p. 431; the cap is slimy when wet.

105. A young *Cortinarius cedretorum* (p. 439), sliced in half to show the cobwebby veil or cortina.

106. *Cortinarius traganus*, p. 447, a common conifer-lover.

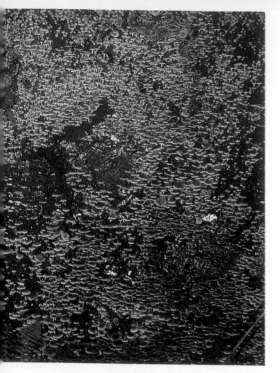

107. Little Brown Mushrooms ("LBM's"), p. 32, on a wet stump.

108. *Cortinarius squamulosus,* p. 445; note the bulbous base.

109. *Cortinarius violaceus,* p. 446, is deep purple to nearly black.

110. Close-up of the forked gills of *Hygrophoropsis aurantiaca* (False Chanterelle), p. 479.

111. *Phylloporus rhodoxanthus* (Gilled Bolete), p. 480; the gills often stain blue or greenish when bruised.

112. *Gomphidius subroseus* (Rosy Gomphidius), p. 483, is associated with Douglas-fir.

113. *Chroogomphus vinicolor* (Pine Spike), p. 485; cap ranges from grayish to orangish to wine-colored or reddish-brown.

114. *Suillus pungens* (Pungent Slippery Jack), p. 503, grows under pine, often with pine spikes.

115. *Suillus brevipes* (Short-Stemmed Slippery Jack), p. 501.

116. *Suillus sibiricus* (Siberian Slippery Jack), p. 498.

117. *Suillus ponderosus*
(see comments under
S. caerulescens, p. 496).

118. *Suillus luteus*
(Slippery Jack),
p. 500.

119. *Suillus fuscotomentosus* (Poor Man's
Slippery Jack), p. 504.

120. Close-up of the glandular-dotted stem
of *Suillus fuscotomentosus*, p. 504.

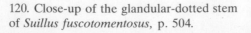

121. *Suillus subolivaceus* (Slippery Jill), p. 499.

122. *Suillus cavipes,* p. 494, is common under larch; note hollow stem.

123. *Fuscoboletinus ochraceoroseus* (Rosy Larch Bolete), p. 506, may be hard to pronounce but is easy to see.

125. *Boletellus russellii*, p. 509, has a long jagged stalk.

124. *Suillus lakei*, p. 495, is common under Douglas-fir.

126. *Fuscoboletinus spectabilis* (Bog Bolete) (see key to *Fuscoboletinus* on p. 506).

127. *Boletus mirabilis*, p. 521; a mature specimen growing on a buried hemlock log.

128. *Boletus zelleri*, p. 518.

129. *Boletus mirabilis*, p. 521; this specimen is younger and darker-capped than #127.

130. *Boletus rubripes* (Red-Stemmed Bitter Bolete), p. 524; young specimens.

131. *Boletus flaviporus*, p. 522. Pores are *brilliant* yellow—this photograph does not do them justice!

132. *Boletus dryophilus*, p. 520, is often shorter and stouter than shown here.

133. *Boletus regius* (Red-Capped Butter Bolete), p. 526; unusually small specimens.

134. *Boletus appendiculatus* (Butter Bolete), p. 525; pores
normally stain blue when bruised.

135. *Boletus erythropus*, p. 526,
is one of many boletes that stain
blue when cut open.

136. *Boletus subvelutipes* (see comments
under *B. erythropus*, p. 526).

137. *Boletus satanas* (Satan's Bolete), p. 527, a poisonous and bulbous oak-lover.

138. *Boletus haematinus* (see comments under *B. pulcherrimus*, p. 528).

139. *Boletus frostii*, p. 528; note the red cap and netted stalk.

140. *Boletus aereus* (Queen Bolete), p. 531. Cap is dark brown beneath a whitish bloom when young, cinnamon-brown or paler in age.

141. *Boletus barrowsii* (White King Bolete), p. 529.

142. A basket of *Boletus edulis* (King Bolete), p. 530.

143. *Boletus edulis* (King Bolete), p. 530. This red-capped variety is abundant in the Rocky Mountains in late summer when many of the wildflowers bloom.

144. *Boletus edulis* (King Bolete), p. 530; the typical brown-capped form.

145. *Leccinum manzanitae* (Manzanita Bolete), p. 539.

146. *Leccinum ponderosum* (see comments under *L. manzanitae*, p. 539).

147. *Leccinum insigne* (Aspen Bolete), p. 540.

148. *Gyroporus castaneus*, p. 510, has a strong, nutty flavor.

149. *Tylopilus gracilis* (see comments under *T. chromapes*, p. 533) is a common eastern bolete.

150. *Tylopilus felleus* (see comments under *T. indecisus*, p. 535), a bitter-tasting eastern bolete.

151. *Polyporus tuberaster* (Stone Fungus), p. 563, fruits from an underground "tuber."

152. *Fistulina hepatica* (Beefsteak Fungus), p. 553, with banana slugs and chinquapin; this is a mature individual.

153. *Phaeolus schweinitzii*, p. 570; young, colorful specimens.

154. *Laetiporus sulphureus* (Sulfur Shelf), p. 572. These juicy specimens are just emerging.

155. *Laetiporus sulphureus*, p. 572. A mature 30-lb. cluster.

156. *Ganoderma tsugae* (see comments under *G. lucidum*, p. 577) is one of several varnished polypores.

157. *Hydnellum zonatum* (see comments under *H. scrobiculatum,* p. 627) has a zoned cap with spines on the underside.

158. *Trametes (= Coriolus) versicolor* (Turkey Tail), p. 594; zoned caps are lined underneath with tiny pores.

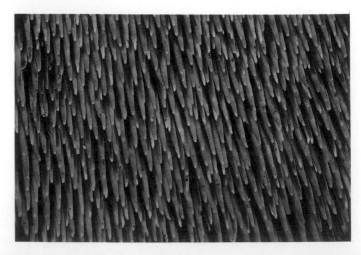

159. Close-up of the spines of *Hydnum imbricatum*, p. 619.

161. *Dentinum (= Hydnum) repandum* (Hedgehog Mushroom), p. 618. In coastal California it can often be found when the wild irises are in bloom.

160. *Hydnellum peckii* (Strawberries and Cream), p. 627, exudes red droplets in wet weather.

162. *Dentinum repandum* (Hedgehog Mushroom), p. 618, has spines underneath the cap instead of gills or pores.

Herb Saylor

163. *Hericium abietis,* p. 614, can weigh 50 lbs. or more!

164. *Hericium ramosum,* p. 615, is more delicate than the above species.

165. *Hydnum scabrosum* group (Bitter Hedgehog), p. 620.

166. *Clavariadelphus truncatus*, p. 634, is a conifer-loving club fungus.

167. *Clavaria purpurea*
(Purple Fairy Club), p. 637.

168. *Ramaria araiospora*
(Red Coral Mushroom),
p. 655.

169. *Clavulina cinerea* (Ashy
Coral Mushroom), p. 641.

170. *Tremella mesenterica*
(Witch's Butter), p. 673.

171. *Clavariadelphus ligula*, p. 633, often grows in troops under conifers.

172. *Sparassis crispa (= S. radicata* or Cauliflower Mushroom), p. 657.

173. *Tremella foliacea* (Brown Witch's Butter), p. 673.

174. *Gomphus floccosus* group (Scaly Chanterelle), p. 661; the caps are not always as brightly colored as these.

175. *Cantharellus cibarius* (Chanterelle), p. 662. This clean, slender-stemmed, yellow form is common in eastern North America.

176. *Gomphus clavatus*
(Pig's Ears), p. 661.

177. The gills of this eastern chanterelle (*Cantharellus cibarius*, p. 662) are unusually thin, deep, and well-developed.

178. *Cantharellus cibarius* (Chanterelle), p. 662; the large orange form common on the west coast.

179. *Cantharellus subalbidus* (White Chanterelle), p. 662, a common western species.

180. *Cantharellus infundibuliformis* group (Funnel Chanterelle, Winter Chanterelle), p. 665, a common northern species; also called *C. tubaeformis.*

181. *Cantharellus cinnabarinus* (Red Chanterelle), p. 664, a common eastern species.

182. *Craterellus cornucopioides* (Horn of Plenty), p. 666, blends in well with its surroundings. The caps are black when moist, brown in dry weather.

183. *Polyozellus multiplex* (Blue Chanterelle), p. 668, a deep blue or violet-tinged conifer-lover.

184. In coastal California, giant puffballs (*Calvatia gigantea* group, p. 682) often grow on poppy-laced hillsides.

185. *Pisolithus tinctorius*, p. 712; a young specimen sliced open to show the numerous spore capsules.

186. *Calvatia booniana* (Western Giant Puffball), p. 684; these rather small specimens are much too old to eat.

187. *Battarrea phalloides*, p. 717, usually grows in deserts or sandy soil.

188. *Calostoma cinnabarina*, p. 718, a colorful stalked puffball.

Michael Fogden

189. *Pisolithus tinctorius* (Dead Man's Foot), p. 712; common in poor soil, along roads, in patios and parking lots.

190. *Scleroderma citrinum* (Common Earthball), p. 708.

Michael Fogden

Phil Sharp

191. *Dictyophora indusiata* (Basket Stinkhorn),
p. 770, is tropical, but a closely related species
occurs in eastern North America.

192. *Lysurus mokusin* (Lantern Stinkhorn), p. 776.
Like other stinkhorns, this species relies on flies for
spore dispersal.

193. *Phallus impudicus*
(Stinkhorn), p. 768;
mature specimen.

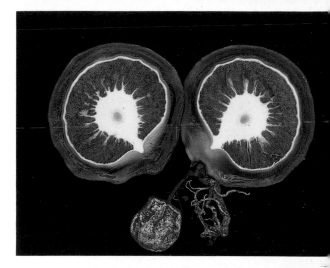

194. A stinkhorn "egg" (*Phallus impudicus*,
p. 768) sliced open to show the olive-colored
spore mass.

195. *Mutinus elegans*
(Dog Stinkhorn) (see
comments under
M. caninus, p. 771).

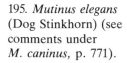

196. *Mutinus elegans*
(see comments under
M. caninus, p. 771);
sometimes called
"Devil's Dipstick."

198. A *Clathrus archeri* (p. 774) "egg" sliced open to show the mucilaginous spore mass.

197. *Clathrus archeri* (Octopus Stinkhorn), p. 774; a mature specimen.

199. In Oregon, black morels (*Morchella elata* group, p. 790) are often found with calypso orchids.

200. *Neolecta irregularis*
(Irregular Earth Tongue),
p. 871.

201. The coastal California version of the white morel (*Morchella deliciosa,* p. 789); ridges are whitish when very young.

202. Black morels (*Morchella elata* group, p. 790) are easily overlooked because they look like fallen pine cones.

203. *Morchella esculenta* (Morel), p. 787.

204. *Helvella compressa* (Compressed Elfin Saddle), p. 811; a pale specimen.

206. *Gyromitra esculenta*
(False Morel, Brain
Mushroom), p. 801.

205. *Verpa bohemica*
(Early Morel),
p. 793.

207. *Peziza domiciliana*, p. 822, a common indoor cup fungus.

208. *Aleuria aurantia* (Orange Peel Fungus), p. 837.

209. *Aleuria rhenana*, p. 836.

210. *Sarcoscypha coccinea* (Scarlet Cup Fungus), p. 836.

211. *Caloscypha fulgens* (Snowbank Orange Peel Fungus), p. 837, is usually greenish- or bluish-stained.

212. Truffles, anyone? These are *Tuber separans*, p. 859.

213. *Pachyphloeus citrinus* (Berry Truffle), p. 856.

214. *Microglossum viride* (Green Earth Tongue), p. 870; young specimens.

215. *Leotia lubrica*
(Jelly Babies), p. 874.

216. *Hypomyces lactifluorum*
(Lobster Mushroom), p. 884.

217. "Too many mushrooms." Overindulgence is a frequent
cause of so-called "mushroom poisoning" (see p. 888).

(usually notched) brown gills, absence of a veil, and radishlike odor (which is usually quite pronounced but sometimes slight) are the important fieldmarks. You will undoubtedly encounter several subtly different species which more or less fit the above description, but they can only be differentiated microscopically. Other species: *H. hiemale* is smaller (cap about 3 or 4 cm broad), with only a slight radish odor and gills not beaded with droplets. *H. sacchariolens* is distinguished by its strong sweet or fruity odor. *H. albidulum* is one of several species with a white or whitish cap and little or no odor. *H. sarcophyllum* is an unusual but distinctive mushroom with a chalk-white cap that may become reddish-gray or brownish-tinged in age, deep flesh-colored gills, and reddish-brown spores. It resembles *Pluteus pellitus*, but has attached gills and grows on the ground. It could also be mistaken for an *Entoloma,* but does not have angular spores. None of the above species should be eaten.

Hebeloma sinapizans group   (Scaly-Stalked Hebeloma)

CAP 4-13 (20) cm broad, convex to broadly convex with an inrolled margin, becoming plane or with an uplifted, often wavy margin; surface slightly viscid when moist, smooth, brown to cinnamon, ochre-brown, pinkish-tan, or dark reddish-brown, but often shaded with gray or overlaid with a pallid sheen toward margin, which is at first minutely cottony. Flesh thick, whitish; odor usually distinctly radishlike. **GILLS** close, usually adnexed or notched, pallid becoming pale brown, then dull brown or dull cinnamon, the edges minutely serrated and often beaded with droplets in wet weather and brownish-dotted when dry. **STALK** 4-13 cm long, 1-3 cm thick, usually swollen at base, whitish, with distinct pallid to brownish flakes or protruding scales; apex usually powdered with small white granules; firm, solid. **VEIL** absent. **SPORE PRINT** dull brown; spores 10-13 × 6-8 microns, elliptical, obscurely roughened. Cystidia present on gill edges.

HABITAT: Scattered to gregarious, sometimes in rings, on ground under both hardwoods and conifers or near planted trees on lawns; widely distributed. This species "complex" is fairly common in our coastal pine forests in the late fall, winter, and early spring. Elsewhere it often grows with oak and a related species (see comments) is fond of aspen.

EDIBILITY: Poisonous—it causes nausea, vomiting, diarrhea, etc.

COMMENTS: This large, sturdy species or species "complex" is best recognized by its brown to reddish-brown cap, thick stem adorned with small scales, absence of a veil, and dull brown, usually notched gills. It is our largest *Hebeloma,* more robust than *H. crustuliniforme,* and not nearly as attractive. It might be mistaken for a dull colored *Cortinarius* were it not for the absence of a cortina even in small buttons. It is sometimes found growing with *Tricholoma imbricatum,* which has a brown cap, white to flesh-colored gills, and white spores. Other species: *H. insigne* is a robust, scaly-stemmed species that occurs commonly under aspen and conifers in the mountains of Colorado and New Mexico.

Hebeloma mesophaeum group   (Veiled Hebeloma)

CAP 2-6.5 cm broad, convex or umbonate, expanding to broadly umbonate, plane, or with margin uplifted in age; surface smooth, viscid when moist but soon dry, color variable but usually dark brown to vinaceous-brown or reddish-brown at the center and paler (grayish, pinkish, buff, or even whitish) at the margin, often fading in age to dingy brown or tan and often adorned with thin filmy or fibrillose patches of veil material at or near the margin, which is inrolled when young. Flesh watery brownish or whitish; odor and taste typically radishlike (but in some variants mild and in others pungent and/or bitter). **GILLS** close, adnate to adnexed or notched, whitish or grayish becoming brown, finally dull brown; edges usually whitish. **STALK** 3-8 cm long, 3-8 (10) mm thick, more or less equal, fibrillose, whitish to dingy brownish, slowly become dark brown from base upward;

Left: *Hebeloma crustuliniforme* (p. 464). Note brown gills and shape of buttons (bottom). **Right:** *Hebeloma mesophaeum* group. The evanescent veil (not visible here) often leaves fibrils on stalk.

apex often mealy. **VEIL** cobwebby-fibrillose, white to grayish or buff-colored, thin, disappearing or forming a slight fibrillose zone on stalk. **SPORE PRINT** brown; spores 7-11 × 5-7 microns, elliptical, minutely roughened. Cystidia present on gill edges.

HABITAT: Widely scattered to densely gregarious on ground in woods and near trees; widely distributed. In the northern United States this species and its numerous look-alikes are common under conifers, especially in cool weather (fall, spring). In our area they occur only sporadically, but I have seen thousands of specimens in a cottonwood-willow woodland adjacent to a small reservoir, in February and March.

EDIBILITY: To be avoided—several Hebelomas are poisonous and all are difficult to identify.

COMMENTS: The presence of a cortina (cobwebby veil) in young specimens distinguishes this species and its close relatives from the larger, more prominent members of the genus such as *H. crustuliniforme* and *H. sinapizans,* but leads to confusion with *Inocybe* and *Cortinarius.* The former, however, usually has a silky to fibrillose or scaly cap, while the latter has a brighter (rusty-brown) spore print and lacks cystidia on the gill edges. There are dozens of equally drab veiled Hebelomas which are best differentiated from *H. mesophaeum* microscopically. (A recent monograph by Alexander Smith, Verna Stucky Evenson, and Duane Mitchell recognizes nearly 100 veiled species in the western United States alone!) Among them are: *H. strophosum,* very similar and widespread, with a thicker, more persistent veil that forms a woolly sheath on the stem below the fibrillose annulus (ring); and *H. fastibile,* one of several larger, fleshier species with a reddish-brown cap with paler inrolled margin and a thicker (1-1.5 cm) stalk.

BOLBITIACEAE

spores

T.WO microscopic features separate this rather small brown-spored family from the Cortinariaceae: the cap cuticle is typically cellular (composed of round to pear-shaped cells) and the spores are smooth and usually furnished with a large germ pore that gives them a truncate (chopped-off) appearance. People not armed with a microscope are better off learning the three genera in the Bolbitiaceae *(Bolbitius, Agrocybe,* and *Conocybe)* individually, rather than trying to devise an unwieldy set of *ifs, buts,* and *ands* for distinguishing them as a unit in the field. The Coprinaceae also have a cellular cap cuticle and spores with a germ pore, but give a much darker (purple-brown to black) spore print.

Like the Coprinaceae, the Bolbitiaceae are mostly frail, saprophytic fungi. They grow in grass, dung, decaying wood, and humus and are among our most common suburban mushrooms. In contrast, the bulk of the Cortinariaceae are mycorrhizal sylvan fungi.

Owing to their fragile consistency, the Bolbitiaceae have little food value. The three common genera are keyed below. If your "LBM" does not key out persuasively, check *Galerina, Tubaria,* & Allies (p. 399), and see comments on p. 32

Key to the Bolbitiaceae

1. Cap viscid when moist and conspicuously striate at maturity, at least at margin; fruiting body soft, sometimes dissolving somewhat in wet weather .... ***Bolbitius,*** p. 473
1. Not as above (cap may be viscid *or* striate but generally not both) 2

2. Cap typically conical to bell-shaped, at least when young; spore print ochre-brown to bright rusty to cinnamon-brown .... ***Conocybe,*** p. 470
2. Spore print dull brown to rich brown ("coffee-brown," "cigar-brown," "earth brown," etc.), *or* if brighter than cap convex to plane 3

3. Gills yellowish to orangish; spore print more or less rusty-orange .. (see ***Gymnopilus,*** p. 407)
3. Not as above; spore print dull brown to rich brown .... ***Agrocybe,*** below

AGROCYBE

Small to medium-sized, saprophytic mushrooms. CAP *convex to plane or broadly umbonate, smooth or cracked but not truly scaly,* dry or slightly tacky, rarely striate. GILLS *typically attached,* brown to rusty-brown at maturity. STALK central, thick or thin, *not markedly fragile (usually pliant).* VEIL present or absent, sometimes forming an annulus (ring) on stalk. VOLVA absent. SPORE PRINT *brown.* Spores smooth, usually with a germ pore. Cap cuticle typically cellular.

AGROCYBE is a difficult genus to characterize. The most common species are best recognized by their smooth or cracked (but not scaly), convex to plane cap, brown spores, and occurrence in grass, manure, wood chips, or cultivated ground. In some types a membranous veil is present and in others it is not. Some species resemble *Hebeloma* and *Pholiota* and have been keyed out with those genera (several Agrocybes were originally placed in *Pholiota*); others resemble *Stropharia,* but can be distinguished by their browner or brighter spore color.

Several Agrocybes are edible, but extreme care must be taken not to confuse them with the metagrobolizing masses of nondescript brown-spored mushrooms—particularly the poisonous Hebelomas. The two species described here are among our most common urban and suburban fungi, but also grow in rural and wooded habitats.

Key to Agrocybe

1. Stalk 5-15 mm thick, with a prominent basal bulb; found in manure, greenhouses, mushroom farms, etc. .... (see ***Conocybe,*** p. 470)
1. Not as above (but may grow in manure) 2

2. Partial veil present and membranous or kleenexlike (check young specimens if unsure!), often (but not always!) forming a ring on stalk or leaving pieces of tissue on cap margin 3
2. Not as above; partial veil absent or rudimentary and evanescent 8

3. Growing on ground in woods; cap dark brown to rusty-brown, often viscid or slimy when moist; cap margin often striate in age, gills often decurrent ***A. erebia*** (see *A. praecox* group, p. 469)
3. Not as above 4

4. Cap margin striate when moist; cap usually less than 5 cm broad and usually hygrophanous (fading markedly as it dries) .... (see ***Galerina, Tubaria,*** & Allies, p. 399)
4. Not as above 5

5. Found in swamps and other wet places; cap typically 3 (4) cm broad or less; stalk 2-4 mm thick .... ***A. paludosa***
5. Not as above 6

6. Growing on hardwoods, often clustered; southern ***A. aegerita*** (see *A. praecox* group, p. 469)
6. Not as above 7

7. Cap olive-brown to yellow-brown, tan, creamy, or whitish; veil membranous or kleenexlike (not fibrillose); stalk lacking scales; cap not normally viscid; found in grassy or cultivated areas or on wood chips or sometimes in woods .... *A. praecox* group, p. 469
7. Not as above .... (see *Pholiota*, p. 384)
8. Fresh cap dark brown to grayish-brown; in groups or clusters on dead hardwoods . *A. firma*
8. Not as above 9
9. Cap medium-sized (5 cm or more); stalk usually 4 mm thick or more ....
    .... *A. sororia* (see *A. pediades* group, below)
9. Cap small; stalk generally less than 5 mm thick 10
10. Either cap translucent-striate when moist *or* cap hygrophanous (brown fading to buff or whitish as it loses moisture) and gills adnate to decurrent .. (see *Galerina, Tubaria,* & Allies, p. 399)
10. Not as above 11
11. Spore print rusty-brown or brighter; stalk white and/ or cap 3-6 cm broad; found on dung and manure; not common .... (see *Conocybe,* p. 470)
11. Not as above; found in grassy or open places, also in dung 12
12. Cap wrinkled or pitted, white to pale tan; subtropical and tropical (on lawns, etc.) *A. retigera*
12. Not as above; widespread and common .... *A. pediades* group, below

Agrocybe pediades group   (Common Agrocybe)

CAP 1-3 (4) cm broad, hemispherical (rounded) to convex, or sometimes broadly convex to plane in age; surface dry or slightly viscid, smooth or sometimes fissured (with cracks) in age, usually ochre to golden-brown or yellow-brown, but varying to yellowish-buff, creamy, or even rusty-brown; margin not striate, but sometimes with whitish veil remnants. Flesh thin, pallid; odor mild or farinaceous. **GILLS** close, at first adnate but often seceding; pallid, soon becoming brown to rusty-brown or cinnamon-brown. **STALK** 2-5 (7) cm long, 1.5-3 (6) mm thick, more or less equal, dry, pallid or buff to yellow-brown (often paler at apex, darker below); often longitudinally striate. **VEIL** absent, or if present then evanescent and fibrillose (*not* membranous) and either disappearing or leaving slight remnants on the cap margin and/ or stalk. **SPORE PRINT** brown; spores 9-13 × 6.5-8 microns, elliptical, smooth, truncate from an apical germ pore.

HABITAT: Scattered to gregarious in grass and cultivated ground, dung or manure, or in sand; widely distributed and common, flourishing in our area whenever conditions are conducive. In warm weather it can often be found mingling with *Marasmius oreades,*

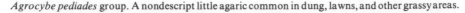

Agrocybe pediades group. A nondescript little agaric common in dung, lawns, and other grassy areas.

Panaeolus foenisecii, and *Conocybe lactea;* in cool weather it associates with *Psilocybe coprophila, Stropharia semiglobata,* and the *Panaeolus campanulatus* group.

EDIBILITY: Edible, but not recommended; it is easily confused with poisonous "LBM's."

COMMENTS: The yellowish cap, brown spores, absence of a ring, and small size typify this commonplace, lawn-loving "LBM" and its close relatives. The cap is not conical as in *Conocybe* and *Panaeolus,* and the spores are neither purple-brown nor purple-black as in *Stropharia,* nor white as in *Marasmius.* Small *Hebeloma* species are similar, but generally grow near trees or in the woods and have a filamentous cap cuticle. Other species: *A. pediades var. platysperma* has larger spores and frequently has watery spots or streaks on the cap; *A. semiorbicularis (=Naucoria semiorbicularis)* is said to have a slightly viscid cap and broader spores, but is otherwise identical; *A. arvalis(=A. tuberosa)* is a rare species that arises from a small, black, easily-overlooked "tuber" (sclerotium); *A. amara* has a bitter taste and grows in dense groups in manure or greenhouses; *A. sororia* is a much larger species (cap 5-15 cm broad, stalk 0.5-1.2 cm thick) with a tan to tawny cap and bitter taste. It resembles *A. praecox* but lacks a veil. I have found it fruiting abundantly on wood chips in a garden. For another photo of the *A. pediades* group, see p. 43.

Agrocybe praecox group   Color Plate 102
(Spring Agrocybe; Early Agrocybe)

CAP (2) 3-10 cm or more broad, convex or broadly umbonate to plane, or at times with an uplifted margin in age; surface dry or slightly tacky, smooth or in age sometimes cracked or fissured; creamy to creamy-ochre, yellow-brown, tan, hazel-brown, or olive-brown; margin often with veil remnants, not striate. Flesh white or pale ochre, soft; odor mild to farinaceous; taste mild or more often somewhat bitter to farinaceous. **GILLS** close, broad, adnate but often seceding; at first pallid, then light brown to grayish, finally dull brown or ochre-brown. **STALK** 3-13 cm long, 0.3-1 or more cm thick, equal or tapered below (or occasionally enlarged below), white or pallid, but often brownish-stained below in age; fibrillose-striate, the base usually with white mycelial threads. **VEIL** membranous but thin; forming a fragile, superior, skirtlike ring on stalk or leaving remnants on cap margin or gills or disappearing entirely. **SPORE PRINT** rich brown ("cigar-brown"); spores 8-12 × 5-7 microns, elliptical, smooth, truncate from the apical germ pore. Cap cuticle cellular.

HABITAT: Solitary, scattered, or gregarious in grassy or cultivated areas, along roads, in wood chips, gardens, woods, etc.; widely distributed, fruiting mainly in the spring. In our area this species "complex" is among the very common early spring mushrooms mushrooms (February-April), but occurs at other times also. The largest fruitings I've seen were in iceplant along a freeway offramp and on a parkway mulched with wood chips. It also occurs fairly commonly under mountain conifers in the spring.

EDIBILITY: Edible, but not recommended. The taste is mediocre at best, disgusting at worst, and there are a number of variants in this group whose edibility is unknown.

COMMENTS: Originally called *Pholiota praecox,* this attractive fungus can be found almost anywhere but is especially common in towns and along rural roads. The creamy to brownish cap, membranous veil (check young specimens!), brown spores, and habitat are the telltate traits. Both slender and fairly robust forms occur, and both are quite attractive in their prime. Specimens with a well-formed ring are reminiscent of *Agaricus,* but the gills are adnate (at least when young) and *never* pink or chocolate-brown. *Stropharia* species (e.g., *Stropharia semiglobata* and relatives) are also very similar, but have a darker (purple-brown to black) spore print and stickier cap. Confusion with *Pholiota* and *Rozites* is also possible, but the habitat and absence of scales or fibrils on the cap and stem should distinguish it. In my experience, the veil forms a distinct ring on the stalk less than 50% of the

Left: This member of the *Agrocybe praecox* group is darker and more robust than the typical form shown in the color plate. It usually grows in wood chips and may well be a distinct species. **Right:** *Agrocybe erebia* (see comments below) has an even darker (brown) cap and slightly decurrent gills.

time, but usually leaves at least some vestiges on the cap margin or gills. The "typical" form has a creamy cap (see color plate) and ranges from slender to fairly robust. Another common form (probably a distinct species) favors wood chips, is usually gregarious or clustered, has a darker (hazel-brown to olive-brown) cap when young, and is fairly robust (see photo above). Also similar are: *A. acericola*, with a yellow-brown cap when young and better-formed ring, growing on decayed wood; and *A. dura* (formerly **Pholiota vermiflua**), with a whitish to ochre-tinged, often cracked cap and slightly larger spores, found in cultivated and disturbed ground. Other species with a membranous veil include: *A. erebia*, a terrestrial woodland species with a dark brown, often viscid cap and/or slightly decurrent gills (see above photo); and *A. aegerita (=A. cylindracea),* a medium-sized to large edible southern species that usually grows in clusters on hardwoods (often living) such as willow, poplar, and box elder. The latter species is prized in Europe and grown there commercially. It has a yellowish to grayish-brown to fulvous-tinged, often wrinkled cap and typically has a well-formed annulus (ring) on the stalk.

CONOCYBE (Cone Heads)

Mostly fragile, little brown mushrooms saprophytic on grass, dung, moss, and humus. CAP sometimes convex but *usually conical or bell-shaped.* GILLS attached or free, *usually rusty-brown to cinnamon-brown at maturity.* STALK central, *usually thin and fragile,* often hollow. VEIL present or absent, sometimes forming an annulus (ring) on stalk. VOLVA absent. SPORE PRINT *rusty-brown to cinnamon-brown to ochraceous.* Spores smooth or rough, typically with a germ pore. Cap cuticle cellular.

CONOCYBES are often called "dunce caps" or "cone heads" because they usually have a conical or bell-shaped cap. They are mostly fragile, often ephemeral, *Mycena*-like mushrooms with a slender stem and rusty-brown to ochre-brown spores. They are sometimes confused with *Psilocybe,* but have brighter brown spores. Among the brown-spored mushrooms, they are apt to be confused with *Bolbitius,* which usually has a distinctly viscid, striate cap, and *Galerina,* which has a filamentous rather than cellular cap cuticle and an often viscid and/or translucent-striate cap. One prominent exception to the above is *C. intrusa* (keyed below), a rather fleshy mushroom with a thick stalk and convex cap. It resembles a *Hebeloma* and was originally classified as a *Cortinarius,* but microscopic characteristics plus its rusty-colored spore print relate it to *Conocybe.*

The "cone heads" are partial to warm weather and fruit in great abundance on watered lawns. Some, such as *C. lactea,* are so frail that they shrivel up or topple over a few hours after appearing. The edibility of most Conocybes is unknown, but they are much too small to be of food value and at least one species, *C. filaris,* is dangerously poisonous. *Conocybe* is a fairly large genus with over 50 species in North America. As these are largely differentiated on microscopic characters, only three are described here.

Key to Conocybe

1. Veil present, usually forming a distinct ring (annulus) on stalk 2
1. Veil absent or evanescent, not forming an annulus 4

2. Growing on dung or manure .... *C. stercoraria* (see *C. filaris* group, below)
2. Growing in grass, woody chips, moss, etc. 3

3. Cap tawny-brown to orange-brown to brown; annulus often prominent and movable ......
   .... *C. filaris* group, below
3. Cap reddish-brown to reddish-cinnamon when moist or sometimes yellower, fading as it dries; annulus not movable, usually small or obscure .... (see *Galerina, Tubaria,* & Allies, p. 399)

4. Stalk 0.5-1.5 cm thick, usually with a bulbous or thickened base; cap convex to nearly plane, 2.5-9 cm broad, viscid when moist, whitish to tawny to ochre-brown or even reddish-brown; growing in hothouses, mushroom farms, compost, etc. .... *C. intrusa*
4. Not as above 5

5. Base of stalk aging or slowly bruising blue to bluish-green ....
   .... *C. smithii* & *C. cyanopus* (see *C. tenera* group, p. 472)
5. Not as above 6

6. Gills crisped, with prominent veins in between; found on lawns *C. crispa* (see *C. lactea,* p. 472)
6. Not as above 7

7. Cap white to buff (or often darker at center), often wrinkled; fruiting body lasting only a few hours before toppling over or withering .... *C. lactea,* p. 472
7. Cap brown to tan, yellow-brown, or cinnamon, but often fading to buff, more or less smooth 8

8. Cap 3-6 cm broad, broadly conical to bell-shaped or convex; growing in compost in a garden
   .... *C. sp.* (unidentified)
8. Cap typically smaller and/or habitat different 9

9. Cap convex; growing in dung or manure .... *C. coprophila* (see *C. tenera* group, p. 472)
9. Cap conical to bell-shaped; growing in many habitats 10

10. Cap translucent-striate when moist, *not* growing in dung or manure ....
    .... (see *Galerina, Tubaria,* & Allies, p. 399)
10. Not as above; growing in many habitats .... *C. tenera* group, p. 472

Conocybe filaris *group*   (Ringed Cone Head)

CAP 0.5-2.5 cm broad, conical or bell-shaped, expanding to convex or plane but usually retaining an umbo; surface smooth, not viscid, tawny-brown to orange-brown or brown; margin faintly striate when moist. Flesh thin, brown. GILLS pallid soon becoming brownish to rusty-brown, close, adnexed or notched, sometimes seceding. STALK 1-6 cm long, 1-3 mm thick, more or less equal, fragile, yellow-brown to brown; smooth. VEIL membranous, brown, forming a prominent but delicate, movable, median to superior ring on stalk; ring often stained by spores and sometimes falling off. SPORE PRINT cinnamon-brown to rusty-brown; spores 7.5-13 × 3.5-6.5 microns, elliptical, smooth, with germ pore.

HABITAT: Scattered to gregarious on lawns and other grassy places, also in moss, on decayed wood and wood chips, etc.; widely distributed, but most common in the Pacific Northwest. In our area I have found it only once, on a lawn in October, but a large number of fruiting bodies were present.

EDIBILITY: POISONOUS—potentially fatal! It will not tempt most mushroom hunters because of its small size, but is dangerous to toddlers in the so-called "grazing stage," particularly because it grows on lawns. Analysis has revealed the presence of amanita-toxins, and 20-30 caps are equivalent to about half of one cap of *Amanita phalloides*—a compelling reason not to eat "LBM's!"

COMMENTS: Also known as ***Pholiotina filaris,*** this species and its close relatives are distinguished from other Conocybes by the membranous ring on the stalk. In age, however, the ring may fall off or be otherwise obliterated. Several poisonous *Galerina* species have a ring, but they usually grow on wood and/or have a somewhat viscid, translucent-striate cap (see *G. autumnalis*). There are also several Conocybes with an annulus (ring) that grow on dung and manure (especially of horses), including *C. **stercoraria,*** fairly common, with a yellow-brown to buff-colored cap.

Conocybe tenera group   (Brown Dunce Cap; Common Cone Head)

CAP 1-2.5 cm broad and tall, conical to bell-shaped, sometimes expanding slightly in age; surface smooth, not viscid, brown to cinnamon-brown or yellow-brown, usually fading in age or as it dries to tan, yellowish, or buff; margin finely striate when moist. Flesh thin, brownish. **GILLS** adnate to adnexed or free, close, pallid soon becoming pale brown, then cinnamon-brown or rusty-brown. **STALK** 4-9 cm long, 1-4 mm thick, equal or with a swollen base, smooth or minutely mealy, colored like cap or paler, fragile. **VEIL** absent. **SPORE PRINT** rusty-brown or cinnamon-brown; spores 8-14 × 5-7 microns, elliptical, smooth, with an apical germ pore.

HABITAT: Scattered to densely gregarious on lawns, in gardens, fields, rich soil, dung, and humus; widely distributed. This species and its close relatives occur practically year-round in our area in a variety of habitats (including the woods), but are not particularly numerous. I've seen more than 100 fruiting bodies crowded into a small planter box filled with lettuce seedlings.

EDIBILITY: Unknown, but do not experiment—the similar *C. filaris* is poisonous!

COMMENTS: The above description will fit a large number of brownish Conocybes with a cute (but not necessarily acute) conical to bell-shaped cap, thin fragile stem, and no veil or ring (annulus). They are reminiscent of *Mycena* and *Panaeolus* (with which they may grow), but have rusty-brown gills at maturity and rusty-brown spores. The cap is darker than that of *C. lactea* and not as wrinkled, and is not viscid as in *Bolbitius*. Other species: *C. coprophila* has a slightly viscid convex cap and grows in dung or manure; *C. **smithii*** and *C. cyanopus* are two rather rare "pupil-dilating" species with a blue to blue-green stem base, at least in age. The former grows in bogs, the latter in grass or sometimes moss. Neither should be eaten by "magic mushroom" hunters because of their close resemblance to ringless representatives of the poisonous *C. filaris.*

Conocybe lactea   (White Dunce Cap; White Cone Head)

CAP 1-2.5 cm broad and/or tall, narrowly to bluntly conical (like a dunce cap), then bell-shaped or thimble-shaped, the margin usually flaring; surface typically not viscid, radially wrinkled or striate when moist, white to creamy or tinged buff to pale cinnamon, especially at the center. Flesh very thin, whitish. **GILLS** very narrow, adnexed or free, close, pallid soon becoming cinnamon-brown to tawny-ochre or brown. **STALK** 3-11 cm long, 1-2 mm thick, very thin and fragile, equal or with a small basal bulb, hollow, white or whitish, often powdery or scurfy above. **VEIL** absent. **SPORE PRINT** reddish-brown or cinnamon; spores 11-16 × 6-10 microns, elliptical, smooth, with an apical germ pore.

HABITAT: Scattered or in troops on lawns, baseball fields, and other grassy places; widely distributed and common in muggy weather. It usually appears in the early morning and topples over or shrivels up by afternoon; it is one of our common summer lawn mushrooms.

Conocybe lactea is abundant on lawns in warm weather. These are rather mature specimens; younger ones can have even narrower caps. The fruiting bodies collapse within a few hours.

EDIBILITY: Unknown, but worthless as food—to say it lacks substance is a gross understatement. As Alexander Smith points out, toddlers would be the only ones tempted to eat it, and if it were poisonous, we would *probably* know by now.

COMMENTS: This frail mushroom is easily told by its pallid "dunce cap" and thin, fragile stem. The striate or wrinkled cap is not as viscid as in *Bolbitius,* and is paler than that of the *C. tenera* group. The stem is so feeble that it is practically impossible to bring home specimens without breaking them. *C. crispa* is a somewhat similar widespread species with an ochre-tinged cap and crisped, veined gills.

BOLBITIUS

Small, fragile, rapidly-decaying mushrooms found on *dung, humus, grass, or wood.* CAP *viscid when moist and usually conspicuously striate at maturity;* usually yellow, white, or purple-gray. GILLS typically adnexed or free at maturity, *often dissolving somewhat in wet weather.* STALK *slender, fragile, hollow.* VEIL and VOLVA *absent.* SPORE PRINT *rusty-brown to ochraceous.* Spores smooth, with a germ pore. Cap cuticle cellular.

THIS small genus of fragile, flaccid, putrescent fungi is reminiscent of *Coprinus* (particularly in the striate cap and tendency of the gills to liquefy somewhat), but has rusty-brown to bright yellow-brown rather than black spores. The viscid striate cap separates *Bolbitius* from *Conocybe.* The cap is rarely brown as in *Galerina* and the habitat is quite different.

Bolbitius species are worthless as food. The most common types fruit during wet weather in dung, manure heaps, straw, tall grass, and other vegetable matter; a few grow on wood. Two species are described here.

Key to Bolbitius

1. Cap bright yellow to pale yellow, at least when fresh .... *B. vitellinus,* p. 474
1. Cap differently colored 2
2. Usually growing on lawns or dung, rarely in woods 3
2. Growing in woods or on rotten wood 6

473

3. Cap pinkish-grayish-cinnamon in age; on dung ..... **B. coprophilus** (see *B. vitellinus,* below)
3. Not as above; growing on lawns or in grass 4
4. Cap milky white or with a tinge of buff, especially in age . **B. lacteus** (see *B. vitellinus,* below)
4. Not as above 5
5. Cap conical or oval, very soft, some shade of brown; gills veined and often fused together; stalk
    very fragile, often bending over; fruiting body ephemeral; found in hot, humid weather ....
    .... **Gastrocybe lateritia**
5. Not as above .... (see **Galerina, Tubaria,** & Allies, p. 399)
6. Cap white; growing on wood or wood chips .... **B. sordidus**
6. Cap not white 7
7. Cap chocolate-brown to rusty-brown; stalk 3-10 mm thick and brown over lower portion ...
    .... (see **Agrocybe,** p. 467)
7. Not as above; stalk white or yellow, thin and fragile .... **B. aleuriatus** & others, p. 475

Bolbitius vitellinus   (Sunny Side Up)

CAP 1.5-7 cm broad when expanded, oval to conical or bell-shaped when young, often
becoming plane in old age; surface smooth, viscid or slimy when moist, bright yellow to
pale yellow (the center sometimes yellow-orange), but often fading in age or as it dries to
whitish, brownish, grayish, etc.; margin striate, at times conspicuously grooved nearly to
the center. Flesh very thin, soft, yellowish. **GILLS** adnate to adnexed or free, close, soft,
dissolving somewhat in wet weather; pallid or pale yellow to pale brown, becoming rusty-
ochre to cinnamon-brown in age. **STALK** (3) 5-12 cm long, 2-8 (10) mm thick, equal or
thicker below, hollow, very fragile when thin (readily collapsing), whitish to pale yellow;
often delicately powdered or scurfy. **SPORE PRINT** rusty-orange to rusty-brown; spores
10-16 × 6-9 microns, elliptical, smooth, truncate from the large apical germ pore.

HABITAT: Solitary, scattered, or gregarious (or even tufted) on dung, manure, straw,
lawns, in tall grass, cultivated ground, etc.; widely distributed and ubiquitous, fruiting
throughout the mushroom season in our area. The most luxuriant fruiting I've seen was in
a horse corral where dozens of large, robust specimens were interspersed with massive

Bolbitius vitellinus varies considerably in size and shape, as evidenced by these photographs and
the one on p. 475. However, it is always viscid, yellow (when fresh), and very fragile. **Left:** Slender,
long-stemmed specimens such as these are common in lawns and other grassy areas. **Right:** Relatively
robust specimens found on horse dung. In old age the cap often becomes plane (flat).

clusters of *Peziza vesiculosa* (a cup fungus). Greg Wright, who coined the common name, says that in southern California it commonly occurs on decayed wood.

EDIBILITY: Harmless, but fleshless and flavorless.

COMMENTS: This fragile ephemeral mushroom varies greatly in size, shape, and habitat but can generally be recognized by its viscid, yellow, striate cap, plus its rust-colored gills (in age) and soft, fragile texture. The stems often collapse of their own accord, and in wet weather the cap and gills tend to dissolve, much as in *Coprinus* (but not as the result of an autodigesting enzyme). As a rule, those growing in grass tend to be quite slender and fragile (see photo), while those growing in a nitrogen-rich environment such as horse manure are larger and more robust. Other species include: *B. coprophilus,* with a grayish-pinkish-cinnamon cap in age, usually found on dung; and *B. lacteus*, with a small whitish or buff-tinged cap, found in dung, grass, straw, etc., but not nearly as common as *B. vitellinus.*

Bolbitius aleuriatus

CAP 1.5-4 cm broad, soon broadly convex to broadly umbonate or plane; surface smooth, viscid when moist, gray to grayish-brown, often with a purplish or lilac tint (center sometimes darker); sometimes fading in age; conspicuously striate, at least at margin. Flesh *very* thin, whitish. **GILLS** adnexed to free, narrow, close, pallid soon becoming pinkish to brown or cinnamon-brown. **STALK** 2.5-7 cm long, 2-3 mm thick, equal or thicker at base, very fragile, smooth or minutely scurfy or powdery; entirely white or tinged yellow at base. **SPORE PRINT** rusty-brown; spores 9-12 × 4-6 microns, elliptical, smooth, with an apical germ pore.

HABITAT: Solitary or in small groups (rarely more than three) on rotting wood, sawdust, and humus. It is fairly common in our area in the fall and winter, mainly under oak and madrone, and probably has a much wider distribution.

EDIBILITY: Unknown.

COMMENTS: Also known as *Pluteolus aleuriatus,* this species is a fairly frequent but oft-overlooked fungal feature of our oak wondlands. When the gills are free and pinkish, confusion with *Pluteus* is likely. However, its extreme fragility and viscid, striate cap are good fieldmarks. A form with a yellow stem—possibly distinct—also occurs in our area. Other woodland species include: *B. reticulatus,* similarly colored but with a reticulate (veined or netted) cap; and *B. (=Pluteolus) callisteus,* with a yellow stalk and a yellow-olive to rusty-orange cap.

Left: *Bolbitius vitellinus,* young specimens with rounded yellow caps. (Nancy Burnett) **Right:** *Bolbitius aleuriatus* is a fragile, slender-stemmed woodland species with a grayish to purple-gray cap that is prominently striate (lined).

PAXILLACEAE

spores

> Fleshy medium-sized mushrooms growing on wood or ground. CAP convex to plane or depressed, margin sometimes strongly inrolled. GILLS *typically decurrent, often forked or veined,* yellow to orange or dull-colored; often peeling easily from cap *(Paxillus).* STALK fleshy;, central or off-center to lateral or even absent. VEIL and VOLVA *absent.* SPORE PRINT *yellowish to brown(Paxillus & Phylloporus),* or white *(Hygrophoropsis).* Spores mostly elliptical, smooth, without a germ pore, often dextrinoid.

THIS is a small family of medium-sized mushrooms with decurrent gills and yellowish to brown (or in *Hygrophoropsis,* white) spores. The flesh is not dry, white, and brittle as in *Russula,* and there is no latex as in *Lactarius.* The spore print is not rusty-orange as in *Gymnopilus,* the stem is much fleshier than in *Tubaria,* and other brown-spored agarics without a veil do not normally have decurrent gills.

There are three genera in the Paxillaceae. In the central genus, *Paxillus,* the margin of the cap is usually inrolled when young and the gills are often veined or poroid near the stem and peel rather easily from the cap. Like the dark-spored genera *Chroogomphus* and *Gomphidius, Paxillus* is thought to be closely allied to the boletes. The second genus, *Phylloporus,* includes several species which, despite the presence of gills, are considered true boletes by many mycologists. The gills are bright yellow and often bruise slightly bluish like the tubes in many species of *Boletus,* and the spores are "boletoid" (narrowly elliptical to spindle-shaped). The third genus, *Hygrophoropsis,* resembles *Paxillus* but has unpigmented (white) spores. It includes the fungus popularly known as the false chanterelle, and has traditionally been placed in the Tricholomataceae (it is also keyed out there).

The Paxillaceae are woodland fungi like the boletes, but are not nearly as numerous or conspicuous. In South America the situation is reversed—*Paxillus* is a fairly prominent group while the boletes are few and far between. In case you hadn't noticed, this is not a field guide to the fleshy fungi of South America, so only five members of the Paxillaceae are described. They are not particularly good eating and at least one species, *Paxillus involutus,* can be *very* poisonous.

Key to the Paxillaceae

1. Gills yellow and rather thick, broad, and widely spaced, often staining blue or greenish (sometimes slowly) when bruised . . . . *Phylloporus rhodoxanthus,* p. 480
1. Not as above; gills usually close or crowded 2

2. Stalk velvety from a dense coating of rusty-brown to brown or blackish-brown hairs, typically 1-3 cm thick; growing on or around conifer stumps, logs, etc. *Paxillus atrotomentosus,* p. 478
2. Not as above; stalk not brown and velvety 3

3. Odor fragrant (reminiscent of root beer); fruiting body pinkish to whitish . . . .
. . . . *Hygrophoropsis olida* (see *H. aurantiaca,* p. 479)
3. Odor mild or at least not fragrant 4

4. Spore print white or creamy; stalk normally present; gills usually orange or orangish (but sometimes differently colored) . . . . *Hygrophoropsis aurantiaca,* p. 479
4. Spore print yellowish to brown; stalk present *or* absent; gills not orange 5

5. Stalk lateral to absent; found on wood or lignin-rich humus . . . . . *Paxillus panuoides,* below
5. Stalk present, central to off-center; usually terrestrial . . . *Paxillus involutus* & others, p. 477

Paxillus panuoides   (Fan Pax)

CAP 1.5-7 (10) cm broad, petal-shaped to mussel- or fan-shaped, attached laterally to substrate or with a stemlike base; surface minutely hairy or downy becoming smooth, not viscid; buff to dingy yellowish, yellow-brown, olive-yellow, or dingy ochre; margin often lobed and at first incurved. Flesh thin, soft, whitish to ochraceous. **GILLS** radiating from

Paxillus panuoides grows on wood and has little or no stalk. Note how gills are poroid near base of cap.

base of cap, close, pale or dingy yellowish to ochre or pinkish-buff, often crimped and forked or connected by cross-veins, especially toward base. **STALK** absent or present only as a small, narrowed, lateral base. **SPORE PRINT** yellowish-buff to brown or dingy ochraceous; spores 4-6 × 3-4 microns, elliptical, smooth, many of them dextrinoid.

HABITAT: Solitary or in groups or clumps on coniferous logs, stumps, debris, and humus rich in lignin; widely distributed but not particularly common. In our area I have found it several times on dead pine in the fall and winter. It is also said to occur on mine timbers, causing a bright yellow discoloration in its host.

EDIBILITY: Unknown—do not experiment!

COMMENTS: The fan-shaped cap and absence or near absence of a stalk rescue this listless little brown mushroom from the obscurity it so richly deserves. It might be mistaken for a *Crepidotus* or *Phyllotopsis,* but the gills are usually forked or veined. The latter feature can lead to confusion with the chanterelles, but the growth on wood, dingy color, and poorly-developed stalk distinguish it. The false chanterelle *(Hygrophoropsis aurantiaca)* is also somewhat similar, but has oranger gills, white spores, and a stalk.

Paxillus involutus   (Poison Pax; Inrolled Pax)

CAP 4-15 (20) cm broad, at first broadly convex with a strongly inrolled margin, then plane or centrally depressed with the margin eventually unfurled; surface viscid when moist but otherwise dry; smooth or with soft matted hairs that wear away, sometimes cracked (areolate) in age; brown to dingy yellow-brown, olive-brown, or dingy reddish-brown, often with darker brown stains; margin often slightly velvety or obscurely ribbed. Flesh thick, firm, pale buff to yellowish, but usually staining reddish to brown when cut. **GILLS** usually decurrent, close or crowded, pallid to pale yellowish becoming dingy yellow to olive, brownish, or yellow-brown, staining dark brown or reddish-brown when bruised or in old

Paxillus involutus is easily told by the strongly inrolled cap margin when young and the tendency of the entire mushroom (especially the gills and flesh) to stain dark brown. Birch is one of its favorite tree associates. See p. 41 for another photo, and see p. 478 for close-up of gills.

Left: The close relationship of *Paxillus* to the boletes is suggested by this close-up of the gills (and "pores") of *Paxillus involutus*. **Right:** *Paxillus atrotomentosus* is a wood-inhabiting species with decurrent gills and a dark velvety stalk.

age; often forking and/or forming pores near the stalk. **STALK** 2-7 (10) cm long, 0.5-4 cm thick, usually shorter than width of mature cap, equal or tapered at either end, central to somewhat off-center, solid, firm, dry, smooth; colored like cap or paler, often with dingy reddish to dark brown stains. **SPORE PRINT** brown to yellow-brown; spores 7-10 × 4-6 microns, elliptical, smooth.

HABITAT: Usually scattered to densely gregarious on ground in woods, around the edges of bogs, and in tree-studded parks or lawns; widely distributed, and by far the most common member of the genus. Two possibly distinct forms occur in our area: the typical one (shown in photo at bottom of p. 477) fruits mainly with planted birch trees in the fall, while a larger form occurs with oak and pine in the late fall and winter.

EDIBILITY: Dangerous! It is often eaten in Europe and by transplanted Europeans in America, but it can cause hemolysis (destruction of red blood cells) and kidney failure if eaten raw, and sometimes even when thoroughly cooked! (Apparently the human body can develop a devastating sensitivity to it.) It often has a sour taste anyway, so it is best avoided and should *never* be eaten raw. The one recorded instance of a bonafide mycologist dying of mushroom-poisoning was attributed to this fungus!

COMMENTS: The dingy brownish cap with inrolled margin when young plus the decurrent gills that stain brown and are usually forked or veined near the stem are the unappealing features of this unappealing fungus. Actually, its symmetry and compactness are quite intriguing, but its choice of color is downright disastrous—an unbecoming blend of dingy browns and murky yellows. The stature is reminiscent of a *Lactarius* or *Russula,* but the gills do not exude a latex when broken and the flesh is not chalky or exceptionally brittle. Other species: *P. vernalis* is similar but larger, with a paler cap when young and reddish-brown to chocolate-brown spores. It grows mainly with aspen.

Paxillus atrotomentosus   (Velvet Pax)

CAP 4-15 (20) cm broad, convex becoming plane or centrally depressed; surface dry, unpolished, velvety or with matted hairs, yellow-brown to rusty-brown to dingy reddish-brown becoming dull brown, dark brown, or blackish-brown in age; margin at first inrolled. Flesh thick, rather firm or tough, pallid to ochre or buff. **GILLS** close or crowded, usually decurrent, tan to dull ochre, dingy yellowish, or paler; often forked or veined near

478

stalk. **STALK** 2-9 (12) cm long, 1-3 (5) cm thick, often short and usually off-center or even lateral; solid, tough, densely velvety from a coating of brown to dark brown or blackish-brown, often matted hairs; apex often paler or yellowish. **SPORE PRINT** yellowish to brownish; spores 5-6.5 × 3-4.5 microns, elliptical, smooth, many of them dextrinoid.

HABITAT: Solitary or in groups or tufts on or around conifers (usually dead) or madrone; northern in distribution. It is fairly common in the Pacific Northwest in the late summer and fall, but rare in our area. It causes a carbonizing decay (brown rot) in its host.

EDIBILITY: Not recommended. Some people apparently collect it for the table, but the same can be said of *P. involutus,* which has caused several deaths! (See comments on edibility of that species.)

COMMENTS: The combination of brown cap, dark brown velvety stem, decurrent gills, and growth on rotting conifers makes this one of the Pacific Northwest's most distinctive agarics. It does not discolor as much as *P. involutus,* and is quite attractive when fresh.

Hygrophoropsis aurantiaca   (False Chanterelle)   **Color Plate 110**

CAP 2-8 (14) cm broad, convex becoming plane or somewhat depressed, the margin at first inrolled; surface dry, often somewhat felty or velvety, typically some shade of dark orange, brownish-orange, brownish-yellow, yellowish-brown, olive-brown, or dark brown (often darker or browner at center and orange to yellowish-orange at margin), but in one form whitish and in another blackish. Flesh thin, pallid or tinged orange or cap color; odor mild. **GILLS** decurrent, close, fairly thin and narrow at maturity (may be blunt when young); usually forked dichotomously; typically deep orange to bright orange, but sometimes pale orange or in one form yellowish. **STALK** 2-10 cm long, (0.2) 0.5-1 (2) cm thick, central or off-center, equal or enlarged toward base, often curved; dry, yellowish to orange or brownish-orange or colored more or less like cap. **SPORE PRINT** white to creamy; spores 5-8 × 2.5-4.5 microns, elliptical, smooth, often dextrinoid.

HABITAT: Solitary, scattered, or in groups or tufts in humus and on rotting wood, usually under conifers; widely distributed. In our area it fruits from the fall through early spring, but seems most abundant in cool, dry weather when there are few other fleshy fungi out and about. I usually find it under pine or redwood.

EDIBILITY: To be avoided. In my experience it is edible—but far from incredible. Some sources, however, list it as mildly poisonous (whether or not this is a result of confusion with *Omphalotus,* as has been suggested, is unclear).

Hygrophoropsis aurantiaca. The cap is usually brown at the center and lighter or brighter toward the margin, but in some forms it is completely brown (or even black) and in others it is whitish. For a close-up of the forked gills, see color plate.

COMMENTS: The typically bright orange, decurrent, dichotomously forked gills (see color plate) and white spore print are the principal fieldmarks of this attractive but variable fungus. In spite of its common name, it is difficult to confuse with the true chanterelle *(Cantharellus cibarius)* if the following is kept in mind: the gills are thinner, crowded, and bladelike at maturity (but often blunter when young), and are usually oranger than those of the chanterelle. In other words, the false chanterelle has "true" gills while the true chanterelle has "false" gills. Also, *Hygrophoropsis* has flimsier flesh, a browner cap, is less robust, differently shaped (not as wavy or frilled) and sometimes grows on rotten wood. The jack-o-lantern mushrooms *(Omphalotus* species) are also similar, but have brightly colored flesh, typically grow in clusters on or under hardwoods, and do *not* have forked gills. The false chanterelle was originally placed in *Cantharellus,* and is listed in many mushroom books as **Clitocybe aurantiaca,** but the forked gills, frequently off-center stalk, and dextrinoid spores connote a closer kinship to *Paxillus.* Other species: **H. olida (=Clitocybe morganii)** is a small (cap 1-4 cm) species with a pinkish cap and stalk (that may fade to buff), whitish to pinkish gills, and a flagrantly fragrant odor that is reminiscent of root beer or cinnamon candy. At least some of its gills are forked and/or have cross-veins, its stalk can be central or off-center, and its spores are white and dextrinoid. Like *H. aurantiaca,* it favors conifers and is widely distributed, but seems to be rather rare. I have seen it only once in California, near Mount Shasta, in June.

Phylloporus rhodoxanthus   (Gilled Bolete)   **Color Plate 111**

CAP 2-5 (12) cm broad, broadly convex to plane or with uplifted margin in age; surface dry, minutely velvety to nearly smooth, dull brown to olive-brown or yellow-brown in one form, red to reddish-brown in another; sometimes developing pallid to yellowish cracks or fissures. Flesh thick, pallid to yellowish. **GILLS** widely spaced, broad, fairly thick, adnate to decurrent but sometimes seceding; sometimes forked or with cross-veins; bright yellow to ochre, typically bruising green or blue (but often slowly, sometimes not at all), sometimes also staining brownish. **STALK** 3-10 cm long, 0.4-1 (1.5) cm thick, equal or tapered below, dry, solid, yellow to dingy yellowish-buff or reddish, often stained dingy brown or reddish-brown below; smooth. **SPORE PRINT** brown to yellowish- or olive-tinged; spores 9-15 × 3-6 microns, narrowly elliptical to spindle-shaped, smooth.

HABITAT: Solitary to widely scattered or in small groups (usually twos or threes) on ground in woods, widely distributed. The brown-capped form can be found nearly every year in our area, but seldom in quantity. Like the related *Boletus subtomentosus,* it fruits throughout the mushroom season, often along roads and trails.

EDIBILITY: Edible, but rarely found in sufficient numbers to be more than just a curiosity. Some "bolete-your-meat'ers" rate it highly, but I've found it to be slimy and insipid.

COMMENTS: This evolutionary oddity is an adamant nonconformist. The dry, minutely velvety brownish to reddish cap is an almost exact replica of *Boletus subtomentosus* and relatives, yet the underside of the cap has gills! These are thicker than the gills of many agarics, and are bright yellow and often bruise blue like the tubes of many boletes (it may take several minutes, however, for the color change to show). What's more, the spores are "boletoid"—long, narrow, and spindle-shaped—causing some taxonomists to place it in the Boletaceae. Thus *Phylloporus,* like many other mushrooms, has been denied a stable family and is at the complete mercy of the "authorities." Like baseball's itinerant tobacco-chewing utility infielders of southern rural origin, its group affiliation is in a state of constant flux as it is systematically shuttled back and forth between one "team" and another. Other species: **P. arenicola** is a very similar species with an olive-brown to olive-yellow cap and adnexed (not decurrent) gills which don't blue when bruised. It was originally described from coastal Oregon under pine, but I have not seen it.

GOMPHIDIACEAE

spores

THIS is a small but prominent family of fleshy-stemmed mushrooms with decurrent gills and smoky-olive to black spores. No other group of black-spored mushrooms has consistently decurrent gills. *Gomphos*, meaning "peg" or "stake," refers to the shape of the young mushrooms, which look like tent stakes with their long stems and small, rounded to conical caps. The spores are typically "boletoid"—i.e., long, narrow, and more or less spindle-shaped, and the Gomphidiaceae are thought to be more closely related to the boletes than to other gilled fungi. They are mycorrhizal exclusively with conifers and often occur with the bolete genus *Suillus*. This may be purely coincidental, since *Suillus* is also mycorrhizal with conifers, or it may be that the fungal mycelia are in some way associated with each other as well as with their mutual host.

As far as is known the family is completely safe from an edibility standpoint. There are two common genera, *Gomphidius* and *Chroogomphus*, but the latter is listed under *Gomphidius* in many older books. A closely related gastroid genus, *Brauniellula* (see p. 732) also occurs, but it often grows underground and does not have true gills.

Key to the Gomphidiaceae

1. Immature gills and flesh in the cap white to grayish .... *Gomphidius,* below
1. Immature gills and flesh in the cap some shade of yellow, orange, buff, salmon, reddish, etc. . ... *Chroogomphus,* p. 484

GOMPHIDIUS

Fairly small to medium-large terrestrial mushrooms *associated with conifers.* CAP *viscid or slimy when moist,* usually smooth. *Flesh white to grayish.* GILLS *decurrent,* soft or somewhat waxy, fairly close to well-spaced, *pallid or white when young* but blackening in age. STALK fleshy, more or less central; *lower portion or base usually bright yellow* (especially within). VEIL usually present, but often disappearing or leaving only a slight ring (annulus) on stalk. VOLVA absent. SPORE PRINT *smoky-gray to black.* Spores narrowly elliptical, smooth. Cap tissue not amyloid.

THIS genus is readily recognized by its soft, somewhat waxy decurrent gills, slimy-viscid cap, white or pallid flesh, and smoky-black spores. In addition, the base or lower portion of the stem is brilliant yellow in most species—a very striking and telltale feature. *Gomphidius* is most likely to be confused with *Chroogomphus,* which has orange to yellowish or pinkish flesh, and the waxy caps (Hygrophoraceae), which have white spores.

Gomphidius species are mycorrhizal exclusively with conifers—particularly fir, spruce, Douglas-fir, hemlock, and larch—but in pine forests they are largely supplanted by *Chroogomphus.* November rains can mean *Gomphidius* galore in our area, with large numbers of the glutinous fruiting bodies littering our Douglas-fir forests, along with maggoty *Suillus* species and assorted Russulas.

Gomphidius species are edible but not often collected—perhaps because they are soft, sluglike, insipid, and putrescent. Also, they do not dry nicely like *Chroogomphus,* and often blacken when cooked—but some people relish them nonetheless (maybe they can't find anything better!). About ten species occur in North America, all of them rather similar in appearance. Two are described here and several others are keyed out.

Key to Gomphidius

1. Slimy veil present in young specimens; stalk usually viscid when moist (from veil remains); associated with various conifers 2
1. Veil absent; stalk *not* viscid; associated primarily (but not exclusively) with larch or tamarack; found in northern North America 5

2.   Cap pink to rosy-red or red, not large (up to 7 cm broad); stalk usually 1.5 cm thick or less   ..
      .... *G. subroseus,* p. 483
2.   Cap sometimes pinkish or salmon, but usually dingier or darker in color (whitish, grayish,
      brownish, vinaceous, purplish, etc.); small to large 3
3.   Stalk with little or no yellow at base; fruiting body medium-sized to fairly small ....
      .... *G. smithii* (see *G. subroseus,* p. 483)
3.   Base or lower portion of stalk typically bright yellow; medium-sized to fairly large 4
4.   Often growing in small clumps, stalk often rooted deeply in soil; associated primarily with
      Douglas-fir; spores typically less than 14 microns long .... *G. oregonensis,* below
4.   Not as above; found with various conifers (including Douglas-fir); spores longer than 14
      microns; widely distributed .... *G. glutinosus* & others (see *G. oregonensis,* below)
5.   Cap whitish to yellowish when young; found in eastern North America; rare ... *G. nigricans*
5.   Cap pale cinnamon to brownish, etc., when young; widely distributed in northern latitudes
      .... *G. maculatus* (see *G. oregonensis,* below)

Gomphidius oregonensis   (Insidious Gomphidius)

CAP 2-15 (18) cm broad, at first peglike, then broadly convex to plane or depressed; surface viscid or slimy when moist, smooth, color variable: whitish to salmon-buff to ochraceous-salmon to dull pinkish when young, becoming dingier (brownish to purplish- or vinaceous-gray to dark reddish-brown) in age, often spotted or stained smoky-gray to black. Flesh rather soft, white or grayish (or tinged cap color under cuticle), but brilliant yellow in lower part of stalk or at base. **GILLS** decurrent, soft and rather waxy, close to fairly well-spaced, white or pallid, then gray and finally blackish as spores ripen. **STALK** 5-15 cm long, 1-5 cm thick, equal or tapered below (or occasionally swollen), solid, rather firm or even tough, dry and white above the veil, whitish or dingy below and viscid when wet, sometimes with darker streaks; lower portion bright yellow. **VEIL** whitish and fibrillose beneath a layer of slime; disappearing or forming a slight hairy-slimy superior ring on stalk which is subsequently blackened by falling spores. **SPORE PRINT** smoky-gray to black; spores 10-14 (16) × 4.5-8 microns, spindle-shaped to narrowly elliptical, smooth.

HABITAT: Solitary, scattered, or gregarious on ground under Douglas-fir and other conifers, often in small clumps which originate deep in the soil and may include one or more aborted fruiting bodies; known only from western North America and very common, (along with *G. glutinosus*—see comments) on the Pacific Coast. In our area it grows with Douglas-fir, often mingling with *Suillus* species *(S. caerulescens, S. lakei, S. ponderosus),* from the fall through early spring.

EDIBILITY: Edible, but its sliminess makes it undesirable—except as a possible *escargot* substitute. To clean the cap, simply peel off the slimy pellicle (skin).

COMMENTS: The intensely yellow flesh in the lower stem plus the whitish flesh elsewhere, viscid-slimy cap, soft decurrent gills, and smoky-black spores immediately identify this mushroom as a *Gomphidius.* However, distinguishing it from its brethren (see below), especially when it is not growing in clumps, can be difficult unless the spores—the shortest in the genus—are measured. In age the fruiting body is often quite murky and insidious-looking, hence its common name. The cap is quite variable in color but is usually dingier and dirtier than that of *G. subroseus,* and the stalk is usually thicker—but not necessarily slicker. Other species: The "Glutinous Gomphidius," *G. glutinosus,* is probably the most common and widespread member of the genus. It is mycorrhizal with a variety of conifers, including spruce, fir, and in our area, Douglas-fir. It is quite similar to *G. oregonensis,* but has longer spores (over 14 microns), does not often grow in clumps, and usually has a darker cap when young (purplish to purple-brown, brownish-gray, purplish-gray, etc.). Another western species, *G. largus,* is essentially a giant version of *G. glutinosus,* and also grows with spruce and fir, mainly at higher elevations. Finally, there is the "Hideous Gomphidius," *G. maculatus,* which has a pale cinnamon to reddish-brown to murky brown

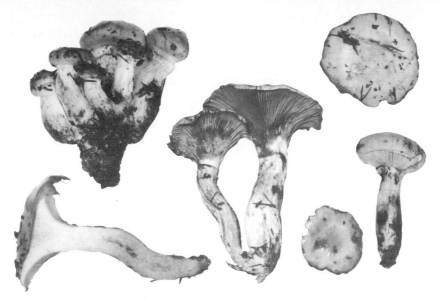

Gomphidius oregonensis, a common associate of Douglas-fir. Gills are decurrent and darken in old age, a veil is present when young (visible in specimen on right), and cap is slimy when moist. Specimen at lower left has been sliced open to show the flesh, which is white in the cap and bright yellow in lower part of stalk. Buttons (top left) of this species are often found in tight clumps.

or blackish-spotted cap and dark-fibered stalk. It lacks the slimy veil characteristic of other Gomphidii and features both robust and very slender forms. It is a northern species that grows mainly with larch, often in the company of *Suillus cavipes* and *S. grevillei*.

Gomphidius subroseus   (Rosy Gomphidius)   Color Plate 112

CAP 2.5-7.5 cm broad, at first peglike, then broadly convex becoming plane or broadly depressed; surface viscid or slimy when moist, smooth, dull to bright pink to rosy-red or even red, often spotted grayish in old age. Flesh white (or tinged pink under cuticle), but yellow in base of stalk and sometimes pinkish in *extreme* base. **GILLS** typically decurrent, soft and rather waxy, fairly well-spaced, white becoming pale gray to smoky-gray or even blackish as spores ripen. **STALK** 3-7.5 cm long, 0.5-1.5 cm thick, equal or narrowed slightly at base, dry and white above the veil, usually viscid below and white to dingy-colored with a pale yellow to bright yellow base; smooth or fibrillose, solid. **VEIL** white and fibrillose beneath a layer of slime; disappearing or forming an obscure hairy-slimy superior ring on stalk which is blackened by spores. **SPORE PRINT** smoky-gray to black; spores (11) 15-21 × 4.5-7 microns, spindle-shaped to narrowly elliptical, smooth.

HABITAT: Widely scattered to gregarious or occasionally tufted on ground under conifers (especially Douglas-fir), northern North America. It can be found throughout the range of Douglas-fir and is fairly common in our area in the fall and winter. It is also said to occur with spruce, fir, and hemlock.

EDIBILITY: Edible, but rather bland; peel off the slimy pellicle (skin) before cooking it.

COMMENTS: The beautiful rosy-red to pink cap and modest size separate this fungus from other *Gomphidius* species, and make it the most attractive member of its clan. It typically shows less yellow in the stem than either *G. oregonensis* or *G. glutinosus,* but more than *G. smithii* (see below). It is often mistaken for a waxy cap (Hygrophoraceae) because of its brightly colored cap and soft waxy gills. However, the smoky-gray to black spore print distinguishes it. Other species: *G. roseus* has a redder cap and longer spores, but has not yet been recorded in North America; *G. smithii* can also be rather small, but shows little or no yellow in the base of the stalk and has a grayish to vinaceous-gray or brownish cap. The latter occurs with Douglas-fir in coastal California but grows with other conifers as well.

CHROOGOMPHUS   (Pine Spikes)

Medium-sized terrestrial mushrooms *associated with conifers.* CAP convex to conical, umbonate, plane, or depressed; smooth or woolly, viscid or dry. Flesh *colored (pale orange to yellow-orange, buff, salmon, pinkish, reddish, etc.).* GILLS *usually decurrent,* fairly well-spaced, *dull orange to yellow-orange or ochraceous when young,* blackening in age. STALK fleshy, more or less central, dry. VEIL present but usually disappearing, fibrillose. VOLVA absent. SPORE PRINT *olive to smoky-gray to black.* Spores long and narrow (narrowly elliptical to spindle-shaped). Gills with prominent cystidia. Cap tissue amyloid.

A FLAGRANT fungal feature of our coastal pine forests, *Chroogomphus* (crow-oh-gom-fus) is easily recognized by its decurrent gills, ochraceous to orange, salmon, buff-colored, or reddish flesh, and smoky-olive to black spores. The fruiting bodies, in contrast to *Gomphidius,* are very attractive: brightly colored with a lean, clean look and a tendency to become reddish or wine-colored in old age. The stem and gills are typically yellow-orange to orange to orange-buff, but the latter blacken as the spores ripen.

Chroogomphus species grow exclusively with pine in our area and usually mingle with slippery jacks (*Suillus pungens, S. fuscotomentosus,* etc.). They have an unusually long growing season—from the onset of the fall rains through the spring—and may even turn up in the summertime. They often seem quite abundant when other fleshy fungi are scarce. They do not decay as rapidly as their cousins the Gomphidii and are usually free of maggots. However, they are susceptible to attack by a greenish mold.

Chroogomphus is an excellent genus for beginners because all of its species are edible and highly distinctive. True, fresh specimens are slimy and insipid when cooked (not to mention *purple*), but they acquire a pleasant, chewy texture that is perfect for tomato sauces if they are chopped up finely and then dried. At least that is the gospel according to "Saint Ciro," a wondrously demented and exceptionally resourceful toadstool-tester of Sicilian extraction. With a large family to feed and an immense forest of unpicked *Chroogomphus* across the street, he isn't the least bit deterred by their lack of renown. Every year he gathers trunkloads and dries them for summer use! About ten species of *Chroogomphus* occur in North America; most of these are keyed out here. Dr. Frank Dutra of Los Gatos, California, reports that eating them can turn one's urine red (just like beets!). This phenomenon is harmless, but can cause anxiety if one mistakes the red pigment for blood.

Chroogomphus vinicolor is a common sight in pine forests and on lawns where pines have been planted. Note the shape, which is very characteristic.

Key to Chroogomphus

1. Fruiting body robust (stalk at least 2 cm thick); cap and/or stalk usually decorated with orange to reddish to wine-colored fibrillose or felty patches or zones; cap convex (not umbonate), not viscid or only very slightly so; spore print often olive-tinged . *C. pseudovinicolor,* p. 486
1. Fruiting body not so robust, or if so then not as above 2

2. Cap not viscid or only slightly so, covered with flattened woolly or felty fibrils or fibrillose scales; often associated with conifers other than pine (but also with pine) 3
2. Cap usually smooth and viscid when moist, often lustrous when dry (but sometimes minutely scaly or faintly fibrillose); associated principally with pine 4

3. Cap pale to dull orange, yellow-orange, or ochraceous . . . . *C. tomentosus,* p. 487
3. Cap usually grayish, at least at margin . . . . *C. leptocystis* (see *C. tomentosus,* p. 487)

4. Cap small (1-4 cm broad) and usually pinkish- or vinaceous-tinged when young; restricted to eastern North America; rare . . . . *C. flavipes*
4. Not as above; widespread and common 5

5. Cap often brightly colored (buff, yellow-orange, orange, ochre) until fairly mature; cystidia thin-walled . . . . *C. ochraceus* (see *C. vinicolor,* below)
5. Not as above (but cap may be orangish when very young); cystidia thin- or thick-walled . . 6

6. Cap often vinaceous (wine-colored) in old age; cystidia thick-walled . . . . *C. vinicolor,* below
6. Cap usually dull reddish-brown in age; cystidia thin-walled *C. rutilus* (see *C. vinicolor,* below)

Chroogomphus vinicolor   (Pine Spike)   Color Plate 113

CAP (1) 2-10 (13) cm broad, peglike to nearly conical or convex when young, becoming somewhat top-shaped to plane or even shallowly depressed (but often with a slight umbo); surface smooth to finely fibrillose (or sometimes scaly in age), viscid when wet, often lustrous or silky when dry; color variable: dull orange to ochraceous to grayish, brownish, yellow-brown, olive-brown, bister, or reddish-brown, often wine-red to dark vinaceous-brown in age and tending to developing wine-red stains where rotten or when dried; margin at first incurved. Flesh thick, usually pale orange but varying to buff, salmon, or ochraceous-buff (often somewhat yellower in lower portion of stalk); often becoming wine-red in old age or where injured. GILLS broad, fairly well-spaced, decurrent or occasionally adnate, pale to dull orange or ochraceous when young, then clouded gray with spores and finally blackish in old age. STALK typically long, slender, and sometimes sinuous, but at other times short and rather thick; 2.5-15 (20) cm long, 0.5-2 (6) cm thick; equal or more often tapered downward; dry, solid, firm; pale orange to yellow-orange to orange-buff, ochraceous, or reddish-tinted when young, often more vinaceous-red in age and/or with reddish fibrils. VEIL dry, fibrillose, ochraceous to orange, soon disappearing or leaving a slight hairy ring on stalk. SPORE PRINT smoky-gray to smoky-black, sometimes with an olive tinge; spores (14) 17-23 × 4.5-7.5 microns, more or less spindle-shaped, smooth. Cystidia on gills large, thick-walled.

HABITAT: Solitary to scattered or gregarious on ground under conifers, particularly pine; widely distributed, but especially common in northern and western North America. Along with *C. rutilus* (see comments), it is a prominent fungal facet of our pine forests and is also frequent in yards and lawns where pines have been planted. The major crop is usually in the late fall or winter and is often accompanied by masses of mushy slippery jacks (*Suillus* species), but it can be found most any time. It is quite sporadic in its fruiting habits— overwhelmingly abundant some years, very sparse during others.

EDIBILITY: Edible, but like all pine spikes, better dried than fresh (see comments on p. 484). It is usually very clean, and unlike *Suillus,* is shunned by maggots.

COMMENTS: Anyone who hunts our coastal pine forests will come across this common mushroom and its equally edible look-alike, *C. rutilus* (see next page). The variable color or mixture of colors (orange, gray, brown, and/or wine-red) plus the pale orange flesh,

Chroogomphus rutilus is easily confused with *C. vinicolor* in the field. The critical difference is microscopic (see coments below).

decurrent gills, and smoky-black spores make it as distinctive as it is attractive. The fruiting body is often quite slender and long-stemmed, but robust forms also occur. *C. rutilus* (formerly *Gomphidius viscidus*) is a very similar, widespread species with thin-walled rather than thick-walled cystidia. It can often be told in the field by its slightly duller or less vinaceous color and broader or flatter cap, but both species are so variable in shape and color that a microscopic examination is often necessary to positively distinguish them. Both differ from *C. tomentosus* and *C. pseudovinicolor* in their smoother cap that is viscid or slimy when wet and often shiny when dry. Other smooth, viscid-capped species: *C. ochraceus,* widely distributed, is closely related to *C. rutilus* (i.e., it has thin-walled cystidia), but is smaller, with a yellow-orange to buff, ochre, orange, or grayish cap, at least when young; it also occurs in our area with pine. See also *C. flavipes* (in the key).

Chroogomphus pseudovinicolor   (Robust Pine Spike)

CAP 5-15 cm broad, convex, sometimes becoming plane in age; surface dry to very slightly viscid, orange-buff to dull orange to ochraceous-orange, sometimes flushed red in places or often mottled with reddish or orangish patches of felty or woolly material, and often becoming entirely dark dull red in old age; margin often fringed with veil remnants at first. Flesh thick, quite firm, pale to dull orange or orange-buff (but often brighter or yellower in base of stalk), reddening in old age or around maggot tunnels. **GILLS** fairly well-spaced, decurrent or sometimes adnate, often forked; pale orange to dingy ochraceous, then clouded olive to gray or black by spores. **STALK** 6-12 cm long, 2-5 cm thick at (or near) apex; usually tapered below, solid, firm, dull orange to orange-buff or ochraceous, usually with zones or patches of reddish to wine-colored fibrils or woolly-felty material; often dingier, darker, or redder overall in age. **VEIL** scanty, fibrillose, soon disappearing or leaving a slight zone of hairs at thickest part of stalk (just below apex). **SPORE PRINT** greenish to smoky-olive to blackish (occasionally lacking olive tinge); spores 15-20 × 5-7.5 microns, spindle-shaped to narrowly elliptical, smooth. Cystidia on gills thick-walled.

HABITAT: Solitary, scattered, or in groups (often in small clumps) on ground under conifers; known only from western North America, not common. In our area it seems to occur only with ponderosa pine—usually in the fall and early winter.

EDIBILITY: Edible. It is firmer and meatier than other *Chroogomphus* species, but unfortunately, not as numerous.

Left: *Chroogomphus pseudovinicolor*, a robust orange to reddish species with patches of fibrils on the cap and stalk. **Right:** *Chroogomphus tomentosus* has a dry fibrillose, orangish to ochraceous cap. It is very common under northern conifers.

COMMENTS: This robust but relatively rare *Chroogomphus* is a very beautiful mushroom when fresh—easily distinguished from other *Chroogomphus* species by its dry, convex (never conical or umbonate) cap and thick, woolly-scaly stalk. Also, the spore print is usually greener than that of its brethren, and it often grows in small clumps of 2-4 individuals rather than in the scattered fashion typical of other pine spikes.

Chroogomphus tomentosus   (Woolly Pine Spike)

CAP 2-9 cm broad, peglike becoming broadly conical to convex, umbonate, or plane; surface dry to very slightly viscid, covered with flattened woolly or felty fibrils or fibrillose scales; pale buffy-orange to pale or bright ochraceous to ochraceous-orange, the fibrils sometimes vinaceous-tinged; sometimes purple-stained in age. Flesh yellow-orange to dull orange or pale orange-buff. **GILLS** well-spaced, usually decurrent but sometimes adnate, yellow-orange to ochraceous or colored like cap, becoming smoky-gray to smoky-brown as spores mature. **STALK** 4-18 cm long, (0.3) 0.7-1.5 (2) cm thick; equal or more often narrowed below; solid, dry and somewhat fibrillose, colored more or less like cap. **VEIL** dry, fibrillose, scanty, colored like cap; disappearing or leaving slight hairy remnants on stalk near apex. **SPORE PRINT** smoky-gray to blackish; spores 15-25 × 6-9 microns, narrowly elliptical to spindle-shaped, smooth. Cystidia on gills with fairly thick walls.

HABITAT: Solitary to widely scattered or gregarious on ground under conifers (hemlock, fir, Douglas-fir, pine); common in the mixed coniferous forests of the Pacific Northwest and northern Rocky Mountains from late summer through early winter; also common at times in northern California. It does not seem to occur in our area, but has been found on the University of California campus in Los Angeles.

EDIBILITY: Edible, but better dried than fresh (see comments on p. 484).

COMMENTS: The dry, woolly-fibrillose cap, overall dull orange to ochraceous color, decurrent gills, smoky-black spores, and growth with conifers form a distinctive set of characteristics. The cap is drier and woollier than that of *C. vinicolor* and *C. rutilus,* and not as variable in color, while the stalk is not as thick as that of *C. pseudovinicolor.* Also, it seems to favor mixed coniferous forests, whereas the others grow almost exclusively with pine. Other species: *C. leptocystis* of western North America tends to have a grayer cap and thin-walled cystidia, but is otherwise quite similar.

487

Boletes

BOLETACEAE

spores

BOLETES have a spongelike layer of **tubes** on the underside of the cap instead of gills. Otherwise they resemble agarics: medium-sized to large, mostly terrestrial fungi with a cap and central stalk. Polypores also possess a tube layer, but are tough or leathery and usually grow on wood. The few polypores that are fleshy and terrestrial typically have an off-center stalk and tubes which do not peel easily from the cap.

The tube layer in the boletes, on the other hand, usually peels away cleanly from the cap. The spores are produced on basidia which line the inner surfaces of the tubes, which are vertically arranged so that the spores, when discharged, drop into the air. The mouths of the tubes are known as **pores.** They are sometimes stuffed with a pith when young, making the pore surface appear smooth. In some species the pores are **radially arranged** (arranged in rows which radiate from the stalk), giving a somewhat gill-like effect. This is carried to an extreme in *Phylloporus rhodoxanthus* (p. 480), a "bolete" with true gills, which serves to illustrate why boletes are thought to have more in common with gilled mushrooms than with, say, the coral fungi, teeth fungi, or polypores.

Boletes used to be lumped together in one giant and unwieldy genus, *Boletus.* Several genera are now recognized, the most prominent being *Suillus, Leccinum, Tylopilus,* and *Boletus.* The common term "bolete," however, is still applicable to *any* member of the Boletaceae, not just those still retained in *Boletus.*

The salient characters for identification are much the same as for agarics. It is particularly important, however, to note color changes on the pore surface, cap, stalk, and flesh. Many boletes will stain blue or greenish-blue when bruised (see Color Plate 135); a few will stain brown or some other color, while others will not stain at all. In a number of boletes the cap is typically **areolate**, i.e., it develops an extensive system of shallow cracks or fissures as it matures, exposing the flesh beneath.

Another important feature to note is the type of stalk ornamentation, if any. In *Boletus* and *Tylopilus,* the upper portion or sometimes the entire stalk is frequently **reticulate** or "netted": covered with a network of veins. Some species are coarsely reticulate (see Color Plate 139), others are very finely so (see photo on p. 489). In *Leccinum,* the stalk is

Close-up of the pores (tube mouths) in *Boletus subtomentosus.* Boletes are one of two groups of fungi that produce their spores in tubes. The other group is the polypores (p. 549).

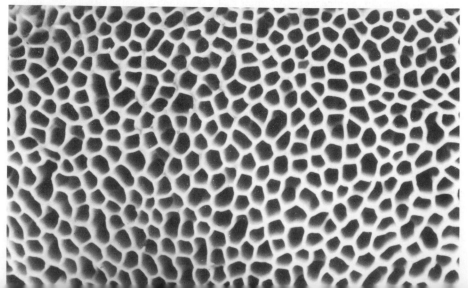

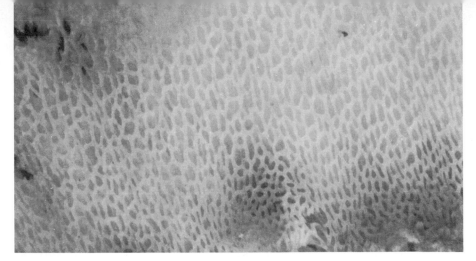
Close-up of the finely reticulate (netted) stalk of a young *Boletus edulis.*

always **scabrous:** decorated with tufted hairs or rough, scurfy scales **(scabers)** which typically darken at maturity (see Color Plates 145-147). In *Suillus,* on the other hand, the stalk is often speckled or smeared with **glandular dots** (see Color Plate 120). These pinkish to brown or blackish spots exude a resinous substance that will stain your fingers brown.

Spore color ranges from yellow to olive, brown, reddish-brown, chocolate-brown, or black. A spore print is easily obtained, but pigmentation in the tubes may stain the paper, making the spore color look brighter (usually yellower or greener) than it actually is. The spore shape is characteristically "boletoid": long and narrowly elliptical or spindle-shaped in face view, inequilateral in profile. Viewed through the microscope, a bolete spore is an unusually large, handsome, and healthy-looking specimen, somewhat reminiscent of a surfer with a good tan.

The boletes are one of the safest groups for the table, as well as one of the most substantial and rewarding. A few members of the genus *Boletus* are poisonous, but the majority are edible (though some are distinctly better than others). Since they are large and fleshy, boletes are a popular picnicground for maggots. These may enter in the usual manner, via the stem, or surreptitiously, through the tubes. Always check for maggots in the field, and cut away soggy or infested areas (assuming you already know the species) before popping them in your basket. That way you are less likely to have a maggot-infested mush two days later. Some people peel off the tubes—when old and spongy they are mostly water anyway. Also, boletes are vulnerable to attack by a variety of moldy-looking parasites. Discard all such individuals when picking for the table. For more on hunting boletes, see p. 546.

The boletes attain their greatest diversity in the deciduous forests of eastern North America, but are also a conspicuous and colorful component of the West's woodland fungi. The cap can be very hefty and the pores are often brightly colored. The vast majority are mycorrhizal. Several hundred species occur in North America, more than 80 on the west coast. Though not complete (what book is?) and more than a trifle expensive (what book isn't?), Harry Thiers' *California Mushrooms: A Field Guide to the Boletes* is a definitive must for the serious and self-respecting bolete buff. (Thiers has also co-authored, with Alexander Smith, the bolete bible for easterners, entitled *The Boletes of Michigan.*)

Key to the Boletaceae

1. Fruiting body often appearing aborted or misshapen and often buried or partially buried in the humus; tubes at least somewhat disoriented (i.e., not necessarily arranged vertically); spore print unobtainable; not common .... ***Gastroboletus,*** p. 544
1. Not as above; tubes vertically arranged; spore print usually obtainable; usually found above the ground when mature 2

489

2. Cap and stalk shaggy or coarsely scaly, the scales gray to brown or black; pores whitish becoming gray or black in age; veil present when young; spore print dark brown to black; spores spiny, warty, or reticulate; found in eastern North America and Southwest . *Strobilomyces,* p. 543
2. Not as above 3
3. Stalk roughened by small tufted hairs or scales (scabers) which are usually gray to dark brown or black by maturity (but differently colored or remaining pallid in a few species); stalk *not* glandular-dotted and typically *not* reticulate; pores rarely yellow .... *Leccinum,* p. 536
3. Not as above; stalk typically without scabers 4
4. Bright yellow, dry, cottony-powdery or cobwebby veil present when young, usually leaving cottony remnants on cap and/or stalk; pores and flesh usually blueing (sometimes slowly) when bruised .... *Pulveroboletus,* p. 509
4. Veil absent, or if present then not as above 5
5. Veil present when young, sometimes forming a ring (annulus) on stalk, at other times clinging to the margin of cap, and in still other cases disappearing in age 9
5. Veil absent (check young specimens if unsure), but cap sometimes fringed with sterile tissue 6
6. Stalk with glandular dots or smears at least when mature (the dots often slightly resinous to the touch); stalk usually more or less equal or at least not markedly bulbous; cap often (but not always) viscid; spore print olive to brown or dull cinnamon .... *Suillus,* p. 491
6. Stalk not glandular-dotted (or very rarely so, but then with much darker spores) 7
7. Pores often radially elongated, sinuous, arranged in rows, and/or veined; tubes often (but not always) decurrent; taste of cap *not* peppery 8
7. Pores not as above, or if so then taste peppery 10
8. Veil absent when young; tube layer often difficult to separate from cap; stalk often short, curved, and/or off-center; spores broadly elliptical to nearly round; associated with hardwoods (mainly ash and alder); common in eastern North America, rare in the West 14
8. Not with above features; veil present or absent; spores elliptical to spindle-shaped; associated with conifers or less commonly hardwoods; widespread and common 9
9. Spore print dark grayish-brown to dark reddish-brown, chocolate-brown, or vinaceous-brown; veil typically present (at least when young) and glandular dots typically absent (rarely present); associated with larch and other northern conifers .... *Fuscoboletinus,* p. 505
9. Spore print olive-brown to brown, yellowish, or dull cinnamon; veil present or absent; stalk glandular-dotted or not; found with many kinds of trees (including larch) .. *Suillus,* p. 491
10. Cap viscid to very slimy when moist; stalk glandular-dotted or if not then stalk typically white to yellow and not reticulate; associated with conifers .... *Suillus,* p. 491
10. Not with above features 11
11. Spore print flesh-colored to pinkish-brown, reddish-brown, vinaceous, or chocolate-brown; pores typically white to pinkish, vinaceous, gray, or brown but not red and only rarely yellow (pores if pallid often but not always staining brown when bruised); common in eastern North America but rather infrequent in the West .... *Tylopilus,* p. 532
11. Not as above; spore print yellow to yellow-brown, olive-brown, olive, brown, or more rarely cinnamon-brown; pores white, yellow, olive, red, orange, or sometimes gray or brown (but if pallid then not typically staining brown when bruised) 12
12. Spore print pale yellow to yellow; stalk often hollow or partially hollow at maturity (especially near the base); common in eastern North America, rare in the West ..... *Gyroporus,* p. 510
12. Spore print olive-brown to olive or brown, or occasionally yellow-brown or cinnamon-brown; stalk not normally hollow; common and widespread 13
13. Spores ornamented with ridges, grooves, pits, etc.; stalk usually relatively long (7 cm or more) and slender and *coarsely* reticulate or conspicuously lined with jagged ridges for most of its length; pores yellow to greenish-yellow, not usually blueing when bruised; typically terrestrial; found in eastern North America and the Southwest .... *Boletellus & Austroboletus,* p. 508
13. Not with above combination of features; spores typically smooth; widespread *Boletus,* p. 511
14. Common under hardwoods (especially ash) in eastern North America; cap olive to dingy yellow-brown or brown; flesh not blueing or blueing only slightly; stalk often quite dark in old age, off-center or lateral .... *Boletinellus (=Gyrodon) merulioides*
14. Associated with alder; known from southern California (and Europe), apparently rare; cap yellowish to reddish-brown; flesh usually blueing when bruised .... *Gyrodon lividus*

SUILLUS   (Slippery Jacks)

Medium-sized to fairly large, fleshy, terrestrial boletes *associated almost exclusively with conifers.* CAP *usually viscid or slimy,* but sometimes dry and scaly. PORES and tubes typically white to yellow, sometimes radially arranged, rarely bruising blue. STALK typically more or less equal (rarely bulbous), rarely reticulate; *often with resinous brown to pinkish glandular dots and smears.* VEIL *frequently present and forming an annulus (ring) on stalk, but often absent.* SPORE PRINT *brown* (usually olive-brown becoming dull cinnamon-brown as moisture escapes). Spores typically elliptical to spindle-shaped, smooth. Bundles of cystidia on inner surfaces of tubes staining dark brown to black in potassium hydroxide (KOH).

MOST slippery jacks have a veil or a glandular-dotted stem (see Color Plate 120) or both. In addition, the cap is often "slippery" (viscid or slimy), the pores are radially elongated in some species (that is, in rows radiating from the stalk toward the cap margin), the stalk is neither conspicuously bulbous nor reticulate as is commonly the case in *Boletus,* and they are mycorrhizal almost exclusively with conifers. *Suillus* is a genus of exceptions, however: several species are not slippery, others lack a veil, still others are not glandular-dotted, many do not have radially-arranged pores, and two or three species in eastern North America occur with hardwoods. However, by using the above features *in combination,* you should have little trouble distinguishing *Suillus* from other bolete genera with the exception of *Fuscoboletinus,* which has darker spores and doesn't occur in California.

Like other boletes, *Suillus* can be found in older mushroom books under *Boletus.* It is perhaps the most primitive genus in the Boletaceae, leading to *Boletus* on the one hand and to the Gomphidiaceae (via *Fuscoboletinus*) on the other. *Suillus* splits neatly into two groups. In the first (the genus *Boletinus* of some authors), a veil is *always* present and the cap is frequently dry and fibrillose, thus belying the moniker "slippery jack." Also, the stalk is *not* glandular-dotted and the pores are often elongated and/or radially arranged. In our area this group occurs only with Douglas-fir, where it is often accompanied by species of *Gomphidius.* Elsewhere it also occurs with larch and various pines. The second group includes the "true" slippery jacks—or "slippery jills" as they are sometimes called. They have a slimy cap or glandular-dotted stem or both, and a veil may or may not be present. They occur principally with pines, often mingling with species of *Chroogomphus.*

Slippery jacks are remarkable for their fecundity. Under favorable conditions they fruit in quantities that must be seen to be believed. Bushels can be harvested from a single row of pine trees and truckloads from a pine plantation. They are indisputably among the most prominent fungal facets of temperate coniferous forests, and in fact, wherever there are conifers of the pine family (Pinaceae)—be it in a forest or park, on a lawn or by a freeway—there are bound to be one or more *Suillus* species also. They are partial to cool weather and generally fruit later than *Boletus* and *Tylopilus* species, which like it warmer. In our area they can be found most any time but peak in the late fall and winter.

None of the slippery jacks is known to be poisonous, but a few have caused "allergic" reactions. In my experience even the so-called "choice" species are insipid and slimy when cooked, but one novel solution is to *take advantage* of their inherent wetness by using them as an *escargot* substitute. (You can boil them, flavor them with herbs, stuff them into empty snail shells, and call it "Parsley, Sage, Rosemary, and Slime.") Slippery jacks also tend to be wormy and do not dry well in the tasty tradition of *Boletus edulis.* They are indisputably *available,* however, so I heartily recommend that you try them. If you find a species that pleases you, you'll never have to worry about a mushroom shortage again!

Nearly half of the 70+ North American species of *Suillus* have been found in California. As they pose a threat to those who don't watch where they step (they are at least as dangerous as banana peels!) and inevitably arouse the curiosity of fungophiles and other weird elements (for their prodigious numbers as well as their slipperiness), I'm offering a generous selection here. Fourteen species are described and many others are keyed out.

Key to Suillus

1. Veil present (check young specimens if possible!), either completely covering the pores when young (and often forming an annulus or ring on stalk after rupturing) or present only as a roll of cottony tissue on cap margin or patches on cap (and not forming an annulus) 2
1. Veil and annulus absent; margin of cap naked, even when young 28

2. Glandular dots or smears present on stalk and conspicuous at least in age; associated mainly with pine, sometimes with spruce 11
2. Stalk lacking glandular dots (or occasionally with a few visible in old age); found with Douglas-fir, larch, hemlock, or sometimes pine in the West and with larch, pine, and oak in East .. 3

3. Stalk usually hollow or partially hollow, at least at base; cap *not* viscid; associated only with larch .... *S. cavipes,* p. 494
3. Not as above; stalk not normally hollow 4

4. Cap with tawny to orangish, red, reddish-brown, pink, or purple-red scales or fibrils, dry or with a viscid layer beneath the scales (i.e., can be viscid in wet weather if fibrils are obliterated) 5
4. Cap more or less smooth (but sometimes streaked) and viscid to slimy when moist 7

5. Found in eastern North America, mainly with pine 43
5. Found in western North America, principally with Douglas-fir 6

6. Cap fibrillose or fibrillose-scaly, the fibrils reddish-brown to brick-red or pinkish, or sometimes orange-buff or tawny; surface dry, or viscid only in wet weather or old age .. *S. lakei,* p. 495
6. Cap with only scattered fibrils, but often streaked; surface usually viscid when moist, *not* reddish but often cinnamon-brown, orangish, greenish, etc. .... *S. caerulescens* & others, p. 496

7. Associated with hardwoods (e.g., oak) in Great Lakes region; veil thick, tough, often gelatinous; spore print olive-yellow to yellow-brown; spores more or less round ..... *S. sphaerosporus*
7. Not as above 8

8. Base or lower portion of stalk often (but not always) staining blue or green when cut (often staining slowly or weakly); associated primarily with Douglas-fir 9
8. Not as above; stalk not blueing and/or associated with larch or pine 10

9. Veil viscid and yellow to bright orange while still covering the pores; cap more or less smooth .... *S. ponderosus* & others (see *S. caerulescens,* p. 496)
9. Veil dry and whitish while covering the pores; cap smooth or fibrillose *S. caerulescens,* p. 496

10. Cap yellow to golden-yellow to dark red or bay-red; flesh usually yellow when young; associated with larch .... *S. grevillei,* p. 497
10. Cap duller (honey-colored to dingy yellow-brown, brown, etc.); flesh white becoming pale yellow in age; associated primarily with pine 11

11. Cap at first covered with a whitish veil that usually leaves small patches or scales on the surface of cap, especially near margin; found mostly with white (5-needle) pines in northern Rocky Mountains (particularly common in northern Idaho) .... *S. albivelatus*
11. Not as above 12

12. Pores large (many of them typically 1 mm or more broad at maturity), often (but not always!) elongated and at least somewhat radially arranged (in radiating rows) 13
12. Pores mostly less than 1 mm broad and *not* radially arranged 15

13. Veil typically forming a gelatinous annulus (ring) on stalk; cap often umbonate, dingy-colored, and rather small; found mainly with lodgepole and beach pines ..... *S. umbonatus,* p. 498
13. Not as above; veil not forming a gelatinous annulus or forming only a very slight one 14

14. Stalk often short, poorly developed, and/or off-center; pores often elongated and irregular; associated with pines and other conifers in the Sierra Nevada and other mountains .... *S. riparius* & *S. megaporinus* (see *S. sibiricus,* p. 498)
14. Not as above; stalk typically well-developed; found with 5-needle (white) pines; widespread 45

15. Veil usually forming a distinct annulus (ring) on stalk 16
15. Veil typically attached only to margin of cap *or* if attached to stalk when young then typically forming only a slight or very obscure annulus 20

16. Annulus usually thin and/or fibrillose; glandular dots on stalk often inconspicuous until old age .... *S. pseudobrevipes,* p. 500
16. Annulus usually well-formed; stalk with conspicuous glandular dots (unless very young) . 17

17.   Stalk and/or underside of veil (annulus) often with purplish or dull lilac shades; cap reddish-brown to dark reddish-brown or sometimes yellow-brown .... *S. luteus,* p. 500
17.   Not as above; purplish shades absent; cap white to dingy yellowish, olive-brown, etc. 18

18.   Found in eastern North America (including the Gulf Coast) 19
18.   Found in the West (especially common in Pacific Northwest) .... *S. subolivaceus,* p. 499

19.   Annulus tall (i.e., broad) and baggy or bandlike (both lower and upper edges often flaring out) .... *S. subluteus* & others (see *S. subolivaceus,* p. 499)
19.   Not as above; annulus not baggy; stalk slender; cap whitish to pale yellow, pale cinnamon, etc., the slime strongly acidic-tasting .... *S. acidus*

20.   Veil purplish to lilac-brown on its underside; cap soon brown to dark brown; associated mainly with western white pine .... *S. borealis* (see *S. pseudobrevipes,* p. 500)
20.   Not as above 21

21.   Cap white becoming olive-gray or olive, then spotted, streaked, or becoming entirely yellow, cinnamon, orange, etc.; common with various pines in central and southern California, and in regions where Monterey pine has been planted .... *S. pungens,* p. 503
21.   Not as above; cap not olive or grayish-olive when young 22

22.   Veil rudimentary (not well-developed); cap yellow to dingy yellowish, mustard-yellow, or orange-buff, often with brown to reddish spots, streaks, or scales or with well-developed brown to grayish fibrils; known only from eastern North America and the Southwest 34
22.   Not as above; if similarly colored then veil well-developed at least when young (as a roll of cottony tissue extending from cap margin toward stalk, or as a true partial veil covering the pores) 23

23.   Cap yellow to dingy olive-yellow or mustard-yellow (even in age), often with reddish to brown scales, spots, or streaks; stalk markedly glandular-dotted; found with 5-needle pines ... 45
23.   Not with above features 24

24.   Young specimens with yellowish to pale tan caps and very dense, conspicuous glandular dots on stalk .... *S. glandulosipes* (see *S. pseudobrevipes,* p. 500)
24.   Not as above; young specimens differently colored or with inconspicuous glandular dots . 25

25.   Slime on cap becoming chocolate-brown in age; associated mainly with sugar pine on the west coast, lodgepole pine in Rocky Mountains . *S. brunnescens* (see *S. pseudobrevipes,* p. 500)
25.   Not as above 26

26.   Growing in volcanic soil, often buried; cap yellow to tawny, often fibrillose or streaked and often developing pinkish to reddish-brown areas *S. volcanalis* (see *S. pseudobrevipes,* p. 500)
26.   Not as above; habitat different 27

27.   Veil material present only on margin of cap, never attached to stalk; cap whitish to buff or at times brown to tawny or cinnamon .... *S. albidipes* (see *S. pseudobrevipes,* p. 500)
27.   Veil attached to the stalk when young; cap dingy yellowish to honey-colored or brown, but not whitish .... *S. pseudobrevipes,* p. 500

28.   Pores large (many of them 1-3 mm in widest dimension), often elongated radially or arranged in radiating rows and often decurrent 29
28.   Not as above 30

29.   Cap hairy or velvety and brown to vinaceous-brown; found under hardwoods in southeastern United States .... *S. castanellus*
29.   Not as above 46

30.   Cap dry to viscid but not thickly slimy, with hairs, fibrils, or scattered fibrillose scales when young (but these often absent in age or rainy weather); pores typically dark brown to yellow-brown, pale cinnamon, or orangish when young (but often dingy yellowish in age) 31
30.   Cap viscid to very slimy when moist, smooth (bald) or streaked; fibrils typically absent on cap (occasionally present, but if so then pores typically white to pale yellow when young) ... 35

31.   Flesh and/or pores staining blue when bruised or cut (but sometimes staining slowly or only slightly—check several specimens if unsure!) .... *S. tomentosus,* p. 504
31.   Not as above 32

32.   Found in eastern North America and the Southwest; cap sometimes yellow or orangish .. 33
32.   Found in California and the Pacific Northwest; cap not yellow . *S. fuscotomentosus,* p. 504

33.   Pores dark brown when young; cap soon more or less bald; associated with conifers, often in boggy areas; odor often fragrant when several specimens are put together .... *S. punctipes*
33.   Not as above; cap usually with scales or colored fibrils; found with hardwoods and conifers 34

34. Associated with conifers (mainly pine); cap scales or fibrils usually conspicuous and glandular dots on stalk prominent at maturity .... *S. hirtellus*
34. Favoring hardwoods; scales on cap usually small and glandular dots obscure . *S. subaureus*
35. Glandular dots or smears present on stalk and usually conspicuous by maturity 36
35. Glandular dots absent or obscure (in old age sometimes visible at apex) 40
36. Cap white when young, whitish to pale yellowish in age (or with darker slime); associated with white pines in eastern North America; common ..... *S. placidus* (see *S. granulatus,* p. 502)
36. Not as above; cap darker, at least in age, and/ or found in the West 37
37. Cap smooth, at first white but soon becoming olive to olive-gray and then orange, cinnamon, yellow, brown, etc. (or splashed with these colors); found with various pines in coastal central and southern California, or in regions where Monterey pine is planted .. *S. pungens,* p. 503
37. Not with above features; cap variously colored but lacking an olive or olive-gray phase .. 38
38. Cap with grayish to dark brown streaks, fibrils, or scales; found with pine (especially Monterey pine) in California; taste harsh, unpleasant .... *S. acerbus* (see *S. fuscotomentosus,* p. 504)
38. Not as above 39
39. Cap buff to yellowish or pale cinnamon at maturity; common with ponderosa pine in the Southwest .... *S. kaibabensis* (see *S. granulatus,* p. 502)
39. Not as above; either found elsewhere or cap darker at maturity *S. granulatus* & others, p. 502
40. Taste peppery (chew on a small piece of cap) and base of stalk usually bright yellow *or* pores brilliant yellow to intense greenish-yellow and associated with hardwoods (see *Boletus,* p. 511)
40. Not as above 41
41. Stalk typically white to yellow (not brown or red) and not viscid or slimy; associated with conifers (mainly pine); pores *not* blueing when bruised 42
41. Stalk viscid or slimy and cap sometimes yellow, or if not then not as above (see *Boletus,* p. 511)
42. Cap dark brown to reddish-brown to dull cinnamon when young .... *S. brevipes,* p. 501
42. Cap whitish or at least paler than above when young ....
.... *S. pallidiceps* & *S. occidentalis* (see *S. brevipes,* p. 501)
43. Pores and flesh typically blueing when bruised; spores ridged or otherwise ornamented .....
.... (see *Boletellus* & *Austroboletus,* p. 508)
43. Pores and flesh not blueing; spores smooth 44
44. Cap fibrils pink to reddish to purple-red; growing with white pine *S. pictus* (see *S. lakei,* p. 495)
44. Not as above; cap dull yellowish to orangish or pinkish-orange; especially common along the Gulf Coast .... *S. decipiens* (see *S. lakei,* p. 495)
45. Veil sometimes forming a slight annulus (ring) on stalk; base of stalk often vinaceous-stained; common in western North America, including the Southwest .... *S. sibiricus,* p. 498
45. Not as above; veil typically *not* forming an annulus; common in eastern North America and also found in the Southwest .... *S. americanus* (see *S. sibiricus,* p. 498)
46. Cap small (up to 5 cm broad) and stalk slender .... *S. helenae* (see *S. umbonatus,* p. 498)
46. Cap medium-sized to large; stalk usually quite thick *S. punctatipes* (see *S. granulatus,* p. 502)

Suillus cavipes   (Hollow-Foot)   Color Plate 122

CAP 3-12 cm broad, convex or broadly umbonate becoming plane or slightly uplifted; surface dry (*not* viscid), densely hairy-fibrillose, often with a suede-like or felty texture; typically dark brown to reddish-brown to orange-brown or tawny (rarely paler), the margin and/ or tips of the fibrils often paler. Flesh firm, white or pale yellow, not blueing when bruised. **PORES** arranged more or less radially, large, angular (elongated); pale yellow becoming dark yellow or greenish-yellow, not blueing; tubes same color, usually decurrent. **STALK** 4-9 cm long, 0.5-2 cm thick, equal or swollen below, dry, lemon-yellow above the ring, colored like cap or paler below; glandular dots absent; lower portion hollow inside, at least in age; not blueing when cut. **VEIL** fibrillose-cottony, white, forming a thin, fragile, fibrillose ring on stalk and/ or leaving remnants on cap margin. **SPORE PRINT** dark olive to brown; spores 7-10 × 3.5-4 microns, elliptical to spindle-shaped, smooth.

HABITAT: Scattered to gregarious on ground in late summer and fall, associated with larch trees; common throughout the northern hemisphere wherever larch occurs. I have seen it in Oregon and Idaho, but not in California.

EDIBILITY: Edible and "choice" according to some sources, but I find it bland. Its one saving grace is that it isn't as slimy as many slippery jacks.

COMMENTS: Also known as *Boletinus cavipes*, this is the only *Suillus* with a consistently hollow stem and dry, densely hairy, brown to tawny cap. Its growth with larch is also diagnostic. Fresh specimens have an unusually clean, pristine appearance that is readily noticeable but difficult to describe (see color plate); the pale pores contrast strikingly with the darker cap, making it one of our most beautiful boletes. Since larch does not occur in California, *S. cavipes* probably doesn't either. However, it is a prominent feature of the coniferous forests of Idaho, Montana, and the Cascades, as well as the northeastern United States and Canada. For other larch-loving boletes, see *S. grevillei* and the genus *Fuscoboletinus* (particularly *F. ochraceoroseus* and *F. aeruginascens*).

Suillus lakei   (Western Painted Suillus)   Color Plate 124

CAP 5-15 cm broad, convex becoming plane or shallowly depressed; surface covered with reddish-brown to brick-red to pinkish or tawny fibrils or fibrillose scales on a yellow to dingy orange or tan background; dry, but in one form viscid in wet weather beneath the fibrils; margin sometimes hung with veil remnants. Flesh thick, yellow, often discoloring pinkish when bruised. **PORES** large (1-3 mm in length), sometimes radially arranged, yellow when young becoming dingy yellow or ochre and usually discoloring reddish to dingy reddish-brown where bruised or in age; tubes same color as pores, adnate to decurrent. **STALK** 3-8 (12) cm long, 1-3 (4) cm thick, more or less equal, firm, dry, solid; yellow above the ring, usually with reddish to brown streaks below; glandular dots absent; interior usually staining weakly blue or green when cut, at least near base. **VEIL** white or becoming cap color, dry, usually forming a fibrillose ring or ragged zone on stalk, but sometimes disappearing. **SPORE PRINT** brown to dull cinnamon; spores 8-11 × 3-4 microns, spindle-shaped to elliptical, smooth.

HABITAT: Scattered to gregarious on ground in woods; associated with Douglas-fir and common throughout the West where its host occurs. In the Rocky Mountains it fruits in the summer and early fall and is the most common bolete associated with Douglas-fir. In our area it is frequent in the fall and winter, but is not as numerous as *S. caerulescens* and *S. ponderosus*, which also grow with Douglas-fir. It prefers poor, exposed soil and often fruits on roadbanks or in campgrounds.

EDIBILITY: Edible, but not choice.

COMMENTS: This is one "slippery jack" that is not slippery except in very wet weather. The dark red to reddish-brown fibrils or scales on the cap plus the presence of a veil and association with Douglas-fir set it apart. When young it is quite attractive, the scales often being more pink then red. Heavy rain will wash them off, however, and gelatinize the cap surface. Then the yellow to dingy orange-brown background color predominates, making it resemble *S. caerulescens* and *S. ponderosus*. In the Pacific Northwest it can also be confused with two larch-lovers—*S. cavipes*, which has a hollow stalk, and *Fuscoboletinus ochraceoroseus*, which is more brightly colored and has darker spores. The "Painted *Suillus*" of eastern North America, *S. pictus*, is a similar but more brightly colored species associated with white pine. It has a dry cap with pink to red fibrillose scales, a somewhat longer stalk with similarly colored fibrils, and a soft, white, cottony annulus (ring). It is edible and highly touted by some, mediocre according to others. Another dry-capped eastern species, *S. decipiens,* has an orangish to dingy ochraceous to pinkish-orange, fibrillose or fibrillose-scaly cap.

Suillus caerulescens commonly occurs with Douglas-fir. Note the presence of a veil which may or may not form a distinct annulus (ring) on the stalk. A closely related and equally common species, *S. ponderosus,* is shown in color plate.

Suillus caerulescens   (Douglas-Fir Suillus)

CAP 5-18 cm broad, convex becoming plane or shallowly depressed; surface viscid when moist, with scattered fibrils, scales, or streaks, color variable: dull cinnamon to yellowish, but most often dull reddish-brown, orange-brown, or yellow-brown toward the center and yellowish to buff near the margin, with dingy greenish stains sometimes developing in cold weather; margin sometimes hung with veil remnants. Flesh thick, pale yellow or discoloring pinkish to vinaceous. **PORES** yellow when young, dingier in age, fairly large (1 mm or more in diameter at maturity) and sometimes radially arranged, not blueing when bruised but often discoloring dingy reddish-brown or brown; tubes same color, adnate to decurrent. **STALK** 2.5-10 cm long, 1.5-3 (4) cm thick, equal or tapered at either end, firm, solid, dry, yellow above the ring, fibrillose and dingier below and often mottled with reddish or brown stains and usually staining brownish after handling; sometimes spotted but not glandular-dotted; flesh near or at base usually staining blue or green when cut (sometimes slowly). **VEIL** white or pallid when young, becoming colored like the cap in age, not slimy, usually forming a slight to distinct ring on stalk; ring fibrillose or bandlike, median to superior. **SPORE PRINT** brown to dull cinnamon; spores 8-11 × 3-5 microns, elliptical to spindle-shaped, smooth.

HABITAT: Scattered to densely gregarious on ground under or near Douglas-fir; known only from the Pacific Northwest and California, very common. In our area it is sporadically abundant, along with *S. ponderosus* (see comments), from the fall through early spring. The largest fruiting is usually around Thanksgiving.

EDIBILITY: Edible. It is generally listed as mediocre, but one collection I sampled had a rather pleasing lemony flavor.

COMMENTS: If you hunt Douglas-fir forests regularly, you will soon tire of finding this mundane mushroom. The dull cinnamon to orange-brown or yellowish cap, presence of a veil, tendency of the stalk base to stain blue when cut, and association with Douglas-fir are key characteristics. A very similar species, *S. ponderosus* (COLOR PLATE 117), is just as common in our area and just as mediocre an addition to a meal. It differs in having a viscid, yellow to orange veil (before breaking) and a smoother, viscid to slimy cap which ranges from reddish-brown to dull yellowish. Its name refers to its sometimes ponderous size, and not to the fact that it isn't associated with ponderosa pine. In fact, it is associated with Douglas-fir just like *S. caerulescens,* and the two sometimes mix company. A third very similar species, *S. imitatus,* has a smooth viscid cap and shorter spores; it occurs under conifers in the Pacific Northwest. A dark or dingy greenish variety *(S. imitatus var. viridescens)* also occurs, and both *S. caerulescens* and *S. ponderosus* can develop greenish stains on the cap and stalk in cold weather. This in no way affects their edibility, however.

Suillus grevillei   (Tamarack Jack)

CAP 3-15 cm broad, convex to nearly plane; surface smooth, viscid or slimy when moist, often shiny when dry; color variable: deep red to reddish-brown with a yellow margin, varying to golden-yellow throughout or yellow with a cinnamon- or rusty-tinged center; margin sometimes with veil remnants. Flesh thick, yellow but often aging or staining pinkish to reddish; not blueing. **PORES** creamy when young becoming yellow and finally dingy ochre or olive-yellow, not blueing but usually turning brownish or rusty where bruised; often somewhat radially arranged; tubes same color, adnate to decurrent. **STALK** 4-10 cm long, 1-3 cm thick, equal or slightly thicker below, solid, firm, pale yellow when young but soon mottled reddish to deep brown or cinnamon below the veil and often staining brown when handled; glandular dots absent. **VEIL** yellowish before breaking, the underside often viscid in wet weather; usually forming a median to superior, cottony, whitish ring on stalk. **SPORE PRINT** olive-brown to dull cinnamon; spores 8-10 × 3-3.5 microns, elliptical to spindle-shaped, smooth.

HABITAT: Scattered to gregarious or in small clusters in woods or bogs, associated exclusively with larch (tamarack); common from late summer through early winter throughout the northern half of the northern hemisphere wherever larch occurs. It is very common in Idaho and Montana and also grows on the eastern slope of the Cascades (frequently in the company of *S. cavipes*), but is not known from California.

EDIBILITY: Edible, but like most slippery jacks, far from incredible.

COMMENTS: Also known as *S. elegans,* this attractive slippery jack is easily told by its smooth golden-yellow to deep red cap, presence of a veil, absence of glandular dots on the stem, and monogamy with larch. In the West the cap tends to be dark red to cinnamon, while in Europe and eastern North America it is yellower. *S. ponderosus* and *S. caerulescens* are somewhat similar but prefer Douglas-fir. Other species: *S. proximus* of eastern North America is closely related, but its spore print is chestnut-brown when moist and its stalk often stains greenish when cut open (at least below); it also favors larch.

Suillus grevillei, a slippery jack that grows only with larch. Note slimy cap and prominent annulus (ring). This is the eastern form with a golden cap; in the West the cap is usually reddish-brown.

Suillus umbonatus is a small slippery jack that favors lodgepole pine. Note the glutinous (slimy) ring on the stalk and the rather large pores.

Suillus umbonatus   (Umbonate Slippery Jack)

CAP 3-9 cm broad, obtusely convex becoming convex or plane, often with a low umbo, the margin sometimes uplifted in age; surface smooth, viscid or slimy when moist, often streaked, olive-buff to dingy tan or dull yellowish-brown, often more olive in age, especially toward margin. Flesh soft, pale yellow to buff, bruising dingy pinkish-brown. **PORES** large, more or less radially arranged, pale yellow when young, dingy greenish-yellow in old age; sometimes staining dingy pinkish-brown when bruised; tubes same color. **STALK** 3-8 cm long, 4-9 mm thick, more or less equal, solid, usually slender; pale yellow or tan above the ring, same color or paler below, with pallid to yellow-brown, often obscure glandular dots; often brownish-stained where handled. **VEIL** viscid-gelatinous, pale brownish, forming a median to superior gelatinous ring on stalk. **SPORE PRINT** olive-brown to dull cinnamon-brown; spores 7-11 × 3.5-4.5 microns, elliptical to spindle-shaped, smooth.

HABITAT: Scattered to gregarious or in small clumps under conifers, particularly lodgepole and beach pines *(Pinus contorta)*, often in large numbers; fruiting in the late summer and fall or early winter, western North America. It is common in the Sierra Nevada and on the California coast north of San Francisco; it has also been taken in eastern Canada.

EDIBILITY: Too gooey and gluey to be of value.

COMMENTS: This species is one of our most distinctive slippery jacks, easily recognized by its dingy yellow-brown to olive-buff color, frequently umbonate cap, viscid annulus (ring), and rather small, slender stature. The glandular dots are quite obscure when young but typically darken as the fungus matures. It is closely related to *S. sibiricus* and *S. americanus,* which, however, are more brightly colored and have reddish scales on the cap. Other species: *S. flavidus* is slightly yellower but otherwise similar if not the same; *S. helenae*, found under lodgepole (beach) pine in coastal Oregon, is also very similar but lacks a veil.

Suillus sibiricus   (Siberian Slippery Jack)   Color Plate 116

CAP 3-10 cm broad, convex to broadly convex or broadly umbonate; surface viscid or slimy when moist, bright yellow to dull yellowish, olive-yellow, ochre-yellow, or dark yellow, with scattered reddish to cinnamon-brown or dark brown spots, scales, streaks, or patches of fibrils, especially toward the margin (but these sometimes washed off by rain); margin incurved and usually hung with cottony veil remnants. Flesh soft, thin, yellow, not blueing but often staining vinaceous or pinkish when bruised. **PORES** large (1-2 mm wide) when mature, often somewhat radially arranged, mustard-yellow becoming duller or darker (ochre to yellow-brown) in age, not blueing when bruised (but may stain vinaceous or pinkish); tubes same color or dingier. **STALK** 3-11 cm long, 0.5-1.5 cm thick, more or less equal, often curved, solid or sometimes hollow in age, yellow to ochre-yellow, often staining vinaceous or brown when handled and often vinaceous-stained at or in the base; glandular dots and smears present, often inconspicuous when young but typically brown to cinnamon and quite conspicuous in age. **VEIL** cottony, whitish to pale yellow or in age sometimes brownish, usually remaining attached to margin of cap, but sometimes forming a slight ring on stalk. **SPORE PRINT** brown to dull cinnamon-brown; spores 8-12 × 3.5-4.5 microns, elliptical to spindle-shaped, smooth.

HABITAT: Scattered to gregarious or clustered under conifers, associated with white (5-needle) pines; widely distributed, but especially common in the summer and fall with western white pine in the Pacific Northwest and northern California. It is also found in the Southwest with southwestern white pine and limber pine. The very similar *S. americanus* (see comments) replaces it in eastern North America.

EDIBILITY: Edible but thin-fleshed, insipid, and slimy. Alexander Smith relates the case of one person who cannot even touch it without suffering an acute allergic reaction.

COMMENTS: The combination of viscid yellow cap with reddish to brown spots or plaques, cottony veil, and glandular-dotted stalk make this one of the most distinctive of all the slippery jacks. The cap is usually bright yellow in dry weather and duller or dingier when moist. The closely related *S. americanus* is often abundant with eastern white pine and also occurs in the Southwest. Its cap is often more streaked than spotted and its stalk is usually thinner (3-10 mm), plus it lacks an annulus because its veil normally doesn't come into contact with the stalk (instead it extends in from the cap margin as a roll of cottony tissue). In the Southwest the two species seem to intergrade; the color plate may represent one such "hybrid." In the Sierra Nevada two somewhat similar species, *S. riparius* and *S. megaporinus,* are often common. Both have very large, irregular or even gill-like pores (up to 5 mm long and 3 mm broad) and yellow to brownish caps with brown to rusty-brown scales, streaks, and/or fibrils. *S. riparius* usually has a well-developed stalk while in *S. megaporinus* the stalk is often poorly developed and off-center and the cap is usually small.

Suillus subolivaceus   (Slippery Jill)   **Color Plate 121**

CAP 4-15 cm broad, convex or obtusely umbonate becoming plane or broadly umbonate; surface smooth but often streaked or with flattened darker fibrils, viscid or slimy when moist; dull olive to olive-brown, grayish-olive-brown, dingy yellow-brown, or dingy tan. Flesh pallid to yellowish or tinged olive-gray, not bruising blue. **PORES** fairly large at maturity, grayish-olive to grayish-buff or olive-buff when young and often beaded with clear droplets in wet weather, becoming dingy yellowish to brownish-yellow in age (and droplets often drying blackish); not bruising blue; tubes same color or yellower. **STALK** 6-12 (17) cm long, 0.8-2 cm thick, equal or thicker below, solid, white or yellowish above the ring, whitish to brownish below; densely covered throughout with prominent pinkish-brown to blackish glandular dots and smears. **VEIL** membranous, whitish, with a gelatinous olive to brownish underside; forming a large bandlike, median to superior ring on stalk which may shrink considerably in age. **SPORE PRINT** brown to dingy cinnamon; spores 8-11 × 3-4 microns, elliptical to spindle-shaped, smooth.

HABITAT: Scattered to gregarious under conifers, particularly western white pine; known only from western North America. In the Pacific Northwest it is often common in the late summer and fall. It probably occurs in California, but I have not seen it south of the Cascades. Similar species occur in eastern North America (see comments).

EDIBILITY: Edible, but of poor quality.

COMMENTS: The large bandlike ring, overall dingy olive-brown to yellow-brown color, slimy cap, and very conspicuous glandular dots on the stem which become blackish in age make this slippery jack easier to identify than it is to grip! The glandular dots are often quite sticky and resinous and will stain your fingers brown more quickly than those of other slippery jacks. *S. umbonatus* is somewhat similar but grows with lodgepole pine and has a gelatinous annulus (ring) and pale yellowish pores when young. *S. luteus* is easily distinguished by its reddish-brown cap and sheathing, purplish-tinged veil. In eastern North America there is a group of "slippery jills" with a yellowish to olive-brown cap and strikingly broad (1-2 cm high), thick annulus that is baggy (flares at both the bottom and top edges) when young and bandlike in age. The most common northern member of this group is *S. subluteus*. Several species are common in the South, including *S. cothurnatus, S. salmonicolor* (with salmon-tinged pores), and *S. pinorigidus*. None are worth eating.

Suillus luteus   (Slippery Jack)   Color Plate 118

CAP 5-12 cm broad, obtuse or convex becoming plane; surface smooth, viscid or slimy when moist, often shiny when dry; typically reddish-brown to dark reddish-brown or chestnut-brown, but ranging sometimes toward yellow-brown or rusty-brown, especially in age, and sometimes streaked; margin often hung with veil remnants. Flesh thick, white to pale yellow, not bruising. **PORES** minute, white at first, becoming yellow and finally dark yellowish, often brownish-spotted in age; not bruising blue; tubes same color. **STALK** 3-10 cm long, 1-3 cm thick, more or less equal, solid, white to pale yellow above, often purplish below or with a purplish zone; prominent pinkish to brown glandular dots and smears soon developing both above and below. **VEIL** sheathing the stalk, membranous, persistent, white above, purplish to dull lilac below and viscid when wet; usually forming a large, often flaring or sleevelike ring that turns brown from falling spores. **SPORE PRINT** brown to dull cinnamon; spores 6-10 × 2.5-3.5 microns, spindle-shaped to elliptical, smooth.

HABITAT: Scattered to gregarious in cool wet weather under conifers, particularly pine but also spruce; widely distributed. It is abundant in pine plantations in northeastern North America, and has also been reported from the Pacific Northwest and Southwest. A favored associate is Scots pine *(Pinus sylvestris)*, and it may very well turn up in California where Scots pine has been planted.

EDIBILITY: Edible and widely collected, though some people are apparently "allergic" to it. According to one source it is "the best of the slippery jacks"—a classic case of damning with faint praise, if you ask me. As in most slippery jacks, the slimy skin peels off easily.

COMMENTS: This species is the "original" slippery jack, i.e., the one from which all the others take their name. It can be told by its reddish-brown cap, glandular-dotted stem, and sheathing veil which typically forms a prominent annulus (ring) on the stalk. The dull purplish color of the lower stalk or underside of the ring is also distinctive. *S. borealis* (see comments under *S. pseudobrevipes*) is a closely related westerner with no annulus.

Suillus pseudobrevipes   (Veiled Short-Stemmed Slippery Jack)

CAP 5-15 cm broad, convex becoming broadly convex or plane; surface slimy or viscid when moist, smooth or with a few patches of veil tissue near margin, but often appearing fibrillose or streaked; dingy yellowish to yellow-brown, dark yellow-brown, or honey-colored, often becoming darker or more cinnamon in age; margin often hung with whitish veil remnants. Flesh thick, white to pale yellow, not blueing. **PORES** and tubes pale yellow to yellow becoming dingy yellowish in age, not blueing. **STALK** 2-8 cm long, 1-3 cm

Suillus pseudobrevipes, a widespread slippery jack with a thin veil and inconspicuous glandular dots. The veil may form a slight ring on the stalk and/or leave remnants on the margin of the cap, or even disappear entirely. These are rather small specimens.

thick, often rather short and thick, equal or tapered downward, solid, firm, white when young, yellowish in age; glandular dots absent or obscure when young but tending to become brown and more visible in age. **VEIL** white to dingy lavender or lavender-brown, usually forming a median fibrillose ring on stalk, but the ring often slight or collapsed; sometimes forming a sheath over the base of stalk or merely leaving remnants on cap margin. **SPORE PRINT** brown to pale brownish; spores 7-9 × 2.5-4 microns, spindle-shaped to elliptical, smooth.

HABITAT: Scattered to gregarious under pines (mainly ponderosa and lodgepole); known only from western North America. In our area it fruits only in scattered localities with ponderosa pine, in the fall and winter. In the Southwest and Rocky Mountains it is quite abundant, however, and I've also seen it in the Sierra Nevada.

EDIBILITY: Edible.

COMMENTS: This slippery jack tends to have a short, stocky stem like its namesake, *S. brevipes,* but the presence of a veil and paler cap color distinguish it. Sometimes the veil persists as a roll of cottony tissue on the cap margin instead of forming a distinct annulus (ring), leading to confusion with several species that typically boast veil remnants on the margin of the cap but not the stalk. These species include: *S. volcanalis,* glandular dots obscure, cap yellowish or tawny (or often with pinkish to reddish-brown areas), growing in volcanic soil (often buried) with Jeffrey pine in Mt. Lassen National Park, California; *S. brunnescens,* cap white with the slime becoming chocolate-brown, glandular dots small, associated with sugar and lodgepole pines; *S. borealis,* cap dark brown, veil dull lavender or purplish (as in *S. luteus*), stalk glandular-dotted in age, associated with western white pine; *S. glandulosipes,* cap fibrillose and pale ochre to cinnamon, glandular dots *very* conspicuous, locally common in northern coastal California; and *S. albidipes,* whose cap is whitish to buff, brownish, or dull cinnnamon and has a roll of cottony veil tissue on the margin when young. I have seen enormous fruitings of the latter species under lodgepole pine in Oregon and Washington.

Suillus brevipes   (Short-Stemmed Slippery Jack)   **Color Plate 115**

CAP 5-13 cm broad, convex becoming broadly convex to plane; surface smooth, viscid or very slimy when moist, often shiny when dry; dark vinaceous-brown to dark brown when young, often fading in age to reddish-brown, dull cinnamon, or even tan (or in one form yellow-brown) and sometimes appearing streaked; margin naked. Flesh thick, white or often becoming yellow in age, soft, not blueing when bruised. **PORES** and tubes pale when young, becoming darker or dingier yellow in age and finally olive-yellow; not blueing. **STALK** 2-7 cm long, 1-2 (3) cm thick, equal or slightly thicker below and often quite short; firm, solid, white becoming pale yellow in age; glandular dots absent or sometimes barely visible in age (never prominent). **VEIL** absent. **SPORE PRINT** brown to dull cinnamon; spores 7-10 × 3-4 microns, elliptical to spindle-shaped, smooth.

HABITAT: Scattered to densely gregarious under conifers (particularly 2- and 3-needle

501

pines, also spruce); very widespread and common. In the West it is especially abundant with lodgepole and bishop pines. In our area it fruits in the fall and winter but is encountered infrequently, perhaps due to the superabundance of *S. pungens.*

EDIBILITY: Edible and perhaps the best of our local slippery jacks—which is hardly a compliment. Peel the slimy pellicle (skin) before cooking it.

COMMENTS: This widespread and often abundant slippery jack is best recognized by its smooth, slimy, dark brown to reddish-brown cap (which may fade in age), absence of a veil, and absence of obvious glandular dots on the stem. True to its name, the stalk is often rather short, but it is a variable species and longer-stemmed individuals are not unusual. The viscid cap and non-reticulate, more or less equal stalk distinguish it from *Boletus;* the cap is never olive-gray as in *S. pungens,* and there is no veil as in *S. pseudobrevipes,* while *S. granulatus* and its relatives have a distinctly glandular-dotted stem. Other species with a smooth viscid cap, no veil, and absent or inconspicuous glandular dots include: *S. pallidiceps* of the Rocky Mountains, with a white cap when young that becomes pale cinnamon or yellowish in age; and *S. occidentalis,* common under ponderosa pine in the Southwest, with a paler (buff to light brown or pinkish-brown) cap that is not white when young.

Suillus granulatus   (Granulated Slippery Jack)

CAP 3-15 cm broad, convex to plane or slightly wavy; surface smooth, viscid or very slimy when moist, often shiny when dry; cinnamon-brown to brown, orange-brown, rusty-cinnamon, or yellow-brown (but in some forms whitish when young), often streaked or mottled and paler or duller in old age; margin naked. Flesh thick, soft, white when young but soon pale yellow. **PORES** and tubes whitish when very young and often beaded with milky droplets, soon becoming buff or yellow and eventually dingy yellowish or brownish-spotted when mature. **STALK** 3-8 cm long, 0.7-2.5 cm thick, more or less equal, firm, solid, whitish when young, becoming yellow above and dingy cinnamon- or reddish-stained toward base; covered with pinkish to reddish-tan or brown glandular dots and smears in age. **VEIL** absent. **SPORE PRINT** brown to dull cinnamon or ochre; spores 7-10 × 2.5-4 microns, spindle-shaped to elliptical, smooth.

HABITAT: Scattered to densely gregarious on ground under pines in the summer, fall, and early winter; very widespread and common, but replaced in our area by *S. pungens.*

EDIBILITY: Edible, but bland (see *S. pungens*).

COMMENTS: This widespread slippery jack can be found almost anywhere where there are pines. It is closely related to *S. pungens,* but lacks the olive-gray tones so characteristic of that species. The combination of glandular-dotted stem, slimy or shiny cap, and absence of a veil are quite distinctive, but there are a number of closely related species which are rather difficult to distinguish (fortunately, all are apparently edible). These variants, which are especially numerous and confusing in the pine forests of the Rocky Mountains and Southwest, include: *S. flavogranulatus* of the Rockies, with a yellowish to dingy ochre cap at maturity and larger pores; *S. kaibabensis,* common in the Southwest, with a buff to pale cinnamon or yellowish cap; and *S. wasatchicus,* with reddish pores when young that become yellowish in age. In the Pacific Northwest another species, *S. punctatipes,* is quite common under various conifers. It also has glandular dots or smears and lacks a veil, but is larger and thicker-stemmed with larger yellow pores that are often somewhat radially elongated and/or decurrent. Its flesh is white and its cap color variable: orange-brown to dark brown, dingy tan, pinkish-brown, gray, vinaceous-gray, or purple-tinged. *S. monticolus* of the Sierra Nevada resembles *S. granulatus* but has a swollen or bulbous stem. *S. placidus* of eastern North America is very similar, but its cap remains whitish for a long time before darkening or turning yellowish. Also see *S. albidipes* (under *S. pseudobrevipes),* which differs in having veil tissue on the cap margin when young.

Suillus pungens. Note slimy cap and presence of milky droplets on pore surface of young specimens, plus the glandular-dotted stem. Cap color is variable (see color plate and description). (Joel Leivick)

Suillus pungens (Pungent Slippery Jack) Color Plate 114

CAP 4-18 cm broad, convex becoming broadly convex or plane; surface smooth, slimy or viscid when moist, often shiny when dry; color extremely variable: usually white when very young but soon olive to grayish-olive and then becoming yellow, tawny-cinnamon, rusty-brown, orange, reddish-brown, or often mottled and streaked with various combinations of the above colors; margin typically naked or with a slight roll of cottony tissue when young. Flesh thick, white when young, soon becoming lemon-yellow and soft, not blueing; odor pungent or pleasant (somewhat banana-like). **PORES** white when young and often beaded with white to pinkish droplets in wet weather, becoming yellow and finally dark yellow-brown or dingy ochre in age, not blueing when bruised; tubes same color. **STALK** 2-10 cm long, 1-2 (3) cm thick, equal or with a slightly tapered or swollen base, firm, solid, white when young becoming yellow in age, with conspicuous reddish to brown glandular dots and smears. **VEIL** absent. **SPORE PRINT** olive-brown to dull cinnamon-brown; spores 9-10 × 3-3.5 microns, spindle-shaped to elliptical, smooth.

HABITAT: Scattered to densely gregarious or clustered on ground near or under Monterey pine; apparently endemic to our area, but also occurring where Monterey pines have been planted (it is a common "lawn mushroom" in our area, often growing with *Chroogomphus vinicolor* and *Helvella lacunosa*). It is most abundant in the fall and winter but can be found whenever it's damp enough. It is also quite common under knobcone and ponderosa pines—but apparently only *within* the geographical range of Monterey pine.

EDIBILITY: Edible; disdained by some due to its "harsh, unpleasant, subnauseous" taste, but prized by others. In my opinion its soft flesh and slimy texture are much bigger drawbacks than its flavor. Since it occurs in such large quantities and is easy to identify, it is definitely worth experimenting with. To remove the slime, simply peel off the skin.

COMMENTS: Locally this is our most prolific bolete, appearing in colossal numbers wherever there are pines. The remarkable series of color changes it undergoes is apt to baffle the color-conscious beginner. On the other hand, when several individuals are present it affords an instant means of recognition! It is closely related to *S. granulatus,* but that species is never grayish-olive when young. The glandular-dotted stem and milky droplets on the pore surface of young specimens are also distinctive.

S. *pungens* vies with *Hygrophorus eburneus* and the "Insidious Gomphidius" *(Gomphidius oregonensis)* for the title of "Slipperiest and Slimiest Fungus Among Us." An organic version of hockey or caroms can be played on a wet surface, using the caps of young *S. pungens* for pucks. Overmature individuals, on the other hand, are more aptly called *S. "spongens"* than *S. pungens*—they literally seethe with fat, agitated maggots and sag with so much excess moisture that they practically demand to be wrung out like a sponge!

Suillus fuscotomentosus   Color Plates 119, 120
(Poor Man's Slippery Jack)

CAP 4-15 cm broad, convex to plane; surface dry, or viscid in age or wet weather, at first covered with olive-brown to dark brown or fuscous (deep grayish-brown) fibrils or small fibrillose scales (in this stage usually dry), the scales often sparser in age, revealing the paler (dull ochre to buff) background, sometimes more cinnamon-colored overall at maturity or appearing streaked; margin naked. Flesh thick, yellow to pale orange or orange-buff, not blueing when bruised; taste mild to slightly unpleasant. **PORES** typically orange-buff when young and sometimes beaded with droplets, becoming yellowish-buff to dark dingy yellow or olive-yellow in old age, not blueing when bruised; tubes same color. **STALK** 4-12 cm long, 1-3 cm thick, equal or somewhat thicker below, solid, firm, pallid to yellow or brownish-buff or often pinkish-orange to pale orange, especially toward base; glandular dots and smears present and usually conspicuous at least in age, often elongated, same color as stalk or darker (brown to cinnamon). **VEIL** absent. **SPORE PRINT** olive-brown to dull cinnamon-brown; spores 9-12 × 3-4 microns, spindle-shaped to elliptical, smooth.

HABITAT: Scattered to densely gregarious on ground under pines, often in large numbersr; originally described from Santa Cruz County, California, but occurring in the Sierra Nevada and Cascades as well. In our area it is often abundant in the late fall and winter under ponderosa, knobcone, and digger pine, while the very similar *S. acerbus* (see comments) favors Monterey pine.

EDIBILITY: Edible, but slimy and insipid. In blind tastings of local edible boletes, it has consistently placed last. Asked to rate each bolete on a scale of 1-10, several tasters gave it a zero and one, a minus five!

COMMENTS: This common bolete resembles *S. tomentosus* but does not stain blue. The pores are usually orange-buff when young, rather than white as in *S. pungens* and *S. granulatus,* and the brownish to grayish-brown cap fibrils and absence of a veil are also distinctive. The "Starving Man's Slippery Jack," *S. acerbus,* is a very similar species with a more viscid, less fibrillose-scaly cap. It is supposedly associated exclusively with Monterey pine, but I wonder whether it is really distinct from *S. fuscotomentosus.* The degree of viscidity seems to depend on age and weather conditions, and I've found specimens under Monterey, knobcone, and digger pines which could be referred to either species.

Suillus tomentosus   (Blue-Staining Slippery Jack)

CAP 5-15 (20) cm broad, convex to plane; surface dry, but in wet weather or age often viscid, at first covered with a dense coating of minute hairs (fibrils) which break up to form scattered scales and are often quite sparse (or even absent) in age; fibrils or scales usually grayish-brown to brown (reddish in one form), sometimes pallid or buff when young; background usually pale yellow to yellow-orange, orange-buff, or dark dull orange; margin naked. Flesh thick, pallid to yellow, usually blueing (but sometimes slightly or slowly) when bruised; taste mild or acidic. **PORES** dingy brown to dark cinnamon when young, becoming dingy yellow to buff or olive-yellow in age; usually blueing at least slightly when bruised (at least in age); tubes yellowish to olive-yellow. **STALK** 3-11 (15) cm long, 1-3 cm thick, equal or thicker below, firm, solid, colored like background of cap or oranger, with numerous small glandular dots which may be same color or browner and are often obvious in age; base sometimes reddish-stained. **VEIL** absent. **SPORE PRINT** dark olive-brown to dull cinnamon; spores 7-11 × 3.5-4.5 microns, elliptical to spindle-shaped, smooth.

HABITAT: Scattered to gregarious or in troops on ground under pines and other conifers, late summer through early winter; widely distributed, but especially common in the West, where it is one of the most common of all the slippery jacks. I have seen enormous fruitings

Suillus tomentosus is a common and widespread species easily told by the tendency of its pores and/ or flesh to stain bluish when bruised. Note presence of fibrillose scales on cap. Older or rain-battered specimens, however, may be practically bald (without scales or hairs).

in Idaho, Washington, northern California, and the Sierra Nevada, especially with lodgepole pine and in mixed forests of aspen and pine. In our area, however, it is largely replaced by the similar *S. fuscotomentosus.*

EDIBILITY: Edible, and every bit the equal of *S. fuscotomentosus* (see comments on the edibility of that species).

COMMENTS: The fibrillose to scaly cap, brownish pores when young, absence of a veil, glandular-dotted stem, and tendency of the pores and/ or the flesh to stain blue or greenish-blue when bruised form a distinctive combination of characters. It is one of the commonest and most variable slippery jacks, with several slightly different color forms. Other species: *S. variegatus* of Europe is very similar if not the same; *S. reticulatus* is a rare, reticulate-stalked species that stains blue.

FUSCOBOLETINUS   (Larch Boletes)

Medium-sized, fleshy, terrestrial boletes *associated mainly with larch.* CAP viscid, or if dry then fibrillose. PORES pallid to yellowish, gray, or grayish-brown, often large and radially arranged. STALK *not usually reticulate or glandular-dotted;* not scabrous. VEIL *present, frequently forming an annulus (ring) on stalk.* SPORE PRINT *dark grayish-brown to dark reddish-brown to chocolate-brown or chocolate-gray.* Spores elliptical to spindle-shaped, smooth. Cystidia on inner surfaces of tubes staining dark brown in KOH (potassium hydroxide).

THIS small genus is an off-shoot of *Suillus.* It differs principally in its darker spore color, and is included in *Suillus* by some boletologists. A veil is present in all species but the stalk usually lacks glandular dots. The cap varies from smooth and slimy to dry and fibrillose.
   Fuscoboletinus species are mycorrhizal mainly with larch or tamarack *(Larix).* This is a helpful feature in identification, though several *Suillus* species (e.g., *S. grevillei* and *S. cavipes*) also grow with larch. Some *Fuscoboletinus* species grow with other northern conifers, but their geographical range still corresponds to that of larch. As larch does not occur in California, *Fuscoboletinus* probably doesn't either. However, it could conceivably appear where larches have been planted. Only two species are known from the western United States, and they are described here; most of the other species occur in the Northeast and Canada. Like slippery jacks, they are edible but of mediocre quality.

505

Key to Fuscoboletinus

1. Stalk distinctly glandular-dotted (the dots dark); known from Minnesota; rare . *F. weaverae*
1. Stalk not glandular-dotted (but may have reddish spots when young); common 2
2. Tubes and pores yellow to yellow-brown or yellow-olive when fresh 3
2. Tubes and pores whitish to grayish or grayish-brown, not yellowish 10
3. Cap 4-10 cm broad, at first covered by cottony grayish to yellow or red veil material which breaks up to form coarse scales, streaks, or plaques; cap viscid beneath the veil material and usually dark red to brownish or red-and-yellow in age; partial veil with both a cottony and a viscid or gelatinous layer; stalk with veil remnants (like cap), usually a mixture of red, yellow, and/or gray; pores and tubes yellow; associated with larch (tamarack), usually growing in bogs; found in the northeastern United States and Canada (a very striking mushroom—see Color Plate 126) .... *F. spectabilis*
3. Not as above 4
4. Cap viscid or slimy when moist and more or less smooth 5
4. Cap dry (not viscid) and fibrillose or scaly 8
5. Flesh in lower stalk staining greenish when cut (see *Suillus proximus* under *S. grevillei*, p. 497)
5. Not as above 6
6. Very common with larch in both eastern and western North America; spore print olive-brown to dull cinnamon .... (see *Suillus grevillei*, p. 497)
6. Found in northeastern U.S., Canada, and Alaska, usually with conifers other than larch; spore print darker than above 7
7. Veil (and stalk below veil) viscid; cap red-brown to mahogany, the stalk often with reddish spots when young; found under conifers in Alaska, Canada, and northeastern U.S. *F. glandulosus*
7. Not as above; stalk dry (not viscid) .... *F. sinuspaulianus*
8. Stalk typically becoming hollow in lower portion; cap color variable: dark brown to reddish-brown, tawny, etc.; spore print dark olive-brown to brown .... (see *Suillus cavipes*, p. 494)
8. Stalk not normally hollow; cap rosy to dark red (but sometimes overlaid with white) when fresh; spore print dark reddish-brown to dark vinaceous-brown 9
9. Stalk typically 1-3 cm thick and cap 6-20 cm broad; known only from western North America .... *F. ochraceoroseus,* below
9. Not as above; typically smaller and eastern ...... *F. paluster* (see *F. ochraceoroseus,* below)
10. Cap only slightly viscid; flesh not blueing when bruised; pores usually elongated near the stalk or forming gill-like rows; found in the eastern United States and Canada, usually in bogs .... *F. grisellus* (see *F. aeruginascens,* p. 507)
10. Not as above; cap viscid to slimy when moist; flesh often (but not always) staining bluish or bluish-green when exposed; found in both eastern and western North America 11
11. Cap more or less chocolate-brown when fresh; known from the eastern United States and Canada .... *F. serotinus* (see *F. aeruginascens,* p. 507)
11. Not as above; widespread .... *F. aeruginascens,* p. 507

Fuscoboletinus ochraceoroseus   Color Plate 123
(Rosy Larch Bolete)

CAP 7-20 (25) cm broad, convex to plane or slightly umbonate, or in age the margin sometimes uplifted; surface dry, densely fibrillose or fibrillose-scaly and often uneven or pitted; rosy-red to bright pink, but often overlaid with whitish hairs (fibrils), and often darker and duller (brick-red) in age; margin sometimes yellowish and hung with veil remnants. Flesh thick, yellow or tinged pink under cap cuticle, bruising slightly greenish-blue or not at all; taste often slightly bitter or acrid. **PORES** large, elongated (2-5 mm long) and usually arranged radially; yellow becoming dark yellow to olive-ochre and finally dingy brown in old age, not bruising blue; tubes shallow, same color, adnate to decurrent. **STALK** 3-7 (10) cm long, 1-3 cm thick, equal or with an enlarged base (and often a flaring apex); solid, firm, yellow (colored like pores) or with reddish to brownish stains, base often whitish and fibrillose; apex often slightly reticulate from tubes; glandular dots absent. **VEIL** membranous, white or yellowish, thin, sometimes forming a slight ring on stalk but more often clinging to margin of cap. **SPORE PRINT** dark reddish-brown to dark vinaceous-brown; spores 7.5-9.5 × 2.5-3 microns, elliptical to spindle-shaped, smooth.

HABITAT: Solitary to scattered or in groups on ground under conifers, associated with larch; known only from the northern Rocky Mountains and Pacific Northwest. I have seen large fruitings in Idaho in September, but apparently it isn't common every year.

EDIBILITY: Edible, but of very poor quality due to the slightly bitter taste.

COMMENTS: The beautiful rosy-red to pink cap, radially arranged pores, presence of a veil, and association with larch are the most distinctive characters of this most distinctive bolete. Its color is somewhat reminiscent of *Suillus pictus,* which grows with eastern white pine, and *S. lakei,* which grows with Douglas-fir, but the spore print is darker and the stalk usually thicker. Other species: *F. paluster* of northeastern North America is somewhat similar (cap deep red) but much smaller and more slender, with even larger pores; it grows in swamps and bogs. Another very striking bog-lover, *F. spectabilis* (**COLOR PLATE 126**), is also restricted to the Northeast and Canada. One glance is usually sufficient to identify it (as evidenced by the color plate), but see the key to *Fuscoboletinus* for details.

Fuscoboletinus aeruginascens   (Grayish Larch Bolete)

CAP 3-12 cm broad, convex becoming plane or slightly umbonate; surface viscid or slimy when moist, smooth or streaked with flattened fibrils, or at times with darker fibrillose scales, sometimes cracking in dry weather; grayish-white to gray, grayish-brown, olive-gray, or dingy yellowish, sometimes with darker scales or developing grayish to dingy olive spots; margin usually hung with veil remnants. Flesh soft in age, white to yellowish, typically turning bluish-green when bruised (but often slowly or only very slightly), at least in the stalk. **PORES** small and round when young but often rather large and somewhat radially arranged in age; white at first becoming gray to grayish-brown at maturity, usually staining blue to dingy greenish when bruised (but often only slightly or slowly); tubes same color. **STALK** 4-6 (9) cm long, 0.7-1.5 cm thick, equal or tapering upward or sometimes pinched at base; solid, pallid or tinged olive above the ring, gray to olive-grayish to brownish below and usually somewhat viscid; glandular dots absent, apex often reticulate from tubes. **VEIL** membranous, white to gray or yellowish, usually forming a median to superior ring on stalk. **SPORE PRINT** dark reddish-brown or vinaceous-brown; spores 8-12 × 3.5-5 microns, spindle-shaped to elliptical, smooth.

HABITAT: Scattered to densely gregarious in woods and around the edges of bogs, associated with larch (or tamarack) and common wherever larch occurs—the most widespread member of the genus. In the Pacific Northwest it is often common in the spring and again in the late summer and fall; it probably does not occur in California.

EDIBILITY: Edible, but unappealing because of its sliminess, thin flesh, and dingy color.

COMMENTS: The whitish to grayish to yellowish cap, white (never yellow!) pores that become grayish in age, presence of a veil, and tendency to bruise weakly bluish-green are the distinguishing fieldmarks of this undistinguished larch-lover. It is likely to be mistaken for a *Suillus* (and is even placed in that genus by some mycologists), but the spore print is vinaceous-brown. In eastern North America it often grows with *F. spectabilis,* a very striking species described in the key to *Fuscoboletinus* and shown in Color Plate 126. *F. serotinus* is a similar species with chocolate-brown slime on the cap; *F. grisellus* is colored like *F. aeruginascens* but does not normally bruise bluish-green, has a more conical and less viscid cap, and more elongated pores. Both species are associated with tamarack (eastern larch) and are fairly common in the northeastern U.S. and Canada.

BOLETELLUS & AUSTROBOLETUS

Medium-sized woodland boletes. CAP fleshy, dry or viscid, sometimes areolate or scaly. PORES and tubes usually (but not always) yellow, sometimes blueing when bruised. STALK fleshy *but usually long, slender, and more or less equal; lacking glandular dots or dark scabers but often longitudinally ridged or jaggedly reticulate.* VEIL absent (except in *B. ananas*). SPORE PRINT olive to brown or cinnamon-brown. Spores elliptical to spindle-shaped, longitudinally wrinkled, ridged, grooved, or "winged" *(Boletellus)* or minutely pitted *(Austroboletus).*

THESE are gracile (slender-stemmed) boletes with ornamented spores (ridged or wrin-kled in *Boletellus,* pitted in *Austroboletus*). The stalk is usually long in relation to the cap and not bulbous as in many species of *Boletus.* In addition, it is often raggedly or jaggedly reticulate or ridged. Both genera are centered in the tropics. In the United States they are most prevalent in the southeastern sector and along the Gulf of Mexico. A few species range north to Canada and west to southern Arizona, but none have been found in Cali-fornia. One *Boletellus* is depicted here; several other species are keyed out.

Key to Boletellus & Austroboletus

1. Stalk smooth to fibrillose or scurfy but not reticulate, jagged, or markedly ridged; pores and flesh usually blueing when bruised; often growing at or near the bases of trees 2
1. Stalk markedly or jaggedly reticulate or ridged; flesh and pores not blueing; terrestrial 3

2. Veil present when young, usually leaving white flaps on cap margin; cap coarsely scaly or fibril-lose-scaly; usually growing on or near pine trees .... *B. ananas* (see *B. russellii,* below)
2. Veil absent; cap often areolate but not fibrillose-scaly or shaggy .... *B. chrysenteroides*

3. Cap and stalk whitish to pale yellow or buff; stalk prominently and coarsely reticulate; taste bitter; usually found in sandy soil in southern U.S. (subtropical) .... *A. subflavidus*
3. Not as above; cap and stalk darker or brighter 4

4. Cap smooth and viscid when fresh and moist ..... *A. betula* (see *Boletellus russellii,* below)
4. Cap not viscid, often areolate (cracked) in age .... *B. russellii,* below

Boletellus russellii   (Jagged-Stemmed Bolete)   Color Plate 125

CAP 3-9 (13) cm broad, convex to broadly convex or rarely plane; surface dry, yellow-brown to buffy-brown or olive-gray, varying to brownish, reddish, or cinnamon-brown; minutely velvety to obscurely scaly, often becoming areolate (breaking up into small scales) in age, revealing the flesh beneath; margin at first incurved. Flesh yellow, not blueing when bruised. **PORES** rather large (1 mm broad or more), yellow when young, greenish-yellow in age, not blueing; tubes same color. **STALK** 10-20 cm long, 0.8-2 cm thick, equal or slightly thickened downward, typically long and slender, often curved at base; coarsely reticulate-lacerate (ragged and deeply ridged) more or less throughout; dull reddish to reddish-brown or cinnamon; solid, dry or with a viscid base when fresh. **VEIL** absent. **SPORE PRINT** dark olive to olive-brown; spores 15-20 × 7-11 microns, elliptical to spindle-shaped, deeply ridged or wrinkled longitudinally, with a cleft at apex.

HABITAT: Solitary, scattered, or in small groups on ground under hardwoods (espe-cially oak) or occasionally conifers; fairly common in the summer and early fall in eastern North America, but rarely fruiting in large numbers. It also occurs in southern Arizona, like many other "eastern" species.

EDIBILITY: Edible, but rather soft and bland.

COMMENTS: This distinctive bolete is easily recognized by its long, slender, lacerated (ragged or jagged) stalk (see the color plate), dry cap that is frequently areolate, and yellow to greenish-yellow pores that do not stain blue. Another species with a long, slim, lacerated stalk, *Austroboletus (=Boletus, Boletellus) betula,* differs in having a smooth, viscid, yellow-orange to reddish-brown cap, a more southern distribution, and pitted spores. Another southerner, **Boletellus ananas,** is quite different. It has a fibrillose to coarsely scaly or shaggy, purplish to reddish (or sometimes yellowish) cap plus a whitish veil when young. The veil usually leaves remnants on the cap margin rather than forming a ring. The stalk is whitish to tan and smooth to slightly fibrillose, and the flesh and pores turn blue when bruised or cut. Neither of the above species is worth eating.

PULVEROBOLETUS

AS DEFINED here, *Pulveroboletus* contains an awesome total of one common species. The prefix *pulvero* means **pulverulent,** which in turns means "powdery." It pertains to the powdery to somewhat cottony or cobwebby consistency of the bright yellow veil that covers the cap and stalk of the young fruiting body—a unique feature among the boletes. Some mycologists, however, broaden the genus concept to include veil-less species such as *P. auriporus* and *P. hemichrysus* (treated in this book under *Boletus*).

Pulveroboletus ravenelii   (Veiled Sulfur Bolete)

CAP 2-8 (12) cm broad, convex or nearly round when young, becoming plane in age; surface dry to slightly viscid if wet, bright yellow with orange, pinkish, or reddish-brown tones developing at the center, especially in age; covered with yellow cottony or powdery veil material, at least when young; margin often covered with veil remnants also. Flesh thick, white to yellow, changing slowly to blue when exposed, then eventually yellowish to dingy brown. **PORES** and tubes bright yellow when fresh, becoming dingier or olive-yellow in age; usually bruising bluish-green, and then sometimes turning dark brownish. **STALK** 6-15 cm long, 0.5-1.5 cm thick, equal to slightly irregular, often rather long; dry, bright yellow, powdery or cottony at least when young, except at apex. **VEIL** bright yellow, cottony-powdery to cobwebby, leaving remnants on cap and cap margin and stalk (and sometimes a slight ring on stalk). **SPORE PRINT** dark olive-brown to brown; spores 8-12 × 4-6.5 microns, spindle-shaped to elliptical, smooth.

HABITAT: Solitary or in small groups in woods; widely distributed, but rather rare on the west coast. In our area I have found it in the fall and early winter in mixed woods of oak and pine and under manzanita.

EDIBILITY: Like myself, not firmly established.

COMMENTS: I was dumbfounded by the sight of my first *Pulveroboletus*. As boletes go it is an odd creature—a gaudy misfit, an audacious anomaly. The brilliant yellow cottony veil which covers the cap and stalk when young sets it apart from all other boletes. In old age or rainy weather the veil remnants may be obliterated, but the color and relatively long, slender stalk help identify it. Though rare in California, it is included here because it is so bizarre. If you should be (un)fortunate enough to fall on top of one, as I did, I would like you to know what it was.

An unusually gracile (long and slender) example of *Pulveroboletus ravenelii*. The distinctive yellow veil is not clearly visible in this picture.

GYROPORUS

Small to medium-sized, terrestrial boletes. CAP typically dry. PORES *pallid or whitish when young, pale yellow to yellow in age.* STALK lacking scabers and glandular dots and typically *not* reticulate; *usually hollow at maturity, at least toward the base.* VEIL *absent or rudimentary.* SPORE PRINT *pale yellow to yellow.* Spores elliptical to spindle-shaped, smooth.

THE pale yellow spore print and hollow or partially hollow stem are the distinctive features of this small genus. Four species—all choice edibles—are included in the key, but only one, *G. castaneus,* has been found in California. In addition to the generic characteristics, *G. castaneus* can be told by its chestnut-brown to orange-brown cap, white to pale yellow pores, and often slender, uneven, similarly-colored, non-reticulate stalk plus its "failure" to stain blue when bruised.

Key to Gyroporus

1. Flesh and pores quickly turning deep indigo-blue or violet-black when bruised or cut; cap yellowish to buff, dry, often somewhat fibrillose; pores whitish becoming yellowish; stalk colored like cap, often thicker toward base, smooth or fibrillose; common in sandy soil and woods in eastern North America (also said to occur rarely in Pacific Northwest) *G. cyanescens*
1. Flesh and pores not blueing when bruised 2
2. Cap and stalk white to yellowish or tinged pink or apricot, sometimes spotted with cinnamon or brown in age; common under oak and pine in southeastern North America . *G. subalbellus*
2. Cap and stalk darker (tawny to orange-brown, chestnut-brown, wine-red, etc.) 3
3. Cap more or less wine-red to burgundy; found under hardwoods in eastern North America . . . . *G. purpurinus*
3. Cap brown to chestnut-brown to orange-brown or tawny; widespread . *G. castaneus,* below

Gyroporus castaneus   (Chestnut Bolete)   Color Plate 148

CAP 2.5-7 (10) cm broad, convex to plane or shallowly depressed; surface dry, minutely hairy to smooth, sometimes with a delicate whitish bloom when young; color variable: chestnut-brown to brown, cinnamon-brown, orange-brown, or rusty-tawny. Flesh thick, firm, white, not blueing when bruised. **PORES** and tubes white, becoming pale yellowish in old age, not blueing. **STALK** 3-9 cm long, 0.5-1 (3) cm thick, often rather slender and more or less equal (but a robust form also occurs), sometimes thicker or swollen below; surface dry, uneven, brown to tawny (colored more or less like cap or slightly paler); hollow or partially hollow at least at maturity (especially toward base). **SPORE PRINT** pale yellow to yellow; spores 8-12 × 4.5-6 microns, elliptical-oblong, smooth.

HABITAT: Solitary or in groups under oaks and other hardwoods; fairly common in Europe and eastern North America, but extremely rare in the West. I have found it only twice in our area, under a tanoak in July (while truffle-hunting) and under a large live oak in October. The slender form is the common one in North America.

EDIBILITY: Edible and highly esteemed in Europe, but alas, hard to come by in our area. One day I was picking fairy ring mushrooms *(Marasmius oreades)* on a lawn. A round Polish woman came out and said the mushrooms I was picking were no good. Then she invited me inside, where she proceeded to stuff me with sour cream cookies (she thought I was too skinny). She showed me a picture of her son, who was in the navy. Then she gave me a necklace of dried *Gyroporus castaneus.* They were her last ones, she said, and they came from "the old country, where everything tastes better." They made a fabulous soup. So did the *Marasmius.*

COMMENTS: In the West it is unlikely to be confused with anything, as it is unlikely to be found.

BOLETUS

Medium-sized to very large, mostly terrestrial and mycorrhizal, woodland boletes. CAP fleshy, *usually dry* (but sometimes viscid), sometimes areolate. PORES and tubes typically white, yellow, orange, red, brown, or gray, often blueing when bruised. STALK fleshy, sometimes bulbous, thick or slender, *often reticulate but not glandular-dotted and without dark scabers*. VEIL *absent*. SPORE PRINT *usually olive to olive-brown or brown*, but sometimes yellow-brown or cinnamon-brown. Spores spindle-shaped to elliptical, smooth.

IF YOUR bolete is not a *Suillus, Fuscoboletinus, Leccinum, Tylopilus, Pulveroboletus, Austroboletus, Boletellus, Gyroporus, Strobilomyces,* or *Gastroboletus,* then it's probably a *Boletus.* In other words, it is much easier to characterize what a *Boletus* isn't than what it is. There is neither a veil nor glandular dots as in *Suillus,* the stem does not have the dark scabers characteristic of *Leccinum,* the spores are neither yellow as in *Gyroporus* nor reddish as in *Tylopilus,* nor are they ornamented as in *Boletellus* and *Austroboletus.*

As such *Boletus* is a large and varied genus that splits nicely into two natural groups. The first group includes the hefty familiar forms like *B. edulis,* and can be subdivided into those with white pores when young (e.g., *B. edulis*), those with red pores (e.g., *B. satanas*), and those with yellow pores (e.g., *B. appendiculatus*). The stem is usually thick, sometimes bulbous, and often reticulate, the pores small (up to 1 mm broad), and the different species are *relatively* easy to distinguish in the field. They are fairly finicky as to their mycorrhizal mates and have exceedingly fickle fruiting habits. For instance, in our area heavy early rains may elicit a bumper crop of nearly every species, whereas if the rains are delayed until December, they may not fruit at all!

Species belonging to the second group (the genus *Xerocomus* of some books) tend to be smaller, with a dry, minutely velvety (**subtomentose**) cap. The pores are often large (1-2 mm or more in diameter), and the stem is relatively slender (or at least not bulbous) and not reticulate or only very coarsely so from decurrent tube walls. Members of this group are more cosmopolitan than those in the first group—that is, they occur in a wider range of habitats and may even grow on rotting wood. They tend to have a much longer growing season and more dependable fruiting habits, but seldom appear in large numbers. Identifying them is often difficult without resorting to chemical and microscopic characters, but few species in the group are good edibles.

Boletus boasts some of the finest and most flavorful of all the fleshy fungi (see photo on p. 512). First and foremost, of course, is that fabulous fungus cherished by the Europeans above all others—the king bolete or cepe, *B. edulis.* Its sister species, *B. barrowsii* and *B. aereus,* are also excellent, and the butter boletes (*B. appendiculatus* and *B. regius*) are little-known local treasures. On the other hand, some species are bitter-tasting (see comments on the edibility of *B. "marshii"*!) and others, such as *B. satanas,* are so poisonous that their consumption can result in hospitalization. A general rule of thumb is to avoid all those species whose pores bruise blue. But this rule eliminates roughly half of all *Boletus* species (including some very good ones such as *B. appendiculatus*), and at least on the west coast it can be amended to: avoid all those blue-staining species with pink, red, or orange pores. However, this does *not* mean that you should munch on those without red pores indiscriminately. The edibility of some species has yet to be established and many (even *B. edulis*) can cause stomach upsets if not thoroughly cooked.

Even in its modern "restricted" sense, *Boletus* is *the* major genus of boletes. It is especially numerous and diverse (over 100 species) during the humid summer months in the hardwood forests of eastern North America. On the west coast, however, the rains come during the cooler months of the year and the *Boletus* flora is much smaller, with only about 40 known species. Our oak-madrone woodlands boast a modest but unique bolete bounty. Prominent species include *B. satanas, B. erythropus, B. amygdalinus, B. appendiculatus, B. regius, B. aereus, B. barrowsii,* and *B. flaviporus* (along with *Leccinum*

Five delectable species of *Boletus* (gathered the same day!): *B. regius,* three at top left; *B. barrowsii,* three at top right; *B. aereus,* three at bottom right; *B. appendiculatus,* small button at bottom and the one just above it; and *B. edulis,* two at bottom left. The smallest bolete in this picture is about 10 cm (4 inches) high! All of these species favor hardwoods (at least in our area) except *B. edulis.*

manzanitae). Pine forests, on the other hand, are favored by *B. edulis,* which can be harvested by the bushel under favorable conditions. *Boletus* species are easily preserved by slicing and drying, but remember: if you should stumble into a "*Boletus* bonanza," take only as many as you can use. This will allow others (namely me) to share your luck, and aside from the good social practice it builds, it will test your will power to its limits!

More than 30 species of *Boletus* occur in California. Nineteen are described here, plus three extralimital species. Don't be intimidated by the apparent length of the key—it is actually composed of two separate keys that diverge at couplet #7: one key to western species and another to eastern ones. (Since *Boletus* is such a prominent group in eastern North America, I have keyed out many of the more distinctive and common species that occur there.) If you happen to be in southern Arizona, southern New Mexico, Mexico, or other regions where the eastern and western fungal floras overlap, try the key to western species first; if that doesn't work, then try the key to eastern ones (couplet #42).

Key to Boletus

1. Growing on earthballs (*Scleroderma* species); fruiting body small to medium-sized, usually rather soft; especially common in southern latitudes .... ***B. parasiticus***
1. Growing on ground or wood, not on earthballs 2

2. Taste distinctly acrid (peppery) when a small piece of the cap is chewed; fruiting body typically rather small, with bright yellow mycelium at base of stalk .... ***B. piperatus*** & others, p. 517
2. Not as above (but taste may be bitter or sour) 3

3. Growing on or near wood or in sawdust (especially coniferous) 4
3. Growing on ground 7

4. Cap yellow to ochre (or developing rusty-orange tones toward center in age); pores and flesh blueing when bruised; often in groups or clusters; spores short-elliptical 5
4. Not as above 6

5. Pores reddish to reddish-brown when young; cap at first powdery .... ***B. hemichrysus***
5. Not as above; pores typically yellow .... ***B. sphaerocephalus***

6. Cap and stalk dark reddish-brown to maroon-brown or chocolate-brown; cap plushlike or fibrillose-scaly; stalk usually long and often streaked or lined; pores yellow to olive-yellow, *not* blueing when bruised; found on or near northern conifers .... ***B. mirabilis***, p. 521
6. Not as above 7

512

7.   Found in western North America (Rocky Mountains, Southwest, and westward)* 8
7.   Found in eastern North America (east of the Rockies, also southern Arizona)* 42

8.   Pore surface red, pink, vinaceous, or orange when fresh (but may be yellowish or dingier in age, or in one case yellow when young becoming red in age); pore surface and flesh usually turning blue or blue-black when bruised or cut 9
8.   Pore surface yellow to olive, brown, white, etc., but not red or orange; pore surface may or may not turn blue when bruised 15

9.   Stalk distinctly reticulate (netted), at least over upper portion 10
9.   Stalk not reticulate, or only very slightly so from decurrent tube walls 13

10.   Stalk equal to bulbous, usually *at least* 2.5 cm thick at apex; reticulation on stalk usually red or pink or vinaceous (but may be stained olive or brown by spores); widespread in West   11
10.   Stalk equal or slightly thicker below, typically 1-3 cm thick at apex; reticulation red or not red; found only in parts of the Southwest (e.g., southern Arizona) 46

11.   Stalk with an exaggerated basal bulb; cap whitish to grayish to olive-buff, or sometimes suffused with pink in age; associated with oak .... ***B. satanas,*** p. 527
11.   Stalk equal or swollen below, but without an exaggerated bulb; cap some shade of brown or red (or in one variety, pallid); associated mainly if not exclusively with conifers 12

12.   Cap brown to olive-brown or yellow-brown (not often reddish); pores often yellow when very young; found in mountains .... ***B. haematinus*** (see *B. pulcherrimus,* p. 528)
12.   Cap brown to reddish-brown; pores never yellow .... ***B. pulcherrimus*** & others, p. 528

13.   Cap intensely yellow, quite large; rare .... ***B. orovillus*** (see *B. erythropus,* p. 526)
13.   Not as above; common 14

14.   Flesh changing to blue slowly or not at all when mushroom is cut open; cap usually areolate in age; not common .... ***B. mendocinensis*** (see *B. chrysenteron,* p. 519)
14.   Flesh blueing quickly when cut; cap not normally areolate ... ***B. erythropus*** & others, p. 526

15.   Stalk typically long, relatively slender, and scurfy or minutely scaly (as in *Leccinum*); cap yellow to ochre, reddish-orange, or cinnamon; pores yellow, not normally blueing when bruised; rare (reported from northern Idaho and the Southwest) 67
15.   Not with above features; common 16

16.   Cap viscid or slimy when moist, cinnamon-brown to chestnut- or reddish-brown; pores an intense, brilliant yellow or greenish-yellow, *not* blueing when bruised; stalk not reticulate or only slightly so (but sometimes ridged) .... ***B. flaviporus,*** p. 522
16.   Not as above; cap not viscid, or if viscid then not as above 17

17.   Stalk entirely or partially pink to red to dark red (or with red flakes or granules) 18
17.   Stalk white to buff, yellow, or brown (i.e., lacking obvious red tones, but may have reddish to cinnamon-brown stains, especially near or at base) 27

18.   Cap red to purple-red, pink, or at times reddish-brown (but sometimes overlaid at first with olive-yellow to olive-gray to olive-brown hairs or fibrils) 19
18.   Cap not markedly reddish (but may be dark brown or black with a slight reddish cast) ... 22

19.   Cap small (2-5 cm broad), red to dark red to purple-red; stalk less than 1 cm thick; pores and flesh typically *not* blueing when bruised; rare .... ***B. coccyginus*** (see *B. bicolor,* p. 521)
19.   Not as above 20

20.   Cap and stalk red or reddish; known only from the Southwest ... ***B. bicolor*** & others, p. 521
20.   Not as above; common in the Pacific Northwest and California, but not in the Southwest; stalk usually only partially red 21

21.   Associated with live oak in central and southern California .... ***B. dryophilus,*** p. 520
21.   Associated with northern conifers .... ***B. smithii*** (see *B. dryophilus,* p. 520)

22.   Taste distinctly bitter (chew on a small piece of the cap), even when cooked; stalk at least 2 cm thick at apex 23
22.   Not as above (but taste may be acidic or sour) 24

23.   Upper part of stalk typically reticulate (often finely so) .... ***B. calopus,*** p. 523
23.   Stalk not reticulate or only obscurely so at the very apex .... ***B. rubripes,*** p. 524

24.   Cap black to blackish-brown to deep olive-gray when fresh, often developing a slight reddish cast toward the margin (especially when wet) and sometimes fading in age; surface of cap *not* typically areolate, but occasionally becoming so in age .... ***B. zelleri,*** p. 518
24.   Cap dark olive-brown to olive-gray to brownish, tan, etc.; surface often areolate 25

*In regions where the eastern and western fungal floras overlap (e.g., southern Arizona), try both choices!

25. Fruiting body medium-sized to large (cap 6-18 cm broad); flesh in base of stalk *not* normally reddish; spores *not* truncate; found under mountain conifers .... ***B. fragrans***

25. Not as above; fruiting body small to medium-sized, or if larger than more common along the coast and flesh in base of stalk often reddish; spores truncate or not truncate; common   . 26

26. Cap when areolate usually showing pink or reddish tints in the cracks; stalk typically rather slender (less than 2 cm thick); cap medium-sized (usually less than 9 cm broad); spores *not* truncate .... ***B. chrysenteron***, p. 519

26. Not as above; cap when areolate *may or may not* show pink or reddish tints; size and shape of fruiting body variable, but sometimes more robust than above species (stalk 1-3 cm thick, cap 5-15 cm broad); spores often amyloid or truncate (appearing chopped off at one end) when viewed under the microscope .... ***B. truncatus*** & others (see *B. chrysenteron*, p. 519)

27. Fruiting body medium-sized to large; stalk 2 cm thick or more, often bulbous or thicker below (but often not); pores small (mostly less than 1 mm broad) 28

27. Fruiting body small to medium-sized; stalk usually 2 cm thick or less at apex, or if thicker then pores at least 1 mm broad; stalk usually equal or tapered below (not bulbous) 38

28. Pore surface whitish when young, becoming yellowish, greenish, or brown in age and *not* blueing when bruised; stalk distinctly reticulate (netted) at least over upper portion; flesh white or tinged reddish, *not* blueing (or sometimes blueing slightly when tubes are peeled off) ... 29

28. Not as above; if stalk reticulate then pores *not* white when young *and/or* pores blueing when bruised 31

29. Cap white to pale grayish or dingy buff, usually dry; associated mainly with oak in central California, with pine and other conifers in the Southwest .... ***B. barrowsii***, p. 529

29. Not as above; cap typically darker (but may be whitish while still under the duff) 30

30. Cap dark brown to blackish-brown beneath a thin whitish bloom when young (but often fading to cinnamon-brown or paler in age); associated with hardwoods .... ***B. aereus***, p. 531

30. Not as above; cap biscuit-brown to yellow-brown, cinnamon-brown, brown, or dark red; associated mainly with conifers but also with oak and other trees . ***B. edulis*** & others, p. 530

31. Taste distinctly bitter (chew on a small piece of the cap), even when cooked; stalk typically at least 2 cm thick at apex 32

31. Not as above (but may taste slightly sour or acidic) 33

32. Found under northern conifers; stalk usually reticulate ***B. coniferarum*** (see *B. calopus*, p. 523)

32. Found mostly with oak in California; stalk not reticulate .... ***B. "marshii,"*** p. 524

33. Stalk distinctly reticulate, at least over upper portion 34

33. Stalk typically *not* reticulate; found under mountain conifers .... ***B. fragrans***

34. Stalk yellow to buff (at least when fresh), sometimes with reddish stains below 35

34. Stalk brown, or white with brown fibrils (apex sometimes yellow) ..... ***B. fibrillosus***, p. 523

35. Cap dark brown when fresh; pores and flesh *not* blueing when bruised ....
.... ***B. sp.*** **(unidentified)** (see *B. fibrillosus*, p. 523)

35. Not as above; pores normally turning blue when bruised (but may not blue in button stage); flesh in very base of stalk often tinged pinkish or vinaceous; cap pink, red, brown, tan, or yellowish, but not normally dark brown 36

36. Cap with distinct fibrillose scales, especially in age; associated with mountain conifers (particularly fir) .... ***B. abieticola*** (see *B. appendiculatus*, p. 525)

36. Associated with hardwoods, or if with conifers then cap lacking obvious scales 37

37. Cap usually brown to reddish-brown, sometimes yellowish .... ***B. appendiculatus***, p. 525

37. Cap reddish to pink or rose-colored, sometimes with yellow .... ***B. regius***, p. 526

38. Flesh, stalk, and pores staining dark blue almost instantly when exposed; cap dull brown to dark brown; stalk usually yellow at apex and brown or reddish-brown at base; pores yellow to olive; rare in California, more common in the Pacific Northwest .... ***B. pulverulentus***

38. Not as above 39

39. Spore print amber-brown to bright yellow-brown; cap vinaceous-brown to dark brown, yellow-brown, tan, etc., often with paler spots; pores whitish to buff or pale tan; reported from the Southwest (but mainly eastern in distribution) .... ***B. affinis***

39. Not as above; spore print typically brown to olive or olive-brown 40

40. Cap brown to dark brown or blackish; stalk pallid to buff (or yellowish at apex); pores brilliant yellow, *not* blueing when bruised .... *B. citriniporus* (see *B. fibrillosus*, p. 523)
40. Not as above; pores typically yellow but not as brilliantly colored as above 41

41. Cap usually extensively fissured (areolate) by maturity .... *B. chrysenteron* & others, p. 519
41. Cap not usually areolate, but sometimes showing a few cracks in age, especially near margin .... *B. subtomentosus* & others, p. 517

42. Pore surface dark red to bright red, orange, orange-brown, red-brown, or dark yellow-brown *when fresh* (may fade to yellow in age) 43
42. Pore surface white to yellow, olive, grayish, etc. when fresh (but sometimes brown in age) 49

43. Pore surface reddish-brown to dark yellow-brown or orange-brown when young, or if not then fruiting body rather small and not blueing when bruised 73
43. Pore surface dark red to red, brick-red, or orange; medium-sized to large, usually blueing 44

44. Cap whitish to buff, grayish-olive, or sometimes tinged pink; stalk neither bulbous nor markedly reticulate; found under southern hardwoods .... *B. piedmontensis*
44. Not as above; cap typically darker or brighter than above 45

45. Stalk typically reticulate, at least over upper portion 46
45. Stalk not reticulate or sometimes slightly so at apex, but often scurfy from numerous minute granules or dots 48

46. Cap yellow to olive-brown, brown, or reddish-brown   *B. luridus* (see *B. pulcherrimus*, p. 528)
46. Cap deep red to bright red to rosy-red, at least when fresh 47

47. Cap viscid when moist; pores dark red, sometimes with yellow droplets; stalk very coarsely reticulate nearly to the base .... *B. frostii*, p. 528
47. Not as above; cap not viscid or only slightly so; stalk only finely reticulate .... *B. flammans* & *B. rubroflammeus* (see *B. frostii*, p. 528)

48. Cap red or rosy-red .... *B. bicolor* var. *borealis* (see *B. bicolor*, p. 521)
48. Cap yellowish to brown or reddish-brown .... *B. erythropus, B. subvelutipes,* & others (see *B. erythropus*, p. 526)

49. Cap white or pallid, at least until old age 50
49. Cap more highly or deeply colored, at least when young 51

50. Stalk white or whitish, usually not reticulate (but sometimes slightly so); very common under hardwoods; taste mild to bitter .... *B. pallidus*
50. Stalk at least partially red or with bright red to pink reticulation; taste bitter .... *B. inedulis*

51. Pore surface *and/or* flesh typically staining blue, blue-green, or blue-black when bruised or cut (but sometimes slowly or weakly) 52
51. Neither the pore surface nor the flesh blueing when bruised 60

52. Fresh fruiting body bright yellow to lemon-yellow (but cap may be yellow-brown at center or streaked with red), quickly staining blue or blue-black when bruised or cut; stalk sometimes reticulate but usually not .... *B. pseudosulphureus*
52. Not as above 53

53. Cap bright red to rose, pink, or rusty-red when fresh (may fade in age) 54
53. Not as above (but cap may show slight red or pinkish tints) 56

54. Stalk slender (usually less than 1 cm thick) ..... *B. rubellus* & others (see *B. bicolor*, p. 521)
54. Stalk usually 0.8-3 cm or more thick 55

55. Stalk distinctly reticulate, at least over upper portion ..... *B. speciosus* (see *B. regius*, p. 526)
55. Stalk not reticulate or only slightly so from decurrent tube walls   . *B. bicolor* & others, p. 521

56. Taste decidedly bitter (chew on a small piece of the cap), even when cooked; stalk typically at least 2 cm thick; not common .... *B. calopus*, p. 523
56. Not as above (but taste may be slightly acidic or sour) 57

57. Flesh, stalk, and pores staining dark blue almost instantly when exposed; cap dull brown to dark brown; stalk usually yellow at apex and brown or reddish-brown toward base; pores fairly small, yellow to olive-yellow; stalk typically not reticulate .... *B. pulverulentus*
57. Not as above (but may stain blue to blue-green) 58

58. Cap typically areolate (extensively fissured or cracked) at maturity; stalk entirely or partially red to dark red (or with red fibrils or granules) 59
58. Not as above; cap not areolate or only occasionally so *and/or* stalk lacking red 66

59.   Stalk scurfy from small rough scales; spores striate under the microscope; often growing at
      the bases of trees .... (see *Boletellus* & *Austroboletus,* p. 508)
59.   Not as above; spores not striate; usually found on ground .. *B. chrysenteron* & others, p. 519

60.   Cap rosy or pink when fresh (but often fading to tan in age); stalk *not* reticulate; stalk base bright
      yellow inside and out; pores whitish when young .... (see *Tylopilus,* p. 532)
60.   Not as above 61

61.   Pore surface white to grayish when young (but usually yellowish, brownish, grayish, or greenish
      in age); stalk distinctly reticulate (netted), at least over upper portion; flesh white or tinged
      reddish; spore print olive to brown but *not* yellow-brown 62
61.   Not as above; pores differently colored and/ or stalk not reticulate or only faintly so at apex 64

62.   Cap grayish (or with darker fibrils) when fresh; pores white becoming gray to grayish-brown
      in age; common under hardwoods .... *B. griseus* (see *B. ornatipes,* p. 522)
62.   Not with above features; pores not typically grayish in age 63

63.   Cap tan to gray-brown, dark brown, or yellow-brown, dry, often areolate in age; cap and stalk
      without purplish tones; associated with hardwoods ...... *B. variipes* (see *B. aereus,* p. 531)
63.   Not with above combination of features (but may have some of them); found with hardwoods or
      conifers .... *B. edulis* & others, p. 530

64.   Cap and stalk bright orange-brown to golden-orange; fairly common in North Carolina and
      adjacent areas, but exact distribution uncertain .. *B. auriflammeus* (see *B. ornatipes,* p. 522)
64.   Not as above 65

65.   Cap and stalk viscid or slimy and yellow, or if not then pores brilliant yellow to *bright* greenish-
      yellow; not large 75
65.   Not as above (pores often yellow but not intensely so) 66

66.   Stalk typically long, relatively slender, and scurfy or minutely scaly as in *Leccinum;* cap yellow
      to ochre, reddish-orange, or cinnamon; pores yellow, not normally blueing when bruised 67
66.   Not with above features 68

67.   Dots or minute scales on stalk reddish; cap usually viscid (sometimes thinly so) when moist
      .... *B. longicurvipes* & *B. rubropunctus*
67.   Not as above; stalk yellow .... *B. subglabripes*

68.   Stalk largely yellow to buff when fresh (but may develop brownish to cinnamon stains from
      the base upward, especially in age) 69
68.   Stalk typically brownish to reddish, tan, etc., or white with brown tones, even when fresh   71

69.   Stalk prominently reticulate throughout or nearly throughout; pores small (about 2 per mm);
      taste mild or bitter; found under hardwoods .... *B. ornatipes* & others, p. 522
69.   Not as above; stalk not reticulate and/ or pores larger; taste usually mild 70

70.   Cap often with brownish to reddish hairs or small scales on a yellowish to orange-buff back-
      ground, usually viscid beneath hairs when wet (hairs may be absent in age) (see *Suillus,* p. 491)
70.   Not as above .... *B. subtomentosus* & others, p. 517

71.   Spore print yellow-brown or amber-brown; cap yellow-brown to reddish-brown or dark brown,
      often with paler spots; pores white to buff or pale tan or yellow .... *B. affinis*
71.   Not as above; spore print olive to brown or olive-brown; pores yellow to olive 72

72.   Stalk with prominent raised ridges and/ or coarsely reticulate, often long and slender in relation
      to cap; cap grayish or olive-tinged when young but usually reddish-brown to dark reddish-
      brown or bay-red in age; pores not blueing .... *B. projectellus* (see *B. mirabilis,* p. 521)
72.   Not as above; cap slightly viscid when young or wet, usually reddish-brown to bay-red to
      chestnut-brown, but at times yellow-brown or olive-tinged; pores may or may not stain blue
      when bruised; stalk not coarsely reticulate or strongly ridged *B. badius* (see *B. zelleri,* p. 518)

73.   Pores and/ or flesh not blueing when bruised; under hardwoods or conifers ... *B. rubinellus*
73.   Pores and/ or flesh staining blue to blue-black when cut; usually under hardwoods 74

74.   Pore surface dark reddish-brown to orange-brown when young; stalk typically *not* reticulate
      .... *B. vermiculosus* & others
74.   Pore surface dark yellow-brown when fresh; stalk often reticulate .... *B. fagicola*

75.   Fruiting body mostly bright yellow to lemon-yellow (including cap); cap and stalk viscid or slimy;
      stalk usually with a white cottony base; common in South .... *B. curtisii*
75.   Not as above; pores brilliant yellow when fresh; cap and stalk viscid or dry ....
      .... *B. viridiflavus* & *B. auriporus* (see *B. flaviporus,* p. 522)

Boletus piperatus is our smallest species of *Boletus*. It has a peppery taste and bright yellow mycelium at the base of the stalk.

Boletus piperatus   (Peppery Bolete)

CAP 2-8 cm broad (but usually less than 5 cm), convex to plane; surface slightly viscid to dry, usually smooth, yellow-brown to buff, rusty-cinnamon, orange-brown, or reddish-brown. Flesh thin, yellowish-buff to pinkish in cap, bright yellow in base of stalk; taste distinctly acrid (peppery). **PORES** rather large (1-2 mm in diameter), yellow-brown to cinnamon, reddish-brown, coppery or brick-red, darkening slightly when bruised but not blueing; tubes tawny to reddish-yellow. **STALK** 2-8 (12) cm long, 0.4-1 (2) cm thick, equal or tapered downward, often slender; smooth, solid, colored more or less like the cap; base coated with bright yellow mycelium. **SPORE PRINT** brown to dull cinnamon; spores 8-12 × 3-5 microns, spindle-shaped to elliptical, smooth.

HABITAT: Solitary to scattered or gregarious on ground in woods, associated mainly with conifers; widely distributed. In our area it fruits in the fall and winter with pine and Douglas-fir, but is not common (I find it only once or twice each year). Farther north, however, it is sometimes abundant.

EDIBILITY: Questionable. It's not as bitter as some of the peppery *Russula* and *Lactarius* species, and therefore could be useful as a spice. However, according to one source the peppery taste disappears when thoroughly cooked, and according to another it is mildly poisonous *unless* thoroughly cooked.

COMMENTS: Our smallest *Boletus,* this species is suggestive of a *Suillus* (and was once placed in that genus), but has neither glandular dots nor a veil. Some view it as a possible "missing link" between *Suillus* and *Boletus*. If you have any doubt as to its identity, just chew a small piece of the cap for a couple minutes! The bright yellow mycelium and yellow flesh at the base of the stalk are also characteristic. *B. piperatoides* is a similar widespread peppery species with blue-staining pores.

Boletus subtomentosus   (Boring Brown Bolete)

CAP 5-15 (20) cm broad, convex to plane; surface dry, minutely velvety (subtomentose), but may appear smooth in age; yellow-brown to drab olive-brown to dull brown, or occasionally yellowish, but in age or especially in wet weather often redder (cinnamon-

517

Boletus subtomentosus is a boring brown and yellow bolete with large yellow pores (see p. 488 for a close-up of the pores).

brown); sometimes with pallid, yellow, or reddish-tinged cracks, especially near margin. Flesh pallid to pale yellow, sometimes blueing slightly when exposed. **PORES** large (1-3 mm in diameter), dull yellow to bright yellow, blueing weakly or not at all when bruised. **STALK** 4-14 cm long, 1-2 (3) cm thick, equal or tapered either way, smooth or scurfy, the apex often coarsely reticulate from downward-extending tube walls; firm, yellow to buff, or often stained brown to dull cinnamon (especially below), but never red. **SPORE PRINT** olive-brown; spores 10-16 × 3.5-5 microns, elliptical to spindle-shaped, smooth.

HABITAT: Solitary to widely scattered or in small groups on ground in woods; widely distributed and very common, but not often occurring in large numbers. In our area it fruits throughout the mushroom season with a variety of tree hosts (oak, conifers, etc.).

EDIBILITY: Edible, but definitely not choice. In blind tastings of local boletes it has consistently placed near the bottom.

COMMENTS: This boring but ubiquitous bolete with the boring brown cap does not show red on the stem, in contrast to *B. zelleri* and *B. chrysenteron*. In addition, the cap is not often as conspicuously areolate as that of the latter species, and the pores are larger. *Xerocomus subtomentosus* is a synonym. *B. spadiceus* is an equally boring but ubiqitous bolete. It has the same general appearance and grows in the same places at about the same time. True, its cap may be more reddish-brown and its pores may blue more readily, but the only way to separate it with certainty is by applying a drop of ammonium hydroxide to the cap surface. A fleeting blue-green to blue-black reaction means *B. spadiceus;* otherwise, it's *B subtomentosus.* Since neither is worth eating, the distinction is academic. Inundating boring brown boletes with ammonia may be your idea of fun, but it's not mine!

Boletus zelleri (Zeller's Bolete) Color Plate 128

CAP 3-16 cm broad, convex to plane; surface dry, often wrinkled or uneven and with a frosted or finely powdered appearance when young; black to dark gray or dark olive-brown when fresh, often reddening somewhat in age or wet weather, especially toward the margin; surface cracking only slightly or not at all. Flesh thick, white to pale yellow, sometimes blueing erratically when exposed. **PORES** and tubes yellow to dark yellow or olive-yellow, often blueing when bruised (but often not blueing). **STALK** 4-12 cm long, 0.5-3(5) cm thick, equal or slightly thicker at either end; firm, yellow to tan with delicate red granules when young, usually dark red above or throughout in age. **SPORE PRINT** olive-brown; spores 12-16 × 4-5.5 microns, spindle-shaped to elliptical, smooth.

HABITAT: Solitary, scattered, or in groups on ground or rotten wood; very common along the Pacific Coast in woods of all kinds. In our area it fruits in the fall, winter, and early spring and is one of the few boletes with a tolerance for redwood (I have even found it growing on a redwood stump!). It is also abundant under oak and farther north, under alder.

EDIBILITY: Edible and highly rated by some sources. In my experience, however, it cooks up slimy and tasteless.

COMMENTS: The sensational combination of black, yellow, red, and often blue makes this our most colorful bolete. The color plate shows specimens whose caps have begun to develop a reddish tinge, but perfectly fresh caps can be coal-black or deep gray. *B. chrysenteron* and *B. truncatus* often have a more copiously cracked (areolate) cap that is usually paler or duller in color. *B. citriniporus* has a dark cap, but its stem is not red. In eastern North America, *B. zelleri* is supplanted by ***B. badius,*** a common edible species with a dark yellow-brown to bay-red cap. Like *B. zelleri,* it grows on rotten wood (as well as on the ground), usually under conifers.

Boletus chrysenteron   (Cracked-Cap Bolete)

CAP 3-11 cm broad, convex to plane; surface dry, minutely velvety when young, in age usually conspicuously cracked or fissured (areolate), especially toward the margin; color variable: dark olive-brown to dark grayish-olive, grayish-brown, or brown; often paler (tan to olive-buff) in age, with pink to reddish tints usually visible in at least some of the cracks (especially those toward the margin). Flesh fairly thick, whitish to yellow, usually blueing slowly when exposed. **PORES** rather large (about 1 mm in diameter), yellow to greenish-yellow or sometimes dingy brownish, usually (but not always) bruising blue or greenish (sometimes slowly!); tubes also yellow. **STALK** 4-13 cm long, 0.5-1.5 cm thick, equal or tapering downward, smooth or minutely scurfy and often longitudinally ridged or striate, firm; color variable, but typically a mixture of yellow and red (yellow

Boletus chrysenteron is one of several species that has a conspicuously cracked (areolate) cap at maturity. Others include *B. truncatus*, *B. porosporus,* and *Tylopilus amylosporus* (not illustrated).

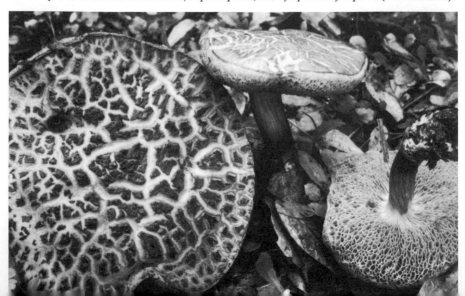

with reddish fibrils or yellow above and red to dark rhubarb-red below). **SPORE PRINT** olive-brown; spores 10-15 × 3.5-6 microns, spindle-shaped to elliptical, smooth.

HABITAT: Solitary or in small groups under trees and in wooded areas, often near trails or on roadbanks; widely distributed and ubiquitous, but rarely fruiting in large numbers. In our area it can be found throughout the mushroom season.

EDIBILITY: Edible, but not choice. It is rather mushy and insipid when cooked.

COMMENTS: The conspicuously fissured (areolate) cap with pinkish-tinted cracks plus the yellow pores that usually bruise blue are the hallmarks of this cosmopolitan bolete. Also known as *Xerocomus chrysenteron,* it is sometimes confused with *B. zelleri,* which has a darker cap that is not usually areolate. There are several very similar "cracked-cap" boletes that are difficult to distinguish in the field, including: *Tylopilus amylosporus,* with dark reddish-brown, erratically amyloid spores and pallid (not pink or red) flesh in the cracks; *B. mendocinensis,* with pinkish to reddish pores; *B. truncatus,* with truncate ("chopped off") spores; and *B. porosporus,* with truncate spores and no pinkish tints in the cracks. The latter two species are particularly widespread and common, and seem to intergrade. A robust form (cap 5-15 cm broad, stalk 1-3.5 cm thick) of *B. truncatus* bears special mention because it occurs commonly in our coastal forests. Its cap is often only slightly areolate and shows little or no pink in the cracks. Since none of the above species are worth eating, it hardly matters whether or not you identify them correctly, unless you aspire to be a professional boletologist (or plan on entertaining a professional boletologist). All of them, along with *B. zelleri* and *B. dryophilus,* are prone to attack by a powdery white to bright yellow parasitic fungus *(Hypomyces chrysospermum)* which eventually engulfs the entire fruiting body, making it look like a very sick puffball.

Boletus dryophilus   (Oak-Loving Bolete)   **Color Plate 132**

CAP 3-10 cm broad, convex to plane or somewhat irregular; surface dry, minutely velvety when young, often cracked or fissured in age, usually reddish to reddish-brown or pinkish, but sometimes overlaid with minute olive-brown to olive-gray hairs. Flesh thick, yellow, usually blueing when exposed. **PORES** often somewhat irregularly shaped, yellow to olive-yellow and usually blueing when bruised; tubes same color. **STALK** 3-8 (12) cm long, 1-2 (2.5) cm thick, often rather short and stout, usually pinched or narrowed at the base and often slightly swollen above it; sometimes nearly equal; firm, solid, usually distinctly yellow at apex and red to dark reddish below. **SPORE PRINT** brown to olive-brown; spores 12-16 × 5.5-8 microns, spindle-shaped to elliptical, smooth.

HABITAT: Solitary or in groups in humus under live oak, fall through early spring, known only from California. It is rather rare in our area, but sometimes abundant in southern California. In fact, it is one of the most common boletes in Los Angeles and San Diego counties.

EDIBILITY: Edible and fairly popular in the above-mentioned region. Collections I have sampled were mediocre, and in our area they are more readily parasitized by *Hypomyces chrysospermum* than any other bolete!

COMMENTS: This interesting endemic species is closely related to *B. chrysenteron,* but is somewhat stouter and apparently grows only with oak. The reddish tones usually present in the cap plus the distinct yellow and red zones on the stalk help identify it. The abruptly tapered or "pinched" stem base is also distinctive when present. Other species: *B. smithii* has a similarly colored cap, but is larger and stouter. Its stalk may be narrowed at the base but is usually pallid or yellow with a bright red band near or at the apex. It seems to grow only with conifers and is quite common in northern California and the Pacific Northwest.

Boletus bicolor   (Red and Yellow Bolete)

CAP 5-15 cm broad, convex to plane or somewhat irregular; surface dry, smooth or minutely velvety, sometimes cracking in age; deep red to pinkish-red or rose, the margin sometimes yellowish in age. Flesh thick, pale yellow, blueing erratically or not at all when exposed. PORES yellow to bright yellow, sometimes with reddish areas in old age; typically staining blue or blue-green slowly when bruised; tubes also yellow. STALK 5-10 cm long, 1-3 cm thick, equal or thicker below or with a tapered base; smooth, firm, solid, not reticulate or only slightly so at apex; deep red to rosy except for yellow apex. SPORE PRINT olive-brown; spores 8-12 × 3.5-5 microns, spindle-shaped to elliptical, smooth.

HABITAT: Scattered to gregarious on ground in woods, associated mainly with oak and perhaps aspen; common in the summer and early fall in eastern North America and also occurring in the Southwest, but apparently not found on the west coast. A related species, B. coccyginus (see comments), occurs in California and the Pacific Northwest, but is rare.

EDIBILITY: Edible, and according to some, excellent. However, gastrointestinal upsets have been attributed to closely related species, so exercise caution.

COMMENTS: The rosy to red cap and stalk contrast vividly with the yellow pores, making this a most beautiful bolete. There are a number of similar red-capped, yellow-pored, blue-staining species that form a close-knit group within Boletus. They are difficult to distinguish in the field and are largely confined to the humid forests of eastern and southern North America and the tropics. They include: B. sensibilis, somewhat larger, with a brick-red cap and yellow stalk; B. miniato-pallescens, cap rosy-red fading to yellow and stalk yellow to reddish-brown, common; and several smaller, slimmer (stalk less than 1 cm thick) species such as B. rubellus, widespread, B. campestris, especially common on shaded lawns, and B. fraternus, with an areolate cap at maturity. All of these species bruise blue (often quickly) and do not have the robust, reticulate stalk of B. regius. Other species: B. coccyginus is a rare, small, slender western species with a red to rose or purple-red cap and yellow pores and flesh that do not stain blue; B. bicolor var. borealis resembles the typical variety but has red to orange pores; it occurs mainly in the Southeast.

Boletus mirabilis   (Admirable Bolete)   Color Plates 127, 129

CAP 5-15 (20) cm broad, convex to plane; surface moist to dry, granulose to plushlike (roughened by numerous small, often erect, fibrillose scales); dark reddish-brown to maroon-brown, bay-brown, or chocolate-brown; margin often hung with fragments of tissue. Flesh thick, white to dingy pinkish or yellow, rarely blueing when bruised. PORES fairly large (1-2 mm in diameter), pale yellow becoming mustard-yellow and finally greenish-yellow, not blueing when bruised but sometimes staining darker yellow; tubes also yellow. STALK 7-20 cm long, 1-3.5 cm thick, usually club-shaped (thicker toward base); dark brown to maroon-brown or reddish-brown, sometimes with yellow, buff, or beige streaks; firm and often roughened, pitted, or longitudinally ridged, the apex often coarsely reticulate; base frequently with yellow mycelium. SPORE PRINT olive-brown; spores 18-24 × 7-9 microns, spindle-shaped to elliptical, smooth.

HABITAT: Solitary or in small groups on or near rotting conifers (especially hemlock), but sometimes appearing terrestrial; common in the fall in the Pacific Northwest and northern California, also reported (rarely) from Michigan. It does not seem to venture south of San Francisco. (Neither does hemlock, and neither would I if I could help it.)

EDIBILITY: Edible and delicious—some collections have a distinct lemony flavor. Don't use specimens attacked by a whitish mold (Hypomyces chrysospermum).

COMMENTS: This beautiful northern bolete is practically unmistakable. The plush-like maroon-brown cap and long roughened, similarly-colored stalk plus the yellow pores that do *not* stain blue and growth on or near rotting conifers form a unique set of characters. Other species: **B. projectellus** of eastern North America is a closely related, terrestrial species that favors sandy situations under pine.

Boletus flaviporus   (Viscid Boletus)   **Color Plate 131**

CAP 4-15 cm broad, convex to plane; surface viscid or slimy when wet, smooth or fibrillose, reddish-brown to cinnamon-brown or chestnut-brown, not normally cracking into scales. Flesh thick, white to pale pinkish, not blueing when bruised. **PORES** and tubes brilliantly, intensely yellow, becoming slightly greenish-yellow in age; *not* bruising blue. **STALK** 6-15 cm long, 0.6-2 (3) cm thick, equal to slightly thicker below or often with a short, tapered, rooting base; smooth, viscid in wet weather but otherwise dry; color variable, but usually yellow at apex and pallid to reddish-brown or dark brown below; typically not reticulate or only slightly so at apex. **SPORE PRINT** dark olive-brown; spores 11-17 × 4-6 microns, spindle-shaped to elliptical, smooth.

HABITAT: Solitary, scattered, or in small groups in mixed woods and under hardwoods; endemic to the west coast. In our area it is fairly common in the fall and winter under oak, madrone, and manzanita, but seldom fruits in large numbers.

EDIBILITY: Edible, but mediocre.

COMMENTS: This is our only *Boletus* with a distinctly viscid cap and intensely yellow pores that do *not* stain blue. It might be mistaken for a *Suillus,* but it isn't associated with conifers and has neither glandular dots nor a veil. The stem is not truly reticulate like that of *B. edulis* and *B. fibrillosus,* but the tubes may form a slight raised network at the stalk apex. Its eastern counterparts with brilliant yellow, non-blueing pores are **B. viridiflavus**, with a reddish- or pinkish-tinged stalk and sometimes viscid cap; and **B. auriporus** (=**B. caespitosus**?), with a yellower stalk and somewhat more velvety cap.

Boletus ornatipes   (Ornate-Stalked Bolete)

CAP 4-20 cm broad, convex to nearly plane; surface dull, often velvety when young, dry or slightly viscid when wet, color variable: gray, purple-gray, olive, olive-brown, yellow, or mixtures thereof. Flesh firm, thick, yellow, not blueing; taste mild to somewhat bitter. **PORES** and tubes lemon-yellow to bright yellow, usually staining orange-yellow to rusty-brown when bruised, but not blueing; pores small. **STALK** 7-15 cm long, 1-3.5 cm thick, more or less equal but often curved, solid, firm, bright yellow to yellow-orange throughout and prominently reticulate nearly to base (reticulation often coarse); dingier in age. **SPORE PRINT** brown to olive-brown; spores 9-13 × 3-4 microns, spindle-shaped to elliptical, smooth.

HABITAT: Solitary, scattered, gregarious or in small clusters on ground under oaks and other hardwoods in eastern North America (particularly the northern half); it is common in the summer and early fall.

EDIBILITY: Non-bitter specimens are said to be edible; I haven't tried it.

COMMENTS: This beautiful eastern bolete is readily told by its yellow pores and flesh and its beautifully patterned (reticulate) golden-yellow stalk. The cap color varies considerably but is usually some shade of gray, olive, or yellow-olive. **B. retipes** is a bitter-tasting southern version (probably a form of the same species); **B. auriflammeus**, also southern, is similar but has a brownish-orange to brilliant gold cap and creamy or pink-tinged flesh. All of these stain your fingers yellow when handled. Another closely related species, **B. griseus,** is grayish but often shows some yellow (at least at the base of the stalk) as it ages. It has softer, mild-tasting, pallid, edible (but often wormy!) flesh, white to gray or grayish-brown (never yellow!) pores that may stain brown, and a reticulate stalk. It is common under hardwoods in eastern North America and also occurs in southern Arizona.

Boletus fibrillosus

CAP 6-17 cm broad, convex to more or less plane; surface dry, minutely velvety becoming fibrillose or sometimes fibrillose-scaly at the center; brown to dark brown to cinnamon-brown, sometimes with paler blotches or paler at the margin (but not yellow). Flesh thick, white or buff, not blueing when bruised; taste mild. **PORES** and tubes pale yellow to yellow or dingy olive-yellow (occasionally pallid when very young), *not* bruising blue. **STALK** 8-16 cm long, 2-4 cm thick at apex, equal or thicker below (but sometimes pinched at base); solid, firm; light brown to brown (often paler than cap); apex often yellow and/or base whitish; usually lined or fibrillose, reticulate at least at the top. **SPORE PRINT** dark olive-brown; spores 13-17.5 × 3.5-5.5 microns, spindle-shaped to elliptical, smooth.

HABITAT: Solitary to gregarious on ground in mixed woods and under conifers; known only from the west coast, sometimes common in the fall in coastal northern California and the Cascades. A similar species (see below) occurs in our area.

EDIBILITY: Edible and fairly good, but certainly not the equal of *B. edulis* or *B. aereus*.

COMMENTS: The dark brown frequently fibrillose cap, yellow pores, brown reticulate stalk, and "failure" of all parts to turn blue when bruised are the telltale traits of this northern bolete (see photo on p. 529). It resembles *B. edulis* and *B. aereus,* but its cap is more fibrillose and its pores are often yellow even when young. *Tylopilus pseudoscaber* is also somewhat similar, but does not have yellow pores. Both *T. pseudoscaber* and *B. fibrillosus* have passed under the defunct name *B. olivaceobrunneus.* A similar, unidentified *Boletus* occurs rarely in our area under live oak, but differs in having a yellow stalk. *B. citriniporus* is a smaller Californian species with a brown to blackish cap, *brilliant* yellow pores that do not bruise blue, a whitish to buff (or yellow at apex) stalk that is not reticulate or only slightly so, and mild taste. I have found it in mixed woods and under oak.

Boletus calopus   (Bitter Bolete)

CAP 10-30 cm broad, convex to plane or somewhat irregular; surface dry, dull, smooth to minutely hairy or fibrillose, often cracking (areolate) in age; dull brown to olive-brown, grayish-brown, or olive-buff, sometimes yellow-brown or in age darker brown. Flesh very thick, pale yellow or whitish, quickly blueing when exposed; taste distinctly and persistently bitter. **PORES** and tubes pale yellow becoming darker or dingier yellow in age; typically bruising blue or blue-green quickly. **STALK** 6-15 (20) cm long, 3-7 cm thick at apex, equal or bulbous; solid, firm; yellow, but usually with pink to red zones or discolorations also present, or sometimes entirely reddish; blueing when bruised; apex or upper half finely reticulate. **SPORE PRINT** dark olive-brown; spores 13-19 × 4-6 microns, spindle-shaped to elliptical, smooth.

HABITAT: Solitary, scattered, or in groups on ground in mixed woods and under conifers in the late summer and fall (or occasionally spring); common in western North America, rare in the East. It is particularly prominent at higher elevations (e.g., in the Sierra Nevada), but I have also seen large fruitings on the Olympic Peninsula in Washington. In our area it is apparently supplanted by another bitter bolete, *B. "marshii."*

EDIBILITY: Bitter-tasting (see comments on the edibility of *B. "marshii"*).

COMMENTS: This sometimes massive bolete is easily recognized by its bitter taste, blueing of all parts when injured, yellow pores, and reticulate stalk. The bitter taste always comes as a disappointment to those seduced by its large size. Our other large "bitter boletes," *B. "marshii"* and *B. rubripes,* have non-reticulate stalks. Still another bitter bolete, *B. coniferarum,* has a yellow reticulate stalk (without any red or pink), a darker (olive-gray to deep brown) cap, and a northern distribution. I have seen it under conifers in northern Idaho and British Columbia.

Boletus rubripes   (Red-Stemmed Bitter Bolete)   **Color Plate 130**

CAP 6-15 (25) cm broad, convex to broadly convex or sometimes plane; surface dry, dull, smooth or velvety, often becoming cracked or furrowed in age; pale buff to olive-buff to tan. Flesh thick, firm, buff to pale yellow or even whitish, but blueing when exposed; taste usually bitter. **PORES** and tubes pale yellow becoming darker or duller yellow in age; blueing when bruised. **STALK** 7-13 (20) cm long, 1-4 cm thick, sometimes swollen below (especially when young), but often more or less equal in age; firm, solid, *not* reticulate but often longitudinally striate in age; color variable, but usually yellow at apex with bright pink to reddish areas below, becoming dark red throughout in age. **SPORE PRINT** olive-brown; spores 12.5-17.5 × 4-5 microns, spindle-shaped to elliptical, smooth.

HABITAT: Solitary to scattered or in groups, associated primarily with conifers; known only from western North America and Mexico. I have not seen it in our area, but it is fairly common in northern California under Sitka spruce and also occurs in the Sierra Nevada. It fruits in the summer and fall. The color photograph was taken in New Mexico.

EDIBILITY: Bitter-tasting (see comments on the edibility of *B. "marshii"*).

COMMENTS: This beautiful but bitter bolete can be told from *B. calopus* by its non-reticulate stalk (use a hand lens if unsure!), and from *B. "marshii"* by the red to pink stalk which beomes dark red or dull purple-red in age. Several *Tylopilus* species are also bitter-tasting, but do not have yellow pores.

Boletus "marshii"   (Ben's Bitter Bolete)

CAP 5-25 cm broad, convex to nearly plane or somewhat irregular; surface dry, smooth or cracking in age, whitish to gray, pale grayish-brown or buff when young, becoming dull brownish and developing darker brown stains in age or upon handling. Flesh thick, dense, white or grayish, turning blue to bluish-gray erratically when exposed; taste bitter, at least latently. **PORES** and tubes pale yellow becoming dingy yellow or olive-yellow, bruising blue quickly, then eventually turning dingy brown. **STALK** 4-15 cm long, 2-5 cm thick at apex, usually enlarged below but sometimes tapered at the base, smooth, *not* reticulate; pallid to buff or yellow above (sometimes with a very slight reddish zone), dingy brown below, darker brown throughout in old age; turning blue when bruised. **SPORE PRINT** olive-brown; spores 11-14 × 4.5-6 microns, spindle-shaped to elliptical, smooth.

HABITAT: Solitary to gregarious on ground near or under live oak in the summer and early fall (*before* the onset of the fall rainy season); common in the vicinity of Santa Cruz, California, and also reported by several collectors from the foothills of the Sierra Nevada.

This bitter-tasting bolete is known locally as *Boletus "marshii"* (or the "Shucks Bolete," because that's what you say when you taste it!). Note pale cap, blue-staining pores, and non-reticulate stalk.

EDIBILITY: Sauteed delicately in butter with a pinch of pepper and a clove of garlic, served steaming hot on toast with cream cheese and celery, broiled belligerently on a skewer with spiced lamb and bell peppers, or layered lovingly in a casserole with parmesan cheese, egg noodles, and onions, *Boletus "marshii"* is still inedible.

COMMENTS: This bulky bolete with the bitter taste appears to be unnamed. However, it is known to local yokels as *B. "marshii"* because it bears an uncanny resemblance to its discoverer, Ben Marsh, who is also dense, bulky, bitter, and bulbous, and who spent many fruitless hours (and ruined many otherwise marvelous meals) in a highly commendable if ill-conceived attempt to make it palatable. It is easily distinguished from *B. calopus* by the non-reticulate stalk, from *B. appendiculatus* by the paler cap, non-reticulate stalk, and bitter taste, from *B. rubripes* by the paler cap and absence of red on the stalk, and from *B. barrowsii* by the bitter taste, blue-staining pores that are yellow when young, and non-reticulate stalk. Its habit of fruiting *before* the rains arrive is odd, but quite reliable. A European species, **B. albidus (=B. radicans)**, is similar in color and taste, but is usually described as having a reticulate stalk with a rooting base. Although *B. "marshii"* never seems to have a reticulate stalk, it often has a rooting base, and a critical comparison may reveal that the two species are one and the same.

Boletus appendiculatus   (Butter Bolete)   Color Plate 134

CAP 6-20 (30) cm broad, convex becoming broadly convex or even plane; surface dry or slightly viscid, smooth or with a fine bloom when young, sometimes cracked in age; brown to cinnamon-brown or yellow-brown, or sometimes buff to yellowish, often with reddish stains or blushes. Flesh thick, very firm and dense, pale yellow to yellow, changing slowly and erratically to blue when cut, or not at all; taste mild. **PORES** and tubes lemon-yellow to bright butter-yellow or in old age olive-yellow; usually blueing when bruised, but sometimes not (especially in button stage). **STALK** 5-15 cm long, 2-6 cm thick at apex, usually bulbous or thicker below, but sometimes equal or tapered at base; solid, very firm; uniformly yellow or butter-yellow, sometimes with brownish or reddish stains; upper portion finely reticulate; flesh in base usually pinkish or vinaceous. **SPORE PRINT** dark olive-brown; spores 12-15 × 3.5-5 microns, spindle-shaped to elliptical, smooth.

HABITAT: Solitary to scattered or gregarious on ground under hardwoods, especially live oak and tanoak; fruiting shortly after the first fall rains and apparently more common in our area than anywhere else in North America. It is abundant some years, practically absent others; it often grows with *B. satanas, B. regius,* and/or *B. aereus.*

EDIBILITY: Edible and popular (although some people are "allergic" to it). It lacks the nuttiness of *B. edulis* but is remarkable for its firmness—a joy to find as well as to eat, something you can really sink your teeth into. When cooked it will often turn blue, then gray, then back to yellow,

COMMENTS: The butter bolete is easily recognized by its heaviness (it is the densest of our boletes—the young buttons are sometimes as hard as rocks), mild (never bitter!) taste, yellow pores that *usually* bruise blue quickly, and finely reticulate, butter-yellow stalk. Don't be misled by the variation in cap color—it is usually brown, but ranges from yellowish to somewhat reddish or rusty-brown. Sometimes it seems to intergrade with *B. regius,* which is also edible. The reticulation is usually the same color as the stalk (unless stained brown by spores), and thus can be difficult to see (a hand lens will help). Specimens of the poisonous *B. satanas* which have lost their red pore color are easily distinguished because they do not have a yellow stalk. *B. "marshii"* differs in its paler cap, white flesh, non-reticulate stalk, and bitter taste. Other species: **B. abieticola**, sometimes common under fir in the Siskiyou Mountains of northern California and southern Oregon, is quite similar, but has a more fibrillose, tan to rose-colored cap.

Boletus regius   (Red-Capped Butter Bolete)   **Color Plate 133**

CAP 6-20 cm broad or more, convex to plane or somewhat irregular; surface smooth to uneven or pitted, sometimes minutely hairy when young, dry or slightly viscid; pink to rosy-red or dark red, sometimes also with brownish tones or yellow areas. Flesh very thick, firm, yellow, slowly and erratically or irregularly blueing when exposed; taste mild. **PORES** and tubes bright yellow becoming darker yellow or olive-yellow in old age; usually blueing when bruised. **STALK** 4-14 cm long, 2.5-6 cm thick at apex, usually bulbous or thicker below, often rather short in relation to cap; solid, firm, pale to bright yellow but usually bruising blue and often reddish-stained below; finely reticulate over at least the upper half; flesh in base often pinkish or vinaceous. **SPORE PRINT** olive-brown; spores 12.5-16.5 × 3.5-5 microns, spindle-shaped to elliptical, smooth.

HABITAT: Solitary to gregarious on ground in woods; known from California, Oregon, and Washington. In our area it fruits in the fall under live oak, often with *B. barrowsii, B. satanas,* and/or *B. appendiculatus.* Like those species it does not seem to fruit every year. In the Sierra Nevada and Cascades it grows under conifers in the spring, summer, and fall.

EDIBILITY: Edible and very similar to *B. appendiculatus* in flavor, but not as firm and therefore not as good; it is often riddled with maggots.

COMMENTS: This sometimes massive bolete can be told by its blue-staining yellow pores, yellow flesh, pink to reddish cap, and large, thick, reticulate yellow stem. It closely mimics *B. appendiculatus* except for cap color, but since both are edible it hardly matters if you confuse them. Small specimens of *B. regius* are shown on p. 512. *B. speciosus* is a similar eastern species with a longer, slimmer stalk and a tendency to stain blue more readily.

Boletus erythropus   **Color Plate 135**

CAP 5-15 (20) cm broad, convex becoming nearly plane to slightly depressed or pitted in age; surface dry, minutely velvety when young, yellow-brown to cinnamon-brown, dark brown, or reddish-brown, quickly staining bluish-black when bruised. Flesh thick, firm, yellow or yellowish but blueing *very* quickly when exposed; often reddish-brown or reddish in base of stalk. **PORES** red to brick-red to orange-red, orange, rusty-orange, or burnt sienna (but sometimes yellowish in old age), blueing quickly when bruised; tubes yellow becoming greenish-yellow in age. **STALK** 4-15 cm long, (1.5) 2-4 (5) cm thick, equal or thicker at either end but not bulbous; yellowish or more often masked by a coating of minute red to orange dots or granules; blueing quickly when bruised and often dingier (darker or browner) in age; solid, firm, often curved; *not* reticulate or only very slightly so at apex from decurrent tube walls. **SPORE PRINT** olive-brown to ochraceous-brown; spores 12-16 × 4-6 microns, elliptical to spindle-shaped, smooth.

HABITAT: Solitary, scattered, or gregarious on ground in woods; widely distributed. It is fairly common under conifers (especially spruce) from northern California to Alaska, and under both hardwoods and conifers in eastern North America; also reported from the Rocky Mountains. In the oak and madrone woodlands of California it is replaced by *B. amygdalinus* (see comments).

EDIBILITY: Poisonous to some people—to be avoided. Two books which list *B. satanas* as poisonous state flatly that *B. erythropus* is edible. However, Bill Everson (an intrepid Californian toadstool-tester) was unaffected by cooked *B. satanas* but ate a small portion of sauteed *B. erythropus* and vomited soon after—an explicit example of why you should be cautious when trying any mushroom for the first time, even a so-called "edible" one!

COMMENTS: The red to orange pores, non-reticulate and non-bulbous stalk, and brown to dark brown cap typify this beautiful bolete. It certainly has substance, but is puny in comparison to the bulky hulk (hulking bulk?) of *B. satanas.* It bruises blue so rapidly that

Boletus erythropus, a poisonous species with orange or reddish pores that quickly stain blue. Flesh also stains quickly when cut (see color plate). Note that the stalk is not bulbous as in *B. satanas*.

a few minutes frenzied handling will render it unrecognizable. A very similar species that is apparently unique to California, *B. amygdalinus*, has oranger pores and grows under hardwoods. Another look-alike, *B. subvelutipes* (COLOR PLATE 136), is quite common in the Southwest as well as eastern North America. It can have a somewhat yellower cap and often has dark red hairs at the base of the stalk. *B. hypocarycinus* is a very similar southeastern species with a dark red to carmine stalk base and shorter spores. *B. orovillus* is a rare but striking bolete with a large brilliant yellow cap, bright red pores that may or may not bruise blue, and a thick, non-reticulate stalk. It has been taken in a few widely scattered localities in northern California and the Pacific Northwest. None of these should be eaten.

Boletus satanas   (Satan's Bolete)   Color Plate 137

CAP 7-30 cm broad, convex to broadly convex or nearly plane in age; surface dry, smooth or sometimes cracked; pallid to gray or olive-buff, becoming suffused with pink in age, especially toward margin (but yellowish where slug-eaten). Flesh white to yellow, blueing when bruised, especially when young or near the tubes. **PORES** deep red becoming red to pink or in old age orange to yellowish, turning blue to blue-black when bruised; tubes yellow to greenish. **STALK** 6-15 cm long, 2-6 cm thick at apex, with a massive abrupt bulb at base which is up to 15 cm broad and is especially prominent when young; solid, firm; upper portion colored more or less like the cap and finely but distinctly reticulate (reticulation usually reddish); bulb usually pink or pinkish-red, but the pink tones usually more evident in youth and tending to disappear in age. **SPORE PRINT** brown to olive-brown; spores 11-15 × 4-6 microns, spindle-shaped to elliptical, smooth.

HABITAT: Solitary or in groups under oaks (often at the edges of pastures); known only from California and Europe. In our area it is sometimes abundant in the fall or early winter. It is usually one of the very first boletes to appear.

EDIBILITY: Poisonous, at least raw! It causes vomiting, diarrhea, and severe cramps. Thorough cooking reputedly destroys the toxins, and some people eat it regularly. Its voluminous avoirdupois is certainly inviting—but when so many more delectable and less dangerous mushrooms abound, why tempt fate?

COMMENTS: The red pores, pallid to olive-buff cap, and bulbous reticulate stem are diagnostic. Young specimens can be told by their obesity alone—the bulb can be several inches broad, often larger than the cap and just as round! The red color of the pores

527

sometimes fades in old age, but the pronounced bulb and blue-staining tubes signify *B. satanas.* I found one specimen at the edge of a meadow that weighed six pounds, and half of it had been eaten by a cow! *B. pulcherrimus* is the only other red-pored, reticulate-stalked bolete in coastal California, but it is not so bulbous and has a brown to reddish-brown cap.

Boletus pulcherrimus   (Red-Pored Bolete)

CAP 8-25 cm broad, convex to broadly convex or nearly plane; surface dry, smooth or minutely velvety when young, sometimes breaking up into small scales in age; brown to pale olive-brown or reddish-brown (often redder at margin). Flesh thick, yellow, blueing when exposed. **PORES** deep red to bright scarlet-red, often duller (reddish-brown) or paler in old age; quickly staining blue or blue-black when bruised; tubes yellow. **STALK** 7-20 cm long, 2-5 cm thick at apex and up to 10 cm thick at base; sometimes equal, more often club-shaped (thicker below), but not abruptly bulbous; solid, firm, yellowish to pale reddish-brown with dark red reticulation over at least the upper half. **SPORE PRINT** olive-brown; spores 13-16 × 5.5-6.5 microns, spindle-shaped to elliptical, smooth.

HABITAT: Solitary to gregarious in mixed woods and under conifers in the summer and fall; endemic to western North America, rather rare but sometimes fruiting in quantity. I've seen it in northern California and New Mexico, but have yet to find it in our area.

EDIBILITY: Poisonous! Like *B. satanas,* it is especially dangerous raw.

COMMENTS: Listed in older books as *B. eastwoodiae,* this large and beautiful bolete is easily told by its red to dark red pores, reticulate stalk, and brown to reddish-brown cap. It is the principal red-pored bolete of the Pacific Northwest, but does not fruit every year. In our area it is replaced by *B. satanas,* which is differently colored and much more obese. A very similar species, *B. haematinus* **(COLOR PLATE 138)** is common under western mountain conifers. It has a yellow-brown to olive-brown or brown cap, sometimes has a yellower stalk, and often has yellow pores when very young (the pores at the cap margin of older specimens can also be yellow). A similar but unidentified species with a pallid cap has been found by Greg Wright under conifers in southern California. Another reticulate-stalked species, *B. luridus,* grows under hardwoods and conifers in eastern North America and the Southwest. It has a slimmer (1-3 cm) stalk, red to orange pores, and a yellowish to olive-brown, brown, or reddish-brown cap. All of the above species are poisonous and all lack the exaggerated bulb of *B. satanas.* For species with bright red caps, see *B. frostii.*

Boletus frostii   (Apple Bolete; Frost's Bolete)   **Color Plate 139**

CAP 5-15 cm broad, convex to broadly convex, often becoming plane or with an uplifted margin in age; surface smooth, viscid when moist and often shiny, dark red (or even blackish-red) when young, becoming blood-red or apple-red in age, and often developing yellowish (faded) areas when old. Flesh thick, pallid to yellow, quickly blueing when cut. **PORES** small, dark red when fresh and often beaded with yellow droplets when young, often fading to paler red in age; typically staining dark or dingy bluish when bruised; tubes yellow to olive. **STALK** 4-12 cm long, 1-3.5 cm thick, equal or thicker below, dry, solid, coarsely and deeply reticulate throughout; dark red to red, sometimes with yellowish areas; reticulation same color or paler; base red to yellow or whitish. **SPORE PRINT** brown to olive-brown; spores 11-17 × 4-6 microns, spindle-shaped to elliptical, smooth.

HABITAT: Solitary, scattered, or in groups under hardwoods; common during the summer throughout eastern North America, also in Texas, southern Arizona, Mexico, and Costa Rica. It favors oak, but is said to be frequent under madrone in Mexico.

EDIBILITY: Edible. In Mexico it is often sold in farmer's markets, but since it has red pores, be careful! Cook it thoroughly and be *sure* of your identification!

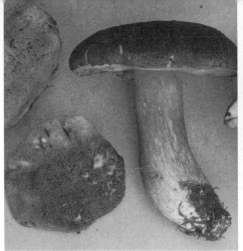

Left: *Boletus fibrillosus* (see p. 523) can be confused with *B. edulis* and *B. aereus,* but has a hairier (fibrillose) cap in age and usually has yellow pores. **Right:** *Boletus barrowsii,* oak-loving form (also shown on p. 512). Note whitish cap. See color plate for pine-loving variety from the Southwest.

COMMENTS: One of the most beautiful and memorable of all boletes, this species is easily told by its viscid red to dark red cap, dark red pores, and coarsely reticulate stalk. The stalk may be thickened downward, but is never bulbous as in *B. satanas.* The tendency of the young pores to exude yellow droplets is also distinctive. In Mexico it is sometimes called *panza agria,* which means "sour belly." Other species: ***B. flammans*** and ***B. rubro-flammeus,*** both poisonous, have a non-viscid red to dark red cap and finely to scarcely reticulate stalk and red pores (at least when young). The first favors eastern conifers, the second grows with hardwoods in eastern North America and southern Arizona.

Boletus barrowsii   (White King Bolete)   Color Plate 141

CAP 6-25 cm broad, convex becoming broadly convex to plane; surface smooth or very slightly velvety, dry, dull white to buff or grayish. Flesh thick, white, not blueing when exposed (or blueing only very slightly near the tubes). **PORES** and tubes white or pallid when young (and appearing stuffed with a pith), becoming yellow to olive-yellow in age; *not* blueing when bruised. **STALK** 6-20 cm long, 2-6 cm thick at apex, equal or thicker below (especially when young), solid, firm, whitish to buff or colored like cap, sometimes with brownish stains; finely but distinctly reticulate at least over the upper portion. **SPORE PRINT** dark olive-brown; spores 13-15 × 4-5 microns, spindle-shaped to elliptical, smooth.

HABITAT: Solitary or scattered to gregarious under both hardwoods and conifers, fruiting mainly in the summer and fall; apparently endemic to the drier parts of western North America and Mexico. It is a warm weather species and more abundant in Arizona and New Mexico than anywhere else. There it is associated almost exclusively with ponderosa pine, while *B. edulis* favors spruce. In Colorado, however, I have found it under spruce and fir, and in coastal California it is usually associated—quite curiously—with live oak, while *B. edulis* grows with pine. In our area it fruits in the fall and winter, but is abundant only after warm early (September-November) rains.

EDIBILITY: Delectably delicious. It is a favorite of collectors in the Southwest, and in the completely objective opinion of its "discoverer," Chuck Barrows, it is the best of the boletes for the table. Unfortunately, the maggots share his opinion, for they attack it even more voraciously than they do *B. edulis!*

COMMENTS: For many years this handsome, meaty bolete has passed as a "white" *B. edulis,* and is listed as such in the first edition of *Mushrooms Demystified.* However, it does not intergrade with *B. edulis* and lacks the clearly differentiated, often viscid cap cuticle (skin) of that species. Result: it is now recognized as a distinct species.

Boletus edulis   Color Plates 142, 143, 144
(King Bolete; Cep; Steinpilz; Porcini; etc.)

CAP 8-30 cm broad or more, convex to broadly convex or bun-shaped, or in age becoming plane; surface dry or more often viscid when moist, smooth to somewhat pitted or in dry weather sometimes cracking into plaques; color variable: biscuit-brown or warm brown to yellow-brown, cinnamon-brown, reddish-brown, or dark red (but often with whitish or pinkish tints while still under the duff); margin sometimes paler or yellower. Flesh thick, firm, white or sometimes tinged yellowish or dingy reddish in age; not blueing when bruised or only blueing slightly near the tubes; odor and taste pleasant. PORES at first white or pallid and appearing stuffed with a pith, then becoming yellow, olive-yellow, or brown in age, *not* blueing when bruised; tubes whitish becoming yellow, then olive-yellow. STALK (3) 8-25 cm long, 2-7 (12) cm thick (often large in relation to cap); usually enlarged below when young, often becoming more or less equal in age; firm, solid, entirely white, or whitish at the base and brownish above; finely reticulate over upper portion or throughout; flesh in base sometimes with dark yellow areas (a parasite?). SPORE PRINT more or less olive-brown; spores 13-19 × 4-7 microns, spindle-shaped to elliptical, smooth.

HABITAT: Solitary, scattered, or in groups or sometimes troups on ground in woods; found throughout the world and very common in western North America. It favors conifers (pine, spruce, hemlock, fir) but also grows with hardwoods such as oak and birch. It is often abundant in our coastal pine forests in the fall and winter (usually 2-4 weeks after the first substantial fall rains) as well as under oak, and a smaller flush may appear in the spring. In coastal northern California and Oregon it grows with pine and spruce in the fall; in the Sierra Nevada a stout form fruits in large quantities under conifers in the spring, summer, and fall; and in the high Rockies the variety with a dark red cap fruits in colossal quantities in August and September, under spruce.

EDIBILITY: One of the finest of fleshy fungi and certainly the best-loved and most sought-after in Europe, where it has more common names than there are languages. If any mushroom deserves the dubious title of "king," this is the one. It is a consummate creation, the peerless epitome of earthbound substance, a bald bulbous pillar of thick white flesh—the one aristocrat the peasantry can eat!

The entire fruiting body is exceptionally delicious, even the tubes if they're firm enough. It is delicious raw, but can cause stomach upsets because it's difficult to digest, so play it safe and cook it (preferably on high heat in an open pan, so as to drive off the excess moisture). The odor and taste of dried *B. edulis* are marvelous—nutty, earthy, and meaty all at once. But you have to find them before you can eat them, and it isn't always easy. You can't just casually look for them the way you can for chanterelles or blewits—you have to hunt them down and root them up from under the duff before they're visible to others. Timing is of paramount importance, because you face formidable competition from both maggots and boletivores (see p. 546). One source suggests getting up at the crack of dawn and wearing your shirt inside out. If that doesn't work, you can always resort to the dried, imported version found in delicatessens.

COMMENTS: This magnificent mushroom has several color forms, two of which are particularly common. The first has a brown to reddish-brown cap (Color Plates 142 & 144) and is the dominant one in coastal California. The second, also called *B. edulis var. pinicola* (=*B. pinicola, B. pinophilus*), has a dark red to red-brown cap. It is prevalent in the Rocky Mountains and Southwest (Color Plate 143) and Gulf Coast region. Remember: brown to reddish cap, thick reticulate stalk, mild to nutty (not bitter) taste, and white pores when young that do *not* bruise blue. No mushroom is more substantial or satisfying to find! Other edible species: *B. mottii* of the Sierra Nevada has a pitted and/or ridged cap; *B. separans* (reddish to liver-colored cap when young) and *B. pseudoseparans* (dark purple cap when young) usually have a purple or purple-tinged stalk and grow under hardwoods in eastern North America.

Boletus edulis, young specimens with white pores. Some forms have much stouter stalks than these, which have been trimmed to check for maggots. The caps will broaden as they mature. See p. 489 for a close-up of the reticulate stalk. Also see color plates and photo on p. 512. (Joel Leivick)

Tylopilus species can be somewhat similar, but have pinker pores, pinkish- to reddish-brown spores, stain dark brown when bruised, and/or are bitter-tasting. *B. fibrillosus* is another look-alike, but its cap is darker and hairier. See also *B. aereus* and *B. barrowsii.*

Boletus aereus   (Queen Bolete)   Color Plate 140

CAP 5-15 (20) cm broad, convex becoming broadly convex to plane in age; surface dry or moist (or viscid only in age), smooth or somewhat pitted, dark brown to nearly black when young and covered at least partially with a fine whitish bloom, but in age becoming smooth and cinnamon or red-brown or blotched with paler (whitish to tan) areas. Flesh thick, white or tinged reddish, not blueing when exposed (or blueing only slightly near the tubes); taste mild or pleasant. **PORES** and tubes white when young and at first stuffed with a pith, becoming yellow to greenish-yellow in age; not blueing when bruised. **STALK** 5-15 cm long, 2-5 cm thick at apex, usually enlarged below when young but often equal in age; firm, solid, white or often brown in age, finely reticulate at least over upper part. **SPORE PRINT** dark olive-brown; spores 12-14 × 4-5 microns, spindle-shaped to elliptical, smooth.

HABITAT: Solitary, scattered, or gregarious in mixed woods and under hardwoods (especially oaks); found in California, but originally described from Europe. In our area it is fairly common in the fall in the coastal mountains, usually under tanoak, madrone, and/or chinquapin. It also grows with live oak, but not as commonly as *B. barrowsii.*

EDIBILITY: Delicious! It is every bit as good as *B. edulis,* and less apt to be wormy!

COMMENTS: This relative of *B. edulis* often grows with *B. appendiculatus,* but has white pores when young, a darker cap, and does not stain blue. The hoary bloom on young caps (see photos on pp. 50 & 512) disappears as the cap becomes more cinnamon-colored, and in age there is little to distinguish it from *B. edulis* other than habitat. Fortunately, both species are so delectably delicious that it hardly matters if you unwittingly confuse them! *B. fibrillosus* is also similar, but has a hairier cap *without* a whitish bloom. Another cousin of *B. edulis,* **B. variipes,** has a dry, tan to grayish-brown, yellow-brown, or dark brown cap that lacks the whitish bloom of *B. aereus* and often becomes areolate (cracked) in age. It is abundant in the summer under oak and beech in eastern North America, but rare or absent in the West. It is edible, but usually riddled with maggots and sometimes bitterish to boot.

531

TYLOPILUS

Medium-sized to large, fleshy, mostly terrestrial boletes. CAP usually dull colored and typically not viscid. PORES and tubes *typically pallid to pinkish, vinaceous, gray, or brown.* STALK fleshy, sometimes reticulate, *but not glandular-dotted and not usually with dark scabers.* VEIL *absent.* SPORE PRINT *flesh-colored to pinkish-brown, reddish-brown, cinnamon-brown, vinaceous, or chocolate-brown.* Spores elliptical to spindle-shaped, smooth (except *T. gracilis*).

TYLOPILUS closely resembles *Boletus,* but gives a flesh-colored to reddish-brown or chocolate-brown spore print instead of a brown to olive-brown one. In the field the two genera can often be told by the color of their pores—in *Boletus* they are usually orange, red, yellow, or white and frequently stain blue when bruised, while in *Tylopilus* they are typically pallid to flesh-colored, vinaceous, gray, or some shade of brown, and are more apt to stain reddish, brown, or black than blue.

In eastern North America *Tylopilus,* like *Boletus,* is quite diverse and often abundant in the summer and early fall. The West presents a completely different picture, however—a mere handful of species occur and only one, *T. pseudoscaber* (a species "complex"), could be characterized as "common." *Tylopilus* does not offer the plethora of good edibles that *Boletus* does. I can find no mention of any *Tylopilus* being poisonous, but many species (e.g., *T. felleus*) have a very bitter taste—even when cooked. Since *Tylopilus* is such a conspicuous component of the bolete bounty of eastern North America, I have included the more common and distinctive eastern species in the key. Four species are described (three of which occur in California); two additional species are shown in the color plates.

Key to Tylopilus

1. Tubes yellow; pores yellow to brownish or reddish, often blueing when bruised; cap usually areolate (cracked) in age; western ...... *T. amylosporus* (see *Boletus chrysenteron,* p. 519)
1. Not as above; tubes not yellow, or if so then found in eastern North America 2

2. Stalk white or creamy (but staining dark brown); pores white when fresh but quickly staining dark red and then dark brown when bruised; flesh in cap and upper stalk often blueing slightly when cut, then turning reddish or brown; found in eastern North America; rare .. *T. snellii*
2. Not as above; staining reactions different, or if similar then stalk *not* whitish; common ... 3

3. Pores and/or flesh staining blue or blue-black when bruised or at least staining waxed or white paper blue or blue-green when wrapped in it; fruiting body dark *T. pseudoscaber* group, p. 534
3. Not as above; not staining waxed paper blue or blue-green; fruiting body dark or light 4

4. Cap and stalk dark gray to brownish-black or black, but the pores white when young and pinkish in age; pores and flesh staining reddish or vinaceous when bruised, then often blackening; fruiting body usually robust; spore print pinkish or vinaceous; found in eastern North America and the Southwest, usually under hardwoods such as oak; edible .... *T. alboater*
4. Not as above; differently colored or with darker (chocolate-brown, etc.) spores or pores .. 5

5. Pore surface dark gray to dark brown or blackish, even when young; stalk densely scurfy from numerous minute scales or granules; cap and stalk chocolate-brown to dark brown or with a purple or bluish tinge; spore print pinkish-brown to vinaceous; found in eastern North America, mainly under northern conifers .... *T. eximius*
5. Not as above (if dark, then spore print chocolate-brown or stalk not densely scurfy) 6

6. Cap and stalk dark (olive-brown to grayish-brown, dark brown, or blackish); pores usually gray to brown, at least in age; spore print chocolate-brown .. *T. pseudoscaber* group, p. 534
6. Not as above; differently colored and/or spore print pinkish to vinaceous or cinnamon ... 7

7. Found in eastern North America (east of the Rocky Mountains)* 8
7. Found in western North America (from the Rocky Mountains westward)* 18

8. Stalk base bright yellow; cap pink when fresh (but fading to tan or paler) *T. chromapes,* p. 533
8. Not as above 9

9. Cap whitish to buff or pale yellow, sometimes dingy tan in age 21
9. Not as above; cap darker or brighter 10

*In regions where the eastern and western fungal floras overlap (e.g., southern Arizona), try both choices!

10. Cap broadly conical and shaggy-fibrillose (or pitted in age), yellow to bright yellow-brown or tinged pink, margin often fringed; found in southern bottomlands *T. (=Mucilopilus) conicus*
10. Not as above; cap not shaggy 11

11. Cap bright orange to dull orange (but may become cinnamon or browner in age); stalk white to yellow or orange (or aging brownish) .... *T. ballouii* (see *T. chromapes*, below)
11. Not as above; cap yellow-brown to cinnamon, brown, purplish, etc., but not truly orange 12

12. Flesh very bitter-tasting (but some people have trouble detecting the bitterness) 13
12. Taste not bitter or only slightly or sporadically so 16

13. Stalk purple or purple-and-white when fresh (but may fade in age!); cap purplish to brown; pores white when young, dingy pinkish in age; stalk not reticulate or only slightly so at apex; usually found under hardwoods; common .... *T. plumbeoviolaceus*
13. Not as above; cinnamon to brown or dark brown but not purple when fresh 14

14. Stalk clearly reticulate; usually (not always) under conifers *T. felleus* (see *T. indecisus*, p. 535)
14. Stalk not reticulate or only slightly so at apex; usually associated with oak 15

15. Stalk thick (1-5 cm at apex); cap medium-sized to large; widespread ..... *T. rubrobrunneus*
15. Stalk thinner than above; cap often rather small; southern .... *T. minor*

16. Stalk typically long and slender (less than 1 cm thick at apex), often curved at base; cap and stalk reddish-brown to cinnamon or tawny (see Color Plate 149); found in many habitats but partial to hemlock; many spores minutely roughened or pitted *T. gracilis* (see *T. chromapes*, below)
16. Not as above; stalk usually at least 1 cm thick; spores smooth 17

17. Cap more or less maroon-red .... *T. badiceps*
17. Not as above 24

18. Stalk distinctly reticulate over upper portion; associated mainly with oak *T. indecisus*, p. 535
18. Stalk not reticulate or only obscurely so at apex from decurrent tube walls 19

19. Stalk often short, poorly-developed, and sometimes off-center; margin of cap often incurved even at maturity; fruiting body often partly buried in soil .... *T. humilis*, p. 535
19. Not as above 20

20. Cap dark brown .... *T. ferrugineus* (?)
20. Cap tan to dull brown or vinaceous-tinged .... *T. ammiratii* (see *T. indecisus*, p. 535)

21. Stalk distinctly reticulate or reticulate-ridged; mainly southern 22
21. Stalk not reticulate or only slightly so at apex; southern or northern 23

22. Reticulation on stalk coarse and raised; stalk usually long in relation to cap; cap often areolate in age; spores pitted or roughened .... (see *Boletellus & Austroboletus*, p. 508)
22. Not as above; taste bitter or mild; fruiting body often robust .... *T. rhoadsiae*

23. Cap typically wrinkled; northern .... *T. intermedius*
23. Cap not normally wrinkled; southern .... *T. peralbidus*

24. Fruiting body medium-sized to large, tawny to orange-brown to brown (including the pores!); stalk usually reticulate; taste mild to slightly bitter; mainly southern .... *T. tabacinus*
24. Not as above; pores usually paler when fresh .... *T. indecisus* & others, p. 535

Tylopilus chromapes (Chrome-Footed Bolete)

CAP (3)5-15 cm broad, convex, often becoming more or less plane in age; surface smooth or felty, dry to slightly viscid, bright pink to rose-colored when fresh, but often fading (to pinkish-tan, tan, or even whitish) in age. Flesh thick, white or tinged pink, not blueing; taste mild or slightly acidic. **PORES** minute, white when young becoming pinkish to flesh-colored to dingy brownish in age, not staining appreciably when bruised, or staining pinkish; tubes more or less same color or slightly yellower. **STALK** 4-20 cm long, 1-2.5 cm thick, equal or slightly tapered either way, solid, firm; upper portion white or pallid or sometimes splotched with pink (especially when young), base bright yellow inside and out; not reticulate but often scurfy above from small hairs or scales. **SPORE PRINT** vinaceous-to pinkish-brown; spores 10-17 × 4-5.5 microns, elliptical to spindle-shaped, smooth.

HABITAT: Solitary to gregarious on ground under both hardwoods and conifers; locally frequent throughout eastern North America in the summer and fall. It is especially common under birch, aspen, and conifers in northern latitudes in June and July.

EDIBILITY: Edible and quite good, but often riddled with maggots.

COMMENTS: Formerly known as *Boletus chromapes* and *Leccinum chromapes,* this lovely bolete is easily recognized by its pink cap and white pores when young and its scurfy stalk with a bright yellow base. (The color scheme is reminiscent of *Gomphidius subroseus,* a gilled mushroom closely related to the boletes.) It is one of several very striking species of *Tylopilus* that are native to eastern North America. Others include: *T. plumbeoviolaceus,* a beautiful but bitter-tasting violet to purple-brown species; *T. (=Austroboletus) gracilis,* a slender-stemmed species **(COLOR PLATE 149)**; and *T. ballouii,* with an orange, *Leccinum*-like cap when fresh. For more details on these and other distinctive eastern species, see the key to *Tylopilus.*

Tylopilus pseudoscaber group   (Dark Bolete)

CAP 5-16 cm broad, convex to plane or somewhat irregular; surface dry, minutely velvety to smooth in age, dark brown to very dark brown or at times grayish-brown to olive-brown, usually bruising darker brown to blackish; margin sometimes slightly paler. Flesh thick, whitish, typically bruising blue (but sometimes slowly or erratically), then slowly discoloring to reddish-brown or brown (or sometimes staining directly to pinkish); odor often pungent, taste mild or at least not bitter. **PORES** and tubes dark yellow-brown to coffee-brown to grayish-brown, chocolate-brown, or very dark brown (but in some forms pallid to grayish or yellow-gray when young); usually (but not always!) staining blue and then slowly reddish to dark brown when bruised. **STALK** (4) 7-16(20) cm long, 1-3(4) cm thick, equal or more often thicker below, firm, dry, minutely velvety and/or scurfy, often longitudinally lined or streaked; distinctly reticulate above in one form, not reticulate in others; brown or colored like cap or darker, usually bruising darker brown or sometimes blue; base typically whitish. **SPORE PRINT** pale to dark chocolate-brown or reddish-brown; spores 11-19 (25) × 5-9 microns, spindle-shaped to elliptical, smooth.

HABITAT: Solitary, scattered, or in groups on ground or rotten wood, usually under conifers but sometimes with hardwoods; widely distributed but mainly northern. It is fairly common in the late summer and fall in the Pacific Northwest and northern California. It is especially fond of Sitka spruce—which may explain its absence or near absence south of Mendocino County, California (the southernmost limit of Sitka spruce).

EDIBILITY: Edible, but according to most sources, far from incredible. Its somber color is unappetizing and makes it difficult to see in the forest gloom. I haven't tried it.

COMMENTS: Also known as *Porphyrellus pseudoscaber,* this distinctive bolete is easily told by its dark color (see photo on p. 535), reddish-brown spore print, dark brown pores in age, and tendency to stain blue. The latter feature varies considerably in duration and intensity and may even be absent; however, if the fruiting body is wrapped in white or waxed paper it will usually stain the paper blue or dark blue-green. The above description is actually a composite of several closely related "dark boletes" that have passed under the names *T. pseudoscaber, T. porphyrosporus, T. olivaceobrunneus, T. atrofuscus, T. sordidus,* etc. Some mycologists consider them to be forms of one highly variable species. Others consider them distinct—for instance, they reserve the name *T. porphyrosporus* for the form with pallid pores when young, and *T. olivaceobrunneus* for the variety with a reticulate stalk. Other species in the group include: *T. pacificus,* which tends to have a copiously cracked (areolate) cap in age; *T. fumosipes,* an eastern species with an areolate cap in age and narrower spores; and *T. nebulosus,* which typically does not stain blue. Two other dark species, *T. alboater* and *T. eximius,* are described in the key.

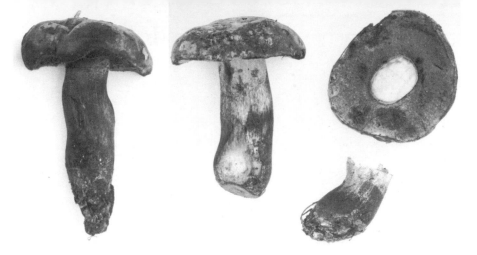

Left: *Tylopilus pseudoscaber* group. Note dark color. **Right:** *Tylopilus indecisus* is apt to be mistaken for a *Boletus*, but like many species of *Tylopilus*, it stains brown when handled.

Tylopilus indecisus   (Indecisive Bolete)

CAP 5-15 (25) cm broad, convex to plane; surface usually dry and smooth, dark brown to brown, dingy brown, or dingy cinnamon (usually darker when young, paler in age), staining dark brown where bruised. Flesh thick, pallid, often staining pale vinaceous or flesh-color when bruised; taste mild. **PORES** and tubes pallid when young, soon becoming dull pinkish or pale flesh-color; slightly darker, dingier, or more vinaceous in age, staining brown where bruised. **STALK** 4-10 cm long, 1-3 (4.5) cm thick, equal or thicker below, dry, firm, buff to brown, the apex often pallid; darker brown where bruised or handled; reticulate at least at the apex (but eastern variety sometimes not reticulate), sometimes nearly throughout. **SPORE PRINT** flesh-colored to reddish-brown or tinged vinaceous; spores 10-15 × 3-5 microns, spindle-shaped to elliptical, smooth.

HABITAT: Solitary to scattered or in small groups in mixed woods and under oaks; fairly common in eastern North America, but rare in the West. I have seen it only a few times in our area, under live oak in the late fall and early winter, and under white oak inland.

EDIBILITY: Edible, according to the literature. I haven't fried it.

COMMENTS: There is nothing particularly indecisive about this drab bolete, but there is nothing particularly decisive about it either. The brown cap and reticulate stalk are reminiscent of *Boletus edulis,* but the pores are pinker (and never olive-yellow), the fruiting body stains dark brown when handled or bruised, and the spore print is pinkish. Eastern material (including *var. subpunctipes*, which differs microscopically) is less apt to be reticulate than the western version. *T. ammiratii* is similar but has a tan or vinaceous-tinged cap and whitish stalk that is not reticulate or only slightly so at the apex; it occurs in California under oak. Similar easterners include: *T. subunicolor*, a small, slim, tawny-ochre southern species with yellowish flesh; *T. tabacinus* and *T. rubrobrunneus*, both medium to large (see key to *Tylopilus*); and *T. felleus* (**COLOR PLATE 150**), with a reticulate stalk and very bitter taste. *T. felleus* is especially common under conifers (e.g., hemlock and pine); it is often mistaken for *Boletus edulis*. For other easterners, see key to *Tylopilus*.

Tylopilus humilus   (Humble Bolete)

CAP 4-12 cm broad, nearly round to convex when young, sometimes expanding to plane in age or often somewhat misshapen or irregular; surface dry or slightly tacky, often minutely velvety, dull brown or sometimes reddish-brown; margin often with vinaceous tints and frequently remaining incurved through maturity or even covering some of the pores. Flesh thick, white, bruising pinkish-brown or vinaceous (rarely blue); taste mild. **PORES** and

Tylopilus humilus is a small, rare, often misshapen bolete. Note how stalk is often poorly developed.

tubes whitish when young, becoming pale flesh-color or dull pale pinkish, staining brown when bruised. **STALK** 2-5 cm long, 1-3 cm thick, usually rather stocky and sometimes off-center; equal or swollen below, but the base often pinched; firm, solid, white at apex, usually brownish- or vinaceous-stained below, especially when handled; not reticulate or only very slightly so at extreme apex. **SPORE PRINT** dull reddish-brown; spores 8-12 × × 3-4 microns, spindle-shaped to elliptical, smooth.

HABITAT: Scattered to gregarious or in small clumps on ground (often partially buried), fruiting in the fall and winter; known only from California, rare. It appears every year in sandy soil in a burned-over area near Santa Cruz, apparently in association with manzanita or live oak, but with knobcone pine also in the general vicinity. It was originally discovered by Harry Thiers in Mendocino County under pines near a destroyed dormitory.

COMMENTS: This bashful bolete is easily separated from other *Tylopilus* species by its semi-underground habit, i.e., it tends not to expose itself until fully mature, and even then may remain half-buried. The stocky or poorly-developed stem is reminiscent of *Gastroboletus,* but the spores are forcibly discharged and the tubes are arranged vertically.

LECCINUM   (Rough-Stemmed Boletes)

Medium-sized to large, fleshy, terrestrial, mycorrhizal, woodland boletes. CAP usually some shade of orange, red, brown, gray, or white; viscid or dry; often rimmed with a flap of sterile tissue when young. PORES and tubes *usually whitish to dingy buff, grayish, grayish-olive, or pale brown,* only very rarely yellow; not staining blue. STALK fleshy, usually rather tough and fibrous, typically not reticulate, *ornamented with scabers which usually darken at maturity.* VEIL *typically absent.* SPORE PRINT *brown.* Spores spindle-shaped to elliptical, smooth.

LECCINUM differs from other boletes by virtue of the tufted hairs or small rough scales **(scabers)** on the stalk. The scabers may be pallid when young, but usually darken to brown or black by maturity. They should not be confused with the resinous or sticky glandular dots found in *Suillus,* which do not protrude as scabers do and are not composed of hairs. Leccinums also have a characteristic appearance which makes them recognizable from a distance: the cap is typically some shade of orange, reddish-brown, brown, or white; the pores are usually white or dingy-colored (not yellow or red) and do not stain blue; and the stalk is usually whitish or pale colored except for the scabers, and usually much tougher and more fibrous than the cap. Also, the stalk sometimes stains blue when cut or handled, and in many species is rather long in relation to the cap.

Leccinums are good boletes for beginners because they are widespread, often abundant (in Alaska they are said to outnumber all other boletes), and edible (though a few, like *L. atrostipitatum,* reportedly cause stomach upsets in *some* people). Nearly all Leccinums have a tendency to blacken when cooked or dried, but in no way does this affect their flavor. They dry quite nicely and are less apt to be maggoty than other boletes, at least in our area. Some species, such as *L. insigne,* are delicious fresh. Others, like *L. manzanitae,* are watery and bland unless dried first. Even the less savory ones, however, have an important role to fulfill—they help stuff the buckets and bellies of boletivores (see p. 546) who might otherwise grow despondent and dangerous!

Like most boletes, Leccinums are mycorrhizal. Their favorite hosts are aspen and birch, but they also occur with conifers and oaks, and in California—where birch is not native and aspen is restricted to the Sierra Nevada—they link up with madrone and manzanita, masquerading under the moniker "manzanita boletes." Over 100 species of *Leccinum* have been described from North America (most of them from the northern part), but many are very similar in appearance and can only be distinguished with a microscope. For instance, *L. aurantiacum* differs from *L. discolor* by the presence of minute pigment globules in the hyphae of the cap cuticle, while *L. discolor* differs from *L. rufescentoides* in its smaller cuticular hyphal end cells! The staining reactions of the flesh upon exposure to air— particularly at the juncture of the cap and stalk—are also significant and should be determined by slicing open the fruiting body lengthwise, rubbing the flesh a couple times, then observing it at 5-minute intervals for a period of ½-1 hour.

Fortunately, *Leccinum* is such a safe genus that boletivorous bipeds such as you and I can leave such dubious and esoteric distinctions to professional boletologists, and concentrate on learning a few common "prototypes"—e.g., the *L. manzanitae* group that is so common with manzanita and madrone; the orange to reddish-brown-capped *L. insigne*- *L. aurantiacum* group that grows profusely wherever there are aspens; and the white-capped *L. holopus* and dingy colored *L. scabrum* groups that favor birch. Three "prototypes" are described here and several others are keyed out.

Key to Leccinum

1. Pores and tubes yellow when fresh 2
1. Pores and tubes not yellow 3
2. Scabers on stalk brown to blackish in age; cap yellow to ochraceous to brown or dark brown; found under hardwoods (mainly oak) in eastern and southern North America 29
2. Not as above; scabers reddish or paler, not darkening appreciably in age   (see *Boletus,* p. 511)
3. Base of stalk bright yellow and cap pink when fresh (but fading!) *or* cap and stalk chocolate-brown to purplish to dark blue-gray and pores chocolate-brown to dark gray to blackish at maturity; restricted to eastern North America .... (see *Tylopilus,* p. 532)
3. Not as above   4
4. Margin of cap *lacking* sterile flap(s) of tissue; cap white to buff, grayish, brown, dingy yellow-brown, olive-tinged, or even black when fresh 5
4. Cap pink to apricot-buff, orange, red, reddish-brown, orange-brown, or rusty-brown, *or* if colored as above then margin of young cap rimmed with a sterile flap of tissue 2-6 mm wide, the flap breaking up into segments as the cap expands, or disappearing 12
5. Cap white or whitish (but sometimes tinged buff, pinkish-buff, brown, or olive in age) 6
5. Cap more highly or deeply colored (pink, orange, red, brown, gray, black, etc.) 8
6. Associated with aspen; known only from California   *L. californicum* (see *L. scabrum,* p. 541)
6. Not as above   7
7. Cap often olive-tinged in age, sometimes viscid; common under birch in northern and eastern North America .... *L. holopus* & others (see *L. scabrum,* p. 541)
7. Not as above; cap not viscid; associated with hardwoods (especially oak) in southern and eastern North America (especially common in the South) *L. albellum* (see *L. scabrum,* p. 541)
8. Cap bluish-black to dark grayish-brown to black when fresh (but often fading in age); associated with aspen and birch in eastern North America 9
8. Not as above; cap dull brown to gray-brown, dingy yellow-brown, olive-tinged, etc. 10

9. Found with birch; flesh in stalk apex reddening when bruised *L. snellii* (see *L. scabrum*, p. 541)
9. Not as above; associated with aspen .... *L. griseonigrum* (see *L. scabrum*, p. 541)
10. Cap yellow-brown when young becoming duller or grayer in age and usually prominently cracked (areolate); associated with oak and other hardwoods; especially common in southeastern North America .... *L. griseum* (see *L. scabrum*, p. 541)
10. Not as above; cap not usually areolate; associated with birch or aspen 11
11. Associated with birch; widely distributed .... *L. scabrum* & others, p. 541
11. Associated with aspen; common in the Sierra Nevada *L. montanum* (see *L. scabrum*, p. 541)
12. Veil present, covering the young cap, then breaking up into whitish patches; associated with aspen in the Great Lakes region .... *L. potteri*
12. Not as above; veil absent (but margin of cap may have sterile flap(s) of tissue) 13
13. Scabers on stalk dense and coal-black even in the button stage and black or dark brown in age; associated mainly with birch in eastern and northern North America, but also found with aspen in Southwest .... *L. atrostipitatum* & *L. testaceoscabrum* (see *L. insigne*, p. 540)
13. Not as above 14
14. Associated with manzanita and madrone (and possibly toyon); cap often (but not always) viscid when moist; common in west coast states, especially Oregon and California 15
14. Associated with aspen, conifers, or other trees; cap viscid or dry; widely distributed but not found in coastal California 18
15. Cap dark red to brown or sometimes orange-brown .... *L. manzanitae* & others, p. 539
15. Cap bright orange to pink or apricot-buff 16
16. Scabers on stalk orange; bruised or cut flesh usually reddening somewhat before darkening .. .. *L. "aurantioscaber"* (see *L. manzanitae*, p. 539)
16. Not as above 17
17. Cap pinkish to apricot; flesh not staining when bruised *L. constans* (see *L. manzanitae*, p. 539)
17. Cap orange; flesh usually reddening or darkening *L. armeniacum* (see *L. manzanitae*, p. 539)
18. Flesh (especially at juncture of cap and stalk) not changing color appreciably when cut or bruised, or staining only very slightly (but lower stalk may stain blue); found with conifers 19
18. Flesh (especially at juncture of cap and stalk) staining purple-gray, bluish-gray, smoky, reddish, vinaceous, etc. when cut or bruised (but often slowly or erratically); lower stalk may stain blue as well 20
19. Fruiting body medium-sized to very large (mature cap 10-30 cm broad); found in the Pacific Northwest .... *L. ponderosum* & others (see *L. manzanitae*, p. 539)
19. Fruiting body smaller (cap usually less than 10 cm broad); known from the Southwest and eastern North America .... *L. vulpinum*
20. Flesh staining purple-gray to bluish-gray or smoky (fuscous) *directly* when bruised or cut (but often staining slowly or erratically) 21
20. Flesh staining reddish, vinaceous, or burgundy when bruised or cut (but often staining purple-gray to bluish-gray or fuscous after that) 23
21. Cap whitish when young but becoming brownish in age; stalk thick and club-shaped (thicker or swollen below); associated with western conifers (mainly in mountains) ...... *L. clavatum*
21. Not as above 22
22. Cap orange to rusty-orange to reddish when young (may become browner in age), or if brown when young then tubes often staining vinaceous or purplish when bruised *L. insigne,* p. 540
22. Cap dark brown to dull reddish-brown or dull brown; tubes *not* staining vinaceous or purplish when bruised .... *L. brunneum* (see *L. manzanitae*, p. 539)
23. Cap liver-colored to liver-brown and fibrillose to fibrillose-scaly; associated with conifers in the Pacific Northwest .... *L. fibrillosum* (see *L. manzanitae*, p. 539)
23. Not as above; cap typically reddish to rusty-orange, brownish, or paler 24
24. Cap pale tan to pale dingy orange-brown or yellow-brown; associated with aspen and birch in northern and eastern North America, and in the Southwest .... *L. cinnamomeum*
24. Not as above; cap usually darker or brighter 25
25. Cap pinkish- or vinaceous-tinged when young; known only from Idaho under whitebark pine .... *L. incarnatum*
25. Not as above 26

26. Stalk more or less equal or narrowed near apex; associated mostly with aspen or pine; very widely distributed and common 27
26. Stalk distinctly and consistently club-shaped (thicker below); associated with conifers (mainly spruce and fir) in the Rocky Mountains and Southwest 28
27. Cap typically reddish to orange to brick-red when young (often somewhat duller in age); associated with aspen and conifers (pine, etc.) .... *L. aurantiacum* (see *L. insigne,* p. 540)
27. Cap somewhat similar in color to above or often duller even when young (brown to orange-brown, pale cinnamon, etc.); associated only with aspen . *L. discolor* (see *L. insigne,* p. 540)
28. Cap fibrillose or fibrillose-scaly .... *L. subalpinum* (see *L. insigne,* p. 540)
28. Not as above .... *L. fallax* (see *L. insigne,* p. 540)
29. Stalk often robust and swollen in the middle or below; cap yellow-brown to dark brown ....
.... *L. crocipodium*
29. Stalk long and more or less equal; cap yellow-brown to ochre, often wrinkled or finely pitted in age .... *L. rugosiceps*

Leccinum manzanitae   (Manzanita Bolete)   Color Plate 145

CAP (5) 7-20 (30) cm broad, rounded becoming convex and then broadly convex to plane or somewhat irregular; surface viscid when moist, practically smooth but usually with flattened fibrils, sometimes pitted in age; dark red to reddish-brown, or at times rusty-brown, rusty-orange, or brown (see comments); margin with flaps of sterile tissue when young, which break up into segments or disappear in age. Flesh thick, firm becoming soft at maturity, white, usually bruising smoky-gray to purple-gray or bluish-gray when cut, but often very slowly or only in certain areas. PORES and tubes whitish to pale olive, olive-gray, or grayish, usually deep dingy olive-buff in age; not blueing when bruised but often discoloring brown. STALK 8-20 cm long, 1.5-4 cm thick at apex, equal or often thicker or swollen below, tough, fibrous, solid, white or whitish when fresh, but roughened by numerous small projecting scabers that are pallid at first but deep brown to black by maturity; lower portion often bruising (or already stained) bright blue to greenish-blue. SPORE PRINT brown; spores 13-19 × 3-5 microns, spindle-shaped to elliptical, smooth.

HABITAT: Solitary to widely scattered or gregarious on ground, associated with manzanita and madrone but often growing in mixed woods; known only from the west coast. It is common in our area in the fall and winter and is easily the most prevalent *Leccinum* in coastal California. It is seldom as prolific as *B. edulis,* but I have seen more than fifty prime fructifications under a patch of stunted madrones in the middle of a pasture.

EDIBILITY: Edible and esteemed by many boletivores, but disappointingly bland in my experience. Mature or waterlogged specimens should be dried in order to concentrate their flavor and get rid of excess moisture. In one blind tasting of several local boletes (all fresh), it placed next to last—slightly behind the boring *Boletus subtomentosus* but ahead of the abominable *Suillus fuscotomentosus.* Michael Cabaniss, who did the line drawings for this book, says: "I have tried only dried slices, a form which is not the most visually stimulating . . . their consistency is rather grainy, giving the gustatory experience of dried earth or decomposed redwood chips. Dried slices are not even chewy; they break up readily when inserted into the oral cavity and are swallowed with great difficulty."

COMMENTS: This large, impressive bolete is easily identified by its reddish to brown cap, hard tough stalk with dark scabers, and growth with manzanita and madrone (see photo on p. 39). It is the central species in a cluster of edible Leccinums that appear to associate only with ericaceous plants such as manzanita and madrone, and are consequently referred to collectively as "manzanita boletes." Their ranks include a common but unidentified (probably unnamed) species in our area with a bright orange to rusty-brown cap, orange scabers, and flesh that tends to stain reddish *before* turning fuscous (it is known locally as *L. "aurantioscaber"*); *L. arbuticola,* with reddish-staining

Leccinum manzanitae is a common associate of manzanita and madrone. Note the dark scabers on the stalk and the white flesh that stains fuscous (smoky to smoky-purple) erratically when cut.

flesh and a buff to pale brown cap; *L. aeneum,* with a non-viscid, oranger cap; *L. largentii,* possibly associated with toyon, with a dry cap, dark olive-gray pores, and very dense scabers; *L. constans,* with a pale pink to apricot-colored cap and unchanging flesh; *L. armeniacum*, a common species with an orangish cap and somewhat paler scabers (especially when young); and *L. arctostaphylos*, an Alaskan species associated with bearberry (a type of manzanita). Other species: *L. ponderosum* (COLOR PLATE 146) is a sometimes massive species that looks like *L. manzanitae,* but typically has unchanging (or only slightly staining) flesh and is associated with conifers in the Pacific Northwest; it is quite tasty. *L. fibrillosum* also grows with conifers in the Pacific Northwest, but it has a fibrillose to fibrillose-scaly, liver-brown cap and its flesh stains reddish or vinaceous when cut. *L. idahoensis* has a roughened, liver-colored cap but does not stain appreciably when bruised, while *L. brunneum* has a dark brown to reddish-brown cap, stains directly to fuscous, and is quite common under aspen in the Sierra Nevada and Southwest. See also *L. insigne* and the species listed under it.

Leccinum insigne   (Aspen Bolete)   Color Plate 147

CAP (4) 6-17 cm broad, round to convex becoming broadly convex to nearly plane; surface dry or only slightly viscid when wet, smooth to minutely fibrillose or sometimes breaking up into small scales, in age sometimes pitted; color variable: bright orange to rusty-orange, reddish-orange, orange-brown, reddish-brown, or cinnamon (or in one form brown), often paler or duller (tan, brown, or dull orange-brown) in age; margin usually with flaps of sterile tissue, at least when young. Flesh thick, often rather soft in age, white, turning bluish-gray to purple-gray or fuscous when bruised or cut, but sometimes very slowly or erratically. **PORES** and tubes pallid or whitish when young, becoming olive-buff to grayish or dull yellowish-buff in age; not blueing when bruised (but may stain yellowish to brown, lavender, or vinaceous). **STALK** 6-15 cm long, 1-2.5 cm thick at apex, equal or swollen below, dry, solid, rather tough and fibrous; white or whitish when fresh, but roughened by numerous small, projecting scabers which are initially pallid but become reddish-orange to brown and finally blackish by maturity; base or lower portion often staining or discoloring blue. **SPORE PRINT** brown to yellow-brown; spores 11-18 × 4-6 microns, spindle-shaped to elongated-elliptical, smooth.

540

HABITAT: Widely scattered to gregarious or in troops on ground in woods and at their edges; associated primarily if not exclusively with aspen, and common in the summer and early fall wherever aspen occurs. It is abundant in the Sierra Nevada (along with other Leccinums—see comments), and I have seen enormous fruitings in the Southwest and Rocky Mountains shortly after the blue columbines bloom. It is also abundant in Alaska.

EDIBILITY: Edible and good. Though not the equal of *Boletus edulis,* it is far better in flavor than its coastal counterpart, *L. manzanitae.* Like most Leccinums it darkens when cooked or dried. If this intimidates you, that's fine with me—I'll be more than happy to take them off your hands!

COMMENTS: This common aspen-lover has many look-alikes in the genus *Leccinum,* most of which have passed under the name *L. aurantiacum.* They are very difficult to distinguish without a microscope but all are apparently edible, though some may cause digestive upsets in sensitive individuals. *L. insigne* is recognized by its rusty-orange to reddish-brown or brown cap, scabers on the stalk which darken in age, and growth with aspen. Some of its more common look-alikes include: *L. aurantiacum,* very similar and widespread under aspen and conifers, but sometimes larger and more robust and with flesh that distinctly stains burgundy-red *before* turning purplish- or bluish-gray; *L. discolor,* very similar to *L. aurantiacum,* but growing only with aspen and with a slightly duller, browner, or more cinnamon-colored cap (see p. 537 for the critical microscopic difference); *L. atrostipitatum* and *L. testaceoscabrum* (the latter more brightly colored), with scabers that are black even in the button stage, associated mainly with birch and poisonous to some people; and *L. fallax* and *L. subalpinum*, with a rusty to dark reddish cap and club-shaped stalk, associated with spruce and fir in the Rocky Mountains and Southwest. Also see *L. manzanitae* and the species discussed under it.

Leccinum scabrum   (Birch Bolete)

CAP 4-10 cm broad, convex to broadly convex to nearly plane or slightly depressed centrally; surface smooth, dry to slightly viscid when wet, dull brown to grayish-brown to dingy yellow-brown or tan (but often developing olive tints in age); margin lacking flaps of sterile tissue. Flesh thick, white, typically *not* staining when cut or discoloring slightly pinkish to brownish in the stalk. **PORES** and tubes dull whitish or pallid becoming dingy brownish in age; not blueing when bruised but sometimes staining slightly yellowish or ochre. **STALK** 7-15 (20) cm long, 0.5-1.5 (3) cm thick at apex, usually thicker below, firm, solid, white to grayish but roughened by numerous projecting tufted hairs (scabers) which soon become brownish to black at least over the upper portion; often staining blue-green below. **SPORE PRINT** brown; spores 14-20 × 5-7 microns, spindle-shaped to elongated-elliptical, smooth.

HABITAT: Solitary, scattered, or in groups near or under birch in the summer and fall; common throughout the northern hemisphere wherever birch occurs. It has not been found in California, but could conceivably appear on lawns where birches have been planted (just like *Lactarius torminosus*).

EDIBILITY: Edible and choice according to some, mediocre according to others; only firm specimens should be used.

COMMENTS: The dingy brownish cap, dark scabers on the stalk, and association with birch are the principal fieldmarks of this bolete. There are several similarly colored species that are best differentiated microscopically, but *L. montanum* of the Sierra Nevada can be distinguished by its different mycorrhizal mate—aspen. Other species: *L. griseonigrum,* associated with birch, is one of several species with a blackish to bluish-gray to dark brown cap (which, however, may bleach out in age); *L. snellii* also has a dark cap but favors aspen; *L. alaskanum*, common with birch in Alaska, has a dark and light, streaked or mottled cap.

Leccinum scabrum is one of many birch-loving Leccinums. Cap is dull colored (brown to tan, gray, or dingy olive) and the stalk is usually rather slender.

There are also several species with a whitish or pallid cap, including: **L. holopus** and **L. rotundifoliae,** both fairly common in northern North America under birch, the former often developing olive tinges on the cap in age and the latter usually growing in bogs; and **L. albellum,** a slender-stemmed, dry-capped eastern species that favors southern oaks. **L. californicum** has a whitish to buff cap, but grows with aspen in the Sierra Nevada, while **L. cretaceum** has a pale buff cap and flesh that stains vinaceous-pink erratically. **L. roseofracta** has a dark brown cap and reddish- or pinkish-staining flesh and is associated with birch. Finally, there is **L. griseum,** an eastern species that somewhat resembles *L. scabrum* but usually has an areolate (cracked) cap in age and favors oak and other hardwoods over birch. All of these species lack the flaps of sterile tissue on the cap margin characteristic of *L. insigne, L. manzanitae,* and the species discussed under them.

Left: *Leccinum holopus,* a white-capped birch-loving species. Note dark scabers on stalk. **Right:** *Strobilomyces confusus* (see comments under *S. floccopus,* p. 543, for details).

"Old Man of the Woods," *Strobilomyces floccopus,* is easily told by its shaggy black or gray fruiting body. Note cottony veil in young specimen at center and dark gray to black pores in older one at left.

STROBILOMYCES

THIS small genus is unique among the boletes in having a shaggy or scaly gray to black fruiting body, gray to black pores when mature, and blackish-brown, reticulate or warty spores. It is so unique, in fact, that some mycologists place it in a separate family. A woolly veil is present when young, and may form a slight annulus (ring) on the stem. Since only one species is described here, a key is not included.

Strobilomyces floccopus   (Old Man of the Woods)

CAP 4-15 cm broad, convex or cushion-shaped becoming broadly convex to plane; surface dry, shaggy from numerous soft, coarse, gray to grayish-brown or black scales on a paler background (but in age sometimes entirely blackish); margin usually hung with pallid to grayish veil remnants. Flesh whitish, usually turning reddish (often slowly) when bruised, then dark brownish to black. PORES fairly large, white or gray becoming darker gray to nearly black in age, usually bruising reddish, then brownish to black; tubes same color. STALK 5-12 cm long, 1-2.5 cm thick, equal or somewhat thicker toward base, solid, firm, dry, sheathed by large gray to grayish-black scales or shaggy material except at apex. VEIL pallid to grayish, woolly-cottony, leaving tissue on cap margin and usually one or more shaggy zones on stalk. SPORE PRINT black to blackish-brown; spores 9.5-15 × 8.5-12 microns, broadly elliptical to nearly round, reticulate.

HABITAT: Solitary to scattered or in groups in humus under hardwoods (especially oak), sometimes also with conifers; common in the summer and fall in eastern North America, and according to Dr. Robert Gilbertson of the University of Arizona, also fairly common in southern Arizona. Its dark color makes it hard to see in the forest gloom.

EDIBILITY: Edible, but not rated highly by most mushroom enthusiasts. I found it rather insipid although the texture was good. Only young specimens should be used.

COMMENTS: This most distinctive of all the boletes is easily told by its shaggy gray to black fruiting body. In contrast to most boletes, it does not decay readily. As a result, it often seems more numerous than it actually is. Other species: *S. strobilaceus* is a synonym. *S. confusus* of eastern North America is quite similar in color and overall aspect, but has more erect scales on the cap (see photo on p. 542) and spiny-warty to only partly reticulate spores; *S. dryophilus* is a southern species that is similarly colored in age but usually paler (cap pinkish-tan to brown) when young.

543

GASTROBOLETUS   (Gastroid Boletes)

Small to medium-sized boletes *usually buried or partially buried in humus.* CAP convex to some-what irregular in shape. PORES variously colored; *tubes often irregularly arranged, i.e., not all vertically oriented.* STALK *usually short, poorly developed, sometimes off-center.* VEIL absent or present as persistent membrane that covers the tubes. SPORE PRINT *unobtainable.* Spores elliptical to spindle-shaped, smooth, brown to golden-brown under the microscope.

THESE funky fungi have been modified or "reduced" from normal, everyday, upright boletes to abnormal, semi-underground, "decadent" ones. They resemble puffballs in that they do not forcibly discharge their spores, yet they retain a modicum of respectability in the form of a rudimentary stalk and definite tube layer arranged rather haphazardly within the misshapen cap.

Gastroboletus is a small genus, with only one species common along the coast. How-ever, several species occur in the Sierra Nevada and Cascades—perhaps in response to the more severe and unpredictable weather. Fruiting underground helps the mushrooms to withstand the pendulum-like hot-and-cold or warm-and-dry spells. None are known to be poisonous, yet none are known to be edible, since none are known to have been tried.

Key to Gastroboletus

1. Tubes and flesh blueing when bruised or cut .... *G. turbinatus* & others, below
1. Not as above 2

2. Stalk relatively long (6-10 cm) and roughened by small scales or scabers; known only from eastern North America .... *G. scabrosus*
2. Not as above 3

3. Cap whitish to buff or pale tan (sometimes slightly darker in age); tubes whitish to buff to olive-brown, or grayish-yellow; stalk averaging 2-5 cm thick .... *G. subalpinus,* p. 545
3. Not as above; cap pale or dark; tubes yellow to olive-yellow; flesh often yellow; stalk typically 1.5 cm thick or less .... *G. suilloides* & *G. amyloideus* (see *G. turbinatus,* below)

Gastroboletus turbinatus   (Bogus Boletus; Gastroid Bolete)

CAP 2-5 (8) cm broad, convex to slightly depressed to irregular (misshapen); surface dry, minutely velvety, often pitted or wrinkled; color variable, but usually brown to reddish-brown to bright yellow with red or brown spots and stains. Flesh soft, yellow (but often reddish in base of stalk), typically bruising blue quickly. PORES pinkish to reddish, orange, or yellow, typically blueing quickly when bruised; tubes greenish-yellow to yellow, irregular and sinuous or disoriented (arranged at various angles rather than exclusively vertical). STALK 1-4 (7) cm long, 1-2.5 cm thick, equal or narrowed at base, often (but not always) protruding only a short distance below the tubes; solid, dry, yellow or with reddish streaks or granules or colored like the cap, sometimes reddish in old age and often reddish at the base. SPORE PRINT unobtainable; spores (9.5) 13.5-18 × 6.5-9.5 microns, elliptical to spindle-shaped, smooth, brown or golden under the microscope.

HABITAT: Solitary to scattered or gregarious on ground in woods or at their edges, often partially buried; widely distributed. It is fairly common in the summer and fall under conifers in the Pacific Northwest, northern California, and the Sierra Nevada. I have found it twice in our area under live oak, in July and October (it is not necessarily rare, however—just easy to overlook).

EDIBILITY: Unknown, like myself, and likely to remain so.

COMMENTS: A curiously wrought fungal afterthought, easily recognized by its irregular shape, semi-underground growth habit, and disoriented tube layer. Since the spores are not forcibly discharged, there is no need to elevate the cap above the ground. Just how the spores are spread is a mystery—perhaps with the help of boletivorous bipeds such as I.

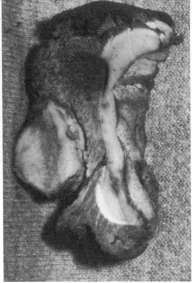

Left: *Gastroboletus turbinatus*. These two specimens have unusually well-developed stems, but the misshapen caps are typical. Right: *Gastroboletus xerocomoides*. This specimen has been sliced open to show the disoriented (haphazardly arranged) tube layer. (Herb Saylor)

Many agarics have been "reduced" to gastroid forms (see Podaxales & Allies, p. 724); this is one instance in which a bolete has. Most of the other members of the genus have limited distributions. Several occur under conifers in the Sierra Nevada, including: *G. xerocomoides,* which blues only slightly and has mostly truncate spores, and three species which do not blue when bruised: *G. suilloides,* with a brown cap; *G. amyloideus,* with a yellow, buff, or reddish-spotted cap and amyloid spores; and *G. subalpinus*(below).

Gastroboletus subalpinus   (Gastroid King Bolete)

CAP 5-12 cm broad, convex becoming plane to depressed or somewhat irregular; surface smooth to slightly velvety, often pitted and/or wrinkled, not viscid but often covered with dirt and debris; whitish to buff or pale tan, sometimes darkening somewhat in age; margin extending downward and over the tubes as a thin white membrane that often disappears or breaks up in age. Flesh thick, soft, white or with yellow to olive discolorations; not blueing when bruised but sometimes staining pinkish. PORES pallid or whitish becoming grayish-yellow to buff and finally darkening to olive-brown, not blueing when bruised; tubes occasionally oriented vertically (especially near stalk) but mostly arranged haphazardly (at various angles); colored more or less like pores. STALK 2-6 cm long, 2-5 cm thick, tapered downward or sometimes equal or even swollen below; solid, dry (but often dirty), white to buff, sometimes darkening to tan in age; base sometimes orange-yellow. SPORE PRINT unobtainable; spores 10-16 (18) × 4.5-6 (8) microns, elliptical to spindle-shaped, smooth, yellowish to brown under the microscope.

HABITAT: Solitary to gregarious in soil or duff under mountain conifers in the late spring and summer; known only from the Sierra Nevada (where it is fairly common) but to be expected in other mountain ranges of the West. It develops underground but is often partially exposed (erumpent) at maturity.

EDIBILITY: Presumably edible; I haven't tried it.

COMMENTS: This gastroid bolete is easily told from its brethren by its pale color and white, non-blueing flesh. It is likely to be mistaken for an aborted button of *Boletus edulis,* or more likely still, *B. barrowsii*. However, the disoriented tube layer, tendency to develop underground, and refusal to give a spore print are distinctive. The thin layer of tissue which at least partially covers the tubes of young specimens may represent a more elaborate version of the whitish pith that plugs the tubes of young *B. edulis* and *B. barrowsii*.

BOLETIVORES

THESE curious creatures, as their name suggests, are a major predator of *Boletus edulis.* Since they are likely to be encountered by anyone who looks for mushrooms, it seems appropriate to describe them and include a few observations on their habits.

Boletivores appear shortly after the onset of the rainy season, not coincidentally, at the same time *Boletus edulis* appears. As a group they share a number of telltale traits which the experienced observer can discern at a glance. Their overall appearance is rather dingy and disheveled, their color drab. The cap is quite variable, ranging from depressed or deflated to uplifted, ecstatic, or inflated; its surface may be dry and distinctly tomentose or smooth and somewhat viscid. The flesh is normally pallid, becoming blue when bruised and exuding a red latex when cut, but it is most often obscured by a ragged woolly cuticle which is typically appressed above and baggy below. The stalk may be fleshy or cartilaginous, and more often than not, bent. The odor varies considerably from individual to individual; the taste is unpleasant, and a spore print is not obtainable, at least with normal methods. Boletivores are strictly terrestrial, and are invariably equipped with a long pointed stick, rapacious eye, and capacious bucket or "universal pail." At least two distinct species occur in our area; these are keyed below.

Key to the Boletivores

1. Retreating furtively when approached .... ***Boletivorus clandestinus,*** below
1. Advancing boldly when approached .... ***Boletivorus brutalosipes,*** below

Boletivorus clandestinus is the more common of the two species. In addition to its secretive nature, it can be recognized (if you can get close enough!) by its tough, wizened stalk with a persistent, even permanent, stoop. The gait is also highly distinctive: curiously hitched and truncated, marvelously efficient, yet flailing and disjointed. The overall impression is that of a creature completely immersed in, yet not designed for, its element—like a kayaker trying to negotiate a rapids without wetting her back. When one is crept up on unawares, it can be heard alternately cooing to and swearing at its prospective prey in a vaguely familiar, yet unintelligible, tongue. This species occurs solitary to scattered or in small groups in the woods, but always near roads. There is usually a rusty pick-up truck, station wagon, or '65 Dodge Dart nearby, parked on the side of the road. Near the vehicle will be found one or more telltale "middens"—neat piles of discarded tubes and wormy stalks. These resemble the feathers that remain from a freshly and systematically disemboweled bird—positive proof of boletivores' singular lust and unbridled craving for their quarry.

Boletivorus brutalosipes (the "Brutal-Footed Boletivore"), on the other hand, can be instantly recognized by its inquisitiveness, its bold advance and cavalier stance. In addition, the stalk is more fleshy than that of *Boletivorus clandestinus*—often swollen in the middle or even bulbous. The gait is decidedly more compact, the stride purposeful, yet completely arbitrary. When one comes rushing toward you, the net effect is that of a rapids intent upon wetting *your* back. Its insatiable greed for *Boletus edulis* may be cleverly disguised by an amicable disposition and disarmingly friendly, fibrillose smile. But it will stop at nothing to achieve its ends, so never leave your basket unattended in the woods! If there are *Boletus edulis* in it, they will be gone, and other species will be stepped on, masticated and regurgitated, or otherwise obliterated. Trampled fly amanitas *(Amanita muscaria),* incidentally, are a sure sign that brutal-footed boletivores are in the vicinity! *Boletivorus brutalosipes* is found in roughly the same habitats as *Boletivorus clandestinus,*

A telltale "midden" left by *Boletivorus clandestinus.*

but fortunately, is not as common. Invariably there is a rusty pick-up truck, station wagon, or '65 Dodge Dart nearby, parked in the *middle* of the road, and the discarded tubes and wormy stalks of its quarry are apt to be haphazardly strewn about rather than stacked in the "middens" characteristic of *B. clandestinus.*

Boletivorus clandestinus is probably harmless, though I haven't been able to get close enough to find out. *Boletivorus brutalosipes,* on the other hand, has a well-deserved reputation for unprovoked acts of aggression. If you have some *Boletus edulis* in your basket, watch out!

Between these extremes, as might be expected, "hybrids" occur, in addition to possibly autonomous species* (I have even seen a skinny, blonde, bare-stemmed variety). However, further study—and preferably, a critical comparison with European specimens—is necessary before a definitive and natural classification can be worked out. Both species occur throughout the range of *Boletus edulis.* On weekends they are particularly abundant on lower Empire Grade above Santa Cruz, California, where their crafty cousin, *Pseudoboletivorus incognitus,* is also in evidence. (The latter closely mimics *Boletivorus clandestinus,* but only *pretends* to be stalking *Boletus edulis*—the true object of its desires being *Amanita calyptrata.*)

As a final note, I might add that what boletivores do during the summer is a mystery. It has been suggested by one colleague of mine that they hibernate in the trunks of their cars. Another possibility is that they follow the bolete season up and down the coast. A third is that they migrate east—or west. But perhaps the most reasonable hypothesis is that they are like squirrels—they stash away what they gather in hedges, holes, nests, or sheds, for when the hunting is good, they have neither the time nor the inclination to eat. Then, in the summer, when they have nothing *but* time, they do nothing *but* eat.

*In the newsletter of the Mycological Association of Washington, D.C., Anne Dow announces the discovery of a third species, *Boletivorus europaeus,* in the Tetons of Wyoming. Like *B. brutalosipes,* it "advances boldly when approached," but it "fills [rather than empties] the baskets [sic] of other boletivores" (or does she mean other bolete hunters?). I have never encountered such a species in California, but would welcome its introduction!

APHYLLOPHORALES

THIS large and diverse assemblage of fleshy—and not-so-fleshy—fungi includes all the Hymenomycetes (p. 57) with the exception of the agarics, boletes, and jelly fungi. The fruiting body is not normally gelatinous (as in the jelly fungi), there is no veil (as in many agarics and boletes), and the basidia are typically simple and clublike. The hymenium (spore-bearing surface) can be completely smooth and unspecialized as in the crust and parchment fungi, or it can take the form of tubes (the polypores), spines or "teeth" (the teeth fungi), upright clubs or branches (the coral fungi), or veins, wrinkles, or even rudimentary gills (the chanterelles).

More than twenty families in the Aphyllophorales are now recognized, but many of the distinguishing criteria are esoteric. To facilitate field identification, the Aphyllophorales have been divided into five major groups (plus one aberrant species), keyed below. The coral fungi and chanterelles (Clavariaceae and Cantharellaceae) are placed in a separate order, the Cantharellales, by some taxonomists.

Key to the Aphyllophorales

1. Fruiting body with a layer of downward-pointing spines or "teeth" on underside of cap or with icicle-like spines hanging from branches or a cushion of tissue .... **Hydnaceae**, p. 611
1. Not as above (but fruiting body may have pores whose walls become torn or toothlike in age) 2

2. Fruiting body with a layer of tubes (usually on underside of cap); tube mouths (pores) large and sinuous to small or very minute .... **Polyporaceae & Allies**, p. 549
2. Not as above; tubes and pores absent (but spore-bearing surface may have veins that give it a somewhat poroid appearance) 3

3. Fruiting body consisting of a small (up to 2.5 cm) hollow, bladderlike cap mounted on a thin stalk; surface of cap smooth or wavy, whitish to yellowish or pale brownish; found on ground or wood in eastern North America, often in groups or clusters .... *Physalacria inflata*
3. Not as above; common and widely distributed 4

4. Fruiting body usually bracketlike, crustlike, or sheetlike (i.e., usually without a stalk and often without a cap); usually growing on wood or herbaceous stems .. **Stereaceae & Allies**, p. 604
4. Not as above; fruiting body upright, either simple and clublike *or* branched (coral- or bushlike) *or* with a cap (often vase-shaped) and stalk; usually on ground, sometimes on wood 5

5. Fruiting body with a cap (i.e., a clearly differentiated, usually flattened or depressed upper sterile surface) and stalk .... **Cantharellaceae**, p. 658
5. Fruiting body clublike or branched or a rosette-like mass of flattened, white to creamy or tan lobes or segments (i.e., without a well-defined cap) .... **Clavariaceae**, p. 630

Bondarzewia montana (see description on p. 565). When these compound fruiting bodies were younger, they were whitish and resembled coral fungi (e.g., *Sparassis*) but for the presence of pores. As they mature, however, the upper lobes or "branches" broadened into tan or brown caps. This species is also capable of producing simple fruiting bodies (with just one cap and stalk).

Polypores and Bracket Fungi

spores

POLYPORACEAE & Allies

THIS large and exceedingly diverse group comes in a mind-boggling multiplicity of shapes and sizes and a pleasing array of gay decorator colors. As the name "bracket fungi" implies, the fruiting body is most often bracketlike or shelflike. "Polypore" (meaning "many-pored") describes the spore-producing tube layer which lines the underside of the cap. However, the tube mouths or **pores** are sometimes so minute that they're virtually invisible without a hand lens, and in some bracket fungi the pores are replaced by large mazelike pockets or even gills and in others the tube walls break up to form "teeth." The boletes are also equipped with tubes, but they are rapidly-decaying, mostly terrestrial fungi with a soft fleshy cap, central stem, and tubes that usually peel away easily from the cap.

Polypores, in contrast, grow mostly on wood and have a tube layer which is scarcely detachable. The most familiar types fruit on logs and stumps in dense, shelving masses (or in ascending tiers if the tree is still standing). Others, called **conks**, are those hard, woody, hooflike growths you see on living trees. Still others are **resupinate**: with neither cap nor stem, they lie flat on the wood or incrust it. Those that grow on the ground often originate from roots or buried wood, and their stem is usually off-center or even lateral.

Polypores are usually described as being tough, leathery, or woody. This is true of old specimens, but fresh individuals of some species, such as *Phaeolus schweinitzii,* are so soft and watery they can be wrung out like a sponge. Others, like the beefsteak fungus *(Fistulina hepatica)* and the sulfur shelf *(Laetiporus sulphureus)* "weep" in wet weather, exuding colored water droplets. In contrast, *Fomitopsis* and *Ganoderma* can be woodier than the wood they feed on!

Like most fungi, polypores fruit during the rainy season. However, mature fruiting bodies are so tough they tend to dry out instead of decaying. They may thus persist for months, even years. Some (e.g., *Ganoderma, Fomitopsis, Phellinus*) are actually perennial: rather than going to the trouble of manufacturing new fruiting bodies every year, they quite sensibly make use of existing resources by adding a new growth layer onto the old one. In this manner they may attain gargantuan dimensions—a 5 ft. × 3 ft. specimen of *Oxyporus nobilissimus* from the Pacific Northwest weighed 300 pounds, undoubtedly a record for fleshy fungal fructifications!

There is much more to polypores than their fruiting bodies, however. One cannot fully appreciate them without appreciating their work—that is, the chemical and structural changes they produce in their hosts. Polypores are absolutely indispensable to the forests of this world. They are *the* major group of wood-rotting fungi. Though they wreak economic havoc (15-20% of all standing timber in this country is said to be defective or unusable because of fungal decay, more than 90% of it caused by bracket fungi; structural timber is also destroyed—in ships, mines, houses, bridges, etc.), the polypores should not be seen as enemies. Without them there would be no logging industry in the first place; every cut stump, felled log, and lopped-off limb would lie indefinitely on the forest floor, the woods would quickly become impenetrable, and new trees would have neither room nor nutrients to grow.

In living trees the dead tissue (heartwood) in the center of the trunk is more susceptible to attack than the living tissue (sapwood) beneath the bark. Trees infected by **heart rot** may look outwardly sound, even when they are completely hollowed out inside. Sometimes the only clue to the pernicious presence of the rot is the fruiting body of the fungus! Since heartwood is largely dead, heart rot fungi are technically not parasites, but they *can* be deleterious—the infected trees are steadily weakened until they come crashing down at the slightest provocation. (So remember not to breathe too hard when picking polypores from a standing tree!) A few species, such as *Heterobasidion annosum*, are virulent parasites, destroying both the sapwood and the heartwood.

Spores generally gain access to living trees through wounds in the bark. Trees bruised or battered by wind, nearby construction, careless machinery, or mindless vandalism ("Jody loves Judy") are especially vulnerable. Sometimes spores enter through holes bored by grubs or beetles, which may in turn munch on mycelium growing in their tunnels.

Heart rot continues after the tree dries, and **sap rot** sets in. On fallen logs and branches you will find large clusters of colorful bracket fungi (e.g., *Trametes versicolor* and *Trichaptum abietinus*) which are digesting the sapwood. Several varieties may simultaneously or successively inhabit the same log. Recently felled timber is also susceptible to attack. Only complete immersion in water will prevent infection.

Wood is composed largely of dead, empty cells from which the fungus extracts nutrients. The cell walls have two principal constituents: cellulose, which makes the wood soft and tough, and lignin, which makes the wood hard and brittle. Fungi which digest the cellulose and leave the lignin behind are called **carbonizing decays** or **brown rots** because they render the wood dry, brittle, and darker than normal wood. Fungi which digest cellulose *and* lignin are called **delignifying decays** or **white rots** because they make the wood soft, spongy, and whiter than normal wood. Some species may completely hollow out the trunk or produce small, hollow pockets called **pocket rot.** The mycelial threads of bracket fungi can often be seen if the wood is broken open and examined closely. But if more than one species inhabit the same piece of wood, interpretation and diagnosis become complicated. Only repeated observation can tell you which ones produce which types of rot.

The part of the tree infected should also be noted. **Butt rots** such as *Phaeolus schweinitzii* are confined to the roots and base of the tree and the fruiting bodies are often found on the ground. **Trunk rots** like *Phellinus pini* infect the entire trunk, while **top rots** such as *Fomitopsis rosea* inhabit the top of the trunk.

Practically all of the polypores were originally lumped together in one giant genus, *Polyporus.* Dozens of families and genera have subsequently been proposed in a laudable effort to break the group down. As usual, however, there are widely divergent opinions on exactly how the task should be accomplished. The result is scholarly chaos, and the goal of arriving at one name for each kind of organism is still a long way off. To give you an idea of the difficulties faced by a prospective polyporologist (or any toadstool taxonomist) when she sets about consulting the literature on her pet polypore, here is a list of names given to *Trametes occidentalis* by various investigators:

> *Polyporus occidentalis, Coriolus occidentalis, Polystictus occidentalis, Microporus occidentalis, Coriolopsis occidentalis, Boletus sericeus, Trametes lanata, Polystictus lanatus, Microporus lanatus, Polyporus lanatus, Polyporus lenis, Polystictus lenis, Microporus lenis, Trametes wahlenbergii, Trametes scalaris, Polystictus scalaris, Polystictus cyclodes* var. *homoporus, Polystictus scorteus, Microporus scorteus, Polyporus gourliaei, Fomes gourliaei, Scindalma gourliaei, Trametes hispidula, Trametes devexa, Polystictus malachodermus, Polystictus substrigosus, Polyporus illotus, Polystictus illotus, Microporus illotus, Daedalea subcongener, Polystictus subcongener, Trametes heteromalla, Polystictus extensus, Polyporus badiolutescens* . . . (I'm suddenly very grateful that I'm *not* a taxonomist!)

Several hundred different polypores are known from North America; perhaps half of them occur in California. Unfortunately, the vast majority are too tough or too bitter to eat (or too tough *and* too bitter) to eat. Two happy exceptions are the sulfur shelf *(Laetiporus sulphureus),* a succulent and unmistakable treat, and the beefsteak fungus *(Fistulina hepatica),* which looks like a slab of steak. The hen of the woods, *Grifola frondosa,* and its look-alike, *G. umbellata,* are also excellent, but do not seem to occur in our area. A few polypores, however, may actually be poisonous (e.g., *Phaeolus schweinitzii*), and even the edible species are difficult to digest unless cooked thoroughly.

Since so few bracket fungi are edible, they are ignored by most people and barely mentioned in many popular mushroom books. As I am also ignored by most people (and have yet to be mentioned in any kind of book), I feel I have something vital in common with these unheralded but indispensable organisms. Therefore, a fairly extensive (but by no means comprehensive) treatment is offered here, in hopes that readers will at least learn to notice polypores, if not identify them. In the following key they are broken down into several groups based on field characters. If you have trouble with some of the choices, you can take solace in the fact that almost *everyone* finds the polypores difficult to identify!

Key to the Polyporaceae & Allies

1. Spore-bearing surface composed of closely-packed but discrete (separate) tubes that look like small pipes (use a hand lens if unsure and see photo at top of p. 553); flesh usually marbled or streaked when sectioned; fruiting body fleshy (*not* hard and woody) and often exuding a bloodlike juice when moist; found on or around hardwoods .... **Fistulina,** p. 553
1. Not as above; spore-bearing surface not composed of tubes, or if composed of tubes then the tubes forming a united layer (i.e., not discrete); fruiting body fleshy, tough, woody, etc. .. 2

2. Fruiting body with a stalk(s) and cap(s) (but may appear fingerlike when emerging); found on ground, or if on wood then stalk usually central or off-center; cap and stalk not varnished   3
2. Fruiting body knoblike, hooflike, bracketlike, shelflike, or crustlike; stalk absent, rudimentary, or attached to side or top of cap (or if central, then varnished); growing on wood or roots   4

3. Flesh yellow-brown to tawny, rusty-brown, or brown (or sometimes dark yellow, but if so then associated with conifers); flesh darkening (staining red to vinaceous-brown or black) in potassium hydroxide (KOH); pores often colored like flesh; usually growing on ground (often around stumps or trees or on roots) .... **Phaeolus, Inonotus, Coltricia,** & **Allies,** p. 566
3. Not as above; flesh usually white or pallid to pale yellow when fresh (in one case yellow to olive-yellow, but if so then associated with hardwoods in eastern North America), but sometimes pale brownish or otherwise discolored in age or where bruised; pores usually colored like flesh when fresh; on wood or ground .... **Polyporus, Albatrellus,** & **Allies,** p. 554

4. Pore surface completely enclosed (hidden) by a tough membrane (i.e., fruiting body with a tube-lined internal cavity) *or* fruiting body growing on birch and possessing a blunt projecting margin that forms a "curb" around the pore surface .... **Piptoporus & Cryptoporus,** p. 584
4. Pore surface exposed; not growing on birch, or if on birch then margin not curblike 5

5. Tube layer elastic and waxy to somewhat gelatinous when fresh, often separating easily from rest of fruiting body; pores usually pinkish to reddish-purple to blackish-purple (but occasionally white to buff); cap white and hairy (when present) 19
5. Not as above; pore surface differently colored and/or not separable and elastic-gelatinous   6

6. Fruiting body typically resupinate (i.e., lying flat on the wood or incrusting it), consisting of a simple layer of tubes; cap (sterile surface) absent or rudimentary .... **Poria** & **Allies,** p. 602
6. Fruiting body normally with a cap (upper sterile surface), but sometimes resupinate, especially if growing on the undersides of logs 7

7. Spore bearing surface composed of *shallow* veins which may form very broad "pores" or pits .... (see **Stereaceae & Allies,** p. 604)
7. Spore-bearing surface comprised of a true tube layer which forms minute to fairly large pores *or* spore-bearing surface with deep, elongated, mazelike pockets or even gills or "teeth" .. 8

8. Underside of cap (spore-bearing surface) with gills 9
8. Spore-bearing surface with tubes (pores), but the pores sometimes elongated or mazelike or breaking up to form small "teeth" 10

9. Gills appearing to be split lengthwise (along their edges), but actually composed of two adjacent plates which curl back in dry weather (see photos on p. 591) .... **Schizophyllum,** p. 590
9. Not as above .... **Lenzites, Daedalea,** & **Allies,** p. 586

10. Fruiting body annual (with only one tube layer), the cap soon dark brown to blackish or bluish-black with a resinous, often roughened (sandpaper-like) and/or radially wrinkled surface; fruiting body at first watery and often beaded with droplets, but tougher in age; pore surface whitish at first but aging or bruising darker (brownish); found on dead trees (logs, stumps, etc.); widespread but not common .... **Ischnoderma,** p. 573
10. Not with above features (but may have some of them); very common 11

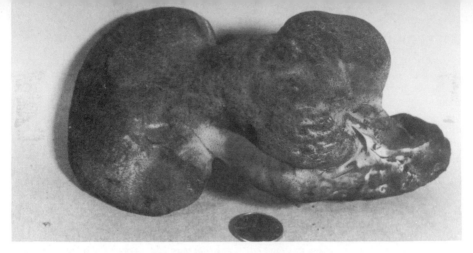

Fistulina hepatica, a young specimen in which the cap is still velvety. It's easy to see why it is sometimes called the "Ox Tongue." A mature specimen is shown in the color plate. (Ralph Buchsbaum)

11. Fruiting body punky, corky, or hard and woody even when fresh, medium-sized to very large and often thick; usually perennial, i.e., with more than one tube layer, but if annual then often with a varnished surface crust .... ***Ganoderma, Fomitopsis, Phellinus,* & Allies,** p. 574

11. Not as above; fruiting body usually annual, small to medium-sized or if large then usually fleshy or spongy when young and fresh; if tough when fresh then cap usually fairly thin and lacking a highly varnished surface crust 12

12. Flesh soon red to orange, rusty-brown, tawny, yellow-brown, or dark brown *and* darkening (staining red to vinaceous-brown to black) in potassium hydroxide (KOH); pores variously colored (including red, orange, and mustard-yellow), but not rosy, purple, or bright sulfur-yellow)* .... ***Phaeolus, Inonotus, Coltricia,* & Allies,** p. 566

12. Not as above; flesh white to yellow, yellowish, beige, or sometimes light brown or salmon-tinged, *not* darkening in KOH* 13

13. Underside of cap with elongated pocketlike or mazelike pores (see photos on pp. 586, 588) .... ***Lenzites, Daedalea,* & Allies,** p. 586

13. Not as above (but pores may break up to form "teeth") 14

14. Fruiting body usually preceded by a small, often zoned cup- or saucerlike "nest" (0.5-2 cm broad) from which small whitish to yellowish shelves (fruiting bodies) develop; common on dead hardwoods, especially elm, in eastern North America .... ***Poronidulus conchifer***

14. Not as above 15

15. Cap only 1-5 mm broad, with a stubby or knoblike stalk attached to the top; fruiting body white to brownish; common on hardwoods (rarely conifers) in eastern states ***Porodisculus pendulus***

15. Not as above; usually larger, widely distributed 16

16. Pore surface white to buff to yellowish or bright sulfur-yellow when fresh (but may stain blackish or rusty-reddish in age or when bruised) 17

16. Pore surface red to orange, orange-yellow, lavender- or violet-tinged, rosy, gray, black, or brown, or soon becoming these colors .... ***Trametes* & Allies,** p. 592

17. Pore surface bright sulfur-yellow when fresh (or rarely white, and if so then cap salmon-colored); cap orange-red to bright orange, bright yellow, or salmon-colored (but fading), usually at least 6 cm broad when mature .... ***Laetiporus,*** p. 572

17. Not as above 18

18. Cap thin and tough (but pliant) even when fresh, typically tearing easily in a radial direction; cap surface usually hairy or velvety and often concentrically zoned; often fruiting in large masses; very common, especially on hardwoods .... ***Trametes* & Allies,** p. 592

18. Cap soft, watery, spongy, or fleshy when fresh (but often tough in age), or if tough when fresh then not as above; cap often but not always thick, the surface sometimes hairy but not usually zoned; usually (but not always) found in small numbers ...... ***Tyromyces* & Allies,** p. 597

19. Pores averaging 4-8 per mm; found mainly on hardwoods ***Caloporus (=Gloeoporus) dichrous***

19. Pores averaging 2-4 per mm; tubes very shallow; found on dead conifers ***Skeletocutis amorpha***

---

*Several household cleaning agents (e.g., Drano, 1 teaspoon per ¼ cup water) can be substituted for KOH, but unless the flesh is borderline in color (i.e., dark yellow or light brown) the test is usually unnecessary. If there's any doubt, simply try both choices!

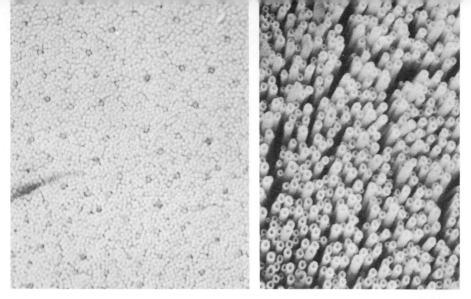

Left: A close-up of the tubes in *Fistulina hepatica.* Note how bruised area is darker. **Right:** An even closer view, in which the tubes can be seen as discrete units or "pipes"—a unique feature of *Fistulina.*

FISTULINA

THIS genus contains only one common species, the beefsteak fungus, *F. hepatica.* It is not a "true" polypore because the tubes, though packed together closely, are discrete units arranged like the bristles of a brush (see above photos). This character, which is best observed with a hand lens, has led polyporologists to give *Fistulina* a family of its own.

Fistulina hepatica   (Beefsteak Fungus; Ox Tongue)   Color Plate 152

FRUITING BODY annual, at first knoblike, then cushion- or shelflike; juicy when fresh, soft to fairly firm. **CAP** 7-30 cm broad and 2-6 cm thick when mature, tongue-shaped or fan-shaped; surface minutely roughened and suedelike or velvety, but often gelatinous when wet, firm or very soft and spongy, sometimes knobby; orange-buff to reddish-orange to pinkish, reddish-brown, dark red, or liver-colored; margin often lobed. Flesh thick, soft and watery when fresh and oozing with dark reddish juice; whitish to grayish, olive-brown, or reddish, streaked or marbled horizontally with paler veins (usually reddening after standing); taste rather sour or strongly acidic. **PORES** 1-3 per mm, circular, whitish to yellowish, buff, or pale flesh-color, becoming dark reddish-brown when bruised or in old age; tubes 10-15 mm long, closely packed but discrete (their walls not joined). **STALK** absent, or if present then lateral and rather short, 2-6 (10) cm long, 1-4 cm thick; tougher than the cap but continous with it and colored similarly, firm. **SPORE PRINT** pinkish-salmon to pale ochre or pale rusty-brown; spores 4-6 × 3-4 microns, elliptical, smooth.

HABITAT: Solitary or sometimes several together at the bases of hardwoods and on stumps, especially of chestnut and oak; widely distributed, but most common in Europe and eastern North America. In coastal California it seems to grow exclusively on chinquapin (a close relative of chestnut) in the fall and winter. In my experience it is not common, but *Chroogomphus*-king Ciro Milazzo discovered one large local fruiting of over thirty individuals on some recently cut chinquapin stumps. It causes a serious carbonizing decay in its host. The rich brown hue it imparts to the wood is prized by cabinetmakers.

EDIBILITY: Edible. The strong sour taste is displeasing to some, esteemed by others. Parboiling may remove some of the sourness as well as some of the nutrients (it is rich in vitamin C). I like to marinate thin, raw slices in seasoned vinegar and olive oil.

Fistulina hepatica, sliced open to show the streaked, meatlike flesh. Fresh specimens even exude a bloodlike juice. For other views of this photogenic fungus, see color plate and photos on pp. 552-553.

COMMENTS: This remarkable fungus looks like a slab of raw meat, especially after it is sliced open and starts oozing "blood." The streaked or marbled flesh is very distinctive, plus it never becomes as tough as other polypores and bracket fungi. The fruiting bodies seem to develop very slowly and are rarely attacked by maggots. Other species: ***Pseudo-fistulina radicata (=F. pallida)*** of eastern North America and Mexico also has discrete tubes, but the cap is grayish-brown to pale reddish and the stalk is usually longer (to 15 cm) and somewhat rooting. It is edible also.

POLYPORUS, ALBATRELLUS, & Allies
(Stalked Polypores)

Fruiting body annual, fleshy to tough, *with a cap and stalk or sometimes compound* (with many caps and stalks); growing on ground or on wood. CAP convex to plane, depressed, or misshapen. PORES large to very minute, often decurrent on the stalk; tubes usually not cleanly separable from the cap. STALK *central to off-center or sometimes lateral, usually well-developed but sometimes stubby.* SPORE PRINT usually white (when obtainable), but sometimes brownish in *Boletopsis.* Spores smooth, warty, or spiny; amyloid in *Bondarzewia,* otherwise not amyloid.

THESE are fleshy to rather tough polypores with a clearly defined cap, pore surface, and stalk. As such they are easy to recognize—the presence of a stalk separates them from most other polypores, and confusion with the boletes is unlikely because boletes have a soft, fleshy texture, central stalk, and tubes which are usually longer (deeper) and peel away easily from the cap (and often from each other).

The genus *Polyporus* once embraced practically all the polypores or woody pore fungi. The "splitters" have steadily whittled away at it, however, so that it now contains a modest number of stalked polypores with white, cylindrical, smooth, non-amyloid spores. Most of its species grow on dead hardwoods, but some arise from roots or underground "tubers" and thus may appear terrestrial. *Albatrellus,* on the other hand, is one of the few groups of polypores that is *truly* terrestrial, fruiting in sometimes massive numbers under both hardwoods and conifers. Microscopically it differs from *Polyporus* in having elliptical to nearly round spores which may or may not be amyloid.

To facilitate identification, a number of other stalked polypores are treated here as well, though their similarity to *Polyporus* and *Albatrellus* may be entirely superficial. These include: *Boletopsis,* terrestrial like *Albatrellus,* but with warted spores; *Heteroporus,* with a frequently misshapen fruiting body that is often entirely covered with pores; *Grifola,* with a compound, intricately branched fruiting body that gives rise to many small overlapping caps; and *Bondarzewia,* with a frequently compound fruiting body and warted amyloid

554

spores. The latter two genera can be reminiscent of coral fungi (especially *Sparassis*), but are easily differentiated by the presence of pores on the underside of each cap or "lobe."

Tender specimens of *Grifola* are delicious when thoroughly cooked, but most of the other polypores treated here are too tough or chewy to be more than marginally edible. None are known to be dangerously poisonous, however. Fourteen species are described here and several others are keyed out.

Key to Polyporus, Albatrellus, & Allies

1. Fruiting body often misshapen and covered in various areas (or almost entirely) by pores; pore surface usually reddening when handled .... ***Heteroporus biennis,*** p. 566
1. Pores confined to underside of cap (or decurrent on stalk), reddening in only a few cases .. 2

2. Fruiting body wood-inhabiting (on wood or roots or near the bases of trees or stumps) ... 3
2. Fruiting body truly terrestrial (growing on ground) 13

3. Stalk black or with a black base, and only one cap; pores minute (averaging 3-7 per mm) .. 4
3. Stalk not black at base, or if so then pores larger and/or fruiting body compound 5

4. Cap reddish-brown to chestnut-brown to blackish; stalk usually black .... ***P. badius,*** p. 562
4. Cap tan to ochre or paler; stalk often black only at base .... ***P. elegans*** & others, p. 562

5. Flesh very dense and heavy, becoming *bone-hard* as it dries out; cap(s) white to gray, buff, or even brownish, 2-10 cm broad, often in fused clusters but not truly compound; pores white to yellowish; found on dead conifers or rarely birch; not common   ***Osteina obducta (=O. ossea)***
5. Not as above (larger or with different habitat or color, etc.); *not* drying bone-hard 6

6. Fruiting body with one cap, small to medium-sized (cap usually 1-8 cm broad), flesh usually 2 mm thick or less; stalk usually less than 6 mm thick; found on wood above the ground   . 7
6. Not as above; if growing on wood above ground then fruiting body larger or compound   .. 9

7. Margin of cap often ciliate (fringed with hairs); pores averaging about 1 mm broad, usually hexagonal; stalk typically central .... ***P. arcularius,*** p. 563
7. Margin of cap not ciliate; pores usually larger or smaller than above; stalk central to lateral 8

8. Pores large (up to 3 mm), diamond-shaped or hexagonal, usually arranged in radiating rows; stalk very short, often lateral .... ***P. mori*** (see *P. arcularius,* p. 563)
8. Not as above; pores smaller; stalk central to lateral; stalk short or long ....
   .... ***P. brumalis*** and ***Microporellus*** spp. (see *P. arcularius,* p. 563)

9. Pores minute (3-7 per mm); pore surface *and/or* margin of cap staining gray to dark brown or black when bruised or dried; fruiting bodies often quite large; found at bases of hardwoods in eastern North America .... ***Meripilus giganteus*** (see *Bondarzewia montana,* p. 565)
9. Not with above features 10

10. Fruiting body compound, i.e., with many small (mostly 2-7 cm broad) overlapping caps or segments arising from a common, often branched base; fairly common in eastern North America, especially with hardwoods, but rare in the West; spores smooth 11
10. Not as above; fruiting body not compound, or if so then mature caps larger (5 cm broad or more) or fewer or habitat different; spores smooth *or* ornamented 12

11. Fruiting body white or tinged cream; fertile surface smooth, toothed, or with pores; growing in rosettes on hardwood stumps or roots in tropics and Gulf Coast   ***Hydnopolyporus palmatus***
11. Not with above features 28

12. Pores minute (2-4 per mm); pore surface sulfur-yellow, or if not then cap salmon-colored or pinkish *or* found in Southeast and cap buff to pinkish- or red-brown   (see ***Laetiporus,*** p. 572)
12. Pores larger, or if small then not as above 22

13. Cap blue to blue-gray to bluish-green (but often salmon-, ochre-, or rusty-stained in age) . 14
13. Cap not bluish when fresh (but may be greenish) 15

14. Pore surface white when fresh; found in western North America .... ***A. flettii,*** p. 558
14. Pore surface bluish when fresh; found in eastern states ***A. caeruleoporus*** (see *A. flettii,* p. 558)

15. Cap yellowish to yellow-brown to greenish or brownish-stained, with matted hairs or prominent scales in age; pores usually staining greenish, averaging 0.5-2 per mm ***A. ellisii*** & others, p. 559
15. Not as above; differently colored and/or cap not hairy-scaly 16

16. Spore print olive-brown to brown; flesh and pores *not* white; pores yellow to greenish, averaging 3 *or less* per mm; associated with hardwoods (especially ash and alder) (see ***Boletaceae,*** p. 488)
16. Not as above; spore print whitish and/or flesh white before exposure or differently colored 17

17.   Found under eastern hardwoods (occasionally conifers); pores 1-3 per mm, yellow to greenish or
      if white then cap yellowish, ochre, or greenish; spores smooth *A. cristatus* (see *A. ellisii*, p. 559)
17.   Not as above; differently colored or pores smaller or spores not smooth 18
18.   Fruiting body with many yellowish petal-shaped caps arising from a common base; found under
      western conifers; apparently rare .... *A. dispansus*
18.   Not as above 19
19.   Cap and stalk grayish to purple-gray, black, vinaceous, or with olive, pinkish, or brownish tones
      (sometimes whitish when young), but not typically yellow-stained; cap often with a somewhat
      streaked appearance; pores small (1-4 per mm), white when fresh but often discoloring grayish,
      vinaceous, etc., in age and usually staining olive to black in KOH; spores angular-warty under
      the microscope, not amyloid .... *Boletopsis subsquamosa group,* below
19.   Not as above; pores larger and/or cap and stalk differently colored and/or developing yellowish
      tones in age or after handling; spores smooth *and/or* amyloid 20
20.   Pores small (2-5 per mm); cap white to yellow, pinkish-tan, or pinkish-cinnamon (occasionally
      darker), not hairy or scaly (but may be cracked); stalk *not* blackish at base; tuber absent   21
20.   Not as above; pores larger or stalk darker or tuber present or cap hairy or scaly, etc. 22
21.   Cap or entire fruiting body often developing reddish or orangish stains in age or when dried;
      taste usually bitter .... *A. confluens* & others (see *A. ovinus*, p. 557)
21.   Not as above; taste mild or bitter .... *A. ovinus* & others, p. 557
22.   Cap and stalk brown to grayish-brown and covered with small hairs; taste very bitter (chew on a
      small piece of cap); odor sometimes (but not always!) like iodine; stalk with neither a tuber nor
      a gnarly rooting base; spores smooth; found in western North America .... *P. hirtus*, p. 560
22.   Not as above 23
23.   Fruiting body arising from a large, swollen, brown to black underground "tuber" (sclerotium);
      cap often with fibrils or scales; spores smooth .... *P. tuberaster*, p. 563
23.   Not as above (but fruiting body may have a rooting base); spores smooth or warted 24
24.   Found at or near bases of trees or stumps; fruiting bodies often clustered or compound (with one
      to many caps), usually arising from a gnarly, rooting base; caps pallid to tan, ochre, or brown,
      but usually lacking darker scales; spores with amyloid warts or spines 29
24.   Not as above; fruiting body typically *not* compound (but may be clustered); spores smooth 25
25.   Stalk with a black rooting base; pores usually 2-3 per mm *P. radicatus* (see *P. tuberaster*, p. 563)
25.   Not with above features 26
26.   Cap dark brown to brown, reddish-brown, or pinkish-brown; terrestrial *A. pescaprae*, p. 560
26.   Found on wood, or if terrestrial then cap usually paler, oranger, or yellower than above . 27
27.   Cap with brown scales, often large; pores often more than 1 mm in widest dimension; stalk often
      blackish or dark brown at base; found on wood, widespread but especially common in eastern
      North America .... *P. squamosus* (see *P. decurrens*, p. 561)
27.   Not as above; usually medium-sized; if found on wood then pores averaging 1 mm wide or
      smaller; cap smooth or scaly; stalk base not normally blackish *P. decurrens* & others, p. 561
28.   Caps usually spoon- or fan-shaped; stalks off-center to lateral ..... *Grifola frondosa*, p. 564
28.   Caps more or less circular; stalks central .... *Grifola umbellata* (see *G. frondosa*, p. 564)
29.   Each fruiting body with one to many caps; favoring conifers .. *Bondarzewia montana*, p. 565
29.   Favoring hardwoods, with one to several caps *Bondarzewia berkeleyi* (see *B. montana*, p. 565)

Boletopsis subsquamosa group   (Kurokawa)

CAP 4-15 (20) cm broad, broadly convex to plane or slightly depressed; circular in outline
to somewhat irregular; surface dry, smooth to fibrillose or breaking up into small scales,
especially at center; color variable: dingy whitish to gray, purple-gray, bluish-gray,
vinaceous-tinged, or with olive, pinkish, brownish, or even black tones (usually darker
in age), often with a somewhat streaked appearance; margin incurved at first, often wavy or
lobed. Flesh thick, firm, white or tinged variously with cap colors; taste mild to bitter.
PORES small or minute (1-5 per mm), usually white when fresh and young but soon
discolored grayish, vinaceous-buff, brownish, etc.; tubes 2-4 mm long, usually at least

Boletopsis subsquamosa is variable in color, but usually quite dark (gray, black, purplish, etc.) in age.

somewhat decurrent. **STALK** (2) 4-13 cm long, 1-4 cm thick, central or off-center, equal or with a narrowed base, solid, smooth; whitish or colored like the cap or pores. **SPORE PRINT** white to pale brown; spores 4-7 × 3.5-5 microns, elliptical to nearly round, but distinctly angular-warty.

HABITAT: Solitary to scattered or gregarious on ground under both hardwoods and conifers; widely distributed. In our area it is fairly common in the fall and winter under tanoak and madrone; to the north it often fruits in coastal sand dunes under pine.

EDIBILITY: Edible, but often bitter. It is esteemed by the Japanese, who usually soak it in brine to remove the bitterness.

COMMENTS: Like *Albatrellus ovinus,* this polypore is sometimes mistaken for a bolete because of its frequently central, well-developed stalk and terrestrial growth habit. However, the tube layer cannot easily be peeled from the cap and the overall shape is quite different. The color of the fruiting body (especially the cap) is extremely variable, leading to the naming of several different "species," including *B. leucomelas,* with a bluish-black to black cap, and *B. griseus,* with a grayish cap. These color forms appear to intergrade, however, and the common one in our area can be bluish-gray to purplish, pinkish-purple, or blackish, depending on season, temperature, and exposure to sunlight. The warty-angular spores are the critical microscopic feature. *Albatrellus avellaneus* (often listed as a variety of *A. ovinus*) can be similar in color but stains yellow and has smooth spores.

Albatrellus ovinus   (Sheep Polypore)

CAP 4-15 (20) cm broad, circular in outline or irregular; convex becoming plane or slightly depressed; surface dry, unpolished, smooth or breaking into scales in age; usually whitish at first, but often yellow, ochre, buff, pinkish, or tan (especially at center) in age, or in some forms purplish to purple-gray throughout; margin often wavy. Flesh thick, white or yellow, firm and rather tough. **PORES** minute (2-4 per mm), white but often becoming yellow or yellowish in age or when bruised; tubes shallow (1-2 mm long), usually decurrent. **STALK** 3-10 cm long, 1-3 (4) cm thick, central or slightly off-center, equal or enlarged below with a narrowed base; solid, firm, whitish or tinged cap color, but sometimes staining pinkish. **SPORE PRINT** white; spores 3-4.5 × 2.5-3.5 microns, elliptical to nearly round, smooth, not amyloid.

HABITAT: Solitary to scattered or gregarious, sometimes in fused masses, on ground in mixed woods and under conifers; widely distributed. It is very abundant in the Rocky Mountains in the late summer under spruce and I have also seen impressive fruitings in Maine. In our area it is locally common under manzanita in the fall and winter.

EDIBILITY: Edible when cooked well, and according to European sources, fairly good. Large quantities can have laxative effects, however, and material I tested was rather slimy ("okraceous") when cooked.

Two farflung terrestrial polypores. **Left:** *Albatrellus ovinus,* with a white to tan or yellowish cap. **Right:** *Albatrellus pescaprae* (see p. 560), with a scaly brown cap and well-developed stalk.

COMMENTS: This widespread *Albatrellus* is variable in color (the purplish-capped form is often called *A. avellaneus*), but can usually be recognized by the yellow tints that develop in age or after handling. The pores are sometimes so tiny that the pore surface may appear completely smooth. Because of its terrestrial habit this polypore is sometimes mistaken for a bolete, but is much tougher and does not have detachable tubes. Other species: *A. confluens* is a very similar, widespread species which becomes orange to pinkish-cinnamon in age or upon drying and has slightly larger, weakly amyloid spores and a bitter taste; *A. similis,* reported from Arizona, is also similar, but has distinctly amyloid spores; *A. peckianus* is an eastern beech-lover with a yellow to cinnamon-buff cap.

Albatrellus flettii   (Blue-Capped Polypore)

CAP 5-20 or more cm broad, convex becoming plane or centrally depressed; surface dry, blue to blue-gray or sometimes blue-green, but developing ochraceous, salmon, or rusty stains in age; margin at first incurved, often lobed or wavy. Flesh thick, firm and rather tough, white. **PORES** fairly small (1-4 per mm), white, but often becoming torn and developing salmon or ochraceous stains in age or upon drying; tubes 1-7 mm long,

Albatrellus flettii. The blue-gray to blue-green cap, white pore surface (the pores are too tiny to see here), and well-developed stalk distinguish this beautiful species. (Ralph Buchsbaum)

decurrent. **STALK** 5-15 cm long, (1) 1.5-4 cm thick, central or off-center, equal or thicker at either end, white to pale bluish-gray, or aging dingy ochraceous to reddish; solid, firm. **SPORE PRINT** white; spores 3.5-4 × 2.5-3 microns, elliptical to nearly round, smooth, weakly amyloid.

HABITAT: Scattered to gregarious or in fused clusters on ground in mixed woods and under conifers; known only from western North America. I have found it under knobcone pine in the fall and winter, but it seems to be quite rare in our area.

EDIBILITY: Edible. It has been sold commercially in northern California, but I haven't tried it. If it's anything like *A. ovinus,* then it isn't worth collecting.

COMMENTS: The unusual blue to blue-gray cap color contrasts nicely with the white pores to set this beautiful terrestrial polypore apart. Dingy ochraceous to rusty-ochraceous stains develop in age, but there are almost always vestiges of blue-green or blue somewhere on the cap. Other species: *A. caeruleoporus* of eastern North America is similar, but entirely indigo-blue to blue-gray (including the pores).

Albatrellus ellisii   (Greening Goat's Foot)

CAP 8-25 cm or more broad, convex becoming plane, wavy, or depressed; surface dry, at first hairy or plushlike but the hairs often matted in age or grouping to form coarse scales; greenish to sulfur-yellow or yellow-brown (often a mixture of colors), sometimes with darker brown shades or stains in age; margin often wavy and at first inrolled. Flesh thick, firm, white, sometimes bruising greenish slowly. **PORES** 0.5-2 mm in diameter, white, usually staining greenish or yellow-green when bruised or becoming yellowish to dingy greenish in old age; tubes 2-6 mm long, often decurrent. **STALK** 3-12 cm long or more, 2-6 cm thick, usually off-center or lateral, solid, tough, equal or thicker at either end; usually colored more or less like cap. **SPORE PRINT** white; spores 8-9 × 5-7 microns, elliptical, smooth.

HABITAT: Solitary, scattered, gregarious, or in fused clusters on ground in woods; widely distributed. It favors conifers and is especially common in the mountains of western North America. I have seen huge fruitings in the late summer and fall in the Cascades and Sierra Nevada, but in our area it is supplanted by *A. pescaprae.*

EDIBILITY: Edible, with a mild flavor and pleasantly chewy texture. Care must be taken to simmer it slowly and thoroughly or it will toughen.

Albatrellus ellisii. This terrestrial species is especially common under conifers in the Pacific Northwest and Sierra Nevada. Note hairy-scaly cap. The pore surface usually stains greenish when bruised.

COMMENTS: Also known as *Scutiger ellisii*, this impressive conifer-lover is easily told by its yellow-green to yellow-brown color, fairly large size, growth on the ground, and tendency to stain greenish (especially the pores). Its close relative, *A. pescaprae*, is browner or redder in color and more southern in distribution. *A. sylvestris* is a somewhat similar species with a smoky-olive or darker pore surface and roughened spores. *A. cristatus* of eastern North America is similar in color but has a less hairy cap and favors hardwoods.

Albatrellus pescaprae   (Goat's Foot)

CAP (2.5) 5-20 cm broad, convex to plane or centrally depressed; surface dry, covered with fine fibrils which form small, dense scales or sometimes a plush (especially toward center); dark brown to brown, reddish-brown, or pinkish-brown; margin often deeply indented. Flesh thick, fairly tough, white, bruising pinkish slowly; taste mild. PORES large, angular, 1-2 mm or more in diameter (or length); white to yellowish or with greenish stains, sometimes becoming pinkish in age; tubes 2-5 mm long, often decurrent. STALK 2.5-8 cm long, 1-3 (4) cm thick, solid, usually enlarged below (sometimes several arising from a common base); occasionally central, but usually off-center to lateral; white to yellowish or in age brownish. SPORE PRINT whitish; spores 8-11 × 5-6 microns, elliptical, smooth.

HABITAT: Solitary or in groups or clumps on ground in woods; widely distributed, but more southern than *A. ellisii*. In our area it is not uncommon in mixed woods and under tanoak and madrone, especially in the fall and winter.

EDIBILITY: Edible, at least according to European sources; I haven't tried it. Young caps are said to be quite tasty (though chewy) if cooked thoroughly.

COMMENTS: Also known as *Scutiger pescaprae*, this terrestrial polypore can be told by its reddish-brown to brown, densely scaly cap plus its large pores, well-developed stem (see photo at top of p. 558), and mild taste. *Bondarzewia* species are somewhat similar, but have spiny spores and less scaly caps and are not normally yellowish- or greenish-stained. *Polyporus hirtus* is also somewhat similar, but has a bitter taste.

Polyporus hirtus   (Bitter Polypore; Iodine Polypore)

CAP 5-18 cm broad, convex to plane or somewhat irregular; surface dry, covered with minute short, stiff, erect or matted hairs; uniformly brown to dark brown or grayish-brown. Flesh thick, white, firm and rather tough; taste very bitter (but often latently). PORES white to creamy, drying yellowish, 1-2 per mm; tubes 2-6 mm long, adnate to decurrent. STALK 2-10 cm long, 0.7-3 cm thick, usually lateral or off-center; equal or tapering downward, hairy and colored like cap; solid, firm, rather tough. SPORE PRINT white; spores 12-17 × 4.5-6 microns, spindle-shaped or cylindrical ("boletoid"), smooth.

HABITAT: Solitary or in groups on ground, around old stumps and trees (especially conifers), sometimes also on wood (often buried); northern North America. It is not uncommon in our area in the fall and winter, particularly in mixed woods of tanoak and redwood. It is also fairly common in the Pacific Northwest.

EDIBILITY: Inedible. If you boil it for several days and change the water frequently, you *might* remove the bitter taste, but you would remove everything else as well!

COMMENTS: Also known as *Albatrellus hirtus* and *Scutiger hirtus*, this species is the most common terrestrial polypore in our area, but is by no means numerous. The evenly colored brown hairy cap plus the white pores, bitter taste, and presence of a stalk are good fieldmarks (see photos at top of next page). According to mycologist Daniel Stuntz, specimens in the Pacific Northwest emit an iodine odor soon after being picked, but I have not detected such an odor in local material. *Bondarzewia montana* can be similar when it doesn't grow in clusters, but has warted amyloid spores and is not as bitter-tasting.

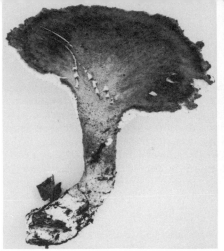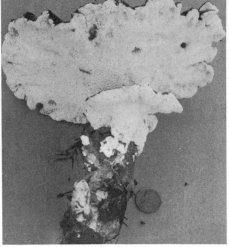

Polyporus hirtus. **Left:** Top view showing hairy brown cap. **Right:** Underside, showing white pores.

Polyporus decurrens

CAP 4-14 cm broad, convex to plane or slightly depressed; surface dry, yellowish to yellow-brown to ochre or dull orange (rarely reddish-brown), with darker (usually brown) erect fibrillose scales which become flattened in age. Flesh thick, white, tough. **PORES** fairly large (0.5-2 mm in diameter), angular or becoming torn or toothlike in age, white, but sometimes discoloring when dried; tubes 2-6 mm long, usually decurrent. **STALK** 2-10 cm long, (0.5) 1-3 cm thick, equal or thicker at either end, solid, tough; white or somewhat brownish, usually reticulate above from the decurrent tube walls. **SPORE PRINT** white; spores 10-18 × 4-6 microns, cylindrical, smooth.

HABITAT: Solitary or in small groups or tufts on the ground (usually originating from buried wood) or sometimes on dead wood, fall through spring; known from California, occasional. It favors hardwoods, but I have found it in a variety of habitats. It is the most common terrestrial polypore of southern California.

EDIBILITY: Unknown. The closely related *P. squamosus* (see comments) is edible when thoroughly cooked, but is thoroughly mediocre.

COMMENTS: The dull orange to yellowish-brown to ochre scaly cap with fairly large decurrent pores distinguish this species from *Albatrellus pescaprae* and others. It is also known as ***P. mcmurphyi.*** The "Dryad Saddle," ***P. squamosus,*** is a better-known, closely

Polyporus decurrens has large pores, a scaly cap, and may or may not be terrestrial. *P. squamosus* (not illustrated) is a similar but larger wood-inhabiting species.

related species. It is larger (cap 6-30 (60) cm broad and stalk up to 5 cm thick) and paler with dense, flattened brown to dark brown scales (but may become darker brown overall in age) and very large pores (1-10 mm each in largest dimension). It usually grows solitary or clustered on hardwoods (often living) rather than on the ground and its stem, when well-developed, usually has a black base. In North America it is most common east of the Rocky Mountains, but occurs occasionally in Washington and California. (I've seen what *might* be a photograph of this species from Fresno.) Another related species, *P. fagicola* *(=P. lentus)* of eastern North America, occurs on hardwoods but is smaller and less scaly.

Polyporus badius   (Black-Leg)

CAP 4-20 cm broad, convex becoming depressed or umbilicate; surface tan to chestnut-brown to dark reddish-brown or often becoming blackish in age; margin usually paler, often lobed. Flesh tough when fresh, rigid when dry, very thin, white becoming brownish. **PORES** minute, 4-7 per mm, white when fresh, brownish in age; tubes shallow, adnate to somewhat decurrent. **STALK** 1-6 cm long, 0.3-1.5 cm thick, central or off-center or lateral, more or less equal; pallid above and black below or more frequently black throughout, tough. **SPORE PRINT** white; spores 5-9 × 3-4 microns, cylindrical to elliptical, smooth.

HABITAT: Solitary or in groups on rotting hardwoods (especially aspen) or occasionally conifers; widely distributed, and occasional in our area in the fall, winter, and spring.

EDIBILITY: Too tough to eat.

COMMENTS: Also known as *P. picipes,* this species is close to *P. elegans* but has a darker and often larger cap (see photo on p. 42). Also, the entire stalk is usually black and it often grows on logs and stumps whereas *P. elegans* is more common on branches and sticks. Both species are retained in the genus *Polyporus* in its residual (narrowest) sense. For similar species, see comments under *P. elegans*.

Polyporus elegans   (Elegant Polypore; Black-Foot)

CAP 1.5-7 cm broad, round to kidney-shaped in outline; convex, soon becoming depressed, vase-shaped, or umbilicate; surface smooth or finely striate, pale tan to tan or ochre, often weathering to white; margin often wavy or lobed. Flesh tough when fresh, rigid when dry, thin, white to pale cinnamon. **PORES** minute, 4-6 per mm, white when fresh but often becoming grayish or brownish in age; tubes shallow, usually decurrent. **STALK** 0.5-5 cm long, 2-6 (10) mm thick, central or off-center or lateral, more or less equal or swollen at base; pallid or tan above, soon becoming black below; tough. **SPORE PRINT** white; spores 6.5-10 × 2.5-4 microns, cylindrical, smooth.

HABITAT: Solitary or several together on decaying hardwoods sticks, branches, and debris (only rarely on conifers); widely distributed. It is fairly common in our area in the

Polyporus elegans, a common wood-inhabitor. Note pale cap and black lower half of stalk. Compare it to the photo of *P. badius* on p. 42.

Polyporus arcularius grows on hardwoods. Note ciliate (hairy) cap margin and stalk that is *not* black.

winter and spring, especially on willow, alder, poplar (cottonwood), and oak.

EDIBILITY: Too tough to eat, but dries nicely for decorative purposes.

COMMENTS: This dainty polypore is easily recognized by its tan to whitish cap, small size, and black "foot." *P. varius* is sometimes listed as a synonym, but is also applied to a form with a radially streaked cap that is intermediate in color between *P. elegans* and *P. badius.* The latter species is usually larger than *P. elegans,* and has a darker cap and entirely black stem. Another species, *P. melanopus,* grows on the ground or from buried wood; it has a velvety stem and velvety-scurfy cap, at least when young.

Polyporus arcularius   (Fringed Polypore)

CAP 1-8 cm broad, round in outline, convex becoming depressed, vase-shaped, or umbilicate; surface dry, golden-brown to dark brown, usually minutely scaly; margin often ciliate (fringed with fine hairs). Flesh thin, white, tough. **PORES** large (1-2 per mm), angular or hexagonal, white or yellowish; tubes shallow, sometimes decurrent. **STALK** 2-6 cm long, 2-4 mm thick, usually central or slightly off-center, more or less equal, dark brown to yellowish-brown, smooth or minutely scaly, the base sometimes hairy. **SPORE PRINT** white; spores 7-11 × 2-3 microns, cylindrical, smooth.

HABITAT: Solitary or in small groups on dead hardwoods; widespread, but rare in our area. I've seen it in the Southwest and Pacific Northwest in the summer, and in eastern North America in the spring. Several closely related species are abundant in the tropics.

EDIBILITY: Much too small and tough to be edible.

COMMENTS: The small size and fairly large pores plus the typically central stem and frequently ciliate cap margin characterize this attractive polypore. *P. mori* (also called *Favolus alveolaris*), similar but without hairs on cap, and with a stubby stalk and large, diamond-shaped or hexagonal pores, occurs east of the Rockies. *P. brumalis* has much smaller pores and its cap is some shade of brown. It is common on birch and other hardwoods in eastern North America, rare in the West. *Microporellus dealbatus* has a concentrically zoned cap and central stalk; *M. obovatus* has a whitish, unzoned cap and central to lateral stalk. Both are common on dead hardwoods in the southeastern United States. None of these have the black or half-black stem of *P. elegans* and *P. badius.*

Polyporus tuberaster   (Stone Fungus)   **Color Plate 151**

CAP 4-15 cm broad, convex to plane with a depressed or umbilicate center or becoming funnel-shaped; surface dry, tan to brown or ochre, darker with age and with scattered darker brown fibrils or fibrillose scales which are often radially arranged; margin often indented or lobed. Flesh thin, pallid, rigid when dry. **PORES** 1-3 per mm, white to pale tan; tubes 1-3 mm long, usually decurrent. **STALK** 2.5-10 (20) cm long, 1-2.5 (4) cm thick, more or less equal, central or off-center, brown or colored like cap; solid, tough, arising from a large brown to black underground sclerotium ("tuber") which is rather rubbery when fresh but rock-hard when dry; exterior of "tuber" rough and irregular; interior often marbled with blackish-brown and paler areas and usually full of dirt and debris. **SPORE PRINT** white; spores 10-16 × 3.5-6 microns, cylindrical, smooth.

563

HABITAT: Solitary or in twos or threes on ground in woods; widely distributed, but not common. I have found it several times in our area in mixed woods and under oak and madrone. It fruits in the fall, winter, and spring.

EDIBILITY: Edible, but tough unless young, fresh, and thoroughly cooked. Native Americans ate an underground sclerotium which they called "tuckahoe." It was once thought to be this species (as reported in the 1st edition of *Mushrooms Demystified*), but is now believed to be the sclerotium of another polypore (see *Poria cocos*, p. 604). The "tubers" of *P. tuberaster* are inedible because they are full of dirt. However, they are sold in southern Italy under the name *pietra fungaia* ("Stone Fungus"). Buyers plant the "tubers" in flower pots, water them regularly, and then eat the fruiting bodies that result.

COMMENTS: The large underground "tuber" immediately identifies this polypore. The "tuber" might be mistaken for a piece of buried wood or a parasitized potato (it's usually the same size or larger), but the texture when fresh is distinctive. It is thought to be a resting stage of the fungus, or a secret storage compartment for nutrients. *Bondarzewia berkeleyi* may have a thick, rooting stem or "tuber," but is much larger. Other species: *P. radicatus* of eastern North America is somewhat similar but larger, and has a long, black, narrowed rooting base instead of a sclerotium; *P. mylittae* of Australia has a yellower cap and produces edible sclerotia weighing up to 40 lbs. each! The sclerotia are known as "Blackfellow's Bread," a reference to their use as food by the aboriginees.

Grifola frondosa   (Hen of the Woods; Sheep's Head)

FRUITING BODY compound, 15-60 cm broad or more, consisting of a mass of numerous small, overlapping caps arising from a common, fleshy, repeatedly branched base. **CAPS** 2-7 (10) cm broad, spoon-shaped, tongue-shaped, or fan-shaped and flattened; surface dry, smooth or rough to fibrillose, gray to brown or grayish-brown; margin often wavy. Flesh white and firm, rather tough; taste mild when young. **PORES** 1-3 per mm, white or yellowish; tubes shallow (2-3 mm long), decurrent. **STALKS** (branches) smooth, fleshy but tough, white or pale grayish, off-center or more often lateral (attached to sides of caps). **SPORE PRINT** white; spores 5-7 × 3.5-5 microns, broadly elliptical, smooth.

HABITAT: At or near the bases of trees and stumps (usually oak), fruiting year after year in the same spots; widely distributed, but common only in eastern North

Left: The "Hen of the Woods," *Grifola frondosa,* is a prized edible mushroom in eastern North America. The often massive fruiting body is composed of numerous petal- or fan-shaped caps arising from a common base. **Right:** *Grifola umbellata* mimics *G. frondosa,* but each of its caps has a more or less central (rather than lateral) stalk, as shown here. It is also delicious.

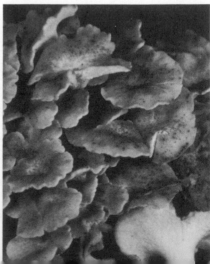

America. It has been reported from Idaho, but is rare in the West and apparently absent in our area. It causes a delignifying butt rot of both the heartwood and sapwood of its host.

EDIBILITY: Edible and choice—along with the sulfur shelf and beefsteak fungus, the best of the polypores for the table. Long, slow cooking is recommended and only the young, tender caps are worth eating. "Allergies" have been reported. It is excellent pickled.

COMMENTS: The numerous overlapping, more or less spoon-shaped caps are reminiscent of a fluffed-up hen, making this one of the safest and most easily recognized of all edible mushrooms. Clusters weighing 100 pounds have been recorded, but 5-10 pound specimens are the norm. There are usually more caps per cluster than in *Bondarzewia*, and the stems are attached to the sides of the caps rather than being central. *Polyporus* or *Polypilus frondosus* are synonyms. *G. umbellata (=P. umbellatus)*, also edible, is a similar species with whitish to gray to smoky-brown, circular caps with central stems. Its fruiting bodies arise from sclerotia ("tubers") that have been used by the Chinese as an immune system stimulant. It occurs across the northern half of the continent, but is not common.

Bondarzewia montana

FRUITING BODY simple (with a cap and stalk) or compound (with numerous caps and stalks arising from a common base), but often fingerlike when first emerging or taking the form of a lumpy, misshapen, or rosette-like mass (when compound); often massive when mature (up to 1 m (3 ft.) broad if compound); usually arising from a rooting base. **CAP(s)** 5-25 cm broad when mature, convex to irregular or depressed; surface dry, finely velvety to fibrillose, or becoming smooth; tan or ochre to brown or dark brown, sometimes paler when immature. Flesh fairly thick, white, firm but brittle when fresh; taste mild or sometimes bitter in age. **PORES** fairly large (0.5-2 mm broad), usually angular or irregular in shape or breaking up to form "teeth"; pore surface often lumpy or nodulose in compound fruiting bodies, white to creamy or buff, dingier when dried or in age; tubes 1-6 mm long, usually decurrent. **STALK(s)** central to lateral, continuous with individual caps and similar in color and texture, usually arising from a tapered, underground, gnarly rooting base that is 4-12 cm long and 2-5 cm thick. **SPORE PRINT** white; spores 5-7 microns, round, with conspicuous amyloid warts.

HABITAT: Solitary or in groups under conifers, usually near stumps or trunks (presumably arising from roots or buried wood); locally common in the late summer and fall in western North America, particularly at higher elevations. I have found it in quantity near Lake Tahoe under fir and white pine, but have not seen it on the coast. The very similar *B. berkeleyi* (see comments) favors living hardwoods, producing a serious delignifying butt rot of the heartwood ("string and ray rot") that completely hollows out its host.

EDIBILITY: Tempting because of its size, but rather tough and sometimes bitter to boot. It is *probably* harmless, but certainly not the equal of *Grifola frondosa*.

COMMENTS: The warted amyloid spores, often compound fruiting body, and gnarled rooting base are characteristic of the genus *Bondarzewia*. The spores are unique among the polypores, leading some investigators to suggest a relationship to the agaric genera, *Russula* and *Lactarius*. When *B. montana* has a massive compound fruiting body (see photo on p. 548) it is reminiscent of the hen of the woods, *Grifola frondosa,* an eastern hardwood-loving species with smooth spores. Fruiting bodies with one to several caps, on the other hand, are more likely to be confused with *Polyporus hirtus,* a bitter-tasting, smooth-spored polypore. Its only close relative, ***B. berkeleyi,*** is quite similar but usually has fewer (one to five), broader, paler (whitish to grayish or yellowish-tan) caps and slightly larger spores, plus it favors hardwoods (especially oak and maple). It seems to intergrade with *B. montana,* and is very widely distributed. Another sometimes gigantic polypore, ***Meripilus giganteus,*** occurs at the bases of hardwoods in eastern North America. It stains, ages, or dries gray to dark brown or black on the pore surface and/ or margin and has much smaller pores (3-7 per mm). Like *Bondarzewia,* it often has a compound fruiting body, but has smooth, non-amyloid spores. The individual caps in all of these species are usually larger than those of *Grifola.*

Heteroporus biennis

FRUITING BODY arising from a poorly developed stalk or fleshy base, with one cap or several fused together in an overlapping rosette, or very distorted with most of the surface covered with pores. **CAP** 3-9 (20) cm broad when well-developed, plane to depressed; surface dry, woolly-hairy or felty, white to tan or aging pinkish to reddish. Flesh tough, duplex, white or pinkish. **PORES** 1-3 per mm, angular, irregular, or mazelike, or becoming toothlike; whitish, but often discoloring or bruising reddish, dingier in age; tubes 2-6 mm long, usually decurrent. **STALK** central to off-center, poorly developed or even absent, continuous with and colored like the cap; hairy. **SPORE PRINT** white; spores 4-8 × 3-5 microns, elliptical, smooth; thick-walled, nearly round spores (chlamydospores) often present also.

HABITAT: Solitary or in groups on ground around hardwood stumps and trees (presumably growing on the roots), occasionally under conifers; widely distributed, but not common. I have found it in our area under oak and pine in the fall and winter.

EDIBILITY: Unknown; too tough to be worthwhile.

COMMENTS: Misshapen fruiting bodies are usually found with pores covering much or most of the mushroom—the best fieldmark of this otherwise unimposing, profoundly forgettable, pitiful excuse for a polypore. The growth habit is sometimes reminiscent of *Phaeolus schweinitzii. Abortiporus biennis* and *Polyporus biennis* are synonyms.

PHAEOLUS, INONOTUS, COLTRICIA, & Allies

Fruiting body usually annual, small to large, often *soft and spongy when fresh* but tough or corky-woody in age; growing shelflike to bracketlike to practically resupinate on wood *or* growing on ground and possessing a stalk and one to several caps. CAP usually hairy, fuzzy, or velvety. Flesh *orange to yellow-brown, rusty-brown, or brown.* PORES *orange to greenish-yellow, rusty-brown, brown, or grayish (but not white) at maturity;* small or large. STALK present or absent. SPORE PRINT whitish to yellow or brown when obtainable. Spores usually elliptical, smooth. Setae (large brown sterile cells) often present among basidia. Cap tissue darkening (turning red to black) in potassium hydroxide (KOH).

THESE orange to brown or rusty-yellow polypores are somewhat intermediate in aspect between the woody perennial polypores (conks) and the smaller and thinner, annual bracket fungi and stalked polypores. *Phaeolus* and *Inonotus* most closely resemble *Phellinus* (a genus of conks), but are usually softer and fleshier when fresh, and have annual rather than perennial fruiting bodies. *Coltricia,* on the other hand, has a central stalk and is reminiscent of a *Polyporus.*

  All three genera have tissue that darkens or blackens dramatically when touched with a 2% aqueous solution of potassium hydroxide (KOH). This feature is also shared by *Phellinus,* leading some mycologists to sequester the four genera in a family of their own. *Phaeolus* and *Inonotus* are wood-inhabiting but appear terrestrial when growing from roots or buried wood, while *Coltricia* is nearly always terrestrial. *Inonotus* species typically produce white rots and *Phaeolus* species cause brown ones (*P. schweinitzii* is a serious pest of standing timber). They are not edible because of their tough texture, and *P. schweinitzii* may actually be poisonous. Five species are described here.

Key to Phaeolus, Inonotus, Coltricia, & Allies

1. Fruiting body with a stalk, typically appearing terrestrial (but may be at or near base of tree) 2
1. Fruiting body growing shelflike or bracketlike on logs, stumps, trees, etc.; stalk absent ... 5

2. Underside of cap with concentrically-arranged plates or with very large pores (0.5-2 mm in diameter); pores whitish to brownish; stalk 2-5 cm long, with only one cap; confined to eastern North America; not common .... *Coltricia montagnei* (see *C. cinnamomea,* p. 568)
2. Not as above 3

3. Stalk well-developed and distinct from cap (only one cap present); flesh corky-tough and thin (usually less than 2 mm thick), never spongy 4
3. Stalk often present only as a narrowed, often rooting base, with one to several caps; flesh usually thicker than above and spongy when fresh (but tough in age) 11

4. Cap shiny or silky, rarely more than 5 cm broad; pores usually not decurrent ....
    .... *Coltricia cinnamomea,* p. 568
4. Cap often velvety but not usually silky or shiny, sometimes small but up to 11 cm broad ....
    .... *Coltricia perennis* (see *C. cinnamomea,* p. 568)

5. Pores small (2-4 per mm) and bright saffron (orange-yellow) to bright orange or red when fresh; fruiting body soft or tough; usually found on hardwoods .. (see *Trametes* & Allies, p. 592)
5. Not with above features; pores differently colored and/or larger 6

6. Flesh and tubes bright orange or soon becoming orange; typically on conifers 7
6. Not as above (but may be rusty-yellow, reddish-brown, etc.) 8

7. Pores large (1 mm or more in diameter); tubes 1 cm or more long; cap rudimentary (usually present only as a free margin) .... *Phaeolus alboluteus,* p. 571
7. Pores smaller; tubes shorter; cap present .... *Phaeolus fibrillosus* (see *P. alboluteus,* p. 571)

8. Typically growing on or under conifers 9
8. Typically growing on hardwoods (but occasionally on conifers) 12

9. Spore-bearing surface taking the form of gills or very long, meandering mazelike pockets *or* if not then fruiting body usually with an anise odor when fresh; flesh corky-tough; pores not yellowish or greenish when fresh; found on wood . (see *Lenzites, Daedalea,* & Allies, p. 586)
9. Not as above; found on wood or ground; flesh often spongy when young and fresh 10

10. Pores minute (3-6 per mm); cap surface soon more or less bald; growing shelflike or bracketlike on wood .... *Inonotus dryadeus* (see *I. tomentosus,* p. 569)
10. Pores usually larger (up to 4 per mm); cap surface usually hairy or velvety, at least when young; growing shelflike *or* on ground 11

11. Pore surface yellow to greenish-yellow when fresh and young; cap often showing bright (orange to yellow) zones when young (but fruiting body entirely reddish-brown to dark brown in old age); stalk (when present) often with more than one cap ..... *Phaeolus schweinitzii,* p. 570
11. Pore surface brown to grayish or hoary when fresh; cap yellow-brown to tan when young; stalk (when present) typically with only one cap; setae present ..... *Inonotus tomentosus,* p. 569

12. Fruiting body a roundish to cylindrical, compact mass of numerous small, thin, tough, closely overlapping caps arising from a solid core or "knot"; pores gray to brown; found on hardwoods in eastern North America (common name: "Sweet Knot," because it is *sometimes* very fragrant) .... *Globifomes graveolens*
12. Not as above 13

13. Fruiting body usually resupinate (i.e., lacking a well-defined cap), either dark brown to black and hard *or* growing in sheets under oak bark 14
13. Not as above; fruiting body usually with a cap (upper sterile surface) 15

14. Fruiting body yellow-brown to dull brown; usually growing under the bark of living oaks ...
    .... *Inonotus andersonii*
14. Fruiting body dark brown to black, hard, often cracked and usually irregular or cankerlike in shape; found mainly on birch .... *Inonotus obliquus*

15. Pores often mazelike or pocketlike (elongated), the walls between them fairly thick; cap thin (less than 1 cm, the flesh only 1-4 mm thick), corky or leathery even when fresh ....
    .... (see *Lenzites, Daedalea,* & Allies, p. 586)
15. Not as above 16

16.   Cap tan to pale brown (or weathering grayish) and distinctly hairy, usually 2 cm thick or less;
       pores whitish to gray, grayish-brown, or even blackish; found mainly on *dead* cottonwood
       (Populus) or willow, often in confluent rows .... ***Funalia hispida*** & relatives
16.   Not as above; if growing on cottonwood or willow, then differently colored, etc. 17
17.   Fruiting body soft and watery when fresh, entirely tawny-yellow to cinnamon or yellow-brown
       and staining red or vinaceous (*not* black) in KOH; typically found on *dead* hardwoods in
       eastern North America (reported rarely from the West on conifers) .. ***Hapalopilus nidulans***
17.   Not as above (but may have some of above features) 18
18.   Fruiting body tough and corky or woody even when fresh; cap surface not hairy or only slightly
       so; usually found on dead trees .... (see ***Phellinus gilvus***, p. 582)
18.   At least the upper layer of fruiting body usually fleshy (watery or spongy) when young; cap
       surface often (but not always) hairy; often on living trees *Inonotus hispidus* & others, p. 569

Coltricia cinnamomea   (Fairy Stool)

FRUITING BODY with a cap and stalk, usually terrestrial. **CAP** 1-5 cm broad, more or
less circular in outline, centrally depressed or umbilicate; surface dry, bright cinnamon
to reddish-brown, yellow-brown, rusty-brown, or darker, with shiny or silky striations and
narrow or inconspicuous concentric zones; margin often fringed or torn. Flesh very thin
(1 mm thick or less), pliant when fresh; rusty-brown. **PORES** 2-3 per mm, yellow-brown
to brown or reddish-brown; tubes shallow (0.5-3 mm long), typically *not* decurrent.
STALK 1-5 cm long, 1-4 mm thick, usually central and more or less equal, brown to
reddish-brown, hairy or velvety, tough. **SPORE PRINT** yellowish-brown; spores 6-10
× 4.5-7 microns, elliptical, smooth. Cap tissue staining black in KOH.

HABITAT: Solitary or in small groups on ground or moss in woods, often along well-
beaten paths, roadbanks, and in clearings (rarely on rotten wood); widespread. In our area
this species and *C. perennis* (see comments) occur year-round, but are not very common.

EDIBILITY: Too tough to eat, but makes an intriguing addition to seedpod arrangements.

COMMENTS: This beautiful little polypore is easily told by its very thin, silky-shiny,
cinnamon to amber-brown cap, brown pore surface, and reddish-brown to dark brown
(not black!) stem. The rusty-brown flesh distinguishes it from *Polyporus* and other stalked
polypores. In similar habitats or on charred ground and just as common is the elegant
C. perennis (see photo below). Its cinnamon to yellow-brown or grayish cap is up to 10 cm
broad and usually strongly zoned and finely velvety (rather than silky). It also tends to have
a thicker stem and brown to grayish-brown pores that are sometimes decurrent. It is widely
distributed, but a third species, *C. montagnei*, is restricted to eastern North America. It
has large, often mazelike pores or even thick, *concentrically* arranged plates or "gills."

Left: *Coltricia cinnamomea* has a small silky cap, brownish pores, and thin rusty-brown flesh. **Right:**
Coltricia perennis is often larger and more strongly zoned. Note tough central stalk in both species.

Inonotus hispidus

FRUITING BODY annual, growing shelflike or bracketlike on wood, at first soft and spongy but tough in age and rigid when dry. **CAP** 5-20 (30) cm broad and 2-10 cm thick, convex or plane; surface at first densely covered with stiff or bristly hairs, the hairs tending to wear away in age; bright reddish-orange to yellowish-brown to rusty-yellow when fresh, becoming brown to dark reddish-brown or even blackish in age. Flesh up to 5 (10) cm thick, at first watery or spongy (but fibrous), tougher in age; rusty to rusty-yellow to dark reddish-brown; odor often rather pleasant when fresh. **PORES** 1-3 (4) per mm, yellowish to brown or rusty-yellow, sometimes with an olive tinge and sometimes beaded with droplets, darkening with age or where bruised; tubes 0.5-3 cm long. **STALK** absent. **SPORE PRINT** ochre- to chestnut-brown; spores 7.5-11 × 6-9 microns, broadly elliptical to nearly round, smooth. Flesh darkening (staining red to black) in KOH.

HABITAT: Usually solitary (occasionally several) on living or recently dead hardwoods or rarely conifers; fruiting mostly in the summer and fall but occurring year-round, widely distributed. It is a destructive parasite of oak and walnut (causing a white rot of the heart-wood), but occurs on a wide range of other trees, including mulberry and willow. The fruiting bodies usually emerge from the wounds of trees, often at a considerable distance from the ground. I have not seen this species locally but it occurs in southern California.

EDIBILITY: Unknown.

COMMENTS: This species and its close relatives (see below) are reminiscent of *I. tomentosus* and *Phaeolus schweinitzii,* but are strictly shelving species with no stalk. They can also be confused with species of *Phellinus,* but are soft and spongy when fresh and are not perennial. Similar species include: *I. cuticularis,* with a woolly-matted to nearly smooth (not bristly) cap, found on various hardwoods such as willow, cottonwood, and pepper trees; *I. dryophilus,* with a thick granular core between the upper fibrous tissue and the tube layer, found on hardwoods (especially oak); *I. texanus,* found on desert legumes (particularly acacia and mesquite); *I. arizonicus,* found on sycamore; *I. dryadeus* (see comments under *I. tomentosus*); and *I. radiatus,* widespread (but especially common in eastern North America) on hardwoods such as birch and alder, with a rather thin, firm, brightly colored (golden to rusty-brown or even yellowish-green) cap that is often zoned and radially fibrillose plus unpigmented spores. See also *I. tomentosus,* which sometimes grows shelflike on conifers.

Inonotus tomentosus

FRUITING BODY usually terrestrial with a cap and a short or rudimentary stalk, but sometimes lacking a stalk and growing shelflike on wood. **CAP** 3-12 (18) cm broad, circular to fan-shaped in outline, convex to plane to centrally depressed; surface dry, soft and hairy (tomentose) or velvety, whitish when very young but soon yellowish-brown to tan, dull ochre, brown, or rusty-brown, sometimes with faint concentric zones. Flesh yellow-brown to brown, duplex: upper layer soft and spongy when fresh, lower layer rather thin, firm, and fibrous. **PORES** round to angular or irregular becoming torn or sometimes toothlike in age, 2-4 per mm, pale buff to grayish or beige or becoming brownish, but often with a hoary sheen or surface covering; darker brown where bruised; tubes 1.5-7 mm long, usually decurrent. **STALK** often rudimentary, when present short (1-5 cm long), 0.5-2 cm thick, central to off-center or lateral, continuous with cap and more or less same color and texture, or darker. **SPORE PRINT** pale yellow to pale brown; spores 4.5-7 × 2.5-4 microns, elliptical, smooth, hyaline under the microscope. Brown sterile cells (setae) abundant among basidia, straight and pointed. Cap tissue blackening in KOH (sometimes with a fleeting red intermediate phase).

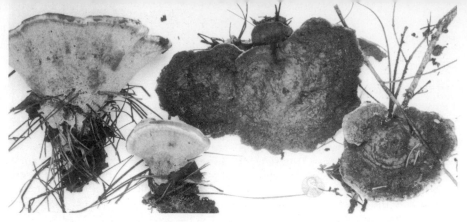

Inonotus tomentosus is reminiscent of *Phaeolus schweinitzii*, but is usually smaller and paler in color. In addition, the fruiting body is usually simple (with one cap rather than several). Note how pine needles and sticks are incorporated into the fruiting bodies.

HABITAT: Solitary or in groups on ground under conifers (presumably arising from roots or buried wood), sometimes also on stumps or bases of trunks; widely distributed. In our area it is fairly common throughout the mushroom season, especially in coastal pine forests. It causes a white pocket rot of the roots and butt of a wide variety of conifers in the pine family, but is especially fond of pine and spruce.

EDIBILITY: Unknown, but too tough to be of value.

COMMENTS: Also known as *Onnia tomentosa* and *Polystictus (*or *Mucronoporus) tomentosus,* this common polypore can be recognized by its soft hairy or velvety cap and overall brown to yellow-brown color. It is thicker and spongier than *Coltricia;* it could be confused with *Phaeolus schweinitzii,* but is duller in color, somewhat smaller, and does not usually have a compound fruiting body. Also, the fresh pore surface is brown to grayish or hoary rather than yellow or greenish. *I. circinatus* is a very similar species with hooked setae; it is also common on or near conifers across North America. *I. dryadeus* favors conifers in the West (but oak in the eastern states); it always grows shelflike on wood, has a bald or nearly bald cap, minute pores, and attains sizes of 30 cm broad or more. For shelflike (stalkless) species of *Inonotus* that favor hardwoods and have more highly pigmented (browner) spores, see *I. hispidus* and the species listed under it.

Phaeolus schweinitzii (Dyer's Polypore) Color Plate 153

FRUITING BODY usually compound, composed of several caps arising in tiers from a common base, but sometimes simple (with one cap) and sometimes growing shelflike on wood. **CAP(s)** 5-30 or more cm broad, circular to fan-shaped in outline; cushion-shaped becoming plane or depressed; soft and spongy when fresh and often knobby when actively growing, tough or corky in age, rigid and brittle when dry; surface covered with a dense felty or woolly mat of hairs, smoother in age; color variable: orange to ochraceous to yellowish or greenish-yellow (especially on the margin) when growing, becoming rusty-brown to dark brown in age (or from the center outward); staining brown to blackish when bruised and becoming entirely dark brown to blackish in old age; sometimes concentrically zoned with several of above colors; margin often wavy. Flesh yellowish to rusty-brown or brown, often appearing zoned. **PORES** 1-3 per mm or fused together to form larger pores; mustard-yellow to greenish when fresh, but quickly becoming brown or blackish when bruised or in age; tubes 2-10 mm long, usually decurrent. **STALK** (if present) 1-6 cm long and 1-5 cm thick, usually tapered downward, often rooting, central or off-center, same color and texture as cap. **SPORE PRINT** white or tinged yellow to green; spores 5-9 × 3.5-5 microns, elliptical, smooth. Setae absent. Cap surface and tissue staining black in KOH, often with a fleeting cherry-red intermediate phase.

Phaeolus schweinitzii, a common conifer-lover. Note compound fruiting body; these are rather small specimens.

HABITAT: Solitary or in groups on or around dead and living conifers, usually but not always appearing terrestrial (originating from the roots); widely distributed and very common throughout the West. In our area it favors pine and Douglas-fir, the fruiting bodies usually appearing after the first fall rains but persisting year-round. The mycelium attacks the roots and heartwood of its host, causing a serious carbonizing decay known as "red-brown butt rot." It fractures the wood into cubical blacks as high as fifteen feet up the trunk, weakening the tree so that it blows over easily. It also occurs on coniferous slash, in which case it is more apt to be shelflike than terrestrial. In the West it is one of the two or three most prevalent brown rot fungi. There is one report of it growing on eucalyptus.

EDIBILITY: Possibly poisonous. It contains a stimulant found in the roots of the kava kava plant, but apparently some toxic substancces as wcll. It is too tough and hairy to bc of food value anyway. However, it is prized by dye-makers for the rich and varied hues it imparts to yarn.

COMMENTS: One of the most common and conspicuous of the larger polypores, this cosmopolitan species is apt to confuse the beginner because of the color and texture changes it undergoes as it ages. However, the yellowish to greenish-yellow pores when young, yellow to rusty-orange tones in the caps of young specimens, and growth at the base of or near conifers are diagnostic. (The color plate shows young specimens that are predominantly orange rather than greenish-yellow.) It grows gradually, engulfing pine needles, sticks, plants, and other debris in its way, as shown in the color plate. Old, weathered fruiting bodies are quite light in weight and entirely dark brown in color. *Inonotus tomentosus* is somewhat similar, but has a tan to dull brown cap and brownish or hoary pore surface. *Hydnellum* species are superficially similar, but have short, blunt spines or "teeth" on the underside of the cap instead of pores.

Phaeolus alboluteus   (Orange Sponge Polypore)

FRUITING BODY resupinate and spreading or bracketlike, soft and spongy when fresh, up to 1 meter or more in length. **CAP** (upper surface) usually present only as a free margin, soft and spongy; surface bright orange but eventually weathering or fading to whitish. Flesh thin, orange. **PORES** large (1-3 mm or more in diameter), angular, the walls often jagged or splitting to form "teeth"; whitish or yellowish soon becoming orange (or tubes orange with white edges); tubes long (1-3 cm). **STALK** absent. **SPORE PRINT** white; spores (7) 10-14 × 3-4 microns, cylindrical, smooth. Flesh reddening in KOH.

571

HABITAT: On fallen logs of conifers, usually on the undersides and often developing in snow; common in the subalpine forests of the West, fruiting mainly in the spring but persisting into summer, fall, or even winter. It is quite common in the southern Rocky Mountains, and I have also seen it in the Sierra Nevada. It produces a brown rot (carbonizing decay) and is said to occasionally occur on aspen.

EDIBILITY: Unknown.

COMMENTS: This species, which has also been placed in *Pycnoporellus* and *Aurantioporellus,* is easily told by its large pores and orange spongelike fruiting body that peels off in one continuous felty layer. *P. fibrillosus* is a somewhat similar species that is shelflike (i.e., with a well-defined fiery red to orange or brownish-orange cap). It has creamy to orange pores and grows on conifer slash or rarely aspen. *Pycnoporus, Aurantioporus,* and several Porias are also brightly colored, but are tougher and/or have much smaller pores.

LAETIPORUS

THIS is a small genus of large, fleshy, shelflike fungi. Only one colorful species—the well-known and edible sulfur shelf—is described here. A key hardly seems necessary.

Laetiporus sulphureus   Color Plates 154, 155
(Sulfur Shelf; Chicken of the Woods)

FRUITING BODY annual, emerging knoblike or fingerlike, soon becoming shelflike; soft and fleshy when young, tough in age. **CAP** 5-50 (70) cm broad and up to 4 cm thick, fan-shaped to elongated or semi-circular in outline; surface smooth to suedelike, often uneven or wrinkled; red-orange to bright orange, yellow-orange, sulfur-yellow, or salmon (the margin usually yellow), fading slowly with age to yellowish, buff, or eventually dull whitish; margin at first thick and blunt. Flesh thick, soft and watery when very fresh, becoming tough and eventually crumbly; white to pale yellow or salmon-tinged; when *very* fresh often exuding yellow or orange droplets and reminiscent of uncooked chicken; odor fungal or rather pungent; taste nearly mild or acidic becoming quite sour or unpleasant in age. **PORES** 2-4 per mm but barely visible when young, bright sulfur-yellow, but often darkening when bruised and fading slowly in age; tubes shallow (1-4 mm long). **STALK** absent or present only as a narrowed base. **SPORE PRINT** white; spores 5-7 × 3.5-5 microns, broadly elliptical to nearly round, smooth.

HABITAT: Solitary or more often in overlapping clusters or shelving masses on dead stumps and logs, sometimes also on living trees or sometimes growing in rosettes from roots or buried wood; usually appearing on the same stumps year after year; widely distributed and common. In our area it fruits mainly in the late summer and fall, but old fruiting bodies may persist for months. It grows on a wide range of hardwoods and conifers. In coastal California it favors eucalyptus (September-October, ushering in the new mushroom season) and to a lesser extent conifers and oaks (November-December). In the Sierra Nevada it is common on fir in the summer; in eastern North America it favors oak. It causes a destructive red-brown carbonizing heart rot that eventually hollows out the tree, producing cracks in the wood in which thin sheets of white mycelium may appear. It is said to have caused considerable damage to British sailing vessels.

EDIBILITY: Edible and delectable when thoroughly cooked. However, there have been several cases of sulfur shelf poisoning on the west coast, so try it cautiously the first time and never eat it raw. When young the succulent flesh has a mild flavor, tofu-like texture, and "Candy Corn"-like color, making it especially attractive and delicious in omelets. Maturing specimens are tougher and develop a strong sour flavor; their texture when cooked is reminiscent of white chicken meat, so they're good in sandwiches. If the specimens you find are mature (but not so old as to be asbestos-like),, you can trim off the tender, rapidly growing margin (about 5 cm), and perhaps return later for more!

COMMENTS: One of the "Foolproof Four"—the brilliant yellow-orange shelving masses are unmistakable. Actually, nothing is *fool*proof, but the sulfur shelf is certainly *intelligence*-proof, and I trust that no one reading this book is a fool! I always experience an element of disbelief when I stumble onto a large cluster. It looks like something out of a Jacques Cousteau movie—you no more expect to find it on an aging eucalyptus stump by the railroad tracks than you do a freight train at the bottom of the sea! Fresh specimens can be so soft that it's difficult to handle them without leaving fingerprints. They are full of water and weigh far more than their size suggests (I have found clusters weighing over 50 pounds!). It seems strange that such a large fungus should require so little moisture to fruit, for in our area it often appears *before* the arrival of the fall rains. The cap color ranges from deep orange to yellow or salmon, but the fresh pore surface is always sulfur-yellow—except in the rare *var.* **semialbinus,** which has a salmon-colored cap, white pores, and a frequently rooting base. *Polyporus sulphureus* and *Grifola sulphurea* are synonyms. Other species: *L. persicinus* is a southeastern species that usually grows in stalked clumps; it has whitish to creamy pores and a buff to pinkish-brown or darker cap.

ISCHNODERMA

THIS genus is easily recognized by its roughened, resinous, dark brown to blackish, shelflike fruiting body. A single species is described here.

Ischnoderma resinosum (Resinous Polypore)

FRUITING BODY annual; shelflike or bracketlike, watery at first and often exuding droplets, especially near margin of cap, becoming tougher and drier in age. **CAP** 5-30 cm broad and 1-3 cm thick, fan- to kidney-shaped or semicircular in outline; surface rough or velvety (like sandpaper) at first (but often nearly smooth in age), often wrinkled radially and saturated or incrusted with a resin (the resinous areas often darker and more metallic-looking and sometimes slightly bluish), often concentrically zoned or ridged in age; color usually dark brown to blackish, but often overlaid with a thin golden or ochre coating when very young; margin usually quite thick. Flesh whitish to beige, tan, or brownish, watery when young becoming tough and corky in age. **PORES** minute (3-6 per mm), white or creamy, but often becoming brownish when bruised and ochre-brown to brownish in age; tubes 1-10 mm long. **STALK** absent. **SPORE PRINT** white; spores 4-7 × 1.5-2.5 microns, sausage-shaped to cylindrical, smooth.

HABITAT: Solitary or several together (often overlapping) on dead hardwoods and conifers, fruiting mostly in the summer and fall; widely distributed but not particularly common, at least in my experience. In northern California and the Pacific Northwest it favors larger (first-growth) conifers. It causes a delignifying decay of both the sapwood and heartwood, and the infected area sometimes has a strong aniselike odor. I have not seen it in our area.

EDIBILITY: Said to be edible when young and watery, but soon tough and corky.

COMMENTS: This distinctive shelving polypore is easily identified by its resinous dark brown to blackish cap that is often radially wrinkled and its tendency to exude droplets of liquid (especially when young or in wet weather). Some mycologists reserve the name *I. resinosum* for the form that grows on hardwoods (particularly elm), while using the name *I.* **benzoinum** for the conifer-loving version with darker (browner) flesh and tubes.

GANODERMA, FOMITOPSIS, PHELLINUS, & Allies
(Conks)

Fruiting body medium-sized to very large; *tough, woody, corky, or punky, usually perennial (but some species annual); often thick; growing on dead or living trees.* CAP knoblike to hooflike to shelflike or bracketlike, *sometimes with a hard surface crust;* often zoned, ridged, or grooved. PORES fairly small to minute or barely visible; *tubes* often stratified (with more than one layer) as seen in longitudinal section. STALK *usually absent or rudimentary,* but sometimes present as a lateral extension of the cap (see *G. lucidum*). SPORE PRINT brown to whitish but difficult to obtain. Spores smooth or minutely prickly.

THESE are the tough, woody, hoof-shaped or shelflike growths you see so often on the trunks of dead trees or in the wounds and crotches of living ones. They are commonly called "conks" because of their woody to corky texture and the frequent presence of a hard surface crust. Conks are larger and thicker than the other tough bracket fungi (e.g., *Trametes*) and usually—but not always—have perennial fruiting bodies (i.e., each year a new tube layer is added onto the already-existing ones). In many species these tube layers are distinctly stratified so that the age of the fruiting body can be determined by counting the number of layers (just like counting the growth rings on a tree). Ages of 50-70 years have been recorded! If the yearly tube layers are not stratified, the age can often be estimated by the number of growth zones (represented by ridges or furrows) on the cap. A new zone is "issued" each year as an outer or downward extension of the cap in conjunction with the new tube layer.

Most conks have at one time or another been classified in the genus *Fomes.* However *Fomes,* like *Polyporus,* has been obliterated by the "splitters," so that it now contains only one common species *(F. fomentarius).* The principal genera recognized here are *Ganoderma,* with a hard, sometimes shiny surface crust, whitish to pallid pore surface when fresh, and minutely prickly, double-walled spores; *Fomitopsis,* with whitish to pale-colored flesh and smooth spores; and *Phellinus,* with rusty-brown to yellow-brown flesh that darkens in potassium hydroxide (KOH), and smooth spores. Several small genera (e.g., *Fomes* in its residual sense and *Heterobasidion*) are also described and/or keyed out.

Conks are responsible for many serious rots. *Phellinus pini, Fomitopsis officinalis,* and *Heterobasidion annosum* are especially destructive to living or standing conifers, *Fomitopsis pinicola* to dead ones, and the *Ganoderma applanatum* group to living and dead hardwoods.

Conks are obviously far too tough or woody to eat (see comments on the edibility of *Fomitopsis pinicola!*), but several enjoy other uses (see in particular *Ganoderma lucidum, G. applanatum,* and *Fomitopsis officinalis*). Only some of the more distinctive and/or common species in North America are treated here.

Key to Ganoderma, Fomitopsis, Phellinus, & Allies

1. Pore surface rosy or pink when fresh (but often duller or darker in age); flesh usually pinkish also (when fresh) .... *Fomitopsis cajanderi* & others, p. 580
1. Not as above; pore surface not rosy 2

2. Cap (and stalk if present) with a surface crust that appears varnished, at least when young (unless covered by spore dust); fruiting body usually annual (with one tube layer); flesh usually light-weight or punky; *fresh* pore surface white, but typically turning brown if rubbed or in age   3
2. Varnished surface crust absent, or if present then not as above; stalk typically absent 6

3. Typically growing on hardwoods 4
3. Typically growing on conifers 5

4. Cap ochre to whitish or only partly red; stalk typically present; mainly southern (southeastern United States), rare northward .... *Ganoderma curtisii* (see *G. lucidum,* p. 577)
4. Cap usually reddish (or mostly reddish), at least in age; stalk present or absent; widespread in temperate zone and tropics .... *Ganoderma lucidum,* p. 577

5. Stalk usually absent; flesh up to 10 cm thick; known only from western North America .....
.... *Ganoderma oregonense* (see *G. lucidum*, p. 577)
5. Stalk usually present; flesh up to 3 cm thick ..... *Ganoderma tsugae* (see *G. lucidum*, p. 577)

6. Flesh bright reddish-orange to rusty-orange; fruiting body often thick (tall) in relation to diameter; restricted to juniper .... *Truncospora demidoffii* (see *Phellinus pini*, p. 582)
6. Not as above 7

7. Flesh bright yellow-brown to rusty-brown, brown, or dark brown when fresh 8
7. Flesh whitish to yellowish or straw-colored (or light brown to dingy yellow-brown in age) 18

8. Pore surface white when fresh but turning brown when scratched (or in old age); cap with a hard surface crust; very common .... *Ganoderma applanatum* group, p. 576
8. Not as above 9

9. Fruiting body annual (though sometimes large), i.e., with only one tube layer; usually soft or spongy or exuding water droplets when fresh, but usually tough in age; if found on hardwoods, the hardwoods usually living .... (see *Phaeolus, Inonotus, Coltricia,* & Allies, p. 566)
9. Not as above; fruiting body very tough (corky or woody) even when fresh, usually with more than one tube layer (perennial), especially if growing on living hardwoods 10

10. Typically growing on conifers .... *Phellinus pini* & others, p. 582
10. Typically growing on hardwoods 11

11. Fruiting body typically without a cap (resupinate) or with only a rudimentary one 12
11. Fruiting body with a cap (upper sterile surface), usually hooflike or shelflike 13

12. Fruiting body hard and woody; free margin ("cap") when present often with a surface crust
.... *Phellinus laevigatus* & *P. pomaceus* (see *P. igniarius*, p. 581)
12. Fruiting body corky or fibrous-tough but not woody; free margin when present not incrusted
.... *Phellinus ferruginosus* & *P. ferreus* (see *P. gilvus*, p. 582)

13. Fruiting body with a clearly differentiated hard surface crust (at least in older specimens), or if not then the older (buried) tube layers showing whitish flecks or streaks when sectioned; cap gray, brown, or black; pore surface brown or gray; especially common in northern regions 14
13. Not with above features 15

14. Older (buried) tube layers typically stuffed with white hyphae that show as white streaks or flecks when broken open; old caps often black and cracked ..... *Phellinus igniarius,* p. 581
14. Not as above .... *Fomes fomentarius* (see *Phellinus igniarius*, p. 581)

15. Typically growing on locust or other legumes ...... *Phellinus rimosus* (see *P. gilvus*, p. 582)
15. Not as above 16

16. Cap ochraceous to rusty-brown (sometimes blackish in old age); flesh usually less than 1.5 cm thick; usually found on dead wood .... *Phellinus gilvus* & others, p. 582
16. Cap differently colored and/or flesh thicker; usually found on living trees 17

17. Flesh bright yellow-brown .... *Phellinus robustus* (see *P. gilvus*, p. 582)
17. Flesh rusty-brown or darker .... *Phellinus everhartii* (see *P. gilvus*, p. 582)

18. Growing on incense cedar; pore surface usually yellow or yellowish when fresh; typically with only one layer of tubes .... (see *Tyromyces amarus,* p. 601)
18. Not as above; usually growing on a different host 19

19. Cap distinctly hairy and white to pale ochraceous (or greenish from algae and moss) 20
19. Not as above 21

20. Fruiting body often massive; cap surface coarsely hairy "like a doormat" (Stuntz); found on conifers in western North America; rare .... *Oxyporus nobilissimus*
20. Not as above; found on hardwoods (especially maple) in eastern states . *Oxyporus populinus*

21. Flesh cheesy when young but soon becoming chalky, very bitter-tasting; cap white to yellowish or becoming grayish or greenish in old age; found on conifers . *Fomitopsis officinalis,* p. 579
21. Not as above 22

22. Underside of cap with mazelike or elongated pores or sometimes even gills; cap 5-15 cm broad, tube walls often thick; found mainly on hardwoods (see *Lenzites, Daedalea,* & Allies, p. 586)
22. Not as above 23

23. Cap surface usually (but not always) partly or entirely reddish to reddish-brown or reddish-black; cap smooth or grooved; tube layers stratified .... *Fomitopsis pinicola,* p. 578
23. Cap surface usually roughened, pitted, knobby, ridged, etc., not normally reddish; tube layers not usually stratified .. *Heterobasidion annosum* & others (see *Fomitopsis pinicola,* p. 578)

Artist's Conk *(Ganoderma applanatum* group), with my wife Judith Mattoon included for scale. This large specimen shows ridges and furrows characteristic of many perennial polypores (conks). The underside is white when fresh and turns brown when scratched (see photo on next page).

Ganoderma applanatum group   (Artist's Conk; Artist's Fungus)

FRUITING BODY emerging whitish and knoblike, becoming hooflike or shelflike to somewhat irregular; very hard and woody; perennial. **CAP** 5-75 cm broad or more, 2-20 cm thick, usually fan-shaped or semi-circular in outline, with a hard surface crust that is usually cracked, furrowed, ridged, and/or lumpy or knobby in age, but *not* varnished; surface gray to brown or grayish-brown, sometimes aging grayish-black, but often covered with brown to cocoa-brown spore powder. Flesh punky or corky, 0.5-5 cm thick, brown or cinnamon-brown, rarely whitish. **PORES** barely visible (4-6 per mm), white or whitish when fresh but instantly turning brown when scratched; often dingy yellowish to brown when dried or in old age; tube layers distinctly stratified and separated by a thin layer of chocolate-brown tissue; each tube layer 4-12 or more mm deep. **STALK** usually absent. **SPORE PRINT** brown or reddish-brown; spores 6-9.5 × 4.5-7 microns, broadly elliptical to slightly truncate at apex, thick-walled, appearing minutely spiny.

HABITAT: Solitary or in groups on hardwood logs and stumps, or growing from wounds in living trees (usually near the ground); also common on conifers in some areas. Very widely distributed, found year-round. Its hosts include virtually every hardwood found in North America, plus numerous conifers. In coastal California it (and close relatives—see comments) is especially common on bay laurel, but can also be found on oak, magnolia, pepper trees, acacia, eucalyptus, elm, and Douglas-fir. Along with *Fomitopsis pinicola,* it is the commonest conk in our area. The only regions where it seems to be absent are those where there aren't any trees! It usually attacks dead or dying trees, but can also be parasitic, producing a whitish, delignifying decay of the sapwood and heartwood. Infected trees blow over easily, so if one appears on a tree near your house—watch out!

EDIBILITY: Much too tough and woody to be worthwhile (see edibility of *Fomitopsis pinicola*); however, sturdy specimens can be made into stools.

COMMENTS: This is the hard, woody growth you see so often at the bases of bay laurels and other hardwoods. Since the brown-staining of the pore surface is reasonably permanent, it makes an excellent medium for etching (or better yet, leaving cryptic messages in the woods)—hence its popular name, artist's conk. It's been calculated that a large specimen (I've found one weighing 26 pounds) liberates 30 billion spores a day, 6 months a

Ganoderma applanatum group, underside (pore surface). It is sometimes called the "Artist's Conk" because the white pore surface turns brown when scratched. This means you can draw pictures on it, or better yet, leave messages such as this one in the woods!

year—or over 5,000,000,000,000 (5 trillion!) spores annually. Millions of these may be borne aloft by air currents and deposited on *top* of the cap, turning it brown. *G. applanatum* as discussed here is actually a "collective" species, embracing at least two others: *G. annularis,* with unstratified tubes and no flesh; and *G. brownii (=G. adspersum? G. europaeum?),* with thicker (3-8 cm), darker flesh. Both have longer spores (9-12 microns) than the "true" *G. applanatum.* The latter is especially common in our area on bay laurel, perhaps more so than *G. applanatum.* All of these conks used to be placed in *Fomes* and have also been put in *Elfvingia.*

Ganoderma lucidum   (Varnished Conk; Ling Chih)

FRUITING BODY annual, corky and tough; often emerging as a whitish or pallid knob but soon becoming shelflike or developing a cap and stalk. **CAP** 2-20 (35) cm broad, 4-8 cm thick, circular to semi-circular to fan-shaped or kidney-shaped in outline; surface usually with a varnished (shiny) surface crust, smooth or often concentrically zoned and grooved; color variable: dark red to reddish-brown, orange-brown, mahogany, or reddish-black, but often ochre or yellowish toward the margin (which is often white when actively growing); surface sometimes covered with brownish spore powder. Flesh ochraceous-brown to dark brown, or pallid near the cap and brownish near the tubes; soft-corky or punky when fresh, tough when dry or old. **PORES** minute (4-7 per mm), whitish or yellowish-white when fresh, usually bruising or aging brown; tubes 2-20 mm long, one layer only (rarely two). **STALK** sometimes absent, but often present; usually attached laterally, but often vertical and well-developed, 3-14 cm long, 0.5-3 (4) cm thick; often gnarled or twisted, equal or enlarged below; dark red to reddish-black and appearing varnished like the cap. **SPORE PRINT** brown; spores 7-13 × 5-8 microns, elliptical, double-walled, appearing minutely roughened.

HABITAT: Solitary or in small groups at bases of living hardwoods or on stumps and roots, very rarely on conifers; fruiting during mushroom season but often persisting year-round. Distribution worldwide—in both the tropical and temperate zones, from Canada to Argentina and Europe to Siberia, China, India, Australia, and Africa. It is fairly common in eastern North America but rather infrequent in California. It occurs on a wide

577

range of hosts, but in North America is partial to maple. The very similar *G. tsugae* and *G. oregonense* occur on conifers (see comments). It is sometimes parasitic, but can also be saprophytic, producing a wet white rot in its host.

EDIBILITY: Too tough to be edible, but in the Orient it is pictured in many classical art works and has been revered for centuries as a symbol of success and well-being (Ling Chih means "marvelous herb" or "mushroom of immortality"). It has been used in the treatment of cancer and various other maladies, and is believed by some to have the power of arousing the dead. One method of preparation is to soak the fruiting body in wine for several months. The resulting liquid "essence" or elixir is then drunk or put into candies. The varnished fruiting bodies are also used for decorative purposes.

COMMENTS: This striking polypore and its close relatives are easily recognized by their shiny ochre to reddish-black caps that look as if they had been artificially varnished or shellacked (see Color Plate 156) unless old or covered by spore powder. Like *Piptoporus betulinus,* the fruiting bodies are annual (with one tube layer) but persist for months. Two extreme growth forms occur: one is fairly large, often sessile (with little or no stalk) and particularly common in North America. The other is smaller (cap rarely over 15 cm broad), has a long slender stalk, and appears to be more common in the tropics and Old World. However, a large number of intermediate forms also occur, plus occasional abnormal ones in which the fruiting body is branched and antler- or tree-like, and these are all currently thought to be forms of one highly variable, farflung species. *G. curtisii* is a similar species with an ochre to whitish or only partly reddish cap that often lacks the varnish in age. It typically has a stalk and grows on hardwoods in eastern (especially southeastern) North America. *G. tsugae* (**COLOR PLATE 156**) is also very similar to *G.lucidum,* but has white flesh and grows only on conifers, particularly hemlock, in northern North America. Still another similar species, *G. oregonense*, also grows on conifers, but is usually larger (cap 5-100 cm broad, 2-20 cm thick!), with larger spores and a somewhat darker or slightly duller cap. It is the most common "Varnished Conk" of Washington, Oregon, and California. All of these can be distinguished from *Fomitopsis pinicola* by their punkier (softer) flesh, annual fruiting body, and brown-staining pore surface (when fresh).

Fomitopsis pinicola   (Red-Belted Conk)

FRUITING BODY emerging as a whitish, pale yellow, or lilac-tinged knob, becoming hoof-shaped or shelflike or sometimes bracketlike; hard and woody in age. **CAP** 5-40 (75) cm broad, 3-22 cm thick, usually fan-shaped to semi-circular in outline; surface developing a thin, hard, resinous crust in age which is sometimes slightly varnished; usually at least partially reddish to dark red, but sometimes brown and often rusty or blackish-brown toward the base and brightly colored (white, yellow, ochraceous, or reddish) at the margin; concentrically furrowed and/or zoned in age; growing margin rounded, thick, and blunt. Flesh corky or woody, very tough, white to pinkish-buff or yellow when young, pale brownish or straw-colored in old age; usually bruising pinkish when actively growing; odor when fresh rather strong and fungal. **PORES** minute (3-5 per mm), white or pale yellow becoming brownish in old age; *not* turning brown when scratched but sometimes bruising yellow or pinkish-lilac; tube layers distinctly stratified, each layer 2-8 mm deep. **STALK** typically absent. **SPORE PRINT** white or pale yellowish; spores 5-8 × 3.5-5 microns, cylindrical to elliptical, smooth. Flesh staining reddish to dark reddish-brown in KOH.

HABITAT: Solitary or in groups on dead trees, logs, and stumps or rarely on living trees, perennial; very common and widely distributed. It has been recorded from a wide range of hosts but is found mainly on conifers (in our area primarily Douglas-fir and redwood). It attacks both the heartwood and sapwood of its host, producing a slow carbonizing rot which fractures the wood and turns it brown ("brown crumbly rot"). Large mycelial mats can often be seen in the fractured wood.

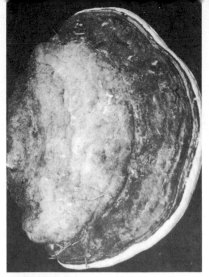

Left: *Fomitopsis pinicola,* one of the most common of all the conks (top view). **Right:** A large, old specimen of *Fomitopsis officinalis* (described below). Note how small the penny is!

EDIBILITY: This woody conk is not often eaten, but in a pinch you can use the following recipe developed by my wife: "Saw into 2-inch cubes, then marinate in olive oil and dandelion wine for at least 48 hours (be sure to use LOTS of garlic!). Roast slowly on skewers over charcoal indefinitely (minimum time: 20 hours). Cool. Pound vigorously with a large mallet between two pieces of thick leather. Pulverize in a meat grinder and then force through a braced sieve (allow several hours for this step). Wrap the resulting mess in several thicknesses of cheesecloth and hang up someplace high and out of the way (on a clothesline or TV antenna). Allow to dangle thus for at least one week. (Aging has a mellowing effect, so you may want to try one year.) Wring periodically, making sure to reserve the drippings for gravy or as a motor oil additive. To eat, boil for twenty-four hours, squeeze thoroughly, garnish with gravel, and serve forth."

COMMENTS: This beautiful but cosmopolitan conk is a major destroyer of dead coniferous timber. The reddish or partially reddish cap with a brightly colored margin (when growing) and pallid pores that do *not* bruise brown are good fieldmarks. Varnished specimens might be confused with *Ganoderma tsugae* or *G. oregonensis* (see *G. lucidum*), but are perennial and much harder and denser. *Fomes pinicola* and *Ungulina marginata* are synonyms. Another common, widespread woody conk with whitish flesh and white to yellowish pores is *Heterobasidion annosum (=Fomes annosus).* It has a bracketlike to shelflike to resupinate or irregularly knobby fruiting body that, when shelflike, is usually thinner and rougher (knobby, pitted, grooved, etc.) than *F. pinicola.* The cap is usually brown to grayish-brown with a pallid growing margin, but is sometimes reddish-brown (and can be whitish when very young) and its flesh does not redden in KOH. It is a frequent parasite of conifers (or occasionally hardwoods) and usually grows *at the base* of the trunk or from its roots. It is especially common on second-growth mountain conifers (e.g., ponderosa and Jeffrey pines). Two other species with whitish to yellow-brown flesh, *F. fraxinophilus* and *F. ellisianus,* somewhat resemble *H. annosum* but have different hosts: the first favors ash trees, while the second grows on buffalo berry (a shrub).

Fomitopsis officinalis   (Quinine Conk; Quinine Fungus)

FRUITING BODY perennial, emerging as a whitish knob, then becoming convex and finally hoof-shaped to cylindrical; hard and tough in age. **CAP** 4-30 cm broad and 5-40 cm or more thick (high); surface with a thin crust, often cracked and/or furrowed in age;

white to yellowish, but aging grayish and sometimes with a greenish covering of algae. Flesh thick, white, cheesy when young but chalky or friable (crumbly) in mature or old specimens; odor farinaceous; taste very bitter. **PORES** 3-4 per mm, white or whitish when fresh, discoloring in age or drying; tube layers each 3-20 mm long, often stratified. **STALK** absent. **SPORE PRINT** whitish; spores 4-5.5 × 3-4 microns, broadly elliptical, smooth.

HABITAT: Solitary or several on living and dead conifers; common year-round throughout much of the West, especially on larch and pine, but also on spruce, fir, hemlock, and Douglas-fir. It apparently does not occur in our area, but I have seen it on sugar pine and ponderosa pine on the slopes of the Sierra Nevada. It is said to be one of the three major destroyers of standing coniferous timber in the West *(Phellinus pini* and *Phaeolus schweinitzii* are the other two). It causes "felted heart rot," an extensive carbonizing trunk rot with thick mycelial mats often several feet long.

EDIBILITY: Friable but not fryable—it is inedible due to the bitter taste and tough texture, but has been used as a laxative and quinine substitute. (It does not have antimalarial properties, but was thought to because of its taste.) Since the fruiting bodies often grow high above the ground, commercial "quinine" collectors used to dislodge them with rifles.

COMMENTS: Also known as *Laricifomes officinalis* and *Fomes officinalis,* this destructive conifer-killer is easily recognized by its pale color, chalky white flesh, bitter taste, and sometimes massive size. New fruiting bodies are convex, but add layers yearly until they are hoof-shaped if growing on slash or vertically elongated (cylindrical) if growing high up on living trees (see photo at top of p. 579). Specimens 60 cm (2 ft.) in height with more than 70 tube layers have been recorded!

Fomitopsis cajanderi   (Rosy Conk)

FRUITING BODY often perennial; shelflike or bracketlike to somewhat hoof-shaped, tough in age. **CAP** 2.5-10 (13) cm broad, 0.3-2 cm thick; surface covered with hairs, becoming nearly bald in age (but often roughened), not incrusted; pinkish-red to pinkish-brown becoming brown to grayish-brown and in old age blackish except for the margin; often zoned or grooved concentrically; margin usually thin, acute. Flesh rosy-pink to reddish-brown or pinkish-brown, rather soft when fresh but corky in age. **PORES** minute (3-5 per mm), rosy to pinkish-red or pinkish-brown when fresh, often duller and darker (reddish-brown) in age; tube layers not distinctly stratified; each layer 1-3 mm deep. **STALK** absent. **SPORE PRINT** whitish; spores 4-8 × 1.5-2.5 microns, cylindrical but slightly curved (sausage-shaped), smooth.

HABITAT: Usually in colonies on dead conifers, but also reported on madrone and various fruit trees; widely distributed. It is fairly common in our area, especially on Douglas-fir, and is usually but not always perennial. I find it most often on the cut ends of recently felled trees. It produces a carbonizing brown pocket rot but is of minor economic importance.

EDIBILITY: Inedible (but see comments on edibility of *F. pinicola).*

COMMENTS: The beautiful rosy pore surface is the outstanding feature of this outstanding polypore, which is better known as *Fomes subroseus.* The upper surface is not as hard and crusty as that of many species of *Fomitopsis* and *Fomes,* and the fruiting bodies are not so obviously perennial—therefore it is apt to be looked for in another genus. Other species: *F. rosea* is practically identical, but has paler (silvery-pink to pale rose) flesh, slightly broader, cylindrical (not curved) spores, and a cap surface that is sometimes incrusted slightly in older specimens. It also favors conifers, causing a top rot of dead trees.

Phellinus igniarius. The black cracked cap is typical of older specimens. The fruiting body can be hoof-shaped as well as shelflike.

Phellinus igniarius   (False Tinder Polypore; False Tinder Conk)

FRUITING BODY perennial, soon shelflike or bracketlike to hoof-shaped; hard and woody. **CAP** 5-20 cm or more broad, 2-12 (20) cm thick; surface usually with a crust, at least in older specimens; brown and finely hairy or velvety when young, soon becoming bald and grayish, then finally black; often cracked, furrowed, and/or knobby in age; margin brown and velvety when actively growing. Flesh hard, woody, rusty-brown to brown; taste sour or bitter. **PORES** minute, 4-5 per mm, grayish-brown to brown; tube layers each 2-5 mm long but not always stratified; tubes and pores of older (buried) layers often stuffed with white mycelial threads that show as streaks or flecks when cut open. **STALK** absent or rudimentary. **SPORE PRINT** whitish; spores 5-7 × 4-6 microns, round or nearly round, smooth. Brown sterile cells (setae) sometimes abundant among the basidia. Cap tissue blackening in potassium hydroxide (KOH).

HABITAT: Solitary or in small groups on hardwood trunks (usually living); widely distributed. It is especially common on birch and aspen in the northern half of North America, and occurs year-round. In our area it grows on alder, madrone, manzanita, maple, and willow, but in my experience is not common. It can be quite destructive, causing an intensive white heart rot that reduces its host to a "soft, spongy, whitened mass."

EDIBILITY: Inedible unless you are fond of wood (see recipe under *Fomitopsis pinicola*).

COMMENTS: This conk is easily recognized by its brown to grayish-black cap and pores, brown to rusty-brown flesh, velvety brown actively-growing margin, and whitish-flecked or streaked tubes when old. A smaller variant, often called *P. tremulae*, is common on aspen. *P. laevigatus* is a closely related resupinate variety that usually grows on birch; *P. pomaceus,* often resupinate, causes extensive heart rot in fruit trees *(Prunus)*, especially wild ones; *P. robustus* (see comments under *P. gilvus*) is similar, but has yellow-brown flesh and favors oak or eucalyptus. Finally, there is the "true" tinder polypore or "Amadou," *Fomes fomentarius.* This hoof-shaped fungus resembles *P. igniarius* in color but has a hard thick surface crust, non-stratified tubes *without* white mycelial threads, and cylindrical spores 14-20 microns long. It grows on dead hardwoods (especially birch and maple) or from wounds in living trees, and is widely distributed. Unlike *Ganoderma*

581

applanatum, its pore surface is not white and does not turn brown when scratched. As its name implies, it has been used for centuries to ignite fires. Since the tubes are not stratified, they are quite long and readily soak up liquids through capillary action. Chunks of the tube layer were soaked in a salt peter solution, then dried and used as "matches."

Phellinus gilvus   (Oak Conk)

FRUITING BODY shelflike or bracketlike, often perennial; tough and corky when fresh. **CAP** 2.5-15 cm broad and 1-3 cm thick; fan-shaped or semi-circular in outline; surface at first velvety (just the growing margin if perennial), becoming rough to nearly smooth and/or somewhat zoned; bright rusty-yellow to ochraceous when young (as is the growing margin of older specimens), dark rusty-brown and finally blackish in age; not incrusted. Flesh tough, bright ochraceous to dark yellow-brown or colored like cap. **PORES** minute (4-8 per mm), grayish-brown becoming reddish-brown or dark brown; tubes in 1-5 layers, each 1-5 mm long. **STALK** absent. **SPORE PRINT** whitish; spores 4-5 × 2.5-3.5 microns (but rarely found), oblong-elliptical, smooth. Large brown sterile cells (setae) abundant among the basidia. Cap tissue blackening in potassium hydroxide (KOH).

HABITAT: Solitary or more often in colonies on dead or occasionally living hardwoods (rarely on conifers); widely distributed. It occurs year-round in our area on oak and tan-oak and is our most common perennial polypore of *dead* hardwoods. It produces a general delignifying decay of the sapwood, rendering it whitish and very brittle.

EDIBILITY: Unknown.

COMMENTS: The dark rusty-brown color and velvety yellow-ochre growing margin are the fallible fieldmarks of this common conk. The upper surface is not crusty as in many conks, and the color and texture are somewhat reminiscent of old *Phaeolus schweinitzii,* which grows with conifers (usually on the ground). A specimen of *P. gilvus* can be seen in the photograph at the bottom of p. 887. Other *Phellinus* species that favor hardwoods include: *P. robustus,* widespread, especially on oaks, large and woody but not incrusted, with a gray-brown to blackish cap, bright yellow-brown flesh, and stratified tubes (it also grows on eucalyptus, pittosporum, walnut, cactus, various other hardwoods, and on conifers); *P. everhartii* (see photo at bottom of p. 583), very hard and woody, up to 40 cm broad, growing mainly on oak, with unstratified tubes, brown spores, and setae; *P. rimosus (=P. robiniae),* with a rich brown cap, growing on locust, acacia, and other legumes; and *P. ferruginosus* and *P. ferreus,* common on dead hardwoods and rather similar to *P. gilvus,* but with a rusty-brown resupinate (capless) fruiting body. Most of these were once placed in *Fomes,* but now belong to *Phellinus* because they darken in KOH, have rusty-brown to tawny flesh, and often possess setae. For conifer-loving species, see *P. pini,* and for a similar but more brightly colored annual species, see *Inonotus radiatus* (under *I. hispidus*).

Phellinus pini   (Pine Conk)

FRUITING BODY perennial; shelflike, hooflike, or bracketlike, very tough or woody. **CAP** 2-20 cm broad and 1-15 cm thick; hoof-shaped to convex or fan-shaped; surface hard, often crusty but not shiny; rough or cracked, minutely hairy or roughened (like sandpaper) at first, often concentrically grooved in age; tawny to rusty-brown becoming brown to reddish-brown or brownish-black in age; margin sometimes brighter. Flesh less than 1 cm thick, tough, tawny to rusty-reddish or ochre. **PORES** 2-5 per mm, round to irregularly sinuous, ochraceous-tawny to rusty-brown becoming brown; tubes 2-5 mm long, in one or several layers. **STALK** absent or rudimentary. **SPORE PRINT** brown; spores 4-6 × 3.5-5 microns, round or nearly round, smooth. Large brown sterile cells (setae) intermingled with basidia. Cap tissue blackening in potassium hydroxide (KOH).

HABITAT: Solitary or more often in rows or columns up and down living or recently

Phellinus pini is a major parasite of conifers, particularly pines. The tough fruiting bodies often form lines along the entire length of the trunk. Note the somewhat sinuous pores.

conifers; widely distributed and common, infecting all important members of the pine family (pine, Douglas-fir, etc.). Perennial, in our area occurring mainly on pine. It attacks the heartwood and sometimes the sapwood of living trees, resulting in a delignifying pocket rot known as "conch rot." The fruiting bodies are largely confined to older trees, partly because years of growth must take place before the mycelium can fruit. As a result, the appearance of a single fruiting body means that extensive heart rot has already taken place 10-15 feet above and below it. This fungus is said to cause more timber loss than any other.

EDIBILITY: Inedible (but see comments on edibility of *Fomitopsis pinicola).*

COMMENTS: The rough unpolished cap, frequently sinuous pores, ochraceous to rust-colored flesh, and growth on conifers are the fallible fieldmarks of this destructive fungus. It varies considerably in size, from large hoof-shaped specimens to moderately sized ones (see photo), to a small thin shelving form that often occurs in fused masses and is considered a distinct species, *P. chrysoloma,* by many investigators. Other species: ***Truncospora demidoffii*** (also known as ***Fulvifomes, Pyrofomes,*** or ***Fomes juniperinus***) is somewhat similar, but has rusty to reddish-orange flesh and grows only on juniper; *P. texanus* resembles a small *P. robustus* (see comments under *P. gilvus)* but also grows on juniper, while a resupinate form of *P. robustus* with bright yellow-brown flesh often grows on conifers; *P. taxodii* causes heart rot in bald cypress; *P. nigrolimitatus,* common on conifers in the Rocky Mountains, usually has a spongy upper layer of tissue and fine black lines running through its flesh; *P. torulosus* looks like *P. gilvus,* but has yellow-brown pores and grows on conifers in Arizona. For hardwood-loving species, see *P. gilvus* and *P. igniarius.*

An old *Phellinus everhartii* (see comments under *P. gilvus).* Bark protects a tree from infection, but nailing a signboard to a tree (as shown here) or otherwise defacing it can allow fungal spores to enter!

PIPTOPORUS & CRYPTOPORUS

THESE two oddball genera are each represented in North America by a single farflung species. *Piptoporus* has a conklike (but annual) fruiting body whose margin projects *below* the pore surface to form a "curb." In *Cryptoporus* this theme is carried to its illogical extreme: the margin of the cap is so ingrown (or overgrown) that it completely covers the pore surface, thus hiding it from view!

Key to Piptoporus & Cryptoporus

1.   Pore surface exposed; cap margin curblike; found on birch .... *P. betulinus,* below
1.   Pore surface completely hidden (at least until old age); found on conifers   *C. volvatus,* p. 585

Piptoporus betulinus   (Birch Conk; Birch Polypore)

FRUITING BODY annual, nearly round becoming shelflike or hooflike at maturity; tough or corky when fresh, rigid and hard when dry. **CAP** (2.5) 5-25 cm broad, 2-6 (10) cm thick, kidney-shaped to nearly round in outline, convex to nearly plane; surface covered by a thin, smooth or suedelike, white to buff, tan, brown, or grayish-brown crust that often breaks up into scales or flat patches or wears away, revealing the whitish undersurface; margin thick, blunt, inrolled, curblike (projecting below the pore surface), sometimes wavy. Flesh punky or corky, thick, white. **PORES** appearing recessed due to curblike margin, 2-4 per mm; white when fresh, in age becoming pale brown or grayish-brown and occasionally torn up or toothlike; tubes 2-10 mm long, one layer only. **STALK** absent or present only as a stubby extension of the cap; lateral or attached to top of cap. **SPORE PRINT** white; spores 3-6 × 1.5-2 microns, cylindrical to sausage-shaped, smooth.

HABITAT: Solitary or in groups or columns on dead or sometimes living birch trees; common year-round in the northern hemisphere throughout the range of birch. I have not seen it in California, undoubtedly because birch does not occur naturally. However, it is found in Washington and Idaho and is very abundant in northeastern North America. It causes a reddish-brown to yellow-brown carbonizing decay.

EDIBILITY: Edible when young (according to McIlvaine), but tough. The fruiting bodies are quite attractive, however, and have enjoyed a variety of non-culinary uses, e.g., as

Piptoporus betulinus. Note the thick blunt margin and recessed pore surface of this unique birch-loving polypore.

tinder, as a razor strop, and as a mounting medium for pinned insect specimens.

COMMENTS: Formerly known as *Polyporus betulinus,* this common and distinctive polypore is easily told by its recessed pore surface with a curblike margin (see photo) plus its white to tan or grayish-tan color and growth on birch. The fruiting bodies are annual in the sense that they possess only one tube layer; however, they remain intact for many months and as a result can be found year-round.

Cryptoporus volvatus   (Cryptic Globe Fungus; Veiled Polypore)

FRUITING BODY annual, tough or corky, more or less round to oval to slightly compressed or hooflike; 1.5-8.5 cm broad, with a hollow interior. **CAP** (upper or outer surface) with a thin, smooth, glazed or resinous crust; whitish to warm tan or yellowish, drying darker (ochre-brown to reddish-brown); margin extending down and under to form "veil" which completely covers pore surface; in age the underside perforated by one (rarely two) holes. Flesh whitish, tough; odor often fragrant (like *Sparassis*). **PORES** hidden by the "veil," 3-4 per mm, white becoming pinkish or brownish in age; tubes 2-5 mm long. **STALK** absent. **SPORE PRINT** pinkish or flesh-colored; spores often collecting in a heap on the inner "floor" of fresh specimens), 8-12 × 3-5 microns, cylindrical to elliptical, smooth.

HABITAT: Solitary or more often in groups on dead or occasionally old living conifers throughout northern North America. Common year-round in our area, especially on pine. It seems to favor standing or recently felled trees and produces a delignifying decay that scarcely damages its host.

EDIBILITY: Too tough to be edible. However, Alexander Smith says that "worms" (insect larvae) found inside the fruiting bodies can be used as fishbait. So can worms found outside the fruiting bodies.

COMMENTS: This bizarre evolutionary anomaly looks like a cross between a confused puffball and a bemused oak gall. The smooth, warmly tanned exterior is quite attractive (often reminding me of a small loaf of bread) and gives no hint of the tube layer within. Slicing it open, however, reveals a hollow interior with a "ceiling" of tubes. The "floor" eventually ruptures and tiny bark-boring beetles enter the "trap door" in search of tasty tube tissue and spores. After feasting they depart to construct brood tunnels in old or dying conifers, and the spores they carry with them gain entry to a new host. Later, fruiting bodies may emerge through the very holes bored by the beetles!

Cryptoporus volvatus can be hoof-shaped (like these specimens here) or spherical. The tube layer is *inside* the fruiting body, as shown in sectioned specimen at right.

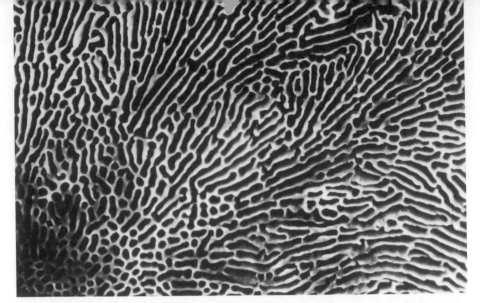

Close-up of the pore surface in *Daedaleopsis confragosa* (p. 588). Note how the pores range from round (bottom left) to elongated or mazelike.

LENZITES, DAEDALEA, & Allies

Fruiting body small to medium-sized, *tough and leathery or sometimes woody,* usually annual (but often perennial in *Daedalea*), growing shelflike or bracketlike on wood. CAP zoned or unzoned, variously colored, hairy to smooth. PORES *usually mazelike, sinuous, or meandering, or forming plate-like gills.* STALK *absent or rudimentary.* SPORE PRINT whitish to brown, but difficult to obtain. Spores usually cylindrical, smooth.

THESE corky or leathery fungi can be told from *Trametes* and other bracketlike polypores by the configuration of their spore-producing surface. In *Daedalea* and *Daedaleopsis* (derived from the Greek word *daedalus,* meaning "maze") the pores are usually long and sinuous or mazelike and have relatively few cross-walls. In *Lenzites* and *Gloeophyllum,* on the other hand, the cross-walls have all but disappeared, leaving plate-like gills instead of pores! However, the configuration varies considerably, especially in *Daedalea* and *Daedaleopsis,* leading to confusion with numerous other polypores (when the pores are not distinctly mazelike), and with species such as *Phellinus pini,* a "conk" with somewhat sinuous pores.

Like most polypores, the species treated here are too tough to eat. If your polypore has gills, it should key out here. If it has mazelike pores, however, it may not key out convincingly, in which case other groups of polypores should be checked.

Key to Lenzites, Daedalea, & Allies

1. Flesh and gills or pores yellow-brown to reddish-brown or brown when fresh; flesh typically at least 1 mm thick and usually darkening or blackening in potassium hydroxide (KOH) . 2
1. Not as above; flesh white or pale-colored when fresh (but may become pale brown in old age or after weathering), or if brown when fresh then typically less than 1 mm thick; pores or gills variable in color (white, violet-tinged, brown, gray, blackish, etc.) 5
2. Typically growing on hardwoods (or occasionally on conifers such as bald cypress); flesh thin .... *Gloeophyllum trabeum* (see *G. saepiarium,* p. 590)
2. Typically growing on conifers (but occasionally on hardwoods) 3
3. Fruiting body usually with an anise-odor when fresh; underside of cap usually with pores ... . *Osmoporus odoratus* (see *Gloeophyllum saepiarium,* p. 590)
3. Not as above 4

4. Underside of cap usually with gills; cap surface often (but not always) with white, yellow, or orange tones or zones when fresh; very common **Gloeophyllum saepiarium** & others, p. 590

4. Underside of cap usually with sinuous or mazelike pores; cap dark or dull brown . . . .
. . . . **Daedalea berkeleyi** (see *D. quercina*, below)

5. Cap surface decidedly hairy, woolly, or velvety when fresh (but may be nearly smooth in old or very weathered specimens) 6

5. Cap surface not hairy or only very slightly so when young 11

6. Underside of cap typically composed of gills or plates (but sometimes varying to mazelike)   7

6. Underside of cap typically composed of elongated pockets or mazelike pores (but sometimes forming gills or regular pores) 9

7. Gills crisped (crimped), usually forked or shallow and veinlike; cap up to 2.5 cm broad, typically *not* zoned concentrically; found on hardwoods in eastern North America . . . .
. . . . (see **Plicaturopsis crispa** under *Schizophyllum commune*, p. 590)

7. Not as above; larger or cap concentrically zoned or found on conifers or in West 8

8. Gills often violet-tinged when fresh; cap often grayish to blackish in old age, zoned or unzoned; typically found on conifers (see **Trichaptum abietinus var. abietis** under *T. abietinus,* p. 593)

8. Not as above; cap usually zoned; usually found on hardwoods . . . . . **Lenzites betulina,** p. 589

9. Fruiting body soft or watery when fresh, discoloring rusty or reddish after handling, or if not, then often quite thick (3-8 cm); fresh pore surface whitish . (see **Tyromyces** & Allies, p. 597)

9. Not as above; fruiting body usually quite thin and tough when fresh, not soft and watery; pore surface often gray to brown or blackish, at least in age 10

10. Cap brown to blackish and velvety when fresh; tubes often unequal in length or appearing slot-like (see photo on p. 596) . . . . (see **Datronia mollis,** p. 595)

10. Cap white to grayish, brown, greenish, etc., usually hairy; tubes often breaking up in age to form "teeth" . . . . **Cerrena unicolor** (see *Daedaleopsis confragosa,* p. 588)

11. Gills when present thick, blunt, and widely spaced (1 mm or more apart); pores when present 1 mm or more broad, the walls thick and blunt 12

11. Gills when present thinner and closer (less than 1 mm) than above; pores when present smaller (averaging 1-3 per mm), the walls thin 14

12. Typically found on hardwoods . . . . **Daedalea quercina,** below

12. Typically found on conifers (occasionally on hardwoods) 13

13. Found on juniper . . . . **Daedalea juniperina** (see *D. quercina,* below)

13. Found on other conifers; cap often poorly developed . . . .
. . . . (see **Coriolellus heteromorpha** under *Poria corticola,* p. 603)

14. Cap white, at least when fresh . . . . **Daedaleopsis ambigua** (see *D. confragosa,* 588)

14. Cap pallid to brown, grayish, reddish-brown, etc. . . . . **Daedaleopsis confragosa,** p. 588

Daedalea quercina   (Thick-Walled Maze Polypore)

FRUITING BODY usually perennial; shelflike, rigid and very tough or corky. **CAP** 4-20 cm broad and 1.5-8 cm thick, more or less fan-shaped; convex or plane; surface uneven or roughened, usually concentrically furrowed or zoned in older specimens (from new growth layers); whitish to tan, ochraceous, grayish, or brown (usually quite pale when fresh), often blackening and cracking in old age; margin often thick. Flesh very tough and corky; white to buff, ochraceous, or pale brown (never dark). **PORES** usually greatly elongated and pocket- or mazelike, sometimes even forming gills; whitish to buff, tan, or dull ochre; tube walls (or "gills") thick (1 mm or more), the spaces between them at least 1 mm broad; tubes 0.5-3 cm long, the layers not distinctly stratified. **STALK** absent or rudimentary. **SPORE PRINT** whitish; spores 5-7.5 × 2-3.5 microns, cylindrical to elliptical, smooth.

HABITAT: Solitary or sometimes in shelving groups on dead or living hardwoods, especially oak, chestnut, and chinquapin; widely distributed, but especially common in eastern North America. I have seen it in northern California and Oregon, but have yet to find it in our area. It causes a brown heart rot, and the tough fruiting bodies occur year-round.

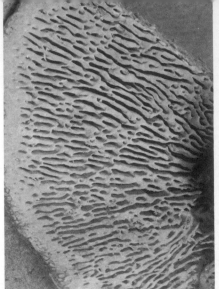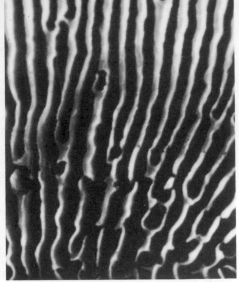

Left: Mazelike pore surface of *Daedalea quercina*. Note the thick walls separating the "pores," and compare them to the thinner walls of *Daedaleopsis confragosa* (photo on p. 586). **Right:** Close-up of the thick gills of *Gloeophyllum saepiarium* (see description on p. 590).

EDIBILITY: Much too tough to be edible.

COMMENTS: The color of this species is rather variable, but the configuration of its spore-bearing surface is relatively constant (i.e., mazelike), and not nearly as polymorphic as that of *Daedaleopsis confragosa*. In addition, the tube walls are thicker and often deeper (see photo) and the fruiting body is usually perennial. *Daedalea berkeleyi* is a southern species that grows on dead conifers and has dark brown to dark rusty-brown flesh. *Daedalea juniperina* is a widespread but smaller pale-fleshed species that grows on juniper. Both of the above tend to have mazelike pores and neither is worth eating.

Daedaleopsis confragosa   (Thin-Walled Maze Polypore)

FRUITING BODY usually annual, shelflike or sometimes bracketlike, leathery or corky when fresh, rigid when dry. **CAP** (3) 5-15 (22) cm broad, fan-shaped to semi-circular in outline, broadly convex to plane; surface dry, smooth or slightly hairy, usually zoned or ridged concentrically, often radially wrinkled or bumpy in age; reddish-brown to brown to grayish, sometimes blackish in old age; margin thin, acute. Flesh white to pinkish or brownish, tough. **PORES** 0.5-1.5 mm in diameter, usually elongated and mazelike with relatively thin walls, but sometimes circular and at other times forming gills or becoming toothlike in age; white to tan or brown, sometimes bruising pinkish or reddish; tubes up to 1.5 cm long. **STALK** absent. **SPORE PRINT** white; spores 7-11 × 2-3 microns, cylindrical, smooth.

HABITAT: Solitary or in groups on dead hardwoods or from wounds in living trees, occurring year-round (sometimes perennial); widely distributed. Its favorite host is willow, but it also occurs on birch and other hardwoods and very rarely on conifers. I have yet to find it in our area, perhaps because it is not fond of oak. However, it occurs in the Pacific Northwest and is very common in eastern North America. It produces a delignifying decay of the sapwood.

EDIBILITY: Not edible.

COMMENTS: The maze-like ("daedaloid") pattern of the pores is characteristic of this species and several others, all of which were originally placed in a single genus, *Daedalea*.

The configuration of the pores is extremely variable, however, leading to confusion with *Lenzites* and *Gloeophyllum* (when they form gills), and dozens of other polypores (when they are not noticeably daedaloid). This species can usually be recognized, however, by its relatively thin, corky, fan-shaped, brown to grayish cap with a bald, often zoned surface. The depth of the tubes and thickness of the cap are also quite variable, depending on how old the fruiting body is, but the tube walls are never as thick as those of *Daedalea quercina* (see photo at top of p. 586). *Daedalea confragosa* is an older name for it. Other species: *Daedaleopsis ambigua* is a similar southern species with a whiter cap. *Cerrena* (=*Daedalea) unicolor* has a thin, hairy, white to grayish (or greenish from a coating of algae) cap and pallid to grayish or even blackish mazelike pores that often break up to form small "teeth." It is common on dead hardwoods in many regions, but not ours. For other "daedaloid" species, see *Daedalea quercina* and *Gloeophyllum saepiarium*.

Lenzites betulina   (Gilled Polypore)

FRUITING BODY usually annual; shelflike, bracketlike, or forming rosettes; tough and leathery. **CAP** 2-13 cm broad, nearly round to fan-shaped in outline; surface dry, velvety or hairy, with narrow concentric zones or grooves of various colors: whitish, tan, buff, gray, brown, yellow-brown, dull orange, etc. (or in old age often greenish from algae). Flesh thin (1-2 mm), tough, white. **GILLS** platelike, often branching toward the margin or forming elongated pockets (especially in young specimens); white or whitish, drying dingy yellowish or darker, often wavy in age. **STALK** absent or rudimentary. **SPORE PRINT** white; spores 4-7 × 1.5-3 microns, cylindrical to sausage-shaped, smooth.

HABITAT: Scattered or more often in overlapping rows, columns, or shelving masses on rotting hardwood logs and stumps (rarely on conifers); widely distributed and very common in our area throughout the year, often sharing logs with *Trametes versicolor* and *Stereum hirsutum*. The species epithet means birch, but it grows on a wide range of hardwoods, especially oak and willow. It produces a delignifying decay of the sapwood.

EDIBILITY: Inedible, but dries nicely. If you are adamant about trying it, see comments on the edibility of *Trametes versicolor* for cooking suggestions.

COMMENTS: This species is an excellent example of convergent evolution—a gilled fungus that is not an agaric. In all respects save the gills it is a typical polypore—the multicolored, hairy, zoned cap bearing an uncanny resemblance to *Trametes* species. In fact, it sometimes can't be distinguished from those mushrooms without examining the underside of the cap. It can also be confused with *Gloeophyllum*, which has brown gills, and *Daedalea* and *Daedaleopsis,* which have "daedaloid" (elongated or mazelike) pores. In eastern North America, *L. betulina* often shows orange tones on the cap, whereas in our area it is usually quite dull in color (grayish, buff, etc.).

Lenzites betulina is a common hardwood-loving polypore with a hairy zoned cap and gills instead of pores. It is often found with *Trametes versicolor*.

Gloeophyllum saepiarium   (Rusty Gilled Polypore)

FRUITING BODY annual; shelflike or bracketlike, leathery when fresh, rigid when dry. **CAP** 2-12 cm broad, more or less fan-shaped in outline; surface dry, hairy to nearly smooth, concentrically zoned or ridged and often radially wrinkled, rusty-brown to dark brown or maroon-brown, often with brighter (yellow, orangish, etc.) zones, but sometimes fading or weathering to grayish; margin orange, yellow, or whitish when actively growing. Flesh 1-5 mm thick, yellow-brown to rusty-brown. **GILLS** close, ochre to yellow-brown or rusty-brown becoming brownish in old age, often fused to form elongated pores or sometimes the underside of cap entirely poroid or even toothlike. **STALK** absent. **SPORE PRINT** white; spores 8-13 × 3-5 microns, cylindrical, smooth. Cap tissue blackening in potassium hydroxide (KOH).

HABITAT: Solitary or in groups or overlapping tiers on dead conifers (or occasionally dead hardwoods, particularly aspen); very widely distributed. I have yet to find it in our area, but it is *very* common in the Southwest, Pacific Northwest, and the mountains of California. It causes a rapid carbonizing decay (brown rot) of both the heartwood and sapwood, and along with *G. trabeum* (see comments), is a pest of telephone poles, structural timber in houses and bridges, etc.

EDIBILITY: Inedible.

COMMENTS: Formerly known as ***Lenzites saepiaria,*** this common conifer-lover, like the hardwood-loving *Lenzites betulina,* typically has gills instead of pores (see photo on p. 588). However, the gills are not white and the flesh is rusty-brown—a typical feature of *Gloeophyllum.* The gills at times join to form mazelike pockets and sometimes there are pores instead of gills (leading to confusion with *Daedalea* and *Daedaleopsis*), but the rusty to orange-yellow colors of the cap are distinctive. A related species, ***G. trabeum,*** has a gray to brownish, unzoned cap. It usually grows on dead hardwoods (but occasionally on conifers), and is more apt to have mazelike pores than gills. It is said to be common on the woodwork of automobiles! Another unzoned species, ***G. striatum,*** occurs only on dead juniper and cypress. Other species: ***Osmoporus odoratus*** is rather similar in color to *G. saepiarium,* but usually has pores on the underside of the cap and often has an aniselike odor when fresh. It is widely distributed and produces a brown rot in dead conifers.

SCHIZOPHYLLUM

SCHIZOPHYLLUM is unique by virtue of its longitudinally split or grooved gills (see description and photographs). It has few close relatives and is placed in its own family, the Schizophyllaceae, by most taxonomists. It looks like a polypore, however, and is treated as one here. One cosmopolitan species is described here.

Schizophyllum commune   (Split-Gill)

FRUITING BODY shelflike or with a narrowed base, tough and leathery both fresh and dry. **CAP** 1-4 cm broad, more or less fan-shaped (or vase-shaped if stalk central); surface dry, densely hairy, white to grayish-white, gray, or sometimes brownish-gray when wet; margin usually lobed and inrolled in dry weather. Flesh tough, leathery, thin, pallid or grayish. **GILLS** radiating from point of attachment, well-spaced, white to grayish; edges appearing split or grooved lengthwise (i.e., cuplike in cross-section), rolling back in dry weather. **STALK** absent or present only as a narrowed basal point of attachment. **SPORE PRINT** white; spores 3-4 (6) × 1-1.5 (3) microns, cylindrical, smooth.

HABITAT: Scattered or in groups, rows, or fused clusters on hardwood sticks, stumps,

Schizophyllum commune. This close-up shows how each gill is actually composed of two adjacent plates. Note hairs on cap.

logs, etc.; distribution worldwide. It survives dry spells by folding back its gills, and hence can be found practically year-round. In our area it is common on oak, producing a white rot (delignifying decay) in its host.

EDIBILITY: Too small and tough to be of value. However, some natives of Madagascar are said to chew them, for reasons unknown.

COMMENTS: The hairy white to grayish cap of this farflung fungus is reminiscent of the common bracket polypores in the genus *Trichaptum*. However, the peculiar manner in which the gills split lengthwise is unique. The "split" gills are actually two adjacent plates which separate and roll up in dry weather, thus protecting the spore-bearing surface (scc photographs). Specimens sealed in a tube in 1911, then moistened 50 years later, unrolled their gills and began shedding spores! *Schizophyllum* has been widely used in genetic studies because it fruits readily in the laboratory. ***Trogia (=Plicaturopsis) crispa*** of eastern North America is a related mushroom with crisped (wavy) gills and a tough, hairy, tan to yellowish cap.

Schizophyllum commune. **Left:** Mature specimens. Cap color ranges from pure white to brownish or grayish. **Right:** A close-up of the "split" gills.

TRAMETES & Allies

Fruiting body *small to medium-sized, usually annual but persistent; tough and leathery even when fresh; growing shelflike or bracketlike on wood* (usually dead wood), often in masses. CAP often hairy or velvety and zoned concentrically, *usually thin.* PORES variously colored, often minute and barely visible, round to angular or becoming toothlike or rarely gill-like; tubes usually shallow, one layer only. STALK *absent or rudimentary.* SPORE PRINT whitish to yellowish (when obtainable). Spores typically oblong to cylindrical or sausage-shaped, smooth.

GROUPED here are a number of smallish tough or leathery bracket fungi. The fruiting bodies are much thinner than those of the perennial polypores or "conks," and are not soft and spongy when fresh as in *Tyromyces.* In age the tube walls may break up to form small "teeth," but do not normally form the mazelike pattern characteristic of *Daedalea* and *Daedaleopsis.*

The various genera treated here are not closely related and can easily be distinguished by pore (not spore!) color. The most common genus, *Trametes,* usually has a zoned cap and white to dingy buff pores. *Trichaptum,* also abundant, has violet to brownish pores, while *Bjerkandera* and *Datronia,* with brownish to gray or black pores, and *Pycnoporus,* with brilliant red to orange pores, are less frequent.

All five genera feed largely on dead wood—logs, branches, even twigs. They are by far the most numerous polypores, often completely smothering their substrate in hundreds of shelving fruiting bodies arranged in attractive clusters or rosettes. Unfortunately, they are not tender enough to eat. Six widespread species are described here.

Key to Trametes & Allies

1. Pore surface bright saffron to orange-red or red . *Pycnoporus cinnabarinus* & others, p. 597
1. Not as above (but fertile surface may be rosy, violet, or rusty-reddish) 2

2. Fertile surface (or pores) typically rosy, violet, or with a distinct violet tinge when fresh . . . 3
2. Fertile surface not rosy or violet-tinged, or rarely with a very faint lavender tinge (and then growing on poplar or willow) 4

3. Cap pinkish to reddish or brown when fresh (never white), often fairly thick; pores rosy when fresh . . . . (see *Ganoderma, Fomitopsis,* & Allies, p. 574)
3. Not as above; pores tinged violet or lavender when fresh; cap thin and differently colored . 8

4. Fruiting body fleshy (watery or spongy) when fresh (but may toughen in age) *and* either staining rusty-reddish when handled or in age *or* with bluish spores and bluish or bluish-gray-tinged pore surface . . . . (see *Tyromyces* & Allies, p. 597)
4. Not as above 5

5. Pore surface brown (or gray from a hoary coating), the tubes often of unequal length or slotlike (see photo on p. 596) but not usually forming "teeth"; cap very thin, brown to blackish, velvety (but not coarsely hairy) when fresh, usually zoned . . . . *Datronia mollis,* p. 595
5. Not as above (but pores may be brown or grayish) 6

6. Pore surface (tube walls) often breaking up in age to form numerous small "teeth"; cap typically hairy 7
6. Not as above 9

7. Pore surface usually grayish to smoky-brown or even blackish in age; cap usually zoned concentrically . . . . *Cerrena unicolor* (see *Bjerkandera adusta,* p. 596)
7. Pore surface violet-tinged, brown, or occasionally pallid; cap zoned or unzoned 8

8. Cap commonly up to 7 cm broad; found mainly on hardwoods, especially common in eastern North America . . . . *Trichaptum biformis* (see *T. abietinus,* p. 593)
8. Cap usually less than 4 cm broad, or if larger then usually with gills on underside; found mainly on conifers; common and widespread . . . . *Trichaptum abietinus,* p. 593

9. Pore surface soon gray to smoky-brown or black; pores small to minute (3-7 per mm), not pallid when young or if so then tubes usually separated from flesh by a narrow dark line . . . . *Bjerkandera adusta* & others, p. 596
9. Not with above features (but pore surface may become beige or slightly grayish in age) . . 10

10. Cap smooth (bald) or nearly so, or if not then growing on living cypress or juniper and causing a brown rot . . . . (see **Tyromyces & Allies,** p. 597)
10. Not as above; cap hairy or velvety; very common, usually on dead wood 11
11. Cap typically *lacking* marked concentric zones, grayish to ochraceous to tan or tawny; found mainly on poplar, willow, or birch; not common . . . . (see **Tyromyces & Allies,** p. 597)
11. Cap usually zoned and/or differently colored; very common on many trees (including poplar, willow, and birch) 12
12. Cap typically 2-7 (10) cm broad, usually with zones of contrasting colors; surface velvety or with velvety zones alternating with silky-smooth zones . . . . **Trametes versicolor,** p. 594
12. Cap 2.5-10 cm broad or more, usually hairy throughout, the zones not sharply contrasting in color . . . . **Trametes hirsuta** & others, p. 595

Trichaptum abietinus (Violet-Pored Bracket Fungus)

FRUITING BODY shelflike or bracketlike, tough and leathery when fresh, rigid when dry. **CAP** 1-4 cm broad, fan-shaped to kidney-shaped or elongated in outline; surface dry, covered with coarse, stiff hairs; usually concentrically zoned or grooved, white to grayish, but old weathered specimens often greenish from algae or blackish; margin often wavy. Flesh very thin (up to 1 mm thick), tough, pale gray to brownish or purplish. **PORES** 2-4 per mm, round to angular but in age often irregularly torn or toothlike; whitish to brownish but usually tinged bright lavender to purplish when fresh, especially toward cap margin; duller or browner in age; tubes very shallow (up to 3 mm long). **STALK** absent. **SPORE PRINT** whitish; spores 4-8 × 2-4 microns, cylindrical or sausage-shaped, smooth.

HABITAT: In groups, shelving masses, or overlapping tiers on decaying conifers; widely distributed and common practically year-round. It is said to be the most important de-lignifier of coniferous slash in North America. In our area it favors Douglas-fir and pine.

EDIBILITY: Boil for 26 hours, squeeze thoroughly, and serve forth.

COMMENTS: Also known as *Hirschioporus abietinus,* this little polypore is roughly to rotting conifers what *Trametes versicolor* and *Lenzites betulina* are to rotting hardwoods —overwhelmingly abundant. The violet-tinged pores (when fresh) that break up into "teeth" are the best field character. The cap is usually zoned, but not multicolored as in *Trametes versicolor. Fomitopsis cajanderi* has rosy pores, but is much thicker and has a reddish to brown cap. Other species: *T. biformis* (better known as *Hirschioporus pargamenus*) is a hardwood-loving version of *T. abietinus* with a cap 2-7 cm broad, violet-tinted pores that quickly become toothlike, and soft, velvety hairs on the cap. It is one of the most common polypores of eastern North America (even more abundant than *Trametes versicolor!*), often growing in huge masses that completely cover rotting logs, but is not as prevalent in the West. *T. abietinus var. abietis* resembles the typical variety but is usually larger, weathers darker, and has "gills" instead of pores on underside of cap.

Trichaptum (=Hirschioporus) abietinus is abundant on rotting conifers. Note hairy white cap. The fertile surface is violet-tinged when fresh and features pores, tiny spines, or even gills!

Trametes (=Coriolus) versicolor. This picture shows why it is called the "Turkey Tail." Multi-colored zones on the cap are distinctive.

Trametes versicolor   Color Plate 158
(Turkey Tail; Many-Colored Polypore)

FRUITING BODY shelflike or bracketlike, thin and leathery when fresh, rigid or slightly flexible when dry. **CAP** 2-7 (10) cm broad, tongue-shaped becoming fan-shaped, or growing in circular rosettes; plane or wavy; surface dry, velvety (covered with fine hairs) or silky, strongly zoned with narrow, concentric bands of contrasting colors (but more uniformly colored in sheltered situations), hairy zones usually alternating with silky-smooth ones; colors extremely variable: a mixture of white, gray, brown, yellowish-buff, bluish, reddish, or black (or even greenish from a coating of algae), or sometimes dark brown with a white margin; margin often wavy and white or creamy when actively growing. Flesh very thin (1-2 mm), tough, white. **PORES** white to dingy yellowish, minute (3-5 per mm) but visible; tubes shallow (up to 2 mm long). **STALK** absent or rudimentary. **SPORE PRINT** white or yellowish; spores 4-6 × 1.5-2.5 microns, cylindrical or sausage-shaped, smooth.

HABITAT: Typically in groups, rows, tiers, shelving masses, or overlapping clusters on logs, stumps, and fallen branches of dead hardwoods (particularly oak), sometimes also on wounds in living trees and rarely on conifers; very common and widely distributed. It is abundant in our oak woodlands year-round, but fruits mainly in the winter and spring—it can be seen on almost any jaunt through the woods. It causes a general delignifying decay of the sapwood, and along with *T. hirsuta,* is sometimes parasitic on wounded fruit trees and lilac bushes.

EDIBILITY: Boil for 62 hours, squeeze thoroughly, and serve forth.

COMMENTS: Our most common polypore, this species is well known—by sight if not by name—to everyone who spends time in the woods. The multicolored, concentrically zoned caps do not decay readily, making beautiful ornaments, brooch clips, earrings, and necklaces. The multiplicity of colors is both its most bewildering and most distinctive characteristic—no two are colored quite alike. Other *Trametes* species are less radically colored and/or hairier (see *T. hirsuta*). *Stereum* species are superficially similar, but lack pores; *Trichaptum* species have violet-tinged to brownish pores. Synonyms include ***Coriolus versicolor, Polystictus versicolor,*** and ***Polyporus versicolor.***

594

Trametes hirsuta. This species resembles *T. versicolor,* but is not as colorful and is often larger.

Trametes hirsuta (Hairy Turkey Tail)

FRUITING BODY shelflike or bracketlike, tough and leathery when fresh, fairly rigid when dry. **CAP** 2.5-15 (30) cm broad, fan-shaped to nearly circular in outline; surface dry, densely and conspicuously hairy to coarsely velvety, usually concentrically zoned or grooved, but the colors of each zone dull and not sharply contrasting; whitish to grayish, yellowish, dull ochre, buff, beige, or pale brownish, the margin sometimes darker and often wavy (but entire surface sometimes greenish from algae). Flesh tough, white (or pale brownish to yellowish in age), 1-5 mm thick. **PORES** (1) 2-4 per mm, white to dingy yellowish or buff when fresh, often tinged brownish or gray in age; tubes 1-3 (5) mm long. **STALK** usually absent. **SPORE PRINT** whitish or pallid; spores 4.5-7.5 × 1.5-3 microns, cylindrical to sausage-shaped, smooth.

HABITAT: Solitary or more often in groups, fused rows, or overlapping clusters on dead hardwoods (or occasionally conifers); widely distributed throughout the northern hemisphere. It occurs year-round in our area on oak, alder, and other hardwoods, but is rather uncommon, at least in comparison to *T. versicolor.* Like the latter species, it causes a general delignifying decay of the sapwood and sometimes parasitizes fruit trees.

EDIBILITY: Tough and hairy (but see *T. versicolor* for cooking suggestions).

COMMENTS: This lesser known cousin of *T. versicolor* is thicker and often larger than that species, but not as brightly colored. Also, the cap is hairier. *T. occidentalis* is a very similar, slightly larger species with a southern and tropical distribution. It is so similar, in fact, that the two may very well be forms of a single wide-ranging species. For a list of synonyms for *T. occidentalis,* see p. 550. *Lenzites betulina* looks quite similar from the top, but has gills on the underside of the cap. Other species: *T. velutina* is very similar, but has a more velvety cap and its pores are often smoky-tinged; *T. pubescens* is smaller and has a white to grayish-yellow, silky-hairy cap with a radially striate margin plus a somewhat fleshier texture when fresh; both occur on hardwoods and are widely distributed.

Datronia mollis

FRUITING BODY annual; bracketlike or at times resupinate; tough when fresh or dry. **CAP** (when present) 1-7 cm broad, shelflike and often wavy; surface umber-brown and velvety when fresh but becoming smooth and dark brown to blackish in age; usually zoned concentrically. Flesh thin, tough, pale brown, often separated from the velvety surface

595

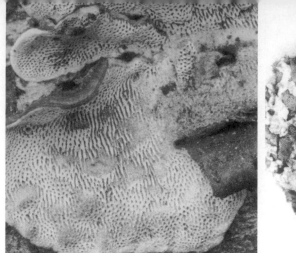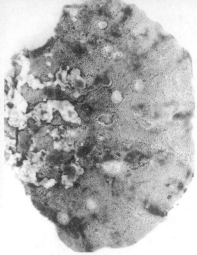

Left: Close-up of slotlike pores of *Datronia mollis*. **Right:** Dark gray pore surface of *Bjerkandera adusta.*

by a thin dark line. **PORES** 1-2 per mm, often becoming elongated and sinuous (slot-like); brown, but covered with a hoary bloom when fresh that gives them a grayish appearance, the bloom rubbing off easily (i.e., surface bruising brown); tubes 0.5-5 mm long. **STALK** absent. **SPORE PRINT** whitish; spores 8-11 × 2.5-4 microns, cylindrical, smooth.

HABITAT: Usually in groups on dead hardwoods, but also reported on conifers; widely distributed and fairly common, but often overlooked because of its small size. In our area I have found it several times in the winter and spring on dead oak. It causes a delignifying decay of the sapwood.

EDIBILITY: Unknown, but like myself, too small, thin, and tough to bother with.

COMMENTS: Formerly known as *Trametes mollis* and *Daedalea mollis,* this small, nondescript hardwood-lover is best recognized by its thin, tough, velvety cap and grayish to brown, slotlike pores (see photograph). It is most likely to be confused with *Bjerkandera* species, but has larger pores and a thinner cap.

Bjerkandera adusta   (Smoky Polypore)

FRUITING BODY shelflike or bracketlike, several often fused together; tough or corky when fresh, rigid when dry. **CAP(s)** 1.5-7 cm broad, plane or wavy, elongated or fan-shaped in outline; surface white to tan, smoky-gray, or grayish-brown, dry, finely hairy (velvety or suede-like) to nearly smooth; margin whitish when young, darkening or blackening in age. Flesh thin (1-6 mm), tough, white becoming grayish or brown; odor fungal, taste often rather sour. **PORES** very minute (5-7 per mm), scarcely visible, at first whitish and darkening where bruised, but soon becoming gray or smoky throughout, and finally blackish; tubes shallow (up to 2 mm long). **STALK** absent. **SPORE PRINT** white or yellowish; spores 4-6 × 2.5-3 microns, elliptical, smooth.

HABITAT: On dead hardwoods (or rarely conifers), usually in dense, overlapping or fused clusters; widely distributed. It is fairly common in our area throughout the mushroom season, but not nearly as abundant as *T. versicolor.* It produces a general delignifying decay of the sapwood, giving it a whitish-flecked appearance.

EDIBILITY: Unequivocally inedible.

COMMENTS: The gray to black pore surface is the principal fieldmark of this small, tough bracket fungus. *Gloeoporus adustus* is a synonym. *B. fumosa* is a very similar species that is somewhat larger and thicker; it has slightly larger, pale gray to smoky-brown (rarely black) pores, larger spores, and a stronger (aniselike to unpleasant) odor.

596

In both species, but particularly *B. fumosa,* a thin dark line separates the pores from the flesh. Intermediate forms occur, however, and it is questionable whether the two are distinct. *Cerrena unicolor* (see comments under *Daedaleopsis confragosa*) is somewhat similar, but its grayish to blackish spore-bearing surface usually breaks up into small flattened "teeth"; it is widely distributed on hardwoods.

Pycnoporus cinnabarinus   (Red Polypore)

FRUITING BODY shelflike, tough when fresh, nearly rigid when dry. **CAP** 2-12 cm broad, nearly round to elongated or fan-shaped in outline; surface dry, smooth or finely hairy, wrinkled or warty, bright orange to orange-red, red, or cinnabar-red, but fading in age. Flesh red to yellowish-red, tough, up to 1.5 cm thick. **PORES** 2-4 per mm, bright orange to orange-red or red, scarcely fading; tubes 1-6 mm long. **STALK** absent. **SPORE PRINT** white; spores 5-6 × 2-2.5 microns, oblong-elliptical, smooth. Cap tissue staining magenta or black in potassium hydroxide (KOH).

HABITAT: Solitary or in groups on dead hardwoods (particularly cherry, but also oak, birch, etc.) or occasionally on conifers; mainly northern in distribution. I have not found it in California, but it occurs in the Southwest and Pacific Northwest and is quite common in eastern North America. The very similar *P. sanguineus* (see comments) is prevalent in the southern United States and tropics.

EDIBILITY: Too tough to be edible, but makes a colorful ornament.

COMMENTS: The beautiful bright red to orange color of the pores makes this one of the easiest of all polypores to recognize. The cap may fade in age, but the pores will retain their color for years if stored properly. The fruiting body is not spongy when fresh as in *Phaeolus,* and has much smaller pores. *P. sanguineus* is a closely related brilliant red to orange-red southern species with a thinner (up to 5 mm) cap. *Aurantioporus croceus* has saffron to bright orange pores when fresh; it is also larger (cap 5-30 cm broad) and softer when fresh, and grows on eastern hardwoods.

TYROMYCES & Allies

Fruiting body typically small to medium-sized (occasionally large) and annual; *usually soft, watery, spongy, or cheesy when fresh,* becoming tougher in age; *growing shelflike or bracketlike on wood.* CAP usually dull colored (white, gray, etc.), cottony or hairy to smooth, usually not zoned. Flesh *usually white, pallid, or pale brown.* PORES minute to fairly large, usually white, pallid, or yellow when fresh, but sometimes bruising darker. STALK *absent or rudimentary or present only as a narrowed, lateral extension of the cap.* SPORE PRINT white or bluish (when obtainable); spores usually smooth and cylindrical or sausage-shaped.

TYROMYCES is an undistinguished genus of fairly forgettable, fan-shaped to hooflike fungal fructifications with a fleshy rather than leathery texture when fresh. The latter feature distinguishes them from *Trametes* and other tough, bracketlike polypores.

Over 20 species of *Tyromyces* occur in North America. Most cause brown rots in dead conifers (in contrast to *Trametes,* which causes white rots in dead hardwoods). However, a few common species (e.g., *T. chioneus*) occur on hardwoods. Despite their fleshiness they are not good edibles, being infrequent, indigestible, and insipid.

Key to Tyromyces & Allies

1. Spore print bluish or bluish-gray; fruiting body often tinged blue or bluish-gray, especially at cap margin .... *Tyromyces caesius* & others, p. 599
1. Not as above 2

2. Fruiting body spongy, watery, or cheesy when fresh and quickly staining rusty-reddish or red-dish-brown when handled or in age (especially the pore surface) or staining yellow and then rusty-reddish; usually found on dead conifers 3
2. Not as above (but pore surface may stain pinkish-red if found on dead hardwoods) 4

3. Pores averaging 3-5 per mm, usually staining yellow before reddening .... *T. fragilis,* p. 600
3. Pores averaging 1-3 per mm, reddening directly ... *T. mollis* & others (see *T. fragilis,* p. 600)

4. Upper layer of cap very soft (marshmallow-like) when fresh (but may toughen in age); cap often elongated and narrow, its margin often projecting below the pore surface; found only on dead mountain conifers (usually above 4,000 ft.) .... *T. leucospongia,* p. 600
4. Not as above 5

5. Tube walls soon breaking up to form spines or "teeth"; cap white to ochre 6
5. Not as above (but tube walls may sometimes be toothlike or partially toothlike in old age) . 7

6. Found on conifers; "teeth" 5-10 mm long *Spongipellis pachyodon* (see *T. leucospongia,* p. 600)
6. Found mainly on hardwoods (rarely conifers); "teeth" 1-5 mm long; cap hairy, white, leathery; flesh very thin .... *Irpex lacteus*

7. Either pore surface staining dark brown to gray or black when bruised or dried and fruiting body often large *or* forming white to creamy rosettes on tropical and subtropical hardwood stumps and roots *or* fruiting body with very dense, heavy flesh that dries *bone-hard*; usually growing in groups or clusters .... (see *Polyporus, Albatrellus,* & Allies, p. 554)
7. Not as above 8

8. Cap small (up to 4 cm), with a narrowed stemlike base *T. floriformis* (see *T. chioneus,* p. 599)
8. Not as above 9

9. Found on incense cedar; fruiting body fairly large (7 cm broad or more when mature) and pore surface usually yellow or yellowish when fresh .... *T. amarus,* p. 601
9. Not as above (usually found on a different host) 10

10. Found on cypress or juniper (usually living); fruiting body very tough or corky, even when fresh .... *T. basilaris* (see *T. chioneus,* p. 599)
10. Not as above; usually found on a different host 11

11. Fruiting body leathery or corky when fresh; cap lacking scales and usually bald or nearly bald, often concentrically zoned or furrowed near margin; pores varying from mazelike to round, sometimes staining pinkish-red when bruised; found almost exclusively on hardwoods ....
      .... (see *Daedaleopsis confragosa* & others, p. 588)
11. Not as above 12

12. Pores fairly large (usually averaging 1 mm broad or more in widest dimension) 13
12. Pores smaller (averaging 1-5 per mm) 15

13. Pores hexagonal or angular and usually arranged in radiating rows, or if not then cap ochre-buff to tan, often with darker scales; rudimentary stalk often present; found on or near hardwoods
      .... (see *Polyporus, Albatrellus,* & Allies, p. 554)
13. Not as above; cap smooth or hairy but not normally with scales 14

14. Cap 4-30 cm broad and 3-8 cm thick, spongy or watery when fresh, the margin usually thick; found on hardwoods .... *T. unicolor* (see *T. leucospongia,* p. 600)
14. Not as above 15

15. Cap often poorly developed or rudimentary; tubes often unequal in length; pores large or small .... (see *Poria* & Allies, p. 602)
15. Not as above; cap usually well-developed and pores small 16

16. Fruiting body tough (corky or leathery) even when fresh; pore surface often grayish to smoky-brown in age or when dry; found mainly on poplar, willow, or birch 17
16. Not as above; fruiting body usually watery, spongy, or cheesy when young (but often toughening in age); found on a variety of hosts 18

17. Odor aniselike when fresh; cap hairy or smooth .... *Coriolellus (=Trametes) suaveolens*
17. Odor not aniselike; cap hairy .... *Funalia trogii*

18. Odor fragrant when fresh and taste not bitter; usually found on hardwoods *T. chioneus,* p. 599
18. Odor not fragrant *and/or* taste bitter; found on hardwoods or conifers 19

19. Taste bitter; usually found on conifers .. *T. guttulatus* & *T. stipticus* (see *T. chioneus*, below)
19. Taste not bitter; on hardwoods or conifers *T. tephroleucus* & others (see *T. chioneus*, below)

Tyromyces caesius   (Blue Cheese Polypore)

FRUITING BODY shelflike or bracketlike, soft and spongy or watery when fresh, tougher when dry. **CAP** 1-5 (8) cm broad, fan-shaped to semi-circular in outline; surface white to gray, but usually tinged or mottled bluish or blue-gray, especially toward the margin; covered with soft, whitish hairs; not zoned. Flesh up to 1 cm thick, white or aging gray to yellowish, spongy when fresh; odor often sweetish. **PORES** 2-4 per mm, white or colored like cap; tubes 2-8 mm long. **STALK** absent. **SPORE PRINT** pale ashy-blue; spores 3-6 × 1-2 microns, cylindrical or sausage-shaped, smooth.

HABITAT: Solitary or in small groups on decaying wood of both hardwoods and conifers; widely distributed and fairly common but rarely occurring in large numbers. In our area it can be found practically year-round. It is associated with a brown carbonizing decay.

EDIBILITY: Unknown.

COMMENTS: The delicate bluish-gray tint to the cap and soft texture when fresh distinguish this unobtrusive polypore. Other species: *T. perdelicatus* also has pale bluish spores, but the cap is smaller (up to 2 cm broad) and not normally bluish, and the taste is bitter. It is fairly common on dead conifers in the Pacific Northwest.

Tyromyces chioneus   (White Cheese Polypore)

FRUITING BODY shelflike or bracketlike; fleshy and rather soft or spongy-watery when fresh, rigid and tough when dry. **CAP** 2-12 cm broad, fan-shaped to semi-circular, broadly convex to plane; surface smooth or slightly hairy, pure white to buff, yellowish-buff, or watery gray, not zoned. Flesh 2-15 mm thick, white; soft and spongy with a fragrant odor when fresh, crumbly when old and dry. **PORES** 3-5 per mm, white to creamy or yellowish; tubes 1.5-3 (7) mm long. **STALK** absent. **SPORE PRINT** white; spores 3.5-5 × 1-2 microns, cylindrical to sausage-shaped, smooth.

HABITAT: Solitary or in groups on dead hardwoods or occasionally on dead conifers; widely distributed and common, but rarely fruiting in large numbers. In our area it is fairly frequent in the fall and winter, especially on oak. It produces a wet, stringy delignifying decay of the sapwood.

EDIBILITY: Unknown.

COMMENTS: The pale color, spongy or cheesy texture when fresh, modest size, fragrant odor, and absence of any staining reactions are the principal characteristics of this polypore, which is better known as *T. albellus.* There are a number of more or less similar white or pallid *Tyromyces* species that are best differentiated microscopically, including: *T. tephroleucus (=T. lacteus)*, with a mild odor, on both hardwoods and conifers; *T. galactinus,* common on hardwoods in eastern North America, with a hairier cap and elliptical spores; *T. balsameus,* small, thin, with a faintly zoned cap, on conifers; *T. spraguei,* with a cap margin that stains greenish or blackish when young and nearly round to elliptical spores, on hardwoods; *T. guttulatus,* often larger (cap up to 16 cm broad), with circular spots, pits, or droplets on the cap, growing mainly on conifers; *T. stipticus (=T. immitis),* even more bitter but lacking spots or pits, on conifers; *T. floriformis,* usually bitter but small (cap up to 4 cm broad) and petal-like (i.e., the cap has a narrowed, stemlike base), on conifers; and *T. basilaris,* tough, attacking living juniper and cypress trees (including Monterey cypress). *T. spumeus* should also be mentioned. In contrast to *T. chioneus,* nearly all of the above species produce brown rots (carbonizing decays) in their hosts.

Tyromyces fragilis, a nondescript polypore that stains rusty-yellow to reddish when handled.

Tyromyces fragilis   (Rusty-Staining Cheese Polypore)

FRUITING BODY shelflike or bracketlike, soft and spongy or watery when fresh; rigid and brittle when dry. **CAP** 2-10 cm broad, fan-shaped to elongated in outline; surface covered with soft white hairs that become matted in age; white, but becoming reddish to pinkish-red in age; staining yellowish, then rusty to reddish when handled. Flesh soft when fresh, white, discoloring like the cap surface. **PORES** 3-5 per mm, white, quickly bruising yellowish, then rusty-reddish; tubes 2-8 mm long. **STALK** absent. **SPORE PRINT** whitish; spores 4-5 × 1-2 microns, cylindrical to sausage-shaped, smooth.

HABITAT: Solitary or in groups or fused clusters on rotting conifers; widely distributed. In our area it is fairly common (along with *T. mollis*—see comments) in the fall and winter on pine and Douglas-fir. It produces a brown carbonizing decay in its host.

EDIBILITY: Unknown, and like myself, likely to remain so.

COMMENTS: The tendency of all parts, especially the pores, to stain rusty-red when handled is the one noteworthy feature of this otherwise unnoteworthy polypore. There are several very similar reddish-staining species with slightly larger pores (1-3 per mm) and thicker flesh (0.5-2 cm) that are best differentiated microscopically, including: *T. mollis,* also common in our area, with a larger, smoother cap that stains rusty-red *directly; T. transmutans,* with dextrinoid spores; and *Amylocystis lapponica,* with amyloid cystidia and spores 7-11 microns long.

Tyromyces leucospongia   (Marshmallow Polypore)

FRUITING BODY annual, shelflike or bracketlike to somewhat irregular; soft and rather watery when fresh, but soon becoming tough and rigid except for the soft upper surface. **CAP** usually elongated (2-10 cm long and 1-5 cm wide), convex; surface white to buff, grayish-buff, pinkish, or even brown or cinnamon; usually velvety and very soft or spongy to the touch, but sometimes papery and fairly firm if the velvety layer wears away; margin usually extending below the pore surface and sometimes partially covering it. Flesh white and rather tough, but covered by a thick (up to 1.5 cm), soft, spongy layer of cottony tissue that is pallid to pinkish-buff or cinnamon. **PORES** 2-3 per mm, angular and often becoming torn and markedly toothlike in age (or interspersed with spines or "teeth"); white when fresh, often discoloring (buff to tan, brownish, or cinnamon) in age or when dried; tubes 2-6 mm long. **STALK** absent or rudimentary. **SPORE PRINT** white; spores 4-6 × 1-1.5 microns, more or less sausage-shaped, smooth.

HABITAT: Solitary to gregarious on dead conifers, usually fruiting from fissures in the bark; common at higher elevations throughout most of western North America. It usually fruits in the spring (often beginning its development under the snow), but the fruiting bodies persist for months without decaying, and can thus be found most anytime. It causes a brown (carbonizing) rot of both the heartwood and sapwood of its host.

EDIBILITY: Not edible.

Tyromyces (=Spongiporus) leucospongia has a unique cottony or marshmallow-like texture. Note how the margin of the cap extends over the pore surface.

COMMENTS: Also known as ***Spongiporus*** (or ***Spongipellis***) ***leucospongia***, this is one of my very favorite polypores because of the unique texture of its skin. The texture is difficult to describe, but it is undeniably *soft*—something like cotton candy (but not sticky) or a chamois-covered cotton ball, and also reminiscent of a "Nerf ball." In age it may toughen and is not so readily identified, but the tendency of the margin to grow over the pore surface is also distinctive. Other species: ***Spongipellis pachydon (=Irpex mollis)*** is a widely distributed species that grows on conifers and has pores which soon become toothlike; it is tough, has a spongy skin only when very young, and lacks the extended cap margin of *T. leucospongia*. *T. (=Spongipellis) unicolor* is also widespread and spongy when fresh, but has larger pores (1 mm or more broad), nearly round spores, a thicker (3-8 cm) and larger cap, and grows on living or dead hardwoods (usually oak).

Tyromyces amarus   (Incense Cedar Polypore)

FRUITING BODY more or less hoof-shaped, annual; soft and watery when very young, hard and rigid in age or when dry. **CAP** 7-30 cm broad and 4-20 cm thick, convex; surface downy when young, usually bald in age but often becoming roughened, wrinkled, or shallowly fissured; whitish or buff to pale brown, sometimes darkening or becoming more ochraceous as it dries; surface usually quite hard (or even crustlike) in age; margin thick and blunt. Flesh thick, soft when fresh but very hard when dry; yellowish to pale brown. **PORES** 2-3 per mm (but may fuse to form larger pores), yellowish to bright lemon-yellow when fresh, often discoloring with age or handling; tubes 5-15 mm or more long. **STALK** absent. **SPORE PRINT** white; spores 6-7.5 × 3.5-5 microns, elliptical, smooth.

HABITAT: Solitary or occasionally several together on incense cedar *(Libocedrus)*; known only from western North America, fruiting mainly in the late summer, fall, and early winter (but persisting year-round) wherever incense cedar occurs. It causes a destructive carbonizing trunk rot that creates large tunnels in the wood. One third of all logged incense cedar is infected by it, and must either be rejected or used as "pecky cedar" for fencing and other purposes.

EDIBILITY: Unknown, but too tough to be worthwhile.

COMMENTS: This interesting polypore is restricted to incense cedar (although there is one report of it from Idaho, on fir). It might be mistaken for a conk *(Fomitopsis* or *Phellinus)* because of its large size and tough texture in age, but the fruiting body is usually annual and quite watery when actively growing. The yellow pore surface and growth on incense cedar are the principal fieldmarks.

601

PORIA & Allies

Fruiting body usually annual but sometimes perennial, *resupinate* (lying flat on the substrate); soft, tough, or chalky; growing on wood. CAP *absent or rudimentary*. PORES fairly large to minute; tube layer always present. STALK *absent*. SPORE PRINT white or yellowish. Spores elliptical to cylindrical, smooth.

THIS large group of polypores is distinguished by its **resupinate** fruiting body: a simple layer of tubes devoid of cap and stem. In several other genera the fruiting body is *sometimes* resupinate—especially when growing on the undersides of logs—but in *Poria* it is *nearly always* resupinate.

Because they lack both a cap and stem, Porias are among the most lackluster of fleshy fungi—they just lie there on their logs, listless and limpetlike, unobtrusively going about their business while boletivores and polypores blatantly go about theirs. They usually grow on dead wood and play an important role in the forest ecosystem. A few are pests of structural timber in mines, cellars, and other dank places where the dismal, abysmal conditions are to their liking.

More than 150 species of *Poria* have been described from North America, but identification is a difficult, tedious, and time-consuming task. Like *Polyporus* and *Fomes,* *Poria* has been split into several smaller genera on the basis of microscopic and chemical characters beyond the scope of this book. Two representative species are described here, and a few others are keyed out plus several resupinate species that belong to other genera. They represent but a fraction of the total number, however, and serious students will want to consult the technical literature on the subject. If the fruiting body is tough and possibly perennial, check the key to the conks *(Ganoderma, Fomitopsis, Phellinus,* & Allies).

Key to Poria & Allies

1. Spore-bearing surface composed of meandering veins that form shallow pits . . . .
   . . . . (see **Stereaceae & Allies,** p. 604)
1. Spore-bearing surface composed of a layer of tubes (pores) or sometimes "teeth" 2

2. Spore-bearing surface soon breaking up into small "teeth" *and/or* violet-tinged when fresh   3
2. Spore-bearing surface not normally violet-tinged or composed of "teeth" 4

3. Spore-bearing surface violet-tinged when fresh . . . . (see *Trametes* & **Allies,** p. 592)
3. Spore-bearing surface white or pallid . . . . (see *Tyromyces* & **Allies,** p. 597)

4. Pore surface *and/or* flesh pink to orange, salmon, or reddish 5
4. Not as above; pore surface white to yellow or some shade of brown or gray 7

5. Pores large (typically at least 1 mm broad); found on conifers (see *Phaeolus alboluteus,* p. 571)
5. Pores minute (3-7 per mm); found on hardwoods or conifers 6

6. Pore surface dark reddish-orange; fruiting body cheesy when fresh; pores scarcely visible (averaging 7-9 per mm); found mainly on hardwoods . . . . *P. spissa*
6. Not as above; pore surface pink to salmon or brick-red . . . . . *Chaetoporus euporus* & others

7. Flesh yellow-brown to rusty-brown, dark brown, etc. 8
7. Flesh white to creamy, straw-colored, yellowish, or beige (at least when fresh) 10

8. Found on charred conifers; pores fairly large (1-2 per mm) and usually hexagonal; fruiting body not normally perennial . . . . *Coriolellus carbonarius* (see *Poria corticola,* p. 603)
8. Not as above 9

9. Fruiting body either dark brown to black, hard, often cracked and cankerlike, and usually found on birch *or* fruiting body growing in sheets under the bark of living oak . . . .
   . . . . (see *Phaeolus, Inonotus, Coltricia,* & **Allies,** p. 566)
9. Not as above; fruiting body usually perennial, quite tough or woody . . . .
   . . . . (see *Ganoderma, Fomitopsis, Phellinus,* & **Allies,** p. 574)

10. Found on living cypress or juniper (causing a brown rot) . . . (see *Tyromyces* & **Allies,** p. 597)
10. Not as above 11

A member of the *Poria corticola* group. Note the absence of both cap and stem—most Porias consist solely of a layer of tubes.

11. Fruiting body hard or woody even when fresh, the margin often incrusted; usually perennial
   .... (see *Ganoderma, Fomitopsis, Phellinus,* & Allies, p. 574)
11. Not as above 12
12. Rudimentary cap often present; tubes often unequal in length 13
12. Not as above 14
13. Pores large (1-3 mm in diameter) .... *Coriolellus heteromorpha* (see *Poria corticola,* below)
13. Pores smaller .... *Coriolellus sepium* & others (see *Poria corticola,* below)
14. Pore surface bright yellow; found on dead conifers ...... *P. xantha* (see *P. corticola,* below)
14. Not as above (pore surface may be yellowish, but not bright yellow) 15
15. Fruiting body sometimes growing from an underground "tuber"; uncommon *P. cocos,* p. 604
15. Not as above; very common 16
16. Pores averaging 4-6 per mm; fruiting body often perennial (with more than one tube layer)
   .... *Perenniporia subacida* (see *Poria corticola,* below)
16. Pores larger than above and/or fruiting body not perennial *P. corticola* & many others, below

Poria corticola   (Boring Poria)

FRUITING BODY usually annual, rather tough; resupinate (consisting of a simple layer of tubes). **CAP** absent or sometimes present as a free margin. Flesh thin, white, fibrous but rather soft when fresh. **PORES** white or creamy, discoloring to pale tan in age or upon drying, 1-4 per mm; tubes 3-10 mm long, rather soft when fresh but drying rigid and tough. **STALK** absent. **SPORE PRINT** white; spores 5-8 × 3-5 microns, broadly elliptical, smooth. Cystidia present among the basidia.

HABITAT: Solitary or more often in rows or fused masses on rotting hardwood logs and branches, less commonly on conifers; widely distributed and fairly common in our area practically year-round. It produces a white rot; many similar species occur on conifers.

EDIBILITY: The entire group is worthless from an edibility standpoint.

COMMENTS: This species typifies a large number of white to yellowish Porias that can only be differentiated microscopically. Another common species, *Perenniporia subacida,* is often perennial (i.e., with more than one tube layer). It is usually creamy or yellowish, has minute pores (4-6 per mm), occurs on both hardwoods and conifers, and is particularly abundant in the Pacific Northwest. There are several similar species that cause brown rots in structural timber as well as in dead conifers and hardwoods, including: *P. vaillantii,* white to yellowish; *P. incrassata,* white to yellowish but drying grayish-brown to black and with yellow spores; *P. xantha,* yellow or yellowish-white with a crumbly or chalky texture, bitter taste, and minute pores; and *P. sequoiae,* which infects the butts of living redwoods. Several *Coriolellus* species should also be mentioned. They were formerly placed in *Trametes* because they often possess a cap, but can also be resupinate or have only a rudimentary cap. *Coriolellus serialis* is a corky or woody, white to creamy, resupinate species that is common on dead conifers and hardwoods; *C. carbonarius* has brownish pores and flesh and grows on charred conifers, including redwood; *C. hetero-morphus* is whitish and has a rudimentary cap or free margin and large pores (1-3 mm each in diameter); *C. sepium* also has a rudimentary cap but its pores are smaller (1-2 per mm); and *C. alaskanus* is a resupinate species with very thin (less than 1 mm) flesh.

Poria cocos   (Tuckahoe)

FRUITING BODY annual, rather tough; resupinate (consisting of a simple tube layer), but sometimes arising from a buried sclerotium ("tuber"). **CAP** absent or rudimentary. Flesh thin, ivory-whitish to tan. **PORES** yellowish-white to tan or pinkish-buff, 1-3 per mm. **STALK** absent. **SCLEROTIUM** when present often large, somewhat resembling an oblong coconut; outer surface brown and scaly, interior white and cheesy when fresh. **SPORE PRINT** white; spores 7-11 × 3-4 microns, cylindrical, smooth. Cystidia absent.

HABITAT: Occasionally growing on conifer logs, but more often infecting tree roots and stumps (of both hardwoods and conifers), and the fruiting bodies appearing terrestrial if they arise from sclerotia; widely distributed, but most common in southeastern North America. It is reported from California, but I have not seen it in our area. It causes a brown cubical root and butt rot in its host.

EDIBILITY: The fruiting bodies, like those of other Porias, are worthless, but the large underground "tuber" was apparently eaten by various tribes of Native Americans (some of whom called it "tuckahoe"). The "tubers" are said to be visually unappetizing, but I've seen "Unnative" Americans foraging for even stranger fare in supermarkets—Spam, for instance, or pickled pig's feet.

COMMENTS: The ability to form large, coconut-sized sclerotia is the only redeeming feature of this otherwise boring *Poria*.

Crust and Parchment Fungi

spores

STEREACEAE & Allies

Fruiting body soft and spongy to thin and tough; *resupinate, bracketlike, shelflike, or occasionally stalked; usually growing on wood.* CAP present or absent, when present usually small. SPORE-PRODUCING SURFACE *smooth to wrinkled, veined, or warty.* STALK usually absent but occasionally present. SPORE PRINT white to yellow or brown, but often not obtainable. Spores variously shaped.

THIS is a very difficult and immense group of very simple, less-than-immense fungi with a smooth to wrinkled, veined, or warty spore-bearing surface. They can be found almost anywhere, at almost any time, and play an integral role in the breakdown of wood and plant material. Some forms, such as *Stereum,* mimic polypores by fruiting in shelving masses on wood, but the majority (*Corticium, Peniophora,* etc.) are resupinate, i.e., they form crust- or parchment-like sheets on logs or stumps. Others form small, tile-like plaques (*Xylobolus*), "whitewash" sticks and branches, or cause dark discolorations on the bark of trees. Still others cause serious damage to structural timber (e.g., rickety bridges and rotting rafters), and a few, such as *Cotylidia,* are stalked or branched and/or grow on the ground. Whatever their shape and growth habit, however, they all differ fundamentally from the polypores by virtue of their relatively unspecialized spore-producing surface—in other words, they lack a layer of tubes and pores.

Most of the crust and parchment fungi were originally classed together in a single family, the Thelephoraceae. On the basis of various chemical and microscopic features several families are now recognized: the Stereaceae, Corticiaceae, Coniophoraceae, Hymenochaetaceae, etc. As relatively few are conspicuous or fleshy enough to warrant attention from the average mushroom hunter and even fewer can be accurately identified without painstaking microscopic examination and still fewer are edible, only a very *very* few are described here.

Key to the Stereaceae & Allies

1. Growing in white to creamy rosettes of flattened segments or lobes on hardwood stumps and roots in tropics and along Gulf Coast .... *Hydnopolyporus palmatus*
1. Not as above 2

2. Fruiting body with a cap and stalk, the stalk central to somewhat off-center; usually growing on ground or on very rotten wood 3
2. Not as above 4

3. Fruiting body chocolate- to purple-brown or dark brown *Thelephora terrestris group,* p. 608
3. Fruiting body paler (hazel to buff, whitish, or yellowish) *Cotylidia diaphana* & others, p. 608

4. Fruiting body hard, black, usually cracked, somewhat irregular in shape or cankerlike; growing on hardwoods, especially birch .... (see *Phaeolus, Inonotus, Coltricia,* & Allies, p. 566)
4. Not as above 5

5. Fruiting body bright to dark blue or purple or with a purple tinge when fresh; on hardwoods 6
5. Not as above 7

6. Fruiting body bright to dark blue, velvety .... *Pulcherricium caeruleum*
6. Fruiting body purple or purple-tinged, at least when fresh ....
.... *Chondrostereum purpureum* (see *Stereum hirsutum* group, below)

7. Spore-bearing surface smooth to slightly uneven or sometimes cracked 8
7. Spore-bearing surface conspicuously warty, lumpy, furrowed, ridged, wrinkled, pitted, or honeycombed with large "pores" 13

8. Fruiting body terrestrial and branched (coral-like), usually dark brown or purple-brown ....
.... *Thelephora palmata,* p. 609
8. Fruiting body usually found on wood, or if on ground then usually with a cap (upper surface) 9

9. Fruiting body purple- to chocolate-brown or dark grayish-brown, usually growing on herbaceous stems or in rosettes or clusters on the ground; spore print brown; spores warty or angular under the microscope .... *Thelephora terrestris group,* p. 608
9. Not as above 10

10. Fruiting body usually "bleeding" when cut or handled (exuding a red juice); found on conifers .... *Stereum sanguinolentum* (see *S. hirsutum* group, below)
10. Not as above; found on hardwoods and/or fruiting body not "bleeding" 11

11. Cap silky-striate, thin; found on hornbeam in eastern North America *Stereum striatum,* p. 607
11. Not as above 12

12. Cap very thin, pliant, typically with long, loosely-arranged white hairs pointing toward the margin .... *Stereum striatum,* p. 607
12. Not as above (but cap usually with fine velvety hairs) *Stereum hirsutum* group & others, below

13. Growing on structural timber (in houses, mines, etc.) 14
13. Growing in the wild (on logs, stumps, etc.) 15

14. Spore-bearing surface-honeycombed with large pits .... *Serpula lacrymans,* p. 610
14. Spore-bearing surface wrinkled or bumpy *Coniophora puteana* (see *Serpula lacrymans,* p. 610)

15. Spore-bearing surface irregularly warty or with radiating or concentric wrinkles or furrows; cap absent .... *Phlebia radiata* & others, p. 610
15. Not as above; spore-bearing surface usually with a honeycombed or veined appearance; cap or free margin sometimes present 16

16. Cap or free margin hairy, white ...... *Merulius tremellosus* (see *Serpula lacrymans,* p. 610)
16. Cap or free margin bright pink to reddish *Merulius incarnatus* (see *Serpula lacrymans,* p. 610)

Stereum hirsutum group   (False Turkey Tail; Hairy Stereum)

FRUITING BODY thin, leathery, pliant when moist, rigid when dry, annual but persistent; bracketlike to shelflike or partially resupinate with a free margin. **CAP** 0.5-4 (5) cm broad but sometimes fused laterally to form larger, lobed shelves 10 cm long or more; plane to folded or wavy (crisped); surface dry, often zoned concentrically, with whitish to brownish or grayish matted hairs (but often smooth toward the margin), the hairs wearing

Stereum hirsutum group. Noted the zoned cap which is finely hairy (under a hand lens) when fresh but often smooth in old age. The underside is golden to orange or buff when fresh (specimen at top right), but often darkens in old age (specimen at top center).

away in narrow zones to reveal the reddish-brown to dark chestnut-brown cap cuticle; margin often orange to golden or tawny, especially when young or growing; overall color thus variable: orange-brown to reddish-brown to cinnamon when moist, but appearing buff to grayish or paler from the hairs when dry; in old age sometimes greenish from algae or even blackish. Flesh thin, tough. **UNDERSIDE** smooth to slightly bumpy or cracked (when dry), sometimes exuding a red or yellow liquid when cut (fresh); color variable: orange to dull orange-buff or tawny to ochraceous, varying to buff or pinkish-buff, sometimes zoned concentrically, often browner or darker toward the base; in old age often dark brown to chestnut-brown. **STALK** absent or present only as a narrowed lateral base. **SPORE PRINT** white; spores 5-8 × 2-3.5 microns, cylindrical, smooth.

HABITAT: In groups, rows, fused masses, or dense overlapping clusters on hardwood sticks, fallen branches, logs, stumps, etc., occasionally on living trees or conifers; widely distributed and extremely common. Found in our area year-round, especially on dead oak. *Trametes, Lenzites,* and/or *Tremella* species often co-inhabit the same piece of wood.

EDIBILITY: Like myself, too thin and too tough to be edible.

COMMENTS: This omnipresent little bracket fungus can be mistaken for the turkey tail *(Trametes versicolor)*, but a close inspection of the spore-bearing surface reveals the complete absence of tubes or pores. The above description is rather lengthy because it encompasses a number of confusing forms that are sometimes recognized as distinct species. For instance, the common North American variety with a bright orange to orange-buff spore-bearing surface has been called *S. complicatum* and *S. rameale.* Another variety, *S. gausapatum,* "bleeds" red when cut, others exude a yellow juice, while typical *S. hirsutum* does not bleed at all. However, these "species" grade into each other, making a neat, clean separation almost impossible. As a group they can be recognized easily by their omnipresence (in the hardwood forests of California they outnumber even *Trametes versicolor*), the reddish-brown cap cuticle (best seen by sectioning the fruiting body) beneath the hairs, the frequent orange to yellowish tones on the cap margin or underside, and the smooth spore-bearing surface.

Other species: *S. fasciatum (=S. ostrea, S. lobatum)* differs microscopically, but can usually be told by its slightly larger caps (1-7 cm broad) that are more prominently zoned (dark reddish and brown) and usually form individual brackets rather than fusing, and buff to cinnamon-buff underside; it, too, is common on hardwoods, especially oak. *S. (= Haematostereum) sangunolentum* is one of the few widespread Stereums to grow on conifers. When fresh its fertile surface "bleeds" dramatically when cut, sometimes even staining one's hand. *Hymenochaete tabacina* resembles the *S. hirsutum* group in shape and color, but its tissue blackens in potassium hydroxide (KOH). *H. rubiginosa* also blackens in KOH but has a velvety, rusty-brown to chestnut-brown to blackish cap and rusty-brown underside. Both of these favor oak. *Chondrostereum purpureum* is purple when fresh and often resupinate; it parasitizes apple and plum trees, causing "silver leaf"

606

disease, but is also frequent on other trees (including oak in our area). *Laxitextum bicolor,* found on hardwoods, has a brown cap and white to pale buff fertile surface. *Peniophora gigantea* forms a paper-thin crust on dead conifers. *Veluticeps berkeleyi* has a minutely bristly fertile surface and grows on both hardwoods and conifers, but favors ponderosa pine. Dozens of other species also occur.

Stereum striatum

FRUITING BODY very thin, leathery and pliant when fresh, annual but persistent; bracketlike to cuplike. **CAP** 0.3-3 (4) cm broad but sometimes fused laterally to form lines 10 cm long or more; flat and circular to fan-shaped in outline *or* if small, often conical (inverted cup-shaped); surface dry, whitish to buff or pale brown, sometimes zoned concentrically when moist; covered with long, loosely-arranged white hairs which usually point toward the margin (*var. ochraceoflavum*) *or* the hairs pressed against the cap to give it a silky-shiny striate appearance (*var. striatum*). Flesh *very* thin, tough. **UNDERSIDE** (fertile surface) smooth, buff to pale brown or in some forms yellow to yellow-brown, sometimes fading in age to whitish, sometimes zoned concentrically. **STALK** absent or present only as a small knob or "umbo" on top of the cap. **SPORE PRINT** whitish; spores 5-8.5 × 2-3.5 microns, cylindrical, smooth.

HABITAT: In groups or masses on dead branches and twigs (rarely logs) of hardwoods; widely distributed. Variety *ochraceoflavum* can be found year-round in our oak woodlands, but is shrivelled up and inconspicuous in dry weather. Variety *striatum* occurs in eastern North America on hornbeam *(Carpinus).*

EDIBILITY: Not edible.

COMMENTS: Like myself, this species is too tough, too thin, and too small to be edible. Unlike myself, it contents itself with unambitious undertakings—decomposing branches, sticks, and lopped-off limbs—while leaving the larger stumps, logs, and standing trees to the polypores and other bracket fungi. Variety *ochraceoflavum (=S. ochraceoflavum)* can be distinguished from the more common *S. hirsutum* group by its duller, paler color and thinner, more pliant cap with long white hairs pointing toward the margin, plus the absence of a red-brown cap cuticle. Variety *striatum (=S. sericeum)* is easily told by its silky-striate cap. At least two growth forms of var. *ochraceoflavum* occur in our area: a small one with a concave spore-bearing surface and a somewhat conical cap less than 1 cm broad that is usually attached by its top to small twigs; and a larger one with a flatter cap that usually inhabits larger sticks and branches and is often partially resupinate.

Stereum striatum is common on sticks and branches of various hardwoods. Note the long hairs that protrude from cap.

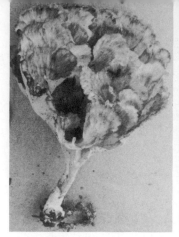

Left: Top view of *Cotylidia diaphana* (this specimen has several caps). **Center:** Underside of *Cotylidia diaphana*. **Right:** *Thelephora terrestris* group. Note how underside is darker than that of *Cotylidia*.

Cotylidia diaphana (Stalked Stereum)

FRUITING BODY annual, erect, thin and tough, with a cap and stalk. **CAP** 0.5-3 cm broad, vase-shaped or funnel-shaped or often split into petal-like lobes; surface dry, with fine radiating silky fibrils, whitish to buff to pale hazel-brown, sometimes with obscure concentric zones. **UNDERSIDE** (fertile surface) smooth or somewhat uneven but without pores, whitish to buff, pinkish-buff, or tinged cap color. **STALK** 0.5-3.5 cm long, 1-2 mm thick, more or less central, solid, smooth, colored like rest of fruiting body; base usually with white mycelial down. **SPORE PRINT** whitish; spores 4-6 (8) × 2.5-4 microns, elliptical, smooth. Long, narrow, projecting cystidia present among the basidia.

HABITAT: Solitary or in groups among humus and debris in woods; widely distributed. Occasional in our area in the fall and winter, but easily overlooked.

EDIBILITY: A worthless, miniscule morsel.

COMMENTS: Formerly known as *Stereum diaphanum,* this species differs from *Stereum* by its well-developed, more or less central stalk, and from *Thelephora terrestris* by its paler color and whitish spores. It is considered by some to be a variety of *C. aurantiaca,* a widely distributed, often larger and yellower species. Other species: *C. decolorans* (*=Stereum burtianum)* of eastern North America is similar but lacks cystidia.

Thelephora terrestris group (Earth Fan; Fiber Vase)

FRUITING BODY annual but persistent, tough; usually vase-shaped to fanlike, often clustered or in confluent masses, sometimes bracketlike or shelflike on plant stems. **CAP** 2-5 cm broad or forming rosettes or clusters up to 12 cm broad; surface dry, with radiating silky fibrils or small scales; brown to reddish- or chocolate-brown to grayish-brown or fuscous, often darker in age; margin usually fringed, splitting, often paler or whitish. Flesh very thin, tough; odor mild or earthy. **UNDERSIDE** smooth or wrinkled, without pores; some shade of brown. **STALK** when present lateral to central, thin and tough, colored like cap or paler, short. **SPORE PRINT** purplish-brown; spores 8-12 × 6-9 microns, elliptical-angular, warted (often minutely so).

HABITAT: Occasionally solitary but more often in groups or clusters in humus, sandy soil, and decomposing vegetable matter; sometimes on old stumps or climbing up herbaceous stems or tree seedlings; widespread and common. In our area it grows year-round, but is easily overlooked. I've seen it several times on potted plants in nurseries.

Left: *Thelephora terrestris* group, top view of a compound fruiting body. **Right:** At top is *Thelephora palmata*—note its flattened branches. At bottom are small specimens of the *Thelephora terrestris* group.

EDIBILITY: I can find no information on it.

COMMENTS: The size and shape of this species is fairly variable but the color is quite constant (quite variable? fairly constant?). When growing on the ground in small, erect clusters it might be mistaken for an emaciated *Craterellus*, but is smaller and differently colored. When growing on herbaceous stems, on the other hand, it looks more like a *Stereum*. *T. laciniata* is apparently a synonym. Other species: *T. multipartita* is a small (1-3 cm high), widespread, terrestrial species with a vase-shaped cap that splits into several lobes or "branches"; *T. vialis* of eastern North America is a terrestrial species with a larger fruiting body and a fetid odor in age, and smaller spores; *T. spiculosa* encrusts twigs and stems, but has spiky protuberances and a whitish growing margin; *Sebacina incrustans* also grows on herbaceous stems (usually at their bases) but is paler in color and has white spores. None of these are worth eating.

Thelephora palmata   (Fetid False Coral)

FRUITING BODY annual, erect, usually profusely branched from a common base; 2-10 cm high and just as broad or broader. **BRANCHES** purplish-brown to chocolate-brown or darker, flattened; tips also flattened (palmlike) and usually paler (whitish) when actively growing. **STALK** present only as a common base or short "trunk" below the branches. Flesh tough, leathery; odor garliclike becoming fetid (unpleasant) in age. **SPORE PRINT** dark reddish-brown; spores 8-11 × 7-8 microns, elliptical-angular, spiny.

HABITAT: Solitary or in groups on moist ground in woods and at their edges; widely distributed. It often grows along woodland paths but blends uncannily into its surroundings. In our area it fruits in the late fall, winter, and spring, but is not particularly common.

EDIBILITY: Unknown.

COMMENTS: This mushroom looks like a coral fungus and is keyed out under that group. However, the flattened branches, dark color, fetid garlic odor, and angular-elliptical spores are distinctive. Other species: *T. vialis* of eastern North America is more variable in color and has smaller spores.

609

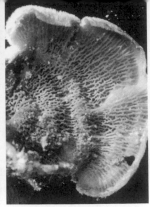

Left: *Radulum orbiculare* (see comments below) forms sheets of irregularly warted or lumpy tissue.
Right: The veined underside of *Merulius tremellosus* (see comments under *Serpula lacrymans*).

Phlebia radiata

FRUITING BODY annual, resupinate (lying flat on substrate), sometimes with a free margin but no cap or stalk; soft when fresh, tough in age. **FERTILE SURFACE** usually fused to form patches 30 cm or more long, but often with smaller, discrete systems of radiating wrinkles or warty veins 1-4 cm broad; flesh-colored to bright orange to pinkish-red, fading to whitish in old age. Underside of margin (if free) with white woolly hairs. Flesh thin, rather soft or slightly gelatinous when fresh, tough in age or when dry. **SPORE PRINT** whitish; spores 3.5-7 × 1-3 microns, sausage-shaped or elliptical, smooth.

HABITAT: On fallen logs and branches of both hardwoods and conifers; widely distributed. I find it occasionally on oak logs in the fall and winter.

EDIBILITY: Inedible. It looks as if it has already been eaten (see comments).

COMMENTS: Also known as *P. merismoides,* this species is unique by virtue of its resupinate, orange to pinkish fruiting body with radiating wrinkles. Its overall appearance is somewhat reminiscent of regurgitated dog food, and a similar species, *Radulum orbiculare,* has an irregularly lumpy or warty spore-producing surface that is *distinctly* reminiscent of regurgitated dog food. Other species: *Punctularia strigoso-zonata* is also similar, but has concentric wrinkles and furrows instead of radiating ones; it grows on dead hardwoods.

Serpula lacrymans   (Dry Rot Fungus)

FRUITING BODY annual, forming widely-spreading, nearly flat, fanlike sheets on horizontal substrates, but sometimes bracketlike if growing on vertical substrates; soft and spongy when fresh, 5 cm-1 m (3 ft.) broad or more. Upper surface (or free margin) silvery-white to gray, hairy. Flesh thin, dingy yellowish; odor often unpleasant, musty. **FERTILE SURFACE** consisting of very shallow (1 mm deep), large, irregular pits or "pores" formed by a honeycomb-like network of folds and ridges rather than by tubes; olive-yellow to brownish-yellow, rusty-brown, orange-brown, or cinnamon. **STALK** typically absent, but white or grayish mycelial strands (by which it spreads) often present. **SPORE PRINT** orange-brown to orange-yellow to brownish; spores 8-12.5 × 4-6 microns, elliptical, smooth, thick-walled.

HABITAT: A serious pest of structural wood in old houses, buildings, ships, etc., usually developing indoors or in poorly ventilated situations, often hiding under floorboards. Bulging wood and a musty odor are telltale signs of its presence. Common in Europe, where the ventilation in many houses was sealed off during the war; less common in North America. It is called dry rot because it extracts water from the wood and cracks it into cubical blocks, eventually reducing it to a fine, dark powder. A related species, *S. himantioides,* turns up occasionally in the wild (on dead conifers), as do *Merulius* species (see comments).

EDIBILITY: Utterly and indisputably inedible.

COMMENTS: One of the few fleshy fungi that lives up to the label *fungus* in its most pejorative sense—odious, insidious, hideous, obnoxious, downright abominable. Once it gains a foothold it is hard to eradicate because the often gigantic, padlike fruiting bodies exude great quantities of water, stimulating further fungal growth. The mycelial strands spread with astonishing rapidity. Like a horde of hungry army ants in search of food, they will overrun anything and everything in their way: bricks, stones, tiles, plaster, drain pipes, wires, leather boots, cement floors, books, tea kettles, even corpses. For a fascinating account (and pictures) of some of its more heroic feats, see John Ramsbottom's mycological treasure trove, *Mushrooms and Toadstools.*

The veined or honeycombed network of large, irregular, shallow "pores" is characteristic of *Serpula* (colored spores) and *Merulius* (white spores). Both are now quite rightfully placed in families of their own, apart from other crust and parchment fungi. Other species: *Coniophora puteana* ("Wet Rot") is another pest of structural timber; it has a similar growth habit but has an irregularly wrinkled to bumpy fertile surface. *Merulius tremellosus* (see photo on p. 610) has an orange to orange-buff to pinkish, veined or honeycombed fertile surface and a hairy white cap or free margin; it grows in the wild, mainly on dead hardwoods. *M. incarnatus* is an eastern species with a bright coral-pink cap and paler, duller, honeycombed or veined fertile surface. Several other species differ microscopically.

Teeth Fungi

HYDNACEAE

spores

As their name implies, the teeth fungi produce their spores on pendant **spines** or "teeth" (see Color Plate 159). The fruiting body is usually stipitate (equipped with a cap and stem), with the spines lining the underside of the cap. *Hericium,* however, grows on wood and has spines which hang like icicles from a rooting base or network of branches, and several others are shelflike, i.e., with a cap but no stalk. Those that grow on the ground can be tough and quite reminiscent of polypores (*Hydnellum* and *Phellodon*) or fleshy, brittle, and agaric-like (*Hydnum* and *Dentinum*).

The teeth fungi include many highly distinctive mushrooms and some strikingly beautiful ones. They are most common and diverse in northern pine and spruce forests, but relatively sparse in our area. *Hericium* and *Dentinum* are excellent eating and have the added advantage of being virtually unmistakable. Most of the others are either too tough or too bitter (or too tough *and* too bitter) to eat.

Just as all the boletes were once lumped together in *Boletus,* so all the teeth fungi were originally placed in a single genus, *Hydnum.* Several families and genera are now recognized based on differences in the shape and texture of the fruiting body and color and ornamentation of the spores. However, to facilitate identification all of the teeth fungi have been retained here in a single family, divided into the five groups keyed below.

Key to the Hydnaceae

1. Growing on decaying conifer cones or sometimes in mats or debris made up partly of cones; cap and stalk hairy, brown to dark brown; stalk only 0.5-3 mm thick . *Auriscalpium,* p. 629
1. Not as above; not growing on cones; stalk if present usually thicker 2

2. Fruiting body rubbery and flexible, small (cap typically 5 cm broad or less), translucent white to watery gray or with a brownish cap; stalk lateral (attached to side of cap); spines minute and very short .... (see *Pseudohydnum gelatinosum,* p. 671)
2. Not as above 3
3. Growing on wood 4
3. Growing on ground (or rarely on very rotten wood) 5
4. Fruiting body a branched framework or unbranched cushion of tissue from which spines are suspended (i.e., icicle-like); lacking a distinct cap .... *Hericium,* p. 613
4. Fruiting body resupinate (i.e, crustlike or sheetlike) or more often with a cap (clearly defined upper sterile surface), the spines lining the underside of the cap *Echinodontium* & Allies, below
5. Fruiting body tough and fibrous or woody *or* if soft and spongy then with a tough, fibrous inner core (especially in stalk); fruiting body sometimes encompassing needles and other debris as it grows; surface of cap often roughened irregularly by projecting spikes, lumps, warts, ridges, pits, etc.; stalk continuous with cap and sometimes not well-defined; flesh sometimes showing zones or lines when fruiting body is sliced open lengthwise   *Hydnellum & Phellodon,* p. 622
5. Not as above; fruiting body fleshy and usually brittle (not spongy or woody), *not* typically absorbing needles and other debris as it grows; cap smooth, cracked, or scaly but not normally roughened by projecting spikes, lumps, pits, or ridges; stalk usually well-defined; flesh not often zoned by lines .... *Hydnum, Dentinum,* & Allies, p. 616

ECHINODONTIUM & Allies

Medium-sized to large, fleshy to very tough or woody fungi *growing on wood.* FRUITING BODY *typically shelflike or hooflike.* CAP clearly defined, smooth to hairy, rough, or cracked. SPINES long or short, variously colored. STALK *typically absent* (except in *Mycorraphium*). SPORE PRINT typically white (when obtainable). Spores smooth or minutely spiny, sometimes amyloid.

THESE shelving mushrooms are more likely to be mistaken for polypores than for other teeth fungi. However, they bear their spores on numerous downward-pointing spines instead of in tubes or pores. They differ from the genus *Hericium* in possessing a clearly defined, sterile upper surface (cap) and fertile lower surface (spine layer). Several diverse genera are keyed below. They are not closely related but share a similar growth habit. Only one species is described, the others being rare or absent in the West. None are worth eating.

Key to Echinodontium & Allies

1. Stalk often present; cap more or less kidney-shaped, less than 8 cm broad, white to tan (but may blacken when handled); spines only 1-3 mm long, whitish becoming pinkish, brown, or cinnamon in age; found on dead hardwoods in eastern North America ....
.... *Mycorraphium (=Steccherinum) adustum*
1. Not as above 2
2. Entire fruiting body white to yellowish or pale ochre or pale gray when fresh (but may discolor in old age), fleshy or tough but not woody; flesh white; growing mostly on hardwoods ... 3
2. Not as above 5
3. Usually growing in overlapping, shelving masses high up on living hardwoods (especially maple); caps usually 10-30 cm broad, several arising from a common base; fairly common in northern North America (see photo at top of p. 613) ... *Climacodon (=Steccherinum) septentrionale*
3. Not as above 4
4. Cap(s) usually roughened or with sterile upright spines; spines on underside usually quite long (up to 2 cm); spores amyloid; usually on hardwood stumps; rare .... *Creolophus cirrhatus*
4. Not with above features; pores often present when very young (the tube walls that form the pores often breaking up in age to form spines); common ..... (see *Polyporaceae* & Allies, p. 549)
5. Fruiting body woody, more or less hoof-shaped; flesh bright orange to rusty-red or cinnamon; growing on conifers in western North America .... *Echinodontium tinctorium,* p. 613
5. Not as above; pores often present when young (their walls breaking up to form spines or "teeth" in age) .... (see *Polyporaceae* & Allies, p. 549)

Climacodon septentrionale looks like a polypore but has spines under the cap instead of pores (see key at bottom of p. 612). It forms large shelving masses on northern hardwoods such as maple.

Echinodontium tinctorium   (Indian Paint Fungus; Toothed Conk)

FRUITING BODY shelflike, very tough or woody, often perennial, up to 15 cm thick. **CAP** 4-25 cm broad, more or less hoof-shaped; surface dry, finely hairy to rough, often covered with moss. Flesh very tough or woody, bright orange to rusty-red or cinnamon, zoned. **SPINES** brittle, blunt, thick, flattened, long (1-3 cm); grayish to pale olive-buff, the tips sometimes darker. **STALK** absent. **SPORE PRINT** white (when obtainable); spores 5.5-8 × 3.5-6 microns, elliptical, minutely spiny, amyloid.

HABITAT: Solitary or several on living or occasionally downed conifers; known only from western North America, where it favors mountain conifers such as fir and hemlock (I haven't seen it on the coast). It causes an extensive white heart rot in its host.

EDIBILITY: Unequivocally inedible, but can be used as a red dye. Its common name is a tribute to its use by Native Americans in the preparation of war paint.

COMMENTS: The woody hoof-shaped fruiting body looks like a conk, but has long spines or "teeth" on its underside. The bright orange to reddish-orange flesh is also distinctive. It is not closely related to other teeth fungi and is now sequestered in a family of its own.

HERICIUM

Medium-sized to very large fleshy fungi *growing on wood.* FRUITING BODY *usually white to yellowish or pale salmon, branched or unbranched* but lacking a distinct cap; *spores borne on clusters or rows of delicate hanging spines.* SPORE PRINT white. Spores more or less round, smooth or minutely roughened, amyloid.

A PRISTINE full-grown *Hericium* is a breathtakingly beautiful sight. The fruiting body is unmistakable: a mass of fragile "icicles" suspended from a branched supporting framework, or in the case of *H. erinaceus,* from a tough, unbranched cushion of tissue.

Hericiums are as delectable as they are beautiful, providing they are not too old or too tough—and providing you can bear to pick them in the first place. They are excellent fresh as well as marinated or pickled, and they have absolutely no poisonous look-alikes. They grow exclusively on wood and are easily cultivated. Their amyloid spores and novel fruiting body have led some taxonomists to reward them with a family of their own.

Four well-known species occur in North America. Three favor hardwoods, while *H. abietis* grows on conifers. In most regions, including California, they are infrequent to rare, so you should consider yourself fortunate to find one.

Key to Hericium

1. Fruiting body unbranched, consisting of a tough cushion of tissue from which long (2-7 cm) spines are suspended (but spines short in one eastern variety); found on hardwoods, especially oak .... *H. erinaceus,* 615
1. ˙ Fruiting body branched, the spines hanging from the branches or branch tips (sometimes scarcely branched and very compact, but if so, then usually growing on conifers) 2
2. Growing on conifers in western North America; fruiting body white to salmon- or yellowish-tinged when fresh .... *H. abietis,* below
2. Growing mainly on hardwoods; fruiting body white when fresh (but may turn yellowish in age); widely distributed 3
3. Spines rather short (3-10 mm), arranged in rows along the branches (like teeth on a comb); branching usually open rather than compact; fruiting body often delicate *H. ramosum,* p.615
3. Not as above; spines often long (up to 4 cm), arranged mostly in tufts or clusters, especially at the branch tips; branching open or compact ...... *H. coralloides* (see *H. abietis,* below)

Hericium abietis   (Conifer Coral Hericium)   Color Plate 163

FRUITING BODY 10-75 cm or more broad and high at maturity, consisting of an open to compact branched framework from which tufts of icicle-like spines hang; branches arising from a thick, tough, rooting base; color variable: white to creamy, yellowish-buff, pale ochraceous, or salmon-buff. **SPINES** up to 25 mm long but usually 5-10 mm; soft but brittle, arranged in tufts or clusters that are mainly grouped at the branch tips. **SPORE PRINT** white; spores 4.5-5.5 × 4-5 microns, round or nearly round, smooth or minutely roughened, amyloid.

HABITAT: Solitary or sometimes several together on dead conifers (especially fir and Douglas-fir); known only from the Pacific Northwest and northern California, fruiting mainly in the fall. It is rather infrequent but locally common, particularly at higher elevations (the largest fruitings I've seen were in the Cascades). It appears year after year on the same logs, causing a conspicuous white pocket rot. The closely related *H. coralloides* (see comments) favors hardwoods and is more widely distributed.

EDIBILITY: Eminently edible, delectably delicious. When thoroughly cooked it is reminiscent of fish and is excellent sauteed, curried, or marinated. Its large size (a 100-lb. specimen was wheeled into one mushroom show!) plus its distinctive appearance make it an excellent mushroom for beginners. However, its breathtaking beauty poses a minor moral dilemma: should one ruthlessly uproot it for the sake of a meal, or leave it for others to see?

COMMENTS: It is hard to believe that this astonishing *Hericium,* with its cascading clusters of pristine "icicles," is a fungus. It is easily distinguished from other Hericiums by its white to salmon-buff color, clustered spines, and growth on conifers. It has long been a favorite among nature-lovers and photographers in the Pacific Northwest, but only recently has it been recognized as distinct from the better known, equally beautiful and delicious *H. coralloides.* Some investigators, in fact, still consider the two to be the same species. Others reserve the name *H. coralloides* for a pure white (when fresh) hardwood-loving form with slightly longer spines (up to 4 cm) and slightly larger spores. This form occurs in eastern North America, but reports of it from the West are based at least in part on *H. abietis.* To complicate matters, it has now been suggested that the name *H. coralloides* is better applied to *H. ramosum,* and that the eastern form traditionally called *H. coralloides* should be given a new name, *H. americanum!* To muddle matters even more, there are several growth forms of *H. abietis,* including a very compact, scarcely branched one (formerly *H. weirii*) that somewhat resembles *H. erinaceus,* and an extensively branched one with very short spines (1-5 mm) that mimics *H. ramosum* (both forms grow on conifers, however). Since all of these Hericiums are equally edible, their exact identities needn't concern you—at least, they don't concern me!

Hericium ramosum   (Comb Hericium)   **Color Plate 164**

FRUITING BODY 8-35 cm broad and 6-15 cm high when mature, comprised of an open framework of rather delicate, toothed (spine-laden) branches arising from a tough, repeatedly branched rooting base or "trunk"; pure white when fresh, discoloring creamy to buff or yellowish-tan in old age. Flesh white. **SPINES** more or less evenly distributed in lines along the branches (like teeth on a comb), sometimes also in small tufts at the branch tips; spines rather short (3-10 mm long or up to 25 mm long in the tufts). **SPORE PRINT** white; spores 3-5 × 3-4 microns, nearly round, smooth or minutely roughened, amyloid.

HABITAT: Solitary or in small groups on fallen hardwood branches, logs, and stumps; widely distributed. It is said to be the most common *Hericium* in North America, but like the others, is uncommon in our area. I have found it in the fall, winter, and early spring on dead oak, and I have seen it in the summer on aspen and poplar in New Mexico.

EDIBILITY: Edible and delicious when cooked slowly, but not as fleshy as the other Hericiums.

COMMENTS: Formerly known as *H. laciniatum,* this lovely species is smaller and more delicate than *H. abietis* and *H. coralloides,* with slightly shorter spines and smaller spores. Also, the branching is more open and the spines are arranged in lines lengthwise along the branches, rather than exclusively in tufts. A compact form of this species occurs, but is rare.

Hericium erinaceus   (Lion's Mane Hericium; Old Man's Beard)

FRUITING BODY an unbranched mass of numerous long, closely-packed, icicle-like spines hanging from a tough, solid, hairy, rooting cushion of tissue; 8-40 cm or more broad and high when mature; entirely white when fresh, discoloring yellowish to tan or dingy ochre in age. Flesh white. **SPINES** (1) 2-5 (7) cm long, soft and pliant when fresh, with pointed tips. **SPORE PRINT** white; spores 5-6.5 × 4-5.5 microns, broadly elliptical to nearly round, smooth to minutely roughened, amyloid.

HABITAT: Solitary (rarely several together) on wounds of living hardwoods or on the cut ends of recently felled logs; widely distributed. In our area it favors oak and is fairly common (for a *Hericium*) in the fall, winter, and spring; farther north it grows on maple.

EDIBILITY: Excellent when fresh, but tougher than other Hericiums and sometimes

Left: *Hericium ramosum* is the most delicate species in its genus. It favors dead hardwoods. **Right:** *Hericium erinaceus* has long spines hanging from an unbranched cushion of tissue. It favors living or recently killed hardwoods.

Hericium erinaceus. Note how long the spines are in this rather small specimen.

developing a rather sour, unpleasant taste in age. Slow cooking is called for and the base should not be eaten—it's so tough that it's difficult to remove from the tree without a knife!

COMMENTS: The unbranched white to yellowish fruiting body and long, slender spines distinguish this impressive fungus from its relatives. Though hardly common, it is more numerous in our area than the branched Hericiums and is one of our most distinctive wood-inhabiting fungi. Gigantic specimens weighing several pounds each are not uncommon. An unbranched version completely covered with small *short* spines occurs on hardwoods in eastern North America, but I have not seen it in the West.

HYDNUM, DENTINUM, & Allies
(Hedgehog Mushrooms)

Medium-sized to fairly large, fleshy, *terrestrial* fungi with a cap and stalk. CAP smooth or cracked to conspicuously scaly. *Flesh firm, usually brittle.* SPINES soft, brittle. STALK central or off-center, *usually well-developed,* fleshy but not woody. SPORE PRINT *white (Dentinum) or brown (Hydnum).* Spores smooth *(Dentinum);* rough to warty or angular *(Hydnum).*

HEDGEHOG mushrooms are the best known and most common of the teeth fungi, and as a group are easy to recognize. The fruiting body has a well-defined cap and stalk and might be mistaken for a gilled mushroom but for the layer of delicate spines or "teeth" on the underside of the cap. Also significant is the texture of the flesh: fleshy or firm but brittle, rather than tough and pliant or leathery or woody as in the other major genera of terrestrial teeth fungi *(Hydnellum* and *Phellodon).*

In the small but common genus *Dentinum,* the spores are white and smooth, the spines white to pale orange, and the cap usually smooth. In the larger genus *Hydnum,* the spores are brown and warty, the spines are variously colored but usually darker than in *Dentinum,* and the cap is often scaly. A third genus, *Bankera,* has a brownish fruiting body and white roughened spores but is relatively rare, at least in the West.

Although there is no doubt as to what hedgehog mushrooms *are,* there is, as usual, considerable controversy as to what hedgehog mushrooms should be *called.* Some mycologists campaign for the use of *Hydnum* instead of *Dentinum,* and the creation of the genus *Sarcodon* to account for the Hydnums of this book. A "correct" classification of the teeth fungi may not be essential to your well-being (it certainly isn't to

616

mine), but please bear in mind that the exacting specialists owe their livelihoods to the resolution of such matters—and that they are doing their best. Giving even tacit approval to one system of classification at the expense of another is thus transformed into an act of inordinate importance, with the taxonomist's professional reputation at stake. Why they can't arrive at a consensus is anyone's guess—but since when do human beings agree on anything? I for one find them (human beings, that is) even more perplexing and mystifying than mushrooms!

Dentinums are delicious and a good choice for beginners since nothing poisonous remotely resembles them. Hydnums, on the other hand, though sometimes beautiful, are mostly bitter-tasting or of unknown edibility.

Both *Dentinum* and *Hydnum* species are strictly woodland fungi with rather erratic fruiting habits. For instance, in February and March of 1975 Dentinums were outrageously abundant on the poison-oak-shrouded hillsides of our coastal pine forests. While barely denting the crop I managed to harvest and can over 200 pounds, which I am still enjoying today. For several years thereafter, however, they were practically absent in the same area, then produced another stupendous crop in 1981. Similarly, I did not find a single *Hydnum fuscoindicum* in our area until 1979, when it fruited by the dozens under tanoak and madrone. Five distinctive hedgehog mushrooms are described here and several others are keyed out.

Key to Hydnum, Dentinum, & Allies

1. Fruiting body (including spines) white to pale orange to dull orange, the cap sometimes slightly darker; spore print white; spores smooth 2
1. Not as above; some part of fruiting body usually darker (gray, brown, etc.) 3

2. Cap convex to plane or depressed, 2-15 cm broad or more; stalk usually at least 1 cm thick .... *D. repandum* & others, p. 618
2. Cap usually umbilicate (with a navel-like central depression) and typically less than 5 cm broad; stalk usually less than 1 cm thick .... *D. umbilicatum* (see *D. repandum*, p. 618)

3. Spore print white; spines usually pale gray or grayish in age; found under conifers (usually pine) in eastern North America, but rare in the West 15
3. Spore print brown; spines variously colored but often dark in age; widespread and common 4

4. Flesh showing distinct violet tints (deep violet to lilac-gray) 5
4. Not as above; flesh white to brownish, vinaceous-buff, etc. (but may stain purple when cut) 6

5. Entire fruiting body (including spines!) violet to deep violet or blackish-violet .... *H. fuscoindicum,* p. 622
5. Not as above; spines brownish to cinnamon-brown with paler tips .... *H. cyanellum* & *H. fuligineo-violaceum* (see *H. fuscoindicum,* p. 622)

6. Base (tip) of stalk and flesh within it black to olive-black, olive-gray, or bluish-green; taste very bitter or bitter-farinaceous (chew on a piece of the cap) 7
6. Not as above; taste mild or bitter 8

7. Flesh becoming pinkish or lilac-tinted when cut; spines whitish or pallid until old age .... *H. subincarnatum* (see *H. scabrosum* group, p. 620)
7. Not as above .... *H. scabrosum group,* p. 620

8. Scales on cap brown to blackish, usually large and conspicuous even when young and often raised or upturned; spines usually brownish at maturity; common .. *H. imbricatum,* p. 619
8. Not as above; cap usually developing scales only in age and/or differently colored 9

9. Cap white and plushlike when young, yellowish or tan in age; spines dull yellowish or brown becoming almost blackish in old age; growing under hardwoods in eastern North America .... *H. cristatum*
9. Not as above 10

10. Base of stalk usually white from coating of mycelium; fruiting body often large, thick-fleshed, buff to tan to yellow-brown, orange-brown, or cinnamon; cap usually with small flattened scales .... *H. calvatum group* & others, p. 621
10. Not as above 11

11. Fruiting body developing greenish-olive tones in old age or when dried; flesh often becoming greenish also; not common ....*H. fumosum*
11. Not as above 12
12. Flesh staining lilac, purplish, purple-gray, pinkish, or vinaceous when cut (quickly *or* quite slowly) *and/ or* fruiting body often with vinaceous or purplish tints; widely distributed but especially common in the West 13
12. Not as above; cap pale to dull brown or dark reddish-brown, quite hard and woody when dried; fairly common under conifers in eastern North America, also reported from the Pacific Northwest and California .... *H. stereosarcinon*
13. Cap smooth to felty or conspicuously cracked but not truly scaly, reddish-brown to grayish-brown, vinaceous, grayish, or darker; taste mild to somewhat farinaceous or occasionally bitter 14
13. Cap usually with scales, brownish-orange to dark brown, grayish-brown, or pinkish-brown; taste usually bitter .... *H. subincarnatum* (see *H. scabrosum* group, p. 620)
14. Cap and stalk often with vinaceous or purplish tints; cap often conspicuously cracked (areolate) in age; common in western North America .. *H. rimosum* (see *H. scabrosum* group, p. 620)
14. Cap grayish to grayish-brown or reddish-brown, smooth to minutely areolate (cracked) in age; widespread but not common (at least in West) *H. laevigatum* (see *H. scabrosum* group, p. 620)
15. Cap some shade of brown at maturity (the margin often paler), not normally scaly but usually with needles and other debris adhering to the surface; found under conifers (usually pine) in eastern North America, but rare in West .... *Bankera fuligineo-alba*
15. Not as above; cap more or less grayish-brown, often cracked or scaly in age; widespread but not common .... *Bankera carnosa (=B. violescens?)*

Dentinum repandum   Color Plates 161, 162
(Hedgehog Mushroom; Pig's Trotter)

CAP 2-17 (25) cm broad, broadly convex to plane or depressed, the margin often wavy or deeply indented and at first inrolled; surface dry, more or less smooth, but sometimes cracking into scales in age; pale flesh-color to pale or dull orange, orange-tan, salmon, tan, or pale cinnamon to reddish-tawny (but white to creamy in *var. album*); bruised areas often darker orange. Flesh thick, occasionally zoned concentrically, firm, brittle, white, usually discoloring yellow to yellow-ochre or orange-brown when bruised; odor mild, taste mild to somewhat bitter or peppery. SPINES 2-7 mm long, whitish to yellowish, salmon-buff, or pale orange, bruising dark orange to ochraceous; slender, brittle but soft, usually decurrent. STALK 3-10 cm long, (0.5) 1-3 (5)) cm thick, central or off-center, equal or enlarged below or occasionally tapered downward, firm, white or colored like cap but usually paler; bruising ochre to dark orange-brown; smooth or downy at base. SPORE PRINT white; spores 6.5-9 × 5.5-8 microns, broadly elliptical to nearly round, smooth.

HABITAT: Solitary, scattered, gregarious, or in troops on ground under both hardwoods and conifers; widely distributed throughout the north temperate zone and probably the most common of all the teeth fungi. In California it fruits throughout the mushroom season but develops slowly and normally does not peak until the late winter or early spring, when the wild irises are in bloom (see Color Plate 161) and most other fungi have long since rotted away. In our area it favors fern, bramble, and poison oak thickets under pine; in the Pacific Northwest and Rocky Mountains it grows under a wide range of conifers, and in eastern North America it is often common under oaks in the summer and fall.

EDIBILITY: Edible and choice! It is comparable to the chanterelle in color, texture, and flavor, and like that species is usually maggot-free. It is easier to clean, however, and is also easier to recognize.The peppery taste (if present) disappears in cooking. It's superb in casseroles, tomato sauces, or sauteed with sour cream, but should be cooked slowly and lengthily to make it tender. Be careful to keep the spines clean while picking them!

COMMENTS: Also known as *Hydnum repandum,* this late bloomer resembles a dull-colored chanterelle when first spotted amongst the needles and humus. However, the white to pale orange spines on the underside of the cap immediately distinguish it, making it one of the safest of all edible mushrooms. The cap color, though variable, is typically some shade of pale orange to pinkish-orange to reddish-tan, with wounded areas darker orange. The white form (var. *album*) does not seem to occur in our area, but is common farther north under conifers, especially Sitka spruce. Other species: *Var. **macrosporum*** has larger spores but is otherwise similar; ***D. umbilicatum*** is a closely related, equally edible conifer-lover with a smaller (usually less than 5 cm) umbilicate cap, a slimmer stalk (typically less than 1 cm), and larger spores. It is similar in color and widely distributed, and sometimes mingles with *D. repandum*. Two edible white or whitish southern species should also be mentioned: ***D. albomagnum*** and ***D. albidum***. The first has a mild taste and non-staining flesh, while the latter has an acrid taste and smaller spores.

Hydnum imbricatum   (Shingled Hedgehog)   **Color Plate 159**

CAP 5-20 cm broad, convex to plane or centrally depressed; surface dry, buff to pale brown or dull reddish-brown, but covered with large, coarse, broad, raised or shingle-like, darker brown to nearly black scales that are often upturned in age; becoming darker brown throughout in age and sometimes cracking, with the scales sometimes wearing off except at the center. Flesh thick, firm but brittle; pallid to grayish, tan, or brownish; odor mild or when dry somewhat smoky or chocolate-like; taste mild to bitter. **SPINES** pale brown or grayish or pallid becoming dark brown in age, 2-15 mm long; soft, brittle, often slightly decurrent. **STALK** 4-10 cm long, 1.5-3.5 (5) cm thick, central or off-center, often enlarged below; some shade of brown, often hollow toward the top in age; usually more or less smooth. **SPORE PRINT** brown; spores 6-8 × 5-7 microns, nearly round but prominently warted (angular-nodulose or shaped like a Maltese cross).

HABITAT: Solitary to gregarious on ground in woods; widely distributed—it is probably the most common *Hydnum* in North America. In many regions it is abundant under conifers in the late spring, summer, and fall. In our area it occurs under hardwoods in the late fall and winter, but is rare. I have seen enormous fruitings under spruce and fir.

EDIBILITY: Edible, but of poor quality. Many collections have a bitter taste and parboiling does not necessarily help, plus it causes indigestion in some people. The European version is apparently better because it is often sold in markets there.

Hydnum imbricatum is easily told by the large brown to blackish scales on the cap, which make it look like a charred macaroon. See color plate for close-up of spines, which range from grayish to brown.

COMMENTS: Also known as *Sarcodon imbricatum,* this arresting hedgehog is easily identified by the prominent brown to blackish scales on its cap. The stalk base is not olive-black as in the *H. scabrosum* group, and the cap is scalier—even when young. In old age the central depression of the cap may become perforated, i.e., join up with the hollow in the upper portion of the stalk. For similar species, see the *H. scabrosum* group.

Hydnum scabrosum group   (Bitter Hedgehog)   Color Plate 165

CAP 4-14 (20) cm broad, convex to plane or slightly depressed; surface dry, at first smooth but soon becoming cracked and scaly, the scales cinnamon-brown to reddish-brown becoming dark chocolate-brown to vinaceous-brown at maturity and the background dingy yellowish-brown or darker. Flesh thick, brittle, firm, white or buff, but olive-gray to olive-black in base of stalk; odor farinaceous or smoky; taste strongly bitter and/ or acrid to farinaceous. **SPINES** often unequal in length, 2-10 mm long, pallid or buff, darkening to tan with paler tips, then darker brown in old age; usually slightly decurrent. **STALK** 2.5-10 cm long, 1-3.5 cm thick, usually tapered below, central or off-center, often curved, flesh-color becoming brown or dark brown in age, the base blackish-olive to grayish-olive, olive, or dark bluish-green; firm, solid. **SPORE PRINT** brown; spores 6-7.5 × 4-5.5 microns, elliptical to round, prominently warted. Cap tissue staining blue-green in KOH.

HABITAT: Solitary to scattered or gregarious on ground under conifers or sometimes hardwoods; widely distributed. It is fairly common in the late summer and fall in the Pacific Northwest and northern California, but rare in our area.

EDIBILITY: Unequivocally and indisputably inedible due to the awful taste.

COMMENTS: The scaly reddish-brown to brown cap, blackish-olive to blue-green stem base, and bitter taste distinguish this species from most other Hydnums. The scales are usually quite conspicuous, but not as large as those of *H. imbricatum.* *H. fennicum* is a very similar species with a scaly reddish-brown to brown cap and blackish stem base. However, it does not stain blue-green in potassium hydroxide (KOH) and it has an even more intensely bitter taste. It grows under both hardwoods and conifers and is widespread. Three other somewhat similar species are quite common in the West: *H. sub-incarnatum* is usually bitter-tasting and similarly colored or more vinaceous, but its flesh usually stains pinkish or lilac when cut, its spines are pallid or whitish becoming buff or pale brown only in old age, and its stalk base may or may not be olive-tinted. In California it grows under both hardwoods and conifers, sometimes in the company of *H. scabrosum.* *H. rimosum* (see photo below) is also similar, but has a more or less mild (or rarely slightly

Hydnum rimosum (see comments under *H. scabrosum* group) is not nearly as scaly as *H. imbricatum,* but the surface of its cap often cracks in age.

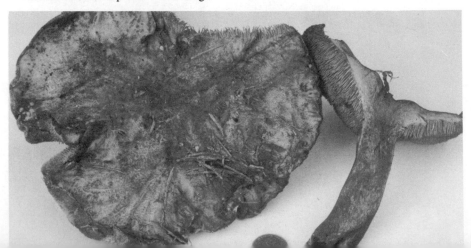

acrid) taste, is not normally olive at the stem base, and has a vinaceous- or purple-tinged cap that usually develops cracks in age but is not scaly when young. It is fairly common under conifers in the Pacific Northwest, and I have found it in our area in the winter and in the Sierra Nevada in the spring. Finally, there is *H. laevigatum*, a widespread but infrequently-encountered species that has a smooth to minutely cracked cap and, like *H. rimosum*, often stains purplish when cut open. None of these are worth eating.

Hydnum calvatum group   (Robust Hedgehog)

CAP (5) 10-25 (35) cm broad, convex to nearly plane or somewhat irregular; surface dry, soon breaking up into small, flattened scales and usually cracking in age; creamy-buff to yellowish-tan or pale cinnamon-brown, the scales slightly darker and becoming brownish where bruised; margin often lobed. Flesh very thick, firm, pallid to pinkish-buff or pale brownish, or grayer near the spines; odor and taste variable: mild to spicy-fragrant to farinaceous. **SPINES** pallid, soon darkening to brown or grayish-brown, the tips usually paler; 2-12 (15) mm long, usually very uneven in length or many aborted or fused together or forked or with small swellings; sometimes decurrent. **STALK** 2-9 cm long, 2-4 (6) cm thick, central or off-center, usually narrowed at the base and sometimes rooting; solid, firm, colored like cap or slightly paler, the base often whitish from mycelium. **SPORE PRINT** brown; spores 4-5.5 × 3-5 microns, nearly round to elliptical but prominently warted (angular-nodulose). Cap surface staining blue-green to olive-black in KOH.

HABITAT: In groups or clumps, often with small or aborted fruiting bodies present, on ground in mixed woods and under conifers; known only from western North America. I've found it in August under spruce in New Mexico; it is also fairly common in Oregon.

EDIBILITY: Unknown—but fleshy enough to warrant *cautious* experimentation.

COMMENTS: The large size, thick flesh, tendency to grow in groups or clumps, whitish stalk base, and presence of small appressed (flattened) scales on the cap distinguish this species from most other Hydnums. The color is never as dark as that of *H. imbricatum*, nor the cap as coarsely scaly; the stalk is not olive or blue at the base as in the *H. scabrosum* group. A similar species, *H. crassum*, is slightly brighter in color, has larger spores, and its flesh may stain yellow-green when cut. Its cap surface stains brownish in KOH and it also occurs under conifers. A somewhat similar and widespread species, *H. martioflavum*, has bright cinnamon-orange spines when young and a cinnamon to tawny cap.

Left: *Hydnum calvatum* group. Note small flattened scales on cap. **Right:** Underside of *Hydnum fuscoindicum* (see description on p. 622). Entire fruiting body is deep purple to black in this beautiful and unmistakable hedgehog mushroom.

Hydnum fuscoindicum   (Violet Hedgehog)

CAP 4-18 cm broad, convex to plane or centrally depressed; surface dry, at first smooth but usually cracking to form scales in age; violet-black to bluish-black, black, or raisin-colored; margin often somewhat paler or purpler and wavy. Flesh thick, firm but brittle, deep slate-purple or violet; odor and taste mild to somewhat farinaceous or cinnamon-like. **SPINES** deep violet to deep bluish-violet to deep lavender, the tips usually paler or lilac; soft, brittle, 2-6 (15) mm long, usually decurrent. **STALK** 2-10 cm long, 1-2 (3.5) cm thick, equal or more often tapered below, central or off-center, firm, colored more or less like the spines (deep purplish). **SPORE PRINT** brown; spores 5-7 × 4.5-6.5 microns, broadly elliptical to nearly round, prominently warted.

HABITAT: Widely scattered to gregarious on ground in woods; locally common but erratic in its fruiting behavior—absent some years and abundant others; known only from western North America. In our area it fruits in the fall and winter under tanoak and madrone at higher elevations in the coastal mountains; in the Pacific Northwest, however, it favors conifers, especially hemlock and pine. It is yet another example of a conifer-lover that crosses over to tanoak-madrone (others include *Armillaria ponderosa, Cantharellus subalbidus, Hygrophorus chrysodon,* and *Tricholoma aurantium*).

EDIBILITY: Not recommended. Although not exactly bitter-tasting, small pieces which I sampled caused a peculiar burning sensation in the back of my throat. It would be interesting to see what color it yields as a dye.

COMMENTS: The striking deep violet color, which is reminiscent of *Cortinarius violaceus,* sets apart this attractive *Hydnum* (see photo at bottom of p. 621). Its color makes it hard to pick out in the forest gloom, and even the most eagle-eyed *Hydnum*-hound is likely to miss it or else mistake it for an old, blackened *Russula albonigra. Phellodon atratus* is somewhat similar, but is bluish-black and much smaller and tougher. Other purplish Hydnums include: *H. cyanellum,* with a similarly colored cap but paler ("lilac-gray") flesh, a bitter taste, and cinnamon-brown spines with whitish tips (it was originally described from California, but I have not seen it); *H. fuligineo-violaceum,* with a vinaceous-brown cap, brown or cinnamon-brown spines, an acrid taste, and gray flesh in the base of the stalk; and *H. rimosum* (see comments under the *H. scabrosum* group), which is vinaceous-brown to vinaceous-buff rather than deep purple.

HYDNELLUM & PHELLODON

Small to medium-large, *terrestrial* fungi with a cap and stalk (or sometimes several caps). CAP columnar to top-shaped becoming plane, depressed, or irregular; often spongy or felty when young and often lumpy, pitted, or ridged in age. Flesh *tough; spongy to fibrous; pliant or woody, often duplex.* SPINES typically short, often blunt; variously colored. STALK central or off-center, sometimes nearly absent; *tough or woody,* continuous with cap. SPORE PRINT *brown (Hydnellum) or white (Phellodon).* Spores roughened by warts or spines.

THESE tough or woody, terrestrial teeth fungi occur primarily in coniferous forests. They have both a cap and stalk. but the latter is sometimes present only as a poorly-defined tapered base. Many species have the general aspect of a polypore, but a closer look reveals the presence of spines rather than pores on the underside of the cap, although the spines are sometimes so short and blunt that they look like minute warts.

The fruiting body develops and decays over a long period of time, frequently engulfing needles, twigs, and other debris in the growth process. If there is a dry spell, growth may cease, then begin anew around the margin of the cap when it is damp again. As a result the cap frequently has an irregular or somewhat misshapen appearance; several caps may fuse together or arise in a rosette from a common stem, and their surfaces are often pitted, ridged, or lumpy. This **indeterminate** growth pattern plus the tough and pliant to fibrous or woody texture separate *Hydnellum* and *Phellodon* from the fleshier, more brittle hedgehog mushrooms *(Hydnum* and *Dentinum).* In many Hydnellums and some Phellodons the cap is spongy or felty to the touch when young or actively growing, and is sometimes beaded with colored droplets. However, beneath or within the spongy-felty outer layer there is a tougher, corky or woody-fibrous core, particularly in the stalk. The flesh thus has two distinct layers of different textures, i.e., it is **duplex.**

Hydnellum contains about 50 species in North America. They are medium-sized to fairly large and have brown spores. *Phellodon* includes only a handful of species, but some of them are quite common. They are distinguished from *Hydnellum* by their white spores and smaller size. Neither genus is common in our area but both are prominent farther north. Because of their indeterminate growth most species vary tremendously in their appearance according to environmental conditions. It is therefore imperative, when using the following key, to have several specimens in hand if at all possible. Seven species are described fully and others are keyed out. All are much too tough and/or bitter to eat.

Key to Hydnellum & Phellodon

1. Flesh in the cap *and/or* stalk black or tinted or zoned (lined) with blue or violet (purple-black, blue-gray, etc.) when fruiting body is cut in half lengthwise 2
1. Not as above; flesh orange, cinnamon, brown, or shades thereof 9

2. Flesh in lower half of stalk reddish-orange to bright rusty-brown or salmon; flesh in cap usually zoned with brown and blue or mauve (sometimes faintly) ... *H. caeruleum* & others, p. 625
2. Not as above 3

3. Spines bright blue to dark blue; stalk with a swollen buried base or "tuber"; found in eastern North America; rare .... *H. scleropodium*
3. Not as above 4

4. Flesh in lower half of stalk bluish-black to purple-black; cap whitish to yellowish to tan or violet-tinged when fresh; odor often fragrant .... *H. suaveolens* & others, p. 624
4. Not as above 5

5. Flesh black to grayish-black (not blue or violet-tinted) in both cap and stalk; spore print white (when obtainable); found mainly from the Rockies eastward *P. niger* (see *P. atratus*, p. 629)
5. Not as above 6

6. Cap rather small (5 cm broad or less); stalk slender (averaging 3-5 mm thick); spore print white 7
6. Cap medium-sized to rather large (usually 4 cm broad or more); stalk usually thicker; spore print brown 8

7. Flesh purple-gray to purple-black; cap typically dark brown to purplish-gray with a pale margin .... *P. melaleucus* (see *P. atratus*, p. 629)
7. Flesh purple-black to bluish-black; cap more or less same color .... *P. atratus*, p. 629

8. Found in eastern North America; fruiting body slate-gray to black ....
   .... *H. nigellum* (see *P. atratus*, p. 629)
8. Found mainly from the Rocky Mountains westward; flesh and/or cap usually with some blue or violet tones .... *H. cyanopodium & H. regium* (see *H. caeruleum*, p. 625)

9. Spines bright yellow at least at the tips; cap usually with bright yellow to olive-yellow tones at least at the margin; found in eastern North America .... *H. geogenium*
9. Not as above 10

10. Flesh yellow-orange to bright orange, rusty-orange, or reddish-orange, at least in the stalk; cap and/or spines often showing same colors 11
10. Not as above 12

11. Flesh in cap and stalk orange to orange-red or rusty-cinnamon; cap surface with cinnamon, orange, or rusty-cinnamon tones, at least in age .... *H. aurantiacum* & others, p. 626
11. Flesh in cap duller (mauve, grayish, brownish, etc.); cap surface also typically duller ....
    .... *H. caeruleum* & others, p. 625

12. Odor sweet *and/ or* taste of flesh very acrid (peppery); cap beaded with bright red to dark red droplets in wet weather .... *H. peckii* & others, p. 627
12. Cap not beaded with red droplets, or if so then taste not typically acrid nor odor sweet ... 13

13. Cap usually small (6 cm or less); stalk often slender (less than 1 cm thick); spore print white 14
13. *Mature* cap often more than 5 cm broad; stalk often thick; spore print brown or brownish 15

14. Spines pale cinnamon to brown at maturity; cap usually zoned ...... *P. tomentosus,* p. 628
14. Spines grayish at maturity .... *P. confluens* (see *P. tomentosus,* p. 628)

15. Stalk exuding yellow juice when broken or crushed (if fresh); cap pale to dark brown, or sometimes also with yellow-brown tones and exuding a brownish juice when fresh .. *H. mirabile*
15. Not as above 16

16. Cap smooth or radially ridged but not lumpy, warty, or spongy, often zoned concentrically (see Color Plate 157) with various shades of brown, cinnamon-brown, pinkish-brown, etc.; stalk *not* spongy or bulbous .... *H. zonatum* (see *H. scrobiculatum,* p. 627)
16. Not as above 17

17. Associated with hardwoods (mainly oak) in eastern North America 18
17. Associated principally with conifers; widely distributed 19

18. Stalk very spongy and swollen or bulbous; cap and stalk brown to rusty-brown or cinnamon-brown and finely hairy or velvety; common .... *H. spongiosipes*
18. Not as above; cap very dark brown to blackish in old age .... *H. piperatum*

19. Cap sometimes beaded with pinkish droplets in wet weather; odor typically mild or faint but not farinaceous; common under pines in northeastern North America .... *H. pineticola* (see *H. peckii,* p. 627)
19. Cap sometimes beaded with red to dark red droplets in wet weather; odor usually farinaceous; widely distributed under various conifers .... *H. scrobiculatum* & others, p. 627

Hydnellum suaveolens   (Fragrant Hydnellum)

CAP (3) 5-15 (30) cm broad when mature, top-shaped becoming plane to somewhat depressed, often with needles or other debris incorporated into it; surface white and thinly felty or velvety at first, often bumpy and/ or pitted in age, soon becoming yellowish to tan, brownish, olive-brown, or violet-gray from the center outward (margin often paler); usually staining brown where bruised. Flesh duplex, whitish to yellowish-buff zoned with blue lines in the cap, entirely deep blue to purple-black in the stalk; odor often strongly fragrant (like anise or peppermint). **SPINES** whitish to creamy when young, becoming grayish-brown with pallid tips in age; short (up to 3 mm long), irregularly decurrent. **STALK** 1-5 cm long, 1-3 cm thick, central or slightly off-center, very tough, the base usually swollen, rooting; grayish-blue to bluish-black. **SPORE PRINT** brown; spores 4-6 × 3-4 microns, elliptical to nearly round, prominently warted.

HABITAT: Solitary to gregarious or in fused clusters under northern and montane conifers, late summer through early winter; widespread but especially common in the Rocky Mountains. I have seen large fruitings under spruce and fir in Idaho and New Mexico, but have yet to find it in our area.

EDIBILITY: Unknown, but much too fibrous to be worthwhile.

COMMENTS: The blue-lined flesh in the cap and blue-black flesh in the stalk plus the flagrantly fragrant (fragrantly flagrant?) odor make this an easy species to identify. In some collections the fragrance is absent, but in others it is overpowering—I once had to remove two specimens from my car because their aroma was so heady! *H. peckii* is often fragrant but much different in color, while *H. caeruleum* shows orange-red flesh in the stalk. Other species: *H. cruentum,* described from Nova Scotia, has a menthol odor, but when young and fresh its cap has red droplets and it spines are lilac- or bluish-tinged.

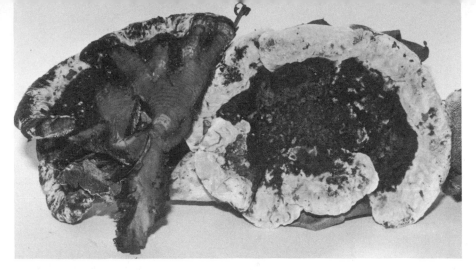

Hydnellum caeruleum. Note the conspicuously zoned flesh in the sliced specimen at left.

Hydnellum caeruleum   (Blue-Gray Hydnellum)

CAP 3-12 (17) cm broad, often with leaves or needles incorporated into it; top-shaped becoming plane or slightly depressed; surface felty or velvety at first, often bumpy and/or pitted in age and with matted hairs; mauve to pale blue, whitish, or tan when young, becoming light to dull dark brown or even blackish from the center outward (often a mixture of these colors), the soft, felty growing margin white or pale blue to bluish-gray. Flesh duplex: upper or outer layer spongy, inner core tough and fibrous; zoned variously with bluish, blue-gray, mauve, and brown in the cap; bright rusty-colored to orange-red in stalk; odor and taste farinaceous. **SPINES** whitish when young or tinged blue, becoming brown to dark brown with pallid tips in age; rather short (1-5 mm long), often decurrent. **STALK** 2-9 (12) cm long, 1-3 cm thick, central or off-center, very tough, often rooting deeply in humus; equal or thicker at either end, buff to brown or orange-brown, but usually covered with debris. **SPORE PRINT** brown; spores 4.5-7 × 3.5-5 microns, nearly round to elliptical and irregularly lobed or warted.

HABITAT: Solitary to gregarious or in fused clusters on ground in woods; widely distributed. It is common in the Pacific Northwest under pines and other conifers, but in our area I find it under oak, tanoak, and madrone, usually early in the fall.

EDIBILITY: Indisputably inedible.

COMMENTS: Like most Hydnellums this species develops over a period of several weeks and undergoes a number of confusing color changes. Young specimens are more or less top-shaped, and have a thick, felty cap margin that shows white and/or bluish tints. In age, however, or after being battered by rain, the cap becomes depressed (wouldn't you?), the margin thins out, and the color becomes darker or duller brown. However, the bluish lines in the flesh of the cap (though sometimes faint) and the bright rusty- to orange-red flesh in the stalk (which distinguishes it from *H. suaveolens*) are fairly constant characters. Other species: *H. ferrugipes* of eastern North America has an orange-red stalk, but favors hardwoods. *H. cyanopodium* has a vinaceous-blue cap shading toward lavender (or whitish at margin) that is often beaded with red droplets in wet weather, flesh that is zoned bluish-black, and spores shaped like jacks. *H. regium* often forms compound fruiting bodies (with several caps); its cap is violet-black with a paler margin and the flesh is brown to pale orange in the stalk and brownish to grayish with violet tones in the cap. The latter two species occur under conifers (mainly spruce and pine) in northern California, the Pacific Northwest, and the Rocky Mountain region.

625

Hydnellum aurantiacum looks like a terrestrial polypore, but has minute spines on underside of cap instead of pores. Lumpy specimen on right is young and actively growing.

Hydnellum aurantiacum   (Orange Hydnellum)

CAP 3-15 cm broad, columnar or somewhat top-shaped becoming plane or depressed in age; often with pine needles and other debris incorporated into it; surface velvety and suedelike when fresh and often roughened by projecting knobs and lumps when mature, or in some forms with radiating ridges; white when young or on actively growing margin, otherwise orange to rusty-orange to rusty-cinnamon (or mixture of these colors), and eventually darker (brown) in old age. Flesh thick, tough and corky except for frequent presence of a spongier outer or upper layer; orange to rusty-cinnamon to orange-red in both the cap and stalk; odor mild, taste bitter to farinaceous. **SPINES** short and blunt (1-4 mm long), whitish to grayish or orange, becoming brown in age with the tips often paler. **STALK** 2-6 cm long, 1-3 cm thick, usually central, very tough or woody, equal or tapered downward or in one form enlarged at base; orange to bright rusty-cinnamon becoming dark brown in age, with a large mat of pine needles and debris usually stuck to the base. **SPORE PRINT** brown; spores 5.5-7.5 × 5-6 microns, nearly round, prominently warted.

HABITAT: Solitary or in groups on ground under pines and other conifers, sometimes also in fused clusters; widely distributed. It is one of the two most common Hydnellums in our area (from late fall through early spring), but does not often occur in large numbers.

EDIBILITY: Unequivocally inedible.

COMMENTS: The tough texture, knobby or lumpy cap surface, and bright orange to rusty-cinnamon colors of the cap and flesh are the fallible fieldmarks of this species and its close relatives. It never exudes the red droplets characteristic of *H. peckii,* but might be casually mistaken for a polypore such as *Phaeolus schweinitzii.* However, a close look at its underside reveals the presence of small "teeth" instead of pores. Closely related species include: *H. complectipes,* usually with many caps fused to form large, complicated masses or rosettes, found in the Pacific Northwest under conifers; *H. conigenum,* with very thin flesh in the cap, also found under conifers; and *H. ferrugipes* and *H. earlianum,* found under hardwoods in eastern North America, the latter with a nearly smooth cap.

626

Hydnellum peckii   Color Plate 160
(Strawberries and Cream; Bleeding Hydnellum)

CAP 2.5-15 cm broad, often with needles and other debris incorporated into it; top-shaped becoming broadly convex to plane or finally depressed; surface felty or velvety and white to pink in young specimens or on actively-growing margin; in age becoming nearly hairless and lumpy or jagged with projecting nodules, and often ridged and/or pitted; darkening to brown, dark brown, or vinaceous-brown from the center outward and almost entirely these colors in old age; beaded with or exuding bright ruby-red to dark red droplets when fresh and moist. Flesh in both cap and stalk tough and fibrous-corky, faintly zoned; pinkish-buff to cinnamon-brown, dark reddish-brown, or dingy brown; odor mild to fragrant or pungent; taste typically extremely acrid (peppery). SPINES rather short (1-6 mm long), dull pinkish becoming brown or purplish-brown, often with paler tips; sometimes decurrent. STALK 0.5-7.5 cm long, 1-2 (3) cm thick, central or off-center, equal or tapered below and sometimes rooting, or occasionally swollen at base; felty or velvety and colored more or less like cap or darker; solid, tough or woody. SPORE PRINT brown; spores 4.5-5.5 × 3.5-4.5 microns, round or nearly round, prominently warted.

HABITAT: Solitary to scattered, gregarious, or in fused clusters on ground under conifers; widely distributed, but particularly common in the Pacific Northwest in the late summer and fall. I've recorded only one questionable collection from our area, but have seen it farther north under pine and fir.

EDIBILITY: Indubitably inedible due to the burning-acrid taste and tough, corky texture.

COMMENTS: The bright red droplets that cling to the surface of the cap in moist weather make this a striking and easily-identified mushroom (see color plate!). As in other Hydnellums, the cap varies considerably in color and texture according to age and environmental conditions. When young, white, and beaded with droplets it looks like a Danish pastry topped with strawberry jam. Older, battered specimens, on the other hand, are scarcely recognizable and easily confused with other species. Intermediate stages are typically brown or dark reddish-brown with a white to pink, beaded margin. Other species exuding red droplets in wet weather include: *H. diabolum,* very similar if not the same, with a hairier cap at maturity and stronger odor; and *H. pineticola,* common under pines in northeastern North America, with a rather unpleasant but not acrid taste and a cap that sometimes has pink droplets. See also *H. scrobiculatum.*

Hydnellum scrobiculatum   (Rough Hydnellum)

CAP 3-10 cm broad, more or less top-shaped when young becoming plane to depressed in age, sometimes with smaller caps on top; surface usually roughened and irregular from numerous pits and projecting warts or blunt spikes, often also with radial ridges; pallid to pale salmon-buff or pinkish and plushlike or felty when young or on growing margin of older specimens; darkening in age from the center outward to buffy-brown, then dull cinnamon and finally darker brown, but lacking conspicuous concentric zones; sometimes beaded with dark red droplets when young and moist, and bruising reddish-black to black when rubbed, especially at margin. Flesh zoned, sometimes also with white dots; usually duplex, the upper layer spongy when fresh and colored like cap, the lower layer or core brown to dark reddish-brown; often exuding dark red juice when squeezed (if moist and fresh); odor and taste mild to farinaceous. SPINES short (1-5 mm long), usually decurrent, sometimes fused; pallid or colored like cap margin, becoming buffy-brown to cinnamon-brown and finally purple- or dark brown in old age. STALK 1-4 cm long, 0.3-1.5 cm thick, usually tapered downward but often with a swollen, buried, spongy base; central or off-center; tough; cinnamon-brown or colored more or less like cap. SPORE PRINT brownish; 4.5-5.5 (7) × 3.5-5 microns, elliptical to nearly round, prominently warted.

HABITAT: Scattered to gregarious or in fused clusters in woods, usually associated with conifers; widely distributed and sporadically common in some regions. It can be found nearly every fall in the Pacific Northwest and extends into northern California, but occurs very rarely if at all in our area.

EDIBILITY: Incomparably inedible.

COMMENTS: This species is rather hard to characterize, as evidenced by the lengthy description. It is best recognized by its irregularly roughened cap with pits, radial ridges, and spikelike projections, plus the overall cinnamon to brown color with a pinkish growing margin that darkens when bruised. When beaded with red droplets it can be confused with *H. peckii,* but is not as peppery-tasting as that species and usually has a buried, swollen "tuber" on the stalk. When older it can be mistaken for various other brownish to reddish Hydnellums. *H. zonatum (=H. scrobiculatum* var. *zonatum, H. concrescens*— COLOR PLATE 157) is a closely related, similarly colored species with a smooth or radially corrugated (but not lumpy) cap that is usually zoned concentrically (at least somewhat). It is quite common under hardwoods in eastern North America but also occurs on the west coast both in its normal form and a diminutive one whose cap is 4 cm broad or less. Both *H. zonatum* and *H. scrobiculatum* are quite distinct in their typical forms, but appear to intergrade. Other similar species: *H. subsuccosum* resembles *H. zonatum,* but is more irregular in shape and juicier when fresh, and exudes a pinkish to reddish juice when squeezed; it also has yellowish-gray to yellow-green mycelium at the base of the stem and its cap is often flecked with yellowish spots or particles. *H. cumulatum* has many caps built on top of one another, and little or no stalk. Both of these occur under conifers.

Phellodon tomentosus   (Zoned Phellodon)

CAP 1.5-4 (5.5) cm broad, often fused with others, plane to depressed or broadly funnel-shaped; surface dry, smooth to ridged or corrugated, minutely hairy (tomentose), white when very young, soon becoming concentrically zoned with yellow-brown, cinnamon-brown, darker brown, etc., the growing margin usually remaining white and felty to the touch, but bruising brownish. Flesh thin, leathery and fibrous, brownish (darker brown in stalk); odor usually fragrant (like fenugreek); taste mild or slightly bitter. **SPINES** short (1-3 (5) mm), crowded, delicate, whitish becoming pale cinnamon to brown with paler tips; slightly decurrent. **STALK** 1-5 cm long, 2-5 (8) mm thick, usually central, equal or tapering downward, colored more or less like cap, arising from spongy pad of brownish mycelium. **SPORE PRINT** white; spores 3-4.5 microns, round or nearly round, minutely spiny.

HABITAT: Scattered to densely gregarious or clustered (several caps often fused together but the stalks usually separate) under conifers; widely distributed and fairly common. I have found it in Mendocino County, California, in the fall and early winter, and it is sometimes abundant in the Pacific Northwest; just how far south it occurs is unclear.

EDIBILITY: Unknown, but too small and too tough to be of value.

COMMENTS: The small size, beautifully zoned yellow-brown to cinnamon-brown to dark brown cap, brown flesh, slender stem, and frequently fragrant odor are good field-marks. It bears an uncanny resemblance to polypores of the genus *Coltricia* (see *C. cinnamomea*), but the underside of the cap features minute spines or "teeth" instead of pores. *Hydnellum zonatum* (see comments under *H. scrobiculatum*) is also somewhat similar, but has darker spines and brown spores. Other species: *P. confluens* is quite similar, but more common under hardwoods. It is more irregular in shape, with an often roughened or pitted, whitish cap that darkens to creamy or dark tan from the center outward, and spines which are grayish at maturity.

Phellodon atratus   (Blue-Black Phellodon)

CAP 1-5 cm broad but often fused with others; plane to depressed or irregular; surface dry, usually at least faintly zoned concentrically; bluish-black to purple-black or black, the margin often slightly paler or purpler. Flesh in both cap and stalk purple-black to bluish-black; thin, tough, fibrous, pliant, sometimes with a thin outer or upper spongy layer; odor mild or faintly fragrant; taste mild. SPINES very short (1-2 mm), irregularly decurrent; gray to dark purplish-gray-brown, darker where bruised. STALK 2-5 cm long, 3-5 mm thick, usually central, sometimes compound or branched; tapering downward but usually thickened at ground level by a felty mycelial layer; rough, often flattened, colored more or less like cap. SPORE PRINT white; spores 4-5 × 3-5 microns, round or nearly round, minutely spiny. Cap tissue staining blue-black in potassium hydroxide (KOH).

HABITAT: Scattered to gregarious, often forming compound or fused clusters, on ground under conifers (particularly Sitka spruce); apparently endemic to the Pacific Northwest and California and quite common in the fall and winter in second-growth forests. It has been found in Big Basin State Park, but is rare south of San Francisco.

EDIBILITY: Unknown, but like myself, too tough and too small to be of value.

COMMENTS: The small size, tough texture, and bluish-black color distinguish this conifer-lover. *Hydnum fuscoindicum* is somewhat similar in color but larger and fleshy-brittle rather than pliant and tough. A closely related species, *P. melaleucus,* has purplish-black to purple-gray flesh, but its cap is dark brown to purplish-gray with a pallid margin, the spines are whitish to gray, and the stalk is very thin, dark brown to black, and sometimes deeply rooted. It also occurs under conifers but has a wider (albeit northern) distribution. In eastern North America a similar species, *P. niger,* is common. It is larger and thicker than *P. atratus,* with a white to brownish, gray, or black cap and black flesh (in both the cap and stem). *Hydnellum nigellum*, another eastern species, is small and slate-gray to black, but has brown spores.

AURISCALPIUM

THIS genus contains a single odd species with a worldwide distribution. It is not closely related to other stalked teeth fungi. In fact, microscopic characters suggest a possible relationship to the agaric genus *Lentinellus.*

Auriscalpium vulgare   (Ear Pick Fungus)

CAP 1-2 (4) cm broad, more or less kidney-shaped in outline, broadly convex to plane or slightly depressed; surface dry, covered with dense fibrils or hairs, brown to dark brown, sometimes blackish in age; margin often fringed and paler. Flesh thin, tough, pliant, white to pale brown. SPINES whitish to flesh-colored, sometimes darkening to brown; very fine and crowded, short (1-3 mm long). STALK 2-10 cm long, 0.5-3 mm thick, very slender, equal or slightly enlarged below, usually attached to side of cap (lateral), densely hairy, especially toward base, rusty-brown to dark brown or blackish. SPORE PRINT white; spores 4.5-6 × 3-3.5 microns, round or nearly round, smooth or minutely spiny, amyloid.

HABITAT: Solitary or in twos and threes on rotting, often buried cones of conifers, or sometimes on thick mats of debris made up partly of decaying cones; widely distributed, but rare in our area (or else frequently overlooked). In my experience it favors Douglas-fir cones, at least on the west coast.

EDIBILITY: Much too small and much too tough to be of value.

Auriscalpium vulgare. This dainty fungus grows on decaying cones (in this case, Douglas-fir). Note the slender stalk and layer of spines on underside of cap.

COMMENTS: The small size and growth on decaying cones plus the thin, hairy, lateral stem and fine spines that line the underside of the cap are the forthright fieldmarks of this unique, petite fungus. The stem, though lateral, may occasionally appear to be central when the cap is deeply indented. The stem is usually longer than the width of the cap and far thinner than that of any other fungus with spines. The dark color and small size make it very difficult to see unless you are specifically looking for it.

Coral and Club Fungi

spores

CLAVARIACEAE

THIS large and lovely group of fleshy fungi includes simple, unbranched, upright clubs and fleshy, intricately branched, coral-like forms. With the exception of *Clavariadelphus* the fruiting body is *not* differentiated into an upper sterile surface (cap) and fertile underside. Instead the spore-bearing basidia line the smooth to occasionally wrinkled surfaces of the upright clubs or branches. A sterile base, stalk, or "trunk" is normally present, however.

Coral fungi are a conspicuous and colorful component of our woodland fungi. They come in every imaginable color, and some of the larger branched forms (notably *Sparassis*) are edible. They are difficult from a taxonomic standpoint. Nearly all the coral fungi were originally lumped together in one unwieldy genus, *Clavaria,* but now more than 30 genera are recognized. These are delimited largely on microscopic and chemical characteristics (e.g., whether or not the spore-bearing surface stains green in ferrous sulfate), so to facilitate identification the family has been divided into five groups, keyed below. The largest and most common group, *Ramaria,* is microscopically similar to *Gomphus* of the chanterelle family (Cantharellaceae), and some taxonomists place them together in the family Gomphaceae in the belief that they arose from a common ancestor. Another genus, *Sparassis,* is usually placed in a family of its own, but is traditionally grouped with the coral fungi because of its branched fruiting body.

Key to the Clavariaceae

1. Fruiting body unbranched or *very* sparsely branched (but often tufted or clustered)   ... 2
1. Fruiting body profusely branched from a stalk or common base 4

630

Left: *Clavaria vermicularis* has unbranched but clustered fruiting bodies (it is also shown on p. 637). **Right:** *Ramaria stricta*, one of many species with a branched fruiting body.

2. Fruiting body entirely brownish-black to black *or* blackish beneath a white powdery coating *or* entirely green to olive or blue-green *or* interior with large chambers or compartments *or* parasitic on insects, spiders, or truffles; spores borne asexually or in asci (see **Ascomycetes,** p. 782)

2. Not as above (may be white, but if so then not powdery); spores borne on basidia 3

3. Fruiting body tough, the flesh pithy, stringy, or punky; apex often enlarged; usually 7 mm thick or more . . . . ***Clavariadelphus,*** p. 632

3. Fruiting body typically fragile or if tough then much smaller; mostly less than 7 mm thick; apex acute or blunt or occasionally enlarged . . . . ***Clavaria* & Allies,** p. 634

4. Fruiting body small and tough with *very* thin, almost hairlike branches, brown to grayish-brown to dark brown or purple-brown; growing on twigs, needles, etc.; rare (mostly tropical) ***Pterula***

4. Not as above; common 5

5. Fruiting body consisting of numerous *flattened*, wavy, ribbonlike, or leafy segments or lobes arising from a common base; rather tough; overall color white to creamy, yellowish, or tan; growing at or near the bases of trees and stumps 6

5. Not as above 7

6. Found on hardwood stumps or roots in tropics and along Gulf Coast; fertile surface usually developing pores, spines, or "teeth" in age . . . (see ***Polyporus, Albatrellus,* & Allies,** p. 554)

6. Not as above; common and widespread . . . . ***Sparassis,*** p. 657

7. Fruiting body *bright* yellow to orange; spore print white, or if not then branches usually viscid; spores smooth; typically growing on wood . . . . ***Clavaria* & Allies,** p. 634

7. Not as above 8

8. Branch tips crownlike (in the form of small fringed cups); spore print white; growing on wood . . . . ***Clavulina* & Allies,** p. 640

8. Not as above 9

9. Fruiting body bright yellow to orange when fresh and small (typically 2-7 (10) cm high) . . 10

9. Not as above 11

10. Stalk slender and not particularly fleshy; branches often hollow and/ or viscid; spore print white or yellowish; extensive mycelial mat typically absent . . . . . . ***Clavaria* & Allies,** p. 634

10. Not as above; stalk thick and fleshy or if not, then an extensive mat of mycelial threads usually present at base and in substrate; spore print buff to tan, yellowish, or ochre ***Ramaria,*** p. 645

11. Branches tough, usually flattened, grayish-brown to dark brown to purple-brown (but tips often pallid when growing); odor typically garliclike or fetid (see ***Thelephora palmata,*** p. 609)

11. Not as above 12

12. Spore print creamy to yellow, tan, yellow-orange, or ochraceous (rarely white); fruiting body medium-sized to fairly large, often brightly colored, or if dull colored then usually with a large fleshy base (stalk); fertile surfaces staining greenish to blue in ferrous sulfate *Ramaria,* p. 645
12. Spore print typically white; fruiting body rather small to medium-sized, white or dull-colored (grayish, brownish, tan, bluish-gray, or tinged purple); base typically *not* large and fleshy; fertile surfaces typically *not* staining green or blue in ferrous sulfate ... *Clavulina* & Allies, p. 640

CLAVARIADELPHUS   (Club Corals)

Medium-sized, terrestrial, woodland fungi. FRUITING BODY *erect, unbranched or occasionally forked, more or less club-shaped or with a flattened top, usually at least 5 mm thick;* surface smooth to wrinkled. Flesh rather tough and fibrous or pithy. SPORE PRINT white to pale yellow, buff, or ochre. Spores typically elliptical, smooth. Spore-bearing surface staining green in ferrous sulfate.

THESE are rather tough, club-shaped fungi with a smooth or somewhat wrinkled spore-bearing surface. They are larger and thicker than most fairy clubs (*Clavaria* & Allies) and not nearly so fragile. In *C. truncatus* and its close relatives the apex of the club is flattened and sterile—in other words, a rudimentary cap—but in the other species it is typically rounded, pointed, or only slightly flattened.

Club corals are harmless but rather tough, stringy, and/or bitter-tasting. They grow only in the woods and in our area fruit mostly during cold weather. Three widespread species are described here. If your "club coral" is small and irregularly shaped, check the earth tongues (on p. 865) as well as *Clavaria* & Allies (p. 634).

Key to Clavariadelphus

1. Fruiting body with a consistently flattened (truncate) or depressed apex or even a rudimentary cap; associated with conifers 2
1. Not as above; apex of fruiting body rounded to obtuse or pointed, or if sometimes flattened then associated with hardwoods 4

2. Upper portion of fruiting body red to reddish-orange   *C. lovejoyae* (see *C. truncatus,* p. 634)
2. Not as above; upper portion of fruiting body orange to yellow, ochre, etc. 3

3. Spore print pale ochre .... *C. truncatus,* p. 634
3. Spore print white .... *C. borealis* (see *C. truncatus,* p. 634)

4. Apex of fruiting body usually with a sharply defined point or "nipple"; associated with conifers .... *C. mucronatus* (see *C. ligula,* p. 633)
4. Not as above; apex rounded to bluntly pointed or somewhat flattened 5

5. Associated mainly with hardwoods; mature fruiting body 1-3 cm thick and 6-20 cm or more high .... *C. pistillaris* & others, below
5. Associated with conifers; mature fruiting body up to 1.5 cm thick and typically 2-10 cm high .... *C. ligula* & others, p. 633

Clavariadelphus pistillaris   (Common Club Coral)

FRUITING BODY simple, erect, unbranched or sometimes forked; club-shaped or tapering downward, the apex rounded or somewhat flattened but not normally depressed; 6-20 (30) cm high, 0.8-4 (6) cm broad; surface smooth or often longitudinally wrinkled or grooved in age; usually pallid at first, but soon dull pinkish-brown to reddish-brown, flesh-colored, or ochraceous-brown; staining brown to vinaceous-brown when handled or bruised; apex often yellowish at first but soon colored like the rest of the fruiting body; base usually pallid, with white hairs. Flesh tough, fibrous or pithy, whitish, bruising brown; taste mild or bitter. SPORE PRINT white or tinged yellow; spores 9-16 × 5-10 microns, elliptical, smooth.

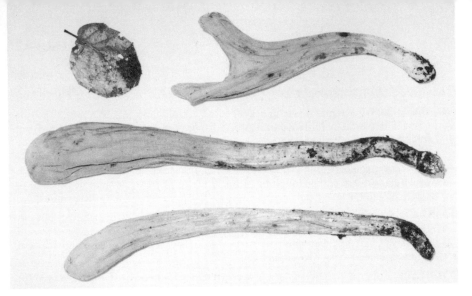

Clavariadelphus pistillaris is a common terrestrial club-shaped species. In our area it favors oak. Forked specimen at top is not unusual.

HABITAT: Solitary, scattered, or in groups on ground under hardwoods and in mixed woods; widely distributed. It is common in our area from the late fall through early spring, especially under live oak, tanoak, and madrone.

EDIBILITY: Harmless. The taste and texture are reminiscent of stale rope.

COMMENTS: The ochre-brown to flesh-colored, club-shaped fruiting body that stains brown when handled is characteristic of this cosmopolitan club coral. It is the commonest *Clavariadelphus* in our area and the only one found under hardwoods. Its apex is sometimes quite broad, but not as flagrantly flattened as that of *C. truncatus.* Other species: *C. subfastigiatus* is a brownish-orange species found under conifers in northern California and elsewhere; it does not discolor as much when handled and turns bright green in potassium hydroxide (KOH).

Clavariadelphus ligula   (Strap Coral)   **Color Plate 171**

FRUITING BODY simple, erect, unbranched or rarely forked, cylindrical to flattened-cylindrical or club-shaped, the apex usually rounded or bluntly pointed; 2-10 cm high and 0.3-1 (1.5) cm broad at apex. Surface smooth to slightly wrinkled, dull-colored (buff to dull yellowish, ochre-buff, pale reddish-brown, or vinaceous-buff); base whitish and hairy, often with white mycelial threads penetrating the surrounding humus. Flesh white, pithy but tough; taste mild or bitter. **SPORE PRINT** white to pale yellowish; spores 8-18 × 3-6 microns, elongated-elliptical, smooth.

HABITAT: Scattered to densely gregarious or tufted in humus under conifers; widely distributed. It is common throughout much of the West in the summer and fall, but absent or very rare in our area.

EDIBILITY: Worthless.

COMMENTS: This species is smaller and slimmer than *C. pistillaris* and typically occurs under conifers rather than hardwoods—often in large troops (see color plate). The apex of the club is not noticeably flattened or depressed as in *C. truncatus,* and is not brightly colored. Other species: *C. sachalinensis* is a macroscopically identical species with larger, buff to ochraceous spores; it also grows gregariously in humus under conifers. Another conifer-lover, *C. mucronatus,* differs in its whitish, sharply nippled apex. Still another, *C. subfastigiatus,* is somewhat thicker and stains green when touched with potassium hydroxide (KOH). None of these are worth eating.

Clavariadelphus truncatus (Truncate Club Coral)   Color Plate 166

FRUITING BODY simple, erect, unbranched or occasionally forked; club-shaped or more often with a broadly flattened or depressed apex (a rudimentary cap); 5-15 (18) cm high and 2.5-8 cm broad at apex. Surface smooth or often wrinkled or veined (especially near apex), more or less pinkish-brown to ochre or brownish-orange, the apex usually brighter (yellow to golden-yellow or yellow-orange), at least when young; base often pallid, with white hairs. Flesh rather tough or pithy, white to ochre; taste mild to sweetish or bittersweet. **SPORE PRINT** pale ochre; spores 9-13 × 5-8 microns, elliptical, smooth.

HABITAT: Scattered to gregarious in duff under conifers; widely distributed, but not nearly as common in our area as *C. pistillaris*. I have seen large fruitings of this species and its look-alike, *C. borealis* (see comments), in the fall and winter in northern California and in the summer in the Rocky Mountains.

EDIBILITY: Edible and delicious when sweet. It can be sauteed by itself and served for dessert!

COMMENTS: The broad, golden, flattened top distinguishes this species from its cousins *C. pistillaris* and *C. ligula*. The color of the fruiting body actually varies considerably from dull pinkish-brown to bright golden-orange, and the apex can be quite inflated so as to resemble a chanterelle *(Gomphus* or *Cantharellus)*. Other species: *C. borealis* is a white-spored version of *C. truncatus* that sometimes shows a lilac tinge to the fruiting body; it is common and widespread under conifers. *C. lovejoyae* of the Rocky Mountains is also similar, but red to reddish-orange in color.

CLAVARIA & Allies   (Fairy Clubs)

Small, mostly fragile fungi found on ground, leaves, or occasionally on wood. FRUITING BODY erect, *usually unbranched or sparingly branched and finger-shaped or clublike, slender,* often tufted or clustered. *Flesh usually fragile.* SPORE PRINT *white.* Spores typically smooth. Basidia typically 4-spored. Clamp connections typically absent *(Clavaria),* present *(Clavulinopsis).* Fertile surface often staining green in ferrous sulfate *(Clavulinopsis),* or not greening *(Clavaria* & others).

THESE are primitive fungi with an erect, relatively unspecialized fruiting body. The most common forms are unbranched and smaller, slimmer, and frailer than the club corals *(Clavariadelphus).* They are sometimes confused with earth tongues—which are tougher, often flattened and velvety, and capitate (with a cap distinct from the stalk)—and which belong to an entirely different group of fungi, the Ascomycetes. Fairy clubs often grow in clumps, but the individual clubs do not usually arise from a fleshy base as in the branched corals *(Clavulina, Ramaria,* etc.). The few branched species are not as fleshy as Ramarias and usually more vividly colored than *Clavulina, Ramariopsis,* and *Clavicorona.*

Clavaria and *Clavulinopsis* are the two most common genera of fairy clubs, but their defining features are esoteric, involving the presence or absence of carotenoid pigments, clamp connections, and the behavior of nuclei in the basidia. For the sake of convenience they are treated together here, along with several other small, miscellaneous genera *(Multiclavula, Macrotyphula,* and *Typhula).*

Fairy clubs are too small and fragile to have any food value, but they are an attractive addition to our woodland decor. They are saprophytic on humus, soil, or occasionally grass and decaying wood. In our area they fruit—as do most of the Clavariaceae—from late fall through early spring. Seven species are described here.

Key to Clavaria & Allies

1. Growing on algae-covered wood or soil; fruiting body minute (up to 1.5 cm high and 1-3 mm thick) .... *Multiclavula mucida* & others, p. 636
1. Not as above (if growing on algae, then larger) 2

2. Fresh fruiting body yellow to orange, red, salmon, or pink 3
2. Not as above (fruiting body white, yellow-brown, grayish, purple, etc.) 10

3. Fresh fruiting body yellow to orange 4
3. Fresh fruiting body rose-pink to red or orange-red 9

4. Fruiting body branched 5
4. Fruiting body unbranched or occasionally forked (but often clustered) 6

5. Usually growing on or near wood, the base often deeply rooted; branches rather tough and usually somewhat viscid .... (see *Calocera viscosa*, p. 674)
5. Not as above; growing on ground or wood; not viscid .... *Clavulinopsis corniculata,* p. 639

6. Fruiting body with a wide, often flattened head *and/or* fruiting body often irregular in shape; texture rather tough; spores borne inside asci .... (see **Helotiales,** p. 865)
6. Not as above; fruiting body typically clublike to spindle-shaped or fingerlike or rarely forked; usually rather fragile; spores borne on basidia 7

7. Fruiting bodies 5-15 cm tall, usually growing in bundles or large clusters, yellow .... *Clavulinopsis fusiformis* (see *C. laeticolor,* p. 638)
7. Fruiting bodies up to 6.5 (10) cm high, solitary to gregarious or tufted; yellow to orange .. 8

8. Fruiting body less than 15 mm high, usually somewhat viscid; typically growing on or near wood .... (see *Calocera cornea* under *C. viscosa,* p. 674)
8. Not as above; usually growing on ground .... *Clavulinopsis laeticolor* & others, p. 638

9. Fruiting body pink to rose-colored when fresh; found mainly in eastern North Ameica; rare . ... *Clavaria rosea* (see *Clavulinopsis laeticolor,* p. 638)
9. Fruiting body orange to orange-red or red; widespread but rare .... *Clavulinopsis aurantio-cinnabarina* & others (see *C. laeticolor,* p. 638)

10. Fruiting body very thin, arising from a small beadlike body (sclerotium) .... *Typhula* spp. (see *Macrotyphula juncea,* p. 636)
10. Not as above; fruiting body not very thin, or if so then not attached to a beadlike body .. 11

11. Fruiting body very thin (up to 2 mm), 3-10 cm tall; pallid to yellowish or brown .... *Macrotyphula juncea,* p. 636
11. Not as above 12

12. Fruiting body lavender to purple, deep purple, or grayish-purple, at least when fresh 13
12. Fruiting body differently colored 15

13. Fruiting bodies unbranched but often clustered; brittle or fragile; found mostly under northern and mountain conifers .... *Clavaria purpurea,* p. 637
13. Not as above 14

14. Fruiting body sparingly or much-branched (the branching often dichotomous); basidia 4-spored; found in eastern North America ..... *Clavaria zollingeri* (see *C. purpurea,* p. 637)
14. Fruiting body much-branched; basidia 2-spored; widespread (see *Clavulina* & **Allies,** p. 640)

15. Fruiting body branched ..... *Clavulinopsis umbrinella* & others (see *C. corniculata,* p. 639)
15. Fruiting body unbranched or occasionally branched very sparingly (but often growing in tufts or clusters) 16

16. Fruiting body small (up to 2 cm tall), white, the apex broadly enlarged; growing on needles, twigs, etc. .... *Clavicorona taxophila* (see *Multiclavula mucida,* p. 636)
16. Not as above 17

17. Fruiting bodies very fragile, crumbling easily, pure white to translucent white or yellow-stained, *not* prominently wrinkled; tip acute or sometimes blunt but not broadly enlarged; often growing in tufts or clusters; very common .... *Clavaria vermicularis,* p. 637
17. Not with above features 18

18. Fruiting body very long (tall) and narrow (7-30 cm high; 2-8 mm thick); yellowish to brown; not normally growing in clusters .... *Macrotyphula fistulosa* (see *M. juncea,* p. 636)
18. Not as above 19

19. Growing on wood; fruiting body more or less clublike, the surface often roughened, 1-4 cm high
     and 0.5-1 cm thick; spores borne inside asci . . . . (see *Podostroma alutaceum,* p. 879)
19. Not as above; usually growing on ground 20
20. Fruiting body 0.3-1.5 cm thick, texture tough; flesh white and pithy or stringy; color variable:
     buff to yellowish or some shade of brown, but not white or gray; typically growing in duff under
     conifers, often in troops . . . . (see *Clavariadelphus,* p. 632)
20. Fruiting body usually 2-7 mm thick and/ or differently colored, usually fragile 21
21. Fruiting body white and sparingly branched; branches thick, hollow, blunt, somewhat gelati-
     nous; found under hardwoods in eastern North America . . (see **Tremellales & Allies,** p. 669)
21. Not as above 22
22. Fruiting body fragile, usually tufted or clustered, grayish to yellowish-gray to pinkish-gray
     or dingy flesh-colored; basidia typically 4-spored . . . . *Clavaria fumosa* & others, p. 638
22. Not as above; fruiting body tough, or if fragile then differently colored (white or tinged gray or
     buff); basidia typically 2-spored . . . . (see *Clavulina* & Allies, p. 640)

Multiclavula mucida   (Scum-Lover)

FRUITING BODY simple, unbranched or sometimes forked, erect; small (only 0.5-1.5
cm high and 1-2 mm thick); cylindrical or tapered; surface smooth, white to creamy, buff,
or yellowish, but often aging salmon to brick-red. Flesh rather waxy and tough, pliant,
white. **SPORE PRINT** white; spores 4.5-7.5 × 2-3 microns, oblong to elliptical, smooth.
Cystidia absent.

HABITAT: In groups—but not clusters—on wet algae-covered wood or occasionally on
soil; northern in distribution. I have seen it in our area in the fall and winter, but it is inconspicuous because of its small size.

EDIBILITY: Utterly inconsequential.

COMMENTS: The small size and association with green algae typify the genus *Multiclavula,* and the whitish color of the fresh fruiting body and tendency to grow on wood are
characteristic of this species. *Clavaria mucida* is an older name for it. Other species:
M. vernalis (=Clavaria phycophila) is similar but pale orange, has cystidia, and grows on
algae-covered soil; *Clavicorona taxophila* is small and whitish but has a widened, flattened
apex and grows on twigs, needles, and other debris rather than with scum. All of these are
smaller than *Clavaria vermicularis* and shorter than *Macrotyphula juncea.*

Macrotyphula juncea   (Fairy Hair)

FRUITING BODY simple, unbranched or rarely forked, erect, very thin (up to 2 mm
thick), 3-10 cm high when mature; cylindrical or tapering upward; leather-colored to
yellowish-buff or pallid, smooth; tip acute or in age sometimes blunt; base somewhat
fibrillose and often creeping horizontally, often with large whitish mycelial threads (rhizomorphs) attached. Flesh very thin; taste sometimes acrid. **SPORE PRINT** white; spores
6-12 × 3.5-5.5 microns, elliptical or almond-shaped, smooth.

HABITAT: Scattered to gregarious in humus and leaf litter, on rotting twigs, etc.; widely
distributed. It occurs in our area on oak and tanoak leaves and redwood needles, but is easy
to overlook. It is fairly common in the fall and winter, especially along streams and in other
dank places.

EDIBILITY: Utterly irrelevant—a couple hundred would be needed for a mouthful!

COMMENTS: This species is so thin that it can be mistaken for the bare stem of an herbaceous plant or a small agaric that has lost its cap. Its slimness alone separates it from our
other common coral and club fungi. Its generic disposition is problematic—some authors
place it in *Typhula* (other *Typhula* species are similar in size and shape but arise from a

Left: *Macrotyphula juncea,* a very thin species. **Right:** Fragile white clusters of *Clavaria vermicularis* look like bunches of bean sprouts. See p. 631 for a photo of younger specimens.

small beadlike body or sclerotium); it has also been placed in *Clavaria* and *Clavariadelphus* and may eventually merit a genus of its own. Other species: *M. fistulosa (=Clavaria-delphus fistulosus)* has a very long (7-30 cm), slender (2-8 mm), hollow fruiting body that is yellowish to brownish; it grows on dead sticks and debris, especially of alder.

Clavaria vermicularis   (Fairy Fingers)

FRUITING BODY simple, unbranched (but usually clustered) or rarely forked at the tip, erect or often curved, slender, soon withering; 3-12 (15) cm tall, 3-5 mm thick; surface pure white to translucent white or often stained yellow and yellowing in age from the tip downward; cylindrical or flattened somewhat, smooth or sometimes grooved, usually tapered toward the tip, which is acute or sometimes blunt and often discolored. Flesh thin, white, very brittle or fragile. **SPORE PRINT** white; spores 5-7 × 3-4 microns, elliptical to nearly round, smooth.

HABITAT: In tufts, clusters, or groups, often with solitary fruiting bodies interspersed, on ground in woods or grassy places; widely distributed. Common in our area throughout the mushroom season, but most prevalent in December and January, especially in dank wooded areas.

EDIBILITY: Edible but insubstantial; the fragile watery flesh has little or no flavor and dissolves when chewed.

COMMENTS: The distinctive clumps of slender white "fingers" make this species most attractive. It is by far our most common *Clavaria,* growing wherever moisture is sufficient. It is not as deeply wrinkled as *C. fumosa* or *Clavulina rugosa* (see comments under the *Clavulina cristata* group), and it crumbles seemingly without provocation.

Clavaria purpurea   (Purple Fairy Club)   **Color Plate 167**

FRUITING BODY simple, unbranched but often clustered, erect, 2.5-12 cm tall, 2-6 mm thick; grayish-purple to deep purple or purple when fresh, fading as it ages to lavender-gray, lavender-buff, purple-brown, smoky-brown, etc.; cylindrical or tapered; tip acute or blunt; base usually paler and/or with white hairs. Flesh white or purplish, brittle. **SPORE PRINT** white; spores 5.5-9 × 3-5 microns, elliptical to oblong, smooth. Cystidia intermingled with basidia.

HABITAT: Scattered to densely gregarious (often in tufts or clusters) on ground in wet areas, usually under or near conifers; common during the summer under spruce and fir in the Rocky Mountains, also found in the Pacific Northwest and California (but not in our area). I have seen large fruitings in New Mexico.

EDIBILITY: Edible, but thin-fleshed and fragile.

COMMENTS: The purple color immediately distinguishes this beautiful species from other fairy clubs. *Clavaria (=Clavulina) zollingeri* is a sparingly to profusely branched purple species occasionally found in eastern North America. For other branched purple coral fungi, see *Clavulina cinerea, C. amethystina,* and *Ramariopsis pulchella* (under *C. cinerea*) and *Ramaria fumigata.*

Clavaria fumosa   (Grayish Fairy Club)

FRUITING BODY simple, unbranched (but usually clustered), erect, 3-10 (14) cm tall, 2-7 mm thick; surface grayish to yellowish-gray (color variable), usually grooved or wrinkled longitudinally, often somewhat flattened or twisted and/or hollow in age; tip usually blunt, sometimes brownish; base often paler or whitish. Flesh brittle, whitish. **SPORE PRINT** white; spores 5-8 × 3-4 microns, elliptical, smooth.

HABITAT: In tufts or clusters on ground in woods or grassy areas; widely distributed but not common. I have found it several times under oak and madrone in the winter and spring.

EDIBILITY: Harmless, fleshless, flavorless.

COMMENTS: The gray to yellowish-gray (or occasionally brownish-gray) color, frequently wrinkled surface, and clustered but unbranched fruiting bodies typify this forgettable fairy club. Other species: *C. rubicundula* is just as forgettable but more fragile, and usually slightly pinker (dingy flesh-colored to pinkish-gray to vinaceous-buff).

Clavulinopsis laeticolor   (Golden Fairy Club)

FRUITING BODY simple (unbranched or occasionally forked once) but often tufted; erect, small, cylindrical or often somewhat flattened, grooved, and/or twisted; 1.5-6.5 (10) cm tall but usually about 3-4 cm, 1-5 (10) mm thick; surface bright orange to yellow (sometimes yellower below) to yellow-ochre or in some forms orange-red, often duller (yellowish-buff) as it dries or fades; extreme base whitish; tip usually acute, often brownish in age or when dry. Flesh thin, somewhat pliant, pallid or yellowish; odor and taste mild. **SPORE PRINT** white; spores 4.5-7 (9) × 3.5-5.5 (6.5) microns, broadly elliptical to nearly round to triangular or pear-shaped, smooth, prominently apiculate.

HABITAT: Solitary, scattered, tufted, or in groups on mossy banks, wet soil, and woodland humus; widespread and common, fruiting in our area in the fall, winter, and spring.

EDIBILITY: Inconsequential.

COMMENTS: Also known as *Clavaria pulchra,* this dainty fairy club can be distinguished from all but a few close relatives (see below) by its bright yellow to orange or sometimes reddish-orange color. It is not viscid like the coral-like jelly fungi *(Calocera),* and does not usually grow on wood. Certain earth tongues (e.g., *Microglossum rufum* and *Neolecta irregularis)* are similarly colored, but have more enlarged fertile "heads" and bear their spores in asci rather than on basidia. There are a number of closely related, brightly colored fairy clubs, including: *Clavulinopsis helvola,* with anguar-warty spores; *C. fusiformis,* taller (5-15 cm) and bright yellow with round spores, usually found in bundles in grass or humus (common in eastern North American, also found in California); *C. gracillima (=C. luteoalba),* with a well-defined sterile base or "stalk"; *C. appalachiensis,* creamy to creamy-yellow with a deep ochre "stalk"; *C. miniata,* pale pinkish-orange or

Two bright yellow to orange club fungi: *Clavulina laeticolor* (**left**) grows singly or in small tufts; *Clavulina fusiformis* (**right**) is usually taller, slimmer, and clustered.

apricot-colored; *C. aurantio-cinnabarina,* a rare but widely distributed striking blood-red to orange or pinkish-orange species with only slightly apiculate spores; and *C. sub-australis,* with a delicate pink fertile portion and dark yellow to orange "stalk." Finally, there is *Clavaria rosea,* a beautiful rose-pink species that occurs occasionally in eastern North America but hasn't yet been found in California. None of these are worth eating.

Clavulinopsis corniculata

FRUITING BODY branched (sometimes sparingly) from a common base or stalk, usually small and delicate-looking; 2-9 cm high and broad (but usually 2-5 cm). **BRANCHES** smooth, mostly hollow, bright yellow to yellow-orange or egg-yellow to deep ochraceous, sometimes fading when dry or in age to creamy-buff; tips acute. **STALK** 2-4 mm thick, colored like branches or duller, often with a coating of downy mycelium at base. Flesh rather tough, thin, whitish; odor mild or farinaceous; taste mild or bitter. **SPORE PRINT** white; spores 4.5-7.5 microns, round or nearly round, smooth, apiculate.

HABITAT: Solitary, scattered, or in groups on ground or dead wood in forests and at their edges, in grassy areas, etc.; widely distributed. I have found it locally under cypress and oak in December and January, accompanied by *C. laeticolor* and numerous waxy caps (*Hygrocybe* and *Camarophyllus* species).

EDIBILITY: Probably edible, but too small to be of value.

Clavulinopsis corniculata is a small brightly colored branched species. Both specimens shown here are elaborately branched, but some specimens have a well-developed "stalk" with only a few branches.

COMMENTS: The bright yellow to ochre color and modest size make this branched coral fungus fairly easy to recognize. Other yellow to orange coral and club fungi are either unbranched (see *C. laeticolor*) or much larger and fleshier (see *Ramaria*). The jelly fungus genus *Calocera* is somewhat similar, but usually has a viscid fruiting body. Other small branched species include: *C. umbrinella (=Clavaria cineroides)*, a dingy buff to grayish or brownish species with round spores and several primary branches that arise from a common base or "trunk," occasional in our area in the fall and winter (usually in forests), but more widely distributed; *C. holmskjoldii,* which resembles *C. umbrinella* but has purplish branch tips and an aniselike odor when fresh; *C. dichotoma* and *C. subtilis,* which are both white to faintly yellowish; and *Clavulina (=Clavaria) ornatipes,* with pallid to pinkish-gray to brownish branches and a hairy brown stalk or "trunk."

CLAVULINA & Allies

Small to medium-sized, coral-like fungi found on ground *(Clavulina, Ramariopsis, Tremellodendropsis)* or wood *(Clavicorona).* FRUITING BODY *usually branched,* sometimes with a fleshy base. BRANCHES mostly erect, *usually pale or dull-colored (white to gray, tan, brownish, dingy yellowish, or sometimes tinged pinkish or purple).* SPORE PRINT *white* (or in *Clavulina* sometimes yellowish after prolonged storage). Spores smooth or spiny. Basidia typically 2-spored *(Clavulina)* or 4-spored (others). Spore-bearing surface typically *not* greening in ferrous sulfate.

THIS is a motley, artificial grouping of mostly white or dingy colored, branched coral fungi. Their overall aspect is intermediate between that of the fairy clubs *(Clavaria & Allies)* and the Ramarias. The pale or dingy color, white spores, and moderate size distinguish them from most Ramarias, while the branched fruiting body separates them from most fairy clubs.

The most common of the genera treated here is *Clavulina,* which has smooth spores and 2-spored basidia. In *Ramariopsis,* also common, the spores are minutely ornamented and the basidia are 4-spored. In *Clavicorona* the fruiting body grows on wood and typically has crownlike branch tips and amyloid spores, while *Tremellodendropsis* has tough, usually flattened branches and basidia which appear partially partitioned.

The species treated here are probably edible, but are not as fleshy or desirable as the Ramarias. In addition, some are quite tough while others are exceedingly fragile. They fruit primarily in the woods. Five species are described and several others are keyed out.

Key to Clavulina & Allies

1. Growing on wood; branch tips usually crownlike .... *Clavicorona pyxidata* & others, p. 642
1. Not as above; usually growing on ground 2
2. Fruiting body tough and pliant, *not* white, typically with a mat of copious white mycelial threads at base or in surrounding humus .... (see *Ramaria,* p. 645)
2. Not as above; mycelial mat absent *and/or* fruiting body brittle 3
3. Fruiting body lavender to purple .... *Clavulina amethystina* & others (see *C. cinerea,* p. 641)
3. Not as above (but branches may have a purple tinge) 4
4. Texture very tough or cartilaginous to slightly gelatinous; branches usually flattened; overall color white or pallid (or sometimes greenish from algae); found mainly under hardwoods in eastern North America (see photo at bottom of p. 644) ....
.... *Tremellodendron pallidum* & others (see *Tremellodendropsis tuberosa,* p. 643)
4. Not as above 5
5. Texture tough; "stalk" usually comprising at least one half the height of fruiting body; branches often flattened, pallid to brownish-gray or tinged dull purplish; base often with whitish down; basidia appearing partially septate under microscope . *Tremellodendropsis tuberosa,* p. 643
5. Texture brittle to fragile, or if tough then not as above 6

6. Fruiting body branched, but stalk usually at least half the total height; color of brownish; taste
    often bitter and/ or stalk with brown hairs; basidia 4-spored . (see *Clavaria* & Allies, p. 634)
6. Not as above 7
7. Fruiting body white or pallid (or tinged buff, gray, or pinkish) 8
7. Fruiting body darker (gray to brownish-gray, bluish-gray, purplish-gray, etc.) ....
    .... *Clavulina cinerea*, below
8. Fruiting body unbranched or sparingly so .. *Clavulina rugosa* (see *C. cristata* group, below)
8. Fruiting body branched 9
9. Fruiting body profusely branched, fragile; branch tips typically neither toothed nor enlarged
    .... *Ramariopsis kunzei*, p. 643
9. Fruiting body branched (but at times rather sparingly so), fragile to rather tough; branch tips
    often toothed and/ or enlarged .... *Clavulina cristata* group, below

Clavulina cinerea   (Ashy Coral Mushroom)   Color Plate 169

FRUITING BODY erect or somewhat spreading, profusely branched, 2-11 cm tall and
broad. **BRANCHES** often irregular in shape, often resulting in a somewhat tangled appearance; pallid soon becoming grayish to ashy-gray, purple-gray, bluish-gray, dark gray,
or even brownish-gray; smooth or wrinkled to somewhat flattened; tips acute or blunt,
often forked. **STALK** present as a short, fleshy sterile base or "trunk"; colored like the
branches, or often whitish at the very base. Flesh white, brittle; taste usually mild. **SPORE
PRINT** white; spores 6.5-11 × 5.5-10 microns, broadly elliptical to nearly round, smooth.
Basidia 2-spored.

HABITAT: Solitary, scattered, or in groups on ground in mixed woods and under
conifers; widely distributed but mainly northern. In California I have seen it in the fall and
winter, but have yet to find it in our area.

EDIBILITY: Edible and highly rated by some authorities, but rather fragile and insipid
in my experience.

COMMENTS: The ash-gray to dark gray or purple-tinged fruiting body separates this
conifer-lover from other branched coral fungi. Microscopically it is distinct by virtue
of its 2-spored basidia with curved spore stalks (sterigmata). The *C. cristata* group is closely
related but paler in color. *C. (=Clavaria) amethystina* is a beautiful purple branched
species; it is found occasionally in eastern North America and I have found it (or something
very similar) near Aptos, California, under redwood and tanoak. *Ramariopsis pulchella*
is a very small, sparingly branched, lavender species. *Clavaria zollingeri* (see comments
under *C. purpurea*) is also small and purple, but has 4-spored basidia.

Clavulina cristata group   (Crested Coral; Wrinkled Coral)

FRUITING BODY erect but extremely variable in shape and form, 2-7 (12) cm high, up to
5 cm broad; "typical" form branched, but other forms sparsely or irregularly branched and
still others unbranched. **BRANCHES** smooth in typical form, uneven or knobby or longitudinally wrinkled or flattened in other forms; tip(s) acute and often finely toothed in
"typical" form, but blunt and often enlarged in others; color usually white, but sometimes tinged gray, buff, yellowish, or pinkish; tip(s) often darkening in age or dry weather.
STALK present as a sterile base; slender, white or darkened by a parasite. Flesh white,
brittle to rather tough. **SPORE PRINT** white; spores 7-11 (14) × 6.5-10 (12) microns,
nearly round, smooth. Basidia 2-spored.

HABITAT: Solitary to scattered or densely gregarious on ground in woods and grassy
areas; widely distributed and common from sea level up to timberline. In our area this
group fruits throughout the mushroom season and is especially prevalent under pine.

EDIBILITY: Edible. Some rate it highly, others do not.

Clavulina cristata, a cosmopolitan white to grayish coral fungus. Note how the branch tips are "crested." In our area it is especially common under pine.

COMMENTS: This species "complex" is extremely polymorphic (variable in shape and form), and as a result is likely to confound the beginner. The "typical" form is distinguished from *Ramariopsis kunzei* (which is also white and branched) by its frequently crested (finely toothed) branch tips, smooth spores, and tendency of the stalk to darken in age when attacked by a fungal parasite. The unbranched to sparingly branched, wrinkled or knobby form, which is often called *C. rugosa,* is also common in our area (see photo at top of next page). It is likely to be confused with species of *Clavaria,* which are differently colored and/or smooth and more regular in appearance. It also seems to intergrade with "typical" *C. cristata.* Another closely related species, *C. cinerea,* is usually darker (see description on p. 641). Also see *Clavulinopsis dichotoma* and *C. subtilis* (under *C. corniculata*).

Clavicorona pyxidata (Crown Coral Mushroom)

FRUITING BODY profusely branched from a common base or "stalk"; 5-12.5 cm high and 2-8 cm wide. **BRANCHES** usually arising in tiers from the enlarged tips of lower branches; whitish to pale yellow when young, becoming dull ochre to tan or tinged pinkish, but lower portion often darkening to brownish or grayish-brown in age; tips usually enlarged and pyxidate (i.e., crownlike or terminating in fringed cups). **STALK** present as a slender, short sterile base or "trunk"; colored like lower branches. Flesh white, tough and rather pliant; taste often somewhat acrid (peppery). **SPORE PRINT** white; spores 3.5-6 × 2-3 microns, elliptical, smooth, amyloid.

HABITAT: Solitary or in small groups on dead hardwoods, especially aspen, willow, and cottonwood; very widely distributed and common in some regions. I have seen it often in the southern Rocky Mountains. It also occurs in northern California, but I have yet to find it in our area.

EDIBILITY: Edible, but rather stringy and tough.

COMMENTS: The crownlike branch tips, pale color, and growth on wood distinguish this widespread coral mushroom. Several *Ramaria* and *Lentaria* species grow on wood (see *R. stricta),* but do not have crownlike branch tips. Other species: *C. avellanea* is grayish-brown with paler crownlike tips; it grows on rotting conifers.

642

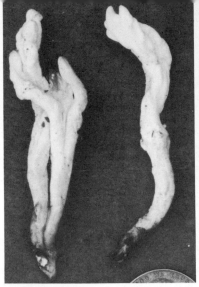

Left: *Clavulina rugosa* (see comments under the *C. cristata* group) is a wrinkled clublike or sparingly branched whitish species. **Right:** *Ramariopsis kunzei* is an elaborately branched whitish coral fungus that is especially common under conifers.

Ramariopsis kunzei (White Coral Mushroom)

FRUITING BODY erect or somewhat spreading, profusely branched, 2.5-10 cm tall and 3-8 cm wide. **BRANCHES** white to creamy-white, often tinged pinkish in age; smooth, not usually compact; tips blunt or acute. **STALK** absent or present only as a short, fragile, sometimes hairy base. Flesh white, fragile; taste mild. **SPORE PRINT** white; spores 3-5.5 × 2.5-4.5 microns, broadly elliptical to round, minutely spiny. Basidia 4-spored.

HABITAT: Scattered to densely gregarious on ground in mixed woods and under conifers, frequently hidden in the duff; widely distributed. It is common in our redwood forests from late fall through early spring, but is not restricted to that habitat.

EDIBILITY: Harmless, fleshless, flavorless.

COMMENTS: Formerly known as *Clavaria kunzei,* this ubiquitous branched coral fungus is best recognized in the field by its white color and marked fragility. It is more profusely branched than typical *Clavulina cristata* and lacks the toothed branch tips often found in that species, and has ornamented spores. Other species: *R. californica* is a rare, smooth-spored, branched, white to yellowish species.

Tremellodendropsis tuberosa

FRUITING BODY rather sparsely branched from a tough base (stalk) or sometimes scarcely branched; 2-7 (10) cm high, 0.5-4 cm broad. **BRANCHES** usually somewhat flattened (especially the lower ones), erect, tough, whitish to buff, brownish, or grayish, sometimes with a purple or pinkish tinge (caused by a parasite?); tips often paler and brighter (whiter) when actively growing. **STALK** well-developed, usually one-third to one-half the height of fruiting body; colored like branches or paler, often with a coating of white mycelial down. Flesh white, tough, not staining. **SPORE PRINT** white; spores 13-20 × 4.5-6.5 microns, elongated-elliptical or spindle-shaped, smooth. Many or all of the basidia appearing partially septate (partitioned) longitudinally at their apices.

HABITAT: Solitary or in groups on ground in woods or clearings; widely distributed. It is fairly common in coastal California in the late fall and winter, especially with bracken fern, redwood, or cypress, but is easily overlooked because of its humble appearance.

Tremellodendropsis tuberosa is best recognized by its tough texture, pale color, and upright branches.

EDIBILITY: I can find no information on it.

COMMENTS: This rather nondescript coral fungus is included here because it represents a possible "bridge" between the coral fungi, which have "normal" club-shaped basidia, and the jelly fungi, some of which are coral-like and have longitudinally septate (partitioned) basidia. In the field it can be told by its rather long stalk or trunk, pale or dingy color, tough texture, and growth on the ground. *Lentaria byssiseda* (see comments under *Ramaria stricta*) is a rather similar whitish to pinkish-tan species that often grows on wood, has less flattened branches, "normal" basidia, and white mycelial threads at the base of the stalk. It is widely distributed, but in California is more common in the Sierra Nevada than on the coast. Other species: *Tremellodendron pallidum (=T. schweinitzii)* of eastern North America is a beautiful whitish branched species that is classified as a jelly fungus because it has longitudinally partitioned basidia. It is common under hardwoods along with a smaller relative, *T. candidum,* and can be distinguished from other whitish coral fungi by its tough or tenacious texture and flattened branches (see photograph). Another eastern hardwood-lover, *Tremella reticulata,* has very blunt, hollow, somewhat gelatinous, white branches.

Tremellodendron pallidum of eastern North America is a coral-like jelly fungus. It may be related to *Tremellodenropsis tuberosa* (see comments under that species for more details).

RAMARIA (Coral Fungi)

Medium-sized to large, coral-like fungi found on wood or ground. FRUITING BODY profusely branched from a common, often fleshy base or stalk. BRANCHES mostly erect, smooth, never ribbonlike; often brightly colored. SPORE PRINT typically yellowish to tan or ochraceous. Spores usually ornamented (warted or spiny), produced on the surfaces of the branches. Spore-bearing surface staining green or bluish in ferrous sulfate.

THESE are brittle or pliant fungi with elaborately branched fruiting bodies. They represent an evolutionary advancement over the simpler coral fungi or fairy clubs (*Clavaria* & Allies) insofar as branching greatly increases the available surface area on which to produce spores. (The same can be said for the teeth fungi (Hydnaceae), except that their "branches" hang downward and are called spines.)

Ramaria can be distinguished from other branched coral fungi (see *Clavulina* & Allies) by its tan to ochraceous or orange-yellow spores (which are usually ornamented with minute warts, spines, or ridges) and frequently colorful fruiting body. Virtually every hue is represented, with yellow, orange, red, pink, and tan predominating. It is easily the most attractive and prominent group of coral fungi, and never fails to attract the attention of collectors. However, it is also the largest and most complex group, with over 35 species in California, many more in the Pacific Northwest, and at least 100 in North America.

To facilitate identification, *Ramaria* can be crudely divided into two groups: medium-sized to large, terrestrial species with a fleshy, gelatinous, or brittle fruiting body (e.g., *R. botrytis*); and wood- or duff-inhabiting species with a fairly small, slender, pliant-tough fruiting body (such as *R. stricta,* and the small, smooth-spored "satellite" genus *Lentaria,* which intergrades with *Ramaria*). However, pinpointing the exact identity of a species within either group is a difficult task, even for a specialist. In part this is because of the nature of the fruiting body—aside from color and texture, there are few criteria by which to separate species in the field. As a result, certain names have been applied indiscriminately to a slew of similar—but autonomous—species. Any attempt to correct this trend short of an exhaustive study of *Ramaria* would only contribute to the confusion. Therefore, the descriptions offered here are rather broad in scope. This may not satisfy *Ramaria*-researchers, but it will enable collectors to refer most of the Ramarias they find to a species group or "complex" without resorting to detailed microscopic study and special chemical tests. Besides, it is not necessary to know the exact identities of these coral fungi to appreciate their beauty. The manner in which they arise from the murky depths of the forest floor is indeed reminiscent of corals.

Ramarias are a popular group for the table, probably because they are so distinctive and none are known to be *dangerously* poisonous. Only the large, fleshy species are worth collecting, but some are bitter or prone to attack by fungal parasites such as *Hypomyces transformans*, and all are apt to be riddled with maggots and are difficult to clean. A few (notably *R. gelatinosa* and the *R. formosa* group) are mildly poisonous, and even the so-called "edible" species have a laxative effect on some individuals. Therefore it is best to sample each type cautiously (if at all) and not to overindulge.

Ramarias are woodland fungi. The slender, pliant forms are saprophytic on humus and wood and are especially common under conifers. The large fleshy types are usually terrestrial and may or may not be mycorrhizal. In the Pacific Northwest they are partial to hemlock and other conifers, in our area they favor tanoak, and in the Rocky Mountains they gravitate toward spruce, but are by no means restricted to these habitats. Only eleven species are described here but a number of others are keyed out. Most of them occur in the Pacific Northwest because that region is particularly rich in Ramarias. (The key is based on my own field experience plus information in *The Ramarias of Western Washington* by Currie Marr and Daniel Stuntz, and *Trial Key to the Species of Ramaria in the Pacific Northwest,* by Kit Scates.)

Key to Ramaria

1. Growing on wood 2
1. Growing on ground (or occasionally on very rotten wood) 5

2. Branches pale yellow to tawny-buff, orangish, or pinkish-tan, the tips yellow at first; spores warty .... *R. stricta*, p. 648
2. Not as above; fruiting body pallid to buffy-tan, pinkish-tan, reddish-brown, brown, or darker, (occasionally yellowish but then spores smooth); branch tips not usually yellow when fresh 3

3. Taste acrid (peppery); branches usually with a pinkish tinge, becoming darker (reddish-brown) in age .... *R. acris* (see *R. stricta*, p. 648)
3. Not as above; branches pallid to pinkish-tan, buffy-tan, brown, greenish-tinged, etc. 4

4. Fruiting body creamy to yellowish or pinkish-tan (branch tips sometimes greenish); spore print whitish and spores smooth ... *Lentaria byssiseda* & *L. pinicola* (see *Ramaria stricta*, p. 648)
4. Fruiting body buffy-tan to pinkish-tan to brown or darker (sometimes greenish-tinged); spore print yellowish to ochraceous and spores ornamented .. *R. apiculata* (see *R. stricta*, p. 648)

5. Fruiting body pliant and rather tough, small or medium-sized (rarely taller than 10 cm); stalk or "trunk" slender to practically absent, with a mat of conspicuous white mycelial threads attached to the base and/or permeating the substrate (see photo on p. 650) 6
5. Fruiting body medium-sized to large; mycelial mat absent, or if present then not as above (sometimes tough and pliant but usually fleshier) 8

6. Spore print whitish; fruiting body creamy to pinkish-tan or yellowish, sometimes with greenish tips; often found near wood or in lignin-rich humus .... *Lentaria byssiseda* & *L. pinicola* (see *Ramaria stricta*, p. 648)
6. Spore print yellowish to ochraceous; fruiting body pallid to yellowish, ochraceous, cinnamon-tan, etc., sometimes with greenish stains; found in duff 7

7. Fruiting body (or at least the lower branches) bruising or aging blue-green to olive-green, especially in cold weather .... *R. abietina*, p. 650
7. Not as above .... *R. myceliosa* & others, p. 649

8. Flesh in base of stalk rusty to rusty-brown when cut open lengthwise (i.e., vertically) 9
8. Not as above 11

9. Branches orange to pinkish or pinkish-red .... *R. amyloidea*
9. Branches creamy to yellow or yellow-orange 10

10. Branches whitish to pale yellow with whitish tips .... *R. velocimutans*
10. Branches yellowish to yellow-orange with yellow tips .... *R. celerivirescens*

11. Branches white to creamy *when fresh*; tips similarly colored *or* pinkish to purplish, brick-red, orangish, brownish, or yellowish-buff (but *not* yellow); fruiting body often (but not always) compact when young; flesh in base *not* gelatinous or semi-gelatinous 40
11. Not as above; branches darker or more brightly colored when young and fresh or tips yellow or flesh gelatinous to semi-gelatinous 12

12. Odor fragrant, like cocoa butter; branches pinkish-lavender to vinaceous or duller; found in eastern North America .... *R. cacao*
12. Not as above 13

13. Branches dingy-colored (yellow-brown to olive-brown or olive-gray, etc.) with a distinct violet tinge to lowermost ones (when fresh) and/or upper stalk; base (stalk) usually well-developed and fleshy (see photo on p. 651) .... *R. fennica*, p. 650
13. Not as above (but branches may be entirely violet or have violet to vinaceous stains) 14

14. Branches entirely violet to lavender when fresh (but may fade or discolor in age) 15
14. Branches differently colored (but may have some vinaceous or purplish stains) 16

15. Spore print white; base not particularly fleshy .... (see *Clavulina* & Allies, p. 640)
15. Spore print yellowish to ochre; base usually fleshy .... *R. fumigata* & others, p. 651

16. Branches red to coral-red, rose-pink, scarlet, or magenta when fresh; branch tips *not* yellow, of if yellow then branches red 17
16. Branches differently colored (pink to orange, salmon, yellow, brown, etc.); tips yellow or not yellow 18

17. Branches usually red to coral-pink when fresh; flesh in base of stalk *not* amyloid . . . .
. . . .***R. araiospora*** & others, p. 655
17. Not as above . . . . ***R. stuntzii*** & others (see *R. araiospora,* p. 655)
18. Branch tips blunt and conspicuously swollen (like the fruiting body of a *Clavariadelphus*)   19
18. Not as above 20
19. Branches white to yellowish or becoming tan; widespread but not common . . ***R. obtusissima***
19. Branches tan to dull grayish-orange; taste often bitter; found in the Pacific Northwest . . . . . .
. . . . ***R. claviramulata***
20. Base or lower branches covered with a cottony white tomentum (fuzz), or if not then branches
or stalk often hollow 21
20. Not as above 23
21. Fruiting body consisting of a dense bundle of upright, slender, sparsely or densely branched
"stems" arising from a common base; branches ("stems") salmon to yellow-orange to peachy
with yellow tips when fresh (sometimes with purplish stains), pliant to somewhat waxy and
often hollow; widely distributed under both hardwoods and conifers . . . . . . ***R. conjunctipes***
21. Not as above 22
22. Branches yellow or pale yellow; odor often fragrant or pungent-sweet . . . .
. . . . ***R. cystidiophora*** & ***R. synaptopoda*** (see *R. sanguinea* group, p. 653)
22. Not as above; differently colored or soon becoming so 23
23. Flesh in base (stalk) gelatinous or showing gelatinous streaks or pockets when sliced open *or*
if not gelatinous then cartilaginous (tough and brittle) 46
23. Flesh in base firm or fibrous or with watery areas, but not gelatinous or cartilaginous   . . . 24
24. Branches brown to dark brown with white or pallid tips when fresh; found in eastern North
America . . . . ***R. grandis***
24. Not as above 25
25. Base or stalk (and sometimes the branches) staining brown to reddish-brown to wine-red,
violet, or blackish when bruised (often already with vinaceous to reddish-brown stains) . 26
25. Not as above; fruiting body not staining appreciably when bruised 30
26. Branches becoming yellow-brown to reddish-brown in age; branches and base staining reddish-
brown to brown when bruised; found under conifers in the Pacific Northwest . . . .
. . . . ***R. testaceoflava*** (see *R. formosa* group, p. 654)
26. Not as above; branches usually lighter or brighter in color, at least when fresh; common   . 27
27. Branches and tips yellow to creamy . . . . ***R. sanguinea*** *group,* p. 653
27. Branches pinkish to orangish or yellow-orange, the tips sometimes yellow 28
28. Fruiting body with a yellow zone or band near ground level (on lower branches or just below),
at least when young and fresh . . . . ***R. sandaracina*** (see *R. gelatinosa,* p. 655)
28. Not as above 29
29. Fruiting body staining brown to blackish, at least at base . . . . ***R. formosa*** *group,* p. 654
29. Fruiting body staining wine-red to burgundy, at least at base . . . .
. . . . ***R. rubribrunnescens*** & ***R. maculatipes*** (see *R. formosa* group, p. 654)
30. Fresh branches *and* branch tips bright orange to yellow-orange with a distinct yellow zone or
band just below them (near ground level) . . ***R. aurantiisiccescens*** (see *R. rasilispora,* p. 652)
30. Not with above features (but may have some of them) 31
31. Branches *and* branch tips bright orange to deep orange to yellow-orange or pinkish-orange 32
31. Not as above (but branches may be pinkish or orangish with yellow tips) 34
32. Branches yellow-orange when young and fresh, becoming pale or dull orange in age 37
32. Branches bright orange to deep orange or pinkish-orange when fresh but not yellow-orange 33
33. Branches bright orange to deep orange in all stages ***R. largentii*** (see *R. formosa* group, p. 654)
33. Branches usually scarlet to pinkish or pinkish-orange, at least when young and fresh . . . .
. . . . ***R. stuntzii, R. subbotrytis,*** & ***R. cyaneigranosa*** (see *R. araiospora,* p. 655)
34. Odor usually fragrant; branches yellow becoming yellow-brown in age; fruiting mainly in the
fall under conifers . . . . ***R. flavobrunnescens*** (see *R. rasilispora,* p. 652)
34. Not as above 35

35. Branches orange to pinkish or pinkish-tan, the tips usually yellow or yellowish when fresh
.... *R. formosa group* & others, p. 654
35. Not as above 36
36. Branches yellow to yellow-orange or whitish with yellow tips (when fresh) 37
36. Branches differently colored (usually dull) 39
37. Found under hardwoods in eastern North America .... *R. aurea* (see *R. rasilispora*, p. 652)
37. Found under both hardwoods and conifers; common in western North America 38
38. Fruiting body with a large, broad, conical or steeply tapered, rooting base; upper branches
never orange when fresh; found under northern and mountain conifers in the spring or
summer .... *R. magnipes* (see *R. rasilispora*, p. 652)
38. Not as above; fruiting body with a large or small base and with yellow to orangish branches;
found in above habitat and season as well as many others .... *R. rasilispora*, p. 652
39. Branches yellow-brown to orangish-brown or duller even when fresh (sometimes with a purplish
tinge); branches typically erect and slender; taste usually bitter when cooked; common under
conifers in the Pacific Northwest .... *R. acrisiccescens* (see *R. fennica*, p. 650)
39. Not as above 40
40. Fruiting body compact, creamy to tan, dingy yellow-brown, or cinnamon-buff; taste mild;
spores *not* striate; found in eastern North America .... *R. caulifloriformis*
40. Not with above features; widespread and common 41
41. Branch tips typically pinkish when young but soon fading so that the mature fruiting body is
entirely white or creamy; spores striate; found under conifers in the Pacific Northwest .....
.... *R. rubrievanescens* (see *R. botrytis*, p. 656)
41. Tips never pinkish, or if pinkish when young then retaining their color longer; spores striate
or not striate 42
42. Base with reddish-brown or vinaceous stains or staining these colors when bruised; branch
tips whitish to yellow; spores *not* striate *R. vinosimaculans* (see *R. sanguinea* group, p. 653)
42. Not as above 43
43. Odor usually pungently sweet; branch tips dull orange to brownish, yellowish-buff, or whitish
when fresh; spores striate .... *R. strasseri* & others (see *R. botrytis*, p. 656)
43. Odor mild, or if sweetish then branch tips differently colored; spores striate or not 44
44. Fruiting body whitish with yellow to creamy-yellow tips, at least while under the duff (often
yellower or brighter in color when exposed); spores *not* striate *R. rasilispora* & others, p. 652
44. Not as above; branch tips usually pinkish to reddish or purplish, at least when young 45
45. Taste typically mild (occasionally bitter or acrid); spores striate; very common ....
.... *R. botrytis* & others, p. 656
45. Taste typically bitter to *very* bitter; spores not striate or only obscurely so; widespread but not
particularly common .... *R. botrytoides* (see *R. botrytis*, p. 656)
46. Fruiting body brownish-yellow to tan or yellowish-tan; flesh cartilaginous ....
.... *R. cartilaginea* (see *R. gelatinosa*, p. 655)
46. Not as above; differently colored and/or flesh in base gelatinous or semi-gelatinous 47
47. Branches usually yellow or yellow-orange ..... *R. flavigelatinosa* (see *R. gelatinosa*, p. 655)
47. Branches creamy to pinkish, orangish, brownish, etc., but not yellow (but fruiting body may
have a yellow zone just below branches) 48
48. Fruiting body usually with a yellow zone or band near ground level (near bases of branches),
at least when young .... *R. gelatiniaurantia* & *R. sandaracina* (see *R. gelatinosa*, p. 655)
48. Fruiting body lacking marked yellow zone .... *R. gelatinosa*, p. 655

Ramaria stricta (Strict Coral Mushroom)

FRUITING BODY profusely branched from a poorly developed base or stalk; 4-12 cm
high and 3-10 cm broad. **BRANCHES** erect, slender, mostly parallel, usually compact
but sometimes openly branched, sometimes grooved or flattened slightly; pale yellow
becoming pinkish-tan to tawny-buff, orangish, or sometimes light brown, usually

staining slowly brown or vinaceous-brown when bruised; branch tips fine, pale yellow to pale greenish-yellow when fresh. **STALK** rudimentary or practically absent, not well-developed. Flesh tough, pliant; odor often faintly sweetish; taste often somewhat metallic or bitter. **SPORE PRINT** cinnamon-buff to yellowish; spores 7-10 × 3.5-5.5 microns, elliptical, minutely roughened.

HABITAT: Solitary or in groups or tufts on rotting logs and branches (mostly of hardwoods but also conifers); widely distributed. It is common in our area in the late fall and winter, but seldom fruits in large numbers.

EDIBILITY: Inedible—there is very little flesh and the flavor is not pleasing.

COMMENTS: This is one of several Ramarias that grow on wood, though it may form "log lines" (i.e., grow in rows originating from buried or extremely decomposed logs). The upright, parallel orientation of the branches (see photo on p. 631) plus the yellowish branch tips are distinctive. The form that grows on conifers is often rather bushy and yellowish when young, while the one that grows on hardwoods is usually oranger. Other wood-inhabiting species: *R. apiculata* favors conifers but is dull buffy-tan to dull orange-brown or vinaceous-cinnamon (often with whitish tips when young). It is widespread and common, also bruises brown, and in one form exhibits green to blue-green tints on the branch tips and/or stalk base. *R. acris (=R. rubella)* is usually acrid-tasting and pinkish-tinged when young with paler tips but becomes reddish-brown or darker throughout in age. *Lentaria pinicola* and *L. byssiseda* (both placed in *Ramaria* by some authorities) grow on logs (mostly coniferous) or lignin-rich humus and have smooth whitish spores. The former is tan to yellowish with spores shorter than 10 microns, while the latter is pallid to pinkish-tan, sometimes has greenish branch tips, and has longer spores. Both species have white mycelial threads or cottony material at the base; *L. byssiseda* is especially common in the Sierra Nevada (but is widespread). None of these are worth eating.

Ramaria myceliosa

FRUITING BODY abruptly and profusely branched from a slender base (stalk); 2-6 (10) cm high and broad. **BRANCHES** slender, spreading, pliant, yellowish to tan, ochraceous, olive-ochre, cinnamon-buff, or dull orange, etc.; tips same color. **STALK** slender, pliant, not very fleshy; same color as branches or paler, with abundant white mycelial threads (rhizomorphs) attached to base and/or permeating the surrounding humus. Flesh thin, whitish, pliant; taste usually bitter. **SPORE PRINT** yellowish or pale ochraceous; spores 3.5-6 × 2-4 microns, elliptical to nearly round, minutely spiny.

HABITAT: Scattered to densely gregarious in duff under conifers; known only from the west coast, but similar species (see comments) are more widely distributed. In our area it is often abundant under redwood in the fall, winter, and early spring.

EDIBILITY: Unknown, but hardly worth experimenting with because of its small size and bitterish taste.

COMMENTS: This is one of several small (usually less than 10 cm high), dull, pliant, terrestrial Ramarias with a conspicuous white mycelial mat and minutely spiny spores. Other members of the club include: *R. pusilla,* an eastern version of *R. myceliosa; R. invalii,* with golden to dull yellowish-orange branches and spores 6-8.5 × 4-6 microns; *R. flaccida,* pale creamy-ochre with paler tips and spores 5-8 microns long; *R. suecica,* pallid to pinkish-tan with spores 8-10 microns long; *R. gracilis,* with minutely roughened rather than spiny spores and a frequent licorice-like odor; and *R. murrillii,* with cinnamon-brown to dark reddish-brown branches and a whitish base plus spiny spores, found in southeastern North America. None of these grow on wood like *R. stricta* and relatives, nor do they develop the greenish or blue-green stains typical of *R. abietina.*

Ramaria abietina (=R. ochraceovirens) is an ochraceous species that develops greenish stains in age or cold weather. Note the extensive mat of white mycelial threads at base of fruiting body.

Ramaria abietina   (Green-Staining Coral Mushroom)

FRUITING BODY profusely branched from a slender base (stalk); 2-10 cm high and broad. **BRANCHES** slender, pliant, ochraceous to pale cinnamon becoming olive-brown to dingy brown in age, slowly staining greenish or blue-green when bruised or as it matures, especially the lower branches. **STALK** up to 2.5 cm long, slender, white to ochraceous or colored like branches, with white mycelial threads (rhizomorphs) at base and/or in surrounding humus. Flesh pliant, often discoloring slightly vinaceous-brown; taste usually bitter. **SPORE PRINT** pale yellowish-tan; spores 5.5-8 × 3-4.5 microns, elliptical, minutely spiny.

HABITAT: Scattered to densely gregarious in duff under conifers (less commonly hardwoods); widely distributed. It is common in our area from the fall through early spring, especially under redwood.

EDIBILITY: Too small, tough, and/or bitter to bother with.

COMMENTS: The tendency of the lower branches to develop greenish stains distinguishes this species, which is also known as *R. ochraceovirens,* from the throngs of other small, pliant, buff to dingy ochraceous, terrestrial Ramarias (see *R. myceliosa*). The green-staining is especially evident in cold weather, and can be enhanced—if there is any doubt—by placing the fruiting body in a freezer. *R. apiculata* (see comments under *R. stricta*) may stain greenish, but normally grows on wood.

Ramaria fennica   (Bitter Coral Mushroom)

FRUITING BODY typically with two to four large primary branches arising from a common base (stalk), and many smaller secondary branches arising from the primary ones; 6-18 cm high and 5-12 cm broad. **BRANCHES** smooth, mostly erect, olive-gray to olive-umber to smoky-yellow, grayish-tan, or yellow-brown, the basal (primary) branches dark olive-brown with a violet tinge when fresh; tips olive-yellow to dingy buff. **STALK** large, fleshy, white below, upper portion colored like the branches and usually tinged violet when fresh. Flesh white, firm but brittle; taste often bitterish. **SPORE PRINT** pale ochraceous; spores 8.5-12 × 3.5-5 microns, elliptical, roughened.

HABITAT: Solitary, scattered, or in groups in humus under both hardwoods and conifers; widely distributed. In our area it is usually associated, like other fleshy Ramarias, with tanoak. It is not normally a common species, but occasionally fruits in great abundance in the fall and early winter.

EDIBILITY: Unknown; the bitterish taste (when present) is a deterrent.

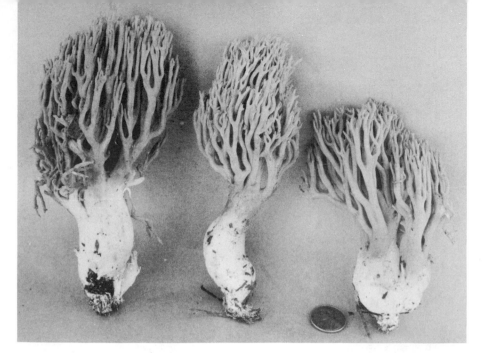

Ramaria fennica is a handsome upright coral fungus. Note the well-developed fleshy base or "trunk." Fresh specimens are usually tinged purple at base of branches.

COMMENTS: One of our most lovely coral fungi, this species is easily told by its overall shape (see photograph) and dull color, plus the subtle violet tinge to the lower branches and top of the trunk (when fresh). In contrast to many of the larger Ramarias, it is typically taller than it is wide. *R. fennnica var. violaceibrunnea* is a more precise name for the west coast variant. Other species: *R. fumosiavellanea* is a rare similarly-colored species with a much smaller base. *R. acrisiccescens,* called the "Blah Ramaria" by Kit Scates in her extensive key to Ramarias of the Pacific Northwest, is a common dull-colored species with slender, erect branches and a slim, tapered base plus an acrid or bitter taste, at least when cooked. It is common under conifers in the Pacific Northwest, but I have not seen it in our area.

Ramaria fumigata (Violet Coral Mushroom)

FRUITING BODY typically with two to four large primary branches arising from a common base (stalk), and many smaller secondary branches arising from the primary ones; 5-12 cm high and 8-12 cm broad. **BRANCHES** smooth, violet or lilac, slowly discoloring dingy yellowish-brown or smoky-yellow in old age; tips also violet. **STALK** distinct as a white fleshy base or "trunk" below the main branches. Flesh white, firm; taste slightly bitter or acrid. **SPORE PRINT** pale ochraceous; spores 8.5-11 × 3-4 microns, elliptical, roughened.

HABITAT: Solitary or in small groups on ground in woods; widely distributed, but apparently very rare in California. I have found it only once in our area, near Boulder Creek, under tanoak and Douglas-fir, in December.

EDIBILITY: Unknown.

COMMENTS: The violet branches and branch tips make this beautiful *Ramaria* unmistakable. It is closely allied to *R. fennica,* but is somewhat smaller in my experience, and more vividly colored. Other distinctly purple coral fungi are more sparsely branched *and/or* have white spores (see comments under *Clavaria purpurea* and *Clavulina cinerea*). Other species: *R. cedretorum,* reported by Kit Scates from northern Idaho, is another rare amethyst-colored species. It does not discolor as much in age and has slightly larger spores.

651

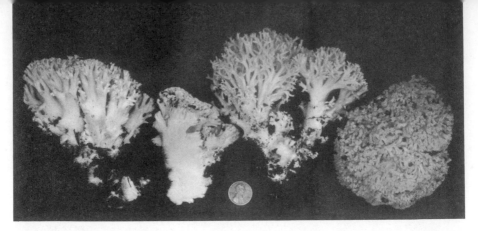

Ramaria rasilispora is the most common of several yellow to pale orange coral fungi. Note that the flesh in sliced specimen is solid and white, *not* gelatinous.

Ramaria rasilispora (Yellow Coral Mushroom)

FRUITING BODY abundantly branched from a fleshy base or stalk, 5-40 cm high and broad. **BRANCHES** smooth, mostly erect, creamy to pale yellow (*var. scatesiana*) or yellow to yellow-orange or pale orange (*var. rasilispora*), but the lower branches often whitish when very young and branches in both varieties often pale or dull orange to ochraceous in old age; tips yellow or same color as branches, sometimes brownish in age. **STALK** fleshy, tapering to a point or rooting; white or whitish. Flesh brittle or fibrous, often with watery areas, white; taste mild. **SPORE PRINT** pale orange-yellow or ochraceous; spores 8-12 × 3-4.5 microns, cylindrical, smooth or slightly roughened.

HABITAT: Solitary to scattered or in groups or rings on ground in woods; common in the West and probably more widely distributed. In our area it fruits in the fall and winter under tanoak or less commonly live oak, but in other regions it is common under conifers, especially fir. In mountain ranges such as the Sierra Nevada it usually fruits in the spring and early summer, as does *R. magnipes* (see comments), which is common from the Siskiyou Mountains northward under hemlock, fir, and other conifers.

EDIBILITY: Edible and quite popular, but difficult to clean and sometimes lacking in flavor. Coral fungus connoisseur Herb Saylor says it's good raw in salads or candied like grapefruit rinds. Some people, however, are adversely affected by it, and *R. magnipes* (see comments) is sometimes bitter when cooked.

COMMENTS: The fleshy, terrestrial, yellowish Ramarias do not lend themselves to glib categorization. As a group, however, they are easily recognized by their shape and color, and none appear to be dangerously poisonous. The names **R. flava** and **R. aurea** have traditionally been applied indiscriminately to practically all yellow coral fungi, but are inadequate for the many variants in the group. *R. rasilispora* is probably the most common yellow species in California. **R. magnipes** is a very similar, characteristic feature of the springtime mushroom flora of western mountains. It differs in having a larger (7-14 × 4-6 cm), rooting, steeply tapered or broadly conical stalk, yellow (never orange) branches, and slightly longer spores. Other fleshy, terrestrial, yellow Ramarias include: **R. flavobrunnescens,** which usually has a sweet odor, yellow branches that become brownish-yellow (or brownish-tipped) in age, and is very common at times in the Pacific Northwest under western hemlock; *R. flavigelatinosa* (see comments under *R. gelatinosa*), with watery-gelatinous or sometimes cartilaginous flesh; **R. aurantiisiccescens,** with yellow-orange to orange branches and a yellow zone near ground level; and **R. aurea,** a pale orange to yellow species that is said to be common under hardwoods in eastern North America. None of these bruise wine-red at the base as in the *R. sanguinea* group (or the "true" *R. flava*), and none are choice edibles.

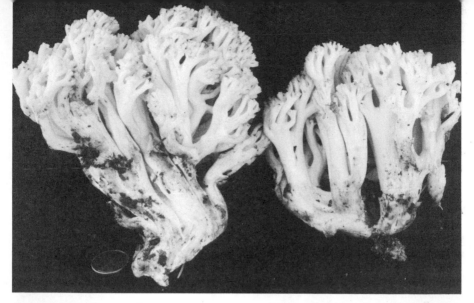

Ramaria sanguinea group. Branches are yellow to nearly white and the base stains reddish or burgundy when bruised.

Ramaria sanguinea group (Bleeding Coral Mushroom)

FRUITING BODY profusely branched from a fleshy base or "stalk"; 4-25 cm or more high and broad. **BRANCHES** smooth, mostly erect; pale or clear yellow, the tips usually slightly brighter yellow; lower (primary) branches turning dark red or vinaceous to reddish- or vinaceous-brown when bruised. **STALK** fleshy, tapered below, often very short; white or whitish, typically staining dark red or wine-red to vinaceous-brown when bruised or already with such stains. Flesh white or yellowish, rather brittle; taste mild. **SPORE PRINT** pale ochraceous; spores 8.5-12 × 3.5-4 microns, oblong or elongated-elliptical, roughened or smooth.

HABITAT: Solitary, scattered, or in groups on ground under both hardwoods and conifers; widely distributed. In our area this species "complex" fruits in the late fall and winter under tanoak, but is not as numerous as *R. rasilispora.* In the Pacific Northwest and northern California several similar species (see comments) are common under conifers in the fall, and I have also seen large fruitings near Lake Tahoe in October.

EDIBILITY: Presumably consumable; I haven't tried it.

COMMENTS: The yellow branches and tendency to stain vinaceous or reddish-brown plus the mild taste and non-gelatinous flesh are the telltale traits of this lovely coral mushroom and its close relatives. *R. flava* is a very similar, longer-spored species that has not yet been reported from the west coast. Actually, there is considerable doubt as to whether the "true" *R. sanguinea* occurs in the West, but there are a number of similar vinaceous-staining species, including: *R. vinosimaculans,* with creamy to light yellow branch tips and vinaceous stains at the base (and sometimes on the branches), and *R. rubiginosa,* a fleshy species with darker yellow tips and prominent vinaceous or purplish-brown stains at the base. Both of these species are common under conifers in California and the Pacific Northwest, particularly in the fall. Two other yellow, vinaceous-staining species, *R. cystidiophora* and *R. synaptopoda,* differ in possessing a coating of cottony white fuzz at the base. The former is often fragrant while the latter resembles a bundle of strings with "ox-blood red" spots at the base and usually has a pungent-sweet odor. For vinaceous-staining species with oranger or pinker branches, see comments under the *R. formosa* group, and also check *R. conjunctipes* (couplet #21 of the key to *Ramaria*) and the species discussed under *R. gelatinosa.*

653

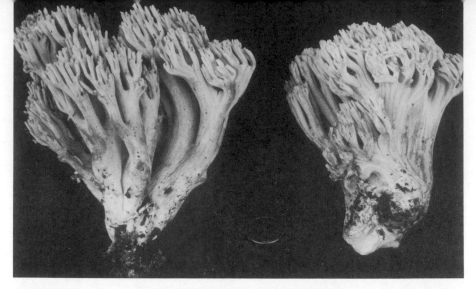

Ramaria formosa group. Branches are pinkish to orangish and tips are usually yellow when fresh.

Ramaria formosa group   (Pinkish Coral Mushroom)

FRUITING BODY profusely branched from a fleshy base (stalk); 5-20 cm high and wide. **BRANCHES** smooth or grooved, mostly erect; pinkish to pinkish-orange, salmon, pinkish-tan, or even reddish, fading to pinkish-buff, tan, or dingy ochraceous in age; tips yellow or colored like branches when fresh, also fading. **STALK** fleshy, usually tapering downward; white at base, otherwise colored like branches (or paler). Flesh brittle to fibrous, whitish or becoming pinkish to orange in age, *not* gelatinous; taste usually mild but often astringent or bitter when cooked. **SPORE PRINT** pale ochraceous; spores 9-12.5 × 3.5-6 microns, elliptical, minutely roughened.

HABITAT: Solitary, widely scattered, or in groups or rings on ground in woods, widely distributed. In the Pacific Northwest and Rocky Mountains it is one of the most common Ramarias, fruiting mainly under conifers, but also hardwoods. In our area it favors tanoak (like just about every other fleshy *Ramaria*) and is fairly common in the fall and winter.

EDIBILITY: Said to have cathartic effects on some (or many) individuals. Until the identity of our variety (or varieties) is definitely established, it is best left alone.

COMMENTS: The above description encompasses several forms that pose great difficulties to the avid coral-categorizer but have collectively passed under the name *R. formosa* in North America. The pinkish to pinkish-orange to pinkish-tan branches and non-gelatinous flesh are the main fieldmarks, but older specimens are easily confused with other species. The European *R. formosa* is said to stain cinnamon to vinaceous-brown or blackish when bruised (at least at the base). Several forms in the Pacific Northwest do, whereas the one in our area does not. Which is the "true" *R. formosa* is for the *Ramaria*-researchers to decide! Similar species which stain include: *R. rubribrunnescens* and *R. maculatipes,* both bruising vinaceous or burgundy (the former often fragrant); the *R. formosa* of the Pacific Northwest already alluded to, which bruises brown; and *R. testaceoflava var. brunnea* (=*R. brunnea)*, a handsome, bitter-tasting conifer-lover of the Pacific Northwest with "apricot-buff" to golden-yellow branches that soon become brownish and quickly stain reddish-brown, plus a brown-staining base. Similarly-colored western species that do *not* stain include: *R. leptoformosa* and *R. rubricarnata,* common in the Pacific Northwest under conifers, the latter with a stout, chunky fruiting body and salmon-tinged flesh inside the branches; *R. longispora,* with a yellow zone or band near or at ground-level when young; *R. cyaneigranosa*, whose fruiting body is often flattened (somewhat fanlike); and *R. largentii,* a beautiful species that is orange to deep orange throughout (never pinkish and not yellow-tipped).

654

Ramaria gelatinosa (Gelatinous Coral Mushroom)

FRUITING BODY profusely branched from a fleshy base or "stalk"; 5-15 cm high and wide. **BRANCHES** mostly smooth and erect, orange to pinkish-orange to orange-buff, pinkish-brown, or occasionally yellow-orange, sometimes developing a faint purple-gray tinge in age. **STALK** or main trunk whitish, rather tough. Flesh colored like branches or paler and showing translucent, often gelatinous pockets or streaks when stalk is sectioned lengthwise; taste often bitter. **SPORE PRINT** pale ochraceous; spores 7-10 × 4.5-6 microns, elliptical, roughened (warted).

HABITAT: Widely scattered to gregarious in duff under conifers (hemlock, fir, etc.); common in the fall in the Pacific Northwest and northern California, but more widely distributed. I have yet to find it south of San Francisco.

EDIBILITY: Poisonous to some people (causing diarrhea and digestive upset); apparently edible for others, but hardly worth the risk.

COMMENTS: The gelatinous interior of the base and overall dingy pinkish to dull orange color are the forthright features of this fungus. It is most likely to be mistaken for the *R. formosa* group, which does not have gelatinous flesh. The above description is for *R. gelatinosa var. oregonensis.* The typical variety, which was originally collected under hardwoods in the southeastern states, is similar but paler (creamy) when young. Other western species with gelatinous or semi-gelatinous flesh include: *R. flavigelatinosa,* which is yellower in color and quite common on the west coast; *R. gelatiniaurantia* and *R. sandaracina,* both with orange to deep orange branches and orange to yellow tips, and a band of yellow just above ground level (near the bases of the branches); and *R. cartilaginea,* a yellowish-tan to tan or brownish-yellow species with cartilaginous to very slightly gelatinous flesh. None of these are worth eating.

Ramaria araiospora (Red Coral Mushroom)   **Color Plate 168**

FRUITING BODY profusely branched from a fleshy base (stalk); 4-13 cm high, 3-10 cm broad or more. **BRANCHES** smooth, brilliant coral-red to red, crimson, or even magenta, but often fading to coral-pink, light red, orange, or even paler in old age; tips red to yellowish, fading slightly in age. **STALK** usually rather short, white or whitish (or discolored) at base, colored like branches (or paler) above. Flesh brittle; taste mild or slightly krautlike. **SPORE PRINT** pale ochraceous; spores 8-13 × 3-5 microns, elliptical to cylindrical, roughened. Flesh in base of stalk *not* amyloid.

HABITAT: Solitary, widely scattered, or in small groups on ground in woods; known only from the west coast. I find it every year in the fall and winter under tanoak, but it is not as common in our area as some of the other fleshy Ramarias. Farther north it frequently occurs with hemlock and other conifers.

EDIBILITY: Presumably edible, but not as fleshy as *R. botrytis* and *R. rasilispora.* Be sure not to confuse faded specimens with the *R. formosa* group!

COMMENTS: The brilliant red color of fresh specimens makes this gorgeous coral mushroom unmistakable. It used to pass under the name *R. subbotrytis,* but that species is apparently eastern in distribution. There are at least two distinct varieties of *R. araiospora:* var. *araiospora,* with yellowish branch tips in age, and *var. rubella,* with red tips. As in other Ramarias, the color tends to fade with age, and old specimens barely hint at their former splendor. *R. cyaneigranosa* is a somewhat similar species that has broader spores and is usually more peach-colored when fresh. *R. stuntzii,* common in northern California and the Pacific Northwest, is similar to *R. araiospora* in color or can show more yellow below, deep red or magenta near the tips, and orange in between, but it has amyloid flesh in the stalk. *R. subbotrytis* is an eastern species with salmon-orange to pink branches and branch tips when fresh, but it becomes duller or slightly yellower in age.

Ramaria botrytis, young specimens. Note how compact the fruiting body is—it looks more like a cauliflower than the cauliflower mushroom *(Sparassis).* Older specimens are not necessarily compact, however. Branch tips are purple-red to pinkish when young, but the color may fade with age.

Ramaria botrytis   (Pink-Tipped Coral Mushroom)

FRUITING BODY profusely branched from a fleshy base or stalk; sometimes massive, 7-20 cm high and 6-30 cm broad or more when mature; usually very compact when young, but less so in age. **BRANCHES** often crowded, white or pallid when young, gradually becoming buff to dull ochraceous or tan in age; tips typically short, clefted, blunt, pink to purple, vinaceous, or red when fresh (but orange to brownish-orange in *var. aurantii-ramosa*), fading in age. **STALK** thick (1.5-6 cm), fleshy, the base often rounded and swollen at or above ground level; white or whitish when fresh, tan to ochraceous in old age. Flesh firm, solid, white; taste mild in some forms, somewhat bitter in others. **SPORE PRINT** pale ochraceous to ochre-buff; spores 11-16 (20) × 4-6 microns, oblong-elliptical to nearly cylindrical, longitudinally lined (striate).

HABITAT: Solitary, scattered, or in groups or rings on ground in woods, often partially buried in the humus; very widely distributed. Along with *R. strasseri* (see comments) it is common in the Pacific Northwest and Sierra Nevada under conifers in the summer, fall, and spring (but is less common in the spring than *R. rubripermanens*—see comments). In our area it favors tanoak and other hardwoods and fruits in the fall and winter.

EDIBILITY: Edible and choice, according to some. However, it has laxative effects on some individuals and bitter look-alikes (see comments) occur. It is one of the largest and fleshiest of all the Ramarias, so it is definitely worth trying.

COMMENTS: This handsome, compact coral mushroom is reminiscent of a cauliflower —more so, in fact, than the cauliflower mushroom *(Sparassis).* Young specimens with stubby pink- or purple-tipped branches are the most readily recognized of all the Ramarias (see photo). In age, however, the fruiting body discolors and often loses its compact form, and it is then easily confused with other Ramarias. The longitudinally striate spores, however, are distinctive. There are several closely related species with longitudinally striate spores. One, *R. strasseri,* is quite common (at least in western North America), but typically has a tapered, rooting stalk and a frequently fragrant or spicy-sweet odor; it is whitish to ochraceous or tan with whitish to ochre, tan, dull orange, or brownish tips. A second, *R. secunda,* closely resembles *R. strasseri,* but has shorter spores. A third, *R. rubripermanens,* is colored like typical *R. botrytis,* but has shorter spores (less than 13 microns long). A fourth, *R. rubrievanescens,* also has short spores, but shows pink on the branch tips only when very young and is often entirely white or creamy at maturity. All of these occur in western North America, mainly (but not exclusively) with conifers. Finally, there is *R. botrytoides,* a widespread species that mimics *R. botrytis* in color (whitish with pink to purplish branch tips) but has an exceedingly acrid taste that can linger in one's mouth for up to twelve hours after tasting it! This species, which occurs occasionally in our area, also differs in having non-striate (or only very faintly striate) spores, suggesting a closer kinship to *R. araiospora* than to *R. botrytis.*

SPARASSIS   (Cauliflower Mushrooms)

Medium-sized to large fungi *parasitic on the roots of trees (especially conifers).* FRUITING BODY *typically a compact mass of flattened, wavy, ribbonlike branches or leafy lobes,* often with a tough rooting base; *white to buff or tan.* SPORE PRINT white. Spores smooth, elliptical.

THE branched fruiting body with wavy, flattened, leafy or ribbonlike lobes is unique to *Sparassis,* and most mycologists now sequester it in a family of its own. Despite the common name, our species looks more like a giant brain than a cauliflower. It fruits at the bases of trees (another distinctive feature), appearing year after year in the same spots.

From a size and edibility standpoint, *Sparassis* is indisputably the king (or queen) of the coral fungi. It is highly esteemed for its fragrance, flavor, and keeping quality, and I found it to be a staple item in the monsoon diet of many Himalayan villagers. Only two species occur in North America, one western and one eastern; a key hardly seems necessary.

Sparassis crispa   (Cauliflower Mushroom)   Color Plate 172

FRUITING BODY a compactly branched mass of flattened, leafy lobes arising from a tough, rooting base or "stalk"; 12-60 cm or more high and broad. **BRANCHES** (lobes) flattened, thin, usually leafy in appearance; wavy, crisped, or crimped, pliant; surfaces of lobes smooth, whitish to creamy or yellowish, often becoming cinnamon-buff or tan in age or dry weather, sometimes with darker brown stains on the edges. **STALK** 5-13 cm long, 2-5 cm thick, usually tapered downward, fleshy but very tough, buried deep in ground (or wood). Flesh firm, white, fairly tough or elastic; odor spicy-fragrant. **SPORE PRINT** white; spores 5-7 × 3-5 microns, broadly elliptical, smooth.

HABITAT: Parasitic on the roots of conifers, usually growing solitary at or near the bases of trunks or stumps; found in the coniferous forests of western North America. It is not uncommon in our coastal pine forests, fruiting in the same spot year after year (or sometimes twice a year), usually in the late fall or winter. It also grows on Douglas-fir and other members of the pine family. It produces a brown or yellow carbonizing rot in its host, and is economically significant in some regions. The best place to look for it is in older forests where there are plenty of mature trees. *S. spathulata* (see below) can grow on hardwoods.

EDIBILITY: Edible and exceptional. The perfect mushroom for a special occasion, it is as elegant as it looks intelligent. Thorough cooking is necessary, however, to render it tender, and it is a royal pain-in-the-posterior to clean! Gentle sauteeing or parboiling followed by baking or stewing is best. Fresh specimens can be stored for a week or two in a cool dry place, but be sure to check first for maggots!

COMMENTS: Better known as *S. radicata,* this fantastic fungus looks more like a sea sponge or "bouquet of egg noodles" (Alexander Smith) than a cauliflower. The flattened, ribbonlike lobes or "branches", plus the overall white to yellowish color and tough, rooting base distinguish it from other coral fungi. The spicy odor is also distinctive, but difficult to characterize. In the Pacific Northwest, 20-30 (or even 50!) pound specimens are not unheard of, but in our area the size range is generally 1-5 pounds—or about the size of a human head (a cross-section, coincidentally, reveals brainlike convolutions or "canals"). Certain cup fungi (e.g., *Peziza proteana* form *sparassoides* and *Wynnea sparassoides*) sometimes mimic it, but are brown or lilac-tinted and/or smaller and bear their spores in asci rather than on basidia (a microscopic distinction). The name *S. crispa* has also been applied to the equally edible cauliflower mushroom of eastern North America, which favors oak and pine, has thicker, more erect, rigid-looking branches or lobes, and lacks a rooting base. However, recent studies indicate that its "correct" name is *S. spathulata* or *S. herbstii,* and that the western species, which has been called *S. radicata,* is the "true" *S. crispa.* (I'm tempted to say that mushroom taxonomy is as intricately twisted and convoluted as a cauliflower mushroom, but I won't!)

Chanterelles

 spores

CANTHARELLACEAE

Fairly small to medium-large, mostly terrestrial woodland fungi. FRUITING BODY *with a cap and stalk.* CAP typically depressed to vase-shaped or trumpet-shaped at maturity, surface not typically viscid. UNDERSIDE *smooth, wrinkled, veined, or with primitive foldlike, forking, shallow, blunt, decurrent gills.* STALK fleshy or hollow, continuous with cap. VEIL and VOLVA absent. SPORE PRINT *white to buff, yellowish, pinkish, salmon-tinged, or tan.* Spores smooth or roughened, not amyloid. Basidia long and narrow.

THESE are colorful and conspicuous mushrooms with a more or less vase-shaped to funnel-shaped fruiting body differentiated into an upper or inner sterile surface (the cap) and lower or outer spore-bearing surface (the hymenium). The latter may be smooth, wrinkled, veined, or gilled, but the gills when present are foldlike (thick, shallow, blunt, and usually joined by connecting veins) rather than bladelike and/or thin-edged as in the true agarics. The long, narrow basidia and relatively unspecialized hymenium suggest a kinship between the chanterelles and coral fungi. In fact, if the flattened apex of a club coral *(Clavariadelphus)* were broadened into a cap and the wrinkles on its sides amplified, you would essentially have a chanterelle!

The three principal genera in the Cantharellaceae are treated together here. In *Craterellus* the fruiting body is more or less trumpet-shaped and dark brown to gray or black, and the spores are smooth; in *Cantharellus* it is broadly convex to plane or vase-shaped and often brightly colored, gills are usually present, and the spores are smooth; and in *Gomphus* the fruiting body is cylindrical to vase-shaped with a veined or wrinkled hymenium, and the spores are warted or wrinkled. A fourth genus, *Polyozellus,* is also included here though it is usually relegated to a family of its own.

The chanterelles are great favorites with fungophiles. With the exception of the *G. floccosus* group, all are apparently edible and many are excellent. The best (and most plentiful!) of the local species are the renowned chanterelle *(Cantharellus cibarius)* and the unheralded but far tastier horn of plenty *(Craterellus cornucopioides).* Both refrigerate well and are usually maggot-free, and since large crops are commonplace, they can be preserved for later use. The horn of plenty is excellent dried. Chanterelles, on the other hand, dry miserably (they become as tough as leather), but can be pickled, canned, or sauteed and frozen. (Enormous quantities of chanterelles are gathered in the Pacific Northwest and shipped in brine to Germany—five million pounds annually according to one estimate!)

The chanterelles are strictly woodland fungi and are saprophytic or mycorrhizal with a broad range of tree hosts. They are not a particularly large group, but are quite conspicuous, easy to identify, and have an interesting distribution pattern. Many species are characteristic components of cold northern and montane coniferous forests—e.g., *Gomphus clavatus, G. floccosus,* and *Cantharellus infundibuliformis.* A slew of others favor the warm hardwood forests of eastern North America—e.g., *Cantharellus cinnabarinus, C. ignicolor,* and *C. lateritius.* But there is a puzzling paucity of species in our area, with only a few of the truly cosmopolitan ones (like the chanterelle) represented. Nine species are depicted here (most of them with color plates) and several others are keyed out.

Key to the Cantharellaceae

1. Fruiting body gelatinous or rubbery and pliant . . . . (see **Tremellales & Allies**, p. 669)
1. Fruiting body fleshy to rather tough and pliant, but not rubbery-pliant or gelatinous 2

The chanterelle, *Cantharellus cibarius,* is a prized edible mushroom. Note shallow, blunt, veined gills. Cap is depressed or funnel-shaped in these mature specimens, but can be convex when very young (see description on p. 662).

2. Fruiting body small or minute (usually less than 2 cm broad), typically white to grayish- or brownish-tinged; stalk absent, or if present often short and off-center to lateral; fertile surface (underside of cap) with meandering veins or shallow gills; growing on sticks, moss, etc. (not often terrestrial); especially common in northern regions (see **Pleurotus & Allies,** p. 132)
2. Fruiting body medium-sized to large, or if small then not as above 3

3. Fruiting body a gilled mushroom (usually *Lactarius* or *Russula*) covered by a layer of minutely pimpled tissue that completely engulfs the gills or makes them look blunt and chanterelle-like; flesh crisp, brittle, usually white; spores borne inside asci ..... (see **Pyrenomycetes,** p. 878)
3. Not as above; surface not finely pimpled; spores borne on basidia 4

4. Cap dark when fresh (dark gray to dark brown, black, violet-black, or bluish-black) 5
4. Cap brighter or lighter than above (including olive-buff, tan, light brown, etc.) 11

5. Entire fruiting body deep blue to blue-black or violet-black, usually compound (a cluster of spoon- or fan-shaped caps); found under northern and montane conifers (especially spruce and fir) ....***Polyozellus multiplex,*** p. 668
5. Not as above (but fruiting body may be black or bluish-gray or more or less funnel- to trumpet-shaped, and may occur in clusters) 6

6. Underside of cap (fertile surface) with distinct gills, the gills often forked 7
6. Underside of cap smooth to wrinkled or with sinuous (squiggly) veins, but lacking gills ... 8

7. Entire fruiting body dark when fresh (colored more or less like cap or the fertile surface sometimes paler) .... ***Craterellus cinereus*** & others, p. 665
7. Not as above; either the stalk or gills (or both) yellowish to orange (or stalk with a yellowish to orange undersurface and darker fibrils) 17

8. Underside of cap (fertile surface) with prominent veins (at least at maturity); odor sickeningly sweet in age, or if not then fertile surface usually developing a strong yellowish tinge as spores mature; not common .... ***Craterellus foetidus*** & *C. sinuosus* (see *C. cinereus,* p. 665)
8. Not as above; underside of cap smooth to slightly wrinkled; common 9

9. Fruiting body typically 4 cm or more high when mature, trumpet-shaped to petunia-shaped or tubular (i.e., hollow nearly to base) .... ***Craterellus cornucopioides*** & others, p. 666
9. Fruiting body smaller and/or differently shaped (not hollow); cap margin sometimes white or pale when actively growing 10

10. Cap typically less than 2 cm broad and stalk 1-3 mm thick; fruiting body grayish to brown or blackish but not chocolate-brown or purple-brown; spores smooth; found in eastern North America (especially the Southeast) ... ***Craterellus calyculus*** (see *C. cornucopioides,* p. 666)
10. Not as above; cap up to 5 cm broad (or larger if compound), usually purple-brown to chocolate-brown (but sometimes grayish-brown) and/or compound (with more than one cap); cap margin often fringed; spores often angular or warty ...... (see **Stereaceae & Allies,** p. 604)

11. Fruiting body more or less vaselike and coarsely scaly at the center (or throughout), at least in age (see Color Plate 174) 12
11. Not as above; cap smooth or breaking up to form small scales, but not coarsely scaly 13

12. Cap reddish to bright orange to orange-buff, at least when fresh (may fade to tan or paler) .. .. *Gomphus floccosus* group, p. 661
12. Cap with tan to brown scales, never brightly colored .... *Gomphus kauffmanii* (see *G. floccosus* group, p. 661)

13. Flesh and gills (or underside of cap) white or whitish, but staining purplish to purple-brown when bruised; cap usually salmon-colored to orange; known only from the Southeast, not common .... *Cantharellus (=Gloeocantharellus) purpurascens*
13. Not as above (but gills or fertile surface may be purple or purple-tinged) 14

14. Fertile surface and/or upper stalk purple or with a purplish tinge *when fresh*; cap tan to olive-buff, olive-brown, yellow-brown, or purplish; stalk thick and solid, firm, continuous with cap (i.e., fruiting body resembling a chopped-off, often lopsided club, at least when young); found under conifers, often (but not always) in fused clusters ... *Gomphus clavatus*, p. 661
14. Not with above features (differently colored or stalk slender and hollow in age, etc.) 15

15. Fertile surface (underside of cap) with gills (or lacking them only in the button stage), the gills often forked or with connecting veins 16
15. Fertile surface (underside of cap) smooth to slightly wrinkled or at times developing shallow veins at maturity, but lacking gills 22

16. Fresh fruiting body bright red to pink to orange-red ..... *Cantharellus cinnabarinus*, p. 664
16. Fruiting body differently colored (white, yellow, orange, salmon-orange, brown, etc.) ... 17

17. Either growing on wood (or lignin-rich humus) and stalk absent or rudimentary *or* gills orange to deep orange and usually dichotomously forked and fairly thin in age and cap often brownish, at least at the center .... (see **Paxillaceae**, p. 476)
17. Not as above; gills usually blunt and thick (at least when young) and often forked irregularly (rather than dichotomously), orange to yellow-orange, white, or differently colored 18

18. Cap grayish to brown or yellow-brown when young (but may break up into scales or fibrils, revealing the yellowish to orangish undersurface); stalk often hollow, at least in age 19
18. Not as above; cap white, yellow, or orange when fresh (unless faded); stalk hollow or solid 20

19. Stalk brown or with brownish fibrils when young; gills yellowish to orangish; found in eastern North America .... *Cantharellus appalachiensis*
19. Stalk usually yellowish to orange when young and fresh; gills often with a grayish, brownish, or purplish tinge; widespread .... *Cantharellus infundibuliformis* group, p. 665

20. Cap white or whitish, sometimes with orange or yellow stains *Cantharellus subalbidus*, p. 662
20. Cap orange-buff to salmon-buff to ochre-tawny to bright yellow or orange, at least when fresh (may fade to whitish in sunlight) 21

21. Stalk slender (usually less than 1 cm thick) and usually hollow in age; flesh very thin; cap smallish; common in eastern North America .... *Cantharellus ignicolor & C. minor* (see *C. infundibuliformis* group, p. 665)
21. Stalk slender to very thick (0.5-2 cm or more), solid (unless maggot-eaten); cap medium-sized to very large or sometimes small; flesh thick; common, widespread *Cantharellus cibarius*, p. 662

22. Mature fruiting body with a flattened top, but without a true cap (i.e., cap margin blunt or non-existent—see Color Plate 166); flesh often punky or stringy .. (see *Clavariadelphus*, p. 632)
22. Not as above; fruiting body normally with a cap that has a well-defined, acute edge (margin) 23

23. Fruiting body whitish or dull-colored; stalk thin (1-3 mm); cap usually small, often rosette-like (with petal-like lobes) in age; flesh very thin and tough ... (see **Stereaceae & Allies,** p. 604)
23. Not as above; fruiting body fleshy though sometimes thin-fleshed, usually brightly colored at least in part 24

24. Cap brown to ochre-brown or yellow-brown, or sometimes more brightly colored; flesh very thin; odor typically mild .. *Cantharellus xanthopus* (see *C. infundibuliformis* group, p. 665)
24. Cap brightly colored (usually yellow to orange or pinkish-orange); flesh fairly thick (at least at center of cap); odor often fruity or fragrant 25

25. Fruiting body more or less trumpet-shaped, usually clustered and fragrant .... *Cantharellus odoratus* (see *C. cibarius,* p. 662)
25. Not as above .... *Cantharellus lateritius & C. confluens* (see *C. cibarius,* p. 662)

Gomphus clavatus   (Pig's Ears)   Color Plate 176

FRUITING BODY 6-20 cm or more high, nearly cylindrical to more or less club-shaped but with a flattened top; often growing in fused or compound clusters. **CAP(s)** 2-10 (15) cm broad, plane or depressed with an uplifted, often wavy or lobed margin that is often more highly developed on one side than the other; surface moist or dry but not viscid, smooth or breaking up into minute scales, light purplish to purplish-tan to olive-brown, olive-buff, tan, or yellowish-buff. Flesh thick, firm, white or buff; odor mild. **UNDER-SIDE** (fertile surface) with numerous shallow, blunt, deeply decurrent, forking veins or wrinkles which occasionally give it a poroid appearance; color variable but usually dull purple to purplish-tan or showing at least some purple tints below (near the stalk), slowly fading to dull ochre, tan, or buff. **STALK(s)** 3-5 (10) cm long, 1-3 cm thick, continuous with cap, central or off-center, often fused together below, equal or narrowed at base (if solitary), often curved; solid, firm, buff to pale purple. **SPORE PRINT** pale tan to pale ochre; spores 10-13 × 4-6.5 microns, elliptical, slightly wrinkled or warted.

HABITAT: Scattered to gregarious, often in fused pairs or clusters, on ground under conifers; widely distributed in northern North America, fruiting mainly in the late summer and fall. It is common in the Pacific Northwest and northern California, but I have not seen it south of San Francisco.

EDIBILITY: Edible and considered choice by some, but I am not particularly fond of it.

COMMENTS: Also known as *Cantharellus clavatus,* this northern conifer-lover is easily distinguished by its overall shape (like a chopped-off, frequently lopsided club), dull purplish veined spore-producing surface, and growth habit in fused clusters. It is aptly characterized in one book as "that funny looking thing which is purplish underneath." The common name, "pig's ears," is also applied to an Ascomycete, *Discina perlata.* Other species: *G. (=Pseudocraterellus) pseudoclavatus* is said to be similar, but has smooth spores and favors hardwoods. It is apparently quite rare; I have not seen it.

Gomphus floccosus group   Color Plate 174
(Scaly Chanterelle; Woolly Chanterelle)

FRUITING BODY nearly cylindrical, soon becoming more or less trumpet- or vase-shaped, 5-20 cm high. **CAP** 3-15 cm broad, depressed and soon hollow in the center; surface reddish to bright orange, buffy-, or yellow-orange (but often fading!), with large cottony or woolly scales, the ones at the center often recurved in age; margin often wavy or lobed. Flesh rather fibrous, whitish; taste mild to sour. **UNDERSIDE** (fertile surface) milky white to creamy, buff, dingy yellowish, or pale ochre; with blunt, shallow, frequently forking, deeply decurrent veins or ridges, sometimes with a somewhat poroid appearance in age. **STALK** 3-10 cm long, 1-3 cm thick, continuous with cap, central or off-center, tapering downward, solid becoming hollow in age, white to orange or yellowish. **SPORE PRINT** ochraceous; spores 10-16 × 5-8 microns, elliptical, slightly wrinkled or warted.

HABITAT: Scattered to gregarious or sometimes clustered on ground under conifers; widely distributed in northern and montane North America. Along with *G. bonari* (see comments), it is common in northern California and the Sierra Nevada in the summer, fall, and even spring, but like *G. clavatus,* it does not seem to occur on the coast south of San Francisco. It is also common in the Southwest, Pacific Northwest, and Rockies.

EDIBILITY: Not recommended. Some people pronounce it delicious, but others are adversely affected by it (they suffer gastric upsets) and the specimens I've sampled had an unpleasant sour taste. Its only asset is its distinctiveness—it would be very difficult to confuse it with anything other than its close relatives (see comments).

COMMENTS: Also known as *Cantharellus floccosus,* this striking mushroom and its close relatives are easily told by their vase-shaped fruiting body with a reddish-orange to orange-buff scaly cap and pallid veined exterior. The name *G. bonari* has been given to the western variety with milky white exterior, smaller spores, tendency to grow in clusters, and duller, paler, or more cinnamon-colored cap; it is often common under mountain conifers. Another western conifer-lover, *G. kauffmanii,* is similar but has a tan to brown or ochre-tawny cap up to 35 cm broad with very prominent, brittle scales. It is sometimes common in northern California and the Sierra Nevada as well as in the Pacific Northwest and Southwest. Like *G. floccosus* and *G. bonari,* it should be eaten very cautiously if at all.

Cantharellus subalbidus   (White Chanterelle)   Color Plate 179

CAP 4-15 cm broad, plane to broadly depressed with an often wavy or irregularly lobed margin; surface smooth or breaking up into small scales in age, moist to dry but not viscid; dull white or whitish, bruising yellowish-orange to orange-brown. Flesh thick, firm, white; odor mild or faintly fragrant. **UNDERSIDE** (fertile surface) with thick, well-spaced, shallow, blunt, deeply decurrent, foldlike gills which are usually forked or veined between; dull white or pinkish-tinged, often staining yellowish to orange in age or where bruised. **STALK** 2-7 cm long, 1-5 cm thick, equal or tapered downward, central or off-center, solid, firm, smooth, dull whitish, discoloring yellowish-orange to orange-brown in age or where bruised. **SPORE PRINT** white; spores 7-9 × 5-5.5 microns, elliptical, smooth.

HABITAT: Solitary, scattered, or gregarious on ground in woods; known only from the Pacific Northwest and California. It is quite abundant under second-growth conifers in the late summer, fall, and winter in Washington, Oregon, and northern California, and I have also seen large fruitings in Idaho. In our area it fruits in the fall and early winter, but is only occasionally common. The best places to look for it are on the high, exposed ridges of our coastal mountains under tanoak and madrone or in manzanita thickets mixed with knobcone pine. It often grows with *Armillaria ponderosa,* another choice edible.

EDIBILITY: Edible and choice—as good as *C. cibarius* or even better, and like that species, rarely attacked by maggots. In the Pacific Northwest the buttons are firm, meaty, and very compact—ideal for marinating or prolonged refrigeration. Unfortunately, it is not nearly so common in our area and the fruiting bodies, when found, are often water-logged and overripe.

COMMENTS: The white chanterelle can be mistaken for a *Clitocybe* or *Hygrophorus,* but the gills are distinctly "chanterellesque"—thick, shallow, blunt, and usually veined. It differs from the common chanterelle *(C. cibarius)* in its dull whitish color and white spores, but in size and shape the two are quite similar and can even be found growing together.

Cantharellus cibarius   (Chanterelle)   Color Plates 175, 177, 178

CAP (2) 3-15 (25) cm broad, broadly convex when young, becoming plane to depressed or vase-shaped in age; surface smooth or occasionally cracked, not viscid; orange to bright golden-orange, egg-yellow, or yellow (but in dry weather or direct sunlight often bleached out and cracked or even greenish from algae); margin usually lobed or wavy, at first in-curved. Flesh thick, firm, whitish or tinged yellow to orange under the cuticle; odor mild or faintly fruity (like pumpkins or apricots); taste mild to somewhat peppery or occasionally bitter. **UNDERSIDE** (fertile surface) with thick, well-spaced to close, shallow (narrow), blunt, foldlike, deeply decurrent gills which are often forked or cross-veined; colored like cap or more often paler (but brighter if cap has faded); not fading, but sometimes developing dingy orange-brown stains in old age. **STALK** 2-10 cm long, 0.5-3 (5) cm thick,

Cantharellus cibarius. Note the white flesh in sliced specimen (upper left) and shallow, decurrent, veined gills. It varies greatly in stature and color (see color plates) and favors oak as well as conifers.

equal or tapered downward or sometimes enlarged at base, solid, dry, firm, colored like cap or paler, often staining ochraceous to orange-brown. **SPORE PRINT** creamy or yellow in most forms, but pinkish in one variety; spores 7-11 × 4-6 microns, elliptical, smooth.

HABITAT: Widely scattered to gregarious on ground in woods, occurring throughout the north temperate zone and very common on the west coast in the fall, winter, and spring. In northern California and the Pacific Northwest it grows mainly with conifers, but in our area it favors live oaks, especially those at the edges of pastures. It has a long growing season and is a joy to find—brilliant splashes of gold against the subdued backdrop of decaying leaves. If you see one, probe around and you'll probably uncover more—they often hide under the humus. If rainfall is sufficient, successive crops are produced over a long period of time, so check your patches regularly (but don't check mine!).

EDIBILITY: Edible and choice—the best known wild mushroom in California, if not in North America. It is so plentiful and popular on the west coast that it is now being harvested commercially and shipped abroad or sold to restaurants and markets. A few tips for the uninitiated: Chanterelles should always be cooked as they are somewhat fibrous raw. Pick selectively—firm specimens will keep in the refrigerator for a week, whereas waterlogged individuals will rot rapidly and cook up slimy. Even firm ones have a high water content and should be sauteed slowly and thoroughly in an open pan. You waste them if you can't taste them, so don't just mix them in with a bunch of other vegetables. They are best in simple dishes that highlight their delicate flavor and fruity fragrance—cream of chanterelle soup is a traditional favorite. On the west coast chanterelles are virtually free of maggots, an asset keenly appreciated by those who hunt them—but they're often full of dirt, a quality deeply detested by those who have to clean them! (If your "chanterelles" are pristine and wormy, they may not be chanterelles!). Curiously, it's just the opposite in the eastern United States, where they are usually quite spotless, but often lunched on by maggots! Imported chanterelles can be purchased in small tins at delicatessens—for an outrageous price, of course. They are low in protein but high in vitamin A. Attempts to cultivate them commercially have not been successful.

COMMENTS: *Cantharellus cibarius* is the proud possessor of a plethora of popular names—more than any other mushroom with the possible exception of *Boletus edulis.* The best known are *chanterelle* and *girolle* (both French) and *Pfifferling* (German). From region to region the chanterelle varies considerably in size, stature, color, and fruiting habits (several forms have been described), but is easily identified by its yellow-orange cap

663

and "primitive" gills (blunt, thick, shallow, veined or forking, and foldlike rather than bladelike). The firm whitish flesh and wavy cap margin are also characteristic. On the west coast it fruits in cool weather and is often very large and thick-stemmed (one pound specimens are not uncommon), with an orange cap, faintly fruity odor, and pale, copiously veined gills (although a smaller, slimmer, cleaner form grows under Sitka spruce). In eastern North America it is commonest in the summer and is usually much smaller (caps average 3-6 cm broad) and often yellower, with a slender, well-developed stalk and little or no odor (see Color Plate 175). Also, the gills may be less "primitive" (deeper, with thinner edges and few if any cross-veins, as shown in Color Plate 177), and are sometimes brightly colored. In the southern Rocky Mountains, on the other hand, the fruiting bodies are often packed closely together and are small and stubby, with bleached-out caps (if growing in sunlight) and brilliant yellow to yellow-orange gills.

Mushrooms most likely to be mistaken for the chanterelle include the false chanterelle *(Hygrophoropsis aurantiaca),* various waxy caps *(Camarophyllus pratensis, Hygrocybe flavescens,* etc.), *Lactarius alnicola, Leucopaxillus albissimus,* the *Gymnopilus spectabilis* group, *Hypomyces lactifluorum,* and the jack-o-lantern mushrooms *(Omphalotus* species*).* The latter are poisonous, but have brightly colored flesh and thin, crowded, unforked, bladelike ("true") gills, and usually grow in clusters on or near wood. Except for the *Hypomyces,* which is a parasitized gilled mushroom, the others also have bladelike gills. (It is sometimes said that true chanterelles have "false" gills, while false chanterelles have "true" gills.) Aside from the many variations within *C. cibarius,* there is *C. formosus,* a northern conifer-lover with a convex, yellow to brownish cap and pinkish-tinged gills, and three similar edible species with a smooth to only slightly wrinkled fertile surface. These are common in eastern North American and are often mistaken for *C. cibarius.* The most common and widespread of the three, *C. lateritius (=Craterellus cantharellus)* has a yellow-orange to pinkish-orange fruiting body; *C. confluens* has a yellower, often compound fruiting body and a tropical-subtropical distribution; *C. odoratus (=Craterellus odoratus)* is a gorgeous trumpet-shaped, yellow-orange-capped southern species that is usually strongly fragrant and cespitose (clustered). *C. lateritius* has also been reported (but not verified) from the Pacific Northwest.

Cantharellus cinnabarinus   (Red Chanterelle)   Color Plate 181

CAP 1-4 (7) cm broad, convex to nearly plane at first, plane to shallowly funnel-shaped in age, the margin often lobed or wavy and often remaining incurved for a long time; surface more or less smooth, dry, bright red to red-orange to flamingo-pink when fresh, often paler (or even whitish) in age or sunlight. Flesh thin, whitish below, usually tinged cap color above; odor typically mild, taste mild to peppery. **UNDERSIDE** (fertile surface) with thick, shallow, blunt, decurrent gills which are fairly close to well-spaced, usually forked or veined, and colored like the fresh cap or slightly paler (or pinker). **STALK** 1-4 (6) cm long, 0.3-1 cm thick, equal or tapered downward, often curved, solid, firm, dull red or colored like fresh cap (or pinker), the base sometimes whitish. **SPORE PRINT** usually pinkish-tinged (pinkish-cream); spores 6-9 (11) × 4-6 microns, elliptical, smooth.

HABITAT: Scattered to gregarious or clustered on ground in woods or at their edges, open areas, in beds of moss, etc.; fairly common in eastern North America (especially southward) in the summer and early fall. I have seen large fruitings under oak and pine in Michigan, Pennsylvania, and North Carolina. It apparently does not occur in the West, but is to be looked for in Arizona (one unverified report places it in the Pacific Northwest).

EDIBILITY: Edible and quite good, though delicate in flavor. It sometimes grows with *C. cibarius,* and the two species make colorful companions in the same dish.

COMMENTS: The modest size and bright pinkish-red to orange-red color make this chanterelle practically unmistakable. It is vaguely reminiscent of a waxy cap *(Hygrocybe),* but the blunt, veined or forked gills distinguish it.

Cantharellus infundibuliformis group   Color Plate 180
(Funnel Chanterelle; Winter Chanterelle)

CAP 1-6 (11)) cm broad, convex to plane with a depressed center or becoming broadly depressed to funnel-shaped (often with a perforated or hollow center) in age; surface smooth to slightly wrinkled or finely fibrillose-scaly, not viscid, dark brown to brown to dingy yellow-brown, sometimes fading to paler brown or the yellowish background color becoming dominant as it dries or ages; margin usually wavy or irregularly lobed. Flesh thin, rather tough and pliant, colored like cap or paler; odor mild or slightly fragrant; taste mild or bitterish. **UNDERSIDE** (fertile surface) with shallow (narrow), thick, blunt, well-spaced, decurrent, forked or veined gills; pale yellow or pale orange-yellow becoming grayish, brownish, or violet-tinted in age. **STALK** 2-8 cm long, 0.3-1 (2) cm thick, equal or tapered at either end, smooth, often grooved and/ or flattened, becoming hollow; orange-yellow to yellow, duller in age (dingy yellow or yellow-brown to grayish-orange). **SPORE PRINT** creamy-buff to yellowish (or white in one form); spores 9-12 × 6.5-8 microns, elliptical, smooth.

HABITAT: Scattered to gregarious in moss and humus and on rotting wood in cold, damp coniferous forests and bogs throughout northern North America; fruiting from late summer through the winter, common. The farthest south I've seen it is in Mendocino County, California, in December, January, and February.

EDIBILITY: Edible, but small and thin-fleshed. Some people relish it nevertheless, and in Finland it is harvested commercially.

COMMENTS: This characteristic feature of northern bogs and cold coniferous forests can be recognized by its modest size, dark brown to dingy yellow-brown cap, yellowish to gray or purple-tinged gills, and slender, hollow, yellow to yellow-orange stalk. The gills have a somewhat waxy look (leading to possible confusion with *Hygrocybe* and *Camarophyllus),* but are characteristically "chanterellesque": thick, blunt, shallow, and conspicuously forked or veined and/ or wavy. The name *C. tubaeformis* has been used for the white-spored variety, but many cantharellologists treat it as a synonym for *C. infundibuliformis.* In eastern North America there are numerous other small, edible, hollow-stemmed chanterelles, including: *C. ignicolor,* very similar but with a brighter cap when fresh (orange to yellow-orange to ochraceous-tawny or salmon-buff), growing under both hardwoods and conifers; *C. minor,* small and entirely yellow to orange (including the gills), favoring hardwoods; and *C. xanthopus* (also called *C. lutescens*), widely distributed (also found in the West), similar to *C. infundibuliformis* in cap color or brighter, but with a smooth to slightly veined (rather than gilled) orangish to yellowish underside.

Craterellus cinereus   (Black Chanterelle)

FRUITING BODY more or less funnel- or vase-shaped but not tubular; 3-12 cm high. **CAP** 1.5-5 cm broad, shallowly to deeply depressed, surface smooth or minutely scaly-scurfy, not viscid; black when moist, becoming dark grayish-brown in age or as it dries out; margin usually wavy, lobed, or torn. Flesh thin, tough, colored more or less like cap. **UNDERSIDE** (fertile surface) with thick, well-spaced, shallow, blunt, foldlike gills which fork frequently near the cap margin and/ or have cross-veins; bluish-black becoming bluish-gray to gray, or paler from spore dust. **STALK** 2-8 cm long, 0.4-1.3 cm thick, nearly equal or tapering downward, central or off-center, tough, hollow except at base, colored like cap or underside. **SPORE PRINT** whitish; spores 8-11 × 5-6 microns, elliptical, smooth.

HABITAT: Solitary to scattered or in groups or clusters in mixed woods and under oaks; widely distributed but not common. In our area it fruits in the winter and early spring, often in the company of *C. cornucopioides,* or a few days before it.

Craterellus cinereus, a bluish-gray to black species with veins or shallow gills on underside of cap.

EDIBILITY: Edible and delectable—as good or even better than *C. cornucopioides.* It is delicious simply sauteed in butter.

COMMENTS: This species, which has also passed under the name *Cantharellus cinereus,* closely mimics its commoner cousin, *Craterellus cornucopioides,* but has primitive gills on the underside of the cap and is not as brown in dry weather. Other species: *C. caeruleofuscus* of eastern North America is similar, but has smaller spores; *C. foetidus,* also eastern, has a sickeningly sweet odor, veined fertile surface, and thicker stalk (1-3 cm thick at apex); *C. (=Pseudocraterellus) sinuosus* has a gray to dark grayish-brown cap, slightly larger spores, and a veined or copiously wrinkled (but not gilled) underside that is grayish at first but acquires a distinct yellowish or ochre tinge as the spores mature; it is edible and widely distributed, but rather rare (Craig Mitchell has found it in mixed woods near Felton, California, in February).

Craterellus cornucopioides   (Horn of Plenty)   Color Plate 182

FRUITING BODY 3-14 cm high, tubular becoming trumpet- or funnel-shaped. **CAP** 2-8 (10) cm broad, hollow at the center with the margin at first decurved (folded down), then spreading out and becoming wavy, split, or lacerated (sometimes with a lacy appearance or convoluted to form small "suction cups"); surface not viscid, usually minutely scaly or scurfy, grayish-black to very dark brown or black when moist, paler (brown to grayish-brown) when dry (but one form sometimes developing a yellowish margin or yellowish blotches). Flesh thin, brittle but tough, colored like cap or paler; odor pleasant. **UNDERSIDE** (fertile surface) smooth to uneven or with slight, deeply decurrent wrinkles (but not gills); colored like cap but usually paler or grayer, in age or dry weather with a whitish to yellowish or buff coating of spore dust. **STALK** 1-5 cm long, 0.5-1 (1.5) cm thick, continuous with cap and tapering downward, central or off-center, hollow except at very base, tough, often twisted; colored like cap or underside. **SPORE PRINT** whitish to buff or pale yellow; spores 8-11 × 5-7 microns, elliptical, smooth.

HABITAT: Scattered or in groups, clumps, or lacy clusters in woods; widely distributed. It is common in our area in the winter and spring under hardwoods (live oak, tanoak, manzanita, madrone, huckleberry), but farther north it fruits in the late fall under conifers. South of San Francisco it rarely appears before Christmas, usually peaks at the same time that the fetid adder's tongue blooms, and may fruit on into April. In eastern North America

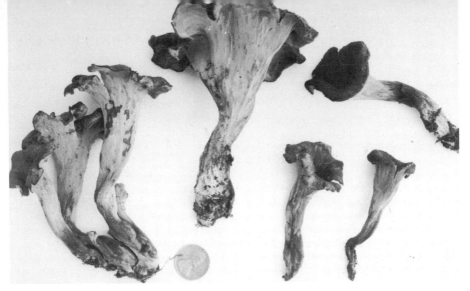

Craterellus cornucopioides. This delicious mushroom is easily recognized by its dark trumpetlike fruiting body. The undersides of these mature specimens are dusted white or buff by spores. Note complete absence of gills.

it is largely replaced by *C. fallax* (see comments), which also occurs in our area. Though terrestrial, it often fruits near fallen branches or at the bases of trees or manzanita burls.

EDIBILITY: One of my fifteen "five favorite flavorful fleshy fungal fructifications." Like myself, it is thin, tough, and dark, and like myself, it goes largely unappreciated. It is forever being passed up in favor of the larger, fleshier, more colorful and impressive types, yet its flavor is superb and its potential unlimited. It is partly to blame for this sad state of affairs, for its appearance is admittedly drab and rather somber, and it shuns attention, blending unobtrusively into the dark, secretive situations where it thrives.

But though it takes an accomplished eye to detect its presence in the woods (from the top it looks like a hole in the ground), anyone can detect its presence in a dish—for it cooks up black, announcing itself with unmistakable earthy authenticity. It can play any culinary position with equal finesse, from first base to second fiddle, enhancing practically any dish, be it soup, souffle, or sauce. Like its more popular cousin the chanterelle, it is not often

Craterellus cornucopioides **(right)** often grows in clusters, blending into its surroundings so uncannily that you have to search for black holes in the ground, as shown on **left** (actually a photo of *Craterellus fallax,* essentially the same but for its salmon-tinged spores).

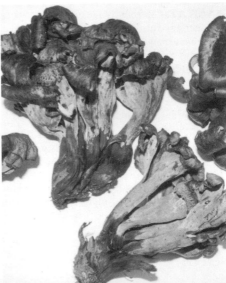

inhabited by maggots. Dried and powdered it has a cheesy odor and is known, quite appropriately, as "Poor People's Truffle."

COMMENTS: This cryptically-colored fungus usually occurs in large groups. It is hard to pick out in the forest gloom, but once you locate one, you're bound to find more. In France it is sometimes called *trumpet de mort* (trumpet of death)—a tribute to its somber appearance, not its edibility. Actually, it is quite lovely when fresh—fully-grown specimens look more like black or brown petunias than anything else, and clusters often have a lacy look. The horn of plenty that is so common in the summer and fall in eastern North America, *C. fallax*, looks and tastes identical but has a salmon- or yellow-tinted underside in age and longer, salmon or ochre-yellow spores; it also occurs in California. Otherwise, there is little that *C. cornucopioides* can be confused with—the dark trumpet-shaped fruiting body with smooth or slightly wrinkled exterior and hollow center is unique. *C. cinereus* (the black chanterelle) is somewhat similar but has primitive gills; *Polyozellus multiplex* is also similar but is dark blue or violet-tinted, while *C. sinuosus* (see comments under *C. cinereus*) has yellowish-ochre spores and a more copiously wrinkled or veined hymenium (underside). Several Ascomycetes are black, but not trumpet-shaped (*Plectania* species are cup-shaped; *Urnula* is urn-shaped). *Thelephora terrestris* is sometimes funnel-shaped but is thinner, smaller, and purple-brown rather than gray or black. Occasionally hollow, deformed, clublike specimens of *C. cornucopioides* occur alongside normal ones. These are identical in color and edible. Other species: *C. calyculus* of eastern North America is one of several similar but much smaller and thinner (cap less than 2 cm broad, stalk 1-3 mm thick), solid-stemmed, mostly southern species.

Polyozellus multiplex   (Blue Chanterelle)   Color Plate 183

FRUITING BODY usually compound—in compact, clustered masses 5-15 cm high and up to 1 meter broad (but usually smaller). CAP(s) 2-10 cm broad, spoon-shaped to fan-shaped, often centrally depressed and lop-sided; surface dull purple to purple-gray to black, deep blue, or frosted blue; margin usually irregularly lobed or wavy. Flesh deep violet to bluish-black, soft and rather brittle; odor mild or fragrant. UNDERSIDE (fertile surface) nearly smooth or more often with shallow, crowded wrinkles or veins or at times nearly poroid (but lacking gills); colored more or less like cap or slightly paler. STALK(s) 2-5 cm long, 0.5-2 cm thick, central to off-center or lateral, usually short and often fused at base; colored more or less like cap or underside, often grooved; solid or becoming hollow. SPORE PRINT white; spores (4) 6-8.5 × 5.5-8 microns, round to broadly elliptical, warted or appearing angular. Cap tissue staining olive-black in KOH.

HABITAT: Clustered on ground under conifers (usually spruce and fir) or aspen in late summer and fall; known only from northern and montane North America. It is usually listed as rare, but is locally common in the Rocky Mountains. I have seen it in the Oregon Cascades and it occurs very rarely in northern California. It was originally described from Maine, and the color photograph was taken near Santa Fe, New Mexico, in an aspen-spruce forest, in August.

EDIBILITY: Edible, and according to some sources, delicious. I have tasted it only once— the texture was reminiscent of *Craterellus*, but the flavor was far inferior.

COMMENTS: This strikingly beautiful mushroom is easily told by its unusual deep violet to bluish-black color and clustered growth habit. It might be mistaken at first glance for a *Craterellus*, but the caps are spoon- to fan-shaped rather than tubular or trumpetlike. The warty-angular spores suggest a closer relationship to the genus *Thelephora* (or perhaps to the teeth fungi) than to other chanterelles, but it has traditionally been treated as a chanterelle because of its superficial resemblance.

Jelly Fungi

TREMELLALES & Allies

spores

Small to medium-sized fungi *found mostly on wood.* FRUITING BODY *typically gelatinous or rubbery when fresh;* variously shaped, but most often lobed, convoluted, or bloblike. SPORE BEARING SURFACE smooth, warty, lobed, or with small spines, covering the outer or lower surfaces of the fruiting body. SPORE PRINT white to yellowish, but difficult to obtain. Spores smooth, sometimes septate. Basidia septate or forked.

JELLY fungi differ fundamentally from other Basidiomycetes in the structure of their basidia, which are partitioned (**septate**) or forked rather than simple and clublike. The rust and smut fungi also have partitioned basidia, but are not considered to be mushrooms. Although the basidia are microscopic, jelly fungi can usually be told in the field by the gelatinous (jellylike) or rubbery texture that gives them their name. The most familiar types are collectively called "witch's butter." They have lobed to convoluted or rather amorphous (shapeless) fruiting bodies that look—at least to the hungry mushroom hunter staggering home under a basketful of boletes—like lumps of melting butter. Other jelly fungi have a cap and stalk, still others are cuplike, and a few are tough, branched, and coral-like. The spore-bearing surface ranges from smooth to veined, lobed, or in one case, toothed, and is always exposed (hence the jelly fungi are classed here as Hymenomycetes).

The design of the basidium (see illustration) forms the basis for dividing the jelly fungi into three separate groups. The Tremellales, which are the most common, have obliquely or longitudinally septate basidia that look like hot cross buns when viewed from above, the Auriculariales have transversely septate basidia (i.e., with cross walls), and the Dacrymy-cetales have Y-shaped basidia that look like tuning forks. As these distinctions are micro-scopic, the three groups are treated as one unit in this book.

Different types of basidia in the jelly fungi. Left to right: longitudinally septate (Tremellales); Y-shaped (Dacrymy-cetales); and transversely septate (Auri-culariales).

Most of the jelly fungi grow on rotten wood. They thrive in cool wet weather, shrinking down to almost nothing when it dries out, then swelling up again as soon as it rains. None are known to be poisonous, but most are very watery and flavorless. Some species, however, can be marinated, candied, or even eaten raw, and in the Orient two types are very popular: a translucent white species called *Tremella fuciformis,* and the brown to black "tree ears" or "cloud ears" (*Auricularia*). The latter can be purchased in dried form in many specialty stores. Eight jelly fungi are described here and several others are keyed out. *Tremella mesenterica* is the most common of the lot, at least in our area.

Key to the Tremellales & Allies

1. Fruiting body brightly colored (yellow, orange, pink, red, or greenish) when fresh, but sometimes losing its color in rainy weather or old age 2
1. Not as above; fruiting body white, grayish, black, reddish-purple, brown, yellow-brown, etc. 9

2. Fruiting body spatula-shaped to funnel-shaped (but with one side usually incised or split), pink to orange or red; stalk usually present .... ***Phlogiotis helvelloides,*** p. 672
2. Not as above 3

3. Fruiting body erect, either simple and clublike (unbranched) or antlerlike or branched, or with a
    cap and stalk 4
3. Not as above; fruiting body cup-shaped to cone-shaped, cushion-shaped, irregularly lobed and
    contorted, or amorphous (bloblike) 6
4. Fruiting body with a yellow-orange to greenish "head" or cap ...... (see **Helotiales,** p. 865)
4. Not as above; fruiting body simple and unbranched or antlerlike to coral-like 5
5. Fruiting body branched, usually more than 15 mm high .... *Calocera viscosa,* p. 674
5. Fruiting body unbranched (but may be clustered), or if sparingly branched then typically less
    than 15 mm high .... *Calocera cornea* (see *C. viscosa,* p. 674)
6. Fruiting body cup-shaped to cone-shaped (with a narrowed base or point of attachment); usually
    found in the spring (often near melting snow) ...... *Guepiniopsis alpinus* & others, p. 674
6. Not as above; fruiting body bloblike to cushion-shaped to irregularly lobed or brainlike .. 7
7. Fruiting body 1-5 mm broad, drop-like, usually growing in large masses or rows ....
    .... *Dacrymyces deliquescens* (see *Tremella mesenterica,* p. 673)
7. Not as above; fruiting body usually larger 8
8. Typically growing on hardwoods; bone-hard when dry; basidia shaped like hot-cross buns in
    top view .... *Tremella mesenterica,* p. 673
8. Usually growing on conifers; collapsing when dry, but with a tough whitish basal point of
    attachment; basidia Y-shaped ... *Dacrymyces palmatus* (see *Tremella mesenterica,* p. 673)
9. Fruiting body translucent to whitish, grayish, or brownish, with a cap (and usually a stalk),
    the underside of the cap lined with tiny spines or "teeth" *Pseudohydnum gelatinosum,* p. 671
9. Not as above; underside of cap lacking minute spines or "teeth" 10
10. Fruiting body tough, erect, and usually branched (coral-like), white or pallid; found mainly on
    ground under hardwoods in eastern North America 11
10. Not as above; usually found on wood or plants 12
11. Branches few and very thick, blunt, hollow, and somewhat gelatinous ... *Tremella reticulata*
11. Not as above; texture very tough, branches often flattened ....
    ...... *Tremellodendron pallidum* & *T. candidum* (see *Tremellodendropsis tuberosa,* p. 643)
12. Fruiting body black (or nearly black) when fresh 13
12. Not as above (but fruiting body may be dark brown and/or may blacken as it dries out) . 14
13. Fruiting body small and cushion-shaped, warty, or lobed, but often fusing to form large con-
    tinuous patches or sheets; growing on dead hardwoods .... *Exidia glandulosa,* p. 672
13. Not as above; fruiting body broadly top-shaped to cuplike and not fusing to form sheets ....
    .... (see **Pezizacaceae & Allies,** p. 817)
14. Fruiting body a gelatinous mass of wavy or leafy lobes up to 20 cm broad (or sometimes larger),
    some shade of brown; spores borne on basidia .... *Tremella foliacea,* p. 673
14. Not as above (if brown and leafy, then fruiting body thin and rubbery instead of gelatinous) 15
15. Fruiting body reddish to brown or dark brown and bracketlike to cuplike, ear-shaped, broadly
    top-shaped, or occasionally forming a cluster of earlike lobes 16
15. Not as above 17
16. Fruiting body cup-shaped to broadly top-shaped, reddish to purplish to brown or dark brown;
    flesh thick or thin; usually on hardwoods; spores borne in asci ..... (see **Helotiales,** p. 865)
16. Fruiting body bracketlike to earlike or cuplike (or with several earlike lobes); flesh thin; on
    hardwoods or conifers; spores borne on basidia ..... *Auricularia auricula* & others, p. 675
17. Fruiting body containing whitish granules; overall color whitish becoming pinkish- to reddish-
    brown or vinaceous-brown .... *Exidia nucleata* (see *E. glandulosa,* p. 672)
17. Not as above 18
18. Fruiting body white or pallid when fresh (may be tinged tan in age) 19
18. Fruiting body not white or pallid 21
19. Fruiting body gelatinous or tough and usually hollow; growing on herbaceous stems and other
    vegetable matter in eastern North America .... *Tremella concrescens*
19. Not as above; usually growing on wood; widespread 20

20. Fruiting body lobed or convoluted, 1.5-7 cm broad (or merging to form larger patches up to 12 cm or more); spores elliptical; found in the southern United States and warmer parts of the world (including the Orient) .... *Tremella fuciformis*
20. Not as above; smaller and/or spores sausage-shaped   . *Exidia alba* (see *E. glandulosa*, p. 672)
21. Fruiting body more or less brownish-yellow and top-shaped or cone-shaped ....
.... *Exidia recisa* (see *Guepiniopsis alpinus*, p. 674)
21. Fruiting body bloblike to lobed, convoluted, or brainlike and flesh-colored to brownish, yellowish, or buff .... *Tremella encephala* & *T. frondosa* (see *T. foliacea*, p. 673)

Pseudohydnum gelatinosum   (Toothed Jelly Fungus)

FRUITING BODY flexible and flabby, rubbery-gelatinous, more or less tongue-shaped to spoon-shaped or fan-shaped (with a cap and usually a stalk). **CAP** 1-6 (7.5) cm broad; surface minutely roughened or downy to nearly smooth, not viscid; translucent white to watery gray or bluish-gray, or sometimes dingy brownish (or spotted brown). Flesh rubbery-gelatinous. **UNDERSIDE** lined with small pallid spines or "teeth" that are 0.5-3 (5) mm long. **STALK** up to 6 cm long, usually lateral, continuous with cap and similar in color and texture. **SPORE PRINT** white; spores borne on the "teeth," 5-8.5 microns, round or nearly round, smooth. Basidia longitudinally septate, elliptical to nearly round.

HABITAT: Solitary, scattered, or gregarious on rotting logs, twigs, and humus under conifers; widely distributed. It is sporadically common in our area in the late fall and winter, especially in dank, dark, damp situations under Douglas-fir.

EDIBILITY: Edible. It is said to be fairly good with honey and cream—but what *isn't*? It can also be marinated for use in salads. The texture is interesting, the flavor nonexistent.

COMMENTS: Also known as *Tremellodon gelatinosum,* this denizen of dank places is one of my fifty "five favorite fleshy fungal fructifications." The rubbery or flabby tongue-shaped fruiting bodies with small "teeth" on the underside are as attractive as they are unique, and look funnier than they do fungal—in fact, it is hard to take them seriously! (They remind me of the "Creepy Crawlers" I used to buy at the dime store with my lunch money.) The small teeth resemble those of the teeth fungi (Hydnaceae), hence the name *Pseudohydnum,* which means "false tooth fungus." Specimens on the west coast tend to have a well-developed, often vertical stalk whereas those in eastern North America tend to have little or no stalk, especially when growing from the sides of logs.

Pseudohydnum gelatinosum. Underside (left) is lined with tiny "teeth." Cap (right) is whitish to grayish or brown. Entire fruiting body is rubbery.

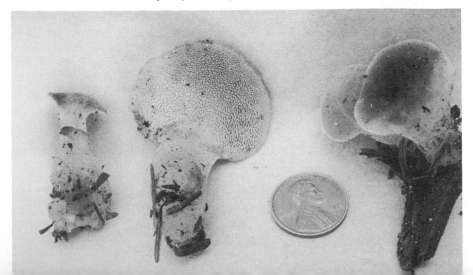

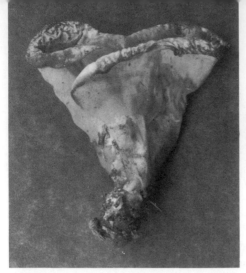

Left: A mature, flabby specimen of *Phlogiotis helvelloides.* Note how it is split down one side. **Right:** *Exidia glandulosa,* whose small fruiting bodies often fuse to form black rubbery or gelatinous sheets.

Phlogiotis helvelloides   (Apricot Jelly Mushroom)

FRUITING BODY flabby or rubbery, with a cap and stalk; spatula-shaped to somewhat funnel-shaped, but usually indented or split down one side; 2-8 (18) cm high. **CAP** 1-7 (10) cm broad, pale to deep rosy-pink to reddish-orange, apricot , or salmon-colored; surface more or less smooth, margin often wavy. Flesh rubbery to somewhat gelatinous. **UNDERSIDE** smooth to faintly veined or coarsely wrinkled, colored like cap or paler. **STALK** 1-6 cm long, off-center or lateral but usually vertical (upright); continuous with cap, with the same color and texture, or the base whitish. **SPORE PRINT** white; spores borne on underside of cap, 9-12 (16) × 4-6.5 microns, oblong-elliptical, smooth. Basidia longitudinally septate.

HABITAT: Scattered to gregarious on ground, debris, and rotten wood under conifers; widely distributed, but uncommon in most regions. In California I have seen impressive fruitings in Yosemite National Park and the Trinity Alps, but have found it only once in our area, near Big Sur. The fruiting bodies do not decay readily and seem to like cool weather (either late in the season or in the spring).

EDIBILITY: Highly prized by some people—presumably for its color and texture—for it has no flavor that I can detect. It can be eaten raw in salads, pickled, or candied, but should not be cooked because of its high water content. Larry Stickney says if it is sucked on a hot day it "cools you right down."

COMMENTS: The pink to orange color, rubbery texture (similar to that of *Pseudohydnum gelatinosum),* and overall shape—which is sometimes reminiscent of a miniature calla lily—identify this fetching little fungus. Actually, the word "little" is not always appropriate, as specimens 7 inches (18 cm) high have been reported by Sam Ristich of New York. These apparently developed over a period of two months and had smaller ones riding "piggy-back" on top of them! The normal size range, however, is 1-3 inches high.

Exidia glandulosa   (Black Witch's Butter; Black Jelly Roll)

FRUITING BODY flabby or gelatinous, beginning as a pallid or translucent blister but soon becoming cushion-shaped to irregularly lobed; reddish-black to olive-black soon becoming jet-black (or black from beginning); 1-2 cm broad but often fusing with others to form rows or masses up to 50 cm long; upper surface smooth to minutely roughened or warty. Flesh gelatinous, black. **STALK** absent. **SPORE PRINT** whitish; spores borne on the lobes, warts, or wrinkles, 10-16 × 3-5 microns, sausage-shaped, smooth. Basidia longitudinally septate.

HABITAT: Scattered to densely gregarious or in fused rows and masses on rotting hardwood logs and branches; widely distributed. It is fairly common in our area throughout the mushroom season, but is usually overlooked because of its dark color.

EDIBILITY: Unknown, and like most of us, likely to remain so.

COMMENTS: This sinister-looking black jelly fungus can be confused with little else. Individual fruiting bodies are quite small but usually fuse to form rows or sheets that look like black gelatin or slime. The frequently roughened spore-bearing surface and sausage-shaped spores help distinguish the genus *Exidia* from *Tremella.* Other species: *E. alba* is small and whitish; *E. nucleata* varies from white or colorless to pinkish- or vinaceous-brown, but is sprinkled with small white granules or "nuclei." Both are widely distributed.

Tremella foliacea (Brown Witch's Butter) Color Plate 173

FRUITING BODY flabby or gelatinous when moist, bone-hard when dry; 2.5-20 cm broad or more, typically consisting of a complicated mass of wavy or leaflike folds, lobes, and convolutions; reddish-cinnamon to brown, vinaceous-brown, or tinged purple; often paler when waterlogged. Flesh gelatinous. **STALK** absent. **SPORE PRINT** white to yellowish; spores produced on the upright lobes, 7-9 (13) × 6-9 microns, round to broadly elliptical, smooth. Basidia longitudinally septate.

HABITAT: Solitary or several on hardwood stumps, logs, and fallen branches (occasionally on conifers as well); widely distributed. In our area it is not uncommon on oak in the late fall, winter, and early spring, but is not nearly as numerous as *T. mesenterica.*

EDIBILITY: Harmless, but mostly water. As for the flavor—well, I once encountered an Asian-American gathering this species and *T. mesenterica* in a live oak woodland. "How do you prepare them?" I said. "With garlic and soy sauce," she said. "And how do they taste?" I said. Slight hesitation. "Like garlic and soy sauce," she said.

COMMENTS: This brown version of *T. mesenterica* is our largest *Tremella* and is quite striking and seaweed-like when fresh. The fruiting body is never cuplike or earlike as in *Auricularia auricula,* and is much more gelatinous. *T. frondosa* is a somewhat similar yellowish to buff species. *T. encephala* is also similar but smaller (1-6 cm) and flesh-colored to brownish. It is thought to be parasitic on *Stereum,* a wood-inhabiting bracket or parchment fungus. Most Tremellas, in fact, are frequently found with *Stereum,* suggesting some kind of relationship between the two. The lobed, brainlike, or convoluted fruiting bodies of Tremellas are reflected in their names: *foliacea* means "leaflike"; *mesenterica* means "middle intestine"; and *encephala* means "brainlike."

Tremella mesenterica (Witch's Butter) Color Plate 170

FRUITING BODY flabby or gelatinous when fresh, bone-hard when dry; 1-10 cm broad, typically consisting of several to many convoluted or brainlike lobes or folds, but in wet weather or old age often bloblike or amorphous; clear yellow to golden-yellow to bright orange, paler (to nearly colorless) when old or waterlogged. Flesh gelatinous. **STALK** absent. **SPORE PRINT** pallid to yellow; spores produced on the upright lobes or folds, 7-18 × 6-14 microns, elliptical to nearly round, smooth. Basidia longitudinally septate.

HABITAT: Solitary or in groups on hardwood sticks, logs, etc.; very common and widespread. In our area it favors dead oak and fruits throughout the mushroom season, often on the same branches as *Stereum* species. Shrivelled up specimens are inconspicuous, but usually revive in wet weather. In colder climates it may appear during winter thaws.

EDIBILITY: Harmless but flavorless (see comments on edibility of *T. foliacea*)—it is mostly water. My one attempt at cooking it was a failure: most of it evaporated!

COMMENTS: Also known as *T. lutescens,* this fungus and its look-alikes (see below) are a familiar sight in rainy weather across the continent. At first the fruiting body is lobed or brainlike and relatively firm, but as it swells with water it often loses its original shape and looks like a dollop of melting butter on a log. Also common is ***Dacrymyces palmatus,*** which closely mimics *T. mesenterica* but is slightly smaller (1-6 cm broad), tends to be oranger in color, favors conifers, and has a small, tough, whitish, basal point of attachment. The resemblance is only superficial, however, for *Dacrymyces* species have Y-shaped basidia and long narrow spores that develop cross walls, while *Tremella* species have longitudinally septate basidia and simple (non-septate) spores. Other species: ***T. frondosa*** is buff to yellowish but larger and leafier (like *T. foliacea*); ***D. deliquescens*** is a small (1-5 mm broad), yellow-orange to reddish species that grows on decaying wood, usually in large masses or rows. Like most jelly fungi, the above species are not worth eating.

Guepiniopsis alpinus   (Alpine Jelly Cone; Poor Man's Gumdrop)

FRUITING BODY flabby or gelatinous; top- to cone-shaped or sometimes cup-shaped, 0.3-2.5 cm broad; yellow to orange or occasionally reddish-orange; surface smooth. Flesh gelatinous. **STALK** present only as a small, narrowed point of attachment. **SPORE PRINT** yellowish; spores borne on the concave surface of the cone, 11-18 × 4-6 microns, sausage-shaped, developing 3-4 cross walls. Basidia Y-shaped.

HABITAT: Scattered to gregarious on coniferous logs, stumps, branches, twigs, etc.; fruiting in the spring (usually as or just after the snow melts) at higher elevations from the Rocky Mountains westward (rarely in the summer or fall); sometimes abundant.

EDIBILITY: Presumably consumable, but of negligible value.

COMMENTS: The golden cone-shaped fruiting body and fondness for dead conifers typify this omnipresent "snowbank" mushroom of western mountains. The gelatinous texture distinguishes it from the cup fungi, and the gumdrop-like fruiting body separates it from *Tremella* and *Dacrymyces.* The Y-shaped basidia place it in the Dacrymycetales, alongside *Dacrymyces* and *Calocera.* ***G. chrysocomus*** is a closely related species with a yellow cup-shaped fruiting body, larger spores, and a preference for hardwoods. ***Exidia recisa*** resembles *G. alpinus* in shape, but has longitudinally septate basidia and is brownish-yellow in color; it grows on hardwoods and is exceptionally abundant in some regions (especially the southern United States).

Calocera viscosa   (Staghorn Jelly Fungus)

FRUITING BODY erect, branched (antler- or coral-like); 2-7 (10) cm high, 1-3 (5) cm wide. **BRANCHES** tough, pliant, usually somewhat viscid, the tips usually forked (with

Left: *Guepiniopsis alpinus* looks like a gumdrop. **Right:** *Calocera viscosa.* (Herb Saylor)

2-3 tips); bright yellow to deep yellow to orange. **STALK** present as a yellow to whitish base or "trunk", often rooting. Flesh tough but often somewhat gelatinous. **SPORE PRINT** dingy yellowish to ochraceous; spores produced on the branches, 9-14 × 3-5 microns, elongated-elliptical to sausage-shaped, smooth, developing one cross wall at maturity. Basidia Y-shaped.

HABITAT: Solitary, scattered, or in small groups on or near coniferous logs, stumps, roots, and debris; widely distributed. I have found it several times in our coastal pine forests in the winter and spring, but never in quantity.

EDIBILITY: Inconsequential.

COMMENTS: This dainty little jelly fungus looks like a coral mushroom (especially *Clavulinopsis corniculata*) and is keyed out as one. However, it is usually viscid and has Y-shaped basidia, plus it typically grows on or near wood. Another widespread species, *C. cornea,* is smaller (up to 15 mm high), yellow, and clublike or very sparingly branched; it grows in groups or clusters on twigs and branches of both hardwoods and conifers. The forked basidia place *Calocera* in the Dacrymycetales.

Auricularia auricula   (Wood Ear; Tree Ear)

FRUITING BODY rubbery to pliant or flabby to somewhat gelatinous when fresh, hard when dry; 2-15 cm broad; cup-shaped to ear-shaped or sometimes with several earlike lobes originating from a central point of attachment. Outer surface sterile, often veined or ribbed, minutely silky or with fine downy hairs, pale brown to brown to liver-brown, drying blackish. Inner (fertile) surface smooth to slightly wrinkled, somewhat gelatinous when wet, tan to yellow-brown, grayish-brown, brown, liver-brown, or tinged purple; blackish when dried. Flesh thin, rubbery. **STALK** absent or rudimentary. **SPORE PRINT** white; spores borne on inner (usually the lower) surface, 12-18 × 4-8 microns, sausage-shaped, smooth. Basidia transversely septate (with 3 cross walls).

HABITAT: Solitary or in groups or clusters on logs, dead branches, stumps, etc. (attached centrally or laterally); very widely distributed on both hardwoods and conifers, and often common in cool weather. I have seen it in the springtime in the Cascades, but it seems to be rare in our area.

EDIBILITY: Edible, but I haven't tried it. A similar but hairier species, *A. polytricha,* is prized in the Orient and can be purchased dried in many stores. Along with the shiitake *(Lentinus edodes)* it is the mushroom most often served in Chinese restaurants, usually under the name Yung Ngo or Muk Ngo. Dried specimens of *A. auricula* and *A. polytricha* are quite dark, hard, and unappetizing, but billow up like clouds when soaked in water, showing off their delicate curves and convolutions to great effect.

COMMENTS: This species can be mistaken for a cup fungus, but has a more rubbery texture, bears its spores on basidia (a microscopic feature), and usually—but not always—grows with its fertile (concave) surface facing downward. Cup fungi, on the other hand, bear their spores in asci with the concave (fertile) surface facing upward, and usually have a brittle or fragile texture. Sometimes *Auricularia* is irregularly lobed rather than cuplike, and I have found old specimens which looked more like pieces of soggy seaweed than anything else. *Hirneola auricula-judae* is a sinister-sounding synonym. It also used to be called "Judas' Ear" because it was believed that when Judas' hanged himself on an elder tree, these ear-shaped "excrescences" were condemned to appear on elders thereafter. *Auricularia* is the most prominent genus in the Auriculariales. It embraces several similar brownish species, including: *A. mesenterica,* which is more or less bracketlike with a hairy, concentrically zoned sterile surface and veined fertile surface; and *A. delicata,* a striking gelatinous tropical species with large honeycomb-like pits or "pores." For a photo of *Auricularia,* see p. 958.

GASTEROMYCETES

THIS large division of the Basidiomycetes includes those fungi better known as puffballs, earthstars, stinkhorns, bird's nest fungi, false truffles, and gastroid agarics. *Gastero-* (meaning "stomach") describes the manner in which the spores are produced—*internally,* rather than externally as in the Hymenomycetes (agarics, boletes, coral fungi, etc.). The Gasteromycetes are also unique in that their spores are not forcibly discharged. Instead the basidia disintegrate and the leftover spore mass is dispersed by wind, rain, and animals.

The most familiar fungi in this group are the puffballs and earthstars. They bear spores in a round to oval "stomach" or **spore case.** The mature spore mass is powdery and easily dispersed. The false truffles are similar, but their spore mass remains intact and does not become powdery, while the gastroid agarics resemble gilled mushrooms that haven't opened. The stinkhorns strike a decidedly different pose—their slimy spore mass is initially enclosed by a membrane, but later it is elevated on a stalk, arms, or latticed ball, with the membrane forming a volva at the base. Last and least, there are the bird's nest fungi, which look like tiny nests with spore-containing eggs or **peridioles.**

Gasteromycetes occur in a wide range of habitats. They are especially prominent in arid regions where there is a selective advantage to producing spores internally (it affords protection from moisture loss and heat). They are notable more for their size range (giant puffballs weigh up to 50 pounds each, bird's nest fungi are only a few millimeters broad) and different strategies for spore dispersal than for their colors. Only the puffballs are of importance to the mushroom-eater. Six major groups are keyed below.*

Key to the Gasteromycetes

1. Fruiting body minute (typically less than 15 mm high), consisting of a "nest" (cup, vase, or bowl) containing one or more "eggs" (peridioles); (older specimens, however, may lack peridioles and young ones often have a covering or "lid" over the top of the "nest") **Nidulariales,** p. 778
1. Fruiting body differently constructed and usually larger than above 2

2. Fruiting body at first enclosed in a membrane, then emerging as a cylindrical, phallic, branched, tentacled, or latticed structure with the membrane forming a volva or sack at the base; mature spore mass slimy or mucilaginous (never powdery) and usually foul-smelling, coating the head, branches, or latticework of the fruiting body . . . . **Phallales,** p. 764
2. Not as above (but fruiting body may be slimy or malodorous at some stage) 3

3. Fruiting body with a stalk below the spore case or "cap" 4
3. Stalk absent or rudimentary 7

4. Stalk penetrating the spore case or "cap" and usually percurrent (i.e., extending to top of fruiting body); stalk *not* normally branched inside spore case . . . **Podaxales & Allies,** p. 724
4. Not as above; stalk not percurrent, sometimes branched 5

5. Spore mass slimy and greenish or olive, divided into several large chambers; fruiting body club-shaped to pear-shaped; usually found on rotten wood **Hymenogastrales & Allies,** p. 739
5. Not with above features 6

6. Stalk well-developed and clearly defined, almost always longer than diameter of spore case; stalk not composed of tough rootlike fibers, or if so then fruiting body usually brightly colored and/ or gelatinous . . . . **Tulostomatales,** p. 715
6. Not as above; stalk a tough mass of "roots" or fibers (and sand or dirt) *or* stalk an elongated, narrowed sterile base which often shows minute chambers when sliced open lengthwise . . 7

7. Spore case rupturing or disintegrating at maturity; spore mass firm and solid when young (chambers if present hardly discernible), powdery or cottony when mature and usually dispersing fairly soon; columella (internal stalk) typically absent; mature fruiting body usually (but not always) above the ground; found in many habitats . . **Lycoperdales & Allies,** p. 677
7. Spore case and spore mass remaining intact for a long time; spore mass often chambered (chambers large to minute), firm, rubbery, spongy, or slimy but *not* cottony or powdery at maturity; columella present or absent; usually growing underground in association with trees or shrubs, but sometimes surfacing at maturity . . . . **Hymenogastrales & Allies,** p. 739

*Some agarics, boletes, or other mushrooms superficially resemble puffballs, especially when young. Examples are *Entoloma abortivum, Amanita* "eggs," parasitized *Boletus* spp., and *Cryptoporus volvatus* (a common polypore).

Puffballs and Earthstars

spores

LYCOPERDALES & Allies

MANY people don't think of puffballs and earthstars as mushrooms, and indeed they have little in common with the cap-gill-and-stem commodity you buy at the grocery store. The fruiting body simply consists of a roundish to oval **spore case**, sometimes with a sterile region beneath it called, quite appropriately, the **sterile base**. The skin **(peridium)** of the spore case is usually composed of an inner and outer layer **(endo-** and **exoperidium)**. In the earthstars, the outer skin separates completely from the spore case and splits into several starlike rays; in the puffballs it does not. The sterile base, which is composed of minute compartments that give it a spongelike appearance, is best seen by slicing open the fruiting body lengthwise. In some species the sterile base is absent or inconspicuous; in others it is as large as the spore case and narrowed below so that it looks like a stem. However, only the stalked puffballs (discussed on p. 715) have a true stem.

The interior of the spore case is called the **spore mass** (or **gleba**). It is usually white and firm when young, but turns yellow, greenish, brown, or purplish as the spores mature, first becoming mushy as moisture is released, then powdery or cottony as the moisture evaporates. The spore color corresponds to that of the mature spore mass, and is usually some shade of brown or purple. Once the spores have matured, the spore case either splits open or ruptures irregularly or disintegrates—thereby exposing the spore mass to the elements—*or* a mouth or slit **(apical pore)** forms at the top, so that the spore case looks and acts like a miniature volcano. It has been suggested that the word "puffball" is a derivative of "puckfist," which in turn was derived from "pixie fart." All of these names testify to the puffballs' distinctive method of spore dispersal. If a mature (ruptured) specimen is poked, tapped, squeezed, or kicked (thereby duplicating the action of a raindrop or gust of wind), a cloud or "puff" of spore dust will emerge.

The true puffballs (Lycoperdales) also have several distinctive microscopic features. The spores are typically borne in a **hymenium** (palisade of basidia) and are usually round, sometimes with a **pedicel** ("tail") attached. They are often warted or spiny like sea urchins, and bounce around like furry bon-bons under the microscope. Usually intermingled with the mature spores are microscopic threadlike cells called **capillitium**. These are the remains of the hyphae on which the basidia form. Their shape and branching pattern are of great significance to the puffball taxonomist, but are not emphasized here.

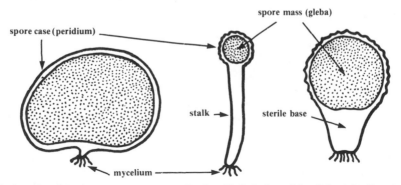

Vertical sections of two puffballs and one stalked puffball. Left to right: *Calvatia* (stalk and sterile base absent), *Tulostoma* (stalk present), and *Lycoperdon* (sterile base present).

Make sure every puffball you eat is firm, white, and solid (homogeneous) inside. These are rather scaly specimens of coastal California's giant puffball (see *Calvatia gigantea* group, p. 682). (Joel Leivick)

Besides the Lycoperdales, which includes the true puffballs such as *Calvatia* and *Lycoperdon* and the earthstar genus *Geastrum,* a superficially similar order, the Sclerodermatales, is traditionally included under the title "puffballs and earthstars." This group differs microscopically in lacking capillitium *and/or* a hymenium. It includes the earthballs *(Scleroderma)*, which can usually be separated from puffballs by their hard, one-layered peridium and purple-black spore mass, certain earthstars (e.g., *Astraeus*), and two dusty monstrosities that bear their spores in small chambers within the spore case *(Pisolithus* and *Dictyocephalos)*. A third group, the Tulostomatales (stalked puffballs), differs in possessing a true stalk. It is treated separately here, although microscopically it shares features of both the Lycoperdales and Sclerodermatales.

It is often said that all puffballs are safe to eat so long as they're firm and white inside. This is *not* necessarily the case, however. Several species of *Scleroderma* are poisonous, some of the so-called edible puffballs have purgative effects on certain individuals, and others don't taste good. Therefore, it behooves you puffball-pickers to identify each puffball before you eat it and to sample it cautiously. Once picked, puffballs must be refrigerated or they will ripen rapidly. Any showing the *slightest* traces of color (usually yellow or green) should be discarded, as they become bitter and indigestible. Also be sure that you don't inadvertently mistake a deadly poisonous *Amanita* "egg" for a puffball. When sliced open lengthwise (i.e., perpendicular to the ground), puffballs are solid and homogeneous within (see above photo), whereas *Amanita* "eggs" or other agaric buttons reveal the outline of cap, gills, and stalk (see photo at top of p. 679). False truffles are sometimes mistaken for puffballs, but they usually grow underground and do not have a powdery spore mass at maturity, while stinkhorn "eggs" are gelatinous within.

Puffballs and earthstars can be found almost anywhere at any time, but are especially prominent in prairies, deserts, and high mountains, where other fungi are not so plentiful. Though they form a distinctive group, identifying the various species can be difficult because you often have to know the characteristics of both mature *and* immature fruiting bodies, as well as microscopic features such as the size and shape of the spores and capillitium. It is tempting to eat puffballs without bothering to identify them, but indiscriminate sampling of *any* mushrooms—even puffballs—is poor practice, *so at least make an attempt* to key out each kind you find, even if you have only young (or old) specimens at hand. In the following key, the various genera in the Lycoperdales and Sclerodermatales have been grouped into several categories.

678

Amanita "eggs" can look like puffballs, but reveal a mushroom "embryo" of cap, gills, and stalk when sliced open *lengthwise* (perpendicular to the ground). Although this *Amanita calyptrata* "egg" is edible, the "eggs" of several Amanitas are deadly poisonous!

Key to the Lycoperdales & Allies

1. Outer layer(s) of fruiting body splitting into several starlike rays which unfold or bend under (at least in wet weather) to expose the inner skin or spore case ....
.... ***Geastrum, Astraeus, & Myriostoma***, p. 699
1. Not as above (fruiting body may rupture in starlike fashion, but if so then there is no separate spore case within) 2

2. Spore mass containing numerous minute peridioles (spore-bearing chambers) which look like particles of sand; fruiting body small to medium-sized 3
2. Spore mass not containing peridioles, or if so then the peridioles considerably larger than grains of sand (usually appearing more like seeds) 4

3. Spore mass penetrated by a columella (internal stalk); fruiting body with an external stalk also, but the stalk often falling off; found in deserts (often mingling with *Endoptychum arizonicum*)
.... ***Araneosa columellata***
3. Fruiting body puffball-like, i.e., lacking a columella and stalk; found mostly in sandy soil or grassy or open places; widespread but not very common ...... ***Arachnion album*** & others

4. Mature fruiting body sticking out of the ground like a dusty root or half-rotted stump (but may be roundish to pear- or club-shaped when young); spores produced inside numerous small chambers or seedlike bodies (peridioles) imbedded in the fruiting body (the peridioles are best seen by slicing open a young fruiting body because they soon disintegrate); usually found in poor soil, along roads, in deserts, etc. .... ***Pisolithus & Dictyocephalos***, p. 711
4. Not as above; peridioles absent; spores produced in a single large chamber (the spore case)   5

5. Spore case typically hard or tough with a thick rindlike skin, at least when young; spore mass white when very young but soon darkening (usually purple-gray to black) *while remaining firm*, eventually becoming dark brown to blackish and powdery; basidia not borne in a hymenium; capillitium absent .... ***Scleroderma***, p. 707
5. Not as above; skin (peridium) thick or thin; spore mass white when young and normally softening or becoming mushy *as it darkens*, then becoming powdery; basidia usually borne in a hymenium; capillitium usually present 6

6. Fruiting body thick-skinned, not rupturing, usually underground; spore mass revealing a thick short columella (internal stalk) when sectioned lengthwise *through the center* (but columella sometimes disintegrating in old age)—see photos on pp. 761-762 ... (see ***Radiigera***, p. 760)
6. Not as above 7

7. Spore mass with prominent veins or cords running through it, the veins seeming to originate from the base or peridium; rare .... ***Scleroderma***, p. 707
7. Not as above 8

8. Sterile base present, often as a narrowed stemlike base beneath the spore case (section fruiting body lengthwise if unsure) 9
8. Sterile base absent or rudimentary 10

9. Fruiting body medium-sized to quite large, rupturing (in old age) irregularly or through radial tears or general disintegration; peridium (skin) thick or thin ...... *Calvatia* & Allies, below
9. Fruiting body small to medium-sized (usually smaller than a baseball), typically rupturing through an apical pore, slit, or large mouth; usually thin-skinned *Lycoperdon* & Allies, p. 690
10. Fruiting body golfball-sized to very large, rupturing (in old age) irregularly or through radial tears or general disintegration; peridium (skin) thick or thin; found in many habitats .... *Calvatia* & Allies, below
10. Fruiting body usually marble- to golfball-sized or occasionally as large as a baseball, usually rupturing through a pore or large mouth (and often blowing about in the wind when old); peridium usually rather thin; found mostly in grassy or open places *Bovista* & *Disciseda*, p. 696

CALVATIA & Allies   (Giant Puffballs)

Medium-sized to very large terrestrial puffballs. FRUITING BODY round to top- or pear-shaped (broader above) to somewhat flattened. Peridium (skin) two-layered, *disintegrating or rupturing irregularly at maturity,* smooth or warty. STERILE BASE present or absent. SPORE MASS firm and white when immature, then slowly darkening to olive-brown, dark brown, or purple *and becoming powdery or cottony.* Spores typically more or less round, smooth to warted or spiny. Capillitium present.

THESE are baseball- to basketball-sized puffballs that disintegrate in old age, i.e., a distinct apical pore is not formed as in *Lycoperdon.* In some species the outer layer of the peridium takes the form of large warts which break up into plates and then flake off, but in other species it is smooth and in many it adheres to the inner layer so that the two are indistinguishable. Calvatias with a thick peridium are sometimes mistaken for earthballs (*Scleroderma* species), but are usually whiter (both inside and out) when immature and not as hard-fleshed. The texture (whether powdery or cottony) and color of the mature spore mass are important features in the identification of Calvatias, as is the presence or absence of a sterile base.

Calvatias are among the most prolific of living organisms. It has been calculated that an average-sized (30 cm) specimen of the giant puffball *(Calvatia gigantea)* contains 7,000,000,000,000 (7 trillion) spores! In these inflationary times that may not sound like much, but consider this: if all 7 trillion spores (each one measuring 1/200 of a millimeter) were lined up in a row, they would circle the earth's equator! If each spore produced a 30 cm offspring, the resulting puffballs would stretch from the earth to the sun and back, and if their spores were equally successful, the formidable puffball mass would weigh 800 times as much as the earth! Each spore is theoretically capable of germinating, yet very few (obviously!) do. It would be interesting to know why so many don't, or conversely, why such a surplus of spores is (needlessly?) produced.

The truly giant puffballs (the *C. gigantea* group and *C. booniana*) are among the best known and most popular of all edible mushrooms. In fact, they are eaten by people who don't know a gill from a gall. Some of the smaller species (e.g., *C. sculpta* and *C. cyathiformis*) are also excellent, but a few (e.g., *C. fumosa*) are bitter. None are known to be poisonous, but the edible species have purgative or laxative effects on some people. Each kind should be tested cautiously and eaten in moderation, and specimens that have begun to ripen should be discarded.

Calvatia species are partial to arid climates, which is ironic considering the gigantic size of some. It is a large and complex genus, especially in the prairies of the Midwest and the sagebrush deserts and mountains of the West, where many endemic species still await classification. (In our area there are several odd species which I haven't been able to identify to my satisfaction.) Only a handful of the more common or easily recognized Calvatias are described here, and three small, superficially similar genera, *Mycenastrum, Calbovista,* and *Abstoma,* are also treated.

Key to Calvatia & Allies

1. Sterile base absent or rudimentary (make a perpendicular section of the fruiting body if unsure) 2
1. Sterile base present (but sometimes small), often but not always chambered 14

2. Fruiting body large (10-40 cm broad or more unless very young); peridium (skin) thick in immature specimens but thinner in age and disintegrating soon after the spore mass matures or turns powdery 3
2. Not as above; fruiting body small to medium-sized (typically less than 12 cm broad), or if larger then peridium persistent (remaining intact), tough or hard, and thick even at maturity ... 4

3. Fruiting body with large warts or plaques when young, usually depressed-globose (broader than it is tall); found in sagebrush flats and mountains of the West ..... *C. booniana,* p. 684
3. Fruiting body smooth or areolate (cracked) when young (or western form often breaking up into plaques in dry weather), round to somewhat flattened; found mostly in grassy or open areas, roadsides, etc.; widely distributed .... *C. gigantea group,* p. 682

4. Peridium (skin) thick and quite tough or hard, even after spore mass has matured (i.e., skin not readily disintegrating); spore mass *not* purple at maturity 5
4. Peridium thin, fragile, and/or disintegrating soon after spores mature (may be thick when young); spore mass variously colored at maturity, including purple 11

5. Fruiting body typically less than 10 cm broad; found under conifers, especially in mountains; spores *not* reticulate 6
5. Fruiting body consistently larger than above, or if not then usually found elsewhere (in open areas, livestock corrals, etc.) *and/or* spores reticulate 7

6. Peridium (skin) smooth when young or sometimes finely cracked (areolate); fruiting body often with a small cord at base .... *C. fumosa,* p. 688
6. Not as above; peridium with whitish to grayish or smoky-brown warts that eventually break up into plaques and flake off .... *C. subcretacea,* p. 688

7. Outer layer of peridium (skin) a thick, white felty coat which turns grayish and fibrillose and wears away in patches, exposing the hard thick (about 2 mm) brown persistent inner layer; capillitium thorny; often (but not always) found in areas where livestock loiter .... *Mycenastrum corium,* p. 689
7. Not with above features 8

8. Peridium (skin) with distinct warts when young; found in arctic and tundra regions .... *C. cretacea* (see *C. subcretacea,* p. 688)
8. Not as above 9

9. Fruiting body averaging 2-8 cm in diameter 10
9. Fruiting body averaging 5-20 cm or more in diameter .... *C. pachyderma & C. lepidophora* (see *C. fumosa,* p. 688)

10. Peridium (skin) rupturing through radial tears *and/or* spores reticulate .... *Abstoma townei & A. reticulatum* (see *Mycenastrum corium,* p. 689)
10. Not as above; fruiting body usually breaking up irregularly at maturity; spores *not* reticulate .... *C. hesperia* (see *C. fumosa,* p. 688)

11. Mature spore mass purple or dull purple; fairly common in grass or other open areas .... *C. fragilis* (see *C. cyathiformis,* p. 687)
11. Not as above; mature spore mass typically some shade of brown or ochre 12

12. Mature spore mass typically persistent (i.e., rather cottony in texture) .... *C. lycoperdoides* & others, p. 687
12. Mature spore mass typically powdery and easily dispersed 13

13. Peridium (skin) often rupturing through radial tears *and/or* spores reticulate .... *Abstoma townei & A. reticulatum* (see *Mycenastrum corium,* p. 689)
13. Not as above; spores not reticulate .... *C. paradoxa, C. owyheensis,* & others

14. Fruiting body with prominent warts, at least when young; warts large (4-20 mm), polygonal or pyramidal and often lined (i.e., adorned by lines); found in western mountains 15
14. Not as above (but fruiting body may have small warts), or if large warts present then habitat different 16

15. Warts usually pyramidal or pointed when young and sometimes very exaggerated, their tips
     sometimes joined; usually found in forests; capillitium not thorny .... *C. sculpta,* p. 684
15. Warts often flattened or truncated but sometimes pointed, not joined at their tips; usually found
     in open places or edges of woods, roadsides, etc.; capillitium thorny ....
     .... *Calbovista subsculpta* (see *Calvatia sculpta,* p. 684)
16. Outer surface of fruiting body staining yellow when bruised or rubbed, at least when young;
     found mainly in cultivated soil .... *C. rubroflava* (see *C. lycoperdoides,* p. 687)
16. Not as above 17
17. Outer surface of fruiting body with red or reddish spots when fresh; mature spore mass pur-
     plish; found under pine and perhaps other trees; not common ....
     .... *C. rubrotincta* (see *C. cyathiformis,* p. 687)
17. Not with above features 18
18. Spore mass distinctly purple when mature (powdery); common in grass or other open areas
     `... *C. cyathiformis,* p. 687
18. Mature spore mass ochre to brown, etc., but not purple 19
19. Spore mass cottony at maturity and persisting for a long time (remaining intact); peridium (skin)
     smooth or granular to very wrinkled but lacking distinct warts, plaques, or spines; fairly com-
     mon in eastern and southern North America . *C. craniiformis* (see *C. lycoperdoides,* p. 687)
19. Not as above 20
20. Sterile base very prominent and elongated (up to 12 cm long) to form a stalklike base; spore
     case typically less than 8 cm broad and usually smaller in height than sterile base ....
     .... *C. elata* & *C. excipuliformis* (see *C. bovista,* p. 686)
20. Sterile base sometimes prominent but not as above, and/or spore case larger 21
21. Typically found in pastures, lawns, roadsides, and other grassy or open places; common at low
     elevations 22
21. Typically found in arid regions (deserts, sagebrush flats, etc.) or under conifers, sometimes also
     in mountain meadows 23
22. Fruiting body rather small, typically less than 5 (rarely 7) cm broad ....
     .... (see *Lycoperdon* & Allies, p. 690)
22. Fruiting body medium-sized to large (5-25 cm broad) unless very young .. *C. bovista,* p. 686
23. Fruiting body typically developing warts or blunt spines (at least on top) when young or as it
     matures; found under western conifers, mainly in mountains *C. lloydii* (see *C. bovista,* p. 686)
23. Not as above; fruiting body lacking distinct warts and/or habitat different 24
24. Fruiting body averaging 5-10 cm broad; base often wrinkled or furrowed and found in arid
     habitats, or if not wrinkled, then found under conifers .. *C. tatrensis* (see *C. bovista,* p. 686)
24. Not with above features; usually smaller; if growing in arid habitats, then base not usually
     wrinkled, and if growing in woods then base usually wrinkled ....
     .... *C. pallida* & *C. candida* (see *C. bovista,* p. 686)

Calvatia gigantea group   (Giant Puffball)   **Color Plate 184**

FRUITING BODY softball- to basketball-sized (8-60 cm or more in diameter), round
or sometimes lobed or occasionally somewhat flattened on top in age. Outer layer of
peridium (skin) thick when young (2-4 mm), at first smooth (with texture of kid glove) and
pure white or brownish-stained (but in the California version often breaking up into
brownish scales even when young), then cracking up into flat scales or plates which
eventually flake off to expose the thin olive-brown inner layer—which soon disintegrates
—or both layers falling off together. **STERILE BASE** absent or rudimentary, but a cordlike "root" often present. **SPORE MASS** at first white and firm or cheesy, becoming
greenish-yellow and mushy with an unpleasant odor (like old urine), finally deep olive-
brown to brown and powdery. Spores 3.5-6 microns, round or nearly round, smooth or
minutely spiny, apiculate.

HABITAT: Solitary, scattered, or in groups or large circles in fields, pastures, open

Immature giant puffballs, *Calvatia gigantea* group. In the fall or spring, a casual jaunt through a "puffball pasture" can yield quite a haul—the skillet in the foreground is one foot in diameter! For scalier specimens, see photo on p. 678. (Bill Everson)

woods, cemeteries, on exposed hillsides, along roads, in drainage ditches, etc.; fairly common in eastern North America and the Midwest. The west coast form, which is slightly different (see comments), is sometimes abundant in our area after the first fall rains and again in the spring. In fact, when conditions are favorable it is not unusual to find 30-40 pounds on a casual jaunt through a "puffball pasture." Because of its preference for open hillsides, it can often be spotted from the road. Large specimens, in fact, have been mistaken by passersby for herds of grazing sheep! (Mushroom hunters, on the other hand, are more likely to mistake grazing sheep for giant puffballs.) Dried specimens found under houses have been mistaken for bleached skulls, while a sinister-looking individual found in England during the war was labelled "Hitler's Secret Weapon" and used for propaganda purposes at an exhibition to raise war funds!

EDIBILITY: Edible and choice when the flesh is firm and white, but with laxative effects on certain individuals. It can be sliced and fried like pancakes, or better yet, cubed like tofu and dropped into clear soups or eaten raw in salads. The tough outer skin should be peeled and those which have begun to ripen should be discarded. Size it *not* necessarily an indication of maturity, so slice them open in the field. This will enable you to check for maggots, which are fanatically fond of them. Infested areas can be trimmed away and the solid portions carried home. Dried giant puffballs have been used as sponges, tinder (before matches were invented), toys, and dyes. They were burned under beehives to stupefy bees and used to squelch bleeding.

COMMENTS: The giant puffball is one of the best known and most familiar of all the fleshy fungi—and it *is* fleshy—5 foot, 50 pound specimens have been recorded! (Alas, the largest one I've found weighed "only" 7 pounds.) The exact identity of our local giant puffball, strangely enough, is a minor mystery. It does not seem to be either the "true" *C. gigantea* of Europe and eastern North America, or *C. booniana* of arid regions. The former is smoother and whiter than our giant puffball, while the latter is wartier and broader. To most people, however, its "true" identity is an academic problem best left to the puffball pundits, who are paid to pore over such matters—after all, any large puffball is a *giant* puffball, and any *giant* puffball is a giant meal! *Langermannia gigantea* and *Calvatia maxima* are synonyms for *C. gigantea*. *C. bovista* can also be quite large, but has a prominent sterile base, while *C. cyathiformis* has purple spores at maturity. Also see *C. pachyderma* and *C. lepidophora* (under *C. fumosa*).

683

Immature western giant puffball, *Calvatia booniana.* Note the shape (longer than it is high) and large warts on the surface. See color plate for mature specimens. (Chuck Barrows)

Calvatia booniana   (Western Giant Puffball)   Color Plate 186

FRUITING BODY 15-60 cm or more broad and 7-30 cm or more high; sometimes round or lobed, but more often somewhat flattened or depressed on top. Outer layer of peridium (skin) thick, white to buff or tan and finally brown; at first sculptured with large warts which soon separate to form flattened scales, plaques, or plates and eventually disintegrate along with the thin inner layer. **STERILE BASE** absent or rudimentary. **SPORE MASS** firm and white at first, then turning yellow or greenish and mushy and stinky, finallly becoming powdery and olive-brown to brown. Spores 4-6.5 × 3-5.5 microns, round or nearly round, smooth or minutely spiny.

HABITAT: Solitary, widely scattered, or in groups or "flocks" in fields, under sagebrush or juniper and in other open areas; confined to the arid and semi-arid regions of western North America, and sometimes common, especially in the late spring and summer. I have seen it in New Mexico and Idaho, and in the mountains of southern California. A similar but less warty species is common along the west coast (see the *C. gigantea* group).

EDIBILITY: Edible and choice when firm and white inside, but with laxative effects on certain individuals. Like *C. gigantea,* it was apparently eaten by pioneers as well as by some Native Americans. Be sure to check for maggots when gathering them, and remember: any specimens whose flesh shows the *slightest* traces of color (yellow, greeen, or brown) are no longer good to eat (unless you're a maggot!).

COMMENTS: This giant puffball is approximately the same size as *C. gigantea* (50 lb. specimens have been reported), but is not normally as round and has larger plaques or warts—even when young. *Calbovista subsculpta* (see comments under *Calvatia sculpta*) is somewhat similar but much smaller (softball- or grapefruit-sized).

Calvatia sculpta   (Sierran Puffball; Sculptured Puffball)

FRUITING BODY more or less egg-shaped, pear-shaped, top-shaped, or somewhat irregular, (4) 7-18 cm high and/or broad. Outer layer of peridium composed of long, pointed, white pyramidal warts (up to 2.5 cm long) which are erect *or* bent and joined at their tips; warts often lined or grooved and arising from angular plaques which soon crack and fall away, exposing a fragile inner layer which soon disintegrates. **STERILE BASE** present, often prominent but sometimes inconspicuous, white to yellowish, often with

684

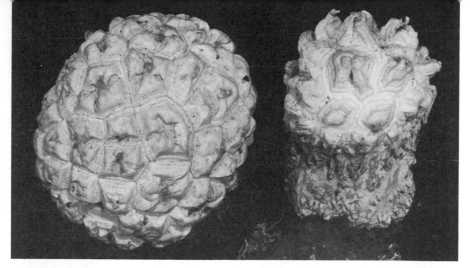

The Sierran puffball, *Calvatia sculpta,* is easily recognized by its enormous pyramidal or polygonal warts. Note sterile base beneath spore case (right); specimen on left is being viewed from the top.

a purplish interior (especially in age). **SPORE MASS** white and firm at first, turning yellow and then deep olive-brown as it ripens, eventually powdery. Spores 3.5-6.5 microns, round or nearly round, minutely spiny.

HABITAT: Solitary or in small groups under conifers or sometimes in the open; known only from the mountains of the West. It is fairly common in the Sierra Nevada in the late spring, summer, and fall; I have not seen it on the coast.

EDIBILITY: Edible when immature, and better than most puffballs.

COMMENTS: This is easily the most spectacular of all the puffballs, and is well known to hikers in the Sierra Nevada. The white pyramidal "peaks" make fresh specimens look like a cross between a geodesic dome and a giant glob of meringue. *Amanita magniverrucata* can sometimes resemble it superficially (especially in the egg stage), but has gills and a stalk. Other species: ***Calbovista subsculpta*** is a common edible western mountain puffball that is similar in size but has flatter (but sometimes pointed), less flagrant warts. It used to be placed in *Calvatia,* but its capillitium have thornlike branches—a momentous enough difference in the eyes of puffball specialists to merit a genus of its own. It is quite common in the spring, summer, and early fall in open and grassy places or at the edges of woods in the Sierra Nevada, Cascades, and other western mountains.

Left: *Calvatia sculpta* in the bush (Bob Winter). **Right:** *Calvatia sculpta* in the hand (Nancy Jarvis).

Calvatia bovista, immature specimens. Note large size and prominent sterile base. Specimen in center is being viewed from above, hence the sterile base is not visible.

Calvatia bovista

FRUITING BODY 10-25 cm high and 5-25 cm broad, top-shaped or pear-shaped with a broad, often flattened top in age and a large, prominent stemlike sterile base. Outer layer of peridium (skin) white to grayish or occasionally yellowish-brown, covered with pointed scurfy warts or soft particles, slowly breaking up into flat scales that slough off, exposing the thin inner layer which soon disintegrates. **STERILE BASE** very large, constituting up to one half the total height of the fruiting body; chambered; exterior white when young, brown in age; smooth, persisting long after the spore case has decomposed. **SPORE MASS** white and cheesy, then yellow or olive and finally olive-gold to olive-brown or dark brown and powdery. Spores 4-6.5 microns, round, minutely warted or spiny.

HABITAT: Solitary, scattered, or in groups in pastures, exposed soil, open woods, etc.; widely distributed. It is fairly common in our coastal pastures from fall through spring, sometimes mingling with the giant puffball *(Calvatia gigantea* group*)*.

EDIBILITY: Edible when immature, but rather soft and in my experience insipid or even bitter. According to puffball scholar William Burk, it has been used to stop up holes in drafty dwellings as well as nosebleeds.

COMMENTS: Also known as *C. utriformis* and *C. caelata,* this is the second largest of our local puffballs and is easily recognized by its flattened top and large sterile base (see photograph). It looks something like a gigantic *Lycoperdon perlatum,* but does not form a pore at the top. The bowl-shaped base persists long after the rest of the spore case has disintegrated, and old bowls are often found filled with rainwater, mosquito larvae, and spores, or they may be windblown and completely empty. In the latter condition you may not recognize them as puffballs, but don't let this faze you—a special genus, *Hippoperdon,* was once erected based on these empty bowls!

    C. excipuliformis (=C. saccata) and *C. elata* are two similar species which have a narrower spore case (3-10 cm broad) that is granular or coated with small spines and pale buff to brownish, plus a narrower and greatly elongated (stemlike) sterile base. Both are widespread in wooded and open areas; the latter has practically smooth spores. *C. tatrensis* is a western species with a purplish to purple-brown sterile base; it occurs under sagebrush and in other arid waste places, or occasionally under conifers, and usually has a wrinkled or furrowed base. *C. pallida* and *C. candida* are two small species (up to 5 cm broad) with a sterile base and powdery spore mass; the former has a wrinkled base and grows in woods and mountain meadows, while the latter prefers more arid habitats. Finally, there is *C. lloydii,* a fairly common species in the dry coniferous forests of the Sierra Nevada and other western mountains. It is also rather small (less than 10 cm broad) and has a sterile base, but features warts or blunt spines on the top, at least in age. It sometimes grows with *C. fumosa and C. subcretacea,* but is easily separated from those species by its thinner skin, and from *C. sculpta* and *Calbovista subsculpta* by its smaller warts.

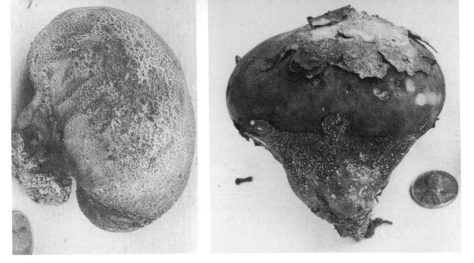

Calvatia cyathiformis. **Left:** Immature specimen (turned sideways) with a well-developed sterile base. **Right:** Mature specimen in which the outer layer of the peridium (spore case) is disintegrating. The purplish color of the mature spores is the most distinctive characteristic of this species.

Calvatia cyathiformis (Purple-Spored Puffball)

FRUITING BODY 5-20 cm high and/or broad; nearly round when young, becoming top-shaped or pear-shaped, or round with a flattened top and narrowed base. Outer layer of peridium (skin) smooth at first, but soon cracking into small, flat scales or patches, at least on top; white to tan or pinkish-tan becoming purplish or purple-brown in age; inner layer dark purple or purple-brown, smooth, thin and delicate; both layers flaking off in old age. **STERILE BASE** usually present (but see comments), chambered, white to dingy yellow or darker, persisting as a deep purplish to purple-brown cuplike structure after the spores have dispersed. **SPORE MASS** firm and white when young, becoming yellowish, then brownish and finally dull purple and powdery. Spores 3.5-7.5 microns, round, spiny or warty to nearly smooth.

HABITAT: Widely scattered to gregarious or in rings in pastures and other grassy places, widely distributed. It is common locally in the fall, occasional at other times.

EDIBILITY: Edible and quite good when firm and white inside. It is not quite as tasty as the giant puffball, but is not as rich either.

COMMENTS: The striking purple color of the mature spore mass sets apart this *Calvatia* from its brethren. It frequents the same habitats as *C. bovista* and our version of *C. gigantea,* but there is normally only one major fruiting (at least locally)—in the fall. It is smaller and firmer than *C. bovista,* and usually darker in color than the *C. gigantea* group. Fairy rings of this species in the prairies of Colorado are estimated to be 420 years old—or older than most trees! *C. cyathiformis form fragilis* (also called *C. fragilis*) is very similar, but has only a rudimentary sterile base; it is also common in grassy areas on the west coast, and is widely distributed. Other species: *C. rubrotincta* has purplish spores, but is pallid with red spots when young; it has been found in Oregon under ponderosa pine.

Calvatia lycoperdoides (Cotton-Spored Puffball)

FRUITING BODY 1.5-5 cm high and broad, more or less round to somewhat cushion-shaped. Outer layer of peridium (skin) white or pallid becoming brownish in age, at first with soft granules, flakes, or spines (or often with small warts on top); adhering to the inner layer and breaking up into large flakes at maturity. **STERILE BASE** absent or rudimentary. **SPORE MASS** firm and white when young, olive-brown to brown at maturity, with a cottony texture that causes it to remain intact for a long time. Spores 4-6.5 microns, round, with small warts.

687

HABITAT: Solitary or in small groups on ground in woods and under trees; known only from western North America, but *C. craniiformis*(see comments) is common in the eastern states. It can be found in our area in the late fall, winter, and spring, but is not common.

EDIBILITY: Presumably edible when immature, but not big or plentiful enough to warrant collecting. I haven't tried it.

COMMENTS: This is one of several Calvatias with a cottony(instead of powdery) mature spore mass. Others include: *C. umbrina,* a dark brown to black species; *C. diguetii,* a smooth-spored westerner with an ochre gleba; and *C. rubroflava*, with a greenish-orange mature spore mass and yellow-staining exterior, found mostly in cultivated soil(especially in southern latitudes). In these species an apical pore is not formed as in *Lycoperdon,* and there is no sterile base(except in *C. rubroflava*). Another species with a cottony spore mass, *C. craniiformis,* is often wrinkled, softball-sized, and white to tan when young, with a well-developed sterile base and yellow-green to yellow-brown spores. It is quite common in southern and eastern North America in open areas and under hardwoods, but to my knowledge does not occur on the west coast. It is edible, like most Calvatias, when firm and white inside. There are several other species with a sterile base that may or may not have a cottony spore mass when mature. These include *C. elata, C. excipuliformis,* and *C. lloydii*(see comments under *C. bovista* for more details).

Calvatia subcretacea   (Small Warted Mountain Puffball)

FRUITING BODY golfball- to baseball-sized, 1.5-5 (7) cm in diameter, round or more often cushion-shaped to somewhat flattened. Peridium(skin) thick and white when fresh, the outer layer composed of white to smoky-gray or grayish-tipped warts which break up into plates or polygons and then flake off, exposing the yellowish to brownish inner layer, which breaks up irregularly. **STERILE BASE** absent or rudimentary. **SPORE MASS** white and firm becoming yellowish, then olive-buff to olive-brown or brown and powdery. Spores 3.5-6.5 microns, round or nearly round, smooth or minutely ornamented.

HABITAT: Solitary to widely scattered or in small groups in duff under mountain conifers, especially spruce and fir; fairly common in the mountains of western North America from late spring through early fall. I have seen it several times in the Sierra Nevada (often with *C. fumosa*), but never on the coast.

EDIBILITY: Edible when young, but difficult to find in sufficient quantity for a meal.

COMMENTS: The modest size, warty peridium, and growth under mountain conifers help to distinguish this commonplace *Calvatia* (see photo at top of p. 689). It is not as smooth-skinned as *C. fumosa,* and is usually much smaller than *Calbovista subsculpta* (see comments under *Calvatia sculpta*). Mature specimens lack the apical pore characteristic of *Lycoperdon,* and their habitat is also distinctive; Sclerodermas are different in color and texture. *C. cretacea* is a similar but larger species of arctic and tundra regions.

Calvatia fumosa

FRUITING BODY golfball- to baseball-sized, round to oval, 3-8 (10) cm broad. Outer and inner layers of peridium (skin) adhering to each other, thick (1-5 mm) and persistent (not disintegrating); at first smooth and white, but soon becoming grayish to brownish and often areolate (cracking to form small scales), the undersurface and cracks white. **STERILE BASE** absent or rudimentary, but a mycelial cord often present. **SPORE MASS** firm and white at first, then yellowish or olive, finally dark brown and powdery (or yellow-brown in one variety); odor often unpleasant ("a combination of sour milk, diesel oil, and pit toilet"—Robert Ramsey). Spores 4.5-8 microns, round or nearly round, spiny.

Calvatia fumosa (two specimens in center) and *Calvatia subcretacea* (specimens at far left and far right) are common under mountain conifers and sometimes grow together. Both have a thick, tough skin. *C. subcretacea* has a warty exterior, while *C. fumosa* is smoother and has a cord at the base.

HABITAT: Solitary to gregarious in duff (sometimes buried) under spruce, fir, and other mountain conifers; common in the western United States, spring through early fall. A similar species occurs in our area under cypress, and others occur in cultivated or hard-packed soil (see comments).

EDIBILITY: Bitter-tasting according to Orson K. Miller; I haven't tried it.

COMMENTS: The thick, tough, persistent skin (peridium) and modest size plus the absence of warts distinguish this *Calvatia* from most others. Old specimens might be mistaken for Sclerodermas, but possess capilliitum and are differently colored. The peridium seldom disintegrates of its own accord, but rodents are apparently fond of chewing on it, thus enabling the spores to escape. There are several similar Calvatias with a thick, persistent peridium, including: *C. hesperia,* similar in size but white to grayish, with smooth spores and growing mostly in open places (farmland, deserts, etc.); *C. pachyderma,* larger (5-15 cm in diameter) and whitish when young, with smooth to minutely warted spores, found in open, cultivated, and arid places; and *C. lepidophora* of the Midwest prairies, which is even larger (15-20 cm) and has densely warted spores. See also *Mycenastrum corium.*

Mycenastrum corium

FRUITING BODY 4-20 cm broad or more, round to somewhat pear-shaped when young, eventually rupturing in irregular fissures to form rays or plates which may bend back somewhat in a star-shaped pattern. Outer layer of peridium (skin) a thick, white, felty coat which becomes areolate (separates into blocklike areas), forming thin, grayish, fibrillose patches which eventually wear away to expose the tough, hard, persistent, smooth inner layer, which is brown to purple-brown and about 2 mm thick. **STERILE BASE** rudimentary or absent, but mycelial fibers often present. **SPORE MASS** firm and white becoming olive-yellow to olive-brown and finally dark brown to purple-brown and powdery. Spores 8-12 microns, round, warted-reticulate. Capillitium branched, thorny.

HABITAT: Scattered to gregarious on ground (sometimes partially buried) in horse corrals, composted areas, and fields where livestock have been grazing; widely distributed, but especially common in the West. In our area it occurs year-round. The tough spore cases persist for months, sometimes breaking loose to blow about in the wind.

EDIBILITY: Presumably edible when firm and white inside; I haven't tried it.

Mycenastrum corium, immature specimens. These were found in a horse corral, a favorite haunt of this species. Note the thick skin and white felty material on exterior.

COMMENTS: This peculiar puffball is easy to recognize but difficult to describe. The thick, tough inner peridium (skin) distinguishes it from *Bovista* and the thin-skinned Calvatias, while the white, felty outer layer separates it from *Scleroderma* and the thick-skinned Calvatias. The presence of capillitium in the spore mass indicates a much closer kinship to the true puffballs *(Calvatia,* etc.) than to the earthballs *(Scleroderma).* Its tendency to fruit in localities where livestock loiter is another helpful (but fallible) field-mark. *A bstoma townei* and *A. reticulatum* are smaller and dirtier puffballs (2-6 cm broad) with reticulate spores and unbranched capillitium. The first is said to occur in old pastures and other waste places; the latter has been found in coastal California under cypress.

LYCOPERDON & Allies   (Common Puffballs)

Small to medium-sized puffballs found mostly on rotten wood or on ground in woods *(Lyco-perdon),* or in grass *(Vascellum).* FRUITING BODY round to pear-shaped or top-shaped; peridium two-layered, the outer layer usually with spines, warts, granules, or particles; *usually rupturing through an apical pore.* STERILE BASE *usually present and often conspicuous or stem-like.* SPORE MASS firm and white when young, becoming powdery and olive-brown to brown or purplish in old age. Spores more or less round, smooth to warted or spiny, sometimes pedicellate. Capillitium present *(Lycoperdon)* or replaced by paracapillitium (septate, colorless hyphae) in *Vascellum.*

THESE are small to medium-sized puffballs that rarely exceed 10 cm (4 inches) in diameter. In contrast to *Calvatia* and *Scleroderma,* the spores are usually released through an **apical pore** (a hole, slit, or mouth that forms at the top of the *mature* spore case). The frequent presence of a well-developed, often stemlike sterile base distinguishes *Lycoperdon* from *Bovista,* but the two genera intergrade through a series of forms with a *slight* sterile base. In most species the spore case is initially coated with a layer of spines, warts, or fine particles, but these eventually fall away to expose the inner layer of the peridium, in which the pore forms.

   Lycoperdons are our most common woodland puffballs, but also grow in open areas, waste places, and sawdust piles. In our area the major fruiting is in the fall and winter, but old weathered specimens can be found at any time. All Lycoperdons are thought to be edible when firm and white inside, but some taste better than others and it is imperative to discard any specimens that have begun to ripen. (In Charles McIlvaine's words, "one ageing *L. pyriforme* will embitter a hundred.") The various species are rather difficult to distinguish—particularly when immature—but it is *always* a good practice to identify each type before eating it. Five representative species are described here, plus one species of *Vascellum,* a small "satellite" genus that differs microscopically. Two other small genera, *Bovistella* and *Morganella,* are included in the key.

Key to Lycoperdon & Allies

1. Exterior of fruiting body covered with dark brown to black spines *when young;* yellow tones often developing in age .... *L. foetidum,* 692
1. Not as above; fruiting body not dark brown when young (but may be pale to medium brown when young and become dark brown in old age) 2

2. Growing on wood, sawdust, or lignin-rich humus (if in humus, then base with white mycelial threads or rhizomorphs and spore case with inconspicuous spines if any) 3
2. Not as above; usually terrestrial 4

3. Fruiting body with cinnamon-buff to brown spines when young, conspicuously pitted at maturity (after spines have worn off); sterile base well-developed to practically absent; found mainly in eastern North America   *Morganella subincarnata (=Lycoperdon subincarnatum)*
3. Not as above; fruiting body never pitted, usually with white mycelial threads (rhizomorphs) at base or in surrounding substrate; sterile base well-developed; common and widespread .... *L. pyriforme* & others, p. 691

4. Fruiting body often broader than it is tall, white or pinkish-tinged when young; outer layer of skin peeling away *in sheets* at maturity (see photo on p. 695) **L. marginatum** & others, p. 694
4. Not as above (if peeling in sheets, then much taller than it is broad) 5

5. Fruiting body 3-12 (14) cm broad, with a narrowed rooting base; outer layer of peridium (skin) composed of spines and granules, the spines often tufted or joined at their tips; usually in open, sandy, or cultivated ground ... **Bovistella radicata** (see *Lycoperdon pulcherrimum*, p. 694)
5. Not with above features; rooting base absent and/or fruiting body considerably smaller .. 6

6. Fruiting body golden-orange to bright yellow when young . (see **Bovista & Disciseda**, p. 696)
6. Not as above; differently colored 7

7. Fruiting body lavender-tinged or with lavender-tinged spines when young; found mainly in eastern North America .... **L. peckii** (see *L. perlatum*, p. 693)
7. Fruiting body not lavender-tinged when young 8

8. Typically growing in grass, prairies, and other open areas; either paracapillitium present *or* capillitium with small round pits 9
8. Not as above; typically growing in woods and at their edges, under trees, on roadsides, etc. 11

9. Sterile base very small or rudimentary; spore case rupturing through an apical pore; usually densely gregarious or in clusters .... **Vascellum curtisii** (see *V. pratense*, p. 695)
9. Not as above 10

10. Fruiting body typically with a large mouth at maturity (the top disintegrating); very common on lawns, golf courses, etc. .... **Vascellum pratense** & others, p. 695
10. Fruiting body typically rupturing through a slit or tear at maturity; widespread but not common .... (see **Bovista & Disciseda**, p. 696)

11. Spines on exterior of young fruiting body 2-6 mm long and often joined at their tips 12
11. Spines absent, or if present not as above (usually shorter and sparser) 13

12. Spines white becoming brown, leaving small scars or pockmarks on the inner peridium when they fall off .... **L. americanum** (see *L. pulcherrimum*, p. 694)
12. Spines remaining white until they fall off, not leaving scars (i.e., inner peridium smooth) .... **L. pulcherrimum**, p. 694

13. Outer layer of peridium (skin) sloughing off in sheets or chunks at maturity; known only from the Pacific Northwest; rare (?) .... **L. nettyana** (see *L. marginatum*, p. 694)
13. Not as above; very common and widespread 14

14. Mature spore mass olive-brown to brown; spines leaving pockmarks on inner peridium when they fall away (but marks may disappear, leaving inner peridium smooth) **L. perlatum**, p. 693
14. Mature spore mass purple-brown; spines not leaving scars .... **L. umbrinum** (see *L. perlatum*, p. 693)

Lycoperdon pyriforme   (Pear-Shaped Puffball)

FRUITING BODY pear-shaped to nearly round, but usually with a stemlike sterile base; 1.5-5 cm high and sometimes almost as broad in the widest part. Peridium (skin) whitish to pale brown when young, yellowish to dark rusty-brown in age; at first smooth or with a few small scattered spines on top, then becoming finely cracked to form small patches or minute granules or particles (making it rough to the touch), this rough outer layer slowly but eventually falling away to expose the smooth inner layer in which an apical pore or tear is very slow to form. **STERILE BASE** small or well-developed, spongy when fresh, occupying the stemlike base (if base is present); chambers very small; conspicuous white mycelial threads (rhizomorphs) usually radiating from the base and connected to others in the surrounding wood or humus. **SPORE MASS** at first firm and white, then yellow to olive and finally deep olive-brown and powdery. Spores 3-4.5 microns, round, smooth.

HABITAT: Scattered to densely gregarious or clustered on stumps, rotting logs, sawdust, and in lignin-rich humus; widely distributed and common, fruiting mostly in the fall and winter in our area, but old bleached-out fruiting bodies can be found most any time. It sometimes forms dense clusters "as large as a loaf of bread" (to borrow a phrase from Alexander Smith).

Lycoperdon pyriforme, immature specimens. Note absence of prominent spines, the narrowed sterile base, plus the clustered growth habit and white mycelial threads.

EDIBILITY: Edible when young, but only worth collecting when it occurs in quantity. In my fickle fungal opinion it is one of the better puffballs, but is not as good as "a loaf of bread" and is apt to be bitter if not absolutely white and firm inside.

COMMENTS: The tendency to grow on rotting wood is a distinctive feature of this pear-shaped puffball, but it often *appears* to be terrestrial (when growing from buried wood or humus rich in lignin). The white rhizomorphs or "roots" that emanate from the base of the fruiting body plus the narrowed or stemlike base and absence of prominent spines are also good fieldmarks. It is one of the few Lycoperdons that occurs in sufficient quantity to merit collecting for the table. Other species: ***L. pedicellatum*** also grows on rotten wood, but has longer spines and ornamented spores.

Lycoperdon foetidum   (Dark Puffball)

FRUITING BODY pear-shaped to nearly round, but usually with a narrowed base; 1.5-6 cm high and 1.5-4 cm broad. Outer layer of peridium (skin) composed of fine pointed black to dark brown spines (especially dense on top) interspersed with granular material, the spines persisting or eventually falling away; background (inner layer) thin, grayish-tan to yellowish (usually distinctly yellow in age), developing an apical pore or tear at maturity. **STERILE BASE** well-developed, chambered and spongy when fresh; exterior usually paler than the rest of fruiting body, at least when young. **SPORE MASS** white when young, then yellow, finally dull cinnamon-brown to dark brown or sepia, and powdery. Spores 4-5 microns, round, minutely spiny.

Lycoperdon foetidum, immature specimens. Note the dark brown to blackish color—its principal fieldmark. A pore forms at the top in old age.

HABITAT: Solitary, scattered, or in groups in humus and debris in deep woods along the west coast (also Europe). Fairly common in our area from fall through early spring—especially under conifers—but easily overlooked because of its dark color.

EDIBILITY: Presumably edible when firm and white inside; I haven't tried it.

COMMENTS: Also known as *L. nigrescens,* this attractive puffball is our only *Lycoperdon* with dark brown to black spines *when immature* (several species may be quite dark in age, however—see *L. perlatum*). The dark spines contrast nicely with the white flesh, and the yellowish background color that develops in age is also distinctive. The species epithet is something of a misnomer, for I have never found it to have an odor other than the usual slightly unpleasant smell that all ripening puffballs develop.

Lycoperdon perlatum   (Common Puffball; Gemmed Puffball)

FRUITING BODY pear-shaped or top-shaped (broader above) or with a flattened top, or at times practically round with a narrowed, often wrinkled stemlike base; 1.5-6 (9) cm broad, 3-7 (10) cm high. Outer layer of peridium (skin) consisting of slender, short, cone-shaped spines interspersed with smaller spines or granules, the larger ones leaving scars or pockmarks when they fall off; spines white to gray or in one form brown. Inner layer at first with scars, but often smooth in old age, white to tan becoming yellowish-brown to grayish-brown or dark brown in old age; eventually rupturing through a pore at the top. **STERILE BASE** large, well-developed, chambered, forming a stalklike base beneath the spore mass; at first white and spongy, then yellow, olive, brown, or chocolate-colored. **SPORE MASS** firm and white but soon becoming soft and turning yellow, then olive, and finally becoming dark olive-brown to chocolate-brown or brown and powdery. Spores 3.5-4.5 microns, round, minutely spiny.

HABITAT: Solitary, scattered, gregarious, or clustered on ground in woods and under trees, along roads, or sometimes in the open; very widely distributed. It is probably the most abundant woodland puffball in North America, and our region is no exception. It can be found most any time in our area, but usually fruits in the fall or winter.

EDIBILITY: Edible when firm and white inside, but specimens showing the slightest traces of yellow should not be eaten. It occasionally occurs in enough quantity to merit collecting, but in my experience it is bland at best and bitter at worst.

Lycoperdon perlatum, maturing specimens. If you look closely you can see some of the spines and the small scars they leave when they fall off. Younger specimens usually lack the apical pore that is beginning to form in these; older ones are often smooth (without scars or pockmarks).

COMMENTS: Also known as *L. gemmatum,* this cosmopolitan puffball is easily told by its white to gray or brownish, slender, cone-shaped spines which leave small scars behind when they fall off. As pointed out in the description, however, the pockmarks may eventually disappear. It is one of the most variable of all puffballs, especially in color and size, but the well-developed sterile base, dark olive-brown mature spore mass, and ubiquitousness help to identify it. Very large specimens are occasionally found, but these can be distinguished from *Calvatia* by the apical pore which forms in old age. There are several similar species, including: *L. umbrinum (=L. molle),* widely distributed and common, with spines that do not leave scars plus a dark brown to purple-brown mature spore mass and a purplish-tinged sterile base at maturity; *L. muscorum,* small and rare, found in deep moss; *L. peckii,* with lavender-tinged spines when young; and *L. rimulatum,* with purplish spores and a nearly smooth peridium. See also *L. pyriforme.*

Lycoperdon pulcherrimum   (Long-Spined Puffball)

FRUITING BODY more or less pear-shaped or rounded above with a narrowed, stem-like, often wrinkled base; 2-5 cm broad and high. Outer layer of peridium (skin) with a dense coating of long (3-6 mm), slender, white spines which typically form numerous fascicles (by uniting at their tips); spines eventually wearing away to expose the smooth brown to dark purple-brown inner layer; pore forming at the top in age. **STERILE BASE** occupying up to one half the height of fruiting body, chambered, white when young, brown to purple-brown in age. **SPORE MASS** white and firm at first, becoming yellow and eventually dark purple-brown and powdery. Spores 4-4.5 microns, round, minutely spiny, with a long pedicel ("tail") 10-13 microns in length.

HABITAT: Solitary or in small groups in humus or on very rotten wood, usually under hardwoods. It is not uncommon in eastern North America (especially the southern states), and I occasionally find it (or something very similar) in our area in the fall and early winter.

EDIBILITY: Edible when young, but too small and infrequent to be of value.

COMMENTS: The long white spines that are frequently joined at their tips distinguish this small puffball from the more common *L. perlatum* and other small species. *L. echinatum* (now called *L. americanum* by some puffball pundits) is a similar but more common species with white spines that soon turn brown and leave small marks or reticulations on the inner layer of the spore case when they fall off. A third puffball that often has united —albeit shorter—spines, *Bovistella radicata,* strikes a much different pose: it is larger (3-10 (14) cm), with a thick, rooting base and preference for disturbed or open ground. Though widespread, it is not particularly common, at least in our area.

Lycoperdon marginatum   (Peeling Puffball)

FRUITING BODY at first round to somewhat flattened, at maturity often broader than it is tall, the underside usually wrinkled (at least in age) and sometimes with a short, rooting base; 1-5 cm in diameter. Peridium (skin) white or tinged pinkish when young, the outer layer composed of short erect warts or spines which peel off *in sheets,* exposing the smooth to slightly scurfy or faintly pitted, pale to dark olive-brown inner layer; an apical pore eventually forming. **STERILE BASE** chambered, usually well-developed but sometimes inconspicuous. **SPORE MASS** white and firm at first, then olive to grayish-brown and eventually powdery. Spores 3.5-5.5 microns, round, minutely ornamented or smooth, sometimes with a broken pedicel ("tail").

HABITAT: Scattered or in groups or clusters on ground, usually in sandy soil or waste places; widely distributed. I find it regularly in sandy soil under oak, pine, madrone, and manzanita, usually in the late fall and winter.

Lycoperdon marginatum, maturing specimens. Note how the outer layer of the spore case is peeling away *in sheets.* Exterior is white or pinkish-tinged when immature, browner in age.

EDIBILITY: Edible when firm and white inside. In Mexico, this species and *L. miztecorum* are used to induce auditory hallucinations. It is known as "gi-i-sa-wa," meaning "fungus of the second quality" (*L. miztecorum* being the fungus of first quality). However, no intoxicating substances were found when these puffballs were analyzed.

COMMENTS: The peculiar manner in which the outer layer of the peridium separates completely from the inner layer and peels off in sheets is the hallmark of this pleasing puffball. *L. candidum* is a synonym. *L. nettyana,* known only from the Pacific Northwest, also peels off in sheets, but is more or less pear-shaped (with a well-developed, stalklike sterile base). *Vascellum (=Lycoperdon) curtisii* is also quite similar, but usually grows in grass and has little or no sterile base. *L. rimulatum* (see comments under *L. perlatum*) sometimes peels off in sheets, but has a smooth (spineless) peridium.

Vascellum pratense   (Field Puffball)

FRUITING BODY typically top-shaped (broader above) or round with a narrowed base, often wrinkled somewhat below; 2-6 cm high and 2-4 cm broad. Outer layer of peridium (skin) scurfy from a coating of particles or fine spines (especially on top) which disappear in age, exposing the smooth inner layer; white when young, yellowish-tan to metallic brown in age (or the inner layer grayish); rupturing at the top to form a large pore that soon widens so that only the base and lower portion of the spore case remain, forming a "bowl." **STERILE BASE** usually well-developed, chambered, white becoming brownish or purplish in age, separated from the spore mass by a thin membrane (but see comments). **SPORE MASS** white but rather soft when young, then greenish-yellow and finally olive-brown and powdery. Spores 3-5 microns, round, minutely warted, more or less apiculate.

HABITAT: Widely scattered to gregarious in grassy places—lawns, golf courses, pastures, etc; widely distributed, but especially common on the west coast (along with *V. lloydianum* —see comments). In our mild climate it occurs year-round, but is most abundant in the fall. I have also seen large fruitings in Oregon.

EDIBILITY: Edible when immature, but of mediocre quality.

Vascellum pratense is a small grassland puffball with a conspicuous sterile base. Specimens at left are immature; bowl-shaped specimen at right is quite old.

COMMENTS: This undistinguished grassland puffball can be distinguished by its modest size, the presence of a sterile base and soft, deciduous, scurfy spines or granules, and the bowl- or urnlike remains of ruptured specimens. It is often found in the company of *Bovista plumbea,* which is rounder and smoother and lacks a sterile base. It used to be called *Lycoperdon hiemale,* but was transferred to the genus *Vascellum* because of microscopic differences. However, now that taxonomists have settled upon its genus name, they cannot seem to agree on a species epithet. Some hold out for *V. pratense,* others for *V. depressum,* while Alexander Smith has described the common western variant as a "new" species, *V. lloydianum,* because the membrane separating the spore mass from the sterile base is not very distinct. It hardly matters what you call it, however, since it belongs to that vast army of fungi that are "better neglected than collected." Other species: *V. (=Lycoperdon) curtisii* is a small (0.5-2 cm) widespread species with numerous spines when young and a distinct apical pore at maturity. It somewhat resembles *Lycoperdon marginatum* when fresh, but usually grows gregariously in grass.

BOVISTA & DISCISEDA   (Tumbling Puffballs)

Small or occasionally medium-sized puffballs *found mostly in grassy or open areas.* FRUITING BODY *more or less round to somewhat flattened* or irregular; peridium two-layered, rupturing to form a large mouth at the top *(Bovista),* or often rupturing basally *(Disciseda); inner layer usually thin and papery when mature; outer layer usually lacking distinct spines or warts, persistent in Disciseda, disappearing in Bovista.* STERILE BASE *typically absent or rudimentary.* SPORE MASS firm and white becoming powdery and brown to yellowish. Spores round to oval or elliptical, minutely warted or spiny or smooth, often pedicellate. Capillitium present, typically branched.

THESE small (marble- to baseball-sized) puffballs usually lack the sterile base characteristic of *Lycoperdon* and are smaller and/or smoother than most *Calvatia* species. In old age they often break loose from the soil so that the thin, lustrous, papery spore cases tumble about freely in the wind.

Bovista species release their spores through an apical pore or large mouth, and are probably the most common puffballs of our lawns and pastures. *Disciseda,* on the other hand, is as rare as it is bizarre. In the most common species, *D. candida,* the tough outer peridium splits around its "equator" and the papery inner layer ruptures basally, then the spore case breaks loose and flips over so as to resemble an acorn in a cup.

Bovistas are edible when firm and white inside, but are bland at best and bitter at worst. Discisedas are too rare to be of food value and are not normally found until they are old and powdery. One representative of each genus is described here.

Key to Bovista & Disciseda

1. Outer layer of peridium (skin) persistent, often incrusted with sand or dirt particles and often forming an acornlike cup around the inner layer; spores typically released through a small pore or tear at maturity; not common .... **D. candida** & others, p. 698
1. Not as above; outer layer of peridium not persistent, or if persistent then typically rupturing radially or forming a large mouth at top of spore case 2

2. Fruiting body bright yellow to golden-orange when young **B. colorata** (see *B. plumbea,* p. 697)
2. Not as above 3

3. Fruiting body typically buried in ground except for the apex or mouth, even at maturity .. 4
3. Not as above (but fruiting body may be buried or half-buried when very young) 5

4. Outer peridium often persisting for a long time; found in grass or prairies in north-central United States and Canada .... **Gastrosporium simplex**
4. Not as above; found mainly in eastern North America under trees, along roads, in bare soil and waste places, etc. .... **B. minor** (see *B. plumbea,* p. 697)

5. Sterile base present ....*B. dakotense* (see *B. plumbea*, below)
5. Sterile base absent or rudimentary 6
6. Fruiting body small, with crowded spines and scurf when young, rupturing through an apical pore at maturity; usually clustered or densely gregarious . (see *Lycoperdon* & **Allies,** p. 690)
6. Not as above; fruiting body smooth or with small warts or granules when young 7
7. Fruiting body small (less than 2 cm broad) and round, but usually with a narrowed or pinched, rooting base; rupturing through an apical pore at maturity ....
    ....*B. pusilla* & *B. longispora* (see *B. plumbea*, below)
7. Not as above; usually rupturing through a large mouth or tear at maturity 8
8. Fruiting body typically 1-3 (6) cm broad and attached to ground by a small patch of dirt-bound fibers when young ....*B. plumbea* & others, below
8. Fruiting body typically 3-9 cm broad, usually attached to ground by a small cord or "root" when young 9
9. Spore case typically dark brown to bronze-brown or coppery when mature; common in North America ....*B. pila* (see *B. plumbea*, below)
9. Spore case typically gray to lead-colored when mature; common in Eurasia, uncommon or absent (?) in North America ....*B. nigrescens* (see *B. plumbea*, below)

Bovista plumbea   (Tumbling Puffball; Tumble Ball)

FRUITING BODY round or slightly flattened, 1-4 (8) cm in diameter, usually with a small patch of dirt (held together by fibers) at base. Outer layer of peridium (skin) white, smooth, felty, or with small flattened scales, peeling away or shrivelling up in age to reveal the smooth, papery inner layer which is blue-gray to purplish-brown or lead-colored in age and usually has a metallic luster; inner layer rupturing at the top to form a large circular "mouth." **STERILE BASE** absent. **SPORE MASS** white and firm at first, then yellow-olive and mushy, finally powdery and olive-brown to deep chocolate-brown. Spores 5-7 × 4-5 microns, oval or broadly elliptical to nearly round, minutely spiny to nearly smooth, with a long pointed pedicel ("tail") 8-14 microns in length.

HABITAT: Solitary, scattered, or gregarious in grassy places; widely distributed and very common. In our area it fruits mainly in the fall and winter in pastures, year-round on lawns, golf courses, and cemeteries. The thin, papery windblown spore cases can be found most any time.

EDIBILITY: Edible when immature, but mediocre. It is bland, and a great many would be needed for a meal.

COMMENTS: This paltry puffball is a common fungal feature of our lawns and pastures. In the immature stage it can be distinguished from *Vascellum pratense* and other grass-loving puffballs by its small size, smooth white skin, and absence of a sterile base, while older specimens are easily told by their lustrous, paper-thin spore cases. Another wide-spread cosmopolitan species, *B. pila,* is slightly larger (3-9 cm in diameter) and attached to the ground by a small cord or "root." It has a dark brown to bronze mature spore case, smooth apedicellate (tail-less) spores, and grows in cultivated and manured soil as well as in grass and open woods. I have found it several times in our area, but it is not nearly as

Bovista plumbea, immature specimens. This common small round puffball of lawns and pastures lacks a sterile base. Note the patch of dirt-bound fibers below. *B. pila* (not illustrated) is similar but slightly larger and has a rootlike cord at the base rather than a patch of fibers.

Bovista plumbea, mature specimens. Note metallic lustre, thin skin, and the large mouth that is starting to form in each one.

numerous as *B. plumbea.* Another larger species with a cordlike root, **B. nigrescens,** is lead-gray when mature and is of questionable occurrence in North America. A cosmopolitan species, **B. pusilla (=Lycoperdon pusillum),** is one of our smallest puffballs. It is marble-sized and more or less round above and pinched below to a cordlike root. It also lacks a sterile base and the spores are not pedicellate. It usually grows in grass, bare ground, vacant lots, sandy soil, etc.; it is also known as **B. californica.** Other small species include: **B. longispora,** similar to *B. pusilla,* but with a slight sterile base and elliptical spores; **B. leucoderma,** reddish- to bronze-brown when mature with a patch of fibers at the base when young; **B. minor,** rare, with a trash- or dirt-covered spore case, found in poor soils and usually developing underground; **B. colorata (=Lycoperdon coloratum),** easily told by its bright yellow to golden-orange color; and **B. dakotense,** one of the few members of the genus to have a sterile base (leading to confusion with *Lycoperdon,* but usually growing in grassy or open areas). None of these are worth eating.

Disciseda candida   (Acorn Puffball)

FRUITING BODY round to somewhat flattened or irregular, 0.8-3 cm broad. Outer layer of peridium (skin) dirt-incrusted, composed of white mycelial matter that is cottony when young but tough in age; inner layer minutely granulose or scurfy, pallid to tan or grayish, forming a basal pore at maturity. Outer layer rupturing around its periphery, the upper portion retaining the inner layer (spore case) within it, and detaching and flipping over so as to resemble an acorn in its cup (thereby causing the pore to appear apical). **STERILE BASE** absent. **SPORE MASS** white when very young, soon yellowish to olive, finally olive-brown to brown and powdery. Spores 3.5-4.5 microns, round, minutely warted, with a short stubby pedicel ("tail").

HABITAT: Solitary or in groups in pastures and other grassy or open areas, along paths, in barnyards, etc; usually developing at or beneath ground level and then flipping over or breaking free to tumble about in the wind. It is widely distributed, but not common. It has been found in southern California on sandy and grassy soil. In eastern North America it is sometimes found with *Arachnion album* (see the key to Lycoperdales). *D. subterranea* (see comments) has been reported from Washington, Colorado, and Wyoming.

EDIBILITY: Worthless—it is rarely found when white inside, and is insubstantial anyway.

COMMENTS: This peculiar puffball is easily told by its small size and acornlike mature fruiting body. Other species: **D. subterranea** is similar, but has a grayish to bluish-gray spore case and larger spores, while **D. pedicellata** has spores with long "tails." Several other species occur in sandy or exposed soil in California and the Southwest, but are rarely collected and difficult to identify because they are not necessarily acornlike. They have very thin spore cases (as in *Bovista*) that usually retain vestiges of the outer peridium. Some of them are: **D. ater,** which has a blackish outer peridium; **D. luteola,** originally described from Lake Tahoe, California, with a yellowish to dull yellow-brown spore case; and **D. brandegeei,** a desert species with smooth spores.

698

GEASTRUM, ASTRAEUS, & MYRIOSTOMA
(Earthstars)

Small to medium-sized, mostly terrestrial fungi. FRUITING BODY round or flattened when young (usually underground in this stage), but at maturity *the outer wall (peridium) splitting into several starlike rays which curl back (and often under) the spore case.* SPORE CASE (inner peridium) smooth or roughened, *rupturing through an apical pore (Geastrum & Astraeus) or through several pores (Myriostoma), or irregularly (Astraeus).* SPORE MASS white when *very* young (underground), brown and powdery when mature. STALK and STERILE BASE absent, but spore case sometimes mounted on a pedicel (short stalk) or pedicels. Spores round and warted or occasionally smooth. Capillitium present, branched in *Astraeus,* unbranched in *Geastrum* and *Myriostoma.*

EARTHSTARS are modified puffballs in which the thick outer skin splits into starlike rays which unfold and often recurve, exposing the spore case (inner skin) to the elements. Some puffballs and earthballs (e.g., *Scleroderma geaster, Mycenastrum corium*) rupture in starlike fashion, but only the earthstars have a discrete spore case intact within the rays.

Some earthstars are **hygroscopic;** the rays open in wet weather to expose the spore case to raindrops, and then close up in dry weather to protect it. This phenomenon can be observed at home by placing a closed-up earthstar in a shallow bowl of water. If hygroscopic, it will open up in a few minutes—even if it is a year or two old! Earthstars are also accomplished acrobats, coming in an amazing assortment of unorthdox and extraordinary poses. *Geastrum fornicatum* in particular is so prodigious a contortionist that it is difficult to find two that are alike.

The earthstars are comprised of three genera: *Geastrum (=Geaster), Astraeus,* and *Myriostoma. Geastrum* embraces a sizable number of small to medium-sized, difficult-to-identify earthstars that rupture through a distinct apical pore and have small spores (typically less than 7 microns long). Some Geastrums are hygroscopic, but most are not. *Astraeus,* on the other hand, is always hygroscopic. It includes two species (one quite large) that rupture irregularly or through a large apical pore or slit, and have relatively large spores (typically more than 7 microns in diameter). The third genus, *Myriostoma,* includes a single rare species whose spore case is perforated by many holes (like a saltshaker) instead of just one. Most puffball pundits place the latter two genera alongside the earthballs in the Sclerodermatales because of certain microscopic features. However, they are grouped here with *Geastrum* because of their superficial similarity.

Fruiting body development in *Geastrum fornicatum*. Earthstars resemble puffballs when very young (specimen at bottom left), but their outer skin splits into rays and unfolds (bottom right), revealing the inner skin or spore case (specimens at top). See p. 701 for another photo of this species.

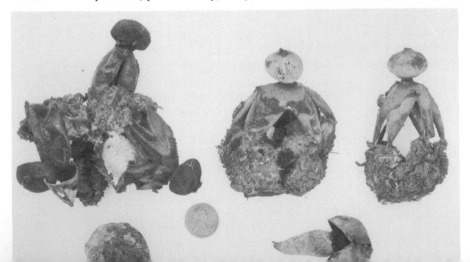

Earthstars occur in a variety of habitats (pastures, woods, waste places, etc.), but like many of the Gasteromycetes, they are especially prevalent and diverse in arid and semi-arid regions. The Southwest, for instance, is unusually rich in *Geastrum* species. They are real crowdpleasers, but have no culinary value because of their tough or woody texture.* Several earthstars are described and/ or keyed out here, but in the case of *Geastrum,* they represent only a fraction of the total number.

Key to Geastrum, Astraeus, & Myriostoma

1. Spore case typically rupturing irregularly or through a single apical pore (mouth at top) .. 2
1. Spore case typically with more than one (usually several) pores at maturity 18

2. Rays distinctly hygroscopic (i.e., folding over the spore case in dry weather to protect it and unfolding when moistened); spore case usually sessile (*not* seated on a short stalk) 3
2. Rays not hygroscopic; spore case may or may not be seated on a short stalk 7

3. Fruiting body 5-15 cm or more broad when expanded; spore case 2.5-5 cm broad, rupturing irregularly (i.e., not forming a distinct pore) .... *Astraeus pteridis,* p. 706
3. Fruiting body smaller and/ or rupturing through an apical pore or slit 4

4. Spore case roughened by numerous particles; apical pore often irregular or poorly defined or merely a slit; underside (exterior) of rays often with blackish fibrils; spores over 7 microns in diameter .... *Astraeus hygrometricus,* p. 705
4. Not with above features; apical pore usually well-defined and spores less than 7 microns .. 5

5. Apical pore (mouth) with distinct radial grooves ....
      .... *Geastrum drummondii* & others (see *Astraeus hygrometricus,* p. 705)
5. Apical pore not grooved (but may be fibrillose) 6

6. Apical pore fibrillose . *Geastrum mammosum* & others (see *Astraeus hygrometricus,* p.705)
6. Apical pore more or less naked .. *Geastrum floriforme* (see *Astraeus hygrometricus,* p. 705)

7. Spore case mounted on a distinct pedicel (short stalk) 8
7. Spore case sessile (stalkless) or nearly so 15

8. Mouth of spore case "beaked" (i.e., forming a raised cone), the beak longitudinally grooved 9
8. Not as above; mouth not beaked, or if beaked, then not grooved 11

9. Pedicel with a distinct collar around base of spore case ....
      .... *Geastrum striatum* (see *G. pectinatum,* p. 702)
9. Not as above 10

10. "Beak" long; pedicel also long (2-6 mm) and slender .... *Geastrum pectinatum,* p. 702
10. Not as above .... *Geastrum nanum* & others (see *G. pectinatum,* p. 702)

11. Fruiting body small, typically less than 3 (4) cm broad when expanded; spore case typically 3-10 mm broad 12
11. Fruiting body larger, typically 2-6 cm or more broad; spore case typically 1-2.5 cm broad   13

12. Fruiting body usually with 8-12 rays .... *Geastrum minimum* (see *G. fornicatum,* p. 701)
12. Fruiting body usually with 4-6 rays ...... *Geastrum quadrifidum* (see *G. fornicatum,* p. 701)

13. Fruiting body with 4-5 (6) rays (but the tips may split) that are usually attached to a "cup" of tissue, mycelium, and debris; fruiting body poised on the tips of its rays when mature; apical pore or mouth more or less same color as rest of spore case ... *Geastrum fornicatum,* p. 701
13. Fruiting body usually with 6-12 rays, or if with fewer rays then not as above 14

14. Mouth area usually delimited (set off from rest of spore case), sometimes paler in color than rest of spore case .... *Geastrum limbatum* (see *G. pectinatum,* p. 702)
14. Mouth area not clearly delimited .... *Geastrum rufescens* (see *G. pectinatum,* p. 702)

---

*Captain Charles McIlvaine, as usual, casts a dissenting opinion, saying of *Astraeus hygrometricus:* "when young** it is, when cooked, soft and creamy inside. The outer part is tough and semi-glutinous but of pleasant texture. It has not a marked flavor, but makes a succulent dish." (Does this make any sense to you?)

**Immature earthstars are not commonly found because they're inconspicuous and often develop underground.

15. Fleshy upper (inner) layer of rays often breaking away to form a broad cup or saucer around the base of spore case .... *Geastrum triplex*, p. 703
15. Not as above; spore case not seated in a saucer 16
16. Mouth area sharply delimited (i.e., with a circular zone around it) 17
16. Mouth not sharply delimited .... *Geastrum fimbriatum* (see *G. saccatum*, p. 703)
17. Mouth with distinct radial grooves *Geastrum campestre* (see *Astraeus hygrometricus*, p. 705)
17. Mouth not grooved, but often silky-fibrillose .... *Geastrum saccatum*, p. 703
18. Spore case mounted on several pedicels (short stalks) .... *Myriostoma coliforme*, p. 704
18. Spore case sessile (stalkless) ..... *Geastrum pluriosteum* (see *Myriostoma coliforme*, p. 704)

Geastrum fornicatum   (Acrobatic Earthstar)

FRUITING BODY round or flattened when young and incrusted with debris, the outer wall (peridium) then splitting into 3-6 (usually 4 or 5) rays which peel back and under until they're more or less erect. Mature fruiting body 3-6 cm broad and 5-10 cm high, usually poised on the pointed tips of its rays, the tips joined to a concave mycelial mat or "cup" which has a grayish interior and debris-incrusted exterior and may be broken into rays itself. **RAYS** not hygroscopic; underside usually smooth and tan to brown; upper surface chocolate-brown to dark brown or nearly blackish, but peeling off in patches to reveal the smooth, paler brown or tan undersurface. **SPORE CASE** 1-2.5 cm broad, round or somewhat flattened to urn-shaped, mounted on a pedicel (short stalk), often with a compressed collar underneath it (at the top of the pedicel); surface brown to dark chocolate brown, finely velvety; rupturing at the top to form a fairly large, irregularly torn or lacerated pore. **SPORE MASS** chocolate-brown to blackish-brown and powdery at maturity. Spores 3-5 microns, round, warted.

HABITAT: Scattered to densely gregarious in humus under trees, on rubbish, around stables and other waste places, etc.; widely distributed, but apparently rare except in southern California and the Southwest, where it is quite common. In our area I have found it in only one locality—under oaks where there is run-off from a bathing area for horses—but it fruits there by the hundreds and can be found every month of the year. In New Mexico it is quite common under juniper.

Geastrum fornicatum, mature specimens perched on the tips of their rays. Note how the rays peel in patches and how the specimens at left are firmly attached to a mat or cup of mycelium and debris. See p. 699 for another photograph of this acrobatic species.

EDIBILITY: Inedible, but makes an extraordinary ornament or conversation piece (it is easily dried).

COMMENTS: One of the most distinctive of all fleshy fungi, this earthstar is a true pleasure to find. Mature individuals defy description, but look something like a cross between a flat tire and a ballet dancer. The tendency of the rays to stand erect on their tips (like a ballet dancer) plus the membranous mycelial cup ("flat tire") to which they're attached are the principal fieldmarks. Because of the likeness to a human figure, it was named *Fungus anthropomorphus* when first described in 1688. Specimens are sometimes found with only one or two "arms" extended, as if they were putting out tentative "feelers" before committing themselves completely (and irrevocably) to exposure. Other species: *G. quadrifidum* (also called *G. coronatum*) is a similar but somewhat smaller species with a well-defined, conical mouth that is paler than the rest of the spore case; it is also equipped with a mycelial "cup" and is widely distributed. *G. minimum,* which may or may not be a diminutive form of *G. coronatum,* has 8-12 rays and usually lacks a mycelial cup; it occurs commonly in the Southwest under pinyon and juniper. For other species with a pedicel beneath the spore case, see *G. pectinatum.*

Geastrum pectinatum   (Beaked Earthstar)

FRUITING BODY round to flattened when young, the outer wall (peridium) then splitting into 5-12 rays (usually 6-10) which peel back and under or stand upright; 3-6 cm broad when expanded. **RAYS** not hygroscopic, copiously incrusted with mycelial matter and debris on underside (exterior); upper surface brown to grayish-brown, fleshy, cracking irregularly or flaking off as it dries, revealing the smooth undersurface. **SPORE CASE** 0.5-2 cm broad, round to somewhat urn-shaped, mounted on a prominent pedicel; surface more or less smooth, lead-colored to purple-brown, but often appearing paler from a hoary sheen; underside sometimes striate or grooved near the pedicel; rupturing to form a prominent apical pore that is beaklike (more or less conical) and strongly grooved or furrowed longitudinally. Pedicel slender, 2-6 mm long, sometimes with a collar or ring around its middle or base. **SPORE MASS** brown and powdery when mature. Spores 4-6 (7) microns, round, warted.

HABITAT: Solitary or in groups in humus under trees, especially conifers such as juniper, or less commonly in the open; widely distributed but not common, fruiting after rains but persisting for weeks or months. I have found it (or something very similar) in New Mexico.

EDIBILITY: Too tough to be edible.

COMMENTS: The prominent pedicel and "beaked" mouth of the spore case distinguish this earthstar from most others. The rays are not hygroscopic, and their tips are not joined to a mycelial "cup" as in *G. fornicatum.* Other earthstars with a pedicel and grooved mouth include: *G. striatum (=G. bryantii),* similar but typically with a prominent collar at the base of the spore case (where it meets the pedicel); *G. nanum (=G. schmidelii),* also similar but much smaller (up to 2.5 cm broad) and more common, with a shorter "beak" and a short, thick pedicel; *G. smithii,* common in the Southwest, with a flattened-conical mouth seated in a well-defined, depressed area; and *G. xerophilum,* also common in the Southwest (but usually growing in the desert), with a short pedicel (if any) and a granular spore case. Two species with a non-grooved mouth should also be mentioned: *G. limbatum* is a medium-sized to fairly large species with a well-defined fibrillose (but not grooved or beaked) mouth and 6-12 rays (more than *G. fornicatum*—see photo on p. 703). It comes in a striking dark brown to black version that is common in the Southwest, and a paler (pinkish-buff to brown) form that is more widely distributed. *G. rufescens,* on the other hand, has a poorly-defined, often fringed or torn mouth and a short or indistinct

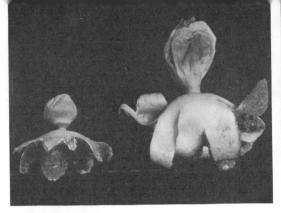

Left: *Geastrum limbatum* (see comments under *G. pectinatum*). Note the numerous rays and the pedicel (short stalk) below spore case. **Right:** *Geastrum triplex,* mature specimen. The "saucer" around the spore case is distinctive.

pedicel. It is fairly large (5-10 cm broad) and is more common in eastern North America than in the West. All of these are better eyed then fried.

Geastrum triplex (Saucered Earthstar)

FRUITING BODY round to flattened or bulblike when young, the outer wall splitting at maturity into 4-8 rays which unfold and then bend under the spore case; 3-10 cm broad when expanded. **RAYS** not hygroscopic, usually free of adhering debris on underside; upper surface with a rather thick, fleshy, pinkish to tan or brown layer that cracks into patches, the central region usually breaking loose to form a broad cup or saucer around the spore case. **SPORE CASE** 1-3 cm broad, round or flattened, sessile (*not* mounted on a stalk); pale to dark tan, grayish, or even reddish-brown, rupturing through an apical pore which is often paler, slightly raised, and radially fibrillose. **SPORE MASS** deep brown to smoky-brown and powdery when mature. Spores 3.5-4.5 microns, round, warted.

HABITAT: Solitary or in groups in forest humus and under trees (usually hardwoods); widely distributed and common in some regions. I have found it only twice in our area (in the fall and winter), but have seen large fruitings under aspen in the Southwest.

EDIBILITY: Supposedly edible when immature (white inside), but like all earthstars, rarely found in that stage, and too tough and fibrous to eat when older.

COMMENTS: Also known as *G. indicum,* this relatively large, attractive earthstar is distinguished by the shallow cup or "saucer" that often forms around the base of the spore case. There seems to be a great deal of variation in the extent to which the fleshy upper surface of the rays cracks, and those which don't crack and lack a saucer can be confused with *G. saccatum* and *G. fimbriatum* (which, however, are often smaller). For other relatively large, non-hygroscopic earthstars, see comments under *G. pectinatum.*

Geastrum saccatum (Sessile Earthstar)

FRUITING BODY round, flattened, or bulblike when young, the outer wall splitting at maturity into 4-8 rays which unfold and then bend back under the spore case; (1.5) 2-5 cm broad when expanded. **RAYS** not hygroscopic, rather rubbery when fresh, relatively clean and buff to pale tan on underside; upper surface fleshy, pallid becoming pale tan, pinkish, or ochre-brown; sometimes cracking. **SPORE CASE** 0.5-2 cm broad, round to somewhat flattened, smooth, sessile (without a stalk or pedicel), papery, pallid to buff, dull grayish, or brownish; rupturing through an apical pore that is often paler in color and set off from the rest of the spore case by a circular line, ridge, or groove; pore usually raised and fibrillose. **SPORE MASS** brown and powdery when mature. Spores 3.5-5 microns, round, warted.

Left: *Geastrum saccatum* lacks a pedicel and "saucer," but features a circular zone around the pore at top. The rays often bend under the spore case as shown in photo at right. **Right:** *Geastrum fimbriatum* resembles *G. saccatum,* but lacks circular zone (see comments under latter species).

HABITAT: Solitary, scattered, or in groups in humus under trees (especially conifers), in woods, or occasionally in the open; widely distributed. This species and *G. fimbriatum* (see comments) seem to be the most common Geastrums in our area. They fruit mainly in the fall and winter, but persist for weeks or months without decaying.

EDIBILITY: Inedible in the mature state in which it is usually found.

COMMENTS: The spore case in this species has neither the pedicel (short stalk) of *G. fornicatum* and *G. pectinatum,* nor the saucer of *G. triplex,* and the rays are not hygroscopic. There are several similar species, including: *G. fimbriatum (=G. sessile)*, often slightly darker, with a torn or fringed apical pore that is not set off by a circle, usually growing under trees and widely distributed; and *G. rufescens* and *G. xerophilum,* both of which may have a slight pedicel (see comments under *G. pectinatum* for more details on these species, and also note the hygroscopic Geastrums discussed under *Astraeus hygrometricus*, most of which lack a pedicel).

Myriostoma coliforme   (Saltshaker Earthstar)

FRUITING BODY round to flattened when young, the outer wall (peridium) splitting into 5-14 pointed rays which open out; 2-12 cm broad when expanded. **RAYS** not hygroscopic; more or less smooth on underside (exterior), but often covered with dirt near base; inner (upper) surface with a brown to cinnamon fleshy layer that weathers or peels away, revealing the paler (buff) undersurface. **SPORE CASE** 1-5 cm broad, round to somewhat flattened, usually mounted on several slender, short, sometimes inconspicuous pedicels (columns); surface minutely roughened, brown to silvery-brown or grayish in age, rupturing through several to many mouths or pores, mostly on the upper portion (the number of mouths often corresponding to the number of pedicels). **SPORE MASS** powdery and brown at maturity. Spores 4-6 microns, round, warted.

HABITAT: Solitary or in groups in sandy soil, sometimes under trees, fruiting after heavy rains but persisting year-round; widely distributed. It is usually listed as rare, but is not uncommon in New Mexico, where I have seen it in the summer and fall.

EDIBILITY: Not edible.

COMMENTS: This fungal rarity can be recognized with clarity by its saltshaker-like spore case mounted on several short columns. Most taxonomists place it alongside *Astraeus* in the Sclerodermatales because of certain microscopic features. *Geastrum pluriosteum* is a rare southwestern earthstar that forms one to several pores on a stalkless spore case.

Astraeus hygrometricus, mature specimens. Ten minutes before this picture was taken, the fully expanded individuals were curled up into tight "fists" like the one in foreground. However, a quick "bath" caused the "fists" to unclench. Note the roughened surface of the spore case.

Astraeus hygrometricus (Hygroscopic Earthstar)

FRUITING BODY round to flattened when young, the outer wall splitting at maturity into 6-15 pointed rays; 1-5 (8) cm broad when fully expanded. **RAYS** hygroscopic (unfolding in wet weather and closing up over the spore case in dry weather), often unequal in length, tough and leathery when moist, hard when dry; underside (exterior) fibrillose, usually with adhering sand and sometimes with black hairlike threads (rhizomorphs) at base; upper surface (interior) at first smooth, but often developing numerous cracks, pale tan darkening to brown, gray, or blackish-brown. **SPORE CASE** 0.5-3 cm broad, round or somewhat flattened, sessile (without a stalk), thin and papery; surface whitish to grayish or brownish and roughened by particles (and/ or finely reticulate); rupturing at maturity through an irregular or poorly defined apical pore or slit. **SPORE MASS** brown to cocoa-brown and powdery when mature. Spores 7-11 microns, round, finely warted.

HABITAT: Solitary, scattered, or gregarious in old fields, sand or sandy soil, pastures, roadsides, waste places, etc.; worldwide in distribution. It grows from sea level to above timberline, from verdant pastures to desert wastelands. In our area it fruits mainly in the fall, but the tough fruiting bodies persist for months without decaying. I have seen hundreds of spiderlike fruiting bodies in Arizona under pinyon pine, but when the weathered, blackened outer skins (rays) are found without their spore cases, they look like old tire patches.

EDIBILITY: Indisputably inedible because of its toughness.

COMMENTS: This veritable barometer is the most theatrical of all the earthstars. A few minutes' immersion in water will open up old, dried-up specimens that seem as tightly closed as clenched fists. It is distinguished from *A. pteridis* by its smaller size (but see comments under that species). There are also several hygroscopic species of *Geastrum* that have spores *less* than 7.5 microns in diameter. They grow mainly in arid climates (e.g., the deserts of the West), but a few are more cosmopolitan. Several of these Geastrums can be distinguished from *A. hygrometricus* by the presence of a prominently grooved apical pore, e.g., *G. drummondii,* with a granular spore case; *G. umbilicatum,* with a smooth spore case; and *G. campestre,* with a roughened spore case and sometimes a short pedicel (stalk). Others have a fibrillose or fringed pore, e.g., *G. mammosum,* especially common in the Southwest under pinyon and juniper; *G. recolligens,* very similar, but found in sand or waste places; and *G. arenarium,* also in sand or deserts, but only slightly hygroscopic. Still others have a naked mouth (merely a puncture or slit) like that of *A. hygrometricus.* One of these, *G. floriforme (=G. delicatus)* is widely distributed and occurs in our area, but can be separated from *A. hygrometricus* by its small (up to 1.5 cm broad), smoother spore case and smaller spores.

705

Astraeus pteridis, mature specimen. This species is much larger than *A. hygrometricus* and often has a more pronounced crazy-quilt pattern on its rays. Note how the spore case disintegrates irregularly. (Joel Leivick)

Astraeus pteridis   (Giant Hygroscopic Earthstar)

FRUITING BODY round or somewhat flattened when young, the outer wall then splitting into 6 or more rays; 5-15 cm or more broad when fully expanded. **RAYS** hygroscopic (unfolding in wet weather, closing up over the spore case in dry weather), thick (3-6 mm), tough and leathery becoming hard or woody when dry; exterior (underside) with a thin coating of matted brownish fibrils which may or may not wear away; upper surface (interior) often conspicuously cracked or fissured transversely to form a checked pattern; tan when fresh (but dark brown in the cracks), darker overall in age (but weathered specimens sometimes whitened). **SPORE CASE** 2.5-5 cm broad, round or somewhat flattened, sessile (not seated on a stalk); at first roughened, veined, or fibrillose, but soon smooth or merely roughened; grayish-brown to brown; rupturing irregularly in age, i.e., not forming a distinct pore. **SPORE MASS** dark brown and powdery when mature. Spores 8-12 microns, round, warted.

HABITAT: Solitary, widely scattered, or in groups on ground, usually along roads, railroad tracks, in waste places, old fields, etc., but sometimes in woods; known only from western North America. In our area it is fairly common year-round.

EDIBILITY: Much too leathery or woody to be edible.

COMMENTS: The thick rays which are frequently checkered like a crooked crossword puzzle plus the large size and irregular rupturing of the spore case distinguish this impressive species from its more mundane cousin, *A. hygrometricus,* and other hygroscopic earthstars. However, the two species seem to intergrade in our area (e.g., you can find fairly large, uncheckered specimens), and consequently some earthstar authorities "reduce" *A. pteridis* to a variety of *A. hygrometricus.* For some inexplicable reason, the fructifications remind me of "tribbles"—those lovable but relentlessly prolific creatures of "Star Trek" renown.

706

SCLERODERMA (Earthballs)

Medium-sized to fairly large fungi found in humus, soil, sand, or on rotten wood. FRUITING BODY more or less round to oval, lobed, or with a stemlike base; rupturing irregularly in old age or splitting into starlike rays or sometimes forming an apical pore or tear. PERIDIUM usually one-layered and typically *thick, tough, and rigid* when fresh; smooth or wrinkled or with scales, typically yellowish to brown but sometimes whitish. STERILE BASE absent, but a stalklike base composed of tough mycelial fibers often present. SPORE MASS initially white and very firm, but in most cases *soon becoming gray to purple-black* (sometimes marbled with paler veins) *and remaining firm,* then eventually becoming powdery and sometimes browner. Spores more or less round, spiny and/ or reticulate, *not* borne in a hymenium. Capillitium absent or rudimentary.

ALSO known as the thick-skinned or hard-skinned puffballs, the earthballs superficially resemble the true puffballs (*Calvatia, Lycoperdon,* etc.), but differ in several fundamental respects. Their peridium (skin) is typically hard (rindlike) when fresh and tough or leathery in age, and the spore mass becomes colored (usually purple-gray to black) at an early age *while remaining firm.* In the puffballs, on the other hand, the peridium is often thin and papery in age and the spore mass becomes soft or mushy as it ripens (darkens). There are other differences as well. In *Scleroderma* there are no specialized threadlike cells (capillitium) in the spore mass, the exterior of the spore case lacks the spines or soft particles found in many puffballs, and there is no sterile base, though a stalklike or rootlike mass of tough mycelial fibers is often present. These differences have led most taxonomists to group the Sclerodermas in a separate order, the Sclerodermatales, along with several other genera such as *Pisolithus.*

As their name implies, earthballs are often buried or partially buried before maturity, leading to possible confusion with the false truffles (Hymenogastrales & Allies). The spore mass, however, becomes powdery in old age and the thick, tough peridium is distinctive. The deer truffles *(Elaphomyces)* also have a thick peridium, but tend to grow deeper in the ground, lack a distinct basal point of attachment, and bear their spores in asci.

Sclerodermas are common and ubiquitous, but are not as conspicuous as the true puffballs. Their nomenclature is very confused. Over 150 species have been described but in a 1970 monograph, Gaston Guzman condenses them into a mere 21 widespread, polymorphic species. However, even with only 21 species to choose from, identification is difficult without a microscopic examination of the spores. The genus is divided into three "sections" based on whether the spores are echinulate (spiny), reticulate (covered with a network of ridges), or a compromise between the two, and species in the different sections resemble each other closely.

Many Sclerodermas are thought to be mycorrhizal, but they can grow almost anywhere—in woods or at their edges, under planted trees and shrubs, in sand and asphalt, along roads and trails, on exposed hillsides, around old stumps, etc. In our mild climate they occur year-round, but seem to favor warm weather. Three representative species are described here (one from each "section") and several others are discussed. *Sedecula,* a rare, smooth-spored, underground genus possibly related to *Scleroderma,* is also keyed out.

Key to Scleroderma

1. Spore mass (interior) divided into large irregular chambers by tough cordlike veins (which may turn into coarse black mycelial cords at base); fruiting body 2-9 cm broad, oval to cushion-shaped; exterior often pitted or split in age, whitish to grayish to olive-buff or yellowish (but often blackening in age); spore mass black and powdery to slightly gelatinous when mature; spores 20-28.5 × 11-16 microns, *smooth*; widespread under western conifers (especially in mountains), but usually rare; usually growing underground .... *Sedecula pulvinata*
1. Not as above; mature spores ornamented; found underground or above; common 2
2. Fruiting body growing underground, typically *without* an obvious basal point of attachment to substrate; peridium (skin) *sometimes* marbled in cross-section; spores borne inside asci (but asci soon disintegrating) .... (see *Elaphomyces,* p. 862)
2. Not as above; fruiting above the ground, or if underground then usually with an obvious base or point of attachment; peridium *not* marbled in cross-section; spores borne on basidia 3

3. Peridium (skin) *very* thick (3-10 mm), rupturing into starlike lobes in old age; fruiting body medium-sized to large; spores partially reticulate .... *S. geaster* & others, p. 710
3. Not as above; either peridium thinner (averaging 1-4 mm) or not rupturing into starlike lobes; fruiting body fairly small to medium-sized (rarely large) 4
4. Fruiting body with a long "stalk" composed of tough strands and fibers that totals *at least half* (and usually three-quarters) of the fruiting body's total height; spores reticulate; found in sand dunes or sandy soil .... *S. macrorhizon* (see *S. citrinum,* below)
4. Not as above ("stalk" if present shorter and/or habitat different) 5
5. Peridium (skin) covered with prominent *inherent* rosette-like scales (i.e., each scale often with a central wart); widespread, but especially common in forests (see Color Plate 190) 6
5. Not as above; peridium typically smooth (at least when young) but often becoming fissured or cracking and peeling to form scales in age 7
6. Peridium (skin) rather thin (typically less than 2 mm), usually rupturing in old age through a pore or slit at top; spores spiny .... *S. verrucosum* & *S. areolatum* (see *S. citrinum,* below)
6. Peridium fairly thick (usually 1-4 mm), rupturing into lobes in age or forming an irregular pore; spores reticulate .... *S. citrinum,* below
7. Spores spiny but not reticulate .... *S. cepa* group, p. 709
7. Spores reticulate or partially reticulate ...... *S. bovista* & others (see *S. cepa* group, p. 709)

Scleroderma citrinum   (Common Earthball)   **Color Plate 190**

FRUITING BODY 2-10 (12) cm broad and 2-6 cm high, round or somewhat flattened, the underside often with a stemlike base composed of ridges, mycelial fibers, and/or adhering debris; eventually cracking into lobes to form an irregular pore, the lobes *not* normally bending outward or unfolding appreciably. **PERIDIUM** (skin) hard and fairly rigid (rindlike), 1-3 mm thick, white when sectioned but usually staining pinkish (if fresh); surface yellow-brown to dingy yellow to ochre or tan and cracked or arranged into prominent, inherent scales which often have a smaller, central wart. **SPORE MASS** white when very young, soon gray to purple-gray with whitish veins often running through it, then dark purple-gray to purple-black or black and still solid and firm; eventually becoming powdery and blackish-brown. Spores 8-13 microns, round, strongly reticulate.

HABITAT: Solitary, scattered, or in groups or clumps on ground and rotten wood, usually in forests but sometimes in gardens or in sandy or disturbed soil; widespread and the most common of all *Scleroderma* species. Although it is more typically a northern than a southern species, I have seen enormous fruitings in the Great Smoky Mountains of Tennessee. In our area I have found it in a garden in October, but it is rare—at least in comparison to other species.

EDIBILITY: Poisonous! If eaten raw or consumed in quantity it causes nausea, vomiting, diarrhea, and even chills or cold sweats. Nevertheless, it has been used by some unscrupulous individuals to adulterate truffles.

COMMENTS: Also known as *S. aurantium* and *S. vulgare,* this earthball is easily recognized by its scaly, yellow-brown, rindlike skin that tends to stain pinkish when cut, and the purple-black spore mass (see color plate). It might be mistaken for an edible puffball, but is only rarely found while still white inside. The scales often looked embossed (i.e., with a central wart), a character that helps distinguish it from scaly specimens of *S. cepa, S. bovista,* etc. (see comments under *S. cepa*). Other species: *S. macrorhizon* has a long "stalk" composed of tough strands and fibers and grows exclusively in sand (often in sand dunes!). It is widely distributed in northern latitudes and has spores similar to those of *S. citrinum,* but slightly larger. *S. verrucosum* and *S. areolatum (=S. lycoperdoides)* are smallish species with a thin (up to 1 mm thick), delicately scaly peridium that usually ruptures in old age through an apical pore or tear. Both have spiny rather than reticulate spores, and a stalklike, fibrous base, and are widely distributed under hardwoods or sometimes conifers. (In *S. verrucosum* the spores are 7-11 microns in diameter; in *S. areolatum* they are 10-18 microns.)

Scleroderma cepa group. Note thick, relatively smooth skin and dark (purple-black) interior when young. Also note how the older specimen at right is beginning to split open. These specimens stained burgundy when rubbed, especially near base (staining is visible in middle specimen).

Scleroderma cepa group   (Smooth Earthball)

FRUITING BODY 1.5-8 (10) cm broad and/or high, more or less round to somewhat flattened or lobed (often broader than it is high), the underside often—but not always—with a rootlike or stalklike base made up of tough mycelial fibers and adhering debris; eventually rupturing irregularly or splitting at the top into lobes which may peel back slightly. **PERIDIUM** (skin) hard and fairly tough (rindlike) when fresh, 1-3 mm thick, white in cross-section but often staining reddish-pink to vinaceous when cut (if fresh). Surface whitish or pallid, soon becoming buff, yellowish, straw-colored, or brownish, often staining reddish or vinaceous when rubbed (especially the underside), then discoloring brownish; smooth when young, often becoming areolate (cracked to form scales) in age or where exposed to light (especially on top). **SPORE MASS** white and very firm when young, soon becoming black or purple-black (sometimes with paler veins) while remaining firm, then eventually becoming powdery and somewhat paler or browner. Spores 7-12 microns, round, spiny but not at all reticulate.

HABITAT: Solitary, scattered, or gregarious under trees (both hardwoods and conifers), in woods or cultivated areas, along roads, in gardens, lawns, etc.; widely distributed. In our area this species and its numerous look-alikes (see comments) are common year-round in a variety of habitats, but seem especially prevalent under pine and/or eucalyptus.

EDIBILITY: Poisonous! Chanterelle picker and *Russula*-kicker Bob Sellers of Santa Cruz, California, ate a small piece of *S. laeve* (see comments) under the impression that all "puffballs" were edible. Twenty minutes later he broke into a cold sweat, felt nauseous, and started vomiting. After expelling the offender, he recovered quickly.

COMMENTS: There are a number of microscopically distinct Sclerodermas that will more or less fit the above description. Collectively they can be recognized by their pallid to buff or yellow-brown peridium that is initially smooth but may crack into scales as it

Scleroderma cepa group. These specimens differ from those in above photo in their more highly developed "stalk" composed of tough fibers. Microscopically, however, they are indistinguishable. **Left:** Fairly young specimens. **Right:** An older individual which has split into lobes or "rays."

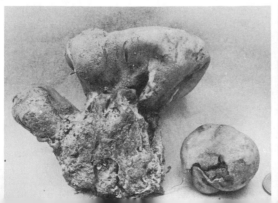
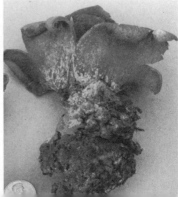

is exposed (the scales are not inherent as in *S. citrinum*), by their tough rindlike skin, and by their firm purple-black spore mass when young. Features such as the tendency of the peridium to stain reddish when rubbed and the presence or absence of a "stalk" composed of mycelial fibers seem to vary according to environmental conditions. Other species: *S. flavidum* is considered by some *Scleroderma*-scholars to be a form of *S. cepa*. It is microscopically identical, but does not usually bruise reddish and normally splits open into starlike lobes which bend back to expose the spore mass. *S. laeve* closely resembles the "true" *S. cepa*, but has slightly larger spores (9-15 microns); it appears to be the most common *Scleroderma* in our area. *S. albidum* is also similar, but has even larger spores (12-17 microns). *S. reae*, which favors arid habitats and ruptures irregularly, can be distinguished microscopically by its partially reticulate spores (9-18 microns). *S. floridanum* also has partially reticulate spores, but is tropical and subtropical and ruptures stellately (i.e., splits into starlike lobes). There are also four farflung Sclerodermas with *completely* reticulate spores: *S. bovista*, which sometimes develops blackish spots or stains in old age and has spores 9.5-16 microns broad; *S. fuscum*, which favors conifers and has spores averaging 13.5-19.5 microns; *S. hypogaeum (=S. arenicola)*, a common vinaceous-staining conifer-lover that often grows underground and has an unusually thick peridium and spores 18-23 microns broad; and *S. michiganense*, which has spores like *S. hypogaeum* and often grows underground, but has a thinner peridium and favors hardwoods. All of the above species are widespread and vary considerably in size, form, and scaliness, and all but the latter species *may* stain reddish or vinaceous when bruised. They should be considered poisonous until proven otherwise. Also see the species listed under *S. citrinum*.

Scleroderma geaster   (Dead Man's Hand)

FRUITING BODY 4-15 cm broad or high when closed, 12-30 cm broad when expanded (after rupturing); at first round to oval (but often somewhat flattened or irregularly lobed), eventually splitting into several (usually 4-8) coarse, irregular, thick rays which curl back in starlike fashion to expose the spore mass. **PERIDIUM** (skin) very thick (3-10 mm) and firm and tough when fresh; exterior whitish becoming yellowish, straw-colored, or brownish, roughened or somewhat hairy, often with adhering debris and often fissured or cracked into scales, especially in age; interior (upper side of rays) brownish becoming blackened and empty in old age; base often attached to the soil by a mycelial "root" of tough fibers. **SPORE MASS** at first firm and pallid, soon dark gray to purple-black or black (and still firm), eventually becoming brown to dark brown or purple-brown

Scleroderma geaster (=S. polyrhizon), mature but rather small specimens that have split into rays. In this stage they might be mistaken for old cup fungi, but usually contain traces of spore powder.

Scleroderma geaster (=S. polyrhizon), young specimens which have yet to split open. Note how thick the skin is! These were found along a road with *Pisolithus tinctorius* (p. 712).

and powdery. Spores (5) 6-11 (12) microns, round, with warts or spines that often form incomplete lines or ridges (i.e., partially reticulate). Peridium containing few if any thick-walled hyphae.

HABITAT: Solitary, scattered, or in groups on hillsides, along roads, in ditches, poor soil, sand, asphalt, gravel, etc.; often buried or partially buried before maturity. It is widely distributed and quite common in our area, especially in sandy soil. It usually appears in the fall, but the thick tough skins take a long time to decompose.

EDIBILITY: Poisonous? This is one fleshy fungus I've never been tempted to eat!

COMMENTS: The large size and tough skin that splits into coarse, thick rays are the telltale traits of this bizarre fungus. It is also known (more correctly) as *S. polyrhizon* or *S. polyrhizum.* It rivals *Pisolithus tinctorius* ("Dead Man's Foot") for grotesqueness. The two thrive in similar milieu—asphalt, sand, poor soil, etc.—and make a well-matched if not exactly charming couple. It might be mistaken for a large cup fungus (e.g., *Sarcosphaera*), but it usually has traces of the powdery spore mass to distinguish it. The rays lack the intact inner spore case characteristic of the earthstars, and there may or may not be a stemlike base of mycelial fibers. It can also be confused with *Mycenastrum corium,* but that puffball has a thick white felty outer peridium and a smooth, purple-brown inner one which splits into lobes at maturity. Other species: *S. texense (=S. fur-furellum)* is a very similar, thick-skinned species with a scalier (often shingled) exterior and thick-walled hyphae in the peridium; it has a more southerly (tropical and subtropical) distribution but also occurs on the west coast. *S. flavidum* (see comments under the *S. cepa* group) also splits in starlike fashion, but is much smaller, with a thinner (about 1 mm thick) peridium that is tawny and more or less smooth.

PISOLITHUS & DICTYOCEPHALOS (Oddballs)

Medium-sized to fairly large terrestrial fungi *found mostly in poor soil.* FRUITING BODY variously shaped at maturity. PERIDIUM variously colored and rupturing irregularly in old age. STERILE BASE absent, but a stalk often present. VOLVA absent *(Pisolithus)* or often present *(Dictyocephalos).* SPORE MASS *either composed of numerous small capsules (peridioles) or divided up into numerous "cells" or chambers;* brown and powdery at maturity, the peridioles or "cell" walls mostly disintegrating. Spores more or less round, warty or spiny. Capillitium absent.

THESE two oddballs are easily distinguished from puffballs and earthballs by the structure of their spore mass. In *Pisolithus* there are hundreds of small spore-containing capsules **(peridioles)** imbedded in the fruiting body. The peridioles disintegrate, however, so they are best seen by making a lengthwise section of a fairly young specimen. The

711

fruiting body is internally sticky when young, but at maturity it becomes crumbly and protrudes from the ground like a dusty stump, half-rotted root, or ball of dried-up dung. *Dictyocephalos* is just as unsightly, but has a long, woody stalk. It is likely to be mistaken for a stalked puffball at first glance, but its spore mass is divided into numerous "cells" or chambers. Most of the chamber walls disintegrate, but the lowermost ones can often be seen after the spore mass has dispersed. In each genus there is a single variable species with a worldwide distribution.

Key to Pisolithus & Dictyocephalos

1. Fruiting body with a long, tough or woody stalk; volva often present at base of stalk; spore mass chambered when young; found mostly in deserts ...... ***Dictyocephalos attenuatus,*** p. 713
1. Fruiting body lacking a volva and true stalk (but often with an elongated stalklike sterile base); spores borne in numerous lentil-like capsules imbedded in upper part of fruiting body (best seen when young); widespread in poor soils, along roads, etc. .. ***Pisolithus tinctorius,*** below

Pisolithus tinctorius   (Dead Man's Foot)   **Color Plates 185, 189**

FRUITING BODY 5-30 cm or more high and 4-20 cm broad, at first round to pear-shaped, then club-shaped (i.e., usually but not always with a narrowed, rooting base or "stalk"), finally breaking up in age or taking on the appearance of a large dusty root or stump; odor mild to aromatic or unpleasant, depending on the age. **SPORE CASE** with a thin, brittle peridium (skin) that is variously colored (but usually yellowish, purplish, olive-black, or brown) and often lustrous and soon ruptures irregularly or flakes away. Upper portion of fruiting body containing hundreds of seedlike peridioles which gradually disintegrate, turning into a crumbly or dusty mass of brown spores (the disintegration process starts at the top of fruiting body and proceeds downward). Peridioles small (2-4 mm long), elongated to oval or circular, whitish to greenish-yellow, yellow, brownish, or vinaceous, at first imbedded in a sticky dark or blackish substance which dries out and becomes brittle or crumbly at maturity. Lower portion of fruiting body typically consisting of a fibrous, stalklike, persistent, sterile, rooting base which may have coarse greenish-yellow mycelial fibers attached. **VOLVA** absent. **SPORE MASS** powdery when mature, dark brown to cinnamon-brown. Spores 7-12 microns, round, warty or spiny.

HABITAT: Solitary, widely scattered, or in small groups on or along roads, in waste places, and in hardpacked, poor, sandy, or gravelly soil; very widely distributed, but especially common in California and the Pacific Northwest (the largest fruitings I've seen were in the Siskiyou Mountains of northern California). In our area it fruits chiefly in the late summer and fall, but is slow to decay and can be found most any time. It forms mycorrhiza with a wide range of trees and shrubs (oak, pine, etc.) and is said to be of value in reforestation projects.

EDIBILITY: Not recommended. In Europe it is known as the "Bohemian Truffle" and used as an aromatic seasoning when unripe, and in China it is employed medicinally. It is scarcely appetizing, however, and should be tried *very* cautiously, if at all. The name *tinctorius* reflects its use as a dye—a variety of rich colors (mostly browns and golds, but also blacks and dark blues) can be obtained, depending on the mordants used and the type of soil in which it is found.

COMMENTS: This dusty monstrosity is among the most distinctive and memorable of all the fleshy fungi. Young specimens are somewhat puffball-like (i.e., pear- or club-shaped), but are easily told by the hundreds of seedlike peridioles they contain (see Color Plate 185). In this stage the peridium often has a beautiful metallic luster. Older specimens, on the other hand, can be recognized from a great distance by their dusty brown stumplike stance. Fresh specimens will stain practically anything they come into

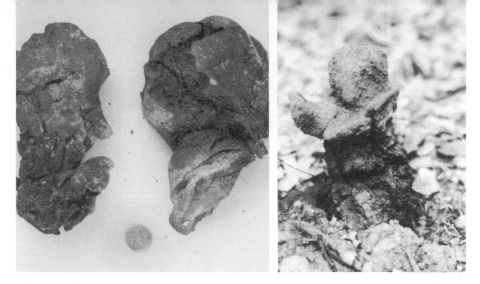

Pisolithus tinctorius. When young this monstrosity is puffball-shaped, but its outer skin ruptures **(left)**, exposing the powdery brown spore mass. Old specimens protrude from the ground like dusty stumps **(right)**, and bear little resemblance to puffballs or any other fleshy fungi. For other views of this distinctive fungus (including a close-up of the lentil-like spore capsules), see the color plates.

contact with, and the dry spore dust coats everything in the vicinity. This species has a penchant for adversity, inevitably fruiting in poor soil, ditches, chaparral, or road cuts (often in the company of "Dead Man's Hand," *Scleroderma geaster*), or even bursting up through asphalt. As might be expected, it is disdained by many mycologists ("It is the most objectionable of all fungi ..."; "This ugly, stinky fungus ..."), but to me it is one of the most enthralling—if not beautiful—of all fungi (I call it "Dead Man's Foot" not to demean it, but because it looks like one). Its wide distribution and variable features have resulted in a plethora of aliases, including *P. arenarius, P. arrhizus,* and *Polysaccum pisocarpium.*

Dictyocephalos attenuatus   (Stalked Oddball)

FRUITING BODY beginning underground as an "egg" completely encased in a membrane, the stalk then elongating rapidly and the ruptured outer membrane forming a volva at its base. Mature fruiting body consisting of a spore case mounted on a tough stalk. **SPORE CASE** round to somewhat flattened or even depressed on top, or irregular (e.g., sometimes with several lobes), 2-6 cm high and 5-13 cm broad, seated on a disclike enlargement of the stalk apex which is white to brownish in old specimens. Outer peridium (skin) fleshy to slightly gelatinous when young, becoming tougher and developing brown to reddish scales or warts which vary from small, flat, and more or less persistent to large, pyramidal and deciduous; lower margin sometimes with a hanging "veil" composed of adhering tissue stripped from the stalk during expansion. Inner layer of peridium 1-2 mm thick, usually paler (whitish to buff) than outer peridium, but sometimes darkening in age. Upper portion of spore case breaking up into irregular pieces which fall away; lower portion tougher and persisting as a more or less bowl-shaped structure. **STALK** (5) 10-40 (52) cm long, 2-5 cm thick at apex, equal or more often tapered downward (rarely thicker below or bulbous), the base sometimes pointed and free of the volva, sometimes with 1-2 rootlike mycelial cords attached; solid (or with insect cavities), sometimes forked (with one spore case on each branch), fleshy-tough becoming woody when dry; at first white, but becoming brown in age; often grooved, ridged, and/or flattened and sometimes twisted; often cracking, peeling, or curling back to form scales or sometimes even an irregular "ring" or "annulus." **VOLVA** at base of stalk saclike, but occasionally poorly developed or absent; white to tan when fresh, 3-11 cm high; tough, hard, or chalky in age.

713

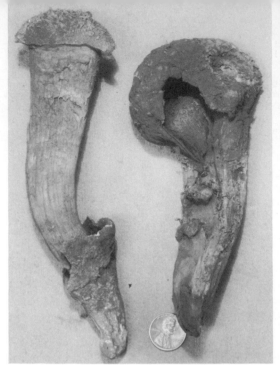
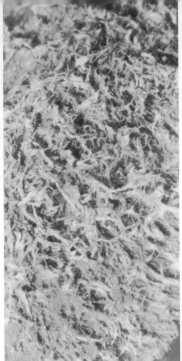

Dictyocephalos attenuatus. **Left:** Two typical dusty, mature specimens. The one on the left has lost most of its spore case but retains its volva, while the one on the right still has its spore case but has lost its volva. Note the well-developed stalk in both specimens. **Right:** A close-up of the lower part of a mature spore mass, showing the hairy or fibrous remains of the walls that originally divided the spore mass into numerous small chambers (a feature which separates it from the stalked puffballs).

SPORE MASS divided up into numerous whitish-walled "cells" (peridioles), but all but the lowermost "cell" walls disintegrating, so that the mature spore mass is brown and powdery, usually with an unpleasant odor (like decaying fish); lowermost "cell" walls often persisting after the spores have been dispersed as flattened, pointed "teeth" on the inner surface of the spore case, giving it a pitted or reticulate appearance. Spores 5-7 microns, round or nearly round, warted or spiny.

HABITAT: Solitary, widely scattered, or in small groups (often in tufts of 2-5 individuals and occasionally two emerging from a single volva) in barren sandy soil or clay, waste places, etc. It is widespread in the arid and semi-arid parts of the West, but is most common in the deserts of southern California, the Four Corners area, and the Southwest—often near or under saltbush *(Atriplex)* or along washes. It fruits after seasonal rains, but the fruiting bodies are practically impervious to bacterial and fungal decay, persisting for as long as 30 years in their natural environment! Mature specimens can protrude from the soil like dusty roots (in the grand tradition of *Pisolithus tinctorius*), or remain underground, depending on the hardness of the soil and weather conditions. W. H. Long, who gathered a gaggle of them near Antelope Valley, California, reports that developing specimens raised hard blocks of soil weighing 15 pounds each!

EDIBILITY: Much too crusty, musty, and dusty to be worthwhile, but possibly useful as a dye (see *Pisolithus tinctorius*).

COMMENTS: As can be surmised from the description, this "oddball," like *Pisolithus tinctorius,* exhibits a great deal of variation in size, shape, color, and degree of scaliness —in other words, it is not difficult to identify but is easy to *misidentify*. Its polymorphism has resulted in the naming of dozens of "new" species, when in fact there is only one highly variable one. It can be distinguished from the stalked puffballs and other desert fungi by

714

its chambered spore mass, the pitted-reticulate upper surface of old (empty) spore cases, the long and tough or woody stalk, plus the frequent presence of a volva and irregular breaking up of the peridium (a distinct apical pore is not formed). Its closest relative, *Pisolithus tinctorius,* lacks the well-developed stalk and volva, and does not usually grow in such desolate places.

Stalked Puffballs

spores

TULOSTOMATALES

Small to medium-sized puffballs often found in arid habitats or in sand or poor soil. FRUITING BODY *with a spore case mounted on a well-developed, differentiated stalk.* SPORE CASE typically with a two-layered peridium, rupturing apically, peripherally, stellately, or irregularly. STERILE BASE absent. STALK *not percurrent; usually tough or woody and often hairy or scaly* (but gelatinous in *Calostoma*). VOLVA present or absent. SPORE MASS *powdery* at maturity and buff to rusty-salmon to cinnamon-brown or dark brown. Spores typically round and ornamented, but occasionally smooth. Capillitium usually present, well-developed.

THESE are puffballs with a clearly differentiated, well-defined stalk. As in the true puffballs (Lycoperdales), the mature spore mass is powdery and the peridium (skin) that encloses it is composed of at least two layers. The presence of a stalk can lead to confusion with the gastroid agarics (Podaxales & Allies, p. 724), but in those fungi the stalk is percurrent, i.e., it extends through the spore mass to the top of the spore case or "cap," just like the stalk of a gilled mushroom. In the stalked puffballs, on the other hand, the stalk terminates at the spore case rather than extending through it.

Tulostoma, with over 30 species, is the largest and most cosmopolitan genus of stalked puffballs, yet it is by no means common. It can aptly be characterized as "a puffball on a stick" because that's just what it looks like, with its slender stalk and its small round spore case with an apical pore. There are six other genera of stalked puffballs, none with more than five North American species and several with only one. *Calostoma* is the most outlandish of the lot, with its brightly colored gelatinous fruiting body. *Battarrea* is also unmistakable, for its outer peridium forms a volva at the base of the stalk and the spore case ruptures *around its rim* or through *several* holes. *Chlamydopus,* like *Tulostoma,* forms an apical pore, but lacks the ball-and-socket relationship of stalk to spore case that characterizes that genus. *Schizostoma* ruptures stellately (that is, splits into several lobes), while *Phellorina* and *Queletia* rupture irregularly. An eighth genus, *Podaxis,* is often treated as a stalked puffball, but in this book is grouped with the gastroid agarics because it has a percurrent stalk.

With the notable exception of *Calostoma,* the stalked puffballs tend to grow in environments that most fleshy fungi find inhospitable—sand, poor soil, waste places, etc. Several like it *hot* and are found strictly in the desert, where they form a fairly large percentage of the fungi in the arid wastelands of the Great Basin and Southwest. *Calostoma,* as already noted, is the exception. It prefers the humid forests of eastern North America and is particularly common in the southern Appalachians.

None of the stalked puffballs are known to be poisonous. None are known to be edible either, because most of them spend their youth underground (a necessary adaptation to their harsh environment). As a result, they are invariably encountered *after* their spore mass is mature and powdery (and thus inedible), and the stalks are much too tough to eat, unless you're starving *and* barefoot (in other words, you're better off chewing on a shoe!). Six stalked puffballs are fully described here, and several others are keyed out.

Key to the Tulostomatales

1. Stalk (and often spore case) with a viscid or gelatinous outer layer when fresh; spore case rupturing through an apical pore at maturity; confined to humid regions, especially southeastern North America 2
1. Not as above 4

2. Spore case red when fresh (but may fade to yellow or buff) .. *Calostoma cinnabarina,* p. 718
2. Spore case yellow to buff when fresh (but mouth may be red) 3

3. Spore case with a viscid or gelatinous outer layer when fresh ....
   .... *Calostoma lutescens* (see *C. cinnabarina,* p. 718)
3. Spore case not viscid ... *Calostoma ravenelii & C. microsporum* (see *C. cinnabarina,* p. 718)

4. Spore case rupturing around its lower periphery, the top falling off like a lid in old age (as shown in photo on p. 717); stalk usually long, often scaly at maturity ....
   .... *Battarrea phalloides* & others, p. 717
4. Not as above; spore case not rupturing around its lower periphery 5

5. Cap cylindrical (shaggy mane-shaped); stalk extending into (and often through) the spore mass .... (see *Podaxis,* p. 725)
5. Not as above 6

6. Spore case typically rupturing through a single small mouth at the top (an apical pore) ... 7
6. Spore case rupturing through *several* pores or by splitting into lobes or by irregular disintegration 11

7. Stalk comprised of a mass of tough rootlike fibers; spore mass often purplish or black when young and firm (but often browner when powdery) .... (see *Scleroderma,* p. 707)
7. Not as above 8

8. Stalk thickest at apex, *not* inserted into the spore case like a ball-and-socket; volva present when fresh; found mostly in deserts .... *Chlamydopus meyenianus,* p. 721
8. Stalk sometimes thicker above but usually equal, with a ball-and-socket-like attachment to the spore case (but not necessarily separating easily); volva present or absent; found in many habitats, including deserts 9

9. Fruiting body medium-sized (spore case typically 1.2-3.5 cm broad; stalk typically 0.5-2 cm thick) .... *Tulostoma macrocephalum* & others, p. 720
9. Fruiting body usually rather small and slender (spore case typically up to 2 cm broad; stalk typically 2-6 mm thick) 10

10. Stalk with a volva at base and one or more "roots" below the volva; spore case white or whitish; found in desert .... *Tulostoma cretaceum* (see *T. brumale* group, p. 719)
10. Not with above features; volva present or absent; found in many habitats (including desert) .... *Tulostoma brumale* group & many others, p. 719

11. Spore case rupturing through *several* holes (pores); volva usually present when fresh; found in deserts .... *Battarrea digueti* (see *B. phalloides,* p. 717)
11. Spore case disintegrating, rupturing irregularly, or splitting into several lobes; volva present or absent; found in many habitats (including deserts) 12

12. Stalk continuous with spore case (i.e., bottom of spore case merely an expansion of stalk)   13
12. Stalk and spore case distinct, with a ball-and-socket-like attachment 15

13. Fruiting body full of spore-containing capsules when young and protruding from ground like a dusty stump in old age *or* spore mass composed of numerous cells or chambers (when young) whose walls usually give a pitted, reticulate, or hairy appearance to inside surface of bottom of spore case in old specimens; volva present or absent (see *Pisolithus & Dictyocephalos,* p. 711)
13. Not as above; spore mass lacking capsules, chambers, or "cells"; volva typically absent .. 14

14. Stalk composed of tough rootlike strands or fibers; spore mass often purplish to black when young and firm (but often browner when powdery); found in many habitats ....
    .... (see *Scleroderma,* p. 707)
14. Not as above; found in deserts .... *Phellorina strobilina,* p. 723

15. Spore case rupturing stellately (i.e., splitting into several starlike rays or lobes) when old; found in deserts .... *Schizostoma laceratum* (see *Chlamydopus meyenianus,* p. 721)
15. Spore case rupturing irregularly or disintegrating; very rare (reported from Pennsylvania) .... *Queletia mirabilis* (see *Chlamydopus meyenianus,* p. 721)

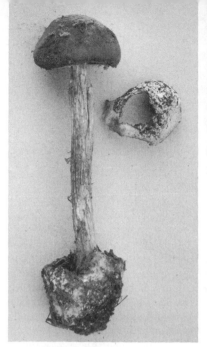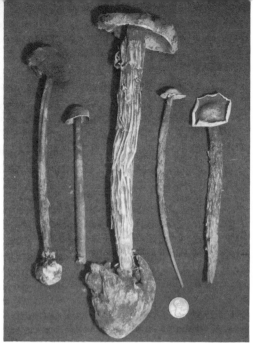

Battarrea phalloides. **Left:** A mature specimen in which the spore case has ruptured around its perimeter and the top has fallen off. Note large volva at base. **Right:** Old specimens, some of which have lost their volvas. Note how stalks vary considerably in thickness and degree of scaliness.

Battarrea phalloides   (Scaly-Stalked Puffball)   Color Plate 187

FRUITING BODY initially underground and completely enclosed by a membrane (outer peridium) which ruptures around its periphery, the stalk then bursting through and rapidly elongating. Mature fruiting body composed of a spore case (the inner peridium) mounted on a long stalk, with the ruptured outer peridium leaving a volva at base of stalk and sometimes a volval patch on top of the spore case. **SPORE CASE** 3-5 (6) cm broad and 1-3.5 cm high, typically rounded (convex) on top, with a flat or even depressed (concave) underside that somewhat resembles a "veil"; surface smooth, pallid or whitish, soon splitting around its lower periphery so that the upper (rounded) portion falls off, thereby exposing the spore mass; lower portion remaining intact as a disc beneath the spore mass and sometimes persisting (after spore mass has dispersed) as a disclike "cap" or sliding down the stalk to form a "ring." **STALK** 10-40 cm long, 0.4-1.5 cm thick, *not* percurrent, equal or tapered downward (but volva-encased base may be swollen); rigid and tough or even woody, hollow, white to brownish or rusty-brown, longitudinally striate or grooved and covered with fibers that often peel or split to form fine to very coarse needlelike, ribbonlike, or shaggy scales. **VOLVA** at base of stalk saclike, whitish or dirt-incrusted, often loose, buried in ground and soon rotting away. **SPORE MASS** rusty-brown and powdery when mature, adhering to everything it comes in contact with; odor sometimes unpleasant in old age. Spores (4) 5-8 (10) microns, round to broadly elliptical, warted. Capillitium of two types: simple, short hyphae up to 10 microns long, and elators (elongated cells with internal spirals) up to 100 microns long.

HABITAT: Solitary, widely scattered, or in groups in sand, poor soil, and waste places (but often under trees in deep shade); very widely distributed, but in the United States apparently confined to the West. It is generally regarded as rare, but in the deserts and sagebrush country of the Great Basin and Southwest it is relatively common for a desert fungus. It also occurs along the California coast and Orson Miller reports it from Alaska. In our area it fruits during the rainy season but persists for months without decaying.

EDIBILITY: Inedible.

717

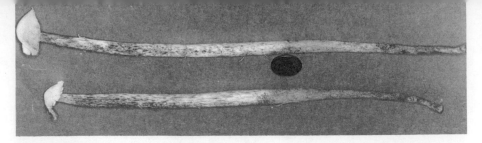

Battarrea phalloides, old specimens bereft of volva and spores. Only the bottom of the spore case remains, forming a thin "cap" on the stem. In this condition they are as light as straws.

COMMENTS: The fibrous-scaly stalk, presence of a volva (which, however, often rots away or can be left behind in the ground), and curious manner in which the spore case ruptures around its periphery are the telltale traits of this intriguing fungus. The flattened or concave, veil-like underside of the spore case is also unique. The upper portion or "lid" usually detaches completely; it can often be found on the ground nearby, but is sometimes blown away by the wind. The volval patch (when present) may tear away from the "lid," leaving a gaping hole as shown in the photograph. The stature of the fruiting body varies considerably according to soil and weather conditions. Some specimens (especially the coastal ones) have relatively thick, coarsely scaly stalks, whereas others (particularly desert dwellers) have longer, thinner stalks. All sorts of intergradations can be found, however. Some authorities claim that the "true" *B. phalloides* of Europe has a gelatinous-layered volva in the egg stage. If this is really the case, then the American species, which lacks the gelatinous layer, is more correctly called *B. stevenii.* (Since the "eggs" develop deep in the ground, they are rarely encountered and it is difficult to know whether or not this distinction is a valid one.) Another questionably distinct species, *B. laciniata,* has a more persistent, multi-layered volva with inner, concentrically-arranged "leaflets" around the stalk base. It usually has a white to buff or reddish volval patch on top of the spore case and tends to be more robust (stalk 2-4 cm thick at apex and spore case 4-8 cm broad, 2-4 cm high). It occurs in southern California and the Southwest, usually in "rich loamy alkaline soil" in the open desert (not under trees). Still another species, *B. digueti,* is *definitely* distinct, for it does not rupture by peripheral cleavage. Instead, its outer peridium ruptures apically (thus never leaving a volval patch) and its spore case *remains intact,* with several holes forming in the upper (convex) portion. It also has longer elators than *B. phalloides,* but in other respects (shape and overall appearance) is quite similar. It appears to be strictly a desert fungus. I have found it in Baja California, and it is also known from southern California, mainland Mexico, and the Southwest.

Calostoma cinnabarina (Red Slimy-Stalked Puffball) Color Plate 188

FRUITING BODY consisting of a spore case mounted on a thick, short, fluted stalk (the spore case often barely protruding above the ground). **SPORE CASE** 0.5-2 cm broad, more or less round to somewhat oval, at first covered by the outer peridium, which is composed of a thick, gelatinous or slimy, translucent layer and a thin, bright red inner layer; outer peridium breaking up into pieces and falling away (the pieces of the inner layer often imbedded in the gelatinous pulp of the outer layer so as to resemble small tomato or pomegranate seeds). Inner peridium persistent, *not* gelatinous, at first red and scurfy or powdery, but often fading to orange, yellow, or buff as the scurf wears away; rupturing through a crosslike apical pore (mouth) formed by 4-5 dark red to scarlet ridges or "teeth." **STALK** 1.5-4 cm long, 0.5-2 cm thick, typically thick and short, more or less equal, spongy, with an outer gelatinous layer when fresh; otherwise composed of firm, branching strands that form a ridged, netted, and/or pitted pattern; ochraceous to red or cinnabar-red to yellow-brown, often fading in age. **VOLVA** absent. **SPORE MASS** white at first, becoming powdery and buff-colored at maturity. Spores 14-28 × 6-11 microns, elliptical to oblong, pitted. Capillitium present when young but soon disintegrating.

Calostoma cinnabarina. Since these mature specimens are dried out, the gelatinous layers evident in the color plate do not show. The bright red color (when fresh) and thick, fibrous stalk are distinctive.

HABITAT: Solitary to gregarious in soil and humus in woods, along roadcuts, under trees, etc.; fairly common in the late summer and fall in the southern and eastern United States, especially at higher elevations. It occurs as far west as Texas, and is to be looked for in southern Arizona and New Mexico, where several "eastern" mushrooms occur.

EDIBILITY: Unknown—but too small and slimy to merit experimentation.

COMMENTS: In a group (the puffballs) not noted for its bright colors, the genus *Calostoma* stands out. The complex structure of the outer and inner peridiums (each has at least two layers) and peculiar microscopic features such as the absence of capillitium in the mature spore mass have made it difficult for taxonomists to relate it to other puffball genera. The thick gelatinous outer layer, which can be likened to a universal veil, distinguishes *Calostoma* from other stalked puffballs, and the powdery spore mass separates it from the stinkhorns, which may also be slimy. *C. cinnabarina* is the only red *Calostoma*. Others—all southeastern in distribution—include: *C. lutescens,* with a longer stalk that usually elevates the spore case above the ground, and a light to bright yellow spore case (except for the red mouth or pore) that usually has a wide collar at its base formed by the outer peridium; *C. ravenelii,* a smaller species with a gelatinous-fibered stalk and non-gelatinous spore case that is tan to yellowish (with a red mouth) and often decorated with warts left by the outer peridium; and *C. microsporum,* rather similar to *C. ravenelii* but slightly larger, with smaller spores (up to 11 microns long).

Tulostoma brumale group   (Common Stalked Puffball)

FRUITING BODY consisting of a small spore case mounted on a stalk. **SPORE CASE** round (but often somewhat collapsed or flattened), 1-2 cm broad, the underside shrinking as it dries and often pulling away from the stalk to form a collar. Outer layer of peridium (skin) brownish, but covered with sand and other debris, peeling away to reveal the inner layer, but often remaining intact around the base. Inner layer smooth or with minute particles, tan to pinkish-buff to buff, gray, or whitish, developing an apical pore at maturity; pore usually encircled by a raised collar or tube. **STALK** 1-6 cm long, 2-4 (6) mm thick, usually slender, equal except for a small bulb at base; tough but usually pliant, covered at first with coarse brown to rusty-brown fibers, but sometimes smoother and/ or paler in age; longitudinally striate, joined to spore case via a ball-and-socket arrangement. **VOLVA** absent or inconspicuous. **SPORE MASS** rusty-salmon and powdery at maturity. Spores 3-5 microns, more or less round, minutely warted. Capillitium well-developed.

HABITAT: Scattered or in groups in sand and gravel, sandy soil, and other waste places (even in the mortar of brick walls!); very widely distributed, but not particularly common.

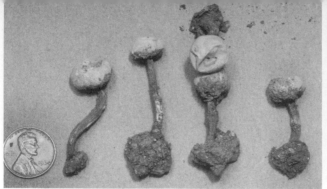

Tulostoma berteroanum (see comments under *T. brumale* group) is probably the most common stalked puffball in coastal California. **Left:** Stocky specimens. **Right:** More slender ones. Note the small pore that forms at the top of each spore case.

In our area this species and its close relatives (see comments) fruit mainly in the fall and winter, but the weathered fruiting bodies can be found most any time.

EDIBILITY: Inedible when mature, and rarely found in the immature stage.

COMMENTS: *Tulostoma* species are easily recognized by their slender stalk, small spore case, and rusty-salmon spores. The fruiting body develops underground as in *Battarrea,* and the stalks of mature specimens are often buried so that only the spore case is visible. The absence of a volva in most species plus the small size and formation of an apical pore distinguish *Tulostoma* from *Battarrea,* and the stalk is not as consistently thickened at the apex as it is in *Chlamydopus.* There are several closely related Tulostomas that will more or less fit the above description, including *T. berteroanum* (see photo), the most common in our area. They are partial to sandy soil and are best differentiated microscopically (in other words, leave them to the *Tulostoma* taxonomists). Some of the more widespread species are: *T. striatum,* which has ridged spores and an outer peridium that leaves an acornlike cup around the lower half of the spore case; *T. simulans,* in which the spore case is persistently covered by adhering sand particles and other debris; *T. campestre,* also sand-incrusted, but lacking a well-defined tube around the apical pore; and *T. fibrillosum,* a fairly common species which also lacks a tube but has practically smooth spores and a longer, coarsely hairy stalk.

There are also several species apparently endemic to the Southwest (or at least to arid regions), including: *T. involucratum,* common, with a membranous (not granular) outer peridium that often forms a frilled cup around the inner peridium plus a tubular apical pore; *T. opacum,* a rare species with large spores (7-11 microns); *T. meristostoma,* small, with a slit or irregular tear instead of a raised apical pore or tube *even when fresh* (in many species the pore becomes slitlike or irregularly torn after weathering); *T. cretaceum,* a common and highly distinctive sand-loving species with a chalk-white spore case and a stalk that tapers downward to a volva beneath which are one or more rootlike processes (at least in most specimens); and *T. excentricum,* with an off-center tubular mouth and inconspicuous volva plus a spore case which is *not* white unless weathered. All of these species differ from *Chlamydopus meyenianus* in the ball-and-socket relationship between the spore case and stalk (though the two may be firmly attached), and all are rather small and slender-stemmed. For larger species, see *T. macrocephalum.*

Tulostoma macrocephalum   (Fat-Headed Tulostoma)

FRUITING BODY consisting of a spore case mounted on a stalk. **SPORE CASE** 1.2-3 cm broad and 0.8-1.5 cm high, nearly round to somewhat flattened (usually broader than it is high), firmly attached to the stalk. Outer layer of peridium (skin) grayish-white to buff and soon falling away to reveal the inner layer, but often remaining intact around the base. Inner layer smooth, dingy whitish, tough, developing a short tubular apical pore (mouth)

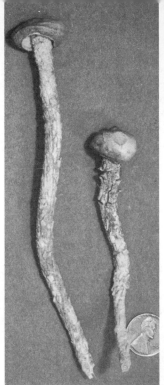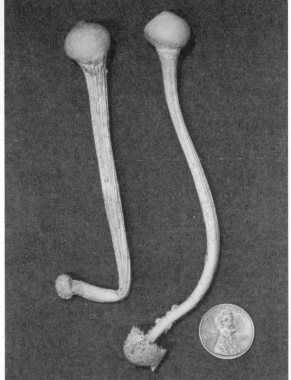

Left: *Tulostoma macrocephalum* has a thicker stem and larger "head" than most Tulostomas. **Right:** *Chlamydopus meyenianus,* rather small specimens in which the rough outer peridium (skin) has completely worn away, leaving the beautiful smooth inner layer. Note volva in specimen on right.

at maturity. **STALK** 5-15 cm long, 0.5-1.5 cm thick, usually rather long, equal or tapered slightly toward the base, which usually has a bulb; tough or woody, usually scaly and/or transversely cracked. **VOLVA** absent or present only as a small fragile dirt-incrusted sack or collar at base of stalk. **SPORE MASS** rusty-salmon to cinnamon-brown and powdery at maturity. Spores 4-5.5 microns, round, warted. Capillitium present.

HABITAT: Solitary to gregarious in sand or sandy soil in arid regions, fruiting most any time if rainfall is sufficient. It was originally found in the gypsum dunes of White Sands National Monument in southern New Mexico (I have seen it there), but has also turned up in southern California and can probably be found throughout the Southwest.

EDIBILITY: Inedible because of its toughness.

COMMENTS: The broad head, scaly stalk, and relatively large size (for a *Tulostoma*) are the hallmarks of this distinctive stalked puffball. The presence of an apical pore distinguishes it from *Battarrea,* and the equal or only slightly tapered stalk separates it from *Chlamydopus.* It is not likely to be found by the average mushroom hunter, but any mushroom willing to grow in gypsum crystals warrants attention. Another robust southwestern species, *T. lysocephalum,* has a coarser, dirtier appearance and its soil-incrusted "head" topples off easily and is often found beside the stalk. It grows under mesquite and other desert shrubs, but is not common.

Chlamydopus meyenianus   (Desert Stalked Puffball)

FRUITING BODY beginning as an underground "egg" encased in an outer peridium, the stalk then elongating rapidly; outer peridium rupturing around its periphery, the lower half forming a volva at base of stalk, the upper portion forming a patch of warty, brittle tissue that breaks up and disintegrates or occasionally clings to the top of the spore

721

case. Mature fruiting body consisting of a spore case (inner peridium) mounted on a stalk. **SPORE CASE** round to somewhat flattened, 1-3.5 cm broad and 0.5-2 cm high, tough, persistent, attached firmly to the stalk; surface smooth or slightly roughened, pallid to buff, pinkish-buff, yellowish, or sometimes cinnamon, developing an apical pore at maturity which may enlarge in age to form a "mouth." **STALK** 4-15 (35) cm long, 0.2-1.5 (3.5) cm thick at apex, tapering downward, usually solid and tough or rather woody, often curved and/or flattened; usually longitudinally striate or grooved, smooth to silky-fibrillose or sometimes fibrillose-scaly; colored more or less like spore case or browner; *not* percurrent. **VOLVA** at base of stalk saclike, two-layered, thick, often rotting away or staying behind in the ground; white to brownish, usually incrusted with dirt or sand. **SPORE MASS** rusty to ochraceous to brown and powdery at maturity. Spores 5.5-9 microns, round, spiny (or rarely smooth). Capillitium present.

HABITAT: Solitary to scattered or in small groups in sandy, gravelly, or volcanic soil, sand dunes, gypsum flats, and other barren places; sometimes also in adobe soil. It is found throughout the arid and semi-arid regions of the world (Australia, North Africa, etc.) and was originally described from Peru. It is widespread in western North America, but is common only in the Southwest—and then only sporadically. I have seen it near the Painted Hills in eastern Oregon and outside San Bernardino in southern California. It fruits after heavy rains but, like other stalked puffballs, persists for months afterward.

EDIBILITY: Worthless—it is much too tough and fibrous, even as a substitute for beef jerky.

COMMENTS: Like other desert-dwelling stalked puffballs, this species is quite variable in size and appearance but still easy to recognize. When still covered by the warty outer peridium it is reminiscent of *Phellorina strobilina,* but the presence of a volva (in most specimens) distinguishes it. Once it has shed its outer coat (as shown in photo on previous page), it resembles the cosmopolitan genus *Tulostoma,* but is slightly larger, lacks the ball-and-socket connection of spore case to stalk typical of that genus, and is characteristically thickened at the apex of the stalk and tapered downward. The presence of a volva is also noteworthy, but it decays more quickly than the rest of the fruiting body and is not always evident in older specimens. The tendency of the spore case to rupture through a single pore distinguishes it from *Battarrea,* which also has a volva. Two other distinctive stalked puffballs should also be mentioned. *Schizostoma laceratum* is a desert dweller whose spore case ruptures along sutures to form rays or lobes which subsequently spread out somewhat (see photo). The spore case is attached firmly to the stalk via a ball-and-socket arrangement (as in *Tulostoma*) and there is no volva, though the base of the stalk may split to form a free rim or collar. *Queletia mirabilis* resembles *Schizostoma,* but has a spore case that ruptures irregularly. It is widely distributed but rare—there is one report of it from Pennsylvania.

Schizostoma laceratum (see comments above) is easily distinguished from other stalked puffballs by the way in which its spore case ruptures into rays in old age.

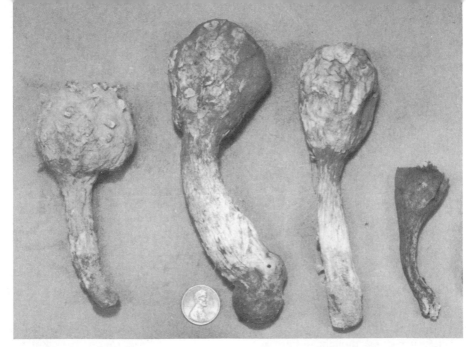

Phellorina strobilina. This desert dweller releases its spores through general disintegration of the spore case, leaving an urnlike structure behind (specimen at far right). Note presence of large warts in the youngest individual (far left).

Phellorina strobilina   (Urnlike Stalked Puffball)

FRUITING BODY consisting of a spore case which narrows downward into a distinct stalk. **SPORE CASE** round to oval or pear-shaped, 1.5-6 cm broad and/ or high. Outer layer of peridium (skin) scaly, the scales erect or pyramidal and often quite large on top, usually smaller and more or less shingled below; white when fresh but soon becoming brown to cinnamon and rupturing irregularly and falling away in pieces, exposing the inner layer. Inner peridium a cup- or urnlike expansion of the stalk apex (i.e., completely continuous with the stalk), pinkish-buff to cinnamon, with a large (2-5 cm) irregular mouth forming at the top in age. **STALK** 2-5 cm long, 0.8-2 cm thick at apex, dilated at apex and tapered downward, but with a bulb at the base; firm, solid, woody, brown or cinnamon and scaly like the outer peridium of the spore case (but scales may wear away in age); *not* percurrent. **VOLVA** typically absent, but the basal bulb sometimes splitting or peeling to form a volva-like collar or rim. **SPORE MASS** powdery when mature and cinnamon to rusty-brown. Spores 5-7 × 4.5-6 microns, round or nearly round, warty or spiny. Capillitium present but often scanty.

HABITAT: Solitary or in small groups (occasionally two or three growing from a common base) in desert soil, usually near shrubs; widely distributed in the warm, arid regions of the world. It occurs throughout the Southwest after heavy rains but is not common and only rarely fruits in large numbers. I have seen it outside Reno, Nevada and near Tucson, Arizona as well as in India.

EDIBILITY: Much too woody to be edible.

COMMENTS: Also known as *P. inquinans,* this peculiar stalked puffball is reminiscent of the common desert fungus *Podaxis pistillaris* when young, but its stalk is not percurrent (i.e., it terminates at the base of the spore case). The scales flake off the spore case or "cap" and the top eventually ruptures to form a large mouth so that old specimens resemble wine goblets. These might be confused with *Dictyocephalos attenuatus,* but the spore mass lacks the chambers characteristic of that species and there is normally no volva at the base of the stalk.

723

spores

Gastroid Agarics

PODAXALES & Allies

THIS motley assortment of bizarre and inelegant fungi combine the features of an agaric (presence of a cap and stalk and a spore mass composed of chambers, plates, or rudimentary gills) with those of a Gasteromycete (lack of forcible spore discharge and a frequently enclosed spore mass). Most of them look like gilled mushrooms that have failed to open or develop properly, and that is precisely what they are most frequently mistaken for. A few, such as *Montagnea arenarius,* have a fully exposed spore mass at maturity, but they lack true gills and do not forcibly discharge their spores. Those with an enclosed spore mass might be mistaken for stalked puffballs, but their stalk is **percurrent** (i.e., it extends all the way through the spore mass to the top of the cap, just as in a gilled mushroom). In the stalked puffballs, on the other hand, the stalk terminates at the bottom of the spore case or "cap."

The gastroid agarics are sometimes called "secotioid" fungi because many of them were originally grouped in a single genus, *Secotium.* To facilitate identification they are retained together here as an artificial subgroup of the Gasteromycetes. It should be realized, however, that they are thought to have evolved independently from each other, and that toadstool taxonomists try to express their true relationships by scattering them among the various families of gilled mushrooms which they resemble. Most of them thrive in severe environments (e.g., deserts or high mountains), suggesting that they are indeed metamorphosed agarics which, through a process of adaptation and specialization, have lost the ability or necessity to forcibly discharge their spores (hence they also go by the unflattering moniker "reduced agarics"). *Podaxis pistillaris,* a common desert fungus, is a good example: it resembles the shaggy mane *(Coprinus comatus),* but its cap remains closed to protect the developing spores from the desert heat, and the mature spores are thick-walled to prevent moisture loss. Taxonomists still argue about its relationship to *Coprinus,* but most of the other gastroid agarics have clearcut counterparts among the gilled mushrooms—e.g., *Longula* is closely related to *Agaricus, Montagnea* to *Coprinus, Brauniellula* to *Chroogomphus, Macowanites* and *Arcangeliella* to *Russula* and *Lactarius,* and so forth. Some gastroid agarics are closely related to the false truffles. As defined here, however, the false truffles lack a stalk and/or grow underground.

The gastroid agarics are particularly prominent and diverse in western North America. They fruit after seasonal rains but the tougher types persist, like the stalked puffballs, for weeks or even months before decaying. Some are said to be edible when very young, but as a rule they are better observed or preserved than served.

Key to the Podaxales & Allies

1. Spore mass composed of tubes or empty tubular chambers 2
1. Spore mass solid, slimy, or powdery or with rudimentary gills, plates, or cavities 3
2. Fruiting body resembling a misshapen or aborted bolete; spore mass (tubes) exposed at maturity, lining the underside of cap .... (see *Gastroboletus,* p. 544)
2. Spore mass enclosed by a peridium (skin); not as above (see **Hymenogastrales & Allies,** p. 739)
3. Fruiting body averaging 3-8 cm high, narrowly to broadly club-shaped or with an oval cap; peridium (skin) of "cap" often wearing away in age; spore mass greenish-brown to brown and often slimy, or becoming powdery in old age; found on living or dead wood (especially pine) and other debris in humid regions (e.g., southeastern U.S.) . ***Rhopalogaster transversarium***
3. Fruiting body not as above, or if similar then habitat different 4
4. Fruiting body shaggy-mane-like; cap oval to cylindrical (much taller than it is wide), usually with shaggy scales or fibrils which may, however, wear off in age; spore mass enclosed, at least until very old age; found in deserts and other hot, arid environments .... ***Podaxis,*** p. 725
4. Not as above; differently shaped (including narrowly conical), etc. 5

5. Fruiting body *Russula*- or *Lactarius*-like: mature spore mass or "gills" white to yellowish, ochraceous, or pinkish; stalk also pale in color; fruiting body crisp, brittle or fragile, the stalk typically snapping open cleanly like a piece of chalk; spores colorless under the microscope, with amyloid ornamentation; found in forests or under trees   ***Macowanites, Arcangeliella,* & Allies,** p. 736
5. Spore mass brown to black at maturity, or if paler then not as above 6
6. Flesh in stalk orange to ochraceous to vinaceous when fresh and mature spore mass smoky to gray or black; found under conifers .... ***Brauniellula,*** p. 732
6. Not as above (but flesh may stain yellow) 7
7. Flesh white in upper part of stalk, bright yellow in base; spore mass minutely chambered; found under conifers .... ***Gomphogaster leucosarx*** (see *Brauniellula nancyae,* p. 732)
7. Not as above 8
8. Growing on hardwoods in Midwest; spore mass or "gills" white; cap usually with small brownish scales .... ***Lentinus tigrinus*** (gastroid form) (see *Lentinus & Lentinellus,* p. 141)
8. Growing on ground, or if on wood then not as above 9
9. Cap narrowly conical and not expanding appreciably (see photo on p. 734, top right); usually found in grass or alpine meadows .... ***Thaxterogaster, Nivatogastrium,* & Allies,** p. 733
9. Not as above 10
10. Fruiting body *Agaricus*-like or *Coprinus*-like; mature spore mass or "gills" dark brown to black; terrestrial .... ***Endoptychum* & Allies,** p. 727
10. Not as above (mature spore mass may be paler brown or grayish); found on wood or ground 11
11. Fruiting body somewhat *Lepiota*- or puffball-like, at least when young; spore mass often white at first but yellowish to brown or grayish and often powdery in age; fruiting body sometimes broadly conical but not narrowly conical; growing mainly in grass, cultivated earth, deserts, open places, etc.; terrestrial .... ***Endoptychum* & Allies,** p. 727
11. Not as above; spore mass soon yellow-brown to cinnamon-brown or reddish-brown; cap very narrowly conical, or if not then usually found on wood or on ground in forests and under trees (including eucalyptus) .... ***Thaxterogaster, Nivatogastrium,* & Allies,** p. 733

PODAXIS

THIS genus occurs throughout the hotter parts of the world. It displays several interesting protective adaptations which enable it to survive in its hostile environment, e.g., the spores are thick-walled to prevent moisture loss and the cap never opens out. The single species described here is easily recognized by its tough, shaggy mane-like fruiting body and powdery mature spore mass. The presence of capillitium (see comments) is also distinctive.

Podaxis pistillaris   (Desert Shaggy Mane)

FRUITING BODY with a cap and stalk. **CAP** 2-15 cm high and 1-4 cm broad, oval to cylindrical or sometimes somewhat twisted; surface dry, pure white to tan, yellow-brown, or brown; typically breaking up to form shaggy fibrils or scales which may eventually peel or wear away to reveal the smooth undersurface; eventually tearing radially or irregularly, the margin usually remaining attached to the stalk for a long time. **SPORE MASS** consisting of contorted plates or rudimentary gills; at first white, soon darkening to yellowish, then reddish, and finally reddish-brown to dark brown to blackish; powdery at maturity (but liquefying slightly under certain conditions). **STALK** 4-15 (26) cm long, 0.2-1 (1.5) cm thick, equal above, the base usually swollen; extending into the spore mass and often percurrent; usually long and slender, rather tough and fibrillose to raggedly scaly (but often smoother in age); solid or hollow, often twisted-striate; white, or in age often discolored like the cap. **VEIL** not clearly differentiated from cap and stalk tissue. **VOLVA** absent. **SPORE PRINT** not obtainable; spores averaging (7) 10-16 (20) × (5) 9-15 microns, but sometimes much larger (to 36 microns!); round to broadly elliptical, pear-shaped, or irregular, smooth, thick-walled, with an apical germ pore. Capillitium present.

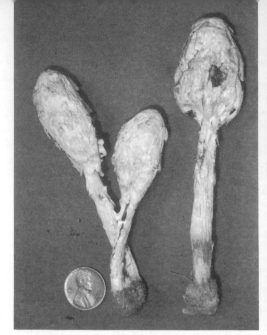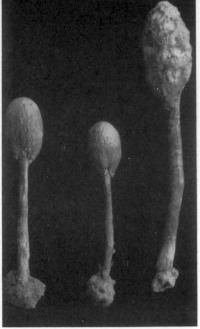

Podaxis pistillaris looks like a shaggy mane, but lacks gills and grows in the desert. Stalk is percurrent (i.e., it extends into the spore case or "cap"). **Left:** Typical specimens with a shaggy or scaly cap. **Right:** Weathered specimens in which the scaly layer has worn away, revealing the smooth undersurface.

HABITAT: Solitary or scattered to gregarious or clustered on ground in deserts, washes, lawns, gardens, irrigated fields, etc.; occurring throughout the warm, dry parts of the world (known from every continent except Antarctica). It is the most prominent desert fungus of the western United States, showing a preference for sandy soil but also coming up in clay or adobe. In California it has been found from below sea level to over 5,000 feet (in the Panamint Mountains). It usually fruits between April and October, but the fruiting bodies persist for months and consequently can be found most any time. I have seen it growing outside the Taj Mahal in India, accompanied by *Agaricus bitorquis*.

EDIBILITY: Said to be edible when very young (white inside), but tough and dusty in age.

COMMENTS: This distinctive denizen of the desert looks like a shaggy mane *(Coprinus comatus)* but does not have gills, does not normally deliquesce, and has a powdery spore mass when mature. To the feverish, sun-fried, dust-incrusted fungophile driving deliriously across the monotonous mushroom-meager desert while dreaming of cool coastal pine forests bulging with boletes and flower-filled mountain meadows overflowing with *Agaricus*, it is likely to be mistaken for a miraculous mirage or wistful hallucination. To the methodical mushroom taxonomist, on the other hand, it is an evolutionary anomaly: the stalk is percurrent as in other gastroid agarics, but the spore mass contains capillitium as in the stalked puffballs. (Some taxonomists consider it to be a "gastroid" relative of *Coprinus,* while others sequester it in an exclusive order of its own, the Podaxales.) The percurrent stalk distinguishes it from the similarly-shaped *Phellorina strobilina* (a stalked puffball), and the shape and habitat separate it from other gastroid agarics and stalked puffballs (but see *Rhopalogaster transversarium* in key on p. 724). It's interesting to note that, like the shaggy mane, it sometimes grows with *Agaricus bitorquis*. Perhaps since the latter fruits underground, it has a greater temperature range than either the shaggy mane (which likes it cool) or the desert shaggy mane (which likes it hot). Other species: *P. argentinus,* with an olive-brown to yellow-brown mature spore mass, and *P. microsporus,* with a reddish-brown mature spore mass, are both very similar but have much smaller spores (5-9 microns long). *P. longii* also has smaller spores, but has a large and robust fruiting body (17-45 cm tall, stalk 1-2.5 cm thick at apex). All of these occur in the Southwest, but are not nearly as common as *P. pistillaris*.

ENDOPTYCHUM & Allies

Medium-sized to fairly large, terrestrial fungi found mostly in arid places. FRUITING BODY *with a cap and stalk.* CAP variously shaped. Flesh *usually white when fresh,* but often staining or discoloring. SPORE MASS *typically composed of plates, irregularly contorted gills that may branch to form cavities, etc.; enclosed or exposed, typically brown to deep brown or black at maturity and sometimes powdery.* STALK *percurrent,* long to very short. VEIL and/ or VOLVA *typically present when fresh,* but sometimes disintegrating or easily overlooked. SPORE PRINT *not obtainable.* Spores yellow-brown to dark brown under the microscope, round to elliptical, smooth. Capillitium absent.

THESE dark-spored fungi are remarkably reminiscent of *Agaricus* and *Coprinus,* but do not forcibly discharge their spores. Five genera are treated here. The central genus, *Endoptychum,* may someday be divided into two genera because some of its species (e.g., *E. depressum*) are obviously related to *Agaricus,* while others (e.g., *E. agaricoides*) may be closer to the parasol mushrooms (*Lepiota* and *Chlorophyllum*). In *Endoptychum* the spore mass is typically enclosed by the cap and becomes rather powdery in old age. A similar genus, *Longula,* usually has an exposed spore mass at maturity. It also resembles *Agaricus,* but has congested or disfigured "gills." *Neosecotium* is a small genus which, like *Endoptychum,* is probably related to the Lepiotas. The other two genera, *Montagnea* and *Gyrophragmium,* have blackish spores and very small, dislike caps with a fully exposed spore mass. They are thought to be related to the genus *Coprinus,* although the name *Gyrophragmium* has also been used for *Agaricus*-like forms.

Most of the above fungi fruit in hot, open, arid or semi-arid habitats (*E. depressum,* however, favors mountain conifers). None are known to be poisonous, but they are usually found after the spore mass has matured and the fruiting body has toughened.

Key to Endoptychum & Allies

1. Cap thin and dislike (less than 4 cm broad), with spore-bearing plates or "gills" hanging from its underside or margin; mature spore mass black 2
1. Spore mass enclosed for a long time, or if exposed then not as above 3
2. Spore-bearing plates or "gills" attached to *underside* of cap; rare .... **Gyrophragmium californicum** (see *Montagnea arenarius,* below)
2. Spore-bearing plates hanging from margin of cap; fairly common **Montagnea arenarius,** below
3. Found with mountain conifers (or aspen); mature spore mass chocolate-brown to black; *young* fruiting body often with a sweet odor when broken open . **Endoptychum depressum,** p. 730
3. Not with above features; usually found in deserts, lawns, gardens, open areas, etc. 4
4. Mature spore mass usually exposed and brown to blackish .... **Longula texensis,** p. 729
4. Spore mass usually enclosed until very old age, white becoming yellowish to brownish (but never chocolate-brown or blackish) 5
5. Spore mass containing minute peridioles (spore-bearing capsules) which look like grains of sand; stalk separating easily from cap .... (see **Lycoperdales & Allies,** p. 677)
5. Not as above .... **Endoptychum agaricoides** & others, p. 731

Montagnea arenarius (Gastroid Coprinus)

FRUITING BODY at first deeply buried and enclosed in a tough membrane, then expanding as the ruptured outer skin forms a volva at base of stalk. Mature fruiting body with a stalk and dislike cap (the cap first appearing oval, but all but the disc (center) soon splitting into spore-bearing plates. **CAP** 1-3.5 cm broad, merely a thin dislike expansion of the stalk; convex becoming plane or depressed, persistent; surface smooth, white to grayish, buff, or occasionally straw-colored, often with a volval patch or remnants; margin often tattered or fringed in age. Flesh (in stalk) white when fresh. **SPORE MASS** exposed at maturity, composed of thin plates which hang from the margin of the dislike cap

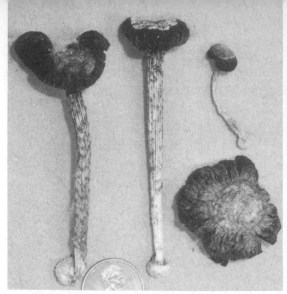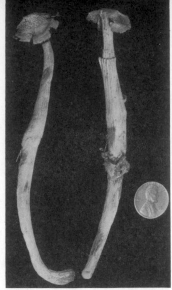

Left: *Montagnea arenarius,* small specimens. Note how central disc or "cap" has "gills" radiating from its edge. **Right:** *Gyrophragmium californicum* (see comments under *Montagnea arenarius*), a rare mushroom whose "gills" are attached to underside of cap. Both species have a volva when fresh.

and are entirely free from the stalk; plates often wavy or curled, up to 3.5 (6) cm long, reddish-black to blackish at maturity, eventually falling off the cap or disintegrating, but not deliquescing. **STALK** (5) 8-30 cm long, 0.2-1.5 (2.5) cm thick, percurrent, more or less equal or often tapering downward, hollow, tough or almost woody when old and dry (but very light); smooth or longitudinally fibrillose-striate, often splitting or cracking into fibrillose or shaggy scales, white to buff or sometimes discoloring darker in old age. **VEIL** absent or rudimentary. **VOLVA** at base of stalk saclike, usually buried in soil, loose (often remaining in ground), two-layered, the outer layer white and ample, the inner layer composed of tough fibers. **SPORE PRINT** not obtainable; spores (7.5) 12-20 (28) × (4.5) 6-11 (14) microns, elliptical to nearly round, smooth, with a germ pore. Capillitium absent.

HABITAT: Solitary, scattered, or gregarious in sandy soil, old fields, and other waste places; widely distributed and fairly common in the arid and semi-arid parts of western North America (from Mexico and Texas to California and eastern Washington). It has been found high up on the slopes of Mount Shasta, and by the hundreds in fields in eastern Oregon. In California it is quite common inland. I have yet to find it in our area, but the very similar *Gyrophragmium californicum* (see comments) does occur rarely in sandy soil.

EDIBILITY: Like myself, too thin and tough to be of value.

COMMENTS: This odd but interesting fungus can be told by its long stalk and thin disclike cap from whose margin the spore-bearing plates are suspended. The presence of a volva is also distinctive, but the volva is deeply buried and easily left behind in the ground (or sometimes rots away). The size and stature vary considerably—those from Oregon and north-central California being rather small and slender (usually less than 20 cm tall), and those from southern California and the Southwest being somewhat taller (20-30 cm) and often thicker. The blackish spores and thin spore-bearing plates suggest that *Montagnea* evolved from *Coprinus,* but the fruiting body does not deliquesce, nor does it have true gills. A similar but much rarer fungus, *Gyrophragmium californicum,* has dark brown to blackish spore-bearing plates that hang from the *underside* of the disclike cap rather than from the margin. It also has a loose volva plus a double-layered partial veil that either disappears *or* forms an annulus (ring) on the stalk. It is known only from the San Francisco Bay region of California, and appears to be quite rare. In both of these

728

species the volva and spore mass disintegrate or rot away, while the light woody stalk and thin disclike cap remain intact and persist for months without decaying. In this condition they can be mistaken for old specimens of *Battarrea phalloides* (a stalked puff-ball) whose spore mass has dispersed. In the latter species, however, the disc at the top of the stalk represents the underside of the old spore case rather than the cap.

Longula texensis   (Gastroid Agaricus)

FRUITING BODY with a cap and stalk. **CAP** 3-9 cm broad, oval to round or broadly convex or somewhat flattened; surface dry, white to buff becoming tan to ochre or brownish in old age, smooth or often breaking up to form white to brown fibrillose scales or warts that may wear away in age; becoming fragile as it dries out, often rupturing longi-tudinally in age (especially near margin); margin at first joined to cap by a veil. Flesh firm, white, but bruised areas often stained yellowish (or sometimes pinkish-stained in stalk). **SPORE MASS** composed of crowded, convoluted, folded or branched plates and/or cavities; free from the stalk (at least in age); brownish becoming deep chocolate-brown to blackish at maturity and usually exposed or partially exposed in age. **STALK** 2-10 cm long, 1.5-3.5 (5) cm thick, equal or thicker at base, smooth or longitudinally striate, firm and solid when fresh, becoming tougher or even woody as it dries out; white or colored like cap; percurrent. **VEIL** two-layered, at first continuous with the cap margin and stalk, but in age usually pulling away from the stalk or separating from the cap to form a superior ring on stalk. **VOLVA** present or absent (usually intergrown with stalk and therefore inconspi-cuous). **SPORE PRINT** not obtainable; spores 6-7.5 × 5-6.5 microns, nearly round, smooth. Capillitium absent.

HABITAT: Solitary, scattered, or gregarious in poor soil, waste places, irrigated fields, lawns, disturbed areas, etc.; apparently widespread in the arid and semi-arid areas of the West (from Texas to California and north at least to Oregon), fruiting after rains or artificial watering. It used to be common in the San Francisco Bay area, but has apparently been pushed out by development. It can still be found with regularity in the Sacramento and San Joaquin Valleys, and has also been found in Santa Clara and San Mateo counties.

EDIBIITY: Unknown. It has probably been eaten as an *Agaricus* button, but I can find no specific information on it. According to Rick Kerrigan, it is easily cultivated.

Longula texensis looks like an aborted or deformed *Agaricus*. Note wide variation in size and shape. At left a stout button has been sliced open to show the dark spore mass inside; at far right are two slim buttons; at center are three small, mature, dried-up individuals (two with a fully exposed spore mass).

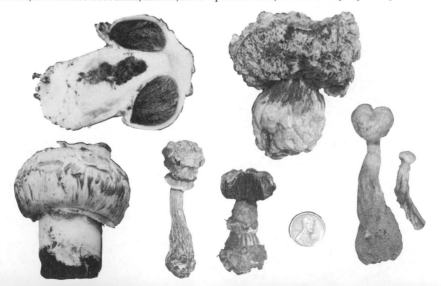

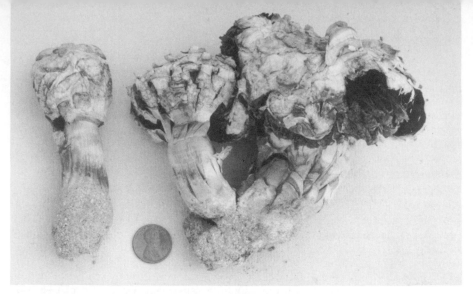

Longula texensis. These large specimens were found near Coalinga, California. The cap can be scaly or smooth (see photo on previous page).

COMMENTS: Originally called *Longia* (or *Gyrophragmium*) *texensis,* this mushroom is closely allied to *Agaricus,* and is likely to be mistaken for that genus because of the veil (which sometimes forms an annulus or ring), cap, stalk, and dark spore mass. However, it lacks true gills and the spores are not forcibly discharged. When the veil breaks or pulls away from the stalk or the cap splits radially, the spore mass is exposed to the elements, whereas in *Endoptychum* the spore mass usually remains enclosed. A giant version of this species, *L. texensis var. major,* occurs throughout the range of the typical variety. Its cap is 6-12 or more cm broad and often quite scaly, and the stalk is 10-25 cm long and 2.5 cm or more thick (see photograph above).

Endoptychum depressum   Color Plate 84
(Mountain Gastroid Agaricus)

FRUITING BODY with a cap and stalk. **CAP** 3-15 cm broad, rounded or convex or flattened on top, often centrally depressed in age; surface smooth or occasionally scaly, white to dingy whitish or buff, often staining yellow or amber when bruised and in age (and sometimes staining vinaceous-brown near bottom); margin depressed around the stalk and joined to it by a veil, sometimes pulling away from stalk in age or the cap rupturing irregularly. Flesh white, firm becoming tough in age, often bruising yellow; odor usually sweet (like almond extract or anise) when young but often unpleasant or musty at maturity. **SPORE MASS** composed of branched or contorted, crowded plates (rudimentary gills) or elongated chambers, pallid soon becoming chocolate-brown to blackish-brown, usually powdery at maturity and not normally exposed except in old age. **STALK** 1-4 cm or more long, 0.8-3 cm thick, equal or tapered, usually short (but see comments),

Endoptychum depressum, the widespread short-stalked form. Specimen at right is being viewed from top, the specimen next to it from the bottom. Specimen at far left has been cut open to show the dark mature spore mass. Specimen next to it is being viewed from side. See color plate for long-stalked form.

firm, white or stained like the cap, smooth, percurrent. **VEIL** rather tough, covering the juncture of cap margin and stalk, not normally rupturing but sometimes breaking away from the stalk in age. **VOLVA** typically absent. **SPORE PRINT** not obtainable; spores variable in size but usually averaging 6-10 × 5.5-8 microns, round to broadly elliptical, smooth, thick-walled. Capillitium absent.

HABITAT: Solitary to gregarious or clustered in duff or soil under conifers (or sometimes aspen); common in the mountains of western North America, especially in the late summer and fall. The short-stemmed form typical of the Sierra Nevada often develops underground but usually pokes through the surface by maturity. A long-stemmed version (see comments) is sometimes abundant under ponderosa pine in the Southwest.

EDIBILITY: Said to be edible (use only firm specimens); I haven't tried it.

COMMENTS: This mushroom is closely related to the section Arvenses of the genus *Agaricus* by virtue of its sweet odor when young, tendency to stain or age yellow, and dark spores. However, the veil doesn't normally break until *after* the spores are mature (if it breaks at all) and the gills are misshapen to practically absent. The stalk is usually quite short (as shown in the black-and-white photo), but the long-stemmed variety (which may or may not be distinct) shown in the color plate is the prevalent one in the Southwest. Several other species may occur in the West, but have not been formally described. The names *Secotium* and *Gyrophragmium* are used by some authors instead of *Endoptychum*.

Endoptychum agaricoides   (Gastroid Lepiota)

FRUITING BODY with a round to oval to pear-shaped, broadly conical, or heart-shaped (narrowed at top) cap and a short stalk. **CAP** (1) 2-7 (10) cm broad, 2.5-10 (12) cm high; surface dry, white when young and fresh, but discoloring buff to tan, ochre, or brownish with age; fibrillose, but often breaking up to form scales which give it a shredded appearance; margin typically joined to the stalk. Flesh (in stalk and cap) white when fresh, but in one form staining vinaceous-brown when exposed. **SPORE MASS** composed of distorted chambers and/or plates (rudimentary gills) crowded within the cap; at first white and fleshy, then turning yellowish and finally becoming brown to yellow-brown in old age, and sometimes powdery. **STALK** percurrent, sometimes entirely internal (a vertical column that extends to the top of cap), but usually extending slightly below the cap (up to 2 cm), usually thickest at base of cap; white when fresh, but often discoloring like the cap in age; usually with a mycelial cord at base. **VEIL** not clearly differentiated from

This immature *Endoptychum agaricoides* has been sliced open to show the percurrent stalk and white spore mass. Later the spores turn yellowish or brown and may become powdery. Note the *Lepiota*-like scales on cap.

cap (i.e., underside or "margin" of cap joined to stalk). **VOLVA** absent. **SPORE PRINT** not obtainable; spores 6-9 (11) × 5-7 microns, elliptical, smooth, with a thick inner wall that has an apical pore. Capillitium absent.

HABITAT: Solitary, scattered, or in groups or clusters in lawns, fields, flower beds, waste places, etc.; widely distributed. It is fairly common in the Southwest and Rocky Mountain region in the summer and early fall, but I have not seen it in coastal California. It is one of the few gastroid agarics to be found in eastern North America.

EDIBILITY: Said to be edible when young; I haven't tried it.

COMMENTS: Also known as *Secotium agaricoides,* this unsung, farflung fungus looks like an unexpanded *Agaricus* or *Lepiota,* but has only rudimentary gills and never opens out. The rather pale spore color and thick inner spore wall with a pore suggests a closer relationship to *Lepiota* or *Chlorophyllum* than to *Agaricus.* It does not appear to have much in common with other Endoptychums (e.g., *E. depressum*), but is the "type" species of the genus. Other species: *E. arizonicum* of the Southwest is very similar, but has much larger spores; *Neosecotium macrosporum* of the Midwest, Southwest, and southern California also has larger spores, but they are ornamented rather than smooth and the fruiting body is usually slightly smaller (1-4 cm high).

BRAUNIELLULA

EASILY recognized by its orangish to vinaceous flesh and gastroid appearance, this small genus is clearly related to the pine spikes *(Chroogomphus).* Like the latter, it has grayish-black spores and amyloid flesh and grows exclusively with conifers. As there is only one common species, a key hardly seems necessary.

Brauniellula nancyae   (Gastroid Pine Spike)

FRUITING BODY with a cap and short stalk. **CAP** (0.5) 1-6 cm broad, rounded to convex or lobed, or sometimes flattened on top; surface dry to very slightly viscid, fibrillose, ochre to pale or dull orange beneath a layer of flattened grayish to brownish-gray fibrils, often becoming vinaceous in age; margin often lobed, at first attached to the stalk by the veil, separating from the veil in age but remaining incurved. Flesh thick, firm, pale orange to ochraceous when fresh. **SPORE MASS** composed of crowded, convoluted plates and/ or cavities; ochraceous to pale orange, becoming darker (grayish to nearly black) as spores mature; usually remaining covered by the veil, but sometimes partially exposed. **STALK** 0.3-2 cm long, 0.5-1 cm thick, equal or narrowed below, often so short that it scarcely protrudes below the cap; solid, firm, ochraceous to dull orange, usually fibrillose, often becoming vinaceous or reddish-stained in age and/ or at base. **VEIL** fibrillose, ochraceous or becoming vinaceous. **VOLVA** absent. **SPORE PRINT** not obtainable; spores 16-20 × 6.5-9 microns, narrowly elliptical to spindle-shaped, smooth.

HABITAT: Solitary or scattered to densely gregarious or clustered in duff under conifers (especially fir and pine) in the summer and early fall; known only from western North

Brauniellula nancyae (= B. albipes) is common under mountain conifers throughout the West. Note short stalk. It is closely related to *Chroogomphus.* (Herb Saylor)

America. It is common in the Sierra Nevada and Cascades as well as in Idaho; it is also said to occur in northern Arizona.

EDIBILITY: Unknown, but probably edible.

COMMENTS: This curious mountain fungus is an almost exact replica of a *Chroogomphus,* except that it has only rudimentary gills (if any) and does not open out, and is often buried or half-buried in the humus layer. Note that it is the same color as *Chroogomphus* (dull or pale orange to ochraceous, becoming vinaceous in age) and has the same spore color (gray to black) and habitat. The grayish fibrils on the cap are suggestive of *C. leptocystis* (see comments under *C. tomentosus*). According to Eric Gerry, *B. albipes* is an earlier (more correct) name for *B. nancyae.* Other gastroid members of the *Chroogomphus*-family (Gomphidiaceae) include: an undescribed species (or perhaps just a "freak") found under a pine in southern California (larger than *B. nancyae,* and lacking the grayish fibrils on the cap); and *Gomphogaster leucosarx,* described from Idaho, which has white flesh in the upper stalk and a bright yellow stem base (like the agaric genus *Gomphidius*), but which has a minutely chambered spore mass instead of gills.

THAXTEROGASTER, NIVATOGASTRIUM, & Allies

Medium-sized to small fungi found on wood *(Nivatogastrium)* or ground *(Thaxterogaster* & others). FRUITING BODY *with a cap and stalk.* CAP narrowly conical to convex, plane, or centrally depressed, viscid or dry. SPORE MASS enclosed by the cap but occasionally exposed at maturity, *composed of rudimentary gills or plates that sometimes branch to form cavities, ochre to brown, rusty-brown, or reddish-brown at maturity,* not normally becoming powdery. STALK usually well-developed (but sometimes very short), percurrent. VEIL absent or if present then thin and fibrillose. VOLVA typically absent. SPORE PRINT *not obtainable.* Spores tawny to brown under the microscope, mostly elliptical, smooth or roughened. Sphaerocysts and capillitium absent.

SIX miscellaneous genera of gastroid agarics are treated here. All have brown spores and are thought to have evolved independently from various brown-spored gilled mushrooms. *Nivatogastrium,* for instance, is closely related to *Pholiota;* like that genus, it grows on wood. *Thaxterogaster,* on the other hand, is terrestrial and allied to *Cortinarius; Weraroa* and *Galeropsis,* with their distinctive pointed caps, may be derived from *Psilocybe* or possibly *Conocybe; Setchelliogaster* and *Gastrocybe* are both suggestive of the Bolbitiaceae. All six are small genera and none are worth eating. Three characteristic species of our western mountains are described here.

Key to Thaxterogaster, Nivatogastrium, & Allies

1. Growing on wood (on mountain conifers) . . . . ***Nivatogastrium nubigenum,*** p. 735
1. Growing on ground 2

2. Cap very narrowly conical (like a dunce cap); usually found in grass or mountains 3
2. Not as above 4

3. Found on lawns; fruiting body soon dissolving or collapsing . . . .
. . . . ***Gastrocybe lateritia*** (see *Weraroa cucullata,* p. 734)
3. Usually found in mountains (but often among grasses); fruiting body not dissolving quickly
. . . . ***Weraroa cucullata*** & others, p. 734

4. Associated principally (if not exclusively) with eucalyptus; stalk slender . . . .
. . . . ***Setchelliogaster tenuipes*** (see *Weraroa cucullata,* p. 734)
4. Not as above; associated mainly with conifers 5

5. Veil thick and/or persistent; gills present; spore print obtainable . . . (see ***Cortinarius,*** p. 417)
5. Veil fibrillose or not well-developed, persisting or disappearing; spore mass composed of contorted plates or cavities; spore print not obtainable . . . . . ***Thaxterogaster pingue,*** p. 734

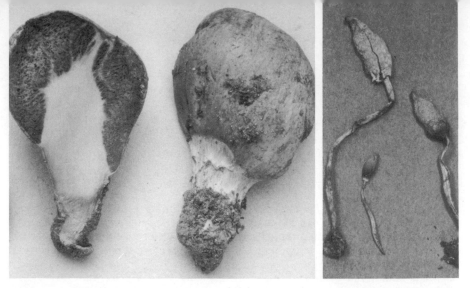

Left: *Thaxterogaster pingue*. Note percurrent stalk in sliced specimen on left. The stalk can be very short or fairly long depending on environmental conditions. **Right:** *Weraroa cucullata* is easily told by its small size and narrow, pointed "head."

Thaxterogaster pingue (Gastroid Cortinarius)

FRUITING BODY with a cap and short stalk. **CAP** 1-5 cm broad, rounded or obtuse to convex or lobed, the top often flattened somewhat in age; surface slightly viscid to slimy when moist, smooth, buff to olive-yellow to dingy yellow-brown or dark brown. Flesh firm, white or tinged cap color. **SPORE MASS** composed of crowded, contorted plates and/or small cavities; yellowish becoming dull brown to pale or dark cinnamon-brown. **STALK** 0.5-4.5 cm long, 0.8-2.5 cm thick, very short or rudimentary in some forms, well-developed in others; equal or swollen at base; percurrent (the portion in the spore mass usually quite narrow and unbranched); smooth, often viscid near base when moist, dull yellowish to buff or colored like the cap, often purple or lilac where exposed to light. **VEIL** fibrillose or cobwebby, persisting or disappearing. **VOLVA** absent. **SPORE PRINT** not obtainable; spores (12) 14-16.5 × 8-9.5 microns, elliptical to somewhat oblong, warty and/or wrinkled.

HABITAT: Solitary to gregarious in duff under conifers in western North America; fairly common in the summer and fall, especially at higher elevations. It is one of the most frequently encountered gastroid agarics of the Rockies (usually under spruce). In the Sierra Nevada and Cascades it favors fir. In our area it is rare—one specimen of local origin was brought to me, but its collector did not note the habitat.

EDIBILITY: Unknown—but hardly tempting.

COMMENTS: This species is best recognized by its viscid cap, dingy yellowish-brown to brown overall color, and frequently violet-tinged stalk. It is a variable species insofar as shape is concerned—specimens growing in regions of high rainfall are apt to fruit above the ground and have a well-developed stalk, while those growing in drier habitats are more likely to remain buried in the duff and have a short stalk that barely protrudes from the cap (i.e., the fruiting body can be roundish and *Rhizopogon*-like). *Thaxterogaster* is thought to be closely related to the agaric genus *Cortinarius*. Several other species of *Thaxterogaster* occur in the United States, some of them undescribed.

Weraroa cucullata (Gastroid Liberty Cap)

FRUITING BODY with a cap and stalk. **CAP** 1-4 cm high, 0.4-1.5 cm broad, narrowly conical with a pointed apex; surface dry to slightly viscid, smooth or fibrillose, often wrinkled, yellow to yellowish-buff to tawny or yellow-brown; margin often splitting in age and retracting slightly from the stalk. **SPORE MASS** composed of crisped or contorted gills that may or may not branch to form cavities; brown to reddish-brown. **STALK** long and thin (5-12 cm long, 1-4 mm thick), equal or with a small bulb at base, hollow or stuffed, rather tough, smooth or fibrillose, colored like cap, dry. **VEIL** (partial) absent or evanescent. **VOLVA** absent. **SPORE PRINT** not obtainable; spores 11-14 (16) × 6-8 (10) microns, elliptical, smooth, with a germ pore; tawny to brown under the microscope.

HABITAT: Scattered to gregarious on ground, usually among grasses in wet alpine meadows; widely distributed in the mountains of western North America. It occurs in the Sierra Nevada and I have found it in the southern Rockies in August.

EDIBILITY: Utterly and irrefutably inconsequential.

COMMENTS: The narrowly conical, pointed cap perched on a long thin stalk plus the rudimentary gills make this a most distinctive mushroom. It is reminiscent of a liberty cap *(Psilocybe semilanceata)*, but does not have true gills. *Galeropsis cucullata* is an older name for it. *Galeropsis polytrichoides,* found in the same habitats, may be a smaller-spored version of the same species. *G. angusticeps* is an eastern species. *Gastrocybe lateritia* is also similar but grows on lawns and, like *Conocybe,* collapses or dissolves very quickly (within a few hours). Also deserving mention is *Setchelliogaster tenuipes,* a small species probably related to the Bolbitiaceae. It has a convex to round or cylindrical, yellow-brown to brown or reddish-brown, non-viscid cap with a chambered or gill-like, exposed spore mass and a short or long but slender stalk. It is associated with eucalyptus and is either especially fond of erudite settings or is only sought for by "academia nuts." (It has been found on the Stanford and University of California campuses.)

Nivatogastrium nubigenum   (Gastroid Pholiota)

FRUITING BODY with a cap and stalk. **CAP** (1) 2-8 cm broad, nearly round to convex, sometimes expanding in age, the center often flattened or slightly depressed; surface smooth or slightly fibrillose, ochraceous or tawny to dingy yellowish, buff, or white (usually fading to whitish in age), slightly viscid when wet, sometimes pitted; margin at first incurved and often lobed, remaining attached to the stalk or sometimes pulling away in old age. Flesh white in the cap and columella (that portion of the stalk surrounded by the spore mass), usually brown in the stalk; rather soft but tough (especially in the stalk); odor mild or fragrant (like bubble gum). **SPORE MASS** brown to cinnamon-brown when

Nivatogastrium nubigenum grows on dead mountain conifers in the spring. Spore mass is composed of contorted gills and is sometimes exposed in old age (large specimen). Note pale color.

mature, composed of irregularly contorted gills that often form chambers; exposed only in old age if at all. **STALK** 0.5-2.5 cm long, 0.5-2 cm thick, equal or thicker at either end, percurrent, usually short and stout, quite tough (difficult to section); white or stained brownish to rusty-brown, not viscid. **VEIL** present as whitish fibrillose tissue that extends from the cap margin to the stalk; often disappearing in age, *not* forming an annulus (ring). **VOLVA** absent. **SPORE PRINT** not obtainable; spores (3) 7-9 (12) × (3) 5.5-6.5 microns, elliptical, smooth, brown in mass, honey-colored under the microscope.

HABITAT: Solitary to gregarious or in small clusters on rotting conifers, often near melting snow or shortly after the snow disappears; fairly common in the mountains of the West in the spring and early summer, especially on fir and lodgepole pine. I have seen large fruitings in the Sierra Nevada and Cascades, but have yet to find it on the coast.

EDIBILITY: I can find no information on it.

COMMENTS: This distinctive member of the "snowbank" mushroom flora is easily identified by its pale color, growth on wood, and irregularly contorted or chambered brown "gills" which remain enclosed by the cap or are exposed only in old age. Like other gastroid or "reduced" agarics, it is apt to be mistaken for an unopened gilled mushroom. Its occurrence on wood and brown "gills" plus the color and shape of the spores relate it to the agaric genus *Pholiota*. Another *Pholiota*-relative, *N. wrightii*, has been found in the mountains of southern California.

MACOWANITES, ARCANGELIELLA, & Allies

Small to medium-sized, terrestrial woodland fungi *with a crisp and brittle or fragile texture.* FRUITING BODY *with a cap and stalk.* CAP variously shaped but never conical, dull or brightly colored; margin frequently joined to or touching the stalk, at least when young. Flesh white or pallid. SPORE MASS *white to yellow, pinkish, or ochraceous; usually composed of contorted gills which may branch to form cavities;* enclosed or exposed. Latex present in *Arcangeliella,* otherwise absent. STALK often short, percurrent, white or pale-colored, *brittle* (typically snapping open cleanly like a piece of chalk). VEIL and VOLVA *absent.* SPORE PRINT *not obtainable.* Spores hyaline (colorless) under the microscope, round to elliptical, ornamented with amyloid warts, spines, or ridges. Sphaerocysts usually present in some part of tissue. Capillitium absent.

THESE curious fungi look like disfigured specimens of *Russula* and *Lactarius,* which is essentially what they are. Like those genera, they have white to ochre gills, a crisp or brittle texture, and ornamented amyloid spores. However, as in other gastroid agarics, the cap does not expand fully and the gills are irregularly contorted or aborted and have lost the ability to forcibly discharge spores. Furthermore, while they are not truly hypogeous (subterranean), they exhibit a *tendency* in that direction, i.e., they are frequently only partially exposed at maturity. In other words, they are intermediate in aspect between the Russulaceae *(Russula* and *Lactarius)* and certain genera of false truffles *(Martellia, Gymnomyces,* and *Zelleromyces).* (The latter are microscopically similar but have gone completely underground, lost their stalk, and typically have a chambered rather than gilled spore mass.)

In *Macowanites* the fruiting body is *Russula*-like (i.e., the cap is sometimes brightly colored and there is no latex). In *Arcangeliella,* on the other hand, a latex is usually present as in *Lactarius* (the latex is best seen by slicing open a *fresh* fruiting body). A third genus, *Elasmomyces,* closely resembles *Macowanites,* but has spores which are not modified for forcible discharge (thus it is a step closer to the false truffles) and lacks sphaerocysts in the gills. (*Macowanites* has sphaerocysts in the gills and its spores are modified for forcible discharge though they are not actually discharged.)

Like *Russula* and *Lactarius,* these mushrooms are strictly mycorrhizal. They are especially common and diverse under conifers in the mountains of western North America, but several (e.g., *Macowanites magnus*) occur in our area. They are not often collected for food and those with an acrid (peppery) taste or unpleasant odor should be avoided. Over 25 species of *Macowanites* have been described from the United States. Most of them are difficult to identify without a microscope. *Arcangeliella* and *Elasmomyces* have fewer species, but are still not easy to identify. One *Macowanites* and one *Arcangeliella* are described here.

Key to Macowanites, Arcangeliella, & Allies

1. Taste very acrid (peppery), but sometimes latently so 2
1. Taste mild or distinctive but not acrid 4

2. Fruiting body exuding a latex (milk or juice) when cut open, at least when fresh and moist (but latex sometimes very scant) 3
2. Latex absent 9

3. Stalk usually well-developed; latex usually copious; found mainly in coastal forests .... *Arcangeliella variegata* (see *A. crassa*, p. 738)
3. Stalk usually short, often poorly developed; latex copious or scanty; found mainly under mountain conifers (in Sierra Nevada, Cascades, etc.) ...... *Arcangeliella crassa* & others, p. 738

4. Fruiting body medium-sized to fairly large (cap 3-14 cm broad, stalk 3-7 cm long and often 3 cm thick); cap dull-colored (cinnamon-buff to pale tan to dark brown); odor usually unpleasant in age .... *Macowanites magnus* (see *M. americanus*, below)
4. Not as above 5

5. Odor antiseptic (like iodoform) or chlorine-like, at least at maturity 6
5. Odor mild or distinctive but not as above 7

6. Odor chlorine-like; taste unpleasant   *Macowanites chlorinosmus* (see *M. americanus*, below)
6. Odor antiseptic; taste mild .... *Macowanites iodiolens* (see *M. americanus*, below)

7. Odor sweet, at least in age or when dried 8
7. Odor mild or unpleasant but not sweet 9

8. Cap whitish .... *Elasmomyces odoratus* (see *Macowanites americanus*, below)
8. Cap tan to rusty-brown .... *Arcangeliella camphorata* (see *A. crassa*, p. 738)

9. Associated primarily with hardwoods (oak, etc.); cap whitish to bright red or reddish 10
9. Associated primarily with conifers; cap whitish to yellow, blue, olive, vinaceous, salmon, rosy, purplish, vinaceous, or mixtures thereof 11

10. Spore mass composed of contorted or rudimentary gills .... *Elasmomyces russuloides* & others (see *Macowanites americanus*, below)
10. Spore mass minutely chambered, not normally exposed .... (see *Martellia* & Allies, p. 742)

11. Taste acrid (peppery) .... *Macowanites luteolus* & others (see *M. americanus*, below)
11. Taste mild or at least not acrid .... *Macowanites americanus* & many others, below

Macowanites americanus   (Gastroid Russula)

FRUITING BODY with a cap and stalk. **CAP** 1-5 cm broad, convex or irregularly lobed, but often expanding in age to plane or even centrally depressed; surface viscid when wet, smooth or often breaking up into scales (especially at center); color variable: lilac, purplish, vinaceous, olive, blue, yellow, or even whitish, or often mixtures thereof; margin not typically joined to the stalk but often touching it when young. Flesh brittle, crisp, white; odor and taste mild. **SPORE MASS** consisting of irregular or distorted, often veined or even chambered gills that are whitish at first but soon become ochre to dull yellow-orange; usually attached to stalk. **STALK** 1-3 cm long, 0.3-1.5 cm thick, more or less equal, firm, brittle, smooth, white, percurrent. **VEIL** and **VOLVA** absent. **SPORE PRINT** not obtainable; spores 8.5-13.5 × 8-12 microns, broadly elliptical to round, with amyloid warts and ridges, modified for forcible discharge but not actually discharged.

HABITAT: Solitary to gregarious or clustered in duff under conifers (sometimes buried); fairly common and widespread in western North America, especially under spruce, fir, Douglas-fir, and pine. It fruits mainly in the summer and fall and is said to be the most common member of the genus. I have seen it in northern California.

EDIBILITY: Said to be edible—but be sure of your identification.

COMMENTS: This conifer-lover looks like a deformed or aborted *Russula*, which is essentially what it is! The spores are not forcibly discharged and hence a spore print is unobtainable. There are more than 20 other members of the genus in North America, including: *M. iodiolens,* very similar but with a distinct iodoform (antiseptic) odor, found under conifers; *M. subolivaceus,* a mild-tasting species wtih a white or olive-tinged cap and smaller spores, found in the Sierra Nevada under conifers; *M. chlorinosmus,* with a whitish to yellow cap, deep ochre spore mass, chlorine odor, and unpleasant taste, found under conifers; *M. luteolus,* a dull or pale yellowish, acrid-tasting species found in coastal California and the Pacific Northwest under conifers; *M. subrosaceus,* a pinkish-capped, ochraceous-staining species; *M. (=Elasmomyces) roseipes,* with a pinkish to salmon or vinaceous cap whose margin never touches the stalk, favoring spruce and fir; *M. alpinus,* a white to creamy-buff montane species; and *M. magnus,* a large species (cap 3-14 cm broad, stalk 3-7 cm long and up to 3 cm thick) with a viscid tan to dark reddish-brown cap and an unpleasant odor, originally collected in Santa Clara County, California. The closely related genus *Elasmomyces* has spores which are not modified for forcible discharge and seems to be more common under oak than conifers, at least in California. Its members include: *E. russuloides,* with a red to reddish-brown cap (at least in age), widely distributed; *E. pilosus,* which tends to grow underground in association with oak; and *E. odoratus* of Washington, with a strong fragrant odor. None of these are worth eating.

Arcangeliella crassa   (Gastroid Milk Cap)

FRUITING BODY with a cap and stalk. **CAP** 2-8 cm broad, convex when young, becoming plane to shallowly depressed or irregular in age; surface pale buff to pinkish-buff, sometimes darkening slightly (to cinnamon-buff) in old age, dry or slightly viscid, smooth; margin typically remaining attached to the stalk, but often not quite reaching it in some places, and in age sometimes separating slightly from it. Flesh thick, crisp, brittle, white to pinkish-buff; odor often unpleasant in age; taste very acrid (peppery). Latex present but often scanty (especially in dry weather), white, unchanging. **SPORE MASS** comprised of contorted plates or rudimentary, often chambered gills; pale pinkish-buff to ochre-buff, usually completely enclosed by the cap but occasionally exposed when old, especially near the stalk. **STALK** 0.5-2.5 cm long, 0.5-1.5 (2) cm thick, usually stubby and poorly-developed, often off-center; solid or hollow, crisp, brittle, dry, smooth, percurrent; whitish or colored like the cap. **VEIL** and **VOLVA** absent. **SPORE PRINT** not obtainable; spores 7.5-11 × 5.5-8 microns, elliptical, with strongly amyloid ridges (reticulate).

Left: *Macowanites* sp. (perhaps *Macowanites magnus*), a *Russula*-like fungus with deformed gills.
Right: *Arcangeliella crassa* has short stalk plus white latex which is not visible here. (Herb Saylor)

HABITAT: Solitary to gregarious in duff under conifers and in mixed woods; known only from the mountains of western North America. It is sometimes common (along with *A. tenax*—see comments) in the Sierra Nevada in the spring, summer, and fall. Like many gastroid agarics, it does not seem to occur on the coast.

EDIBILITY: Unknown, but the acrid taste is a deterrent.

COMMENTS: The genus *Arcangeliella* is closely related to *Lactarius,* and this species looks like an aborted milk cap. The overall buff to pinkish-buff color and white acrid latex are important field characters. *A. tenax* and *A. lactarioides* (the latter with a gelatinous cap cuticle) have also been described from the mountains of Oregon and California; they are very similar if not the same. Other species: *A. variegata* is a coastal species with a grayish-buff to olive-buff cap that is often yellowish-spotted in age plus a copious white to almost clear, acrid latex and well-developed white stalk. It occurs in our area but is more common northward. Still another species, *A. (=Elasmomyces) camphorata* of Washington, is remarkable for its strong fragrant odor like that of candy caps (*Lactarius fragilis* and relatives), especially when dry. As in candy caps, its latex is often absent, but the odor and overall tan to rusty-brown color distinguish it. See also *Zelleromyces* (under *Martellia* & Allies).

False Truffles

spores

HYMENOGASTRALES & Allies

THE false truffles, deprecatingly dubbed "squirrel food" by at least one jaded fungophile I know, are easy-to-overlook, difficult-to-identify, subterranean, tuberlike fungi. They are neither delicious (unless you're a squirrel) nor conspicuous (unless you're a squirrel) nor attractive (unless you're a squirrel), but they *are* interesting from an evolutionary standpoint (unless you're a squirrel) because many are thought to have evolved from the Hymenomycetes (e.g., agarics and boletes) or perhaps to have given rise to them.

The fruiting body of a false truffle is typically potatolike: round to oval or knobby with a tough or cartilaginous to rubbery or gelatinous interior. The spore mass or **gleba** (interior) is often composed of small chambers that give it a spongelike appearance. The spores are borne on the walls of the chambers or sometimes within the chambers themselves. The gleba usually remains intact through maturity rather than becoming powdery and dispersing. These features, plus the fact that most false truffles are **hypogeous** (subterranean) or **erumpent** (i.e., they "erupt" through the surface of the ground at maturity) distinguish them from the puffballs and earthballs, which have a powdery spore mass in old age and usually fruit above the ground.

Some false truffles, such as *Truncocolumella,* possess an internal stalk or **columella**— a branched or unbranched central column which penetrates the spore mass and sometimes protrudes slightly beneath it as a rudimentary stalk or "stump." The columella can be quite slender, so you should always make several sections of the fruiting body so as not to overlook it. The columella or "stalk" is not usually percurrent as in the gastroid agarics, and in many false truffles (including the most common genus, *Rhizopogon*) it is altogether lacking.

The false truffles can also be confused with the "true" truffles, which are not as common, usually have a channelled or marbled or hollow interior, and lack a columella (see lengthy footnote on p. 844). As might be expected, there are some exceptions (e.g., false truffles such as *Melanogaster* and *Alpova* have veins in the spore mass, and some truffles have a

False truffles look something like puffballs, but usually grow underground and do not have powdery spores at maturity. The interior is usually divided into chambers, but the chambers can be so minute that a hand lens is required to see them. This species is *Rhizopogon ochraceorubens* (p. 755).

chambered interior), and it is sometimes necessary to make a microscopic determination as to whether the spores are borne inside asci (the true truffles) or on basidia (the false truffles).*

As defined here, the false truffles are an artificial and heterogeneous group. Several genera are related to the agarics (e.g., *Hymenogaster* to the Cortinariaceae, *Hydnangium* to *Laccaria, Martellia* to *Russula,* and *Zelleromyces* to *Lactarius*). A few, like *Truncocolumella,* show affinities to the boletes, while the origins and relationships of others (e.g., *Rhizopogon* and *Alpova*) are obscure.

Western North America harbors more than its share of false truffles—more, in fact, than any other region in the world. Most if not all are mycorrhizal. They are *very* common, but since they tend to grow underground, they're unlikely to be seen unless sought for. The best way to find them is to take a handrake and sift gently through the humus or needle duff and upper soil layer (see chapter on truffles for more details), giving special attention to areas where squirrels, deer, or wild pigs have been digging. You face fierce competition, however, for those not devoured by animals are eagerly snatched up by mycologists (it is only in the last fifty years that the false truffles have been studied closely—dozens have yet to be named and many of the named ones are known only from a single locality). Relatively few are left in the ground to rot in peace, which may actually be to their advantage—since their spore mass is not powdery and easily dispersed, they rely on rodents, insects, rain, and mycologists to spread their spores for them. Many have strong odors (e.g., like marzipan, garlic, or vanilla) that help to broadcast their presence.

The false truffles are of more interest to the evolutionary taxonomist than the most revolutionary gastronomist. Some have an unpleasant taste or texture; a few, I am told, are edible. The overwhelming majority, however, have not been tested (by humans), in part because field identification can be *very* difficult. A microscope is often needed to determine the genus and family as well as the species.

The false truffles are a vast group—there are over 150 species of *Rhizopogon* alone! Only a few representatives are depicted here. The keys, which unavoidably resort to microscopic characteristics, are greatly modified versions of those in *How to Know the Non-Gilled Mushrooms* by Alexander Smith, Helen Smith, and Nancy Weber. Those of you who wish to know more about the false truffles should consult the extensive keys in that book, the technical literature on the subject (see bibliography), and/or the squirrel representative nearest you.

Key to the Hymenogastrales & Allies

1. Spore borne inside asci* .... (see **Tuberales,** p. 841)
1. Spores borne on basidia* 2

2. Columella present, branched or unbranched, percurrent (extending all the way through spore mass) or stumplike, sometimes very thin (be sure you slice the fruiting body perpendicular to the ground and through its center—you may want to make several thin slices so as not to overlook the columella) 3
2. Columella absent (but a stalklike base of tough fibers sometimes present) 7

*For some fairly reliable ways to distinguish the Tuberales (truffles) from the false truffles *in the field,* see the extensive footnote at bottom of p. 844.

3. Columella short and stumplike *or* branched, usually prominent; peridium (skin) typically yellow to olive; spore mass composed of empty tubular chambers, *not* blueing when cut; spores *smooth*, yellowish to pale brown under the microscope; associated mainly if not exclusively with Douglas-fir .... ***Truncocolumella,*** p. 752

3. Not as above 4

4. Outline of gills and cap present in longitudinal (perpendicular) section (see **Agaricales,** p. 58)

4. Not as above 5

5. Peridium (skin) 2-5 mm thick; columella fairly thick (sometimes rounded or cushion-shaped), *not* percurrent; spore mass lacking chambers but with radiating lines or plates when young; white becoming brown or black; firm to gooey or in old age powdery .... ***Radiigera,*** p. 760

5. Not as above 6

6. Spore mass usually (but not always) greenish to olive-gray or olive-brown, typically tough or cartilaginous when young but often slimy in old age; columella often translucent (but sometimes whitish), branched or unbranched; peridium (skin) usually well-developed; spores smooth or enclosed in a wrinkled outer coat .... ***Hysterangium*** & **Allies,** p. 762

6. Not as above 11

7. Spore mass white even in age, the chambers usually gel-filled or exuding a white latex when cut; spores with a gelatinous sheath, *not* amyloid .... ***Leucophleps*** & ***Leucogaster,*** p. 759

7. Mature spore mass darker (may be white when young), or if white then not as above 8

8. Peridium (skin) thick (occasionally 1 mm, usually 2-5) and tough or leathery; spore mass at first white but soon black, purplish, or dark brown, remaining firm for a long time but finally powdery, sometimes with veins but *lacking* gel-filled chambers ...... (see ***Scleroderma,*** p. 707)

8. Not as above; peridium usually thinner; spore mass only rarely powdery 9

9. Spore mass typically marbled with paler veins, the chambers usually gelatinous or gel-filled; spores smooth .... ***Alpova*** & ***Melanogaster,*** p. 756

9. Not as above (but may have some of above features) 10

10. Spores smooth (*not* ornamented), elliptical to oblong, sometimes amyloid but not dextrinoid, spore mass composed of minute chambers; outer surface often (but not always) with mycelial threads; *very* common, especially under conifers .... ***Rhizopogon,*** p. 753

10. Not as above; spores typically ornamented with lines, warts, wrinkles, spines, etc., or showing dextrinoid "stripes" 11

11. Spores round, elliptical, or elongated, colorless to yellowish or brown under microscope, *not* amyloid; spore mass variously colored (sometimes dark); peridium absent or present ... 12

11. Spores round to elliptical with amyloid or partially amyloid warts, rods, ridges, or spines; spores usually colorless or nearly so under the microscope; spore mass variously colored but often white at first and not typically dark brown to blackish; peridium usually present .. 16

12. Spores round to broadly elliptical and ornamented with large warts, cones, or spines; spores usually colorless under the microscope but sometimes brown; found in many habitats but especially common under eucalyptus .... ***Hydnangium*** & **Allies,** p. 744

12. Spores nearly round to elliptical, spindle-shaped, or elongated and typically wrinkled, warted, or longitudinally lined (rarely smooth or honeycombed), yellow-brown to rusty-brown or brown under the microscope; common in many habitats (including eucalyptus) 13

13. Spores appearing smooth except in iodine solution (and then showing prominent brown bands or stripes, i.e., dextrinoid); spore mass white to yellow or yellowish 14

13. Not as above; spores typically ornamented (rarely smooth, but if so then not as above) .. 15

14. Peridium white to brownish; found in Sierra Nevada .... ***Protogautieria substriata***

14. Peridium absent or practically so; known from eastern Washington ... ***Protogautieria lutea***

15. Peridium (skin) often thin and soon wearing away; spore mass quite tough and crisp when young; spores ornamented with longitudinal lines .... ***Gautieria,*** p. 746

15. Not as above; peridium usually well-developed, persistent ... ***Hymenogaster*** & **Allies,** p. 748

16. Outer layer of each spore consisting of amyloid rods imbedded in non-amyloid material; sphaerocysts absent; fruiting body 1-4 cm broad, usually irregular in shape, white becoming yellowish to olive-yellow or olive-gray; spore mass white and very dry, becoming olive in age; found under conifers, not common but sometimes fruiting prolifically ***Mycolevis siccigleba***

16. Not as above; spores ornamented with free spines, warts, and/or ridges; sphaerocysts usually (but not always) present somewhere in fruiting body; mature spore mass variously colored but not often olive; widespread and fairly common .... ***Martellia*** & **Allies,** p. 742

MARTELLIA & Allies

Small to medium-sized *woodland fungi usually found underground.* FRUITING BODY round to knobby or lobed (potato-like), variable in color. SPORE MASS (interior) typically composed of large to small or minute chambers; *usually rather crisp when fresh, white to pinkish, ochre, ochre-brown, or cinnamon-brown.* Latex often present in *Zelleromyces,* otherwise absent. STALK absent except in a few species; columella present or absent. SPORES round to elliptical, *ornamented with amyloid warts, spines, and/or ridges;* hyaline (colorless) under the microscope. Basidia forming a hymenium that lines the walls of the chambers. Sphaerocysts usually present somewhere in the fruiting body (but sometimes absent in *Martellia*). Capillitium absent.

THESE are poorly known, difficult-to-identify false truffles with a white to yellow- or cinnamon-brown spore mass and amyloid spores. Like the gastroid agarics *Macowanites, Arcangeliella,* and *Elasmomyces,** they are thought to be closely related to *Russula* and *Lactarius,* but are one step farther removed (i.e., they lack the cap and stalk of those genera, grow underground, and typically have a chambered rather than plated or gilled spore mass).

The most common genus, *Martellia,* lacks a latex and is thought to be related to *Russula.* A second genus, *Gymnomyces,* mimics *Martellia* but differs microscopically (the central strata of tissue in the walls between the chambers contain sphaerocysts; in *Martellia* they do not). A third genus, *Zelleromyces,* is probably closer to *Lactarius* than to *Russula* because the fresh fruiting body often exudes a latex (juice) when cut. (A *Zelleromyces* recently discovered in Yosemite National Park has a scanty dark red latex and stains greenish—just like *Lactarius rubrilacteus!*)

Unless the spores are examined, all three genera are easily confused with other false truffles such as *Hymenogaster* and *Octavianina.* However, they are often crisper or fleshier (or not as tough) as many false truffles and often paler in color, at least when young.

Martellia and *Gymnomyces,* and to a lesser extent, *Zelleromyces,* are fairly common in our area under oak, madrone, and various conifers. However, they do not normally fruit in the large numbers characteristic of some of the other false truffles (e.g., *Rhizopogon* and *Gautieria*). All three genera need critical study, especially *Martellia,* which is fairly sizable. As the various species are differentiated almost exclusively on microscopic characteristics, only two are described here.

Key to Martellia & Allies

1. Fruiting body with a scanty dark red latex; wounded tissue usually staining greenish (within several hours); columella present but stalk absent; spore mass chambered; peridium (skin) practically absent; known only from Yosemite National Park .... *Zelleromyces* sp. (name soon to be published by Harry Thiers and Herb Saylor)
1. Not as above 2

2. Columella and/or stalk present, usually percurrent (extending clear through spore mass) . 3
2. Stalk absent; columella absent or rudimentary 5

3. Fruiting body exuding a latex when cut; columella thin, branching; found mainly under oak along the west coast .... *Zelleromyces gardneri*
3. Latex absent 4

4. Spore mass rosy to pinkish to pinkish-buff .... *Gymnomyces socialis,* p. 743
4. Mature spore mass more or less cinnamon-buff *Martellia cremea* (see *M. brunnescens,* p. 743)

5. Fruiting body exuding a white latex when cut; exterior cinnamon to reddish; spore mass (interior of fruiting body) pale cinnamon-buff; widespread in eastern North America, especially under pine .... *Zelleromyces cinnabarinus*
5. Not as above; latex absent 6

6. Exterior tinged or spotted rose or red . *Gymnomyces roseomaculatus* (see *G. socialis,* p. 743)
6. Not as above 7

---

* *Macowanites, Arcangeliella, Elasmomyces, Zelleromyces, Gymnomyces,* and *Martellia* constitute the so-called "astrogastraceous series." In modern classifications they are usually placed alongside *Russula* and *Lactarius* in the Russulales and/or Russulaceae.

Martellia fallax (see comments under *M. brunnescens*). Specimen at upper right is immature, as evidenced by white interior. Specimen at upper left is mature and distinctly fragrant. Since many species share these features, a microscope is usually needed to distinguish them.

7. Walls of the spore-bearing chambers containing sphaerocysts in their central layers (a microscopic feature); uncommon . *Gymnomyces ferruginascens* & others (see *G. socialis,* below)
7. Walls of the spore-bearing chambers lacking sphaerocysts in their central layers (sphaerocysts may or may not be present elsewhere) .. *Martellia brunnescens* & many others, below

Martellia brunnescens

FRUITING BODY 1-3 cm broad, round to oval or irregularly knobby (potato-like). Outer surface white when young, developing ochre- to rusty-brown or brown stains in age or after handling, and usually entirely brownish in old age; smooth or pitted. **SPORE MASS** (interior) composed of small chambers, crisp and white when young, but discoloring brown to rusty-brown when cut (often slowly) and becoming ochre-brown to brown at maturity; odor often sweet (somewhat reminiscent of vanilla), at least in age. **STALK** and columella absent. **SPORES** 8-11 × 8-9 microns, broadly elliptical to round, with strongly amyloid warts and spines; hyaline (colorless) under the microscope. Cystidia absent. Walls of tissue between the chambers lacking sphaerocysts in their central layers.

HABITAT: Solitary, scattered, or in groups in soil or humus (buried) in woods; known only from western North America. It favors conifers and is one of the most common Martellias of Oregon and California. I have found it in mixed woods and under Douglas-fir in April, July, and October.

EDIBILITY: I can find no information on it.

COMMENTS: There are many Martellias that will more or less fit the macroscopic features of the above description. The sweet odor which develops in age might seem to be a distinctive feature, but is found in a number of other species (e.g., *M. fragrans* of Idaho, which has a strong vanilla-like odor). Another common western species, *M. fallax,* smells like *Russula fragrantissima* (at least to me). It is quite common in our area under hardwoods, and differs microscopically from *M. brunnescens* in having cystidia which are golden in potassium hydroxide (KOH). Also worth mentioning are *M. foetens,* an Idaho species that smells like *M. fallax; M. cremea,* which has a percurrent columella; and *M. californica, M. boozeri, M. parksii, M. ellipsospora,* et al, which differ microscopically.

Gymnomyces socialis

FRUITING BODY 1-2 cm broad (but several sometimes tightly clustered to form larger masses), roundish to somewhat irregular. Outer surface pallid to creamy or pinkish, usually staining ochraceous to rusty-brown when handled or dried; more or less smooth.

SPORE MASS (interior) composed of minute chambers, rosy-pinkish to pinkish-buff at maturity. **STALK** present as a short sterile base (at least in larger specimens) which is tucked into the depression formed by the lower portion of the peridium (skin), giving rise to a columella which penetrates the spore mass and is usually percurrent; whitish to pinkish-buff. **SPORES** 9-14 × 8-11 microns, round or nearly round, ornamented with amyloid warts or spines, hyaline (colorless) under the microscope. Walls of tissue between the chambers containing sphaerocysts in their central layers.

HABITAT: Solitary to gregarious or clustered in soil or humus under oak; known only from California. According to Herb Saylor, it is fairly common in our area in the late fall, winter, and spring.

EDIBILITY: Unknown.

COMMENTS: This species can be told by its small size, pinkish-tinged peridium (skin) and spore mass, and the presence of a columella which usually extends below the spore mass as a short stalk. The latter feature is unusual for a *Gymnomyces,* and serves to distinguish it from *G. roseomaculatus,* which also develops pinkish or reddish spots or tinges and grows under oak. Specimens with a well-developed stalk might be mistaken for a *Macowanites* or *Elasmomyces,* but do not have the rudimentary gills and exposed spore mass of those genera. Other species: *G. cinnamomeus* and *G. ferruginascens* both lack a columella and have a brownish to rusty-ochre or cinnamon exterior, at least in age. The first occurs in California under oak; the latter is recorded from Idaho under conifers.

HYDNANGIUM & Allies

Small to medium-sized fungi *found mostly underground.* FRUITING BODY round to irregularly lobed (potato-like), variously colored. SPORE MASS (interior) typically composed of chambers or elongated cavities; *pinkish* in *Hydnangium, yellow-orange at maturity* in *Sclerogaster,* variously (but usually differently) colored in *Octavianina.* Latex present or absent. STALK absent or present as a short sterile base; columella present or absent. SPORES *usually round or nearly round and warted or spiny but not amyloid;* hyaline (colorless) or pale under the microscope (or sometimes brown in *Octavianina*). Capillitium absent.

THREE miscellaneous genera with spiny or warted, non-amyloid spores are treated here. The most common genus, *Hydnangium,* is easily told by its pinkish to flesh-colored spore mass and frequent association with eucalyptus. *Sclerogaster* also has a distinctively colored spore mass: bright golden-yellow to orange-yellow at maturity. It favors conifers. *Octavianina,* on the other hand, is variable in color and cannot reliably be distinguished from other false truffles without examining its spores. Its spore mass, however, is not normally pinkish or orange-yellow.

    Hydnangium is thought to be closely related to the agaric genus *Laccaria.* The affinities of *Sclerogaster* and *Octavianina* are unclear.

Key to Hydnangium & Allies

1. Spore mass pinkish to flesh-colored; peridium (skin) pinkish or whitish; usually (but not always) associated with eucalyptus .... ***Hydnangium carneum,*** p. 745
1. Not as above 2

2. Spore mass typically golden-yellow to orange-yellow when mature (may be whitish at first)   3
2. Not as above .... ***Octavianina asterosperma*** group & others, p. 745

3. Fruiting body 0.5-2.5 cm broad, white to pale yellow; spore mass sometimes powdery in age; odor often strong at maturity but not like peanut butter; spores 8-11 microns broad; not uncommon in Southwest and Rocky Mtns., especially under pine .. ***Sclerogaster xerophilum***
3. Not as above; odor resembling peanut butter when mature; known from southern California .... ***Sclerogaster columnatus***

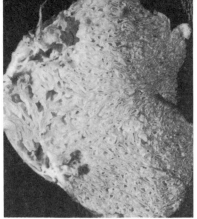

Hydnangium carneum, a pinkish false truffle that favors eucalyptus. **Left:** Specimens with a well-developed sterile base or "stalk." **Right:** A sectioned specimen lacking a sterile base. (Herb Saylor)

Hydnangium carneum   (Rosy Eucalyptus False Truffle)

FRUITING BODY 0.5-4 cm broad, round to irregularly lobed and/or flattened. Outer surface often pitted or wrinkled, dry, fibrillose or minutely velvety, pale rose to pinkish, pinkish-cinnamon, pinkish-buff, or flesh-colored (but often whitish when young and/or retaining white areas in age), sometimes also with darker spots. **SPORE MASS** (interior) composed of small, often elongated or mazelike chambers that give it a spongy or marbled appearance; pinkish or rosy (like exterior or darker); odor usually mild. **STALK** absent or often present as a fragile whitish to pinkish sterile base which may extend into the spore mass as a thin columella. **SPORES** 10-18 microns, round or nearly so, with large non-amyloid warts or spines; hyaline (colorless) or tinged yellow under the microscope.

HABITAT: Solitary to gregarious in soil or humus under eucalyptus and possibly other exotic trees and shrubs (e.g., *Leptospermum*); widely distributed wherever eucalyptus is planted or naturalized. In California it fruits from late fall through the spring, but is particularly common from January through April. I often find it with *Hymenogaster albus* and *Hysterangium fuscum*. Harry Thicrs reports finding a large-spored variant under pine.

EDIBILITY: Edible according to one source; I haven't tried it.

COMMENTS: The pinkish color (at least of the spore mass) and round spiny spores of this false truffle are strongly suggestive of the agaric genus *Laccaria,* and it is very likely that the two are closely related. It is interesting to note that pinkish Laccarias are sometimes common in our area under acacia, eucalyptus, and other Australian imports. The common *Hydnangium* of California has long gone under the names **H. roseum** and **H. soederstroemii**, but they are probably synonyms for the wide-ranging and variable *H. carneum.* One source of confusion is the presence or absence of a sterile base or "stalk." Though often present, it is very easily broken and consequently apt to be overlooked.

Octavianina asterosperma group

FRUITING BODY 1-3 cm broad, round or more often knobby and pitted (potato-like), firm, often with white mycelial threads (rhizomorphs) at the base. Outer surface whitish at first from a thin layer of cottony tissue which wears off in patches, revealing the dingy brown to grayish to yellowish-olive undersurface; discoloring pale greenish to faintly bluish or grayish-olive where bruised or rubbed. Peridium (skin) often wavy, whitish to buff, discoloring like the surface when cut. **SPORE MASS** (interior) composed of minute chambers, very firm, dark brown to blackish when mature, with whitish plates of sterile tissue sometimes visible; latex (fluid) present or absent (see comments). **STALK** absent, but a slight sterile base often present as a thickening of the basal peridium, and in some specimens this sterile base apparently giving rise to a thin columella. **SPORES** 13-18 microns, round, covered with large, blunt spines; yellowish to brown under the microscope, not amyloid.

745

HABITAT: Solitary to gregarious in soil or humus under trees in woods (usually buried); widely distributed but rare. Specimens tentatively identified as this species have been found near Santa Cruz, California, under redwood and Douglas-fir in June and July.

EDIBILITY: Unknown.

COMMENTS: Most species of *Octavianina* are rare, and this one is no exception. It was originally collected in Europe and described as exuding a latex when cut. Local material seems nearly identical to the European version, but shows no sign of a latex. *Var. potteri,* described from Michigan, also lacks a latex, but has a paler spore mass and blackens when bruised. The large, relatively few, non-amyloid warts or spines on the spores are the outstanding feature of the species. *Hymenogaster utriculatus* (see comments under *H. sublilacinus*) looks quite similar in the field, but has ridged or honeycombed spores. There are several Octavianinas with a paler (pallid to cinnamon-brown) spore mass, including: *O. papyracea,* with an abundant cream-colored latex and a papery peridium when dry, originally found in northern California under redwood; *O. rogersii,* with smaller warts on its spores and no latex; and *O. macrospora,* with elliptical spores. All of these are rare.

GAUTIERIA

Small to medium-large *woodland fungi usually found underground.* FRUITING BODY roundish to oval to lobed or knobby (potato-like). Peridium (outer skin) *absent or poorly-developed in many species.* SPORE MASS (interior) composed of small to large cavities or chambers, usually ochre-yellow *to yellow-brown, cinnmamon-brown, or dull brown;* firm and tough or rubbery when fresh, often rubbery-gelatinous and strong-smelling in age. STALK absent or rudimentary; *columella often present, well-developed, and branched.* SPORES typically elliptical and yellowish to rusty-cinnamon under the microscope, *longitudinally striate or ridged (or even "winged"), not amyloid.* Basidia borne in a hymenium lining the walls of the chambers. Capillitium absent.

THE ochre to cinnamon to dull brown overall color and tendency of the outer skin to wear away (thereby exposing the inner cavities) are the main field characters of this distinctive genus of false truffles. Those few species with a persistent peridium (skin) can be distinguished from *Hymenogaster* and other false truffles by their brownish, longitudinally lined spores. *Gautieria* is a common genus (if you're looking for truffles), but the various species are difficult to distinguish from each other and several are still undescribed. One representative is described here.

Key to Gautieria

1. Peridium (skin) well-developed and persistent, i.e., the spore mass not exposed 2
1. Peridium absent or present only as a thin layer of tissue that soon wears away, i.e., the spore mass (spore-bearing chambers or cavities) usually exposed at maturity 3

2. Peridium whitish when young .... *G. gautierioides* (see *G. monticola,* p. 747)
2. Not as above .... *G. parksiana* (see *G. monticola,* p. 747)

3. Spore-bearing cavities quite large (2-10 mm broad); columella typically absent or rudimentary; found under hardwoods .... *G. morchelliformis* (see *G. monticola,* p. 747)
3. Not as above; spore-bearing cavities often smaller and columella usually present; very common under conifers, but also occurring with hardwoods 4

4. Spore mass yellowish to bright yellow-brown or ochre; spores appearing "winged" under the microscope .... *G. pterosperma* (see *G. monticola,* p. 747)
4. Spore mass typically rusty-brown or darker or duller; spores longitudinally lined but not "winged" .... *G. monticola* & others, p. 747

Gautieria monticola. **Left:** Exterior; note how there is practically no skin, i.e., the chambered spore mass is visible. **Right:** A specimen sliced open to show the interior. Note the thin branching columella.

Gautieria monticola

FRUITING BODY (1) 2-9 cm broad, round to elongated or irregularly knobby (potato-like), often with a mycelial cord (rhizomorph) at base. Outer surface pallid when very young, soon becoming light brown to ochre to rusty-brown or dull brown. Peridium (skin) very thin, soon wearing away to expose the spore-bearing cavities. **SPORE MASS** (interior) composed of numerous small (0.5-3 mm) round or elongated chambers, firm and rubbery (*not* fragile) and crisp when fresh, ochre-brown to rusty-brown to dull cinnamon to dull brown at maturity; odor often strong in age (sometimes reminiscent of decaying onions). **STALK** absent, but a distinct columella usually present and often penetrating to the center of the spore mass; columella whitish or translucent, thin, often branched. **SPORES** 10-15 × 6-9 microns, elliptical, longitudinally grooved, rusty-brown to ochre-yellow under the microscope.

HABITAT: Solitary, scattered, or in groups or clusters in duff under conifers (sometimes partially exposed at maturity); common in western North America, particularly under mountain conifers in the late spring, summer, and early fall. I have seen large numbers in the Sierra Nevada and Cascades in June, and a similar species occurs in our area under hardwoods. The strong odor that develops in age presumably aids animals in locating them. The animals then help to disperse the spores.

EDIBILITY: Edible, I am told, but rather rubbery.

COMMENTS: The rubbery brown spore mass composed of numerous irregularly-shaped cavities plus the absence or near-absence of an enveloping peridium (skin) are characteristic of this species and its close relatives. *Hysterangium* species are somewhat similar, but have a persistent skin and are differently colored. Similar species include: *G. graveolens,* a larger-spored conifer-lover; *G. pterosperma,* fairly common in Oregon and California, with "winged" spores and yellow-brown to ochre cavities; and *G. candida,* with melon-like spores. Other species: *G. morchelliformis* is a widespread hardwood-loving species with very large (up to 10 mm broad) spore-bearing cavities, no peridium, and little or no columella. I have sniffed it out under oak because, like *G. monticola,* it develops a strong odor in age. Two small Gautierias with a *persistent* peridium should also be mentioned: *G. parksiana,* whose peridium is yellowish-orange to ochre at first; and *G. gautierioides,* whose peridium is whitish, at least initially. Both of these species occur on the west coast and are likely to be mistaken for a *Hymenogaster* or *Martellia* until their spores are examined.

747

HYMENOGASTER & Allies

Small to medium-sized *woodland fungi usually found underground.* FRUITING BODY roundish to pear- or cushion-shaped to knobby (potato-like), variously colored. Peridium (skin) *present, usually well-developed and persistent.* SPORE MASS (interior) composed of numerous irregular chambers or tubelike cavities or at times nearly solid (i.e., the chambers very minute); *typically some shade of brown (rusty- to blackish-brown) when mature;* fragile to firm, rubbery, or slightly gelatinous. STALK absent or rudimentary, *but a well-developed columella often present.* SPORES tawny to brown under the microscope, elliptical to nodulose, *usually warted or wrinkled, sometimes longitudinally ridged (Chamonixia)* or in one species coarsely netted; *not amyloid.* Basidia borne in a hymenium (except in *Destuntzia*); hymenium lining the chamber walls. Capillitium absent.

THESE are small to medium-sized false truffles with an ochre to brown or blackish mature spore mass, a persistent peridium (skin), and warted or wrinkled, non-amyloid spores. *Hymenogaster* is a diverse genus that is gradually being split into smaller units such as *Destuntzia.** A small "satellite" genus, *Chamonixia,* is also treated here; it intergrades with *Hymenogaster.* All of these are thought to be related to the Cortinariaceae or possibly the Boletaceae. None are significant from an edibility standpoint unless you're a squirrel or aspire to be one. Three species are described; several others are keyed out.

Key to Hymenogaster & Allies

1. Peridium (skin) white when fresh but staining vinaceous to vinaceous-brown or blackish after handling (sometimes also with blue-green to yellow-green stains); columella and/or short stalk usually present, often percurrent; mature spore mass dark brown to blackish; spores longitudinally ridged; known from California under oak ..... *Chamonixia ambigua,* p. 751
1. Not with above features 2

2. Peridium typically whitish when young but becoming pale lilac or dull bluish, then yellowish to orangish and finally brownish in age; spores warted, not beaked; common under mountain and northern conifers (e.g., fir, spruce, lodgepole pine) .... *H. sublilacinus,* p. 749
2. Not with above features 3

3. Fruiting body staining blue or bluish-green when bruised or cut in half 4
3. Not as above 6

4. Peridium yellowish, often with reddish stains or becoming reddish to reddish-brown in age; spores smooth .... (see *Truncocolumella rubra* under *T. citrina,* p. 752)
4. Not as above; spores longitudinally lined or ridged 5

5. Columella short and thick, occupying only the base of the fruiting body ....
.... *Chamonixia brevicolumna* (see *C. ambigua,* p. 751)
5. Not as abvoe .... *Chamonixia caespitosa* & others (see *C. ambigua,* p. 751)

6. Peridium usually becoming pink, reddish, or vinaceous in age or when handled; chambers of spore mass often (not always!) gel-filled .... *Destuntzia rubra* & others, p. 750
6. Not as above 7

7. Spore mass (interior) dark brown at maturity (see photo on p. 749); fruiting body often 1.5 cm or more broad; mature spores somewhat honeycombed (i.e., with ridges and "pits"); found mainly under conifers, especially Douglas-fir; rare *H. utriculatus* (see *H. sublilacinus,* p.749)
7. Not as above; common 8

8. Spores longitudinally ridged, grooved, or "striped" 9
8. Spores wrinkled or warted, but not as above 10

9. Spores 9-15 microns long, with an apical pore; peridium light brown to dark yellow-brown or maroon-brown .... *Chamonixia caudata* (see *C. ambigua,* p. 751)
9. Not as above; spores usually larger .... (see *Gautieria,* p. 746)

10. Associated with eucalyptus 11
10. Not as above; found with oak and other trees *H. parksii* & others (see *H. sublilacinus,* p. 749)

11. Peridium and spore mass containing dirt; spore mass dark brown and gelatinous at maturity; rare .... *Chondrogaster*
11. Not as above; peridium white to yellowish; common .. *H. albus* (see *H. sublilacinus,* p. 749)

*The genus *Destuntzia* and its species (see comments under *D. rubra*) are described in an article by Robert Fogel and James Trappe, now in press. The names are used here with their permission.

Hymenogaster sublilacinus

FRUITING BODY 1-3 (5.5) cm broad, round to irregularly lobed (potato-like). Outer surface often fibrillose, variable in color but typically whitish to lilac or dull bluish when young, becoming yellowish and then orangish and finally brown in age; often staining dingy ochraceous when handled (if older) or bluish (if young). Peridium (skin) often separating rather easily from spore mass. **SPORE MASS** (interior) composed of small or minute chambers that give it a sponge-like appearance; pallid when very young but brown to cinnamon-brown at maturity; odor variable (mild, sweetish, farinaceous, etc.). **STALK** absent, but sterile tissue usually present as an expansion or thickening of the basal peridium, which may or may not give rise to a thin, short, branching columella. **SPORES** 8-13 (15) × 5-8 (9.5) microns, elliptical, warted-wrinkled, brown under the microscope.

HABITAT: Solitary to gregarious in soil or duff under mountain conifers, especially fir, spruce, and lodgepole or ponderosa pine; fairly common in the Sierra Nevada and other mountain ranges of the West in the spring, summer, and early fall.

EDIBILITY: Unknown.

COMMENTS: This species appears to be the most common conifer-loving *Hymenogaster,* at least in California. The confusing color changes it undergoes as it ages and dries out has led to the naming of several "new" species (e.g., *H. brunnescens, H. diabolus, H. subochraceus)* where only one exists. The brown color of the mature spore mass plus the warted brown spores separate it from other genera of false truffles. Other species: *H. parksii* is very common in our area under oak and other trees. It is small (5-15 mm broad) and whitish when fresh and has a grayish to ochre-brown or cinnamon-brown spore mass and prominently beaked spores (see photo at bottom of p. 750). *H. gilkeyae* closely resembles *H. parksii* but differs microcopically. *H. mcmurphyi* also grows under oak, but yellows in age or when bruised. *H. albus (=H. albellus, H. luteus)* is a white or yellow-staining species that grows under eucalyptus (often with *Hydnangium carneum*). *H. utriculatus* (see photo below) is a conifer-lover with distinctive pitted-reticulate spores at maturity. Macroscopically it rather resembles the description of *Octavianina asterosperma* (dark spore mass, etc.). I have found it in large numbers under Douglas-fir and redwood in June and July, but it probably fruits throughout the mushroom season. Because of its distinctive spores it may deserve a genus of its own. Many other Hymenogasters occur, but most can only be identified with the aid of a microscope.

Hymenogaster utriculatus has very unusual spores (see comments under *H. sublilacinus*). Note the dark spore mass (interior).

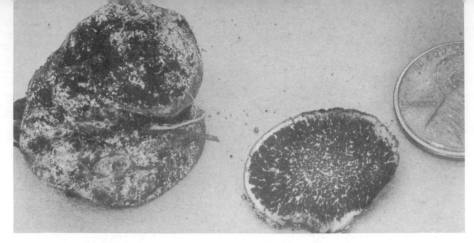

Destuntzia rubra is best told by its tendency to develop reddish or pinkish tones as it ages or is handled, plus its dark, sometimes gelatinous spore mass (interior).

Destuntzia rubra

FRUITING BODY 1-4 cm broad, roundish to oblong or lobed (potato-like). Outer surface white when young, but staining pink, reddish, or vinaceous (or even bluish) in age or after handling. Peridium (skin) fairly thick, white when sectioned. **SPORE MASS** (interior) olive-brown to grayish-brown becoming darker (sometimes nearly black) in age; often speckled with white sterile tissue that separates clusters of minute chambers (use hand lens!); firm or gelatinous (chambers may or may not be gel-filled). **STALK** and columella absent, but vinaceous mycelium sometimes attached to base or visible in humus. **SPORES** 8-12 × 6.5-9 microns, elliptical, with striate conical warts, brownish under the microscope. Basidia borne in chambers but not typically forming a hymenium.

HABITAT: Solitary to gregarious in soil or duff (usually buried) in mixed woods and under conifers; fairly common in our area in the late winter, spring, and early summer (but probably more widely distributed). It seems to favor Douglas-fir. A similar species, *D. saylorii*, is fairly common in the Sierra Nevada.

EDIBILITY: I can find no information on it.

COMMENTS: Formerly known as *Hymenogaster ruber*, this species and its relatives can be recognized by their dark, often somewhat gelatinous spore mass and pink- or reddish-staining exterior. The specimen shown in the photograph was pure white when collected and entirely vinaceous when I brought it home. Several Rhizopogons stain reddish, but usually have a more fibrillose exterior, paler interior, and/or smooth spores. The spore mass of *Destuntzia* may gelatinize somewhat, leading to confusion with *Melanogaster,* but the chambers are much smaller than in that genus. Other Destuntzias include: *D. saylorii* (with 1-spored basidia like *D. rubra*) and *D. fusca* of California; *D. subborealis* of Idaho; and *D. solstitialis*, an eastern species (see footnote on p. 748).

Hymenogaster parksii (see comments under *H. sublilacinus* on previous page). This small whitish false truffle is very common in our area, especially under oak. The distinctly chambered spore mass is grayish to brownish and usually lacks a well-developed columella. There are a number of closely related look-alikes, including *H. albus,* which commonly grows under eucalyptus.

Chamonixia ambigua is a rare oak-loving species with a dark spore mass and whitish exterior that darkens when handled. Note presence of columella in sectioned specimen at center.

Chamonixia ambigua

FRUITING BODY (0.5) 1-5 cm broad, more or less pear-shaped to cushion-shaped (i.e., round or oval with a narrowed base) to somewhat lobed (especially if growing in clusters). Outer surface white when fresh, but discoloring vinaceous to dark vinaceous-brown to blackish-brown after handling or standing, and either staining yellow-green to bluish-green when bruised *or* showing such stains, especially near the base; smooth but rather soft or cottony. Peridium (skin) very thin. **SPORE MASS** dark brown to vinaceous-brown to deep grayish-brown when fresh, minutely chambered, the chambers empty so that they resemble small tubes. **STALK** absent or present as a white or greenish-stained sterile base which usually gives rise to a columella; columella branched or unbranched, usually percurrent. **SPORES** (9) 11-15 (18) × 8-12 microns, elliptical with a prominent "beak" at one end, longitudinally ridged and slightly warted, the inner wall with an apical pore; brown to dark brown under the microscope.

HABITAT: Solitary, scattered, or in groups or small clusters in humus and soil (buried) under live oak; known only from California. Like many subterranean fungi, it is described as being "rare," but may very well be common once you know when and where to look for it. I have found it in June and July in Santa Cruz County.

EDIBILITY: Unknown.

COMMENTS: The genus *Chamonixia* intergrades with *Hymenogaster,* but most of its species can be told in the field by their tendency to stain bluish or to exhibit natural bluish-green stains. The distinctly tubulose (spongelike) spore mass is reminiscent of a bolete, and it has been suggested that *Chamonixia* is related to the boletes (along with *Truncocolumella*). *C. ambigua* may or may not stain bluish when bruised, but typically stains vinaceous to black when rubbed (within a few minutes) and usually has a percurrent columella. *C. caespitosa* is a widespread species whose skin is white when fresh and rapidly stains indigo-blue when bruised; it also has larger spores. *C. brevicolumna* has an olive-ochre peridium that blues when bruised, plus a short broad rounded columella; it has been found under conifers in Idaho. *Hymenogaster pyriformis* is a small blue-staining species originally collected under oak in California. *C. caudata* has a yellowish-brown to brown or maroon peridium and a short whitish stalk, but does not stain blue; it occurs under oak and other hardwoods in California and Oregon.

751

TRUNCOCOLUMELLA

THIS small but common genus is distinguished by its smooth spores, branched or stump-like columella, and yellowish to olive color. The columella may extend below the spore mass to form a short stalk (see photo), leading to possible confusion with the gastroid agarics, but it is not usually percurrent. *Rhizopogon* is somewhat similar, but lacks a colu-mella, while *Hymenogaster* and its allies have wrinkled, striate, or warted spores. *Trunco-columella* is thought to be closely related to the boletes, but it is not nearly as bolete-looking as *Gastroboletus* (p. 544), which has a cap, exposed tube layer, and percurrent stalk. At least two species of *Truncocolumella* occur on the west coast. One stains blue when bruised and is perhaps better placed in *Chamonixia* (see key to *Hymenogaster* & Allies); the other, described below, does not stain blue.

Truncocolumella citrina

FRUITING BODY 2-5 (7) cm broad, 1.5-5 cm high, irregularly rounded to oval, bulblike, lobed, or kidney-shaped, often with a stubby stalk. Outer surface sometimes whitish when very young but soon becoming yellow to ochre, greenish-yellow, or grayish-olive, smooth or sometimes cracking to form a few scales or patches. Flesh(in peridium and stalk) yellow to buff, firm. **SPORE MASS** (interior) yellow-brown to olive-gray, becoming darker (sometimes blackish) in age; composed of small empty tubular chambers, completely enclosed by the peridium; firm when young but often quite gelatinous in old age. **STALK** often (but not always) present as a short, thick, narrowed base; columella present also, either stumplike (i.e., a broad basal region of sterile tissue) or branched and extending into the spore mass for a considerable distance; yellow to buff. **SPORES** 6-10 × 3.5-5 microns, elliptical, smooth, yellowish to light brown under the microscope.

HABITAT: Solitary to gregarious under Douglas-fir (and possibly other western coni-fers); common, especially in the summer and fall, but fruiting practically year-round in our area. Like most false truffles it develops underground, but often surfaces at maturity or is dug up by rodents.

EDIBILITY: I can find no published information on it, but ethnomycologist Jim Jacobs had a bizarre experience after putting some in an omelet: although they had little or no

Truncocolumella citrina is particularly common under Douglas-fir. Note narrowed stalklike base in specimen at left and stout columella in sectioned specimen at right.

flavor at the time of eating, a strong licorice-like flavor began to permeate his mouth an hour or two later, and lingered for several hours!

COMMENTS: The yellow to buff, stumplike or branched columella plus the yellowish to olive exterior make this a relatively easy species to identify. Other species: *T. rubra* has a reddish or reddish-staining exterior, a dingy yellowish spore mass that stains blue when cut, and larger spores. It was originally described from Washington but has been found in the San Bernardino Mountains of southern California. For other blue-staining false truffles, see comments under *Chamonixia ambigua.*

RHIZOPOGON

Small to medium-large *underground or erumpent fungi found mainly under conifers.* FRUITING BODY round to oval or irregular (potato-like), variously colored. Peridium (skin) present, often fibrillose, felty, or overlaid with rhizomorphs (mycelial strands). SPORE MASS (interior) sponge-like, i.e., composed of minute chambers; *firm, crisp, rubbery, or cartilaginous when young,* sometimes becoming soft or gelatinous in age; *usually cinnamon-brown to dingy olive-brown or grayish at maturity* (but often white when young). STALK *and columella typically absent.* SPORES hyaline (colorless) to yellowish or brownish under the microscope, *typically smooth and elliptical to cylindrical,* sometimes amyloid. Basidia forming a hymenium that lines the walls of the chambers. Capillitium absent.

THESE dingy, unattractive, potato-like fungi are the Russulas of the underworld—unappreciated except by squirrels, but ubiquitous.* The 100+ known species are differentiated primarily on chemical and microscopic features such as whether or not the spores are pronged and what color the hyphae of the peridium stain when mounted in potassium hydroxide. However, the sameness and mundaneness of the Rhizopogons make them relatively easy to recognize as a group. The fruiting body usually has a tough or rubbery ("better bounced than trounced") consistency and the interior is composed of tiny chambers that give it a spongelike appearance (use a hand lens!). Also, the exterior is often overlaid with mycelial strands (rhizomorphs), there is no stalk or columella (or rarely a rudimentary columella) inside the spore mass, and the spores are typically smooth. Finally, nearly all Rhizopogons are mycorrhizal with members of the pine family. (One unidentified local species seems to grow only *beneath* cow patties or "meadow muffins," but may still be mycorrhizal.) In some species the spore mass becomes soft or gelatinous in old age, but the chambers are never filled with a gel as in *Alpova* and *Melanogaster,* nor are they separated by pallid veins, nor does the spore mass become powdery as in the puffballs and earthballs.

Rhizopogons are not only the most ubiquitous of all the hypogeous (underground) fungi, they are also among the most visible. Many are erumpent (i.e., they burst through the surface of the ground at maturity); others are excavated by squirrels. A few species, (e.g., *R. occidentalis, R. smithii*) are edible, but most have not been tested, and as already pointed out, identification is very difficult. My own experience with them is limited. Not only do I have an "allergy" to microscopes, but I just can't seem to get excited about these dingy, dumpy, dirty "small potatoes" of the mushroom microcosm when there are so many boletes to be picked and Russulas to be kicked. Over 200 species of *Rhizopogon* have been described, most of them from western North America, but my carefully cultivated ignorance of the subject permits only a *very* meager treatment here: a grand total of three species are described and several others are keyed out. For a much more extensive treatment, see *How to Know the Non-Gilled Mushrooms* by Alexander Smith (who has "the world's largest collection of unidentified Rhizopogons"), Helen Smith, and Nancy Weber.

*Whereas the Russulas' brittle flesh is irresistible to those who like to trounce things, the Rhizopogons' rubbery texture is a blessing to those who like to *bounce* things.

Key to Rhizopogon

1. Peridium (skin) yellow to ochre or tawny beneath a coating of brown to reddish-brown rhizomorphs (mycelial threads), but often developing reddish to brownish stains in age; common under pine . . . . *R. ochraceorubens,* p. 755
1. Not with above features; peridium often white or whitish when young 2

2. Fruiting body small (0.5-1.5 cm), bright yellow . . . *R. cokeri* (see *R. ochraceorubens,* p. 755)
2. Not as above 3

3. Peridium *and/or* spore mass (interior) staining pinkish, reddish, or distinctly vinaceous when cut or bruised (and often in age) . . . . *R. rubescens* & many others, below
3. Not as above (but may stain purplish-gray, bluish, fuscous, or black) 4

4. Common on west coast under conifers other than pine (e.g., Sitka spruce, Douglas-fir); exterior staining blue-black or purplish when bruised or in age . . . *R. parksii* (see *R. ellenae,* p. 755)
4. Not as above 5

5. Exterior usually staining or aging bluish . . . . *R. subcaerulescens* (see *R. ellenae,* p. 755)
5. Not as above 6

6. Peridium staining yellow when cut open . . . . *R. pinyonensis* (see *R. ochraceorubens,* p. 755)
6. Not as above 7

7. Fruiting body rather small (averaging 1-2 cm), vinaceous-tinged at least in age; spores *not* amyloid; found under coastal pines . . . . *R. maculatus* (see *R. ochraceorubens,* p. 755)
7. Not as above; fruiting body usually larger, often staining brown, gray, pinkish-gray, purple-gray, or black with age or handling; spores amyloid or not   *R. ellenae* & many others, p. 755

Rhizopogon rubescens   (Blushing False Truffle)

FRUITING BODY 1-6 cm broad, round to oval or somewhat irregular (potato-like). Outer surface somewhat cottony or fibrillose, often sparsely overlaid with mycelial strands (rhizomorphs); white when young but staining pinkish to red or vinaceous when bruised or cut, and becoming yellow to greenish-yellow or reddish-stained in age. **SPORE MASS** (interior) minutely chambered, firm or rubbery to spongy or slightly gelatinous in age; white when young, but staining pinkish when cut and becoming tawny-olive to olive-buff to dark olive-brown or brown in age. **STALK** and columella absent. **SPORES** 5-10 × 3-4.5 microns, elliptical to oblong, smooth, usually tinged yellow under the microscope.

HABITAT: Scattered to gregarious or clustered in duff or soil under conifers (usually buried or half-buried), especially pine; widespread and common. In our area this species and/or its numerous look-alikes (see comments) can be found whenever it is damp.

EDIBILITY: Delectable—if you're a squirrel!

COMMENTS: The reddish-staining peridium (skin) is the outstanding feature of this *Rhizopogon,* but there are many other "blushing" species that can only be separated with certainty by using a microscope. Their ranks include: *R. smithii,* an aromatic species that grows with *Boletus edulis* in pine forests and is eaten by Italian-Americans as a white truffle substitute (it is whitish when young, reddens when bruised, and darkens considerably with age); *R. vinicolor,* a common associate of Douglas-fir that is white when young and becomes vinaceous-red in age; *R. roseolus,* which becomes pinkish in age; *R. subaustralis,* which develops black stains in age; *R. succosus,* with a strong rotten egg odor in age; *R. occidentalis,* a common white to yellow, edible western species said to have a sweetish taste and skin that stains orange to reddish-brown when rubbed and reddish when sectioned; *R. couchii,* widespread, with few or no rhizomorphs and a pink-staining spore mass when fresh; *R. evadens,* similar to *R. smithii* but with a disagreeable "metallic" (Smith) odor in age, very common with ponderosa pine; *R. atroviolaceus,* with strongly amyloid spores and a white fruiting body that stains pinkish, then develops grayish-purple to dark brownish tones in age; and *R. subsalmonius,* a diffcult-to-identify species "complex" that is white to salmon-pink with salmon- or vinaceous-tinged rhizomorphs on its exterior, especially common under mountain conifers (e.g., fir and spruce).

Rhizopogon ochraceorubens is one of many Rhizopogons with prominent mycelial threads (rhizomorphs) on its exterior (specimen at right). The interior (left) is white when young, darker in age. It is also shown on p. 740. (Herb Saylor)

Rhizopogon ochraceorubens

FRUITING BODY 2-8 cm broad, round to oval or somewhat flattened or irregular (potato-like). Outer surface yellow to tawny or ochraceous when young, but usually overlaid with conspicuous brown to reddish-brown rhizomorphs (mycelial strands), often mottled with red or reddish-brown areas (especially in age), sometimes reddening somewhat when bruised or discoloring brown to rusty-brown. **SPORE MASS** (interior) pallid to grayish becoming olive or olive-brown in age, firm or spongy becoming tough as it dries; minutely chambered. **STALK** and columella absent. **SPORES** 6-8 × 2-3 microns, oblong, smooth, hyaline or tinged yellow or greenish under the microscope.

HABITAT: Scattered to gregarious or clustered (usually buried, sometimes erumpent) under pine; widely distributed in western North America and very common. It is often abundant in the Sierra Nevada as well as along the coast whenever it is damp enough.

EDIBILITY: Unknown.

COMMENTS: The conspicuous brown to reddish-brown rhizomorphs that cover the exterior of the fruiting body are a distinctive though by no means unique feature of this species. It doesn't redden as much as *R. rubescens,* and is not white when young; there are several similar species which can only be differentiated microscopically. Other species: *R. maculatus* is a small (1-2 cm) vinaceous-tinged species with small rhizomorphs on its outer surface and a grayish mature spore mass that becomes quite hard as it dries; it occurs under coastal pines. *R. cokeri (=R. truncatus)* is an infrequent but widely distributed species with a small (0.5-1.5 cm), bright yellow fruiting body that does not stain; it grows under conifers also, particularly white pines. *R. pinyonensis* has a peridium that stains yellowish when broken; it is one of several Rhizopogons associated with pinyon pine.

Rhizopogon ellenae

FRUITING BODY 1-9 cm broad or long, oval to elongated to roundish, often somewhat flattened. Outer surface white when young but developing dingy pinkish-gray to purplish-lilac to fuscous (dark purple-gray or smoky) tones in age or after handling; smooth to fibrillose or with a few rhizomorphs (mycelial strands). **SPORE MASS** (interior) minutely chambered, firm and white when young, becoming gray to olive-gray in age. **STALK** and columella absent. **SPORES** 7-9 × 3-4 microns, elliptical to oblong, smooth, hyaline (colorless) or tinged yellow under the microscope, weakly amyloid at maturity.

Rhizopogon ellenae is very common under pine in many parts of the West. Note relatively large size (for a false truffle) and dark stains.

HABITAT: Scattered to gregarious or clustered in duff or soil under conifers (particularly pine), often erumpent (breaking through the ground); common in western North America. I have seen large fruitings under pine in the Sierra Nevada and along the coast.

EDIBILITY: Unknown.

COMMENTS: This is one of several medium-sized to rather large Rhizopogons that stain dingy purplish- or pinkish-gray (not reddish as in *R. rubescens!*) to fuscous when handled and/or in age. Other species: *R. idahoensis* also has amyloid spores (a rather unusual feature for a *Rhizopogon*); it stains brownish to grayish-lilac when bruised and is especially common in the northern Rockies under Douglas-fir. *R. amyloideus* has strongly amyloid spores, but is more prevalent in the coastal forests of northern California. *R. colossus* is a large species that is whitish when fresh but stains brown to blackish when handled or dried; it has non-amyloid spores and, like the others, occurs under western conifers. *R. parksii* is a very common coastal species, particularly under Sitka spruce and Douglas-fir. It is pallid to brown or pinkish, but stains dingy purple to bluish-black on its surface when bruised or with age. *R. subcaerulescens* tends to develop bluish or greenish stains when handled or in age and is widely distributed. It is white at first, dark vinaceous-brown in age. Also see the species listed under *R. rubescens*.

ALPOVA & MELANOGASTER

Small to medium-sized *underground* fungi. FRUITING BODY roundish to oval or knobby (potato-like), variously colored but not normally white. SPORE MASS (interior) *yellow to greenish, brown, or black at maturity; composed of numerous chambers filled with spore-containing gel, the chambers separated by paler, meandering veins.* STALK *and columella typically absent.* SPORES elliptical to cylindrical and often truncate or cupped at one end, *smooth, non-amyloid;* thick-walled and brown to dark brown or purple-black, with a pore *(Melanogaster),* thin-walled, pore-less, and hyaline (colorless) to medium brown under the microscope *(Alpova).* Basidia and spores contained *inside* the chambers, not forming a hymenium. Capillitium absent.

THESE false truffles, which comprise the family Melanogastraceae, are easily told by their chambered spore mass that is gelatinous and colored at maturity and marbled with pallid or yellow veins. Microscopically they are also distinctive by virtue of their smooth, lemon-shaped to cylindrical spores. In *Alpova,* the spores are colorless to brownish under the microscope and the spore mass ranges in color from yellow to brown when mature. In *Melanogaster,* on the other hand, the spores are darker and the most common species have a dark brown to black mature spore mass. The marbled interior can lead to confusion with

756

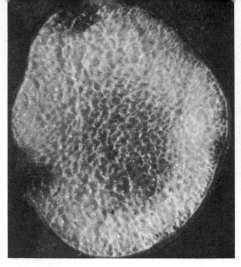 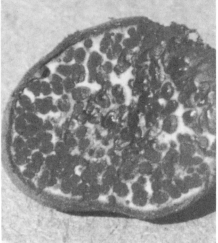

Left: Close-up of an *Alpova* (*A. olivaceotinctus*), showing the numerous gel-filled chambers that comprise the spore mass. (Herb Saylor) **Right:** *Melanogaster* species are easily told by their blackish interior marbled with white to yellow-orange veins. This is probably *M. variegatus*.

the true truffles (e.g., *Tuber*), which, however, have a harder, non-gelatinous interior and bear their spores inside asci rather than on basidia. *Hysterangium* is also somewhat similar, but usually has a translucent columella instead of veins, while *Leucogaster* and *Leucophleps* have a whitish spore mass even when mature.

Both *Alpova* and *Melanogaster* are widely distributed. Neither is worth eating (unless you're a squirrel or wood rat!), although the latter has been used in Europe as a cheap substitute for truffles. One representative of each genus is described here and several others are keyed out.

Key to Alpova & Melanogaster

1. Peridium (skin) bright yellow when young (but often mottled with reddish, orange, or brown in age); interior (spore mass) brownish-yellow to brownish-black but not truly black; associated with western conifers .... ***A. trappei*** (see *A. diplophloeus,* below)
1. Not as above 2

2. Interior (spore mass) typically black, at least at maturity; spores strongly brown to dark brown under the microscope, usually with rather thick walls and a pore ....
.... ***Melanogaster euryspermus*** & others, p. 758
2. Not as above (spore mass sometimes brownish-black, but usually paler and spores paler under the microscope) 3

3. Typically associated with alder .... ***A. diplophloeus,*** below
3. Not as above 4

4. Found in western North America ..... ***A. alexsmithii*** & others (see *A. diplophloeus,* below)
4. Found in eastern North America 5

5. Peridium (skin) darkening when handled or bruised; associated with conifers ....
.... ***A. olivaceoniger*** (see *A. diplophloeus,* below)
5. Not as above; found mostly with hardwoods .... ***A. nauseosus*** (see *A. diplophloeus,* below)

Alpova diplophloeus   (Alder False Truffle; Red Gravel)

FRUITING BODY 0.5-3 (5) cm broad, oblong or oval to nearly round, or less commonly irregularly lobed. Outer surface pallid to pinkish-yellowish-buff when young, soon becoming yellow-brown to cinnamon, reddish-brown, or brown, usually staining dark reddish-orange to reddish-brown where bruised; smooth or with only a very few rhizomorphs (mycelial threads); odor usually fruity at maturity. **SPORE MASS** (interior) composed of small chambers separated by pale yellow or whitish veins; chambers 0.5-3 mm broad, filled with a gelatinous substance, pallid to pale yellowish or olive when young,

Alpova diplophloeus. Note somewhat gelatinous interior (specimen on left) and smooth exterior.

becoming orange-brown to reddish-brown to dark vinaceous-brown at maturity. **STALK** and columella absent. **SPORES** 4-6 × 1.5-3 microns, elliptical to oblong, smooth, hyaline (colorless) under the microscope.

HABITAT: Widely scattered to gregarious in humus or soil (usually buried) under or near alder; widely distributed, but particularly common in western North America. I have found it in the fall and winter in coastal California and in the late summer, fall, and spring in the Sierra Nevada and Cascades.

EDIBILITY: I can find no information on it.

COMMENTS: Also known as *A. cinnamomeus* and *Rhizopogon diplophloeus,* this false truffle can be distinguished by its cinnamon-colored exterior (when mature), fondness for alder (a tree not noted for its diversity of false truffles), and its spore mass composed of small, gel-filled chambers and pallid veins. At maturity the interior (spore mass) is reddish-brown and somewhat reminiscent of "red gravel." It can even turn dark vinaceous-brown, but never becomes black as in *Melanogaster.* Other species: *A. trappei(=A. luteus)* is a brilliant yellow species that becomes dull yellow to orange- or reddish-brown (or mottled with these colors) in age. It has a brownish-yellow to brownish-black spore mass and a garlicky to slightly fruity odor when mature. It grows with fir, Douglas-fir, and other western conifers; I have found it locally in the winter and spring. *A. olivaceotinctus* is a yellowish-brown or olive-tinged western species that is apparently rare. *A. alexsmithii* is yellowish-brown to dark brown and grows with mountain conifers (especially hemlock). *Melanogaster parksii* of California is said to have a brown interior. Alpovas in eastern North America include: *A. olivaceoniger,* a brownish-olive conifer-lover that darkens when handled and has a fruity odor at maturity; *A. nauseosus,* brown to reddish-brown, with a yellow-brown to blackish-brown interior and strong aroma of rotten fruit at maturity, favoring hardwoods; and *A. superdubius,* whose stalklike rooting base causes false truffle expert James Trappe to be "superdubious" as to whether it is really an *Alpova.*

Melanogaster euryspermus   (Black Veined False Truffle)

FRUITING BODY 1.5-4 (6) cm broad, round to elongated or lobed (potato-like), often with rhizomorphs (mycelial threads) at base. Outer surface slightly fibrillose to downy or felty, rusty-brown to reddish-brown, warm brown, or dark brown (but often yellower when young), often blackening where bruised. **SPORE MASS** (interior) composed of large (up to 5 mm broad) blackish gel-filled chambers separated by whitish to yellow meandering veins; odor usually strong in age (sometimes like sewer gas, at other times with a pleasant citrus component). **STALK** and columella absent. **SPORES** 9-18 × 7-11 microns, lemon-shaped (but with two small prongs at one end), thick-walled, smooth; dark brown under the microscope.

HABITAT: Solitary, widely scattered, or in small groups in duff or soil under both hardwoods and conifers; widely distributed and fairly common (for a false truffle), but rarely fruiting in large numbers. In California this species and *M. variegatus* occur throughout the mushroom season. I have found them under oak, tanoak, and madrone, and in the Sierra Nevada, under conifers.

Melanogaster variegatus (see comments under *M. euryspermus*). Note brown exterior (right) and gelatinous black interior that is divided into chambers by paler veins (left). (Herb Saylor)

EDIBILITY: According to European sources, *M. variegatus* (see comments) is edible when young. I can find no information on American material.

COMMENTS: This species and its close relatives are easily distinguished from other false truffles by their blackish gelatinous spore mass composed of large chambers separated by pallid to yellow veins. They might be mistaken for true truffles because of the veins, but they bear their spores on basidia and are much softer and stickier inside. The name *M. variegatus* has traditionally been used for our common species. However, the genus is being critically studied and species concepts may be revised. It is very similar to *M. euryspermus*, but has elliptical spores that are usually less than 11 microns long. *M. intermedius* is also similar, but has spores intermediate in size between those of *M. euryspermus* and *M. variegatus*. *M. ambiguus* has large, elongated spores and a spongy, more olive-colored exterior plus a somewhat metallic odor in age; I usually find it under Douglas-fir. *M. macrocarpus* is a large (up to 12 cm!) conifer-loving species with large spores. *M. parksii* of California is said to have paler brown spores and a brown rather than black spore mass. See also *Alpova trappei* and *A. nauseosus* (under *A. diplophloeus*).

LEUCOPHLEPS & LEUCOGASTER

Small to medium-sized *underground woodland fungi.* FRUITING BODY roundish to oval or knobby (potato-like), *usually pallid in Leucophleps and usually brightly colored in Leucogaster.* Peridium (skin) sometimes overlaid with rhizomorphs. SPORE MASS (interior) *white or pallid even when mature; composed of numerous chambers that are filled with a clear spore-containing gel and/or exude a white latex when cut* (unless old). STALK and columella absent or rudimentary. SPORES hyaline (colorless) under the microscope, round to elliptical, *pitted-reticulate or spiny within a clear gelatinous outer wall, not amyloid.* Basidia and spores contained *inside* the chambers, not forming a hymenium. Capillitium absent.

THESE potato-like fungi can be told by their white or pallid interior which is composed of gel-filled chambers (which may, however, be very small). Fresh specimens often exude a white latex when cut, but the spores are not amyloid as in *Zelleromyces,* and there is no stalk or columella. The interior of most other false truffles is darker when mature and/or has a different texture.

In *Leucophleps* the exterior of the fruiting body is usually whitish, the chambers are tiny, and the spores are spiny (within their outer wall), while in *Leucogaster* the exterior is often brightly colored, the chambers are usually larger, and the spores are pitted or reticulate. Their evolutionary affinities are unclear, and modern taxonomists isolate them in their own order, the Leucogastrales. Both are fairly small genera, but relatively common, especially under conifers. As is commonly the case with false truffles, many of the species are poorly known and/or difficult to identify. One representative is described here and several others are keyed out.

759

Key to Leucophleps & Leucogaster

1. Peridium (skin) white or at times pale yellowish, ochraceous, or grayish; spore mass chambers minute; spores prickly inside a clear gelatinous outer wall 2
1. Peridium usually darker or more brightly colored than above; spore-bearing chambers relatively large (easily visible); spores reticulate or pitted within clear outer wall 3
2. Peridium white or whitish (or grayish in age), usually covered with conspicuous mycelial threads, at least near base . . . . *Leucophleps spinispora,* below
2. Not as above . . . . *Leucophleps magnata* (see *L. spinispora,* below)
3. Peridium pinkish to reddish or reddish-brown (often browner in age); found under western conifers . . . . *Leucogaster rubescens* (see *Leucophleps spinispora,* below)
3. Not as above . . . . *Leucogaster odoratus* & others (see *Leucophleps spinispora,* below)

Leucophleps spinospora   (White Jellied False Truffle)

FRUITING BODY 1-3 cm broad, round to oval or irregularly lobed (potato-like). Outer surface white or whitish, often becoming grayish in age, usually wrinkled and typically covered with rhizomorphs (mycelial threads), but young specimens usually showing rhizomorphs only at base. **SPORE MASS** (interior) white, often exuding a white latex when cut, composed of many tiny (less than 1 mm broad), clear, gel-filled chambers (unless old or late to develop); sterile veins absent. **STALK** and columella absent. **SPORES** 10-13 × 10-11 microns, round or nearly so, with small spines inside a clear gelatinous outer wall.

HABITAT: Solitary to gregarious in soil or duff (usually buried) under western conifers. In the Sierra Nevada it is fairly common from spring through fall. The duff of our coastal forests is more apt to cough up species of *Leucogaster* (see below) when ruffled for truffles.

EDIBILITY: Unknown, but the texture is hardly appealing.

COMMENTS: This common whitish false truffle might be mistaken for a *Rhizopogon* because of its rhizomorphs, but the white gel-filled chambers and/or presence of a latex distinguish it. Specimens which mature slowly, however, may lack a gel and latex or be hard and *Tuber*-like. Other species of *Leucophleps* are best differentiated microscopically. One, *L. magnata,* can usually be told by its whitish to pale yellowish or ochraceous exterior with less conspicuous rhizomorphs, plus its larger spores with larger warts or spines. *Leucogaster* is also fairly common. Its ranks include: *Leucogaster rubescens,* with *large* gel-filled chambers and an exterior that becomes pinkish to brick-red to brownish as it ages; *L. odoratus,* less common, with a yellowish-orange exterior, large chambers, and a pungent odor in age; and *L. carolinianus,* one of several species in eastern North America.

RADIIGERA

Medium-sized woodland fungi *usually found underground.* FRUITING BODY typically round or slightly flattened. Peridium (skin) *thick,* composed of two or three layers of tissue. SPORE MASS (interior) with threads or plates that radiate from the columella, at first white and fleshy, *then becoming gooey and finally powdery; brown to black at maturity.* STALK *absent, but a columella present; columella fairly thick, often rounded,* white, unbranched, *not* percurrent. SPORES *warted or spiny, round, brown under the microscope.* Capillitium present.

THIS small genus is easily recognized by its thick skin and rounded, unbranched columella (see photo). The multi-layered peridium (skin), warted or spiny round spores, powdery spore mass in old age, presence of capillitium, and rounded columella (or "pseudocolumella," to be technically accurate) indicate a close relationship to the earthstar genus *Geastrum,* and it is placed in the Lycoperdales by most taxonomists. However, the outer skin never splits into rays as it does in *Geastrum,* and the underground growth habit is also distinctive. It is not a particularly common genus. Three species are keyed below, but undescribed species also occur, at least in California.

The interior of an immature *Radiigera*, showing the thick rounded columella and thick skin. This species was never determined because the spores had not yet matured; it was growing under an oak.

Key to Radiigera

1. Outer peridium (skin) splitting into rays when mature (look around for older specimens); exterior usually smooth (but often dirty); sometimes growing just below the ground, sometimes on top .... (see **Geastrum, Astraeus, & Myriostoma,** p. 699)
1. Peridium not normally rupturing or splitting; exterior smooth or felty; usually underground 2

2. Mature spore mass blackish; outer layer of peridium rather heavy and felty **R. atrogleba,** below
2. Not as above; mature spore mass brown to dark brown 3

3. Fruiting body typically 2-4 cm broad; spores warted ... **R. taylorii** (see *R. atrogleba,* below)
3. Fruiting body typically 3-8 cm broad; spores spiny .. **R. fuscogleba** (see *R. atrogleba,* below)

Radiigera atrogleba

FRUITING BODY 2.5-5 cm broad, round to slightly flattened or depressed, often imbedded in white mycelial threads (rhizomorphs). Peridium (skin) three-layered. Outer surface tough and felty or fibrous-cottony, rough, white to grayish, often developing buff, pinkish, vinaceous, or brownish tints in age; usually separating readily from the two inner layers. Inner layers closely adhering to each other, whitish to rosy or pinkish when fresh (but may turn olive or buff when bruised), 3-5 mm thick, white when sectioned (or with a greenish tinge). **SPORE MASS** (interior) composed of plate-like bundles of hyphae which radiate from the columella to the peridium; white, fleshy, and rather soft at first, becoming gooey or inky and blackish, and eventually powdery (and blackish); odor often rather unpleasant in the inky stage. **STALK** absent, but the basal portion of peridium usually thickened slightly and giving rise to a prominent columella which is nearly round, cushion-shaped, or narrowed at the base and enlarged (rounded) at the apex; columella usually 0.8 cm thick or more at the widest portion, usually penetrating the spore mass at least half way (never percurrent); white or tinged gray, sometimes disintegrating in old age.

Radiigera atrogleba, maturing specimen. Interior (shown at right) is black and gooey in this stage, but eventually turns powdery. The top of the columella can be seen at the center, but its base has been obscured by the gooey spore mass. Note how thick the skin is. (Herb Saylor)

Radiigera fuscogleba, immature specimens. Note how lines radiate from the columella in sliced specimen. See comments below for more details.

SPORES 5.5-6.5 microns, round, minutely warted, deep brown under the microscope.

HABITAT: Solitary or more often in groups or large, mycelium-incrusted clusters in soil or duff in woods and along old roads through the woods, usually buried or only partially exposed; not uncommon in the Pacific Northwest and California, but probably more widespread. Radiigeras occur mainly with conifers in the summer and fall, but I have found *R. taylorii* (see comments) locally under oak and pine in the winter.

EDIBILITY: Unknown; the gooey black spore "mess" of older specimens isn't appetizing!

COMMENTS: The thick, often rounded or cushion-shaped columella plus the thick skin and blackish spore mass at maturity make this a most distinctive fungus. *Scleroderma* species also have a thick skin and dark spore mass, but lack a columella. The spore mass passes through a gooey stage before becoming powdery. In this stage (see photo) it looks like the "ink" from a shaggy mane and stains everything it comes into contact with. Other species: *R. fuscogleba* is quite similar, but has an olive-brown to brown mature spore mass and spiny spores (see photo above). *R. taylorii*, on the other hand, is smaller, with a smoother brownish to drab exterior, paler brown spore mass, and warted spores. Both of these species occur in California and the Pacific Northwest but may be more widespread.

HYSTERANGIUM & Allies

Small to medium-sized, mostly woodland fungi *usually found underground.* FRUITING BODY typically round to irreguarly lobed (potato-like), often with one or more mycelial threads or cords at base. SPORE MASS (interior) composed of small chambers, usually greenish to olive-brown to olive-gray (but not always); typically firm and tough when young *but usually gelatinous or muci-laginous in old age;* odor sometimes fetid at maturity. STALK *typically absent, but a columella usually present; columella whitish or translucent,* branched or unbranched, rarely percurrent. SPORES spindle-shaped to oblong or elliptical, *smooth or encased in a wrinkled utricle (outer sac);* hyaline (colorless) to greenish-brown under the microscope. Capillitium absent.

HYSTERANGIUM is closely related to the stinkhorns (Phallales), but is treated here with the false truffles because of its hypogeous (underground) tuberlike fruiting body. Like the stinkhorns, most Hysterangiums have a greenish to olive-brown spore mass that becomes slimy and stinky in old age. Those species with mycelial cords at the base are reminiscent of stinkhorn "eggs," but never hatch, i.e., a specialized spore-bearing structure never emerges. *Hysterangium* is a fairly sizable genus. Most of its 30+ species occur in the woods and are best differentiated microscopically. One very common and farflung species is described here. A related genus, *Phallogaster,* is included in the key.

Key to Hysterangium & Allies

1. Fruiting body with a stalk or distinctly narrowed and elongated base, pear-shaped or developing lobes on top; spore mass divided into distinct chambers; found on rotten wood, mainly in eastern North America .... ***Phallogaster saccatus*** (see *Hysterangium separabile,* below)
1. Not as above 2

Hysterangium separabile, immature specimens. In this stage the spore mass (shown in specimens on left) is odorless and quite firm or cartilaginous. Note presence of a branched, translucent columella.

2. Fruiting body showing a thick gelatinous layer just under the peridium (skin) when sliced open, eventually "hatching" if old enough (i.e., a specialized spore-bearing structure emerging from it); one or more thick mycelial cords usually emanating from the base; often found in lawns, gardens, and other urban areas (but also in woods) .... (see **Phallales**, p. 764)
2. Not with above features; usually found in woods 3
3. Columella typically percurrent (extending all the way through spore mass); often fruiting above the ground .... *H. darkeri* (see *H. separabile,* below)
3. Columella not typically percurrent; usually underground or only partially exposed 4
4. Associated with eucalyptus .... *H. fuscum* (see *H. separabile,* below)
4. Not as above 5
5. Spore mass (interior) tough and pale pinkish when young; found under oak .. *H. occidentale*
5. Not as above 6
6. Fruiting bodies enmeshed in a copious mass of white mycelial threads; usually growing in rocky or gravelly soil .... *H. crassum* (see *H. separabile,* below)
6. Not as above .... *H. separabile* & many others, below

Hysterangium separabile

FRUITING BODY round to somewhat flattened or lobed, 0.5-3 cm broad, sometimes with mycelial fibers at the base. Outer surface white to pinkish or sometimes buff, often becoming pinkish where bruised or handled. Peridium (skin) easily cracking or peeling away from the spore mass, at least at maturity. **SPORE MASS** (interior) olive-brown to greenish, composed of small chambers; at first firm and tough (cartilaginous), but becoming slimy and stinky (putrid) at maturity or in old age. **STALK** absent, but a columella present; columella thin, translucent or whitish, arising from a rudimentary sterile base, often branched and typically extending about half way into the spore mass. **SPORES** 12-19 × 6-8 microns, more or less spindle-shaped, smooth within a wrinkled utricle (sac), hyaline (colorless) to pale greenish-brown under the microscope.

HABITAT: Solitary to gregarious in humus or soil (usually buried) under both hardwoods and conifers; very widely distributed, but especially common in western North America. In our area it is abundant under oak and other trees practically year-round. In the Sierra Nevada and Cascades it's common in the late spring, summer, and fall.

EDIBILITY: Edible and choice—if you're a squirrel or chipmunk.

COMMENTS: Once called *H. clathroides* and also known as *H. coriaceum,* this subterranean fungus look like a stinkhorn "egg," but never hatches. The greenish spore mass and translucent, branched columella distinguish it from most other false truffles, and the tendency of the peridium to crack and peel easily from the spore mass is also distinctive. There are a number of similar species that differ microscopically. A few can be recognized in the field, including: *H. setchellii* and *H. crassirhachis,* with thick whitish septa (partitions) emanating from a scarcely branched columella (the latter especially common); *H. aureum,* with a golden-brown exterior; *H. crassum,* whose fruiting bodies are enmeshed in a copious mass of white mycelial threads, fairly common under western hardwoods and conifers, particularly in rocky or gravelly soil; *H. darkeri,* a slightly larger (1-5

Hysterangium separabile. Spore mass is putrid and gelatinous in old age (specimen on right), but rubbery or tough when young, and skin is easily separable. Note mycelial cords emanating from base.

cm) species that has a percurrent columella and often grows above the ground; *H. fuscum (=H. fischeri)*, associated with eucalyptus; and *H. stoloniferum,* fruiting body anchored by a thick mycelial cord or "stolon" and spore mass sometimes bluish-tinged.

Another stinkhorn relative, *Phallogaster saccatus,* is quite different. It usually grows on decayed wood and has a more or less pear-shaped fruiting body with a tapered, stalklike sterile base. The peridium is white to pinkish or lilac, and typically develops lobes and shallow depressions at the top which rupture irregularly at maturity in order to expose the spores. The spore mass is mucilaginous, olive-colored, and fetid at maturity, and separated into several large chambers within the fruiting body. It is fairly common in eastern North America but absent in our area. Also see *Rhopalogaster* (in key on p. 724).

Stinkhorns

PHALLALES

spores

STINKHORNS are among the most fascinating and highly specialized of the fleshy fungi. They differ from other Gasteromycetes in having a slimy or sticky, putrid spore mass **(gleba)** that is borne aloft at maturity. However, they begin as puffball-like "eggs" completely encased in a skin **(peridium)**, and usually anchored by a thick mycelial cord(s). A sectioned stinkhorn "egg" reveals a differentiated interior with a gelatinous substance beneath the outer skin, a compressed stalk (and/or arms), and a brown or olive-colored spore mass (see Color Plates 191-198). The superficial resemblance to a puffball is soon rudely—and forever—dispelled, however, as the enclosed stalk or other spore-bearing structure elongates rapidly, bursting out of its "shell" and into the world.*

In the common stinkhorns (Phallaceae), the mature fruiting body is unbranched and explicitly phallic, with the spore slime coating the apex of the stalk or "head." In the ornate stinkhorns (Clathraceae), the fruiting body is branched or latticed and the spore mass generally coats the inside surfaces of the branches or latticework. In both families the ruptured peridium forms a sack **(volva)** at the base of the fruiting body, much as in *Amanita.***

*This rapid elongation process, which can take as little as one hour, is possible because all of the stinkhorn's parts are fully formed (though greatly compressed) within the "egg." Its emergence is mainly an act of *expansion* (elongation of cells) rather than of growth or development. As a result, stinkhorn "eggs" can be "hatched" at home by keeping them in a humid environment.

**Though their developmental stages are superficially similar, the stinkhorns differ from the Amanitas in several fundamental respects. Stinkhorns lack gills, and like other Gasteromycetes, do not forcibly discharge their spores. The spores, although borne aloft at maturity, are produced *internally* (within the "egg"). Amanitas, on the other hand, do not form their spores until the cap has expanded. Their spores are produced on gills and are forcibly discharged when ripe.

The stinkhorns' most outlandish feature, however, is the unpleasant or provocative odor of the mature spore slime, which has been variously characterized as "foul," "fetid," "evil," "odious," "obnoxious," "cadaverous," "putrid," "maddening," "aggravating," "compelling," "intolerable," "filthy," "vile," "disgusting," "distressing," "disconcerting," "spermatic," "garbageous," "nauseating," "like rotting carrion," "like spent incense," "like the damp earthy smell we meet with in some of our churches on Sundays," "enough to cause one to think that all the bad smells in the world had been turned loose," and most apt and understated of all: "indiscreet." Lured from afar by the stench, flies and carrion beetles come to feast on the slime, and if the day is hot, roll around in it. With their eventual reluctant departure, spore dissemination is accomplished (some spores stick to their feet, others are presumably passed through their digestive tracts). All in all, it is a rather ingenious method of spore dispersal more typical of the so-called "higher" plants than of the "lowly" fungi.

Opinions on the edibility of stinkhorns range from ill-disguised disgust to idle specu-lation to passionate praise. The "intolerable" odor of mature specimens would seem to be enough to discourage even the most ardent and confirmed toadstool-tester from sampling them. However, the odorless stinkhorn "eggs" are considered a delicacy in parts of China and Europe, where they are pickled raw and even sold in the markets (sometimes under the name "devil's eggs"). Captain Charles McIlvaine, of course, pro-nounces them delicious (see comments on edibility of *Phallus impudicus*), suggesting they be sliced and fried like a Wiener schnitzel. My own experience with them has left me nonplussed (see comments on the edibility of *Clathrus archeri*). None are known to be dangerously poisonous, but it is reported that a young person of twenty-two, "having eaten a morsel, was seized with violent convulsions, lost the use of his speech, and ulti-mately fell into a stupor which lasted forty-eight hours; prompt attention was given to him, but it appears to have been some months before he was perfectly cured." (Note: one source described the victim as a "young English girl.")

Stinkhorns are notoriously spontaneous and unpredictable in their habits—they are liable to pop up at almost anytime, anywhere, providing conditions are to their liking. They occur throughout the world but attain their maximum diversity in the tropics. Several of our species were probably introduced accidentally—surely no one would do it on purpose!—along with soil or plant material from exotic lands. They are not common in comparison to other fleshy fungi, but they command a degree of attention far dispro-portionate to their numbers because of their fantastic shapes and repugnant odors. For this reason I have keyed out the common North American species, plus several "aliens" that may turn up in greenhouses, botanical gardens, the backyards of stinkhorn specialists, or other hospitable places. They are, after all, among the most beautiful of all mushrooms—providing you hold your nose!

Key to the Phallales

1. Spore slime exposed or borne aloft at maturity; fruiting body variously shaped, emerging from an "egg" whose peridium (skin) forms a sack or volva 2
1. Not as above; peridium either remaining intact or rupturing to form irregular holes at the top; volva absent .... (see *Hysterangium* & Allies, p. 762)
2. Fruiting body unbranched (but sometimes with a drooping lacy skirt or "veil"); spore slime coating the outer surface of the apex or "head" (unless washed off) ..... **Phallaceae,** p. 766
2. Fruiting body branched to form several arms (which may or may not be fused at their tips), columns, or a lattice-like framework; spore slime borne on the inner surfaces of the arms or latticework (but if arms unfold, the inner surfaces become the upper surfaces) ....
   .... **Clathraceae,** p. 772

PHALLACEAE　(Common Stinkhorns)

Medium-sized, *foul-smelling* (at maturity) fungi found on ground or rotten wood. FRUITING BODY *emerging from a round to oval "egg" that has an inner gelatinous layer; roughly cylindrical and unbranched* when fully expanded, with or without a swollen head. STALK *present as a column by which the spore slime is elevated;* hollow, spongy, usually perforated at the tip. VEIL absent, or present as a drooping, netlike "skirt" (indusium) below the "head." VOLVA *present as a membranous sack or pouch at base of stalk.* SPORE MASS usually greenish to dark olive-brown, *mucilaginous or slimy at maturity, coating the exterior of the "head" or top of the stalk.* Spores more or less oblong (bacillus-like), smooth. Capillitium absent.

THE common stinkhorns are unmistakable by virtue of their explicitly phallic fruiting body which is simple (that is, unbranched), but may terminate in an enlarged "head" on which the foul-smelling spore slime resides. Three genera are common in North America: *Phallus,* which has a well-defined cap or "head"; *Dictyophora,* which resembles *Phallus* but boasts a netlike veil or **indusium** that hangs from the lower edge of the "head"; and *Mutinus* (the so-called "dog stinkhorns"), which lacks a differentiated "head." (*Dictyophora* is incorporated into *Phallus* by some stinkhorn specialists, and *Ithyphallus* is an outdated synonym for *Phallus.*)

Whether or not stinkhorns are handsome or repulsive has been the subject of considerable debate. The verdict seems to rest largely on personal prejudice and one's ability to overlook their obnoxious odor and the swarms of blowflies that come to wallow in their spore slime. They are undeniably *phallic,* however, and as might be expected, their suggestiveness has given rise to a veritable "mother lode" of stinkhorn lore. For instance, German hunters believed they grew where stags rutted, and it is said that their putrid carcasses are burned outside houses in Thailand to discourage unwanted guests (a rather drastic practice that might discourage *wanted* guests as well!). They've been used in countless ointments and potions, e.g., as a cure for gout, epilepsy, and gangrenous ulcers. They've been blamed for cancer and prescribed as a sure-fire remedy for it. And of course, they've been employed as aphrodisiacs, and are supposedly still given to cattle for that purpose in some parts of Europe. Alexander Smith tells of a remarkable encounter with an elderly gentleman who carried a mammoth specimen under his hat in the certain belief that it would cure his rheumatism! And, in an otherwise tedious book of Victorian reminiscences by Gwen Raverat, there is this astonishing passage:

> In our native woods there grows a kind of toadstool, called in the vernacular The Stinkhorn, though in Latin it bears a grosser name. This name is justified, for the fungus can be hunted by the scent alone; and this was Aunt Etty's greatest invention: armed with a basket and a pointed stick, and wearing special hunting cloak and gloves, she would sniff her way round the wood, pausing here and there, her nostrils twitching, when she caught a whiff of her prey; then at last, with a deadly pounce, she would fall upon her victim, and then poke his putrid carcass into her basket. At the end of the day's sport, the catch was brought back and burnt in the deepest secrecy on the drawing-room fire, with the door locked, *because of the morals of the maids.*

One wonders how such an innocuous, splendid, upright organism could be so ruthlessly and unjustly maligned, and I submit that this particularly perverse brand of fungophobia be christened *phallophobia.* ("Aunt Etty," incidentally, was Charles Darwin's daughter!)

Attitudes change, however, and in the twentieth century the stinkhorns' impudence and imprudence have begun to be appreciated. Mycological literature is replete with wonderfully provocative accounts of close encounters (of the casual kind) with stinkhorns —as riddled with them, in fact, as an old *Suillus pungens* is with maggots. When stinkhorns are discussed, the language makes a startling and unprecedented qualitative leap, from monotonous minutiae to half-baked hyperbole, as if the authors were suddenly taking

Fruiting body development in the tropical basket stinkhorn, *Dictyophora indusiata* (p. 770). This sequence was photographed within a time period of thirty minutes! Note how the ridges on the cap become more visible in age (as the spore slime is carried away by flies and/or rain). In old specimens completely bereft of spore slime the cap is whitish. (Keith Muscutt)

an interest in what they were saying. They are lavish in their praise as they tread the fine line between double-entendre and forthright fungal fact: "The mischief-maker is a handsome specimen, as its plate shows" ... "This is a highly specialized type of fungous fruiting body" ... "It is one of the seven unnatural wonders of the natural world" ... "Never have I seen such intricate lacework as on a *Phallus*" ... "It undoubtedly emerges from the depths for a single noble and grand purpose—that of disseminating its spores. All of its parts have been developed to accomplish this function in the most effectual manner possible" ... "It is inconspicuous while encased in its skin, and unlikely to be noticed by anyone who is not looking for it. When fully elongated, however, it is pointedly, inescapably prominent" ... "By extraordinary growth and expansion it carries the banquet of spores into the light—a natural Jack-in-the-box" ... "When its expansion is complete, it begins immediately to become limp, bent, sinks down, and undergoes putrefaction" ... "Once while collecting fungi with other students we smelled a phalloid and tried to trace it down in order to learn the species, but it seemed to keep moving. Finallly, we noticed that an old man was also in the woods collecting fungi, and as I worked over toward him I realized that a phalloid of some kind was very close" ... "One is curious to learn the mechanism by which so much is accomplished in apparently so short a time, and find in this instance, as in all others where great things are accomplished with ease, that many forces have been slowly at work to insure everything being in readiness for the success of the final flourish" ... "The banquet is prepared underground and the table, with its viands ready, is pushed into the light while the invitation to guests is wafted swiftly on the breeze" ... "It is a glorious fungus—an admirable specimen of herculean proportions pokes up periodically in my front yard."

The common stinkhorns are more prevalent in temperate regions than their ornate cousins, the Clathraceae, but still attain their greatest diversity in the tropics. Species of *Phallus* are especially fond of populated areas, where they lurk in lawns, gardens, flower beds, along roads, in ditches, under bushes, hedges, porches, and houses, in the vicinities of churches and lumberyards, trash heaps, and old sawdust piles. *Mutinus* and *Dictyophora* species, on the other hand, are more frequent in forests and are largely restricted to eastern North America and the tropics. All of these fruit whenever conditions are conducive to development—that is, warm and moist—and none are known to be poisonous. One species from each genus is described here, and several others are mentioned and/or keyed out.

Key to the Phallaceae

1. Stalk with a latticed head, the spore slime coating the *inside* of the latticework . . . .
   . . . . (see **Clathraceae,** p. 772)
1. Not as above; "head" absent or if present, then spore slime coating its exterior 2

2. Lacelike or netlike skirt (indusium) present and usually prominent (hanging from the lower
   margin of cap) 3
2. Skirt (indusium) absent or rudimentary 5

3. Fruiting body, or at least the skirt, yellow to orange-yellow or even orange-red; primarily
   tropical . . . . **Dictyophora multicolor** (see *D. indusiata,* p. 770)
3. Fruiting body typically white (excluding the spore slime and volva); temperate or tropical . 4

4. Skirt (indusium) large, often reaching the ground; tropical . . . **Dictyophora indusiata,** p. 770
4. Skirt smaller (3-6 cm long); fairly common in forests of eastern North America . . . .
   . . . . **Dictyophora duplicata** (see *D. indusiata,* p. 770)

5. Spore slime borne on a ring or "collar" *beneath* the tip of the fruiting body; fruiting body white
   except for the slime; tropical . . . . **Staheliomyces cinctus**
5. Not as above; spore slime not borne on a sharply defined ring or collar 6

6. Spore slime borne on the upper portion of the stalk, i.e., a sharply defined "head" *not* present;
   stalk typically slender (up to 1.5 cm thick) 7
6. Spore slime borne on a differentiated (sharply defined), often swollen "head"; stalk sometimes
   slender, but usually at least 1.5 cm thick 9

7. Fruiting body entirely or partially red to pink or orange 8
7. Fruiting body white except for spore slime **Mutinus caninus var. albus** (see *M. caninus,* p. 771)

8. Fruiting body 9-17 cm long, usually slightly thicker in the middle and tapered gradually toward
   the tip . . . . **Mutinus elegans** (see *M. caninus,* p. 771)
8. Fruiting body 5-10 cm long, usually more or less equal except for the very tip, which is bluntly
   and rather abruptly narrowed . . . . **Mutinus caninus,** p. 771

9. Fruiting body red to scarlet (at least in part), usually quite slender . . . .
   . . . . **Phallus rubicundus** (see *P. impudicus,* below)
9. Fruiting body white to pale pinkish (except for spore slime), but volva may be pink to purple 10

10. Cap or "head" reticulate (pitted and ridged) beneath the spore slime; widely distributed . . . . .
    . . . . **Phallus impudicus** & others, below
10. Cap or "head" not reticulate, pitted, or ridged, but often granular; found mainly in eastern
    North America . . . . **Phallus ravenelii** (see *P. impudicus,* below)

Phallus impudicus   (Stinkhorn)   Color Plates 193, 194

FRUITING BODY beginning as an "egg" up to 6 cm high. **PERIDIUM** (skin) white to
yellowish-white in one variety, lurid pinkish to purple in another (see comments), with a
gelatinous layer beneath; rupturing to form a volva as the stalk elongates and thrusts the
slimy, swollen "head" upward. **CAP** ("head") 1.5-4 cm broad, coated with the putrid,
copious spore slime which eventually drips off or is carried or washed away, revealing
the whitish reticulate (pitted and ridged) surface beneath; top with a hole which is sometimes covered by a clinging piece of the peridium (volva) and/or by numerous flies.
STALK 1.5-3 cm thick, equal or tapered at both ends, entirely white or sometimes pinkish
below; minutely honeycombed (spongelike), hollow, fragile. **INDUSIUM** ("veil") absent
or rudimentary. **VOLVA** present at base of stalk as a white to pinkish or purple, loose,
lobed sack formed by the ruptured peridium; base usually with a thick, similarly-colored
mycelial cord(s). **SPORE MASS** slimy or mucilaginous, olive-green to olive-brown,
with an obnoxious odor at maturity (see p. 765). Spores 3-5 × 1.5-2.5 microns, elliptical
or oblong, smooth.

HABITAT: Solitary or in groups or clusters in lawns, gardens, sandy or cultivated soil,
under trees or shrubs, in rich humus, etc.; widely distributed and especially common in
the West, fruiting whenever conditions are favorable. McIlvaine states that, "its favorite

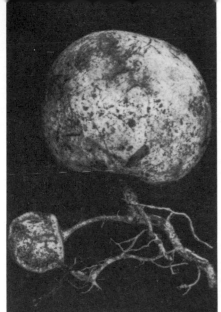 

This purple-egged variety of *Phallus impudicus* is considered a distinct species, *P. hadriani,* by some stinkhorn specialists. **Left:** A round "egg" with telltale mycelial cord. **Right:** An "egg" just prior to hatching, and a malodorous full-grown (but rather small) specimen. Note how the spore slime coats the reticulate "head"; after it has been dispersed the "head" is whitish. See color plates for another full-grown specimen and a beautiful sectioned "egg."

abode is in kitchen yards and under wooden steps where, when mature, it will compel the household to seek it out in self-defense," and his contemporary Nina Marshall says, "the distracted housewife searches in vain for a solution to the difficulty and the odor disappears as mysteriously as it came. If she is one of the initiated, however, she will search until she finds the haunt of the offender and then destroy it on the spot to avoid further repitition of the nuisance." The largest fruitings I have seen were on the lawn of a school in Los Angeles, and in front of an old cathedral in Santa Fe, New Mexico. Stinkhorn specialist William Burk recalls seeing massive clusters that apparently sprung from the foundation of a house in Salt Lake City, Utah. If the "eggs" are carried home and transplanted in cool, damp earth, more often than not they will continue to develop so that the fascinating elongation process can be observed firsthand.

EDIBILITY: Not poisonous (see comments on p. 765), but as Alexander Smith says, "who would want to eat even the eggs?" Captain Charles McIlvaine (the plenipotentiary extraordinaire of turn-of-the-century toadstool testers), for one. He says, "(the eggs) are semigelatinous, tenacious, and elastic, like bubbles of some thick substance. In this condition, they demand to be eaten . . . cut in slices and fried or stewed, they make a most tender, agreeable food."

COMMENTS: This is one "wild" mushroom that is truly unmistakable. In shape it resembles nothing so much as a slimy cigar, leaky pipe, malodorous thumb, putrid horn, hollow baton, spongy candle, or putrescent pencil. But the most distinctive thing about the stinkhorn is its utter spontaneity. In the words of one founding member of the New York Mycological Society, "you never know when one is going to pop up right in front of you, so fast you can actually see it growing. And there is no way to predict or control them. It's the one thing in the world you can't push or hurry." The stinkhorn also has a highly refined sense of poetic justice. One "going strong" is said to have appeared in the concrete floor of the newly built house greeting a newly married couple. To accomplish this instructive feat, the mycelial cords had to force their way into the foundation from an old stump in the garden!

The stinkhorn is sometimes mistaken inexplicably for an old morel, perhaps because

769

of the pitted head and somewhat similar shape. However, the "indiscreet" odor and slimy spore mass (both of which may be gone in old age) and presence of a volva (and frequently, a squadron of flies) leaves no doubt as to its identity. The common variety in California has a pinkish to purple peridium (see Color Plate 194) and is considered a distinct species, *P. hadriani (=P. imperialis, P. iosmus)* but some authorities on the subject. Other species: *P. ravenelii* is very similar, but has a smooth to granular rather than reticulate "head." It is quite common in eastern North America, especially in lignin-rich humus and on old sawdust piles. The tropical "Devil's Stinkhorn," *P. rubicundus,* is a rather long, slim, red to scarlet species. It might be mistaken at first glance for a dog stinkhorn *(Mutinus),* but has a clearly differentiated, detachable, more or less conical or bell-shaped "head." It occurs in southern and eastern North America (at least as far west as New Mexico), but is not common. See also *Dictyophora duplicata* (under *D. indusiata).*

Dictyophora indusiata   (Basket Stinkhorn)   Color Plate 191

FRUITING BODY beginning as an "egg" up to 6 cm high. **PERIDIUM** (skin) white or sometimes tinged buff, gray, or reddish-brown, with an inner gelatinous layer; rupturing to form a volva as the stalk elongates. Mature fruiting body 7-25 cm tall, unbranched, consisting of a more or less conical to bell-shaped cap on a stalk, with a large lacelike veil or "skirt" flaring out from beneath the cap. **CAP** ("head") 1.5-4 cm broad, coated at first with the putrid olive-green to brown spore slime, which eventually drips off or is carried away by insects, revealing a white or yellowish, ridged and pitted (reticulate) surface beneath; top with a hole in it. **STALK** 1.5-3 cm thick, equal or tapered at either end, white, hollow, fragile, sometimes curved, minutely chambered and porous (spongelike). **INDUSIUM** ("veil") white, initially tucked under the cap margin but soon unfurling; skirtlike or basketlike when fully expanded and touching or nearly touching the ground (6 cm or more high); attached to the top of the stalk or just under the margin of the cap; composed of white strands that form an intricate chainlike or lacelike net whose "holes" are large and polygonal. **VOLVA** present at base of stalk as a white or pallid, loose, often lobed sack formed by the ruptured peridium; base usually with a thick mycelial cord(s) attached. **SPORE MASS** slimy or mucilaginous, olive-green to olive-brown or brown, with a fetid odor at maturity that attracts flies, beetles, and other insects (but odor not always strong, and in one form somewhat sweetish). Spores 3.5-4.5 × 1.5-2 microns, elliptical to oblong, smooth.

HABITAT: Solitary or in groups in humus or on rotten wood in tropical forests and at their edges; fruiting in wet weather, widely distributed. It is quite common in Central America and South America, as well as in Australia, the South Pacific, Africa, India, and Japan.

EDIBILITY: Edible in the egg stage, but probably not choice (see comments on edibility of *Clathrus archeri*)—in other words, it is better eyed than fried. In many parts of the world (e.g., New Guinea) it is worshipped for its beauty and/ or used as an aphrodisiac.

COMMENTS: What's an indisputably tropical fungus doing in a reputedly topical field guide like this? Well, I could justify its presence on the flimy pretext that it just *might* show up in somebody's hot house, sauna, or backyard bamboo forest. Or I could cite the popular but shamefully arrogant political notion that Central America is our "backyard," and that any stinkhorn found in one's backyard merits mention. However, the truth is that I've included it because I couldn't *resist* including it. In my opinion, such a *piece de resistance* of nature deserves a full-fledged description no matter *where* it grows (and I offer the color plate as proof). The lacy or netlike "veil" which is responsible for the name "basket stinkhorn" is initially hidden (greatly compressed), but swells out soon after the stalk elongates (see photographs on p. 767). In its prime the veil touches—or nearly touches—the ground, and may be globelike or broadly skirtlike. Later, however, it shrinks or collpases somewhat and begins to droop. Aside from its spectacular skirt, *D. indusiata* resembles our common stinkhorn *(Phallus impudicus)* in shape, size, color, and texture (including the pitted head), and some phallologists consider it "a good *Phallus*." Other species: *D. duplicata* (the "Veiled Stinkhorn") is a robust temperate zone version that is fairly common in the hardwood forests of eastern North America. It resembles *D. indusiata,* but is not so spectacular, the skirt being smaller-meshed and only 3-6 cm long (it looks somewhat like the second photo on p. 767). *D. multicolor* is a colorful (yellow to orange-red) tropical species. *D. rubrovolvata* of China has a reddish volva.

Mutinus caninus   (Dog Stinkhorn)

FRUITING BODY beginning as an "egg" up to 2.5 (4) cm high. **PERIDIUM** (skin) white or occasionally with a faint pinkish or yellowish tinge, with an inner gelatinous layer; rupturing to form a volva at base of stalk. Mature fruiting body 5-10 cm high and 0.5-1.2 cm thick, unbranched, slender, erect or curved slightly, roughly cylindrical (equal) or thicker near top; lacking a differentiated cap, but with a blunt, rounded or abruptly narrowed, often perforated tip. **FERTILE PORTION** covering the upper 2-3 cm of fruiting body (except the very tip), bright orange-red to red to pink, but covered at first with olive to olive-brown spore slime. **STALK** more or less equal, colored like cap (orange-red to orange or pink) or often paler or even white toward the base; hollow, fragile, spongy (minutely chambered). **INDUSIUM** ("veil") absent. **VOLVA** present at base of stalk as a white, lobed sack or pouch formed by the ruptured peridium; base usually with one or more white mycelial cords attached. **SPORE MASS** mucilaginous, olive to deep olive-brown, coating the upper portion of stalk; odor fetid at maturity (but not as malodorous as some stinkhorns). Spores 3-7 × 1.5-2.5 microns, elliptical or oblong, smooth.

HABITAT: Solitary to gregarious or clustered on ground and rotten wood in gardens, roadsides, woods, etc.; widely distributed and fairly common in eastern North America in the late summer and fall (along with *M. elegans*—see comments). It is apparently absent in the West, but a white variety (see comments) has been found in Oregon, and it seems only a matter of time before the typical form and its longer look-alike, *M. elegans,* show up on the west coast, as several other erotic exotics have.

EDIBILITY: Nonpoisonous. Mature specimens, of course, are hardly tempting, but Bill Roody of Elkins, West Virginia, says the "eggs" are excellent "peeled and rolled in flour seasoned with garlic salt and pepper, dipped into beaten egg and then once again in the flour mix before frying in butter or oil. They're great as is or served with crackers and cream cheese."

COMMENTS: The slim, cylindrical, pink to orange-red fruiting body capped with olive-brown spore slime is a unique and memorable—if suggestive—sight, and it is easy to see how this "phalloid" got its common name (see color plates of *M. elegans*). The odor of mature specimens can be rather feeble compared to that of *Phallus impudicus* or *Clathrus ruber,* but is obnoxious nevertheless. The stalk varies from orange or pink to white, and a variety that is *entirely* white (except for the spore slime) has been found in Oregon and Michigan. (It has been called *M. caninus var. albus,* but may very well be a distinct species.) Another dog stinkhorn, *M. elegans (=M. bovinus, M. curtisii)* (COLOR PLATES 195, 196) is sometimes called "Devil's Dipstick." It is also common in eastern North America, and is frequently mistaken for *M. caninus.* It can be distinguished, however, by its longer (9-18 cm) fruiting body that is usually thickest in the middle and tapered *gradually* toward the tip. Also, a larger area of the fruiting body (up to 6 cm) is smeared with spore slime. In my experience it favors the same habitats as *M. caninus,* but is more common.

CLATHRACEAE   (Ornate Stinkhorns)

Medium-sized, *foul-smelling* (at maturity) fungi found on ground or rotten wood. FRUITING BODY *emrging from a round to oval or flattened "egg" that has an inner gelatinous layer; usually branched to form arms, columns, tentacles, or a latticed framework;* often brightly colored. STALK absent or if present then hollow, spongy, fragile, and often perforated or open at the top. VEIL (indusium) absent. VOLVA *present as a membranous sack or pouch at base of fruiting body.* SPORE MASS usually greenish to dark olive-brown (or drying blackish), *mucilaginous or slimy at maturity, coating the inner surfaces of the arms, tentacles, or latticed framework.* Spores more or less oblong (bacillus-like), smooth. Capillitium absent.

THESE stinkhorns are more fantastic than—but not nearly as phallic as—the common or "true" stinkhorns (Phallaceae). Some have long "tentacles" that make them look like starfish, others have stubby arms, still others boast two or more thick columns or an elaborate, lovely lattice-like framework. None look like "horns," but all of them *stink* (to a greater or lesser degree), and flies seem to find them just as desirable as members of the Phallaceae. Thus the label "stinkhorn," though not completely accurate, is amply deserved. In most cases the spore slime coats the inward-facing surfaces of the latticework, columns, branches, or tentacles (or the upper surfaces of the tentacles if they unfurl). As in the Phallaceae, the stalk (if present) is tubular, spongy, and fragile, and emerges from a membranous peridium (volva) which has an inner gelatinous layer.

Most of the stinkhorns in this family are tropical, but several have established themselves as "resident aliens" in the United States. The three most common genera are: *Clathrus,* whose fruiting body is composed of columns, a latticed ball, or four or more tentacles (with or without a short stalk); *Pseudocolus,* with three or four tentacles and a short stalk; and *Lysurus,* which has several thick arms or a latticed "head" atop a long stalk or column.

Most of the ornate stinkhorns are edible in the egg stage, but in the words of Alexander Smith, "are unlikely to become popular as food" (for obvious reasons!). One species, *Clathrus ruber,* is said to be poisonous raw, but this reputation may rest on a single bizarre incident (see p. 765). Five species are described here and several others are keyed out.

Key to the Clathraceae

1. Stalk typically much longer than the arms or fertile "head"; arms (if present) typically short (less than 3 cm long) and thick 2
1. Stalk absent (but two or more fused columns may be present), or if stalk present then relatively long (2.5 cm or more) and slender arms or "tentacles" also present; stalk when present usually short but sometimes fairly long 4

2. Fruiting body consisting of a rounded to somewhat flattened latticed "head" or netlike framework mounted on a stalk . . . . *Lysurus periphragmoides,* p. 776
2. Not as above; stalk with several short arms that may or may not be fused at their tips 3

3. Arms usually bright red and fused at their tips to form a "spire" (but sometimes breaking free from each other in age); stalk usually fluted or several-sided . . . . . *Lysurus mokusin,* p. 776
3. Arms red to pinkish, flesh-colored, brownish, or white, their tips sometimes touching at first but not fused into a "spire," and usually separating in age; stalk more or less cylindrical (i.e., round in cross-section) . . . . *Lysurus cruciatus,* p. 777

4. Fruiting body with three or more slender arms or "tentacles" which arise from stalk (but stalk may be very short and hidden by the volva, or occasionally practically absent) 5
4. Fruiting body either a latticed ball or netlike framework *or* comprised of two or more thick columns which are usually fused at their summit; stalk absent or rudimentary 7

5. Fruiting body typically with 3-4 "tentacles" that often remain joined at their tips . . . . *Pseudocolus fusiformis* (see *Clathrus archeri,* p. 774)
5. Fruiting body typically with 4 or more "tentacles" that may initially be joined at their tips but which usually break free and sometimes bend back in age 6

6. Arms or "tentacles" diverging from a flat disclike expansion of the stalk apex, often appearing divided or paired; spore slime mostly in center of fruiting body or coating bases of the arms; tropical . . . . *Aseroe rubra* (see *Clathrus archeri,* p. 774)
6. Not as above; spore slime borne on inner or upper surfaces of arms . *Clathrus archeri,* p. 774

7. Fruiting body lantern-like, i.e., the spore slime borne in a specialized structure slung between 2-4 red to orange or pink columns that merge at the top to form an arch; tropical (reported from Miami, Florida) . . . . **Laternea triscapa**
7. Not as above; fruiting body with columns or a latticed framework, but lacking a separate spore-bearing structure 8
8. Fruiting body consisting of a latticed network of branches and "windows" (somewhat like a "whiffle ball"—see photo on p. 774) . . . . **Clathrus ruber** & others, below
8. Fruiting body composed of 2-5 thick columns joined at their summit (the columns occasionally with one or two transverse branches, but not enough to form a lattice) 9
9. Fruiting body composed of two creamy to yellow to orangish columns at first pressed closely together but then bowing at the center; rare (native to Japan) . . . . . . **Clathrus bicolumnatus**
9. Fruiting body typically with (2) 3-5 red to pink or orange columns; widely distributed and not uncommon (especially in eastern North America) **Clathrus columnatus** (see C. ruber, below)

Clathrus ruber   (Latticed Stinkhorn)

FRUITING BODY beginning as a round to somewhat flattened or knobby "egg" up to 6 cm broad. **PERIDIUM** (skin) white and membranous, smooth becoming wrinkled and grooved as it gets larger, with an inner gelatinous layer; rupturing to form a volva. Mature fruiting body 5-14 cm high, consisting of a round to oval latticed ball or netlike framework with large polygonal or elongated "windows." **BRANCHES** of the framework bright pink to red to orange or pale orange, and often paler toward the base; flattened, hollow, very fragile, minutely chambered (like a sponge) and transversely ribbed or wrinkled on the outer surfaces; inner surfaces covered with the sticky, putrid spore slime. **STALK** absent or rudimentary. **VOLVA** present at base of fruiting body as a thick, loose, white sack or pouch, usually with thick mycelial cord(s) attached. **SPORE MASS** coating the inside surfaces of the latticework; mucilaginous, olive to olive-brown or drying blackish, with an *extremely* obnoxious stench at maturity. Spores 5-6 × 1.5-2.5 microns, oblong, smooth.

HABITAT: Solitary to densely gregarious or clustered in soil, wood chips, rich humus, etc.; widely but erratically distributed, apparently fruiting most any time (providing conditions are favorable). It is said to be a native of southern Europe (its likeness appears on more than one postage stamp from that region), but has turned up in many localities in North America (Florida, Virginia, North Carolina, etc.), usually in landscaped areas or other places where exotic plants have been introduced. It is a common sight in the parks of San Francisco (especially in the late fall), and I have also seen it in Santa Clara County.

EDIBILITY: Nonpoisonous according to some sources, harmful (at least raw) according to others (see the quote on stinkhorn poisoning on p. 765).

COMMENTS: Also known as *C. cancellatus,* this "wild" mushroom looks like a red or orange "whiffle ball" (see photograph). The white volva from which it emerges and the intolerable stench that develops at maturity are also distinctive. The color of the branches varies considerably—from bright red to very pale orange—apparently depending on the temperature and humidity. The putrid perfume attracts flies in large numbers and is, in my humble fungal opinion, the vilest of any stinkhorn. It must be smelled to be believed! (According to Ramsbottom, the smell is "so fetid that more than one artist has related that it was impossible to paint it without discomfort . . . when the color photograph was taken it would have been pleasing to have shown one or two flies at work, but they settled in such swarms that they had to be driven off."). The "eggs," on the other hand, are quite odorless and very knobby or furrowed just prior to bursting. In this stage they are easily "hatched" at home. *C. crispus* is a similar but even more spectacular neotropical species whose "windows" are circumscribed by "coronas"; it has been found in Florida. A species of the Old World tropics, *C. crispatus,* is also quite similar, but the upper part of its latticework regularly breaks up into fragments in age. *C. columnatus (=Laternea columnata)* has 2-5 thick, orange to red columns which arise independently but are fused at the top and

Clathrus ruber is easily recognized by its bright red to pink to pale orange latticework. **Left:** Mature specimens; note absence of a stalk. **Right:** Note how lumpy this "egg" is!

may branch horizontally (but not enough to form a latticework). It is widely distributed and has been found in many localities in eastern North America (particularly the South), plus Hawaii. *C. preussii* and *Ileodictyon cibarium (=C. cibarius)* are but two of many southern and/or tropical stinkhorns with a delicate *white* latticework and large "windows." The first is a native of Africa; the latter is common in the southern hemisphere.

Clathrus archeri   (Octopus Stinkhorn)   Color Plates 197, 198

FRUITING BODY beginning as a round to somewhat flattened "egg" up to 6 cm in diameter. **PERIDIUM** (skin) membranous, white to dingy buff, pinkish, or tinged purple, smooth or scurfy, with a gelatinous inner layer; rupturing to form a volva at base of stalk. Mature fruiting body 5-12 cm high, consisting of (4) 5-7 (8) arms or "tentacles" arising from a common stalk or "tube," or sometimes with up to 12 arms arising from two fused stalks. **ARMS** long and slender, 3-9 (12) cm long, tapered toward their tips, at first upright and roughly parallel and joined at their tips (often in pairs), soon opening outward like the petals of a flower and eventually curling back and under so that the tips often touch the ground (but sometimes one or more arms branched, and at other times the tips remaining joined to others). Inner (or upper) surfaces of arms bright red to pinkish-red, usually paler toward base, ribbed or reticulate and coated with spore slime. Outer surfaces (undersides) pale pink, longitudinally grooved. **STALK** 1-3 (5) cm long, typically short and sometimes hidden by the volva (but a long-stalked form is also known); hollow and tubular (open at the top), detaching easily from the volva, white or pallid below, pinkish above. **VOLVA** present at base of stalk as a loose sack formed by the ruptured peridium; base usually with one or more mycelial cords (rhizomorphs). **SPORE MASS** coating the inner (upper) surfaces of the "tentacles," mucilaginous, olive to dark olive-brown when fresh, but usually blackening as it dries; odor obnoxious at maturity (somewhat reminiscent of rotting crab). Spores 4-5.5 (7) × 2-2.5 microns, elliptical, smooth.

HABITAT: Solitary to gregarious or clustered on ground (especially sandy soil) or rotten wood in various habitats; widely distributed (Tasmania, Russia, etc.), but rare in North America. It is fairly common in the riparian woodlands along the San Lorenzo River in Santa Cruz County, California (especially in the spring, but fruiting most any time). I have also found it under a rose bush growing on a mixture of compost and rice hulls.

EDIBILITY: Nonpoisonous, and like most stinkhorns, edible in the egg stage. One spring evening a former friend and I decided to sample two "eggs," following a recipe for French-fried stinkhorn eggs (see comments on edibility of *Mutinus caninus*). The flavor wasn't bad, but we neglected to strip away the gelatinous outer layer. It proved to be so slippery that the "eggs" slid down our throats before we could savor them, leaving behind only the sticky spore mucilage (see Color Plate 198), which clung to our throats and tongues so tenaciously that we were still trying to wash it away several hours later!

Clathrus archeri is aptly called the "Octopus Stinkhorn" or "Stinkopus" because of its long, tentacle-like arms. They are intially joined at their tips (central specimen), but then separate and recurve (specimen at left). Note cords emanating from the "egg" in foreground. Also see the color plates.

COMMENTS: The octopus stinkhorn—or "stinkopus," as I am fond of calling it—is one of my five favorite fleshy fungal fructifications. It is a most attractive fungus when fresh, easily told by its long red "tentacles" which are initially joined at their tips but then unfold to form a slender-armed "star." It definitely looks more like a sea creature than a mushroom, but its obnoxious aroma—which attracts flies and jaded fungophiles from afar—identifies it as a member of the stinkhorn tribe. The arms are longer than those in *Lysurus,* and the stalk is proportionately shorter (or at times practically absent). It was formerly placed in the genus *Anthurus* (as *A. archeri* or *A. aseroeformis*), but since the "tentacles" are sometimes branched to form a rudimentary latticework (when young), it is now classified as a *Clathrus*. A common tropical stinkhorn, *Aseroe rubra* (see photo below) is sometimes confused with *C. archeri*, but looks more like a starfish or sea anenome than an octopus. It has five to ten or more "tentacles" which are deeply divided lengthwise (so that they may seem to occur in pairs) and arise from a flattened, disclike expansion of the stalk apex. The spore slime rests in the center of the disc or coats only the very bases

Top view of the tropical stinkhorn *Aseroe rubra* (see comments above). Note how arms are deeply divided and spore slime is concentrated at center. Stalks are not visible in this photo. (Michael Fogden)

of the "tentacles." Although strictly tropical, it may turn up, like so many other exotic fungi, in a greenhouse or similarly sultry environment. Also similar to *C. archeri* is the "Stinky Squid," *Pseudocolus fusiformis (=P. schellenbergiae, P. javanicus)*. It is common in scattered localities in eastern North America (especially Chapel Hill, North Carolina, the home of stinkhorn specialist William Burk), as well as in Europe, Australia, Asia, and Tierra del Fuego. It has a white to brown volva and only three to four "tentacles" which are red-orange to orange or pinkish, arise from a common stalk, and are usually fused at their tips (see photo on p. 777) but sometimes break free in age.

Lysurus periphragmoides   (Stalked Lattice Stinkhorn)

FRUITING BODY beginning as a round to oval "egg" up to 5 cm broad. **PERIDIUM** (skin) white to buff, with a gelatinous inner layer, rupturing to form a volva at base of stalk. Mature fruiting body 6-16 cm high, composed of a small rounded latticed "head" on a long stalk. **CAP** or "head" 1.5-3.5 cm broad, round to somewhat flattened, comprised of a latticework of red to orange (or sometimes yellowish or white) branches which form rather small meshes or "windows"; outer edges of branches keeled. **STALK** 5-13 cm long, 0.8-3 cm thick, hollow, fragile, rather spongy, equal or tapered, usually red above and paler below, but yellow or white in some forms. **VOLVA** present at base of stalk as a loose, lobed sack or pouch formed by the ruptured peridium, usually with whitish mycelial cords attached to base. **SPORE MASS** coating the inside surfaces of the latticed "head" and sometimes spilling out, mucilaginous, dark olive to olive-brown or drying blackish; fetid at maturity. Spores 3.5-4.5 × 1.5-2.5 microns, elliptical to oblong, smooth.

HABITAT: Solitary to gregarious (occasionally two arising from the same volva) in rich soil, lawns, gardens, open woods, on rotten wood, etc.; widely distributed. It is not uncommon in mild wet weather in the southern United States and Midwest (North Carolina, Texas, Nebraska, New Mexico, even New York), but has yet to be found in California.

EDIBILITY: Presumably edible in the egg stage, but who wants to eat it?

COMMENTS: Also known as *Simblum sphaerocephalum* and *S. texense*, this southern stinkhorn looks like a miniature *Clathrus ruber* mounted on a long stalk. In some specimens the transverse (horizontal) branches of the network are reduced or even absent, making its relationship to other species of *Lysurus* more obvious. White and yellow variants are known, but the red one is the most common in North America.

Lysurus mokusin   (Lantern Stinkhorn)   **Color Plate 192**

FRUITING BODY beginning as a round to oval "egg" up to 6 cm high. **PERIDIUM** (skin) white and membranous, with an inner gelatinous layer; rupturing to form a volva at base of stalk and sometimes also leaving a piece of tissue stuck to the top of fruiting body. Mature fruiting body (3) 5-12 (16) cm high, consisting of a stalk that is branched above into 4-6 (usually 5, rarely 7) arms which normally remain joined at their tips to form a "spire," but which sometimes break free from each other. **ARMS** 0.8-3 (4) cm long, short and thick, usually bright red, erect and only sightly separated, or bowed to form a lantern-like structure; 3-sided; outer surfaces roughened, with a central longitudinal ridge; sides at first coated with spore slime; "spire" long or short (1-20 mm long), erect or bent downward. **STALK** 6-13 cm long, 0.5-2 cm thick, equal or more often tapered downward, flesh-pink to pink or reddish-pink above, paler below and usually white at base; hollow, tubular, fragile, minutely chambered and marked by longitudinal rows of V-shaped depressions; cross-section usually showing 4-6 (7) angles (same number as arms). **VOLVA** present at base of stalk as a loose, lobed, white sack or pouch formed by the ruptured peridium, usually with one or more white mycelial cords attached to base. **SPORE MASS** borne in the vertical slits between the arms and coating their sides, mucilaginous, light brown to olive-brown becoming darker (blackish) as it dries, with an unpleasant odor at maturity. Spores 3.5-4.5 × 1.5-2 microns, oblong, smooth.

HABITAT: Solitary to densely gregarious or clustered in lawns, gardens, hard-packed soil, etc.; common in southern California and fruiting there year-round, but partial to warm weather. It is probably native to Asia, but is well established in California at least as far north as Fresno. It has also been found in Texas and Washington, D.C.

EDIBILITY: Edible in the egg stage, but in my opinion, no better than *Clathrus archeri* (I have fried it). However, it is considered a great delicacy in China.

COMMENTS: This gaudy stinkhorn is easily recognized by its red to pink color and distinctive shape. In the words of Paul Rea (who wrote an extensive analysis of the developmental stages of this species), "the form of the expanded plant may be likened to a moat (the volval cup) surrounding the base of a tower (the polygonal stipe) bearing a belfry or lantern (the arms) surmounted by a spire"—a very apt description, except that the "belfry" is usually besieged by flies (see color plate)! Specimens growing in hard ground are apt to be larger than those growing in rich or loose soil, perhaps because they take longer to develop. The arms sometimes break free from each other, but the several-sided stalk and often brighter color distinguish it from *L. cruciatus*.

Lysurus cruciatus   (Lizard's Claw Stinkhorn)

FRUITING BODY beginning as a round to oval "egg" up to 6 cm high. **PERIDIUM** (skin) white and membranous, with a gelatinous inner layer; rupturing to form a volva at base of stalk. Mature fruiting body 6-12 (16) cm high, composed of a stalk that is branched at the top to form 4-7 (usually 5) stubby arms. **ARMS** initially incurved and touching (but not permanently fused) at their tips, then separating at least slightly and remaining more or less erect (i.e., not unfolding); short and thick (1-2.5 cm long), hollow, 3-sided; outer surfaces pallid to brownish, flesh-colored, pinkish, orange, or red, with a longitudinal groove. Inner surfaces wrinkled and irregularly roughened or knobby, at first covered with spore slime. **STALK** 6-10 cm long, 1-2 cm thick, usually tapered downward, hollow, fragile, minutely chambered and faintly striate longitudinally; entirely white or tinged yellowish above and white below. **VOLVA** present at base of stalk as a thick, loose, lobed, white sack formed by the ruptured peridium; usually with one or more mycelial cords attached to base. **SPORE MASS** supported within the arms and coating their inner surfaces, mucilaginous, olive to olive-brown becoming blackish as it dries, with an unpleasant (fetid) odor at maturity. Spores 3-4 × 1-2 microns, elliptical to oblong, smooth.

Left: *Pseudocolus fusiformis* (see comments at top of p. 776) has three or four long arms which are normally fused at their tips but may separate in age; stalk is not visible in this picture. (Joan Zeller)
Right: *Lysurus cruciatus* (=*Anthurus borealis*) has several short, thick arms at apex. (Bob Tally)

HABITAT: Solitary or in groups or clusters in lawns, gardens, under trees, in rich soil, on rotten wood, etc.; apparently native to Australia and New Zealand, but now widely distributed and well established in various parts of the United States. It is not uncommon in southern California, but is not nearly as numerous as *L. mokusin.* It favors warm weather but can be found most anytime. I have not seen it north of Santa Barbara.

EDIBILITY: Presumably edible, but see comments on the edibility of *Clathrus archeri.*

COMMENTS: Better known as *L. (=Anthurus) borealis,* this stinkhorn is easily told by its short, thick arms which separate slightly but do not unfold a la *Clathrus archeri.* It can be dull or brightly colored, and the tips of its arms are never fused into a spire as in *L. mokusin.* It has often been confused with *L. gardneri,* a tropical species whose arms have sterile bases.

Bird's Nest Fungi

spores

NIDULARIALES

Tiny fungi found on soil, wood, dung, and vegetable debris. FRUITING BODY usually rather tough, at first round to cylindrical or cushion-shaped, *usually becoming cup- to mug-shaped (nest-like) at maturity, with several "eggs" enclosed* (only one "egg" in *Sphaerobolus*). PERIDIUM (wall of "nest") conposed of 1-4 layers. PERIDIOLES ("eggs") usually flattened and lens- or lentil-shaped, white to gray, brown or black, sometimes with minute cords attached. STALK absent (but base of "nest" may be narrowed). SPORES *borne inside the peridioles,* typically smooth and hyaline. Capillitium absent.

THESE minute Gasteromycetes look like miniature bird's nests. The fruiting body typically consists of a tiny vase or "nest" **(peridium)** furnished with several spore capsules or "eggs" **(peridioles),** and in most cases a protective membrane or "lid" **(epiphragm)** that initially covers the top of the nest. As such there is very little that the bird's nest fungi can be confused with, though when very young they might be mistaken for minute puff-balls, and when old and bereft of eggs they look like tiny, tough cup fungi.

The spores are dispersed with the aid of raindrops and animals. The force of a single raindrop will splash the eggs out of the nest and as much as seven feet away (hence they are sometimes called "splash cups"). The outer wall of the egg then decays or is eaten away by insects and the spores within are exposed. Whereas most fungi produce millions of spores, the bird's nest fungi need only a few—each egg contains the correct mating strains, so a fertile secondary (dikaryotic) mycelium develops directly, obviating the need for large numbers of spores.

There are four common genera of bird's nest fungi. In *Cyathus* and *Crucibulum,* each egg is anchored to the nest by a minute cord or "stalk" **(funiculus),** while in *Nidula* and *Nidularia* the eggs are imbedded in a sticky gel. Both the cords and the sticky mucilage help the splashed eggs to adhere to whatever they land on. Also treated in this chapter is *Sphaerobolus,* a dynamic relative of the bird's nest fungi that is sometimes called the "cannon fungus." It features only one egg that is shot out with terrific force, accompanied by an audible "pop." (The Latin name literally means "sphere-thrower.")

Bird's nest fungi are found on various types of organic or vegetable matter—rotting wood and sticks, herbaceous stems, humus, dung, and manure (the spores of dung-inhabiting individuals presumably pass through—and out of—grazing animals). They are gregarious creatures, but as they rarely attain heights of more than 15 mm, they are difficult to see and worthless as food. Most of the common North American species are keyed out here, but additional exotic types may turn up in greenhouses and flower pots.

Key to the Nidulariales

1. Fruiting body containing only one peridiole ("egg"), round at first, then splitting open to form 4-9 orange rays around the central "egg" .... *Sphaerobolus stellatus,* p. 781
1. Not as above; fruiting body cylindrical to mug- or cup-shaped when mature, containing more than one egg (unless all but one have been expelled) 2

2. Fruiting body lacking a "lid" when young, at first round to cushion-shaped and more or less covered with a felty or powdery material that ruptures irregularly; nest soon disintegrating; eggs imbedded in a sticky mucilage which eventually dries out 3
2. Not as above; fruiting body typically with a "lid" when very young, the nest usually well formed and persistent; eggs may or may not be imbedded in a mucilage 4

3. Eggs brown to reddish, round and flattened (lens-shaped); not common .... *Nidularia farcta*
3. Eggs grayish-brown and often irregularly-shaped; fairly common ..... *Nidularia pulvinata*

4. Eggs gray to brown or dark brown, imbedded in a sticky mucilage or jellylike substance which eventually dries out; eggs *not* attached to the nest by a small cord or stalk; sides of nest usually vertical (i.e., fruiting body more or less mug-shaped) 5
4. Eggs white to gray, brown, or black, often (but not always) attached to side of nest by a minute cord or short stalk, *not* imbedded in a mucilage; sides of nest vertical to tapered 6

5. Exterior of nest white and shaggy when fresh *Nidula niveotomentosa* (see *N. candida,* p. 780)
5. Exterior of nest scurfy or hairy, brown to cinnamon or at times grayish *Nidula candida,* p. 780

6. Interior of nest longitudinally striate (with distinct radial grooves) ....
.... *Cyathus striatus* & others (see *C. stercoreus,* p. 780)
6. Interior of nest smooth or at least not striate or grooved 7

7. Eggs typically white to buff; interior of nest not black ... *Crucibulum laeve* & others, below
7. Eggs typically gray to brown or black; interior of nest variously colored (including black) . 8

8. Fruiting body typically less than 5 mm high; exterior of nest smooth to fibrillose but not shaggy; found on sticks, dung, etc., in arid regions ... *Cyathus pygmaeus* (see *C. stercoreus,* p. 780)
8. Fruiting body typically 5-15 mm high; exterior smooth or shaggy; cosmopolitan 9

9. Exterior of nest shaggy (with long hairs), at least when young; fruiting body averaging 5-10 mm high, the rim more or less circular at maturity .... *Cyathus stercoreus,* p. 780
9. Exterior of nest fibrillose to smooth even when young (without long hairs); fruiting body averaging 10-15 mm high, the rim often wavy in age *Cyathus olla* (see *C. stercoreus,* p. 780)

Crucibulum laeve (Common Bird's Nest Fungus)

FRUITING BODY tiny, at first nearly round, becoming cylindrical and then deeply cup-shaped (i.e., with a wide flaring mouth); 5-12 mm high and broad (at the top) when mature, the rim more or less circular and covered at first by a hairy lid. Peridium (wall of nest) one-layered, tough, persistent. Exterior velvety or shaggy, yellowish or tawny to cinnamon-brown, becoming nearly smooth in age and often darker or whiter. Interior of nest smooth, somewhat shiny, white to silvery, gray, or pale cinnamon. **PERIDIOLES** (eggs) 1-2 mm in diameter, several, whitish to buff or with a very slight brownish tinge, circular but flattened (lens- or disclike), usually attached to nest by long thin cords. Spores (4) 7-10 × 3-6 microns, elliptical, thick-walled, smooth, hyaline.

HABITAT: Scattered to densely gregarious on sticks, wood chips, nut shells, vegetable debris, humus, and manure; widely distributed. It is not as common locally as some of the other bird's nest fungi, but occurs year-round, sometimes in the company of *Cyathus.*

EDIBILITY: Much too miniscule to merit being munched on.

COMMENTS: Also known as *C. levis* and *C. vulgare,* this fetching and farflung little fungus is easily told by its whitish eggs that are initially attached to the nest by long, thin cords. It lacks the sticky mucilage of *Nidula,* and the eggs are never lead-colored or black as in *Cyathus.* Other species: *Crucibulum parvulum* is a similar but even more minute (2-4 mm high) species with a white to grayish or buff exterior. It grows on dead juniper and other organic matter in arid habitats.

 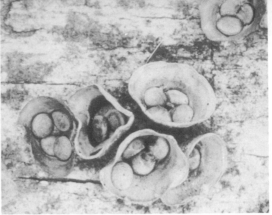

Left: *Cyathus stercoreus* growing on dung. The camera's strobe has made the "eggs" appear paler than they actually are. Note how rim of nest is circular. **Right:** *Cyathus olla* growing on wood. Note how rim of nest is often wavy. (Nancy Burnett)

Cyathus stercoreus   (Dung-Loving Bird's Nest Fungus)

FRUITING BODY tiny, goblet- to vase-shaped or resembling an inverted cone, 5-10 (15) mm high and 4-10 mm broad at the top, the rim more or less circular when mature and covered at first with a thin pallid or whitish lid that soon disappears. Peridium (wall of nest) tough, persistent, 3-layered (but the layers not always distinguishable). Exterior tan to golden-brown, brown, grayish-brown, or reddish-brown and shaggy (but often smooth and blackish in old age); base often with a pad of brown to reddish-brown mycelium. Interior of the nest smooth, pale gray becoming dark gray or lead-colored and often blackish in age. **PERIDIOLES** (eggs) several, 1-2 mm in diameter, dark gray to black, flattened or lentil-like, hard, smooth, often with a short cord or "stalk" attached (especially the lower eggs in the nest). Spores large: 22-40 × 18-30 microns, variable in size and shape but mostly round to oval, thick-walled, smooth, hyaline.

HABITAT: Densely gregarious on dung, manure, and other organic debris; widely distributed and common, but seldom noticed. In our area it can be found most any time. The largest fruiting I've seen was on a well-manured lawn in March.

EDIBILITY: A meager morsel, much too puny to be of value.

COMMENTS: The dark gray to black eggs, shaggy exterior (at least when young), and smooth interior are the principal fieldmarks of this cosmopolitan bird's nest fungus. The cords by which the eggs are attached are not always evident, at least in my experience. Other species: *C. olla* is a similar but slightly larger species. Its nest is gray to brown with an often wavy (rather than circular) rim and a smooth to finely hairy (fibrillose) but not shaggy exterior. Its eggs are also larger (2-3 mm) and gray to brown or blackish, and its spores are much smaller (8-15 microns). It is widely distributed and common, but not as numerous in our area as *C. stercoreus*. *C. pygmaeus* is a miniature version of *C. olla* (fruiting body only 4-5 mm high) that commonly grows on sticks, dung, etc., in the drier parts of the West. Two other species are worth mentioning because of their beautifully pleated or grooved (radially striate) interiors: *C. striatus*, widely distributed and quite common, with a shaggy cinnamon-brown to grayish-brown or dark brown exterior plus slightly triangular eggs; and *C. helenae*, found mostly in arid or alpine habitats, with a grayer, thicker-walled nest that has tufted hairs on its exterior. The latter two species can grow in manure but are more common on sticks and other vegetable matter.

Nidula candida   (Jellied Bird's Nest Fungus)

FRUITING BODY tiny, cylindrical or cushion-shaped becoming mug- or flower-pot-shaped at maturity (with vertical sides and a flaring mouth); 5-15 (20) mm high and 3-8 mm broad at the top when mature, the rim more or less circular and covered at first by a

Nidula niveotomentosa (see comments under *N. candida*), young specimens in which the nest is still covered by the epiphragm ("lid"). Note white exterior and mug-shaped fruiting body. (Herb Saylor)

lid. Peridium (wall of the nest) tough, persistent. Exterior whitish beneath a gray to brown or dull cinnamon scurfy or shaggy layer that covers at least the basal portion and the lid. Interior of the nest smooth, white to yellowish-brown or brown. **PERIDIOLES** (eggs) several, 1-2 mm in diameter, pallid to gray or brown (but often darker on underside), flattened, without cords, instead imbedded in a sticky mucilage or gel which eventually dries out. Spores 6-10 × 4-8 microns, elliptical to nearly round, smooth, hyaline.

HABITAT: In groups on rotting wood, berry canes, and herbaceous debris in gardens, woods, along streams, etc.; widely distributed. It is common in our area in the late fall and winter or even spring, but the empty nests persist for months without decaying and sometimes give rise to new ones.

EDIBILITY: Academic; you'd have to be slightly looney to bother with something so puny.

COMMENTS: A common but frequently overlooked little fungus, easily told by the brownish to cinnamon scurfy exterior and the sticky gel in which the eggs are imbedded, plus the presence of a covering or lid when young. The species name, *candida,* which means "shining white," is a misnomer since the fruiting body is neither shiny nor white. However, a similar species with a shaggy *white* exterior (and numerous small brown eggs), *N. niveotomentosa,* is common in California and the Pacific Northwest, usually on sticks or in moss. See also the genus *Nidularia* (in the key) and *Cyathus stercoreus.*

Sphaerobolus stellatus   (Cannon Fungus; Sphere Thrower)

FRUITING BODY minute, 1-3 mm broad, at first more or less round and white to dull yellow-orange or ochraceous, the outer wall (peridium) then splitting into 4-9 bright orange starlike rays or "teeth," exposing the single spore-containing peridiole ("egg"), which is then shot out like a cannonball as the entire structure turns inside out (leaving behind a translucent whitish ball perched on the rays). **PERIDIOLE** ("egg") chestnut-brown to olive-black, sticky or slippery, smooth, more or less round. Spores 7-10 × 3.5-5 microns, oblong, smooth, hyaline.

HABITAT: Gregarious on rotting wood, sawdust, plant debris, and dung or manure; widely distributed, but easily overlooked. I have seen it several times in nursery flats; it fruits whenever moisture is sufficient.

EDIBILITY: Unknown. Several hundred would be needed for a mouthful!

COMMENTS: This cousin of the bird's nest fungi is easily recognized by its diminutive dimensions, bright orange star-shaped "catapult," and single central "egg" or spore ball. It makes a marvelous—albeit liliputian—laboratory pet: a small *pop* is said to accompany the ejection of the spore ball, which sticks to whatever it lands on. Its flight path has been measured at 14 ft. high and 17 ft. long—or more than 1000 times the size of the fruiting body! To equal such a prodigious feat, we puny humans would have to throw a discus over one mile high and far! It is also interesting to note that the "sphere thrower" does not "do its thing" in complete darkness—light is apparently needed to trigger the "cannon," as well as a sufficient supply of moisture.

781

Ascomycetes

ASCOMYCOTINA

THE Ascomycetes are the largest subdivision of the true fungi. With over 15,000 known species, they pose a challenge, in the words of one noted mycologist, to "an army" of students. They are also an exceedingly diverse group, ranging from the most economically important of all the fungi, the single-celled yeasts, to powdery mildews to bread molds like *Penicillium* (the original source of penicillin) to prized edible mushrooms such as the morels and truffles. Their common bond is fundamental but microscopic: the sexually-produced spores are formed inside saclike mother cells called **asci** (singular: **ascus**). The ascus is roughly analagous to the basidium of a Basidiomycete, although fusion of the parent nuclei typically occurs in a separate cell, the ascogonium. Each ascus contains one to several thousand spores depending on the species. The most frequent number is eight.

Only a small fraction of the Ascomycetes have fruiting bodies large enough to merit mention in this book. In the field they can usually be distinguished from Basidiomycetes by a process of elimination, i.e., if your fleshy fungal fructification does not fit one of the common categories of Basidiomycetes pictured on pp. 52-54 (agarics, boletes, puffballs, etc.), then chances are it's an Ascomycete. The fleshy Ascomycetes treated in this book fall into two broad categories, keyed below.

Key to the Ascomycetes

1. Fruiting body more or less round (spherical) to oval or knobby (potato-like), growing underground or *inside* very rotten wood .... **Discomycetes,** p. 783
1. Not as above; growing *on* wood or on ground, on insects, other mushrooms, plants, etc. .. 2

2. Growing on wood (but wood sometimes buried) 3
2. Growing on ground or on insects, herbaceous plants, or other mushrooms 4

3. Fruiting body usually black or very dark brown (but often covered with white or grayish powder), rounded to irregularly knobby and charcoal-like *or* fingerlike to clublike or antlerlike and very tough or hard; asci borne in flasklike nests (perithecia) which often give the fertile area of fruiting body a minutely pimpled appearance .... **Pyrenomycetes,** p. 878
3. Fruiting body cuplike or variously shaped but not as above, *or* if colored as above then texture usually different (fragile, fleshy, rubbery, gelatinous, etc.); asci typically borne in a palisade (hymenium), not in perithecia .... **Discomycetes,** p. 783

4. Growing on insects (adults, pupae, larvae), spiders, or on other mushrooms (but hosts are often buried, so dig up carefully!); asci typically borne in flasklike "nests" (perithecia) ....
.... **Pyrenomycetes,** p. 878
4. Growing on ground or plant material, but not on insects or other mushrooms; asci typically borne in a palisade (hymenium), not in perithecia .... **Discomycetes,** p. 783

Left: Fruiting body of a cup fungus (a common type of Discomycete), showing how the paraphyses (sterile cells) and asci are arranged in a palisade on the upper (inner) surface. In reality there are thousands more asci than shown here. **Right:** An ascus containing eight spores (the most common number).

DISCOMYCETES

DISCOMYCETES are not Ascomycetes that like to dance, nor are they necessarily disc-shaped. Discomycetes are Ascomycetes in which a palisade of asci line an exposed surface of the fruiting body much as a palisade of basidia line the gills of an agaric. This palisade of asci (see drawing at bottom of p. 782) is called the **hymenium**, and the fruiting body of a Discomycete is called an **apothecium** (as opposed to the perithecium of a Pyreno-mycete). All but a handful of the Ascomycetes treated in this book are Discomycetes. Examples are the morels, false morels, elfin saddles, cup fungi, and earth tongues. Truffles are also Discomycetes, but they have specialized underground fruiting bodies in which the hymenium is infolded and internalized, or in some cases, non-existent (see p. 841 for more details). The Discomycetes have been divided here into three orders (large groups), keyed below. To facilitate identification, the key is based largely on the shape and size of the fruiting body rather than on more critical (i.e., microscopic) characteristics. Exceptions to the generalizations put forth in the key are then keyed out individually under the various orders, families, and genera.

Key to the Discomycetes

1. Fruiting body nearly always underground (or *inside* very rotten wood), round to oval or knobby (potato-like), i.e., stalk absent or extremely rudimentary; spores nearly always borne *inside* the fruiting body, the interior usually (but not always) with channels, veins, or one or more cavities .... **Tuberales,** p. 841
1. Not as above; fruiting body occasionally buried but usually above the ground at maturity or on wood, moss, etc.; spore-bearing surface exposed (external) at maturity 2
2. Fruiting body cup- to ear-shaped, spoon-shaped, disclike (flat), cushion-like, top-shaped, or sometimes contorted; stalk absent or present only as a narrowed base (but fruiting body some-times growing erect like a rabbit's ear); asci operculate (i.e., with "lid" at tip) **Pezizales,** below
2. Fruiting body erect, with a stalk and often a cap, cup, or "head" 3
3. Fruiting body with a well-defined cap *or* splitting into rays at maturity; cap cuplike, disclike, wrinkled, brainlike, saddle-shaped, pitted, honeycombed, or thimble-like (i.e., usually with a sterile underside or sterile hollow interior) 4
3. Not as above; fruiting body clublike or with an enlarged or flattened "head," but lacking a distinct cap with a sterile underside; "head" *not* cuplike, brainlike, saddle-shaped, disclike, or honeycombed, and not splitting into rays; fruiting body usually small; asci inoperculate (i.e., tip usually thickened, with a pore but no "lid") .... **Helotiales,** p. 865
4. Flesh gelatinous or semi-gelatinous (slice open fruiting body lengthwise); surface of fruiting body often viscid; cap rounded or wrinkled, often brightly colored but not dark brown to black; asci inoperculate .... **Helotiales,** p. 865
4. Not as above; asci operculate .... **Pezizales,** below

Morels, Elfin Saddles, and Cup Fungi

PEZIZALES

THIS large order includes most of the familiar fleshy Ascomycetes and nearly all of the prominent ones: the fabulous morels, grotesque false morels, elegant elfin saddles, and lowly cup fungi. All of these fungi typically have an exposed spore-bearing surface (hymenium) and their asci have "lids" which open when the spores are forcibly expelled. The asci usually discharge their spores *simultaneously*. The discharge can easily be triggered by tapping, breathing on, or otherwise disturbing the fruiting body, with the result that it will sometimes "smoke" (spew out clouds of spores) when picked!

The fertile surface ranges from concave or flat in the cup fungi to convex or convoluted in the false morels and elfin saddles to deeply pitted in the true morels. A stalk is present in the latter forms but often absent in the cup fungi.

Members of the Pezizales can be found almost anywhere at any time, but are most numerous and diverse in forests, in the spring. This means that in regions with mild winters they usually peak *after* most of the Basidiomycetes have fruited, while in regions with cold winters they are among the first fungi to appear after the snow has melted. Excluding the truffles (which are treated separately in this book), there are seven major families in the Pezizales, five of which are cup fungi. These are keyed below.

Key to the Pezizales

1. Fruiting body an irregularly cabbage-like, cauliflower-like, contorted, brainlike, or pitted mass of tissue, with or without a stalk; flesh *not* gelatinous 2
1. Not as above 3

2. Stalk absent or rudimentary, *or* if stalk present then stalk long (15-30 cm!) and solid and brown and often buried *and* restricted to eastern North America; fertile portion of fruiting body whitish to yellowish-brown, beige, pinkish, or lilac-tinged; rare . **Pezizaceae & Allies,** p. 817
2. Not as above; stalk present, long or short, usually hollow or partially hollow or complexly folded or chambered in cross-section; fertile portion of fruiting body colored as above or often darker; very common and widely distributed 5

3. Fruiting body cup-shaped (concave) to disclike (flat), cushion-shaped, or sometimes top-shaped or splitting into rays; stalk present or absent; flesh gelatinous, fragile, or tough ... 4
3. Fruiting body with a cap and stalk; cap convex to conical to bell-shaped, round, lobed, brainlike, saddle-shaped, or pitted but not cuplike (or cuplike only when *very* young); flesh fragile or tough but not gelatinous 5

4. Stalk absent, or if present then often (but not always!) short or merely a narrowed, downward extension of the cup; stalk when present usually *lacking* distinct ribs; fruiting body fleshy, fragile, rubbery, or gelatinous, *sometimes* brightly colored, large to minute; tips of asci amyloid or not amyloid .... **Pezizaceae & Allies,** p. 817
4. Stalk present, clearly differentiated from cap; stalk usually longer than width of cup *or* if shorter then usually ribbed or fluted; fruiting body fleshy or fragile but *not* gelatinous and *not* brightly colored, large to fairly small but *not* minute; tips of asci *not* amyloid .. **Helvellaceae,** p. 796

5. Cap honeycombed with ridges and pits, the pits usually fairly deep, or if not then the ridges with a strong vertical orientation; cap intergrown with stalk (or in one case, only upper part of cap intergrown with stalk) .... **Morchellaceae,** below
5. Not as above 6

6. Flesh gelatinous; fruiting body often viscid and brightly colored .... (see **Helotiales,** p. 865)
6. Not as above; flesh not gelatinous 7

7. Cap roughly conical to bell-shaped, like a thimble on a finger (i.e., attached only to very top of stalk, the sides hanging down freely like a skirt) .... **Morchellaceae,** below
7. Not as above; cap lobed, brainlike, saddle-shaped, etc. *and/or* attached to stalk differently .... **Helvellaceae,** p. 796

Morels and Allies

spores

MORCHELLACEAE

THIS is a small but famous family with only three genera: *Morchella, Verpa,* and *Disciotis. Morchella* (the true morels or "sponge mushrooms") has a pitted cap that is intergrown with the stalk (see p. 785 for more details). *Verpa* also has a well-developed stalk, but differs in having a smooth to wrinkled or somewhat pitted, thimble-like cap. *Disciotis* lacks an obvious stalk and is likely to be mistaken for a veined *Peziza;* consequently it is keyed out under the cup fungi. All three genera have non-amyloid asci and smooth elliptical spores. The spores lack large oil droplets but typically have "crowns" of minute droplets at both ends.

Key to the Morchellaceae

1. Fruiting body cuplike or spreading (flat); stalk very short or absent . . . . *Disciotis,* p. 796
1. Not as above; fruiting body with a cap or "head" and well-developed stalk 2

2. Cap with pits and ridges; at least the upper part of cap intergrown with stalk *Morchella*, below
2. Cap smooth to wrinkled or shallowly pitted, sitting on the stalk like a thimble (i.e., attached only to very top of stalk, the sides hanging free like a skirt) . . . . *Verpa*, p. 793

MORCHELLA (Morels)

Medium-sized to large, mostly terrestrial fungi. FRUITING BODY *with a cap and stalk; interior hollow.* CAP *honeycombed with ridges and pits (chambers), attached to the stalk for all of its length (or in one case, for 1/3-2/3 of its length).* STALK well-developed, smooth to scurfy or wrinkled below. SPORES typically elliptical, smooth, without large oil droplets. Asci lining inside surfaces of pits; 8-spored, operculate, not amyloid.

THESE prized delicacies are among the best-known wild mushrooms in North America, and they are so esteemed in Europe that people used to set fire to their own forests in hopes of eliciting a bountiful morel crop the next spring! Luckily, morels are among the most unmistakable of all fungi by virtue of their hollow, pitted or honeycombed "heads" (they are also known as "sponge mushrooms"). False morels *(Gyromitra)* are vaguely similar, but have a wrinkled or convoluted (not pitted) cap, while thimble morels *(Verpa)* have a smooth to wrinkled or shallowly pitted cap with free (skirtlike) sides. Morels have also been confused inexplicably with stinkhorns *(Phallus),* perhaps because of their similar shape. The head of a fresh *Phallus,* however, is coated with a sticky, stinky spore slime and there is a sack or volva at the base of the stalk.*

Much has been said about where and when to find morels, but no self-respecting morel hunter will divulge any truly crucial information unless he or she is planning to permanently leave the country (and then only for a stiff price!). Morel hunters are so protective of their favorite "patches," in fact, that they regularly disseminate misleading —if not downright erroneous—information, and they practice a presidential evasiveness when asked: "Where did all those morels come from?" Thus any "tips" or "secrets" you manage to squeeze out of morel hunters should be taken with a grain (better make that a bucket!) of salt. Having said this, I will now tell you where and when to find morels . . .

Morels *usually* grow outdoors: in forests (under both hardwoods and conifers) and open ground, in abandoned orchards, gardens, landscaped areas, under hedges, on roadcuts and driveways, near melting snow, in gravel, around wood piles or tree trunks, and in sandy soil along streams. In other words, morels grow wherever they please! They will even fruit in barbecue pits and on scorched ground in the wake of forest fires—sometimes in awesome quantities, *but only if conditions are favorable!*

Morels are almost universally associated with the spring ("May is morel month in Michigan"), but occasionally appear in the summer, fall, and winter. In the Midwest, where they are particularly abundant, the various species appear in a definite succession from late April through early June, and the annual morel festivals are a major tourist attraction. The state of Minnesota, in fact, recently declared the morel its official "state· mushroom." Morels seem to respond to a warming trend following a cold spell. This explains why they are particularly abundant in regions with cold winters—which is only right, because the people who survive those winters deserve a delicious spring! Alas, in tepid coastal California morels are not nearly as common as one would wish. In our area, in fact, finding them is largely a matter of luck, i.e., being in the right place at the right

*Don't feel too embarrassed should you manage to confuse them— *Morchella esculenta* was originally classified as a *Phallus* by Linnaeus!

Springtime bounty: A basket of black morels (*Morchella elata* group, p. 790).

time. In the Sierra Nevada and Cascades, however, they can actually be *hunted*—and sometimes harvested in large quantities, particularly if there are spring rains. Although timing is of critical importance because of competition from other morel hunters, it has been shown that morels develop and age more slowly than most mushrooms, over a period of three to four weeks! They have resisted all attempts to raise them commercially, but one ambitious entrepeneur tells me that success is just around the corner. The spores germinate readily in culture and the mycelium flourishes, but getting it to fruit has been the major stumbling block (perhaps it needs to be chilled to simulate a cold winter?).

Although the genus *Morchella* is unerringly distinct, the disposition of species is largely a matter of opinion. Dozens have been described based on differences in color, shape, and orientation of the pits and ridges. However, each "species" seems to intergrade with the next, leading some morelizers to recognize only three or four species. To most people the "true" identity of a morel is academic anyway—what counts is that it is edible, and incredible!

Raw morels often cause digestive upsets, so to stay on their good side, *always* cook them. They are delicious sauteed (providing you don't drown them in butter and herbs), and you can serve them over toast or in soup. Because they are hollow, morels are easy to stuff, but it is even easier to stuff yourself with morels! Before cooking, always split them length-wise to check for millipedes, slugs, and other critters that like to hide inside. Transfer all such tenants to the compost pile or some other comfortable spot (just because you're evicting them from their home doesn't mean you have to kill them!) and carefully remove all grit or sand, using water if necessary. If you are lucky enough to stumble on a large batch of morels and are unwilling to share the surplus with me, you can preserve them by canning or sauteeing and freezing. Or you can string them into beautiful necklaces and hang them up to dry (I will accept them in any form).

Four "species" of *Morchella* are described here, and several others are discussed. As I am meticulously mapping their variation and distribution, I request you, generous reader, to deliver any and all morels you find to my doorstep. Then, while I am savoring the superb flavor of *croutes aux morilles a la normande* (having first, of course, studied them thoroughly), you can bask in the altruistic satisfaction that comes from contributing to science.*

*The least you can do is invite me over for dinner.

786

Key to Morchella

1. Volva (sack) present at base of stalk (dig up carefully!); odor often fetid (see **Phallales,** p. 764)
1. Volva absent 2

2. Cap usually brainlike and irregularly lobed (as well as pitted), brown to reddish-brown; stalk complex (i.e., cross-section half way up stalk revealing numerous internal folds), often massive (up to 20 cm thick!) . . . . (see ***Gyromitra,*** p. 799)
2. Not as above; cap not conspicuously lobed; upper stalk or mid-portion more or less simple in cross-section (but base often folded or complex) 3

3. Lower 1/3-2/3 of cap free from stalk (only the upper part intergrown) ***M. semilibera,*** p. 791
3. All or nearly all of cap intergrown with stalk (lower edge may be creased) 4

4. Ridges of cap dark (dark gray to olive-brown to black) at maturity and sometimes dark when young . . . . ***M. elata*** group, p. 790
4. Not as above (but ridges may blacken when they dry out and shrivel up) 5

5. Ridges white or markedly paler than the pits *when young,* but often becoming same color as pits in age; pits usually large and often elongated vertically; fruiting body small to medium-sized (*not* large); common in suburbia, orchards, etc., less often in woods ***M. deliciosa,*** p. 789
5. Not with above features (but may have some of them) 6

6. Fruiting body pale (whitish to buff), not darkening in age; pits usually vertically elongated; mostly found in woods (especially montane) ***M. sp. (unidentified)*** (see *M. deliciosa,* p. 789)
6. Not as above; if pale then pits roundish to irregular 7

7. Fruiting body medium-sized to large (11-30 cm tall or more), *not* reddish-tinged; stalk often swollen at base and sometimes massive in age; base usually wrinkled, folded, or buttressed (like a tree trunk); found mainly with hardwoods . . . ***M. crassipes*** (see *M. esculenta,* below)
7. Not as above; fruiting body usually small to medium-sized or reddish-tinged; stalk often somewhat wrinkled at base but not normally buttressed; found in many habitats 8

8. Fruiting body reddish to reddish-brown or with a reddish tinge *and/or* pits arranged in definite rows . . . . ***M. elata*** group, p. 790
8. Fruiting body not reddish-tinged; pits usually rounded or irregular . . . . . ***M. esculenta,*** below

Morchella esculenta   (Morel; Yellow Morel)   Color Plate 203

CAP 3-11 cm high, 2-6 cm broad, round to oval to bluntly conical or irregular; margin attached to the stalk (but often with a crease at point of attachment); overall color tan to yellow-brown to warm buff or even buff. Pits roundish to irregular in shape and not normally arranged in well-defined rows; often quite small but sometimes large, yellowish to brown or tan. Ridges typically meandering rather than in lines, usually quite narrow, same color as pits or paler (or occasionally slightly darker). Interior hollow, whitish, roughened. Flesh often rather thin. **STALK** 1-5 (10) cm long, 1-2.5 (3.5) cm thick, usually minutely granular or scurfy, equal or enlarged at the base, usually relatively short and narrower than cap, the base often somewhat wrinkled or pitted; white to buff, sometimes with brownish or cinnamon stains; typically hollow in cross-section. **SPORES** 16-25 × 9-14 microns, elliptical, smooth, without oil droplets.

HABITAT: Solitary to widely scattered, gregarious, or clustered in a variety of habitats (woods, streamsides, old orchards, cultivated or disturbed ground, burned areas, etc.); very widely distributed, but especially common in eastern North America and the Midwest. Like other morels, it fruits in the spring but will occasionally turn up at other times. In eastern North America it is most often found under oak, maple, beech, hickory, elm, ash, fruit trees, and other hardwoods, especially in May (or in the words of Ingrid Bartelli, "when the oak leaves are as big as squirrel's ears"). Large crops can also be found around the bases of dying (but not quite dead) elms attacked by Dutch elm disease. In the West I look for it under deciduous oaks and in sandy soil or riverbottoms where there is willow, cottonwood, or alder, from February to May or June depending on the elevation and climate. It also occurs in burned areas, but not as commonly as the black morels, and it is not as frequent as the latter at higher elevations.

Left: *Morchella crassipes* (see comments below). This morel varies considerably in shape but is always large and has a wrinkled, buttressed stem. A sliced specimen is shown at top of p. 792. **Right:** *Morchella deliciosa* (p. 789), young specimens from eastern North America with very white ridges. The form in coastal California is not usually as white (see photo on p. 789). (Alan Bessette)

EDIBILITY: Edible and one of the most avidly hunted of all wild mushrooms (see comments on pp. 785-786).

COMMENTS: This, the common morel or "sponge mushroom," is one of the most readily recognized of all edible fungi. It can be told from other morels by its modest size, warm brown to tan or yellowish color, and irregularly-arranged pits (see color plate), and from the poisonous false morels *(Gyromitra* species) by its pitted or honeycombed rather than brainlike or lobed cap. It could also be confused with the stinkhorn *(Phallus impudicus),* but does not have a volva and lacks the foul-smelling spore slime of that species. It is typically a denizen of low hardwood forests, riparian woodlands, and fruit orchards, whereas the black morels are more common in northern latitudes or at higher elevations under conifers. The ranges of the two overlap, however, and they can sometimes be found growing in the same area. Like other morels, *M. esculenta* varies considerably in shape and appearance. In most cases the pits are quite irregular-looking, but in some forms the transverse ridges are not as well-developed as the vertical ones, giving the pits an elongated, slotlike appearance. The ridges never blacken as in the *M. elata* group, but conical forms intergrade with *M. deliciosa* (see comments under that species). Pale forms can also be confused with *M. deliciosa,* but usually have smaller, paler pits that are not as vertically elongated. A giant morel with irregularly arranged pits occurs in the same habitats as *M. esculenta,* but usually a little later. This is *M. crassipes* (or *M. esculenta* var. *crassipes*). It measures 6-20 or more cm high and usually has a massive stalk with an enlarged, wrinkled and folded or buttressed base (hence its nicknames, "Thick-Footed Morel," "Big-Foot," and "Tree Trunk Morel"). It also tends to have grayer pits when young with paler or even whitish ridges, but like *M. esculenta,* becomes tan in age. Many morel connoisseurs in eastern North America prize it above all others. In the West, unfortunately, it is relatively rare. I have found it under cottonwood in Oregon, among forget-me-nots in an old orchard in California (see photo above), and in landscaped areas.

Morchella deliciosa   (White Morel)   **Color Plate 201**

CAP 1.5-6 cm high, 1-3.5 cm broad, round to oval or conical, the margin attached to the stalk but sometimes creased at its juncture. Pits usually vertically elongated and quite large at maturity, but not necessarily arranged in rows; dark gray to dark brown to grayish-tan, brown, or tan (usually darker when young and paler in age). Ridges usually widely spaced and mostly vertical (the horizontal ones often few or not as prominent), white to creamy (lighter than pits) when young, usually becoming tan or same color as pits in age but not blackening unless shrivelled up. Interior hollow, the surface pallid or whitish, roughened. Flesh rather thin. **STALK** 1.5-6 cm long, 0.5-2 (3.5) cm thick, equal or thicker below, the base often somewhat wrinkled, pitted, or irregular; white to creamy or buff, usually scurfy or minutely warted; typically hollow in cross-section. **SPORES** 18-25 × 10-15 microns, elliptical, smooth, without oil droplets.

HABITAT: Solitary to gregarious in gardens and other suburban habitats, under fruit trees, in old orchards, in woods and at their edges; etc.; widely distributed, fruiting mainly in the spring. In eastern North America its appearance marks the beginning of the end of the morel season. In coastal California it is not uncommon in the spring, especially in sandy soil, but occurs practically year-round. One couple I know gets a small crop under their plum tree every Christmas! Another "white morel" occurs in the Sierra Nevada and Cascades (see comments).

EDIBILITY: Delectably delicious, as the species epithet implies. Like all morels, it should be cooked (see comments on pp. 785-786 for more details).

COMMENTS: As already pointed out, morels are perplexingly polymorphic and resist our obtrusive attempts to categorize them. The name *M. deliciosa* has been applied to more than one kind of morel, and the above description has been broadened to include a number of confusing and intergrading, small to medium-sized morels with white or pallid ridges *at least when young* and pits which are usually large and often vertically elongated. In age the pits and ridges often become the same color (yellowish-tan to brown) as shown in the color plate, but the shape and size of the pits helps distinguish it from *M. esculenta.* The name *M. conica* has also been used for this species, but is more properly applied to morels with blackening ridges (see the *M. elata* group). Other species: Another "white morel" occurs in the Sierra Nevada and Cascades about the same time as the black morels. It has a whitish to buff cap that does not darken appreciably in age. The pits are usually elongated vertically but the ridges often form vertical lines as in the black morels. I do not know the identity of this morel, but I *do* know that it is delicious!

The coastal Californian form of *Morchella deliciosa* (or what I have identified as that species). Note how the ridges are whitish when very young but become colored like the pits in age. Also note how large and elongated the pits are, and how the ridges on dried-up specimen at left have darkened. Two mature specimens are shown in the color plate.

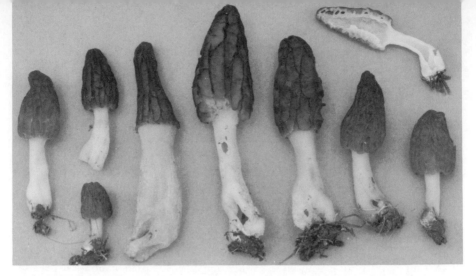

Black morels *(Morchella elata* group). This narrow-headed variety has also been called *M. angusti-ceps* and *M. conica* (see comments on p. 791). Note the conical head and strong vertical orientation of the ridges. They were found in a mountain meadow—a favorite haunt of the narrow-headed variety.

Morchella elata group (Black Morel) Color Plates 199, 202

CAP 2-6 (10) cm or more broad and 2-10 (18) cm or more high; usually conical to oval or somewhat irregular in shape; margin joined to the stalk but often with a crease at its point of attachment; overall color usually quite dark, especially in age. Pits usually vertically elongated *and/or* arranged in vertical rows, but in some forms meandering (see comments); yellow-brown to brown, grayish, olive-brown, or even reddish-brown, sometimes becoming blackish in age. Ridges vertically aligned in some forms, colored like the pits at first, usually darker (olive-brown to smoky-brown to black) before or by maturity (and sometimes dark from the beginning). Interior hollow, the surface pallid or whitish or tinged cap color, roughened. Flesh fragile in some forms. **STALK** 1.5-10 (20) cm or more long, (0.5) 1-4 cm or more thick, slender in some forms, as thick as the cap in others (often proportionately longer and thicker in age); equal or enlarged at either end, often grooved and/or wrinkled, especially toward base; sometimes nearly smooth, more often entirely or partially scurfy from a coating of small granules or warts; white to creamy, buff, pinkish-tan, or sometimes with a reddish tinge; typically hollow in cross-section. **SPORES** (18) × 20-25 (30) × 11-15 microns, elliptical, smooth, without oil droplets.

HABITAT: Solitary, scattered, or in groups, clusters, or troops on ground in woods and at their edges (especially under conifers and aspen) and in burned areas, less commonly in urban and suburban settings; very widely distributed (Japan to the Himalayas, Europe, etc.), but especially common in northern and western North America. Like other morels, its season is the spring (or early summer at higher altitudes), but "freak" fruitings can occur in the fall, particularly if a warm wet spell follows cold weather. In coastal California, where the seasons are not strongly defined, black morels occur year-round, but seldom in quantity. In colder regions they are often abundant, usually two to five weeks after the snow melts. Look for them in the Sierra Nevada when the snow plants (bright red sapro-phytic plants) have fully emerged, and in the Cascades when the calypso orchids are in bloom (see Color Plate 199). Don't expect to find them everywhere, however. For one thing, you face fierce competition from other collectors; for another, they are devilishly difficult to see, because they look just like fallen pine or fir cones (see Color Plate 202). Also, they show a definite preference for semi-disturbed areas, e.g., campgrounds, along roads, and in logged and burned areas. Bushels can be harvested one to two years after a forest fire, providing the spring weather is favorable.

EDIBILITY: Edible and delectable (see comments on pp. 785-786). However, some people are apparently "allergic" to it. Like all morels, it should never be eaten raw.

Black morels (*Morchella elata* group). **Left:** This fat-headed specimen has the radially-arranged pits of the narrow-headed variety (see comments below). **Right:** Typical examples of the fat-headed form.

COMMENTS: The above description covers a number of black morels (the so-called "*M. elata-M. angusticeps-M. conica*" complex). Many of the morels in this category are black or at least dark in age, but some are brown, reddish, or pinkish-tinged and/or have pits that are usually more or less radially arranged (that is, arranged in rows). Others are quite dark and may have pits that are *not* arranged in obvious rows. Excluded are those morels with a "half-free" cap (see *M. semilibera*), brown or tan to yellowish-capped morels with irregularly arranged pits (see *M. esculenta*), and morels with dark pits and pale ridges when young (see *M. deliciosa*). There are at least two common kinds of black morels in North America. The first variety, sometimes called the "Fat-Headed Black Morel," is probably the "true" *M. elata,* though it has gone under the name *M. angusticeps* and has also been called *M. conica*. It has a roundish to oval or conical "head" that is relatively large in comparison to the stalk, pits which are often elongated but not necessarily radially arranged, and ridges that are gray to black before maturity (and are often dark while very young). This variety is shown above and in the color plates. The second variety, sometimes called the "Narrow-Headed Black Morel," is shown in the photo on p. 790. Its cap is often more conical or elongated and often spongier or more fragile than the first variety. Its pits are usually arranged in rows and its ridges usually blacken, but not as quickly or as consistently as the fat-headed variety. Its stalk is often long and relatively thick (often as thick as the base of the cap), and is typically more furrowed or wrinkled toward the base than the fat-headed variety, and also wartier or scurfier. According to "morelizers," this second variety, if distinct from the first, is probably the "true" *M. angusticeps,* but it has also been called *M. conica*. A giant form of this variety is shown on the back cover of the book.

These two black morels grow in the same regions, but the fat-headed one seems to be the commoner of the two. Sometimes they seem distinct, at other times they intergrade or "hybridize." A third variety, often called the "Red Morel," is shown on p. 31. It resembles the previous two (especially the narrow-headed one), but a distinct reddish or pinkish tinge pervades its stalk and/or cap. This variety, which may simply be a growth form, has a particularly fine flavor. Still another variety has a whitish to buff cap. It may be an albino form, or a separate species entirely (see comments under *M. deliciosa*). Whew!

Morchella semilibera   (Half-Free Morel)

CAP 1.5-5 cm high and broad, bluntly conical to round or oval when young, usually conical in age; margin (lower edge) free from stalk for about one half (1/3-2/3) the distance to the apex of the cap, often flared outward away from the stalk in age. Pits large and elongated, yellowish-brown to brown or grayish-brown. Ridges vertically oriented (with few if

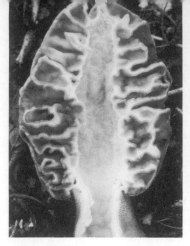
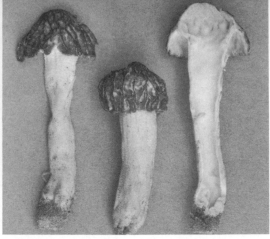

Left: In most species of *Morchella* the cap is completely intergrown with the stalk, as shown in this sliced *M. crassipes*. **Right:** In *Morchella semilibera,* however, the lower part of the cap hangs free from the stalk. These are mature specimens.

any transverse ribs), yellowish-brown to brown or olive-brown (usually slightly darker than the pits), often becoming blackish in old age or as they dry out. Underside (interior) whitish or pallid, roughened. Flesh rather thin and fragile. **STALK** 3-10 cm long, 1-2.5 cm thick at apex, equal or thicker below in age, fragile, white to yellowish; surface usually noticeably rough or with scurfy granules that may form ribs (as in *Verpa*); usually hollow in cross-section. **SPORES** 22-34 × 15-21 microns, elliptical, smooth, without oil droplets.

HABITAT: Solitary to gregarious on ground in woods and under trees (mainly hardwoods); widely distributed, fruiting in the spring. In our area it occurs in sandy soil along streams with cottonwood and alder, but is not as common as *Verpa bohemica,* which favors the same habitats. Farther north (e.g., in Idaho) it is more frequent.

EDIBILITY: Edible, but more fragile and not quite as flavorful as other Morchellas.

COMMENTS: Also known as *M. hybrida* and *Mitrophora semilibera,* this is a distinctive species because of its "half-free" cap that is intermediate between *Verpa bohemica* (whose cap is attached to the stalk only at its apex) and other Morchellas (whose cap is completely intergrown with the stalk). It is also intermediate in its fruiting behavior, for it typically appears shortly *after* the Verpas and shortly *before* the other Morchellas. The stalk may be quite short when young and the "head" relatively large, but in age the "head" shrinks and the stalk lengthens considerably, giving it the aspect of a *Verpa*.

Morchella semilibera, prime specimens. Note "half-free" cap and roughened stalk. Compare the cap to the completely free, thimble-like caps of the Verpas (pp. 793-795).

The roughened stalk surface is, in the words of mycologist Orson K. Miller, "the texture of a cow's tongue." Typical *M. semilibera* is unlikely to be confused with other morels, but it sometimes intergrades with the black morel *(M. elata* group).

VERPA   (Thimble Morels)

Medium-sized to fairly large terrestrial fungi. FRUITING BODY *with a cap and stalk.* CAP *bell-shaped to conical and thimble-like, i.e., free from the stalk except at the very top; surface smooth to deeply wrinkled or sometimes pitted.* Flesh rather fragile and thin. STALK well-developed, smooth or roughened by granules but not fluted; *usually stuffed with a pith.* SPORES elliptical, smooth, without large oil droplets. Asci lining outer surface of cap; asci 8-spored or 2-spored, operculate, not amyloid.

A *VERPA* looks like a thimble stuck on a finger, i.e., its smooth to wrinkled or pitted cap is attached only to the very top of the stalk so that its sides hang free like a skirt. The true morels *(Morchella)*, in contrast, featured a pitted cap that is entirely or partially inter-grown with the stalk, while the false morels and elfin saddles (*Gyromitra* and *Helvella*) have lobed, brainlike, or saddle-shaped caps.

Verpas are edible, but are more fragile than the true morels and not as flavorful. At least one species, *V. bohemica,* is poisonous to some people, so it is advisable to eat Verpas sparingly if at all. You can find them in woods and under trees and shrubs, particularly in well-drained soil along rivers and creeks. They are most abundant in the spring about a week to a month before the true morels (they are sometimes called "early morels"), but in coastal California they also occur in the winter. Two farflung species are depicted here.

Key to Verpa

1. Upper portion of cap intergrown with stalk; cap pitted   ... (see *Morchella semilibera,* p. 792)
1. Not as above; cap free from stalk except at very top (like a thimble on a finger) 2

2. Cap strongly wrinkled, the wrinkles usually oriented vertically and sometimes branched to form pits; asci 2-spored .... *V. bohemica,* below
2. Cap smooth or at times irregularly wrinkled; asci 8-spored 3

3. Stalk often rather spongy and usually stuffed with a cottony pith (but may be hollow in age); spores without oil droplets .... *V. conica,* p. 794
3. Not as above; spores usually with one oil droplet at maturity .... (see *Helvella,* p. 805)

Verpa bohemica   Color Plate 205
(Early Morel; Wrinkled Thimble Morel)

CAP 1-5 cm broad and 2-5 cm high (but sometimes much larger—see comments), bluntly conical to somewhat bell-shaped, squarish, or irregular, attached to the stalk only at the apex, the sides hanging down freely like a skirt; margin often touching (but not joined to!) stalk when young, sometimes flaring or upturned in old age. Fertile surface pale to dark yellow-brown or tan, often becoming darker brown in age; deeply wrinkled by branching folds or ribs which are often vertically oriented and sometimes branch to form pits. Underside whitish to brownish. Flesh rather thin and fragile. STALK 6-15 cm long, 0.8-3 thick (or sometimes much larger), equal or tapered slightly in either direction; whitish to creamy or becoming tan or ochre in age, smooth or often roughened by small, orangish to brownish granules which may form transverse belts or "ribs"; more or less round in cross-section, usually stuffed loosely with a cottony white pith. SPORES huge: 54-80 × 15-18 microns, elliptical-elongated, smooth, without oil droplets. Asci 2-spored.

HABITAT: Widely scattered to gregarious in woods, thickets, and forest edges, etc., especially in sandy or well-drained soil in stream valleys and ravines; usually fruiting in the early spring, widely distributed. It typically appears one to three weeks before the true

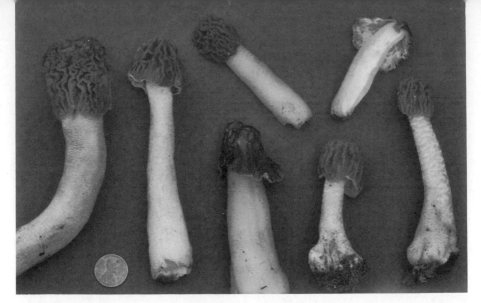

Verpa bohemica. Note how the cap is strongly wrinkled vertically, and may even be pitted (as shown in old specimen at center). Note also how cap is attached only to very apex of stalk, and how the stalk is stuffed with a cottony white pith, at least when young.

morels (*Morchella* species), sometimes in the same places. It is abundant in parts of the Pacific Northwest. In our area I have seen large fruitings under cottonwood and alder.

EDIBILITY: Edible with caution. Although eaten by many people, it can cause severe stomach cramps and loss of muscular coordination, particularly when consumed in large amounts or on several successive days. The flavor is strong but not on a par with the true morels. *Always* cook it, and beware: some of the "morels" sold in markets are Verpas!

COMMENTS: Also known as *Ptychoverpa bohemica* because of its two-spored asci, the early morel is easily told by its conspicuously wrinkled or pitted, thimble-like cap and smooth to roughened but not fluted stem. The cap is not as lobed as that of *Gyromitra* and *Helvella,* and the sides of the cap are not intergrown with the stalk as in *Morchella.* Medium-sized specimens are the norm, but as in some of true morels, a giant form also occurs. Typical *V. bohemica* is most likely to be confused with *Morchella semilibera,* whose cap is "half-free" (intergrown with the stalk over its upper portion), and with *V. conica,* which has a smoother or more irregularly wrinkled cap and 8-spored asci. When in doubt, you can quickly resolve the issue by examining the spores. Those of *V. bohemica* are the largest of any stalked Discomycete!

Verpa conica   (Thimble Morel; Bell Morel)

CAP 1-4 cm broad and high, usually more or less thimble-shaped (broadly conical to bell-shaped), but sometimes lobed (see comments) and sometimes developing a depression at the top in age; attached to stalk only at its apex, the sides free like a skirt; margin inrolled or incurved at first and sometimes touching the stalk (but not joined to it!), often lobed, often flaring out or turning up in old age. Surface smooth to slightly wrinkled, or in one form irregularly wrinkled (see comments), ochre-brown to brown or dark brown. Underside pallid. Flesh thin and rather brittle or fragile. **STALK** (2.5) 4-12 cm long, (0.4) 0.5-1.5 cm thick, equal or tapered upward (or occasionally downward), white to yellowish, tan, or tinged ochre-orange; smooth or with granular or minutely scaly transverse bands or ribs (granules often browner or oranger than background); often rather spongy and usually stuffed with a loose cottony pith (but often becoming hollow in age); more or less round in cross-section. **SPORES** (20) 22-30 (34) × 12-17 (19) microns, elliptical, smooth, without oil droplets. Asci 8-spored.

Verpa conica, typical specimens. Cap is relatively smooth and thimble-like, i.e., it is attached only to the apex of the stalk and it pops off easily.

HABITAT: Solitary, widely scattered, or gregarious in soil or humus in forests, riparian woodlands, under shrubbery and fruit trees, etc., usually in the spring; widely distributed but not particularly common. In our area I have found it as early as February and as late as May, but it usually peaks a week or two after *V. bohemica.* I have seen fairly large fruitings under redwood, oak, and cottonwood; in southern California it sometimes fruits in great quantity under chaparral shrubs. I have also seen it in Yosemite.

EDIBILITY: Edible when cooked, but rather fragile and not often occurring in enough quantity to warrant collecting.

COMMENTS: The thimble-like cap with a free margin is the principal fieldmark of this far-flung fungus. A rather strongly wrinkled, irregularly lobed form sometimes occurs, however (see photo below). The shape of the typical form is somewhat reminiscent of a stinkhorn, but it lacks a volva and its cap is never coated with putrid spore slime. Since the cap is attached to the stalk only at the very top, it pops off easily, and the entire mushroom is rather fragile. The strongly wrinkled form does *not* have the vertical ridges characteristic of *V. bohemica,* and is actually reminiscent of a small *Gyromitra.* The spongy stalk with a cottony pith inside plus the free cap margin that is inrolled when young are sufficient to distinguish it. *Morchella semilibera* is also somewhat similar, but has a pitted cap that is partially intergrown with the stalk.

Verpa conica, a gyrose (lobed and wrinkled) form (see comments above). The cap is reminiscent of a small *Gyromitra,* but is attached only to the apex of the stalk and has a strongly inrolled margin at first. Note how stalk is stuffed with a cottony pith. This form is quite different from the typical form shown at top of this page, but is microscopically identical.

DISCIOTIS

THIS genus includes one farflung species, described below, with a brown, more or less cuplike fruiting body. Consequently, it is keyed out under the cup fungi. Microscopic features, however, relate it to the morels.

Disciotis venosa   (Veined Brown Cup Fungus)

FRUITING BODY (3) 5-20 cm broad, at first more or less cup-shaped, but usually flattening out in age. Fertile (upper or inner) surface reddish-brown to brown, dark brown, or at times ochre-brown, sometimes smooth when young but usually becoming radially wrinkled or veined to reticulate, corrugated, or pebbled by maturity (at least toward the center); margin often wavy and/or splitting. Exterior scurfy, roughened, or minutely warty, whitish to buff or tinged brown, often fluted or wrinkled at base (i.e., appearing "gathered") to form a short stalk. Flesh fairly thick and brittle. **STALK** when present short, thick, usually buried. **SPORES** 19-25 (30) × 12-15 (17) microns, elliptical, smooth, without large oil droplets; asci lining upper surface of fruiting body, *not* amyloid.

HABITAT: Solitary, scattered, or in small groups in damp soil or humus under or near trees or occasionally in the open; widely distributed, fruiting mainly in the spring. I usually find it on well-drained soil in riparian woodlands (cottonwood, willow, etc.), but in my experience it is not very common.

EDIBILITY: Edible, but not recommended. Although the flavor is said to be like that of morels, people without a microscope can easily confuse it with other brown cup fungi (*Peziza, Discina,* etc.) which may or may not be edible. One source lists it as poisonous unless cooked.

COMMENTS: Until it is seen several times, this prominent brown cup fungus can be difficult to identify in the field. It is often quite large and conspicuously veined or reticulate, but several other brown cup fungi can mimic it. Microscopically it is quite distinct, however, for the smooth, elliptical spores lack oil droplets and the asci do not blue in iodine. It used to be called *Peziza venosa,* but because of these microscopic differences it has been rewarded with its own genus and placed in the morel family.

False Morels and Elfin Saddles

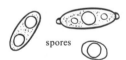

spores

HELVELLACEAE

MOST of the fungi in this family have a well-developed cap and stalk, but a few are cup-like or disclike. In some species of *Helvella* (the largest genus in the family), the cap is concave (cup-shaped), with the hymenium (fertile tissue) lining the upper or inner surface of the cup. In other Helvellas the cup turns inside-out as it matures and becomes lobed or saddle-shaped, while in still others it is saddle-shaped or irregularly lobed from the beginning. In *Gyromitra* the cap is often brainlike, but not deeply pitted as in the morels. Microscopic features that unite the family include the non-amyloid asci and smooth to roughened spores that typically contain one to three large oil droplets.

Several species of *Gyromitra* contain a deadly poisonous rocket fuel called MMH (see p. 893). Although these species are usually harmless when cooked or dried, it is safer to avoid them and to treat *all* members of the family with extreme caution (i.e., *always* cook them, don't eat large amounts, and be certain of your identification).

Key to the Helvellaceae

1. Fruiting body flattened and spreading (often undulating), attached to substrate by numerous "roots" (rhizomes); stalk absent; fertile (upper) surface dark reddish-brown to brown or even blackish, the margin often paler; growing under conifers, especially in recently burned areas; widespread but not very common .... *Rhizina undulata (=R. inflata)*

1. Not as above; fruiting body differently shaped, or if similar then not attached by "roots" .. 2

2. Fruiting body shallowly cup-shaped to flattened, disclike, umbilicate, or with a slightly down-curved margin; stalk absent, or if present then short and rather thick; fertile (upper) surface smooth or wrinkled but not brainlike, yellowish to tan, brown, or reddish-brown but not normally gray, grayish-brown, or black; underside lacking prominent hairs or ribs (but stalk may be ribbed or base of fruiting body may appear "gathered"); found mainly in spring and early summer when or soon after snow melts; spores often apiculate at maturity .. *Discina*, below

2. Not with above features (but may have some of them) 3

3. Fruiting body an erect club or column that lacks a distinct cap, up to 12 cm tall; interior of fruiting body with large chambers or compartments; exterior white to creamy or brownish, usually fluted or furrowed lengthwise; found in eastern North America and Midwest, usually under hardwoods; not common .... *Underwoodia columnaris*

3. Not as above; fruiting body with a cap (or cup) and stalk 4

4. Fruiting body small (cap up to 3 cm broad but usually less than 2 cm); cap usually round or convex, but sometimes compressed or broadly saddle-shaped, the margin usually tucked in toward stalk; usually found in groups or clusters; spores often needle-like .. (see **Helotiales**, p. 865)

4. Not with above features; usually larger or differently shaped; spores not needle-like 5

5. Stalk very thin (less than 3 mm); cap hollow and bladderlike, whitish to yellowish; found in eastern North America, usually clustered; spores borne on basidia (see **Aphyllophorales**, p. 548)

5. Not as above; spores borne on asci; widespread 6

6. Cap (or "cup") gray to black *or* white 9

6. Cap (or "cup") some shade of brown or tan 7

7. Cap brainlike to irregularly convoluted or wrinkled (at least at maturity), or if saddle-shaped then spores with two oil droplets .... *Gyromitra*, p. 799

7. Cap saddle-shaped, lobed, cup-shaped, plane, etc., but the surface only slightly wrinkled if at all; spores typically with one oil droplet 8

8. Stalk arising from a swollen tuberlike structure (sclerotium); cap cup-shaped to disclike; fruiting body small .... (see **Helotiales**, p. 865)

8. Not as above; stalk not normally arising from a sclerotium 9

9. Fruiting body minute; cap usually cup-shaped or disclike and less than 1.5 cm broad; stalk very thin (usually 1-2 mm thick); growing mostly on plant tissues (living or dead); asci *not* operculate (i.e., without "lids") .... (see **Helotiales**, p. 865)

9. Not as above; fruiting body small to fairly large, *not* minute; cap variously shaped (including cuplike or disclike); growing in soil, humus, or rotten wood; asci operculate *Helvella,* p. 805

DISCINA   (Pig's Ears)

Medium-sized fungi found on ground or rotten wood, *usually in the spring soon after the snow melts.* FRUITING BODY *shallowly cup-shaped to disc-shaped (flat), umbilicate, or slightly convex, with or without a short stalk. Fertile (upper) surface yellowish to tan or brown* (or less commonly reddish-brown), smooth to somewhat wrinkled; underside *not* normally ribbed. STALK when present short and thick, usually ribbed or appearing "gathered." SPORES elliptical to spindle-shaped, smooth to roughened or reticulate, the ends typically pointed or with short projections (apiculate). Asci lining upper surface of fruiting body, typically 8-spored, operculate, *not* amyloid.

THESE are springtime cup- or disc-shaped fungi with a yellowish to brown fertile surface and a thick, short stalk or none at all. As such they are likely to be mistaken for cup fungi (and are keyed out under that group), but microscopic features such as the non-amyloid asci and apiculate spores relate them closely to *Gyromitra* of the Helvellaceae.

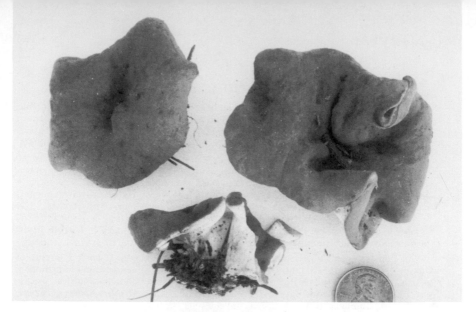

Discina perlata. This common "snowbank" mushroom looks like a cup fungus, but usually has a short stalk and is often umbilicate (i.e., with a "navel," as shown here). The fertile surface can be smooth or quite wrinkled. Microscopic features relate it to *Gyromitra.*

Pig's ears (a common name also applied to *Gomphus clavatus,* a member of the chanterelle family) are a common feature of our western springtime or "snowbank" mushroom flora. They are said to be edible, but because of their close relationship to the poisonous Gyromitras, it is best to eat them in moderation (if at all) and *always cook them.* The dozen or so species of *Discina* are distinguished primarily on microscopic features such as the shape and ornamentation of the spores. As only one is described here, a key hardly seems necessary.

Discina perlata (Pig's Ears)

FRUITING BODY (2) 3-10 (20) cm broad, at first cup-shaped but soon becoming saucer-like, then wavy or flattened (disclike) to very broadly convex (with the margin turned down) or umbonate, *or* umbilicate (with a central depression); outline round to irregular or somewhat angular in old age. Fertile (inner or upper) surface some shade of brown: tan to yellow-brown to brown, cinnamon-brown, or dark brown (usually paler or yellower when growing in or near snow); usually wrinkled, veined, or convoluted, especially toward the center, but sometimes smooth. Exterior (underside) more or less smooth, paler or whitish or somewhat translucent when moist. Flesh rather thick but brittle. **STALK** sometimes absent but more often present as a short (up to 1 cm long), thick (1-3 cm), narrowed base; white or tinged tan to brown, usually appearing "gathered" (i.e., with broad ribs and/ or pits). **SPORES** 25-35 × (8) 11-16 microns, spindle-shaped with an apiculus (knob or short projection) at each end, smooth or becoming minutely roughened at maturity, with one large central oil droplet and two or more smaller ones at the ends.

HABITAT: Solitary to gregarious or clustered on ground or around old stumps or occasionally on rotten wood in forests; widely distributed and common, fruiting mainly in the spring or early summer under conifers. It is one of the characteristic "snowbank" fungi of western mountains and often occurs with *Gyromitra gigas, G. esculenta,* and morels. Sometimes it begins developing while still under the snow!

EDIBILITY: Not recommended. Some people eat it, but care must be taken to cook it thoroughly and identify it correctly.

COMMENTS: This common springtime fungus is best told by its saucerlike to flattened, umbilicate, or slightly downturned, yellowish to brown fruiting body that frequently has a short, thick stalk. Like many of the larger Discomycetes, it develops quite slowly, and since spores are not produced until the fruiting body is fully grown, "sterile" specimens are often encountered. *D. ancilis* is apparently a synonym. There are several similar species of *Discina* that are best differentiated microscopically. These include: *D. apiculatula* and *D. macrospora* (the latter often with a reddish-brown fruiting body when mature); *D. olympiana,* a smaller species found in the Pacific Northwest; and *D. leucoxantha,* a yellowish, sometimes stalkless, truncate-spored species found under both hardwoods and conifers, particularly in eastern North America. Also very similar is *Gyromitra melaleucoides (=Peziza melaleucoides, Paxina recurvum),* with a somewhat waxy-looking underside when moist, a sometimes slightly longer and narrower stalk, and much shorter (10-14 microns), non-apiculate spores. See also *Disciotis venosa* and *Peziza* species.

GYROMITRA (False Morels)

Medium-sized to large fungi found on ground or rotten wood. FRUITING BODY *with a cap and stalk.* CAP *usually contorted, lobed, wrinkled, or brainlike (but sometimes saddle-shaped);* margin free from stalk or attached. *Fertile surface some shade of brown.* Flesh often rather brittle. STALK smooth or ribbed, simple (round) or complex (folded or chambered) in cross-section. SPORES typically elliptical or elongated, smooth or roughened, usually with two oil droplets. Asci lining outer surface of cap; typically 8-spored, operculate, non-amyloid.

THIS is a small but prominent and infamous genus whose grotesque fruiting bodies have inspired a number of fanciful nicknames: brain mushrooms, elephant ears, lorchels, beefsteak morels, and of course, false morels. Despite the latter name, Gyromitras are unlikely to be mistaken for morels (*Morchella* and *Verpa*) because their cap is neither deeply pitted nor conical or thimble-like. They are easily confused with Helvellas, however, and several species have been shuffled back and forth between *Gyromitra* and *Helvella* because of their "in-betweenness." As currently defined, forms with a brown brainlike cap belong to *Gyromitra,* while those with a differently colored and/or saddle-shaped cap (with the exception of *G. infula* and its look-alikes) are placed in *Helvella.*

Gyromitras are particularly common under northern and mountain conifers in the spring, often at the same time as morels. *G. infula,* however, fruits in the summer and fall (or the winter in coastal California). Several species, notably *G. esculenta, G. infula,* and *G. ambigua,* are dangerously poisonous (even deadly!) raw, yet are eaten without ill effect by many people. This apparent contradiction can be explained by two facts: (1) The toxin, MMH (monomethylhydrazine), is extremely volatile, i.e., it is usually removed by cooking or drying, and (2) there is a very narrow threshold between the amount of MMH the human body can "safely" absorb and the amount that will cause acute poisoning and even death. It seems foolhardy to risk eating the MMH-containing species, yet many people do and dried *G. esculenta* is sold in many European markets. If you *must* try them, then *always* either dry them out (then rehydrate and cook them) *or* parboil them first and throw out the water, being sure not to inhale the cooking vapors (which contain MMH), then saute them. *Never eat them raw* and never eat a large amount. MMH, incidentally, is also carcinogenic and is used as a rocket fuel. See p. 893 for more details on its effects.

Gyromitra is a much smaller genus than *Helvella,* but the fruiting bodies are usually larger. Four species are fully described here and several others are discussed.

Key to Gyromitra

1. Cap shallowly cup-shaped to flattened, disclike, or with the edges turned down slightly; underside of cap *not* ribbed .... *G. melaleucoides* (see *Discina perlata,* p. 798)
1. Not as above 2

2. Stalk smooth or indistinctly grooved; cross-section of mid-stalk round to somewhat flattened (i.e., with one round hollow or two flattened ones) 3
2. Stalk with very distinct vertical ribs *and/or* complex (i.e., folded, chambered, or convoluted in cross-section—but make the cross-section *above* the base); stalk usually thick 6
3. Cap usually saddle-shaped or "winged" (with 2-3 lobes), the surface usually wrinkled in central portion (directly above stalk), but usually smooth toward edges; lower edge of cap usually free from stalk; stalk stout, thick; typically found under hardwoods in eastern North America, mostly in the spring .... *G. brunnea* (see *G. infula,* p. 802)
3. Not as above; either more conspicuously brainlike at maturity or found elsewhere or surface not wrinkled; stalk often relatively slender; common and widespread 4
4. Stalk whitish or pale-colored, often rather spongy and usually stuffed with a cottony pith (but may be hollow in age); cap joined to stalk only at very top, the margin usually inrolled when young; spores without oil droplets .... (see *Verpa,* p. 793)
4. Not with above features 5
5. Cap lobed and brainlike (intricately folded and wrinkled) at maturity, but surface usually smoother when young; found in spring or early summer .... *G. esculenta,* p. 801
5. Cap lobed (saddle-shaped, hoodlike, etc.) but not brainlike, the surface smooth or only slightly wrinkled, even in age; found mostly in late summer, fall, and winter *G. infula* & others, p. 802
6. Stalk with deep, widely-spaced ribs that continue onto underside of cap; cap thin and spreading (umbrella-like) or appearing "puffed up"; margin usually free from stalk 7
6. Not as above 8
7. Stalk often reddish- or purplish-tinged near base; spores elliptical; found in western North America .... *G. californica,* p. 804
7. Not as above; spores round; eastern .... *G. sphaerospora* (see *G. californica,* p. 804)
8. Cap usually yellow-brown to ochre to tan or occasionally darker brown; stalk massive (nearly as thick as cap), often short; entire fruiting body with a chunky, almost cubical stature; found in spring and early summer when or soon after the snow melts ... *G. gigas* & others, below
8. Not as above; cap usually brown to reddish-brown 9
9. Cap brainlike to irregularly wrinkled or even pitted at maturity, not strongly lobed .... *G. caroliniana* (see *G. esculenta,* p. 801)
9. Cap often saddle-shaped or "winged" (with 2-3 large lobes), surface not wrinkled or wrinkled mainly at center (in area directly above stalk) .... *G. brunnea* (see *G. infula,* p, 802)

Gyromitra gigas
(Snow Mushroom; Snowbank False Morel; Bull Nose; Walnut)

CAP 3-10 (25) cm broad, 3-6 (15) cm high, brainlike or strongly and deeply convoluted and wrinkled, but typically compact (i.e., without strongly projecting lobes); fertile surface typically yellow-brown to butterscotch-brown or tan when fresh, but at times darker brown or even reddish-brown (especially in age); attached to the stalk at or near the margin. Interior chambered; underside (sterile surface) usually whitish. Flesh thin and fairly brittle. **STALK** massive (usually as thick or almost as thick as the cap) and usually short (sometimes completely hidden by the cap margin!), 2-10 cm long and thick, white or whitish, often rather irregular in shape or thicker at base; ribbed or wrinkled and grooved; strongly channelled or folded in cross-section. **SPORES** 24-36 × 10-15 microns, elliptical, the ends lacking projections or with only very short, blunt ones, smooth or finely roughened, typically with one large central oil droplet and smaller ones at the ends.

HABITAT: Solitary, scattered, or in groups on ground or rotten wood in coniferous forests (occasionally under hardwoods); common throughout the mountains and colder parts of western North America in the spring and early summer, typically near melting snow or soon after the snow disappears. In our area, where the winters are mild, it seems to be absent. However, I have seen enormous fruitings in the Sierra Nevada and Cascades.

EDIBILITY: Edible and popular. Apparently it can be eaten safely without parboiling (e.g., sauteed like a morel), but should *never* be eaten raw. Some people prefer it to the true morels—but be sure you identify it correctly!

Gyromitra gigas. This edible "snowbank" species does not have the bad reputation of other Gyromitras. Note the chunky, almost cubical stature and thick stalk that is folded or chambered within.

COMMENTS: The chunky or almost cubical stature (see photo), strongly wrinkled cap, and internally folded stalk are the telltale traits of this common "snowbank" fungus. *G. montana* and *Neogyromitra gigas* are synonyms. It sometimes grows with the dangerous *G. esculenta,* but can be distinguished by its yellower cap color, chunkier stature, and complex stalk. *G. californica* has a ribbed stalk, but usually has a strongly spreading (umbrella-like) or puffed-up cap. *G. gigas* shows considerable variation in size. Specimens weighing more than two pounds have been reported, but the normal size range is walnut- or softball-sized. *G. korfii* (called *G. fastigiata* in some books) is a very similar vernal species. It is more numerous in eastern North America than in the West (under both hardwoods and conifers) and has spindle-shaped spores with prominent projections at both ends. For other complex-stalked species, see *G. caroliniana* (under *G. esculenta*) and *G. brunnea* (under *G. infula*).

Gyromitra esculenta (False Morel; Brain Mushroom) Color Plate 206

CAP 3-12 cm high and broad or occasionally larger, sometimes cup-shaped when *very* young but soon becoming lobed or even saddle-shaped and at maturity intricately wrinkled and folded (brainlike) but not pitted. Fertile surface nearly smooth when very young but becoming more wrinkled as it matures, deep reddish-brown or purplish-brown to bay-brown, dark brown, or brown, or in some forms yellowish-brown; typically attached to the stalk at several points. Underside usually paler; interior hollow or chambered. Flesh rather thin and brittle. **STALK** 2-10 (15) cm long, 1-2.5 (4) cm thick, white to tan, flesh-colored, reddish, or sometimes colored like the cap (but often paler); equal or sometimes thicker at either end, smooth or grooved vertically but not ribbed; stuffed or hollow (or with two narrow hollows) in cross-section. **SPORES** 17-28 × (7) 9-13 (16) microns, elliptical, smooth, usually with two oil droplets.

HABITAT: Solitary, scattered, or in groups under both hardwoods and conifers; fruiting in the spring (or early summer in colder climates) and very widely distributed, but especially common in northern and montane coniferous forests. It is not very common in our area (perhaps the winters aren't cold enough), but is often abundant in the Sierra Nevada, Cascades, and other mountain ranges of the West. It is often encountered while morel hunting. The fruiting bodies persist for several weeks before decaying.

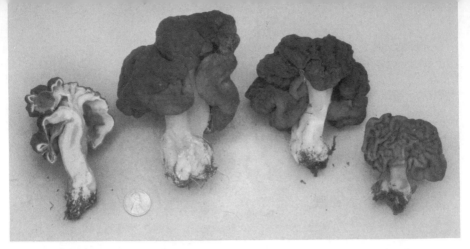

Gyromitra esculenta. Note how younger specimens at center have a relatively smooth cap, and older (but smaller) one at right is intricately convoluted or brainlike.

EDIBILITY: Dangerously poisonous, at least raw! Although eaten without ill effect by many people, this species has caused numerous deaths in Europe (see p. 799 for more details). Far fewer cases of poisoning have been reported from North America, suggesting that the American version is safer than the European one, or that certain ecological variants are less toxic. However, the paucity of poisonings may also indicate that fewer Americans eat it, or that those who do treat it with more caution. I certainly don't think it's worth the risk. In the words of mycophagist Charles McIlvaine, who was famous for his willingness to eat almost *anything*: "It is not probable that in our great food-giving country anyone will be narrowed to *G. esculenta* for a meal. Until such emergency arrives, the species would be better left alone."

COMMENTS: The brown cerebral cap and smooth to slightly grooved stalk plus its occurrence in the spring are the hallmarks of this infamous fungus. Despite the moniker "false morel," it can scarcely be confused with the true morels, for the brainlike cap is never deeply pitted or honeycombed. California specimens are more often reddish-brown than yellowish-brown, but regional variation can be expected. The stalk is often slightly grooved or compressed, but does not show the intricate folds in cross-section that are typical of *G. gigas*. Other species: ***G. caroliniana*** is a springtime species that favors low hardwood forests in eastern North America, but has also been reported from scattered localities in the West. It has a brainlike to irregularly wrinkled or even pitted, brown to reddish-brown cap, but has a massive stalk (up to 20 cm thick!) that is deeply furrowed and complex (folded) in cross-section. It can attain weights of several pounds each, and is said to be edible if prepared properly (parboiled, etc.). Like *G. esculenta,* however, it is best avoided, particularly if there are morels out and about!

Gyromitra infula   (Hooded False Morel)

CAP 3-12 (15) cm broad and high, sometimes cup-shaped when *very* young but soon developing projecting lobes; at maturity with 2-3 (rarely 4) lobes, saddle-shaped to hood-shaped to irregularly lobed but *not* intricately wrinkled or brainlike; fertile surface usually reddish-brown to dark brown, but in some forms yellow-brown, smooth to uneven but not intricately folded; cap typically attached to stalk at several points, the margin usually incurved (toward stalk). Underside (sterile surface) paler, minutely velvety; interior hollow or chambered. Flesh rather thin and brittle. **STALK** 1-8 (12) cm long, 0.8-3 cm thick, equal or tapering upward, smooth or minutely velvety, sometimes indistinctly grooved but *not* ribbed; colored like the cap or paler (in some forms whitish); hollow (or with two narrow hollows) in cross-section. **SPORES** (15)17-23(26) × (6)7-10(12) microns, oblong-elliptical, smooth, typically with two large oil droplets.

Gyromitra infula, rather old specimens. Note how stalk darkens with age and how the shape of the cap is quite variable but never brainlike. See next page for a distinctly saddle-shaped specimen.

HABITAT: Solitary to gregarious under both hardwoods and conifers, usually on rotten wood but also on soil, humus, along roads, in burned areas, etc.; widely distributed. In most regions it fruits in the late summer and fall—a feature that helps distinguish it from *G. esculenta* and other vernal species. In coastal California, however, it fruits in the winter and early spring and is the most common *Gyromitra* (especially under oak and pine).

EDIBILITY: Poisonous! It contains MMH and should *never* be eaten raw! For more details, see comments on p. 799 and p. 893.

COMMENTS: The saddle-shaped to irregularly lobed cap which is usually reddish-brown to dark brown plus the more or less smooth (non-fluted) stem are the distinguishing features of this false morel, which is also known as *Helvella infula*. Its penchant for growing on rotten wood is also distinctive, but by no means consistent or definitive. Saddle-shaped specimens are apt to be confused with *Helvella,* and it is keyed out under that genus. The color is quite distinctive, however, and the stalk is thicker than that of most non-fluted Helvellas. The surface of the cap can be smooth to uneven or slightly wrinkled, but is not pitted as in the true morels nor intricately folded or brainlike as in *G. esculenta.* Smooth specimens of the latter species can usually be distinguished by their growth in the spring soon after the snow melts. Other species: *G. ambigua* is a "dead ringer " for *G. infula.* Although it sometimes has a slight lilac or purplish tinge to the fruiting body, it can only be differentiated with certainty by its spores, which are longer (22-33 microns) and have blunt projections at both ends. It has a northern distribution, fruits at the same time, and is also poisonous. *G. brunnea* (=*G. underwoodii* and now called *G. fastigiata,* a name which has also been applied to *G. korfii*) is a springtime species found principally under hardwoods in eastern North America. It has a pale to dark reddish-brown cap that is usually saddle-shaped or "winged" (divided into 2-3 lobes—hence its common name, "Elephant Ears"). The cap surface is often wrinkled above the stalk but relatively smooth near the margin, and the undersides of the lobes are free from the stalk but often fused to

Gyromitra infula. Note the relatively smooth spore-bearing surface. It can be confused with young specimens of *G. esculenta,* but fruits at a different time, often on rotten wood.

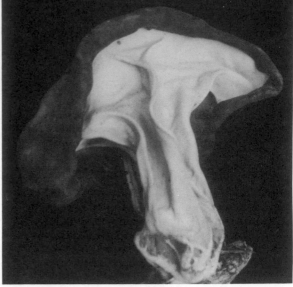

Left: A dark, saddle-shaped example of *Gyromitra infula*. Right: *Gyromitra californica*. Although dried out, this specimen shows the widely spaced ribs on the stalk which fan out on underside of cap.

each other. The stalk is usually white and fairly thick (2-5 cm) and ranges from hollow to complex (folded) in cross-section. It has finely warted spores and is edible *with caution*.

Gyromitra californica   (Umbrella False Morel)

CAP 4-12 (25) cm broad, convex to umbrella-like or undulating or occasionally broadly saddle-shaped (usually strongly spreading and often appearing "puffed-up"); fertile surface uneven to somewhat convoluted but not brainlike, tan to brown, olive-brown, or light to dark grayish-brown; margin typically free from the stalk and incurved toward it. Underside (sterile surface) whitish or creamy, ribbed, minutely hairy to smooth. Flesh very thin and fragile, almost papery when dried. **STALK** 3-8 (12) cm long, 2-5 (6) cm thick, equal or narrowed at base, often rather short, deeply and irregularly fluted or with several widely spaced ribs that extend onto the underside of the cap, where they fan out and continue to or nearly to the margin; white or sometimes aging grayish, buff, or yellowish, the base (or sometimes the entire stalk) usually with a pinkish, rosy, vinaceous, or purplish tinge; typically *not* internally chambered in cross-section. **SPORES** 13-19 × 7-10 microns, elliptical, smooth; ends often apiculate at maturity, with small oil droplets.

HABITAT: Solitary, scattered, or gregarious in woods and at their edges, along streams, or often in somewhat disturbed soil (e.g., along old skid roads); widespread in western North America, fruiting mainly in the late spring and summer, but sometimes in the fall or even winter. I have not found it in our area, but it occurs in the Sierra Nevada. As the species epithet implies, it was originally collected in California, but seems to be more common in the Pacific Northwest and Rocky Mountains. Alexander Smith says he could have collected "several bushel baskets" of it on the Olympic Peninsula one year.

EDIBILITY: To be avoided. It is edible according to some reports, poisonous according to others. The presence of MMH (see p. 799) would account for this contradiction.

COMMENTS: Also known as *Helvella californica*, this species can be told from other Gyromitras by its thin, strongly spreading or umbrella-like cap and distinctly ribbed stalk, and from *Helvella maculata* by its pinkish- or vinaceous-tinged stalk (a fairly reliable but not foolproof feature). The cap color is rather variable, but is not typically as yellow as that of *G. gigas,* nor as red as that of *G. esculenta.* Other species: *G. sphaerospora* is its eastern North American counterpart. It is quite similar but has round spores, and occurs at least as far west as Montana.

804

HELVELLA   (Elfin Saddles)

Small to medium-large fungi found on ground or rotten wood. FRUITING BODY *with a cap and nearly always a stalk.* CAP *cuplike to plane, saddle-shaped, or irregularly lobed but not usually brainlike;* margin usually free from stalk (except *H. lacunosa*). Fertile surface *white to brown, gray, or black.* Flesh often rather brittle and thin. STALK *usually well-developed* (occasionally practically absent), *smooth or ribbed or deeply fluted and pitted;* simple or complex (folded) in cross-section, fragile or tough. SPORES typically elliptical and smooth with one large central oil drop-let. Asci lining exterior of saddle or interior of cup, typically 8-spored, operculate, not amyloid.

HELVELLA is a common, distinctive, and attractive genus with a wide range of shapes and sizes. Some species are cuplike (and are more aptly called "elfin cups"); others look like miniature saddles (hence the popular name "elfin saddles"); still others are irregularly lobed. The cap, however, is not brown and brainlike as in *Gyromitra,* nor thimble-like as in *Verpa,* nor prominently pitted as in *Morchella.*

Helvella (spelled *Elvela* in some older books) splits nicely into four groups: those species with a cuplike cap and a ribbed (or practically absent) stalk (e.g., *H. acetabulum*); those with a more or less saddle-shaped cap and smooth or non-ribbed stalk (e.g., *H. compressa*); those with a cup-shaped cap and smooth, well-developed stalk (e.g., *H. macropus*); and those with a saddle-shaped to irregularly lobed cap and a deeply ridged or fluted stalk (e.g., *H. lacunosa*). Some authorites elevate some of these groups to genus rank (e.g., *Leptopodia, Cyathipodia, Macropodia, Macroscyphus, Paxina*), but their characters intergrade to some extent and microscopically they are very similar.

The cup-shaped Helvellas are apt to be mistaken for cup fungi, but tend to have more markedly ribbed and/ or longer stems. The convoluted species, on the other hand, can be confused with *Gyromitra,* but are usually smaller and/ or differently colored.

Helvellas do not have the sinister reputation of the Gyromitras, but should be treated with extreme caution. *H. lacunosa* is definitely edible, but many of the species have not been adequately tested. Others are too small or rare to be of value. If you really must eat Helvellas, *always* cook them thoroughly (in case they contain MMH) or dry them out.

Helvellas, like other Ascomycetes, are most numerous in the spring, but a few species fruit in the summer and fall, and in mild climates such as ours they can be abundant in the winter. They grow under trees and shrubs, along streams, and in the woods, or less commonly in the open. Over 30 species have been reported from North America. Most of these are keyed out or mentioned in this chapter; eleven are fully described.

Key to Helvella

1. Cap typically concave (cup-shaped) to plane, occasionally becoming convex or saddle-shaped in old age (examine several specimens if possible) 2
1. Cap typically convex (umbrella-like) to miter-shaped, hoodlike, saddle-shaped, irregularly lobed, or brainlike, but sometimes with an upturned or inrolled margin when young that makes it appear somewhat cup-shaped (examine several specimens if possible) 14

2. Fertile (upper or inner) surface of cup black or blackish; stalk and base of cup *not* whitish . 3
2. Fertile surface not black, or if so then stalk or base of cup white or significantly paler than fertile surface 5

3. Stalk and exterior (underside) of cup or cap black or very dark, sometimes with ribs; fertile (upper) surface not fading appreciably 4
3. Stalk usually grayish (paler than fresh fertile surface); fertile surface often fading to grayish-brown in age .... *H. villosa* (see *H. macropus,* p. 810)

4. Stalk well-developed (1-4 cm long, 2-5 (7) mm thick); fertile (upper) surface black or sometimes narrowly margined with white; cup fleshy or brittle; spores elliptical, with one large oil droplet; found in many habitats but not with melting snow ... *H. corium* (see *H. macropus,* p. 810)
4. Not with above features; stalk merely a narrowed base *or* if well-developed then fruiting body tough or at least not breaking easily *and/or* growing in or near melting snow ....
.... (see *Sarcosoma* & **Allies,** p. 826)

5. Stalk well-developed and slender, without prominent ribs, usually less than 6 mm thick (or if thicker than fertile surface yellow-brown to olive-brown and underside blackish *or* fertile surface more or less pinkish-brown with a scalloped or lobed margin) 6
5. Not as above; stalk thicker and stouter than above *and/or* with prominent longitudinal ribs 9

6. Upper stalk and exterior or underside of cup usually hairy or densely scurfy; stalk well-defined, usually 1 cm or more long; fertile surface of cup grayish-tan to gray or grayish-brown or dark brown; exterior same color or paler (*not* black) 7
6. Not with above features .... (see **Pezizaceae & Allies**, p. 817)

7. Underside of cup (sterile surface) often whitish, at least toward base; stalk often whitish also; rather rare .... ***H. cupuliformis*** (see *H. macropus,* p. 810)
7. Not as above; upper stalk and underside of cup usually grayish or colored like fertile surface (but base of stalk may be whitish); fairly common 8

8. Cap often becoming plane or even convex in age, the margin sometimes splitting to form lobes; upper surface quite dark (grayish-brown); spores elliptical *H. villosa* (see *H. macropus,* p. 810)
8. Cap usually remaining cup-shaped (not often plane in age); upper or inner surface gray to grayish-brown to grayish-tan or paler; spores often elongated   ***H. macropus*** & others, p. 810

9. Cap often spreading or flattened in age and/or center often wrinkled; fertile (upper) surface yellowish to tan or brown (not normally gray or black); stalk usually short; usually found in spring shortly after snow melts .... (see ***Discina***, p. 797)
9. Not with above features 10

10. Ribs on stalk extending onto underside of cup, often nearly to the margin 11
10. Ribs on stalk terminating at or near base of cup 12

11. Cap gray to grayish-brown .... ***H. griseoalba*** & ***H. costifera*** (see *H. acetabulum,* p. 807)
11. Cap light to dark brown or occasionally tinged violet .... ***H. acetabulum***, p. 807

12. Stalk (1) 3-7 cm or more long, usually well-developed, markedly ribbed; cap shallowly cup-shaped with opposite sides rolled up when young, often nearly plane in age ***H. queletii,*** p. 809
12. Stalk short (usually less than 2.5 cm long) or poorly developed; cup usually deeper or more regularly shaped than above 13

13. Interior of cup usually tan to pale brown to yellow-brown; exterior with brown hairs; texture rather tough .... (see **Pezizaceae & Allies,** p. 817)
13. Not as above .... ***H. leucomelaena*** & others, p. 808

14. Stalk distinctly fluted, ridged, or ribbed longitudinally, often showing folds or chambers in cross-section 15
14. Stalk not as above, but may show a few indistinct grooves, especially near base; stalk simple and round to slightly compressed in cross-section 21

15. Cap typically spreading, undulating, or appearing puffed-up or umbrella-like, tan to brown or grayish-brown, dark brown, or brownish-black (but *not* truly gray or black); margin of cap usually free from stalk; stalk with widely spaced ribs that fan out on underside of cap and often extend nearly to margin; stalk at least 1 cm thick (usually 2 cm or more), sometimes pinkish- or reddish-stained at base .... (see ***Gyromitra***, p. 799)
15. Not with above features (but may have some of them); base of stalk not normally reddish- or pinkish-stained; cap usually differently shaped or colored or stalk thinner 16

16. Cap white to buff, creamy, or *pale* tan .... ***H. crispa*** & others, p. 816
16. Cap darker (dark tan to brown, gray, or black) or sometimes pale gray or covered with a whitish fungal parasite 17

17. Cap typically gray to black (occasionally dark grayish-brown) 18
17. Cap typically some shade of tan, brown, or grayish-brown 20

18. Cap usually convex or umbrella-like; stalk slender, usually lacking external holes or pits ....
.... ***H. phlebophora*** (see *H. lacunosa,* p. 815)
18. Cap usually saddle-shaped to irregularly lobed or brainlike; stalk often with external holes 19

19. Cap irregularly lobed or brainlike to saddle-shaped, the surface often wrinkled; stalk with external pits and/or chambered in cross-section; very common, especially in western North America .... ***H. lacunosa***, p. 815
19. Cap usually more or less saddle-shaped, surface not wrinkled; stalk ribbed but usually lacking pits; more common in eastern North America than West ***H. sulcata*** (see *H. lacunosa,* p. 815)

20. Cap brown to pale or dark reddish-brown or chocolate-brown; spores finely warted at maturity; found mainly under hardwoods in eastern North America .... (see *Gyromitra,* p. 799)
20. Cap brown to grayish-brown but not reddish-brown; spores smooth; found in western North America .... *H. maculata* & others, p. 814

21. Stalk white to creamy or buff (significantly paler than cap unless cap is also pale-colored) 22
21. Stalk darker (brown, dark gray, or blackish), at least above (may be whitish toward base) 27

22. Stalk typically 0.8-2.5 cm thick; cap reddish-brown to dark brown or yellow-brown but not grayish-brown or black; found on ground or rotten wood; spores with two oil droplets ....
.... (see *Gyromitra,* p. 799)
22. Stalk typically less than 1 cm thick, or if thicker then cap grayish-brown to blackish but not yellow-brown or reddish-brown; spores with one oil droplet; usually on ground 23

23. Sterile surface (underside of cap) without hairs (use hand lens if unsure!) 24
23. Sterile surface minutely hairy, at least when fresh 26

24. Fertile surface of cap not usually very dark (i.e., yellow-brown to tan to brown or sometimes grayish-brown or violet-tinged) .... *H. elastica,* p. 813
24. Fertile surface of cap darker than above (grayish-brown to very dark brown to black) ... 25

25. Cap 3-7 cm broad and stalk 0.5-2 cm thick; margin of cap *not* significantly upturned when young .... *H. leucopus,* p. 812
25. Cap 1-2.5 cm broad and stalk 2-6 mm thick; margin of cap often upturned when very young
.... *H. albella* (see *H. compressa,* p. 811)

26. Cap buff to tan, light brown, or cinnamon-buff ..... *H. stevensii* (see *H. compressa,* p. 811)
26. Cap typically darker than above, at least when fresh .... *H. compressa* & others, p. 811

27. Cap irregularly lobed, or if saddle-shaped then stalk typically at least 8 mm thick; cap 3-15 cm broad, yellow-brown to reddish-brown or dark brown but not gray-brown or black; spores typically with two oil droplets .... (see *Gyromitra,* p. 799)
27. Not as above; cap usually rather small, 1.5-4 (5) cm broad, usually more or less saddle-shaped, gray-brown to dark brown to black; stalk typically 2-7 mm thick (occasionally thicker) . 28

28. Underside of cap (sterile surface) minutely hairy, at least when young and fresh 29
28. Underside of cap hairless (use hand lens if unsure!) 30

29. Cap grayish .... *H. ephippium* (see *H. atra,* p. 813)
29. Cap dark brown to grayish-brown to blackish .... *H. pezizoides* (see *H. atra,* p. 813)

30. Cap dark sooty-gray to black .... *H. atra,* p. 813
30. Cap usually paler (brown to gray but not black) .... *H. subglabra* (see *H. atra,* p. 813)

Helvella acetabulum   (Brown Ribbed Elfin Cup)

CAP 2-8 cm broad and up to 4 cm deep, cup- or bowl-shaped. Fertile surface (interior) light to dark brown (or in one form tinged violet), smooth; margin even or irregularly split in age. Exterior (underside) brown above, paler (white or creamy) at base, conspicuously ribbed, the ribs blunt or sharp-edged, branching, usually creamy and typically extending at least half way up the cup and often nearly to the margin; smooth or minutely hairy. Flesh thin, rather brittle. STALK sometimes absent but usually present as a stout, deeply ribbed base, 1-5 (9) cm long and 0.5-3 cm thick; equal or thicker below, white or creamy (or sometimes brown if very short); convoluted or chambered in cross-section. SPORES 16-20 × 11-14 microns, elliptical, smooth, with a central oil droplet.

HABITAT: Solitary to gregarious on ground in woods and at their edges; widspread and fairly common. It is usually listed as a spring and early summer species, but in our area it fruits in the winter and early spring. I find it once or twice a year, usually under oak.

EDIBILITY: I can find no information on it.

COMMENTS: Formerly known as *Paxina acetabulum,* this species looks like a cross between a cup fungus and an elfin saddle. (If the cup were inverted, you'd have an elfin saddle!) The well-developed ribs that extend far up the underside of the cup help to separate it from *H. leucomelaena* and *H. queletii.* Other species with ribs extending up the

Helvella acetabulum. Note how prominent whitish ribs on stalk extend onto underside of cup (left), and how stalk is complex in cross-section (bottom); upper surface of cup is brown or tan.

underside of the cup include: ***H. griseoalba,*** similar in size but with a gray cup; and ***H. costifera,*** smaller and grayish-brown.

Helvella leucomelaena   (White-Footed Elfin Cup)

FRUITING BODY cup- or bowl-shaped, with a short stalk or base, (1) 2-5 (7) cm broad and high. Fertile surface (interior of cup or "cap") smooth, dark gray to dark brown, dark grayish-brown, or blackish, the margin often finely scalloped, lobed, or split, especially in age. Exterior colored like the interior above, shading into white or creamy below, minutely roughened, *not* ribbed or only ribbed basally (i.e., appearing "gathered"). Flesh thin, rather brittle. **STALK** short and stout to practically absent, 0.5-2 (4) long, 0.5-1.5 (3) cm thick at apex, with distinct, broad, low, rounded, whitish ribs that terminate at the base of the cup; usually white but at times dingy grayish; channelled in cross-section (especially the upper portion). **SPORES** 20-23 (25) × 10-14 microns, elliptical to somewhat oblong, smooth, with a single central oil droplet.

HABITAT: Scattered to densely gregarious or clustered (occasionally solitary) on ground, usually near or under pine and other conifers; widespread, but particularly common in western North America. In our area it fruits in the winter and early spring, but is infrequently encountered (or easily overlooked because of its dark color). Farther north and at higher elevations it is quite common in the spring and early summer. In all regions it favors bare, grassy, or hard-packed soil along roads and paths. The fruiting bodies begin their development underground, and their stalks are often buried until maturity.

Helvella leucomelaena. **Left:** Side view showing the short whitish ribbed stalk. **Right:** View of dark inner surface and finely scalloped margin; note how short stalk is covered by dirt.

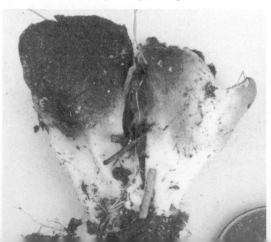
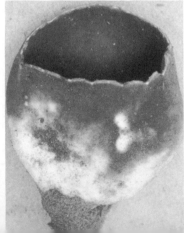

EDIBILITY: I can find no information on it.

COMMENTS: Formerly known as *Paxina leucomelas,* this species is likely to be mistaken for a cup fungus. However, the dark "bowl" and whitish, distinctly ribbed stalk plus the fairly fragile flesh place it in *Helvella.* The stalk is sometimes so short as to be non-existent, in which case the base of the bowl appears "gathered" (folded or ribbed). *H. crassitunicata* and *H. solitaria* are two very similar western species that differ microscopically. The first, however, usually fruits in the late summer and fall and has a northern distribution (the Pacific Northwest and Alaska), while the latter has a more distinct stalk and usually develops above the ground. None of these species have ribs which extend up the exterior of the cup as in *H. acetabulum* and relatives, nor is their stalk as long as that of *H. queletii,* nor as slender as that of *H. macropus* and relatives.

Helvella queletii (Ribbed Elfin Saucer)

CAP 2-8 (12) cm broad, usually concave with opposite margins rolled up and in (incurved) at first, becoming shallowly cup- or saucer-shaped to plane or even with a downcurved margin in age. Fertile (upper) surface blackish to grayish-brown to brown (often darker when young) or in one form pale brown to buffy-brown, smooth; margin often splitting in age. Exterior (underside) same color or more often paler or grayer, minutely roughened (granulose) to nearly smooth, *without* ribs. Flesh thin, rather brittle. **STALK** (1) 2.5-7 (12) cm long, 0.5-2 (5) cm thick, equal or thicker below, deeply ribbed longitudinally, the ribs extending only to the base of the cup (cap); white (including the ribs) or sometimes tinged buff, tan, or ochre. **SPORES** (17) 19-22 × 11-14 microns, elliptical to oblong, smooth, with a large central oil droplet.

HABITAT: Solitary, scattered, or in small groups in forest humus or occasionally on rotting wood, usually under hardwoods; widely distributed but not particularly common. In most of North America it is a late spring and summer species, but in coastal California it fruits, like most Helvellas, in the winter and early spring.

EDIBILITY: I can find no information on it.

COMMENTS: This is a most distinctive species, as the photograph shows. The concave to plane, grayish-brown to nearly black fertile surface plus the prominently ribbed or fluted stalk are the principal fieldmarks. The stalk is much longer than that of *H. leucomelaena* and *H. solitaria,* and the ribs do not extend onto the underside of the cup ("cap")

Helvella queletii. This beautiful species has a well-developed stalk whose ribs do *not* extend to margin of cup. Note how opposite sides of "cup" are inrolled when young (specimen at right).

as in *H. acetabulum, H. griseoalba,* and *H. costifera.* The peculiar way in which opposite sides of the "cup" are often rolled up when young (see photo on previous page) is suggestive of a young *H. compressa,* but that species does not have a ribbed stalk.

Helvella macropus   (Scurfy Elfin Cup)

CAP 0.8-3 (6) cm broad, sometimes closed up when young but soon opening to become shallowly cup-shaped, rarely becoming plane in old age; margin at first incurved. Fertile (upper or inside) surface usually gray to grayish-brown, varying to grayish-tan or grayish-olive or grayish-buff, smooth. Exterior (sterile surface) colored like the interior or slightly paler or grayer, hairy or minutely fibrillose-scaly or dandruffy. Flesh thin. **STALK** 1-5 (7) cm long, 2-5 mm thick, equal or often thickened at the base, not ribbed but sometimes with wrinkles or pits at the base; colored like the exterior of cup above and hairy or scaly-dandruffy, usually paler or whitish at the base; round or flattened but not chambered in cross-section. **SPORES** (18) 20-25 × 10-12.5 microns, more or less spindle-shaped, finely roughened or smooth, with one large oil droplet and a smaller one at each end.

HABITAT: Solitary, scattered, or gregarious on ground or rotten wood under both hardwoods and conifers; widely distributed. It is one of the commoner summertime Helvellas of eastern North America, but in our area it fruits in the winter and spring and is not very numerous (or at least not often collected because it is so inconspicuous).

EDIBILITY: Unknown, but too small to be of value.

COMMENTS: This cute little elfin cup is best told by its overall grayish color and hairy or scurfy-scaly sterile surface and stalk. The stalk is not ribbed as in *H. leucomelaena* or *H. acetabulum,* and the elongated spores are unusual for a *Helvella.* Both short- and relatively long-stemmed varieties are known. *H. villosa* is a very similar and widespread, slightly darker species with a slightly less hairy stalk and a shallower cup that usually becomes plane or even slightly convex in age (see photo below). I have found it several times in our area, also in the winter and spring. *H. cupuliformis* is quite similar in shape, but has a paler (often whitish) exterior or underside. *H. corium* is similar in shape but is entirely black, and the lower portion of its stalk is often somewhat ribbed. It seems to have a circumpolar (northern) distribution and often occurs in sandy soil or debris. *H. pallidula* of eastern North America is a pale tan to pale grayish-brown, cup-shaped species. None of these have the markedly ribbed stalks of *H. queletii* and *H. acetabulum,* and all have the elliptical spores typical of most Helvellas, rather than the elongated spores of *H. macropus.*

Left: *Helvella macropus,* a young specimen with hairy-scurfy underside. **Center:** *Helvella macropus,* mature specimen with a broader cup. **Right:** *Helvella villosa* (see comments above) resembles *H. macropus,* but its cap or "cup" is often plane in age, as shown here.

  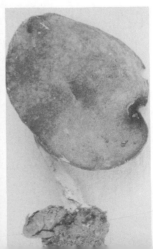

Helvella compressa. **Left:** Young specimens in which opposite sides of the cap are rolled up and in, making the cap somewhat cuplike and hiding the fertile surface; sterile surface is whitish and minutely hairy. **Right:** A mature saddle-shaped specimen which has faded somewhat (to brown).

Helvella compressa (Compressed Elfin Saddle) Color Plate 204

CAP 1.5-5 cm high and broad, saddle-shaped to somewhat irregularly lobed (with 2-3 lobes) when mature, typically with a well-developed sinus (cleft), but opposite margins rolled up and over the fertile surface when young; margin unrolling and often flaring in age, free from the stalk. Fertile surface brown to dark brown or grayish-brown, smooth. Underside (sterile surface) white or creamy to grayish-white, minutely hairy. Flesh thin, rather brittle. **STALK** 3-10 cm long, 0.3-1.2 (1.7) cm thick, equal or thicker below, smooth or finely downy, *not* ribbed; white to pale cream-colored; round or somewhat flattened in cross-section, but not chambered; base sometimes pitted. **SPORES** 19-22 (25) × 12-15 microns, broadly elliptical, smooth, with a large central oil droplet.

HABITAT: Solitary to gregarious on ground in woods and at their edges, under trees, etc.; known only from western North America. In our area it usually fruits in the late winter and spring under redwood, oak, and various other trees. It seems to be the most common of our non-fluted Helvellas.

EDIBILITY: Unknown.

Helvella compressa, mature specimens. Note deep cleft (sinus) in the saddle-shaped cap. These small, dark specimens could just as easily be referred to *H. albella* (see comments on next page).

COMMENTS: This attractive elfin saddle is best identified by its combination of brown to grayish-brown cap with a finely hairy underside and slender, white, non-ribbed stem. The transformation in the shape of the cap as it matures is apt to cause some confusion unless intervening stages are found. When young the margin is curled over the fertile surface so that the whitish underside is most visible and the cap is almost cuplike (see photo at top of p. 811). In age, however, the margin unfurls and the saddle shape is more apparent (see other photos on p. 811). **H. albella** is a very similar but smaller (cap 1-2.5 cm broad and high, stalk up to 5 cm long and 2-6 mm thick) species with a grayish-brown to blackish cap. Some authors describe its underside as hairless while others maintain it is minutely hairy when young. If the latter is true then its small size is the principal point of departure from *H. compressa*. A small form meeting this description is not uncommon in California in the winter and spring, usually under oak. Another closely related species, **H. stevensii** **(=H. connivens)** is small to medium-sized but has a paler (tan to yellowish- or cinnamon-buff) cap and slightly smaller spores. It has a wide distribution and also occurs in our area. Both *H. compressa* and *H. stevensii* are distinguished from *H. elastica* by their upturned cap margin when young, deeper and narrower cleft (sinus), and minutely hairy sterile surface (see color plate), and from *H. atra* and relatives by their whitish stalk.

Helvella leucopus   (Elfin Miter)

CAP 3-7 cm broad and high, often with three or more lobes but sometimes with only two; shape variable: miter-shaped or appearing saddle-shaped from three different angles, but sometimes irregularly contorted or occasionally suggestive of an elephant head (one longer lobe with two "ears"); margin nearly straight when young (not conspicuously inrolled), the lower edge of each lobe typically joined to the stalk at one point. Fertile surface typically more or less smooth and very dark (dark brown to dark grayish-brown to blackish), occasionally mottled with lighter brown areas. Underside (sterile surface) white or tinged faintly with the cap color, *not* minutely hairy. Flesh thin, rather brittle. **STALK** 4-12 cm long, 0.7-2 cm thick, equal or thicker below, often flattened or compressed but not ribbed or deeply fluted; smooth (not hairy), often with hollows or holes at the base; white or sometimes developing slight smoky-brown stains in age, often curved (giving it the appearance of a bleached rib); hollow in cross-section. **SPORES** 20-23 × 14-15 microns, elliptical or slightly oblong, smooth, with a large central oil droplet.

Helvella leucopus. Note dark cap which often has three or more distinct lobes. Stalk is whitish, usually hollow, and fairly thick (over 5 mm).

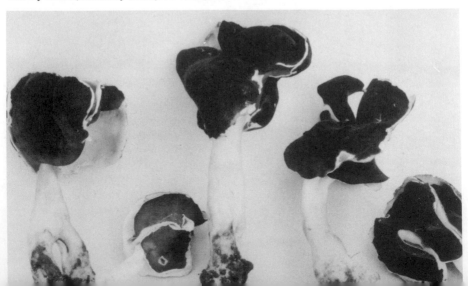

HABITAT: Scattered to gregarious on ground under trees, usually in the spring; widely distributed. In our area I have found it under cottonwood in the late winter and spring, but it grows in other habitats as well.

EDIBILITY: Not recommended. The European version is said to be edible, but I can find no information on North American material.

COMMENTS: The above description is based on several dozen specimens that grew in sandy soil under a cottonwood near Santa Cruz, California. It agrees perfectly with European descriptions of *H. monachella*, a name now considered synonymous with *H. leucopus; H. albipes* is yet another synonym. No matter what you call it, the dark cap contrasts nicely with the white, unribbed stalk, making it a most attractive elfin saddle. It is most apt to be confused with *H. compressa* and *H. albella,* but is usually larger and stouter, with a more irregular, multi-lobed cap whose underside is hairless.

Helvella elastica   (Brown Elfin Saddle)

CAP 0.5-4 (6) cm broad and high, typically saddle-shaped or miter-shaped with a broad, shallow sinus (cleft) between the lobes, but sometimes convex (with little or no sinus). Fertile surface smooth, brown to tan or grayish-tan, the margin straight or somewhat incurved toward the stalk; lobes free from the stalk or sometimes partially attached to it or to each other. Underside (sterile surface) whitish to buff or pale tan, smooth, *not* hairy. Flesh thin, rather brittle. **STALK** (2) 4-8 (14) cm long, (0.2) 0.4-1.2 cm thick, equal or tapered upward, sometimes curved, smooth or indistinctly grooved at the base but not fluted or ribbed; white to buff or pale yellowish-buff; round to slightly flattened in cross-section, not chambered. **SPORES** 18-22 (24) × 10-14 microns, elliptical or oblong, smooth or warted, with one central oil droplet.

HABITAT: Solitary to widely scattered or gregarious in woods or at their edges, particularly near streams and paths; widely distributed. It occurs in our area in the fall, winter, and early spring, but is not as common as *H. compressa.*

EDIBILITY: Unknown.

COMMENTS: Many authors (myself included) have applied the name *H. elastica* indiscriminately to a slew of smallish Helvellas with slender white, non-ribbed stems and brownish, saddle-shaped caps. However, in the "true" *H. elastica* the cap is smooth (hairless) on its underside and usually has a broad, shallow cleft and a straight to slightly downcurved margin, whereas in species such as *H. compressa* and *H. stevensii* the cleft is narrower and the lobes more upright, the underside is minutely hairy, and the margin is often curled up and in (thus hiding the fertile surface) when young.

Helvella atra   (Dark Elfin Saddle)

CAP 0.7-2 cm high and broad, distinctly saddle-shaped or occasionally 3-lobed when mature, with a narrow sinus (cleft); margin free from the stalk or occasionally attached. Fertile surface deep sooty-gray to black, smooth. Underside (sterile surface) pale gray to blackish, hairless. Flesh thin, fairly brittle. **STALK** 1-3.5 cm long, 2-5 mm thick, equal or slightly thicker below, not fluted but occasionally with shallow pits at the base; dark sooty-gray to black (often palest at base), round or flattened in cross-section but not chambered. **SPORES** 17.5-20 × 10.5-12.5 microns, oblong-elliptical, smooth, with a large central oil droplet.

HABITAT: Solitary, scattered, or in small groups on ground or rotten wood under both hardwoods and conifers; widely distributed but rare (at least in North America), fruiting mainly in the summer and fall. I have found it only once—in New Mexico under pine. It is also reported from Alaska, Montana, and Washington.

EDIBILITY: Unknown.

COMMENTS: This diminutive elfin saddle is easily recognized by its black, distinctly saddle-shaped cap and non-ribbed, blackish stalk. *H. lacunosa* is similar in color but larger and much more common and it has a fluted and pitted (lacunose) stalk; *H. corium* (see comments under *H. macropus*) is black but has a cuplike cap. *H. atra* has several close relatives with a dark saddle-shaped cap and dark, non-ribbed stalk, including: *H. subglabra* of eastern North America, which has a brownish to gray (never black) cap, and two widespread species with a minutely hairy or downy underside (sterile surface): *H. ephippium,* which is the same size as *H. atra,* but grayish; and *H. pezizoides,* which is slightly larger and dark brown to grayish-brown or blackish. None of these are worth eating.

Helvella maculata   (Fluted Brown Elfin Saddle)

CAP 1.2-6 cm broad and high, saddle-shaped or irregularly lobed at maturity, the margin free from the stalk and at first rolled up (thus obscuring the fertile surface), but unfurling or flaring out with age and sometimes splitting. Fertile surface brown to grayish-brown or buffy-brown, often mottled with darker and lighter shades, smooth or slightly wrinkled. Underside (fertile surface) creamy to yellowish, buff, or grayish, minutely hairy, sometimes with a few ribs extending a little way up from the stalk. Flesh thin, rather brittle. **STALK** 2-10 cm long, 0.5-3 cm thick, equal or tapered above, deeply ribbed and sometimes lacunose (pitted), white to pale buff, sometimes with grayish or brownish stains in age; convoluted or chambered in cross-section. **SPORES** (18) 20-23 × 12-13.5 microns, bluntly elliptical or oblong, smooth, with one central oil droplet.

HABITAT: Solitary to gregarious in mixed woods and under conifers; widespread in western North America. I've seen large fruitings locally in the winter and early spring, but it is not nearly as common as *H. lacunosa*. In many regions it fruits in the summer and fall.

EDIBILITY: Unknown.

COMMENTS: This showy elfin saddle can be identified by its grayish-brown to brown mottled cap and deeply ribbed stalk. It is reminiscent of *H. crispa* and *H. lacunosa*, but is darker than the former and paler than the latter. It might also be confused with *Gyromitra californica*, but is less fragile and lacks the strongly spreading or umbrella-like cap of that species. *H. fusca* is said to be quite similar, but has distinct ribs on the underside of the cap and lacks a flaring margin. I have not seen it.

Helvella maculata. Cap is brown to grayish-brown but never black. Note deeply fluted stalk.

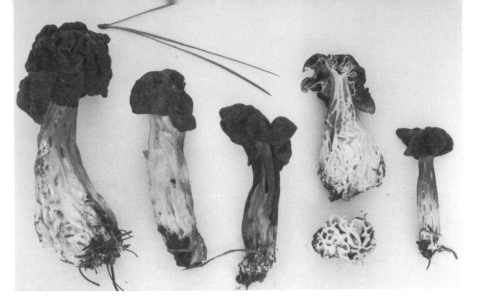

Helvella lacunosa is our most common elfin saddle. The cap is black or gray and usually lobed or wrinkled, while the stalk is white to dark gray, *always* deeply fluted, and usually lacunose (i.e., with visible pits). Note how cross-section of stalk (bottom right) is chambered. In our area this species occurs in many habitats but is especially abundant under pine.

Helvella lacunosa   (Fluted Black Elfin Saddle)

CAP 1-10 cm or more broad and high (but averaging 2-5 cm), saddle-shaped or more often irregularly lobed or convoluted. Fertile surface smooth or wrinkled, black to grayish-black or gray (or rarely whitish); margin typically attached to the stalk at several points or intervals, often incurved (toward the stalk) when young. Underside (sterile surface) grayish to grayish-brown or black (rarely white), smooth (*not* hairy), ribbed or unribbed. Flesh thin, rather brittle. **STALK** 3-15 cm long, (0.5) 1-3 cm thick, equal or tapering upward, deeply ribbed and typically lacunose (i.e., the ribs branching to form elongated holes or pockets); ribs often sharp and double-edged; white to gray or sometimes black (often white when young and darker in age), the base often paler; convoluted and chambered in cross-section. **SPORES** (12) 15-21 × (9) 11-14 microns, broadly elliptical to nearly round to oblong, smooth or slightly roughened at maturity, with one central oil droplet.

HABITAT: Solitary, scattered, or in groups in woods and under trees; very widely distributed, but especially common in western North America. In our area it is usually abundant in the winter and early spring in pine forests or on lawns or roadsides where pines have been planted, but it also occurs with oak and Douglas-fir. In other regions it usually fruits in the late summer and fall.

EDIBILITY: Edible when cooked and rated highly by some people. It can be sauteed, but in my experience is rather chewy and bland. The tough stalks should be cooked separately or discarded. It can also be dried and powdered for use as a seasoning.

COMMENTS: This species is by far the most common of our Helvellas, often appearing in large groups or troops after winter rains. When fresh it is virtually unmistakable because of its gray to black cap and deeply fluted stem, but it is frequently disfigured by a white, moldy-looking parasite *(Hypomyces cervinigenus)*. Such specimens should not be collected for eating. The cap is not brown as in *H. maculata,* and pale individuals can be told from *H. crispa* by the partially attached rather than free cap margin. *H. mitra* is an obsolete synonym. There are several similar species, including: *H. palustris (=H. philonotis)* of eastern North America, a very similar but smaller species that favors damp or swampy places and has a deeply ribbed but not lacunose stalk; *H. sulcata,* which has a

Left: This fluted black elfin saddle *(Helvella lacunosa)* shows its lacunose (pitted) stalk to good advantage. (Rick Kerrigan) **Center:** Another example of *Helvella lacunosa* (others can be seen in photo on p. 815). **Right:** *Helvella crispa,* a common species in eastern North America, but infrequent in California. The pale (white to buff) cap and fluted stalk are its trademarks.

small, pale gray to black cap that is usually saddle-shaped with a deep, well-defined cleft plus a ribbed but not lacunose stalk (it often grows on rotten wood); and *H. phlebophora*, a small, slender, rare species with a more or less convex or umbrella-like cap. All three of the latter species have been reported from the Midwest and/or eastern North America by Nancy Weber, and may occur in the West. I can find no information on their edibility.

Helvella crispa   (Fluted White Elfin Saddle)

CAP 1-5 cm broad and high, saddle-shaped to irregularly lobed at maturity; margin typically at first rolled up, then unfurling and flaring, typically free from the stalk. Fertile surface smooth or slightly wrinkled, white to creamy, buff, pale pinkish-buff, or tinged yellowish. Underside (sterile surface) colored like the cap or slightly darker or grayer, scurfy or minutely hairy. Flesh thin, rather brittle. **STALK** 3-10 cm long, 0.8-3 cm thick, equal or tapering upward, deeply ribbed and often lacunose, white to pinkish-buff or colored like the cap, sometimes darkening slightly in age; convoluted or chambered in cross-section. **SPORES** (14) 17-21 (24) × 10-14 microns, elliptical or oblong, smooth, with a large central oil droplet.

HABITAT: Solitary to gregarious on ground or very rotten wood under both hardwoods and conifers; widely distributed and fairly common (especially in eastern North America), but almost totally supplanted in our area by *H. lacunosa.* I have found it only twice—in October and March under redwood and tanoak. In other areas it fruits in summer and fall.

EDIBILITY: To be avoided. Some sources report it as edible, but it should be tried cautiously if at all—and only after cooking it thoroughly.

COMMENTS: The white to buff cap and deeply ribbed or fluted stalk are characteristic of this beautiful elfin saddle, which is the type species of the genus *Helvella.* The cap is not brown as in *H. maculata,* nor gray or black as in *H. lacunosa.* Pallid specimens of the latter occasionally occur, but can be distinguished by their partially attached rather than free cap margin. Other species: *H. lactea* has a white to creamy or buff cap and a fluted stalk, but the underside of its cap is hairless.

Cup Fungi

spores

PEZIZACEAE & Allies

IN THESE ubiquitous but inconspicuous fungi the fruiting body is usually concave (cup-shaped) to disclike (flat) or cushion-shaped (very slightly convex) and may or may not have a stalk. In a few species it is ear-shaped, top-shaped, cabbage-like, or irregular. The spore-bearing asci line the upper surface of the fruiting body or inner surface of the cup. The spores are shot out of the asci with such terrific force that cup fungi, like other members of the Pezizales, will often "smoke" visibly (spew out clouds of spores) when handled.

The cup fungi are a vast group embracing five families and nearly one hundred genera —or more than twice that number if you count the minute cuplike members of the earth tongue order (Helotiales). Obviously, they can't be covered thoroughly in a book of this kind, and no attempt is made to do so. Most cup fungi are difficult to identify anyway and are too small to be worth eating. A few, such as the Pezizas, can be quite large, but are still unlikely to show up on anyone's list of best edibles.

Cup fungi occur in a wide range of habitats—on dung and manure, in grass, humus, soil, and moss, on wood and foliage, and in recently burned areas. As the characters used to delineate the five families (Pezizaceae, Sarcosomataceae, Sarcoscyphaceae, Pyronemataceae, Ascobolaceae) and numerous genera are almost entirely microscopic, I have rather arbitrarily divided the cup fungi into five groups based on color, size, and texture. These are keyed below.

Key to the Pezizaceae & Allies

1. Fruiting body merely the bowl- or cup-shaped remains of a puffball, usually with traces of spore powder inside (or if not, then usually very dry and light as a straw) (see **Gasteromycetes,** p. 676)
1. Not as above 2
2. Fruiting body beginning as a hollow underground ball, then splitting at the top into several starlike rays; fertile (inner) surface sometimes grayish or whitish but more often pinkish or purplish or lilac; exterior *without* brown hairs .... ***Sarcosphaera,*** p. 825
2. Not with above features (but may have some of them) 3
3. Fruiting body an erect hollow club or closed urn that splits lengthwise from the top to form several rays; found on or near dead hardwoods .... ***Sarcosoma*** **& Allies,** p. 826
3. Not as above 4
4. Fruiting body an irregular mass of cabbage-like, cauliflower-like, brainlike, or contorted tissue, with or without a stalk 5
4. Not as above; fruiting body cuplike, earlike, disclike, etc. (but sometimes contorted, especially if growing in clusters) 6
5. Stalk present, long (15-30 cm), brown, solid; fertile "head" brainlike, pitted, or cabbage-like, usually beige or yellowish-brown; found under hardwoods in eastern North America; rare .... ***Wynnea sparassoides***
5. Stalk absent or rudimentary; widely distributed but not common ... ***Peziza*** **& Allies,** p. 818
6. Fruiting bodies consistently slit down one side, usually erect or semi-erect (often standing on one end like a rabbit's ear); usually medium-sized, *not* huge or minute, often clustered or in contorted masses; *not* growing on manure or dung; tips of asci *not* amyloid . ***Otidea,*** p. 831
6. Not as above; fruiting bodies sometimes slit down one side but not consistently so, and not usually growing erect; sometimes growing on dung 7
7. Fertile (upper or inner) surface of fruiting body brightly colored (red, orange, yellow, blue, or green but not violet) .... ***Aleuria*** **& Allies,** p. 833
7. Fertile surface some shade of brown, black, tan, dingy yellowish, or violet, or sometimes with a pinkish or lilac tinge 8
8. Flesh gelatinous to rubbery-gelatinous, or if not then flesh rather tough (not breaking easily) and fertile surface dark brown to black .... ***Sarcosoma*** **& Allies,** p. 826
8. Not as above; flesh fragile or brittle to slightly rubbery, but usually breaking easily 9

9. Stalk usually present and flesh rather tough; either fertile surface more or less pinkish-brown with a scalloped or stellate margin *or* fertile surface yellow-brown to olive-brown and underside (sterile surface) blackish .... ***Sarcosoma* & Allies,** p. 826
9. Not as above 10

10. Exterior or underside (sterile surface) of fruiting body clothed with brown or black hairs (the hairs sometimes sparse); tips of asci *not* amyloid .... ***Aleuria* & Allies,** p. 833
10. Not as above 11

11. Fruiting body large to fairly small but not minute, usually (0.5) 1-10 cm broad, often flattened (disclike) or spreading at maturity; fertile surface yellow-buff to brown to dark brown, or violet, or occasionally whitish or tinged pinkish; asci with amyloid tips; usually growing solitary, scattered, or in small groups (a few species in clusters) .... ***Peziza* & Allies,** below
11. Fruiting body fairly small to minute, 0.5-2(3) cm broad, often remaining cup-shaped at maturity; fertile surface variously colored but not often dark brown or violet unless very small; sometimes growing in large swarms; tips of asci *not* amyloid .... ***Aleuria* & Allies,** p. 833

PEZIZA & Allies

Small to fairly large fungi found in a wide variety of habitats. FRUITING BODY *typically cup-shaped to flattened (disclike) at maturity,* but in a few cases truffle-like or irregularly contorted. Fertile (upper or inner) surface *typically some shade of yellow-brown, brown, black, or purple* (but sometimes paler or tinged pinkish. Exterior (underside) smooth, scurfy, or roughened by warts but not usually hairy. *Flesh often rather brittle or at least breaking easily.* STALK *absent or if present usually short and continuous with the cup and not normally ribbed.* SPORES elliptical to round, smooth or roughened at maturity, with or without oil droplets. Asci lining upper surface of cup or disc, typically 8-spored, operculate, with amyloid tips.

THIS is the largest and most common genus of cup fungi. It is also the most mundane. The majority of its species are dull or dark colored. A short stem is sometimes present but more often absent. Although most species are cup- or disclike, a few are truffle-like (e.g., *P. ellipsospora*) and one, *P. proteana,* sometimes produces compound fruiting bodies that look like heads of cabbage.

Identification of Pezizas is difficult even with a microscope. Most species grow in soil, humus, or rotten wood, but some have more distinctive milieu. For instance, *P. vesiculosa* grows on dung, *P. violacea* on burnt ground, and *P. domiciliana* in bathrooms, cellars, rugs, and other domestic settings.

Little is known about the edibility of Pezizas, and considering how difficult they are to identify, the entire group is best avoided. Only a few of the many species are described here. Two closely related genera, *Plicaria* and *Pachyella,* are also treated.

Key to Peziza & Allies

1. Fruiting body flattened or spreading, attached to substrate by numerous "roots" (rhizomes); stalk absent; usually under conifers .... (see **Helvellaceae,** p. 796)
1. Not as above; fruiting body if flattened *not* attached by numerous "roots" 2

2. Growing on dung or manure .... ***P. vesiculosa* & others,** p. 823
2. Not as above 3

3. Fruiting body compound, somewhat reminiscent of a head of cabbage (i.e., composed of many spoon-shaped or distorted lobes or "cups"); fertile surface whitish to pinkish, creamy, tan, or lilac-tinged; often but not always growing in burned areas .... ***P. proteana,*** p. 824
3. Not as above (but fruiting bodies can be clustered and somewhat distorted by mutual pressure) 4

4. Growing in charcoal or recently burned areas 5
4. Not as above 8

5. Fertile (upper) surface violet- or lilac-tinged when fresh .... ***P. violacea* & others,** p. 824
5. Fertile surface creamy to pinkish, tan, yellowish, reddish, brown, or black, occasionally with a lilac tinge, but if so then overall color quite pale 6

6. Fertile surface dark brown to black .... *Plicaria endocarpoides* & others, p. 820
6. Fertile surface paler than above 7

7. Fertile surface creamy to pinkish or faintly lilac-tinged; tips of asci amyloid *P. proteana,* p. 824
7. Fertile surface reddish to yellowish to tan to brown, grayish-brown, or sometimes distinctly pink; tips of asci *not* amyloid .... (see *Aleuria* & Allies, p. 833)

8. Growing in bathrooms, on rugs, in cars, and other domestic situations *P. domiciliana,* p. 822
8. Not as above; usually growing outdoors 9

9. Growing underground or under the humus (but sometimes partially exposed); fruiting body round, lobed, or saucer-shaped with a permanently inrolled margin; interior hollow or with canals *or* solid and marbled with veins .... *P. sp.* (unidentified) & others, p. 824
9. Not as above 10

10. Fruiting body developing underground with only the mouth at ground level (like a hole in the ground); interior brown to dark brown; margin often lobed or scalloped at maturity; growing in sand, sandy soil, or sand dunes .... *P. ammophila* (see *Sarcosphaera crassa,* p. 825)
10. Not as above 11

11. Fruiting body shallowly cup-shaped to umbilicate (with a central depression) to flat or even with a downcurved margin; fertile surface usually yellow-brown to tan, but sometimes darker brown or reddish-brown; short stalk often present; nearly always found in the spring when or soon after the snow melts, especially under mountain conifers; mature spores often apiculate, with oil droplets; tips of asci *not* amyloid .... (see *Discina,* p. 797)
11. Not with above features (but may have some of them) 12

12. Fruiting body somewhat rubbery or rubbery-gelatinous, almost the entire underside attached to substrate (not just the center); usually found on wet wood 13
12. Not as above; fruiting body rather brittle and centrally attached, on wood or ground 14

13. Fruiting body small (less than 5 mm broad) *Pachyella babingtonii* (see *Peziza repanda,* p. 821)
13. Fruiting body larger (typically 2-8 cm broad) *Pachyella clypeata* (see *Peziza repanda,* p. 821)

14. Fruiting body small (usually less than 2.5 cm broad) and stalkless, dark brown to nearly black, usually becoming shallowly saucer-shaped or disclike (flat) at maturity *P. brunneoatra,* p. 820
14. Not as above; fruiting body usually larger, or if small then paler or more deeply cup-shaped or with a stalk 15

15. Fruiting bodies usually slit down one side and often clustered, causing them to be somewhat contorted; tips of asci *not* amyloid .... (see *Otidea,* p. 831)
15. Not as above; fruiting bodies often round (spherical) when very young and flattened or spreading in age, or at least not as above 16

16. Fruiting body 1-3 cm broad, brown to grayish-brown, often with a short stalk or narrowed base (especially when young), *not* normally flattening out in age; tips of asci *not* amyloid .... (see *Aleuria* & Allies, p. 833)
16. Not as above 17

17. Flesh staining yellow or exuding a yellowish juice when squeezed or cut .... *P. succosa* (see *P. sylvestris,* p. 821)
17. Not as above 18

18. Fertile (upper or inner) surface grayish to dark grayish-brown to nearly black; base of fruiting body often with obscure ribs .... (see *Helvella,* p. 805)
18. Not as above; fertile surface usually tan to brown, reddish-brown, etc., but not grayish or blackish 19

19. Exterior (underside) of fruiting body brown to reddish-brown even in dry weather .... *P. badia* & others (see *P. sylvestris,* p. 821)
19. Exterior usually white to pale tan, at least toward base (but often brown if wet) 20

20. Fruiting body usually flattening out in age, the fertile surface usually strongly veined, wrinkled, or pebbled (especially toward center); usually found on well-drained or sandy soil (e.g., along streams) in spring; tips of asci *not* amyloid; spores smooth, *without* oil droplets .... (see *Disciotis,* p. 796)
20. Not with above features (but may have some of them); tips of asci amyloid 21

21. Usually found on rotten wood, wood chip mulch, or humus rich in lignin; fruiting body often quite large and spreading (flattened) in age; spores smooth .... *P. repanda,* p. 821
21. Usually found on ground, sometimes on very rotten wood; fruiting body usually medium-sized, the spores often roughened at maturity .... *P. sylvestris* & many others, p. 821

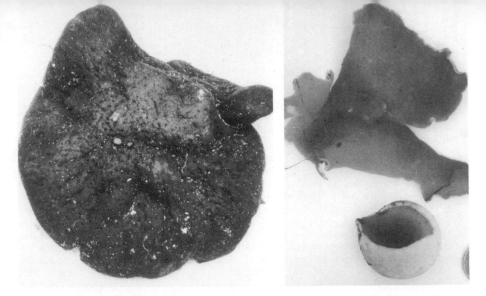

Left: *Plicaria endocarpoides,* a blackish charcoal-loving disclike fungus. **Right:** *Peziza sylvestris.* Young specimen at bottom is cuplike; older one at top resembles a torn piece of discarded rubber.

Plicaria endocarpoides

FRUITING BODY 1-7 cm broad, at first cup-shaped but soon becoming flattened or undulating. Fertile (upper) surface dark brown to black, smooth or roughened, sometimes also wrinkled. Exterior (underside) same color or paler, smooth or roughened. Flesh rather brittle, usually paler or browner than fertile surface. **STALK** absent or rudimentary. **SPORES** 8-10 microns, round, smooth, typically with several small oil droplets.

HABITAT: Solitary to gregarious on burnt ground, old campfire sites, etc.; usually but not always fruiting in the spring; widely distributed.

EDIBILITY: Who knows?

COMMENTS: Also known as *P. leiocarpa,* this charcoal-lover is easily told by its dark brown to black fruiting body that is usually stalkless and flattened at maturity (see photo). *P. trachycarpa* is a very similar but smaller (up to 2.5 cm broad) warted species with larger, roughened spores. *P. carbonaria* should also be mentioned. All three of these grow in ashes, have round spores, and are placed in *Peziza* by some taxonomists.

Peziza brunneoatra

FRUITING BODY 0.5-2.5 cm broad, cup-shaped becoming flattened (disclike) in age. Fertile (upper or inner) surface brown to brownish-black, sometimes with an olive tinge, smooth. Exterior (underside) same color or slightly paler or redder (red-brown). Flesh thin, brownish, brittle. **STALK** absent or present only as a very short, narrowed base. **SPORES** 16-22 × 8-12 microns, elliptical, smooth becoming roughened (warty) or partially reticulate at maturity, with one or two oil droplets.

HABITAT: Scattered to gregarious or clustered on damp soil along roads and paths through the woods, near streams, under trees, etc.; widely distributed but not particularly common. I have found it several times in our area in the winter and spring.

EDIBILITY: Unquestionably inconsequential.

COMMENTS: The small size, dark color, and tendency to become disclike in old age help to identify this species. It is a "cup fungus" in name only, for mature specimens are usually disclike or even slightly convex. *Plectania* species are also quite dark, but are tougher and more deeply cup-shaped, while *Pachyella* species usually grow on wood and *Plicaria* species favor ashes.

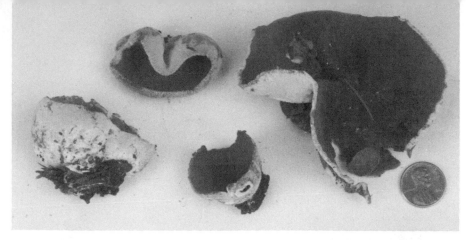

Peziza sylvestris. This common woodland cup fungus resembles many others in its genus. Fertile surface is brown; exterior is whitish or buff, at least when dry. See photo on p. 820 for more specimens.

Peziza sylvestris   (Boring Brown Cup Fungus; Fairy Tub)

FRUITING BODY 2.5-8 cm broad, deeply cup-shaped when young, becoming shallowly cup-shaped or sometimes flattened in age. Fertile (upper or inner) surface brown, smooth or often somewhat wrinkled; margin sometimes wavy or scalloped. Exterior finely warted or roughened to nearly smooth, whitish to pale tan when fresh. Flesh rather fragile. **STALK** absent or rudimentary. **SPORES** 15-20 × 9-10 microns, elliptical, smooth or becoming finely roughened at maturity, without oil droplets.

HABITAT: Solitary, scattered, or in groups or small clusters on ground, usually in woods; widely distributed. It is common in our area, especially in the winter and spring, under both hardwoods and conifers.

EDIBILITY: Unknown.

COMMENTS: This is one of many terrestrial brown Pezizas that are best differentiated microscopically. As a group they can be recognized by their brown color, more or less cup-shaped fruiting body, fragile texture, and growth (usually) on the ground in the woods. *P. sylvestris* seems to be the most common of the lot in coastal California, but may not be so abundant in other regions. Other boring brown cup fungi include: *P. echinospora (=P. pustulata?),* with a pale brown fertile surface and coarsely roughened or pimpled whitish exterior; *P. succosa,* light brown with a whitish exterior and turning yellow (or exuding a yellow juice) when broken; *P. cerea,* a yellowish-buff species; and three species that often have a reddish tinge when fresh (i.e., they range from brown to reddish-brown to brownish-orange): *P. badia* and *P. badioconfusa,* both medium-sized to fairly large with a brown to reddish-brown exterior (the former with partially reticulate spores, the latter with warty spores); and *P. petersii,* a smaller species (up to 4 cm broad) with a smooth exterior and smaller spores (10-12 microns long). All of these Pezizas are cuplike at first and are sometimes called "fairy tubs" because they look like miniature bathtubs when filled with rainwater. In age, however, they can be quite flat or irregular in shape, or in the words of Gary Lincoff, "may resemble discarded torn pieces of rubber" (see photo on p. 820). None are worth eating.

Peziza repanda   (Spreading Brown Cup Fungus)

FRUITING BODY 4-13 cm broad or occasionally larger, cup-shaped when young but expanding to nearly flat or often wavy (undulating), the margin often splitting. Fertile (upper or inner) surface pale brown to medium brown, tan, or in age somewhat darker, smooth to somewhat wrinkled or convoluted at the center. Exterior (underside) pallid. Flesh fairly brittle. **STALK** absent or present only as a short, narrowed base. **SPORES** 14-18 × 8-10 microns, elliptical, smooth, without oil droplets.

821

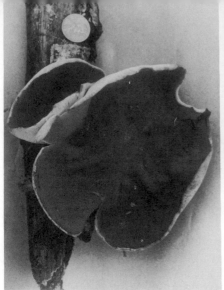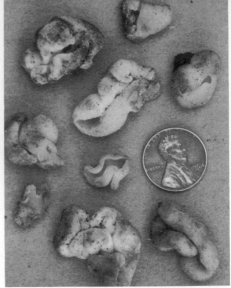

Left: *Peziza repanda*, fairly young specimens which have not yet flattened out; note growth on wood. **Right:** This unidentified *Peziza* (see description on pp. 824-825) is one of several truffle-like species that grow underground or on the surface of the soil beneath the humus. Note irregular shape.

HABITAT: Solitary, gregarious, or in clusters on logs and branches (especially of hardwoods), lignin-rich humus, etc. (often in nurseries where wood chip mulch is used); widely distributed. It is common in our area in the winter but fruits practically year-round.

EDIBILITY: Said to be edible, but easily confused with look-alikes of unknown edibility.

COMMENTS: The salient features of this brown cup fungus are its medium to large size, growth on wood or wood chips, light brown color, and smooth spores. There are several similar species, including: *P. varia,* with minutely roughened spores and a slightly grayer fertile surface, widespread on rotting wood or occasionally in basements; *P. emileia,* with minutely roughened spores and an ochre-brown fertile surface; and *P. badioconfusa* (see comments under *P. sylvestris*), often found on or near rotting conifers. See also *P. domiciliana,* which prefers to grow in domestic situations, and *P. sylvestris,* which usually grows on the ground. Other species: *Pachyella* (formerly *Peziza*) grows on soggy logs or stumps, but has a somewhat gelatinous-rubbery fruiting body that is broadly attached to the substrate (rather than just at the center) and is usually flattened (disclike) or shallowly cuplike. The most common species, *Pachyella clypeata,* grows 2-8 cm broad and has a brown to chestnut-brown fertile surface. *Pachyella babingtonii* is minute (1-5 mm) and cushion-shaped, reddish-brown to purplish-brown, and often somewhat translucent.

Peziza domiciliana   (Domicile Cup Fungus)   Color Plate 207

FRUITING BODY 2-10 cm broad or occasionally larger; cup-shaped when young and often with a short stalk, becoming flattened or wavy in age but often retaining a central depression (i.e., umbilicate); outline frequently irregular or somewhat angular in old age. Fertile (upper) surface at first white or buff, but often darkening in age to tan or brown; smooth to slightly wrinkled. Exterior (underside) same color or paler. Flesh rather fragile, at times slightly waxy, sometimes turning yellow when broken. **STALK** often prominent when young and inconspicuous or absent in age, short (when present) and stout, up to 1 cm long, usually whitish. **SPORES** 11-15 × 6-10 microns, elliptical, smooth to very slightly roughened, at times with two small oil droplets.

HABITAT: Solitary to gregarious or clustered on a wide range of domestic materials: plaster, cement, sand, gravel, coal dust, carpets, fireplace ashes, etc. (It is said to favor strongly alkaline substrates.) Hugo Sloane, Santa Cruz County's incorrigible *enfante* *terrible*-in-residence, has had specimens grow out of the wall above his bathtub—in two different houses! It also grows in cellars, greenhouses, shower stalls, damp closets, under porches, on wet rugs, behind refrigerators, around leaky water beds—and in my car!

EDIBILITY: If it were poisonous we would *probably* know by now, but I can find no specific information on it.

COMMENTS: This white to tan or brown fungus is best identified by its propensity for appearing in unexpected places. Several other cup fungi will occasionally grow indoors (e.g., *P. varia, P. petersii*), but this one makes a habit of it. The color plate shows specimens photographed *in situ*—in the carpeted "romper room" of a nursery school! It has been shown that the fruiting bodies develop quite slowly, taking 3-5 weeks to mature. During this time their shape, size, and color can change considerably.

Peziza vesiculosa   (Common Dung Cup)

FRUITING BODY 2-8 cm broad, at first more or less round (spherical), then opening to become cup-shaped with an inrolled (but often crimped or convoluted) margin. Fertile (inner or upper) surface yellowish to yellow-brown to pale brown or buff, smooth or wrinkled toward the center, sometimes darker brown in age. Exterior (underside) whitish to buff or pale tan, minutely roughened or scurfy. Flesh rather fragile and soft. **STALK** absent or present only as a narrowed basal point of attachment. **SPORES** 18-24 × 10-14 microns, elliptical, smooth, without oil droplets.

HABITAT: Solitary or more often gregarious (sometimes in large clusters) on manure, dung, rotting straw, in corrals, around stables, gardens, and other fertilized areas, etc.; widely distributed and common. In our area it fruits practically year-round, whenever it is damp enough. I have seen massive clusters growing with *Bolbitius vitellinus* in a corral.

EDIBILITY: Not recommended. One source says it is poisonous unless well-cooked.

COMMENTS: This is one of our characteristic dung-inhabiting fungi. The habitat is its most distinctive feature, and when it grows in fertilized soil it may be necessary to examine it microscopically in order to distinguish it from *P. repanda* and other species. *P. fimeti* also grows in dung, but is smaller (up to 2 cm) and dull brown, seldom grows in clusters, and has smaller spores. For even more miniscule "dung cups," see *Cheilymenia coprinaria*.

Peziza vesiculosa, a common dung lover. These are young, almst spherical specimens. As they mature they will open out or even become flat; those growing in clusters will often be distorted.

Peziza violacea   (Violet Cup Fungus)

FRUITING BODY 1-3 (4) cm broad, at first nearly round but soon becoming cup-shaped, then expanding to shallowly cup-shaped or disclike; often somewhat irregular in old age with the margin splitting. Fertile (upper) surface smooth or slightly wrinkled and often depressed at the center, violet to reddish-violet, often darker in age. Exterior (underside) pallid to grayish or tinged violet, delicately powdered at least near margin. Flesh brittle, thin, tinged violet. **STALK** absent or present as a short, narrowed base (especially when young). **SPORES** 16-17 × 8-10 microns, elliptical, smooth, without oil droplets.

HABITAT: Solitary to gregarious on burnt ground (forest fire and campfire sites, etc.); widely distributed, but not very common. In our area this species and *P. praetervisa* (see comments) fruit in the winter and spring; elsewhere I've seen them in spring and summer.

EDIBILITY: Unknown, or at least I can find no information on it.

COMMENTS: This is one of several cup fungi that grow almost exclusively on burnt ground. The violet color is distinctive, although another charcoal-lover, *P. praetervisa*, is also violet (but usually darker: deep purple to purple-brown), with smaller, roughened spores. *P. proteana* also likes ashes, but is paler (white to pinkish). For smaller charcoal-lovers, see comments under *Cheilymenia coprinaria* and *Geopyxis vulcanalis*.

Peziza proteana   (False Sparassis)

FRUITING BODY 1-6 cm broad and cup-shaped to disclike in the typical form, but forming large cabbage-like clumps 10-30 or more cm in diameter in *form sparassoides,* the "cups" in these fruiting bodies much distorted by mutual pressure and usually lopsided, spoon-shaped, or completely misshapen (especially at center of clump). Fertile surface(s) smooth or wrinkled, whitish to tan, often with a pinkish or lilac tinge; margin(s) often wavy. Exterior same color or slightly paler or darker, or often lilac-tinged at or toward the base; often slightly scurfy. Flesh thin, brittle. **STALK** absent or rudimentary. **SPORES** 10-13 × 4.5-7 microns, elliptical, minutely roughened, usually with two small oil droplets.

HABITAT: Solitary to gregarious or clustered on ground in woods, usually in burned areas; widely distributed but not particularly common (form *sparassoides* is rare). I have found it only twice in our area, in the winter and spring.

EDIBILITY: Edible if cooked thoroughly. One book lists form *sparassoides* as "choice."

COMMENTS: The typical form of this cup fungus can be recognized by its pale color and preference for burnt ground (it is not normally as dark or as purple as *P. violacea* and *P. praetervisa*). The cabbage-like form was originally placed in a different genus, but is now thought to be a growth form of *P. proteana* in which the "cups" are "bent out of shape" by overcrowding (wouldn't you be?). This form superficially resembles the cauliflower mushroom *(Sparassis)*, but is differently colored, more brittle in texture, and bears its spores in asci rather than on basidia. It has been suggested that "sparassioid" Pezizas such as *P. proteana* link the hordes of regular cup-shaped species to the rare, convoluted, truffle-like forms (e.g., *P. ellipsospora*).

Peziza sp. (unidentified)   (Truffle-Like Peziza)

FRUITING BODY 0.5-5 or more cm broad, usually buried in ground, but sometimes partly exposed; ranging from simple and cuplike with a strongly inrolled margin to complex and lobed (a loosely joined mass of folded chambers). Exterior smooth or finely hairy, white to creamy when fresh, yellowing when handled and developing ochre to orangish to rusty-yellow stains in age (and old or dried areas often becoming brown or reddish-brown). Interior (fertile surface) sometimes intricately folded, with one to several large chambers which open to the exterior; colored like exterior. Flesh fragile; odor usually sweet in age. **STALK** absent. **SPORES** 10-17 × 9-14 microns, broadly elliptical, minutely roughened at maturity. Asci 8-spored, forming a palisade that lines the interior.

HABITAT: Solitary, scattered, or in groups under oak and other trees, usually growing on the surface of the soil *beneath* the humus layer, but sometimes partially exposed; fairly common in our area in the winter and spring.

EDIBILITY: Unknown, but too fragile to bother with.

COMMENTS: This is one of several truffle-like Pezizas. The internalizing of the fertile surface is an obvious adaptation to growing underground (see chapter on truffles). The above description is based on material collected near Santa Cruz, California, where it is a common species (see photo on p. 822). It is apparently unnamed, although it approaches *P.* (formerly *Hydnotrya*) *ellipsospora*, a sometimes large Californian species described as being "purplish to brown" with slightly larger spores. Other truffle-like Pezizas include: *P. stuntzii,* discovered under conifers in Washington, an aromatic species whose solid interior is marbled with brown veins; and *P. gautierioides,* also found in the Pacific Northwest, but growing quite deep in the soil (up to 6 inches down!), usually in the spring. All of these can be confused with other underground Ascomycetes (e.g., *Hydnotrya*), but their amyloid or amyloid-tipped asci place them in *Peziza.*

SARCOSPHAERA

THIS distinctive genus includes a single widespread species, described below. It usually develops underground as a hollow ball, but when it surfaces the wall usually splits into several lobes at the top to form a crown. In the underground stage it is apt to be mistaken for a truffle, and was once given its own truffle genus, *Caulocarpa*. It is now placed alongside *Peziza* in the Pezizaceae because its asci have amyloid tips.

Sarcosphaera crassa   (Crown Fungus)

FRUITING BODY beginning as a hollow, round to flattened or lobed ball 3-10 cm broad, usually with a "soft spot" or slight depression at the top; wall usually splitting at maturity from this spot downward to form several (usually 6-10) pointed segments or rays which open up part way to form a deep crownlike cup; 5-20 cm broad when expanded. Fertile (inner) surface smooth or breaking into fine scales, usually whitish to grayish at first, but becoming grayish-pink to pinkish, lilac, purple, or purple-brown in age (especially after splitting open). Exterior whitish to creamy, scurfy or roughened, usually dirt-incrusted. Flesh rather thick but brittle, whitish. **STALK** usually (but not always) present as a short, thick, narrowed base up to 3 cm long. **SPORES** 14-22 × 7-9 microns, elliptical with blunt or truncate ends, smooth or very slightly roughened, with 1-3 (usually 2) oil droplets.

Sarcosphaera crassa, rather small specimens. This distinctive fungus begins as an underground ball (right) which is hollow inside (center). Eventually, however, it splits at the top into starlike rays (left).

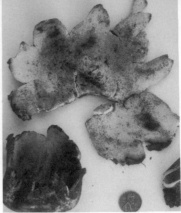

Sarcosphaera crassa. **Left:** Large, young specimens that are just beginning to split into rays. **Right:** Older specimens. Large one at top has flattened out more than is usual. Flesh is thick but fragile.

HABITAT: Solitary to gregarious or in clusters of 2-5 individuals, developing at or below ground level but usually exposed or partly exposed at maturity; widespread, but especially common under western conifers. It is often found in the spring (while morel hunting), but also fruits in the summer and fall. It is common under pine and other conifers in the Sierra Nevada, Cascades, and Rocky Mountains, but I have yet to find it in our area.

EDIBILITY: Not recommended. It is rated highly by some but is difficult to clean and a few people are adversely affected by it. Cook it thoroughly if you decide to try it. O.K. Miller describes the texture as "a little like a rubber eraser that's been softened by time."

COMMENTS: Also known as *S. eximia* and *S. coronaria,* this is a curious and highly distinctive fungus. Specimens which haven't split open are likely to be mistaken for truffles (especially if they're growing underground), but are easily distinguished by their completely hollow interior. Older individuals, on the other hand, might be confused with *Scleroderma geaster,* but can be told by their pinkish to purplish color. Other species: *Peziza ammophila* is somewhat similar in shape when mature, but has a brown to dark brown interior and is cup-shaped when young. It grows in sand dunes or sandy soil around the world, but is not common. *Geopora* species can also be similar, but have fuzzy brown hairs on the exterior and are not pinkish- or purple-tinged. *Neournula pouchetii* has a pinkish fertile surface, but is smaller and always cuplike (see key to *Sarcosoma* & Allies).

SARCOSOMA & Allies

Small to medium-large fungi found on ground or rotten wood. FRUITING BODY *variously shaped but usually cuplike to urnlike or top-shaped.* Fertile (upper or inner) surface *usually dark brown to black,* sometimes paler but not brightly colored. Exterior (sterile surface) often but not always minutely hairy. *Flesh gelatinous or very rubbery, or if not then rather tough* (i.e., not breaking easily). STALK present or absent. SPORES elliptical to round, smooth or roughened, with or without oil droplets. Asci lining upper surface of fruiting body or interior of cup, mostly 8-spored, operculate, with non-amyloid tips.

THESE dark, tough to rubbery or gelatinous cup fungi are not necessarily cup-shaped. The paler, non-gelatinous types can usually be separated from *Peziza* and *Helvella* by their tougher texture. The thickly gelatinous ones are more apt to be mistaken for jelly fungi, but produce their spores in asci and are usually darker or differently shaped.

Sarcosoma, with gelatinous flesh, and *Plectania* and *Urnula,* with tough flesh, are the three principal genera treated here. Along with a few other genera they comprise the family Sarcosomataceae. They are saprophytic on humus and dead wood, and like most Ascomycetes, are especially prevalent in the spring. Although distinctive, they are unpalatable except in an emergency because of their texture. Four species are described here and others are keyed out, including gelatinous relatives of the earth tongues (Helotiales).

Key to Sarcosoma & Allies

1. Fruiting body an erect, hollow club or closed urn that splits lengthwise from the top to form several rays; fertile (inner) surface pallid to yellowish, exterior brown and hairy; found on or near dead hardwoods; southern (fairly common in Texas) .... *Choriactis (=Urnula) geaster*
1. Not as above 2

2. Fruiting body shallowly cup-shaped or ear-shaped or like a piece of seaweed, dark brown to reddish-brown or purplish but not normally black unless it dries out, 2-10 cm broad; flesh *thin* and rubbery or rubbery-gelatinous; sterile surface (exterior) minutely hairy; spores borne on basidia which often (but not always) line the lower surface of the fruiting body ....
   .... (see *Auricularia auricula,* p. 675)
2. Not as above; spores borne in asci which line the upper surface of fruiting body 3

3. Fruiting body rounded and nearly closed becoming shallowly cup-shaped in age; stalk short or absent; fertile (upper or inner) surface brown to reddish-brown; exterior with a dense covering of dark hairs; flesh *thick* and rubbery or gelatinous; found in groups or clusters on hardwood sticks and branches in eastern North America and tropics (one report from California) ....
   .... *Galiella (=Bulgaria) rufa*
3. Not as above 4

4. Fruiting body 1-5 mm broad *or* if larger then shallowly cup-shaped to disclike (flat) at maturity (*not* top-shaped), stalkless, and broadly attached to the substrate (i.e., only the margin free); flesh rather waxy; found on wet logs; asci with amyloid tips ... (see *Peziza* & Allies, p. 818)
4. Not as above 5

5. Fruiting body top-shaped to cup-shaped or irregular and very rubbery or gelatinous; flesh usually gelatinous and sometimes translucent, at least when young and fresh 6
5. Not as above; fruiting body usually cup-shaped or urnlike or occasionally disclike, sometimes with a long stalk; flesh tough to slightly rubbery, but not usually gelatinous 9

6. Growing on hardwoods (usually dead); fruiting body cylindrical to top-shaped to shallowly cup-shaped and usually less than 4 cm broad *or* if larger then fruiting body lobed or irregular (like a jelly fungus) and not black; asci *not* operculate .... (see **Helotiales,** p. 865)
6. Not as above; growing on ground or dead wood, usually under conifers; fruiting body round to cylindrical to top-shaped or shallowly cup-shaped, dark brown to purple-brown to black, 2-12 cm broad when mature; asci operculate (i.e., with "lids") 7

7. Found in eastern North America (rarely in West?); base of fruiting body usually as broad as the top .... *Sarcosoma globosum* (see *S. mexicana,* 828)
7. Found in western North America; base usually tapered (narrower than top) 8

8. Fruiting body 4-12 cm or more broad when mature; flesh *thickly* gelatinous; mature spores with 1-3 oil droplets .... *Sarcosoma mexicana,* p. 828
8. Fruiting body 2-5 (7) cm broad; flesh gelatinous when young but often less so in age; mature spores lacking obvious oil droplets .... *Sarcosoma latahensis* (see *S. mexicana,* p. 828)

9. Stalk present 10
9. Stalk absent or rudimentary 16

10. Fertile (inner or upper) surface more or less pinkish-brown; margin of cup scalloped or lobed; stalk whitish; found in northern North America, usually under conifers; rare ....
    .... *Neournula (=Urnula) pouchetii*
10. Not as above; fertile surface usually brown to black, at least in age 11

11. Fertile surface tan to brown, yellow-brown, or olive-brown (but may blacken in age) 12
11. Fertile surface dark brown to black, even when young 13

12. Exterior (underside of cup) wtih sparse brown hairs; stalk usually ribbed or appearing "gathered" .... (see *Aleuria* & Allies, p. 833)
12. Underside black or nearly so; not as above ... *Plectania melaena* (see *P. nannfeldtii,* p. 830)

13. Fruiting body urn-shaped to deeply cup-shaped, 4-12 cm high and 3-10 cm broad; found mainly on hardwood sticks and logs in eastern North America (also reported from Alaska) ....
    .... *Urnula craterium* & others, p. 829
13. Not as above; fruiting body shallowly cup-shaped or smaller than above; widespread 14

14. Stalk 2-6 cm long, 2-4 mm thick, not ribbed; fruiting under mountain conifers in the spring, when or shortly after the snow melts .... *Plectania nannfeldtii* & others, p. 830
14. Not with above features 15

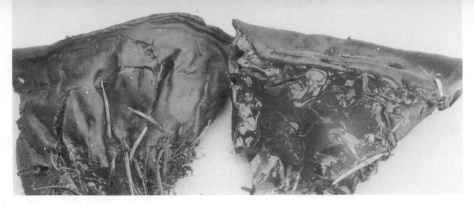

Sarcosoma mexicana. Specimen on right has been sliced open to show the thickly gelatinous interior. Specimen on left shows the wrinkled blackish exterior.

15. Stalk well-developed, 2-5 (7) mm thick; cup fleshy or fragile; exterior of cup *lacking* orangish granules; spores with one large central oil droplet .... (see *Helvella,* p. 805)
15. Not as above; fruiting body tough, not breaking easily; stalk usually present as a narrowed base beneath cup; exterior usually black, *sometimes* with orangish granules ....
.... *Plectania melastoma,* p. 829

16. Fruiting body small and shallowly cup-shaped to disclike, usually less than 1 cm broad; usually found on ground in association with a minute orange cup fungus called *Byssonectria aggregata* .... *Nannfeldtiella aggregata*
16. Not as above 17

17. Found on burnt ground or in dung, or if not, then flesh fragile 18
17. Not as above; flesh usually tough .... *Plectania melastoma* & others, p. 829

18. Fruiting body small to minute; found on dung or burnt ground, sometimes in swarms ......
.... (see *Aleuria* & Allies, p. 833)
18. Fruiting body usually larger than 1 cm broad; found on ground or in scorched areas but not in dung .... (see *Peziza* & Allies, p. 818)

Sarcosoma mexicana   (Starving Man's Licorice; Giant Gel Cup)

FRUITING BODY 5-10 (20) cm broad and 3-10 (15) cm high, rubbery, sometimes cup-shaped but more often more or less top-shaped (i.e., with a narrowed base and broader, flattened to slightly concave "cap"); margin often lobed. Fertile (upper) surface black or sometimes dark brown. Exterior dark gray to black, finely velvety (especially above), usually narrowed below to form a thick "stalk" which is often deeply wrinkled or ribbed or has large pockets. Flesh (interior of fruiting body) a thick, watery-gelatinous mass, clear gray to black or brownish. **STALK** usually present as a narrowed base (see above). **SPORES** 23-34 × 10-14 microns, elliptical to somewhat sausage-shaped, smooth, with one to three oil droplets.

HABITAT: Solitary to gregarious or clustered on rotting wood or duff under conifers; fruiting in the late winter, spring, summer, and early fall. It is known only from western North America and Mexico, and has been characterized as "rare." However, it is sometimes very common (along with *S. latahensis*—see comments) in the mountains of Oregon and northern California during the morel season (May and June) or shortly after the snow melts. I have not seen it in our area, but *S. latahensis* occurs occasionally.

EDIBILITY: Unknown. As its common name implies, you would really have to be hungry to be tempted by it.

COMMENTS: Originally known as ***Bulgaria mexicana,*** this is one of my fifty "five favorite fleshy fungal fructifications." Its black color, thickly gelatinous flesh (best seen by slicing it open lengthwise as shown in above photo), and growth with conifers make it virtually unmistakable. It is much larger than "Poor Man's Licorice" *(Bulgaria inquinans),* and is associated with conifers rather than hardwoods (at least in my experience). *S. lata-hensis* is a similar but slightly smaller (2-7.5 cm broad), purple-brown to black species

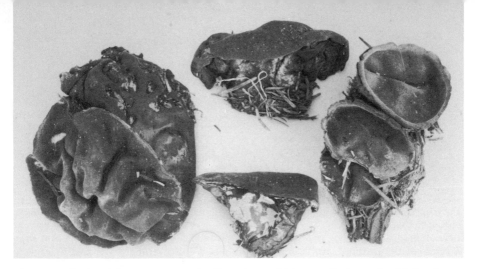

Sarcosoma latahensis (see comments under *S. mexicana,* which resembles it very closely). Specimens on right and left are being viewed from the top; those in center, from the side (the one at the bottom has been sliced open to show the gelatinous flesh).

that is common under western conifers, particularly at higher elevations. Its flesh is gelatinous when young but often becomes tougher and less gelatinous in age, plus its spores lack oil droplets at maturity. Another species, *S. globosum,* may occur in the West but is much more common in eastern North America. It is massive, black, and gelatinous like *S. mexicana,* but its water-and-gel-filled base is very broad (not tapered) and may "leak" when collected!

Urnula craterium   (Crater Cup; Devil's Urn)

FRUITING BODY 4-12 cm high and 3-6 (10) cm broad, usually urn-shaped (i.e., with a narrowed base or stalk); upper portion at first closed, then opening to form a deep cup or "urn"; margin remaining incurved for some time, usually scalloped, sometimes torn at maturity. Fertile (inner) surface smooth or scurfy, dark brown to black. Exterior variable in color: dull pinkish-gray to dark brown and scurfy at first, often smoother and blacker in age. Flesh tough and fibrous or leathery, black or very dark. **STALK** 2-4 cm long, 5-10 mm thick, continuous with and colored like exterior of cup, or darker; usually with a patch of dark brown to black mycelial hairs emanating from the base. **SPORES** 24-36 × 10-15 microns, elliptical to spindle-shaped, smooth.

HABITAT: Solitary or more often in groups or clusters on or near rotting hardwood sticks and logs (the wood often buried); common in eastern North America in the spring.

EDIBILITY: Too tough to be worth eating.

COMMENTS: This dark brown to black urn-shaped fungus is one of the first fungi to appear each spring in the hardwood forests of eastern North America. It differs from *Plectania melastoma* by its deeper cup and better developed stalk, but the two species might just as well be merged into one genus. I can find no record of it from the west coast, but a similar species, *U. hiemalis,* has been found in Alberta and Alaska. It might conceivably be mistaken for a *Craterellus,* but the growth in the spring plus the fertile upper or inner (rather than lower or outer) surface and slightly different shape distinguish it.

Plectania melastoma   (Black Cup Fungus)

FRUITING BODY 1-2.5 cm broad and 1-3 (4) cm high; at first nearly round (or with a stemlike base), then opening slowly at the top and eventually becoming more or less cup-shaped; margin usually remaining incurved for a long time but sometimes splitting in places. Fertile (upper or inner) surface smooth or sometimes with veinlike markings when

Plectania melastoma. This small black cup fungus can be recognized by its color, tough texture, and frequently wrinkled underside. Note how young specimens (right) are nearly closed at the top.

dry, often glistening when wet, black or deep brownish-black. Exterior tough (but with a semi-gelatinous inner layer when wet), minutely hairy, strongly wrinkled or veined, black, but often with a rusty-orange tinge near the margin from the presence of minute orange granules. Flesh tough or cartilaginous, not brittle. **STALK** absent or present as a short, stout, narrowed base; continuous with and colored like exterior of cup, the base often with wiry black mycelial threads that extend into the substrate. **SPORES** 20-28 × 8-12 microns, elliptical or spindle-shaped, smooth, with oil droplets when immature.

HABITAT: Solitary or more often in small groups or clusters on decaying sticks and other debris of both hardwoods and conifers; widely distributed and not uncommon, but easily overlooked, fruiting mainly in the spring. In our area this species or something very similar (see comments) occurs under oak in the late winter and early spring. I usually find it when I'm foraging for *Craterellus cornucopioides,* perhaps because both are black.

EDIBILITY: Unknown.

COMMENTS: The black color, wrinkled exterior (or underside), small size, and tough texture are the principal fieldmarks of this attractive cup fungus. The typical form with rusty-orange granules usually grows under conifers. Our local oak-loving version usually lacks visible granules, but orange granules can be seen under the microscope. *Bulgaria melastoma* is an older name for it. Two very similar black cup fungi with little or no stalk and no orange granules are also worth mentioning: *P. milleri* has a stellate margin (i.e., with starlike points) and elliptical spores, while *P. (=Pseudoplectania) nigrella* has round spores. Both usually occur under conifers. See also *Urnula craterium,* which has a larger, deeper (urnlike) fruiting body.

Plectania nannfeldtii   (Black Snowbank Cup Fungus)

FRUITING BODY consisting of a shallow cup mounted on a well-developed, slender stalk. Cup 0.5-2 (3) cm broad, the margin at first incurved. Fertile (inner or upper) surface smooth, black. Exterior also blackish but delicately hairy. Flesh black, rather tough or cartilaginous. **STALK** always present and typically rather long and slender, 2-6 cm long, 2-4 mm thick, more or less equal, not ribbed, black, often with black mycelium at base. **SPORES** 21-30 (35) × 10-14 microns, elliptical, smooth or slightly roughened, without oil droplets.

HABITAT: Solitary to gregarious in duff and debris or on buried or rotten wood under conifers (particularly fir and spruce); fruiting in the spring shortly after the snow melts or even developing under the snow. It is known only from western North America and is fairly common at higher elevations (but easily overlooked). I have not seen it in our area.

EDIBILITY: Unknown.

COMMENTS: This attractive black cup fungus can be distinguished in the field by its color, small size, well-developed stalk, and growth in the spring under conifers. *Helvella*

Plectania nannfeldtii. This small black cup fungus has a long thin stalk and often grows near snow.

corium (see comments under *H. macropus*) is somewhat similar but larger, not as tough, and often shows ribs on the stalk. *P. (=Pseudoplectania) melaena* also grows in the spring under conifers, but is slightly larger and has a yellow-brown to olive-brown cap or cup when young (but blackens in age) and a short or long stalk. For species with little or no stalk, see *Plectania melastoma.*

OTIDEA

Medium-sized, mostly terrestrial fungi. FRUITING BODY *usually erect or semi-erect* (often standing on one end), *usually lopsided and open or slit on one side, sometimes shaped like a rabbit's ear, at other times more cuplike.* Fertile (inner) surface variously colored, usually smooth. Exterior smooth or scurfy but not usually hairy. *Flesh usually rather brittle, not gelatinous.* STALK absent or present as a short, thick base. SPORES elliptical, smooth, usually with two oil droplets. Asci lining inside or upper surface of fruiting body, typically 8-spored, operculate, tips *not* amyloid.

THIS is a small but common group of lopsided or earlike cup cungi. The fruiting bodies are usually erect (i.e., they stand on end) and slit down one side, and often occur in groups or contorted clusters. Many other cup fungi can be open on one side or lopsided, but *Otidea* is the only common genus that is *consistently* slit or lopsided and erect or semi-erect.

The dozen or so North American Otideas are differentiated primarily on microscopic characters, but can be divided into two groups based on their shape. One group has an elongated, erect, rabbit-ear-like fruiting body; the other is more open and cuplike and often truncate (chopped off) at the top. To avoid otidealogical debate, only one species from each group is described here. *Otidea* belongs to the Pyronemataceae.

Key to Otidea

1. Fruiting bodies arising in groups or clusters from an underground "tuber" (sclerotium), shaped more or less like rabbit ears, 5-15 cm tall; fertile (inner) surface orange to pinkish to reddish-brown, often blackening in age; exterior dark brown to blackish; found under hardwoods in eastern North America, usually in the summer; rare, but often occurring in spectacular numbers when it fruits .... ***Wynnea americana***
1. Not as above; fruiting bodies not arising from an underground "tuber" 2

2. Fruiting body bright yellow to orange but usually developing dark bluish or greenish stains; found mainly in spring under northern and mountain conifers (see *Caloscypha fulgens,* p. 837)
2. Not as above 3

3. Fertile (inner) surface of fruiting body pale yellow to orange, ochraceous, or pinkish 4
3. Fertile surface predominantly tan to brown, sometimes with brighter areas 7

4. Fruiting body pale to bright yellow, usually with a truncate (broad, flattened) apex ....
.... *O. concinna* & *O. cantharella* (see *O. onotica*, below)
4. Not as above 5

5. Fertile surface *bright* orange; underside usually whitish; fruiting bodies only sometimes lop-sided or slit down one side (particularly when clustered) ..... (see *Aleuria* & Allies, p. 833)
5. Not as above 6

6. Fertile surface dull orange to yellowish or tinged pinkish .... *O. onotica,* below
6. Fertile surface yellowish, the exterior usually brownish .. *O. leporina* (see *O. onotica*, below)

7. Fruiting bodies typically elongated and erect (standing on end like rabbit ears) 8
7. Fruiting bodies typically broadened and flattened somewhat on top (truncate) or cuplike or distorted (especially when clustered); usually growing semi-erect 9

8. Fertile surface yellowish-brown .... *O. leporina* (see *O. onotica*, below)
8. Fertile surface usually deep reddish-brown to vinaceous-brown ....
.... *O. smithii* & *Wynnella silvicola* (see *O. alutacea*, below)

9. Fruiting bodies often nearly round (spherical) when young and flattened in age, usually solitary or in small groups but not often clustered; tips of asci amyloid . (see *Peziza* & Allies, p. 818)
9. Not as above; fruiting bodies often clustered and distorted by mutual pressure, not normally flattening out in age nor spherical when young; asci *not* amyloid *O. alutacea* & others, below

Otidea alutacea   (Brown Clustered Ear Cup)

FRUITING BODY 2-6 cm high and 2-4 cm broad, shape variable: usually cup-shaped but lop-sided (the shorter side split lengthwise or open) and semi-erect, but often irregularly wavy or contorted when growing in clusters; apex often truncate when growing erect. Fertile surface (interior) smooth, tan or light brown to grayish-brown or brown. Exterior often slightly scurfy, pale to dull brown (or yellowish in one variety). Flesh brittle. **STALK** absent or present as a narrowed, whitish, downy base. **SPORES** 14-16 × 7-9 microns (or smaller in one variety), elliptical, smooth, typically with two oil droplets.

HABITAT: Scattered to densely clustered in forest humus, usually under conifers; fairly common in western North America. In our area this species and its look-alikes (see comments) is frequent in the fall and winter, particularly under Douglas-fir and oak.

EDIBILITY: Unknown—better chucked than plucked.

COMMENTS: The brownish color, lopsided fruiting bodies that are often contorted, broadened at the apex, and split down one side, plus the tendency to grow in dense clusters are the distinguishing features of this undistinguished fungus. There are a number of similar brownish Otideas, including: *O. bufonia,* dark brown, usually growing in clusters; *O. rainierensis,* found in the Pacific Northwest and normally not clustered; *O. abietina,* medium to dark brown and often cup-shaped, but with much larger spores (18-22 microns long); and *O. grandis,* brown but frequently with reddish-orange patches on the fertile surface and a thicker, more prominent stalk. There are also several brown Otideas with an elongated, erect, spoon-shaped to earlike (not truncate or cup-shaped) fruiting body like that of *O. onotica.* These include: *O. smithii,* deep reddish-brown to vinaceous-brown and fairly large (up to 8 cm high), with a thick stalklike base, fairly common under conifers in the Pacific Northwest and northern California; and *Wynnella silvicola (=O. auricula),* a northern species that is colored like *O. smithii* but has larger spores (22-25 microns long).

Otidea onotica   (Donkey Ears)

FRUITING BODY (3) 5-10 cm high and (2) 4-6 cm broad (more if completely expanded); shape variable but usually spoon-shaped or like an elongated ear (standing erect on one end with one side open or slit); margin at first inrolled, but expanding somewhat in age. Fertile surface (interior) smooth, ochraceous to dull orange to orange-buff or yellowish, often with a pinkish or rosy tint when fresh. Exterior similarly colored but without a

Left: *Otidea alutacea,* rather pale (tan) specimens. Note rather erect growth habit. **Right:** *Otidea onotica.* Note erect growth habit and also how the fruiting body is slit down to the base on one side.

pinkish tinge, often slightly scurfy. Flesh thin, pallid, brittle. **STALK** present as a whitish, narrowed, hairy or downy base that arises from a litter-binding mycelium. **SPORES** 12-14 × 6-8 microns, elliptical, smooth, with two oil droplets. Paraphyses (sterile cells) strongly hooked.

HABITAT: Scattered or more often in groups or clusters under both hardwoods and conifers; widely distributed. In our area it fruits in the winter and spring, but is not as numerous as *O. alutacea.*

EDIBILITY: Edible according to some, but one study revealed the presence of the toxin MMH. In other words, it is better chucked than plucked.

COMMENTS: The erect growth habit and earlike fruiting body plus the ochraceous or orangish to pinkish-tinged fertile surface separate this species from most other cup fungi. The apex of the fruiting body is not broadly flattened or truncate as is typical of *O. alutacea.* Other species: *O. leporina* ("Rabbit Ears") has a yellowish-brown interior and brownish exterior, but is otherwise quite similar. *O. concinna* and *O. cantharella* are both pale to bright yellow, but often have a broadened or truncate apex. All of these species are widely distributed. For duller or browner Otideas, see comments under *O. alutacea.*

ALEURIA & Allies

Mostly small to minute fungi found on dung, soil, moss, wood, foliage, and ashes. FRUITING BODY usually cup-shaped to disclike (flattened) or cushion-shaped, often gregarious. Fertile (upper) surface *often brightly colored,* usually smooth. Sterile surface (underside) *often hairy,* but sometimes bald. *Flesh usually fragile, but sometimes rather tough.* STALK absent or present. SPORES round to elliptical or elongated, smooth or roughened, with or without oil droplets, thick-walled when immature in Ascobolaceae, otherwise thin-walled. Asci lining upper surface of fruiting body, typically 8-spored, operculate, *not* amyloid; thin-walled in Pyronemataceae and Ascobolaceae, thick-walled in Sarcoscyphaceae.

THIS is a motley multitude of miniscule to medium-sized cup- and disclike fungi with non-amyloid, operculate asci and fragile to fairly tough but not gelatinous flesh. In contrast to *Peziza,* many of the species are brightly colored, and those that aren't are usually small. A few, such as the common orange peel fungus, *Aleuria aurantia,* are conspicuous, but most are so minute or mundane that only the most avid devotees of diminutiveness (see p. 224) will notice them.

833

These cup fungi, like others, grow in a wide range of habitats. Many digest dung; others occur in swarms on scorched earth and were among the first organisms to colonize the ash-covered wasteland around Mount St. Helens after that volcano erupted. Space permits descriptions of only a few species, nearly all of which belong to the Pyronemataceae (not to be confused with the Pyrenomycetes or flask fungi). Exceptions are the scarlet cup fungus, *Sarcoscypha coccinea,* and its relatives, which are placed in a separate family, the Sarcoscyphaceae, because of their tougher texture and a number of microscopic differences. A third family, the Ascobolaceae, is only briefly mentioned here. Most of its constituents are minute, dark-spored dung addicts of interest only to scatologists, asco-mycologists, students of Fungi 112B, and other assorted eccentrics who deal with dung on a daily basis.

Key to Aleuria & Allies

1. Growing on dung or manure 2
1. Not as above 3
2. Fruiting body typically at least 1 cm broad when mature, nearly round (spherical) when young becoming cup-shaped to nearly flat in age, yellowish to brown, without prominent hairs; asci with amyloid tips .... (see *Peziza* & Allies, p. 818)
2. Not as above; fruiting body small or minute, often with hairs ....
.... *Cheilymenia coprinaria* & many others, p. 838
3. Exterior (underside) of cup or disc clothed with brown to black hairs; margin often fringed with dark hairs also 4
3. Not as above; exterior either hairless or with white or pale hairs 12
4. Most or all of fruiting body immersed in the ground with only the top (mouth) showing, making it look like a hole in the ground .... (see *Geopora,* p. 846)
4. Not as above 5
5. Fertile (upper or inner) surface yellow, orange, or red 6
5. Fertile surface white to creamy, tan, gray, brown, etc. 9
6. Fruiting body small or minute, growing on burnt soil or charred wood ....
.... *Anthracobia melaloma* & others (see *Cheilymenia coprinaria,* p. 838)
6. Not as above; growing in soil, humus, or on wood but not usually in burned areas 7
7. Fruiting body tough or corky and thick-fleshed; fertile surface orange to red or sometimes yellowish; exterior dark brown to black; found in eastern North America *Wolfina aurantiopsis*
7. Not as above 8
8. Underside of fruiting body with minute brown hairs; fertile surface bright orange ....
.... *Melastiza chateri* (see *Scutellinia scutellata,* p. 839)
8. Underside of fruiting body with fairly obvious hairs which fringe the margin like eyelashes; fertile surface bright red to orange-red or orange .... *Scutellinia scutellata* & others, p. 839
9. Fertile (upper or inner) surface white to grayish; stalk absent or rudimentary 10
9. Fertile surface white to grayish, brownish, or darker; stalk usually present (but often short) 11
10. Fruiting body typically 1-3 cm broad .... *Humaria hemispherica,* p. 839
10. Fruiting body typically less than 1 cm broad ....
.... *Trichophaea boudieri* & others (see *Humaria hemispherica,* p. 839)
11. Stalk ribbed, often short; fertile surface white to tan or brownish but not gray ....
.... *Jafnea semitosta* (see *Humaria hemispherica,* p. 839)
11. Stalk not ribbed, or if so then fertile surface grayish or blackish .... (see *Helvella,* p. 805)
12. Growing on recently fallen branches or foliage of conifers; fruiting body small or miniscule, stalkless, often brightly colored; asci operculate ....
.... *Pithya vulgaris* & others (see *Sarcoscypha coccinea,* p. 836)
12. Not as above 13
13. Fertile (upper or inner) surface bright red to scarlet; exterior usually whitish or with white hairs; usually growing on wood or buried sticks 14
13. Not as above 16
14. Fruiting body very small (less than 1 cm broad); exterior of cup with long hairs 15
14. Fruiting body larger and/ or hairs on exterior short .. *Sarcoscypha coccinea* & others, p. 836

15. Fruiting body arising from an elongated rootlike structure (several often arising together from the same "root") . . . . *Microstoma protacta* (see *Sarcoscypha coccinea*, p. 836)
15. Not as above . . . . *Microstoma floccosa* (see *Sarcoscypha coccinea*, p. 836)

16. Fruiting body yellow to orange but soon developing dark bluish to greenish or olive stains; found under northern and mountain conifers in spring and early summer *Caloscypha fulgens*, p. 837
16. Not as above (but fruiting body may be yellow or orange) 17

17. Fruiting body bright yellow, minute (less than 5 mm broad), occurring in swarms on wood; asci inoperculate (without "lids") . . . . (see **Helotiales,** p. 865)
17. Not as above 18

18. Fruiting body minute (1-6 mm broad) and disclike to cushion-shaped; growing mainly in burned areas, often in swarms or masses 19
18. Not as above; either larger or distinctly cuplike or consistently growing in other habitats . 20

19. Fertile (upper or inner) surface orange to yellow-orange to reddish . . . . *Pyronema omphalodes* & others (see *Cheilymenia coprinaria*, p. 838)
19. Not as above; fertile surface usually darker . . . . *Ascobolus carbonarius* & others (see *Cheilymenia coprinaria*, p. 838)

20. Fertile (upper or inner) surface *bright* orange to yellow-orange 21
20. Not as above (but fertile surface may have a dull yellowish or pale orange tinge) 23

21. Stalk absent or rudimentary; very common . . . . *Aleuria aurantia*, p. 837
21. Stalk present, at least in many specimens 22

22. Fertile surface bright orange to bright yellow-orange . . . . . *Aleuria rhenana* & others, p. 836
22. Fertile surface pale orange to pale or dingy yellowish . . . . *Geopyxis vulcanalis*, p. 840

23. Growing in recently burned areas 24
23. Not as above 26

24. Fertile surface pink to reddish; stalk absent or rudimentary . . . . *Tarzetta rosea* (see *Geopyxis vulcanalis*, p. 840)
24. Not as above; fertile surface differently colored (including brick-red) 25

25. Fertile surface more or less brick-red . . . . *Geopyxis carbonaria* (see *G. vulcanalis*, p. 840)
25. Not as above 26

26. Fruiting body turquoise to blue-green or at least tinged those colors; found on wood . . . . (see **Helotiales,** p. 865)
26. Not as above; differently colored 27

27. Fruiting body with a stalk that arises from a swollen tuberlike structure (sclerotium) immersed in the substrate; asci inoperculate . . . . (see **Helotiales,** p. 865)
27. Not as above 28

28. Fruiting body very small (typically less than 7 mm broad); stalk when present very thin; often found on living plant parts (leaves, stems, etc.); asci inoperculate . . . (see **Helotiales,** p. 865)
28. Not as above; usually found on ground or dead wood; asci operculate (with "lids") 29

29. Fruiting body tough or leathery, the flesh very thin; exterior usually with white or grayish hairs; common on dead hardwood sticks, branches, etc; spores borne on basidia . . . . (see **Stereaceae & Allies,** p. 604)
29. Not as above 30

30. Fertile surface pale orange to pale or dingy yellowish; fruiting body usually less than 1.5 cm broad . . . . *Geopyxis vulcanalis* & others, p. 840
30. Not as above; fertile surface grayish-tan to tan or brown and/or fruiting body larger 31

31. Stalk present (at least in most specimens) 32
31. Stalk absent or rudimentary 34

32. Exterior of cup and upper stalk usually hairy or densely scurfy when fresh; fertile surface often dark; stalk well-developed though sometimes short . . . . (see *Helvella*, p. 805)
32. Not as above 33

33. Fruiting body 1-3 cm broad . . . . *Tarzetta catinus* (see *Geopyxis vulcanalis*, p. 840)
33. Fruiting body typically 1.5 cm broad or less *Tarzetta cupularis* (see *Geopyxis vulcanalis*, p. 840)

34. Fertile surface dark brown to blackish; fruiting body usually saucer-shaped or disclike (flat) in age; tips of asci amyloid . . . . (see *Peziza* & **Allies,** p. 818)
34. Fertile surface brown to yellowish; tips of asci not amyloid . . . . *Tarzetta bronca* (see *Geopyxis vulcanalis*, p. 840)

Sarcoscypha coccinea   (Scarlet Cup Fungus)   Color Plate 210

FRUITING BODY 2-5 (6) cm broad when mature, more or less cup-shaped, the margin usually incurved, often tattered in old age. Fertile (inner or upper) surface bright red to scarlet, sometimes fading to reddish-orange in age, smooth. Exterior whitish, covered with minute hairs. Flesh thin but not particularly brittle. **STALK** absent or more often present, up to 4 cm long, 3-7 mm thick; minutely hairy and white, tapered downward. **SPORES** 24-40 × 10-14 microns, elliptical, smooth.

HABITAT: Solitary or in groups on buried or fallen hardwood sticks or branches; widely distributed and fairly common in the winter and early spring (or late fall in some regions). In our area I have found it in abundance in a riparian woodland composed of willow, alder, buckeye, and cottonwood.

EDIBILITY: Said to be edible; I haven't tried it.

COMMENTS: Formerly known as *Plectania coccinea,* this beautiful cold weather cup fungus is easily told by its bright red fertile surface. The margin of the cup is not fringed with dark hairs as in *Scutellinia scutellata,* and the exterior or underside is whitish. The length of the stem seems to depend partly on how deep its food source (stick) is buried. Other species: *S. occidentalis* is a similar but smaller (up to 1.5 cm broad) eastern species with a well-developed (1-4 cm long) stalk and smaller spores. *Microstoma (=Anthopeziza, Sarcoscypha) floccosa* is a minute (up to 1.5 cm high and 1 cm broad) bright red species whose exterior is clothed with *long* white hairs. It is fairly common on downed sticks in eastern North America, but I have not seen it in the West. *Microstoma protacta (=Plectania hiemalis)* is somewhat similar to *M. floccosa,* but its stalk arises from a hard, elongated rootlike structure and is often branched above, giving rise to up to a dozen bright red cups. It is widely distributed but rather rare. Several closely related genera of bright red to orange or pink cup fungi occur in the tropics, including *Cookeina* and *Phillipsia,* with striate or banded spores. Finally, there are several small stalkless, disclike species that grow on the recently fallen branches or foliage of conifers. These species include: *Pithya vulgaris,* yellow to orange or reddish-orange, semi-gelatinous, and up to 1 cm broad, usually found in the spring on branches of fir and other conifers; *P. cupressina,* similar but not gelatinous, found on cedar or cypress; and *Pseudopithyella miniscula,* a miniscule (1-2 mm) scarlet species.

Aleuria rhenana   (Stalked Orange Peel Fungus)   Color Plate 209

FRUITING BODY 1-2 cm broad, cup-shaped with a stalk. Fertile (inner or upper) surface bright orange to yellow-orange, smooth. Exterior white or whitish and minutely hairy or downy. Flesh thin, brittle. **STALK** present, 1-3 cm long, 2-5 mm thick, slender, equal or tapered downward, colored like the exterior; base often arising from a dense mass of white mycelium that may bind several stalks together. **SPORES** 20-23 × 11-13 microns, ellip-

Aleuria rhenana is a small orange or yellow-orange cup fungus with a well-developed stalk. It often grows in clusters, as shown here (at left) and in the color plate.

tical, coarsely reticulate at maturity.

HABITAT: Gregarious, often in small clusters, on ground or moss in woods (usually under conifers); widely distributed but infrequent. I have found it only twice—in Mt. Rainier National Park in Washington, in September, and near San Francisco in January.

EDIBILITY: Presumably consumable, but much too small and rare to be of value.

COMMENTS: This petite cup fungus is the same color as its cosmopolitan cousin, *A. aurantia,* but is much rarer and usually smaller, possesses a stalk, and likes to grow in dainty clusters (see color plate). *Leucoscypha (=Neottiella, Aleuria) rutilans* is a similar moss-inhabiting species with slightly larger spores and longer hairs on its underside.

Aleuria aurantia   (Orange Peel Fungus)   **Color Plate 208**

FRUITING BODY 1-10 cm broad, sometimes nearly round at first but soon becoming cup-shaped to saucer-shaped to flattened or wavy, or sometimes irregularly contorted (especially if clustered). Fertile (upper or inner) surface bright orange to golden-orange, fading somewhat in age, more or less smooth; margin often wavy or lobed. Exterior (underside) pallid or at least paler, smooth or minutely downy. Flesh thin, brittle or fragile. **STALK** absent or rudimentary. **SPORES** 18-24 × 9-11 microns, elliptical, coarsely reticulate or ridged at maturity, typically with two oil droplets.

HABITAT: Scattered to gregarious or in fused clusters on ground, fruiting mainly in the fall and winter in our area; widely distributed and very common. It seems to prefer bare soil or sand along roads, paths, landslides, etc., but also grows in grass or moss.

EDIBILITY: Edible and highly rated by one authority, but bland according to others. One of my colleagues uses it raw in salads, but it is so thin-fleshed and fragile that it hardly seems worth the trouble to collect it.

COMMENTS: The orange peel fungus is most likely to be mistaken for one of the old orange peels that frequently litter our woods and roadsides. It is much more fragile, however, and less common. Its size and shape vary considerably depending on environmental conditions, but the "aleuring" bright orange color and absence of a stalk are constant. Some species of *Otidea* are orangish, but have a more erect rather than prostrate growth habit, while *A. rhenana* is smaller and has a stalk. A variety of *A. aurantia* with smaller spores (13-15 microns long) occurs in our area. See also *Melastiza chateri* (under *Scutellinia scutellata*), a somewhat similar but smaller species with brown hairs on its exterior.

Caloscypha fulgens   **Color Plate 211**
(Snowbank Orange Peel Fungus)

FRUITING BODY 1-4 (6) cm broad, sometimes nearly spherical when young but becoming cup-shaped or flatter in age, sometimes slit down one side and appearing lopsided; margin inrolled when young. Fertile (upper or inner) surface smooth or slightly wrinkled, bright yellow-orange to orange, sometimes with dark bluish to olive-green stains. Exterior hairless, colored like interior but usually with more pronounced blue or greenish stains, especially toward margin. Flesh thin, brittle. **STALK** absent or present only as a short, narrowed whitish base. **SPORES** 6-8 microns, round, smooth.

HABITAT: Scattered to gregarious or clustered in damp soil or duff under conifers, fruiting in the spring and early summer shortly after the snow melts; widely distributed, but especially common in the mountains of western North America. It is one of the characteristic spring mushrooms of the Sierra Nevada, Cascades, and Rocky Mountains.

EDIBILITY: I can find no information on it.

COMMENTS: The yellow-orange color and dark blue or greenish stains that make it look like a moldy orange peel are the hallmarks of this springtime cup fungus. The mycelium apparently parasitizes the seeds of conifers (mainly spruce and fir) and clusters of fruiting bodies often arise where squirrels stash their seed-containing cones. An albino form of this species with bluish stains has been found in Idaho.

Cheilymenia coprinaria   (Eyelash Dung Cup)

FRUITING BODY (1) 3-7 (10) mm broad, at first closed but soon opening to become shallowly cup-shaped to disclike or somewhat cushion-shaped. Fertile (upper) surface orange to pale orange becoming yellow or brownish in age, smooth; margin fringed with minute dark brown hairs, often wavy. Exterior (underside) paler, also clothed with dark hairs. Flesh thin. STALK absent. SPORES (14) 17-22 (25) × 8-12 microns, elliptical, smooth, without oil droplets.

HABITAT: Solitary to densely gregarious on dung and manure or compost, etc.; fruiting in wet weather, cosmopolitan (along with its numerous look-alikes) but seldom noticed because of its small size.

EDIBILITY: Who knows? Who cares?

COMMENTS: This species is one of several small, difficult-to-distinguish, dung-loving cup fungi. Although the hairs on the exterior are quite conspicuous, in the words of one specialist, "they may be overlooked in the field, where the nature of the substrate discourages close scrutiny." Similar yellow to orange dung-lovers include: *C. theleboides,* with paler hairs (sometimes also growing on soil, humus, or "spent hops"); *C. stercorea*, with branched dark brown hairs; *Coprobia granulata,* minute (1-2 mm), orange, and hairless. Several similar disclike to cushionlike species occur in burned areas, often in vast numbers, or in the heated (sterilized) soil in greenhouses. These include: *Anthracobia macrocystis,* with a reddish fertile surface and brown hairs on its exterior; *A. melaloma,* with a yellowish-brown to ochre-orange fertile surface and brown hairs on its exterior; *Trichophaea abundans,* minute and whitish with pale brown hairs, growing on plaster as well as burnt ground; *Pyronema omphalodes,* very common in confluent masses, with a pale orange to reddish-orange, minute (1-3 mm), hairless fruiting body and elliptical sores; *Pulvinula carbonaria,* also minute and pale to bright orange and hairless, but with round spores; and *Pulvinula archeri,* similar to the previous species but with smaller (7-9 microns), round spores. Also worth mentioning is *Ascobolus,* which usually has a minute dark (greenish to dark brown or black), disclike fruiting body. Most of its species grow on dung, but one, *A. carbonarius,* grows in swarms on burnt ground. For small cup fungi that do *not* grow in dung or burned areas, see *Scutellinia scutellata,* and for larger and more deeply cup-shaped, ash-loving or terrestrial species, see *Geopyxis vulcanalis.* Better yet, go get some exercise!

Cheilymenia coprinaria in its favorite milieu—a "road apple" (piece of horse dung). Note small size, disclike fruiting body, and the long hairs or bristles protruding from its margin.

Scutellinia scutellata typically grows in groups **(left)** and is easily recognized by its orange to red color and the dark hairs that fringe its margin **(right)**. Unfortunately, the hairs do not show up as well in these black-and-whites as they do in the original color photographs. (Ray Gipson, Dan Harper)

Scutellinia scutellata   (Eyelash Pixie Cup)

FRUITING BODY 0.2-1.5 cm broad, at first nearly round (spherical), but soon opening to form a shallow cup and eventually disclike (flattened). Fertile (upper) surface smooth, bright red to scarlet to orange (or rarely paler with a pinkish cast); margin conspicuously ciliate (fringed with dark brown or blackish hairs up to 1 mm long). Exterior (underside) also clothed with dark hairs. Flesh very thin. **STALK** absent. **SPORES** (15) 17-19 (23) × (9) 11-14 (17) microns, elliptical, minutely warted, with one or more oil droplets.

HABITAT: Gregarious on rotten wood or damp soil (or occasionally on ashes, wet leaves, or conks); widely distributed and common, but easily overlooked because of its small size. In our area it fruits in the winter and spring.

EDIBILITY: Unknown, but much too puny to be of importance.

COMMENTS: This is another easily-recognized cup fungus. The bright red to orange fertile surface and ciliate (eyelash-like) margin are good field characters. *S. umbrarum* is a very similar, widespread, *terrestrial* species with a slightly larger fruiting body (up to 2 cm broad), larger spores, and shorter, less conspicuous hairs. *S. erinaceus* is also similar, but is orange to yellow and smaller (2-5 mm broad), has smooth spores, and grows on wood. *Cheilymenia crucipila* is a minute (1-4 mm) orange to orange-red, terrestrial species with paler, shorter hairs and smooth spores that lack oil droplets. *Lamprospora* species are minute and hairless. Finally there is *Melastiza chateri,* a bright orange terrestrial species that is 0.5-2 cm broad and has minute brown hairs on its exterior, especially near the margin. For similar dung- and ash-lovers, see comments under *Cheilymenia coprinaria*.

Humaria hemispherica   (Hairy Fairy Cup)

FRUITING BODY 1-3 cm broad, at first nearly round (spherical), gradually opening up to become cup-shaped. Fertile surface (interior) white or whitish to grayish, smooth; margin fringed with brown hairs. Exterior densely clothed with stiff brown hairs. Flesh thin. **STALK** absent or present only as a rather abruptly narrowed base. **SPORES** 20-24 × 10-12 microns, elliptical, smooth or minutely warted, with 2 or sometimes 3 oil droplets.

HABITAT: Solitary, scattered, or in groups on ground or occasionally rotten wood; widely distributed and fairly common under both hardwoods and conifers, usually fruiting in the summer and fall. I have yet to find it in our area, but it may well occur.

EDIBILITY: Who knows? Who cares? Do you?

COMMENTS: The combination of pallid fertile surface and brown hairy exterior make this an easy cup fungus to recognize. *Jafnea semitosta* is a larger (2-5 cm broad) species with a creamy-white to tan or brown interior, a brown exterior clothed with scattered soft brown hairs, plus a short ribbed stalk; it is fairly common in eastern North America. Other hairy species: *Trichophaea boudieri* and *T. bullata* have a pale gray to whitish interior and brown hairy exterior, but are much smaller (1-6 mm broad) and grow on wet soil under conifers; *T. abundans* is a minute whitish species that grows in burned areas. For more colorful hairy or ciliate species, see *Scutellinia scutellata* and *Cheilymenia coprinaria*.

Geopyxis vulcanalis   (Vulcan Pixie Cup)

FRUITING BODY 0.3-1 (2) cm broad, nearly round (spherical) when young, becoming deeply cup-shaped and then often flattening out in age. Fertile (upper or inner) surface smooth, pale orange to pale or dingy yellowish, the margin usually finely scalloped. Exterior paler or whitish, usually powdery or downy when young but often entirely smooth in age. Flesh thin, fragile. **STALK** usually present, up to 5 mm long and 1-3 mm thick, sometimes so short as to be practically absent, equal or tapered downward, colored like exterior of cup. **SPORES** 14-21 × 8-11 microns, elliptical, smooth, without oil droplets.

HABITAT: Scattered to densely gregarious in duff or moss under conifers, or in burned areas; widely distributed. I have seen large fruitings locally in the fall, winter, and spring.

EDIBILITY: Unknown, but much too puny to be of value.

COMMENTS: This pixieish cup fungus and its close relatives are easily told by their small, deeply cup-shaped (at least when young) fruiting body that often has a finely scalloped margin. In *G. vulcanalis* a short stalk is usually present, but in some of the other species (see below) it is lacking. *G. carbonaria* is a similar species that grows only in burned areas. It has a brick-red fertile surface and normally remains cup-shaped rather than expanding. *Tarzetta* species closely resemble *Geopyxis* in size and appearance, but as currently defined, have spores with two prominent oil droplets. Their ranks include: *Tarzetta cupularis,* widespread under conifers, in moss, on burnt ground, etc., which resembles a miniature goblet with its dainty stalk and grayish-tan to tan or brownish cup; *T. rosea*, found in burned areas, with a pink to reddish cup and little or no stalk; and two somewhat larger (1-3 cm broad), yellowish to brownish, woodland species: *T. bronca* of eastern North America, with little or no stalk, and *T. catinus*, widespread (including California), with a stalk. Most of these species have a scalloped margin or "lip" as in *G. vulcanalis,* and were originally placed in *Geopyxis, Pustularia,* and/or *Peziza*.

Geopyxis vulcanalis. Note small size, gregarious nature, and the small stalk below the cup which is often covered by dirt (visible in specimen at top).

Truffles

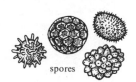

spores

TUBERALES

TRUFFLES are seldom seen because they grow underground.* Most of them look like tiny potatoes or rocks, but reveal a system of canals, veins, and/or cavities when sliced open. They are likely to be confused only with the aptly named false truffles (Hymenogastrales), which also grow underground but bear their spores on basidia rather than in asci and typically have a minutely chambered rather than marbled, channelled, or hollow interior. (For a more detailed comparison, see footnote at bottom of p. 844.)

The truffles are thought to be cup fungi which have gone underground. The evolutionary pathway leading from a cup fungus to a truffle is exquisitely illustrated by the genus *Geopora*. Some Geoporas are truffle-like and subterranean, while others are cuplike and partly exposed (i.e., hollow with a large mouth at ground level that opens to the air as shown on p. 935). Some of the underground Geoporas are also hollow, but the mouth is oriented randomly (at the side, bottom, etc.) and others have become greatly infolded so that the interior is channelled or chambered (see photo on p. 847) rather than hollow. The infolding of the tissue is advantageous because it greatly increases the surface area for producing spores (the spore-bearing asci line the canals or chambers inside the fruiting body). In the "true" truffles (e.g., *Tuber*), this trend is carried to its logical extreme—the interior of the fruiting body is marbled but solid (it presumably evolved through the fusing or merging of folded tissue) and the asci are imbedded randomly in the tissue rather than lining the canals or chambers.

As might be expected, truffles exhibit several other special adaptations to their underground lifestyle. In *Geopora* the spores are forcibly discharged as in the cup fungi, but other truffles have lost the ability—and necessity—to discharge them because they do not depend on the wind for spore dispersal. Instead, their spores are spread by various truffle-eating animals (rodents, deer, pigs, insects, slugs, etc.). The spores pass through the animals' digestive systems unscathed, and a microscopic analysis of the spore content of animal droppings can give you a pretty good idea of which truffles and false truffles grow in your area! Some animals, such as the California red-backed vole (a sort of burrowing mouse) tunnel through the soil eating nothing *but* truffles and are thus restricted to the coastal fog belt, where truffles and false truffles occur year-round.

To attract attention and make themselves desirable, most truffles have developed distinctive odors and flavors. However, the odor and flavor do not normally become strong until well after the truffle is mature, thereby insuring that the eater of the truffle will ingest a large number of viable spores. Furthermore, truffles do not need to develop as rapidly as epigeous (above-ground) fungi because they are insulated from sudden changes in the weather. Instead, the maturation process takes place *gradually,* over a period of several weeks or months, although a few spores often develop earlier coupled with a *slight* scent, perhaps as a safeguard against a prolonged cold or hot spell that would inhibit further development of the fruiting body.

Truffles, particularly species of *Tuber*, have been eaten for centuries. Unfortunately, the fabled truffles of France and Italy have become a fetish of the rich. Due to their rarity and the difficulty in finding them, they have acquired considerable snob appeal and retail for more than $500 a pound! Their flavor and aroma are so powerful that a little goes a long way, but to a person of modest means such as myself, nothing edible is worth that much!

Since truffles grow underground, we humans, with our underdeveloped noses, have

*When speaking of truffles and false truffles, the terms "underground," "subterranean," "hypogeous," and "buried" are used rather loosely to mean beneath the surface of the ground, either in the humus layer *or* in the soil itself *or* in the interface between the two.

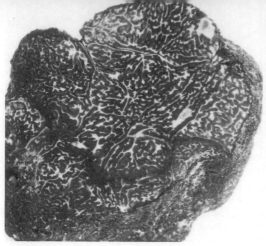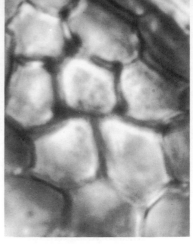

Left: The famous "Black Diamond" of France, *Tuber melanosporum,* sliced open to show the marbled interior (see pp. 854-855 for more details). **Right:** Microscopic view of the surface of a *Tuber* spore, showing the alveolate (pitted-reticulate) pattern typical of many species. (Herb Saylor)

trouble finding them without "hired hounds." Goats have been used to track down truffles in Sardinia and bear cubs have been employed in Russia, but pigs and dogs are the most accomplished truffle hunters. Some truffles contain pig sex hormones, meaning that pigs have a natural nose (and lust!) for truffles. They require little or no training, but must be physically restrained from devouring their quarry, and are hard to control even when there are no truffles about. Acorns are sometimes given as nutritional recompense for finding a truffle (a pitiful substitute, if you ask me) or the pig is muzzled and pulled away just as it begins to dig up the truffle with its exceptional snout. Another problem with pigs is that they tire easily, and must be carted to and from the truffle grounds if they are distant.

Dogs, on the other hand, are tireless and devoted, and care more for humans than truffles. In fact, most dogs *loathe* truffles and must be painstakingly trained to seek them out. There are schools in Italy devoted exclusively to this purpose, and a seasoned truffle hound commands a steep price. Short-legged breeds are traditionally popular, presumably because they're closer to the ground. Both pigs and dogs, incidentally, can detect truffles from as far away as 50 yards, and there is one case on record of a dog that jumped a hedge, crossed a field, and "secured his prize" under a beech tree at least 100 yards away!

Perhaps you are now convinced that you need canine or porcine companions to find truffles. Well, let me state, unequivocally, that *you don't.* True, the odds against casually bumping into a truffle are great, but you *can* find truffles *by making a concerted effort to find them.* This means getting down on your hands and knees and systematically sifting through the forest humus and soil, paying special attention to "truffle tracks": squirrel diggings, small cracks in the soil caused by those that develop close to the surface, strange and compelling odors (some seasoned trufflers claim they can smell them out!), and an occasional cloud of "truffle flies" hovering over the buried object of their affections. Truffles, in fact, are far easier to find than most people realize. Looking for them is both challenging and fun, like hunting for buried treasure or panning for gold, but without the monetary incentive. (Entrepeneurs, are you listening? The expensive truffles of Europe are not known to occur in the United States and most of our native species are not as richly flavored.)

Why, then, do so few mushroom hunters look for truffles? Perhaps because they don't know how, when, or where. (As evidence of this assertion, I offer the chapter on truffles in the first edition of *Mushrooms Demystified!*) The how, the when, and the where are delightfully described by Harold E. Parks in the following excerpt from a 1921 article in the scientific journal, *Mycologia.* A resident of San Jose, California, Parks was one of California's earliest and most avid trufflers, and he has had numerous species of truffles and false truffles named after him.

Even when one knows the ground thoroughly it is surprising how little of it may be covered on a day of good collecting. Frequently two or three hours will be spent in working over the ground under a single large oak, and on several occasions an entire afternoon has been spent in one place . . .

The equipment of the truffle hunter is important. I use a wheel on many trips, as the roads are excellent and the stops are very frequent in some places. It is easily hidden in the brush when I leave the roadways and take to the high hills, and it makes accessible places otherwise out of one's reach. To the wheel is strapped a small combination rake and hoe with a four-foot handle. This implement is very useful in climbing, raking and digging and furnishes good protection in a snake country, as I well know. A short-handled hoe useful for work in thick brush, a trowel, knife, tweezers, lens, kodak, plenty of newspapers and a large number of small pasteboard cartridge boxes obtained from a shooting gallery. These small boxes are very useful in handling the many small specimens or single individual specimens, while large collections are wrapped in the paper. Lunch and thermos bottle complete the outfit, and all are packed compactly in the large canvas bags used by newsboys. These bags ride comfortably with a large load evenly distributed over the shoulders.

In the earlier parts of the season the edges of the forests and the small groups of trees are usually the best places for operations, although frequently the dense forest will yield good specimens. Late in the season the best places are to be found deep in the forest, where the ground retains more moisture. When the collector finds a favorable place for operations the rake comes into use and a small area is raked free of leaves and humus. Watch must be kept in the leaves for certain species . . . Other species will appear entirely exposed on the surface of the earth [under the leaves] and some will be just beneath the surface and out of sight. Excavation may be continued to a depth of a foot, at which depth most species will cease to be found. Care should be taken at all stages, especially near the surface, to avoid injury to specimens, but they will often be injured in spite of it, and many of the dark-colored species will require very careful search and sifting of the soil. The rewards are more often blistered hands and an aching back than truffles, but there are also some intensely exciting moments . . .

To these remarks I would add only this: Digging up the forest can be unsightly as well as destructive, so do it *on a small scale,* in scattered places over a large area, don't go truffling in locales traditionally frequented by mushroom hunters (they have a right to undisturbed duff!), and *always cover up the soil you expose,* leaving the environment as close to its original state as possible.

Truffles are mycorrhizal. This means they can be found wherever there are trees and shrubs, but like the false truffles, they are especially abundant and diverse along the west coast. They are normally terrestrial, but can also occur *inside* very rotten wood that has been permeated by tree rootlets. This is particularly true in dry areas like the Sierra Nevada, where the rotten logs are a major source of moisture for both the trees and the truffles. In our area they seem to favor evergreen oaks (live oak and tanoak) and conifers such as Douglas-fir. Because truffles develop slowly, they are usually found at the end of the mushroom season (February-July in our area). Some, such as *Tuber gibbosum,* are excellent esculents; others are mediocre and still others have yet to be tried.

Although a microscope is often required, truffles are not as difficult to identify as false truffles (for one thing, there are far fewer species). A fairly extensive—but by no means comprehensive—selection is offered here in the hope that it will stimulate mushroom hunters to start looking for these clandestine denizens of our forests. Since I am by no means an expert on truffles, I have gleaned much of the information in this chapter from articles by Helen Gilkey and James Trappe, the past and present authorities on the subject. Anyone who takes truffles seriously should consult these articles (see Suggested Readings and References) or join the North American Truffling Society.

Modern taxonomists try to show the truffles' relationships to the cup fungi by scattering them among several families in the Pezizales, just as they place many false truffles with the Agaricales. The Tuberales, in other words, is a defunct and artificial—but convenient

—category that is used here because it facilitates identification. In the following key, the truffles have been divided into several natural groups. An attempt has been made to use field characters, but microscopic features have unavoidably come into play— particularly the shape and ornamentation of the *mature* spores (you must have at least one mature or partially mature specimen!). Truffle spores, incidentally, are exceptionally ornate. Some are spiny like porcupines or pitted like golf balls, others are covered with a geometrical network of ridges (see photo on p. 842), still others are warted, pegged, or smooth.

Key to the Tuberales

1. Spores borne on basidia* .... (see **Hymenogastrales & Allies,** p. 739)
1. Spores borne inside asci* 2

2. Found in the deserts of the Southwest; fruiting body more or less round and somewhat puffball-like, i.e., developing underground but sometimes emerging at maturity, then drying out and blowing about in the wind; asci brown under the microscope .... *Carbomyces*
2. Not as above 3

3. Fruiting body earthball-like, i.e., consisting of a thick (2-5 mm) tough outer wall and a single large inner cavity which is soon filled with tissue; internal tissue at first white and cottony, becoming divided into several chambers by white sterile bands, then becoming dark brown to blackish *and powdery* when mature (the asci disintegrating quickly and the sterile bands not evident in age) .... *Elaphomyces,* p. 862
3. Not as above; mature spore mass not powdery 4

4. Spores borne inside the fruiting body (or on inside surfaces); common 5
4. Spores borne externally (on outside surface of fruiting body); rare 23

5. Interior of fruiting body either hollow *or* with empty chambers *or* with open veins or empty canals formed by infolding of the fruiting body wall, *or* occasionally with chambers that are loosely stuffed with cottony hyphae 6
5. Interior solid, with pockets or zones of fertile tissue and meandering sterile veins 16

6. Fruiting body hollow inside (with one or sometimes two large empty chambers which may be round *or* convoluted or canal-like due to infolding of the outer wall) 7
6. Interior of fruiting body with separate canals or several separate chambers 9

7. Exterior of fruiting body with small rounded to angular warts, typically with one or more openings to the interior .... *Genea & Genabea,* p. 849
7. Exterior not finely warted; opening(s) present or absent 8

*As already pointed out, this fundamental difference can only be seen with a microscope (and then only after the basidia or asci have formed and before they disintegrate). However, the false truffles (underground Basidiomycetes) and truffles (underground Ascomycetes) can often be differentiated in the field by the following characters:

   If the interior is gelatinous, it is a false truffle.
   If the interior has a columella (e.g., a branched or unbranched internal stalk or well-developed sterile base), it is probably a false truffle (exception: *Fischerula subcaulis,* an Ascomycete).
   If the interior is solid and marbled with veins, it is probably a truffle.
   If the interior is composed of numerous minute holes or empty chambers (giving it a sponge-like appearance) it is probably a false truffle.
   If the interior shows the embryonic beginning of cap, gills, and stalk, it might be a young *Amanita* or other gilled mushroom!
   If the fruiting body is very hard with a solid interior that flakes or chips off like wax, it is probably a truffle.
   If the interior is completely hollow or has several large hollows or is composed of one mazelike hollow, it is probably a truffle.
   If the wall (peridium) of the fruiting body is very thick (several mm) and the interior is not hollow and the fruiting body has a distinct base, it is probably a false truffle or earthball.
   If the outer wall of the fruiting body is very thick and the interior is cottony or powdery and the fruiting body lacks an obvious base, it is probably a truffle.
   If the wall of the fruiting body is infolded to form numerous empty canals or veins or cavities that often open to the exterior, it is probably a truffle.
   If the exterior is covered with rootlike mycelial threads (rhizomorphs) it is probably a false truffle.
   If the exterior is covered with warts (often small or large), it is probably a truffle.

There are also several mycorrhizal Zygomycetes (e.g., *Endogone* and *Glomus*) with truffle-like fruiting bodies. These fungi are not treated in this book because they have neither asci nor basidia. Instead sexual spores are formed by the conjugation of "mother cells" (gametangia), and asexual spores are often formed on hyphae. Most species have gigantic spores (100-200 microns!) and some, such as the common *Endogone lactiflua,* exude a latex when cut.

8. Fruiting body a large hollow ball (but often lobed or flattened); inside surface white to grayish, pinkish, or purplish, *not* warted; outer wall often splitting into lobes (at top) in age; common, especially under northern and mountain conifers .... (see *Sarcosphaera,* p. 825)

8. Not as above 11

9. Exterior of fruiting body with rounded to angular warts 10

9. Exterior of fruiting body smooth or hairy, but not distinctly warted 11

10. Fruiting body black to brownish to reddish or orange, often (but not always) with a tuft of mycelium at the base; spores smooth even at maturity .... *Balsamia* & Allies, p. 852

10. Fruiting body white to yellowish or yellow-gray, or if not then interior warted like exterior; basal tuft present or absent; spores ornamented *at maturity* .... *Genea* & *Genabea,* p. 849

11. Exterior of fruiting body with fuzzy brown hairs (i.e., tomentose); interior a hollow chamber *or* with open canals formed by complex infolding of the fruiting body wall; spores forcibly discharged, smooth at maturity; fairly common, especially under conifers . *Geopora,* p. 846

11. Not as above; exterior not brown *and* tomentose 12

12. Spores smooth *and* round at maturity; chambers of the fruiting body typically stuffed with cottony hyphae; tips of asci *not* amyloid; rare (at least on west coast) .... *Stephensia*

12. Not with above combination of features 13

13. Asci with amyloid tips (i.e., tips staining bluish in iodine solution) (see *Peziza* & Allies, p. 818)

13. Asci not amyloid 14

14. Interior of fruiting body hollow or with open veins, canals, or chambers formed by complex infolding of the fruiting body wall; spores ornamented at maturity; especially common under northern or mountain conifers (often inside rotten wood), but also found in other habitats .... *Hydnotrya,* p. 848

14. Not as above 15

15. Spores elliptical and *smooth* at maturity; channels or canals inside fruiting body usually empty .... *Balsamia* & Allies, p. 852

15. Not as above; spores ornamented (at least at maturity) .... *Tuber* & Allies, p. 854

16. Columella (sterile column or base or rudimentary internal stalk) present, or if not then a very distinct basal pad of mycelium present; exterior of fruiting body pallid to pinkish-gray to brownish; interior pinkish-gray to grayish-purple to nearly black with narrow white veins; spores very large (60-100 microns), elliptical, brown at maturity and ornamented with obscure spines; found under conifers in the Pacific Northwest .... *Fischerula subcaulis*

16. Not as above; columella absent 17

17. Asci with amyloid tips or weakly amyloid throughout; asci arranged in a distinct palisade (hymenium); spores elliptical 18

17. Asci not amyloid; asci arranged in a palisade *or* randomly imbedded in tissue; common .. 21

18. Spores round, ornamented with spines or pegs at maturity; rare ...... *Tuber* & Allies, p. 854

18. Not as above; occasional 19

19. Asci weakly amyloid throughout; fruiting body white to yellowish or yellow-brown; rare (known from coastal California) 20

19. Asci amyloid mainly or only at their tips; fruiting body often darker than above; widespread .... (see *Peziza* & Allies, p. 818)

20. Spores less than 20 microns long .... *Hydnotryopsis setchellii*

20. Spores averaging 20 microns or more long .... *Hydnotryopsis compactus*

21. Fruiting body with a fatty or gristle-like consistency (especially the interior), usually whitish or buff when fresh; spores round and ornamented at maturity ...... *Tuber* & Allies, p. 854

21. Not as above; texture not gristle-like 22

22. Spores smooth even when mature; exterior of fruiting body often warted; interior pallid when mature or grayish to olive with pallid veins (rarely brown) ...... *Balsamia* & Allies, p. 852

22. Spores ornamented at maturity; exterior of fruiting body warted in some species, but more often smooth; interior pallid when immature but usually brown or reddish-brown with pallid veins when mature (but sometimes grayish or olive) .... *Tuber* & Allies, p. 854

23. Fruiting body pale brown to brown or purplish; asci faintly amyloid; spores mostly 20 microns or more broad, hyaline (colorless) under the microscope .... *Sphaerosoma*

23. Fruiting body yellowish to olive or brown; asci not amyloid; spores 8-25 microns broad, hyaline to yellowish or brown under the microscope .... *Sphaerozone*

GEOPORA   (Fuzzy Truffles)

Small to medium-sized fungi growing underground or at ground level. FRUITING BODY *usually more or less round to cuplike, with one (or sometimes more) openings to the interior. Exterior typically brown and hairy or fuzzy* (tomentose). INTERIOR *complexly folded or channelled in one species, hollow in the others.* STALK, columella, and basal mycelial tuft typically absent. SPORES elliptical to nearly round, *smooth,* hyaline (colorless) under the microscope, *forcibly discharged.* Asci arranged in a distinct palisade (hymenium) lining the inside surface of the fruiting body or the internal folds (if present), mostly 8-spored, not amyloid.

THIS genus is best recognized by its fuzzy brown exterior. Some species, such as *G. arenicola*, are traditionally grouped with the cup fungi (in genus *Sepultaria*) because of their hollow fruiting body that grows just below the soil surface with only the large opening or "mouth" at the top exposed to the air. *G. cooperi,* on the other hand, has traditionally been treated with the truffles because it grows underground and has a complexly folded interior and one or more irregularly oriented openings. These apparently disparate fungi are linked by species such as *G. clausa,* which is hollow like *G. arenicola,* but grows underground and has an apical, basal, or lateral opening.

Geopora is currently placed in the Pyronemataceae, alongside *Aleuria, Otidea, Geopyxis,* and many other cup fungi. About a dozen species are known, most of them in the mode of *G. arenicola.* However, *G. cooperi* seems to be the most common species in California. It is said to be a good edible, but I can find no information on other members of the genus.

Key to Geopora

1. Wall of fruiting body complexly infolded to create numerous canals or chambers inside the fruiting body 2
1. Not as above; interior of fruiting body with a simple hollow 3
2. Exterior of fruiting body only slightly hairy if at all; odor often sweet or garlicky *when fully mature;* spores ornamented at maturity; often (but not always) growing inside rotten wood . . . . (see ***Hydnotrya,*** p. 848)
2. Not as above; exterior distinctly hairy or fuzzy; spores smooth . . . . ***G. cooperi,*** below
3. Opening or "mouth" of fruiting body irregularly oriented (at top, base, or side); usually growing underground . . . . ***G. clausa*** (see *G. arenicola,* p. 847)
3. Opening or "mouth" always at top; fruiting body immersed or partly immersed in the ground with the mouth exposed to the air 4
4. Fertile surface (interior) orangish to reddish or sometimes yellowish . . . . ***G. aurantia*** & ***G. pellita*** (see *G. arenicola,* p. 847)
4. Fertile surface white to pale brownish or sometimes drying yellowish 5
5. Fruiting body up to 1 cm broad . . . . ***G. arenosa*** (see *G. arenicola,* p. 847)
5. Fruiting body 1-4 cm broad . . . . ***G. arenicola*** & others, p. 847

Geopora cooperi   (Fuzzy Truffle)

FRUITING BODY usually buried or partially buried, round or nearly round (but often squirrel-eaten), 2-7 (10) cm broad. Exterior fuzzy or velvety from a coating of light brown to dark brown hairs, usually furrowed. **INTERIOR** white to creamy or yellowish-tan, usually streaked with tan or brown, deeply convoluted, the folds often touching each other but leaving at least some open spaces or "canals" between them. Odor usually mild, but in one form resembling fermented cider. **SPORES** 18-27 (30) × (10) 13-21 microns, broadly elliptical in one form, round or nearly round in another; hyaline (colorless) under the microscope, smooth, with one oil droplet. Asci usually 8-spored, forming a distinct palisade (hymenium) that lines the open surfaces of the folds.

HABITAT: Solitary, scattered, or gregarious on or in the ground under both hardwoods

Geopora cooperi. Note the convoluted interior (specimen on left) and fuzzy exterior (specimen on right). It is usually found under conifers, often in sandy soil.

and conifers (but especially the latter); widely distributed in western North America and locally common, especially under mountain conifers during the spring, summer, and fall. It favors pine in coastal California and pine, fir, or spruce in the Sierra Nevada and elsewhere; in Alaska it has been found under willow and aspen. It develops underground but may surface (or be dug up by squirrels) in age, and so is often seen by casual collectors.

EDIBILITY: Edible. Rodents are very fond of it and so are some humans.

COMMENTS: This is one of our largest truffles and also one of the more distinctive. Its telltale traits are the fuzzy brown exterior and convoluted interior. The latter is simply a mass of folded tissue (see photograph), with the spore-bearing asci lining the empty spaces or "canals" between the folds. The spores are shot out of the asci as in the cup fungi. It has numerous synonyms, including *G. harknessii* and *G. magnata.*

Geopora arenicola   (Hole In The Ground)

FRUITING BODY at first closed and buried in ground, then opening at the top and becoming more or less cup-shaped at maturity, the margin remaining incurved or often splitting stellately (in starlike lobes) in age; 1-4 cm broad. Exterior brown and densely clothed with flexible brown hairs that bind surrounding dirt or sand. **INTERIOR** (fertile surface) pallid to creamy or grayish, often becoming yellowish, tan, or brownish in age; smooth. Flesh brittle to rather tough. **STALK** absent or rudimentary. **SPORES** 23-30 × 12-17 microns, elliptical to spindle-shaped, smooth, usually with one (rarely two) large oil droplets. Asci lining inner surface of cup, typically 8-spored, not amyloid.

HABITAT: Scattered to densely gregarious or clustered in sand or silt, disturbed ground, etc., usually immersed in the soil with only the mouth showing; widespread but not common, or at least not often noticed. I have found it once in our area, in the spring.

EDIBILITY: Academic—it is practically impossible to get rid of the sand or dirt!

COMMENTS: Better known as *Sepultaria arenicola*, this species looks like a hole in the ground or an insect burrow or worm tunnel (see photo on p. 935). In old age it is clearly cup-like, however, and often splits into lobes (as shown in photo). The hairy brown exterior distinguishes it from *Sarcosphaera crassa* and other cup fungi that develop in the ground, and relates it to *G. cooperi,* which has a complexly folded rather than hollow interior. The fuzzy brown hairs can be seen with a hand lens by gently brushing away some of the dirt. Other species: *G. longii* of the Southwest is similar but has nearly round spores; *G. arenosa* is also similar but much smaller; *G. aurantia* is similar but has a reddish to orange or egg-yellow interior and rigid hairs on its exterior; *G. pellita* has a yellowish to pale orange interior, but its hairs are not rigid; *G. clausa (=Hydnocystis californica)* has a fuzzy brown exterior and its mouth is oriented randomly with respect to the fruiting body (i.e., at the top, bottom, or side). It usually grows underground, but is rare in our area.

847

HYDNOTRYA   (Wood Truffles)

Small to medium-sized woodland fungi found in soil or very rotten wood. FRUITING BODY roundish to lobed or brainlike. Exterior variously colored, smooth to scurfy or occasionally slightly fuzzy, *but not conspicuously hairy or warted.* INTERIOR *variable in structure, sometimes composed of a single chamber, but more often with several to many chambers or canals formed by complex infolding and fusing of the outer wall.* STALK and columella absent; mycelial tuft also absent. SPORES round to elliptical or cubical, smooth at first but becoming roughened, warted, ridged, pitted, or spiny at maturity; yellowish to brown under the microscope. Asci typically borne in a palisade (hymenium) lining the chamber(s) or canals, *not* amyloid.

THE fruiting bodies of this genus are extremely variable in size and shape. Some species are hollow inside, others are complexly folded. All have open canals or chambers, non-amyloid asci, and spores which suggest a kinship to the elfin saddles and false morels (Helvellaceae). *Geopora* exhibits a similar range of variation, but has a hairier or fuzzier exterior and smooth spores. *Genea* and *Genabea* differ in having a distinctly warted exterior and interior.

Hydnotryas are sometimes called "wood truffles" because several species can frequently be found *inside* rotten wood (they also grow in soil). As pointed out earlier, the fact that they grow inside wood does not necessarily mean they are wood-rotters. Rather, they could be associated with tree rootlets that penetrate the wood in search of moisture.

A dozen species of *Hydnotrya* are known; half occur in North America. They are fairly common in the Sierra Nevada and other mountain ranges, but rare or absent in coastal California. A single species is described here and several others are discussed.

Key to Hydnotrya

1.   Exterior of fruiting body brown to dark brown to dark reddish-brown; interior complexly folded or convoluted; odor often strong (sweet or garlicky) *when fully mature*; spores round (but often knobby) .... *H. cerebriformis & H. tulasnei* (see *H. variiformis,* below)
1.   Exterior whitish to buff, cinnamon-buff, pinkish-cinnamon, or sometimes brown; interior complexly folded *or* a simple hollow; odor not usually as above; spores elliptical 2

2.   Interior usually a simple hollow; spores often cubical; usually found in ground .... *H. cubispora* (see *H. variiformis,* below)
2.   Interior ranging from a simple hollow to complexly folded; spores not cubical; found in ground or inside rotten wood 3

3.   Asci with amyloid tips; common in coastal California .... (see *Peziza* & Allies, p. 818)
3.   Not as above; asci not amyloid .... *H. variiformis* & others, below

Hydnotrya variiformis

FRUITING BODY 0.7-4 cm broad, round to somewhat flattened, depressed, or lobed. Exterior minutely velvety, whitish to creamy to buff or yellowish to cinnamon-buff or brownish, not warted. **INTERIOR** variable in configuration, but small specimens often containing a simple cavity with a prominent opening, and larger ones usually with several chambers or narrow, branching canals formed by crowding and infolding of the outer wall; canals usually empty but their sides often fused; white or pallid, but the hymenium (fertile tissue) often brownish to pinkish-orange at maturity. **SPORES** 32-36 × 24-28 microns, elliptical, smooth becoming minutely pitted and wrinkled at maturity, yellowish-brown under the microscope. Asci borne in a palisade (hymenium) that lines the canals or cavity, typically 8-spored.

HABITAT: Solitary or in groups in soil or inside very rotten wood under conifers; occasional (along with *H. cerebriformis*—see comments) in the Sierra Nevada, Cascades, and other western mountains. It fruits in the spring, summer, and early fall.

EDIBILITY: Edible? I can find no specific information on it.

COMMENTS: This species and *H. cerebriformis* (see below) can usually be told in the field by their fondness for growing inside rotten wood plus their complexly folded interior (at least in large specimens) and non-amyloid asci. The exterior lacks the warts of *Genabea cerebriformis* and the brown hairs of *Geopora cooperi*. *H. cerebriformis* is similar to *H. variiformis* and grows in similar habitats. It has a more consistently complex or convoluted interior (not unlike that of *Geopora cooperi,* shown on p. 847), is usually slightly darker than *H. variiformis* (dull reddish-brown to dark purple-brown), typically has a strong garlicky odor *when mature,* and has round, minutely spiny spores. *H. tulasnei* is a widespread odoriferous species that is very similar to *H. cerebriformis;* it is also brown to reddish-brown, but has coarsely warted spores. *H. cubispora* is a widely distributed, brownish to pinkish-cinnamon, usually terrestrial species with a more or less hollow (but lobed) interior and spores which are often cubical. *H. michaelis (=H. yukonensis)* is a rather rare northern species with elliptical warted spores and a convoluted interior.

GENEA & GENABEA   (Geode Truffles)

Small underground woodland fungi. FRUITING BODY *round to strikingly lobed or brainlike, usually with one or more openings to the interior.* Exterior minutely warted, variously colored, bald or with hairs. INTERIOR *usually warted; hollow or with empty, mazelike canals formed by infolding or inward projections of the outer wall.* STALK and columella absent, but basal tuft of mycelium often present in *Genea.* SPORES round to elliptical, *warted or spiny at maturity* (but ornamentation may dissolve in KOH or Melzer's reagent!), pale or colorless *(Genea)* to brown *(Genabea)* under the microscope. Asci arranged in a palisade (hymenium) lining the surfaces of the inner cavity or canals; typically 8-spored, not amyloid.

THESE small truffles are easily recognized by their finely warted, often lobed fruiting bodies with a hollow or partly hollow, warted, geode-like interior. Geneas typically have a single, often irregularly shaped cavity and warted spores, while Genabeas usually have a more complex or mazelike interior and spiny spores. In addition, the western species differ in color: reddish to brown or black in *Genea,* white or creamy in *Genabea.*

The origins and affinities of *Genea* and *Genabea* are unknown; together they form the family Geneaceae. Both are fairly common, at least in California. They seem to fruit closer to the surface of the ground than many truffles and also have an earlier season, appearing in late November in our area and continuing on into the spring. I can find no information on their edibility. Their small size is hardly an asset, but they seem to be very popular with our local wild pigs. (You can sometimes find them where pigs have been foraging.) *Genea* has over 20 known species, whereas *Genabea* includes only a handful. Three Geneas and one *Genabea* are described here and several others are keyed out.

Key to Genea & Genabea

1. Exterior of fruiting body or at least the "mouth" with brown hairs 2
1. Not as above 3

2. "Mouth" of fruiting body fringed with stiff hairs *Genea kraspedestoma* (see *G. arenaria,* p. 851)
2. Not as above .... *Genea arenaria* & others, p. 851

3. Exterior distinctly reddish; interior often reddish- or pinkish-tinged *Genea intermedia,* p. 851
3. Not as above (but exterior may be very dark reddish-brown) 4

4. Exterior of fruiting body dark reddish-brown to dark brown to almost black when fresh .. 5
4. Exterior white to buff, yellowish, light brown, or medium brown when fresh 6

5. Found in West; spores warted .... *Genea harknessii* & others, p. 850
5. Found in eastern North America; spores spiny *Genabea fragilis* (see *G. cerebriformis,* p. 851)

6. Exterior of fruiting body white to creamy or pale yellow; interior usually complex (mazelike or with many separate chambers); spores spiny ...... *Genabea cerebriformis* & others, p. 851
6. Exterior light yellow-brown to light brown or brown; interior usually a single cavity (but the cavity often convoluted); spores warted .... *Genea compacta* (see *G.harknessii,* p. 850)

Many geode truffles look like bits of knobby coral. This one is *Genea harknessii*. (Herb Saylor)

Genea harknessii (Dark Geode Truffle)

FRUITING BODY underground, 0.5-2.5 cm broad, round to flattened to very knobby and irregularly lobed, usually with an apical opening to the interior and a tuft of mycelium at the base. Exterior dark reddish-brown to dark brown to dark gray or black (but often with a thin whitish covering of hyphae when very young), divided into small, often pyramidal warts. **INTERIOR** basically hollow, but often interrupted by irregular projections of sterile tissue from the outer wall; inner surface of wall warted and dark brown to blackish or bluish-gray; sterile tissue white to grayish. **SPORES** averaging 24-28 × 22-27 microns, elliptical to nearly round, hyaline (colorless) under the microscope, smooth at first but finely warted at maturity. Asci 8-spored, arranged in a palisade (hymenium).

HABITAT: Solitary to gregarious in humus or soil under oak, manzanita, coyote bush *(Baccharis)* and other trees and shrubs; known only from the west coast. It is fairly common (for a truffle) in California in the winter and spring, especially in February-March. I've found this species, *G. gardneri,* and *G. compacta* (see comments) under live oak. It also occurs in the Sierra Nevada. Like other Geneas, it often grows just below the ground or on the soil surface beneath the humus, and is difficult to see because of its color.

EDIBILITY: Prized by pigs, but I can find no mention of humans eating it.

COMMENTS: The hollow interior and dark warted exterior separate this common species from most other truffles. The shape ranges from nearly round to elaborately lobed or cerebriform (brainlike). Round specimens are reminiscent of miniature geodes because of their warted internal cavity, while the more knobby specimens look like bits of coral or piles of intertwined worms (see above photo). Other species: *G. gardneri* is very similar but has larger, coarsely warted spores; it grows under oak in coastal California. *G. compacta* is a similar but much paler (brown to yellowish-brown), knobby species with a hollow, convoluted interior (see photo below). I have found it several times under oak in the spring. It is edible but tasteless.

Left: *Genea gardneri* (see comments above). Note black hollow interior (below) and warted exterior. (Herb Saylor) **Right:** *Genea compacta* (see comments above), a common yellowish to brown, oak-loving species. Note knobby fruiting body, large "mouth" (specimen at center), and hollow interior.

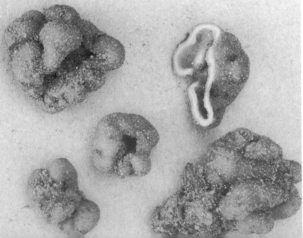

Genea intermedia   (Red Geode Truffle)

FRUITING BODY underground, 0.5-2.5 cm broad, sometimes roundish (especially when young), but more often lobed or knobby, with or without an apical opening to the interior and a basal mycelial tuft. Exterior reddish to reddish-brown, vinaceous, or vinaceous-purple, the lobes or protuberances superimposed with minute warts. **INTERIOR** basically hollow, but often interrupted by irregular projections of sterile tissue from the outer wall; inner surface warted, pink to whitish; sterile tissue also white to pinkish. **SPORES** averaging 36-40 microns, round, finely warted at maturity. Asci arranged in a palisade (hymenium), typically 8-spored.

HABITAT: Solitary to gregarious in soil in woods; known only from Oregon and California. It is not uncommon under conifers in the Sierra Nevada in the spring; I have yet to find it on the coast.

EDIBILITY: Unknown.

COMMENTS: The beautiful reddish, lobed and warted exterior plus the hollow interior make this one of the few truffles that can be recognized instantaneously in the field. No other *Genea* is as red, at least in California. (*G. harknessii* can be dark reddish-brown when fresh but soon blackens after picking.)

Genea arenaria   (Hairy Geode Truffle)

FRUITING BODY underground, 1-3 cm broad, usually irregularly lobed or coarsely knobby (at least at maturity), with or without an opening to the interior and a basal tuft of mycelium. Exterior brown to pale brown, divided into small, often pyramidal warts and covered with scattered long brown to dark brown hairs. **INTERIOR** basically hollow, but the cavity usually irregular in shape due to infolding of the outer wall; inner surface finely warted and colored like the exterior or paler; sterile tissue whitish. **SPORES** averaging 22-32 × (16) 20-24 microns, mostly elliptical, with minute scattered warts. Asci arranged in a palisade (hymenium), typically 8-spored.

HABITAT: Solitary, scattered, or in small groups in soil under trees (mainly live oak); known only from California and Oregon. It typically fruits from the late fall through the early spring (November-April) and is not uncommon in our area. However, it seldom occurs in quantity and is easily overlooked because of its brown color.

EDIBILITY: Unknown.

COMMENTS: This species is best recognized by the highly irregular shape at maturity, the hollow (but usually folded) interior, and the presence of hairs on the warted brownish exterior. Other species: *G. hispidula* is a similar species known from eastern North America; *G. kraspedestoma* is a reddish-brown to brownish Californian with a circular apical opening that is fringed by stiff incurved hairs; it has small spores (only 12-20 microns long) and was originally collected near Almaden, California, under oak. *G. compacta* (see comments under *G. harknessii*) is also similar, but lacks obvious hairs on its exterior.

Genabea cerebriformis   (White Geode Truffle)

FRUITING BODY underground, 0.4-1.5 (2.5) cm broad (but usually under 1 cm); smaller specimens often roundish, larger ones usually quite irregular (knobby, lobed, or brain-like) in shape and typically with several openings to the interior; lacking a basal tuft of mycelium. Exterior covered by small, more or less conical warts superimposed on each knob; white to yellowish-white or yellow-gray. **INTERIOR** often a single cavity in small specimens, but in larger ones usually consisting of a mazelike system of canals formed by infolding and inward projections of the wall; colored more or less like exterior. Odor mild or strong. **SPORES** 28-44 microns, round, smooth at first but covered with long, slender spines at maturity; hyaline (colorless) to grayish-yellow (in age) under the microscope. Asci arranged in a palisade (hymenium), typically 8-spored.

HABITAT: Solitary to gregarious in soil or humus in woods and under trees; fairly common (for a truffle) in western North America under various trees, but especially fond of Douglas-fir. In California it fruits, like other truffles, in the winter, spring, and early summer. Although small, its light color makes it fairly conspicuous.

EDIBILITY: I can find no information on it; too small to be of much value.

COMMENTS: Formerly known as *Genea (*or *Myrmecocystis) cerebriformis,* this is one of our most distinctive truffles. The yellowish-gray to white color plus the irregularly convoluted and warted exterior and complex interior with open canals (in larger specimens) distinguish it. The interior is vaguely reminiscent of *Geopora cooperi,* but that species is much larger and has a fuzzy brown exterior. Other species: *Genabea fragilis* (the type species of the genus) is a blackish species reported from Europe and Quebec; *G. spinospora* is a whitish species that has been found in Virginia.

BALSAMIA & Allies   (Smooth-Spored Truffles)

Small to medium-sized underground woodland fungi. FRUITING BODY round to somewhat flattened or irregular, often with a depression or cavity leading to the interior. Exterior variously colored, warted in *Balsamia,* warted or smooth in *Barssia* and *Picoa.* INTERIOR *solid with sterile veins and pockets of fertile tissue (Picoa), or with open or hyphae-stuffed veins or canals (Barssia & Picoa);* variously colored. STALK and columella absent, but a basal tuft of mycelium sometimes present in *Balsamia.* SPORES *smooth at maturity,* usually elliptical or elongated but sometimes round; colorless under the microscope in *Barssia* and *Balsamia,* sometimes brown in *Picoa.* Asci typically 8-spored, not amyloid, arranged in a palisade or randomly imbedded in the tissue.

THREE genera are treated here: *Balsamia, Barssia,* and *Picoa.* They are intermediate in aspect between the geode truffles *(Genea* and *Genabea)* and the true truffles *(Tuber* and allies). As a unit they are difficult to distinguish in the field because their unifying feature is microscopic: the spores are smooth even at maturity. On an individual basis, however, the three genera are easier to recognize. For instance, *Balsamia* can be told by the frequent presence of a basal mycelial tuft plus its warted exterior and pale (white to yellowish) marbled interior. *Barssia* typically has a broad depression or "mouth" at the top of the fruiting body and several open canals which empty into it. *Picoa,* on the other hand, has a solid interior plus a brown to blackish exterior. It is easily confused with the true truffles, but has a greener or grayer interior and smooth spores.

Balsamia, Barssia, and *Picoa* constitute the family Balsamiaceae, but their relationships to other families are unclear. They occur in a variety of habitats and seem to fruit relatively early for truffles, at least in our area. All three genera are small; one species from each is described here.

Key to Balsamia & Allies

1. Interior of fruiting body solid and usually gray to greenish-gray to greenish-blue at maturity; exterior slate-violet to black, minutely warted; basal mycelial tuft absent ....
    .... ***Picoa carthusiana,*** p. 854
1. Not as above; interior of fruiting body not solid, or if solid then remaining pallid 2

2. Interior of fruiting body with open canals which empty into a broad central depression at or near the top of fruiting body; exterior warted or not warted; basal mycelial tuft absent ....
    .... ***Barssia oregonensis,*** p. 853
2. Not as above; internal veins or canals empty *or* stuffed with hyphae; exterior finely warted; tuft of mycelium often present at base .... ***Balsamia magnata,*** p. 853

Balsamia magnata. Note marbled or veined interior and finely warted exterior. Unlike *Tuber*, the interior does not darken appreciably at maturity. (Herb Saylor)

Balsamia magnata

FRUITING BODY underground, 0.5-2 cm broad, round to somewhat compressed or flattened, the apex usually infolded and the base often with a tuft of mycelium. Exterior divided into numerous rounded to pointed warts, occasionally with small depressions; color variable: bright orange to reddish-brown to brownish-pink or occasionally black (but may be whitish when very young). **INTERIOR** white to pale yellowish, even when mature; composed of crowded folds which form mazelike canals, the canals united or separated into several chambers and either open or filled loosely with cottony hyphae; canals usually converging at the apex or sometimes at several points. **SPORES** 20-24 × 12-14 microns, variable in shape (cylindrical to elliptical to nearly round), smooth at maturity, hyaline (colorless) under the microscope, usually with three oil droplets. Asci mostly imbedded in the tissue between the veins or canals; typically 8-spored.

HABITAT: Solitary to gregarious (usually the latter) in soil under various trees and shrubs (oak, pine, madrone, etc.); common (for a truffle) in California and Oregon in the winter and spring; also reported from Arizona.

EDIBILITY: I can find no information on it.

COMMENTS: The prominently warted orange to reddish-brown exterior plus the pallid interior composed of open or stuffed, often united or converging canals are characteristic of this rather common truffle. It is most likely to be confused in the field with *Pachyphloeus citrinus* (which can also be bright orange), but the smooth spores and pale interior distinguish it. *Pseudobalsamia magnata* is an older alias, and the names *P. alba* and *P. nigrens* have been used for the whitish and black forms (species?), respectively.

Barssia oregonensis   (Depressed Truffle)

FRUITING BODY 1-2.5 cm broad, roundish but usually more or less flattened or slightly lobed, typically with a prominent depression or cavity at the top; firm but not hard, without a mycelial tuft at base. Exterior smooth to roughened or finely warted, pale ochre-buff to orange-cinnamon to brick-red or reddish-brown. **INTERIOR** composed of empty, unconnected, broad or narrow canals, many of which empty into the depression (i.e., more or less solid except for the canals); white to pale gray, even at maturity. **SPORES** 24-36 × 12-21 microns, oblong-elliptical, smooth, colorless under the microscope. Asci mostly 8-spored, forming a palisade (hymenium) that lines the canals.

HABITAT: Solitary or in small groups in soil in woods, associated with Douglas-fir and possibly other trees; known only from the West (California, Oregon, Idaho). Like most truffles, it is commonest in the spring and early summer. I have yet to find it in our area, but see comments.

EDIBILITY: Presumably edible; I haven't tried it.

853

COMMENTS: This truffle is best recognized by its color, the frequent presence of a prominent depression into which several canals empty, and the white or pallid interior. It might be mistaken for a *Genea* or *Genabea,* but is differently colored, not as warted, and has smooth spores at maturity. An unidentified yellowish to pale orange *Barssia* with a prominent broad apical depression has been found recently in the Guadelupe Mines area near Almaden, California (an area as richly endowed with truffles as it is with mercury).

Picoa carthusiana   (Oregon Black Truffle)

FRUITING BODY 0.5-4.5 (8) cm broad, round or nearly round to slightly irregular, without a mycelial tuft at the base. Exterior minutely warted, black to dusky slate-violet. INTERIOR more or less solid, composed of large pockets of fertile tissue marbled with paler (whitish to buff) sterile veins; fertile tissue whitish to buff when young but becoming grayish-green to greenish-blue in age; sometimes exuding a clear latex when fresh which slowly (overnight) stains white paper pale violet. SPORES (56) 74-84 × 20-35 microns, lemon- or spindle-shaped, smooth, typically with one giant oil droplet at maturity; pallid to greenish-yellow becoming brown at maturity (under the microscope). Asci typically 8-spored, imbedded in the tissue (not forming a palisade).

HABITAT: Solitary, scattered, or in small groups in soil and humus in woods; known from Europe and the western United States. In Oregon and California it favors Douglas-fir. In our area it fruits in the spring and summer, but is more common in Oregon, where it occurs earlier. It has been found at Point Reyes by Herb Saylor and Dennis Desjardin, in July.

EDIBILITY: Edible and delicious raw, according to truffle connoisseur Gary Menser.

COMMENTS: In the words of Herb Saylor (*Mycena News,* May 1983), "The general aspect of this fungus is that of a piece of animal dung, with which we frequently confused it while making the collection. It is possible that this may be one reason why it is infrequently collected, as animal dung of similar size and color was common in the area." The solid interior might lead to confusion with *Tuber* and allies, but its greenish to grayish color at maturity plus the large, smooth, spindle-shaped spores and minutely warted, blackish exterior form a distinctive set of characters.

TUBER & Allies   (True Truffles)

Small to fairly large, underground mycorrhizal fungi. FRUITING BODY round to copiously lobed, often hard (especially in *Tuber*). Exterior sometimes warted in *Tuber* and *Pachyphloeus,* otherwise not; variously colored. INTERIOR *typically solid and firm, usually marbled, often waxy, usually white or pallid when young but usually with dark fertile tissue at maturity or pockets of fertile tissue separated by paler walls.* STALK and columella absent; basal mycelial tuft also absent (except in some species of *Pachyphloeus*). SPORES *ornamented with spines, warts, pegs, pits, or ridges at maturity* (but smooth when young), round to elliptical in *Tuber,* round in other genera; *light to dark brown at maturity.* Asci typically 1-6-spored in *Tuber,* 4-8-spored in other genera, randomly imbedded in the tissue between the veins (or sometimes forming a palisade or hymenium in *Pachyphloeus* and *Choiromyces*); *not* amyloid.

THE "true" truffles or "earth nuts," as they are sometimes called, can be told from other truffles by their solid, marbled interior and ornamented spores. There are two families, the Tuberaceae (with one principal genus, *Tuber*) and the Terfeziaceae. *Tuber* is the largest and most famous genus of truffles. It includes the fabled black truffle (*T. melanosporum* —see photo on p. 842) and white truffle *(T. magnatum)* of Europe as well as a number of species endemic to North America. *Tuber* is an easy genus to recognize. The fruiting body is hard and easily mistaken for a small rock or acorn. The interior is solid and marbled (typically whitish when young but become brown or black with white veins at maturity), and has the consistency of wax, i.e., it flakes or chips like a candle. The exterior of the fruiting body is smooth in some species and warted in others, and may or may not be lobed. Microscopically, *Tuber* is distinct by virtue of its relatively large, round to elliptical, geometrically-patterned spores and one- to six-spored asci that are imbedded randomly in the tissue between the veins. However, most species of *Tuber* are practically indistinguishable from each other when young (i.e., without mature spores) and not much easier to differentiate at maturity. A few are distinctive in color, odor, and habitat, but most can only be identified by examining the spores under the microscope. Even then it isn't easy, because the ornamentation of the spores changes as they mature and the size is notoriously variable (a one-spored ascus tends to produce significantly larger spores than a two- or four-spored ascus in the same fruiting body). In other words, *Tuber* may be an easy genus to recognize, but the identification of its species often requires the services of a specialist.

The second family of "true" truffles, the Terfeziaceae, encompasses five genera. Microscopically these genera differ from *Tuber* in several respects (see the key), but they can often be told in the field on an individual basis. *Pachyphloeus,* for instance, has a more or less round, warted fruiting body with a grayish-olive to blackish-olive interior marbled with paler veins, and it often has a tuft of mycelium at its base; *Delastria,* on the other hand, is often pink- or reddish-tinged; *Terfezia* is partial to sandy soil in arid or semi-arid regions; *Hydnobolites* has a very distinctive gristly or fatty texture and pale color, while *Choiromyces* is even harder than *Tuber* and tends to be rougher and more copiously lobed.

Tuber is a fairly sizable genus, with about 60 known species and an equal number of synonyms. Roughly half of these species occur in California and Oregon, making the west coast the best truffle territory in North America. Tubers take an inordinately long time to mature—several weeks or even months. In our area they typically begin developing in the winter, which means they mature in the spring (March-June), after most other mushrooms have long departed. They are mycorrhizal with both hardwoods and conifers, but are particularly abundant under oak and Douglas-fir. Some of our species (e.g., *T. gibbosum*) are good edibles, though not as distinctively flavored, perhaps, as their European counterparts. Many other North American species have yet to be tried. As already mentioned, differentiating the various species can be extremely difficult. Fortunately, none are known to be poisonous. Alas, the famous European truffles *(T. magnatum, T. melanosporum, T. aestivum)* do not seem to occur here, though special truffle hounds have been flown in from Italy to look for them.

Terfezia is the largest genus in the Terfeziaceae. However, its dozen or so species occur mostly in southern and/or arid regions, and have yet to be found in California. The other four genera in the Terfeziaceae are very small. Little is known of the edibility of the North American representatives, but **Terfezia arenaria**, a large (5-12 cm) Mediterranean species that grows in sandy soil (often with rock rose or *Cistus*) is prized in Islamic countries and was a favorite with the Romans and Greeks. Four common Tubers and three members of the Terfeziaceae are described here, and several others are keyed out.

Key to Tuber & Allies

1. Fruiting body white or pale-colored when fresh; interior with a very distinctive gristle-like texture .... *Hydnobolites californicus,* p. 857
1. Not as above; interior not gristle-like 2

2. Exterior of fruiting body black to slate-violet or greenish-black, usually warted; interior usually greenish, grayish, or blackish at maturity (with paler veins) 3
2. Not with above features 4

3. Spores smooth; widely distributed, but on west coast occurring mainly with Douglas-fir ....
   .... (see *Balsamia* & Allies, p. 852)
3. Spores ornamented at maturity; found mainly in eastern North America and Europe ....
   .... *Pachyphloeus melanoxanthus* (see *P. citrinus,* p. 856)

4. Exterior of fruiting body usually warted and often brightly colored; interior either remaining pallid at maturity or becoming olive to grayish to blackish with paler veins; mycelial tuft often present at base of fruiting body; spores smooth or ornamented with spines or pegs 5

4. Not as above; mature interior usually brown to reddish with white veins (but usually pallid when young); exterior warted or not; mycelial tuft usually absent; mature spores ornamented .. 6

5. Spores smooth; interior whitish or pallid even in age .... (see *Balsamia* & Allies, p. 852)

5. Spores ornamented with pegs at maturity; interior pallid when young but becoming olive, grayish, or darker at maturity .... *Pachyphloeus citrinus* & others, below

6. Exterior of fruiting body covered with warts (warts often small) 7

6. Exterior of fruiting body smooth, cracked, downy, pitted, etc., but not warted 8

7. Found in eastern North America; exterior tawny becoming distinctly reddish or brown at maturity; interior usually brick-red or reddish-brown with paler veins (at maturity) ....
.... *T. canaliculatum* (see *T. gibbosum*, p. 858)

7. Not as above; found in western North America *T. murinum* & others (see *T. gibbosum*, p. 858)

8. All of the spores round at maturity and alveolate (pitted-reticulate) 9

8. Spores spiny or alveolate at maturity, at least some of them elliptical or broadly elliptical   10

9. Fruiting body 1-10 cm broad, often lobed and very hard; spores with numerous small pits like those on a golf ball .... *Choiromyces alveolatus*, p. 858

9. Not as above; fruiting body 1-3 (5) cm broad; exterior often with minute white hairs or patches of hairs (i.e., pubescent) .... *T. californicum* & others, p. 860

10. Associated with Douglas-fir; odor often garlicky *when mature* and the peridium (skin) often cracking in age; spores alveolate (pitted-reticulate) .... *T. gibbosum*, p. 858

10. Not as above 11

11. Found in Texas and along the Gulf Coast .... *T. texensis* (see *T. gibbosum*, p. 858)

11. Not as above 12

12. Spores spiny; exterior of fruiting body brown to cinnamon-colored when mature (but usually paler when young) .... *T. rufum*, p. 861

12. Spores alveolate (pitted-reticulate); exterior of fruiting body usually some shade of brown or yellowish-brown when mature, but sometimes reddish-brown (especially when old) ....
.... *T. separans* & many others, p. 859

Pachyphloeus citrinus   (Berry Truffle)   Color Plate 213

FRUITING BODY underground, 0.5-3 cm broad, more or less round (spherical) to slightly lobed or tapered below, firm but not hard; apex often with a depression, circular furrow, or cluster of furrows; base often with a tuft of mycelium. Exterior usually divided into polygonal warts; color variable, but usually bright to dull orange to brown. **INTERIOR** more or less solid, composed of sterile veins which often converge toward the apex or depression and form elongated pockets of fertile tissue between them; entirely whitish when young, becoming grayish to grayish-olive with paler (white to yellowish) veins, and eventually becoming dark olive to blackish with yellow to pale olive veins. Odor of mature specimens sometimes pungent ("like rotting weeds"—Herb Saylor) after collecting, at other times mild. **SPORES** (11) 13-21 microns, round, smooth and hyaline (colorless under the microscope) at first, becoming spiny, and at maturity the spines enlarging into broader, conical to truncate warts or "pegs" that look like miniature golf tees; usually yellowish at maturity. Asci typically 8-spored, forming an irregular palisade (hymenium) along the sterile veins and/or randomly imbedded in the surrounding tissue.

HABITAT: Solitary to gregarious in soil under both hardwoods and conifers; very widely distributed (throughout most of North America and Europe). In our area it fruits, like other truffles, in the winter, spring, and early summer. I have found it several times under tanoak and madrone in June and July, and it has turned up repeatedly under live oak in the Guadelupe Mines area near Almaden, California.

EDIBILITY: Specimens I sampled had little taste, but were immature.

COMMENTS: This species appears to be one of the most widespread of all the truffles. The combination of orange to brown, warted exterior and solid, grayish-olive to blackish interior is distinctive. The "pegs" on the spores are also unusual—at maturity each is usually tipped with a small depression that makes it look like a golf tee. The fruiting bodies are not nearly as hard as Tubers, and are usually rounder. Local material is usually bright orange when immature and has the aspect of a madrone berry (see color plate) or the fruit from a strawberry tree (a European madrone). Other species: *P. virescens* is said to be similar, but has a dull green exterior and yellower interior. It was originally collected in Los Gatos, California, but is also reported from Nebraska! *P. melanoxanthus* of eastern North America and Europe has a black to greenish-black, warted exterior and a grayish to blackish interior marbled with hollow or greenish veins (at maturity) and sometimes has a short "stalk" of mycelial fibers. *P. conglomeratus* has slightly amyloid asci.

Hydnobolites californicus   (Gristly Truffle)

FRUITING BODY underground, 0.5-4 cm broad, roundish to oval, usually lobed or folded but quite compact, gristly or fatty and rubbery in texture (especially the interior), *not* hard. Exterior often roughened but lacking warts, whitish to buff or dingy yellowish, often becoming ochre-tan or dingy brown in age or as it dries. **INTERIOR** more or less solid, with several narrow, meandering sterile veins or canals; white to slightly grayish, but often discoloring brown when cut and dried or in old age. Odor mild or sometimes musty at maturity. **SPORES** 14-18 (24) microns, round, very coarsely alveolate at maturity, the ridges of the alveoli (pits) projecting like needles from the edges of the spore; hyaline (colorless) to yellowish or pale brown under microscope. Asci mostly 8-spored, imbedded in the tissue between the veins, not arranged in a palisade (hymenium).

HABITAT: Solitary, scattered, or in groups or pockets of several individuals in soil and humus under trees and in woods. As its name implies, it was originally discovered in California, but appears to be widely distributed in North America. In our area it is fairly common from January to July under many trees, but especially oak.

EDIBILITY: Prized by slugs, despised by humans. The texture is fatty but the flavor is not.

COMMENTS: This truffle is easily recognized by its pale color and gristly or fatty consistency. The latter feature is particularly striking and almost without parallel among the truffles. The interior is whitish except in old age and the fruiting body is never as hard as a *Tuber*. The *very* coarsely alveolate spores are also diagnostic. Other species: *H. cerebriformis* of Europe is said to have larger spores; it has also been found in Iowa.

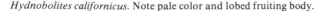

Hydnobolites californicus. Note pale color and lobed fruiting body.

Choiromyces alveolatus   (Hard Truffle)

FRUITING BODY usually underground, (0.5) 1-10 cm broad, nearly round or more often lobed or knobby (potato-like); very hard, without a mycelial tuft at the base. Exterior usually roughened or minutely downy, whitish when very young becoming yellowish to tawny, brown, or rusty-brown in age. **INTERIOR** more or less solid and very firm, consistency rather like hard wax; at first white or pallid, but becoming marbled with darker (yellowish to orange, yellow-brown, or rusty-brown) veins or chambers which are usually solid but occasionally empty. Odor mild or distinctive (see comments). **SPORES** 20-30 (36) microns, round, smooth at first but covered with numerous rounded pits (alveolate) like a golf ball at maturity; yellowish to brown under the microscope. Asci mostly 8-spored (but many appearing 1-4-spored in younger specimens), arranged in a palisade (hymenium) lining the veins or chambers (but see comments).

HABITAT: Solitary to gregarious in soil under trees and in the woods; known only from western North America, occasional (if you're looking for truffles) in the late winter, spring, and early summer. It is particularly numerous in the Sierra Nevada, but also occurs along the coast. Like the Tubers, it takes several weeks or months to mature.

EDIBILITY: Tempting and probably edible, but I can find no mention of anyone eating it.

COMMENTS: This is a variable species as evidenced by its plethora of pseudonyms (e.g., *C. cookei, Piersonia alveolata, P. bispora*). It is likely to be mistaken for a *Tuber* because of its solidity, but microscopic examination reveals that the spores are finely pitted like golf balls and the spore-bearing cells (asci) are arranged in nests or a palisade rather than being randomly imbedded in the tissue. (Apparently the spores tend to form first in "nests" which represent the inner termination of the veins; it is only in older specimens that they line the entire lengths of the veins.) It is one of our largest truffles, capable of attaining the size of a fist! The odor is sometimes distinctive. Helen Gilkey says of one collection: "Odor at first resembling desiccated coconut, changing as [it] dries to that of strong cream cheese."

Tuber gibbosum   (Oregon White Truffle)

FRUITING BODY usually underground, 1.5-5 (8) cm broad, nearly round to irregularly knobby or potato-like; firm or hard, without a basal mycelial tuft. Exterior minutely downy or irregularly roughened but not warted, whitish when young becoming pale buff to tan or brown, then usually developing darker (reddish to purple-brown) areas when fully mature; often cracking in age. **INTERIOR** solid, marbled, crisp; whitish when young, the fertile tissue becoming brown to dark brown to brick-red when mature, the meandering sterile veins remaining whitish. Odor usually strong and garlicky *when fully mature*. **SPORES** 35-52 × 17-40 microns, elliptical or elongated, reticulate-alveolate (ridged and pitted with shallow depressions) when mature but smooth when very young; dark yellow-brown to brown under the microscope when mature. Asci mostly 1- to 6-spored, randomly imbedded in the tissue between the veins.

HABITAT: Solitary, scattered, or gregarious in woods and at their edges, associated mainly if not exclusively with Douglas-fir (usually trees between the ages of 8 and 65 years); found from California to British Columbia, but especially common in Oregon. Although it normally grows underground, I have found specimens on the surface. (They were probably dug up by squirrels, then rejected for reasons known only to squirrels.)

EDIBILITY: Edible and choice, but widespread collecting can be destructive! It smells like the white truffle of Europe *(T. magnatum)*, and some people proclaim it just as good. It can now be bought—for a *slightly* more reasonable price than *T. magnatum*.

COMMENTS: Species of *Tuber* are often difficult to identify in the field, but this one can be told by its growth with Douglas-fir, relatively large size (when mature), tendency

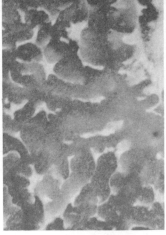

Left: *Choiromyces alveolatus* (p. 858). This specimen is over 5 cm broad! (Herb Saylor) **Right:** Close-up of the marbled interior of a mature *Tuber separans*. When young the interior is whitish. Other Tubers have very similar interiors. See next page and color plate for views of exterior.

to develop cracks in age, and strong garlicky odor (when present). Like other Tubers, it has a marbled white-and-brown (or reddish-brown) interior when mature. The narrowly elliptical spores are also very distinctive, providing you have a microscope. Other species: *T. besseyi* is similar, but has an "olive-buff" exterior and slightly longer spores. *T. canaliculatum* is a sizable choice edible with a distinctly warted, brown to reddish or tawny exterior. It is found in the summer and fall in eastern North America (a region not known for its truffles). Several western truffles also have a warted exterior (e.g., *T. murinum, T. linsdalei, T. gardneri,* and *T. harknessii*), but they are difficult to distinguish without a microscope. *T. texensis* (of Texas, naturally) should also be mentioned. For other species with elliptical, alveolate spores, see *T. separans.*

Tuber separans (Acorn Truffle)   Color Plate 212

FRUITING BODY underground, very firm or hard; (0.7) 1-4.5 cm broad, round to oval or sometimes lobed (potato-like), but usually with no more than three major lobes; lacking a basal mycelial tuft. Exterior smooth to very minutely roughened but not warted; whitish when very young but soon becoming uniformly pale brown to dingy yellow-brown (about the color of an oak gall), then eventually developing dark reddish-brown to dark brown areas (or darkening overall) as it reaches full maturity. **INTERIOR** solid, very firm and crisp, flaking or chipping like wax, marbled; white when young, the fertile tissue becoming light brown and then dark brown with age and the meandering sterile veins remaining whitish. Odor at maturity slight, difficult to describe. **SPORES** (30) 34-58 × 28-50 (56) microns, broadly elliptical to round, brown and alveolate (pitted-reticulate) at maturity, with few to many pits. Asci 1-4 (6)-spored, imbedded randomly in tissue between veins.

HABITAT: Widely scattered to gregarious in soil under oak and other hardwoods; known only from the west coast, fruiting mainly in the spring and early summer (at least in our area). I have found more than sixty specimens growing together in loose soil, associated with tanoak or possibly madrone, in June and July.

EDIBILITY: Edible. It is mild or slightly nutty like *Boletus edulis,* but much crisper. Slice it thinly and saute *very* briefly (about one minute) or the flavor will be lost.

COMMENTS: The above description is drawn from a single large collection made near Santa Cruz, California, and thus may not represent the full range of variation within the species. The yellowish to brown color of the hard, marble- to walnut-sized specimens was quite constant, and they were frequently confused with acorns buried in the duff (see photo on p. 860). The growth with hardwoods and lack of a strong odor at maturity distinguish it from *T. gibbosum,* while the exterior is not pubescent as in *T. californicum*

859

Can you find the acorn among these prime examples of *Tuber separans?* If not, you aren't alone—
I didn't discover it until I'd brought home all of these truffles, which were found under a single tanoak.
Many other objects or "pseudotubers" (rocks, animal dung, etc.) can also be mistaken for truffles!

and the different color and alveolate spores separate it from *T. rufum.* However, there are
many very similar species with elliptical to nearly round, alveolate spores that can only be
differentiated with great difficulty. Part of the problem is that there are no up-to-date
keys available for the North American species, and the spore sizes (an important
feature) are useful only when correlated with the number of spores in each ascus. In other
words, the identification of most Tubers is best left to truffle experts such as James Trappe
(who identified the above-mentioned collection as *T. separans*). Among the many other
Tubers with elliptical, alveolate spores are: *T. monticola,* a rare species found under
conifers in the Sierra Nevada; *T. citrinum,* also rare, with a smooth, pale yellow exterior
in youth; *T. dryophilum,* a very widely distributed species with smaller, coarsely alveolate
spores and a yellowish-brown exterior at maturity; *T. levissimum,* a thicker-skinned
species that is also widespread; *T. shearii,* with large, coarsely alveolate, broadly elliptical
spores; and *T. irradians,* with many nearly round, coarsely alveolate spores. All of these
species have spores about the same size as *T. separans* or smaller. See also *T. gibbosum.*

Tuber californicum   (California Truffle)

FRUITING BODY underground, (1) 1.5-3 (5) cm broad, round to oval or more often
irregularly lobed and/or pitted; very firm, without a basal mycelial tuft. Exterior minutely
but evenly pubescent (covered with tiny hairs) and whitish when young, becoming mottled with darker (olive to dingy ochre to brown) areas as it matures, but usually retaining
patches of the white pubescence; sometimes cracked in age. **INTERIOR** solid, firm,
marbled, chipping like wax, whitish when young, becoming dark brown with large
meandering white sterile veins at maturity. Odor often distinctive when old (rather cheesy
or like "gourmet" crackers). **SPORES** (30) 39-52 microns, round, alveolate (reticulate-
pitted), brown at maturity. Asci 1-4 (6)-spored, imbedded randomly in flesh between veins.

HABITAT: Solitary to gregarious in or on soil (but under humus) in woods and at their
edges; common (for a truffle) in California and Oregon, also reported from Idaho and
Ohio. In our area it is sometimes abundant under oak in the late winter and spring, but I
have also collected it on numerous occasions under Douglas-fir as late as July. The fruiting
bodies are often attacked by slugs, nematodes, and fly larvae.

860

Tuber californicum. This common species always has round (spherical) spores, but can usually be recognized in the field by the patches of white pubescence on its exterior.

EDIBILITY: Edible. Some people detect a bitter taste, but the specimens I sampled had a very strong mushroomy flavor that would go well in sauces or gravies.

COMMENTS: The critical feature of this *Tuber* is its uniformly *round* (not round to elliptical), alveolate spores. However, it can usually be told in the field by its pubescent or downy exterior (use a hand lens!) and tendency to be quite knobby. *T. sphaerosporum* also has uniformly round spores, but it lacks the pubescent exterior of *T. californicum* and has fewer and larger pits on its spores. It occurs in eastern North America and Gary Menser has found it under willow in Colorado. For alveolate-spored species with at least some elliptical spores, see *T. separans* and *T. gibbosum.*

Tuber rufum   (Cinnamon Truffle)

FRUITING BODY underground, 0.5-2.5 (3.5) cm broad, round to somewhat irregular or potato-like, with or without one or more furrows; very firm or hard, without a basal tuft of mycelium. Exterior smooth or broken up into patches or "eyes" but not conspicuously warted, color variable: brown to cinnamon when mature (brown to orange- or reddish-brown), usually duller and lighter (whitish to light brown) when young; sometimes with whitish, pinkish, or golden areas mixed with darker shades. **INTERIOR** solid, firm, chipping like wax, marbled; whitish when young becoming grayish-brown to brown in age; sterile veins large, remaining whitish. Odor not very distinctive. **SPORES** 20-48 × 17-32 microns, elliptical to nearly round, brown and covered with spines at maturity. Asci 1-4 (7)-spored, randomly imbedded in the tissue between the veins.

HABITAT: Solitary to gregarious or clustered beneath the soil under oaks and other trees; widely distributed and common. In California it occurs principally with live oak and is our most common *Tuber.* I have collected hundreds of marble-sized specimens under oaks in the spring. It also fruits in the late winter, but takes several weeks to mature.

Tuber rufum (also known as *T. candidum*) is our most common truffle. The paler specimens in this photo are younger, the darker ones older. Note the relatively smooth exterior and marbled interior.

EDIBILITY: Edible and fairly good; it has a faintly nutty flavor.

COMMENTS: Also known as *T. candidum* and more exactly called *T. rufum var. nitidum,* this common truffle is best told in the field by its modest size and brown to cinnamon-colored peridium (at maturity), and in the laboratory by its spiny rather than pitted or reticulate spores. The latter feature sets it apart from most other North American Tubers. As in other Tubers, the interior is solid and marbled. Other species: *T. harknessii* is an oak-loving western species with spiny spores and a distinctly warted exterior.

ELAPHOMYCES   (Deer Truffles)

Small to medium-sized, underground, mycorrhizal fungi *frequently incrusted with dirt.* FRUITING BODY round to oval, *usually hard and lacking an obvious base.* Exterior smooth or more often covered with small warts. *Peridium (skin) thick (usually 2-5 mm), tough and usually hard.* INTERIOR *a single large cavity soon filled with cottony white tissue that becomes dark (brown to black) and powdery at maturity;* sterile white bands of tissue often visible in intermediate stages. STALK and columella absent; sterile base and mycelial tuft also absent. SPORES *usually dark brown to blackish at maturity,* often so dark under the microscope that the ornamentation is difficult to make out; round, ornamented with warts or spines. Asci typically round at maturity (often irregular when young), 8-spored, not forming a hymenium, soon disintegrating, not amyloid.

THESE thick-skinned truffles are unique among the Ascomycetes in having a dark powdery spore mass at maturity. The powdery texture, which results from early disintegration of the asci,* plus the thick skin can lead to confusion with the earthballs *(Scleroderma).* However, earthballs usually grow near the surface of the ground and have a distinct base or point of attachment to their substrate, and are not usually incrusted with rootlets (mycorrhizae), whereas *Elaphomyces* species grow up to 12 inches under the soil, lack a distinct base, and are usually incrusted with dirt and mycorrhizae.

Elaphomyces differs from other truffle genera in several additional respects. The wall of the fruiting body, in addition to being thick and hard, is marbled in some species (when sectioned). Also, the interior, although apparently homogeneous in old age, is actually divided into several large chambers by thick white bands of sterile tissue when younger. Finally, *Elaphomyces* is the only genus of fungi to be parasitized by certain species of *Cordyceps* (see p. 879). For these reasons, among others, *Elaphomyces* has traditionally been separated from other truffles and placed in its own order and family. However, it is now thought to be less dissimilar to other truffles than once believed.

Elaphomyces is mycorrhizal with both hardwoods and conifers. In our area it fruits throughout the mushroom season and can even be found in the summer if you look hard enough (or dig deep enough!). It is among the most abundant of all underground mushrooms, but is seldom seen or collected, perhaps because the soil-incrusted fruiting bodies look like balls of dirt. In Europe, species of *Elaphomyces* have been used as aphrodesiacs and cheap truffle substitutes, but their rindlike skin is too tough and the mature spore mass too powdery to be worth eating. Over 30 species have been described, based largely on microscopic criteria. Two common and farflung representatives are depicted here.

Key to Elaphomyces

1. Fruiting body with a distinct base or point of attachment to its substrate; growing underground or above it; spores borne on basidia .... (see *Scleroderma,* p. 707)
1. Fruiting body usually lacking an obvious base, but exterior often incrusted with dirt and mycorrhizal rootlets; growing underground; spores borne inside asci 2
2. Peridium (skin) marbled when sliced open (as shown on next page) *E. muricatus* group, p. 863
2. Peridium not marbled when sectioned .... *E. granulatus* group, p. 864

*The early disintegration of the asci is a unique and puzzling feature of *Elaphomyces* that has caused even mycologists to confuse it with *Scleroderma.* The spores are often highly compressed and irregular in shape while inside the asci, presumably as a result of pushing against the ascus wall. It may very well be that this pressure causes the asci to pop like balloons!

Elaphomyces muricatus group, mature specimens. Note how spore mass is dark and powdery in the two sectioned specimens (but the central core is still cottony in one of them), and how the exterior of central specimen is incrusted with dirt and mycorrhizal rootlets.

Elaphomyces muricatus group   (Marbled Deer Truffle)

FRUITING BODY underground, 2-5 cm broad, round to oval or somewhat lobed, very firm. Exterior yellow-brown to ochre-brown and covered with minute, hard, pointed warts that give it a pimpled appearance, but the warts usually obscured by a crust of soil and tiny rootlets (mycorrhizae) that is easily stripped or brushed away. Peridium (skin) thick (2-5 mm), hard and rindlike, marbled when sectioned (dark brown to purplish-brown with whitish to vinaceous-tinged veins). **INTERIOR** at first hollow, soon stuffed with cottony white hyphae, then darkening to grayish, lilac, or purplish and divided into chambers by sterile bands, finally become brownish-black to black and uniformly powdery. Odor not very distinctive. **SPORES** 18-30 × microns, round, warted or warted-spiny, dark brown to black under the microscope. Asci round to pear-shaped or irregular, mostly 8-spored, not forming a palisade (hymenium) and disintegrating soon after the spores form (and before they are completely mature).

Elaphomyces muricatus group. Close-up of a sectioned specimen, showing the thick marbled peridium (skin). The spore mass (interior) is still cottony and divided into chambers by white sterile tissue.

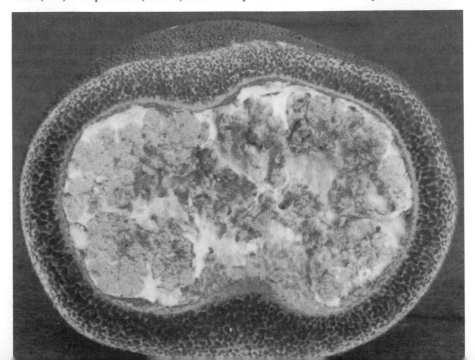

Elaphomyces muricatus group. Note how the thick skin is marbled in sectioned specimen at left, and how the exterior is finely warted in specimen on right. *E. granulatus* and its close relatives (not illustrated) look quite similar, but do not have a marbled peridium.

HABITAT: Solitary to gregarious in soil or duff in woods; widely distributed and fairly common (if you're digging for truffles). It is said to prefer pine woods, but in our area I have found it as early as October under knobcone pine and manzanita and as late as July under tanoak and madrone.

EDIBILITY: Unknown (but see comments on edibility of *E. granulatus*).

COMMENTS: This lesser-known cousin of *E. granulatus* is best told by its marbled peridium (see photographs). Like other *Elaphomyces* species, it is sometimes parasitized by *Cordyceps,* and is easily told from other truffles by its thick rindlike skin and dark, cottony to powdery mature spore mass (interior). The latter features can lead to confusion with the earthballs *(Scleroderma),* which, however, do *not* have a marbled peridium. Other species of *Elaphomyces* with a marbled peridium (e.g., **E. verrucosum, E. variegatus**) differ from *E. muricatus* microscopically.

Elaphomyces granulatus group (Common Deer Truffle)

FRUITING BODY underground, 2-5 cm broad, round to somewhat oval, very firm. Exterior usually covered with small hard warts (but these often hidden by a crust of soil, yellowish mycelium, and/or mycorrhizal rootlets); sometimes pallid when young but usually pale to dingy ochraceous or yellow-brown. Peridium (skin) rindlike, very firm and thick (2-5 mm), showing a very thin yellowish outer layer (when sectioned) and a thick white to grayish inner layer that may feature darker (brown) zones, but *not* marbled. **INTERIOR** at first hollow, soon stuffed with cottony tissue, eventually becoming powdery when spores mature; white at first, soon grayish to purplish (and often separated into chambers by whitish bands), finally becoming blackish and powdery. Odor not very distinctive. **SPORES** 24-45 (65) microns, round, thick-walled, blackish-brown to very dark reddish-brown under the microscope, ornamented with short spines or warts. Asci mostly 8-spored, round to pear-shaped, not forming a hymenium, disintegrating before the spores are fully mature.

HABITAT: Solitary, scattered, or gregarious in soil or humus under conifers or less commonly hardwoods; widely distributed and very common, but seldom seen by the average mushroom hunter because it grows underground. It is usually found 2-3 inches (5-8 cm) below the surface, often imbedded in or resting on clay soil at the point where it meets the humus layer. In many regions it is parasitized by species of *Cordyceps*—which serve as an above-ground indicator of its presence—but I have yet to observe this in our area. It fruits throughout the mushroom season and can sometimes be gathered by the bushel. I have found it under hemlock and pine in northern California, and Herb Saylor reports prolific fruitings from Mendocino County. Alexander Smith calls it "perhaps the most common hypogeous [underground] fungus in North America"—and he should know, since he has collected so many of them!

EDIBILITY: Edible according to some reports, but not choice. In Europe it has been used for centuries as an aphrodesiac and truffle-substitute.

COMMENTS: The thick rindlike skin and purple-gray to black, cottony to powdery interior distinguish *Elaphomyces* from all other underground Ascomycetes. *E. granulatus* and its close relatives differ from the *E. muricatus* group in having a non-marbled peridium (as seen in sectioned specimens). The earthball genus *Scleroderma* is similar, but produces spores on basidia, usually has a distinct base or point of attachment to the substrate, lacks the small hard warts and outer crust of soil and mycorrhizae so frequently found in *E. granulatus,* and usually grows nearer to the ground surface (or often above it). Other species of *Elaphomyces* with a non-marbled peridium are best differentiated microscopically. One, *E. subviscidus,* has a smooth skin, dark brown mature spore mass, and smaller spores.

Earth Tongues

HELOTIALES

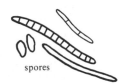

spores

THIS large order includes hundreds of small stalked or cuplike Discomycetes with inoperculate asci (i.e., each ascus has a pore at its tip through which the spores are expelled, but no operculum or "lid" as in the Pezizales). The most conspicuous members of this order are called earth tongues because of their clublike to tongue-shaped fruiting bodies. Many others are cup-shaped or disclike, with or without a stalk.

The larger members of this order are saprophytic on soil, humus, and wood, while most of the smaller types are parasitic or saprophytic on plant stems, leaves, and other tissues. None are prized edibles, being too small or too tough or too small *and* too tough to bother eating. There are several families and over 150 genera in the Helotiales. These are defined largely on the basis of microscopic features, and only a smattering of the larger or more colorful species are described here. These have been divided into five groups, keyed below, based on the shape and texture of the fruiting body.

Key to the Helotiales

1. Fruiting body with a cap and stalk, the cap rounded to convex or wrinkled but *not* cuplike (concave) or disclike 2
1. Fruiting body variously shaped, sometimes with an enlarged "head," but without a clearly differentiated, rounded to convex or wrinkled cap 4

2. Stalk very thin (usually less than 2 mm); cap bladderlike (hollow), whitish to yellowish; found in eastern North America, often clustered; spores borne on basidia (see **Aphyllophorales,** p. 548)
2. Not as above; spores borne inside asci 3

3. Cap with a sterile underside and an abrupt edge or margin (but the margin usually inrolled or tucked in toward the stalk) .... *Leotia* & *Cudonia,* p. 872
3. Not as above; cap merely an enlarged, differentiated "head" that lacks an abrupt margin and sterile underside 4

4. Flesh gelatinous or rubbery-gelatinous; fruiting body variously shaped but not clublike, pinkish to reddish, purplish, brown, or black; growing on wood .... *Bulgaria* & Allies, p. 875
4. Not as above 5

5. Fruiting body cuplike or disclike, with or without a stalk .... *Ciboria* & Allies, p. 877
5. Fruiting bod erect, clublike or with an enlarged or flattened "head" 6

6. Fruiting body with large internal chambers or compartments ..... (see **Helvellaceae,** p. 796)
6. Not as above 7

7. Entire fruiting body black or sometimes dark brown; at least some of the spores brown under the microscope .... *Geoglossum* & *Trichoglossum,* p. 866
7. Not as above; fruiting body usually lighter or brighter than above; spores hyaline (colorless) under the microscope or tinged yellow .... *Microglossum, Spathularia,* & Allies, p. 868

GEOGLOSSUM & TRICHOGLOSSUM
(Black Earth Tongues)

Small fungi usually found on ground or moss, sometimes on rotten wood. FRUITING BODY
upright, usually clublike, with or without an enlarged "head"; dark brown to black. Flesh thin,
tough, *not* gelatinous. STALK present, *usually slender; smooth or velvety.* SPORES very long
and narrow, usually septate, smooth, *brown under the microscope* (at least some of them). Asci
typically 8-spored, lining the upper portion of the fruiting body or the "head" if one is present;
inoperculate and not amyloid, with a pore at the tip.

THESE attractive little Ascomycetes are easily recognized by their black (or occasionally
dark brown) clublike fruiting bodies. Their color separates them from the fairy clubs
(Clavariaceae) and other earth tongues, and they lack the pimpled surface or white spore
powder so often found in *Xylaria.* Their very long (up to 250 microns!) brown partitioned
(**septate**) spores are also distinctive.

Both *Geoglossum* and *Trichoglossum* are charter members of the family Geoglossaceae.
In the more common of the two, *Trichoglossum,* brown lance-shaped cells called **setae**
protrude from the surface of the fruiting body, giving it a velvety texture. In *Geoglossum,*
the setae are absent (at least in the fertile portion), and the texture varies from smooth
to viscid to only slightly velvety.

The black earth tongues are saprophytic on humus, soil, moss, or occasionally rotten
wood. Several of the more than two dozen North American species are common, but all
are difficult to distinguish from their surroundings because of their dark color and small
size. They are also difficult to distinguish from each other, even with a microscope. For
this reason, only one representative from each genus is described here. Neither is worth
eating.

Key to Geoglossum & Trichoglossum

1.   Flesh usually white; exterior of fruiting body often roughened or minutely pimpled but not hairy
     or velvety; growing on wood (but wood often buried); asci borne in flasklike "nests" (peri-
     thecia) imbedded in the fruiting body .... (see **Pyrenomycetes,** p. 878)
1.   Not as above 2
2.   Surface of fruiting body distinctly viscid when moist, often glistening, *not* velvety ....
     .... **G. glutinosum** & others, below
2.   Fruiting body not viscid 3
3.   Surface of fruiting body (especially stalk) distinctly velvety; fruiting body often (but not always!)
     with a spade-shaped or flattened "head" .... **T. hirsutum** & others, p. 867
3.   Not as above; fruiting body not velvety or only very slightly so, usually clublike or twisted, but
     sometimes with a distinct "head" ... **G. nigritum** & many others (see *G. glutinosum,* below)

Geoglossum glutinosum   (Viscid Black Earth Tongue)

FRUITING BODY 1.5-6 cm tall, cylindrical to club-shaped. Upper (fertile) portion 3-6
mm wide, often flattened or slightly twisted but otherwise not sharply differentiated from
lower portion (stalk); surface black, smooth, viscid (at least when moist). Flesh tough,
usually brownish, not gelatinous. **STALK** occupying lower 1/3-2/3 of fruiting body, 2-3
mm thick, dark brown to black, viscid when moist, usually smooth. **SPORES** 60-90 × 4-5
microns, greatly elongated, smooth, brown under the microscope, with 0-7 (usually 3 or 7)
septa (partitions).

HABITAT: Solitary, scattered, or in small groups in humus, moss, or sometimes rotten
wood, usually in the woods; widely distributed but infrequently encountered. I've found
it several times in our area in the late fall, winter, and early spring, but never in quantity.

EDIBILITY: I know of no one who bothers collecting Geoglossums, but Captain Charles
McIlvaine says of this species: "Over a quart found in one patch. Stewed it is delicious."

Left: *Geoglossum glutinosum.* Note all the debris clinging to the viscid fruiting bodies. **Right:** *Geoglossum nigritum* (see comments under *G. glutinosum*) is a fairly common species with a clublike or twisted fruiting body that is *not* flagrantly velvety.

COMMENTS: This dark earth tongue has many look-alikes (see below), but is one of the few species with a distinctly viscid, glistening surface when moist. Similar viscid species include: *G. affine,* rare, with shorter spores; and *G. difforme,* larger (3-12 cm tall), whose spores have 8-15 septa. The genus *Geoglossum* also includes many similar dark brown to black, non-viscid earth tongues that can only be differentiated microscopically. These species are not as velvety as *Trichoglossum* and do not often have a well-defined "head." Some of the more common and widespread ones are: *G. glabrum,* a smooth-stalked species up to 10 cm tall; *G. simile,* the most common species in eastern North America, stalk often scurfy or minutely scaly; and *G. nigritum* (see photo), the most common species in our area (but more widely distributed), with strongly curved paraphyses (sterile cells) whose tips are scarcely enlarged. All of these have brown spores, but some Geoglossums have both brown and hyaline (colorless) spores, including: *G. fallax,* whose hyaline spores are non-septate; and *G. alveolatum* and *G. intermedium,* whose hyaline spores are septate. There are also several similar, dark *Microglossum* species whose spores are all hyaline, including: *Microglossum atropurpureum,* fruiting body dark brown to purplish or black; *M. fumosum,* yellow-brown to brown; and *M. olivaceum* (see comments under *M. viride*), greenish-brown to dark brown with very short spores (10-18 microns long).

Trichoglossum hirsutum (Velvety Black Earth Tongue)

FRUITING BODY 2-8 cm tall, cylindrical to club-shaped or more often with a distinct "head" (i.e., fruiting body shaped more or less like the tongue of a bell). Fertile "head" when distinct 3-8 mm broad, usually flattened laterally or compressed, oval or elongated to spade- or arrowhead-shaped; surface dry, minutely hairy or velvety, sometimes wrinkled, black. Flesh thin, tough, usually brownish. **STALK** thin (1-4 mm thick), more or less equal, densely velvety, often twisted or curved, tough, black. **SPORES** 80-195 (210) × 5-7 microns, greatly elongated, smooth, brown under the microscope, typically with 15 septa (partitions) when mature, but some varieties with consistently fewer or more septa. Both stalk and head lined with long brown sterile cells (setae). Asci 8-spored.

HABITAT: Solitary, scattered, gregarious, or tufted in humus, moss, or soil (or occasionally on rotten wood), usually in woods; very widely distributed and common. It is the most abundant earth tongue in our area, sometimes carpeting large tracts of humus with

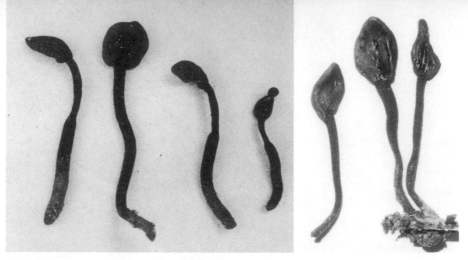

Trichoglossum hirsutum. Note the often spade-shaped "head" of this common black earth tongue. **Left:** Typical fruiting bodies. **Right:** These specimens were photographed with a strobe in an effort to highlight some of the minute hairs that give them a velvety texture.

its little black clubs. In seems to favor habitats shunned by other mushrooms (e.g., redwood), and usually fruits in the winter or spring. Like other black fungi, it is difficult to see.

EDIBILITY: Supposedly edible, but much too tough to be worthwhile.

COMMENTS: This dainty earth tongue is easily recognized by its black velvety fruiting body. In California specimens the fertile portion is usually (but not always) set off from the stem as a thickened or flattened, often spade-shaped "head." The wonderful velvety texture is caused by hundreds of minute projecting hairs or spines (setae) and is most evident on the stalk, especially in dry weather. *Geoglossum* species are very similar, but are not as velvety and do not normally have such a well-defined "head." Other species of *Trichoglossum* can only be differentiated microscopically. They include: *T. velutipes,* with 4-spored asci and mostly 7-11-septate spores; and *T. farlowii,* with 8-spored asci and 0-5 (usually 3)-septate spores.

MICROGLOSSUM, SPATHULARIA, & Allies
(Colorful Earth Tongues)

Small fungi found on ground, moss, or rotten wood. FRUITING BODY *upright, usually clublike to spatula- or tongue-shaped, or with an enlarged fertile "head";* variously colored *but not dark brown to black.* Flesh often rather tough, *not* gelatinous. STALK usually present, often slender; smooth, mealy, minutely scaly, or velvety. SPORES round to elliptical, spindle-shaped, or needle-like, smooth, *hyaline (colorless) under the microscope,* sometimes septate. Asci lining upper part of fruiting body or confined to the "head" if one is present; typically 8-spored, not amyloid, with a pore at the tip but not operculate.

THESE earth tongues are more cheerfully colored than *Geoglossum* and *Trichoglossum,* and their spores are colorless when viewed under the microscope. They may superficially resemble the unbranched coral fungi or fairy clubs (Clavariaceae), but bear their spores in asci rather than on basidia and usually have a swollen or flattened, fertile "head." The "head," when present, lacks the abrupt edge and sterile undersurface of a *Leotia, Cudonia,* or *Helvella.*

Four common genera, all members of the Geoglossaceae, are treated here. Among these, *Spathularia* stands out because of its peculiar flattened, fanlike "head" that runs down opposite sides of the stalk. *Neolecta* is also distinctive because of its highly irregular shape, while *Microglossum* and *Mitrula* have more or less clublike fruiting bodies, the latter with a clearly differentiated "head."

These earth tongues, like *Geoglossum* and *Trichoglossum,* inhabit humus, soil, moss, and rotten wood, but do *not* grow on insects or truffles. None are large enough or tasty enough to collect for the table, but several are quite beautiful, and are worth getting to know for this reason if no other. One species from each of the four genera is described here; several others are keyed out.

Key to Microglossum, Spathularia, & Allies

1. Fruiting body green or greenish 2
1. Not as above 3
2. Fruiting body olive-green to green to dark green or blue-green; stalk often minutely scurfy or scaly .... ***M. viride,*** p. 870
2. Fruiting body with only a slight greenish tinge; stalk smooth ....
   .... ***Microglossum olivaceum*** (see *M. viride,* p. 870)
3. Fruiting body with a flattened (paddle-like or fanlike) fertile "head" that extends down the the stalk on opposite sides (i.e., stalk appears to be wedged into the cap) 4
3. Not as above; "head" absent, or if present and flattened then not running down stalk on opposite sides 5
4. Stalk reddish-brown to dark brown and velvety; found mainly in eastern North America .... ***Spathularia velutipes*** (see *S. flavida,* p. 871)
4. Not as above; widespread .... ***Spathularia flavida*** & others, p. 871
5. Fruiting body with a small oval or rounded "head" that is sharply differentiated (and often differently colored) from stalk; *sometimes* growing in water 6
5. Fruiting body lacking a sharply differentiated "head" (but apex may be swollen, flattened, or broadened); *not* growing in water 8
6. Typically growing on submerged sticks in running water (often in cold mountain streams); stalk usually brownish .... ***Vibressea truncorum*** (see *Mitrula abietis,* p. 870)
6. Not as above; growing in still water or not in water at all 7
7. "Head" pinkish-buff to pale flesh-colored to light brown; stalk light to dark brown; found in duff under conifers .... ***Mitrula abietis,*** p. 870
7. "Head" differently colored (often yellow or orange) and stalk often white or pinkish-tinged *and/or* growing in shallow pools or on very wet soil or leaves ....
   .... ***Mitrula elegans*** & others (see *M. abietis,* p. 870)
8. Fruiting body pallid to pale ochre or dingy yellowish, always growing on wood; exterior usually roughed or minutely pimpled by the projecting tips of perithecia (flasklike nests of asci); rare .... (see **Pyrenomycetes,** p. 878)
8. Fruiting body differently colored, or if similar in color then lacking perithecia; usually terrestrial or in moss, sometimes on rotten wood 9
9. Fruiting body pale yellow to bright yellow to orange 10
9. Fruiting body differently colored 12
10. Fruiting body clublike and slender, without a broadened fertile "head"; spores borne on basidia .... (see **Clavariaceae,** p. 630)
10. Fruiting body usually with a broadened, often flattened fertile "head" or if not, then extremely variable in shape (often lobed, forked, flattened, twisted, etc., and usually at least 0.5-2.5 cm broad at the top); spores borne inside asci 11
11. Fruiting body very irregular in shape (often lobed, forked, flattened, twisted, etc.), usually yellow and usually found under conifers; spores usually less than 10 microns long, elliptical to nearly round; widely distributed .... *Neolecta irregularis* & others, p. 871
11. Fruiting body regularly clublike (but usually with a broad or flattened "head"), yellow to orange; found mainly in eastern North America; spores long and thin ....
    .... ***Microglossum rufum*** (see *Neolecta irregularis,* p. 871)
12. Fruiting body yellow-brown to dark brown, purplish, or even black; usually slender; spores borne in asci . ***Microglossum atropurpureum*** & others (see *Geoglossum glutinosum,* p. 866)
12. Fruiting body differently colored (usually paler or brighter than above) and/or thick; spores borne on basidia .... (see **Clavariaceae,** p. 630)

Microglossum viride   (Green Earth Tongue)   **Color Plate 214**

FRUITING BODY 1-5.5 cm tall, clublike to tongue- or spatula-shaped (i.e., with an enlarged "head" at maturity). Fertile "head" 4-12 mm broad, dark green to pea-green, green, olive-green, or even bluish-green, smooth or furrowed, often flattened or compressed in older specimens. Flesh greenish, rather tough. **STALK** 2-5 mm thick, colored like the head or paler green, usually thinner; surface minutely scurfy or scaly, but sometimes smooth in age. **SPORES** 14-22 × 4-6 microns, sausage-shaped to spindle-shaped, smooth, hyaline (colorless) under the microscope, not septate or septate only when old.

HABITAT: Solitary, scattered, or in groups or tufts in soil, moss, and duff under both hardwoods and conifers; widely distributed. It is not uncommon in coastal California in the winter and early spring, but is often overlooked because of its small size and green color. I find it most often under redwood and tanoak.

EDIBILITY: Too small to be worthwhile.

COMMENTS: The striking green color sets apart this petite, farflung earth tongue. The specimens in the color plate are rather young and club-shaped, but as they grow older their "heads" will become flatter (laterally) and more distinct. The minutely scaly or scurfy stalk is also distinctive. *M. olivaceum* is a somewhat similar but smoother species with an olive-buff to greenish-brown to dark brown (not truly green) fruiting body and shorter spores. I have found it twice in our area, but it is more widely distributed. Other species of *Microglossum* are either yellow to orange (see *M. rufum* under *Neolecta irregularis*) or brown to blackish (see *M. atropurpureum* and *M. fumosum* under *Geoglossum glutinosum*).

Mitrula abietis   (Miniature Earth Tongue)

FRUITING BODY 0.5-4 (5) cm tall, with a stalk and sharply differentiated "head." Fertile "head' 1-7 (10) mm broad and high, roundish to cylindrical (elongated), with a smooth surface; pinkish-buff to pale flesh-colored to light brown. Flesh thin. **STALK** 0.5-3 (4) cm long, 1-4 (7) mm thick, equal or tapered slightly, thin, light to dark brown (usually darker than cap), smooth or slightly powdered above, often with brown hairs at base. **SPORES** 10-14 × 2-2.5 microns, elongated, smooth, not septate, hyaline (colorless) under the microscope.

HABITAT: Scattered to densely gregarious in needle duff under northern and mountain conifers; common in western North America in the spring, summer, and fall, but absent in our area.

EDIBILITY: Fleshless and probably flavorless.

COMMENTS: This little mushroom sometimes grows in large carpets on the forest floor. The sharply differentiated "head" distinguishes it from most earth tongues and fairy clubs, and it is usually smaller than *Cudonia* and lacks the abrupt cap margin and sterile underside of that genus. *M. elegans,* sometimes called the "Swamp Beacon," is a similar, widely distributed and common species that fruits in very wet humus or soil or on leaves in shallow pools. It has a creamy to bright yellow or pale orange "head" and a white or pinkish-tinged, sometimes viscid stalk and can often be found under mountain conifers in the spring and early summer. It has long passed under the name *M. paludosa,* a similar European species that differs microscopically. Other species: *M. gracilis* is a northern and montane species with an ochraceous to orange-buff "head"; it usually grows in moss. *M. borealis*, also northern, has a golden-yellow "head" like that of *M. elegans,* but has elliptical or crescent-shaped spores. *M. lunulatospora* is an eastern springtime, water-loving species with a flesh-colored to yellowish "head" and crescent-shaped spores. *Vibressea truncorum* looks like a *Mitrula* or a miniature *Leotia,* but usually grows on sticks in running water (often in cold mountain streams). It is even smaller than *Mitrula* and somewhat gelatinous.

Neolecta irregularis   (Irregular Earth Tongue)   **Color Plate 200**

FRUITING BODY 1-7 cm tall, clublike to very irregular (lobed, sparingly branched, grooved and twisted, etc.), usually flattened or compressed, 0.4-2.5 cm broad at apex. Fertile surface pale yellow to bright yellow or orange-yellow. Flesh white or yellowish, rather tough. **STALK** sometimes absent but usually present as a sterile, pale yellow to white base beneath the fertile portion; 1-6 (10) mm thick (usually thinner than fertile area). **SPORES** 5.5-10 × 3.5-5 microns, elliptical to nearly round, smooth, not septate, hyaline (colorless) under the microscope. Asci lining at least the upper part of (and sometimes the entire) fruiting body.

HABITAT: Widely scattered to gregarious or occasionally tufted on ground, moss, or duff, usually under conifers; widely distributed. Along with *N. vitellina* (see comments), it occurs throughout much of the West, but is apparently absent in our area. I have seen large fruitings under spruce and fir in New Mexico and Oregon in the summer and fall.

EDIBILITY: In my opinion, not worth collecting. McIlvaine, as usual, dissents: "Those fortunate enough to find this species will hunt for it again assiduously. Even raw, when cut in strips, it makes a picturesque and delicious salad."

COMMENTS: The bright yellow color and highly irregular shape are usually enough to identify this species, which is also known as *Spragueola* (or *Mitrula*) *irregularis*. Club-shaped specimens can be mistaken for fairy clubs *(Clavulinopsis)*, but bear their spores in asci rather than on basidia, are usually broader, and often have irregularly-shaped fruiting bodies growing nearby. *Microglossum rufum* is a farflung earth tongue with a bright yellow to orange fruiting body. It is much more uniform in shape (usually with a slender stalk 2-4 mm thick and a wider, flattened "head") and has much longer spores (20-40 microns). It is especially common in eastern North America in the summer and early fall. Other species: *N. vitellina* closely resembles *N. irregularis,* but is slightly smaller and paler, and differs microscopically.

Spathularia flavida   (Fairy Fan)

FRUITING BODY 1-10 cm tall, with a stalk and fertile "head." Fertile "head" very compressed or flattened laterally, spatula- or fanlike, 1-3 cm broad, decurrent (running down) on *opposite sides* of the stalk; surface (sides) smooth or wrinkled, sometimes lobed or contorted or with a notched apex, pallid when young becoming pale yellow to yellow, buff, or cinnamon-buff to brownish (or occasionally pale orangish). Flesh white, *not* gelatinous. **STALK** (1) 2-8 cm long, 2-10 mm thick, variable in shape but often thicker or swollen at base, usually hollow; surface smooth to finely mealy but not velvety, white to yellowish or colored like the "head" but usually paler, with white to pale yellow mycelium at base. **SPORES** 30-75 (95) × 1.5-3 microns, very long and narrow (needle-like), smooth, with one to several septa (partitions) or none at all; hyaline (colorless) under the microscope but often yellow-brown in mass, especially when dry.

HABITAT: Scattered to gregarious or even clustered, sometimes in lines or circles, on humus or rotten wood under conifers (especially pine) or sometimes hardwoods; widespread. It is common in the summer and fall in the Pacific Northwest and Southwest and occurs in northern California in the fall, winter, and spring, but I have yet to find it locally.

EDIBILITY: Said to be edible, but rather tough. Captain Charles McIlvaine describes it as "tenacious but tender."

COMMENTS: The peculiar flattened, paddle- or fanlike "head" that extends down opposite sides of the stalk is unique to this little mushroom and its close relatives (see next page). It might possibly be confused with *Neolecta irregularis* or a *Microglossum*, but is not as

Left: *Spathularia flavida* is easily told by its flattened, paddle-like "head." Note how stalk of central specimen appears to be wedged into the "head," and how fertile surface can be wrinkled or smooth. **Right:** *Cudonia circinans* (see p. 873). Note incurved cap margin and clustered growth habit.

brightly colored and has a more consistently compressed "head." *S. clavata* is a synonym. Other species: *S. spathulata* is said to be similar but has smaller spores and a somewhat darker (yellow-brown to reddish-brown) fruiting body. Its cap ranges from flattened to rounded (as in *Cudonia*) but is fertile over its entire surface rather than just at the top. It was originally collected in Big Basin State Park, California, but I have not seen it there. *S. (=Spathulariopsis) velutipes* of eastern North America is a common and distinctive "fairy fan" with orange mycelium and a velvety dark brown to reddish-brown stalk that may be thicker at the bottom or top, plus a yellow to yellow-brown flattened "head" that is covered by a "veil" when very young and often retains "veil" remnants at maturity.

LEOTIA & CUDONIA

Small to medium-sized fungi found on ground or rotten wood. FRUITING BODY *with a cap and stalk*. CAP *rounded to convex or lobed, the margin often inrolled;* surface smooth or wrinkled, *viscid when moist in Leotia, not viscid in Cudonia;* variously colored. *Flesh gelatinous in Leotia,* not gelatinous in *Cudonia*. STALK smooth or scurfy or minutely scaly, *well-developed,* variously colored. SPORES spindle-shaped or needle-like, hyaline (colorless) under the microscope; sometimes septate, smooth. Asci lining the upper surface of the cap, typically 8-spored, not amyloid, inoperculate, with a pore at the tip.

THESE two genera are unique among the earth tongues in possessing a well-developed and clearly differentiated cap and stalk. The cap is not just a swollen "head" as in *Mitrula* or other earth tongues; instead, it has a sterile underside and an abrupt margin that sets it off from the stalk. In addition, *Leotia* is easily distinguished by its gelatinous to semi-gelatinous tissue. *Cudonia,* in contrast, is fleshy or tough but not gelatinous. It is apt to be mistaken for a small elfin saddle *(Helvella)*, but its cap is usually convex or rounded and not as dramatically lobed as in that genus.

    Leotia and *Cudonia* used to be classified with other earth tongues in the Geoglossaceae, but are now placed alongside a number of genera (e.g., *Bulgaria*) in a larger family, the Leotiaceae. Both are widely distributed, but only *Leotia* is common in our area. They grow on the ground or on rotten wood, often in groups or clusters. Neither genus is worth eating, although *Leotia* might be useful as a lubricant! Two species of *Leotia* and one *Cudonia* are described here.

Key to Leotia & Cudonia

1. Flesh gelatinous or semi-gelatinous; surface of fruiting body viscid or slimy when wet 2
1. Flesh not gelatinous; surface not normally viscid 6
2. Growing on submerged sticks in running water; small ....
   .... (see *Vibressea truncorum* under *Mitrula abietis*, p. 870)
2. Not as above 3
3. Fruiting body small (0.5-3 cm tall), whitish to pinkish or brownish, sometimes with an ochraceous or lavender tinge; found on rotting hardwoods in eastern North America; rare ......
   .... *Neocudoniella (=Leotia) albiceps*
3. Not as above; usually larger; often with yellow, ochre, or greenish shades; widespread 4
4. Cap consistently and distinctly green or greenish 5
4. Cap some shade of yellow, buff, ochre, or cinnamon, at times with a greenish tinge (olive-brown
   or olive-ochre, etc.) .... *L. lubrica*, p. 874
5. Stalk white to yellow or orange .... *L. viscosa*, p. 874
5. Stalk pale green to green .... *L. atrovirens* (see *L. viscosa*, p. 874)
6. Cap yellowish to olive-buff, sometimes with small "veil" fragments on margin; found in eastern
   North America, usually under hardwoods .... *C. lutea* (see *C. circinans*, below)
6. Not as above; differently colored; widespread 7
7. Cap creamy to pinkish-buff, cinnamon-buff, vinaceous-buff, pale brown, or occasionally
   darker; found mainly in the late summer and fall; widespread .... *C. circinans*, below
7. Cap pinkish-buff to pinkish-cinnamon to grayish-brown or dark grayish-brown; found in
   western North America under conifers, usually in the spring and summer ....
   .... *C. monticola* & *C. grisea* (see *C. circinans*, below)

Cudonia circinans   (Common Cudonia)

FRUITING BODY with a cap and stalk. **CAP** 0.5-2 cm broad, usually rounded or convex,
sometimes with a central depression and sometimes convoluted; surface wrinkled or
smooth, creamy to pinkish-buff, cinnamon-buff, vinaceous-buff, pale brown, or occasionally darker; margin usually curved down and in toward the stalk. Underside sterile,
often with radiating veins that extend up from the stalk. Flesh thin but firm, *not* gelatinous, rather tough or leathery when dry. **STALK** 1.5-7 cm long, 2-12 mm thick (but
usually less than 6 mm at apex), equal or more often thicker below, stuffed or sometimes
hollow in age; drab to dark brown (usually darker than cap), usually minutely scurfy, often
longitudinally striate or ridged, especially above. **SPORES** (28) 32-40 (46) × 2 microns,
very long and thin (needle-like), smooth, sometimes septate (partitioned), but usually not;
hyaline (colorless) under the microscope.

HABITAT: Scattered to gregarious or often in dense clusters in humus, soil, and on rotting
wood; particularly common under conifers, but also found with hardwoods; widely
distributed. It is said to be the most common member of its genus, but I have yet to find it
in our area. In the Pacific Northwest it is fairly common in the late summer and fall.

EDIBILITY: Poisonous, at least raw. It is said to contain high concentrations of MMH
(see pp. 799 and 893 for details).

COMMENTS: The small size, convex to somewhat convoluted cap, and non-gelatinous
flesh are the hallmarks of this species. It is reminiscent of a dry *Leotia*, but is not as brightly
colored. It can also be mistaken for an elfin saddle *(Helvella)*, but the shape is different
and it often forms dense groups or clusters untypical of *Helvella* (see photo on p. 872).
C. monticola is a similar, pinkish-cinnamon to pinkish-buff to grayish-brown westerner.
It is the largest *Cudonia* (up to 10 cm high, cap 1-3 cm broad), but has much smaller spores
than *C. circinans* (only 18-25 microns long) and a frequently compressed or even somewhat
saddle-shaped cap. It is common under conifers in northern California and the Pacific
Northwest, usually in the spring and summer (whereas *C. circinans* is more frequent in
the fall). *C. grisea* also occurs under conifers in the Pacific Northwest, but has a gray to

dark grayish-brown or fuscous fruiting body and usually fruits in the spring. *C. lutea* is a yellowish to olive-buff eastern species that sometimes shows "veil" fragments on the margin of the cap. It grows scattered to gregarious, usually under hardwoods.

Leotia lubrica   (Jelly Babies)   Color Plate 215

FRUITING BODY with a cap and stalk. **CAP** 0.5-4 cm broad, round to convex to slightly lobed or knobby; surface smooth or wrinkled, viscid to slimy when moist (but sometimes drying out), buff to yellow, ochre, olive-ochre, or sometimes cinnamon or greenish-brown; margin usually curved in toward stalk, often lobed or wavy. Underside sterile, paler. Flesh gelatinous (at least the central core), often translucent. **STALK** 2-8 cm long, 0.3-1 cm thick, equal or somewhat thicker below, smooth or scurfy (with minute granules), hollow or more often filled with a gel; surface viscid when moist, colored like cap or sometimes yellower. **SPORES** 16-25 × 4-6 microns, spindle-shaped and sometimes curved, smooth, hyaline (colorless) under the microscope, septate (partitioned) at maturity.

HABITAT: Solitary to gregarious or clustered in duff, soil, or very rotten wood under both hardwoods and conifers; widely distributed. It is the most common *Leotia* in North America, but is not as frequent in our area as *L. viscosa*. I usually find it in the winter and spring, but in other regions it fruits in the summer and fall. Alexander Smith describes finding massive clusters buried in sand dunes!

EDIBILITY: Harmless but glutinous. It might be more useful as a lubricant than a condiment!

COMMENTS: This farflung fungus is unlikely to be mistaken for any other. The combination of rounded or wrinkled cap, viscid fruiting body with gelatinous flesh, and overall yellowish to ochre-buff color set it apart. Greenish-tinged forms approach *L. viscosa,* and the two species may intergrade.

Leotia viscosa   (Chicken Lips)

FRUITING BODY with a cap and stalk. **CAP** 0.5-3 cm broad, round to convex to slightly lobed or knobby; surface smooth or slightly wrinkled, viscid or slimy when moist, dark green to olive-green; margin usually incurved toward the stalk, often lobed or wavy. Underside sterile, usually pallid or paler. Flesh (at least the central core) gelatinous, often translucent. **STALK** 2-9 cm long, 0.3-1 cm thick, equal or tapered slightly upward,

Leotia viscosa, young specimens. Note the viscid-gelatinous stalk and flesh. As they mature the caps will grow larger. The greenish cap distinguishes this species from *L. lubrica* (shown in color plate).

smooth, hollow or filled with a gel, viscid to slimy when moist; white to yellow or orange, sometimes with minute green dots or particles, especially above. **SPORES** 16-28 × 4-6 microns, spindle-shaped and often slightly curved, smooth, usually septate (partitioned) at maturity, hyaline (colorless) under the microscope.

HABITAT: Solitary, scattered, or in groups or clusters in humus or on rotten wood; widely distributed. This is the most common *Leotia* in our area. It fruits in the winter and early spring under oak and various other trees.

EDIBILITY: Harmless but gelatinous.

COMMENTS: This gelatinous Ascomycete with the green head and white to orange stalk can hardly be confused with any other. In dry weather the surface of the cap and stalk may not be obviously viscid, but slicing open the fruiting body will usually reveal gelatinous tissue within. The nickname "Chicken Lips" was obviously given to it by the same person who dubbed *Tricholoma flavovirens* the "Man On Horseback!" Other species: *L. atrovirens (=L. chlorocephala)* is a similar but smaller eastern species with a greenish to dark green cap and green to pale green stalk. *L. lubrica* (see description) is usually yellower.

BULGARIA & Allies

Small to medium-sized fungi *usually growing on wood or bark of hardwoods.* FRUITING BODY *top-shaped to cup-shaped to nearly round to irregularly lobed; pinkish to reddish, purplish, dark brown, or black. Flesh gelatinous or rubbery-gelatinous.* STALK *absent or present only as a short narrowed base.* SPORES elliptical to elongated, smooth or ribbed, septate or not septate; *brown (or many of them brown) or black in Bulgaria,* otherwise hyaline (colorless) under the microscope. Asci lining upper surface of fruiting body, typically 8-spored, inoperculate, with amyloid pore in *Bulgaria,* otherwise not amyloid.

THESE are gelatinous or rubbery-gelatinous, hardwood-inhabiting fungi with little or no stalk and a top-shaped to cup-shaped to irregular fruiting body. The gelatinous texture can lead to confusion with the jelly fungi, which bear their spores on basidia, and with *Sarcosoma* and allies, which have operculate asci. Since these differences are microscopic, the species treated here are also keyed out under those groups. One species, *Bulgaria inquinans,* is described here, and three others are discussed. They are placed with *Leotia* in the Leotiaceae, but lack the well-developed stalk of that genus. Because of their texture they are not worth eating except in dire emergencies.

Key to Bulgaria & Allies

1. Fruiting body usually irregularly lobed or brainlike, often forming confluent masses; found on beech in eastern North America .. *Ascotremella faginacea* (see *Bulgaria inquinans,* p. 876)
1. Not as above 2

2. Fruiting body shallowly cup-shaped to earlike or seaweed-like, reddish-brown to dark brown but not usually black (unless dried out), typically 2 cm broad or more; spores borne on basidia which usually line the lower (but sometimes the upper) surface of fruiting body; sterile surface minutely hairy or downy .... (see *Auricularia auricula,* p. 675)
2. Not as above; spores borne in asci which line the upper surface of fruiting body 3

3. Upper (fertile) surface of fruiting body dark brown to black 4
3. Upper surface of fruiting body pinkish to reddish, reddish-brown, purplish, or tan 5

4. Found on dead hardwoods; asci not operculate .... *Bulgaria inquinans,* p. 876
4. Found on ground or on dead conifers (rarely hardwoods); asci operculate (i.e., with "lids") .... (see *Sarcosoma* & Allies, p. 826)

5. Fruiting body cup-shaped at maturity; fertile surface reddish-brown to brown; exterior with minute dark hairs; asci operculate (with "lids") .... (see *Sarcosoma* & Allies, p. 826)
5. Fruiting body usually more or less top-shaped, pinkish to reddish to purplish; asci not operculate .... *Ascocoryne sarcoides* & *Neobulgaria pura* (see *Bulgaria inquinans,* p. 876)

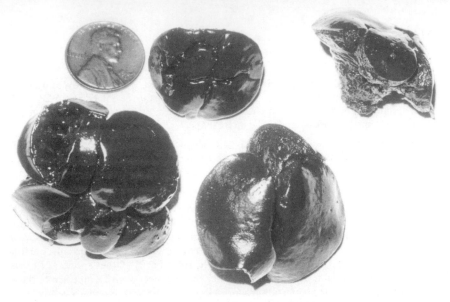

Bulgaria inquinans. These gelatinous specimens look like licorice drops. (If only they tasted like them!) The three on left are being viewed from the top, the one on the upper right, from the side.

Bulgaria inquinans (Poor Man's Licorice; Black Jelly Drops)

FRUITING BODY 1-4 cm broad and/or high, rubbery; at first rounded to somewhat cylindrical to shallow cup-shaped or top-shaped, the top then broadening into a flattened or broadly convex, flabby "cap." Fertile (upper) surface blackish and often shiny when wet. Exterior or underside brown to blackish and roughened. Flesh gelatinous or at least very rubbery; pliant and tough, not brittle. **STALK** absent or present as a narrowed base, continuous with and colored like the exterior or underside of the "cap." **SPORES** 11-14 × 6-7 microns, more or less kidney-shaped, smooth, not septate; brown or black in mass, but only the upper four in each ascus brown under the microscope, the other four hyaline (colorless); asci typically 8-spored, their tips blueing in iodine.

HABITAT: Solitary, scattered, in rows, or densely gregarious or clustered on dead hardwood logs and branches, especially of oak; widely distributed and common. In our area it fruits throughout the mushroom season, especially on live oak and tanoak, and can be seen on almost any wintertime trek through the woods.

EDIBILITY: Unknown, but as the common name implies, not worth eating except in desperation!

COMMENTS: Also known as *Phaeobulgaria inquinans,* the flabby, funky fruiting bodies of this fungus look something like licorice drops, or in the concise words of Judith Scott Mattoon, "They remind me of rubber parts that fit into things by squeezing through and then popping back into shape—you know what I mean?" The texture is reminiscent of India rubber, or can be gelatinous, leading to confusion with the jelly fungi. However, it bears its spores in asci rather than on basidia, and can be told in the field by its distinctive shape and color. Its growth on dead hardwoods and smaller size distinguish it from the

Bulgaria inquinans. These specimens were collected in dry weather and consequently are rubbery rather than gelatinous. Note the narrowed base beneath the fertile portion.

conifer-loving "Starving Man's Licorice" (*Sarcosoma mexicana* and *S. latahensis*). Despite its uncanny resemblance to *Sarcosoma,* it is classified with the earth tongues because of its inoperculate (lidless) asci. Other gelatinous species: *Ascocoryne (=Coryne) sarcoides* is a widespread, more or less top-shaped, gelatinous, flesh-colored to dark purplish or reddish-brown species. It is smaller (up to 1 cm broad), has septate spores, and grows on dead wood. *Neobulgaria pura (=Ascotremella turbinata)* is a flesh-colored to reddish, top-shaped eastern species with non-septate spores. *Ascotremella faginacea* is a larger, raisin-colored, gelatinous species that is lobed or brainlike and often forms continuous masses on beech trunks in eastern North America. All of these look like jelly fungi and are keyed out under that group.

CIBORIA & Allies

Small to minute fungi found on rotten wood, soil, or more often on plant stems, leaves, nuts, and other vegetable matter. FRUITING BODY *usually cup-shaped to disclike or cushion-shaped, with or without a stalk.* Fertile (upper) surface variously colored, usually smooth. Underside sterile, with or without hairs. STALK if present long or short, usually very thin, sometimes arising from a swollen "tuber." SPORES variously shaped, usually but not always smooth and hyaline (colorless) under the microscope. Asci lining the upper surface of the fruiting body, usually 8-spored, inoperculate, with a pore at the tip which may or may not be amyloid.

THE above synopsis covers a vast group of minute but cute cuplike and disclike fungi that are seldom collected except by professional specialists. They differ from the many cup fungi in the Pezizales in having inoperculate (lidless) asci. Some, such as *Ciboria,* have long stalks attached to the substrate. Others, such as *Sclerotinia, Whetzelinia,* and *Myriosclerotinia,* have long stalks that arise from a tuberlike mass of tissue **(sclerotium)**. Still others, like *Dasyscyphus,* have shorter, sometimes hairy stalks. Finally, there are many (e.g., *Mollisia*) that have no stalk at all. A large number of these fungi are parasitic on plant parts, but some are saprophytes. Without exception they are too small to interest the average mushroom hunter and are certainly too small to eat. The two genera described here, *Ciboria* and *Chlorociboria,* belong to different families (the Sclerotiniaceae and Leotiaceae, respectively). Several other related families are not treated here.

Key to Ciboria & Allies

1. Fruiting body blue to green or pallid with a blue or greenish tinge; growing on wood . . . . *Chlorociboria aeruginascens* & others, p. 878
1. Not as above 2

2. Fruiting body minute, bright yellow to orange-yellow; found on rotten wood, usually in swarms; widely distributed . . . . *Bisporella citrina*
2. Not as above 3

3. Stalk arising from a swollen "tuber" (sclerotium) *Sclerotinia, Whetzelinia, & Myriosclerotinia*
3. Not as above 4

4. Fruiting body yellow-brown to brown, with a stalk; growing on willow and alder catkins . . . . *Ciboria amentacea,* below
4. Not as above . . . . *Ciboria, Dasyscyphus, Mollisia,* & many others

Ciboria amentacea   (Catkin Cup)

FRUITING BODY with a stalk and cuplike cap. "Cap" 0.5-1.2 cm broad, at first shallowly cup-shaped, expanding to nearly flat in age. Fertile (upper or inner) surface light brown to yellow-brown, smooth. Exterior of cup similar in color, also smooth. **STALK** 1-5 cm long, 1-2 mm thick, light brown to yellow-brown, equal, smooth, often curved, *not* arising from a sclerotium. **SPORES** 7.5-13 × 4-6 microns, elliptical, smooth.

HABITAT: Solitary or in small groups on old, fallen alder and willow catkins; widely distributed but seldom collected, usually fruiting in the spring.

EDIBILITY: Who knows?

COMMENTS: The growth on alder or willow catkins rescues this little brown cup fungus from the anonymity it so richly deserves.

Chlorociboria aeruginascens   (Blue Stain)

FRUITING BODY 3-7(10) mm broad, at first cup-shaped, then becoming flat or disclike or with a slightly elevated margin. Fertile (upper) surface bright to pale blue-green or turquoise, sometimes with a yellowish or orange-yellow tint developing in age; smooth or slightly wrinkled. Exterior (underside) similarly colored (but not yellowish). Flesh thin, also bluish-green. **STALK** usually present as a short, narrowed, typically off-center base; up to 3 (6) mm long and 1-2 mm thick, same color as rest of fruiting body. **SPORES** 6-10 × 1.5-2 microns, spindle-shaped or elongated, smooth, with an oil droplet at each end.

HABITAT: Gregarious (several often arising from a common base) on dead or barkless wood (usually oak); widely distributed and quite common, but easily overlooked because of its diminutive dimensions. In our area it fruits mainly in the winter and early spring.

EDIBILITY: Indisputably inconsequential.

COMMENTS: This petite Ascomycete merits mention in this book because of its unusual color. *C. aeruginosa* is a very similar, widely distributed species that is smaller (less than 5 mm broad), has a shorter, more or less central stalk, and orange-yellow flesh. Both species can be detected when they are not fruiting because their mycelium stains the host blue-green. The stained wood was once used in the manufacture of inlaid wooden objects known as "Tunbridge Ware." Both species have also been placed in the genus *Chlorosplenium.*

Flask Fungi

spores

PYRENOMYCETES

THE Pyrenomycetes differ fundamentally from the Discomycetes (morels, cup fungi, earth tongues, etc.) because they bear their asci in flask-shaped "nests" called **perithecia.** The perithecia are usually imbedded in the fruiting body, but their necks or mouths often protrude like small pimples.

The flask fungi are a varied lot, but only a few of them are conspicuous enough to be considered in this tome. The most common types, *Xylaria* and *Daldinia,* are tough, usually black, and grow on wood. Another distinctive group, *Cordyceps,* is parasitic on insects and truffles. Still another, *Hypomyces,* engulfs other mushrooms in a pimpled or powdery weft of tissue. The flask fungi treated here belong to a single order (Sphaeriales) within the Pyrenomycetes. They are unpalatable except for one species of *Hypomyces.* They fruit in moist weather, but the tougher types persist year-round.

Key to the Pyrenomycetes

1. Growing on wood (but wood sometimes buried) 2
1. Growing on insects (including pupae or larvae), spiders, truffles, or other mushrooms (but the host sometimes buried inside rotten wood!) 3

2. Fruiting body whitish to yellowish to pale ochre; rare .... *Podostroma,* below
2. Fruiting body gray to dark brown or black, but sometimes covered with a white powder; common .... *Xylaria & Daldinia,* p. 885
3. Fruiting body clublike, threadlike, or with a cap and stalk; growing on insects, spiders, or certain truffles .... *Cordyceps,* below
3. Fruiting body a pimpled or powdery layer of tissue that covers or partially covers its host; growing on other mushrooms .... *Hypomyces,* p. 882

PODOSTROMA

THIS rare, wood-inhabiting genus is represented by a single boring species in our area, described below.

Podostroma alutaceum

FRUITING BODY 1-5 cm tall and 0.5-1 cm thick, cylindrical to club-shaped, without a well-defined cap. Surface dry, minutely roughened by the slightly protruding perithecia (flasklike nests of asci), whitish to yellowish to pale ochre, usually paler (white) at the base. **SPORES** elongated, hyaline (colorless) under the microscope, finely warted and septate (with one partition), breaking up into one-celled, round to elliptical segments averaging 4-4.5 × 3-4 microns. Asci 8-spored, but each spore breaking in two to make 16.

HABITAT: Solitary or in small groups on rotting wood; widely distributed but rare. I have found it only once in our area, on dead oak in the late winter.

EDIBILITY: Who cares?

COMMENTS: This forgettable clublike fungus can be recognized by its yellowish color, growth on wood, and presence of perithecia (flasklike "nests" of asci) on the upper portion of the fruiting body. It is most likely to be mistaken for a fairy club (*Clavaria* or *Clavulinopsis*), but the above-mentioned features distinguish it.

CORDYCEPS

Small fungi *found on insects (pupae, larvae, and adults), spiders, and certain truffles.* FRUITING BODY *threadlike to club-shaped or with a differentiated "head" and stalk;* often brightly colored but sometimes dull or dark; *surface often roughened or minutely pimpled by projecting perithecia.* STALK present, usually slender or very thin, *arising from the host insect or truffle* (which is often buried). SPORES threadlike, typically hyaline (colorless) under the microscope and smooth, multiseptate, but usually breaking up quickly into one-celled, barrel-shaped or elongated segments. Asci borne in perithecia (flasklike structures) imbedded in or projecting from the "head" (if present) or upper portion of the fruiting body.

THESE small but fascinating clublike fungi are obligate parasites of insects, spiders, and certain truffles. As such they are easy to recognize, *providing you dig them up carefully* so they can be traced to their host, which is often buried in humus, soil, or rotten wood. When the host is overlooked or left behind, many species of *Cordyceps* can still be distinguished from other clublike fungi by their minutely roughened or pimpled fertile surface.

Cordyceps is a fairly large genus and only a few species can be treated here. Most of them parasitize insect larvae, pupae, and adults. The mycelium develops inside the insect, killing it and devouring it. After the insect is completely mummified and emptied of nutrients, the mycelium fruits and then dies. As the insect is the sole source of food for the fungus, the size of the fruiting body is often dependent on the size of the host. Since insects are more abundant in warm, humid weather, it's not surprising that *Cordyceps* is particularly prominent and diverse in eastern North America (where there are summer rains) and the tropics. In California and the Pacific Northwest, where the summers are drier, insect-eating species are comparatively rare, but those that parasitize truffles are more common. It should be emphasized, however, that even the "common" species of *Cordyceps* are rare in relation to other mushrooms. Most mycologists consider one or two fruiting bodies of *Cordyceps* a real find!

Cordyceps are worthless as food because of their small size and infrequent occurrence. Their unique diet, however, makes them a fascinating group to study. Perhaps some day we will find a practical use for them in the control of certain insect pests. Three species of *Cordyceps* are described here, and several others are keyed out. Also worth mentioning is the closely related genus, *Claviceps,* which parasitizes plants rather than insects or truffles. The most potent hallucinogenic compound known, LSD, was derived from ***Claviceps purpurea***, better known as wheat ergot..

Key to Cordyceps

1. Growing on truffles (species of *Elaphomyces*) 2
1. Growing on insects (adults, pupae, or larvae) or spiders 3

2. Fruiting body with a cap and stalk .... ***C. capitata*** & others, below
2. Fruiting body clublike (lacking a distinct cap), usually with yellow mycelial threads at base or permeating the host .... ***C. ophioglossoides*** (see *C. capitata,* below)

3. Fruiting body with a cap or "head" which is usually clearly delimited from the stalk or sterile portion of fruiting body 4
3. Fruiting body lacking a differentiated "head" or cap, but often thicker toward the top 7

4. Growing on beetles, moths, or butterflies (or their larvae or pupae) 5
4. Growing on ants or wasps 6

5. Stalk typically yellow; cap ochre to mahogany .... ***C. gracilis*** (see *C. myrmecophila,* p. 881)
5. Not as above; fruiting body brownish or tinged vinaceous; "head" warty .... ***C. entomorrhiza*** (see *C. myrmecophila,* p. 881)

6. Growing on ants .... ***C. myrmecophila,*** p. 881
6. Growing on wasps .... ***C. sphecocephala*** (see *C. myrmecophila,* p. 881)

7. Fruiting body threadlike (less than 2 mm thick) .... ***C. unilateralis, C. clavulata,*** & others (see *C. militaris,* p. 882)
7. Fruiting body not threadlike 8

8. Fruiting body brown to purple-brown or blackish ..... ***C. ravenelii*** (see *C. militaris,* p. 882)
8. Fruiting body white to yellow, orange, or orange-red 9

9. Growing on beetles (usually the larvae or pupae); fruiting body with sterile tip .... ***C. melolanthae*** (see *C. militaris,* p. 882)
9. Growing on butterflies and moths (usually larvae or pupae); tip not sterile 10

10. Fruiting body orange-buff to orange to orange-red .... ***C. militaris,*** p. 882
10. Fruiting body whitish to yellow .... ***C. washingtonensis*** (see *C. militaris,* p. 882)

Cordyceps capitata   (Truffle Eater)

FRUITING BODY arising from certain underground truffles *(Elaphomyces),* 2-8 (12) cm tall, with a well-defined cap or "head" and stalk. Fertile "head" 0.5-2 cm broad and high, nearly round to convex or slightly conical; surface dark reddish-brown to brown, dark olive-brown, or even blackish, roughened or minutely pimpled by protruding perithecia (flasklike nests of asci). Flesh white. **STALK** 1.5-8 cm long, (0.2) 0.4-1.5 cm thick, more or less equal, sometimes slightly flattened, often bent or curved, occasionally forked (with two "heads"), rather tough; surface usually fibrillose of fibrillose-scaly, yellow to yellow-ochre to yellow-olive, sometimes darker (olive to olive-black) in age; base often whitish. **SPORES** threadlike, hyaline (colorless) and smooth under the microscope, usually breaking up into one-celled segments averaging (8) 12-27 (32) × 1.5-3 microns.

HABITAT: Solitary, tufted, or gregarious on ground, but arising from underground deer truffles *(Elaphomyces* species); widely distributed and one of the more common members of the genus. Scattered fruiting bodies are the norm, but sometimes it fruits prolifically. I have not found it in our area, but it may well occur (*Elaphomyces* certainly does). In the mixed coastal forests of northern California it can be found in the fall and winter.

EDIBILITY: Possibly worth trying since it is larger than most species of *Cordyceps,* but I can find no information on it.

COMMENTS: This is one of several *Cordyceps* species that grow only on *Elaphomyces.* The latter can occur several inches deep in the soil, but specimens close to the surface are more apt to be parasitized. The reddish-brown or darker cap which is sharply differentiated from the yellow to olive stalk are the principal fieldmarks. *C. canadensis* is a very similar truffle-eater with much larger spore segments; it is known from eastern North America and Europe. *C. ophioglossoides* is another species that parasitizes *Elaphomyces,* but it has a clublike fruiting body that lacks a sharply defined "head." The club is sometimes yellow when very young but soon becomes reddish-brown to olive-brown to nearly black except for a yellow base and yellow mycelial threads that extend into the host. It is said to be the most common *Cordyceps* in eastern North America, but is rather rare in California.

Cordyceps myrmecophila   (Ant Fungus; Ant Eater)

FRUITING BODY arising from an ant (often buried), 0.8-5 (10) cm tall, with a thin stalk and small "head." Fertile "head" 2-8 mm broad, usually oval; surface ochre to ochraceous-salmon, sometimes with short longitudinal ridges or furrows in dry weather, minutely pimpled from the slightly projecting perithecia (flasklike nests of asci). **STALK** 0.8-9.5 cm long, 0.5-1 (2) mm thick, very thin and more or less equal, colored like the "head" or often paler (pale yellow, sometimes shading to white near base, or entirely white if not exposed to light). **SPORES** threadlike and multiseptate, smooth, hyaline (colorless) under the microscope, breaking into one-celled segments averaging 8-10 × 1.5 microns.

HABITAT: Scattered to gregarious on the mummified, often buried carcasses of ants (usually one per ant); widely distributed, fruiting in damp weather, usually around ant nests in the woods. Although rare, it is sometimes prolific when it fruits. It has been found in British Columbia, Washington, and Oregon (as well as Europe, China, and Brazil), and may well occur in California; there is certainly no shortage of potential hosts!

EDIBILITY: Unknown.

COMMENTS: This is one of several *Cordyceps* species that are capitate (i.e., that have a differentiated "head") and parasitize insects. The ochre to yellow color and growth on ants distinguish it. As in other species, the length of the stalk depends largely on whether or not the host is buried, and if so, how deep. Other capitate species include: *C. sphecocephala,* a very similar southeastern species that grows on wasps and has a very thin, creamy to yellow or yellow-brown fruiting body; *C. entomorrhiza,* brownish or vinaceous-tinged with a very warty "head," growing on beetle larvae in the Pacific Northwest, rare; and *C. gracilis,* growing on larvae of beetles, moths, and butterflies in eastern North America, with a yellow stalk and ochre to mahogany-colored "head," also rare. For capitate species that grow on truffles, see *C. capitata,* and for the clublike (non-capitate) types that grow on insects, see *C. militaris.*

Cordyceps militaris   (Caterpillar Fungus)

FRUITING BODY arising from buried moth and butterfly larvae or pupae, 2-8 cm tall, cylindrical to spindle- or club-shaped (i.e., with a slightly swollen upper fertile region, but lacking a well-defined "head"); often with a longitudinal furrow. Upper (fertile) portion of club 2-6 mm broad, orange to orange-buff to orange-red, finely roughened or pimpled by the slightly protruding perithecia (flasklike nests of asci). **STALK** (sterile lower region) smooth, usually paler, often curved or wavy. **SPORES** threadlike and multiseptate, smooth, hyaline (colorless) under the microscope, breaking up into one-celled, barrel-shaped segments averaging 2-6 × 1-1.5 microns.

HABITAT: Solitary to gregarious or clustered on buried pupae or less commonly larvae (caterpillars) of moths and butterflies; widely distributed. It is one of the more common species in the genus, but is rare in California. In eastern North America I have found it several times in the summer and fall.

EDIBILITY: Unknown.

COMMENTS: The orange to orange-buff, club-shaped fruiting body that arises from mummified pupae or larvae is most distinctive. It might be confused with other clublike fungi (e.g., *Clavulinopsis, Microglossum*) if it is dug up carelessly or the host is overlooked, but the pimpled upper portion (see photo on p. 883) will still identify it. Other species that parasitize insects and do *not* have a clearly defined "head" or cap include: *C. washingtonensis* of the Pacific Northwest, also growing on moth and butterfly pupae or larvae, and very similar, but with a whitish to yellow fruiting body; *C. melolanthae,* growing on beetles (usually buried grubs), sometimes in huge numbers, in eastern North America, often larger than *C. militaris,* with a whitish to yellow to orange fruiting body that has a sterile tip; *C. ravenelii*, a rare eastern species with a brown to chocolate-brown, purple-brown, or blackish, clublike fruiting body, also growing on beetle grubs; *C. unilateralis,* a minute (up to 3 cm tall and less than 1 mm thick) brown eastern and southern species that feeds on ants, bees, and wasps; and *C. clavulata,* growing on scale insects. There are many other species, particularly in eastern North America and the tropics; some can only be identified with a microscope. For species that parasitize insects and have a well-defined "head," see *C. myrmecophila.*

HYPOMYCES

Ubiquitous fungi *that parasitize other mushrooms, partially or completely covering them in a layer of pimpled or powdery tissue.* FRUITING BODY taking on the shape of its host, but disfiguring it; composed of a layer of tissue in which numerous perithecia are imbedded (but asexual stages lack perithecia). SPORES (sexual) elongated, smooth or warted, often 1-septate, hyaline (colorless) under the microscope. Asexual spores of various shapes and sizes also produced by some species. Asci borne in perithecia (flasklike structures).

THIS distinctive genus is parasitic on other mushrooms (mostly agarics and boletes), engulfing them in a pimpled or powdery layer of tissue. The actual fruiting bodies (perithecia) of *Hypomyces* are not large enough to qualify as mushrooms, but are given shape and substance by the mushrooms they grow on. *Hypomyces* can be found wherever suitable hosts occur, sometimes in epidemic proportions. Apparently the mycelium lives with or on the mycelium of its host and fruits at the same time. *H. lactifluorum*, sometimes called the "Lobster Mushroom" because of its bright red to orange color, is eaten by many people. Other species of *Hypomyces* are unpalatable and potentially dangerous, particularly if the host is a poisonous mushroom disfigured beyond recognition. Two common species of *Hypomyces* are described here and several others are keyed out.

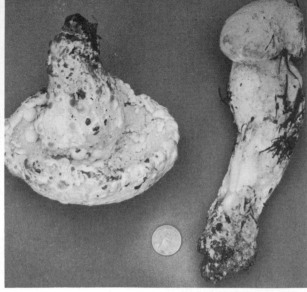

Left: *Cordyceps militaris* (see p. 882). Note the pimpled surface caused by protruding perithecia, and the insect on which it is growing (partly visible on left). (Alan Bessette) **Right:** *Hypomyces chrysospermum* (see below) engulfs boletes in a white to bright yellow mass of tissue.

Key to Hypomyces

1. Growing on boletes, covering them in a white to bright yellow mass of tissue . . . .
   . . . . ***H. chrysospermum,*** below
1. Not as above 2

2. Growing on fruiting bodies of *Russula* and *Lactarius,* covering them in a minutely pimpled layer of bright orange to red or magenta tissue . . . . ***H. lactifluorum,*** p. 884
2. Not as above 3

3. Growing on gilled mushrooms (mostly *Amanita, Lactarius,* and *Russula*) 4
3. Growing on elfin saddles or coral fungi 5

4. Growing on fruiting bodies of *Amanita,* covering them in a white to flesh-colored or pinkish layer of tissue . . . . ***H. hyalinus*** (see *H. lactifluorum,* p. 884)
4. Growing mostly on fruiting bodies of *Lactarius* and *Russula* (usually on the gills and upper stalk), covering them in yellow to greenish tissue ***H. luteovirens*** (see *H. lactifluorum,* p. 884)

5. Growing on coral fungi (e.g., *Ramaria*) . . . ***H. transformans*** (see *H. chrysospermum,* below)
5. Growing on elfin saddles (e.g., *Helvella*) . . . ***H. cervinigenus*** (see *H. chrysospermum,* below)

Hypomyces chrysospermum   (Bolete Eater)

FRUITING BODY beginning as a white moldy-looking growth that attacks and quickly engulfs boletes, then turns bright yellow and powdery, then finally becomes reddish-brown (but this last stage rarely seen) and becomes pimpled. Flesh of the host often soft or mushy. **SPORES** in white stage 10-30 × 5-12 microns, elliptical, smooth; in yellow stage, 10-25 microns, round, thick-walled, and warted; in final (sexual) stage, 25-30 × 5-6 microns, spindle-shaped, hyaline (colorless) under the microscope and usually 1-septate (partitioned). Perithecia only present in final stage.

HABITAT: Solitary, scattered, or gregarious on boletes; widely distributed and very common in our area whenever boletes are out; also reported on *Paxillus* and *Rhizopogon.*

EDIBILITY: Not edible, possibly poisonous. It is often associated with bacterial decay.

COMMENTS: This is the white to bright yellow fungus you see so often on boletes, and which you've probably cursed a hundred times for depriving you of your meal. The spores produced in the white and yellow stages are asexual; only those in the final stage are perfect spores (sexually produced inside asci). This final stage is seldom seen, however, because it occurs only after the host is decayed beyond recognition and is very unpleasant

883

Hypomyces hyalinus (specimens on right) disfigures various species of *Amanita*, in this case *A. rubescens* (shown on left). See comments under *H. lactifluorum* for more details.

to handle. The different stages of the fungus have been given different names, e.g., **Sepedonium chrysospermum** was originally applied to the yellow stage. Other species: In our area, **H. cervinigenus** commonly attacks the black fluted elfin saddle *(Helvella lacunosa),* covering it with white or pinkish tissue. **H. transformans** performs a similar transformation on several species of coral fungi (particularly *Ramaria*). Neither of these should be eaten. For species that attack gilled mushrooms, see *H. lactifluorum.*

Hypomyces lactifluorum   (Lobster Mushroom)   Color Plate 216

FRUITING BODY growing on and engulfing gilled mushrooms (species of *Russula* and *Lactarius*) in a layer of roughened or pimpled, bright orange to orange-red to purple-red or occasionally yellow-orange tissue which is firm to the touch. Overall shape of the host mushroom and parasite often like an inverted pyramid. Flesh (of the host) crisp, white.
SPORES 30-50 × 4.5-8 microns, spindle-shaped or shaped like caraway seeds, hyaline (colorless) under the microscope, septate (with one partition), warted. Perithecia imbedded in the tissue that covers the host, but protruding as small pimples.

HABITAT: Solitary, scattered, or gregarious in woods, often partially buried in the duff, usually on the fruiting bodies of *Lactarius* and *Russula* (especially the large white species like *R. brevipes*); widely distributed. It is common in some regions, but rare in our area. The largest fruitings I've seen were under ponderosa pine in the Southwest.

EDIBILITY: Rated highly by many people and sold in markets in Mexico. There is no absolute assurance that the host species is edible, but I can find no mention of poisonings by this species. Perhaps it only attacks edible species! Material I sampled was fairly good.

COMMENTS: This fungus is best recognized by its bright orange to reddish color and minutely pimpled surface. The gills of the host are often reduced to blunt, chanterelle-like ridges, but the pimpled appearance caused by the numerous perithecia and crisper texture distinguish it. The bright fluorescent color of *H. lactifluorum* make it the most spectacular member of its clan. **H. luteovirens** is a somewhat similar yellow-green to greenish species that covers the gills and upper stalk of various *Lactarius* and *Russula* species. It is more common in our area than *L. lactifluorum,* but not nearly as conspicuous. **H. hyalinus** is a white to pinkish or flesh-colored species that attacks species of *Amanita* (particularly *A. rubescens*), turning them into pimpled or warty upright clubs (see photo above). It should not be eaten because its host might be deadly poisonous! It is common in eastern North America, but I have yet to find it on the west coast.

XYLARIA & DALDINIA

Very tough to hard or charcoal-like, wood-inhabiting fungi. FRUITING BODY *erect and clublike or branched* in *Xylaria, or stalkless and hemispherical* in *Daldinia; usually black when mature,* but sometimes covered with a white or grayish or brown powdery coating of asexual spores (conidia). *Flesh tough or charcoal-like, usually white* in *Xylaria, concentrically zoned* in *Daldinia.* STALK present in *Xylaria,* absent in *Daldinia.* SPORES dark brown to black, usually spindle-shaped or elliptical, smooth; asexual spores (conidia) hyaline (colorless) under the microscope, smooth. Asci borne in flasklike structures (perithecia) imbedded in the fruiting body (usually upper portion).

THESE tough, mostly black, wood-inhabiting fungi are very distinctive. The erect, club-like forms *(Xylaria)* might be confused with the black earth tongues (*Geoglossum* and *Trichoglossum),* but are much tougher or harder and have paler flesh. *Daldinia,* on the other hand, has a sessile (stalkless) fruiting body. It might be mistaken for a charred polypore or crust fungus, but its concentrically zoned interior, pimpled exterior, and brittle, charcoal-like consistency are distinctive. Both *Xylaria* and *Daldinia* frequently produce spores asexually. These asexual spores (conidia) take the form of a white to gray or brown powder that coats the surface of the young fruiting body. The powder may disguise the black undersurface, but is easily rubbed or licked off. In mature specimens, numerous asci-containing flasks (perithecia) can be seen if the fruiting body is sliced open.

Xylaria and *Daldinia* are very common, but much too tough to be edible. Identification of species is based largely on microscopic characteristics, but the genera are easily learned. Two widespread Xylarias and one *Daldinia* are described here.

Key to Xylaria & Daldinia

1. Fruiting body round to hemispherical or lumpy, with little or no stalk (not growing erect); flesh charcoal-like (hard but usually brittle) . . . . ***Daldinia grandis*** & others, p. 887
1. Not as above; fruiting body erect, clublike (unbranched) or antlerlike (branched) 2

2. Fruiting body slender, often branched and/or covered with a whitish powder (at least over upper portion) . . . . ***Xylaria hypoxylon,*** below
2. Not as above 3

3. Fruiting body very tough or hard, up to 3 cm thick; flesh inside usually white or pallid; surface often minutely warted or cracked . . . . ***Xylaria polymorpha*** & others, p. 886
3. Not with above features . . . . (see ***Geoglossum & Trichoglossum,*** p. 866)

Xylaria hypoxylon (Candlesnuff Fungus)

FRUITING BODY 2-8 cm high, very tough, erect, slender, cylindrical or narrowly club-like when young but usually becoming antlerlike (branched sparsely or forked at the tip) in age. Upper portion or tip (or occasionally entire surface) white and powdery when young, eventually becoming black and minutely roughened (use hand lens!). Flesh very tough, white or pallid. **STALK** (lower sterile portion of fruiting body) thin, usually 1-3 (5) mm thick, black, minutely hairy, very tough or wiry. **SPORES** (sexual) 10-14 × 4-6 microns, bean-shaped, smooth, black in mass but brown under the microscope; asexual spores (conidia) smooth and elliptical or elongated, white in mass but hyaline (colorless) under the microscope. Perithecia imbedded in the upper part of fully mature fruiting body.

HABITAT: Scattered to densely gregarious or clustered on rotting logs, stumps, buried sticks, etc.; very widely distributed and common. In our area it occurs year-round in many habitats, but especially on oak and tanoak in the fall and winter.

EDIBILITY: Much too tough to be of value.

Left: *Xylaria hypoxylon.* Antlerlike specimens such as these are typical, but unbranched ones also occur. At least the upper portion of black fruiting body is usually covered with white spore powder until fully mature. **Right:** "Dead Man's Fingers," *Xylaria polymorpha.* Note thick fruiting body with blunt tip and sterile base or stalk. These two specimens are black, but the picture is overexposed.

COMMENTS: Also known as *Xylosphaera hypoxylon,* the candlesnuff fungus is easily told by its very tough, slender, antlerlike fruiting body that is black below and dusted with white powder above. The powdered appearance of young specimens is caused by masses of asexual spores (conidia) that form directly on hyphae instead of in asci. Later on, the asci form inside "flasks" (perithecia) at the top of the fruiting body. There are many Xylarias that more or less resemble this species (e.g., *X. cornu-damae*), but they are best differentiated microscopically. Some are short and cylindrical, others are branched or clustered. All grow on wood (sometimes buried) and are very tough. For thicker club-shaped species, see *X. polymorpha.*

Xylaria polymorpha   (Dead Man's Fingers)

FRUITING BODY 2-8 cm tall, 0.5-3 cm thick, very tough and hard or carbonaceous; erect, club- or finger-shaped to somewhat irregular or twisted, the tip usually blunt or rounded and occasionally lobed. Outer surface hard and crustlike, usually wrinkled, roughened, and/or cracked, black when mature but often covered with a whitish to grayish or brownish powder when very young. Flesh (interior) hard or corky, white or pallid. **STALK** present as a short sterile base, usually well-defined and narrower than fertile part; black. **SPORES** (sexual) 20-32 × 5-10 (12) microns, spindle-shaped, smooth, dark brown to black; asexual spores (conidia) when present smaller, elongated or elliptical, smooth, hyaline under the microscope. Perithecia imbedded in upper portion of fruiting body.

HABITAT: In groups or clusters on hardwood stumps, logs, etc., but often appearing terrestrial if the wood is buried; widely distributed and common, but apparently absent or very rare in our area. In eastern North America, where it favors beech and maple, the fruiting bodies usually appear in the spring and mature (blacken) by the summer. They last for months without decaying.

EDIBILITY: Much too tough and rough to be edible.

COMMENTS: Also known as *Xylosphaera polymorpha,* this fungus is easily recognized by its hard, dark fingerlike fruiting bodies. They are much thicker than those of *X. hypoxylon, Geoglossum,* and *Trichoglossum,* and the white or pallid interior is also distinctive. There are several very similar temperate and tropical species that are collectively called "Dead Man's Fingers." These are differentiated primarily on microscopic characteristics. One, *X. longipes,* is similar but consistently slimmer (0.3-1 cm thick) than *X. polymorpha.* It occurs across the continent, usually on hardwoods.

Daldinia grandis. Note the concentrically zoned interior (shown in sectioned specimen at upper right) and minutely pimpled exterior. Charcoal-like consistency and growth on wood are also distinctive.

Daldinia grandis   (Crampballs; Carbon Balls; King Alfred's Cakes)

FRUITING BODY very tough and woody or charcoal-like, 1-6 cm broad or sometimes larger, hemispherical to nearly round to somewhat lumpy or irregular, stalkless. Exterior black (or sometimes dark brown when young), roughened or pimpled by the perithecia, often cracked in age. Flesh (interior) brown to grayish-black, somewhat lustrous, with lighter and darker concentric zones; very brittle and charcoal-like. **STALK** absent. **SPORES** (sexual) 14-17 (27) × 6.5-11 microns, elliptical or elongated, smooth, dark brown to black; asexual spores (conidia) minute, smooth, and hyaline (colorless) when present.

HABITAT: Scattered to gregarious or in masses on dead logs, branches, or bark of hardwoods; widely distributed. It is abundant year-round in our area, especially on oak.

EDIBILITY: Unequivocally inedible—but perhaps useful as a substitute for charcoal!

COMMENTS: The distinctive charcoal-like fruiting bodies of this fungus can be found on almost any walk through the woods. The pimpled surface of mature specimens (see photo) is caused by the protruding tips of the perithecia (flasks of asci). As in *Xylaria,* younger specimens are sometimes coated with pale, asexual spores (conidia). The nickname "Crampballs," incidentally, can be attributed to the old folk belief that carrying one around under your armpits would cure cramps! Other species: ***D. concentrica*** is the common crampball of eastern North America. It is very similar to *D. grandis,* but has slightly smaller spores and is more apt to be dark brown to bronze-black when young (and black in age). ***D. vernicosa*** is also similar, but usually has a narrowed base beneath the fertile portion and an interior zoned with dark brown and white or gray. Other species can only be differentiated microscopically.

"King Alfred's Cakes," *Daldinia grandis,* sharing a juicy log with *Phellinus gilvus* (see p. 582).

MUSHROOM COOKERY

The abundance boneless
Without husk or scale or thorn,
Granting us this festival of all-embracing freshness

Pablo Neruda's tribute to the tomato is also a tribute to immediacy and vitality, that incomparable freshness that distinguishes a homegrown tomato or cucumber or wild mushroom from its flavorless, mass-produced counterpart. The challenge in cooking wild mushrooms is to maximize their freshness and earthy essence while highlighting their individuality. After all, they are not one vegetable, but many—a pleasant surprise to people who are conditioned to mushrooms that smell and taste "mushroomy." The major constraint is that you must make do with what you have. Obviously, you can't cook boletus broth when you have a basketful of blewits, but you *can* make blewit burgers, or blewit biscuits, or a three-bean blewit salad.

The most important thing to remember is that you can't expect wild mushrooms to be special unless you take the time to make them special. They are ephemeral, temperamental, delicate. It is a relatively simple task to render the most marvelous mushroom tasteless. Likewise, many "mediocre" mushrooms are delightful when cooked with care and imagination. If you make a concerted effort to seek out, gather, identify, and eat wild mushrooms, it only makes sense to do them justice in the kitchen. Don't just throw them into the pot with a bunch of other vegetables—unless, of course, you want them to taste like a bunch of other vegetables. Different mushrooms call for different treatment; only then will they respond with their full measure of flavor.

With each type you will go through a period of discovery and experimentation—succulent successes and unforgettable failures—followed by a process of adjustment and subtle refinement. Each kind of mushroom will gradually acquire its own culinary identity and cease to be a mushroom except in the botanical sense that broccoli is a plant. After all, when you're having broccoli, you don't say, "We're having steamed plants for dinner." Similarly, it will no longer be "mushrooms" for dinner, but "chanterelles," or "poor people's truffle," or in the case of the cultivated mushroom, "*Agaricus bisporus*" (to be pronounced with a subtle insinuation of distaste).

Strive for a marriage (but not a compromise!) between elegance and simplicity. Successful mushroom cookery doesn't require exotic foods, or a bottomless bank account, or idle afternoons, or a degree in gastronomical mechanics. It *does* require patience, sensitivity, enthusiasm, and imagination. There are no rigorous rules, but some basic do's and don'ts are summarized below.

HELPFUL HINTS

1. *Don't* eat a mushroom unless you're *absolutely sure* it's edible. In other words: "When in doubt, throw it out."

2. You wouldn't eat a rotten egg, so *don't* eat a rotten mushroom. Food poisoning is a frequent cause of so-called "mushroom poisoning."

3. You wouldn't eat five pounds of asparagus, so *don't* eat five pounds of mushrooms, no matter how delicious they are. Overindulgence **(COLOR PLATE 217)** is another common cause of so-called "mushroom poisoning," particularly for those who

"What was desire in the hills becomes fulfillment in the kitchen," says Angelo Pellegrini in *The Savory Wild Mushroom*. At left is fresh *Agaricus campestris*. At right, a delicious cream sauce with white onions and sliced *A. campestris*.

don't eat mushrooms regularly. On the other hand, don't be stingy—most wild mushrooms cook down more than the commercial variety, so use them generously.

4. When trying a species for the first time, eat only a small amount. Then wait for a few hours to see if you have an adverse reaction to it. Just as some people are allergic to eggs or chocolate or scallops or strawberries, some are adversely affected by certain kinds of mushrooms. Species to which many people are "allergic" (e.g., *Laetiporus sulphureus, Lepiota* species) should *not* be served to large groups.

5. In the event of an "allergy," you'll want to pinpoint the culprit, so *don't* mix two or more species together unless you've eaten them before.

6. As a rule, maggot-riddled mushrooms (see photo at bottom of p. 277) *should not* be eaten, especially when uninfested specimens are available. However, in the case of certain choice species (e.g., *Agaricus augustus*), you may wish to remove the maggots with a knife (if there are only a few) or even leave them in (they're just a little extra protein). Use your own judgment on this matter.

7. Use as little water as possible when cleaning mushrooms. They absorb it so readily that it dilutes their flavor and causes them to cook up slimy. On the other hand, there's nothing worse than gritty wild mushrooms (unless it's gritty domesticated mushrooms), so don't hesitate to use water if nothing else works. If you wash them, drain them on a paper towel before cooking. The best place to clean mushrooms *is in the field* (providing you already know their identity). Trim away all dirt with a knife or small brush, and don't mix dirty specimens with clean ones. Also, check for maggots and remove any that are present so they won't multiply and spread (even though maggots are the larvae of gnats or flies, they are able to reproduce partheno-genetically).

8. Use mushrooms as soon as possible after picking them. Prolonged refrigeration deprives all vegetables (including mushrooms) of their freshness and flavor. Species of *Coprinus* should be eaten the day they're picked, or they will digest themselves. There's an old saying to the effect that you should boil the water before harvesting the corn. Well, it's not a bad idea to melt the butter before picking the shaggy manes!

Left: One day's catch: *Boletus edulis, Amanita calyptrata, Ramaria* spp., and others. **Right:** Enjoying the day's catch (in this case, shaggy manes fried in egg batter and bread crumbs).

9. During periods of heavy rainfall, most mushrooms will be waterlogged. These are apt to cook up insipid and slimy, but can be sliced and dried for later use (thereby concentrating their flavor).

10. *Don't* steam or pressure-cook mushrooms. You want to get rid of excess moisture, not add to it. Pressure-cooked mushrooms bear an uncanny resemblance to slugs.

11. *Don't* drown mushrooms in spices, butter, salt, garlic, or olive oil. All of these complement mushrooms nicely when used in moderation. Mushrooms and onions, for instance, are practically made for each other, but the mushrooms must always dominate in quantity because their flavor is more delicate.

12. If you don't like a "choice" mushroom, give it a second and third chance. After all, a lot depends on how you cook it and in what condition it was found. The environment can also have an influence. For instance, blewits that grow under cypress are often bitter-tasting. No one agrees on which kinds are best, but a species does not acquire a widespread reputation unless it has something special to offer. On the other hand, some relatively unknown mushrooms (e.g., *Chroogomphus*) are quite good. Improvise!

PRESERVING MUSHROOMS FOR CONSUMPTION

Most mushrooms have fickle fruiting habits, appearing in large numbers for one or two weeks, then disappearing for the rest of the year. To take advantage of their fleeting abundance, you have to harvest them while you can and then preserve them for later use.

DRYING

Drying mushrooms is the easiest and most satisfactory way to preserve them. Fleshy types like *Boletus* must be cut in thin slices; smaller species like *Marasmius oreades* can be dried whole. Don't use an oven. Circulation is more important than heat—you want moisture to be carried away. Spread out the mushrooms on screens and use a light bulb or hot plate as a heat source (unless it's arid enough to sun-dry them). Or string them up

890

Dried mushrooms make a marvelous addition to sauces, soups, and gravies. They should be stored in airtight jars to protect them from insects. This array of dehydrated delicacies includes *Agaricus*, *Boletus*, *Chroogomphus*, *Craterellus*, *Clitocybe*, and *Marasmius*.

on thread and hang them in a warm, dry place. Remove all maggots before drying mushrooms, and try to get them as clean as possible without washing them. If necessary, they can be cleaned before use by placing them in a strainer and scalding them with water.

When thoroughly dried (brittle), the mushrooms should be stored in an airtight jar to protect them from insects and mold. They will keep for months in this state. They can also be pulverized or powdered and used as a condiment. Certain mushrooms, such as *Cantharellus cibarius* and *Laetiporus sulphureus* do *not* dry well (they become too tough and leathery). Others, like *Marasmius, Boletus, Leccinum, Craterellus, Chroogomphus, Morchella*, and *Agaricus* are excellent.

There are several methods for reconstituting dried mushrooms, depending on the type of mushroom and kind of dish. They can be crumbled directly into soups or sauces, but should be soaked first if they're to be put in drier foods. An excellent method is to place dried pieces between wet paper towels overnight. This allows them to absorb moisture gradually while retaining their flavor.

FREEZING

Freezing is another excellent and easy way to preserve mushrooms, *providing you saute them briefly beforehand.* (If you freeze raw mushrooms, they will decompose as soon as they thaw out.) Practically all mushrooms freeze well—I have stored *Amanita calyptrata* for over one year without any noticeable change in flavor! The mushrooms should be stored in an airtight container (e.g., a ziplock bag). Mushroom sauces and stocks can also be frozen for later use.

CANNING

Canning is a big undertaking, and is only worthwhile when you have an enormous amount of mushrooms. A pressure canner, mason jars, and lids are required. Wash the mushrooms thoroughly and let them cook in their own juices for a while. Then pack them in sterilized pint jars, cover with boiling water, seal, and process at 10-15 pounds pressure for 30 minutes (a *very necessary* step because they can easily become tainted with botulism). A little ascorbic acid (vitamin C) can be added to retain color. Mushrooms will keep for years if canned properly, but failure to observe sterile procedures can be disastrous. In my experience, *Cantharellus, Tricholoma, Dentinum, Boletus, Hericium*, and *Agaricus* can well.

PICKLING

Pickling is just marinating on a larger and more elaborate scale. In Europe, mushrooms are often preserved under oil and vinegar. It is a fairly simple procedure, but obviously, the mushroom flavor is masked. Jars should be sterilized, but pressure cooking isn't necessary because of the acidic medium. I have successfully pickled *Cantharellus, Clitocybe nuda,* and *Amanita calyptrata.*

SALTING

This is another technique that is popular in some parts of Europe. The mushrooms are cooked briefly and then packed in rock salt and stored in airtight jars in the refrigerator. Naturally they are salty this way, but the salt is easily dissipated by adding the mushrooms to soups or other watery dishes.

MUSHROOM TOXINS

MOST CASES of "mushroom poisoning" are the result of allergies* (idiosyncratic reactions or hypersensitivity), overindulgence (especially of raw mushrooms), or food poisoning (ingestion of rotten mushrooms). All three usually result in nausea, vomiting, and/or diarrhea. Another common type of "mushroom poisoning" is imaginary—people who have lingering doubts about the safety of their meal are apt to experience discomfort whether or not there is a physiological basis for it. All the more reason not to eat a mushroom unless you're absolutely certain it's edible!

The two most common causes of mushroom poisoning are carelessness and ignorance. Despite what people say, mushroom experts do not die from mushroom poisoning. But of course, it is much more sensational for newspapers to say they do. Relatively few species are poisonous, but some of the most dangerous ones are exceedingly common, and almost any poisonous mushroom can be fatal to a small child or a person in poor health. Therefore it is useful to learn about the different kinds of mushroom poisoning, should you or a friend experience discomfort. A brief rundown of the major groups of mushroom toxins is presented here. "Mushroom toxin," as defined here, is a compound which produces an abnormal effect on the human body. This definition encompasses mind-altering drugs such as psilocybin, whether or not they are ingested deliberately.

As with any kind of poisoning, the two most important things to do are to seek immediate medical attention and identify the agent responsible. Idiosyncratic reactions to edible mushrooms are generally not serious enough to warrant a trip to the hospital, but if there is any doubt, consult a physician.

AMANITA-TOXINS (AMATOXINS)

> **Mushrooms:** *Amanita phalloides, A. ocreata, A. verna, A. virosa, A. bisporigera; Conocybe filaris; Galerina autumnalis, G. marginata, G. venenata; Lepiota castanea, L. helveola, L. josserandii* & close relatives.

Poisoning by amatoxins is extremely serious, with a fatality rate of about 50%. It is doubly dangerous because the symptoms are delayed for 6-24 hours after ingestion of the mushroom, by which time the toxins have been absorbed by the body.

There are several groups of amatoxins, at least in the Amanitas. Phallolysin was the first toxin discovered. It destroys red blood cells when injected into rats and has a very

*In this book the term "allergy" is used loosely to describe an adverse reaction to a normally harmless mushroom even though it may not be a genuine hypersensitivity.

high mortality rate. However, it is unstable and apparently destroyed by cooking and/ or the human digestive tract. A group of complex cyclic polypeptides called phallotoxins comprise the second group. They are also fatal when injected intravenously (they attack the liver), but are apparently destroyed by the digestive tract. It is another group of polypeptides, the amanitins, that are the culprits. They are twenty times more lethal than the phallotoxins. Their concentration varies tremendously from individual mushroom to individual mushroom, but an average fatal dose is about 2 ounces (fresh weight) of *Amanita phalloides.*

All recent mushroom-induced fatalities in California have been caused by *Amanita phalloides* and *A. ocreata.* Both are large, handsome, tempting mushrooms, whereas the other deadly types (such as *Galerina*) are small and nondescript and therefore unlikely to be eaten. Of course, there is no excuse for eating *any* of these mushrooms if you take the time to learn about them.

Symptoms and Treatment

Amanitin poisoning usually manifests itself in four stages: (1) a latency period of 6-24 hours after ingestion, during which time the toxin is actively working on the liver and kidneys, but the victim experiences no discomfort; (2) a period of about one day characterized by violent vomiting, bloody diarrhea, and severe abdominal cramps; (3) a period of about one day during which the victim appears to be recovering (if hospitalized, the patient is sometimes released!); (4) a relapse, during which liver and kidney failure often leads to death. There is sometimes more than one relapse.

The effects of the toxin are centered on the liver and kidneys, but amanitin damages tissue throughout the body by inhibiting RNA synthesis within each cell. To make matters worse, the kidneys are apparently unable to eliminate amanitin from the body. The pancreas, adrenal glands, heart, lungs, muscles, intestines, and brain may be damaged, not only by the amanitin, but indirectly because of liver and kidney failure. It is a slow and painful way to die.

There is no known antidote to amatoxin poisoning. Treatment is largely supportive and symptomatic—maintaining blood sugar and salts, eliminating urea by dialysis, and helping the body to get rid of the toxins. If you have any reason to think someone has eaten a deadly *Amanita* (or an amanitin-containing mushroom), *don't wait for the symptoms to appear!* If the person is taken to the hospital soon after ingesting the mushrooms, at least some of the toxins can be removed before they are absorbed.

GYROMITRIN (MONOMETHYLHYDRAZINE)

Mushrooms: Several *Gyromitra* species (especially *G. esculenta* & *G. infula*), also many related Ascomycetes (e.g., *Verpa, Cudonia, Helvella* spp.)

Gyromitrin's product of hydrolysis, monomethylhydrazine (MMH), is a very toxic carcinogenic compound used in the manufacture of rocket fuel. Gyromitrin poisoning has puzzled scientists for many years because of the very narrow threshold between complete absence of discomfort and severe poisoning or even death. This is due to the volatile nature of gyromitrin, which is removed by the process of cooking or drying *providing it has a chance to escape.* Gyromitras cooked in a closed pan can cause severe poisoning, and inhalation of the vapors is dangerous. Raw Gyromitras, of course, pose the greatest threat.

The situation is complicated by differences in the toxicity of different geographical strains. *Gyromitra esculenta* has caused numerous fatalities in Europe, but not a single one in California.

Symptoms and Treatment

The symptoms, which appear 2-24 hours after ingestion, include headaches, abdominal distress, severe diarrhea, and vomiting. In severe cases the liver, kidney, and red blood cells are damaged (much as in poisoning by amatoxins), which may result in death. Treatment is largely supportive; a physician should be consulted.

MUSCARINE

Mushrooms: *Inocybe* species; *Clitocybe dealbata* and seveal relatives; *Omphalotus* species, and certain red-pored species of *Boletus.*

Muscarine was originally isolated in *Amanita muscaria,* but occurs in that mushroom in insignificant amounts. However, many Inocybes contain large amounts of muscarine—enough so that they can be fatal in large quantities.

Symptoms and Treatment

The effects, which manifest themselves 15-30 minutes after ingestion, are focused on the parasympathetic (involuntary) nervous system. They include excessive salivation, perspiration, tears, and lactation (in pregnant women), plus severe vomiting and diarrhea. These symptoms may be accompanied by visual disturbances, irregular pulse rate, decreased blood pressure, and difficulty in breathing. The victim normally recovers within 24 hours, but in severe cases, death may result from respiratory failure. Atropine is a specific antidote, but must be administered by a physician.

IBOTENIC ACID/MUSCIMOL

Mushrooms: *Amanita muscaria, A. pantherina, A. gemmata.*

There are many contradictions in the literature regarding the principal toxins of *Amanita muscaria* and *A. pantherina.* Muscarine was originally believed to be the toxin, and then bufotenine was put forth as a candidate. It turns out, however, that the main active principle is ibotenic acid, which is converted by the human body into muscimol, a more powerful form that passes out in the urine.

Amanita muscaria is apparently one of the oldest intoxicants known. Its use by certain Siberian peoples has been extensively documented and R. Gordon Wasson, in his book *SOMA: The Divine Mushroom of Immortality,* makes a convincing case for it being the mystical Soma plant of the RgVedas (sacred Hindu texts). It may have been used throughout Eurasia in ancient times, but if so, its use has been suppressed. Curiously, many Europeans fear *A. muscaria* more than its deadly cousin, *A. phalloides.* John Allegro's attempt to link it to the origins of Christianity (*The Sacred Mushroom and the Cross*) is far-fetched and abstruse.

Amanita muscaria is erroneously listed in older books as deadly poisonous. It can be fatal in large doses, but so can practically any poisonous mushroom. According to most sources, *A. pantherina* is somewhat more dangerous, while *A. gemmata* is less so. Deliberate ingestion of these mushrooms has increased now that it has been shown that they are not as dangerous as once believed. However, their effects vary greatly from person to person. As people's metabolisms are different and the concentrations of the toxins vary from mushroom to mushroom, there is no way of predicting what one's reaction will be. Some people experience *extreme* discomfort, others have vivid dreams, still others experience no effects whatsoever. The ingestion of these mushrooms is definitely *not* recommended here. Incidentally, not all of the toxins have been identified. For instance, neither pure ibotenic acid nor muscimol produce the nausea and vomiting that frequently occurs after eating *A. muscaria.*

Symptoms and Treatment

Symptoms normally appear 30 minutes to 2 hours after ingestion, and last for several hours. Nausea and vomiting are common, but the principal effects are on the central nervous system: confusion, mild euphoria, loss of muscular coordination, profuse sweating, chills, visual distortions, a feeling of greater strength, and sometimes hallucinations, delusions, or convulsions. (An inordinate number of "trippers" mistake themselves for Christ). Drowsiness is also a common phenomenon. In fact, those who ingest

A. muscaria frequently fall asleep ("swoon"), to awaken hours later with little or no memory of their experiences. There is no "hangover" effect as with alcohol, but most people who try it (including myself) do not wish to repeat the experience. Since muscimol passes out through the urine, Siberian users "recycled" their *A. muscaria* by drinking their own urine. I know of no one in the United States who has tested this approach.

Treatment of muscimol poisoning is largely supportive—reassuring the victim that the effects are temporary. In the mistaken belief that muscarine was the principal toxin, older texts prescribe atropine as an antidote. Atropine, however, is likely to exacerbate the effects of ibotenic acid/muscimol.

PSILOCYBIN/PSILOCIN

Mushrooms: *Psilocybe baeocystis, P. caerulescens, P. cubensis, P. cyanescens, P. semilanceata* and many others; also certain *Panaeolus* species (e.g., *P. cyanescens, P. subbalteatus*); certain *Conocybe* and *Gymnopilus* species; possibly *Pluteus salicinus, P. cyanopus*, and others.

These indole derivatives are well known for their psychedelic properties, and "psilocybin mushrooms" are often consumed for recreational purposes. Several hallucinogenic mushrooms played an important role in the religious and medicinal rites of Native Americans in Mexico and Central America. But the Spaniards suppressed their use to such an extent that their existence was seriously doubted by early 20th century botanists. The mushrooms were "rediscovered" in Oaxaca in the 1930's, and 20 years later they were identified as belonging principally to the genus *Psilocybe*. Since then their properties have been so publicized that Oaxaca has been inundated by pleasure-seeking gringos. It has subsequently been discovered that many psilocybin-containing mushrooms grow in the United States as well.

Psilocin (a dephosphorylated version of psilocybin) is about ten times as active as psilocybin. Most psilocybin-containing mushrooms have only a trace of psilocin, but the human body coverts most of the psilocybin into psilocin. A blueing reaction associated with the presence of psilocybin and psilocin is caused by an enzyme that oxidizes psilocin. However, not all mushrooms that stain blue contain psilocybin or psilocin, and not all "psilocybin mushrooms" stain blue.

Symptoms and Treatment

Symptoms are similar to those of LSD. Shortly after ingestion, and for a duration of several hours, the "victim" experiences heightened color perception, visual distortions, rapidly shifting shapes and images, a "kaleidoscope effect" with eyes closed, elation or hilarity, and hallucinations or delusions. Nausea and vomiting are rare. Some people report the sensation of leaving their bodies, or of traveling into the future or past, or other highly subjective experiences. Others experience profound anxiety.

In case of accidental ingestion or a "bad trip," the victim should be repeatedly assured that the effects are temporary. A factor to bear in mind is that transferring the person to an unfamiliar environment can be frightening, and that sedatives may worsen the effects, especially if administered forcibly. LSD, incidentally, was derived from another fungus, *Claviceps purpurea* (wheat ergot).

GASTROINTESTINAL IRRITANTS

Mushrooms: Many, including: *Agaricus californicus, A. hondensis, A. placomyces, A. praeclaresquamosus, A. xanthodermus; Boletus satanas, B. erythropus, B. haematinus, B. pulcherrimus, B. subvelutipes; Chlorophyllum molybdites; Entoloma* species; *Gomphus floccosus; Hebeloma* species; many acrid and/or purple- or yellow-staining *Lactarius* species; *Laetiporus* *sulphureus* and *Lepiota rachodes* & *L. naucina* (sometimes); *Naematoloma fasciculare; Omphalotus* species; *Ramaria formosa* and relatives; many acrid *Russula* species; *Scleroderma* species; and several *Tricholoma* species (notably *T. pardinum* and *T. pessundatum*).

As evidenced by the list, this is by far the largest and most prevalent group of mushroom toxins. Very few of the active principles have been identified, however, perhaps because they are rarely fatal. The most frequent culprits in gastrointestinal poisoning are the phenol-smelling *Agaricus* species and *Chlorophyllum molybdites,* undoubtedly because they closely resemble edible types. The most dangerous are *Entoloma* species, *Hebeloma* species, *Tricholoma pardinum, Boletus satanas* and close relatives (raw), *Naematoloma fasciculare,* and *Chlorophyllum molybdites.* "Allergic" reactions to edible mushrooms normally take the form of gastrointestinal upset.

Symptoms and Treatment

Symptoms usually appears shortly after ingestion (20 minutes-4 hours). They include nausea, vomiting, cramps, and diarrhea,* which normally pass after the irritant is expelled. Severe cases, however, may require hospitalization. Treatment is largely supportive—helping the body to eliminate that which it is not equipped to handle. Though not as serious as other types of mushroom poisoning, gastrointestinal upsets are not to be taken lightly, as evidenced by the fact that many people acquire a lingering distaste for mushrooms after an all-night bout with nausea and diarrhea.

MISCELLANEOUS TOXINS

Coprinus atramentarius contains a disulfram-like compound (coprine) that reacts with alcohol in the body to produce acetaldehyde, which in turn produces a peculiar but transitory set of symptoms: reddening of the ears and nose, a metallic taste in the mouth, lightheadedness, rapid heart beat, a throbbing sensation, and sometimes nausea and vomiting. Recovery is normally spontaneous and complete. The alcohol needn't be consumed simultaneously with the mushrooms to have a synergistic effect—therefore anyone who indulges in alcohol regularly should not eat *C. atramentarius.* Individual reactions vary, suggesting that some people may be more sensitive or the mushrooms themselves may differ in coprine content. It has also been suggested that raw *C. atramentarius* does not actually contain coprine, but that it is formed in the process of cooking. Similar alcohol-related effects have been reported for *Clitocybe clavipes.*

Paxillus involutus is said to be toxic raw, but is eaten by many people (especially Europeans) after being pickled or parboiled. However, the human body apparently develops a sensitivity to it, manufacturing antibodies that destroy red blood cells. Thus, someone who has eaten it for years can suddenly be poisoned (even fatally!).

Cortinarius orellanus, C. gentilis, and some close relatives are commonly fatal. They contain toxins which, like the amatoxins, attack the liver and kidneys. Symptoms do not appear for *up to three weeks* after ingestion of the mushrooms, making diagnosis and treatment much more difficult. Some of these Cortinarii occur in the United States. As they are difficult to identify, all dry-capped *Cortinarius* species are best avoided.

Naematoloma fasciculare, a common wood-loving mushroom, can also cause liver and kidney damage. Fortunately, it is rarely eaten because of its bitter taste. There is also evidence to suggest that the substrate on which a mushroom is growing can affect its edibility. The edible honey mushroom *(Armillariella mellea),* for instance, will often cause digestive upsets when growing on buckeye.

Last but not least, there is the danger of contamination by pesticides and other environmental poisons. Always be aware of this possibility, especially when picking mushrooms in towns, along well-traveled roads, and in forests, fields, or range land where herbicides and pesticides are used.

---

*The literal translation of the Japanese name for certain poisonous Tricholomas is "Unable to Cross the Valley"— presumably because one is unable to cross the valley after eating them!

Russula cyanoxantha, a robust mushroom with brittle flesh and white gills. The cap color is extremely variable (see description on p. 94). Maggot-free specimens such as these are unusual.

Lactarius uvidus (see p. 75) is one of several purple-staining milk caps. The cap is usually sticky or slimy, but in one mountain-loving variety it is practically dry. Note how the stalk snaps open cleanly like a piece of chalk—a characteristic feature of *Lactarius* and *Russula*.

Melanoleuca evenosa group (see p. 171) is reminiscent of a *Russula* or *Tricholoma* but has a fibrous stalk and amyloid white spores. Look for it in the spring under mountain conifers soon after the snow melts. Note how the gills are crowded and whitish.

Two agarics that typically grow in dense clusters. **Top:** *Clitocybula familia* (see comments under *Collybia acervata,* p. 215) grows on rotting wood. These are young specimens; as they mature the caps will broaden somewhat. **Below:** *Lyophyllum decastes* group (see p. 174) is usually but not always terrestrial. This cluster is rather unusual because the caps have watery spots.

Abnormalities are common in mushrooms. They are the result of environmental influence or improper development; they are *not* inherited or passed on. One widespread abnormality is the development of a small rosette of gills ("rose gill") on the cap surface. In extreme cases, an entire mushroom may ride "piggy-back" on another, like these Russulas **(left)**. Albino individuals also occur. In this cluster of *Craterellus cornucopioides* **(right)**, half is normally colored and half is whitish! Other abnormalities include aborted, parasitized, or sterile specimens and failure of the cap or gills to form.

WHAT IT ALL MEANS
(A Short Dictionary of Scientific Names)

Otidea, Ramaria, Tricholomopsis, Volvariella
Exidia, Stropharia, Hygrophoropsis, Arcangeliella

You may groan at the sight or sound of scientific names, but the paucity of common names for mushrooms forces us to use them. As explained in the chapter on classification (p. 8), many people are intimidated by scientific names because they are derived from Latin and Greek. "How do you remember all those names?" I am repeatedly asked. Yet posers of this question have usually mastered many scientific names without realizing it:

Asparagus, Magnolia, Sassafras, Eucalyptus
Rhinoceros, Hippopotamus, Chrysanthemum, Citrus

Since that which is understood is more likely to be remembered than that which is not, learning the meanings of scientific names can increase your ability to remember them. By refusing to be intimidated, you can demystify the names of mushrooms while you are demystifying the mushrooms themselves.

Scientific names can be divided into three categories: **descriptive, honorary,** and **geographical.** (A fourth possible category, nonsensical, needs no elaboration.)

Descriptive names are the most numerous as well as the most helpful. They can aid in the identification process because they usually tell us something significant about the mushroom, its habitat, or time of growth. For example:

Flav- means yellow, and it is a safe bet that *Hygrocybe flavescens, Tricholoma flavovirens, Boletus flaviporus,* and *Russula claroflava* are at least partially yellow. (They are.) Similarly, *Macrocystidia cucumis* (*macro*=large, *cucumis* =cucumber) has giant cystidia (sterile cells) on its gills and smells like cucumber, and *Hypsizygus tessulatus* (*hyps*=high up, *zygus*=yoked or joined, *tessulatus*=mosaic-like) grows high up on elms and develops mosaic-like scales on its cap as it matures.

You will soon discover that many of the longest and most "difficult" names are actually composed of shorter, more familiar elements—familiar because the English language is a thicket with many Latin roots. For instance:

Climacodon is derived from *climac-* (ladder, as in the English word climax) and *odon* (tooth or teeth, as in the English word orthodontist), and the Greek element *derm* (skin) appears in many mushroom names (*xanthodermus, calyptroderma, Dermocybe,* etc.) as well as English words like dermatologist, epidermis, and pachyderm.

A few descriptive names, unfortunately, are apparent misnomers, for they bear no obvious relation to the mushroom and may even contradict it:

Appendiculat- means with a fringe or small appendage, but *Boletus appendiculatus* has neither. Although *alnicol-* means alder-dweller, *Lactarius alnicola* favors conifers and oaks (but may have originally been collected in a mixed forest containing alder).

Honorary names usually commemorate a mycologist, mycologist's best friend, or mushroom collector. In most cases an *-ii, -i,* or *-ae* is added to the person's name if it is a species (e.g., *kauffmanii, barrowsii, smithii*), or an *-a, -ea,* or *-ia* if it is a genus (e.g., *Barssia*). Other suffixes, particularly diminutives or those meaning "pertaining to," can also be added (e.g., *booniana, Longula, Lenzites*).

Honorary names are usually recognizable as such because they don't sound Latin. Some, however, may not be obvious, particularly those that honor Europeans. Examples:

> *Hohenbuehelia* is named after the Austrian botanist Hohenbuhel, *Arcangeliella* and *Battarrea* after the Italian mycologists Arcangeli and Battarra, and *Galiella* and *Rozites* after the French mycologists Le Gal and Roze.

One major drawback of an honorary name is that it tells us nothing about the mushroom itself. On the other hand, precisely because it tells us nothing it cannot be misleading!

Geographical names are formed by adding an adjective suffix to the name of a particular region or locality. Most are self-evident (e.g., *californicus, marylandensis, cubensis, mexicana, olympiana*). Some are Latinized, however (e.g., *novaboracensis,* of New York, *suecica,* of Sweden) and others are obscure (e.g., *hondensis,* named after bustling La Honda, California, and *pitkinensis,* which immortalizes populous Pitkin, Colorado). Geographical names can be deceptive because they usually indicate where a species was discovered rather than where it occurs. For instance, *Longula texensis,* originally collected in Texas, and *Suillus sibiricus,* first found in Siberia, are both widely distributed in western North America.

SUFFIXES

Suffixes can have specific meanings, vague meanings, multiple meanings, or no meaning at all. Some suffixes simply transform nouns into adjectives. Others indicate possession, likeness, size, action, or degree. Still others designate gender—a concept familiar to anyone who has studied French or Spanish. Thus a single root word can have several forms depending on the gender and spelling of the word(s) with which it is used. Example:

> *Brunne-*, the root word for brown, becomes *brunneus* (masculine form) in *Boletus olivaceobrunneus, brunnea* (feminine form) in *Leptonia vinaceobrunnea, brunneum* (neuter form) in *Tricholoma albobrunneunm,* and *brunneo-, brunnea-,* or *brunnei* (combining forms), as in *Clitocybe brunneocephala.* Suffixes with specific meanings can also be attached, as in *brunnescens* ("becoming brown") or *brunneola* ("less than brown, brownish").

As you can see, it is not necessary to know all the quirks and nuances of Latin and Greek grammar in order to recognize root words such as *brunne* in their various manifestations.

Since suffixes are sundry and sometimes difficult to recognize, the more common ones are listed separately. However, common prefixes as well as most suffixes with very specific meanings (e.g., *odon*=tooth, *pes*=foot or stem, *osm-*=odor) can be found in the dictionary of word elements.

The dictionary is for descriptive names only. It does *not* include people or places, self-evident words such as *giganteus* (gigantic), *caninus* (canine), *parasitica* (parasitic), or *fragilis* (fragile). Nor does it include names whose meanings are obscure, debatable, or unknown. A hyphen indicates that the root word can have various endings (see discussion of suffixes).

Space does not permit listing *all* the root words used in naming mushrooms, but with the help of the dictionary and the material preceding it, you should be able to make sense out of most of the scientific names in this book—and enrich your English vocabulary in the process! Examples:

> (1) *cortin*=curtain, *arius*=with or pertaining to, thus *Cortinarius* means "with a curtain."
>
> (2) *cybe*=head, *con*=cone, *derm*=skin, *ino*=fiber, *clit*=sloping, *psil*=smooth, thus *Conocybe* means "cone head," *Dermocybe* means "skin head," *Inocybe* means "fiber head," *Clitocybe* means "sloping head," and *Psilocybe* means "smooth head."

(3) *lyc*=wolf, *perdon*=flatulence, thus *Lycoperdon* means "wolf fart."

(4) *vermicul*=little worm, *osus*=full of, thus *vermiculosus* means "full of little worms" (i.e., maggoty).

(5) *canthar*=drinking cup, *ellus* is a diminutive suffix, thus *Cantharellus* means "little drinking cup."

(6) *paxill*=small stick or stake and *-us* is a masculine gender ending, thus *Paxillus* means "small stick" or "small stake."

(7) *leuc*=white, thus *Leucopaxillus* means "white *Paxillus*."

(8) *a* as a prefix means without, *phyll*=leaves or gills, *phor*=bearing or bearer, *ales* denotes an order of fungi, thus *Aphyllophorales* means "order of fungi not bearing gills."

(9) *orth*=straight, *odont*=teeth, *ist* is an English suffix meaning "one who," thus orthodontist means "one who straightens teeth."

(10) *succ*=juice or sap, *ulent*=full of, thus succulent means "full of juice."

Common Suffixes

Note: Some suffixes, particularly those with feminine endings, may not be in alphabetical order because they are listed under their masculine form. For instance, *-ata* is listed under *-atus*, and *-aria* is listed under *-arius*.

-a: most common feminine gender ending

-abilis: see *-bilis*

-abulum, -aculum: see *-bulum*

-aceae: denotes a family of fungi or plants

-aceus, -acea, -aceum: color of, made of, with quality of, closely resembling, relating to (e.g., *olivaceus*, olive-colored)

-acius, -acia, -acium: pertaining to, possessing

-ae: named after, pertaining to (e.g., *annae*)

-aeus, -aea, -aeum: belonging to

-ago: (1) resembling, similar to (2) made of, color of

-ales: denotes an order of fungi or plants

-alis, -ale: belonging to, pertaining to, relating to, characteristic of

-aneus, -anea, -aneum: made of, color of, resembling

-ans: implies action (like -ing in English, e.g., *radicans*, rooting)

-anus, -ana, -anum: pertaining to, characteristic of (often connotes position, e.g., *montanus*, of the mountains)

-arion: diminutive

-aris, -are: belonging to, pertaining to

-arius, -aria, -arium, -arum: pertaining to, possessing, furnished with, relating to, (e.g., *Cortinarius*, with a curtain)

-ascens: becoming, almost, somewhat (e.g., *purpurascens*, becoming purple)

-aticus, -atica, -aticum: denotes place of growth or origin (e.g., *silvaticus*, growing in the woods)

-atilis, -atile: denotes place of growth (e.g., *saxatilis*, growing among rocks)

-ator: see *-tor*

-atus, -ata, -atum: (1) possessing, furnished with (e.g., *armillatus*, ringed or with a ring) (2) resembling, similar to (e.g., *ovatus*, egglike)

-ax: apt or tending to

-bilis: capable of, able to, tending to (e.g., *mutabilis*, able to change)

-bulum: instrument, container, agent, means

-bundus, -bunda, -bundum: implies action

-ceae: see *-aceae*

-cellus, -cella, -cellum: diminutive

-cillus, -cilla, -cillum: diminutive

-colus, -cola, -colum: inhabitant, dweller (e.g., *corticola*, bark dweller)

-culus, -cula, -culum: diminutive (e.g., *auricula*, small ear)

-cundus, -cunda, -cundum: able to, tending to (e.g., *rubicundus*, tending to redden)

-e: (1) gender ending (usually feminine) (2) see *-ae*

-ea: see *-eus*

-ellus, -ella, -ellum: diminutive (e.g., *Cantharellus*, small cup)

-ens: see *-ans*

-ensis, -ense: indicates place of growth or origin (e.g., *arvensis*, growing in fields)

-enus: belonging to

-er, -eres: provided with

-escens: becoming, almost, somewhat (e.g., *rubescens*, becoming red or almost red)

-estris, -ester, -estre: indicates place of growth or origin (e.g., *campestris*, growing in fields)

-etes: (1) denotes class or larger grouping of fungi (2) one who is

-ettus, -etta, -ettum: diminutive

-etum: denotes collective place of growth

-eus, -ea, -eum: (1) pertaining or belonging to (2) color of, made of, similar to (e.g., *melleus,* honey-colored)

-formis, -forme: resembling, especially in shape or form (e.g., *strobiliformis,* shaped like a pine cone)

-genus, -gena, -genum: born of, originating from

-i: see *-ii*

-ia: (1) see *-ius* (2) see *-a* (3) see *-ae*

-iacus, -iaca, -iacum: pertaining to

-iae: see *-ae*

-ianus, -iana, -ianum: see *-anus*

-ibilis: see *-bilis*

-icans: becoming, almost, closely resembling (e.g., *nigricans,* becoming black)

-icius, -icia, -icium: (1) made of, pertaining to (2) implies or denotes action

-icos: pertaining to

-icus, -ica, -icum: belonging to, pertaining to (e.g., *californicus,* of California)

-ides: resembling, similar to

-idion, -idius, -idia, -idium: diminutive (e.g., *Gomphidius,* little peg)

-idus, -ida, -idum: (1) resembling, similar to (2) becoming, almost (e.g., *albida,* whitish)

-iellus, -iella, -iellum: diminutive

-iensis, -iense: see *-ensis*

-ii: named after (e.g., *kauffmanii*)

-ilis, -ile: able to, capable of, with property of (e.g., *fragilis,* able to be broken)

-illus, -illa, -illum: diminutive

-imos: pertaining or relating to

-imus, -ima, -imum: superlative

-inans: becoming, almost

-ineus, -inea, -ineum: color of, made of, similar to (e.g., *ferruginea,* color of rust)

-inius, -inia, -inium: named after, relating to

-inos: (1) made of (2) see *-inius*

-inus, -ina, -inum: (1) pertaining or belonging to, possessing, resembling (2) diminutive

-ioides: see *-oides*

-ion: (1) diminutive (2) denotes occurrence

-ior: (1) comparative (e.g., *strictior,* very straight) (2) see *-ios*

-ios: pertaining or relating to

-is: pertaining or relating to

-isce, -iscus, -iscos: diminutive

-issimus, -issima, -issimum: superlative (e.g., *speciosissimus,* showiest)

-istos: comparative or superlative

-itas: see *-itius*

-ites, -itis: pertaining to, named after

-iticus, -itica, -iticum: (1) possessing, furnished with (2) capable of, able to

-itius, -itia, -itium: (1) with quality or color of, similar to (2) implies action or result

-ius, -ia, -ium: (1) pertaining to, characteristic of, resembling (e.g., *regius,* pertaining to kings) (2) occasionally used as diminutive or comparative

-ivus, -iva, -ivum: (1) possessing, furnished with (2) pertaining to (3) capable of, able to

-izans: becoming like, resembling

-limus: see *-rimus*

-ma, mus: indicates action or result

-nellus, -nella, -nellum: diminutive

-odes: see *-oides*

-oides, -oideus, -oidea, -oideum: resembling, similar to (e.g., *phalloides,* like a phallus)

-olentus, -olenta, -olentum: see *-ulentus*

-olus, -ola, -ole, -olum: (1) diminutive (2) less than, somewhat (e.g., *luteolus,* yellowish)

-on: neuter gender ending

-onius, -onia, -onium: pertaining or related to

-opsis: resembling, similar to (e.g., *Tricholomopsis,* like *Tricholoma*)

-orius, -oria, -orium: capable of, able to (e.g., *tinctorius,* able to dye)

-orum: pertaining to

-os: masculine or feminine gender ending

-osus, -osa, -osum: denotes fullness, abundance, marked degree of development (e.g., *succosa,* full of juice)

-otus, -ota, -otium, -otum: (1) resembling, similar to, possessing, furnished with (2) see *ot-* in dictionary

-rimus, -rima, -rimum: superlative (e.g., *pulcherrimus,* most beautiful)

-ros: pertaining to

-tatos, -tate, -taton: superlative

-ter: agent or means

-teros, -tera, -teron: comparative

-tes, -tis, -tor, -tra, -tria, -tron, -tros, -trum, -trus: denotes agent, means, tool, or object (e.g., *necator,* killer)

-ua: see *-uus*

-ugo: (1) made of (2) able to, capable of (3) disease or rust

-ulentus, -ulenta, -ulentum: denotes fullness, abundance, marked degree of development

-ulus, -ula, -ulum, -ullus: (1) pertaining to (2) diminutive (3) denotes lesser degree of development (e.g., *dulcidulus,* slightly sweet)

-um: most common neuter gender ending

-unculus, -uncula, -unculum: diminutive

-urus, -ura: (1) implies result or action (2) pertaining to

-us: most common masculine gender ending

-usculus, -uscula, -usculum: diminutive

-ustris, -uster, -ustre: see *-estris*

-utus, -uta, -utum: possessing, furnished with

-uus, -ua, -uum: implies result or possibility of action

-ydrion, -yllion: diminutive

-ys: comparative or superlative

Dictionary of Selected
Latin and Greek Word Elements

A

a-, ab-, ad-, etc.: prefix meaning without, absent, or away from
abies, abie-, abiet-: fir, firs
acanth-: thorn, prickle
acer-: (1) maple (2) acrid, sharp
acerb-: bitter, sour
acerv-: heap
acet-: vinegar
acetabulum: vinegar cruet
acicul-: bristle, needle, splinter
acr-: acrid
acumin-, acumen-: sharp point
acut-: sharp, pointed, acute
adust-: scorched
aegerit-: pertaining to poplar
aereus: copper, bronze
aerug-: blue-green, green, or deep green
aest-: summer
affin-: related; adjacent
agaric-: ancient term for mushroom
agath-: pleasant, good, excellent
agglutin-: glued
agr-: field
alb-: white
aleur-: wheat flour
alli-: garlic
aln-: alder
alutace-: leather-colored (yellowish-tan)
amab-: pretty
amanita-: ancient term for mushroom
amar-: bitter
amentace-: pertaining to catkins
amethyst-: violet (amethyst)
amianth-: unpolluted, spotless
ammo-: sand
amoen-: pleasant, lovely
amygdal-: almond
amylo-: amyloid (containing starch)
ananas: pineapple
andros-: a tiny herb or plant
angi-: vessel
angust-: small, narrow
annul-: ring, annulus
anth-: flower
anthrac-: coal, charcoal
apothec-: storehouse
appendiculat-: with a small appendage, appendiculate
apt-: fastened, bound
aqu-: watery
arachn-: spider or spider web
arai-, arae-: thin, narrow
arane-: spider or spider web
arcularius: maker of small chests
aren-: sand
areol-: a small open space
areolat-: areolate (cracked like dry mud)
argent-: silver
argill-: clay
argillace-: clay-colored (dull yellow-brown)
argyr-: silver
armen-: apricot, apricot-color
armill-: bracelet, ring
arquat-: (1) bowed, arched (2) rainbow-colored
arv-: field(s)
asc-: bladder, bag, wineskin
asper-, aspr-: rough
aster-, astr-: star
astragalin-: pertaining to goldfinch
ater-, atr-: black, very dark
atrament-: ink
atrat-: clothed in black
attenuat-: narrowed, tapered, reduced
augean-: pertaining to Augeos, who maintained an uncleaned stable for 30 years
august-: majestic, august
aur-: see *aure-* and *auri-*
auranti-: orange
aurat-: gilded, ornamented with gold
aure-: gold, golden
auri-, auric-: ear
auriscalpium: ear scrape
austr-, austral-: southern
avellan-: avellaneous (pale gray tinged with pink)
azur-: azure, blue

B

baccat-: (1) set with pearls or berries (2) soft, pulpy
badi-: reddish-brown
bae-: slim, small
balsam-: balsam, fir
balte-: belt, girdle
basidi-: basidium or basidia (derived from word for little base or pedestal)
(derived from little base
basilar-: fixed at the base of something, basal
bell-: pretty, handsome
benzoin-: resinous
betul-: birch
bi-: prefix meaning two, double, twice
biennis: lasting two years
bol-: throw, thrower
bolaris: with little lumps of paint
bolbit: cow dung
bolet-, bolites: ancient words for a superior kind of mushroom
bombyc-: silky
boreal-: northern
botry-, botryt-: bunch of grapes
bovista: ancient term for puffball (probably derived from word for ox)

brev-: short, brief
brum-: winter
brunne-: brown, dark brown
buf-: toad
bulb-: bulb
bulbos-: with a bulb, bulbous
bulg-: wine skin, leather bag
butyrace-: buttery
byssised-: seated in or pertaining to a mass
    of fine threads or filaments

C

caelat-: chiselled, carved
caerul-: blue, deep blue
caes-: glaucous, light gray, blue-gray
caesar-: caesarean, regal
caespitos-: cespitose (tufted or clustered)
calamistrat-: curled
calig-: boot
callist-: beautiful
calo-: beautiful
calv-, calvat-: bald, hairless
calyc-: cup, bud, calyx
calyptr-: protective cap or hood
camar-: arched, curved, vaulted
camp-: fields, plains, countryside
campan-: bell
campanul-: little bell
can-: hoary, pale gray, whitish
cancell-: lattice-work
candid-: shing, bright, white, clear
canthar-, canth-: drinking cup
caperat-: wrinkled, folded
capill-: hair
capn-: smoke
capr-: goat
capreol-: (1) wild goat (2) tendril
carbon-: coal, charcoal
carchar-: rough, jagged
carn-, carne-: flesh, flesh-colored
carp-: fruit, fruiting body
cary-, caryc-: nut tree or nut (especially walnut
    or hickory)
castan-: chestnut, chestnut-colored
catathelasma: running down
catin-: small bowl, basin, hollow
caud-: tail
caul-: stalk, stem, stipe
cav-, cavi-: hollow, cavernous
cedr-: cedar (Old World cedar, *Cedrus*)
cedret-: cedar woods, cedars
celer-: swift
centuncul-: a patchwork cloth
cep-: (1) onion (usually *cepa*) (2) head (see
    ceps)
cephal-: head
ceps: heads
cer-: horn
cerace-: waxy
cerea-: wax
cerebr-: brain

cerrusat-: as if painted with white lead
cervin-: pertaining to deer; fawn-colored
chaet-: long hair or bristle
chamal-: false
chamonixia-: on the ground (?)
cheil-: lip, brim, margin
cheim-: winter
chelidon-: russet (colored like a swallow's
    throat)
chelis: (1) hoof or claw (2) cloven
chil-: lip, brim, margin
chion-: snowy
chlamyd-: mantle, cloak
chlor-: green, greenish, greenish-yellow
choir-: sow, pig
chondr-: (1) cartilaginous (2) granular
chro-: (1) color (2) superficial appearance,
    skin
chrom-: bright color, e.g., chrome-yellow
chrys-: golden, gold
cibar-: food
cibor-: drinking vessel or cup
ciliat-: ciliate (fringed with hairs)
cilic-: cloth made of goat's hair
cincinnat-: curly, with curled hairs
cinere-: ash-colored, smoky, of cinders
cingul-: girdle, belt, collar
cinnabarin-: red, cinnabar-colored
cinnamom-: cinnamon-colored
circell-: small ring
circin-: circular, or bent like the head of a
    crosier
cirrhat-: frayed, fringed, curled
citrin-: lemon-yellow
clar-: clear, light, brilliant
clathr-: iron grating, latticework
claud-: lame, closed
clav-: club
clavat-: clavate (club-shaped)
clavicul-: small key
climac-: ladder
clit-: sloping
clype-: round shield
coccin-: scarlet, red
cochleat-: like a snail shell
cocos-: coconut
cognat-: related, kindred
col-: dweller, inhabitant (usually a suffix)
cole-: sheath
coli-: sieve, colander
collinit-: smeared or covered with slime
collyb-: small coin or coin-shaped pastry
columb-: dove
com-: hair
comat-: covered with hair
con-: (1) cone (2) prefix meaning with
conchat-: shell-shaped, shell-like
concinn-: neatly arranged or joined
concrescens: congealed
confluens: confluent; running together
confragos-: rough, scaly

conigen-: growing on or born from cones
connat-: born together
connex-: connected
connivens-: blended together, converging
consobrin-: cousin
controvers-: turned against, in opposite
  directions
copr-: dung
cor-: see *cori-*
cordy-: club, cudgel, swelling
cori-: leather
corn-: horn
corniculat-: with little horns
cornucopi-: horn of plenty
coron-: crown
coronill-: little crown, garland
corrug-: wrinkle, wrinkles
cortic-: (1) bark or hard crust (2) cork
cortin-: curtain
corvin-: shiny black (like a raven)
coss-: wood worm
cost-: rib
cotone-: color of wild olives or quince
cotyl-: cup, cup-shaped
crani-: skull, cranium
crass-: fat, thick, big, heavy
crater-: cup, crater
cren-: notch
crenulat-: notched, scalloped
crep-, crepid-: shoe, slipper
cretace-: chalky
crin-: hair
crinitus: long-haired, very hairy
crisp-: curly, crisped
crist-: crest
croc-: saffron-colored
crocolit-: wearing saffron
cruent-: gory, blood-colored
crust-: crust, crusty
crustulin-: thin bread crust
crypt-: hidden, secret
cucull-: hood, cap
cucum-: cucumber
cudonia: pertaining to a leather helmet
cumatil-: colored like a wave or sea water
cuspidat-: with a sharp tip, pointed
cyan-: blue, deep blue
cyath-: cup
cyb-, cybe: head
cycl-: circle, wheel
cypt-: stooped forward
cyst-: blister, bladder, cyst

D

dacry-: a tear (as in weeping)
daedal-: pertaining to maker of labyrinth,
  beautifully wrought
dal-: fire-brand, charred wood
dasy-: shaggy, hairy
dauc-: carrot
dealbat-: whitewashed
decastes: by tens

deceptiv-, decipiens: deceiving
decolorans: fading
decor-: beautiful, befitting, elegant
decurrens: decurrent (running down)
delibut-: grease, greasy
delica: weaned, without milk
delicat-: dainty, pleasing
delph-: womb
delphus: brother
dendr-, dendron: tree
dens-: crowded, thick, compact
dent-: tooth, teeth
derm-: skin
destruens: destroying
detonsus: sheared, clipped
di-: prefix meaning two, twice
dibaphus: dyed twice
dicty-: net
difform-: deformed
dilatat-: dilated, enlarged
dipl-: prefix meaning double
disc-: disc
discised-: seated in a disc
discoid-: (1) disc-shaped (2) descriptive term
  for one color at the center of another color
dissimulans: one who is disguised
don-, dont-: tooth, teeth
dry-: oak
dryad-: (1) wood nymph (2) oak
dulc-: sweet
dur-: hard, resistant, durable
duracin-: hard-fruited

E

eburne-: ivory
eccilia: rolled up
echin-: hedgehog, sea-urchin (hence spiny,
  prickly)
edodes: food
edulis: edible
elaph-: deer
elasm-: thin plate or gill
elat-: raised, tall
elegan-, elegans: elegant, choice
encephal-: brain
end-, endo-: within, inner, inside (prefix)
ent-: rolled inward
enteron-: internal
ep-: see *epi-*
ephippi-: mounted as on a horse
epi-: prefix meaning on or upon
epipterygia: surmounted by a small wing
ereb-: dark
eric-: heath
ericet-: heaths
erinace-: pertaining to hedgehog
erub-: red, reddish
erythr-: red, reddish
esculent-: edible, good to eat
eury-: large, broad, wide
evanescens: evanescent, withering away
evenos-: without veins

ex-: prefix meaning without, beyond, on the outside, or possessed of
excels-: lofty
excor-: peeled, flayed
exid-: staining, exuding, perspiring
exim-: extraordinary, distinguished

F

fag-: beech
fall-, fallax: deceptive
farin-: flour, meal
farinace-: farinaceous (like fresh meal)
fascicul-: bundle
fastigiat-: pointed, gabled
fav-: honeycomb
fell-: very bitter (bile)
fenn-: Finland
fer-: bearer, carrier, bearing (usually a suffix)
ferr-: iron
ferrug-: rust, rusty
fibr-: fiber
fibril-: fibril (fine fiber or hair)
fibros-: fibrous
fibul-: pin or clasp
fid-: divided, forked
fil-: thread
fim-: dung
fimbriat-: fringed
fimet-: dung heap
firm-: firm, strong
fistul-: tube or pipe
flabell-: small fan
flamm-: flame
flav-: yellow
flocc-: woolly
flu-: secreting fluid, letting flow (usually a suffix)
foc-: fire
foenisecii-: pertaining to hay-making or lawn-mowing
foetens, foetid-: fetid, stinking
foli-: leaf or leaves (gills)
fomes, foment-, fomit-: tinder
form-: shape, form, appearance
formos-: lovely, shapely, beautifully formed
fornicat-: arched or vaulted
fract-, frag-: break
frond-: leaf, frond
fruct-: fruit
fulg-: shine
fulig-: soot
fulmin-: lightning
fulv-: fulvous (fox-colored, reddish-yellow, tawny)
fum-: smoke, smoke-color (gray, usually mixed with brown)
funalia: made of rope
furc-: fork, forked
furfur-: bran, scurf
fus-: spindle
fusc-, fusco-: dark, dusky, fuscous

G

gal-, galact-: milk
gale-, galer-: helmet or fur cap
gambos-: large-hocked, with a swollen base
gan-: shiny, lustrous
gaster-, gastr-: stomach, belly
gausapat-: woollen cloth
ge-: see *geo-*
gemmat-: gemmed
gen-: (1) born of, originating from (2) race
genea: derived from *gen-*, or from ancient word for compact, knobby bodies
gentil-: of the same race
geo-: earth
geotrop-: positively geotropic, i.e., erect
ger-: bearer, carrier, bearing (often a suffix)
gibb-: humped, rounded
gigas-: giant
gilv-: pale yellow, dull yellowish
glaber, glabr-: hairless, smooth
glandul-: gland
glauc-: glaucous (pale grayish-bluish-green) or silvery
gleb-: gleba (derived from word for clod)
gli-: glue
glischr-: glutinous, sticky
gloeo-: glutinous, sticky
gloss-: tongue
glyc-, glycy-: sugar, sweet
gomph-: bolt, peg, stake, nail
gon-: joint, angle
gracil-: gracile (slender, thin)
gramin-: grass, grain
gramm-: sign, mark, line
gran-: seed, grain
granul-: granule (fine grain)
grav-: heavy, strong, pregnant with
grifola: braided fungus
gris-, grise-: gray
gumm-: gum
gutt-: drop, droplet
gymn-: naked
gyr-: round; a circle

H

haem-, haemat-: blood, blood-red
haemorrhoid-: bleeding
hal-: salt, sea
hapal-: soft-boiled, tender
hebe-: blunt, obtuse
helv-: see *helveol-*
helvella: ancient term for an aromatic herb
helveol-, helvol-: light bay, pale reddish, blonde, honey-colored, "the dingy color of oxen"
hemi-: half
hepat-: liver
hericium-: pertaining to hedgehog
hetero-: prefix meaning irregular, different, other
hiem-: winter
hirn-: jug

hirsut-: hirsute (hairy)
hirt-: bristly or shaggy with weak hairs
hispid-: hispid (roughened by stiff hairs),
   shaggy, spiny
hol-, holo-: entire, whole
hort-: garden, gardens
humil-: low, stunted, humble, dwarfish
hyal-: clear, transparent, colorless
hydn-: ancient term for an edible mushroom or tuber (truffle)
hydr-: water
hygr-: moisture, humidity
hymen-: membrane, hymenium
hyp-: see *hypo-*
hyper-: prefix meaning above, beyond, over
hyph-: fringed with tissue, webbed, woven
hypn-: tree-moss
hypo-: prefix meaning below, under,
   beneath, not quite, less than normal
hyps-: high up
hysgin-: red vegetable dye
hyster-: womb
hystrix, hystric-: porcupine

IJK

ianth-: violet
icterin-: jaundiced (pale yellow or greenish-
   yellow)
ign-: fire
illin-: smeared
illot-: dirty, unwashed
illudens: (1) deceiving (2) emitting light
im-: see *in-*
imbricat-: covered with tiles or scales,
   shingled
impatiens: quickly ripening
impolit-: unpolished (matt), rough
impudic-: shameless, immodest
in-: prefix meaning not, lacking, on, toward,
   intensely, to cause to become
inaurat-: gilded, covered with gold
incarnat-: flesh-colored
indic-: pertaining to India or indigo
indigo: dark blue or purple-blue
indusi-: garment, overall
infract-: humble, subdued, uniform
inful-: priest's cap or hood
infundibul-: funnel
ino-: fiber
inquinans: polluting, staining
insign-: illustrious, distinctive, well-marked
insuls-: tasteless, unattractive
integr-: entire, whole, renewed,
intyb-: chicory, endive
invers-: inverted, upside down
involut-: rolled up, inrolled
io-, iod-: violet
irin-: pertaining to iris
irpex: harrow, wolf's teeth
irradians: radiant, radiating
ischn-: slender, thin, weak
ithy-, ithys-: straight, erect

jub-: mane, crest
junonius: pertaining to Juno, wife of Jupiter

L

lacc- paint
lacer-: tear, lacerate
lachn-: wool, woolly
lachrim-: tears (as in weeping)
laciniat-: torn, frayed
lacrim-, lacrym-: tears (as in weeping)
lact-: milk
lacun-: cavity, pit
laet-: bright, gay, pleasing, abundant
laev-: smooth
laevigat-: polished, smooth
lag-: hare, rabbit
lamell-: plate or plates (gills)
lamin-, lamn-: thin plate
lan-: wool
lani-: (1) wool (2) to tear into pieces
lanuginos-: downy
larg-: large, wide
laric-: larch
lat-: broad, large
later-: brick
laterit-: brick-colored, bricklike
laurocerasi: pertaining to English laurel
leccinum: Italian term for fungus
lei-: smooth
lent-: (1) pliable, tenacious (2) sticky, viscid
lenticul-: (1) lentil (2) small lens
lentig-: freckle, speck
leo-: lion
leot-: smooth
lepi-: scale
lepid-: agreeable, pretty, neat
lepide-: scaly
lepis-: scale
lepist-: wine pitcher or goblet
lepor-: rabbit or hare
lepr-: rough, scurfy, scaly
lept-: thin, fine, delicate
leptonia: fine like a small coin
leuc-: white, light
levi-, levigat-: see *laev-, laevigat-*
liber-, libr-: free
lign-: wood
ligny-: ash, soot
ligul-: little tongue
limac-, limax: slug, slime
limb-: border, edge
liquiriti-: licorice
lith-: stone
livid-: livid, lead-colored (an indefinite gray
   or bluish-gray color)
loma-: margin, edge
loricat-: armored
lubric-: lubricous, slippery, smooth
lucid-: glossy, polished, clear
luculent-: clear, beautiful
lup-: wolf

lurid-:   lurid, unclean, dingy, or an indefinite
    color "between purple, yellow, and gray"
    —McIlvaine
lute-:   yellow
lyc-, lyco-:   wolf
lyo-, lyso-:   prefix meaning loose, free,
    dispersed, or dissolving

M

macr-:   large, long
macul-:   spot, stain, blotch
madid-:   moist, soaked
magn-:   great
magnat-:   of magnates (dignitaries)
mamm-:   nipple, teat, breast
mappa: napkin
marasm-:   withered, emaciated
marg-, margin-:   margin, edge, border
marzuol-:   pertaining to March
mastruc-:   sheepskin
maur-:   dark, obscure
maxim-:   largest, greatest
med-:   middle, medium
meg-:   prefix meaning large or great
mela-, melan-, melaen-, melas: black
meleagr-:   speckled
melle-:   pertaining to honey, honey-colored
merd-:   dung
mer-, meri-:   a part
mesenter-:   middle intestine
met-, meta-:   (1) prefix denoting change (2)
    prefix meaning between, next to, etc.
metric-:   measure, measuring
micr-:   small
militar-:   soldierlike, occurring in troops
miniat-:   painted with red lead
minut-:   minute
mir-, mirab-:   admirable, marvelous
mit-, mitis: mild, tender, harmless
mitr-:   miter, headdress, cap
mokusin: a region in China
moll-:   soft, tender
molyb-:   lead or pertaining to lead
monach-:   monk or nun
mont-:   mountain(s)
morb-:   disease
morchell-:   German term for edible fungus
    or morel
morph-:   form, shape
muc-:   slime, mucus
mucid-:   slimy
mucronat-:   with a sharp tip, pointed
multi-:   prefix meaning many
multiplex: many pieces
mund, mundul-:   neat, clean
mur-:   mouse
muricat-:   (1) roughened by many short
    spines (2) pointed
musca-:   a fly
musci-, muscu-, musco-:   moss
mut-, muta-:   change
mutinus-:   a phallic deity worshipped by
    Roman brides, hence a penis

my-:   see *myo-*
myc-:   fungus, mushroom
mycena: ancient term for mushroom
myo-, myos-:   mouse
myri-:   myriad, countless, many
myrm-, myrmec-:   ant
myx-:   mucus, slime

N

naemat-:   with threads
nan-:   dwarf
nap-:   turnip
nasc-, nascens: arising, beginning
nauc-:   nut shell
nebul-:   vapor, cloud, fog, mist, smoke
necator: killer
nemor-:   of the woods, sylvan
neo-:   prefix meaning new
neolecta: new selection (?)
nid-:   nest
nidor-:   fume
nidoros-:   reeking, with a bad or burnt smell
nidul-:   little nest
nidulans: nesting
nidus: nest
nigell-:   blackish, dark
niger, nigr-:   black
nit-, nitel-:   shining, bright, trim
nitid-:   glossy, spotless
niv-:   snow
nola-:   little bell
not-:   back, buttock
nub-:   cloud, clouds
nud-:   naked, bare

O

oblectabil-:   beguiling, delightful
obscur-:   dark, dusky
occidental-:   western
ochr-, ochrac-:   ochre, ochraceous
ocrea-:   sheath, boot, greave
ocul-:   eye
odon-, odont-:   tooth, teeth
officinal-:   with pharmaceutical use
olea-:   the olive tree
olens-:   smelling (usually a suffix)
olid-:   fragrant, smelly
olig-:   little, small, few
olivace-:   olive-colored
oll-:   pot or jar
omphal-:   navel (umbilicus)
ono-:   donkey
onust-:   full, loaden down
onych-:   onyx
ophi-:   snake, serpent
orbicul-:   flat and round (disclike), circular
orcell-:   small vase
oreades: pertaining to mountain nymphs
    or fairies
orellan-:   pertaining to mountains
ori-:   of the mountains
orichalc-:   copper-colored
oriental-:   eastern

orth-: straight
os-: see *oss-*
osecanus-: whitened like bone
osm-: smell, scent, perfume, odor
oss-: bone
ostre-: oyster
ot-: (1) ear (used as suffix or prefix) (2) see list of suffixes
ov-: egg
ovat-: egg-shaped, oval
ovin-: ovine (pertaining to sheep)
ox-, oxy-: sharp
oxyd-: oxide, oxidizing
oz-: branch

P

pachy-: thick
paleace-: chaffy (with small weak scales)
pallens, pallid-: pale, pallid
palm-: palm (of a hand)
palud-: swamp, marsh, bog
palustr-: marshy, pertaining to swamps
panaeolus: variegated, all-variegated
pann-: rag
panus: tumor
papilio-: butterfly
papill-: nipple, papilla, pimple
par-, para-: prefix meaning near, beside, related to
parc-, parci-: stingy, spare, sparse
pard-: leopard
pargamen-: pertaining to parchment
parv-: small, little, pretty, dainty
patel-: dish
pauc-: few
pauper-: poor
paxill-: small stake, small stick, peg
pectin-: comb
ped-, pedes: foot or base
pediades: of the plains, of the soil
pelargon-: pertaining to geranium
peli-: livid, dark
pell-, pelli-: skin
pelliculos-: with a well-developed pellicle
pellucid-: shining through, pellucid, glossy
penetrans: penetrating
penio-: fabric
pepl-: (1) skirt, robe, coat (2) gilded
per-: prefix meaning intensely or very
percomis: very courteous or elegant
perd-, perdon: flatulence
perenn-: perennial, lasting, year-round
perfor-: pierce
peri-: prefix meaning around, all around, enclosing
perlat-: widespread, enduring (may also be a corruption of word for pearl)
peronat-: sheathed, booted, rough-booted
persic-: peach
personat-: masked
pes: foot or base (stalk)
pessundat-: ruined

petal-: petal or leaf
petasat-: with a wide-brimmed hat, having a cap on
peziza: ancient term for a mushroom with little or no stalk
phae-: dark, dusky
phall-: swollen, puffed up, phallic
phallus: rod, phallus
phan-, phaner: appearing, visible
phell-: cork
phil-: loving, fond of, friend of
phleb-, phlep-: vein(s)
phlegm-: mucus
phloe-: rind, bark
phlog-: flame, flame-colored, reddish
phoenic-: red, purple-red, crimson
phol-, pholid-: scale
phor-: bearer or carrier, bearing (usually a suffix)
phragm-: palisade, hedge, fence
phyc-: algae, seaweed
phyl-: tribe, race
phyll-: leaf or leaves (gills)
phys-: (1) small bladder, bubble (2) growth
pic-: magpie
picea-: spruce
pici-: pitch-black (pertaining to pitch)
picoa: coated with pitch or tar (black)
picr-: bitter, pungent
pict-: painted, streaked, embroidered
pil-, pile-: (1) usually means pileus or cap (2) *pile-* can also mean hair (3) *pila* can mean ball or bullet
pin-: pine
pinet-: pine wood
pingu-: (1) fat or grease (2) fat or stout
piperat-: peppery
pipt-: easily detachable, falling off
piri-: see *pyr-*
pis-: pea
pistil-, pistill-: pestle
pithy-: wide-mouthed jar
plac-: a flat round plate, i.e., flat
placid-: mild, gentle
placit-: pleasing
platy-: broad, wide, flat
plect-: plaited, twined, twisted
pleur-: side, beside, on the side
plex: (1) pieces (2) a knitting or interweaving
plicat-: folded, pleated
plinth-: brick, block
plum-: feather or long hair
plumb-: lead or lead-color
plute-: bracket, shed, penthouse, parapet
pocillator: cup-bearer
pod-, podi-: foot, base (stalk)
pog-, pogon: beard
poli-: gray, hoary
poly-: prefix meaning several, many, much
pom-: fruit tree
ponderos-: heavy, ponderous

popul-:   poplar, cottonwood, aspen
por-:   pore, pores
porphyr-:   (1) purple (2) reddish-brown or russet
porrigens: extending forward, projecting horizontally, spreading, stretching out
portentos-:   portentous, prodigious
prae-:   prefix meaning before or very
praeclare-:   very clearly
praecox: early, premature
pras-:   leek, leek-green
prat-:   meadow, meadows
pre-:   see *prae-*
privignus: step son
pro-:   prefix meaning before
procer-:   lofty
prolix-:   stretched, elongated
prote-:   a sea god who could change shape at will, i.e., protean
proto-:   prefix meaning primary or giving rise to
proxim-:   nearest, next
pruinos-:   with a fine bloom, frosted
prun-:   plum
psall-:   ring or collar, fringed or curb
psamm-:   sand
psathyr-:   fragile, brittle
pseud-:   prefix meaning deceptive, similar to, easily confused with, false (often used before the name of a similar mushroom)
psil-:   naked, bare, hairless, smooth
psitt-:   parrot
pter-:   wing, membrane
pterid-:   fern
pteryg-:   wing, fin, membrane
ptych-:   fold or layer, leaf
pub-:   hair
pubescens: downy, pubescent
pudic-, pudor-:   modest, rosy
puellaris: pertaining to girl, i.e., pretty
pulch-, pulcher-, pulchr-:   beautiful
pulver-:   powder, dust
punct-:   fine spot, dot, or puncture
punctat-:   punctate (finely spotted or dotted)
punic-:   red or purple-red, pomegranate-red, garnet-red
pur-:   clean, pure
purpur-:   purple
pus-:   (1) foot, base, or stalk (2) also see *pusil-*
pusil-, pusill-:   very small, little, weak
pustul-:   pustule, pimple
pycn-:   thick, dense, compact, strong
pyr-:   (1) pear (especially *pyri-*) (2) fire (especially *pyro-*)
pyram-, pyramid-:   pyramid, pyramidal
pyren-:   kernel, pit (of a fruit)
pyrr-, pyrrh-:   flame-colored, reddish
pyx-, pyxid-:   a small box

QR

quadr-:   four, fourfold (usually a prefix)
querc-:   oak

racem-:   branched like a grape stalk
rach-:   misspelling of *rhac-*
radi-:   ray, spoke, plate
radians: (1) with rays or spokes (2) radiant
radic-:   root
radul-:   scraper
ram-:   branch
raphan-:   radish
rasil-:   shaved, scraped, worn smooth
recis-:   cut off, cut back
reg-:   king, royalty
renidens: shiny
repand-:   folded backward, turned up
resim-:   turned up or bent back
ret-:   net
reticulat-:   reticulate (netted)
rhac-:   rag
rhach-:   spine or backbone
rhacodes: ragged, shaggy
rhe-:   flow, let flow, exuding, flowing
rhen-:   Rhine
rhiz-:   root
rhod-:   rose, rose-colored (pink or red)
rhopal-:   club, stick
rim-:   crack, fissure
ringens: gaping
ripa-:   stream bank
rivul-:   channel, groove, stream
robinia: locust tree
robust-:   robust, stout, strong
ros-:   rose
rostr-:   beak, snout
rot-:   wheel
rotul-:   little wheel, pinwheel
rub-, ruber: red
rubel-:   almost red, reddish
rubig-:   rusty, reddish-brown
rubr-:   red
rud-:   uncombed, made of untreated wool, rough
ruf-:   rufous (reddish)
rug-:   wrinkle
russ-:   (1) reddish (2) Russian
rutil-:   ruddy or warm red

S

sacc-:   sac, sack
sacchar-:   sugar, sugary, sweet
saep-:   hedge
saev-:   savage
salic-, salix: willow
salmon-:   salmon-colored
sandarac-:   flesh-color
sangui-, sanguin-:   bloody, blood-red
sapine-:   fir or pine
sapon-:   soap
sarc-:   flesh, fleshy
sardoni-:   very bitter, acrid
sarx: flesh
sax-:   stone, rock
scaber, scabr-:   rough, scurfy, pertaining to scaber(s)

scamb-: bow-legged (i.e., curved or bent)
scaur-: with swollen ankle or foot
schiz-: split
sciss-: cut
scler-: hard
sclerotoid-: pertaining to a sclerotium (tuber or knot of tissue)
scolec-: worm
scord-, scorodon-: garlic
scrobiculat-: scrobiculate (pitted)
scut-: shield
scutel-: dish
scyph-: cup
seb-, sebac-: tallow, grease
sec-: an enclosure or nest
secund-: conforming, following
sejunct-: disjoined, separated, severed
semi-: prefix meaning half or less than
semital-: of footpaths, byways
semot-: remote
sep-: see *saep-*
separans: distinctive
separat-: separated, separate, apart
septentrional-: northern (from a northern constellation)
sepulchr-: tomb
sepult: buried
sequoia-: redwood
ser-: (1) whey, serum (2) to fasten or bind
seri-: silk
serial-: in rows
serotin-: late
serp-: snake
serrul-: finely serrate (toothed like a saw)
sibir-: Siberia
sicc-: dry
silv-: woodland, forest
sim-: flattened or concave
simbl-: beehive
sin-: curve, bay
sinap-: mustard
sindon-: muslin
sinopic-: reddish-ochre
sinu-, sinus: curve, bend, bay
sinuat-: sinuate (notched gills)
sistr-: rattle
smaragd-: emerald, green
som-: body
sordid-: sordid (dingy, filthy, foul)
soror-: pertaining to sister
spad-, spadic-: (1) bright brown (the color of a fresh date) (2) shaped like a palm frond or spade
sparassis: torn to pieces, lacerated
spathul-: little spade or blade, spatula
specios-: beautiful, showy
spectabil-: notable, remarkable, visible
sperm-: seed (spore)
sphaer-: sphere, round
sphagn-: sphagnum, moss
sphec-: (1) wasp (2) lichen
sphinct-: a tightening, tightly bound
spin-, spinul-: spine, thorn

spiss-: thick, massive, compact
splendens: polished, glittering, resplendent
spong-: spongy
spor-: spore, seed
spret-: despised, scorned, rejected
spum-: foam, froth
squam-: a scale
squamul-: small scale
squarros-: squarrose (with upright scales, rough, scurfy)
stell-: star
ster-: hard
sterquil-: dung heap, dung pit
stict-: marked, lined, dappled, dotted
stip-: stalk or stem (stipe)
stiptic-: close, dense (also see *styptic-*)
stom-: mouth
stramin-: straw-colored
striat-: striate (finely furrowed or lined)
strict-: very straight, erect (usually narrow)
strigos-: strigose (with coarse flattened, rigid hairs or bristles)
strobil-: pertaining to pine cone, or to something twisted
strom-: bed, mattress
stroph-: (1) belt or pectoral band (2) a turning or twisting
styptic-: astringent (also see *stiptic-*)
su-: pig, swine
suav-: sweet, agreeable
sub-: common suffix meaning (1) somewhat or almost (2) below or under
subtil-: fine, delicate, minute
succ-: juice, sap
sudor-: sweat
sudorific-: sweat-producing
suec-: Sweden
suillus: an ancient term for fungus (derived from word for swine)
sulc-: furrow or groove
summ-: summit, top
surrect-: rising, erect
sylv-: woodland, forest
synaps-, synapt-: joined together

T

tabac-: tobacco (color, odor, etc.)
tabescens: decomposing
tabid-: decomposed
tard-: slow
tect-: roof, cover
telamon-: belt or supporting band; web
tenac-, tenax: tenacious
tener-: tender, delicate
tenu-, tenui-: thin
tephr-: ashes, ash-colored, gray
terfez-: Arabic word for truffle
terr-: earth
tessulat-: mosaic-like, checkered
testace-: testaceous (brick-colored)
thec-: case or box
thej-, thei-: sulfur, brimstone

thraust-:   easily crumbled, fragile, brittle
thrix: hair
thuj-, thy-:   cedar (arbor-vitae, not *Cedrus*)
tigr-:   tiger
tinct-:   tinted, dyed
toga-:   toga (robe)
toment-:   wool or hair
torminos-:   causing dysentery
toros-:   muscular, i.e., bulging or swollen
torqu-:   necklace, collar
tortil-:   twisting, winding
torul-:   (1) tuft of hairs (2) bulge or swelling
torv-:   wild, savage
tost-:   toasted
trabe-:   beam, timber
trachy-:   rough
trag-:   goat
tram-, trama: (1) the woof (as in weaving)
   (2) something thin
trametes: one who is thin
tremell-:   trembling
tri-:   prefix meaning three, thrice
trich-:   hair(s), fiber(s)
trivial-:   common, vulgar
trullisat-:   plastered
trunc-:   trunk (as of a tree)
truncat-:   truncate (appearing chopped off)
tsuga-:   hemlock
tub-, tuba-:   trumpet or tube
tuber: (1) a tuber (2) a bump or knob
tul-, tulo-:   see *tyl-*
tumid-:   swollen
tunic-:   tunic, garment
turbinat-:   shaped like a top
turmal-:   a troop
turp-:   vile, ugly
tyl-, tylo-:   (1) swelling, bump, cushion,
   pillow (2) knob or callus on a club
typh-:   (1) smoke, cloud (2) a plant used for
   stuffing beds
tyr-:   cheese

U

ud-:   wet, moist, damp
uligi-:   moisture
uliginos-:   full of moisture; boggy, swampy
ulm-:   elm
umbell-:   umbrella or sunshade
umbilic-:   umbilicus (navel)
umbon-:   knob (derived from a word for
   shield)
umbr-:   dark, shade
ungu-:   grease
ungul-:   hoof
uni-:   prefix meaning one, single
urbic-:   urban, of the city
und-, undul-:   wave
urens-:   burning
urs-:   bear
ustal-:   russet (derived from word for burnt)
uter-, utri-:   wine skin, womb
uvid-:   damp, humid

V

vacc-:   cow
vaccini-:   huckleberry
vagin-:   sheath, scabbard
valens: powerful, strong, robust
valid-:   powerful, robust
vaporari-:   pertaining to humid places
vari-:   change, changing
variat-:   variable
variegat-:   variegated, multicolored
vas-:   vessel, vase
vel-:   .veil
vell-:   fleece, wool
vellere-:   covered with fleece, woolly
velut-:   velvet, velvety
ven-:   vein
venen-:   poison
venet-:   colored like sea water
ventr-:   belly
ventricos-:   ventricose (swollen, bulging)
venust-:   handsome, elegant
verm-:   worm
vermicul-:   little worm
vern-:   spring, of spring
vernicos-:   varnished
verp-:   rod, penis
verruc-:   wart
vers-:   changing
versicolor: multicolored
vesc-:   (1) edible (2) feeble
vesicul-:   small bladder, blister
vestit-:   dress, attire
veternos-:   weak, lethargic
via-:   way, biway
vibec-, vibic-:   bruise
viet-:   wrinkled, wilted, shrunken
vill-:   shaggy hair
villatic-:   rustic, rural
vin-, vini-:   wine (especially red wine)
viol-:   violet
vir-, virens, virid-:   green
virg-:   stripe, streak
virgin-:   virgin (i.e., white or pure)
virid-:   greenish
viros-:   poisonous
visc-, viscid-:   viscid (sticky)
vitell-:   egg yolk
vitil-:   coiling, twining
volemus: filling the palm of the hand, or
   flowing enough to fill the hand
volva-:   wrapper, womb (volva)
vor-:   devour, devourer, eater
vulgar-:   common, known, usual
vulp-:   fox

WXYZ

xanth-:   yellow
xer-:   dry
xerampelin-:   colored like a dry vine leaf
xyl-, xylon: wood, wooden, woody
zanth-:   see *xanth-*
zon-:   band or zone (usually concentric)
zyg-:   yoked (connected, joined, attached)

GLOSSARY

acanthocytes needle-like crystalline deposits on mycelium of certain mushrooms

acrid taste burning or peppery

acute pointed or sharp

adnate gills attached broadly to stalk (p. 17)

adnexed gills attached narrowly to stalk (p. 17)

agaric a mushroom with gills

alveolate surface of spore or cap with broad pits (p. 842)

amyloid staining blue, gray, or black in iodine solution (e.g., Melzer's reagent)

anastomosing gills connected by cross-veins

angular spores 4- to 6-sided, with corners or angles

annulus ring or collar of tissue on stalk formed by ruptured veil

apex top

apical at or near the top

apical pore in certain Gasteromycetes, the mouth at the top of the spore case through which spores are released; in spores, a germ pore

apiculate furnished with an apiculus

apiculus short projection on either end of a spore

apothecium the fruiting body of an Ascomycete in which the asci are arranged in an exposed hymenium

appendiculate margin of cap fringed or adorned with veil remnants or other tissue

appressed flattened down or pressed against

areolate cap surface cracked into plaques or blocks (like dried mud)

ascocarp fruiting body of an Ascomycete

ascus (pl., **asci**) saclike mother cell in which spores of Ascomycetes are formed (pp. 4, 782)

autodigestion self-digestion

basal at or near the base

basidiocarp fruiting body of a Basidiomycete

basidium (pl., **basidia**) a cell, usually club-shaped, on which spores of Basidiomycetes are formed (p. 4)

bolete a fleshy mushroom with tubes on underside of cap

boletivore one with an inordinate fondness for eating boletes (p. 546)

brown rot carbonizing decay

buff an indefinite pale color: pale dull yellow or very pale tan

bulbous stalk with an enlarged base (p. 17)

BUM Boring Ubiquitous Mushroom

button a young fruiting body before it has opened up

butt rot a rot confined to the base or roots of the tree

cap the caplike part of the fruiting body which supports the spore-bearing surface

capillitium modified threadlike, often-branched hyphae enmeshed in the spore mass of many Gasteromycetes

capitate furnished with a rudimentary cap or head

carbonizing decay a cellulose-decomposing decay

carminophilous see siderophilous

cartilaginous with the texture of cartilage; tough or rindlike, not fleshy

cellular tissue composed of round or pear-shaped cells (p. 19)

cellulose a compound composed of glucose units; it is a major constituent of wood and of plants' cell walls

cespitose (caespitose) tufted or clustered

chlamydospores thick-walled asexual spores formed by breaking up of hyphae

chrysocystidia cystidia with a highly refractive golden content as seen in KOH

clamp connection a small bump, loop, or swollen area formed at the cross-wall between hyphae in the process of cell division

clamps mycological jargon for clamp connections

class a grouping of related orders

clavate club-shaped

club-shaped stalk thickened noticeably toward base (p. 17); basidia thickened toward spore-bearing end

collarlike type of volva formed when the universal veil breaks around the circumference of the cap, forming a collar or free rim near base of stalk (p. 264)

columella internal stalk; a sterile column of tissue projecting into the spore mass of certain Gasteromycetes

complex stalk folded or convoluted in cross-section; a group of closely related species or forms

compound fruiting body with more than one cap and/or stalk arising from a common base

conidia asexual spores formed by the pinching off of hyphae

conifer a cone-bearing tree; one that has needles like a pine or redwood, or scales like a cypress

conk a woody, usually perennial polypore

context the flesh of the cap or stalk

convergent gill hyphae projecting upward toward the cap as seen in cross-section; converging toward the center (p. 19)

convex cap rounded or domed (p. 17)

convoluted wrinkled like a brain

coprophilous dung-loving; growing on dung

cortina a veil with a silky or cobwebby texture

cuticle the skin or surface layer of the cap, more specifically one differentiated from the flesh

cystidium (pl., **cystidia**) specialized sterile cells projecting from the gills, tubes, cap, or stalk (p. 19)

daedaloid pores elongated or meandering like a labyrinth; mazelike

decurrent gills running down the stalk (p. 17)

decurved margin of cap curved downward

delignifying decay a lignin and cellulose-decomposing rot

deliquescing gills dissolving into a fluid; liquefying

depressed cap concave, i.e., the center sunken below the level of the margin (p. 17)

dextrinoid staining brown or red-brown in iodine solution (e.g., Melzer's reagent)

dichotomous forking or dividing in pairs

differentiated different; not the same throughout.

disc center of the cap

distant gills widely spaced

divergent gill hyphae projecting downward (away from cap) as seen in cross-section; diverging from the center (p. 19)

division a group of related classes

dry cap or stalk neither viscid nor hygrophanous

duplex flesh of two distinct textures

eccentric stalk off-center

echinulate spores spiny like a sea urchin

egg the button stage of a mushroom with a universal veil (e.g., *Amanita*); also, spore-containing capsule in a bird's nest fungus

elliptical spores rounded at both ends with sides curved slightly

endemic native to a region and generally restricted to it

endoperidium the inner layer of the spore case in puffballs and their allies

entire gills with even edges; not serrated or toothed

epigeous growing on or above the ground

epiphragm the thin membrane covering the nest of many young bird's nest fungi

equal stalk of more or less uniform thickness throughout (p. 17)

ericaceous pertaining to the heath family

erumpent fruiting body erupting through the ground but scarcely rising above it

evanescent transitory; disappearing

exoperidium the outer layer of the spore case in puffballs and their allies

expanded cap fully developed; spread out

fairy ring a circle or arc of mushrooms (p. 7)

family a group of related genera

farinaceous odor like fresh meal or cucumber

fetid (foetid) ill-smelling

fibril a fine fiber or hair

fibrillose covered with or composed of fibrils

fibrous composed of tough, stringy tissue

filamentous tissue composed of thread-like cells or filaments (p. 19)

flaccid lacking firmness; flabby

flesh the tissue of the cap or stalk; the meaty portion

fleshy having substance (e.g., a fleshy fungal fructification) *or* soft as opposed to tough

floccose woolly or cottony; dry and loosely arranged

fluted stalk ribbed; with longitudinal ridges

free gills not attached to stalk (p. 17)

friable easily crumbling

Friesian pertaining to Elias Fries (the "father" of mushroom taxonomy) and to his classification system based on macroscopic features

fructification fruiting body

fruiting body the reproductive structure of a fungus; a mushroom

fulvous fox-colored

fungophile one who loves fungi

fungophobe one who fears fungi

fungi (pl., **fungi**) any member of a large group of organisms that lack chlorophyll, reproduce by means of spores, and have a filamentous vegetative phase

funiculus the cord attaching peridiole to the "nest" of certain bird's nest fungi

furfuraceous scurfy; roughened by minute particles

fuscous an indefinite dark smoky color, sometimes with a slight violet tinge

fusiform spores spindle-shaped; elongated and tapering from the middle to both ends

gastroid like a Gasteromycete, i.e., not forcibly discharging its spores

gelatinous jelly-like in consistency or appearance

generic pertaining to genus

genus (pl., **genera**) a group of closely related species

germ pore a soft spot in wall of certain spores, through which initial germination takes place

gills spore-producing blades on underside of an agaric

glabrous bald

glandular dots resinous spots or smears on stalk of certain boletes (Color Plate 120)

gleba the spore mass of a Gasteromycete

globose spherical

glutinous slimy or very sticky

granulose covered with granules

gregarious growing close together but not in clusters

habitat natural place of growth

hardwood used here in a broad sense to denote any tree that is not a conifer

heart rot a rot of the heartwood

homogeneous the same throughout; not differentiated

humus decaying organic material in or on soil

hyaline spores colorless under the microscope

hygrophanous cap surface changing color markedly as it loses moisture; usually watery-looking when wet and opaque when dry

hygroscopic sensitive to moisture

hymenium a layer or palisade of spore-bearing cells

hypha (pl., **hyphae**) a threadlike fungal cell, the basic structural unit of any mushroom

hypogeous fruiting underground

incurved cap margin curved inward toward stalk

indeterminate growth that ceases and begins anew according to weather conditions, frequently creating a zoned or lumpy cap

indusium netlike skirt in certain stinkhorns

inferior annulus located near base of stalk

innate a part of; not superficial or easily removed

inner veil see partial veil

inrolled cap margin curved in toward gills and rolled up; tucked under

interwoven gill hyphae entwined or tangled as seen in cross-section; not forming a regular pattern (p. 19)

JAR Just Another Russula

KOH chemical symbol for potassium hydroxide (potash)

lamellae the gills

lamellulae short gills that extend only part way to stalk

lateral stalk attached to side of cap

latex a juice or milk, as in *Lactarius*

LBJ Little Brown Job (p. 33)

LBM Little Brown Mushroom (pp. 32-33)

leathery tough, pliant, not easily breaking

lignicolous wood-inhabiting

lignin   a major constituent of wood

lubricous   somewhat slippery or greasy to the touch but not viscid or slimy

lumper   a taxonomist with a broad species or genus concept

lutescent   becoming yellow

macroscopic   discernible without a microscope

margin   the edge of the cap or gills

marginate   gills with edges differently colored than faces

marginate-depressed   bulb with a raised rim

mealy   having a granular appearance *or* smelling like fresh meal

median   ring at or near middle of stalk

Melzer's reagent   an iodine test solution: 44 parts water (by weight), 3 parts potassium iodide, 1 part iodine, and 40 parts chloral hydrate (poisonous!)

membranous   membranelike; skinlike

metagrobolizing   puzzling, bewildering, confounding, confusing, perplexing

micrometer   see micron

micron   microscopic unit of measure equaling 0.001 mm

microscopic   discernible only with a microscope

montane   of the mountains

mushroom   the fruiting body of a fungus, especially one which has gills; also, a fungus that produces a fleshy fruiting body

mushrump   a hump in the humus caused by a developing mushroom

mycelium   a complex network of hyphae; the vegetative portion of a fungus

mycologist   one who studies fungi

mycology   the study of fungi

mycophagist   one who eats mushrooms

mycophile   one who loves mushrooms

mycophobe   one who despises or fears mushrooms

mycorrhiza   a mutually beneficial relationship between a fungus and the rootlets of a plant (especially trees), in which nutrients are exchanged

naked   without hairs or other tissue

nodulose   spores covered with small bumps or nodules

notched   gills abruptly adnexed, as though a small wedge-shaped piece of tissue had been removed at juncture to stalk (p. 17)

oblong   spores elongated, with approximately parallel sides

obtuse   blunt; not pointed

ochraceous   yellow or yellow-orange with a brownish tinge

order   a group of related families

ornamentation   any discontinuity (warts, ridges, etc.) on the surface of the cap, stalk, or spore

outer veil   see universal veil

oval   spores or cap egg-shaped

pallid   very pale; an indefinite whitish color

parallel   gill hyphae arranged more or less parallel to each other as seen in cross-section (p. 19)

paraphyses   unspecialized sterile cells (p. 782)

parasitic   feeding on another living organism

partial veil   protective tissue stretching from the stalk to the cap margin in many young mushrooms

pedicel   a slender stalk

pellicle   a viscid skin that usually peels easily

pendant   hanging down; also, annulus skirtlike

percurrent   columella or stalk extending through the spore mass to the top of the fruiting body

peridiole   small spore capsule in certain Gasteromycetes

peridium   the wall of the spore case in many Gasteromycetes

perithecium   a flask-shaped "nest" of asci (in the Pyrenomycetes)

peronate   sheathlike; also, a sheathing annulus or veil

persistent   persisting; not evanescent

phenolic   odor reminiscent of phenol

pileate   with a pileus or cap

pileus   the cap of a mushroom

pip-shaped   shaped like an apple seed

plane   cap having a flat surface (p. 17)

pocket rot   a rot producing hollow pockets in a tree

polymorphic   with many shapes or forms

polypore   any of a large group of mostly tough, wood-inhabiting fungi which bear their spores in pores

pores   the mouths of the tubes in boletes and polypores

poroid   resembling pores

potato chip conditions see footnote on p. 35

pruinose powdery or appearing finely powdered; dandruffy

pseudocarp something that resembles a mushroom but isn't

pseudorhiza a rootlike extension of the stalk; a "tap root"

pubescent minutely hairy; downy

pulverulent appearing powdered

punctate dotted with minute scales or points

putrescent readily decaying

radially arranged pores or pits arranged in rows which normally radiate like the spokes of a wheel

recurved curved up and back

resupinate lying flat on substrate; without a stalk or well-defined cap

reticulate stalk marked with lines crossed like the meshes of a net; netted (p. 489); spores with a network of ridges

rhizomorph a rootlike or stringlike bundle of mycelial hyphae

ring see annulus

rivulose marked by riverlike lines

rufescent becoming reddish

rufous brownish-red

rugose wrinkled

saclike volva shaped like a sack, pouch, or cup (p. 264)

saprophytic feeding on dead or decaying matter

sap rot decay of the sapwood

scabers tufted hairs or short projecting scales on stalk (Color Plates 145-147)

scabrous roughened by scabers

scales pieces of differentiated tissue on the cap or stalk, often of a different color than background

scaly furnished with scales; volva with concentric rings or scales at base of stalk (p. 264)

sclerotium a knot or tuber of hyphae, usually underground, in certain mushrooms; a resting stage that fruits periodically

scrobiculate stalk pitted with conspicuous spots

seceding gills breaking away from stalk as the cap expands

secotioid resembling *Secotium*, a genus of gastroid agarics; reminiscent of an undeveloped or aborted agaric

septate partitioned; with one or more cross-walls

serrated gill edges toothed like a saw

sessile lacking a stalk

setae pointed, elongated, thick-walled sterile cells

sheathlike annulus or veil sheathing the stalk (p. 312)

siderophilous basidia with granules that darken when heated in acetocarmine

simple fruiting body unbranched, not compound; stalk not complex

sinuate gills notched (p. 19)

sinuous crooked or curved

skirtlike annulus hanging down like a skirt (p. 312)

sordid dingy or dirty in appearance

spawn the mycelium

species (sp.) a particular kind of organism; the fundamental unit of taxonomy

sphaerocyst a round or swollen cell in certain mushrooms (e.g., *Russula, Lactarius*)

spines pendant spore-bearing "teeth" in the teeth fungi

splitter a taxonomist with a narrow genus or species concept

spore the reproductive unit of a fungus, usually a single cell

spore case the large chamber that holds the spores in puffballs and related fungi (Gasteromycetes)

squamose furnished with scales; scaly

squamulose furnished with small scales

stalk the stemlike structure that supports the cap in most mushrooms

stellate star-shaped; with rays

sterigma (pl., sterigmata) prong on the basidium on which a spore forms

sterile not producing spores; infertile

sterile base the sterile chambered base beneath the spore mass in many puffballs

stipe the stalk

stipitate furnished with a stalk or stipe

striate marked by lines; finely furrowed

stuffed stalk stuffed with a pith

subdistant gills fairly well spaced

subperonate somewhat or slightly peronate

substrate the material to which a fruiting body is attached

subtomentose finely or somewhat tomentose

sulcate   conspicuously furrowed; deeply striate

superficial   scales or warts easily removable; not innate

superior   ring located at or near top of stalk

tacky   slightly sticky but not truly viscid

tawny   the color of a lion

taxonomy   the classification of organisms to reflect their natural relationships

tenacious   tough

terrestrial   growing on the ground

tidepool conditions   see footnote on p. 35

toadstool   a mushroom, especially one that is poisonous

tomentose   covered with soft hairs

top rot   a rot confined to the top of a tree

translucent-striate   cap with the gills showing through to give a striate (lined) appearance

truncate   appearing chopped off or flattened at one end

trunk rot   decay found throughout the trunk of a tree

tuberculate   with low bumps

tuberculate-striate   striate with small bumps

tubes   the tubelike structures on the underside of boletes and polypores, in which spores are produced

turbinate   top-shaped

umber   a deep dull brown

umbilicate   cap with a navel-like central depression (p. 17)

umbo   a knob or bump at center of cap (p. 5)

umbonate   cap furnished with an umbo (p. 17)

universal veil   a protective layer of tissue that envelops all or most of the young fruiting body of certain mushrooms

veil   a protective tissue (see partial and universal veils)

ventricose   stalk swollen at or near the middle

vinaceous   the color of fine red wine or slightly paler

viscid   slimy or sticky to the touch, at least when moist

volva   remnants of the universal veil at the base of the stalk in certain mushrooms, usually in the form of a sack, collar, or series of concentric rings or scales (p. 264)

volval patch   a large patch of universal veil tissue on the cap

warts   small pieces of universal veil tissue deposited on the cap; also, any wartlike scale or protuberance

white rot   a delignifying decay

YAM   Yet Another Mycena

zonate   cap or flesh concentrically zoned

Leucopaxillus amarus, a common but bitter-tasting brown-capped agaric. Note how some specimens have a ribbed cap margin. Gills and flesh are white. See p. 168 for details.

BIBLIOGRAPHY:
Suggested Readings
and Primary References

Non-Technical Literature

(Note: Publications dealing with specific groups of fungi are listed under "Technical Literature," even though they may be written for amateurs)

Field Guides

Biek, David (1984). *The Mushrooms of Northern California.* Redding, Calif: Spore Prints.

Bowers, Jeannette & David Arora (1984). *Mushrooms of the World Coloring Book.* New York: Dover.

Guzman, Gaston (1978). *Hongos.* Mexico City: Limusa.

Kauffman, C.H. (1918). *The Gilled Mushrooms (Agaricaceae) of Michigan and the Great Lakes Region.* New York: Dover (1971 reprint).

Krieger, Louis C. (1936). *The Mushroom Handbook.* New York: Dover (1967 reprint).

Lange, Morton & F.B. Hora (1963). *A Guide to Mushrooms and Toadstools.* New York: E.P. Dutton.

Largent, David, Harry Thiers, Roy Watling, & Daniel Stuntz (1973-). *How to Identify Mushrooms to Genus* (5 vols.). Eureka: Mad River Press.

Lincoff, Gary (1981). *The Audubon Society Field Guide to North American Mushrooms.* New York: Alfred Knopf.

McIlvaine, Charles & Robert Macadam (1902). *One Thousand American Fungi.* New York: Dover (1973 reprint).

McKenny, Margaret & Daniel Stuntz (1971). *The Savory Wild Mushroom.* Seattle: University of Washington Press.

Miller, Orson K. (1972). *Mushrooms of North America.* New York: E.P. Dutton.

Moser, Meinhard (1983). *Keys to Agarics and Boleti.* England: Roger Phillips.

Phillips, Roger (1981). *Mushrooms and Other Fungi of Great Britain and Europe.* London: Pan Books.

Ramsbottom, John (1953). *Mushrooms and Toadstools.* London: Collins.

Rinaldi, Augusto & Vassili Tyndalo (1974). *The Complete Book of Mushrooms.* New York: Crown.

Smith, Alexander (1949). *Mushrooms in Their Natural Habitats.* New York: Hafner (1973 reprint).

Smith, Alexander (1975). *A Field Guide to Western Mushrooms.* Ann Arbor: University of Michigan Press.

Smith, Alexander, Helen V. Smith, & Nancy Smith Weber (1979). *How to Know the Gilled Mushrooms.* Dubuque: William Brown.

Smith, Alexander, Helen V. Smith, & Nancy Smith Weber (1981). *How to Know the Non-Gilled Mushrooms,* 2nd edition. Dubuque: William Brown.

Smith, Alexander & Nancy Smith Weber (1980). *The Mushroom Hunter's Field Guide* (revised). Ann Arbor: University of Michigan Press.

Weber, Nancy Smith & Alexander Smith (1985). *A Field Guide to Southern Mushrooms.* Ann Arbor: University of Michigan Press.

Wright, Greg (1981). *Key to the Gilled Mushrooms and Boletes of Southern California.* Published by the author.

Cookbooks

Grigson, Jane (1975). *The Mushroom Feast*. New York: Alfred Knopf.

Puget Sound Mycological Society (1973). *Oft Told Mushroom Recipes* (republished as *Wild Mushroom Recipes*). Seattle: Pacific Search.

Poisonous & Hallucinogenic Mushrooms

Duffy, Tom & Paul Vergeer (1977). *California Toxic Fungi*. Mycological Society of San Francisco.

Lincoff, Gary & D.H. Mitchel (1977). *Toxic and Hallucinogenic Mushroom Poisoning*. New York: Van Nostrand Reinhold.

Stamets, Paul (1978). *Psilocybe Mushrooms and Their Allies*. Seattle: Homestead Books.

Mushroom Cultivation

Harris, Bob (1976). *Growing Wild Mushrooms*. Berkeley: Wingbow Press.

Stamets, Paul & J.S. Chilton (1983). *The Mushroom Cultivator: A Practical Guide to Growing Mushrooms at Home*. Olympia: Agarikon Press.

Miscellaneous

Bo, Liu & Bau Yun-sun (1980). *Fungi Pharmacopoeia (Sinica)*. Oakland, Calif: Kinoko Company.

Jaeger, Edmund C. (1955). *A Sourcebook of Biological Names and Terms*. Springfield, Illinois: Charles C. Thomas.

Rice, Miriam & Dorothy Beebee (1980). *Mushrooms for Color*. Eureka: Mad River Press.

Wasson, R. Gordon (1968). *Soma: The Divine Mushroom of Immortality*. New York: Harcourt Brace Jovanovich.

Wasson, Valentina & R. Gordon Wasson (1957). *Mushrooms, Russia and History* (2 vols). New York: Pantheon.

Technical Literature

General

Bigelow, H. & H. Thiers, editors (1975). *Studies on Higher Fungi*. Germany: J. Cramer.

Miller, O.K. & D.F. Farr (1975). *An Index of the Common Fungi of North America*. Liechtenstein: J. Cramer, Bib. Myco. 44.

Petersen, R., editor (1971). *Evolution in the Higher Basidiomycetes: An International Symposium*. Knoxville: Univ. of Tennessee Press.

Richardson, M. & R. Watling (1968). *Keys to Fungi on Dung*. Bull. of the Brit. Myco. Soc. 2: 18-43.

Saylor, H., P. Vergeer, D. Desjardin, & T. Duffy. *California Mushrooms 1970-1980: Fungus Fair and Foray Collections*. Myco. Soc. of San Francisco.

Shaffer, R.L. (1968). *Keys to Genera of Higher Fungi*, 2nd edition. Ann Arbor: Univ. of Michigan Biological Station.

Singer, R. (1975). *The Agaricales in Modern Taxonomy*, 3rd edition. Germany: J. Cramer.

Snell, W.H. & E.A. Dick (1957). *A Glossary of Mycology*. Cambridge, Mass: Harvard Univ. Press.

Watling, R. & A.E. Watling (1980). *A Literature Guide for Identifying Mushrooms*. Eureka: Mad River Press.

Russulaceae & Hygrophoraceae

Hesler, L.R. & A.H. Smith (1963). *North American Species of Hygrophorus*. Knoxville: Univ. of Tennessee Press.

Hesler, L.R. & A.H. Smith (1979). *North American Species of Lactarius.* Ann Arbor: Univ. of Michigan Press.

Largent, D. (1985). *The Agaricales of California, 5: Hygrophoraceae.* Eureka: Mad River.

Methven, A. (1985). *New and Interesting Species of Lactarius from California.* Mycologia 77: 472-482.

Shaffer, R. (1962). *The Subsection Compactae of Russula.* Brittonia 14: 254-284.

Shaffer, R. (1964). *The Subsection Lactarioideae of Russula.* Mycologia 56: 202-231.

Shaffer, R. (1970). *Notes on Subsection Crassotunicatinae of Russula.* Lloydia 33: 49-96.

Shaffer, R. (1972). *North American Russulas of Subsection Foetentinae.* Mycologia 64: 1008-1053.

Tricholomataceae

Baroni, T. (1983). *Tricholoma manzanitae— A New Species from California.* Mycotaxon 18: 299-302.

Bigelow, H.E. (1970). *Omphalina in North America.* Mycologia 62: 1-32.

Bigelow, H.E. (1982). *North American Species of Clitocybe,* Part 1. Liechtenstein: J. Cramer.

Desjardin, D.E. (1985). *New Marasmioid Fungi from California.* Mycologia 77: 894-902.

Gilliam, M.S. (1975). *New North American Species of Marasmius.* Mycologia 67: 817-844.

Gilman, L. & O.K. Miller (1977). *A Study of the Boreal, Alpine, and Arctic Species of Melanoleuca.* Mycologia 69:927-951.

Halling, R.E. (1983). *The Genus Collybia in the Northeastern United States and Adjacent Canada.* New York Bot. Gar. Myco. Mem. 8 (Germany: J. Cramer).

Lennox, J.W. (1979). *Collybioid Genera in the Pacific Northwest.* Mycotaxon 9: 117-231.

Miller, O.K. & L. Stewart (1971). *The Genus Lentinellus.* Mycologia 63: 333-369.

Mueller, G. (1984). *New North American Species of Laccaria.* Mycotaxon 20: 101-116.

Singer, R. & A.H. Smith (1943). *A Monograph of the Genus Leucopaxillus.* Pap. Mich. Acad. Sci. 28: 85-132.

Smith, A.H. (1947). *North American Species of Mycena.* Ann Arbor: Univ. of Mich. Press.

Smith, A.H. (1960). *Tricholomopsis in the Western Hemisphere.* Brittonia 12: 41-76.

Smith, A.H. (1979). *The Stirps Caligata of Armillaria in North America.* Sydowia 8: 368-377.

Smith, A.H. & R. Singer (1945). *A Monograph of the Genus Cystoderma.* Pap. Mich. Acad. Sci. 30: 71-124.

Thiers, H.D. & W.J. Sundberg (1976). *Armillaria in the Western United States.* Madrono 23: 448-453.

Entolomataceae & Pluteaceae

Hesler, L.R. (1967). *Entoloma in Southeastern North America.* Germany: J. Cramer.

Homola, R.L. (1972). *Section Celluloderma of the Genus Pluteus in North America.* Mycologia 64: 1211-1247.

Largent, D. (1970-1974). *Studies in the Rhodophylloid Fungi I:* Madrono 21: 32-39; *II:* Mycologia 62: 437-452.

Largent, D. (1971-). *Rhodophylloid Fungi of the Pacific Coast I:* Brittonia 23: 238-245; *II:* Northwest Sci. 46: 32-39; *III:* Mycologia 66: 987-1021.

Largent, D. (1977). *The Genus Leptonia on the Pacific Coast of the United States.* Germany: J. Cramer, Bib. Myco. 55.

Shaffer, R. (1957). *Volvariella in North America.* Mycologia 49: 545-579.

Amanitaceae, Lepiotaceae, & Agaricaceae

Bas, C. (1969). *Morphology and Subdivision of Amanita and a Monograph on its Section Lepidella.* Persoonia 5: 285-579.

Freeman, A.E.H. (1979). *Agaricus in Southeastern United States.* Mycotaxon 8: 50-118.

Isaacs, B.F. (1963). *A Survey of Agaricus in Washington, Oregon, and California.* Master's Thesis, University of Washington; unpublished.

Jenkins, D. (1977). *A Taxonomic and Nomenclatural Monograph of the Genus Amanita, Section Amanita for North America.* Germany: J. Cramer, Bib. Myco. 57.

Kerrigan, R. (1979-1985). *Studies in Agaricus I:* Mycologia 71: 612-620; *II:* Mycologia 77: 137-141; *III:* Mycotaxon 22: 419-434.

Kerrigan, R. (1982). *The Genus Agaricus in Coastal California.* Master's Thesis, San Francisco State University; unpublished.

Smith, H.V. (1944). *The Genus Limacella in North America.* Pap. Mich. Acad. Sci. 30: 125-147.

Sundberg, W. *The Family Lepiotaceae in California.* Master's Thesis, San Francisco State University; unpublished.

Thiers, H.D. (1982). *The Agaricales (Gilled Fungi) of California: Amanitaceae.* Eureka: Mad River Press.

Coprinaceae & Strophariaceae

Guzman, G. (1983). *The Genus Psilocybe.* Liechtenstein: J. Cramer.

Smith, A.H. (1951). *The North American Species of Naematoloma.* Mycologia 43: 467-521.

Smith, A.H. (1972). *The North American Species of Psathyrella.* N.Y. Bot. Gar. Myco. Mem. 24.

Smith, A.H. & L.R. Hesler (1968). *The North American Species of Pholiota.* New York: Hafner.

Van de Bogart, F. (1976-1979). *The Genus Coprinus in Western North America I:* Mycotaxon 4: 233-275; *II:* Mycotaxon 8: 243-291; *III:* Mycotaxon 10: 155-174.

Cortinariaceae & Gomphidiaceae

Ammirati, J.F. & A.H. Smith (1977). *Studies in the Genus Cortinarius II: Section Dermocybe, New North American Species.* Mycotaxon 5: 381-397.

Hesler, L.R. (1969). *North American Species of Gymnopilus.* New York: Hafner, Myco. Mem. 3.

Hesler, L.R. & A.H. Smith (1965). *North American Species of Crepidotus.* New York: Hafner.

Miller, O.K. (1964). *Monograph of Chroogomphus.* Mycologia 56: 526-549.

Miller, O.K. (1971). *The Genus Gomphidius.* Mycologia 63: 1129-1163.

Smith, A.H. (1939). *Studies on the Genus Cortinarius I.* Contr. Mich. Univ. Herb. 2: 1-42.

Smith, A.H. (1957). *A Contribution Towards a Monograph of Phaeocollybia.* Brittonia 9: 195-217.

Smith, A.H., V.S. Evenson, & D.H. Mitchel (1983). *The Veiled Species of Hebeloma in the Western United States.* Ann Arbor: Univ. of Michigan Press.

Smith, A.H. & J. Trappe (1972). *The Higher Fungi of Oregon's Cascade Head Experimental Forest and Vicinity I.* Mycologia 64: 1138-1153.

Stuntz, D. *Macroscopic Field Key to Some of the More Common Species of Cortinarius Found in Washington.* Pacific Northwest Key Council.

Stuntz, D. *Interim Skeleton Key to Some Common Species of Inocybe.* Pacific Northwest Key Council.

Thiers, H.D. & A.H. Smith (1969). *Hypogeous Cortinarii.* Mycologia 61: 526-536.

Boletaceae

Singer, R. (1977). *The Boletineae of Florida.* Liechtenstein: J. Cramer, Bib. Myco. 58.

Smith, A.H. & H.D. Thiers (1964). *A Contribution Towards a Monograph of North American Species of Suillus.* Published by the authors.

Smith, A.H. & H.D. Thiers (1971). *The Boletes of Michigan.* Ann Arbor: Univ. of Michigan Press.

Smith, A.H., H.D. Thiers, & O.K. Miller (1965). *The Species of Suillus and Fuscoboletinus of Priest River Experimental Forest and Vicinity, Priest River, Idaho.* Lloydia 28: 120-138.

Snell, W.H. & E.A. Dick (1970). *The Boleti of Northeastern North America.* Germany: J. Cramer.

Thiers, H.D. (1975). *California Mushrooms: A Field Guide to the Boletes.* New York: Hafner.

Thiers, H.D. (1976). *Boletes of the Southwestern United States.* Mycotaxon 3: 261-273.

Thiers, H.D. (1979). *The Genus Suillus in the Western United States.* Mycotaxon 9: 285-296.

Wolfe, C.B. (1979). *Austroboletus and Tylopilus Subgenus Porphyrellus with Emphasis on North American Species.* Liechtenstein: J. Cramer, Bib. Myco. 69.

Polyporaceae, Stereaceae, & Allies

Canfield, E.,R. & R.L. Gilbertson (1971). *Notes on the Genus Albatrellus in Arizona.* Mycologia 63: 964-971.

Gilbertson, R.L. (1976). *The Genus Inonotus in Arizona.* N.Y. Bot. Gar. Mem. 28: 67-85.

Gilbertson, R.L. (1981). *North American Wood-Rotting Fungi That Cause Brown Rots.* Mycotaxon 12: 372-416.

Overholts, L.O. (1953). *The Polyporaceae of the United States, Alaska, and Canada.* Ann Arbor: Univ. of Michigan Press.

Stuntz, D. (1980). *Trial Field Key to the Pacific Northwest Polypores.* Pacific Northwest Key Council.

Welden, A.L. (1971). *An Essay on Stereum.* Mycologia 63: 790-799.

Hydnaceae

Hall, D. & D.E. Stuntz (1971-72). *Pileate Hydnaceae of the Puget Sound Area I:* Mycologia 63: 1099-1128; *II:* Mycologia 64: 15-37; *III:* Mycologia 64: 560-590.

Harrison, K.A. (1964). *New or Little Known North American Stipitate Hydnums.* Canad. Jour. Bot. 42: 1205-1234.

Clavariaceae & Cantharellaceae

Bigelow, H.E. (1978). *The Cantharelloid Fungi of New England and Adjacent Areas.* Mycologia 70: 707-756.

Marr, C. & D. Stuntz (1973). *Ramaria in Western Washington.* Germany: J. Cramer, Bib. Myco. 38.

Petersen, R.H. (1968). *The Genus Clavulinopsis in North America.* N.Y. Bot. Gar. Myco. Mem. 2.

Petersen, R.H. (1971). *Notes on Clavarioid Fungi IX.* Persoonia 6: 219-229.

Petersen, R.H. (1971). *The Genera Gomphus and Gloeocantharellus in North America.* Nova Hedw. 21: 1-118.

Petersen, R.H. (1975). *Ramaria subgenus Lentoramaria.* Leichtenstein: J. Cramer, Bib. Myco. 43.

Petersen, R.H. (1981). *Ramaria subgenus Echinoramaria.* Leichstenstein: J. Cramer, Bib. Myco. 79.

Scates, C. (1982). *Trial Key to the Species of Ramaria in the Pacific Northwest.* Pacific Northwest Key Council.

Wells, V.L. & P.E. Kempton (1968). *A Preliminary Study of Clavariadelphus in North America.* Mich. Botanist 7: 35-57.

Lycoperdales & Sclerodermatales

Coker, W.C. & J.N. Couch (1928). *The Gasteromycetes of the Eastern United States and Canada.* Chapel Hill: Univ. of North Carolina Press.

Guzman, G. (1970). *Monografia del Genero Scleroderma.* Darwiniana 16: 233-407.

Long, W.H. & D.J. Stouffer (1948). *Studies in the Gasteromycetes XVI: The Geastraceae of the Southwestern United States.* Mycologia 40: 547-586.

Ramsey, R. (1980). *Lycoperdon nettyana, a New Puffball from Western Washington State.* Mycotaxon 16: 185-188.

Smith, A.H. (1951). *Puffballs and Their Allies in Michigan.* Ann Arbor: Univ. of Michigan Press.

Zeller, S.M. (1947). *More Notes on Gasteromycetes.* Mycologia 39: 282-312.

Tulostomatales, Podaxales, & Allies

Long, W.H. (1940-1946). *Studies in the Gasteromycetes I: Dictyocephalos* (with O.A. Plunkett): Mycologia 32: 696-709; *VII: Schizostoma* (with D. Stouffer): Mycologia 35: 21-32; *VIII: Battarrea laciniata:* Mycologia 35: 546-556; *X: Tulostoma:* Mycologia 36: 318-339; *XIV: Chlamydopus* (with D. Stouffer): Mycologia 38: 619-629.

Long, W.H. (1946). *The Genus Phellorina.* Lloydia 9: 132-138.

McKnight, K. (1985). *The Small-Spored Species of Podaxis.* Mycologia 77: 24-35.

Rea, P.M. (1942). *Fungi of California I: Battarrea.* Mycologia 34: 563-573.

Thiers, H.D. & R. Watling (1971). *Secotiaceous Fungi from the Western United States.* Madrono 21: 1-9.

Hymenogastrales & Allies

Fogel, R. (1985). *Studies on Hymenogaster.* Mycologia 77: 72-82.

Singer, R. & A.H. Smith (1960). *Studies in Secotiaceous Fungi IX: The Astrogastraceous Series.* Mem. of Torrey Bot. Club 21: 1-112.

Smith, A.H. & R. Singer (1959). *Studies in Secotiaceous Fungi IV: Gastroboletus, Truncocolumella, Chamonixia.* Brittonia 11: 205-223.

Thiers, H.D. (1979). *New and Interesting Hypogeous and Secotioid Fungi from California.* Sydowia Ann. Myc. 2:8: 361-390.

Trappe, J. (1975). *A Revision of the Genus Alpova with Notes on Rhizopogon and the Melanogastraceae.* Beih. Nova Hedw. 6: 279-309.

Zeller, S.M. (1939). *New and Noteworthy Gasteromycetes.* Mycologia 31: 1-31.

Zeller, S.M. (1941). *Further Notes on Fungi.* Mycologia 33: 196-214.

Zeller, S.M. (1944). *Representatives of the Mesophelliaceae in North America.* Mycologia 36: 627-637.

Phallales & Nidulariales

Arora, D. & W. Burk (1982). *Clathrus archeri, a Stinkhorn New to North America.* Mycologia 74: 501-504.

Blanton, R.L. & W. Burk (1980). *Notes on Pseudocolus fusiformis.* Mycotaxon 12: 225-234.

Brodie, H.J. (1975). *The Bird's Nest Fungi.* Toronto: Univ. of Toronto Press.

Burk, W.R. (1979). *Clathrus ruber in California and Worldwide Distributional Records.* Mycotaxon 8: 463-468.

Dring, D.M. (1980). *Contributions Toward a Rational Arrangement of the Clathraceae.* Kew Bulletin 35: 1-96.

Rea, P.M. (1955). *The Genus Lysurus.* Pap. Mich. Acad. Sci. 40: 49-66.

Pezizales

Denison, W.C. (1961). *Some Species of the Genus Scutellinia.* Mycologia 51: 605-635.

Dissing, H. (1966). *The Genus Helvella in Europe.* Dansk. Bot. Ark. 25: 1-172.

Kanouse, B.B. (1948). *Studies in the Genus Otidea.* Mycologia 41: 660-677.

Kempton, P.E. & V.L. Wells (1970). *Studies on the Fleshy Fungi of Alaska IV: A Preliminary Account of the Genus Helvella.* Mycologia 62: 940-959.

Korf, R. (1973). *Discomycetes and Tuberales* (Chapter 9, Vol. IVA of *The Fungi: An Advanced Treatise,* edited by Ainsworth, Sparrow, and Sussman). New York: Academic Press.

Larsen, H.J. & W.C. Denison (1978). *A Checklist of the Operculate Cup Fungi (Pezizales) of North America West of the Great Plains.* Mycotaxon 7: 68-90.

McKnight, K.H. (1969). *A Note on Discina.* Mycologia 61: 614-630.

Seaver, F.J. (1928). *The North American Cup Fungi: Operculates.* New York: Hafner (1961 reprint).

Seaver, F.J. (1942). *The North American Cup Fungi: Inoperculates.* New York: Hafner.

Tylutki, E.E. (1979). *Mushrooms of Idaho and the Pacific Northwest: Discomycetes.* Moscow, Idaho: Univ. of Idaho Press.

Weber, N.S. (1972). *The Genus Helvella in Michigan.* Mich. Botanist 11: 147-201.

Tuberales

Burdsall, H.H. (1968). *A Revision of the Genus Hydnocystis (Tuberales) and of the Hypogeous Species of Geopora (Pezizales).* Mycologia 60: 496-525.

Gilkey, H.M. (1916). *A Revision of the Tuberales of California.* Univ. of Calif. Publ. Bot. 6: 275-356.

Gilkey, H.M. (1954). *Tuberales.* North American Flora 2, 1: 1-36.

Parks, H.E. (1921). *California Hypogeous Fungi—Tuberaceae.* Mycologia 13: 301-314.

States, J.S. (1983). *New Records of Hypogeous Ascomycetes in Arizona.* Mycotaxon 16: 396-402.

Trappe, J. M. (1975). *The Genus Fischerula (Tuberales).* Mycologia 67: 934-941.

Trappe, J.M. (1979). *The Orders, Families, and Genera of Hypogeous Ascomycotina (Truffles and Their Relatives).* Mycotaxon 9: 297-340.

Helotiales

Mains, E.B. (1954). *North American Species of Geoglossum and Trichoglossum.* Mycologia 46: 586-631.

Mains, E.B. (1955). *North American Hyaline-Spored Species of the Geoglossaceae.* Mycologia 47: 846-877.

Mains, E.B. (1956). *North American Species of the Geoglossaceae, Tribe Cudonieae.* Mycologia 48: 694-710.

Pyrenomycetes

Child, M. (1932). *The Genus Daldinia.* Ann. Missouri Bot. Garden 16: 411-486.

Mains, E.B. (1940). *Species of Cordyceps.* Mycologia 32: 310-320.

Mains, E.B. (1947). *New and Interesting Species of Cordyceps.* Mycologia 39: 535-545.

GENERAL INDEX

Note: This index includes topics, persons mentioned or quoted in the text, scientific names other than those listed in the "Genus and Species Index," and *selected* common names—those which are best known or most likely to be remembered.

Bold face numbers indicated detailed treatment of the subject; if the treatment covers several consecutive pages, only the first is given.

*—Indicates a photograph or illustration *without* accompanying text other than the caption.

To avoid the confusion that results from using two sets of numbers, color plates are not included; they are listed on the page where the subject is described.

A

Abnormalities 898*
Acacia 40
Admirable Bolete **521**
Agaricaceae **310**
Agaricales 57, **58**
Agarics 52*, 57, **58**
Agaricus (section) 313
Alder 39, **42**
Alder False Truffle **757**
Allegro, John 894
Allergies to Mushrooms 892, 896
Almond Mushroom **336**
Alpine Jelly Cone **674**
Amadou 581
Amanitaceae **262**
Amanitas **263**
  egg stage 262*
Amanita-Toxins **892**
Amanitins 893
Amatoxins **892**
Amethyst Laccaria **172**
Ammirati, Joseph 430, 455
Angel Wings **135**
Anise Agaricus **340**
Anise Mushroom **162**
Annulus **16**
  different types *(Agaricus)* 312
Anonymous Amanita **275**
Ant Fungus **881**
Ants 30, 31
Anywhere and Everywhere, Mush-
  rooms Found **47**
Aphyllophorales 57, **548**
Apple Bolete **528**
Apricot Jelly Mushroom **672**
Artist's Conk **576**
Artist's Fungus **576**
Arvenses 313, 731

Asci 4*, 782
Ascobolaceae 817, 834
Ascomycetes 4, 55*, **782**
Ascomycotina—*see* Ascomycetes
Ascus—*see* Asci
Aspen **42**
Aspen Bolete **540**
Astrogastraceous Series 742
Auriculariales 669
Autodigestion 342, 343*

B

Balsamiaceae 852
Banana Slug 28*, 29*
Banded Agaricus **321**
Barrows, Chuck 314, 331, 340, 529
Bartelli, Ingrid 787
Basidia 4*, 6*
  in jelly fungi 669*
Basidioles 5, 6*
Basidiomycetes 4, 52-54*, **57**
Basidiomycotina—*see* Basidiomycetes
Basidium—*see* Basidia
Basket Stinkhorn **770**
Bay Laurel 40
Bearded Milk Cap **73**
Beefsteak Fungus **553**
Beefsteak Morels **799**
Bell Morel **794**
Bell-Shaped Panaeolus **356**
Ben's Bitter Bolete **524**
Bigelow, Howard 154, 207
Big-Foot 788
Big Laughing Mushroom **410**
Bigwood, Jeremy 294
Birch **41**
Birch Bolete **541**
Birch Conk **584**

Birch Polypore **584**
Bird's Nest Fungi 54*, **778**
Bitorques 313
Bitter Bolete **523**
Bitter Hedgehog **620**
Bitter Polypore **560**
Black Chanterelle **665**
Black Cup Fungus **829**
Black Diamond 842*
Black Earth Tongues **866**
Blackening Russula **89**
Blackfellow's Bread 564
Black-Foot **562**
Black Jelly Drops **876**
Black Jelly Roll **672**
Black-Leg **562**
Black Morel 786*, **790**, 959*
Black Truffle 854
Black Witch's Butter **672**
Blah Ramaria 651
Bleeding Agaricus **325**
Bleeding Coral Mushroom **653**
Bleeding Milk Cap **68**
Bleeding Mycena **231**
Blewit 9, 10, 11, **153**
Blue-Capped Polypore **558**
Blue Chanterelle **668**
Blue-Green Anise Mushroom **161**
Blue Leg 154
Blue Stain **878**
Blue-Staining Slippery Jack **504**
Blusher, The **276**
Blushing Amanita **276**
Bohemian Truffle 712
Bolbitiaceae **466**
Boletaceae **488**
Bolete Eater **883**
Boletes 52*, **488**
Boletivores 84, **546**
Booted Amanita **279**
Boring Brown Bolete **517**
Boring Brown Cup Fungus **821**
Boring Ubiquitous Mushrooms 209, 417, 451, 455, 463
Bracket Fungi 52*, **549**
Brain Mushroom **801**
Bread Molds 782
Brown Crumbly Rot 578
Brown Field Mushroom **319**
Brown Rot 550
Brown Witch's Butter **673**
Bufotenine 894
Bulbopodium 418, 441

Bull Nose **800**
BUM's 209, 417, 451, 455, 463
Burk, William 686, 769, 776
Burned Areas, Mushrooms of 45
Butter Bolete **525**
Buttery Collybia **216**
Button Mushroom **319**
Butt Rot 550

C

Cabaniss, Michael 539
Caesar's Amanita **284**
Caldwell, Katy 306
California Agaricus **327**
Candlesnuff Fungus **885**
Candy Cap **80**
Canning Mushrooms 891
Cannon Fungus **781**
Cantharellaceae **658**
Cantharellales 548
Cap **15**
  cuticle 19*
  shape of 17*
Carbon Balls **887**
Carbonizing Decay 550
Caterpillar Fungus **882**
Cauliflower Mushroom(s) **657**
Cedar **41**
Cep **530**
Champignon **318**
Chanterelle **662**
Chanterelles 52*, **658**
Chaplin, Charlie 168
Chemical Characteristics of Mushrooms **20**
Chestnut Bolete **510**
Chicken Lips **874**
Chicken of the Woods **572**
Chinquapin 38
Chrome-Footed Bolete **533**
Classification **8**
Clavariaceae **630**
Cleaning Mushrooms 889
Cloud Ears 669
Club Corals **632**
Club Fungi 53*, **630**
Coccoli **284**
Cocconi 285
Coccora 265, **284**
Cole, C. 185
Collecting Mushrooms **11**, 26
Colorful Earth Tongues **868**
Comb Hericium **615**

Commercial Collecting   26
Commercial Mushroom—*see* Culti-
   vated Mushroom
Common Collybia   **215**
Common Dung Cup   **823**
Common Earthball   **708**
Common Puffball   **693**
Common Puffballs   **690**
Common Stinkhorns   **766**
Compactae   84
Compost, Shalom   325
Conch Rot   583
Cone Heads   **470**
Conical Waxy Cap   **116**
Conifer Coral Hericium   **614**
Coniophoraceae   604
Conks   **574**
Constricted Grisette   **289**
Cookery, Mushroom   **888**
Coprinaceae   **341**
Coprine   896
Coral Fungi   53*, **630, 645**
Corn Silk Inocybe   **457**
Corticiaceae   604
Cortinariaceae   **396**
Cortinarius (subgenus)   418, 446
Cosmopolitan Mushrooms   47
Cottonwood   39, **42**
Cousteau, Jacques   573
Cracked Cap Bolete   **519**
Crampballs   **887**
Crater Cup   **829**
Crepidotaceae   396, 405
Crested Coral   **641**
Crocodile Agaricus   334
Crown Coral Mushroom   **642**
Crown Fungus   **825**
Crust Fungi   53*, **604**
Cryptic Globe Fungus   **585**
Cultivated Mushroom   **319**
Cultivation of Mushrooms   30
Cup Fungi   55*, 783, **817**
Cypress   **36**
Cystidia   5, 6*, 19*, 20
Czarnecki Family   176

D

Dacrymycetales   669
Dark Bolete   **534**
Darwin, Charles   766
Deadly Cortinarius   **444**
Deadly Galerina   **401**
Deadly Parasol   **308**
Dead Man's Fingers   **886**
Dead Man's Foot   **712**

Dead Man's Hand   **710**
Death Angel—*see* Destroying Angel
Death Cap   **269**
Deer Mushroom   **255**
Deer Truffles   **862**
Delicious Milk Cap   **68**
Delignifying Decay   550
Dermocybe (subgenus)   418
Desert Shaggy Mane   **725**
Deserts, Mushrooms of   **46**
Desjardin, Dennis   207, 208, 210, 214,
   854
Destroying Angel   **271**
Devil's Dipstick   771
Devil's Eggs   765
Devil's Stinkhorn   770
Devil's Urn   **829**
Dickenson, Emily   3
Dictionary of Scientific Names (Latin
   and Greek Word Elements)   **899, 903**
Discomycetes   **783**
Disturbed Ground, Mushrooms of   **45**
Dog Stinkhorn   **771**
Domicile Cup Fungus   **822**
Donkey Ears   **832**
Douglas-Fir   **35**
Dow, Anne   547
Doyle, Sir Arthur Conan   2
Dryad Saddle   **561**
Drying Mushrooms   20, 890
Dry Rot Rungus   **610**
Dunce Caps   **470**
Dung Mushrooms   **42**
Dutra, Frank   484
Dyeing   418, 454, 455, 571, 712
Dyer's Polypore   **570**

E

Early Agrocybe   **469**
Early Morel   **793**
Ear Pick Fungus   **629**
Earthballs   **707**
Earth Fan   **608**
Earth Nuts   854
Earthstars   54*, 677, **699**
Earth Tongues   55*, **865**
Edibility of Mushrooms   **23**, 265
Elephant Ears   799, 803
Elfin Cups   805
Elfin Saddles   55*, 783, 796, **805**
Emetic Russula   **97**
Enokitake   220
Entolomataceae   **238**
Erving, Julius   8
Eucalyptus   **39**

Evenson, Verna Stucky   466
Everson, Bill   526
Eyelash Pixie Cup   **839**

F

Fairy Bonnet   **352**
Fairy Clubs   **634**
Fairy Fan   **871**
Fairy Fingers   **637**
Fairy Hair   **636**
Fairy Ring Mushroom   7*, **208,** 510
Fairy Rings   **6,** 209, 687
Fairy Stool   **568**
Fairy Tub   **821**
False Chanterelle   **479**
False Morel   **801**
False Morels   796, **799**
False Sparassis   **824**
False Tinder Polypore   **581**
False Truffles   54*, **739,** 844
False Turkey Tail   **605**
Fan Pax   **476**
Fat-Headed Black Morel   791
Felted Heart Rot   580
Felt-Ringed Agaricus   **326**
Fetid Russula   **92**
Field Mushroom—*see* Meadow Mushroom
Field Puffball   **695**
Fir   **41**
Fires   45
Flask Fungi   55*, **878**
Flesh   **16**
Flower Pot Parasol   **302**
Fluted Black Elfin Saddle   **815**
Fluted Brown Elfin Saddle   **814**
Fluted White Elfin Saddle   **816**
Fly Agaric   265, **282**
Fly Amanita   **282**
Fogel, Robert   748
Foxfire   196
Fragrant Russula   **92**
Freezing Mushrooms   891
Fried Chicken Mushroom   **174**
Fries, Elias   58, 418
Frost's Bolete   **528**
Fungi   **4**
  role in environment   6
Fungophobia   **1**
Funnel Chanterelle   **665**
Fuzzy Truffle(s)   **846**

G

Garden Mushrooms   43

Garlic Mushroom   **207**
Gasteromycetes   57, **676**
Gastroid Agarics   53*, **724**
Gastroid Agaricus   **729**
Gastroid Boletes   **544**
Gastroid Coprinus   **727**
Gastroid Cortinarius   **734**
Gastroid Lepiota   **731**
Gastroid Liberty Cap   **734**
Gastroid Milk Cap   **738**
Gastroid Pholiota   **735**
Gastroid Pine Spike   **732**
Gastroid Russula   **737**
Gastrointestinal Irritants   **895**
Gelatinous Coral Mushroom   **655**
Gemmed Amanita   **281**
Gemmed Puffball   **693**
Geneaceae   849
Geode Truffles   **849**
Geoglossaceae   866, 868, 872
Gerry, Eric   733
Giant Gel Cup   **828**
Giant Gymnopilus   **410**
Giant Horse Mushroom   **333**
Giant Puffball   678*, **682**
Giant Puffballs   **680**
Gilbertson, Robert   543
Gilkey, Helen   843, 858
Gilled Bolete   **480**
Gilled Mushrooms   52*, 57, **58**
Gilled Polypore   **589**
Gills   **16**
  arrangement of hyphae in   19*
  attachment to stalk   17*
  cross-section of   6*
Girolle   663
Glistening Inky Cap   **348**
Glutinous Gomphidius   **482**
Goat's Foot   **560**
Golden Fairy Club   **638**
Golden Pholiota   **390**
Golden Waxy Cap   **115**
Gomphaceae   630
Gomphidiaceae   **481**
Granulated Slippery Jack   **502**
Greek   899
Green Earth Tongue   **870**
Greening Goat's Foot   **559**
Green-Spored Parasol   **295**
Grisette   **288**
Gristly Truffle   **857**
Guzman, Gaston   707
Gypsy Mushroom   **412**
Gyromitrin   **893**

H

Habitats   34
Half-Free Morel   791
Hall, Joseph   29
Hallucinogenic Mushrooms   29, **31**,
   265, 282, 354, 358, 359, 368, 370-374,
   409, 410, 894, 895
Harding, Paul   69
Hard-Skinned Puffballs   707
Haymaker's Panaeolus   360
Hays, William Delisle   1
Heart Rot   549
Hedgehog Mushroom   **618**
Hedgehog Mushrooms   **616**
Helotiales   **865**
Helvellaceae   **796,** 848
Hemlock   **40**
Henis, M.   185
Hen of the Woods   **564**
Hesler, L.R.   64, 385
Hideous Gomphidius   482
Hole in the Ground   **847**
Honey Mushroom   34*, **196**
Hongo, Tsuguo   90
Hooded False Morel   **802**
Horn of Plenty   **666**
Horsehair Fungus   **208**
Horse Mushroom   **332**
Hortenses   **313**
Huckleberry   40
Hydnaceae   **611**
Hygrophoraceae   **103**
Hygroscopic Earthstar   **705**
Hymenium   57, 783
Hymenochaetaceae   604
Hymenogastrales   **739**
Hymenomycetes   **57**
Hyphae   5, 19*

IJK

Ibotenic Acid   **894**
Identifying Mushrooms   14
Incense Cedar   **41**
Indecisive Bolete   **535**
Indian Paint Fungus   **613**
Indigo Milk Cap   **69**
Indoor Mushrooms   **44**
Inky Cap   9, **347**
Inky Caps   **342**
Inocybium   455
Inrolled Pax   **477**
Insects   28, 29
Insidious Gomphidius   **482**

Iodine Polypore   **560**
Irregular Earth Tongue   **871**
Isaacs, Bill   314, 331
Ivory Waxy Cap   **119**
Jack-O-Lantern Mushroom(s)   **147**
Jacobs, Jim   752
JAR's   84
Jelly Babies   **874**
Jelly Fungi   53*, **669**
Jenkins, David   265
Jochelson, Vladimir   29
Judas' Ear   675
Juniper   **41**
Just Another Russula   84
Kerrigan, Rick   314, 323, 324, 332,
   379, 729
Keys   **21**
   dichotomous (initial key)   52
   pictorial key to gilled mushrooms   **61**
   pictorial key to major groups of
      fleshy fungi   **52**
King Alfred's Cakes   **887**
King Bolete   **530**
Kurokawa   **556**

L

Laboulbeniomycetes   47
Lackluster Laccaria   **172**
Landslide Mushroom   374
Larch   **40**
Larch Boletes   **505**
Lantern Stinkhorn   **776**
Largent, David   242, 244, 248, 249, 251
Latex   63, 64
Latin   9, 899
Latticed Stinkhorn   **773**
Lawn Mushrooms   **43**
Lawrence, D.H.   2
LBJ's   33*
LBM's   **32**
Lenin, V.S.   3
Leotiaceae   872, 875, 877
Lepidellas   265, 275
Lepiotaceae   **293**
Leprocybe   418
Leucogastrales   **759**
Liberty Cap   **370**
Lichen Agaric   **223**
Lilac Inocybe   **461**
Lincoff, Gary   373, 821
Ling Chih   **577**
Linnaeus   785
Lion's Mane Hericium   **615**
Little Brown Jobs   33*

Little Brown Mushrooms 32
Little Helmet 352
Lizard's Claw Stinkhorn 777
Lobster Mushroom 884
Long, W.H. 714
Lorchels 799
LSD 895
Luminescence 146, 148, 196
Lumpers 10
Lupine, Bush 40
Lycoperdales 677

M

Macroscopic Characteristics 14
Madrone 38
Maggots 28, 29, 889
Magic Mushroom 373
Magpie Mushroom 346
Man On Horseback 179
Mantis, Praying 28*
Manure Mushrooms 42
Manzanita 38
Manzanita Bolete 539
Maple 39, 41
Marr, Currie 645
Marsh, Ben 525
Marshall, Nina 769
Marshmallow Polypore 600
Matsutake 49*, 189, **191**, 192
Mattoon, Judith Scott 576*, 876
McIlvaine, Charles 113, 123, 141, 355, 376, 419, 445, 584, 690, 700, 765, 768, 769, 802, 866, 871
Meadow Mushroom 318
Medicinal Value of Mushrooms 30
Melanogastraceae 756
Melzer's Reagent 20
Menser, Gary 854, 861
Methven, Andrew 79
Mica Cap 348
Microscopic Characteristics 19
Milazzo, Ciro 145, 553
Milk Caps 64
Miller, Luen 123
Miller, Orson K. 355, 689, 717, 793, 826
Miniature Waxy Cap 113
Minores 313
Miscellaneous Toxins 896
Mitchel, D.H. 466
Mitchell, Craig 414, 666
MMH—see Monomethylhydrazine
Monomethylhydrazine 799, **893**

Morchellaceae **784**
Morel **787**
Morels 31*, 44*, 55*, 783, 784, **785**
Moser, Meinhard 419
Mountain Gastroid Agaricus **730**
Muk Ngo 675
Muscarine 455, **894**
Muscimol **894**
Mushroom(s) **1-959**
  allergies to 892, 896
  carnivorous 29
  chemical characteristics of **20**
  classification of **8**
  collecting **11**, 26
  color **14**
  color changes **15**
  cookery **888**
  definition of 4
  distinctive **48**
  dyeing with 418, 454, 455, 571, 712
  edibility of 23, 265
  evolution 57, 724, 739, 841
  fear of **1**
  growing 30
  hallucinogenic 29, **31**, 265, 282, 354, 358, 359, 368, 370-374, 409, 410, 894, 895
  handling 27
  hunters 84, 224, 546-547
  hunting **11, 25**
    equipment **12**
    field notes **13**
  identification of **14**
  macroscopic characteristics **14**
  medicinal value **30**
  microscopic characteristics **19**
  names **8, 899**
  nutritional value **30**
  odor **15**
  parts of a 5*, **14**
  picking **26**
    dangers of **27***
  poisoning **892**
  preservation of **20, 890**
  role in the environment 6-8
  seventy distinctive **48**
  sexuality 5-6
  size **14**
  succession of 8
  taste **15**, 63
  texture **15**
  toxins **892**
  weeds 47
  when and where they grow **25**

Mycelium   5, **18**, 43*
Mycorrhiza   7
Mycorrhizal Fungi   7, 34
Myxacium   418, 429

NO

Narrow-Headed Black Morel   791
Niche   7
Nidulariales   **778**
Neruda, Pablo   888
Not So Tedious Tubaria   **403**
Nutritional Value of Mushrooms   **30**
Oak   37
Oak Conk   **582**
Oak-Loving Collybia   **215**
Oak Root Fungus   196
Octopus Stinkhorn   **774**
Oddballs   **711**
Old Man of the Woods   **543**
Old Man's Beard   **615**
Orange Peel Fungus   **837**
Orange Sponge Polypore   **571**
Oregon Black Truffle   **854**
Oregon White Truffle   **858**
Ornate-Stalked Bolete   **522**
Ornate Stinkhorns   **772**
Ox Tongue   552*, **553**
Oyster Mushroom   29, **134**

P

Paddy Straw Mushroom   258, 262
Panther Amanita   **280**
Paraphyses   5
Parasitic Fungi   **6**
Parasol Mushroom   **298**
Parasol Mushrooms   **293**
Parchment Fungi   53*, **604**
Parks, Harold   842
Pasture Mushrooms   **43**
Paxillaceae   **476**
Pear-Shaped Puffball   **691**
Peele, Stephen   294
Pellegrini, Angelo   889
Peppery Bolete   **517**
Peppery White Milk Cap   **71**
Pezizaceae   **817**
Pezizales   **783**
Pfifferling   663
Phallaceae   **766**
Phallales   **764**
Phallolysin   892
Phallophobia   766
Phallotoxins   893
Phlegmacium   418
Pickling Mushrooms   892

Piercy, Marge   31
Pigs   29, 842
Pigs   29, 842
Pig's Ears   **661, 797, 798**
Pine   **35**
Pine Conk   **582**
Pine Spike   **485**
Pine Spikes   **484**
Pink-Bottom   318
Pinkish Coral Mushroom   **654**
Pink-Tipped Coral Mushroom   **656**
Pleated Marasmius   **209**
Pluteaceae   **253**
Pocket Rot   550
Podaxales   **724**
Poisoning, Mushroom—*see* Toxins
Poison Oak & Ivy   27*
Poison Pax   **477**
Poison Pie   **464**
Polyporaceae   **549**
Polypores   52*, **549**
Poor Man's Candle   123
Poor Man's Gumdrop   **674**
Poor Man's Licorice   **876**
Poor Man's Slippery Jack   **504**
Poor People's Truffle   668
Poplar   **42**
Poplar Tricholoma   **185**
Porcini   **530**
Pores   488, 549
Potato Chip Conditions   35
Powdery Mildews   782
Preserving Mushrooms
    for study   **20**
    for consumption   **890**
Prince, The   **337**
Psilocin   368, **895**
Psilocybin   368, **895**
Psilocybin Mushrooms   **31**, 895
Psychedelic Mushrooms—*see* Hallu-
    cinogenic Mushrooms
Puffballs   54*, **677**
    parts of   677*
Pungent Slippery Jack   **503**
Purple Fairy Club   **637**
Purple-Spored Puffball   **687**
Purple-Staining Milk Cap   **75**
Pyrenomycetes   **878**
Pyronemataceae   817, 831, 834, 846

QR

Queen Bolete   **531**
Questionable Stropharia   **377**
Quinine Conk   **579**
Quinine Fungus   **579**

Rabbit Ears 833
Ramsbottom, John 611, 773
Ramsey, Robert 688
Rattlesnakes 28*
Raverat, Gwen 766
Rea, Paul 777
Red-Belted Conk **578**
Red-Brown Butt Rot 571
Red-Capped Butter Bolete **526**
Red Chanterelle **664**
Red Coral Mushroom **655**
Red Gravel **757**
Red Morel 31*, 791
Red-Pored Bolete **528**
Red-Stemmed Bitter Bolete **524**
Reduced Agarics 724
Redwood **36**
Redwood Rooter **218**
Reindeer 29
Rhodophyllaceae 238
Ring(s)—*see* Annulus
Riparian Woodland **39**
Ristich, Sam 672
Roody, Bill 771
Rosy Conk **580**
Rosy Gomphidius **483**
Rosy Larch Bolete **506**
Rosy Russula **99**
Rots 549, 550
Rough-Stemmed Boletes **536**
Rufous Candy Cap **82**
Russia 3
Russulaceae **63**
Russulales 63
Russulas **83**
Rusts 57
Rusty Gilled Polypore **590**

S

Saint Ciro 484
Salting Mushrooms 892
Saltshaker Earthstar **704**
Sand Amanita **273**
Sand-Loving Mushrooms **46**
Sandy, The **185**
Sanguinolenti 313
Saprophytic Fungi 6, 7
Sap Rot 550
Sarcoscyphaceae 817, 834
Sarcosomataceae 817, 826
Satan's Bolete **527**
Saylor, Herb 652, 742, 744, 854, 856, 864
Scaly Chanterelle **661**

Scaly Pholiota **389**
Scarlet Cup Fungus **836**
Scarlet Waxy Cap **114**
Scates, Catherine (Kit) 645, 651
Schizophyllaceae 590
Sclerodermatales 678, 699, 707
Sclerotiniaceae 877
Scotch Bonnet 202
Sculptured Puffball **684**
Sellers, Bob 708
Sericeocybe 418, 448
Sex 5, 6
Shaggy Mane 9, 56*, 343*, **345**
Shaggy Parasol **297**
Shaggy-Stalked Parasol **309**
Sheep Polypore **557**
Sheep's Head **564**
Shelley, P.B. 2, 342
Shiitake 31*, 141
Shingled Hedgehog **619**
Shoehorn Oyster Mushroom **136**
Shoe String Root Rot 196
Short-Stemmed Russula **87**
Short-Stemmed Slippery Jack **501**
Shrimp Russula **102**
Siberian Slippery Jack **498**
Sickener, The **97**
Sierran Puffball **684**
Silver-Leaf Disease 606, 607
Singer, Rolf 207
Slippery Jack **500**
Slippery Jacks **491**
Slippery Jill **499**
Sloane, Hugo 822, 823
Slugs 29
Smith, Alexander 7, 64, 118, 138, 224, 235, 261, 345, 360, 361, 384, 385, 419, 434, 435, 455, 466, 473, 489, 499, 585, 657, 691, 696, 740, 753, 754, 766, 769, 772, 804, 864, 874
Smith, Helen 740, 753
Smooth Parasol **299**
Smooth-Spored Truffles **852**
Smuts 57
Snitow, Alan 1
Snowbank False Morel **800**
Snowbank Mushrooms **46**
Snow Mushroom **800**
Snow-Puff Mushroom 220
Soapy Tricholoma **184**
Soma 894
Sordid Waxy Cap **122**
Spawn—*see* Mycelium
Specialized Habitats, Mushrooms with **47**

Sphaeriales 878
Sphaerocysts 63
Sphere Thrower **781**
Split-Gill **590**
Splitters 10
Sponge Mushrooms 785
Spore Color **18**
Spore Prints **18**
Spores 4, 5, 18, 19
  developing 6*
  numbers of 576-577, 680
  truffle 842*
Spreading Brown Cup Fungus **821**
Spring Agrocybe **469**
Spring Mushrooms 46
Springtime Amanita **286**
Spruce **40**
Squirrels 29
Staghorn Jelly Fungus **674**
Stalk **16**
  ornamentation (boletes) 488-489
  position of 17*
  shape of 17*
Stalked Latticed Stinkhorn **776**
Stalked Oddball **713**
Stalked Polypores **554**
Stalked Puffballs 54*, **715**
Stamets, Paul 368
Starving Man's Licorice **828**
Starving Man's Slippery Jack 504
Steinpilz **530**
Stem—see Stalk
Stereaceae **604**
Stickney, Larry 117, 672
Stinker, The **179**
Stinkhorn **768**
Stinkhorns 54*, **764**, 766, 772
Stinkopus 775
Stinky Squid 776
Stone Fungus **563**
Strawberries and Cream **627**
String and Ray Rot 565
Strophariaceae **367**
Stuntz, Daniel 373, 419, 434, 455,
  458, 560, 645
Stuntz's Blue Legs **372**
Suburban Psathyrella **363**
Suffixes, Latin and Greek **900, 901**
Sulfur Shelf **572**
Sulfur Tuft **382**
Sunny Side Up **474**
Swamp Beacon 870
Sweat-Producing Clitocybe **163**
Sweetbread Mushroom **240**

T

Tamarack Jack **497**
Tanoak 37
Tawny Grisette **287**
Taxonomy **8**
Teeth Fungi 53*, **611**
Telamonia 418, 450, 451
Terfeziaceae 854-855
Terminology **14**
Termites 30
Terrestrial Pholiota **389**
Thelephoraceae 604
Thick-Footed Morel 788
Thick-Skinned Puffballs 707
Thick-Walled Maze Polypore **587**
Thiers, Harry 285, 489, 536, 742, 745
Thimble Morel **794**
Thimble Morels **793**
Thin-Walled Maze Polypore **588**
Ticks 27*
Tidepool Conditions 35
Tinder Polypore 581
Toadstool, Definition of 25
Toadstool Tester 24*
Toothed Jelly Fungus **671**
Top Rot 550
Totally Tedious Tubaria **402**
Toxins, Mushroom **892**
Trappe, James 748, 758, 843, 860
Tree Ears **675**
Tree Oyster—see Oyster Mushroom
Trees 34-42
Tree Trunk Morel 788
Tremellales **669**
Tricholomataceae **129**
True Truffles **854**
Truffle-Hounds 842, 855
Truffle Hunting 842-843
Truffle-Like Peziza 822*, **824**
Truffles 55*, 739, **841, 854**
Truly Trivial and Totally Tedious
  Tubaria 403
Truly Trivial Tubaria 403
Trumpet of Death 668
Trunk Rot 550
Tuberaceae 854, 855
Tuberales **841**
Tubes 488, 549
Tuckahoe 564, **604**
Tulostomatales 678, **715**
Tumbling Puffball **697**
Tumbling Puffballs **696**
Tunbridge Ware 878
Turkey Tail **594**

UV

Umbrella False Morel **804**
Van de Bogart, Fred   346
Varnished Conk **577**
Veil(s) **16**
  different types (*Agaricus*)   312
  different types (*Amanita*)   263-264
Veiled Polypore **585**
Veiled Stinkhorn   771
Veined Brown Cup Fungus **796**
Velvet Foot **220**
Velvet Pax **478**
Velvet Stem **220**
Velvety Black Earth Tongue **867**
Violet Cortinarius **446**
Violet Cup Fungus **824**
Violet Hedgehog **622**
Viscid Black Earth Tongue **866**
Vole, California Red-Backed   841
Volva **18**
  different types (*Amanita*)   263-264
Volvariaceae   253
Vulcan Pixie Cup **840**

W

Walnut **800**
Wasson, R. Gordon   29, 894
Wasson, Valentina   320
Waxy Caps **103**
Weber, Nancy S.   740, 753, 816
Western Giant Puffball **684**
Western Grisette **290**
Wet Rot   611
Wheat Ergot   895
White Chanterelle **662**
White King Bolete **529**

White Matsutake   49*, **191**
White Morel   788*, **789**
White Rot   550
White Truffle   854
Willow   39, **42**
Wine-Colored Agaricus **326**
Wine-Red Stropharia **378**
Winter, Bob   143
Winter Chanterelle **665**
Winter Mushroom   220
Witch's Butter **673**
Witch's Hat **116**
Woman on Motorcycle **299**
Wood Blewit   154
Wood Ear **675**
Woodland Agaricus **334**
Wood Truffles **848**
Woolly Chanterelle **661**
Worms—*see* Maggots
Wright, Greg   69, 161, 241, 346, 350,
  404, 475, 528
Wrinkled Thimble Morel **793**

XYZ

Xanthodermati   313
YAM's   224
Yeasts   782
Yellow Coral Mushroom **652**
Yellow Morel **787**
Yellow-Staining Agaricus **329**
Yellow-Staining Milk Cap **74**
Yellow-Veiled Stinkhorn   771
Yet Another Mycena   224-225, **234**
Yung Ngo   675
Zeller's Bolete **518**
Zygomycetes   844

Geopora arenicola (see description on p. 847) is half cup fungus, half truffle. When young it looks like a hole in the ground because all but the mouth is immersed in the soil. As it matures, however, the hole broadens, and in old age the margin of the cup often splits into lobes as shown on left.

GENUS AND SPECIES INDEX

Note: This index includes only genera and species so that it can also serve as a check-list. See "General Index" for common names, topics, persons, etc.

Bold face numbers indicate detailed treatment of the subject; if treatment covers several consecutive pages, only the first is given.

*—Indicates a photograph or illustration *without* accompanying text other than the caption.

To avoid the confusion that results from using two sets of numbers, color plates are not included in the index; they are listed on the pages where subjects are described.

A

Abortiporus
  biennis 566
Abstoma 680
  reticulatum 681, 690
  townei 681, 690
Agaricus 58*, **310**, 896
  abruptibulbus 335
  albolutescens **335**
  albosanguineus 325
  altipes 317
  amethystina 341
  amicosus 316, 326
  argenteus 319
  arorae 311*, **325**
  arvensis 12*, 312*, **332**
  augustus 311*, **337**
  "barrowsii" 331
  benesi 316, 325
  bernardii **322**
  bisporus **319**
  bitorquis 312*, **321**
  blandianus 320
  brunnescens 320
  californicus 312*, **327**, 895
  campestris 312*, **318**, 889*
  chionodermus 315, 317, 335
  chlamydopus 322
  comtulus 315, 340
  cretacellus 315
  crocodilinus 315, 317, 334
  cupreobrunneus **319**
  diminutivus **340**
  dulcidulus 341
  edulis 322
  fissuratus 333
  fuscofibrillosus **325**
  fuscovelatus **324**
  haemorrhoidarius 325
  halophilus 322
  hondensis **326**, 895
  hortensis 320
  lanipes 316

Agaricus (cont.)
  lilaceps **323**
  "luteovelatus" 324
  macrosporus 334
  maritimus 322
  meleagris 329
  micromegathus **340**
  nivescens 334
  osecanus **333**
  pattersonae 316, 324
  perobscurus **339**
  perrarus 338
  pinyonensis **331**
  placomyces 314, 329, 895
  pocillator 314, 329
  porphyrocephalus 319
  praeclaresquamosus 312*, **329**, 895
  purpurellus 341
  rhoadsii 317
  rodmani 322
  rubronanus 325
  rutilescens 317, 319
  semotus 315, 339*, 340
  sequoiae 317, 332
  silvaticus 325
  silvicola 312*, **334**
  smithii 315, 338
  solidipes 318
  spissicaulis 326
  spp. (unidentified) 317, 318, 325
  subfloccosus 320
  subrufescens **336**
  subrutilescens **326**
  summensis 315, 339
  sylvicola 335
  urinescens 334
  vaporarius 317
  villaticus 334
  vinaceo-umbrinus 326
  vinaceovirens 322
  xanthodermus **329**, 895
Agrocybe **467**
  acericola 470
  aegerita 467, 470
  amara 469

Agrocybe (cont.)
  arvalis 469
  cylindracea 470
  dura 470
  erebia 467, 470
  firma 468
  paludosa 467
  pediades 43*, **468**
  praecox **469**
  retigera 468
  semiorbicularis 469
  sororia 468, 469
  tuberosa 469
Albatrellus **554**
  avellaneus 557, 558
  caeruleoporus 555, 559
  confluens 556, 558
  cristatus 556, 560
  dispansus 556
  ellisii **559**
  flettii **558**
  hirtus 560
  ovinus **557**
  peckianus 558
  pescaprae 558*, **560**
  similis 558
  sylvestris 560
Alboleptonia 249
  adnatifolia 253
  ochracea 253
  sericella **252**
Aleuria **833**
  aurantia **837**
  rhenana **836**
  rutilans 837
Alnicola 399
  escharoides 400
  melinoides 400
  scolecina 400
Alpova **756**
  alexsmithii 757, 758
  cinnamomeus 758
  diplophloeus **757**
  luteus 758
  nauseosus 757, 758
  olivaceoniger 757, 758
  olivaceotinctus 757*, 758

Alpova (cont.)
  superdubius 758
  trappei 757, 758
Amanita 262, **263**
  abrupta 275
  alba 266, 288
  aspera **278**
  atkinsoniana 275
  baccata **273**
  bisporigera 273, 892
  boudieri 274
  breckonii 281
  brunnescens 267, 279
  caesarea **284**
  calyptrata 265, **284**,
    679*, **890***
  calyptratoides 269, 286
  calyptroderma 285
  ceciliae 289
  chlorinosma 275
  cinereoconia 269, 276
  cinereopannosa 275
  citrina 267, 279
  cokeri 268, 275
  constricta **289**
  cothurnata 268, 280
  crocea 287, 288
  daucipes 275
  farinosa 267, 276
  flavoconia **278**
  flavorubens 278
  flavorubescens 269, 278
  frostiana 279
  fulva **287**
  gemmata **281**, 894
  hemibapha 284
  hesleri 269
  inaurata 268, 289
  junquillea 281
  longipes 275
  magniverrucata **274**
  mappa 279
  muscaria 15, 29, 264*,
    265, **282**, 894, 895
  mutabilis 273
  ocreata 12*, 266*, **271**,
    286*, 892, 893
  onusta 269, 276
  pachycolea **290**
  pantherina 264*, **280**,
    894
  parcivolvata 267, 284
  peckiana 267
  phalloides 264*, **269**,
    892, 893
  polypyramis 275
  porphyria **279**
  praegraveolens 275
  ravenelii 275
  rhopalopus 275
  roseitincta 269

Amanita (cont.)
  rubescens 264*, 266*,
    276, 884*
  russuloides 281
  salmonea 269
  silvicola **273**, 274*
  smithiana 275
  solitaria 275
  spissa 269, 280
  spp. (unidentified) 268,
    275, 277
  spreta 266
  strangulata 289
  strobiliformis 275
  thiersii 275
  umbonata 284
  umbrinolutea 290
  vaginata **288**, 289*
  velosa 266*, **286**
  verna 273, 892
  virosa 267, 273, 892
  volvata 266
  wellsii 267
Amanitopsis 263
  (also see *Amanita*)
  vaginata 288
Amylocystis
  lapponica 600
Anellaria
  separata 355
Anthopeziza
  floccosa 836
Anthracobia
  macrocystis 838
  melaloma 834, 838
Anthurus
  archeri 775
  aseroeformis 775
  borealis 778
Arachnion
  album 679
Araneosa
  columellata 679
Arcangeliella **736**
  camphorata 737, 739
  crassa **738**
  lactarioides 739
  tenax 739
  variegata 737, 739
Armillaria **189**
  albolanaripes **194**
  bulbosa 197
  caligata **192**
  decorosa 189
  dryina 136
  fusca 190, 194
  luteovirens 194
  mellea 196
  olida **193**

Armillaria (cont.)
  pitkinensis 194
  ponderosa 49*, **191**
  straminea 190, 194
  tabescens 197
  viscidipes 190
  zelleri 189
Armillariella 189
  bulbosa 190, 197
  mellea 34*, **196**, 896
  tabescens 150, 190, 197
Ascobolus 838
  carbonarius 835, 838
Ascocoryne
  sarcoides 875, 877
Ascotremella
  faginacea 875, 877
  turbinata 877
Aseroe
  rubra 772, 775
Asterophora **200**
  lycoperdoides **201**
  parasitica 201
Astraeus **699**
  hygrometricus 700, **705**
  pteridis **706**
Aurantioporellus 572
Aurantioporus
  croceus 597
Auricularia 669
  auricula **675**, 958*
  delicata 675
  mesenterica 675
  polytricha 675
Auriscalpium **629**
  vulgare **629**
Austroboletus **508**
  betula 508
  gracilis 534
  subflavidus 508

B

Baeospora
  myosura 202, 212
  myriadophylla 202
Balsamia **852**
  magnata **853**
Bankera 616
  carnosa 618
  fuligineo-alba 618
  violescens 618
Barssia 852
  oregonensis **853**
  sp. (unidentified) 854
Battarrea 715
  digueti 716, 718
  laciniata 718
  phalloides **717**
  stevenii 718

Bisporella
  citrina 877
Bjerkandera 592
  adusta **596**
  fumosa 596
Bolbitius **473**
  aleuriatus **475**
  callisteus 475
  coprophilus 474, 475
  lacteus 474, 475
  reticulatus 475
  sordidus 474
  vitellinus **474**
Boletellus **508**
  ananas 508, 509
  betula 508
  chrysenteroides 508
  russellii **508**
Boletinellus
  merulioides 490
Boletinus 491
  (also see *Suillus* &
  Fuscoboletinus)
  cavipes 495
Boletopsis 554
  griseus 557
  leucomelas 557
  subsquamosa **556**
Boletus **511**, 894
  (also see *Suillus,*
  Leccinum, Tylopilus,
  etc.)
  abieticola 514, 525
  aereus 50*, 512*, **531**
  affinis 514, 516
  albidus 525
  amygdalinus 527
  appendiculatus 512*,
  525
  auriflammeus 516, 522
  auriporus 516, 522
  badius 516, 519
  barrowsii 512*, **529**
  betula 508
  bicolor **521**
  caespitosus 522
  calopus **523**
  campestris 521
  chromapes 534
  chrysenteron **519**
  citriniporus 515, 523
  coccyginus 513, 521
  coniferarum 514, 523
  curtisii 516
  dryophilus **520**
  eastwoodiae 528
  edulis 489*, 512*, **530**,
  546, 547, 890*
  erythropus **526,** 895
  fagicola 516

Boletus (cont.)
  fibrillosus **523,** 529*
  flammans 515, 529
  flaviporus **522**
  fragrans 514
  fraternus 521
  frostii **528**
  griseus 516, 522
  haematinus 513, 528,
  895
  hemichrysus 512
  hypocarycinus 527
  inedulis 515
  longicurvipes 516
  luridus 515, 528
  "marshii" **524**
  mendocinensis 513, 520
  miniato-pallescens 521
  mirabilis **521**
  mottii 530
  olivaceobrunneus 523
  ornatipes **522**
  orovillus 513, 527
  pallidus 515
  parasiticus 512
  piedmontensis 515
  pinicola 530
  pinophilus 530
  piperatoides 517
  piperatus **517**
  porosporus 520
  projectellus 516, 522
  pseudoseparans 530
  pseudosulphureus 515
  pulcherrimus **528**, 895
  pulverulentus 514, 515
  radicans 525
  regius 512*, **526**
  retipes 522
  rubellus 515, 521
  rubinellus 516
  rubripes **524**
  rubroflammeus 515, 529
  rubropunctus 516
  satanas **527,** 895, 896
  sensibilis 521
  separans 530
  smithii 513, 520
  spadiceus 518
  speciosus 515, 526
  sphaerocephalus 512
  spp. (unidentified) 514,
  523, 528
  subglabripes 516
  subtomentosus 488*,
  517
  subvelutipes 515, 527,
  895
  truncatus 514, 520
  variipes 516, 531

Boletus (cont.)
  vermiculosus 516
  viridiflavus 516, 522
  zelleri **518**
Bondarzewia 554
  berkeleyi 556, 565
  montana 548*, **565**
Bovista **696**
  californica 698
  colorata 696, 698
  dakotense 697, 698
  leucoderma 698
  longispora 697, 698
  minor 696, 698
  nigrescens 697, 698
  pila 697
  plumbea **697**
  pusilla 697, 698
Bovistella 690
  radicata 691, 694
Brauniellula **732**
  albipes 733
  nancyae **732**
Bulgaria **875**
  inquinans **876**
  melastoma 830
  mexicana 828
  rufa 827
Byssonectria
  aggregata 828

C

Calbovista 680
  subsculpta 682, 685
Callistosporium 202
  graminicolor 211
  luteo-olivaceum **211**
Calocera
  cornea 635, 670, 675
  viscosa **674**
Calocybe 173
  carnea **176**
  gambosa 183
  onychina 150, 176
Caloporus
  dichrous 552
Caloscypha
  fulgens **837**
Calostoma 715
  cinnabarina **718**
  lutescens 716, 719
  microsporum 716, 719
  ravenelii 716, 719
Calvatia 677*, **680**
  booniana **684**
  bovista **686**
  caelata 686
  candida 682, 686

Calvatia (cont.)
  craniiformis 682, 688
  cretacea 681, 688
  cyathiformis **687**
  diguetii 688
  elata 682, 686
  excipuliformis 682, 686
  fragilis 681, 687
  fumosa **688**
  gigantea 678*, 680, **682**
  hesperia 681, 689
  lepidophora 681, 689
  lloydii 682, 686
  lycoperdoides **687**
  maxima 683
  owyheensis 681
  pachyderma 681, 689
  pallida 682, 686
  paradoxa 681
  rubroflava 682, 688
  rubrotincta 682, 687
  saccata 686
  sculpta **684**
  subcretacea **688**, 689*
  tatrensis 682, 686
  umbrina 688
  utriformis 686
Camarophyllus **103**
  angelesianus 112
  angustifolius 109
  borealis **109**
  cinereus 112
  colemannianus 108, 112
  cremicolor 109
  graveolens 108
  niveus 105, 109
  pallidus 112
  paupertinus 112
  pratensis **110**
  rainierensis 112
  recurvatus **112**
  russocoriaceus **109**
  subviolaceus **111**
  virgineus 105, 109
Cantharellula 149
  umbonata 151, 165
Cantharellus 658
  appalachiensis 660
  aurantiacus 480
  cibarius 659*, **662**
  cinereus 666
  cinnabarinus **664**
  clavatus 661
  confluens 660, 664
  floccosus 662
  formosus 664
  ignicolor 660, 665
  infundibuliformis **665**
  lateritius 660, 664
  lutescens 665

Cantharellus (cont.)
  minor 660, 665
  odoratus 660, 664
  purpurascens 660
  subalbidus **662**
  tubaeformis 665
  xanthopus 660, 665
Carbomyces 844
Catathelasma 189
  imperialis **195**
  macrospora 195
  singeri 195
  ventricosa 190, 195
Caulorhiza 202
  hygrophoroides 219
  umbonata **218**
Cerrena
  unicolor 587, 589, 592, 597
Chaetoporus
  euporus 602
Chamaeota 253
Chamonixia 748
  ambigua **751**
  brevicolumna 748, 751
  caespitosa 748, 751
  caudata 748, 751
Cheilymenia
  coprinaria **838**
  crucipila 839
  stercorea 838
  theleboides 838
Cheimonophyllum
  candidissimus 132
Chlamydopus 715
  meyenianus **721**
Chlorociboria 877
  aeruginascens **878**
  aeruginosa 878
Chlorophyllum 293
  molybdites **295**, 895, 896
Chlorosplenium 878
Choiromyces 855
  alveolatus **858**, 859*
  cookei 858
Chondrogaster 748
Chondrostereum
  purpureum 605
Choriactis
  geaster 827
Christiansenia
  mycetophila 216
Chroogomphus 20*, **484**
  flavipes 485
  leptocystis 485, 487

Chroogomphus (cont.)
  ochraceus 485, 486
  pseudovinicolor **486**
  rutilus 485, 486
  tomentosus **487**
  vinicolor 20*, 484*, **485**
Ciboria **877**
  amentacea **877**
Clathrus 772
  archeri **774**
  bicolumnatus 773
  cancellatus 773
  cibarius 774
  columnatus 773
  crispatus 773
  crispus 773
  preussii 774
  ruber **773**
Claudopus 238
  byssisedus 238
  depluens 238
  graveolens 238
  nidulans 141
  parasiticus 239
Clavaria 630, **634**
  (also see *Clavulina*, *Clavulinopsis*, *Ramaria*, etc.)
  amethystina 641
  cineroides 640
  fumosa **638**
  juncea 637
  kunzei 643
  mucida 636
  ornatipes 640
  phycophila 636
  pulchra 638
  purpurea **637**
  rosea 635, 639
  rubicundula 638
  vermicularis 631*, **637**
  zollingeri 635, 638
Clavariadelphus **632**
  borealis 632, 634
  fistulosus 637
  ligula **633**
  lovejoyae 632, 634
  mucronatus 632, 633
  pistillaris **632**
  sachalinensis 633
  subfastigiatus 633
  truncatus **634**
Claviceps 880
  purpurea 880, 895
Clavicorona 640
  avellanea 642
  pyxidata **642**
  taxophila 635, 636

Clavulina **640**
  amethystina 640, 641
  cinerea **641**
  cristata **641**
  ornatipes 640
  rugosa 641, 642, 643*
  zollingeri 638
Clavulinopsis 634
  appalachiensis 638
  aurantio-cinnabarina
    635, 639
  corniculata **639**
  dichotoma 640
  fusiformis 635, 638,
    639*
  gracillima 638
  helvola 638
  holmskjoldii 640
  laeticolor **638**
  luteoalba 638
  miniata 638
  subaustralis 639
  subtilis 640
  umbrinella 635, 640
Climacodon
  septentrionale 612,
    613*
Clitocybe **148**
  aeruginosa 151, 162
  alba 160
  albirhiza **161**
  americana 150, 157
  atrialba 165
  atroviridis 151, 162
  augeana 163
  aurantiaca 480
  avellaneialba 152, 161
  brunneocephala **154**
  candicans 161
  candida 151, 159
  cerussata 159, 161
  clavipes **160**, 896
  coniferophila 152, 161
  crassa 151, 160
  cyathiformis **164**
  dealbata 163, 894
  deceptiva **162**
  densifolia 155
  dilatata **159**
  ectypoides 157
  epichysium 151, 166
  fasciculata 155
  flaccida 157
  fragrans 162
  geotropa 158
  gibba **157**
  gigantea **158**
  gilva 157
  glaucocana 149, 152

Clitocybe (cont.)
  graveolens 149, 152
  harperi 152, 160
  illudens 148
  infundibuliformis 157
  inversa **156**
  irina 149, 155
  leopardina 161
  lignatilis 136
  martiorum 149
  maxima **157**
  morbifera 163
  morganii 480
  multiceps 174
  nebularis **159**
  nuda 9, 10, 11, **153**
  obsoleta 163
  odora **161**
  olesonii 154
  oramophila 163
  polygonarum 152, 156
  praemagna 155
  rivulosa 163
  robusta 150, 160
  saeva 149, 154
  sclerotoidea **164**
  septentrionalis 151, 159
  sinopica 157
  socialis 176
  squamulosa 150, 157
  suaveolens 162
  subalpina 156
  subconnexa **155**
  sudorifica 163
  tarda **152**
  variabilis 151, 161
Clitocybula 149
  abundans 150, 204, 215
  atrialba 151, 165
  familia 150, 204, 215,
    898*
Clitopilus **239**
  orcellus 241
  passeckerianus 240
  prunulus **240**
Collybia **201**
  abundans 215
  acervata **215**
  albipilata 212
  alkalivirens 204
  atrata 166
  badiialba 216
  bresadolae 215
  butyracea **216**
  cirrhata 212
  confluens **213**
  conigena 212
  cookei 212
  distorta 216

Collybia (cont.)
  distorta 216
  dryophila **215**
  erythropus 215
  extuberans 205, 217
  familia 215
  fuscopurpurea **214**
  fusipes 204
  iocephala 202
  luxurians 205, 209
  maculata **217**
  marasmioides 205, 215
  oregonensis **218**
  peronata 205, 213
  platyphylla 146
  polyphylla 205, 213
  racemosa 212*, **213**
  radicata 220
  spongiosa 205
  spp. (unidentified) 205,
    213
  strictipes 209
  subsulcatipes 203, 218
  subsulphurea 216
  trullisata 211
  tuberosa **212**
  umbonata 219
  velutipes 220
Coltricia **566**
  cinnamomea **568**
  montagnei 567, 568
  perennis 567, 568
Coniophora
  puteana 605, 611
Conocybe **470**, 895
  coprophila 471, 472
  crispa 471, 473
  cyanopus 471, 472
  filaris **471**, 892
  intrusa 471
  lactea **472**
  smithii 471, 472
  sp. (unidentified) 471
  stercoraria 471, 472
  tenera **472**
Cookeina 836
Coolia
  odorata 198
Copelandia 358
Coprinus **342**
  alnivorus 344, 346
  alopecia 348
  americanus 344
  arenatus 351
  asterophora 344
  asterophoroides 344
  atramentarius 9, **347**,
    896
  bulbilosus 352
  "chevicola" 351

Coprinus (cont.)
  cinereus 351
  colosseus 346
  comatus 9, 56*, 343*,
    345
  disseminatus **352**
  domesticus **349**
  ephemeroides **352**
  ephemerus 344, 352
  fimetarius 344, 351
  impatiens 344, 353
  insignis 348
  lagopides 351
  lagopus **350**
  macrocephalus 351
  macrorhizus 351
  micaceus **348**
  miser 344, 352
  narcoticus 351
  niveus 343, 351
  palmeranus 346
  picaceus **346**
  plicatilis **352**
  pseudoradiatus 351
  quadrifidus 344
  radians 344, 349
  radiatus **351**
  semilanatus 351
  silvaticus 349
  spadiceosporus 346
  spp. (unidentified) 342,
    350, 351
  sterquilinus 343, 346
  sulphureus 344
  tectisporus 351
  umbrinus 346
  variegatus 344
  xerophilus 344
Coprobia
  granulata 838
Cordyceps 878, **879**
  canadensis 881
  capitata **880**
  clavulata 880, 882
  entomorrhiza 880, 881
  gracilis 880, 881
  melolanthae 880, 882
  militaris **882,** 883*
  myrmecophila **881**
  ophioglossoides 880,
    881
  ravenelii 880, 882
  sphecocephala 880, 881
  unilateralis 880, 882
  washingtonensis 880,
    882
Coriolellus 603
  alaskanus 603
  carbonarius 602, 603

Coriolellus (cont.)
  heteromorphus 603
  sepium 603
  serialis 603
  suaveolens 598
Coriolus
  (also see *Trametes*)
  versicolor 594
Corticium 604
Cortinarius **417**
  acutus 427, 452
  adustus 423, 450
  aggregatus 422, 439
  albidus 423
  alboviolaceus **447**
  amethystinus 447
  amoenelens 441
  angulosus 426
  anomalus 425, 448
  argentatus 425, 448
  armeniacus 428, 451
  armillatus **448**
  atkinsonianus 440
  aurantiobasis 454
  aureifolius 425, 454
  aureofulvus 440
  balteato-cumatilis 434
  balteatus **433**
  biformis 428, 451
  bigelowii 443
  bivelus 451
  bolaris 425, 446
  boulderensis 423, 449
  brunneus 423, 428, 450,
    451
  builliardi 450
  bulbosus 451
  cacaocolor 428, 451
  caerulescens 439
  caesiifolius 447
  caesiocyaneus 422, 439
  caesiostramineus 438
  californicus 426, 454
  callisteus 427, 454
  calochrous 422, 441
  calyptratus 421, 444
  calyptrodermus 421,
    444
  camphoratus 425, 447
  caninus 425, 448
  castaneicolor 420, 422,
    431
  cedretorum **439**
  cinnabarinus 426, 455
  cinnamomeo-luteus 453
  cinnamomeus **453**
  citrinifolius 421, 436
  citrinipedes 441
  clandestinus 445

Cortinarius (cont.)
  claricolor 442
  cliduchus 442
  collinitus **431**
  corrugatus 420
  corrugis 423
  cotoneus **445**
  crassus 423, 434
  croceofolius 427, 454
  crocolitus 419*, 423, 432
  crystallinus 428, 429
  cyanites 424
  cyanopus 441
  cylindripes **430**
  damascenus 428, 451
  decipiens 427, 453
  delibutus 420, 431
  dibaphus 438
  dilutus 428, 451
  distans 428, 451
  duracinus 451
  elatior 421, 432
  elegantioides 441
  elegantior 441
  evernius **450**
  fasciatus 452
  flavifolius 426
  flavovirens 441
  fragrans 425, 447
  fulgens 441
  fulmineus **441**
  gentilis **444,** 896
  glaucopus **437**
  griseoluridus 420, 431
  griseoviolaceus 425, 431
  haematochelis 426, 449
  heliotropicus 430
  hemitrichus 428
  hercynicus 446
  herpeticus 441
  humboldtensis 453
  immixtus 436
  impennis 423, 450
  incisus 427, 452
  infractus **435**
  iodes 420, 430
  iodioides 430
  laniger **451**
  largus 434
  latus 442
  lucorum 424, 450
  lustratus 428
  luteoarmillatus 422,
    436
  magnivelatus **442**
  malachius 425, 448
  marylandensis 455
  metarius 441
  michiganensis 422, 439

Cortinarius (cont.)
  miniatopus 426, 450
  montanus 421, 440
  mucifluus 431
  mucosus **429**
  multiformis **442**
  mutabilis **437**
  nigrocuspidatus 427,
    452
  obtusus **452**
  occidentalis 437
  odorifer 422, 436
  olivaceopictus 453
  olympianus **439**
  orellanus 427, 444, 896
  orichalceus 421, 441
  osmophorus 436
  paleaceus 427, 452
  pallidifolius 421, 432
  percomis **436**
  phoeniceus **454**
  pholideus 424, 428,
    446
  pinetorum 428, 451
  plumiger 424, 428, 450
  ponderosus **432**
  prasinus 421, 441
  privignus 428, 451
  psammocephalus 427,
    452
  pseudoarquatus 438
  pseudobolaris 427
  pseudosalor 420, 430
  pulchellus 423, 453
  puniceus 455
  purpurascens 437
  pyriodorus 447
  rainierensis 426, 444
  raphanoides 453
  regalis **443**
  renidens 426
  rubicundulus 427
  rubripes **449**
  rufo-olivaceus 421, 441
  salor 430
  sanguineus **454**
  saturninus 450
  scandens 452
  scaurus **440**
  semisanguineus 426,
    454
  sodagnitus **438**
  sp. (unidentified) 423,
    441
  speciosissimus 427, 444
  sphaerosporus 431
  splendidus 430
  squamulosus **445**
  stemmatus 427, 452

Cortinarius (cont.)
  sterilis 420
  stillatitius 430
  subargentatus 425, 448
  subcuspidatus 452
  subflexipes 423, 453
  subfoetidus **434**
  subpulchrifolius 425,
    448
  subpurpurascens 437
  subpurpureophyllus
    422
  subpurpureus 424, 450
  subtestaceus 426, 449
  superbus 422
  thiersii 453
  torvus 423, 450
  traganus **447**
  triformis 428, 451
  triumphans 432
  trivialis 431
  turmalis 423, 442
  uliginosus 426
  uraceus 427, 428
  urbicus 424, 450
  vanduzerensis **432**,
    433*
  variicolor 422, 442
  varius 422, 442
  velatus 420, 443
  velicopia 422, 439
  verrucisporus 420, 443
  vibratilis **429**
  violaceus **446**
  virentophyllus 421, 441
  volvatus 422, 444
  washingtonensis 427,
    453
  wiebeae 443
  zakii 454
Coryne
  sarcoides 877
Cotylidia 604
  aurantiaca 608
  decolorans 608
  diaphana **608**
Craterellus 658
  caeruleofuscus 666
  calyculus 659, 668
  cantharellus 664
  cinereus **665**
  cornucopioides **666**,
    898*
  fallax 668
  foetidus 659, 666
  odoratus 664
  sinuosus 659, 666
Creolophus
  cirrhatus 612

Crepidotus **405**
  applanatus 405, 406
  calolepis 406
  cinnabarinus 405
  crocophyllus 405, 406
  fulvotomentosus 406
  fusisporus 406
  herbarum **405**
  maculans 406
  mollis **406**
  variabilis 405*, 406
  versutus 406
Crinipellis 202
  campanella 210
  piceae **210**
  stipitaria 210
  zonata 210
Crucibulum 778
  laeve **779**
  levis 779
  parvulum 779
  vulgare 779
Cryptoporus **584**
  volvatus **585**
Cudonia **872**, 893
  circinans 872*, **873**
  grisea 873
  lutea 873, 874
  monticola 873
Cyathipodia 805
Cyathus 778
  helenae 780
  olla 779, 780
  pygmaeus 779, 780
  stercoreus **780**
  striatus 779, 780
Cyptotrama
  chrysopeplum 131
Cystoderma **198**
  ambrosii 199
  amianthinum **200**
  carcharias 199
  cinnabarinum 199, 200
  fallax **199**
  granosum 199
  granulosum 199, 200
  gruberianum 199, 200

D

Dacrymyces
  deliquescens 670, 674
  palmatus 670, 674
Daedalea **586**
  berkeleyi 587, 588
  confragosa 589
  juniperina 587, 588
  mollis 596
  quercina **587**
  unicolor 589

Daedaleopsis 586
  ambigua 587, 589
  confragosa 586*, **588**
Daldinia 878, **885**
  concentrica 887
  grandis **887**
  vernicosa 887
Dasyscyphus 877
Datronia 592
  mollis **595**
Delastria 855
Dentinum **616**
  albidum 619
  albomagnum 619
  repandum **618**
  umbilicatum 617, 619
Dermocybe 418
  (also see *Cortinarius*)
  cinnamomea 453
  sanguinea 455
Destuntzia 748
  fusca 750
  rubra **750**
  saylorii 750
  solstitialis 750
  subborealis 750
Dictyocephalos **711**
  attenuatus **713**
Dictyophora 766
  duplicata 768, 771
  indusiata 767*, **770**
  multicolor 768, 771
  rubrovolvata 771
Discina **797**
  ancilis 799
  apiculatula 799
  leucoxantha 799
  macrospora 799
  olympiana 799
  perlata **798**
Disciotis 784, **796**
  venosa **796**
Disciseda **696**
  ater 698
  brandegeei 698
  candida **698**
  luteola 698
  pedicellata 698
  subterranea 698
Dissoderma
  paradoxum 198

E

Eccilia 245, 246, 249
Echinodontium **612**
  tinctorium **613**
Elaphomyces **862**
  granulatus **864**
  muricatus **863**

Elaphomyces (cont.)
  subviscidus 865
  variegatus 864
  verrucosum 864
Elasmomyces **736**
  camphorata 739
  odoratus 737, 738
  pilosus 738
  roseipes 738
  russuloides 737, 738
Elfvingia 577
Elvela 805
Endogone 844
  lactiflua 844
Endoptychum **727**
  agaricoides **731**
  arizonicum 732
  depressum **730**
Entoloma 238, **242**,
    895, 896
  (also see
    Leptonia, Nolanea)
  abortivum 242
  clypeatum 244
  ferruginans 245
  grayanum 244
  lividum 243, 244
  madidum **243**
  nidorosum **244**
  nitidum 242, 250
  pernitrosum 245
  prunuloides 243, 244
  rhodopolium **243**
  sinuatum 244
  speculum 243, 244
  strictius 246
  trachysporum 242, 250
  vernum 248
  violaceum 249
Exidia
  alba 671, 673
  glandulosa **672**
  nucleata 670, 673
  recisa 671, 674

F

Favolus
  alveolaris 563
Fayodia 151
  anthracobia 166
Fischerula
  subcaulis 845
Fistulina **553**
  hepatica 552*, **553**
  pallida 554
Flammula 407
Flammulina 202
  velutipes **220**
Floccularia 189, 194

Fomes 574
  (also see *Fomitopsis,
    Ganoderma, Phellinus*)
  annosus 579
  fomentarius 575, 581
  juniperinus 583
  officinalis 580
  pinicola 579
  subroseus 580
Fomitopsis **574**
  cajanderi **580**
  ellisianus 579
  fraxinophilus 579
  officinalis **579**
  pinicola **578**
  rosea 580
Fulvifomes
  juniperinus 583
Funalia
  hispida 568
  trogii 598
Fuscoboletinus **505**
  aeruginascens **507**
  glandulosus ·506
  grisellus 506, 507
  ochraceoroseus **506**
  paluster 506, 507
  serotinus 506, 507
  sinuspaulianus 506
  spectabilis 506, 507
  weaverae 506

G

Galera 399
Galerina **399**
  autumnalis **401**, 892
  cedretorum 402
  corneipes 369
  heterocystis **402**
  hypnorum 402
  marginata 400, 401,
    892
  mutabilis 395
  paludosa 400, 401
  semilanceata 402
  tibicystis 402
  triscopa 402
  venenata 400, 401, 892
Galeropsis **733**
  angusticeps 735
  cucullata 735
  polytrichoides 735
Galiella
  rufa 827
Ganoderma **574**
  adspersum 577
  annularis 577
  applanatum **576**
  brownii 577

Ganoderma (cont.)
  curtisii 574, 578
  europaeum 577
  lucidum **577**
  oregonense 575, 578
  tsugae 575, 578
Gastroboletus **544**
  amyloideus 544, 545
  scabrosus 544
  subalpinus **545**
  suilloides 544, 545
  turbinatus **544**
  xerocomoides 545
Gastrocybe 733
  lateritia 474, 733, 735
Gastrosporium
  simplex 696
Gautieria **746**
  candida 747
  gautierioides 746, 747
  graveolens 747
  monticola **747**
  morchelliformis 746,
   747
  parksiana 746, 747
  pterosperma 746, 747
Geaster 699
Geastrum **699**
  arenarium 705
  bryantii 702
  campestre 701, 705
  coronatum 702
  delicatus 705
  drummondii 700, 705
  fimbriatum 701, 704
  floriforme 700, 705
  fornicatum 699*, **701**
  indicum 703
  limbatum 700, 702,
   703*
  mammosum 700, 705
  minimum 700, 702
  nanum 700, 702
  pectinatum **702**
  pluriosteum 701, 704
  quadrifidum 700, 702
  recolligens 705
  rufescens 700, 702
  saccatum **703**
  schmidelii 702
  sessile 704
  smithii 702
  striatum 700, 702
  triplex **703**
  umbilicatum 705
  xerophilum 702
Genabea **849**
  cerebriformis **851**
  fragilis 849, 852
  spinospora 852

Genea **849**
  arenaria **851**
  cerebriformis 852
  compacta 849, 850
  gardneri 850
  harknessii **850**
  hispidula 851
  intermedia **851**
  kraspedestoma 849,
   851
Geoglossum **866**
  affine 867
  alveolatum 867
  difforme 867
  fallax 867
  glabrum 867
  glutinosum **866**
  intermedium 867
  nigritum 866, 867
  simile 867
Geopora 841, **846**
  arenicola **847**, 935*
  arenosa 846, 847
  aurantia 846, 847
  clausa 846, 847
  cooperi **846**
  harknessii 847
  longii 847
  magnata 847
  pellita 846, 847
Geopyxis
  (also see *Tarzetta*)
  carbonaria 835, 840
  vulcanalis **840**
Gerronema 221
Globifomes
  graveolens 567
Gloeocantharellus
  purpurascens 660
Gloeophyllum 586
  saepiarium 588*, **590**
  striatum 590
  trabeum 586, 590
Gloeoporus
  adustus 596
  dichrous 552
Glomus 844
Gomphidius **481**
  glutinosus 482
  largus 482
  maculatus 482
  nigricans 482
  oregonensis **482**
  roseus 483
  smithii 482, 483
  subroseus **483**
  viscidus 486
Gomphogaster
  leucosarx 733

Gomphus 658
  bonari 661, 662
  clavatus **661**
  floccosus **661**, 895
  kauffmanii 660, 662
  pseudoclavatus 661
Grifola 554
  frondosa **564**
  sulphurea 573
  umbellata 556, 564*,
   565
Guepiniopsis
  alpinus **674**
  chrysocomus 674
Gymnomyces 742
  cinnamomeus 744
  ferruginascens 743, 744
  roseomaculatus 742,
   744
  socialis **743**
Gymnopilus **407**, 895
  aeruginosus **409**
  bellulus 407, 408
  flavidellus 407, 408
  fulvosquamulosus 407,
   410
  harmoge 409
  junonius 411
  liquiritiae 407, 408
  luteocarneus 407, 408
  luteofolius **409**
  parvisquamulosus 407,
   410
  penetrans 407, 408
  pulchrifolius 410
  punctifolius 407, 409
  sapineus **408**
  sp. (unidentified) 408
  spectabilis **410**
  subspectabilis 411
  terrestris 407, 408
  validipes 411
  ventricosus 411
Gyrodon
  lividus 490
  merulioides 490
Gyromitra 796, **799**,
  893
  ambigua 803
  brunnea 800, 803
  californica **804**
  caroliniana 800, 802
  esculenta **801**, 893
  fastigiata 801, 803
  gigas **800**
  infula **802**, 893
  korfii 801
  melaleucoides 799
  montana 801
  sphaerospora 800, 804
  underwoodii 803

Gyrophragmium 727
  californicum 727, 728
  texensis 730
Gyroporus **510**
  castaneus **510**
  cyanescens 510
  purpurinus 510
  subalbellus 510

H

Haematostereum
  sanguinolentum 606
Hapalopilus
  nidulans 568
Hebeloma **463**, 895,
  896
  albidulum 465
  crustuliniforme **464**,
    466*
  fastibile 466
  hiemale 465
  insigne 465
  mesophaeum **465**
  sacchariolens 464, 465
  sarcophyllum 463, 465
  sinapizans **465**
  strophosum 463, 466
  syriense 463
Helvella 796, **805**, 893
  (also see *Gyromitra*)
  acetabulum **807**
  albella 807, 811*, 812
  albipes 813
  atra **813**
  californica 804
  compressa **811**
  connivens 812
  corium 805, 810
  costifera 806, 808
  crassitunicata 809
  crispa **816**
  cupuliformis 806, 810
  elastica **813**
  ephippium 807, 814
  fusca 814
  griseoalba 806, 808
  infula 803
  lactea 816
  lacunosa **815**
  leucomelaena **808**
  leucopus **812**
  macropus **810**
  maculata **814**
  mitra 815
  monachella 813
  pallidula 810
  palustris 815
  pezizoides 807, 814
  philonotis 815
  phlebophora 806, 816

Helvella (cont.)
  queletii **809**
  solitaria 809
  stevensii 807, 812
  subglabra 807, 814
  sulcata 806, 815
  villosa 805, 806, 810
Hericium 611, **613**
  abietis **614**
  americanum 614
  coralloides 614
  erinaceus **615**
  laciniatum 615
  ramosum **615**
  weirii 614
Heterobasidion 574
  annosum 575, 579
Heteroporus 554
  biennis **566**
Hirneola
  auricula-judae 675
Hirschioporus
  abietinus 593
  pargamenus 593
Hohenbuehelia 132
  angustatus 137
  atrocaerulea 133, 137
  geogenia 137
  mastrucatus 133, 137
  petaloides **136**
Humaria
  hemispherica **839**
Hydnangium **744**
  carneum **745**
  roseum 745
  soederstroemii 745
Hydnellum 6̣22
  aurantiacum **626**
  caeruleum **625**
  complectipes 626
  concrescens 628
  conigenum 626
  cruentum 624
  cumulatum 628
  cyanopodium 623, 625
  diabolum 627
  earlianum 626
  ferrugipes 625, 626
  geogenium 623
  mirabile 624
  nigellum 623, 629
  peckii **627**
  pineticola 624, 627
  piperatum 624
  regium 623, 625
  scleropodium 623
  scrobiculatum **627**
  spongiosipes 624
  suaveolens **624**
  subsuccosum 628
  zonatum 624, 628

Hydnobolites 855
  californicus **857**
  cerebriformis 857
Hydnocystis
  californica 847
Hydnopolyporus
  palmatus 555, 605
Hydnotrya **848**
  cerebriformis 848, 849
  cubispora 848, 849
  ellipsospora 825
  michaelis 849
  tulasnei 848, 849
  variiformis **848**
  yukonensis 849
Hydnotryopsis
  compactus 845
  setchellii 845
Hydnum 611, **616**
  (also see *Dentinum*)
  calvatum **621**
  crassum 621
  cristatum 617
  cyanellum 617, 622
  fennicum 620
  fuligineo-violaceum
    617, 622
  fumosum 618
  fuscoindicum 621*, **622**
  imbricatum **619**
  laevigatum 618, 621
  martioflavum 621
  repandum 618
  rimosum 618, 620
  scabrosum **620**
  stereosarcinon 618
  subincarnatum 617,
    618, 620
Hygrocybe **103**
  acuta 107, 112
  acutoconica **115**
  albinella 105, 109
  atro-olivacea 112
  aurantiolutescens 116
  aurantiosplendens 115
  caerulescens 112
  caespitosa 105
  calyptraeformis **117**
  cantharellus 106, 113
  ceracea 115
  chlorophana 106, 115
  citrinopallida 115
  coccinea **114**
  conica **116**
  cuspidata 106, 116
  flavescens **115**
  flavifolia 115
  fornicata 109
  laeta 107, 119
  laetissima 106, 115
  langei 116
  marchii 114

Hygrocybe (cont.)
  marginata **112**
  miniata **113**
  minutula 114
  moseri 113
  nigrescens 117
  nitida 106, 115
  nitrata 112
  olivaceoniger 117
  ovina 112
  parvula 106, 115
  persistens 116
  psittacina **118**
  punicea **114**
  pura 104
  purpureofolia 112
  reai 105, 114
  ruber 105
  singeri 117
  splendidissima 115
  squamulosa 113
  subaustralis 105, 109
  subminiata 106, 113
  subminutula 114
  turunda 105, 113
  unguinosa 106, 119
  virescens **118**
Hygrophoropsis 476
  aurantiaca **479**
  olida 476, 480
Hygrophorus **103**
  (also see *Camaro-*
    phyllus, Hygrocybe)
  acutoconicus 116
  agathosmus **128**
  albicastaneus 108, 125
  amarus 108, 124
  bakerensis **126**
  borealis 109
  brunneus 125
  calophyllus **129**
  calyptraeformis 118
  camarophyllus 107, 129
  capreolarius 108, 124
  chrysaspis 121
  chrysodon **119**
  coccineus 114
  conicus 117
  cossus 121
  discoideus 125
  eburneus **119**
  erubescens 108, 124
  flavescens 115
  flavodiscus 121
  fuligineus 107, 127
  fuscoalbus 107, 127
  gliocyclus **120**
  glutinosus 121
  goetzii 108, 125
  hypothejus **126**
  inocybiformis 107, 128
  kauffmanii 108

Hygrophorus (cont.)
  langei 116
  laurae 107, 125
  limacinus 127
  marginatus **113**
  marianae 107
  marzuolus 129
  megasporus 127
  miniatus 113
  monticola 126
  morrisii 128
  nemoreus 108, 111
  occidentalis 127, 128
  odoratus 128
  olivaceoalbus **127**
  pacificus 108, 126
  paludosus 107
  penarius 123
  perfumus 123
  persoonii 128
  piceae 105, 120
  ponderatus 104, 123
  pratensis 110
  psittacinus 119
  pudorinus **124**
  puniceus 114
  purpurascens **124**
  pusillus 110
  pustulatus 128
  pyrophilus 126
  recurvatus 112
  roseibrunneus **125**
  russula **123**
  saxatilis 108, 125
  sordidus **122**
  speciosus **126**
  subalpinus 109*, **121**
  subpungens 125
  subsalmonius 107
  subviolaceus 112
  tennesseensis 125
  tephroleucus 128
  variicolor 107, 126
  vernalis 125
  vinicolor 126
  whiteii 121
Hymenochaete
  rubiginosa 606
  tabacina 606
Hymenogaster **748**
  albellus 749
  albus 748, 749
  brunnescens 749
  diabolus 749
  gilkeyae 749
  luteus 749
  mcmurphyi 749
  parksii 748, 749, 750*
  pyriformis 751
  ruber 750
  sublilacinus **749**

Hymenogaster (cont.)
  subochraceus 749
  utriculatus 748, 749
Hypholoma 361, 381
  (also see *Naematoloma*)
  capnoides 383
  dispersum 384
  fasciculare 383
  incertum 363
Hypomyces 878, **882**
  cervinigenus 883, 884
  chrysospermum **883**
  hyalinus 883, 884
  lactifluorum **884**
  luteovirens 883, 884
  transformans 883, 884
Hypsizygus
  tessulatus 133
Hysterangium **762**
  aureum 763
  clathroides 763
  coriaceum 763
  crassirhachis 763
  crassum 763
  darkeri 763
  fischeri 764
  fuscum 763, 764
  occidentale 763
  separabile **763**
  setchellii 763
  sp. (unidentified) 763
  stoloniferum 763

IJK

Ileodictyon
  cibarium 774
Inocybe **455**, 894
  agardhii 462
  albodisca 456, 459
  bongardii 459
  caesariata 457, 462
  calamistrata **462**
  cookei 457, 459
  corydalina 456, 459
  fastigiata 456, 457
  flocculosa 457, 462
  fuscodisca 456
  geophylla **460**
  godeyi 459
  hirtella 459
  hystrix 457, 462
  jurana **458**
  lacera 457, 462
  laetior 456, 458
  lanatodisca 456, 459
  lanuginosa **462**
  leucoblema 456
  leucomelaena 457
  lilacina **461**
  maculata **458**
  mixtilis 457, 458
  napipes 457

Inocybe (cont.)
  oblectabilis 456, 458
  obscurioides 461
  olympiana 457
  picrosma 457
  pudica **460**
  pyriodora **459**
  serotina 459
  sindonia 459
  sororia **457**
  sp. (unidentified) 457,
    462
  suaveolens 456, 459
  terrigena 457, 462
Inonotus **566**
  andersonii 567
  arizonicus 569
  circinatus 570
  cuticularis 569
  dryadeus 567, 569, 570
  dryophilus 569
  hispidus **569**
  obliquus 567
  radiatus 569
  texanus 569
  tomentosus **569**
Irpex
  lacteus 598
  mollis 601
Ischnoderma **573**
  benzoinum 573
  resinosum **573**
Ithyphallus 766
Jafnea
  semitosta 834, 840
Kuehneromyces
  mutabilis 395

L

Laccaria **171**
  altaica 172
  amethystea 173
  amethysteo-occidentalis
    173
  amethystina **172**
  bicolor 172, 173
  laccata **172**
  ochropurpurea 172, 173
  ohiensis 172
  proxima 172
  striatula 172
  tortilis 172
  trullisata 172, 173
Lacrymaria 366
  velutina 366
Lactarius 63, **64**, 895
  affinis 66
  allardii 67
  alnicola **71**

Lactarius (cont.)
  alpinus 67, 82
  aquifluus 67
  argillaceifolius 63*, **76**
  aspideoides 65, 76
  atrobadius 80
  atroviridis 66, 70
  aurantiacus 79
  barrowsii 65, 69
  caespitosus 76
  californiensis 75
  camphoratus 81
  cascadensis 75
  chelidonium 65, 69
  chrysorheus 65, 74
  cinereus 77
  circellatus 67, 77
  cocosiolens 67, 79
  controversus **70**
  corrugis 66, 78
  croceus 65, 75
  deceptivus 66, 71
  deliciosus **68**
  fallax **77**
  fragilis **80**
  fuliginellus 78
  fumosus 78
  gerardii 78
  glutigriseus 77
  glyciosmus 66
  griseus 67
  helvus 67
  hemicyaneus 69
  hepaticus 80, 82
  herpeticus 82
  hygrophoroides 67, 78
  hysginus 80
  indigo **69**
  insulsus 72
  kauffmanii 67, 77
  lignyotus 65, 78
  luculentus 79
  luteolus 66, 78
  maculatipes 75
  maculatus 75
  manzanitae 67, 80
  mucidus 77
  necator 70
  neuhoffii 71
  occidentalis 67
  oculatus 67, 82
  olivaceoumbrinus **70**
  olympianus 66, 72
  pallescens 65, 75
  pallidiolivaceus 66
  paradoxus 65, 69
  payettensis 66, 72
  peckii 67
  piperatus **71**
  psammicola 66, 72, 73

Lactarius (cont.)
  pseudodeceptivus 71
  pseudodeliciosus 68
  pseudomucidus **77**
  pubescens 41*, **73**
  representaneus **75**
  resimus 65, 74
  riparius 80
  rubrilacteus **68**
  rufulus **82**
  rufus **79**
  salmoneus 65, 68
  sanguifluus 69
  scrobiculatus **73**
  sordidus 70
  speciosus 76
  subdulcis 82
  subflammeus **79**
  subpalustris 75
  subplinthogalus 66
  subpurpureus 65, 69
  subserifluus 67, 82
  substriatus 79
  subvellereus 66, 71
  subvernalis 66
  subvillosus 66, 73
  subviscidus **80**
  thejogalus 67, 82
  thiersii 82
  thyinos 68
  tomentoso-marginatus
    71
  torminosus **73**
  trivialis 77
  uvidus **75**, 897*
  vietus 76
  vinaceorufescens **74**
  volemus **78**, 79*
  xanthogalactus 75
  yazooensis 66, 72
  zonarius 72
Laetiporus **572**
  persicinus 573
  sulphureus **572**, 895,
    896
Lamprospora 839
Langermannia
  gigantea 683
Laricifomes
  officinalis 580
Laternea
  columnata 773
  triscapa 773
Laxitextum
  bicolor 607
Leccinum **536**
  aeneum 540
  alaskanum 541
  albellum 537, 542
  arbuticola 539

Leccinum (cont.)
  arctostaphylos 540
  armeniacum 538, 540
  atrostipitatum 538, 541
  aurantiacum 537, 539
    541
  "aurantioscaber" 538,
    539
  brunneum 538, 540
  californicum 537, 542
  chromapes 534
  cinnamomeum 538
  clavatum 538
  constans 538, 540
  cretaceum 542
  crocipodium 539
  discolor 537, 539, 541
  fallax 539, 541
  fibrillosum 538, 540
  griseonigrum 538, 541
  griseum 538, 542
  holopus 537, 542
  idahoensis 540
  incarnatum 538
  insigne **540**
  largentii 540
  manzanitae **539**
  montanum 538, 541
  ponderosum 538, 540
  potteri 538
  roseofracta 542
  rotundifoliae 542
  rufescentoides 537
  rugosiceps 539
  scabrum **541**
  snellii 538, 541
  subalpinum 539, 541
  testaceoscabrum 538,
    541
  vulpinum 538
Lentaria 645
  byssiseda 644, 646, 649
  pinicola 646, 649
Lentinellus **141**
  cochleatus 142
  crinitis 133
  flabelliformis 144
  montanus 144
  omphalodes 142
  ursinus **144**
  vulpinus 144
Lentinus **141**
  detonsus 141
  edodes 31*, 141
  kauffmanii 142
  lepideus **142**
  ponderosus 43*, 142*,
    143
  sulcatus 142
  tigrinus 141

Lenzites **586**
  (also see *Gloeophyllum*)
  betulina **589**
  saepiaria 590
Leotia **872**
  albiceps 873
  atrovirens 873, 875
  chlorocephala 875
  lubrica **874**
  viscosa **874**
Lepiota **293**
  acutesquamosa 294,
    303
  americana **301**
  asperula 294, 303
  atrodisca **304**
  badhamii 301
  barssii **303**
  birnbaumii 302
  brebissonii 302
  breviramus 302
  brunnescens 305
  bucknallii 295
  castanea **307**, 892
  castaneidisca 307
  cepaestipes **301**
  clypeolaria **309**
  clypeolarioides 310
  cortinarius 309
  cristata **306**
  decorata 307
  eriophora **303**
  excoriata 303
  felina 295, 309
  flammeatincta **304**
  flavescens 302
  fragilissimus 302
  glatfelteri 306
  helveola 309, 892
  hispida 303
  humei 294
  josserandii **308**, 892
  leucothites 300
  lilacinogranulosa 294
  longistriatus 302
  lutea **302**
  luteophylla 295, 302
  molybdites 297
  morgani 297
  naucina **299**, 896
  naucinoides 300
  procera **298**
  pulcherrima 291
  rachodes **297**, 896
  rhacodes 298
  roseatincta 305
  roseifolia 295, 305
  roseilivida 295, 306, 307
  rubrotincta **305**
  sanguiflua 295

Lepiota (cont.)
  scabrivelata 303
  seminuda 306*, **307**
  sequoiarum **307**
  sistrata 307
  spp. (unidentified) 294,
    305, 307
  subincarnata 309
  tinctoria 295, 301
  "tomentodisca" 307
  ventriosospora 310
Lepista 148
  (also see *Clitocybe*)
  inversa 157
  irina 155
  nuda 153*, 154
  saeva 154
  sp. (unidentified) 154
  tarda 152
Leptoglossum 132
Leptonia **248**
  asprella 252
  carnea **250**
  convexa 250
  corvina 251
  cupressa 252
  cyanea 249, 250
  cyaneonita 250
  decolorans 251
  diversa 251
  exalbida 249, 252
  fuligineo-marginata
    249, 252
  gracilipes **252**
  incana 249
  jubata 246
  nigra 249, 252
  nigroviolacea **250**
  occidentalis 249, 250
  parva **251**
  porphyrophaea 249
  rectangula 251
  rosea 249
  sericella 253
  serrulata 249, 252
  undulatella 249, 252
  vinaceobrunnea 252
  violaceonigra 250
  zanthophylla 249, 251
Leptopodia 805
Leucoagaricus 293
  (also see *Lepiota*)
  naucinus 300
  procerus 298
  rachodes 298
Leucocoprinus 293,
    302
  (also see *Lepiota*)
  birnbaumii 302

Leucocoprinus (cont.)
  brebissonii 302
  breviramus 302
  cepaestipes 302
  flavescens 302
  fragilissimus 302
  lilacinogranulosus 294
  longistriatus 302
  luteus 302
Leucogaster **759**
  carolinianus 760
  odoratus 760
  rubescens 760
Leucopaxillus **166**
  albissimus 166*, **167**
  amarus **168**, 918*
  candidus 159
  gentianeus 168
  giganteus 159
  laterarius 168
  paradoxus 168
  septentrionalis 159
  tricolor 167
Leucophleps **759**
  magnata 760
  spinospora **760**
Leucoscypha
  rutilans 837
Limacella 262, **291**
  glioderma **291**
  glischra 291, 292
  guttata 291
  illinita **292**
  kauffmanii 291, 292
  lenticularis 291
  roseicremea 291, 292
  solidipes 291, 292
Longia
  texensis 730
Longula 727
  texensis **729**
Lycoperdon 677*, **690**
  americanum 691, 694
  candidum 695
  coloratum 698
  curtisii 695, 696
  echinatum 694
  foetidum **692**
  gemmatum 694
  hiemale 696
  marginatum **694**
  molle 694
  muscorum 694
  nettyana 691, 695
  nigrescens 693
  peckii 691, 694
  pedicellatum 692
  perlatum **693**
  pulcherrimum **694**

Lycoperdon (cont.)
  pusillum 698
  pyriforme **691**
  rimulatum 694
  subincarnatum 690
  umbrinum 691, 694
Lyophyllum **173**
  atratum 166, 174
  carneum 176
  connatum 175
  decastes **174**, 898*
  infumatum 176
  loricatum 175
  montanum 46*, **175**
  palustre 174
  rancidum 174
  semitale 174, 176
  sp. (unidentified) 175
Lysurus 772
  borealis 778
  cruciatus **777**
  gardneri 778
  mokusin **776**
  periphragmoides **776**

M

Macowanites **736**
  alpinus 738
  americanus **737**
  chlorinosmus 737, 738
  iodiolens 737, 738
  luteolus 737, 738
  magnus 737, 738
  roseipes 738
  subolivaceus 738
  subrosaceus 738
Macrocystidia
  cucumis 131
Macrolepiota 293
  (also see *Lepiota*)
  procera 298
  rachodes 298
Macropodia 805
Macroscyphus 805
Macrotyphula 634
  fistulosa 635, 637
  juncea **636**
Marasmiellus 202
  albuscorticis 206
  candidus **206**
  nigripes 203, 206
Marasmius **201**
  albogriseus 204, 209
  alliaceus 208
  androsaceus **208**
  bellipes 210
  borealis 210
  candidus 206
  capillaris 208
  cohaerens 204

Marasmius (cont.)
  copelandi **207**
  delectans 203, 206
  epiphyllus 206
  foetidus 203 .
  fulvoferrugineus 210
  fuscopurpureus 214
  haematocephalus 210
  iocephalus 202
  magnisporus 206
  nigrodiscus 205, 209
  olidus 208
  oreades 7*, **208**
  pallidocephalus 208
  plicatulus **209**
  prasiosmus 208
  quercophyllus 206
  rotula 203, 206
  scorodonius 202, 208
  siccus 204, 210
  sp. (unidentified) **206**
  strictipes 205, 209
  thujinus 208
  umbilicatus 203
  urens 213
Martellia **742**
  boozeri 743
  brunnescens **743**
  californica 743
  cremea 742, 743
  ellipsospora 743
  fallax 743
  foetens 743
  parksii 743
Melanogaster **756**
  ambiguus 759
  euryspermus **758**
  intermedius 759
  macrocarpus 759
  parksii 758, 759
  variegatus 757*, 759
Melanoleuca **169**
  alboflavida 169
  brevipes 170
  cognata **170**
  evenosa **171**, 897*
  graminicola 169, 170
  lewisii 169
  melaleuca **169**
  polioleuca 170
  subalpina 171
  vulgaris 170
Melanophyllum
  echinatum 317
Melanotus
  textilis 405
Melastiza
  chateri 834, 839
Meripilus
  giganteus 555, 565

Merulius
  incarnatus 605, 611
  tremellosus 605, 610*, 611
Microcollybia 202, 212
Microglossum **868**
  atropurpureum 867, 869
  fumosum 867
  olivaceum 867, 869, 870
  rufum 871
  viride **870**
Micromphale 202
  arbuticola 202, 208
  foetidum 203
  penetrans 208
  sequoiae 204, 208
Microporellus
  dealbatus 563
  obovatus 563
Microstoma
  floccosa 835, 836
  protacta 835, 836
Mitrophora
  semilibera 792
Mitrula **868**
  abietis **870**
  borealis 870
  elegans 869, 870
  gracilis 870
  irregularis 871
  lunulatospora 870
  paludosa 870
Mollisia 877
Montagnea 727
  arenarius **727**
Morchella 784, **785**
  angusticeps 791
  conica 791
  crassipes 787, 788, 792*
  deliciosa 788*, **789**
  elata 31*, 786*. **790**
    959*
  esculenta **787**
  hybrida 792
  semilibera **791**
  sp. (unidentified) 787, 789
Morganella 690
  subincarnata 690
Mucilopilus
  conicus 533
Mucronoporus
  tomentosus 570
Multiclavula 634
  mucida **636**
  vernalis 636
Mutinus 766
  bovinus 771
  caninus **771**
  curtisii 771
  elegans 768, 771

Mycena **224**
  abramsii 235
  acicula **228**
  adonis 228
  albidula 227
  alcalina **234**
  amabilissima 226, 228
  amicta 226, 231
  atroalboides 226, 233, 235
  aurantiomarginata 225, 228
  capillaripes **229**
  capillaris **227**
  citrinomarginata 225, 229
  clavicularis 237
  clavularis **227**, 229*
  corticola 225, 227
  delicatella 227
  elegans 229
  elegantula 226, 230
  epipterygia **237**
  epipterygioides 237
  fibula 221
  filopes 235
  galericulata **235**
  galopus **232**
  griseoviridis 225, 237
  haematopus **231**
  ignobilis 227
  inclinata 235
  iodiolens 234
  juncicola 227
  laevigata 234
  latifolia 235
  leaiana 225, 236
  leptocephala 226, 234
  lilacifolia **236**, 237*
  luteopallens 226, 228
  maculata **235**, 236*
  madronicola 225, 227
  metata 234
  monticola 228
  murina **234**
  occidentalis 226, 234
  olivaceobrunnea 229
  oregonensis 228
  osmundicola 227
  overholtsii 226, 234
  parabolica 234
  paucilamellata 227
  pelianthina 226, 231
  pseudotenax 235
  pura **230**
  purpureofusca **229**
  rorida **237**
  rosella 226, 229
  rubromarginata 230
  rugulosiceps 235
  rutilantiformis 231
  sanguinolenta **232**
  scabripes **233**

Mycena (cont.)
  sanguinolenta **232**
  scabripes **233**
  spp. (unidentified) 204, 209, 226
  stannea 235
  strobilinoides **228**
  stylobates 227
  subcaerulea 226, 231
  subcana **233**
  subsanguinolenta 225, 232
  tenax 237
  tenerrima 227
  viscosa 237
  vitilis 236
  vulgaris 225, 237
Mycenastrum 680
  corium **689**
Mycolevis
  siccigleba 741
Mycorraphium
  adustum 612
Myrioscletorinia 877
Myriostoma 699
  coliforme **704**
Myrmecocystis
  cerebriformis 852
Myxomphalia 149
  maura **165**

N

Naematoloma **381**
  aurantiaca **382**
  capnoides **383**
  dispersum 383, **384**

  elongatum 384
  ericaeum 384
  fasciculare **382**, 896
  myosotis 384
  olivaceotinctum 384
  polytrichi 384
  radicosum 383
  squalidellum 384
  sublateritium 381, 382
  subviride 383
  udum 384
Nannfeldtiella
  aggregata 828
Naucoria 399
  semiorbicularis 469
  vinicolor **404**
Neobulgaria
  pura 875, 877
Neocudoniella
  albiceps 873
Neogyromitra
  gigas 801

Neolecta 868
 irregularis **871**
 vitellina 871
Neosecotium 727
 macrosporum 732
Neottiella
 rutilans 837
Neournula
 pouchetii 827
Nidula 778
 candida **780**
 niveotomentosa 779,
 781
Nidularia 778
 farcta 779
 pulvinata 779
Nivatogastrium **733**
 nubigenum **735**
 wrightii 736
Nolanea **245**
 cuneata 248
 edulis 247
 fructufragrans 245,
 248
 hirtipes 246, 247
 holoconiota 248
 icterina 245, 248
 mammosa 248
 murraii 245
 papillata 248
 salmonea 245
 sericea **246**
 staurospora 246, 247,
 248
 stricta **246**
 verna **247**
Nyctalis 200

O

Octavianina 744
 asterosperma **745**
 macrospora 746
 papyracea 746
 rogersii 746
Omphalia 221
Omphaliaster 151
 asterosporus 166
 borealis 166
Omphalina **221**
 chlorocyanea 162
 chrysophylla 221
 epichysium 166
 ericetorum **223**
 fibula 221
 grossula 162
 hudsoniana 223
 luteicolor 221, 223
 postii 223
 pyxidata 223
 strombodes 221
 umbellifera 223
 wynniae 162, 221

Omphalotus **146**, 894,
 896
 illudens 148
 olearius 147, 148
 olivascens **147**
Onnia
 tomentosa 570
Osmoporus
 odoratus 586, 590
Osteina
 obducta 555
 ossea 555
Otidea **831**
 abietina 832
 alutacea **832**, 833*
 auricula 832
 bufonia 832
 cantharella 832, 833
 concinna 832, 833
 grandis 832
 leporina 832, 833
 onotica **832**
 rainierensis 832
 smithii 832
Oudemansiella 202
 longipes 220
 platyphylla 146
 radicata 218*, **219**
Oxyporus
 nobilissimus 549, 575
 populinus 575

P

Pachyella 818, 822
 babingtonii 819, 822
 clypeata 819, 822
Pachyphloeus 855
 citrinus **856**
 conglomeratus 857
 melanoxanthus 855,
 857
 virescens 857
Panaeolus **353**, 895
 acuminatus 354, 357,
 360
 cambodginensis 358
 campanulatus **356**
 castaneifolius 360
 cyanescens **358**, 895
 fimicola 354, 357
 foenisecii **360**
 papilionaceus 357
 phalaenarum 355
 retirugis 354, 357
 rickenii 357
 semiovatus **355**, 357*
 separatus 355, 357*
 sepulchralis 355

Panaeolus (cont.)
 solidipes **355**
 sphinctrinus 357
 subbalteatus **358**, 895
 tropicalis 358
Panellus 132
 longinquus 133
 mitis 132
 nidulans 141
 ringens 132
 serotinus **137**
 stipticus **138**
Panus 132
 conchatus **138**
 crinitis 133
 dryinus 136
 operculatus 132
 rudis **139**
 strigosus **140**
 torulosus 138
Paxillus 476
 atrotomentosus **478**
 involutus 41*, **477**, 896
 panuoides **476**
 vernalis 478
Paxina 805
 acetabulum 807
 leucomelas 809
 recurvum 799
Penicillium 782
Peniophora 604
 gigantea 607
Perenniporia
 subacida 603
Peziza **818**
 (also see *Pachyella,*
 Plicaria, Geopyxis,
 Tarzetta)
 ammophila 819, 826
 badia 819, 821
 badioconfusa 821, 822
 brunneoatra **820**
 cerea 821
 domiciliana **822**
 echinospora 821
 ellipsospora 822*, 825
 emileia 822
 fimeti 823
 gautierioides 825
 melaleucoides 799
 petersii 821
 praetervisa 824
 proteana **824**
 pustulata 821
 repanda **821**
 sp. (unidentified) 822*,
 824
 stuntzii 825

Peziza (cont.)
  succosa 819, 821
  sylvestris 820*, **821**
  varia 822
  venosa 796
  vesiculosa **823**
  violacea **824**
Phaeobulgaria
  inquinans 876
Phaeocollybia **413**
  attenuata 414, 415
  californica **415**
  christianae 416
  deceptiva 414, 415
  dissiliens 415
  fallax 413, 414
  festiva 413, 414
  gregaria 414, 415
  jennyae 416
  kauffmanii **416**
  laterarius 414, 415
  lilacifolia 414, 415
  olivacea **414**
  oregonensis 416
  piceae 415
  pseudofestiva 413, 414
  radicata 414, 416
  rufipes 416
  scatesiae 414, 415
  similis 414, 415
  sipei 416
  spadicea 414, 416
Phaeolepiota **411**
  aurea **412**
Phaeolus **566**
  alboluteus **571**
  fibrillosus 567, 572
  schweinitzii **570**
Phaeomarasmius
  confragosus 404
  erinaceellus 386
Phallogaster 762
  saccatus 762, 764
Phallus 766
  hadriani 769*, 770
  imperialis 770
  impudicus **768**
  iosmus 770
  ravenelii 768, 770
  rubicundus 768, 770
Phellinus **574**
  chrysoloma 583
  everhartii 575, 582,
    583*
  ferreus 575, 582
  ferruginosus 575, 582
  gilvus **582**
  igniarius **581**
  laevigatus 575, 581

Phellinus (cont.)
  nigrolimitatus 583
  pini **582**
  pomaceus 575, 581
  rimosus 575, 582
  robiniae 582
  robustus 575, 582, 583
  taxodii 583
  texanus 583
  torulosus 583
  tremulae 581
Phellodon **622**
  atratus **629**
  confluens 624, 628
  melaleucus 623, 629
  niger 623, 629
  tomentosus **628**
Phellorina 715
  inquinans 723
  strobilina **723**
Phillipsia 836
Phlebia
  merismoides 610
  radiata **610**
Phlegmacium 418
Phlogiotis
  helvelloides **672**
Pholiota **384**
  abietis 391
  adiposa 391
  albivelata 380, 387
  albocrenulata **392**
  alnicola 388
  astragalina **387**
  aurea 413
  aurivella **390**
  aurivelloides 391
  brunnescens **393**
  caperata 412
  carbonaria 385, 394
  confragosa 404
  connata 391
  decorata 387, 393
  destruens **395**
  elongatipes 384, 387
  erinaceella 386
  ferruginea 393
  ferrugineo-lutescens
    393
  fibrillosipes 388
  filamentosa 391
  flammans **391**
  flavida 388
  fulvosquamosa 381
  fulvozonata 394
  graveolens 394
  hiemalis 391
  highlandensis 385, 394
  lenta 387, 393
  limonella 391

Pholiota (cont.)
  lubrica **392**
  malicola **388**
  multifolia 386, 388
  mutabilis **395**
  myosotis 384, 387
  polychroa 387
  praecox 469
  prolixa 388
  scamba 387, 394
  sp. (unidentified) 385
  spectabilis 411
  spinulifera 388
  spumosa **394**
  squarrosa **389**
  squarroso-adiposa 391
  squarrosoides 385*,
    386, 390
  subangularis 394
  subcaerulea 380, 385
  sublubrica 393
  subochracea 388
  terrestris **389**
  velaglutinosa 387, 393
  vermiflua 470
  vernalis 387, 396
Pholiotina
  filaris 472
Phylloporus 476
  arenicola 480
  rhodoxanthus **480**
Phyllotopsis 132
  nidulans **140**
Physalacria
  inflata 548
Picoa 852
  carthusiana **854**
Piersonia
  alveolata 858
  bispora 858
Piptoporus **584**
  betulinus **584**
Pisolithus **711**
  arenarius 713
  arrhizus 713
  tinctorius **712**
Pithya
  cupressina 836
  vulgaris 834, 836
Plectania 826
  coccinea 836
  hiemalis 836
  melaena 827, 831
  melastoma **829**
  milleri 830
  nannfeldtii **830**
  nigrella 830
Pleurocybella
  porrigens 136

Pleurotellus
  porrigens 136
Pleurotus **132**
  (also see *Hohen-
  buehelia*)
  candidissimus 132
  columbinus 134
  cornucopiae 134
  corticatus 136
  dryinus 135*, **136**
  elongatipes 133
  lignatilis 136
  ostreatus 133*, **134**
  petaloides 137
  porrigens **135**
  sajor-cajou 134
  sapidus 134
  ulmarius 133
Plicaria 818
  carbonaria 820
  endocarpoides **820**
  leiocarpa 820
  trachycarpa 820
Plicaturopsis
  crispa 591
Pluteolus
  aleuriatus 475
  callisteus 475
Pluteus 253, **254**
  admirabilis 254, 257
  atricapillus 256
  atromarginatus 254,
    256
  aurantiorugosus 254,
    257
  californicus 258
  cervinus **255**
  chrysophaeus 258
  coccineus 257
  cyanopus 254, 258, 895
  flavofuligineus **258**
  granularis 258
  leoninus 254, 257
  longistriatus **257**
  lutescens **257**
  magnus 256
  nanus 257
  pellitus 254, 256
  petasatus **255**
  salicinus 254, 258, 895
  seticeps 258
Podaxis 715, **725**
  argentinus 726
  longii 726
  microsporus 726
  pistillaris 724, **725**
Podostroma **879**
  alutaceum 879

Polyozellus 658
  multiplex **668**
Polypilus
  frondosus 565
Polyporus **554**
  (also see *Albatrellus,
  Meripilus, Osteina,
  Ischnoderma*, etc.)
  arcularius **563**
  badius 42*, **562**
  betulinus 585
  biennis 566
  brumalis 555, 563
  decurrens **561**
  elegans **562**
  fagicola 562
  frondosus 565
  hirtus **560**, 561*
  lentus 562
  mcmurphyi 561
  melanopus 563
  mori 555, 563
  mylittae 564
  picipes 562
  radicatus 556, 564
  squamosus 556, 561
  sulphureus 573
  tuberaster **563**
  umbellatus 565
  varius 563
  versicolor 594
Polysaccum
  pisocarpium 713
Polystictus
  tomentosus 570
  versicolor 594
Poria **602**
  cocos **604**
  corticola **603**
  incrassata 603
  sequoiae 603
  spissa 602
  vaillantii 603
  xantha 603
Porodisculus
  pendulus 552
Poronidulus
  conchifer 552
Porphyrellus
  pseudoscaber 534
Pouzarella 238, 239
Protogautieria
  lutea 741
  substriata 741
Psalliota 310
Psathyra 361
Psathyrella **361**
  ammophila 361
  atrofolia 365

Psathyrella (cont.)
  atrofolia 365
  bipellis 362
  candolleana **363**
  canoceps 366
  carbonicola **366**
  circellatipes 364
  conissans 362, 364
  conopilea 365
  elwhaensis 365
  epimyces 361
  fuscofolia 364
  gracilis **365**
  hydrophila **364**
  hymenocephala 363
  incerta 363
  kauffmanii 362
  lacrymabunda 366
  longipes **364**
  longistriata **362**
  maculata 366
  multipedata 361, 364
  sp. (unidentified) 362,
    366
  spadicea 362, 364
  stercoraria 362
  sublateritia 362, 364
  subnuda 365
  uliginicola 362, 363*
  velutina **366**
Pseudobalsamia
  alba 853
  magnata 853
  nigrens 853
Pseudocolus 772
  fusiformis 772, 776,
    777*
  javanicus 776
  schellenbergiae 776
Pseudocoprinus
  disseminatus 353
  impatiens 353
Pseudocraterellus
  pseudoclavatus 661
  sinuosus 666
Pseudofistulina
  radicata 554
Pseudohydnum
  gelatinosum **671**
Pseudopithyella
  miniscula 836
Pseudoplectania
  melaena 831
  nigrella 830
Psilocybe 367, **368**, 895
  angustispora 370
  atrobrunnea 369
  baeocystis 372, 895

Psilocybe (cont.)
  caerulescens 369, 374, 895
  caerulipes 369, 373
  californica 369
  castanella 369
  coprophila **370**
  corneipes 369
  cubensis **373**, 895
  cyanescens **371**, 895
  merdaria 369, 370
  mexicana 373
  montana 369
  pelliculosa 369, 371
  semilanceata **370**, 895
  silvatica 371
  squamosa 381
  strictipes 372
  stuntzii **372**
  tampanensis 374
  umbonatescens 376
Pterula 631
Ptychoverpa
  bohemica 794
Pulcherricium
  caeruleum 605
Pulveroboletus **509**
  auriporus 509
  hemichrysus 509
  ravenelii **509**
Pulvinula
  archeri 838
  carbonaria 838
Punctularia
  strigoso-zonata 610
Pustularia 840
Pycnoporellus 572
Pycnoporus 592
  cinnabarinus **597**
  sanguineus 597
Pyrofomes
  juniperinus 583
Pyronema
  omphalodes 835, 838

QR

Queletia 715
  mirabilis 722
Radiigera **760**
  atrogleba **761**
  fuscogleba 761, 762
  taylorii 761, 762
Radulum
  orbiculare 610
Ramaria **645**, 890*
  abietina **650**
  acris 646, 649

Ramaria (cont.)
  acrisiccescens 648, 651
  amyloidea 646
  apiculata 646, 649
  araiospora **655**
  aurantiisiccescens 647, 652
  aurea 648, 652
  botrytis **656**
  botrytoides 648, 656
  brunnea 654
  cacao 646
  cartilaginea 648, 655
  caulifloriformis 648
  cedretorum 651
  celerivirescens 646
  claviramulata 647
  conjunctipes 647
  cyaneigranosa 647, 654, 655
  cystidiophora 647, 653
  fennica **650**
  flaccida 649
  flava 652, 653
  flavigelatinosa 648, 655
  flavobrunnescens 647, 652
  formosa **654**, 896
  fumigata **651**
  fumosiavellanea 651
  gelatiniaurantia 648, 655
  gelatinosa **655**
  gracilis 649
  grandis 647
  invalii 649
  largentii 647, 654
  leptoformosa 654
  longispora 654
  maculatipes 647, 654
  magnipes 648, 652
  murrillii 649
  myceliosa **649**
  obtusissima 647
  ochraceovirens 650
  pusilla 649
  rasilispora **652**
  rubella 649
  rubiginosa 653
  rubribrunnescens 647, 654
  rubricarnata 654
  rubrievanescens 648, 656
  rubripermanens 656
  sandaracina 647, 648, 655
  sanguinea **653**

Ramaria (cont.)
  secunda 656
  strasseri 648, 656
  stricta 631*, **648**
  stuntzii 647, 655
  subbotrytis 647, 655
  suecica 649
  synaptopoda 647, 653
  testaceoflava 647, 654
  velocimutans 646
  vinosimaculans 648, 653
Ramariopsis 640
  californica 643
  kunzei **643**
  pulchella 641
Resupinatus
  applicatus 132
Rhizina
  inflata 797
  undulata 797
Rhizopogon **753**
  amyloideus 756
  atroviolaceus 754
  cokeri 754, 755
  colossus 756
  couchii 754
  diplophloeus 758
  ellenae **755**
  evadens 754
  idahoensis 756
  maculatus 754, 755
  occidentalis 754
  ochraceorubens 740*, 755
  parksii 754, 756
  pinyonensis 754, 755
  roseolus 754
  rubescens **754**
  smithii 754
  subaustralis 754
  subcaerulescens 754, 756
  subsalmonius 754
  succosus 754
  truncatus 755
  vinicolor 754
Rhodocollybia 202
  butyracea 216
  maculata 217
Rhodocybe **240**
  aureicystidiata 241
  caelata 240, 241
  mundula 240, 241
  nitellina 240, 241
  nuciolens **241**
  roseiavellanea 241
Rhodopaxillus
  nudus 154

Rhodophyllus 238
  (also see *Entoloma,*
    Leptonia, Nolanea)
  sericellus 253
  strictior 246
Rhodotus
  palmatus 130
Rhopalogaster
  transversarium 724
Rickenella
  fibula 221
Ripartites 150
  tricholoma 150
Rozites **411**
  caperata **412**
Russula 63, **83**, 896
  abietina 101
  adusta 85, 91
  aeruginea **95**
  alachuana 96
  albella 96
  albida 96
  albidula **96**
  albonigra **89**
  alutacea **102**
  amoenolens 94
  anomala 96
  aquosa 99
  atrata 89
  atropurpurea 85
  bicolor 86
  brevipes **87**
  brunneola 87, 94
  caerulea 101
  cascadensis 85, 88
  cerolens 94
  chamaeleontina 101
  claroflava **92**
  compacta 85, 91
  crassotunicata 86, 97
  cremoricolor **97**
  crenulata 97
  crustosa 86, 95
  cyanoxantha **94**, 897*
  decolorans **91**
  delica 88
  densifolia **90**
  dissimulans 85, 90
  emetica **97**
  flava 92
  foetens 93
  foetentula 93
  fragilis **98**
  fragrantissima **92**
  gracilis **99**
  gracillima 99
  granulata 94
  grisea 87, 95

Russula (cont.)
  integra **101**
  krombholzii 85
  laurocerasi 93
  lilacea 101
  lutea 86, 92
  maculata **100**
  maculosa 95
  mairei 98
  mariae **96**
  montana 86
  nigricans 85, 90
  obscura 92
  occidentalis 85, 92
  ochroleuca 85, 92
  olivacea 87, 102
  paludosa 92
  parazurea 87, 95
  pectinata 94
  pectinatoides 94
  pelargonia 99
  placita **100**
  polychroma 101
  puellaris 101
  pulverulenta 94
  raoultii 97
  romagnesiana 88
  rosacea **99**
  sanguinea 100
  silvicola 98
  sordida 89
  sororia **93**
  spp. (unidentified) 86,
    102, 103
  subalbidula 96
  subfoetens 93
  subnigricans **90**
  tenuiceps 102
  variata 86, 94
  velenovskyi 100
  ventricosipes 93
  vesca 87, 94
  veternosa 102
  virescens **95**
  xerampelina 83*, **102**

S

Sarcodon 616
  (also see *Hydnum*)
  imbricatum 620
Sarcoscypha 834
  coccinea **836**
  floccosa 836
  occidentalis 836
Sarcosoma **826**
  globosum 827, 829
  latahensis 827, 828
  mexicana **828**

Sarcosphaera **825**
  coronaria 826
  crassa **825**
  eximia 826
Schizophyllum **590**
  commune **590**
Schizostoma 715
  laceratum 716, 722
Scleroderma **707**, 896
  albidum 710
  arenicola 710
  areolatum 708
  aurantium 708
  bovista 708, 710
  cepa **709**
  citrinum **708**
  flavidum 710
  floridanum 710
  furfurellum 711
  fuscum 710
  geaster **710**
  hypogaeum 710
  laeve 710
  lycoperdoides 708
  macrorhizon 708
  michiganense 710
  polyrhizon 711
  polyrhizum 711
  reae 710
  texense 711
  verrucosum 708
  vulgare 708
Sclerogaster 744
  columnatus 744
  xerophilum 744
Sclerotinia 877
Scutellinia
  erinaceus 839
  scutellata **839**
  umbrarum 839
Scutiger
  ellisii 560
  hirtus 560
  pescaprae 560
Sebacina
  incrustans 609
Secotium 724
  agaricoides 732
Sedecula 707
  pulvinata 707
Sepedonium
  chrysospermum 884
Sepultaria 846
  (also see *Geopora*)
  arenicola 847
Serpula
  himantioides 611
  lacrymans **610**

Setchelliogaster 733
 tenuipes 733, 735
Simblum
 sphaerocephalum 776
 texense 776
Simocybe 399
 centunculus 400
Skeletocutis
 amorpha 552
Sparassis **657**
 crispa **657**
 herbstii 657
 radicata 657
 spathulata 657
Spathularia **868**
 clavata 872
 flavida **871**
 spathulata 872
 velutipes 869, 872
Spathulariopsis
 velutipes 872
Sphaerobolus 778
 stellatus **781**
Sphaerosoma 845
Sphaerozone 845
Spongipellis
 leucospongia 601
 pachydon 598, 601
 unicolor 601
Spongiporus
 leucospongia 601
Spragueola
 irregularis 871
Squamanita **197**
 odorata **198**
 paradoxum 198
 umbonata 198
Staheliomyces
 cinctus 768
Steccherinum
 adustum 612
 septentrionale 612
Stephensia 845
Stereum 604
 burtianum 608
 complicatum 606
 diaphanum 608
 fasciatum 606
 gausapatum 606
 hirsutum **605**
 lobatum 606
 ochraceoflavum 607
 ostrea 606
 rameale 606
 sanguinolentum 605, 606
 sericeum 607
 striatum **607**

Strobilomyces **543**
 confusus 542*, 543
 dryophilus 543
 floccopus **543**
 strobilaceus 543
Strobilurus 202
 albipilatus 212
 conigenoides 202, 211
 kemptonae 211
 lignitilis 212
 occidentalis 212
 trullisatus **211**
Stropharia **374**
 aeruginosa **380**
 albocyanea 380
 albonitens 376
 ambigua **377**
 aurantiaca 382
 bilamellata 377
 coronilla **377**
 cubensis 374
 depilata 380
 hardii 375, 377
 hornemannii **379**
 kauffmanii **380**
 magnivelaris 378
 melanosperma 375, 377
 merdaria 370
 riparia 375, 378
 rugoso-annulata **378**
 semiglobata **376**
 siccipes 376
 squamosa 375, 381, 382
 stercoraria 376
 thrausta 382
 umbonatescens 375, 376
Suillus **491**
 acerbus 494, 504
 acidus 493
 albidipes 493, 501
 albivelatus 492
 americanus 494, 499
 borealis 493, 501
 brevipes **501**
 brunnescens 493, 501
 caerulescens **496**
 castanellus 493
 cavipes **494**
 cothurnatus 500
 decipiens 494, 495
 elegans 497
 flavidus 498
 flavogranulatus 502
 fuscotomentosus **504**
 glandulosipes 493, 501
 granulatus **502**
 grevillei **497**
 helenae 494, 498

Suillus (cont.)
 hirtellus 494
 imitatus 497
 kaibabensis 494, 502
 lakei **495**
 luteus **500**
 megaporinus 492, 499
 monticolus 502
 occidentalis 494, 502
 pallidiceps 494, 502
 pictus 494, 495
 pinorigidus 500
 placidus 494, 502
 ponderosus 492, 496
 proximus 497, 506
 pseudobrevipes **500**
 punctatipes 494, 502
 punctipes 493
 pungens **503**
 reticulatus 505
 riparius 492, 499
 salmonicolor 500
 sibiricus **498**
 sphaerosporus 492
 subaureus 494
 subluteus 493, 500
 subolivaceus **499**
 tomentosus **504**
 umbonatus **498**
 variegatus 505
 volcanalis 493, 501
 wasatchicus 502

T

Tarzetta
 (also see *Geopyxis*)
 bronca 835, 840
 catinus 835, 840
 cupularis 835, 840
 rosea 835, 840
Tectella
 patellaris 132
Telamonia 418
Terfezia 855
 arenaria 855
Thaxterogaster **733**
 pingue **734**
Thelephora
 laciniata 609
 multipartita 609
 palmata **609**
 spiculosa 609
 terrestris **608**
 vialis 609
Togaria
 aurea 413
Trametes **592**, 603
 (also see *Coriolellus*)
 hirsuta **595**

Trametes (cont.)
  mollis 596
  occidentalis 550, 595
  pubescens 595
  suaveolens 598
  velutina 595
  versicolor **594**
Tremella
  concrescens 670
  encephala 673
  foliacea **673**
  frondosa 673, 674
  fuciformis 669, 671
  lutescens 674
  mesenterica **673**
  reticulata 644, 670
Tremellodendron
  candidum 644, 670
  pallidum 640, 644, 670
  schweinitzii 644
Tremellodendropsis
  640
  tuberosa **643**
Tremellodon
  gelatinosum 671
Trichaptum 592
  abietinus **593**
  biformis 592, 593
Trichoglossum **866**
  farlowii 868
  hirsutum **867**
  velutipes 868
Tricholoma **176**, 189,
  896
  acerbum 178, 185
  acre 178, 182
  aggregatum 174
  albobrunneum 185
  album 183
  argyraceum 182
  atroviolaceum 178, 181
  aurantio-olivaceum 188
  aurantium **187**
  caligatum 192
  cheilolamnium 180
  cingulatum 177
  columbetta 183
  equestre 180
  flavobrunneum 185
  flavovirens **179**
  focale 189
  fulvum 185
  gambosum 183
  georgii 183
  imbricatum **186**, 188*
  inamoenum 179
  leucophyllum 177, 180
  magnivelare 191
  manzanitae 185

Tricholoma (cont.)
  matsutake 192
  myomyces 182
  niveipes 181
  nudum 154
  orirubens 182
  panaeolum 155
  pardinum **183**, 896
  personatum 154
  pessundatum **185**, 896
  platyphyllum 177, 179
  ponderosum 191
  populinum **185**
  portentosum **180**
  resplendens **183**
  robustum 189
  saponaceum **184**
  scalpturatum 182
  sejunctum **180**
  sordidum 152
  sp. (unidentified) 179,
    182
  squarrulosum 178, 182
  subacutum 181
  sulphurescens 183
  sulphureum **179**
  terreum **182**
  titans 178
  ustale 179, 185
  ustaloides 185
  vaccinum **186**
  venenata 183
  virgatum **181**
  zelleri **188**
Tricholomopsis **144**
  decora 145, 146
  edodes 141
  fallax 146
  flammula 146
  flavissima 146
  platyphylla **146**
  rutilans **145**
  sulfureoides 145, 146
Trichophaea
  abundans 838, 840
  boudieri 834, 840
  bullata 840
Trogia
  crispa 132, 591
Truncocolumella **752**
  citrina **752**
  rubra 748, 753
Truncospora
  demidoffii 575, 583
Tubaria **399**
  confragosa **403**
  furfuracea **402**
  pellucida 403

Tubaria (cont.)
  tenuis 403
Tuber 841, **854**
  aestivum 855
  besseyi 859
  californicum **860**
  canaliculatum 856, 859
  candidum 861*, 862
  citrinum 860
  dryophilum 860
  gardneri 859
  gibbosum **858**
  harknessii 859, 862
  irradians 860
  levissimum 860
  linsdalei 859
  magnatum 854
  melanosporum 842*,
    854
  monticola 860
  murinum 856, 859
  rufum **861**
  separans **859**
  shearii 859
  sphaerosporum 861
  texensis 856, 859
Tulostoma 677*, 715
  berteroanum 720
  brumale **719**
  campestre 720
  cretaceum 716, 720
  excentricum 720
  fibrillosum 720
  involucratum 720
  lysocephalum 721
  macrocephalum **720**
  meristostoma 720
  opacum 720
  simulans 720
  striatum 720
Tylopilus **532**
  alboater 532
  ammiratii 533, 535
  amylosporus 520, 532
  atrofuscus 534
  badiceps 533
  ballouii 533, 534
  chromapes **533**
  conicus 533
  eximius 532
  felleus 533, 535
  ferrugineus 533
  fumosipes 534
  gracilis 533, 534
  humilus **535**
  indecisus **535**
  intermedius 533
  minor 533
  nebulosus 534

Tylopilus (cont.)
  olivaceobrunneus   534
  pacificus   534
  peralbidus   533
  plumbeoviolaceus   533,
    534
  porphyrosporus   534
  pseudoscaber   **534**, 535*
  rhoadsiae   533
  rubrobrunneus   533, 535
  snellii   532
  sordidus   534
  subunicolor   535
  tabacinus   533, 535
Typhula   634, 636
  juncea   636
Tyromyces   **597**
  albellus   599
  amarus   **601**
  balsameus   599
  basilaris   598, 599
  caesius   **599**
  chioneus   **599**
  floriformis   598, 599
  fragilis   **600**
  galactinus   599
  guttulatus   599
  immitis   599
  lacteus   599
  leucospongia   **600**
  mollis   598, 600
  perdelicatus   599
  spraguei   599
  spumeus   599
  stipticus   599
  tephroleucus   599
  transmutans   600
  unicolor   598, 601

UV

Underwoodia
  columnaris   797
Ungulina
  marginata   579
Urnula   826
  craterium   **829**
  geaster   827
  hiemalis   829
  pouchetii   827
Vaginata
  plumbea   288
Vascellum   690
  curtisii   691, 695, 696
  depressum   696
  lloydianum   696
  pratense   **695**
Veluticeps
  berkeleyi   607
Verpa   784, **793**, 893
  bohemica   **793**
  conica   **794**
Vibressea
  truncorum   869, 870,
    873
Volvaria   258
Volvariella   253, **258**
  bombycina   **261**
  hypopithys   **260**
  parvula   260
  pusilla   259, 260
  smithii   **261**
  speciosa   **259**
  surrecta   259, 262
  taylori   259, 261
  villosavolva   259, 261
  volvacea   258, 259, 262

WXYZ

Weraroa   733
  cucullata   **734**
Whetzelinia   877
Wolfina
  aurantiopsis   834
Wynnea
  americana   831
  sparassoides   817
Wynnella
  silvicola   832
Xerocomus   511
  (also see *Boletus*)
  chrysenteron   520
  subtomentosus   518
Xeromphalina   **221**
  campanella   **222**
  cauticinalis   **222**
  fulvipes   221, 222
  kauffmanii   223
  orickiana   223
  picta   222
  tenuipes   221
Xerulina
  chrysopepla   131
Xylaria   878, **885**
  cornu-damae   886
  hypoxylon   **885**
  longipes   886
  polymorpha   **886**
Xylobolus   604
Xylosphaera
  hypoxylon   886
  polymorpha   886
Zelleromyces   742
  cinnabarinus   742
  gardneri   742
  sp. (unidentified)   742

A species of *Auricularia* (probably *A. auricula*) growing on a log. Brown color and rubbery-gelatinous texture are distinctive (see p. 675 for details).